3rd Geoscience and Environmental Management Symposium 2022

IOP Conference Series: Earth and Environmental Science
Volume 1233

Online
7 - 8 September 2022

Editors:

Josaphat Tetuko Sri Sumantyo
Rohayu Che Omar
Ben White
Abraham Cardenas Tristan
Eko Haryono
Wawan Budianta
Sunu Wibirama

ISBN: 978-1-7138-8187-2
ISSN: 1755-1307

Printed from e-media with permission by:

Curran Associates, Inc.
57 Morehouse Lane
Red Hook, NY 12571

Some format issues inherent in the e-media version may also appear in this print version.

Institute of Physics
Dirac House, Temple Back
Bristol BS1 6BE UK

Phone: 44 1 17 929 7481
Fax: 44 1 17 920 0979

techtracking@iop.org

Additional copies of this publication are available from:

Curran Associates, Inc.
57 Morehouse Lane
Red Hook, NY 12571 USA
Phone: 845-758-0400
Fax: 845-758-2633
Email: curran@proceedings.com
Web: www.proceedings.com

TABLE OF CONTENTS

IOP Conf. Series: Earth and Environmental Science 1233 (2023) 011001 doi:10.1088/1755-1315/1233/1/011001

Preface

The 3rd Geoscience and Environmental Management symposium as a part of the 8th International Conference on Science and Technology (ICST) 2022 held by Universitas Gadjah Mada, Indonesia. Following the advice and guidelines from healthcare officials and local authorities, the conference was conducted virtually on 7-8 September 2022 from Universitas Gadjah Mada, Yogyakarta, Indonesia via Zoom Meeting platform. The conference aimed to provide an ideal academic platform for researchers to present the latest research findings and describe emerging technologies and directions in engineering and the natural sciences.

This year, we have invited eight renowned speakers around the world: Prof. Eryk Dutkiewicz from the University of Technology Sydney, Australia; Prof. Lee Wah Lim from Gifu University, Japan; Prof. Tomohiro Kuroda from Kyoto University, Japan; Prof. Taifo Mahmud from Oregon State University, USA; Prof. Abraham Cardenas Tristan from Universidad Autónoma de San Luis Potosí, Mexico; Prof. Jackie Y. Ying from NanoBio Lab, IMRE, A*STAR, Singapore; Prof. Nasir Bin Nayan from Universiti Pendidikan Sultan Idris, Malaysia; and Dr. Vipin Kumar from Oak Ridge National Laboratory, USA. Each speaker delivered 30 minutes of lecture with 10 minutes for question and answer session.

About 161 submissions were received and after a rigorous review process, 74 papers were chosen and presented virtually in the 3rd Geoscience and Environmental Management symposium. The presenters were given 15 minutes to present their paper and answer questions from the audience regarding their findings.

The conference will not have become a reality without the support and assistance of many parties such as the Rector of Universitas Gadjah Mada, BPP, DSSDI officers, all invited speakers and the organizing committee for arranging all needed facilities for the success of this conference.

Wahyono, PhD
Conference chair

IOP Conf. Series: Earth and Environmental Science 1233 (2023) 011001 doi:10.1088/1755-1315/1233/1/011001

Committee

Editors

Prof. Josaphat Tetuko Sri Sumantyo (Chiba University, Japan)
Prof. Rohayu Che Omar (Universiti Tenaga Nasional, Malaysia)
Prof. Ben White (Erasmus University, Netherlands)
Dr. Abraham Cardenas Tristan (Universidad Autónoma de San Luis Potosí, Mexico)
Dr. Eko Haryono (Universitas Gadjah Mada, Indonesia)
Dr.Eng. Wawan Budianta (Universitas Gadjah Mada, Indonesia)
Dr.Eng. Sunu Wibirama (Universitas Gadjah Mada, Indonesia)

Advisory boards

Prof. Rohayu Che Omar (Universiti Tenaga Nasional, Malaysia)
Prof. Ben White (Erasmus University, Netherlands)
Dr. Abraham Cardenas Tristan (Universidad Autónoma de San Luis Potosí, Mexico)
Prof. dr. Ova Emilia, M.Med.Ed.,Sp.OG(K)., Ph.D. (Universitas Gadjah Mada, Indonesia)
Ignatius Susatyo Wijoyo, M.M. (Universitas Gadjah Mada, Indonesia)
Prof. Widodo (Universitas Gadjah Mada, Indonesia)
Dr. I Wayan Mustika (Universitas Gadjah Mada, Indonesia)
Dr. Dyah Rahmawati Hizbaron (Universitas Gadjah Mada, Indonesia)
Dr. Eko Haryono (Universitas Gadjah Mada, Indonesia)

Organizing Committee

Chairman : Wahyono, PhD (Universitas Gadjah Mada, Indonesia)

Co-chairman : Dr.Eng. Wawan Budianta (Universitas Gadjah Mada, Indonesia)

Secretary : Ratih Fitria Putri, PhD (Universitas Gadjah Mada, Indonesia)
Seri Aryati, S.Pd., M.Sc. (Universitas Gadjah Mada, Indonesia)
Muhammad Rafieiy, S.Si. (Universitas Gadjah Mada, Indonesia)

Local Program Committee:

Probo Bhaskoro, S.I.P. (Universitas Gadjah Mada, Indonesia)
Suharman, S.E., M.M. (Universitas Gadjah Mada, Indonesia)
Alif Yuni Alfiani, S.Kom. (Universitas Gadjah Mada, Indonesia)
Fatimah Salsabila Azzahra, S.S. (Universitas Gadjah Mada, Indonesia)
Wildan Ainurrahman, S.T.P. (Universitas Gadjah Mada, Indonesia)
Hana Aulia, S.Sos. (Universitas Gadjah Mada, Indonesia)
Shinta Khoiri F. (Universitas Gadjah Mada, Indonesia)
Dian Rizki Tri Wahyu Nanda (Universitas Gadjah Mada, Indonesia)

Peer review declaration

All papers published in this volume have been reviewed through processes administered by the Editors. Reviews were conducted by expert referees to the professional and scientific standards expected of a proceedings journal published by IOP Publishing.

Type of peer review: Single anonymous
Conference submission management system: Morressier
Number of submissions received: 161
Number of submissions sent for review: 128
Number of submissions accepted: 74
Acceptance Rate (Submissions Accepted / Submissions Received × 100): 45.9 %
Average number of reviews per paper: 3.00
Total number of reviewers involved: 16

Contact:
Name: Dr.Eng. Wawan Budianta
Affiliation: Universitas Gadjah Mada
Email: wbudianta@ugm.ac.id

Adsorption of Cu (II) from Cu Standard Solution Using Expanded Perlite

Abdul Malik Pamasengi[1,a], I Wayan Warmada[1,b*] and Wahyu Wilopo[1,c]
[1]Department of Geological Engineering, Faculty of Engineering, Gadjah Mada University, Yogyakarta 55281, Indonesia
[a]abdul.pamasengi@mail.ugm.ac.id, [b*]warmada@ugm.ac.id, [c]wilopo_w@ugm.ac.id

Keywords: Adsorption, Cu (II), Expanded Perlite

Abstract. The research aimed to determine the ability of expanded perlite to adsorb or remove Cu (II) metal ions in the solution. Furthermore can be used to reduce Cu levels in the aquatic environment caused by industrial and mining waste. Expanded perlite is a product of perlite resulting from heating at temperatures above 1000°C. Perlite is a volcanic product contains high silica content and a lot of water content so when heating occurs will increase the porosity and surface area that can cause adsorption. Expanded perlite sized < 1 mm can be utilized in the remediation process of mining waste because it is effective in the adsorption of ions such as Cu so as not to be in aquatic environtment. Expanded perlites, effectively used in the adsorption process, have a grain size of < 1 mm with a dose of 60 g/l. An experimental batch test method was carried out with the parameters of pH, adsorbent dose, and contact time to determine the ability of expanded perlite to adsorb Cu (II). The optimum pH in the Cu (II) metal ion adsorption process was 7 and reached equilibrium within 240 minutes. The adsorption equilibrium (chemical adsorption due to cation exchange) that can be absorbed by 1 gram of expanded perlite is 6.11 mg Cu (II), while the maximum adsorption capacity (physical adsorption due to the hydroxyl group bonded to silica has a high electronegativity value so that it can attract positively charged ions) that can be absorbed by 1 gram of expanded perlite is 1.47 mg Cu (II).

1. Introduction

Water quality is a significant factor in supporting the health of living things. The presence of pollutants such as heavy metals will adversely affect the biological processes of aquatic organisms that can threaten the sustainability of their lives, including humans, through the food line. Water remediation used to improve water quality as drinking water and to prevent natural contamination of water resources due to industrial waste disposal [3].

Expanded perlite is a perlite product processed through a combustion process at temperatures above 1000°C. Perlite is a pyroclastic rock with a very high silica content of about 70 – 75 wt.% and contains 2 – 6 wt.% water [2]. Through this process, perlite will turn into small grains of white to gray and porous due to loss of moisture content and can expand up to 20 times its previous volume. In general, perlite has a density of about 2400 kg/m^3 while expanded perlite has a density of about 35 kg/m^3 [7]. These characteristics make expanded perlite widely used today in the agricultural, health, and building construction industries. The porosity of expanded perlite causes expanded perlite to absorb water easily. It caused expanded perlite to be resistant to corrosive substances and other harmful substances, so it is very suitable for use as a mixed material in water remediation equipment. In addition, expanded perlite is also an insulator, refractory and inert, or cannot react chemically with other compounds or elements [2].

Copper is a relatively unreactive element but can react with oxygen at high-temperature conditions. Copper can react with halogens and sulfur and can dissolve in ammonia. In addition, copper is a good conductor of electricity and heat, with a soft, malleable, and ductile structure, has a melting point of 1083°C, a boiling point of 2595°C, and a density of 8.96 g/cm^3. The concentration of copper in surface water can range from 20-2200 mg/l [4]. As recommended by the US EPA, the maximum contaminant level for copper is 1.3 mg/l. Therefore, the copper concentration of any industrial effluent must be reduced below 1.3 mg/l before being discharged into the environment. Although copper can exist as Cu(0), Cu(I), and Cu(II), the main species of concern in solution is Cu(II). Cu(II) easily binds to organic and inorganic materials in solution based on the pH of the solution [4]. The Cu (II) threatens aquatic life and makes natural water unsuitable for general use. Copper (Cu) is one of the toxic metal elements and is most widely used in industrial fields such as the metal coating industry, the textile paint and dye industry, and the leather tanning industry [9]. These industries can produce waste containing Cu, which is stable and is carried in industrial waste.

Excessive intake of copper of more than 1.0 mg/l in drinking water can cause hemochromatosis and gastrointestinal catarrh because copper can accumulate in the human and animal liver [8]. In addition, copper (Cu) is one type of metal in tailings in the form of minerals that can be associated with gold (Au), iron (Fe), silver (Ag), zinc (Zn), and various other sulfide minerals such as chalcopyrite ($CuFeS_2$). Mining activities produce much waste in the form of tailings that are dumped on plains or water bodies. Tailings waste may be scattered around the area and can cause environmental pollution. In this case, expanded perlite is one of the suitable adsorbents used to remove or reduce the content of Cu, which is soluble in water. Expanded perlite sized < 1 mm can be utilized in the remediation process of mining waste because it is effective in the adsorption of ions such as Cu so as not to be in the aquatic environment.

2. Experimental Method

2.1. Adsorbent Characterization

The adsorbent was obtained from PT. Perlite Indonesia Abadi, with two kinds of grain sizes < 1 mm and 1-4 mm, has a white color, hygroscopic humidity of 0%, surface pH 6.5 – 7.5, and a fusion point of 1260°C.

The results of the petrographic analysis indicate the presence of several felsic minerals, including plagioclase, opaque, biotite, and base mass of glass. The structure of perlite, that is a concentric fracture structure, can be clearly observed microscopically on PPL observations. X-Ray Fluorescence (XRF) determined the chemical composition of perlite both qualitatively and quantitatively. The results of XRF analysis showed that the pearlite sample had a high silica content of around 80%. Based on these results, plotting was carried out on the Total Alkali-Silica (TAS) diagram to determine the rock origin and series of magma. The results show that the rock samples belong to the rhyolite group. The surface area of perlite and expanded perlite sized 1-4 mm and < 1 mm were analyzed using Brunnauer Emmett Teller (BET) with a surface area of 0.63 m^2/g, 16.85 m^2/g, and 32.53 m^2/g.

The silica atoms present on the surface of expanded perlite tend to maintain their tetrahedral coordination with oxygen. The coordination can be removed at room temperature by attaching to the monovalent hydroxyl group, forming a silanol group. The hydroxyl group that binds to the silicate has a high electronegativity value so that it can attract positively charged ions (physical reaction). It can cause Cu^{2+} ions to be adsorbed on the surface of the expanded perlite. In addition, chemical reactions can occur by exchanging cations between Si and Cu with the following ion exchange reactions (based on experiment):

$SiO_{2\,(s)} + 2Cu(NO_3)_{2\,(aq)} \rightleftarrows 2CuO_{(s)} + Si(NO_3)_{4\,(aq)}$

$Si^{4+} + 4H_2O \rightarrow Si(OH)_4 + 4H^+$

The following groups cause the adsorption properties of perlite:

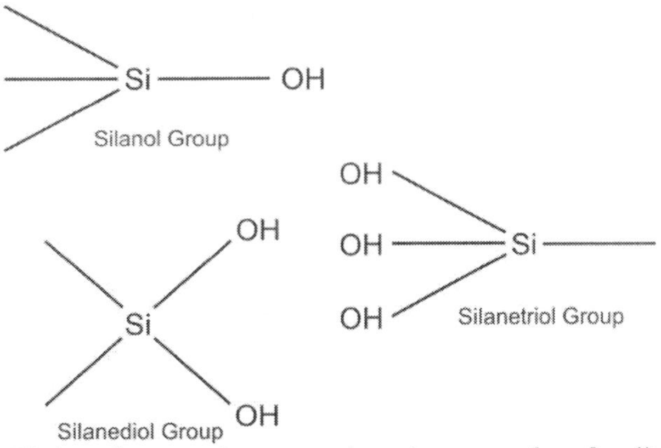

Fig 1. Hydroxyl groups that cause adsorption properties of perlite [4].

2.2. Batch Test Method

Experiments were carried out using the batch test method to obtain effective pH, adsorbent dose, and contact time in the remediation process. A ball pipette takes a standard Cu solution with a specific volume to get a predetermined concentration, then poured into a volumetric flask and filled with aquadest up to 500 ml. The standard Cu solution was dissolved from 1000 ppm to 10 ppm in a 500 ml volumetric flask. The following is a calculation for making a 10 ppm Cu solution.

$$C_1 \times V_1 = C_2 \times V_2 \quad (1)$$

C_1 is the concentration of the standard solution, V_1 is the volume of the standard solution, C_2 is the concentration of the solution to be made, and V_2 is the volume of the solution to be made.

After the 10 ppm Cu solution was prepared, an Atomic Absorption Spectrophotometer (AAS) analysis was performed to determine the initial concentration of the solution to be remedied. The analysis resulted in a Cu solution with a concentration of 10.11 ppm.

To determine the effective pH in the remediation process, used pH variations with values of 3, 5, 7, and 9. The 10 ppm Cu solution was then poured into a 250 ml Erlenmeyer flask as much as 100 ml. The solution was then measured its pH using universal pH paper. The measurement results of the initial solution showed a pH of 3. To make the initial solution with a pH of 5, 7, and 9, NaOH had previously been dissolved in water. After the initial solution with a predetermined pH was successfully made, 3 grams of expanded perlite was added. After that, the Erlenmeyer flask was placed in a shaker and shaken for 60 minutes (110 RPM). After the remediation was carried out, the solution was filtered using paper which AAS then analyzed to determine the concentration of the remaining results.

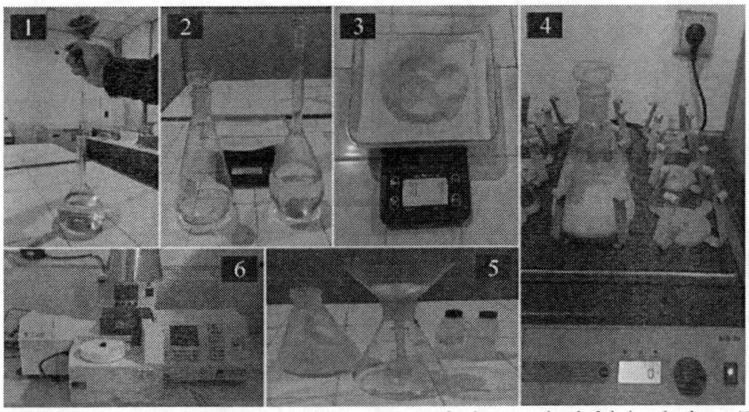

Fig 2. Remediation Process. (1). Preparation of a 10 ppm Cu solution as the initial solution, (2). The solution was put in an Erlenmeyer flask, (3). The adsorbent was weighed and then mixed in an Erlenmeyer flask, (4). Filtering the remedied solution, (5). Measurement of the concentration of the remaining remedied product using AAS.

The adsorbent dose parameter was analyzed with two kinds of expanded perlite grain sizes, namely 1 – 4 mm and < 1 mm. Each grain size was varied again, namely 20, 40, and 60 g/l for grain sizes < 1 mm and 20, 40, 60, and 80 g/l for grain sizes 1 – 4 mm. The contact time parameters were analyzed with five-time variations, namely 30, 60, 120, 240, and 360 minutes. Two calculations were used to analyze the contact time parameters: efficiency or percentage of adsorption and capacity or adsorption equilibrium.
The percentage of metal ion adsorption is calculated by the equation as follows [6]:

$$Adsorpsi\ (\%) = \frac{Ci - Ce}{Ci} \times 100 \quad \text{.......(2)}$$

where Ci and Ce are the initial and final metal ion concentrations, respectively.
The number of metal ions adsorbed per unit mass of adsorbent (qe) is calculated using the equation below [6]:

$$qe = \frac{V}{M}\,(Ci - Ce) \quad \text{.......(3)}$$

where Ci and Ce are initial and equilibrium concentrations, respectively (mg/l). V is the volume of the solution (l), and M is the mass of the adsorbent (g).
Langmuir and Freundlich's adsorption isotherm model was used to determine the maximum adsorption capacity of Cu(II) metal ions by expanded perlite using the adsorbate concentration parameter. The adsorption capacity can be calculated from the trend of the Langmuir and Freundlich graphs using a linear equation. The expanded perlite adsorption capacity for Cu (II) metal ions can be identified from these graphs. The adsorption process by an adsorbent is influenced by many factors and has a specific adsorption isotherm pattern. Each adsorbent that absorbs a substance will not have the same adsorption isotherm pattern as other substances.
The Langmuir adsorption isotherm describes the equilibrium between the adsorbate and the adsorbent system. The adsorbate adsorption is limited to one molecular layer or before the unit relative pressure is reached. Langmuir further assumed that the surface fraction is directly proportional to the rate of desorption from the surface and that adsorption and desorption will be equal in the equilibrium phase [6].

$$K_a . C_e(1-\theta) = K_d . \theta \quad \text{.......(4)}$$

$$\theta = \frac{qe}{a} = \frac{b.C_e}{(1+b.C_e)} \quad \text{.......(5)}$$

$$b = K_a/K_d \quad \text{.......(6)}$$

$$q_e = \frac{a.b.C_e}{1+b.C_e}\ (Nonlinear) \quad \text{.......(7)}$$

$$\frac{C_e}{q_e} = \frac{1}{a.b} + \frac{1}{a} C_e\ (Linear) \quad \text{.......(8)}$$

where Ka is the adsorption rate constant, Kd is the desorption rate constant, is the fractional surface coverage, Ce is the equilibrium concentration in the solution phase (mg/l), qe is the equilibrium concentration in the adsorbent phase (mg/g), a is the adsorption capacity of the coating single (mg/g), and b are constants related to the free energy of adsorption and the reciprocal of the concentration at which half-saturation of the adsorbent is reached.
Freundlich introduced the Freundlich adsorption isotherm in 1906 as an empirical equation consistent with the thermodynamics of adsorption on heterogeneous surfaces [6].

$$qe = K.Ce^{\frac{1}{n}}\ (Nonlinear) \quad \text{.......(9)}$$

$$log\ qe = log\ K + \frac{1}{n} \log Ce\ (Linear) \quad \text{.......(10)}$$

where qe is the equilibrium concentration in the adsorbent phase (mg/g), K is the maximum adsorption capacity (mg/g), n is the adsorption constant, and Ce is the equilibrium concentration in the solution phase (mg/l).

3. Result and Discussion

3.1. Optimum pH

The analysis of pH parameters showed that the remediation process was effective at pH 7, where the lower the pH, the lower the adsorption of Cu metal ions. On the other hand, Cu metal was found in a small amount in solution at a higher pH because Cu ions will be precipitated in the alkaline pH. The analysis was carried out by calculating the efficiency of the adsorption process by calculating the percentage of adsorbed metal ions, as shown in Fig 3.

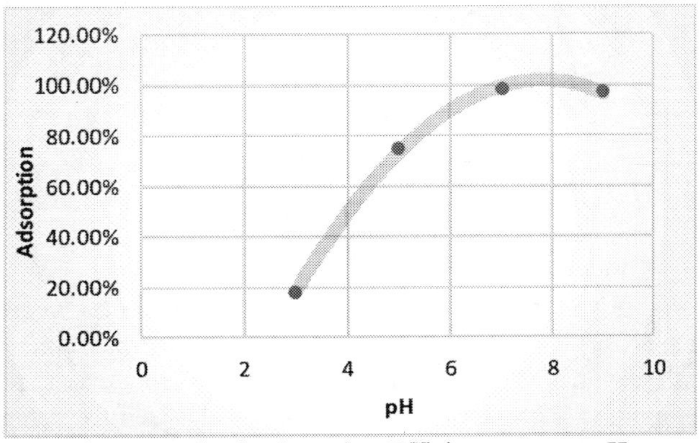

Fig 3. Graph of Adsorption Efficiency versus pH.

3.2. Optimum Adsorbent Grain Size

The adsorbent dose parameter analysis results showed an effective remediation process using an adsorbent with a size of < 1 mm with a dose of 60 g/l. It is caused by the larger surface area of the adsorbent in expanded perlite < 1 mm so that more adsorbate can be adsorbed compared to the size of 1-4 mm. If further interpreted from the trend formed between adsorbents < 1 mm and 1 – 4 mm, adsorbents with sizes < 1 mm have a steeper trend than those with sizes 1 – 4 mm, as shown in Fig. 4. It explains that a larger surface area greatly affects the adsorption process that occurs where the higher the increase in the adsorbent dose, the higher the adsorption of Cu (II) metal ions, while at a size of 1-4 mm, the adsorption process does not increase significantly if the adsorption efficiency has reached the point maximum above 95%.

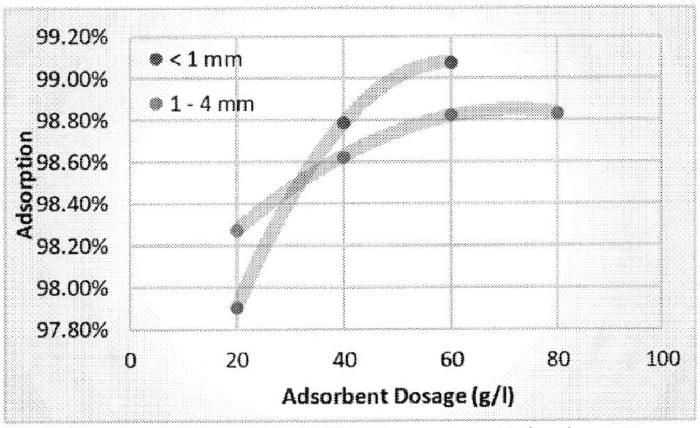

Fig 4. Graph of Adsorption Efficiency versus Adsorbent Dosage.

3.3. Optimum Contact Time

The effective contact time in the adsorption process occurred at 240 minutes with efficiency and adsorption capacity of 99.31% and 0.250 mg Cu per g perlite, as shown in Fig. 5. At times above 240 minutes, the efficiency and adsorption capacity decreased, indicating the starting point for the desorption process.

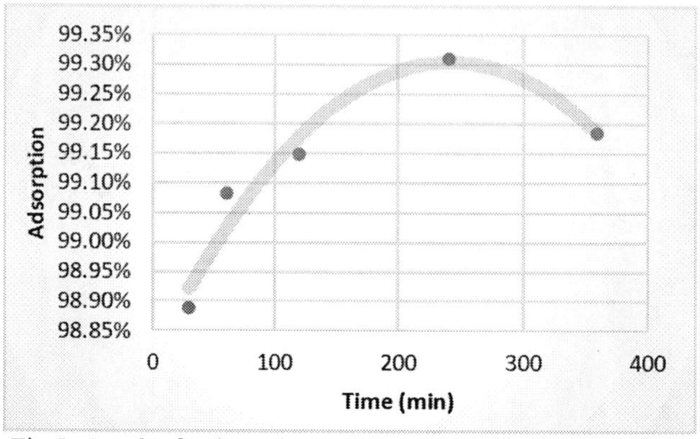

Fig 5. Graph of Adsorption Efficiency versus Contact Time.

3.4. Adsorption Isotherm Model

Langmuir isotherm model is shown by a graph of the relationship between Ce/qe and Ce. Ce is the concentration of residual adsorption in mg/l, and qe is the adsorption capacity in the form of mg/g, as presented in Table 1. Determine the Langmuir isotherm equation was conducted by calculating the values of qe and Ce/qe, making a graph of the relationship between Ce/qe and Ce. The value of Ce or the residual concentration of the adsorption results itself was obtained from the AAS analysis. The Langmuir adsorption equation is proven by an excellent linear graph and has a coefficient of determination ($R^2 = 1$) with the equation ($y = 0.1638x + 0.5831$), as shown in Fig 6.

Table 1. Langmuir Isotherm Data.

Langmuir		
C_e	q_e	C_e / q_e
0.05	0.0845	0.59
0.09	0.1485	0.60
0.30	0.4705	0.63
0.34	0.5372	0.64
0.39	0.6020	0.65
0.41	0.6349	0.65

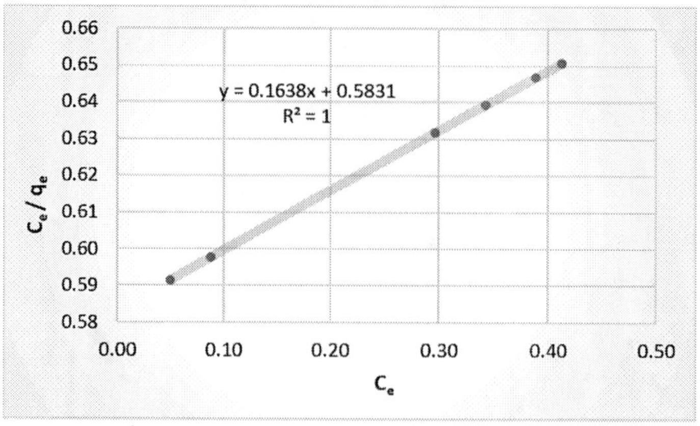

Fig 6. Langmuir Adsorption Isotherm Model.

Freundlich's isotherm model is shown by a graph of the relationship between log qe and log Ce. Determination of the Freundlich isotherm equation calculates the value of log qe and log Ce, as presented in Table 2, to develop a graph of the relationship between log qe and log Ce. The Freundlich adsorption equation is proven by an excellent linear graph and has a coefficient of determination (R^2 = 1) with the equation (y = 0.9331x + 0.1659), as shown in Fig 7.

Table 2. Freundlich Isotherm Data.

Freundlich			
C_e	q_e	Log C_e	Log q_e
0.05	0.0895	-1.30	-1.05
0.09	0.1529	-1.05	-0.82
0.30	0.4724	-0.53	-0.33
0.34	0.5405	-0.46	-0.27
0.39	0.6077	-0.41	-0.22
0.41	0.6423	-0.38	-0.19

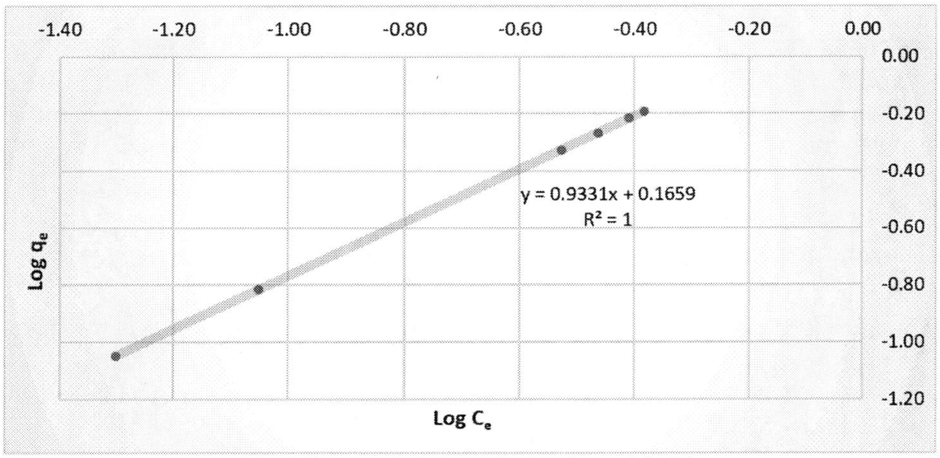

Fig 7. Langmuir Adsorption Isotherm Model.

Tables 1 and 2, indicate that the adsorption capacity of expanded perlite will increase as the Cu concentration of the solution increases until equilibrium is reached. Suppose the initial solution of Cu (II) metal ions has a low concentration. In that case, the adsorption on the surface of expanded perlite is also low due to other reactions in the solution that inhibit the absorption of the adsorbate. On the other hand, if the concentration of Cu(II) metal ions in the solution are initially large, then the metal ions adsorbed in the expanded perlite are also prominent. It is due to the attractive forces between the adsorbates, which results in combined adsorption so that the adsorption process is more optimal.

Table 3. Calculation Results of Adsorption Isotherm Model Constants and Maximum Adsorption Capacity.

Isotherm	Constant	Value
Langmuir	a	6.11
	b	0.28
Freundlich	K	1.47
	n	1.07

The Langmuir and Freundlich adsorption isotherm models have an excellent coefficient of determination (R^2 = 1). This explains that both models can analyze the adsorption isotherm that occurs between the adsorbent (expanded perlite) and the adsorbate (Cu metal ions). The interaction that occurs on the surface of expanded perlite with Cu (II) ions following the Langmuir isotherm model has a strong chemical bond with adsorption of adsorbate limited to one molecular layer on a homogeneous surface. Langmuir isotherm model can be used to identify the equilibrium between adsorbate and adsorbent. In

contrast, the interactions on the expanded perlite surface with Cu (II) ions following the Freundlich isotherm model have relatively weak bonds with adsorbate adsorption occurring in more than one molecular layer [1]. Freundlich's isotherm model is an empirical equation consistent with the thermodynamics of adsorption on heterogeneous surfaces.
The Langmuir isotherm model describes chemical adsorption that occurs due to the exchange of Cu and Si cations contained in expanded perlite. In contrast, the Freundlich isotherm model describes physical adsorption that occurs because the hydroxyl group attached to the silicate has a high electronegativity value so that it can attract charged positive ions. The single layer adsorption capacity/adsorption equilibrium generated from the Langmuir isotherm model was 6.1050 mg/g while the maximum adsorption capacity produced from the Freundlich isotherm model was 1.4652 mg/g.

4. Conclusion

Perlite has a high silica content of about 80%. The surface area of perlite and expanded pearlite sized 1-4 mm and < 1 mm are 0.63 m2/g, 16.85 m2/g, and 32.53 m2/g, respectively. These characteristics cause expanded perlite to be the fine adsorbent for remediation of water contaminated with Cu (II) metal ions. Expanded pearlite was effective in the adsorption of Cu(II) metal ions at pH 7, with grain size < 1mm, adsorbent dose of 60 g/l, and contact time of 240 minutes.
The adsorption equilibrium (chemical adsorption due to cation exchange) that can be absorbed by 1 gram of expanded perlite by 6.11 mg Cu(II), while the maximum adsorption capacity (physical adsorption due to the hydroxyl group bonded to silica has a high electronegativity value so that it can attract positively charged ions) that can be absorbed by 1 gram of pearlite expanded by 1.47 mg Cu(II).

References

[1]Adamson, A., dan Gast, A., 1997, Arthur W. Adamson, Alice P. Gast - Physical chemistry of surfaces.

[2]Angelopoulos, P.M., Maliachova, C., Papakonstantinou, K., Taxiarchou, M., dan Diplas, S., 2016, Structural and physical characteristics of fine perlite expanded with a novel method in a vertical electric furnace: Transactions of the Institutions of Mining and Metallurgy, Section C: Mineral Processing and Extractive Metallurgy, v. 125, p. 71–80, doi:10.1080/03719553.2016.1156244.

[3]Bhattacharya, S., Gupta, A.B., Gupta, A., dan Pandey, A., 2018, Water Remediation Energy, Environment, and Sustainability.

[4]Ghassabzadeh, H., Torab-Mostaedi, M., Mohaddespour, A., Maragheh, M.G., Ahmadi, S.J., dan Zaheri, P., 2010b, Characterizations of Co (II) and Pb (II) removal process from aqueous solutions using expanded perlite: Desalination, v. 261, p. 73–79, doi:10.1016/j.desal.2010.05.028.

[5]Le Bas, M.J., Le Maitre, R.N., Streckeisen, A., dan Zanettin, B., 1986, A chemical classification of volcanic rock based on total silica diagram: Journal Petrology, v. 27, p. 745–750.

[6]Liu, L., Luo, X.B., Ding, L., dan Luo, S.L., 2018, Application of Nanotechnology in the Removal of Heavy Metal From Water: Elsevier Inc., 83–147 p., doi:10.1016/B978-0-12-814837-2.00004-4.

[7]Meriani, S., dan Fontanive, F., 1976, A chemical evaluation of expanding perlitic rocks: Materials Chemistry, v. 1, p. 347–358, doi:10.1016/0390-6035(76)90033-X.

[8]Sari,A., Tuzen, M., Citak, D., dan Soylak, M., 2007, Adsorption characteristics of Cu(II) and Pb(II) onto expanded perlite from aqueous solution: Journal of Hazardous Materials, v. 148, p. 387–394, doi:10.1016/j.jhazmat.2007.02.052.

[9]Sudiarta, I.W., Suarya, P., dan Widya, C.M.P., 2018, Adsorpsi Multi Logam Berat Krom(III), Timbal(II), Dan Tembaga(II) Dalam Sistem Larutan Binary Oleh Silika Gel Terimobilisasi Difenilkarbazida: Jurnal Kimia, p. 159, doi:10.24843/jchem.2018.v12.i02.p11.

Analysis of Consolidation Grouting Effect on The Stability of Portal Slope of The Spillway Tunnel in The Bulango Ulu Dam, Gorontalo, Indonesia

A Purwanto[1,2], I G B Indrawan[1*] and A Setianto[1]

[1]Department of Geological Engineering, Faculty of Engineering, Universitas Gadjah Mada, Yogyakarta, Indonesia
[2]PT Hutama Karya (Persero), Jakarta, Indonesia

*Corresponding author: igbindrawan@ugm.ac.id

Abstract. The problem of portal slope instability often occurs caused by the poor lithological conditions composing the slope. During the portal excavation of the outlet spillway tunnel in the Bulango Ulu Dam, the slope failed. The slope was then reinforced by consolidation grouting. The objectives of this research were to characterize the soil and rock masses composing the slope, evaluate the grouting effectiveness by Limit Equilibrium Method (LEM), and recommend an alternative slope reinforcement method. Surface engineering geological mapping, borehole data evaluation, and laboratory testing were conducted to obtain slope stability analysis parameters. The results showed that the slope consisted of diorite having poor to very poor rock masses. After 7 days of grouting, the triaxial test showed that the soil samples had similar cohesion and increased internal friction angle insignificantly, as compared to the soil samples prior to grouting. The LEM analysis indicated that the portal slope was unstable and consolidation grouting did not increase the slope safety factor (Fs) significantly. As the grouting effect may take significantly long, an alternative slope reinforcement method by soil nailing was proposed. The LEM analysis showed that the soil nailing increased the slope Fs from 0.63 to 1.49.

1. Introduction
The research was conducted at the construction location of the spillway tunnel of the Bulango Ulu dam administratively in Tuloa Village, North Bulango District, Bone Bolango Regency, Gorontalo Province (Figure 1). The Bulango Ulu Dam Development Project is one of the National Strategic Projects that began construction in 2019. Based on The Regional Geological Map of The Kotamobagu Sheet, Sulawesi, the primary rock composed at the research location is the intrusive rock in the Bone Diorite Formation (Figure 2). This unit consists of quartz diorite, granodiorite, and granite aged approximately Late Miocene, which intruded the Bilungala Volcanics and the Tinombo Formation [1].

The problem of instability on the slope of the tunnel portal sometimes often occurs. One of them is caused by the structural geology of rock mass and the poor lithological conditions of the rock compiler on the slopes. Orientation, persistence, roughness, and infilling are all properties of discontinuities in terms of stability [2]. The high strength of the structure quality will be produced by a stable rock mass structure, with the slope geometry, groundwater, and geological condition acting as the controlling variables of the stability [3]. Researchers are concerned about slope stability issues, and as a result, numerous methodologies and methods for assessing slope stability have been developed. Kinematic analysis, limit equilibrium, numerical modeling, and empirical approaches are the four main groups assessing slope stability [4].

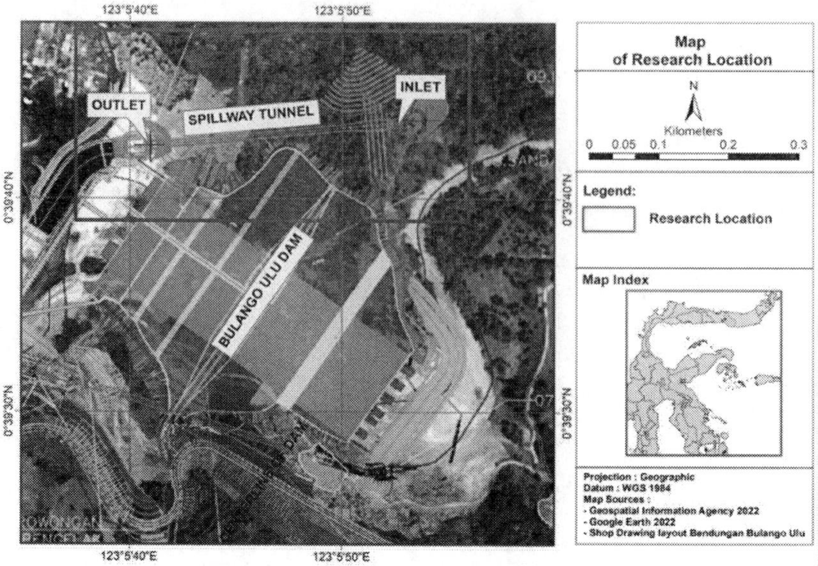

Figure 1. Research location [5] [6] [7]

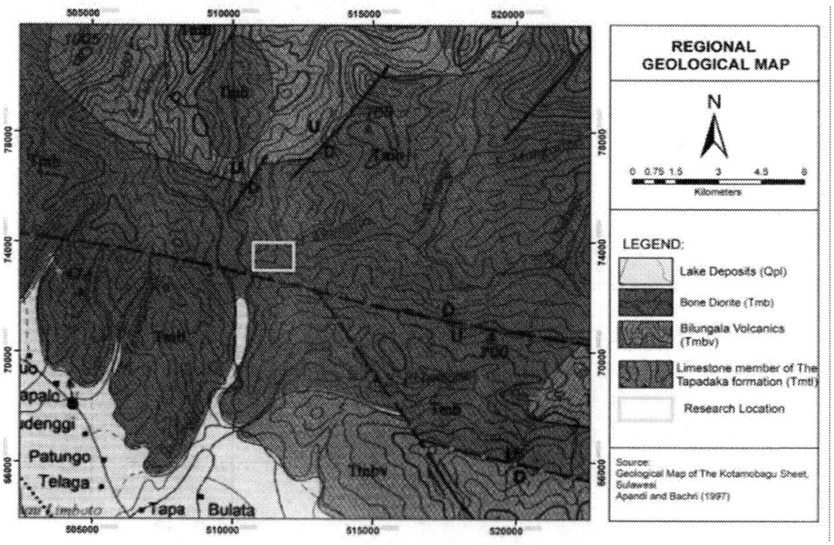

Figure 2. Regional geological map of Kotamobagu, Sulawesi [1]

On the portal slope of the outlet spillway tunnel, the slopes have been excavated with several benches with an inclination of 1:0.5. Slope protection has been carried out by installing wire mesh type M5 combination shotcrete with K-225 concrete quality and 10 cm thickness. However, after the initial mechanical excavation in the outlet tunnel, slope instability occurred, which caused the bench of the slope was collapse on the roof of the tunnel portal. Slope reinforcement after the slope collapse is carried out by consolidation grouting by injecting a liquid grouting material consisting of a mixture of water and cement. Slope stabilization methods generally increase the resisting force of soil, which is increased by chemical treatment (hardening of soil) to increase the shear strength of the ground [8]. Consolidation grouting entails filling open joints, separated bedding planes, faulted zones, voids, and other geological faults up to a certain distance [9], thus hardening the ground and decreasing groundwater flow beyond the excavation [10]. Consolidation grouting was carried out horizontally from the face of the shotcrete slope with 4° inclination, 2.5" diameter, and 3 m grouting spacing point, respectively, until the total grouting depth of 25 m (Figure 3).

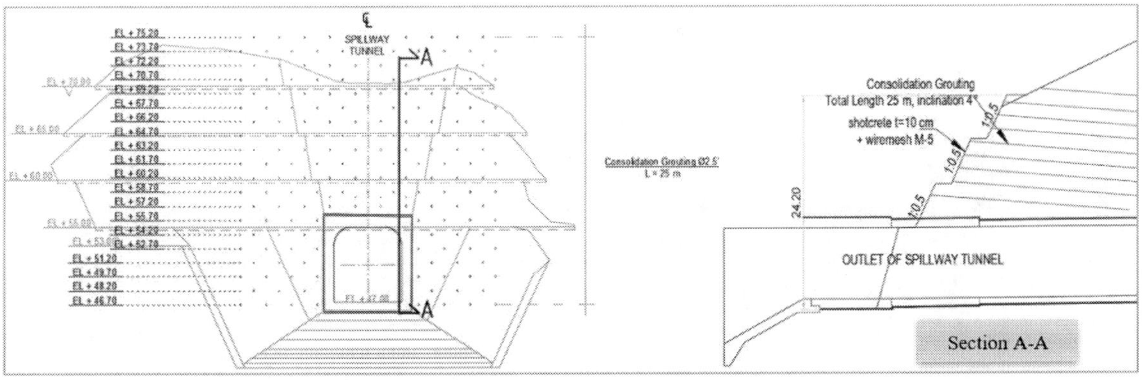

Figure 3. Location of consolidation grouting [6]

This research aims to characterize the soil and rock masses composing the slope, evaluate the effectiveness of grouting based on the shear strength parameters by Limit Equilibrium Method (LEM), and recommend an alternative slope reinforcement method. It is expected that the results of this research contribute to a preliminary investigation to evaluate the consolidation grouting effect on the stability of the portal slope.

2. Method of Research

2.1. Rock Mass Characterization

The surface quality assessment of rock mass on the portal slope of the outlet spillway tunnel area was carried out empirically based on the Rock Mass Rating (RMR) and the Geological Strength Index (GSI). Five outcrop stations were selected in the area of the portal slope that has not been shotcrete, namely STA 01-LP01, STA 01-LP02, STA 01-LP03, STA 02-LP01, and STA 02-LP02. Each observation STA is divided into several zones based on similarities in rock types and differences in the intensity of structural geology and discontinuities, as shown in (Figure 4b).

Unconfined Compressive Strength (UCS) of intact rock, Rock Quality Designation (RQD), discontinuity spacing, and groundwater conditions are the parameters used to assess rock mass quality based on RMR classification by Bieniawski [11]. The UCS value used was obtained from the empirical UCS value estimate in the field based on Hoek et al.[12]. Equations 1 and 2 were used to calculate the RQD value proposed by Priest and Hudson [13].

$$RQD=100e^{-0.1\lambda}(1+0.1\lambda) \quad (1)$$

where,

$$\lambda = \frac{\text{number of discontinuities}}{\text{length of scanline (m)}} \quad (2)$$

where λ is obtained from the joint frequency of the scanline length (m), representing the observed slope area, Equation 3 shows the parameter summation rating in the RMR calculation [11].

$$RMR = R_1+R_2+R_3+R_4+R_5 \quad (3)$$

IOP Conf. Series: Earth and Environmental Science 1233 (2023) 012002 doi:10.1088/1755-1315/1233/1/012002

Figure 4. (a) Existing slope condition of outlet spillway tunnel before shotcrete (b) Slope condition of outlet spillway at the time of field mapping was carried out at locations STA 01-LP01, STA 01-LP02, STA 01-LP03, STA 02-LP01, and STA 02-LP02

where R1 = UCS; R2 = RQD; R3 = discontinuity spacing; R4 = discontinuity conditions (persistence, aperture, roughness, infilling, and weathering); and R5 = groundwater conditions.

In assessing the rock mass quality in the Geological Strength Index (GSI) classification, the parameters considered are rock mass structure and surface conditions of discontinuities and then plotted into a GSI graph by Marinos and Hoek [14] to obtain a range of GSI values.

The GSI values of the core drill data were also determined using a quantitative approach. Hoek et al. (2013) proposed a formula for computing GSI values from core drill data using descriptions and ratings of the drill holes. The correlation between joint condition (Jcond) and RQD is as follows in Equation 4 [15].

$$GSI=1,5JCond_{89}+RQD/2 \tag{4}$$

2.2. *Grouting Samples and UU Triaxial Test*

Grouting is an effective and practical method for improving rock properties [16]. Zakri et al. (2018) analyzed an artificial rock sample made from dental plaster of the type of moldano tara after grouting injection and experienced an increase in cohesion and internal friction angle value using the direct shear test [17]. To evaluate the effectiveness of grouting, coring samples were taken after grouting at the check hole location (CH-01) see Figure 5. As an initial evaluation, coring sampling was carried out seven days after grouting material injection. The sample was then evaluated using the Unconsolidated Undrained (UU) Triaxial method to see how grouting affected the shear strength parameters of cohesiveness and internal friction angle.

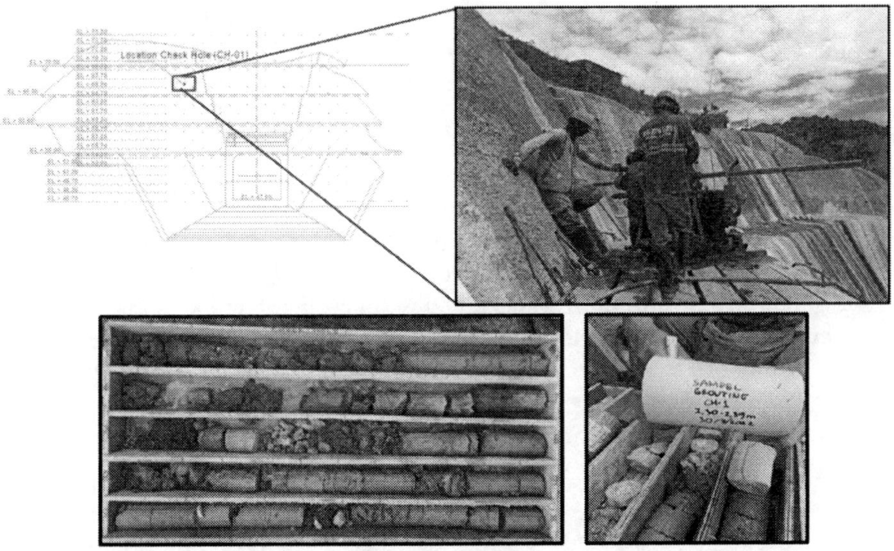

Figure 5. Grouting samples were taken at a location of check hole

The UU Triaxial test before grouting was taken from secondary data from the triaxial test results carried out on the borehole nearest to the check hole location, namely at BP-02 at a depth of 7.50-8.00 m, so that the parameters of shear strength before and after grouting could be compared. In the triaxial test for the sample after grouting, this cylindrical sample is then inserted into triaxial cells with three compression pressures (confinement). For each sample, the compression pressure was the same as for the triaxial test before grouting, namely 1 kg/cm^2, 2 kg/cm^2, and 3 kg/cm^2.

2.3. Slope Stability using Limit Equilibrium Method (LEM)

The LEM approach makes use of the force equilibrium principle [18]. This analysis technique assumes that the failure area could be circular failure or noncircular failure and that the calculating method divides the landslide plane into slices. In this research, slope reinforcement with consolidation grouting was analyzed for stability using LEM (Limit Equilibrium Method) with SLIDE v.6.0 software from Rocscience using 2D modeling analysis (plane strain).

For the condition of the material properties, the soil is modeled with the Mohr Coloumb failure criteria with cohesion parameters (c) and internal friction angle (ϕ), while for rocks using the Generalized Hoek Brown failure criteria with parameters including UCS data, GSI, mi, disturbance factor (D), intact modulus (Ei) with the modulus ratio (MR) approach, the data were then analyzed using the Roclab application so that the Hoek-Brown Criterion output was obtained in the form of constants mb, s, a, and young rock mass modulus (Erm). These parameters were obtained from engineering geological characterization, analysis of rock masses quality, and laboratory test results. Then the material properties are inputted into the modeling on SLIDE software.

The modeling is conducted with the slope condition before and after consolidation grouting. The material properties before grouting are obtained from secondary data at the BP-02 location at a depth of 7.50 - 8.00 m, while the material properties after grouting are obtained from the laboratory test results of the properties index and the triaxial test of the coring sample of CH-01. The results of LEM analysis can simulate the effect of grouting on slope conditions, whether the condition is stable or not.

3. Result and Discussion

3.1. Geological Condition

Surface geological mapping was carried out to determine lithology and geologic features, especially which affected the slope stability. Figure 6 shows three rock units identified in this research location consisting of diorite, andesite, and alluvial deposits. Rock unit dominant is composed of diorite, it is around 85%, including rock unit compose the portal slope of outlet tunnel area. Meanwhile, around 5%

is andesite, and the remaining was an alluvial deposit formed by sedimentation around the Mongiilo River. Diorite has a whitish brown color, phaneritic, holocrystalline, equigranular, and the mineral composition of plagioclase, silica, biotite, oxide minerals, and quartz. Andesite has a gray color, aphanitic, holocrystalline, and equigranular.

The main geological structure found are strike-slip faults which are evidenced by the appearance of features such as slickenside, striation, and stepping in outcrop. This fault consists of dextral strike-slip faults and sinistral strike-slip faults. This fault might control the presence of joints in the rock masses. In addition, discontinuities such as joints allow the passage of surface water into crevices, which will affect slope stability. The joints found were tension joints and shear joints. Tension joints are quite often found in several outcrops, while shear joints are only found in a few locations. Other geologic features such as groundwater conditions might affect slope stability and the result of grouting, but the elevation of groundwater is only obtained from interpretation from bore log data, as shown in Figure 10.

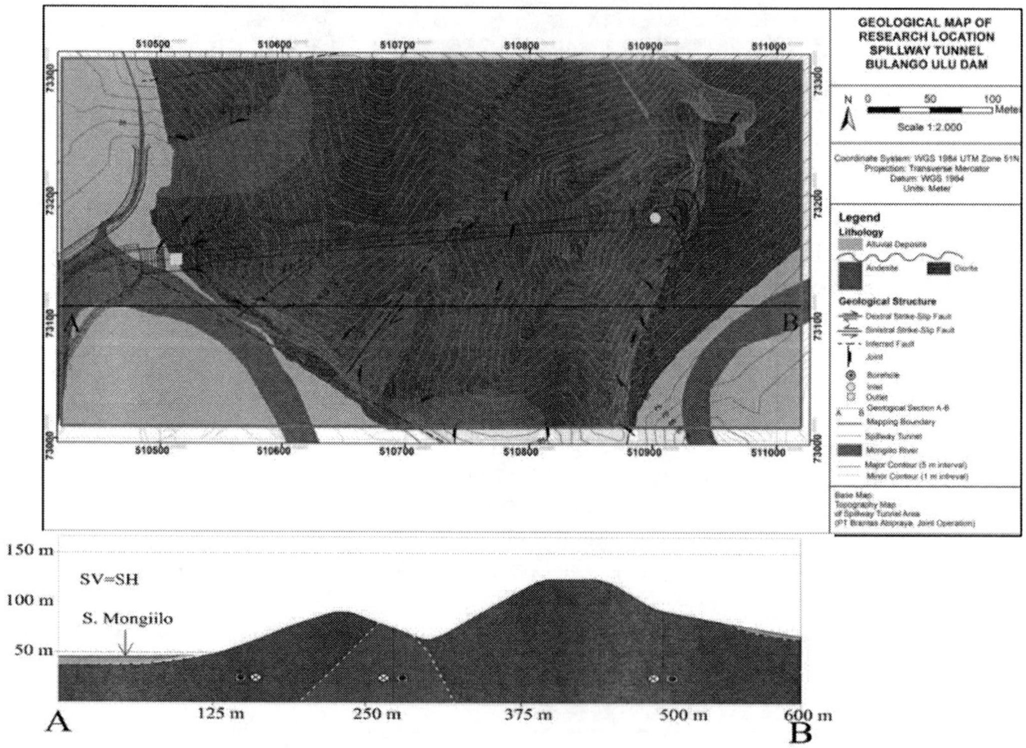

Figure 6. Geological map and geological cross section of research location

3.2. Rock Mass Characterization

Observation zone assessment of surface rock mass quality is carried out based on RMR and GSI. Five outcrop stations were selected in the area of the outlet slope portal that has not been shotcrete, namely STA 01-LP01, STA 01-LP02, STA 01-LP03, STA 02-LP01, and STA 02-LP02 see in (Figure 4b).

Based on field mapping results in the outlet spillway tunnel portal area, the rock lithology that composes the slopes is dominated by highly weathered diorite. According to the RMR and GSI evaluations, the diorite on the slope area has poor to very poor quality, as shown in Table 2, Table 3, and Table 4. The estimated average UCS value for intact rock is 1-5 MPa, some locations, such as in STA 01-LP01, STA 01-LP03, STA 02-LP01, and STA 02 LP02, rocks are very easy to pick up by hand, but core stone can still be found in several locations. The discontinuities in the form of joints are generally spaced tightly to very tight, with an aperture of 1-5 mm having soft clay mineral filling material, average rock roughness are smooth, and dry slope conditions.

The lithological description of the slopes can be seen more clearly at STA 02-LP02 (Figure 7), which is next to ±10 m from the spillway outlet. The lithology of the slopes consists of 1-2 m thick residual soil and highly to completely weathered diorite rock. The estimated UCS value is <1 MPa, while the discontinuity spacing is <0.06 m. The discontinuity condition has a 3-10 m length with a separation of <0.06 m with a soft clay gauge of >5 mm. The roughness of discontinuity is very smooth, and the conditions in the slope area are relatively damp. The RMR value is 20 (very poor rock mass quality).

Based on the GSI, this highly to completely weathered diorite rock has a disintegrated rock structure condition with a very poor surface condition. A GSI value ranging from 10-15 is in the very poor category (<20).

Figure 7. (a) Location of STA 02-LP02 (b) The condition of the weathered diorite becomes soil, but the rock structure and minerals are still visible

The nearest borehole point on the portal slope of the outlet spillway tunnel is BP-02. Evaluation and analysis at the borehole BP-02 were carried out through observations on the core box to determine lithological characteristics and rock mass quality based on GSI by Hoek et al. [15]. The result is that the lithology of the slope of the outlet spillway tunnel consists of residual soil, very poor quality diorite, poor quality diorite, and fair quality diorite seen in Table 1. On the slopes before shotcrete (Figure 4a), the rock lithology conditions were relatively the same as the BP-02 borehole results which can be seen in Figure 8.

Table 1. Rock masses quality at borehole BP-02

Depth (m)	Lithology	J_{cond}	RQD	GSI	Rock Mass Quality
0-4	Residual Soil	0	0	0	-
4-25	Diorite	9	0	14	Very Poor
25-34	Diorite	15	0	22.5	Poor
34-35	Diorite	24	0	36	Fair

Table 2. RQD value calculation on the slope of outlet spillway tunnel area

STA	Lithology	Weathering Degree	The number of joints (n)	Scanline length (L)	λ = n/L	RQD = 100*(0.1λ+1)*e$^{-0.1λ}$
STA 01-LP01	Diorite	Highly Weathered (HW)	34	3	11.33	16
	Diorite	Slightly Weathered (SW)	25	3	8.33	27
STA 01-LP02	Diorite	Highly Weathered (HW)	40	3	13.33	11
STA 01-LP03	Diorite	Highly-Completely (HW-Completely)				
	Diorite	Highly-Moderately (HW-MW)	26	3	8.67	25
STA 02-LP01	Diorite	Highly Weathered (HW)	36	3	12.00	14
STA 02-LP02	Diorite	Highly-Completely (HW-Completely)				

Table 3. Assessment of surfaces rock masses quality on the slope of outlet spillway tunnel area based on RMR [11]

STA	UCS (Mpa)	RQD (%)	Spacing (m)	Joint Discontinuity [11]					GW condition	RMR Total Value	Rock Mass Quality (RMR)
				Persistence (m)	Aperture (mm)	Roughness	Gouge	Weathered			
				Value	Value	Value	Value	Value	Value		
STA 01-LP01	1-5	16	0,06-0.2	1-3 m	1-5	smooth	<5mm, Soft Clay	HW	Dry	36	Poor
	25-50	27	<0.06	1-3 m	0.1-1	rough	None	SW	Dry	56	Fair
STA 01-LP02	1-5	11	0,06-0.2	1-3 m	1-5	smooth	<5mm, Soft Clay	HW	Dry	36	Poor
STA 01-LP03	<1	0	<0,06	3-10 m	>5	very smooth	>5mm, Soft Clay	HW-Completely	Damp	20	Poor
	5-25	25	<0,06	1-3 m	0.1-1	smooth	<5mm, Soft Clay	HW-MW	Damp	39	Poor
STA 02-LP01	1-5	14	0,06-0.2	1-3 m	1-5	smooth	<5mm, Soft Clay	HW	Dry	36	Poor
STA 02-LP02	<1	0	<0,06	3-10 m	>5	very smooth	>5mm, Soft Clay	HW-Completely	Damp	20	Very Poor

Table 4. Assessment of surfaces rock masses quality on the slope of outlet spillway tunnel area based on GSI [14]

STA	GSI [14]		GSI Value	Rock Mass Quality (GSI)
	Structure Condition	Surface Condition		
STA 01-LP01	Blocky/Disturbed-Disintegrated	Poor	20-30	Poor
	Very Blocky - Blocky/Disturbed	Good	45-50	Fair
STA 01-LP02	Blocky/Disturbed-Disintegrated	Poor	20-30	Poor
STA 01-LP03	Disintegrated	Very Poor	10-15	Very Poor
	Blocky/Disturbed	Fair	30-35	Poor
STA 02-LP01	Blocky/Disturbed-Disintegrated	Poor	20-30	Poor
STA 02-LP02	Disintegrated	Very Poor	10-15	Very Poor

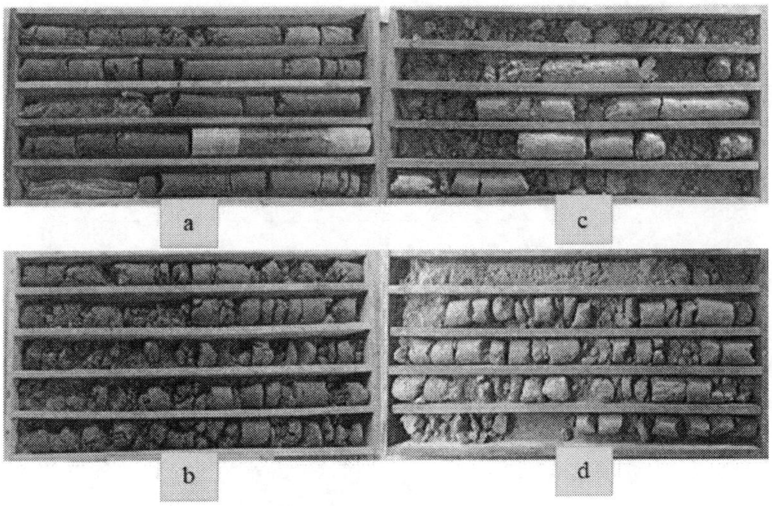

Figure 8. (a) core box conditions at a depth of 0-5 m (residual soil at a depth 0-4 m) (b) core box conditions at a depth of 15-20 m (diorite very poor at a depth 4-25 m) (c) core box conditions at a depth of 26-30 m (diorite poor at a depth 25-34 m) (d) core box conditions at a depth of 30-35 m (diorite fair at a depth 34-35 m)

3.3. Shear Strength Parameters After Grouting

For a preliminary investigation of the effectiveness of grouting implementation, samples at the check hole were taken at the age of 7 days after grouting injection. The number of samples is 3 samples (1 set) were tested for index properties and UU triaxial test, then compared with the test results before grouting from secondary data at the BP-02 location at a depth of 7.50 - 8.00 m to determine the change in shear strength parameters after grouting. The compression pressure (confinement) (σ_3) was used the same as the UU triaxial test of the before grouting sample, namely 1 kg/cm^2, 2 kg/cm^2, and 3 kg/cm^2 for each sample.

Data from the index properties and UU triaxial test shows a change in the value of dry density, and internal friction angle, while the cohesion value is the same before and after grouting. The dry density value increased from 1.819 g/cm^3 to 2.01 g/cm^3 (Table 5), the internal friction angle increased from 15.79° to 20.45°, while the cohesion value is the same (Table 6). The test result showed that there was no significant increase in shear strength parameters (cohesion value and internal friction angle), This shows that the grouting is said to be ineffective and the grouting material injected after seven days indicates that the grouting material is still wet and has not yet hardened. Therefore, sampling needs to wait for the grouting material to have a sufficient age to harden. For example, 28 days after grouting is injected (normal concrete age to achieve the maximum compressive strength) so the sample can be re-evaluated.

Table 5. Index properties test results

Parameters	Sample	
	Before Grouting	**After Grouting**
	BP-02 (Depth: 7.50-8.00 m)	CH-01
Dry Density (ϒdry) (g/cm3)	1.819	2.01
Saturated Density (ϒsat) (g/cm3)	2.158	2.31
Natural Water Content (%)	15.34	8.19
Degree of Saturation (%)	82.20	55.47
Void Ratio (e)	0.514	0.42

Table 6. UU triaxial test results

Type of Sample	Test Piece No.	σ_3 kg/cm^2	σ_1 kg/cm^2	c kg/cm^2	φ degree
Before Grouting	1	1	2.507		
BP-02 (Depth: 7.50-8.00 m)	2	2	4.128	0.181	15.79
	3	3	5.835		
After Grouting	1	1	2.515		
CH-01	2	2	4.591	0.181	20.45
	3	3	6.656		

The comparison of the results of the UU triaxial test before and after grouting can be seen on the triaxial graph in (Figure 9).

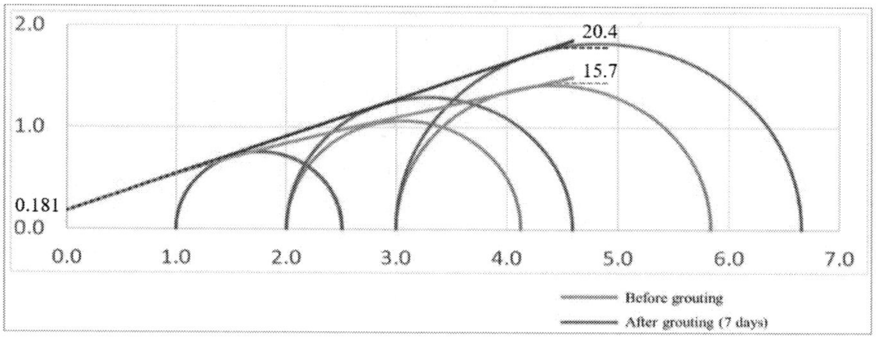

Figure 9. Triaxial graph before and after grouting (7 days)

3.4. Evaluation of slope stability using Limit Equilibrium Method (LEM)

Evaluation for slope stability is conducted with two slope conditions: slope before and after consolidation grouting. The material properties before grouting are obtained from secondary data at the borehole BP-02 location at a depth of 7.50 - 8.00 m seen in Table 7 and Table 8. The material properties after grouting were obtained from the index properties and the UU triaxial test of the CH-01 coring sample seen in Table 5 and Table 6 (the age of 7 days after grouting injection). Then the slope stability is analyzed by the LEM using SLIDE software. The slope modeling of the spillway outlet on SLIDE software is seen in Figure 10.

From the slope stability analysis before grouting, it is obtained safety factor (Fs) = 0.63 <1 (Figure 11a) indicates the slope is still in an unstable condition. It can be concluded that the slope conditions need additional reinforcement so that the slope is in a stable condition.

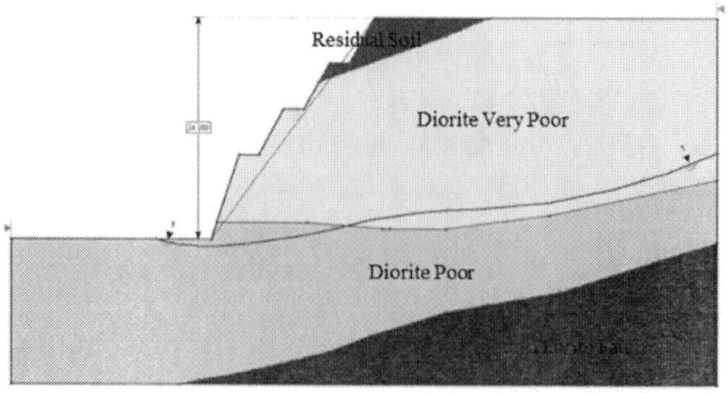

Figure 10. The slope modeling of spillway outlet on SLIDE software

IOP Conf. Series: Earth and Environmental Science 1233 (2023) 012002 doi:10.1088/1755-1315/1233/1/012002

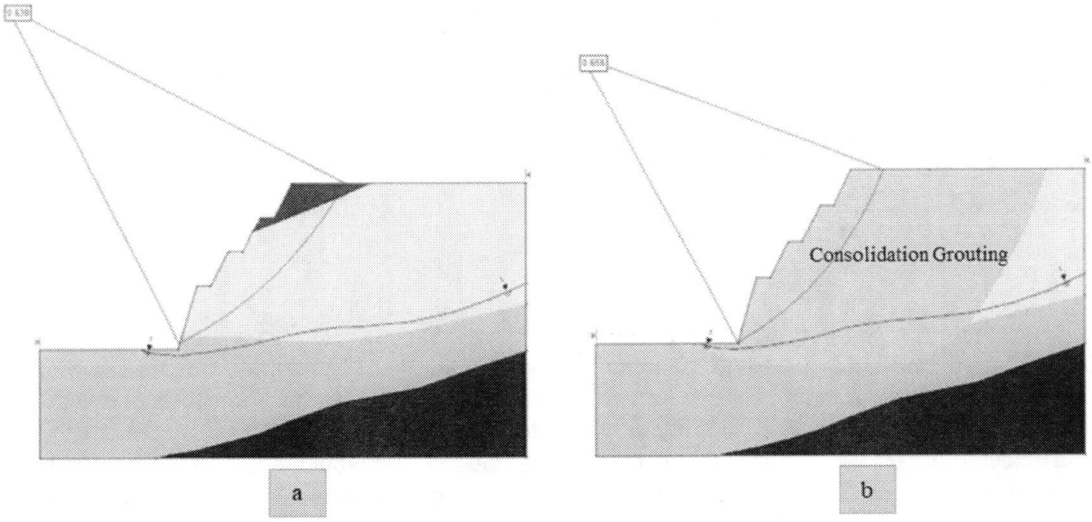

Figure 11. (a) The results of the slope stability before grouting obtained safety factor (Fs) = 0.63 (b) The results of the slope stability after grouting obtained safety factor (Fs) = 0.65

From the slope stability analysis after grouting, it was found that the slope reinforcement with consolidation grouting obtained safety factor (Fs) = 0.65 < 1 (Figure 11b), indicating that the slope was still in an unstable condition. It can be concluded that the insignificant increase in shear strength parameters (cohesion and friction shear angle) at the age of 7 days after grouting injection does not have a significant effect on slope stability.

Zakri et al. [17] evaluate the influence of cement injection to the strength of jointed rock samples using an artificial rock type of moldano tara given the artificial joint with variation Joint Roughness Coefficient (JRC). The different water to cement ratio was injected into the artificial joint. Then, the samples were air dried for 28 days to reach the maximum strength after grouting. The direct shear testing result shows that the grouted samples with low JRC value and cement injection with the ratio of 4W:5C increased cohesion value 398% and internal friction angle 3% from natural samples condition. This result shows the difference in cohesion and internal friction angle after grouting in this research, namely there is no increase in cohesion and internal friction angle increase from 15.79° to 20.45° (increase 29.5%). This is due to differences in sample properties material, grouting samples were taken 7 days after grouting, limitations of tested samples, and geologic features influences such as groundwater conditions.

Table 7. Material properties (Mohr-Coulomb failure criterion)

No.	**Bor Code**	**Depth**	**Lithology**	**Mohr-Coulomb**		
				Unit Weight (γ)	**Cohesion (c)**	**φ**
				kN/m³	**kPa**	**degree**
1	BP-02 (Outlet)	0 - 4	Residual Soil	17.50	30.40	14.54
2		4 - 25	Diorite	17.84	18.10	15.79
3		25 - 34	Diorite	12.06	using Generalized Hoek Brown failure criterion	
4		34 - 35	Diorite	25.57		

Table 8. Material properties (Generalized Hoek-brown failure criterion)

No.	Bor Code	Depth (m)	Lithology	Generalized Hoek-Brown Parameters							Erm
				UCS kPa	GSI	D	mi	mb	s	a	MPa
1	BP-02 (Outlet)	0 - 4	Residual Soil	using the Mohr-Coulomb failure criterion							
2		4 - 25	Diorite								
3		25 - 34	Diorite	4,838.81	22.5	0.7	25	0.353679	1.3245E-05	0.536976	563.44
4		34 - 35	Diorite	19,003.00	36	1	25	0.259	0.00002330	0.515	680.46

3.5. An alternative option of slope reinforcement

Consolidation grouting is a method of strengthening slopes in poor lithological conditions. The results of the grouting sample test taken at the age of seven days after grouting injection showed that the grouting implementation had not significantly affected the slope stability, so it was necessary to wait for a sufficient grout age so that the effectiveness of the grouting implementation that had been carried out could be re-evaluated. As the grouting effect may take significantly long, an alternative to other slope reinforcement that can be applied with similar slope lithological conditions, it can be suggested that slope reinforcement uses the soil nailing method. Based on the results of field mapping, both surface mapping and evaluation of the results of the nearest core drill, show that the lithology of the slopes is dominated by residual soil and diorite rock of very poor to poor rock quality. The slope reinforcement method with soil nailing is suitable for similar conditions.

The reinforcement of the spillway outlet slope with soil nailing was analyzed for stability using the LEM (Limit Equilibrium Method) method with SLIDE v.6.0 software. It was designed according to the SNI 8460:2017 guidelines on geotechnical design (Table 9). The soil nailing method applied is soil nailing with the type of "drilled and grouted soil nailing" where this type of soil nailing is used on permanent structures and is more often applied in the field on relatively high slopes.

Soil nailing is planned to have a nail inclination of 10° with a nail bar length of 25 m and horizontal and vertical spacing of 1.5 m seen in (Figure 12).

Table 9. Soil nailing design based on SNI 8460:2017 [19]

Parameters	Requirements in SNI 8460:2017	Application
Nail bar inclination	10°-20° below the horizontal plane	10° below the horizontal plane
Length of nail bar	0.6 H - 1.2 H, where H = excavation depth or embankment/slope height.	Total slope height (H) = 24.2 m, using length nail bar = 25 m < 1.2 H
Spacing of nail bar	horizontal & vertical spacing 1.5 m for the type of soil nailing "drilled and grouted soil nailing"	horizontal & vertical spacing 1.5 m

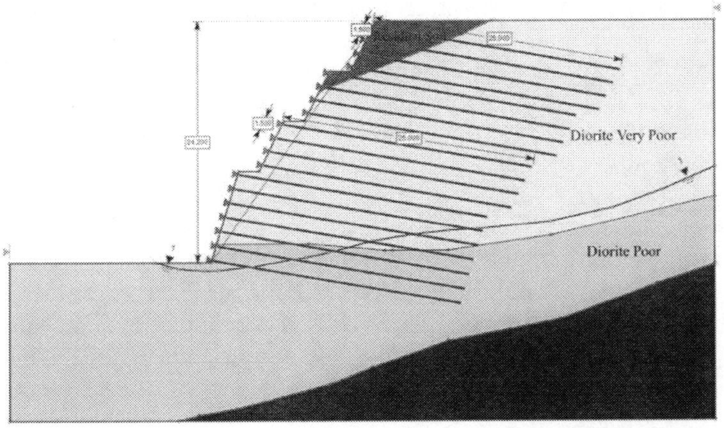

Figure 12. Slope reinforcement modeling with soil nailing on SLIDE software

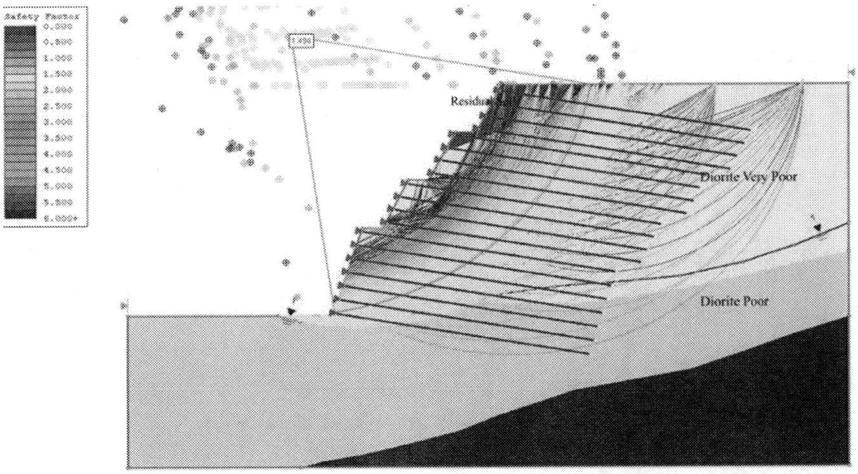

Figure 13. The results of slope stability analysis with the soil nailing using the LEM analysis

From the results of slope stability analysis with soil nailing using the LEM method obtained safety factor (Fs) is 1.49 > 1 (Figure 13), indicating the slope is in stable condition.

4. Conclusion and Recommendations

Based on the results of field mapping, both surface mapping using RMR and GSI parameters and evaluation of the results of the nearest core drill (BP-02) show that material comprising the slope is dominated by residual soil and diorite rock having poor to very poor rock masses.

The analysis results of the slope stability of the outlet spillway tunnel using the LEM on the before grouting condition obtained safety factor (Fs) = 0.63 <1, indicating that the slope is still in an unstable condition so additional slope reinforcement is needed.

The results of the UU triaxial test and the index properties of the after grouting samples were taken at the age of seven days after grouting injection showed that: the same cohesion value results before and after grouting are 0.181 kg/cm^2, while the internal friction angle increased from 15.79° to 20.45°, with a value increase that is not too significant and does not significantly affect slope stability. The results of the slope stability analysis using the LEM (material properties of the age of 7 days after grouting injection) obtained safety factor (Fs) = 0.65 < 1, indicating the slope is still in an unstable condition. Therefore, sampling needs to wait for the grouting material to have a sufficient age to harden namely 28 days after grouting is injected (normal concrete age to achieve the maximum compressive strength). In addition, more samples need to be tested to confirm the change in shear strength after grouting so that the effectiveness of the grouting that has been done can be re-evaluated.

As the grouting effect may take significantly long, as an alternative to other slope reinforcement that can be applied with similar slope lithological conditions, using the soil nailing method, from the results of slope stability analysis using the LEM method obtained safety factor (Fs) is 1.49 > 1, indicating the slope is in stable condition.

Acknowledgments
We would like to express our gratitude to BBWS Sulawesi II for allowing us to conduct research at the Bulango Ulu Dam Project. We would like also to express our deepest gratitude to PT Hutama Karya (Persero) who has given all its support to provide opportunities to gain knowledge and continue education and research through the superspecialist master's program organized by the Ministry of Public Works and Public Housing in 2021.

References

[1] Apandi T and Bachri S 1997 *Peta Geologi Regional Lembar Kotamobagu, Sulawesi (Geological Map of The Kotamobagu Sheet, Sulawesi)* Bandung

[2] Wyllie D C and Mah C W 2017 *Rock slope engineering: Civil and mining 4th edition* p 1–432 doi: 10.1201/9781315274980

[3] Juhari A S and Indrawan I G B 2021 *The Engineering Characteristics and Classifications of Rock Masses along Road Section from Prambanan to Patuk Yogyakarta Indonesia* vol 6 no 2 p 119–127

[4] Basahel H and Mitri H 2017 *Application of rock mass classification systems to rock slope stability assessment: A case study J Rock Mech Geotech Eng* vol 9 no 6 p 993–1009 doi: 10.1016/j.jrmge.2017.07.007

[5] Geospatial Information Agency 2022 *Map of Gorontalo Province* https://tanahair.indonesia.go.id/portal-web/download/perwilayah

[6] PT Brantas Abipraya-PT Bumi Karsa-PT Istaka Karya KSO 2019 *Gambar kerja (Shop Drawing) Pekerjaan Pembangunan Bendungan Bulango Ulu Paket II (MYC) Kab Bone Bulango Provinsi Gorontalo*

[7] Google Earth 2022 *Sulawesi Island & Gorontalo Province region* https://earth.google.com/web/ (accessed Jun. 01, 2022)

[8] Abramson L W, Lee T S, Sharma S and Boyce G M 2001 *Slope Stability and Stabilization Methods* Second Edition vol II United States of America: A Wiley-Interscience Publication John Wiley & Sons Inc

[9] Henn R W 1996 *Practical Guide to Grouting of Underground Structures* 1st ed New York United States of America: ASCE Press American Society of Civil Engineer

[10] Lees D 2006 *The art of grouting and ground treatment* March 2006

[11] Bieniawski Z T 1989 *Engineering Rock Mass Classification: a complete manual for engineers and geologists in mining, civil, and petroleum engineering* A John Wiley & Sons Ltd

[12] Hoek E and Brown E T 1998 *Practical of Estimates or Rock Mass Strength Int J Rock Mech Min*

Sci Geomech Abstr

[13] Priest S D and Hudson J A 1976 *Discontinuity Spacing in Rock Int J Rock Mech Min Sci Geomech* vol 13 p 135–148

[14] Marinos P and Hoek E 2000 *GSI: A geologically friendly tool for rock mass strength estimation ISRM Int Symp 2000*

[15] Hoek E, Carter T G and Diederichs M S 2013 *Quantification of the geological strength index chart 47th US Rock Mech Geomech Symp 2013* vol 3 p 1757–1764

[16] Li Z, Liu J, Xu R, Liu H and Shi W 2021 *Study of grouting effectiveness based on shear strength evaluation with experimental and numerical approaches Acta Geotech* vol 16 no 12 p 3991–4005 doi: 10.1007/s11440-021-01324-4

[17] Zakri R S, Simangunsong G M and Widodo N P 2019 *Effect of Cement Injection on Shear Behaviour of Jointed Rock based on Mohr-Coulomb Criterion* March 2019

[18] Juven D and Herman G 2016 *Analisa Stabilitas Lereng Dengan Limit Equilibrium dan Finite Element Method*

[19] Badan Standarisasi Nasional 2017 *Sni 8640-2017* vol. 8460

IOP Conf. Series: Earth and Environmental Science 1233 (2023) 012003 doi:10.1088/1755-1315/1233/1/012003

Analysis of the Tunnel Face Failure during the Excavation and Evaluation of Excavation Method and Sequential on Ciuyah Tunnel, Lebak Regency, Banten Province

Asyfak Bismoseno[1,2], I Gde Budi Indrawan[1*], and Maulidya Indah Junica[1,2]

[1]Department of Geological Engineering, Faculty of Engineering, Gadjah Mada University, Yogyakarta, Indonesia
[2] Ministry of Public Works and Housing, Jakarta, Indonesia

Abstract. The difference in geological and geotechnical conditions during design and implementation are frequent due to the lack of data collection and diverse rock variations. An important step in designing the excavation method, stages, and support system are determining the rock mass's actual condition. Inaccuracy in evaluating rock mass quality can cause a collapse during tunnel excavation. This study used an example of a collapse case on the inlet side of the Ciuyah tunnel. The classification system used in the design was the JSCE 1975, but collapses occurred at several locations during excavations. An additional assessment of the tunnel stability was carried out by doing surface surveys and evaluations in STA 0+123 using the RMR_{89}, GSI, and JSCE 2016 classification systems. The rock mass in this location consists of tuffaceous claystone and tuffaceous sandstone with very poor quality. The empirical evaluation showed that Daraei's multigraph could be used to determine the collapse pattern for very poor rock. For the excavation method, stages, and support systems, it is preferable to apply the RMR classification system with consideration for the maximum stand-up time. This research aims to provide information that may be used for other projects with similar geological circumstances.

1. Introduction

Cases of collapse in tunnels usually occur due to unstable conditions of the surrounding environment, either due to groundwater seepage [1], geological conditions [2-3], or due to excavation methods and inappropriate support systems [4-5]. The collapse during excavation generally begins with the deformations around the rock excavated and still has no support system. The deformations around the rocks excavated in the construction of tunnels are strongly influenced by the excavation method and the excavation stages [6]. Determination of excavation methods and stages is based on geological conditions, rock mass quality, tunnel dimensions, the level of safety, cost, and implementation time. The tunnel's stability during the excavation and operation execution is important in designing the tunnel [7].

Planning of tunnel structures is usually carried out using empirical and numerical methods. Still, rock engineers and geologists usually only use empirical methods in evaluating construction implementation because of their practical and fast [8]. The Rock Mass Rating (RMR) and Geological Strength Index (GSI) are the rock engineering classification system with empirical methods that have been widely used. The RMR method is commonly used in evaluating the quality of rock masses and support systems; even in 2014, it was recorded that 2,298 tunnels already used this method [9]. These two classification systems have a relationship in assessing the quality of rock masses, so the value of

IOP Conf. Series: Earth and Environmental Science 1233 (2023) 012003 doi:10.1088/1755-1315/1233/1/012003

one can be used as an alternative in determining the other value [10]. Another classification system that is widely used is the Japan Society of Civil Engineers (JSCE), which can accommodate general engineering needs for mountain tunnels (road tunnels, train tunnels, and water tunnels) [11]. Several projects in Indonesia also use the JSCE classification system, including the construction of the Cisumdawu tunnel in West Java [12].

Research on tunnel face failure during excavation has been carried out, and many methods are recommended to overcome this. However, it is still important to continue researching the causes of collapse in poor or very poor-quality rocks and recommend appropriate countermeasures to increase literacy and knowledge. To further explore the problems in tunnel excavation, this paper will examine the occurrence of collapse on the inlet side of the Ciuyah tunnel (Figure 1). The design of the support system and the ciuyah tunnel excavation method were carried out using the JSCE 1975 classification system [13]. [14], [15], and [16] recommend using two or more classification methods of rock masses and at least one equation showing the relationship between them to validate the results. The classification systems used to verify the previous results are GSI, RMR_{89}, and JSCE 2016. The benefits obtained from this study are information and references for evaluation if similar problems arise, as well as providing recommendations for excavation methods and support systems that are safe and optimal in terms of implementation and cost by using data from geological surveys on the surfaces and inside the tunnels.

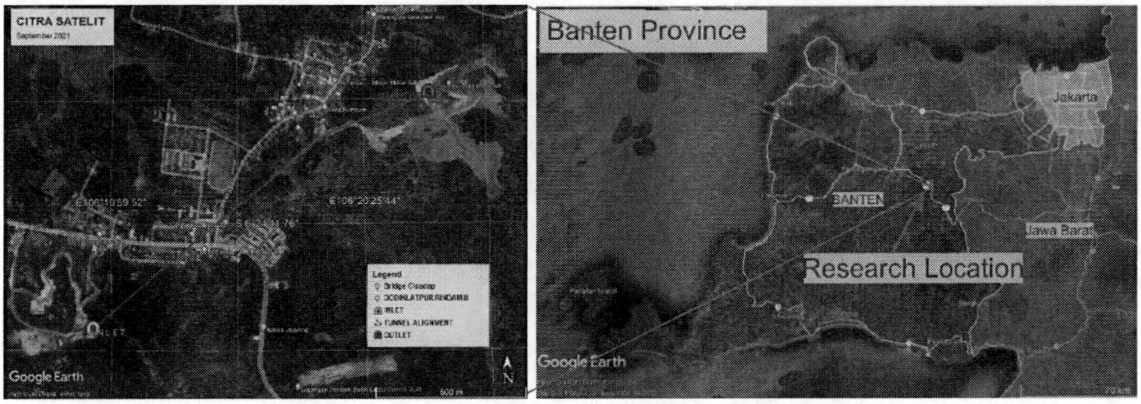

Figure 1. Research Location (source: Google Earth)

2. Ciuyah Tunnel

The construction of the Ciuyah tunnel has been carried out since 2019. It is part of the Karian Dam Development project, which has the function of draining water from the dam to the treatment plant. The completion time is expected before the dam construction is completed. Based on the 1991 Geological Map of Serang, Java by [17] at a scale of 1:100,000, the location of the Ciuyah tunnel consists of sandstone, claystone, and lapilli, which are included in the Genteng Formation. This rock formation was formed between the Late Pliocene and Early Miocene.

Even though the support system was used following the design, a collapse still occurred from the top and front of the tunnel excavation face. It happened after the excavation at STA 0+123 was completed and the steel rib was installed (Figure 2). Subsequent failures occurred at STA 0+123.75 (Figure 3), STA 0+133.5, STA 0+143.1, and STA 0+361.8, where steel supports were installed, but wire mesh and shotcrete had not been implemented [18]. The collapse problem is generally caused by poor rock lithological conditions and excess water pressure [19], geological structures in the rock mass [20], improper excavation systems, or inappropriate support systems [21].

IOP Conf. Series: Earth and Environmental Science 1233 (2023) 012003 doi:10.1088/1755-1315/1233/1/012003

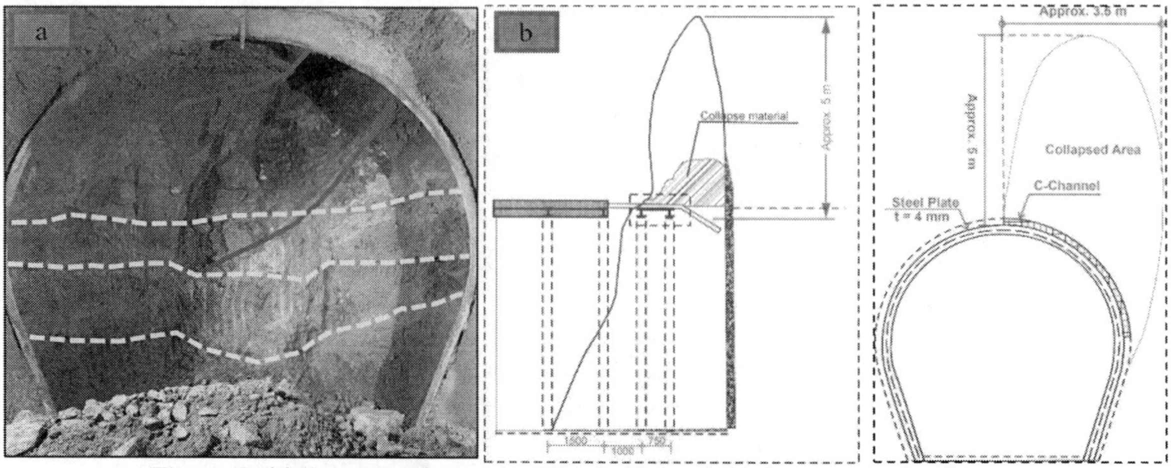

Figure 2. (a) Face Mapping at the STA 0+123 and (b) the pattern of collapse

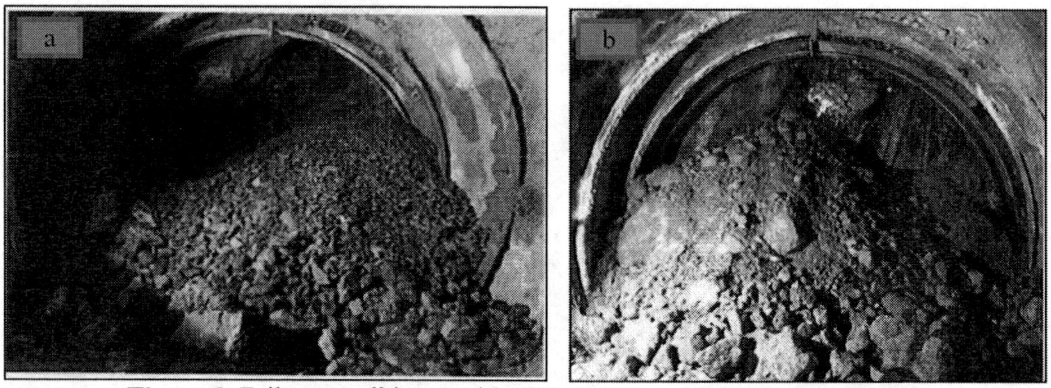

Figure 3. Failure conditions at (a) STA 0+123 and (b) STA 0+123.75

The design data was commonly gained from observations of surface geological conditions and laboratory testing of rock samples. It may not correctly reflect the situations during tunnel construction because once construction starts, we can see a lot of outcrop from surface excavation (example in figure 4), so adjustments for excavation method and tunnel support are still needed. Evaluation of geological conditions to obtain rock mass quality and suitable support systems is necessary to determine the safety and optimal use of costs.

Figure 4. Rock outcrops from excavation; (a) tuffaceous claystone, (b) tuffaceous sandstone

3. Research Methods

3.1. Data Collection and Sampling

Investigations of geological engineering conditions that have been carried out were boring tests and numerical analysis for tunnel design in 2011 [22], and additional investigations that have been carried out in 2019 are 2D Geoelectric surveys and geological surface surveys [23].

Field investigations had been conducted, such as geological surface surveys (Figure 5), analysis of cross sections from borehole data (Figure 6), and diamond coring at three points on the shotcrete-covered tunnel wall (Figure 7). The coring was carried out to determine the actual rock conditions inside the tunnel. Each coring sample was tested in the laboratory to obtain rock engineering parameters and Universal Compressive Strength (UCS) values. All data (primary and secondary data) were evaluated and grouped based on lithology, technical parameters, geological structure, and weathering level to obtain rock mass quality values using the GSI, RMR, and JSCE methods.

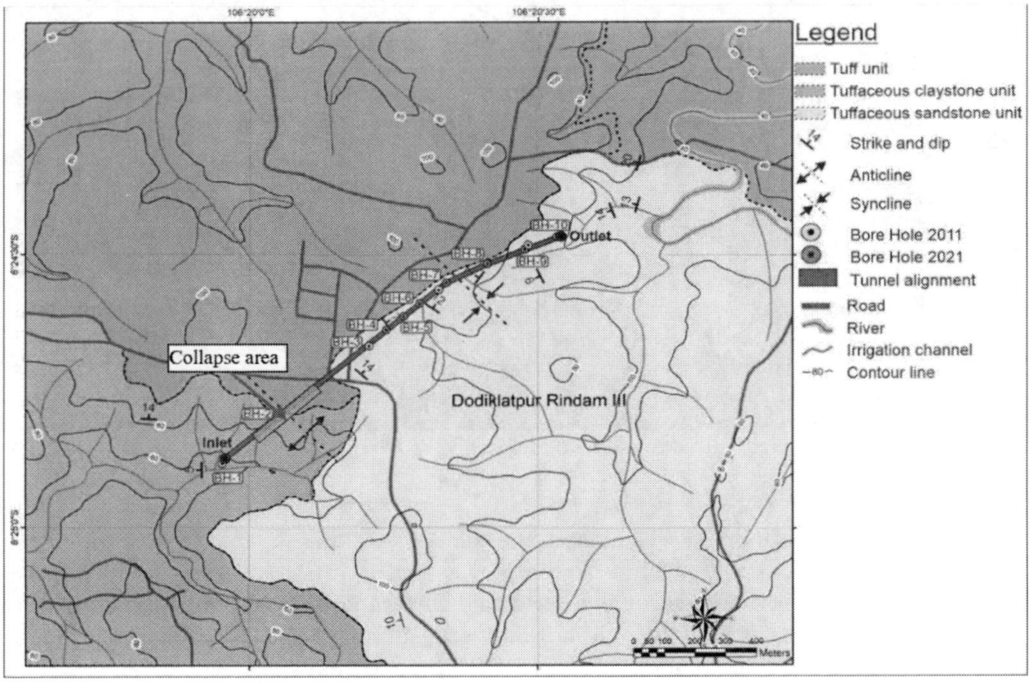

Figure 5. Geological Map from surface surveys

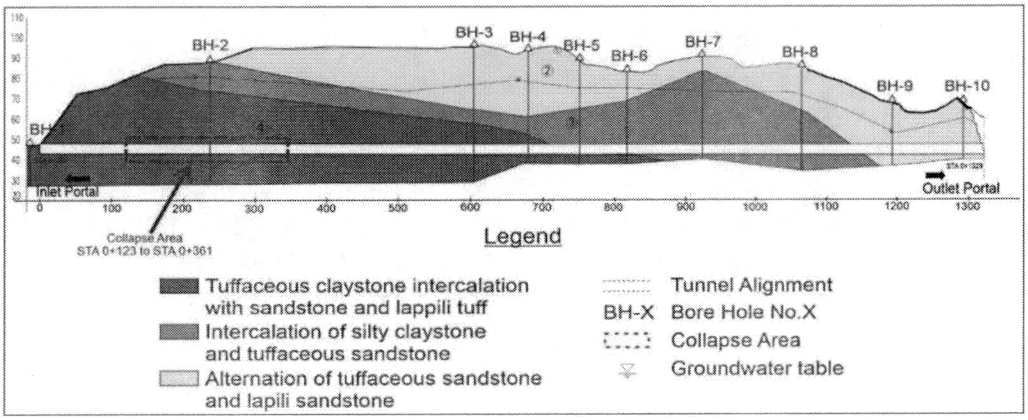

Figure 6. Cross section of Ciuyah Tunnel

The surface survey result showed the presence of folds of rocks that formed anticlines on the side of the inlet and syncline on the side of the outlet. The collapsed tunnel area was in a tuffaceous claystone unit with tuffaceous claystone intercalating with sandstone and lapilli tuff as its lithology. The tunnel condition is below the groundwater level, so it requires additional effort to drain the water out of the tunnel during the excavation. The water that comes from the excavation face is also an indicator of the potential rock mass collapse because water infiltrates through the joint or bedding layered. Besides that, the properties of tuffaceous claystone and sandstone are very easy to absorb and release the water so that the strength can reduce rapidly.

Figure 7. Taking samples with diamond coring (a) STA 0+122, (b) STA 0+128 and (c) STA 0+138

3.2. Rock Mass Quality in GSI, RMR, and JSCE

The Geological Strength Index (GSI) method introduced by [24] is suitable for determining rock mass quality in poor conditions with a high degree of weathering. The rock mass quality is assessed by observing the surface rock outcrop or the excavated surface condition, which includes 2 (two) main parameters, namely the structural condition of the excavation face and its surface conditions. The GSI rating obtained is then classified into five classes according to table 1.

Table 1. Quality of rock mass based on GSI [25]

GSI Rating	95-76	75-56	55-36	35-21	<20
Rock Mass Quality	Very Good	Good	Fair	Poor	Very Poor

GSI assessment can also be done using the formula from [26] as follows:

$$GSI = 1{,}5.JCond_{89} + {}^{RQD}/_{2} \tag{1}$$

Value of $JCond_{89}$ is obtained from observing the joint condition, then the rating value can be determined according to the joint condition (with or without infilling) [26]. The Rock Quality Design (RQD) value is obtained using the equation from [27], by observing and measuring the distance between joints and the number of sets in rock outcrops in the site with the following formula:

$$\text{RQD} = 110\text{-}2.5*\text{Jv} \tag{2}$$

where for joints with random sets using the equation,

$$J_v = \sum_{i=1}^{J} \left(\frac{1}{S_i}\right) + \frac{N_r}{5\sqrt{A}} \tag{3}$$

The value of Si is determined based on the average distance per joint set (meters), Jv is the total number of joint sets excluding the number of random joint sets, Nr is the number of random joint sets, and A is the area in m2.

RMR assessment data is obtained from the surface and inside the tunnel. During the excavation process, the contractor performed an RMR assessment of the excavation face area and obtained approval from the supervising consultant. As additional information, an RMR assessment of surface rocks will also be carried out using the equation:

$$RMR = R1+R2+R3+R4+R5 \tag{4}$$

Where R1 = UCS; R2 = RQD; R3 = discontinuity spacing; R4 = discontinuity conditions and R5 = groundwater conditions. The UCS value is obtained from the results of laboratory tests or from field observations which are equivalent to standard rock strength values [28]

The JSCE method classifies rock masses into five classes: B, C1, C2, D1, and D2. This grouping is categorized based on the value compressive strength (UCS), discontinuity, rock type, elastic wave velocity, geological condition, RQD, rock competence factor, tunnel condition, and rock deformation's standard value. The assessment by using this method can be seen in the ground classification table on Standard Specifications for Tunneling-2016: Mountain Tunnels by JSCE.

3.3. Excavation method and Support system based on RMR and JSCE (2016)
The tunnel excavation method is a crucial construction method to ensure the stability of the excavation face before the support system is installed. The collapse at STA 0+123 is an example of an inappropriate tunnel excavation method because the time required to complete all support systems exceeds the stand-up time limit of the rock around the tunnel. The tunnel excavation method in the design was determined using the excavation system classification by [29], where the evaluation results showed the use of the ripping method in conducting tunnel excavation. The evaluation will be carried out using another method, the excavation chart, revised by [30]. This excavation graph is used to determine the type of excavation equipment and method based on engineering geological parameters such as discontinuity spacing index (If), rock strength class, and point load strength index (Is (50)). An approach was used to determine the stand-up time of rock without support based on a graph of the relationship between excavation distance, support installation time, and rock mass quality (RMR) from [16].

The excavation stages are also part of the tunnel excavation method, where the number of excavation stages and the correct sequence of excavations will increase the stand-up time of the rock. The excavation step method was analyzed using a multigraph from [31] by plotting the Secant Modulus data, strain values, rock strength factors, and excavation lengths in graphs. JSCE (2016) also explains the excavation method, stages, suitable rock conditions, and each method's advantages and disadvantages, which will later become the basis for selection. The classification of RMR_{89} and JSCE (2016) also provides recommendations related to the excavation stage, which is an integral part of the support system based on the rock mass quality value.

4. Analysis and Evaluation Results

4.1. Rock Mass Quality
Rock outcrops on the surface that share the same lithology as the inner rock of the collapsed tunnel underwent quality investigation of the rock mass. Given the position of the rock mass collapse in the tunnel and the fact that the research was concealed by shotcrete and rendered invisible, both investigations are necessary to give a thorough picture of the rock mass's quality. Rock mass quality is evaluated using the GSI (quantitative and qualitative), RMR, and JSCE procedures on the surface and inside the tunnel.

IOP Conf. Series: Earth and Environmental Science 1233 (2023) 012003 doi:10.1088/1755-1315/1233/1/012003

The results of face mapping at STA 0+123 show that the rock at the tunnel elevation has a lithology of claystone and tuffaceous sandstone with a layering angle of N20°E/ 2°. Besides that, there were joints in the direction of N15°E/63°. Looking at the direction and area of joints in the tuffaceous sandstone layer, the main factor causing the rock mass collapse in the STA 0+123 was the joint discontinuity on the weak plane.

The collapsed rock at STA 0+123 is dominated by small and a few large sections, which corresponds to the surface tuffaceous sandstone outcrop of STA 32 (figure 8). Surface tuffaceous claystone outcrops were also discovered at STA 23 (figure 9). These two outcrops are used to qualitatively calculate the RQD and GSI values by calculating the number of joints and joint spacing in a specific area. The more numerous number and dense joints in one area, the lower the RQD value of the rock.

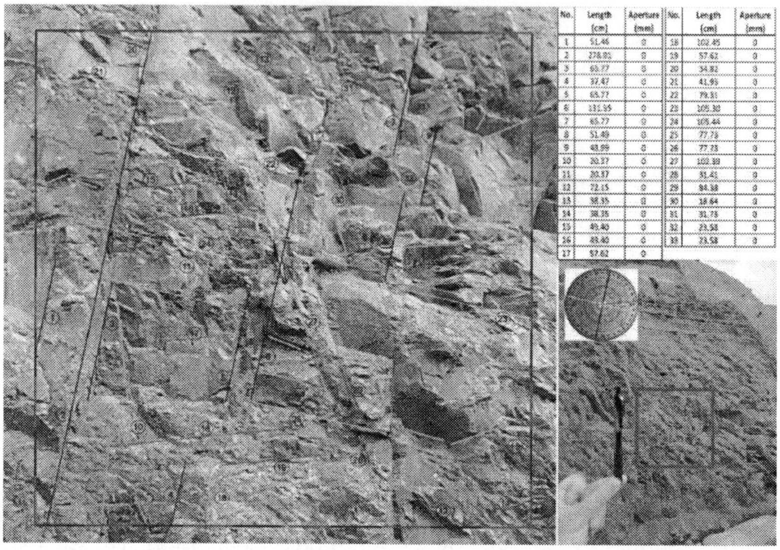

No.	Length (cm)	Aperture (mm)	No.	Length (cm)	Aperture (mm)
1	51.46	0	18	102.45	0
2	278.85	0	19	57.62	0
3	65.77	0	20	34.82	0
4	37.47	0	21	41.95	0
5	65.77	0	22	79.51	0
6	131.95	0	23	105.30	0
7	65.77	0	24	105.44	0
8	51.49	0	25	77.73	0
9	43.99	0	26	77.73	0
10	30.27	0	27	102.38	0
11	30.37	0	28	31.41	0
12	72.15	0	29	94.38	0
13	38.35	0	30	18.64	0
14	38.35	0	31	31.75	0
15	49.40	0	32	23.58	0
16	49.40	0	33	23.58	0
17	57.62	0			

Figure 8. Evaluation of joint sets on surface outcrops on STA 32 (Tuffaceous sandstone)

No.	Length (cm)	Aperture (mm)
1	52.4	0
2	23.7	0
3	75.2	14
4	59.1	0
5	51.6	6
6	58.3	0
7	58.3	10
8	21.4	0
9	71.9	9
10	58.3	10
11	21.5	0
12	37.1	0
13	29.1	0
14	29.1	0
15	55.3	8

Figure 9. Evaluation of joint sets on the surface of STA 23 (Tuffaceous claystone)

IOP Conf. Series: Earth and Environmental Science 1233 (2023) 012003 doi:10.1088/1755-1315/1233/1/012003

To obtain rock mass quality values using the GSI method from [26], it needs to calculate the RQD value for both rock types according to Table 2. An assessment was also carried out using the GSI method from [24]. The analysis showed that tuffaceous sandstone has an RQD value of 7.05 and GSI [25] value is 18.5, but the results of qualitative analysis with the GSI [24] give a value of 20-36. Joint analysis of tuffaceous claystone shows that the rock has an RQD of 5.03 and a GSI value by [26] is 17.5, but the results of qualitative analysis with GSI [24] give a value of 7-19.

Table 2. Assessment of surfaces rock mass quality on the slope of the outlet tunnel is based on the GSI charts proposed by [24] and [26]

Tuffaceous Sandstone						**Tuffaceous Claystone**			
S_i (m)	0.13	0.1	0.22	0.17	0.11	S_i (m)	0.064	0.11	0.08
J_i	7.53	10.23	4.57	5.91	9.14	J_i	7.53	10.23	4.57
Nr			38			Nr		24	
A (m^2)			4			A (m^2)		1	
J_v			41.1795			J_v		41.99	
RQD			7.05			RQD		5.03	
J_{Cond89}			10			J_{Cond89}		10	
GSI (Hoek, 2013)			**18.5 (very poor)**			GSI (Hoek, 2013)		**17.5 (very poor)**	
GSI (Marinos and Hoek, 2000)			**Blocky/Disturbed/Seamy Poor**		**20-36**	GSI (Marinos and Hoek, 2000)		**Disintegrated Very Poor**	**7-19**

Assessment of rock mass quality on the surface was carried out using other methods, namely the RMR method (table 3) and JSCE (table 4). The assessment was carried out based on visual observations, laboratory test results, and tables of relationships between rocks and the classification of compressive strength of rocks in the field from [16].

Table 3. Value of rock mass quality RMR method at STA 32

UCS (Mpa)	RQD (%)	Spacing (m)	Persistence (m)	Aperture (mm)	Roughness	Gouge	Weathered	GW Condition	**RMR Total Value**
1-5 (1)	< 25 (3)	<0.06 (5)	< 1 m (6)	< 0.1 (5)	Smooth (1)	Soft < 5mm (2)	Highly Weathered (1)	Moist (10)	**Poor** **(34)**

Table 4. Value of rock mass quality using JSCE method on STA 32

Class	Sub Class	Rock Type	Rock Name
RQD 10<x<40 **(C II)**	UCS < 20 N/mm^2 **(L)**	The rock which bedding plane or schistosity plane becomes the dominant surface of discontinuity. **(Layered)**	Tertiary sandstone or conglomerate **(Tertiary sandstone)**

Assessment of rock mass quality on the excavation face in the tunnel at STA 0+123 was carried out using the GSI method from [24], RMR, and JSCE. An evaluation was carried out by using observation data from the excavation face (Figures 3 and 4), results of laboratory tests, and the size of the rock mass that collapsed (table 5).

Table 5. Assessment of rock mass quality in the tunnel STA 0+123

GSI (Hoek and Marinos, 2000)

Structure	Surface Condition	Rating
Disintegrated	Poor	**23-38**

RMR (Bienawski, 1989)

UCS (Mpa)	RQD (%)	Spacing (m)	Persistence (m)	Aperture (mm)	Roughness	Gouge	Weathered	GW Condition	Strike Perpendicular to Tunnel Axis	RMR Total Value
5-25 (2)	< 25 (3)	0.06-0.2 (8)	3-10 (2)	0.1-1 (4)	Slickensided (0)	Soft < 5mm (2)	Slightly Weathered (5)	Dripping (4)	P 45°-90° (-12)	**Very Poor** **(18)**

JSCE (2016)

Class	Sub Class	Rock Type	Rock Name
RQD < 10, deformasi 30-60mm	UCS < 20 N/mm^2	The rock which bedding plane or schistosity plane becomes the dominant surface of discontinuity.	Tertiary sandstone or conglomerate
(D II)	**(L)**	**(Layered)**	**Tertiary sandstone**

IOP Conf. Series: Earth and Environmental Science 1233 (2023) 012003 doi:10.1088/1755-1315/1233/1/012003

The quality of rock mass on the surface and in the tunnel using the GSI, RMR, and JSCE methods has differences, where conditions inside the tunnel are worse than conditions on the surface. This difference is due to the poor condition of the excavation surface, tunnel elevation is below the groundwater level, the joint strike direction is parallel to the tunnel excavation direction, and it has a large dip angle. Based on this explanation, the rock mass quality value inside the tunnel was chosen as a reference in determining the support system.

4.2. Rock Mass Excavation Equipment and Tunnel Stand-up Time

Excavations carried out in the field use the ripping method with a twin header tool for the length of the tunnel. The lithology of the excavated face at STA 0+123 consists of tuffaceous claystone (UCS 10.16 Mpa) and tuffaceous sandstone (UCS 13.66 Mpa and 20.34 Mpa). The data result from the Point Load Index test is required to obtain the proper excavation equipment. This value is obtained using the equation of [32], namely UCS = 13.4*Is(50). This equation is used considering the research location, and rock types are identical. The value of the Point Load Index and discontinuity index obtained will be plotted in the excavation graph of Pettifer and Fookes, 1994. The graph of the relationship is shown in Figure 10.

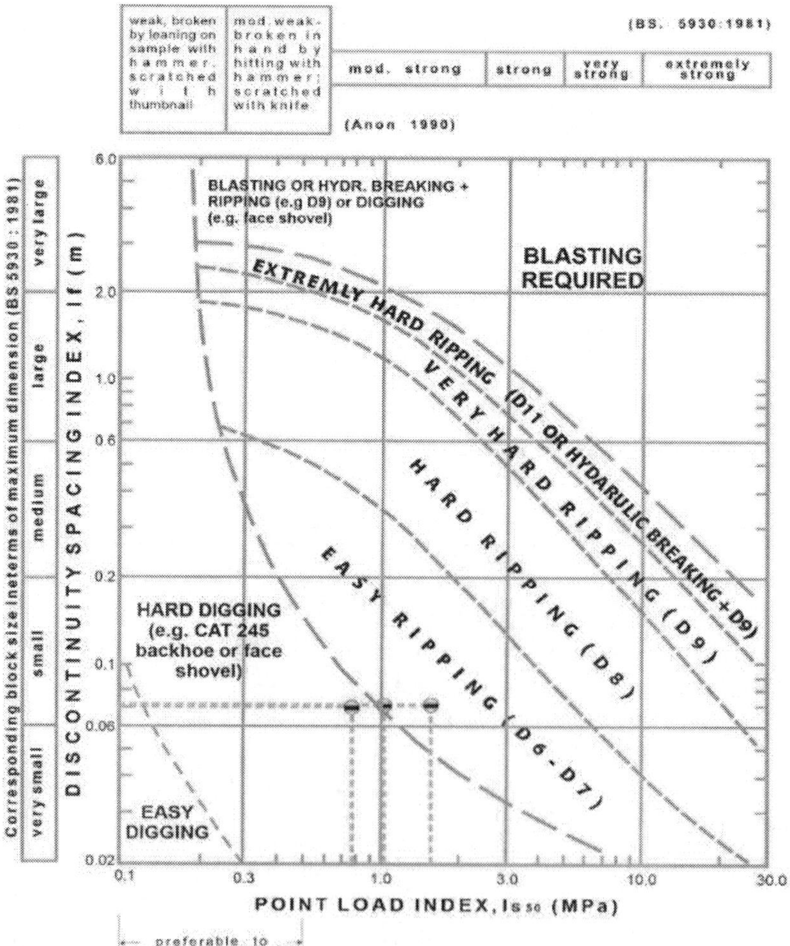

Figure 10. Assessment of the rock mass excavation method from the graph [30]

The plotting results show that rock excavation of tuffaceous sandstone requires an easy ripping method using rippers (D6 or D7), while hard digging for tuffaceous claystone requires hard digging (e.g., CAT 245). The use of inappropriate equipment will reduce the effectiveness of excavation and increase the Disturbance factor. The recommended excavation equipment to extend the rock stand-up time is the equipment that gives a minor impact/disruption to rock stability, namely hard digging equipment in the form of CAT 245.

The excavation span is also essential in determining the potential for collapse. An extensive excavation span will increase the risk of failure. The approach to determining the relationship between excavation distance, support installation time, and rock mass quality (RMR) uses the graph from [16]. The rock mass quality at STA 0+123 is in the very poor category, and the results of plotting this value on the graph of the relationship between RMR and the time of installation of the supports can be seen in Figure 11.

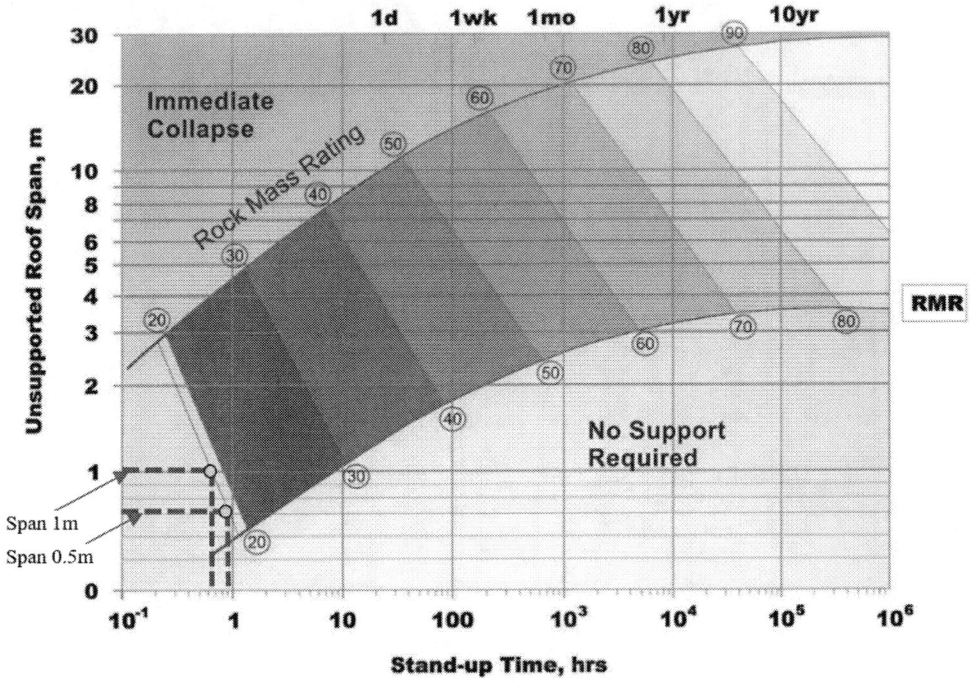

Figure 11. Plots of tunnel excavation distance and stand-up time on the chart proposed by [16]

The plotting graph above shows that for excavations within 1 meter, a support system should be installed within 30 minutes after excavation is complete, while for excavations within 0.5 meters, a support system can be installed within 55 minutes after completion of excavation. The speed of shotcrete installation is also the key to increasing the rock mass stability and extending the tunnel stand-up time.

4.3. Tunnel Excavation Method and Support System

The full face excavation with the face buttress method was chosen as the excavation stage. With this technique, the excavation was done in stages, the first excavation around where the steel ribs were installed, leaving the rock in the core (figure 12). The next step is installing the support system (steel rib, wire mesh, shotcrete, and rock bolt) and excavating the core area.

IOP Conf. Series: Earth and Environmental Science 1233 (2023) 012003 doi:10.1088/1755-1315/1233/1/012003

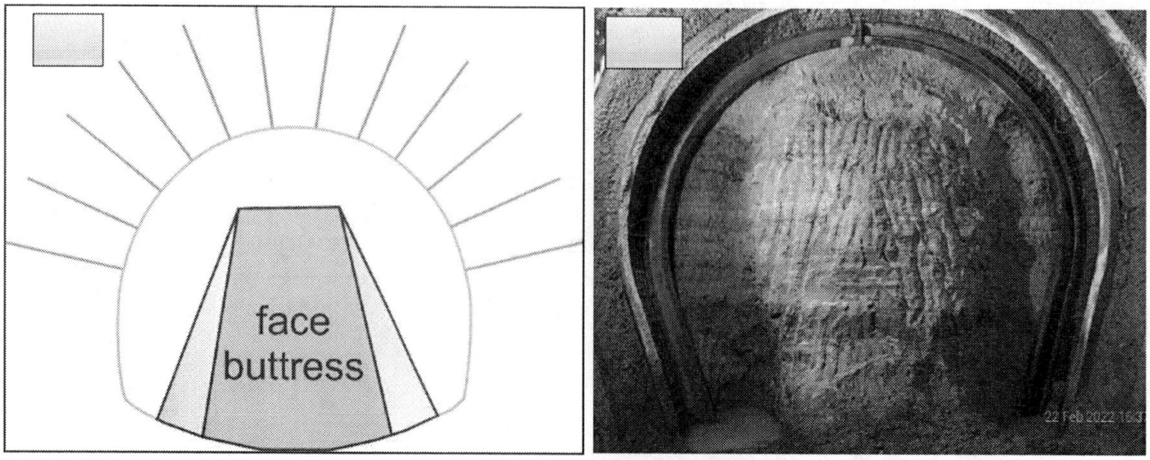

Figure 12. The sketch of the excavation stages (a) Photos of the excavation stages (b)

The design report does not explain the method selection of the full face excavation with face buttress, so analysis is needed to determine the appropriate stages. Analysis of the determination of excavation stages was carried out using a multigraph from [27] by inputting the Modulus Secant (Mod. E_{sec}) data from tuffaceous sandstone (6.71E+04 $^{kg}/_{cm^2}$), and tuffaceous claystone (9.55E+04 $^{kg}/_{cm^2}$) then plotted according to Figure 13.

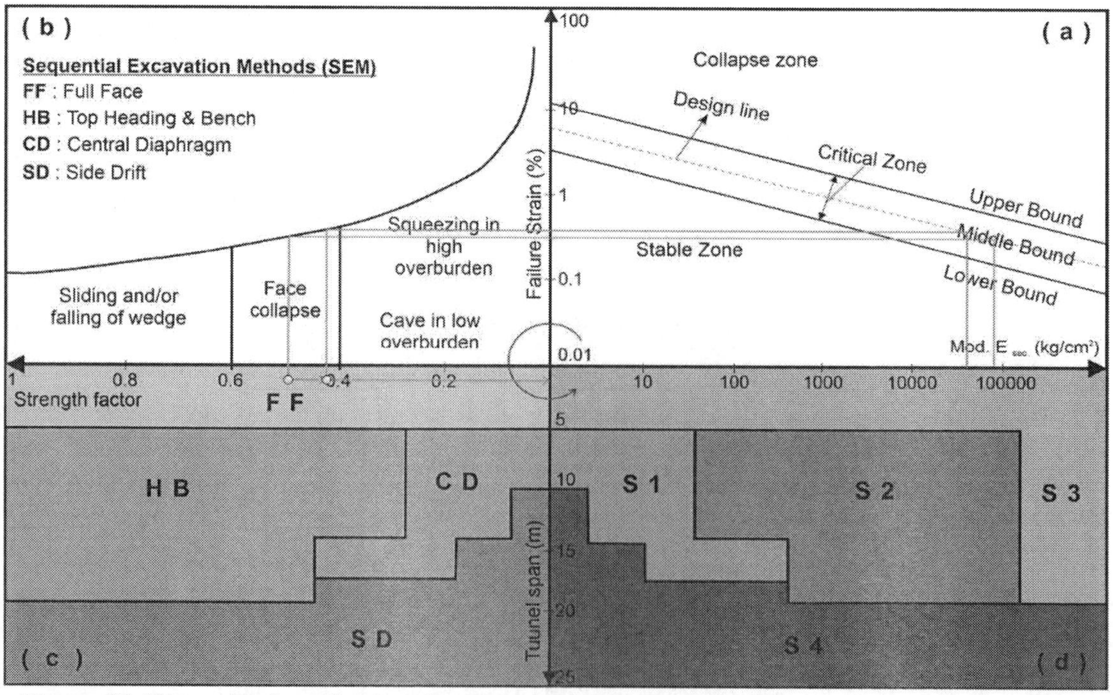

Figure 13. Determination of excavation stages using the multigraph method proposed by [27]

The plotting results of the secant modulus of tuffaceous clay and tuffaceous sandstone indicate that the potential risk of failure is in the form of face collapse. This failure condition corresponds to the collapse at STA 0+123. The strength factor value is then associated with the excavation distance that has been carried out (0.75 m) and shows the results of the full face excavation stage, the same as the method used in the field.

The excavation stages were also analyzed using the RMR and JSCE methods, where the excavation stages were an integral part of the support system. The tunnel support system installed at STA 0+123 (Table 6) is a support system for rock conditions with poor conditions, with details:

Table 6. Typical Support System According to Design

Type	Span Excavation	Steel Rib	Wiremesh	Shotcrete	Rock bolt	Forepolling/ Piperoofing
Type 4	1.0 m	Yes	Yes	Yes	-	Piperoofing Ø73mm, L=9.5m, 14-15 NOS Injeksi Ø60.55mm steel

The implementation of tunnel excavation with a span of 1 meter (type 4 support system) still cannot accommodate the very poor rock mass quality, causing a collapse at STA 0+123. Subsequent installation of the support system was carried out at a smaller distance (0.75 m), but failure still occurred. The second failure indicates that the installed support system still needs to be improved or upgraded in its working method, especially for some instances (e.g., high discontinuity areas). Based on the results from the evaluation of rock mass quality at STA 0+123 using the GSI and RMR, and JSCE methods, the following support systems are recommended (Table 7):

Table 7. Excavation methods and support systems based on RMR_{89} and JSCE (2016)

Methods	Rock Class	Excavation Methods	Support System
RMR_{89}	Very Poor	• Gradual excavation with a distance of 0.5m from the top and immediately installed supports after the excavation is complete.	- Systematic bolts 5-6m long, 1-1.5m spacing at crown and walls with wire mesh and invert - Shotcrete is 150-200mm at the crown, 150mm at the wall, and 50mm at the face of an excavation - Steel sets in the form of medium to Heavy ribs with 0 spacing .75 m with steel lagging and forepolling if required. Installation of close invert.
JSCE (2016)	Class D-II / E	Excavation is carried out every 0.8m	- Shotcrete thickness 20 cm, - Rock bolt length 0.6*De (cm) - Distance between rock bolts in circumferential direction 1m or less - Distance between longitudinal rock bolts 0.8m - Spacing of steel support 0.8m - Lining using reinforced concrete

Note : De = diameter of excavation cross section
For type E, rigid support or allowable deformation should be provided according to ground conditions.

The results of the RMR and JSCE evaluations recommend the same excavation method, namely in stages with a bench cut. The shotcrete installation must be carried out immediately after each stage's excavation. The purpose of the rapid installation of shotcrete is to increase the strength of the rock before all support system is completed. The excavation distance with the RMR method is shorter than the JSCE method. However, the JSCE method uses a face bolt in the face of the excavation to maintain stability during excavation.

5. Conclusion

Rock mass in the Ciuyah tunnel inlet area - Karian Dam is dominated by sandstone, claystone, lapilli, and weathered soil, included in the Genteng Formation. Assessment of rock mass quality using the GSI method shows the rock is in the disintegrated class and has poor surface quality with a rating of 23-38. The assessment with RMR_{89} shows that it is in the very poor class with a rating of 18, and the JSCE (2016) method shows the rock is in the D-II L Layered class.

The recommended excavation equipment based on the chart from Pettifer and Fookes (1994) is CAT 245. The support installation time was analyzed using the Bieniawsky chart (1989), where very poor rock only provides a short stand-up time, which is ± 30 minutes with a 1-meter distance excavation or ± 55 minutes with an excavating distance of 0.5 meters. Hence, it is necessary to install a completed support system immediately.

The excavation method that has been carried out is using full face excavation with a face buttress, and when analyzed using a multigraph from Daraei (2019) shows the same recommendation. However, the excavation method using RMR_{89} and JSCE (2016), where both ways recommended gradual excavation with bench cuts will provide more stability to the tunnel. The excavation distance with the RMR method is 0.5m, and the JSCE method is 0.8m with the installation of face bolts in the face of the excavation.

Based on this study's results, a multigraph from Daraei (2019) can be used to determine the collapse pattern for rocks with very poor conditions. Still, the excavation method and support system are more appropriate using the method from RMR, because it has the shortest steel rib space. In order to prevent collapse, it is crucial to conduct a detailed investigation for condition of the rock on the excavation face at each stage to describe the potential for collapse. For further research, numerical analysis can be carried out to determine the safety of the recommended support system based on the empirical analysis of RMR_{89} and JSCE (2016).

Acknowledgments

We would like to express our gratitude to BBWS Cidanau, Ciujung, and Cidurian for allowing us to conduct research at the Karian Dam Project and Ciuyah tunnel. We would also like to express our deepest gratitude to the Ministry of Public Works Human Resource Development Agency for the financial support.

References

[1] An Q., Qingsong Zhang, Xiao Zhang, Jiaqi Zhang. 2021. Bridging the gap between engineering properties and grouting reinforcement mechanisms for loess in eastern China: taking Jinan loess as an example. *Bulletin of Engineering Geology and the Environment (2021)* **80:4125–4141**, https://doi.org/10.1007/s10064-021-02201-w

[2] Chen Z., Chuan He, Wenbo Yang, Wenqi Guo, Zheng Li, Guowen Xu. 2018. Impacts of geological conditions on instability causes and mechanical behavior of large-scale tunnels: a case study from the Sichuan–Tibet highway, China. *Bulletin of Engineering Geology and the Environment (2020)* **79:3667–3688**, https://doi.org/10.1007/s10064-020-01796-w

[3] Wang K., Shuoshuo Xu, Yujian Zhong, Zhilin Han, Enlin Ma. 2021. Deformation failure characteristics of weathered sandstone strata tunnel: A case study. *Engineering Failure Analysis 127 (2021)* **105565**, https://doi.org/10.1016/j.engfailanal.2021.105565

[4] Ya S., Su Yonghua, Zhao Minghua, Liu Shaofeng, Li Xiang. 2018. A case study of the failure of Liziping tunnel. *Tunnelling and Underground Space Technology (2018)*, https://doi.org/10.1016/j.tust.2018.07.022

[5] Li Z., Liqiong Wang, Bo Feng, Jie Xiao, Qi Zhang, Liying Li, Jie Liang. 2020. Comprehensive collapse investigation and treatment : An engineering case from Qingdao expressway tunnel. *Journal of Cleaner Production*, https://doi.org/10.1016/j.jclepro.2020.121879

[6] Sharifzadeh M., Rahman Daraei, Mohsen Sharifi Broojerdi. 2012. Design Of Sequential Excavation Tunneling In Weak Rocks Through Findings Obtained From Displacements

Based Back Analysis. *Tunneling and Underground Space Technology* **28** (2012) 10–17

[7] Rahimi B., Kourosh Shahriar, Mostafa Sharifzadeh. 2014. Evaluation of rock mass engineering geological properties using statistical analysis and selecting proper tunnel design approach in Qazvin–Rasht railway tunnel. *Tunnelling and Underground Space Technology* **41** (2014) 206–222, http://dx.doi.org/10.1016/j.tust.2013.12.010

[8] H. Basarir, A. Ozsan, M. Karakus. 2005. Analysis of support requirements for a shallow diversion tunnel at Guledar dam site, Turkey. *Engineering Geology* **81** (2005) 131 – 145, doi:10.1016/j.enggeo.2005.07.010

[9] Celada B. and I. Tardaguila, P. Varona, A. Rodriguez, Z.T. Bienawski, 2014. Innovating Tunnel Design by an Improved Experience-based RMR System. *Proceedings of the World Tunnel Congress 2014* – **Tunnels for a better Life**. Foz do Iguaçu, Brazil

[10] Zhang Q., Xianbin Huang, Hehua Zhu, Jianchun Li. 2019. Quantitative assessments of the correlations between rock mass rating (RMR) and geological strength index (GSI). *Tunnelling and Underground Space Technology* **83** (2019) 73–81, https://doi.org/10.1016/j.tust.2018.09.015

[11] Japan Society of Civil Engineer 2018 Standard Specifications for Tunneling – 2016: *Mountain Tunnels*

[12] Suryani A.P., 2019. *Pemodelan Numerik 3d Deformasi Terowongan Cisumdawu Provinsi Jawa Barat*. Tesis Program Studi Magister Teknik Geologi Universitas Gadjah Mada

[13] Committee for Earthquake Resistant Design of Submerged Tunnel (Japan), 1975. *Specifications for earthquake resistant design of submerged tunnels*. Japan Society of Civil Engineering (JSCE), Tokyo

[14] Barton, N., Lien, R., & Lunde, J., 1974. Engineering Classification of Rock Massesfor the Design of Tunnel Support. *Rock Mechanics and Rock Engineering*: **Springer – Verlag**, 189–236, http://dx.doi.org/10.1007/bf01239496

[15] Singh, B., dan Goel, R.K., 1999, *Rock Mass Classification: A Practical Approach in Civil Engineering*, Roorke, India

[16] Bieniawsky, Z. T., 1989. *Engineering Rock Mass Clasification Mining and Mineral Resources Research Institute*. Pennsylvania State University. 272 page

[17] Rusmana, E., Suwitodirdjo, K., Suharsono, 1991. *Peta Geologi Lembar Serang, Jawa*. Pusat Penelitian dan Pengembangan Geologi

[18] Daelim-Wika-Waskita JO. 2019. *Method Statement of Recovery Work From Collapsed Tunnel and Additional Reinforcement Supporting System in Ciuyah Tunnel*. Lebak Distric, Banten Province

[19] Qin Changbing and Siau Chen Chian. 2017. 2D and 3D stability analysis of tunnel roof collapse in stratified rock: A kinematic approach. *International Journal of Rock Mechanics and Mining Sciences* **100** (2017) 269-277

[20] Moussaei N., Mustafa Sharifzadeh, Kourosh Shariar, Mohammad Hossein Khosravi. 2019. A new classification of failure mechanisms at tunnels in stratified rock masses through physical and numerical modeling. *Tunneling and Underground Space Technology* **91** (2019) **103017**

[21] Alija S., F.J. Torrijo, M. Quinta-Ferreira. 2014. Study of the unexpected collapse of the Ampurda´n tunnel (Spain) using a finite elements model. *Bulletin Engineering Geology Environment (2014)* **73:451–463**. DOI 10.1007/s10064-013-0534-z

[22] Directorate General of Water Resources. 2011. Master Planning and Feasibility Study of the Karian Dam ~ Serpong Water Conveyance and Supply System, *Design Note* **Volume III-2 Ciuyah Tunnel for KSCS**

[23] Daelim-Wika-Waskita JO. 2019. *Final Report Geoelectric Survey 2D Main Tunnel Ciuyah*. Lebak District, Banten Province

[24] Marinos, P., & Hoek, E. 2000. GSI: A Geologically Friendly Tool for Rock Mass Strength Estimation. *ISRM International Symposium 2000*, 19–24

[25] Sivakugan, N; Shukla, S.K; Das, B.M. (2013). *Rocks Mechanics an Introduction*. New York:

CRC Press

[26] Hoek, E., Carter, T., dan Diederichs, M. 2013. Quantification of the Geological Strength Index Chart. *US Rock Mechanics /Geomechanics Symposium,* **San Francisco: June 23-26**, 2013

[27] Palmstrom, A. 2005. Measurements of and Correlations between Block Size and Rock Quality Designation (RQD). *Tunnelling and Underground Space Technology* **20**, 362-377

[28] ISRM. 1981. *Rock Characterization Testing and Monitoring*. Brown, E., Ed., Pergamon Press, Oxford, 211 p

[29] Franklin JA, Broch E, Walton G. 1971. Logging the mechanical character of rock. *Trans Inst Min Metall* **80**: A1–A9

[30] Pettifer, GS and Fookes, PG. 1994. "A Revision of The Graphical Method for Assessing The Excavatability of Rock". *Quarterly Journal of Engineering Geology* **27**, 145-164

[31] Daraei A, Zare S. 2019. A new multigraph approach for selecting the sequential excavation method of civil tunnels. *Tunnelling and Underground Space Technology* **91:102999-103010**, DOI: 10.1016/ j.tust.2019.102999

[32] Agustawijaya D.S. 2007. The uniaxial compressive strength of soft rock. *Civ Eng Dimens* **9(1)**: 9–14

IOP Conf. Series: Earth and Environmental Science 1233 (2023) 012004 doi:10.1088/1755-1315/1233/1/012004

Breccia for Diversion Tunnel Construction at Bener DAM in Purworejo, Central Java, Indonesia: Fragments, Matrix, RMR

E R Triristanto[1,2], S Husein[1], and M Y Y Alfarobi[2]

[1]Department of Geological Engineering, University of Gadjah Mada, Yogyakarta, Indonesia
[2]Ministry of Public Works and Housing, Yogyakarta, Indonesia

*Email: eskarutanatriristanto@mail.ugm.ac.id, shddin@ugm.ac.id

Abstract. This study aimed to explain the detailed engineering geology investigation which has breccia geological units on Diversion Tunnel at Bener Dam, Purworejo Regency, Central Java Province. The investigation includes geological conditions both breccia fragment and matrix, to Rock Mass Rating (RMR) values each unit. The data were taken from face mapping tunnel with interval 10 meters at inlet 1-390 and 841-865 meters. The findings showed that geological engineering units consisted of Monomict Breccia. The geological units can be divided into geological engineering; Monomict Breccia Matrix (RMR 72-92), Polymict Breccia Matrix (RMR 82), Andesite Fragmen (RMR 52-62), and Tuff Fragmen (RMR 79-84). The discovery of Andesite and Tuff Fragments needs to be handled differently than Monomict Breccia or Polymict Breccia. Andesite Fragment unit with discontinuity of sheeting joint requires rock bolt reinforcement to be installed perpendicular to the discontinuity plane. While the dominant unit is the Tuff Fragment which has a UCS value of 5-7 MPa, it is not recommended to use the blasting excavation method even though it has an RMR value of 79-84.

1. Introduction

Diversion tunnel is one of the subsurface construction types used to facilitate water passage during dam construction [1]. As a subsurface construction, geological investigations are essential to conduct, starting from the planning, construction, to maintenance phases [2]. Not without cause, it was proposed because the geological conditions up to the time of tunnel excavation during the construction phase could not be known with certainty. Without knowing the geological conditions, the rock mass classification of the tunnel cannot be known, which means that the stand-up time, excavation method, and reinforcement used for excavation cannot be estimated. Therefore, before the construction phase, an approach to estimate the geological conditions and rock mass classification of the tunnel needs to be conducted.

In 2019, in the planning phase, condition of geological tunnel was estimated using geological face mapping [3] [4]. However, it should be noted that the depth of the tunnel from the earth's surface is 60-100 meters so that differences in geological conditions can be found between the earth's surface and tunnel conditions. So that in the same year, five borehole logs were carried out to obtain data on tunnel depth [3] [5]. Although the borehole log data describes the actual geological conditions of the tunnel, the data has very low accuracy to describe the overall condition of the tunnel. By considering the tunnel length of 865 m and 5 borehole logs, the average borehole log must describe the geological conditions and rock mass classification along ±170 m [3] [5]. Based on the data, it was found that the tunnel composition unit consisted of breccia and andesite with RMR values poor-good [4] [5].

The face mapping tunnel in the construction phase is the solution to the above problem. Several studies have been carried out such as lithology Claystone, Breccia, Andesite with RMR values poor to very poor at High-Speed Railways Jakarta-Bandung, West Java, Indonesia [6], meta-clay rock and meta-pelitic rock with RMR values fair to good at Eko-remaja Uranium Exploration Tunnel, Kalan, West Kalimantan, Indonesia [7], and phyllite and slate with RMR poor to good at Hydroelectric Power Plant Conductor Tunnel, Agam, West Sumatra, Indonesia [8]. Based on the studies, an assessment of the

IOP Conf. Series: Earth and Environmental Science 1233 (2023) 012004 doi:10.1088/1755-1315/1233/1/012004

tunnel excavation surface according to the distance of the excavation rate can provide an accurate and comprehensive description of geological conditions and rock mass classification [6] [7] [8].

Considering that the Bener Diversion Tunnel planning phase data is less accurate as data for tunnel excavations and excavations as deep as 400 meters carried out in August 2021-May 2021, an analysis of the lithology and RMR types with a distance of 10 meters according to the distance of the excavation rate needs to be carried out. So that it can be determined the treatment and handling of geological conditions and rock mass classification more accurately and avoid tunnel deformation.

2. Location, Geology, and Geological Engineering Conditions

Bener Dam is located in Guntur Village, Bener District, Purworejo Regency, Central Java Province. It is designed to accommodate 100.94 m3 of water supply, which will reduce the flood discharge by 1.60 m3/second, irrigate 15,069 Ha of land with a flow rate of 210 m3/second, and generate electricity of 6.00 MW [3]. The dam's construction started in 2018 and is estimated to be finished in 2023. The development process of the dam was divided into main and additional constructions. One of additional constructions that will be the subject of this research is the diversion tunnel at the Bener Dam.

The Diversion Tunnel Bener Dam has 865 m in length with a diameter of 9 m, a horseshoe-shaped, and is oriented north-south with an elevation of ±300 m above the sea level [3]. The prediction of the geological condition of the tunnel in the pre-construction phase using surface mapping and borehole logs resulted in geological engineering conditions that affect the RMR value [4] [5].

The regional geology of Bener Dam on a scale of 1:100.000 is composed of the Kebobutak Formation, which consists of breccia, tuff, tuff lapilli, and andesite [9]. The formation can be divided into two sub-formations, Kaligesing and Dukuh [10]. A detailed mapping with scale of 1:16,000, has shown that Breccia units (Kebobutak Formation) existed, which can be further subdivided based on differences of their compositions into Monomict Breccia (Kaligesing) and Polymict Breccia (Dukuh) [4] (Figure 1).

The geological subsurface by the borehole log data shows the possibility of lithology found other than breccia. The lithology is andesite that can be found at the outlet [5]. However, another study stated that andesite is not a separate lithology but is part of a massive breccia fragment so that continuity cannot be found in andesite as a separate lithology [4] (Figure 2).

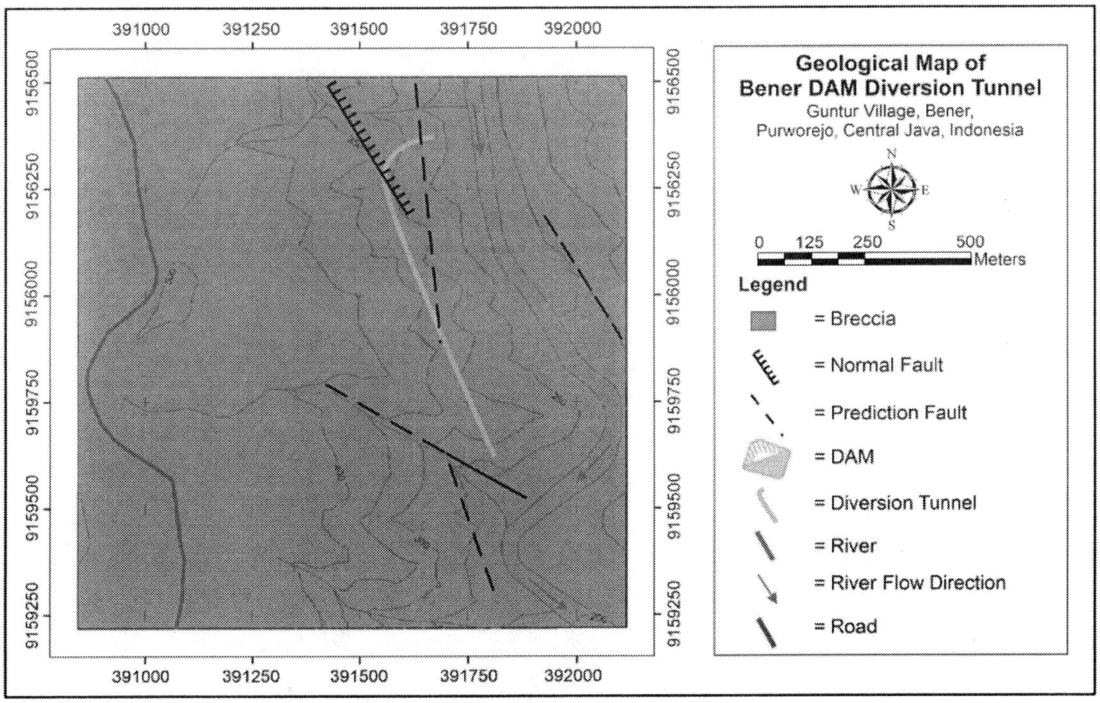

Figure 1. Geology surface mapping

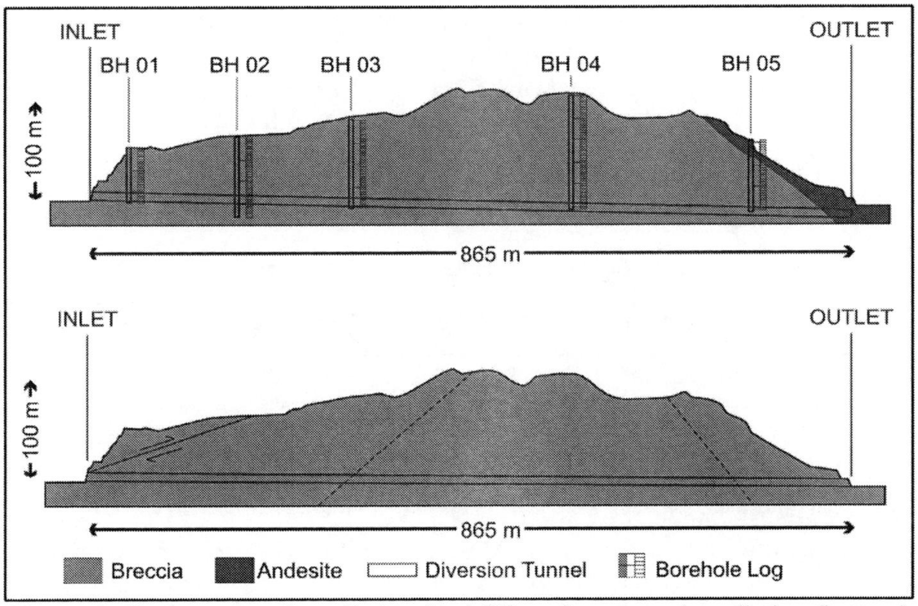

Figure 2. Geology tunnel prediction (top) [4] geology tunnel prediction (bottom)

The Monomict Breccias are made of Andesite Fragment and Tuff Fragment with a matrix of andesite debris. Based on the petrology, the Andesite Fragments and matrix were composed predominantly of plagioclase, clinopyroxenitic, and hornblende [4]. Andesite Fragment has a unique rock structure, namely sheeting joints. Meanwhile, the petrology of Polymict Breccia is composed of Andesite Fragment, Tuff Fragment, Limestone Fragment, and Quartz Fragment, as well as micrite and andesite matrices [4]. The variations in the composition of Monomict Breccia and Polymict Breccia resulted in differences of the compressive strength of rocks, where Monomict Breccia had a higher value than Polymict Breccias.

According to Sasangka (2019) the RMR (Rock Mass Rating) value of the diversion tunnel is 50,55 to 79,19 [4]. Another research found the tunnel's RMR value ranges from 36,5 to 67,76 [5]. (Table 1). The distinction is caused by the differences of the assessment towards geological structure and weathering of rocks.

Table 1. RMR Value Comparison

Depth (Meters from inlet)	Sasangka (2019) [4]		Dharmawan (2019) [5]	
	Zone	RMR Value	Zone	RMR Value
0-180	Zone 1	66,75	Zone 1	76,34
180-300	Zone 2	66,75	Zone 2	75,07
300-350	Zone 3	36,50	Zone 3	79,19
350-525	Zone 4	56,14		
525-650	Zone 5	56,14	Zone 4	72,45
650-865	Zone 6	67,76	Zone 5	55,55

3. Methodology

3.1. Sample Collections and Field Parameters

The data were collected from August 2021 to May 2022, from the face excavation, wall, crown, and invert tunnel lithology part, along the inlet 0–0+400 and inlet 0+765–0+865. On the most dominant and weakest parts, the measurements were taken at intervals of 10 m. Other data were carried out from an additional borehole core of rocks and groundwater levels on March 2022.

Table 2. Measurement parts of sampling

Lithology	Measurements Parts
Monomict Breccia with fragment <9 meter	Matrix
Monomict Breccia with fragment >9 meter	Fragment
Polymict Breccia	Matrix

3.2. Rock Mass Classification

The Rock Mass Classification used was Rock Mass Rating (RMR). The parameter in RMR consisted of Uniaxial Compressive Strength (UCS), Rock Quality Designation (RQD), spacing of discontinuities, condition of discontinuities, and groundwater [11].

UCS was taken by converting Schmidt hammer rebound type-N on each 10 meters interval of the tunnel. The test method was carried out by sampling perpendicularly per intact rock for 10 strokes in each test. Based on those 10 strokes, the five lowest values were eliminated, and the five highest mean values were correlated with the representative values of rock density to obtain the UCS value (figure 3) [12].

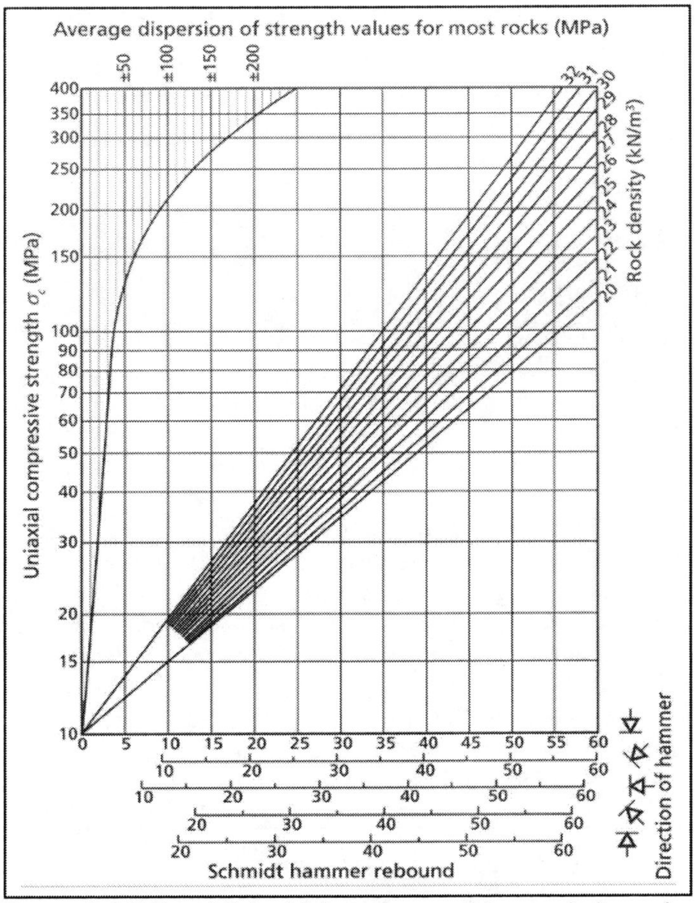

Figure 3. UCS and Schmidt hammer graphic correlation value

RQD values were based on the total joint frequency (λ) in the 1 m^3 (equation 1) area, as can be seen in the following formula.

$RQD=100e^{-0.1}(1+0.1\ \lambda)$ (equation 1) [13]

RQD : Rock quality designation (%)
λ : Total join frequency (/m)

The observation of spacing, condition of discontinuities, and groundwater were carried out directly on the face of the tunnel excavation.

4. Data and Analysis

The geology of the Diversion Tunnel Bener Dam consisted of Monomict Breccia (Figure 4). However, with the condition of the 9m diameter-tunnel, the geological unit doesn't represent the condition of the tunnel, so it is necessary to do a more detailed classification into geological engineering units. The geological engineering of the tunnel can be divided into Monomict Breccia, Polymict Breccia, Andesite Fragments and Tuff Fragments. The distribution of each geological engineering unit can be described as: (a) Monomict Breccia is the dominant unit with a percentage >95%, and is almost found along the tunnel; (b) Monomict Breccia was found as insertion at the outlet; (c) Andesite Fragments and Tuff Fragments are part of the Monomict Breccia fragments with dimensions > 2.5 m which were found at several points in the tunnel.

The 3 groundwater drillings in 2022 found that the tunnel was under the groundwater level. The groundwater flow is directed from the center of the tunnel to the inlet at the outlet at a height of ±40 m above the tunnel (Figure 4).

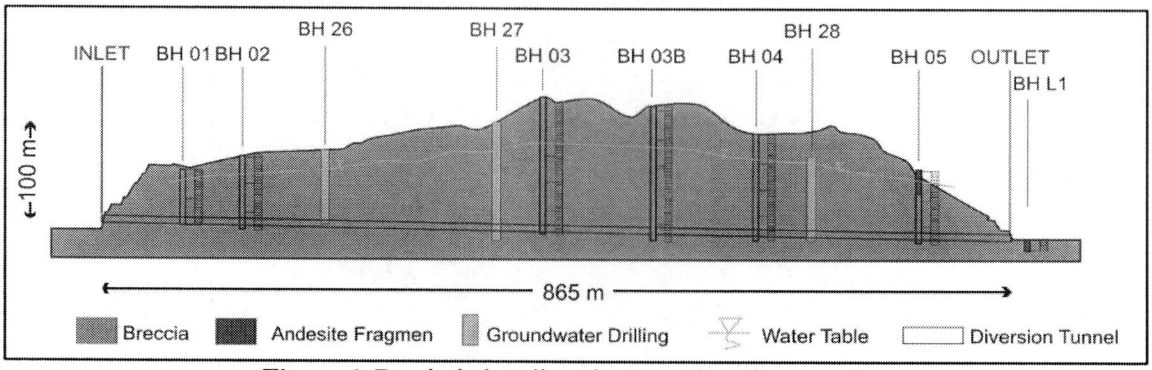

Figure 4. Borehole log diversion tunnel geology section

The tunnel section is presented in 3 illustrations including: (a) geological section (Figure 4); (b) geological engineering section of walls, crowns, and tunnel inverts (Figure 5); and (c) tunnel face excavation section (Figure 6). The geological engineering section of the wall, crown, and invert tunnel was composed of Monomict Breccia with distribution points of Andesite Fragment and Tuff Fragment. Meanwhile, Polymict Breccia was not mapped on the section due to its distribution that was only at the outlet (Figure 5).

The largest Andesite fragment was found at inlet 0+116. The fragment has dimensions of more than 9 m with a depth of 15 m. Tuff fragments were found at inlet 0+100, inlet 0+113, inlet 0+180, and outlet with dimensions of 1 – 2.5 m (Figure 5).

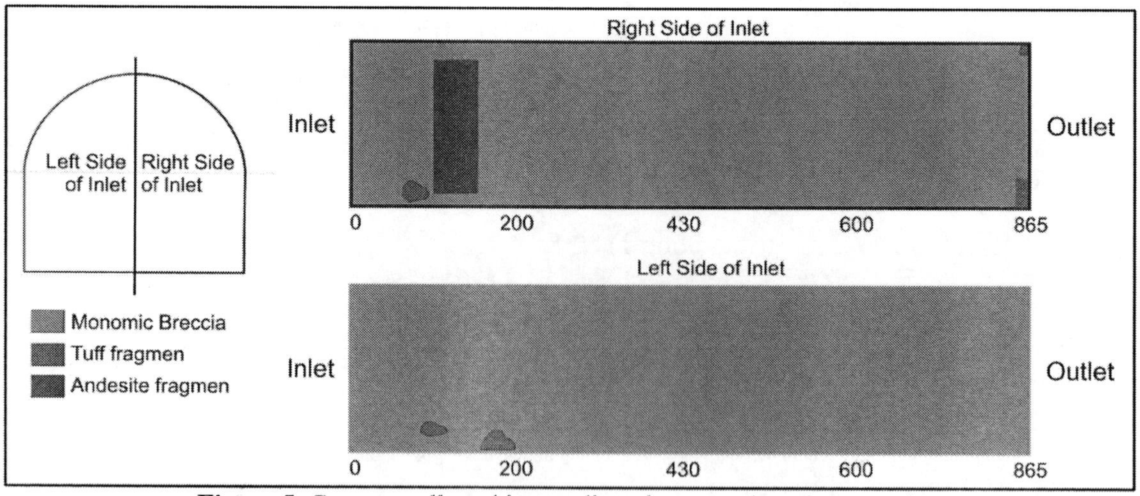

Figure 5. Crown, wall, and invert diversion tunnel geology section

The tunnel face excavation section is a description of rock conditions and discontinuities of the excavation face. The section is a more detailed description of the geological engineering section of walls, crowns, and tunnel invert. The technical geological conditions of the tunnel excavation face can be represented by inlet 0+251, inlet 0+116, and outlet. The excavation face at inlet 0+251 describes most of the geological engineering conditions of the tunnel face, which is composed of Monomict Breccia without any discontinuities. The excavation face at inlet 0+116 is Andesite Fragment with discontinuity in the form of sheeting joints. Meanwhile, the excavation face at the outlet describes the condition of the insertion of Polymict Breccia and Tuff Fragments with discontinuities in the form of geological structures.

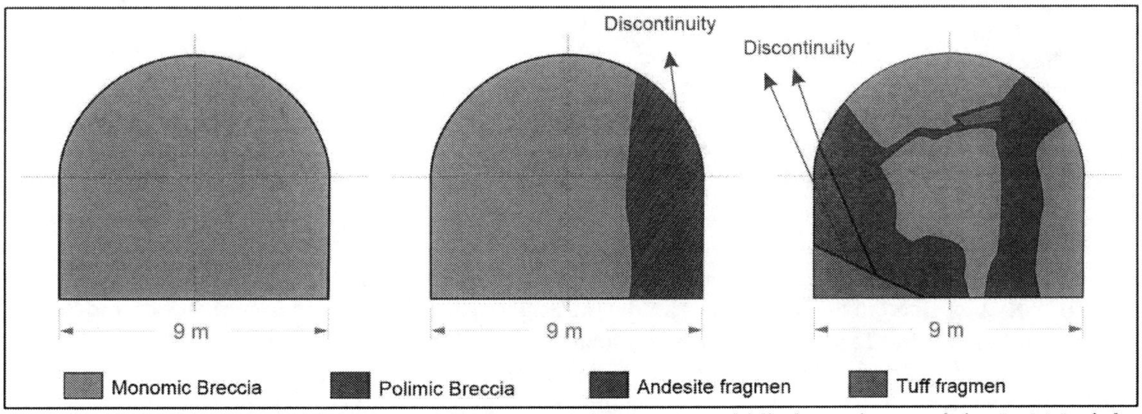

Figure 6. Tunnel face mapping (left is geology at inlet 0+251; middle is geology at inlet 0+116; right is geology at outlet)

The UCS value in each geological engineering unit differs from one another. Based on UCS values, the order of largest to smallest values is Andesite Fragment, Monomict Breccia (matrix), Polymict Breccia (matrix), and Tuff Fragment. The UCS values of Andesite Fragments ranged from 50 – 53 MPa, Monomict Breccia (matrix) between 28 – 51 MPa with an average of 36.7 MPa, while Polymict Breccia (matrix) 19 MPa, and Tuff Fragments 5 – 7 MPa (Figure 7).

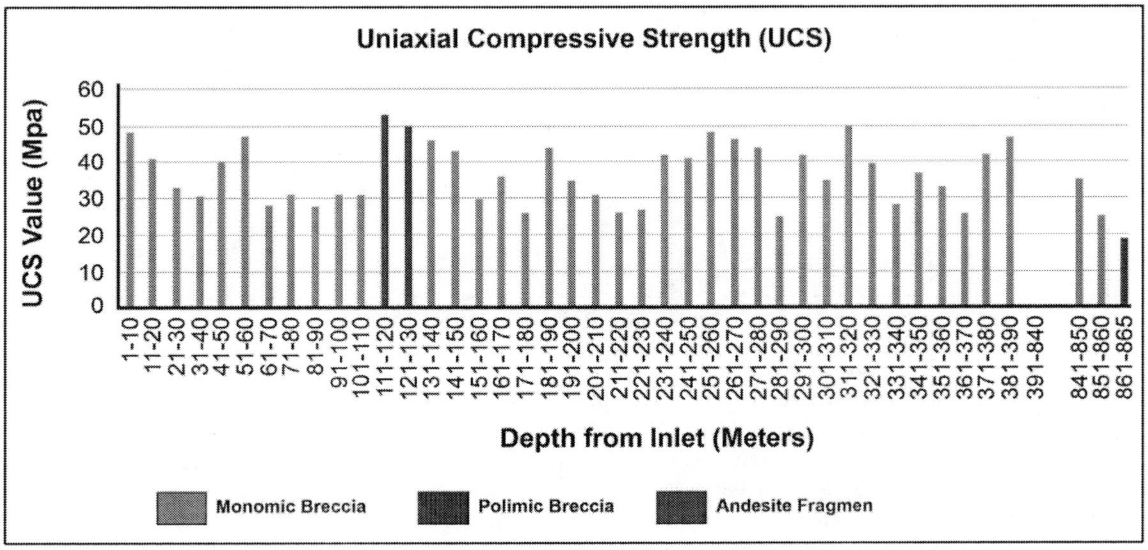

Figure 7. UCS Value

Discontinuity of the tunnel excavation face affected 3 parameters in RMR, namely RQD parameter, spacing, and condition of discontinuity. The discontinuity of this study will be divided into geological structure and sheeting joint. The sheeting joint had a more significant effect on the decrease in the rating of 3 parameters to geological structures.

The geological structure has a discontinuity spacing of more than 2 meters, a length of more than 3 meters, without separation, without filler, and mild to fresh weathered. Meanwhile, the sheeting joint has a discontinuity distance of up to 10 cm, a length of > 1 meter, without filler.

IOP Conf. Series: Earth and Environmental Science 1233 (2023) 012004 doi:10.1088/1755-1315/1233/1/012004

The RQD values at the tunnel excavation face have been significantly affected by Total Join Frequency. Massive rocks have an RQD of almost 100%, where the geological structure has a small effect on the RQD value >80%, and sheeting joints have an RQD value of <50%. The RQD values for Monomict Breccia, Polymict Breccia, and Tuff Fragment ranged from >80%, while the RQD values for Andesite Fragments with sheeting joint were around 37%.

The groundwater conditions in the tunnels are mostly dry. It is because of the petrological conditions of massive and waterproof tunnel rock formation. However, there are some groundwater seepages, qualitatively at (a) inlet 0+190 that has a wet tunnel condition; (b) inlet 0+120, inlet 0+140, and inlet 0+180 that have dripping water conditions.

The data processing of 6 RMR parameters showed that the values of Monomic Breccia, Polymic Breccia, and Tuff Fragment ranged from 75 to 92 (fair to good), while Andesite Fragment ranged from 52 to 62 (good to very good) (Figure 8).

Data processing of 6 parameters have shown RMR value of Monomict Breccia 72-92 (good to very good), Polymict Breccia 88 (very good), and Tuff Fragment 79 – 84 (good to very good) and Andesite Fragment 52 – 62 (fair to good) (Figure 8). Compared to the RMR values in earlier studies 50,55 – 79,19 [4] and 36,5 – 67,76 [5], the values obtained in the tunnel excavation face assessment have several differences in terms of RQD parameters, spacing discontinuity, weathered, and water condition.

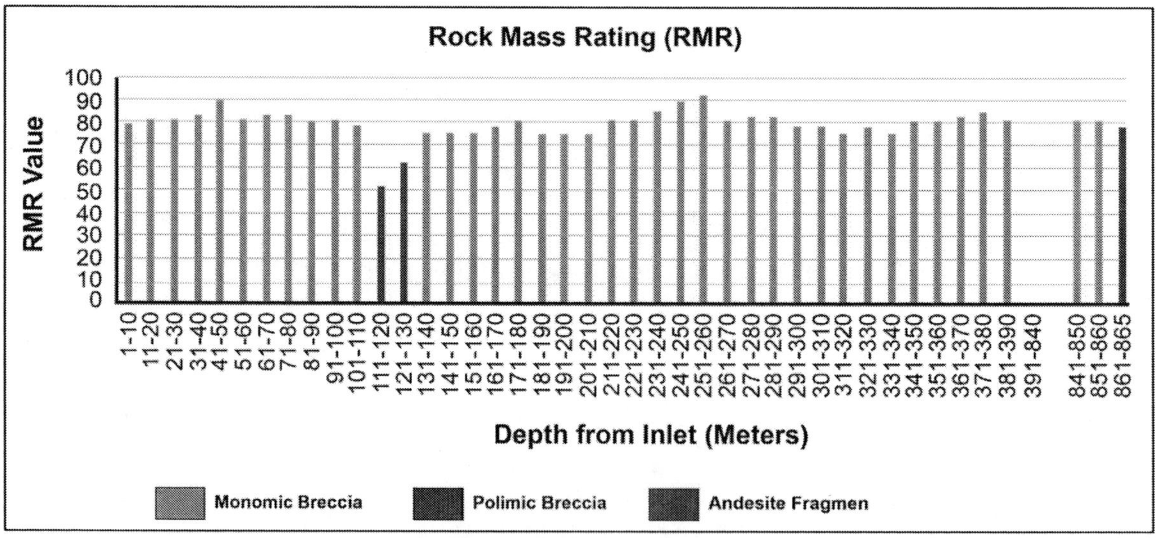

Figure 8. RMR Value

Based on the RMR values, the recommended excavation for Monomict Breccia and Polymict Breccia are blasting and hard ripping. Meanwhile, Ripping is recommended excavation for Andesite and Tuff Fragments. The support tunnel for the whole units is recommended to use rock bolts and shorcrete. The handling of the rock bolts reinforcement on Andesite Fragments needs to be installed perpendicular to the sheeting joint. Although the Tuff Fragment excavation method has an RMR values 79-84 (good to very good), it is not suggested to employ blasting since the UCS value is in the range of 5 – 7 MPa.

5. Conclusion

In conclusion, geology of Diversion Tunnel Bener Dam based on face mapping tunnel is composed of Monomict Breccia. Consider of the tunnel size, the geology Monomict Breccia of the tunnel can be separated into geology engineering: (1) Monomict Breccia Matrix, (2) Polymict Breccia Matrix, (3) Andesite Fragment (of Monomict Breccia), and (4) Tuff Fragment (of Monomict Breccia).

Based on UCS, RQD, discontinuity of rock mass, and groundwater of face mapping observation, RMR values ranged from 52-92. It can be divided into Monomict Breccia RMR 72-92 (good to very good), Polymict Breccia RMR 88 (very good), and Tuff Fragment RMR 79 – 84 (good to very good) and Andesite Fragment RMR 52 – 62 (fair to good).

The discovery of Andesite and Tuff Fragments needs to be handled differently than Monomict Breccia. Andesite Fragment unit at inlet 101-120 with discontinuity in the form of sheeting joint, need is necessary to install rock bolt reinforcemet perpendicular to the discontinuity plane. Meanwhile, in Tuff Fragment unit with UCS values 5-7 MPa is not recommended to use the blasting excavation method eventhough the RMR value is good to very good.

6. Acknowledgement

The author would like to thank Mr. Hendy Setiawan as supervisor from Department of Geological Engineering, Faculty of Engineering, Universitas Gadjah Mada. The author also thanks BBWS Serayu – Opak and BPSDM of Ministry of Public Works and Housing.

References

[1] Purboyo A, Arifin H, Wijaya A, Widodo A, Isnandar, Saptono S, Wiyono B, and Wardhana B K, 2018 *Best Practice Penyelenggaraan Terowongan Jalan dan Underpass* (Jakarta: Directorate of Bridges, Directorate General of Bina Marga, Ministry of Public Works and Housing).

[2] Japan Society of Civil Engineers, 2019, Standard Specifications for Tunneling-2016: Mountain Tunnels 2018 *Tokyo. Journal of System and Computer Engineering.*

[3] Ministry of Public Works and Housing 2017 *Detailed Engineering Design* (unpublished).

[4] Sasangka D J 2019 *Analisis Kestabilan Portal dan Metode Ekskavasi Berdasarkan Kondisi Geologi Teknik pada Terowongan Pengelak Bendungan Bener Kabupaten Purworejo* Thesis (Gadjah Mada University).

[5] Dharmawan A G 2019 *Evaluasi Kondisi Geologi Teknik dan Analisis Kestabilan Terowongan Pengelak Bendungan Bener Kabupaten Purworejo, Propinsi Jawa Tengah* Thesis (Gadjah Mada University).

[6] Malda O 2022 *Analisis Penggalian Dan Kestabilan Sistem Penyangga Terowongan Nomor 6 pada Jalur Kereta Cepat Jakarta-Bandung, di Jawa Barat* Thesis (Gadjah Mada University).

[7] Faizah Y, Cakrabuana W, Kamajati D, and Rahmawati P 2019 Analysis of Rock Mass Quality and Reinforcement of Uranium Exploration Tunnel Eko Remaja Kalan, West Kalimantan Using Rock Mass Rating (RMR) Method *Eksplorium Buletin Pusat Pengembangan Bahan Galian Nuklir* **41** 25–36.

[8] Rakinaung Y E V M 2020 *Analisis klasifikasi massa batuan dengan metode RMR pada terowongan penghantar PLTA, Kecamatan Palupuh dan Palembayan, Kabupaten Agam, Provinsi Sumatera Barat* Thesis (Trisakti University).

[9] Rahardjo W, Sukandarummidi, and Rosidi H M 1995 *Geological Map of the Yogyakarta Sheet* Jawa (Bandung: Geological Research and Development Centre).

[10] Pulunggono A and Martodjojo S 1994 Perubahan tektonik Paleogen-Neogen merupakan peristiwa tektonik terpenting di Jawa *Proc. Geology and Geoengineering Java Island* (Yogyakarta: Gadjah Mada University Press) pp 37-50.

[11] Bieniawski Z T 1989 *Engineering Rock Mass Classification Mining and Mineral Resources Research Institute* (New York: Wiley-Interscience).

[12] Vallejo L I and Ferrer M 2011 Geological Engineering (London: CRC Press).

[13] Priest S D and Hudson J A 1976 Discontinuity Spacings in Rock *International Journal of Rock Mechanism and Mining Sciences & Geomechanics Abstracts* **13** 135–148.

IOP Conf. Series: Earth and Environmental Science 1233 (2023) 012005 doi:10.1088/1755-1315/1233/1/012005

Empirical Approach For Preliminary Support and Excavation Method Design Based On Engineering Geological Study Of Mbay Dam Diversion Tunnel

D W E Purba[1,2], **D H Amijaya**[1*]

[1]Department of Geological Engineering, Faculty of Engineering, Universitas Gadjah Mada, Yogyakarta, Indonesia
[2]The Ministry of Public Works and Housing, Directorate General of Water Resources Jakarta, Indonesia

*email: dwitrawinsaelrispurba@mail.ugm.ac.id; hamijaya@ugm.ac.id;

Abstract. This paper presents excavation methods and support systems in the diversion tunnel of the Mbay Dam, East Nusa Tenggara, Indonesia, based on rock mass quality ratings. The rock mass quality was classified through geological surveys, core drilling evaluations, and laboratory tests. The results showed that the lithology along the tunnel alignment was pyroclastic breccia. Using the Rock Mass Rating (RMR) classification system, which was the single classification at the time of designing the Mbay Dam Diversion Tunnel, the pyroclastic breccia rock mass was classified III and IV (fair and poor). Meanwhile, according to the Q-System classification,the second classification used, pyroclastic breccias were also classified as fair and poor. Therefore, according to the RMR classification, the recommended excavation methods were top heading and bench. Support systems suitable for rock mass quality in the Mbay Dam diversion tunnel were rock bolt, wire mesh, shotcrete, and steel sets, with different variations and types depending on the rock mass class and classification used.

1 Introduction

Tunnels are underground excavations with an excavation width smaller than 17 m and, have a minimal length of numerous hundred meters [1]. The hydraulic tunnel is one type used to divert water in dam construction work. For example, during the construction of the main dam at the Mbay dam, which is sited in the Rendubutowe village in the South Aesesa District of East Nusa Tenggara of Indonesia, a tunnel is being used as a diversion (Figure 1). This tunnel will later function as an intake structure for the dam. The Mbay dam diversion tunnel is on the left side of the Lambo river. With a tunnel length of 355 m, the diversion tunnel of the mbay dam has various excavation diameters. 7 m at the inlet and outlet while 7.5 m at the center of the tunnel. The purpose of Mbay dam is planned to supply 250 liters/second of raw water, irrigation, and flood control downstream of the dam.

At the planning stage, geological mapping and engineering geology benefit civil engineers [2]. Geological map depicting the surface and subsurface geological conditions is based on necessary tunnel excavation investigations and the support system that will be used later.

Since 2002, numerous surveys had performed to plan the Mbay dam. In 2019, as a consultant for planning the Mbay dam construction, PT. Indra Karya (Persero) [3] conducted surveys and investigations, including the alignment and design of tunnel supports using RMR. However, in assessing rock mass quality, it is necessary to use more than one classification system because of the importance of describing rock mass characteristics from various points of view in order to provide an appropriate empirical approach to determining tunnel supports [4]. Using two or more categorization systems in design and rock engineering will typically produce better and more precise results [5]. Several researchers outside Indonesia determine the selection of tunnel supports and excavation methods based on the quality of the rock mass [6,7]. Zekrinaldi et al. [8] and Putera et al. [9] are some researchers who use empirical methods in designing tunnel supports, especially water tunnels in Indonesia. Zekrinaldi et al. implemented it at the pre-construction stage using three classsification, while Putera et al. carried it out during the tunnel excavation work.

Rock Mass Rating [10] and Q-system classification [11] are rock mass classification systems used in this study to determine the quality of rock mass classification, excavation methods, and tunnel supports. Using three drilling data during pre-construction (2002, 2018 and 2019), the condition of the subsurface rock mass along the tunnel route is assessed. Unfortunately, updating the subsurface investigation data has not been carried out due to several social problems at the research site. Despite these limitations, this research is expected to contribute to the initial understanding of tunnel construction, especially regarding selecting tunnel supports and appropriate excavation methods.

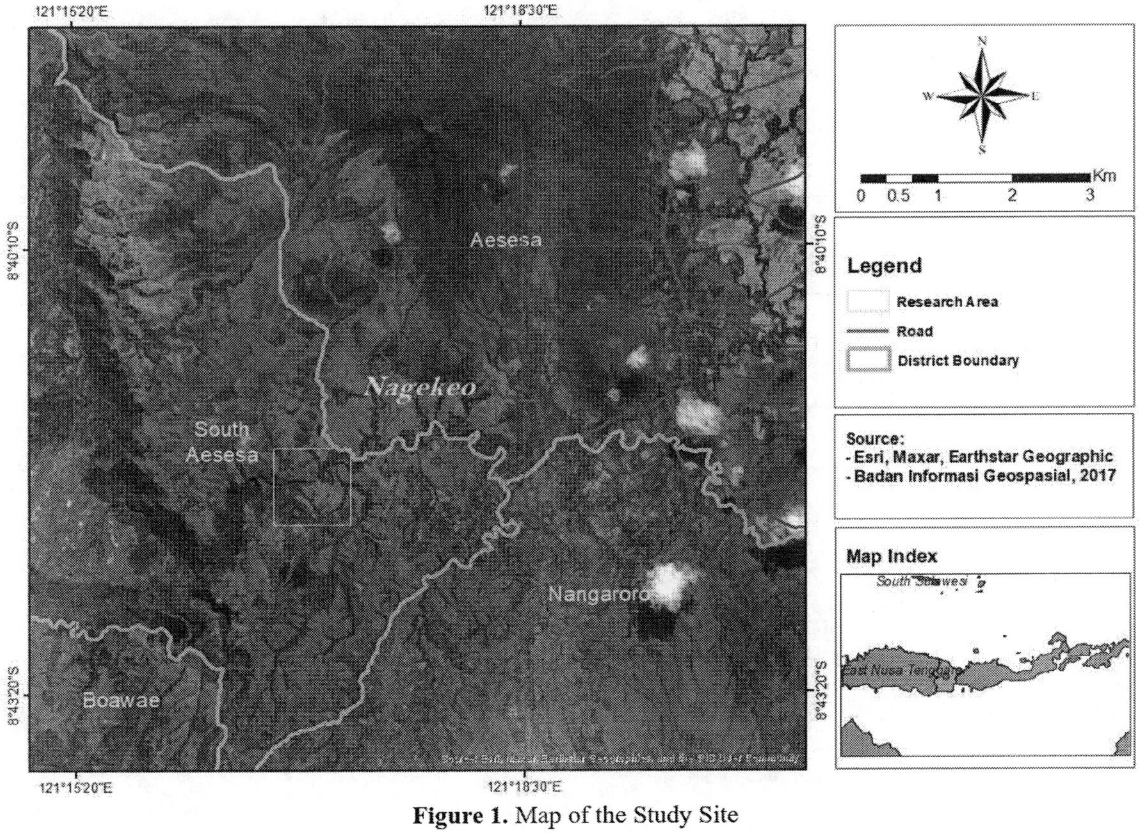

Figure 1. Map of the Study Site

2 Methodology

Surface geological surveying, identifying drilled rock core, and laboratory testing for rock specimens was used to determine rock mass quality, excavation methods, and support systems for tunnels. Surface geology surveying was carried out to collect morphology, lithology, and geologic structural characteristics. Meanwhile, the collected rock core describes the underground lithology, groundwater level, and other factors used to assess the rock mass quality. BWS Nusa Tenggara II carried out the drilling through a consultant PT. Indra Karya along the tunnel alignment. Three drilling data with varying depths were used as data to assess the condition of the subsurface rock mass. In order to obtain mechanical rock properties, laboratory testing was carried out on rock samples taken during the survey

3 Rock Mass Quality

3.1 RMR

Bieniawski[10] proposed the RMR classification system, which was revised in 1989. This classification technique uses five rock mass parameters: Uniaxial compressive strength (UCS) of intact rock, Rock Quantity Designation (RQD) [12], discontinuity spacing, discontinuity conditions, and groundwater. RMR values were derived by adding these ratings. In the subsequent phase, corrections for the discontinuity orientations based on the direction of tunnel excavation were made to the baseline RMR values. Table 1 explains the relationship between RMR values and rock mass quality. While provides empirical results regarding the types of tunnel supports and excavation methods

Table 1. Rock Mass Rating classification [10]

RMR Value	Class no.	Description
81-100	I	Very good rock
61-80	II	Good rock
41-60	III	Fair rock
21-40	IV	Poor rock
<20	V	Very poor rock

Table 2 Guidance for tunnel support and excavation method based on RMR [10]

Rock Mass Class	Excavation	Support		
		Rock Bolts (20-mm Dia, Fully Grouted)	Shotcrete	Steel Sets
Very good rock I	Full face 3-m	Generally, no support required except for occasional spot bolting		
Good rock II	Full face 1-1,5 m advance Complete support 20 m from face	Locally, bolts in crown 3 m long, spaced 2,5 m, with occasional wire mesh	50 mm in crown where required	None
Fair rock III	Top heading and bench 1,5-3 m advance in top heading Commence support after each blast Complete support 10 m from face	Systematic bolts 4 m long, spaced 1,5-2 m in crown and walls with wire mesh in crown	50-100 mm in crown and 30 mm in sides	None
Poor rock IV	Top heading and bench 1-1,5m advance in top heading Insert support concurrently with excavation 10 m from face	Systematic bolts 4-5 m long, spaced 1-1,5 m in crown and walls with wire mesh in crown	100-150 mm in crown and 100 mm in sides	Light to medium ribs spaced 1,5 m where required
Very poor rock V	Multiple drifts 0,5-1,5m advance in top heading Insert support concurrently. Shotcrete as soon as possible after blasting	Systematic bolts 5-6 m long, spaced 1-1,5 m in crown and walls with wire mesh. Bolt invert	150-200 mm in crown, 150 mm in sides, and 50 mm on face	Medium to heavy ribs spaced 0,75 m with steel lagging and fore-poling if required. Closed invert

3.2 Q-system

The Q system, commonly known as the Norwegian Technical Institute (NGI) system, was proposed by Barton et al.[11] in 1974. This system was characterized by the functions of RQD, Jn (joint sets number), Jr (joint roughness) [13], Ja (joint alteration), Jw (water pressure), and stress reduction factor (SRF). The following equation was used to calculate the Q values of the rock masses:

$$Q = \left(\frac{RQD}{Jn}\right)\left(\frac{Jr}{Ja}\right)\left(\frac{Jw}{SRF}\right) \quad (1)$$

4 Result and Discussion

4.1 Geological Surface Condition

Based on the Geological Map of Ruteng Sheet compiled by Koesoemadinata et al. [14], the rock formations around the dam and tunnel locations from old to young consist of the Nangapanda Formation (Tmn), Old Volcanic Results (QTv), and Young Volcanic Rocks (Qhv). Rock outcrops in the research area exhibit three rock units ranging in age from old to young, namely the Pyroclastic Breccia Unit, which intermingles with the tuffaceous sandstone unit, and above it, there was the Sandy Tuff Unit. The geological structure that developed based on the analysis of the straightness of the slopes indicated an indication of a sinistral fault in the study area. Geological maps from mapping at the study can be seen in Figure 2. All rock units in the study area were included in the Quaternary (Holocene) Young Volcanic Rock formation.

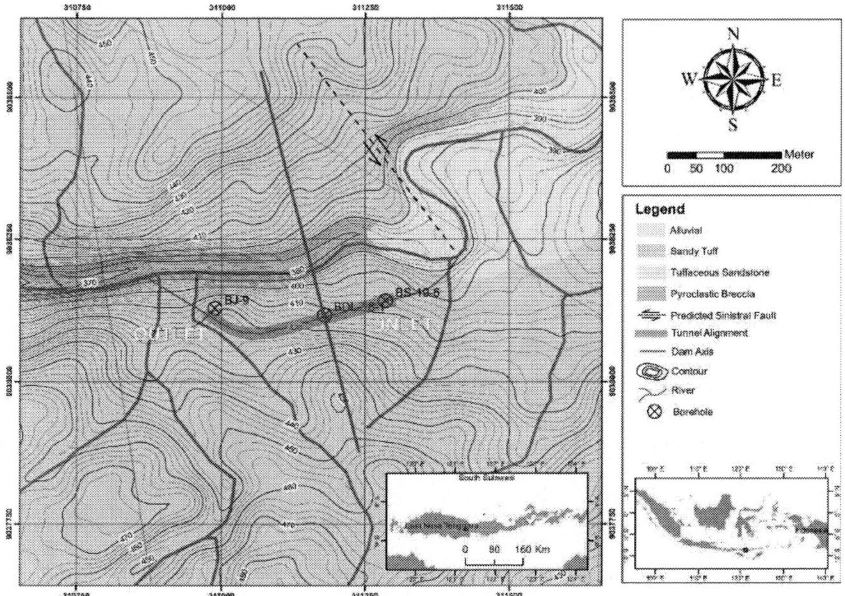

Figure 2. Geological map of study site

4.2 Geological Subsurface Conditions

Based on the drilling point position and the tunnel geometry, the rock mass quality assessment along the tunnel was grouped into three sections (Figure 3). The groupings of these sections are shown in Table 3. In addition, rock mass assessment in each section using the rock core (Figure 4) is presented in Table 4 for the RMR classification and Table 5 for the Q-system classification.

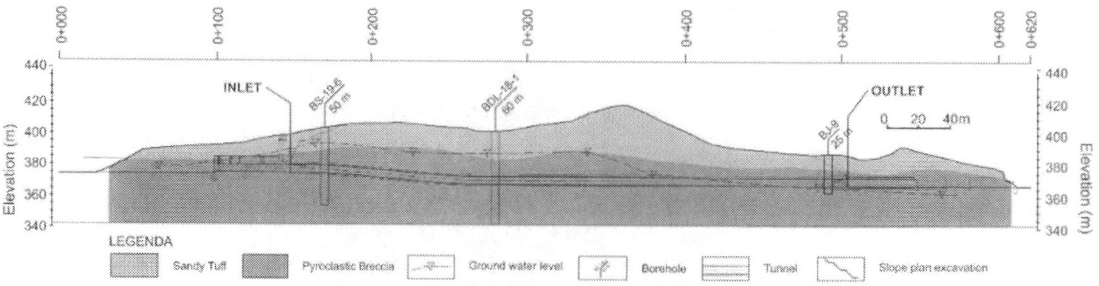

Figure 3. Geological Section in tunnel route

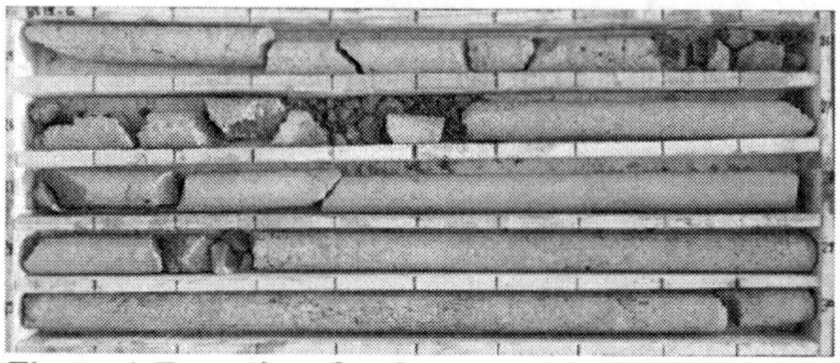

Figure 4. Examples of rock cores analyzed in this study

Table 3. Mbay Dam Diversion Tunnel section definitions

Section	Drill Point	Depth (m)	Tunnel Elevation (m)	Lithology
I	BS-19-6	50	24-31	Pyroclastic Breccia
II	BDL-18-1	60	28-35	Pyroclastic Breccia
III	BJ-9	25	14-21	Pyroclastic Breccia

Table 4. RMR Classification for Section I, II and III

Parameter	Section I	Section II	Section III
		Rating	
UCS	4	2	2
RQD	17	8	3
Spacing of Discontinuities	10	8	8
Condition of Discontinuities	20	10	10
Groundwater	7	7	10
Adjustment for joint orientation	-5	-5	-5
RMR	53	30	28
Quality	Fair	Poor	Poor

Table 5. Q-system Classification for Section I, II and III

Parameter	Section I	Section II	Section III
		Rating	
RQD	80	37.85	25
J_n	6	6	9
J_r	3	3	3
J_a	1	1	1
J_w	1	1	1
SRF	5	5	5
Q Value	8	3.78	1.67
Quality	Fair	Poor	Poor

4.3 Excavation Method

The RMR value in each section was used to determine the method or sequence of tunnel excavation. The results show that sections II and III have the same excavation method: "Top heading and bench 1.5–3 m advance in top heading. Commence support after each blast. Complete support 10 m from face". while *section* I is defined as "Top heading and bench 1–1.5 m advance in top heading. Install support concurrently with excavation 10 m from face" (Table 6).

Table 6. Suggested tunnel excavation method by RMR classification

Section	I	II	III
Excavation Method	Top heading and bench 1–1.5 m advance in top heading. Install support concurrently with excavation 10 m from face	Top heading and bench 1.5–3 m advance in top heading. Commence support after each blast. Complete support 10 m from face	

4.4 Support System

RMR classification provided a rock reinforcement guide for tunnels (Table 2). Therefore, the type of support for the tunnel can be directly determined using the RMR value. The tunnel support was determined using the Q-system graph, which considered the value of Q, the height of the tunnel or span, and the Excavation Support Ratio (ESR) score. In this study, the height of the tunnel or span was 7 and 7.5 meters, and ESR value was 1, resulting in a height to ESR ratio of 7 and 7.5. Figure 5 illustrates how parameter values are plotted on the Q system graph to determine the support system. Table 7 summarises the outcomes of selecting the type of support system for each tunnel section.

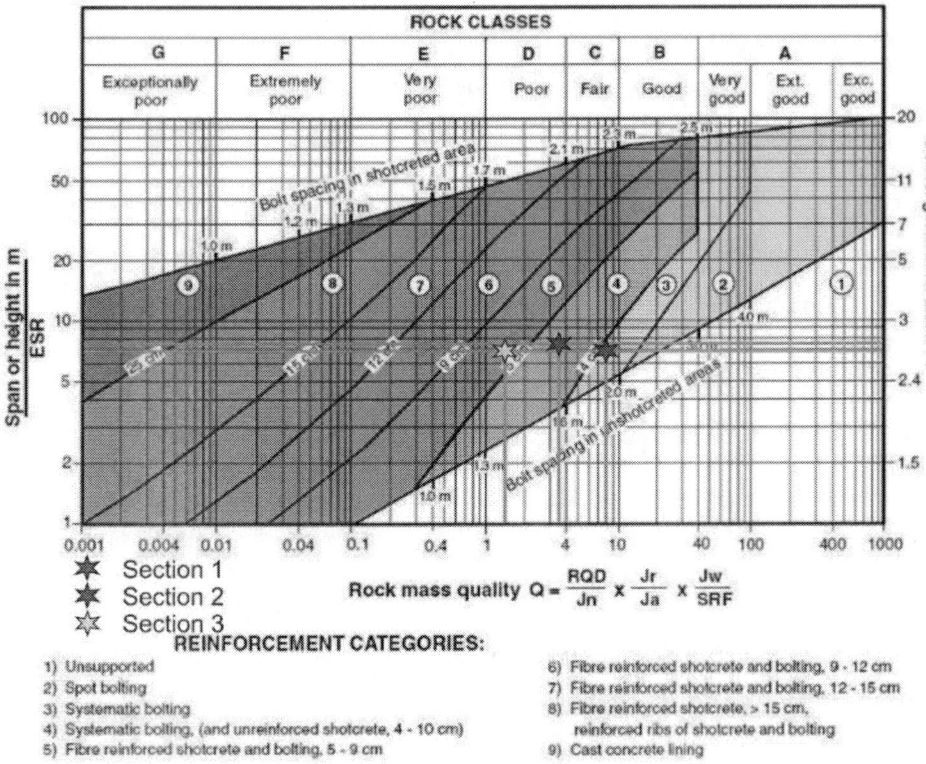

Figure 5. Plotting the Q value and the comparison of tunnel height with ESR to determine the tunnel support [11]

Table 7. The suggested supports are based on the RMR and Q-system

Section	RMR Support			
	Rock Bolt (D20 mm, Fully Grouted	**Shotcrete Thickness**	**Wiremesh**	**Steel Sets**
I	Systematic bolts, L: 4 m, S:1,5-2 m, IP:crown and wall	Crown : 50-100 mm Wall : 30 mm	Crown and Wall	None
II III	Systematic bolts, L: 4-5 m, S: 1-1,5 m IP: crown and wall	Crown : 100-150 mm Wall : 100 mm	Crown and Wall	Light to medium ribs, S: 1.5 m

Section	Q Support	
	Rock Bolt	**Shotcrete Thickness**
I	Systematic bolt, L: 2.7 m, S: 2.3 m	-
II	Systematic bolt, L: 2.8 m, S: 2.1 m	Unreinforced, Crown and wall: 4-10 cm
III	Systematic bolt, L: 2.7 m, S: 1.7 m	Fibre - reinforced, Crown and wall: 5-9 cm

L: Length; S: Spacing; IP: Installation Position

5 Conclusion

The following conclusions can be drawn from the overall data analysis results.

1. Pyroclastic breccia rocks dominate rock units along the tunnel route with fair and poor rock mass quality.
2. The recommended tunnel excavation method was based on the RMR classification: Top heading and bench. "1–1.5 m advance in top heading" in sections I and "1.5–3 m advance in top heading" in sections II and III. And "Install support concurrently with excavation 10 m from face" in sections I and "Commence support after each blast. Complete support 10 m from face" in sections II and III
3. Rock Bolt, Shotcrete, Wiremesh, and Steel Set were the recommended support for mbay dam diversion tunnels. Of course, with the type, size, and amount that adjusted the quality of the rock mass and the classification used
4. The empirical analysis is instrumental in the planning stage of tunnel construction implementation. However, numerical analysis is needed to validate the safety and stability of the proposed support tunnel in the construction of the tunnel

Acknowledgment

The authors would like to thank Mr. Hendy Setiawan as supervisor, the Ministry of Public Works and Public Housing for research funding assistance and BWS Nusa Tenggara II, especially Mr. Davianto Frangky Bernpietson Welkis as mentor, Mr. Nedy Hidayat and Mr. Yohanes Pabi as project leaders who have provided access, permits, and opportunities to us to conduct research in the area of the Mbay Dam construction project.

Reference

[1] Celada B and Bieniawski Z T 2019 Ground Characterization and Structural Analyses for Tunnel Design (CRC Press).
[2] Singh B and Goel R K 2011 Engineering Rock Mass Classification Tunneling, Foundations, and Landslides Bhawani (Elsevier Inc.).
[3] PT. Indra Karya 2019 Lanjutan Penyelidikan Geologi, Pembuatan Model Test Dan Sertifikasi Desain Bendungan Mbay (Lambo) Di Kab. Nagekeo (Balai Wilayah Sungai Nusa Tenggara II).

[4] Genis M, Basarir H, Ozarslan A, Bilir E and Balaban E 2007 Engineering geological appraisal of the rock masses and preliminary support design, Dorukhan Tunnel, Zonguldak, Turkey Eng. Geol. **92** 14–26.
[5] Palmström A 2009 Combining The RMR, Q, AND RMi Classification System Tunn. Undergr. Sp. Technol.
[6] Sopaci E and Akgün H 2008 Engineering geological investigations and the preliminary support design for the proposed Ordu Peripheral Highway Tunnel, Ordu, Turkey Eng. Geol. **96** 43–61
[7] Özsan A, Başarir H, Yüceel S and Cücen Ö 2009 Engineering geological evaluation and preliminary support design for the metro extension tunnel, Ankara, Turkey Bull. Eng. Geol. Environ. **68** 397–408.
[8] Zekrinaldi, Anggara F and Setiawan H 2021 Excavation method and support system in the diversion tunnel of Tiga Dihaji Dam, South Sumatera, Indonesia based on the rock mass quality IOP Conf. Ser. Earth Environ. Sci. **871**.
[9] Putera D A, Hendrayana H and Setiawan H 2021 Analysis of Support Systems and Excavation Methods Based on Rock Mass Quality in the Intake Tunnel of the Jlantah Dam E3S Web Conf. **325** 02009.
[10] Bieniawski Z T 1989 Engineering rock mass classifications: a complete manual for engineers and geologists in mining, civil, and petroleum engineering Eng. rock mass Classif. a Complet. Man. Eng. Geol. mining, civil, Pet. Eng.
[11] Barton N, Lien R, Lunde J, Barton L and Lunde 1974 Engineering Classification of Rockmasses for Design of Tunnel Support Rock Mech. **6** 189–236.
[12] Deere D U 1963 Technical description of rock cores for engineering purposes (Rock Mechanics and Engineering Geology 1).
[13] ISRM 1981 Suggested Methods Rock characterisation, testing and monitoring (Oxford: Pergamon Press).
[14] Koesoemadinata S, Noya Y and Kadarisman D 1994 Peta Geologi Lembar Ruteng, Nusa Tenggara ed N Ratman (Pusat Penelitian dan Pengembangan Geologi, Kementerian ESDM).

IOP Conf. Series: Earth and Environmental Science 1233 (2023) 012006 doi:10.1088/1755-1315/1233/1/012006

Empirical Designs of Excavation Method and Support System of the Intake Tunnel of Meninting Dam, West Nusa Tenggara, Indonesia

Aditya Rasdi Metly[1,2] and Hendarto[2]

[1]Departement of Geological Engineering, Gadjah Mada University, Yogyakarta, Indonesia
[2]Ministry of Public Works and Housing of Indonesia

*email: igbindrawan@ugm.ac.id

Abstract. The intake tunnel of Meninting Dam has been designed based on RMR but with less detailed drill core evaluation in the tunnel area, which will lead to less accuracy in the design. The objective of this study was to evaluate the tunnel's excavation method and support system using the latest data based on rock mass classification of RMR, GSI, and Q–system. The investigation was conducted by surface geological mapping and core drilling evaluation. The results showed that the tunnel area consists of pyroclastic breccia intercalated with lapilli stone. Based on the rock mass quality, the tunnel can be divided into two sections: Section A has fair quality rock, and Section B has poor quality rock. Based on the empirical designs, the recommended rock and tunnel excavation in Section A by hammer and top heading bench, and Section B by ripping and top heading bench. The recommended supports in all sections were using a combination of rockbolts, shotcrete, and wire-mesh. Even though the empirical approach is helpful in the preliminary design, a numerical study is needed to verify the empirical design so the recommended excavation and support system can be safely applied.

1. Introduction

This study is conducted at the Meninting Dam project, located in the Bukit Tinggi village, Lingsar, Lombok Barat, West Nusa Tenggara. A drainage channel from the dam reservoir is needed to drain water downstream to support the fulfillment of raw water needs and irrigation networks. The Meninting Dam is planned to have an intake tunnel with a total length of 463 m and 4.6 m in diameter. Several studies, including geological investigations, core drilling, and tunnel support system design using RMR [1], were done by the planning consultant based on 3 drilling wells that existed in 2017. The investigation results stated that the intake tunnel would be constructed along poor volcanic rocks [2]. However, the investigations carried out have not been detailed on the location of the intake tunnel, and the designs are only calculated using RMR. Based on the insufficient investigation, which will cause less accuracy in the design, it is necessary to evaluate the designs using other methods [3] and add the latest data from 2 additional drilling wells in 2020. The application of various rock mass classification systems is recommended to provide initial estimates of support requirements [4]. A number of previous studies [5,6,7,8] also used RMR, GSI, and Q classification as an empirical method.

In this study, the surface and subsurface mapping is performed, and the results are then correlated into geological section of the tunnel. RMR [1] and Q–system [9] are used to determine the quality of rock mass, while GSI [10] is used as an index for the assessment of excavabilty of rock masses. The recommended excavation method and support system are selected based on the rock mass quality from the core drilling evaluation. The evaluation was carried out by describing the rock cores along the intake tunnel, identifying and calculating the parameters according to the rock mass classification at each depth, and grouping them into engineering geological units. The results of this paper are expected to provide a suitable tunnel excavation and support system with its field condition for the intake tunnel construction at Meninting Dam. However, there are limitations to this study that need to be disclosed, such as: 1. the core samples used are weathered and have been stored since 2017 and 2020, making it difficult to evaluate; 2. the B15-6 drilling well does not cover the tunnel elevation; 3. numerical studies are required to determine which of the two proposed empirical designs is the most stable.

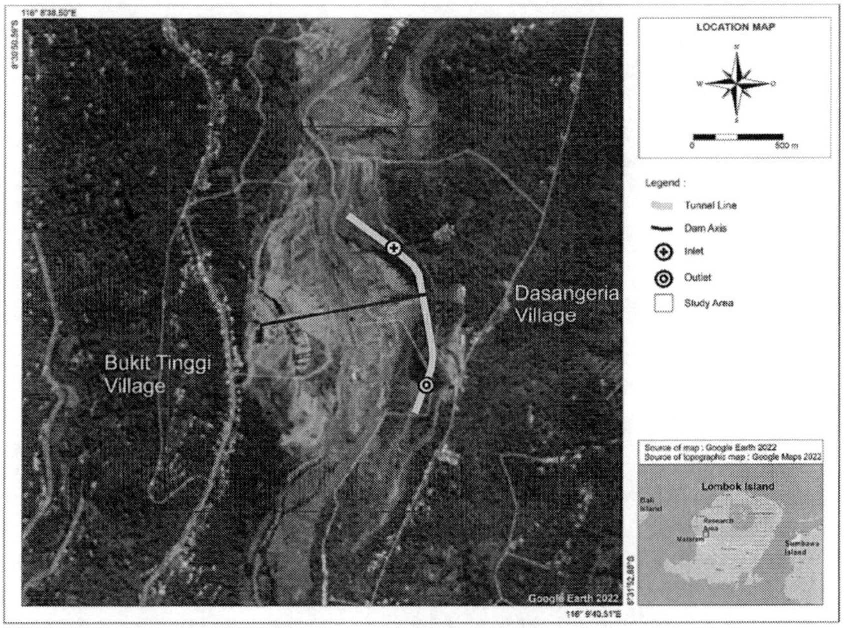

Figure 1. Location of Meninting Dam Intake Tunnel.

2. Methodology

2.1. Geological Investigation

The geological investigations in the study area involved surface and subsurface mapping. Surface mapping consisted of observing and measuring the outcrops, including rock types and their contacts, geological structure elements, textures of rock, and collecting the samples for further laboratory tests and analysis. Engineering aspects can also be described on the rock surface, such as the level of weathering and joint conditions. The mapping was carried out in an area of 1 x 1 km, and the description of rock in the study was based on Fisher's classification [11]. Subsurface geological mapping was carried out by evaluating the core drilling results, including lithology, degree of weathering, rock discontinuities, and rock mass parameters according to RMR, GSI, and Q–system classification. The results then correlated into a rock mass quality profile along the tunnel path. This investigation used data from five core drilling wells along the tunnel path, namely B15–3, B20–5, B14–1, B20–4, and B15–6.

2.2. Geological Strenght Index (GSI)

The GSI index was calculated in this study using the assessment developed by Hoek et al. [9] by considering the factor of rock discontinuity (*Jcond*) and Rock Quality Designation (RQD) value using the Equation 1 [9]. GSI index is then used for the assessment of excavabilty of rock masses according to excavability chart by Tsiambos and Saraglou [12].

$$GSI = 1.5Jcond + {}^{RQD}/_{2} \quad (1)$$

2.3. Rock Mass Rating (RMR)

RMR is a classification using six parameters to determine the rock mass rating by Bieniawski [1]. The parameters are uniaxial compressive strength (UCS) or point load index, RQD, spacing of discontinuities, condition of discontinuities, groundwater conditions, and orientation of discontinuities. The strength of the intact rock used in this paper was obtained from the estimated UCS value of intact rock according to Hoek and Brown [13] and the value of UCS test in the laboratory from five core drilling points at a certain depth (Table 1).

Table 1. UCS value of drill cores from the laboratory test.

No	Hole Number	Depth (m)	UCS Value (Mpa)
1	B15–3	12	21.88
2	B15–3	30	45.49
3	B20–5	21	13.7
4	B20–5	38	148.7
5	B20–5	42	171.7
6	B14–1	23	43
7	B20–4	22	5,7
8	B20–4	24	139,4
9	B20–4	28	24

RQD is a rock mass classification modified from the calculation of core recovery but only considers cores with a length of more than or equal to 100 mm [1]. The entire length of the core pieces longer than 100 mm is added up and then divided by the total length of the rock core in one take to get the RQD value [14]. The gap between each discontinuity plane determines the spacing of rock discontinuities. Discontinuity conditions depend on the joints surface, aperture, persistence, and weathering conditions. Water inflow, water pressure, and groundwater level are parameters for groundwater conditions. The maximum value of the weighting of the RMR parameter is 100, classified into 5 (five) classes by Bieniawski [1]. The guidance for excavation and tunnel support (after Bieniawski, 1989) was used to determine the excavation method and tunnel support systems.

2.4. Q–system

The tunneling Quality Index (Q) is a classification of rock mass using six parameters that expresses the quality of rock mass in the Q value. The parameters are RQD, amount of joint (Jn), the roughness of the joint (Jr), change in stock (Ja), water condition (Ja), and the stress factor (SRF) [10], which are stated in Equation 2.

$$Q = \frac{RQD}{Jn} \cdot \frac{Jr}{Ja} \cdot \frac{Jw}{SRF} \quad (2)$$

In this study, the Q value was calculated by observing rock parameters from the drill core to determine the support design according to the Q–system graph from NGI [15].

3. Results and Discussion

3.1. Surface Geological Conditions

The geology of the study area is in the Lekopiko Formation, composed predominantly of of laharic breccias and lava breccias according to the regional geology map [16]. Based on the geological mapping carried out in the study area, it was found that the lithology consists of two rock units from oldest to youngest, pyroclastic breccia with intercalations of lapilli stone and lapilli tuff (see Fig. 2 and 3). Joints become geological structures that developed in the study area, while no sedimentary layers were found. Breccias unit consisted of pyroclastic breccia and lapilli stone of pleistocene age [16]. These rock outcrops were found on the hillsides and had a thickness of about 60 m. On top of the units, a layer of residual soil and andesitic fragments were widely found.

The majority of the pyroclastic breccia at the ground surface had weathered. The color of highly weathered pyroclastic breccia in the field is dark brown, and the slightly weathered is gray, composed of fragments ranging in size from 6 cm – 4 m (Fig. 4), with < 2 mm sized of a matrix. The texture of pyroclastic breccia is massive, poorly sorted, and fragment supported. The shape of the fragments is angular – rounded, with good interlocking. The composition consists of 70% fragments of andesite and a small portion of lithic tuff, while 30% is a tuffaceous sand matrix.

Intercalated Lapilli stone had a thickness of 2–5 m with a discontinuous depositional geometry. The contact between lapilli stone and pyroclastic breccia can be seen in Fig. 5. The differences between the two breccias in characteristics were based on the size of the fragments. Lapilli stone has a fragment size of 5–15 cm, with a matrix less than 1 mm, well–sorted, matrix–supported, angular–rounded fragments. The composition consists of 40% fragments with the composition of andesite and volcanic rock fragments, while 60% is a matrix with tuffaceous sand.

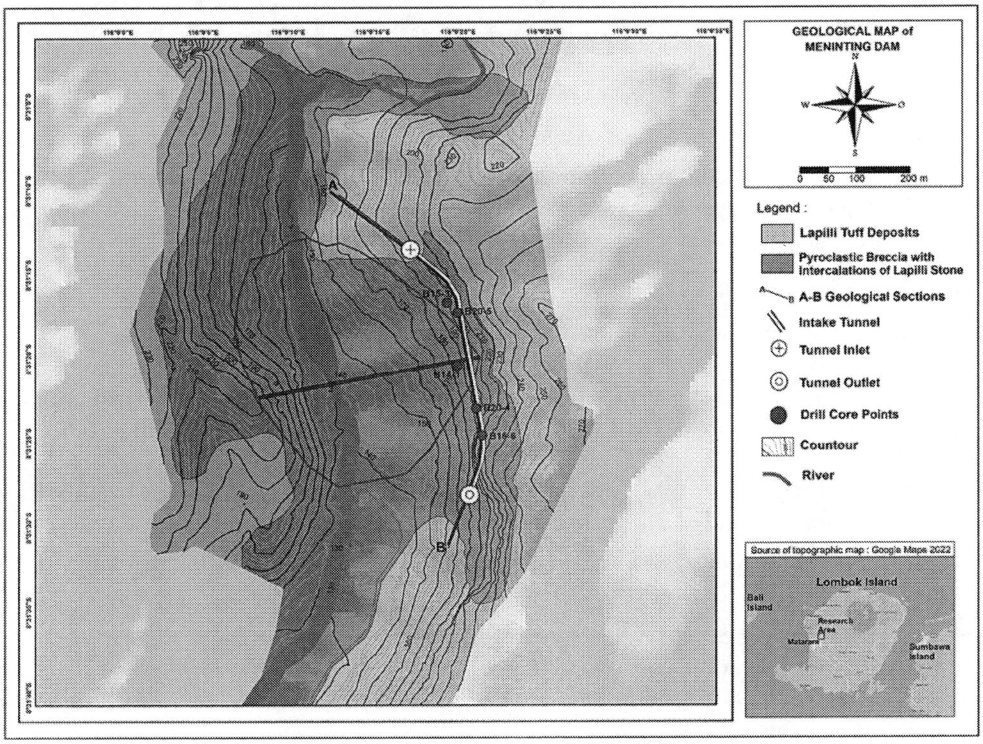

Figure 2. Geological map of the study area.

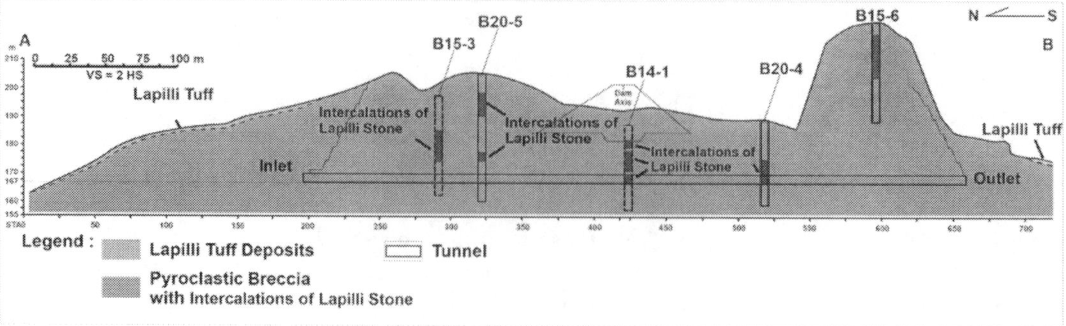

Figure 3. Geological Sections A – B.

Figure 4. Pyroclastic breccia at the study area.

Figure 5. Pyroclastic breccia with intercalations of lapilli stone at the tunnel location.

Lapilli tuff were spread unconformably over the breccia unit, which was controlled by topography. This unit has a gray hue, a layer that is 0.5 to 2 meters deep, has a loose texture, and is formed of 70% pumice fragments that range in size from 0.1 to 4 cm and 30% tuff matrix.

3.2. Subsurface Geological Conditions

The results of surface mapping can be used as a reference for subsurface investigations. In this study, the investigation was carried out by observing five core drilling wells along the tunnel, as shown in Figure 9. The lithology of the subsurface shows similarities to surface rock. The drill core evaluations based on rock mass parameters of RMR, GSI, and Q–system are presented in Tables 2–6.

1. B15–3 borehole is located at the inlet tunnel, with a depth of 35 m, the water level is 10 m from the surface, and the tunnel elevation at 30 m depth.

Table 2. Assessment of rock mass quality of B15–3 drill core.

Depth (m)	Lithology	RMR [1]						Q [3]						GSI [2]
24–35	Pyroclastic Breccia	UCS	RQD	J. spa	J. cond	G.W	Adj	RQD	Jn	Jr	Ja	Jw	SRF	1.5 (J.cond) x (RQD/2)
Rating		4	13	10	20	4	–5	55	4	3	2	0.5	2.5	1.5 (20) x (55/2)
Total Rating		46						4.125						57.5
Quality		Fair						Fair						

IOP Conf. Series: Earth and Environmental Science 1233 (2023) 012006 doi:10.1088/1755-1315/1233/1/012006

2. B20–5 borehole is located at the inlet tunnel, with a depth of 45 m, the water level is 12 m from the surface, and the tunnel elevation at 39 m depth.

Table 3. Assessment of rock mass quality of B20–5 drill core.

Depth (m)	Lithology	RMR [1]						Q [3]						GSI [2]
32–45	Pyroclastic Breccia	UCS	RQD	J. spa	J. cond	G.W	Adj	RQD	Jn	Jr	Ja	Jw	SRF	1.5 (J.cond) x (RQD/2)
Rating		2	13	10	25	4	–5	50	6	3	2	1	3	1.5 (25) x (50/2)
Total Rating				49						4.17				62.5
Quality				Fair						Fair				

3. B14–1 borehole is located in the middle of the tunnel, with a depth of 30 m, the water level is 15,5 m from the surface, and the tunnel elevation at 23 m depth.

Table 4. Assessment of rock mass quality of B14–1 drill core.

Depth (m)	Lithology	RMR [1]						Q [3]						GSI [2]
19–22	Lapilli Stone	UCS	RQD	J. spa	J. cond	G.W	Adj	RQD	Jn	Jr	Ja	Jw	SRF	1.5 (J.cond) x (RQD/2)
Rating		1	3	5	20	4	–5	10	6	3	1	0.6	2.5	1.5 (20) x (10/2)
Total Rating				28						1.32				35
Quality				Poor						Poor				

4. B20–4 borehole is located in the middle of the tunnel, with a depth of 30 m, the water level is 12 m from the surface, and the tunnel elevation at 23 m depth.

Table 5. Assessment of rock mass quality of B20–4 drill core.

Depth (m)	Lithology	RMR [1]						Q [3]						GSI [2]
18–23	Lapilli Stone	UCS	RQD	J. spa	J. cond	G.W	Adj	RQD	Jn	Jr	Ja	Jw	SRF	1.5 (J.cond) x (RQD/2)
Rating		1	3	5	20	4	–5	10	15	2	1	1	2.5	1.5 (20) x (10/2)
Total Rating				28						0.53				35
Quality				Poor						Very Poor				

5. B15–6 is located at the outlet tunnel, with a depth of 25 m, the water level is 21 m from the surface. This drill core did not cover the tunnel elevation.

Table 6. Assessment of rock mass quality of B15–6 drill core.

Depth (m)	Lithology	RMR [1]						Q [3]						GSI [2]
15–25	Pyroclastic Breccia	UCS	RQD	J. spa	J. cond	G.W	Adj	RQD	Jn	Jr	Ja	Jw	SRF	1.5 (J.cond) x (RQD/2)
Rating		4	13	10	25	7	–5	60	4	3	1	0.3	2.5	1.5 (25) x (60/2)
Total Rating				54						5.94				67.5
Quality				Fair						Fair				

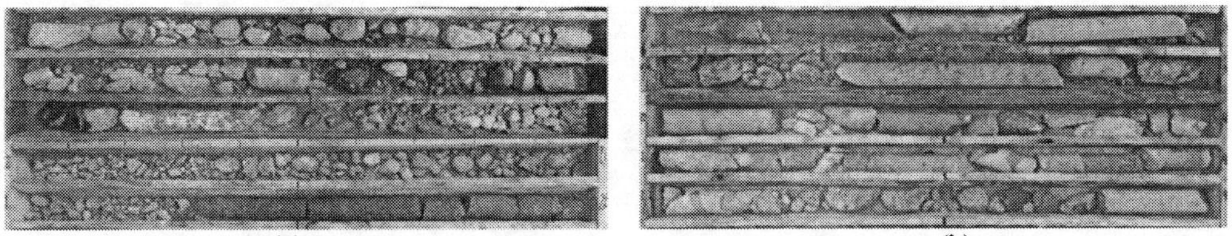

(a) (b)

Figure 6. Cores at the tunnel location,
a) lapilli Stone with poor quality (B20–4); b) Pyroclastic breccia with good quality (B20–5)

IOP Conf. Series: Earth and Environmental Science 1233 (2023) 012006 doi:10.1088/1755-1315/1233/1/012006

The result of rock mass quality assessments based on RMR and Q–system also surface and subsurface geological mapping then correlated into engineering geological profile along the tunnel, as shown in Figure 7. Based on the correlation data, it was discovered that the pyroclastic breccia lithology had relatively better rock quality than the lapilli stone. This is influenced by the discontinuity condition as well as the composition and size of the breccia fragments. Analysis of core drilling shows the tunnel path consists of two types of engineering geological units, that is: pyroclastic breccia with fair quality (RMR: 41–60, Q: 4–10); and lapilli stone with poor quality (RMR: 40–21, Q: 0.5–4.0).

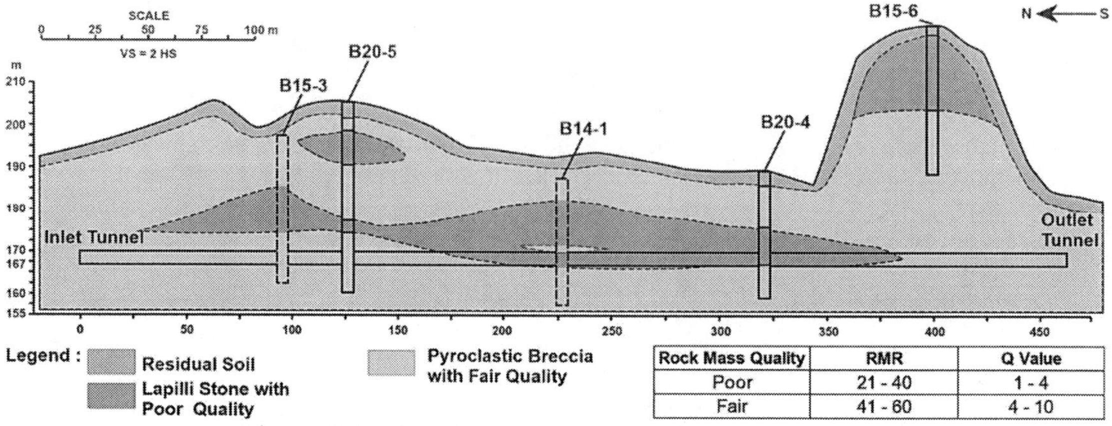

Figure 7. Engineering geological profile along the tunnel.

According to the profile, the tunnel can be divided into two sections: 1.) Section A (St 0–177 and St 370–463) has fair quality rock; 2.) Section B (St 177–370) has poor–quality rock, as shown in Fig. 8.

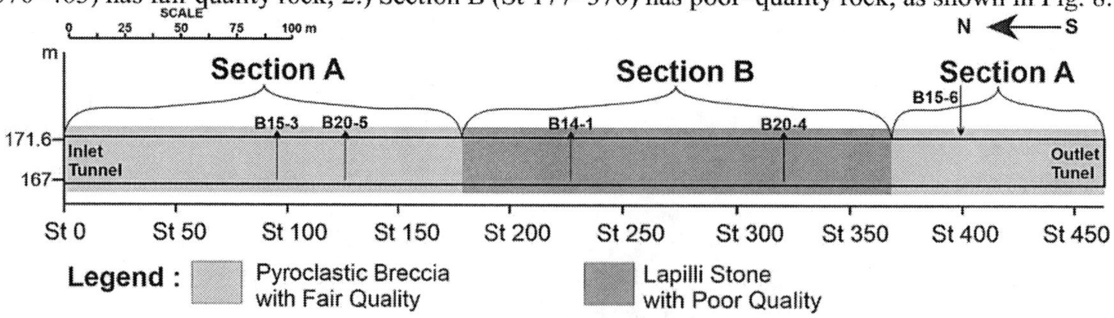

Figure 8. Tunnel section based on rock mass quality.

3.3. *Rock and Tunnel Excavation Method*

In this study, GSI index and point load index (Is50) at the tunnel elevation was used to determine the rock excavation method based on the excavability assessment of a rock mass chart by Tsiambos and Saraglou [12]. GSI index was obtained from rock mass quality assessments as shown in subsection 3.2, whereas the Is50 index was determined by converting the UCS test results in Table 2 with Equation (3) [17].

$$UCS = 24I_{s50} \tag{3}$$

The estimated Is50 index of section A is 1.9 MPa and section B is 0,24 MPa, so the classification chart used was for the rock masses with Is50 value less than 3 MPa. The results of GSI index plotting (Fig. 9) show that the recommended rock excavation in section A is by hammer and blasting, while in section B, it is recommended to use the ripping method.

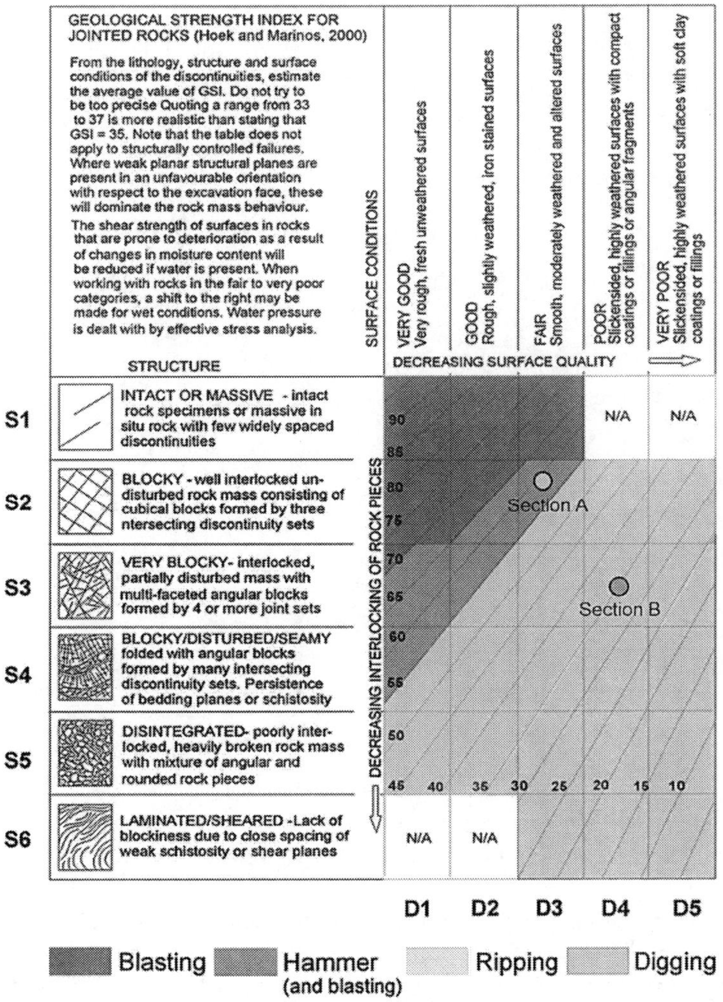

Figure 9. Rock mass excavation based on excavability chart (Is50 <3 MPa) (Tsiambos and Saraglou, 2009).

RMR classification was used in determining the tunnel excavation method by referring to the excavation and tunnel support guides after Bieniawski [1]. It is recommended to apply the top heading and bench methods with an excavation length of 1.5 to 3 meters at the top heading for Section A. While in Section B, top heading and bench with 1 to 1.5 meters advanced on top are also suggested, followed by simultaneously installing support while excavating 10 meters from face. Table 7 summarizes the suggested excavation method for both categories.

Table 7. Rock and Tunnel Excavation Methods based on GSI and RMR classification.

Tunnel Section	Rock Excavation	Tunnel Excavation Method
	GSI	RMR
Section A	Hammer (and blasting)	Top Heading and bench, 1.5 – 3 m advance in top heading
Section B	Ripping	Top Heading and bench, 1.0 - 1.5 m advance in top heading, simultaneously installing support while excavating 10 meters

IOP Conf. Series: Earth and Environmental Science 1233 (2023) 012006 doi:10.1088/1755-1315/1233/1/012006

3.4. Support System

The RMR and Q–system classifications were used as empirical approaches to determine the support system in the tunnel. RMR classification may be applied directly to the excavation and tunnel support guides (Bieniawski, 1989), while Q–system classification uses Q value, tunnel height, and the Excavation Support Ratio (ESR). The tunnel is 4.6 meters in height, and the ESR value used in this study is 1 [15]. The rock support chart proposed by NGI [15] was used to determine the support types based on Q–system classification.

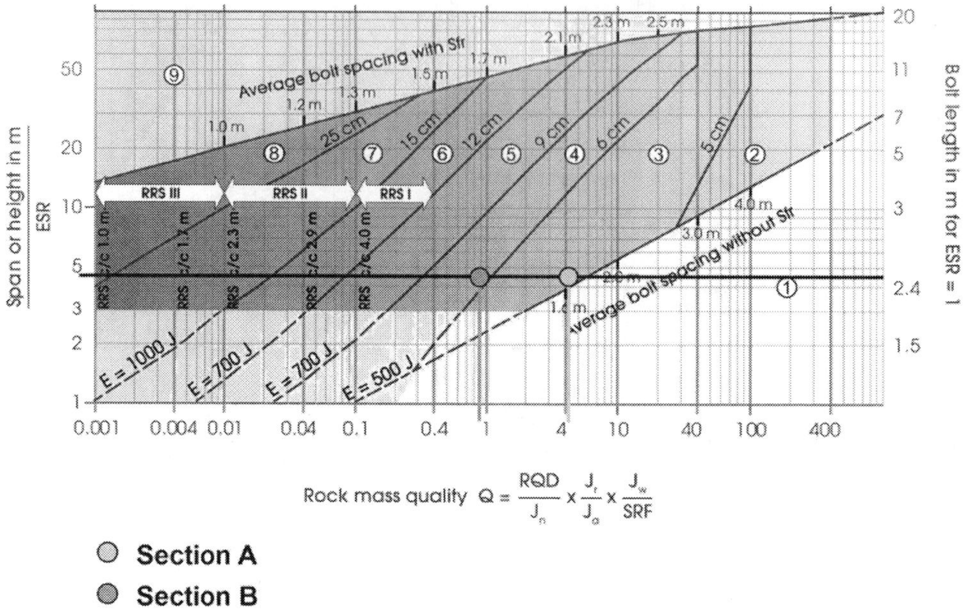

Fig. 10. Plotting Q value on rock support chart (NGI, 2015).

The rock support chart indicates the use of 2.45 m bolt length with support categories 3 and 4. For Section A, systematic bolting and fibre–reinforced shotcrete of 5–6 cm thick are proposed. In Section B, fibre–reinforced shotcrete of 6–9 cm thick and bolting can be used. The recommended support systems based on RMR and Q–system classification are given in Table 8.

Table 8. Proposed support types for the sections along the tunnel.

Tunnel Section	Support System	
	RMR	Q–system
Section A	Rockbolts = systematic bolting, length : 4 m with spaced 1.5–2.0 m in the crown and wall, also adding wire–mesh on the crown. Shotcrete = t : 50–100 mm on the crown and 30 mm on the sides	Rockbolts = 2.1 m space and 2.45 m length Shotcrete = 5–6 cm thick fibre-reinforced sprayed concrete
Section B	Rockbolts = systematic bolting, length : 4–5 m with spaced 1.0–1.5 m in the crown and wall with wire–mesh. Shotcrete = t : 100–150 mm on the crown and 100 mm on the sides	Rockbolts = 1.6 m space and 2.45 m length Shotcrete = 6–9 cm thick fibre-reinforced sprayed concrete

4. Conclusion

The geological unit in the study area consisted of pyroclastic breccia with intercalations of lapilli stone and lapilli tuff. The result of rock mass quality assessments based on RMR and Q–system also geological mapping then correlated into engineering geological profile along the tunnel. It was discovered that the pyroclastic breccia lithology had relatively better rock quality than the lapilli stone. Analysis of core drilling shows the tunnel path consists of two types of engineering geological units, that is: pyroclastic breccia with fair quality (RMR: 41–60, Q: 4–10); and lapilli stone with poor quality (RMR: 40–21, Q: 0.5–4.0). The tunnel can be divided into two sections based on the rock mass quality: 1.) Section A (St 0–177 and St 370–463) has fair quality rock; 2.) Section B (St 177–370) has poor–quality rock.

Based on the GSI and RMR classifications, the recommended rock and tunnel excavation methods are in Section A by hammer with blasting and top heading bench with 1.5–3 m advanced on top heading, and in Section B by ripping and top heading bench with 1.0–1.5 m advanced on top heading, followed by simultaneously installing support while excavating 10 m from the face.

Based on the RMR method, the suitable support system in Section A was systematic bolting in the crown and walls, 4 m long with a spacing of 1.5–2.0 m, and shotcrete 50–100 mm on the crown and 30 mm on the sides, while based on Q–system, 2.45 m bolt length with a 2.1 m space and 5–6 cm thick fiber–reinforced shotcrete. The recommendation in section B based on RMR was to install systematic bolts in the crown and walls, 4–5 m long with spaces of 1.0–1.5 m, wire–mesh, and shotcrete 100–150 mm in thickness on the crown and 100 mm on sides, while on Q–system, 2.45 m bolt length with 1.6 m space and 6–9 cm thick fiber–reinforced shotcrete.

Even though the empirical approach is useful for early design, a numerical analysis is needed to find out about stress distribution, deformation, and support performance so that the empirical design can be used safely on the intake tunnel of Meninting Dam.

Acknowledgements
The authors are grateful to the BWS Nusa Tenggara I, for allowing the study, and the first author wishes to express his gratitude to The Ministry of Public Works and Housing of Indonesia for the 2021 Superspecialist scholarship for the Master Study Program.

References

[1] Bieniawski Z T 1989 *Engineering Rock Mass Classifications: a Complete Manual for Engineers and Geologists in Mining, Civil and Petroleum Engineering* .New York: A Wiley-Interscience Publication.
[2] BWS Nusa Tenggara I 2017 *Laporan Utama Sertifikasi Desain Bendungan Meninting di Kabupaten Lombok Barat.* Unpublished.
[3] Rehman H Ali W Naji A Kim J Abdullah R and Yoo H 2018 *Review of Rock-Mass Rating and Tunneling Quality Index Systems for Tunnel Design: Development, Refinement, Application and Limitation*. Applied Sciences. 8. 27. 10.3390/app8081250.
[4] Hoek E Kaiser P K Bawden W F 1995 *Support of Underground Excavations in Hard Rock.* Balkema, Rotterdam. 215 pp.
[5] Genis M Basarir H Ozarslan A Bilir E and Balaban E 2007 *Engineering geological appraisal of the rock masses and preliminary support design, Dorukhan Tunnel, Zonguldak, Turkey*, Engineering Geology **92** 14–26.
[6] Basarir H 2006 *Engineering geological studies and tunnel support design at Sulakyurt dam site, Turkey*, Engineering Geology **86** 225-237.
[7] Akram M S Zeeshan M Haroon M and Mirza K 2009 *Open Journal of Geology* 9, 11, 809-828
[8] Rasouli M 2009 *Engineering geological studies of the diversion tunnel, focusing on stabilization analysis and support design, Iran*, Engineering Geology **108** 208-224.
[9] Barton N Lien R and Lunde J 1974 *Engineering Classification of Rock Masses for the Design of Tunnel Support* Rock Mechanics Vol. **6/4** by Springer-Verlag p 189 – 236.

[10] Hoek E Carter T G and Diederichs M S 2013 *Quantification of the Geological Strength Index Chart* Proceedings of the 47th US Rock Mechanics / Geomechanics Symposium San Francisco CA USA 23-26 June.
[11] Fisher R V 1966 Earth Sci. Rev. **1**(4), 287-298.
[12] Tsiambaos G and Saroglou H 2009 *Excavability Assessment of Rock Masses Using the Geological Strength Index (GSI)* Bulletin Engineering Geology and Environment p 13-27.
[13] Hoek E Brown E T 1997 *Practical Estimates of Rock Mass Strength*, International Journal of Rock Mechanics and Mining Sciences **34** (8) 1165-1186.
[14] Deere D U and Deere D W 1998 *The rock quality. des.*, J. Rock Class. Syst. For Eng. Purposes, 91-101.
[15] Norwegian Geotechnical Institute (NGI) 2015 *Handbook: Using the Q-system, Rock Mass Classification and Support Design* (Oslo: NGI).
[16] Mangga S A Atmawinata S Hermanto B and Amin T C 2010 *Peta Geologi Lembar Lombok*, West Nusa Tenggara Bandung Pusat Penelitian dan Pengembangan Geologi.
[17] Broch E and Franklin J A 1972 *The Point-Load Strength Test* International Journal of Rock Mechanics and Mining Sciences, 9, 669-697.

IOP Conf. Series: Earth and Environmental Science 1233 (2023) 012007 doi:10.1088/1755-1315/1233/1/012007

Empirical Method for Engineering Geological Evaluation of Pelosika Dam Diversion Tunnel Design, Southeast Sulawesi, Indonesia

J P Diwani[1,2], G I Marliyani[1]

[1]Department of Geological Engineering, Faculty of Engineering, Universitas Gadjah Mada, Yogyakarta, Indonesia

[2]Directorate General of Water Resources, The Ministry of Public Works and Housing, Jakarta, Indonesia

*email: jodanpanrettadiwani@mail.ugm.ac.id, gayatri.marliyani@ugm.ac.id

Abstract. This paper explains a Pelosika Dam Diversion Tunnel in metamorphic rocks (quartz muscovite phyllite) and aims to find out excavation methods, stand-up time, and support systems for tunnel construction. The purpose of this research is that the development of science in tunnel construction with the characteristic of metamorphic rock will increase. Tunnel location in poort rock mass quality based on previous geology investigations. Based on current geology and engineering geological mapping results, diversion tunnel through poor rock mass of metamorphic rocks (phyllite) which have slightly-highly weathered. We classified rock mass quality by GSI, RMR, and Q-System methods which produce the almost same results (poor rock mass quality). The GSI Chart and Pettifer-Fookes Chart which excavation method has the results of the digging and ripping. RMR value determines the stand-up time last 42-240 minutes after excavation. RMR and Q-System values generate the support system which used rock bolts (length= 1.8 m), shotcrete (thick=10-15 cm), and steel sets (spacing=1,5m) if required. A purpose of combination from the different excavation methods and support systems to complement each other and improve the safety of the tunnel without compromising the cost-effectiveness of construction.

1. Introduction

We conducted this study based on a plan to build the Pelosika Dam in the Southeast Sulawesi Province of Indonesia. Pelosika dam has functions for irrigation, raw water, flood control, hydropower, and tourism. Dam construction has an impact on increasing the economy by increasing people's welfare. In the design of the dam structure, the diversion tunnel has the function to drain the river when carrying out the construction of the body of the dam. Tunnel construction requires adequate geological investigation to prevent collapse during construction.

Our investigation is in Asinua Jaya Village, Asinua District, Konawe Regency, Southeast Sulawesi Province as seen in **Figure 1**. Based on the regional geological map, Pelosika Dam stands on the Paleozoic Metamorphic Rocks and alluvial deposits. The research location is in the metamorphic rock which is of poor quality for tunnel construction. The condition of the research area is almost the same as the Athens schist formation (Greece) because they are both located in metamorphic rocks with weak rock conditions [1] [2]. The design of Pelosika Dam has twin diversion tunnels horseshoe-shaped with a vertical diameter of 5,2 m and a length of 596,53 m. The tunnel inlet is planned at +67.25 m elevation and the outlet of the tunnel at +65.50 elevation [3].

Lithology in the research area form of alluvial and metamorphic rock (phyllite) with moderately until hardly weathered affects the level of rock hardness to be low based on previous research [4]. Rock mass quality using the Q-System method [5] had a Q value between 0-0.25 which includes extremely poor until exceptionally poor classification that was influenced by very high weathered conditions [4]. Based on the other past engineering geological investigation [6], the quality of rock mass using the Central Research Institute of Electric Power Industry (CRIEPI) method [7] [8] described the rock condition of the diversion tunnel as low hardness level in D-CL class (D= very soft rock, CL=soft rock). Meanwhile, the rock mass quality using Rock Mass Rating (RMR) [9] had very poor-poor rock results. Based on the above previous research, the main problem at the research site is that the tunnel construction will penetrate metamorphic rocks with poor rock mass classification with high weathering conditions. From the main problem, it is necessary to carry out further research that focuses on engineering geology conditions at the diversion tunnel location using the latest geological investigation data (2022) so that tunnel construction planning will be updated and make it safer during construction. The limitation of the study is investigating geological mapping and engineering geology mapping to determine the rock conditions of the study area in more detail, followed by the classification of rock mass quality using several different methods. The results of the rock mass classification will be used to determine the excavation method according to field conditions, the stand-up time after excavation, and the support system needed for tunnel reinforcement.

We conducted geological mapping, determining rock weathering, and measuring rock mass conditions using the Geological Strength Index (GSI) method to identify the type of surface condition. The analysis of rock mass classification and rock weathering from drilling cores have a function to know the subsurface condition. This research has the purpose to identify the rock mass quality of the research area by using classification methods of the Geological Strength Index (GSI) [1] [10], Rock Mass Rating (RMR) [9], and Q-System [5]. The excavation methods use Pettifer-Fookes empirical method (blasting, ripping, and digging) and the GSI empirical method (blasting, hammer, ripping, and digging). RMR empirical method to know the stand-up time, while RMR empirical method and the Q-system empirical method to know the support system tunnel. The benefit of this research is that there is a new technical justification for complex problems that can be used as advice for job owners to build tunnels in the future to make them safer during construction.

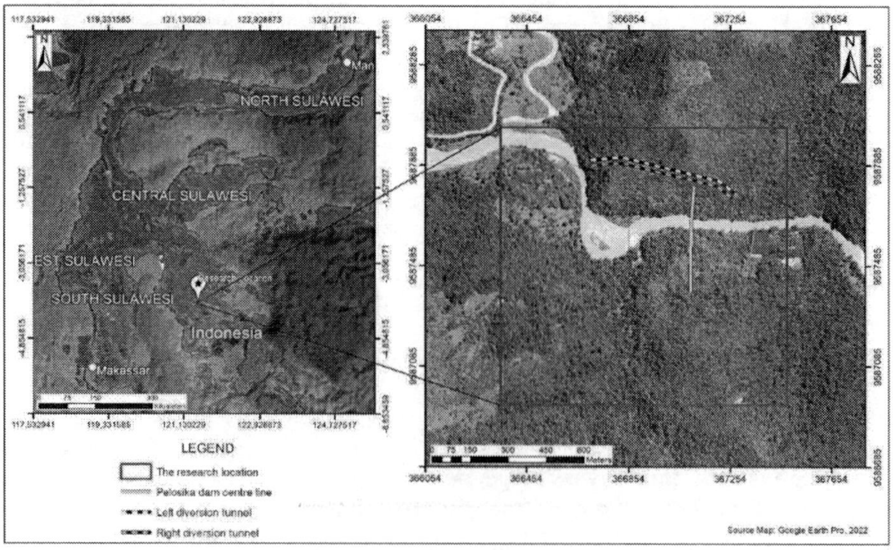

Figure 1. The location of the study area in Pelosika Dam, Southeast Sulawesi Province of Indonesia (Google Earth, accessed in June 2022)

2. Research Method

The flowchart of this research goes through five stages to arrive at the final objective of the research. First, we did a field investigation to know the geological conditions and generate a geological map. The rock identification and mapping by field investigation. In the second stage, we know the quality of rock mass from three steps; field identification, testing of rock samples in the laboratory, and analysis of core drills. We used the GSI method to know field conditions by determining the quality of the rock mass surface, describing discontinuity, and recognizing weathering of rock. The third stage is the analysis of the empirical GSI chart and the empirical method of the Pettifer-Fookes chart to determine the most feasible excavation method. In the fourth stage, we know the tunnel stand-up time after excavation by using the empirical RMR method. The last stage is to determine the tunnel support system with the empirical RMR and Q-System methods. The research method flowchart is shown in **Figure 2.**

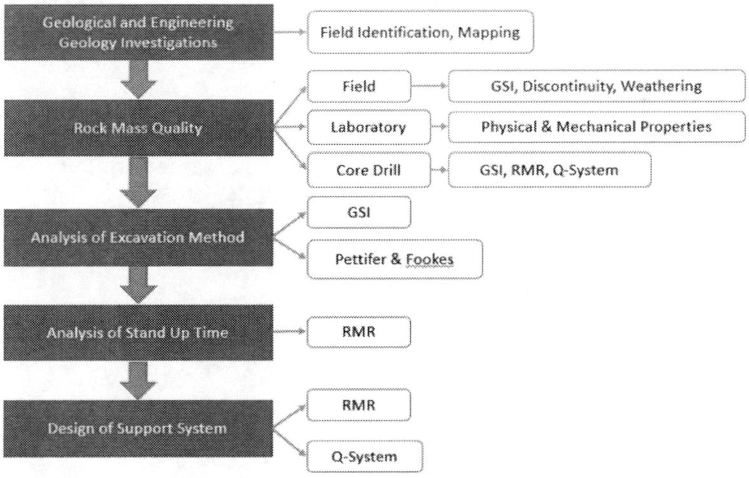

Figure 2. Research flowchart of the Pelosika Dam Diversion Tunnel

2.1. *Geological Strength Index (GSI)*

We know the GSI value by observing the condition of rock outcrops with two main parameters, there are geology structure and surface conditions [11]. This method is suitable for determining the quality of rock in poor conditions and high weathering. The GSI method can be known by extrapolating boreholes to determine subsurface rocks [12]. Drill hole investigations are multiple and inclined to assist the interpretation of rock mass. The Hoek-Brown Failure Criterion in GSI quantifies the rock mass structure and damage conditions above its strength, with the rock assumption not being disturbed. A quantitative GSI assessment can be carried out on the drilled rock core according to the following equation [10]:

$$GSI = 1{,}5\, JCond_{89} + \frac{RQD}{2} \quad (1)$$

Where:
GSI: Geological Strength Index
JCond89: Joint Condition
RQD: Rock Quality Designation

2.2. *Rock Mass Rating (RMR)*

Rock mass quality classification is a method to classify rock masses into groups in a determined relationship [9]. The rock mass quality classification establishes a unique description that can be in the form of a number on the rock mass basis with similar properties so the character of the rock mass can be predicted. Classification of RMR can be used as a basis for designing stand-up time and support of underground structures. The RMR method uses six parameters from the field observations from core drill data. These parameters are Uniaxial Compressive Strength (UCS) values in intact rock materials, Rock Quality Designation (RQD) from a core drill, a distance of discontinuity, conditions of

IOP Conf. Series: Earth and Environmental Science 1233 (2023) 012007 doi:10.1088/1755-1315/1233/1/012007

discontinuity, conditions of groundwater, and orientation of discontinuity. This paper calculates the RMR value on the basis of searching the stand-up time and support system from the tunnel design.

2.3. Q-System

The q-System method has a classification divided into six parameters to determine rock mass, there are RQD, number of joints, the roughness of joints, joint changes, water conditions at joints, and pressure factors [5]. The same approach is used for RMR and Q-System but with different ratings on a log scale. Q-System value is a product of the ratio of the parameter while RMR is the number of parameters [12]. Q index ranges from 0.001 to 1,000 on a logarithmic scale [9]. The geotechnical parameters are indicated by the equalities [13]: 1) Block size relative = RQD/Jn, 2) Frictional strength relative =Jr/Ja, and 3) Number of active pressure = Jw/SRF. Determination of rock mass quality with the Q method is obtained from the following equation [4]:

$$Q = (\frac{RQD}{Jn})(\frac{Jr}{Ja})(\frac{Jw}{SRF}) \quad (2)$$

2.4. Relationship Between RMR and GSI Method

GSI classification was made to improve and be guided by the RMR method in estimating low rock mass quality in the early stages of development [12]. When the rock mass with GSI value of 18 and RMR of 23, then the GSI value can be calculated using the equation [14]:

$$GSI = RMR_{89} - 5 \quad (3)$$

2.5. Relationship Between RMR and Q-System Method

The correlation between RMR and the parameter that can be applied is the relation between RMR and Q-System have the following formula [9]:

$$Q = 10^{\frac{RMR-50}{15}} \quad (4)$$

2.6. Analysis of Excavation Method (Pettifer-Fookes Chart)

Analysis of the excavation method has the goal to obtain the most effective tunnel opening method. The excavation method in this research uses two different methods which have empirical graphic analysis, namely the GSI method and the Pettifer-Fookes method. The different methods will be compared so we can find out which excavation method is the most suitable to apply. According to Pettifer and Fookes [15], the excavation will be determined using the best method among several existing methods such as blasting, extremely hard ripping/hydraulic breaking, very hard ripping, hard ripping, easy ripping, hard digging, and easy digging based on the excavation chart. The excavation method is based on the value of the spacing index of discontinuity or fracture spacing (I_f) and the value of the point load index (Is_{50}). A joint set (Jv) which has defined the sum of frequency fractures that have the same direction pattern and discontinuity spacing index (I_f) can be used in the following equation:

$$Jv = \frac{1}{S1} + \frac{1}{S2} + \frac{1}{S3} \quad (5)$$

$$If = \frac{3}{Jv} \quad (6)$$

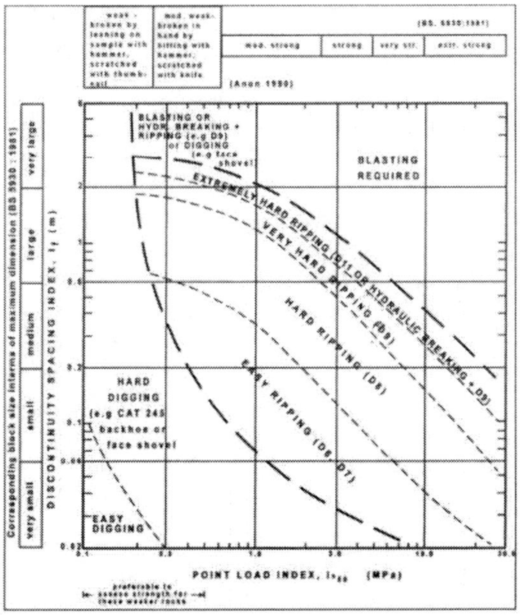

Figure 3 Excavability method using Pettifer-Fookes chart [15]

2.7. Analysis of Excavation Method (GSI Chart)

The excavation method using the GSI chart is divided into two classifications based on the point load index (Is_{50}); $Is_{50} < 3$ MPa chart and $Is_{50} \geq 3$ MPa chart [2].

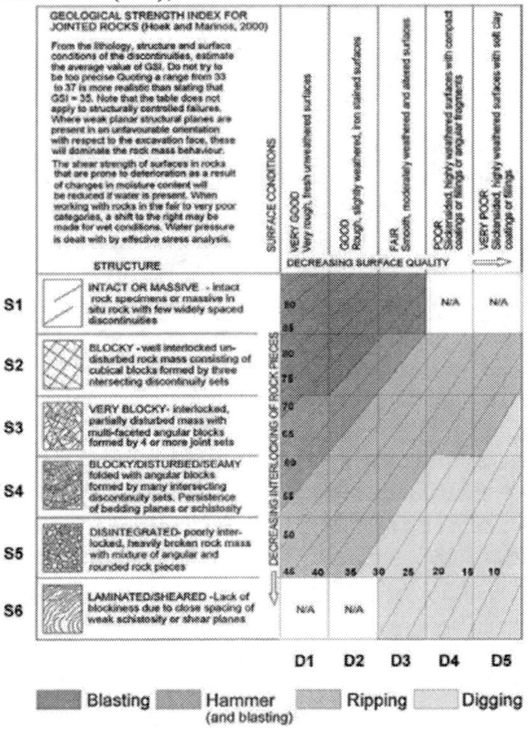

Figure 4 GSI chart for excavatability (Is_{50}) < 3 MPa) [2]

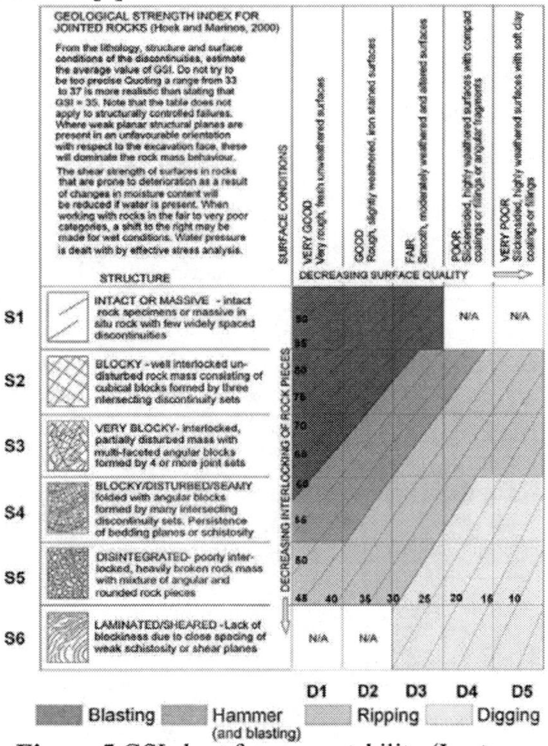

Figure 5 GSI chart for excavatability ($Is_{50} \geq$ MPa) [2]

2.8. *Analysis of Excavation Method and Support Systems (RMR Method)*

The excavation method in this study also uses the RMR method which is equipped with support systems. This method uses rock mass classification based on RMR values of very good rock, fair rock, poor rock and very poor rock. Each RMR value has an excavation method and support system consisting of rock bolts, shotcrete and steel sets [9]. The RMR table with excavation methods and tunnel support systems as shown in **Table 1**.

Table 1 RMR excavation methods and tunnel support systems

Rock Mass Class	Excavation	Support		
		Rock Bolts (20 mm Dia, Fully Grouted)	Shotcrete	Steel Sets
Very Good Rock (I) RMR: 81-100	Full face 3 m advance.	Generally, no support is required except for occasional spot bolting.		
Good rock (II) RMR: 61-80	Full face 1.0-1.5 m advance. Complete support 20 m from the face.	Locally, bolts in crown 3 m long, spaced 2.5 m, with occasional wire mesh.	50 mm in the crown where required.	None.
Fair Rock (III) RMR: 41-60	Top heading and bench 1.5-3 m advance in top heading. Commence support.	Systematic bolts 4 m long spaced 1.5-2 m in crown and walls with wire mesh in crown.	50-100 mm in crown and 30 mm in sides.	None.
Poor Rock (IV) RMR: 21-40	Top heading and bench 1.0-1.5 m advance in top heading. Install support concurrently with excavation 10 m from face.	Systematic bolts 4-5 m long, spaced 1-1.5 m in crown and wall with wire mesh.	100 – 150 mm in crown and 100 mm in sides.	Light to medium ribs spaced 1.5 m where required.
Very poor rock (V) RMR < 20	Multiple drift 0.5 – 1.5 m advance in top heading. Install support concurrently with excavation. Shotcrete as soon as possible after blasting.	Systematic bolts 5-6 m long, spaced 1-1.5 m in crown and walls with wiremesh. Bolt invert.	150 -200 mm in crown, 150 mm in sides, and 50 mm on face.	Medium to heavy ribs spaced 0.75 m with steel lagging and forepolling if required. Close invert.

3. Lithology Conditions

The regional morphology of the research area can be divided into four morphological units, namely mountain, hills, karst, and lowland units. From the regional geological map, the location of Pelosika Dam stands on the Paleozoic Metamorphic Rocks (schist, gneiss, phyllite, and quartzite), and Alluvial deposits (clay, silt, sand, gravel and pebble) [16], as shown in **Figure 6.** The geological structures found in this area are faults, folds, and joints. The faults and lineaments generally have a northwest-southeast trend in the direction of the Lasolo Fault. The Lasolo Fault is a sinistral fault with a northwest-southeast orientation, dividing this area into two parts, the northeast is called the Hailu Arc and the southwest is called the Tinondo Arc. The Hialu Arc generally consists of basaltic rocks which indicates that the rocks are from Oceanic Crust. While the Tinondo Arc consists of acid rocks which indicate that the rocks are from the Continental Shelf. Reverse faults were found in the Wawo area, west of Tampakura, and at Tanjung Labuandala in the south of Lasolo. The types of folds in this rock include closed folds such as recumbent folds, anticlines, and synclines. In contrast, the folds in the Tertiary rocks classify open folds with low-angle slopes. Joints are found in all rock types, for example, joints in limestone have regular patterns and form lineament, while the joints in igneous rocks generally show an irregular pattern [17].

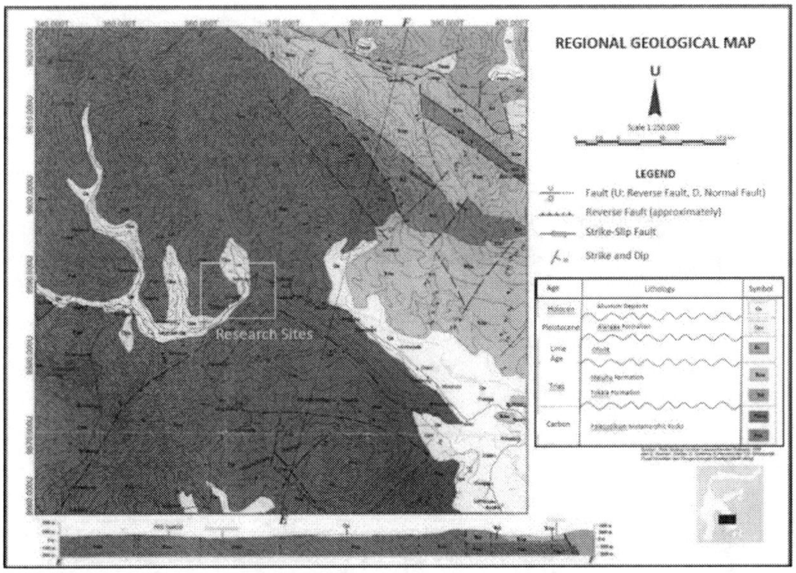

Figure 6. Regional geological map of the Pelosika Dam and its surrounding [16]

The geological condition of the research area consists of metamorphic rocks (quartz muscovite phyllite), fluvial deposits (clay, silt, sand, gravel, and pebble), and alluvial deposits (clay and silt), as shown in **Figure 7.** The geological structure found in this area is a predicted fault that has a northwest-southeast orientation with a value of N333°E/45°. The details of the geological subsurface conditions in the two geological cross-sections (A-A' and B-B') as shown in **Figure 7** (right side) and the geological cross-section of the diversion tunnel location in **Figure 8**.

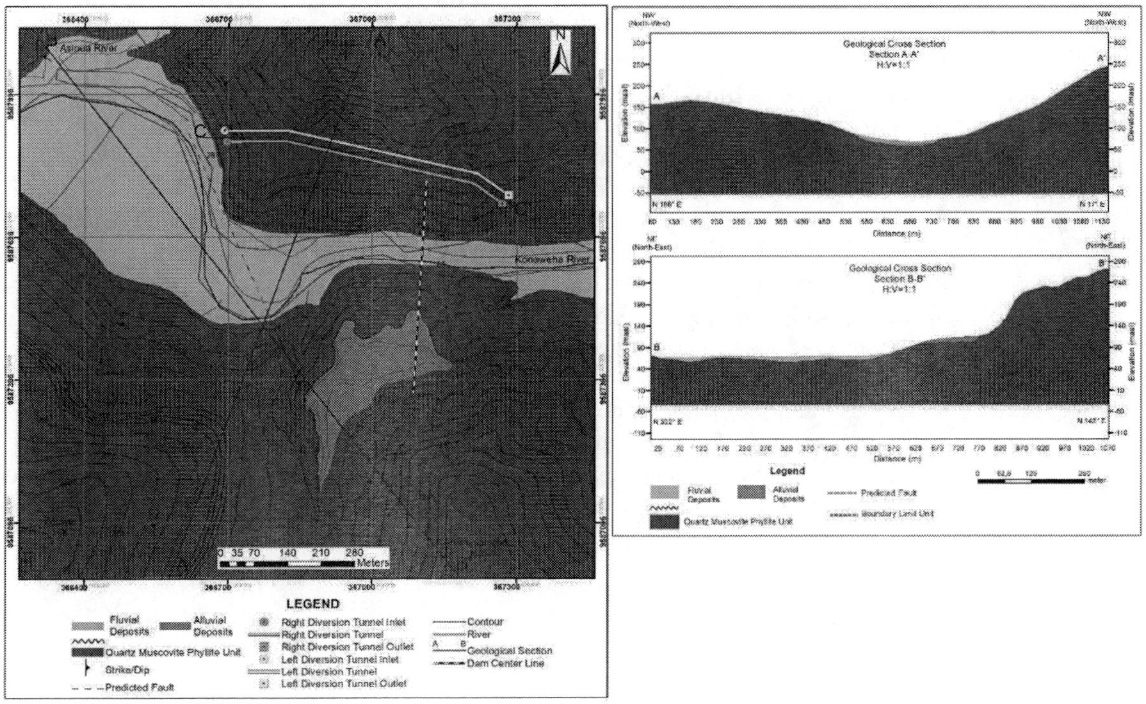

Figure 7. Geological map of the Pelosika Dam Diversion Tunnel area

IOP Conf. Series: Earth and Environmental Science 1233 (2023) 012007 doi:10.1088/1755-1315/1233/1/012007

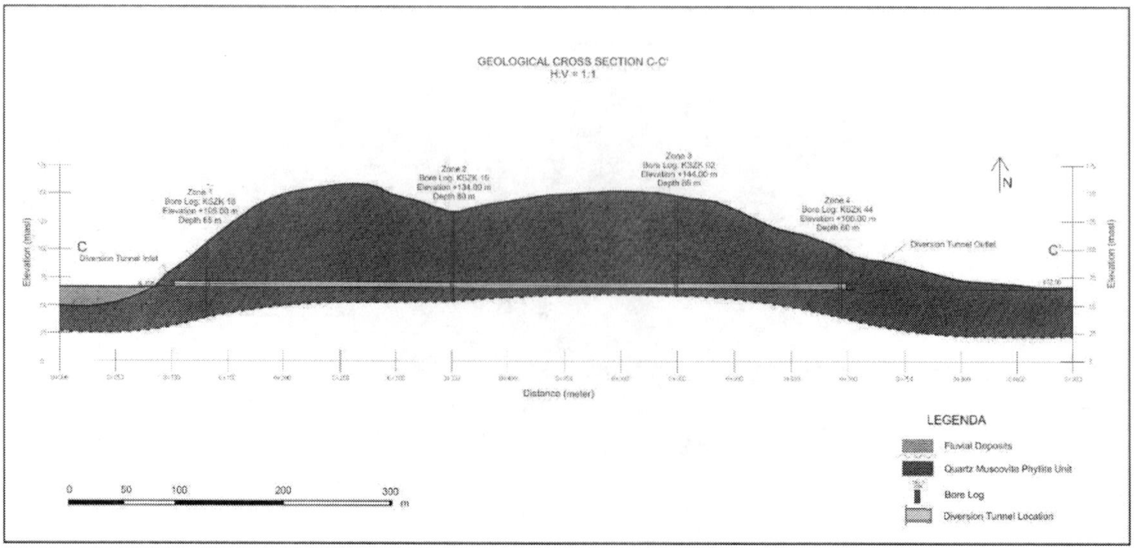

Figure 8. Geological cross-section of the diversion tunnel location (C-C')

Lithology and layering in the geological cross-section in **Figure 8** consist of metamorphic (quartz muscovite phyllite) and fluvial deposits (clay, silt, sand, gravel, and pebble). The lowest and oldest layer consists of quartz muscovite phyllite as a metamorphic rock (purple color). At the top of the layer on the west side, there are fluvial deposits that consist of clay, silt, sand, gravel, and pebble (grey color).

4. Engineering Geological Conditions

We used surface and subsurface GSI methods based on the geological information taken during field investigations to know the rock mass quality. The relation between structure and surface conditions from geological outcrop on the surface area shows surface condition by GSI method chart [9] and produces a technical geological map from the rock mass quality. We obtained surface rock mass quality for the application of slope stability of the tunnel portal. The engineering geology map shows in **Figure 9**. The results of rock mass mapping using the GSI method suggest four rock mass classifications. The category are very poor, poor, fair, and good conditions. The very poor condition has a value of the GSI of less than 20 points which is the weakest level from the rock mass classification and dominates in the research area. Above the very poor conditions, there are poor (21-40) and fair (41-55) from the GSI value which has a middle portion area. At the top, there is a good condition of GSI value (56-75) but has a small area from the research map. The classification of the engineering geological map are fluvial deposits, alluvial deposits, quartz muscovite phyllite with a good quality and slightly weathered unit, quartz muscovite phyllite with a fair quality and moderately weathered unit, quartz muscovite phyllite with poor quality and the highly weathered unit, also quartz muscovite phyllite with very poor quality and completely weathered unit.

We used the rock mass quality of subsurface conditions by analyzing the core drill sample with the GSI subsurface method. A combination of Joint Conditions (JCond89) and Rock Quality Designation (RQD) from the core drill samples will get subsurface rock mass quality data and engineering geological profile [10]. The engineering geological profile shows in **Figure 10**.

IOP Conf. Series: Earth and Environmental Science 1233 (2023) 012007 doi:10.1088/1755-1315/1233/1/012007

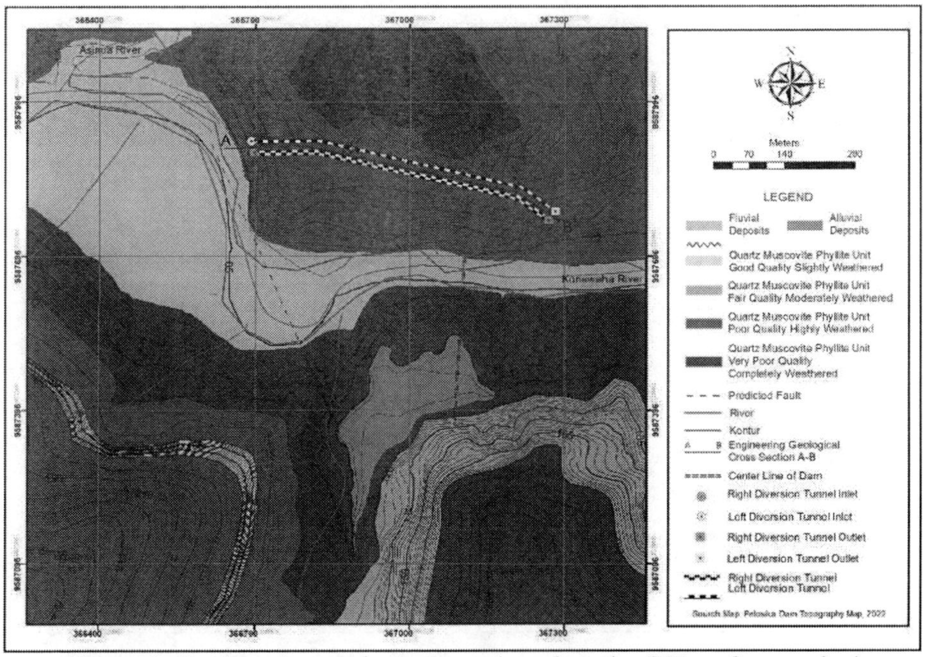

Figure 9. Engineering geological map based on the GSI surface method

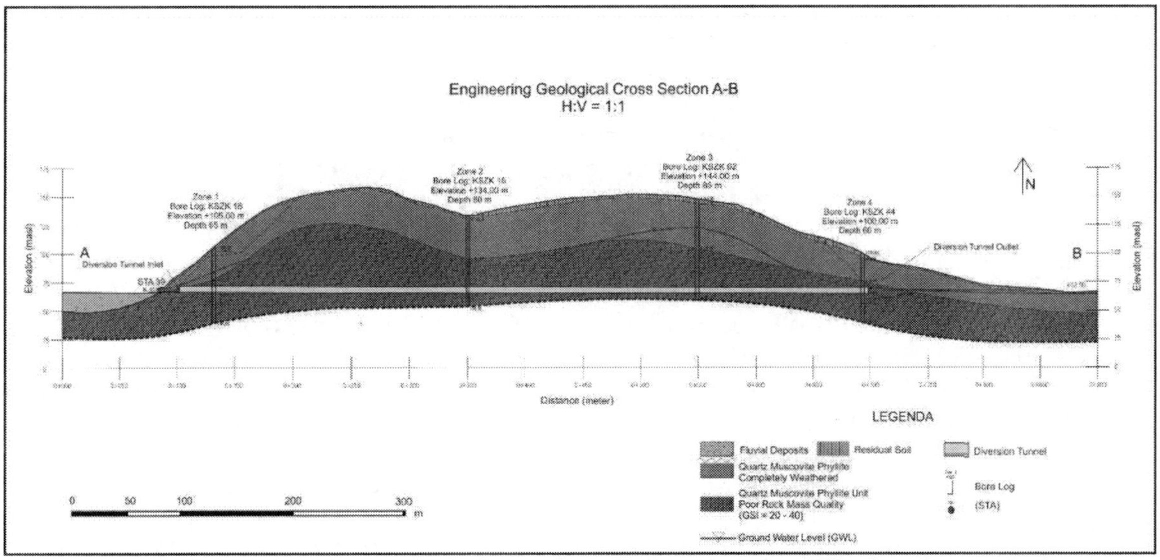

Figure 10. Engineering geological tunnel profile based on the GSI subsurface method

The subsurface tunnel condition as seen in **Figure 10** with the GSI subsurface method uses four core drill points located in the diversion tunnel alignment. The first core drill point is Zone 1 (KSZK 18) in the inlet tunnel section (west side) which has a 65 m depth with poor rock (GSI value 21) condition. The second point is Zone 2 (KSZK 15) with an 80 m depth in the middle section of the tunnel which has poor rock (GSI 28) condition. The third point is Zone 3 (KSZK 02) with 85 m depth in the middle section of the tunnel has the same result as the previous point, there is poor rock (GSI value 35). The last point is Zone 4 (KSZK 44) in the outlet section (east side) has a 60 m depth with poor rock (GSI value 21). In all of the top core drill layers, there is residual soil with a 1-3 m depth from the top elevation.

IOP Conf. Series: Earth and Environmental Science 1233 (2023) 012007 doi:10.1088/1755-1315/1233/1/012007

5. Rock Mass Quality

5.1. Rock Mass Quality Classification (GSI, RMR, and Q-System)

This research uses core drill data to determine rock mass quality using GSI, RMR, and Q-System approach classifications. Rock mass quality from 4 points of core drills data in 4 positions; Zone 1 (inlet), Zone 2 (middle), Zone 3 (middle), and Zone 4 (outlet) tunnel. Analysis data using 3 types of rock mass quality will be compared to each other. Rock mass quality will get value in the following **Table 2**:

Table 2. Rock mass quality values

Zone	Lithology	Weathering Rate	GSI Value	GSI Rock Quality	RMR Value	RMR Rock Quality	Q-System Value	Q-System Rock Quality
1 (Inlet)	Phyllite	Highly Weathered	21	Poor	23	Poor	0,025	Extremely Poor
2 (Middle)	Phyllite	Slightly Weathered	28	Poor	26	Poor	0,227	Very Poor
3 (Middle)	Phyllite	Slightly Weathered	35	Poor	30	Poor	0,322	Very Poor
4 (Outlet)	Phyllite	Moderately Weathered	21	Poor	25	Poor	0,028	Extremely Poor

We did the first rock mass classification using the GSI method at the tunnel elevation point. There were four drill points studied with varying core drill lengths. The material contained in the four drill points studied is a metamorphic rock, namely phyllite. The first is to determine the condition of JCond which consists of core drill conditions with infilling or without infilling in each box with a length of every 1 m. In the tunnel inlet, middle, and outlet, the condition includes infilling because there is filler in the observed core drill results. In the JCond section, there are values of weathering levels and surface conditions of core drilled results. The level of weathering found in the tunnel elevation is dominated by slightly weathered, moderately weathered until highly weathered. Meanwhile, the surface condition of the core drill is dominated by slightly rough. The rating value at the drill point can be given by combining JCond with the RQD value so that it will get the GSI value using equation (1). Furthermore, the available GSI values can be classified based on the previously obtained GSI values. The GSI classification consists of very poor (0-20), poor (20-40), fair (40-55), good (55-75), and very good rock (75-95). After getting the GSI classification value, then the value is entered using equation (3) so that the RMR value will be obtained. The classification of RMR naming based on the RMR value obtained consists of very poor (0-20), poor (20-40), fair (40-60), good (60-80), and very good rock (80-100). If the RMR value has been obtained, then the value is entered into equation (4) so that the Q-System value will be obtained. Q-System naming classification consists of exceptionally poor (0.001-0.1), extremely poor (0.01-0.1), very poor (0.1-1, poor (1-4), fair (4-10), good (10-40), very good (40-100), extremely good (100-400) and exceptionally good rock (400-1000). The GSI value obtained can then be used to determine the appropriate excavation method using the GSI chart. The RMR value is used to determine the stand-up time of the tunnel using the RMR chart. RMR value is also used to determine the support system and excavation method using the RMR table. Q-System value is used to determine the support system using the Q-System chart. Based on its function to determine excavation and support system methods, rock mass classification using GSI, RMR, and Q-System is needed.

6. Analysis of Excavation Method (GSI and Pettifer-Fookes Method)

The range of point load strength and rock mass classification can be estimated to determine the excavation method using the Pettifer-Fookes chart [15]. Phyllite's condition has fracture spacing with

IOP Conf. Series: Earth and Environmental Science 1233 (2023) 012007 doi:10.1088/1755-1315/1233/1/012007

disintegrated and laminated/sheared structures. We did laboratory tests to obtain point load strength (Is_{50}) and discontinuity index (I_f) in phyllite formation as seen in **Table 3**.

Table 3. Empirical rock classification, point load strength, and discontinuity index

Zone	Bore Log	Depth	Lithology	Discontinuity Spacing (S)	Volumetric Joint (Jv)	Is_{50} (MPa)	I_f (Discontinuity Spacing Index)	Excavation Method
	(Code)	(m)		(m)	(each meter)	(MPa)	(m)	
1	STA 25	-	Phyllite	0.137	7,32	0.739	0.41	Easy Ripping
2	KSZK 15	62-68	Phyllite	0.187	5,36	1.066	0.56	Hard Ripping
3	KSZK 02	73-79	Phyllite	0.243	4,11	1.311	0.73	Hard Ripping
4	KSZK 44	29-35	Phyllite	0.127	7,89	0.721	0.38	Easy Ripping

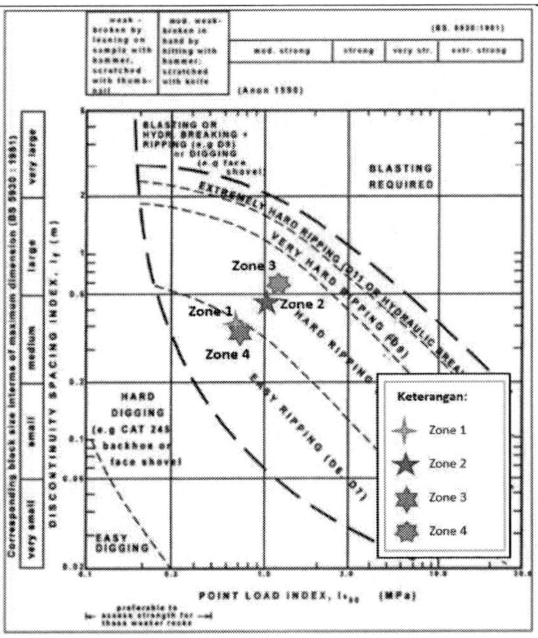

Figure 11. Plotting of The Rock Excavation Method Using Pettifer-Fookes Chart [15]

According to field estimates of UCS in intact rock Marinos and Hoek [11] table, all of the point load index (Is_{50}) is below 3 MPa, so the chart used in this research only uses $Is_{50} < 3$ MPa chart. Plotting of the rock excavation method using the Pettifer-Fookes chart can be seen in **Figure 11** and GSI chart in **Figure 12.** A summary of the plotting data can be seen in **Table 4.**

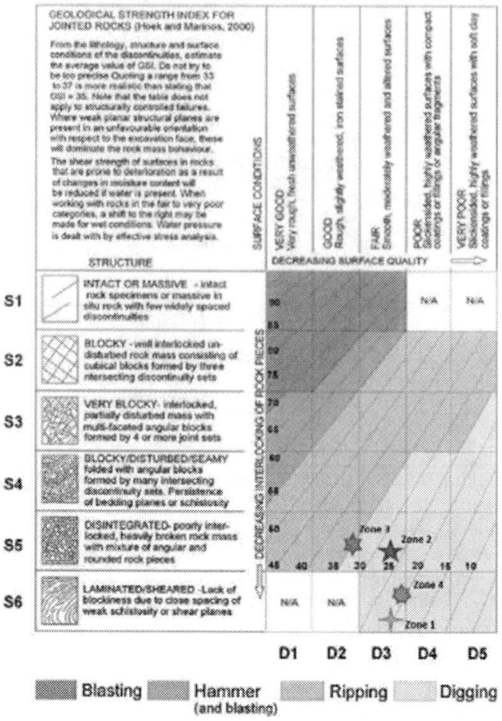

Figure 12. Plotting of Rock Excavation Method Using GSI Chart (Is_{50} < 3 MPa) [2]

Table 4. Summary Data in GSI Chart and Pettifer-Fookes Chart

Zone	Lithology	Rock Class (GSI)	Structure/ Discontinuity	GSI	I_f (m)	Is_{50} (MPa)	Excavation Method (GSI)	Excavation Method (Pettifer- Fookes)
1 (Inlet)	Phyllite	Poor Rock	S2D4	21	0.41	0.739	Digging	Easy Ripping
2 (Middle)	Phyllite	Poor Rock	S4D4	28	0.56	1.066	Digging	Hard Ripping
3 (Middle)	Phyllite	Poor Rock	S4D4	35	0.73	1.311	Digging	Hard Ripping
4 (Outlet)	Phyllite	Poor Rock	S4D4	21	0.38	0.721	Digging	Easy Ripping

Based on the plotting result from the two different methods using GSI and Pettifer-Fookes chart which have different excavation methods that can be applied in the rock-type tunnel, there are digging and ripping methods.

7. Analysis of Stand-up Time (RMR Chart)

Classification of the RMR quality method has a function to analyze a stand-up time value. Stand-up time is the safe time for the tunnel after excavation without any reinforcement. RMR also produces an unsupported span value which is the size of the excavation span without any reinforcement. The larger the stand-up time value, the smaller the unsupported span value because both have an inverse comparison. Meanwhile, the value of stand-up time and unsupported span is directly proportional to the value of rock mass quality. The relationship between stand-up time, span, and RMR correspond in **Table 5** and **Figure 13**.

Table 5. Relationship of The RMR, Span, and Stand-up Time

Zone	Lithology	Rock Type	RMR Value	Span (m)	Stand-up Time (hours)
1 (Inlet)	Phyllite	Poor Rock	23	2.6	42
2 (Middle)	Phyllite	Poor Rock	26	2.6	100
3 (Middle)	Phyllite	Poor Rock	30	2.6	240
4 (Outlet)	Phyllite	Poor Rock	25	2.6	60

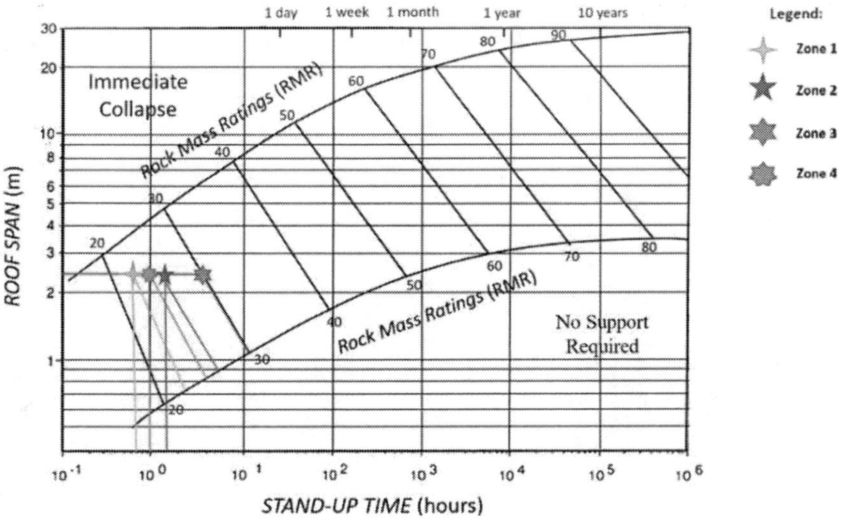

Figure 13. Plotting of The Relationship Chart Between RMR, Span, and Stand-up Time [9]

Based on the plotting result from the stand-up time chart, the RMR value associated with the roof span 2.6 m has the result of stand-up time 42-240 minutes after excavation without support equipment.

8. Excavation and Support System (RMR and Q-System Method)

Evaluation of the tunnel support system analysis is intended to evaluate the results of the analysis by comparing the two methods of determining the tunnel support system used in design planning; the RMR method and the Q-System method. A comparison of tunnel support systems will obtain the most effective tunnel support system. An effective support system can be obtained using rock bolts, shotcrete, or a combination of both with the most effective spacing and quantities produced. After selecting the best reinforcement system formulation, it will be combined with a concrete layer to check the stability of the tunnel as a simulation of the actual tunnel conditions. Recommended support systems using RMR are in **Table 6, Figure 14**, and **Figure 15**. The Q-System methods are in **Figure 16, Table 7,** and **Figure 17**.

Table 6. The Recommended Excavation and Support System Using RMR Method [9]

Zone	RMR	Rock Mass Quality	Excavation	Support		
				Rock Bolts (20 mm Diameter, Fully Grouted)	Shotcrete	Steel Sets
1 (Inlet)	23	Poor Rock (IV) RMR value: 21-40	Excavate at the top heading and bench 1,0 – 1,5 m. After excavating 10 m from the face, support must be concurrently installed	Install the systematic rock bolts with L=4–5 m and spaced for bolts 1,0–1,5 m in the crown and wall. Support crown and wall with wire mesh.	Install shotcrete in the crown t=100–150 mm and t=100 mm in the sides.	If steel sets are required, install light to medium ribs with 1,5 m spaced
2 (Middle)	26					
3 (Middle)	30					
4 (Outlet)	25					

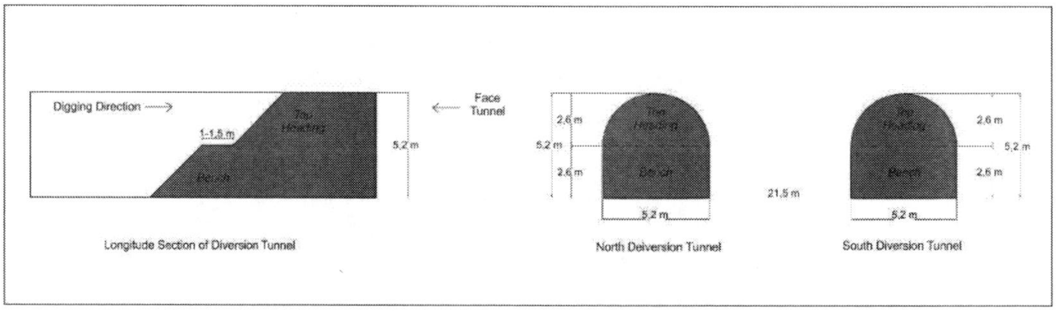

Figure 14 Top heading and bench digging method

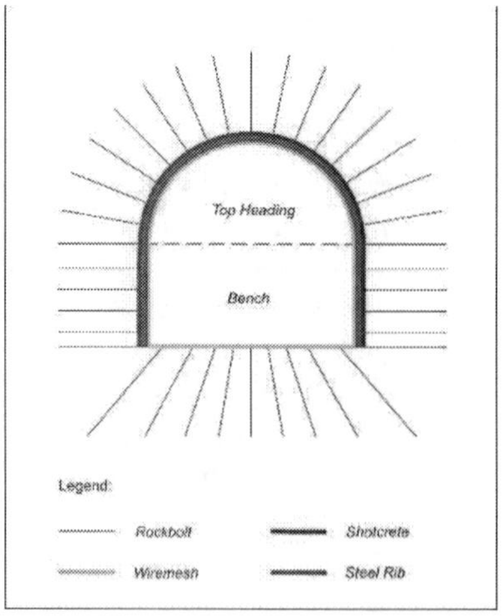

Figure 15 Tunnel support system reinforcement using RMR method

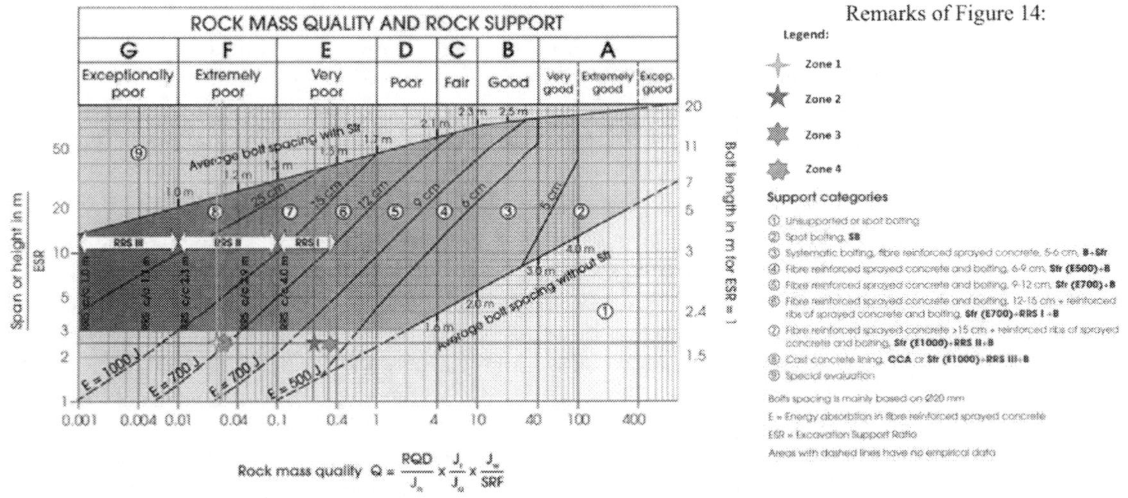

Figure 16. Support System Using Q-System Method [18]

Table 7. The Recommended Support System Using Q-System Method

Zone	Q-System	Rock Mass Quality	Support System
1	0.025	Extremely Poor Rock	- Fiber-reinforced sprayed shotcrete and bolting t=12-15 cm; - Reinforced rib of sprayed concrete (RRS) - Rock bolt L= 1,8 m; RRS= D45/6+2 Ø16-20 (*span* 10 m) D= *double layer of rebars* Ø= *rebar diameter*
2	0.227	Very Poor Rock	- Fiber-reinforced shotcrete and bolting t=9–12 mm;
3	0,322	Very Poor Rock	- Fiber-reinforced shotcrete and bolting t=9–12 mm;
4	0,028	Extremely Poor Rock	- Fiber-reinforced sprayed shotcrete and bolting t=12-15 cm; - Reinforced rib of sprayed concrete (RRS) - Rock bolt L= 1,8 m; RRS= D45/6+2 Ø16-20 (*span* 10 m) D= *double layer of rebars* Ø= *rebar diameter*

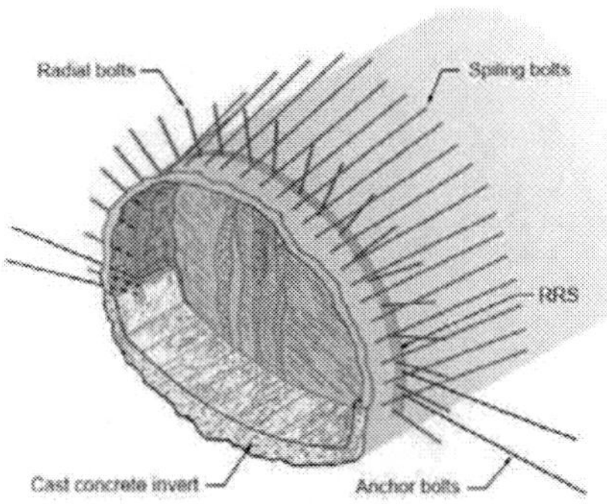

Figure 17 Tunnel support system using Q-System methods [18]

9. Conclusion

Current research shows that the geological conditions in the research area consist of metamorphic rock (quartz muscovite phyllite) and alluvial deposits. Based on the geological conditions in the study area, the location of the diversion tunnel is in the lithology of metamorphic rocks (quartz muscovite phyllite). Rocks in the study area have a slight, medium to the highly weathering. Weathering level affects the strength of the rock that will be excavated for tunnel construction. The GSI value ranges from 21-35 which is categorized as a poor condition. In the RMR classification, the value ranges from 23-30 so it is included in the poor condition. In the Q-System classification, the value ranges from 0.025-0.322 so it is classified into extremely poor and very poor condition. The condition of poor rock mass strength with three methods of classification is probably caused by the high weathering rate and weakening zones that exist in metamorphic rocks. From the rock mass classification value, the excavation method using the GSI chart and Pettifer-Fookes chart has resulted in the GSI chart using the digging method while in Pettifer-Fookes chart uses easy ripping- hard ripping. Analysis of stand-up time using an RMR chart has 42-240 minutes without support after excavation. The RMR method has a recommendation of a support system with 4-5 m rock bolt length with 1-1,5 m spaced in the crown and wall with wire mesh, 100–150 mm thick of shotcrete in the crown, and if steel sets are required, install light to medium ribs with 1,5 m spaced. Meanwhile, the result of the support system using the Q-System method is using fiber-reinforced shotcrete with 12-15 cm spacing, RRS, and 1.8 m bolt spacing in the shotcrete area.

The quality of rock mass with three methods (GSI, RMR, and Q-System) has poor rock conditions so the appropriate excavation method based on the GSI and Pettifer-Fookes methods is a combination of digging and ripping. The tunnel stand-up time after excavation is very shortly after being analyzed by the RMR method, so the tunnel needs to be covered after excavation to increase safety for workers and construction. The recommended protection using a combination of the RMR and Q-System methods is to use rock bolts, shotcrete, and steel sets. The recommended length of rock bolts is 1.8 m. The required shotcrete thickness is 10-15 cm with wire mesh reinforcement. If the reinforcement is still not sufficient, steel sets can be added by installing light to medium ribs with 1.5 m spacing for the last option. A combination of reinforcement methods is used to correct each other methods and improve the safety of tunnel excavations without compromising the cost-effectiveness of construction. In the end, further research needs to be done on the diversion tunnel location such as modeling the stability of the tunnel portal slope or modeling the stability of the tunnel, so that the tunnel will be safer when it is built.

10. Acknowledgment

The author would like to thank Mr. Hendy Setiawan as supervisor from Department of Geological Engineering, Faculty of Engineering, Universitas Gadjah Mada for its guidance in internship and paper writing activities. The author also thanks BWS Sulawesi IV Kendari (Directorate General Water Resources, The Ministry of Public Works and Housing), especially Mr. Arbor Reseda and Mr. Agung Permana as apprentice mentors for the permission to conduct this research in the project location, supplying existing geological data, and publishing the research result for the conference. Thanks to The Ministry of Public Works and Housing for the study scholarships and research funding so that this research can be realized.

IOP Conf. Series: Earth and Environmental Science 1233 (2023) 012007 doi:10.1088/1755-1315/1233/1/012007

References

[1] Hoek E, Marinos P and Benissi M 1998 Applicability of the geological strength index (GSI) classification for very weak and sheared rock masses. The case of the Athens Schist Formation *Bulletin of Engineering Geology and the Environment* **57** 151–60

[2] Tsiambaos G and Saroglou H 2010 Excavatability assessment of rock masses using the Geological Strength Index (GSI) *Bulletin of Engineering Geology and the Environment* **69** 13–27

[3] Anonym 2019 *Laporan Akhir Pekerjaan: Sertifikasi Desain Bendungan Pelosika Tahap III Kab. Konawe* (Kendari)

[4] Nurfarisa M , N 2017 *Kelas Massa Batuan dan Penyangga Terowongan Pengelak Menggunakan Metode Q-System di Desa Asinua Jaya, Kabupaten Konawe, Provinsi Sulawesi Tenggara* (Jatinangor: Universitas Padjajaran)

[5] Barton N, Lien R and Lunde J 1974 Engineering Classification of Rock Masses for the Design of Tunnel Support *Rock Mechanics 6, 189-236 (1974)* 189–236

[6] Anonym 2019 *Laporan Investigasi Geologi Paket Pekerjaan: Sertifikasi Desain Bendungan Pelosika* (Kendari)

[7] Tanaka H 1964 Introduction of Geology for Civil Engineers Sankaidou Japan

[8] Kikuchi K, Saito K and Kusonoki KI 1982 Geotechnically Integrated Evaluation on The Stability of Dam Foundation Rocks *Proceeding of The 14th International Congress on Large Dams. Rio de Janeiro Brazil* 49–74

[9] Bieniawski Z T 1989 *Engineering Rock Mass Classifications* (Canada: John Wiley & Sons, Inc.)

[10] Hoek E, Carter T G and Diendertichs M S 2013 Quantification of the Geological Strength Index Chart *47th US Rock Mechanics / Geomechanics Symposium held in San Francisco, CA, USA, 23-26 June 2013*

[11] Marinos P and Hoek E 2000 GSI: A Geologically Friendly Tool for Rock Mass Strength Estimation *ISRM International Symposium 2000* 19–24

[12] Hoek E 2007 *Practical Rock Engineering* (North Vancouver: Evert Hook Consulting Engineer Inc.)

[13] Grimstad E and Barton N 1993 Updating of The Q-System for NMT *Proceedings of Sarocks '98, 2nd Brazilian Symposium on Sprayed Concrete-Modern Use of Wt Mix Sprayed Concrete for Underground Support*

[14] Hoek E and Brown E T 1997 Practical estimates of rock mass strength *International Journal of Rock Mechanics and Mining Sciences* **34** 1165–86

[15] Pettifer G S and Fookes P G 1994 A revision of the graphical method for assessing the excavatability of rock *Quarterly Journal of Engineering Geology* **27** 145–64

[16] Rusmana E, Sukido, Sukarna D, Haryono E and Simandjuntak T O 1993 Geologi Lembar Lasusua-Kendari, Sulawesi *Pusat Penelitian dan Pengembangan Geologi*

[17] Rusmana E and Sukarna D 1985 Overview Southeast Sulawesi Arm Stratigraphy Compared to The Surrounding Area *Proceeding of Indonesia Association Geologists (IAGI), 14th Annual Convention, h. 61-70.* 61–70

[18] Barton N R, Lien R and Lunde J 2015 *Rock Mass Classification and Support Using The Q-System* (Oslo: NGI)

IOP Conf. Series: Earth and Environmental Science 1233 (2023) 012008 doi:10.1088/1755-1315/1233/1/012008

EXCASS system and Rock Mass Rating (RMR) classification for determining excavation method and tunnel support at Pelosika Dam Spillway Tunnel

H Wisnuaji[1,2*], H Setiawan[1,3*], N I Setiawan[1]

[1]Departemen of Geological Engineering, Faculty of Engineering, Universitas Gadjah Mada, Yogyakarta, 55281, Indonesia
[2]Directorate General of Water Resources, Ministry of Public Work and Housing of Republic Indonesia, Jakarta, 12110, Indonesia
[3]Center for Disaster Mitigation and Technological Innovation (GAMA-InaTEK), Universitas Gadjah Mada, Yogyakarta, 55281, Indonesia

[*]Corresponding author: hidayat.wisnuaji@ugm.ac.id, hendy.setiawan@ugm.ac.id

Abstract. Pelosika dam spillway tunnel was designed to drainage flood discharge and prevent overtopping. In recent investigations, the rock mass classification has not been done specifically in the spillway tunnel. This research aims to obtain technical geological data and information to define excavation methods and tunnel support. For more comprehensive research, engineering geological investigations were carried out regarding the Geological Strength Index (GSI), selecting excavation methods using EXCASS system and Rock Mass Rating (RMR) for tunnel support. The geological mapping showed that the study site was composed of 2 lithological units: quartz muscovite phyllite and pebblely sand deposits. The GSI value in the tunnel is between 14-25, indicating poor and very poor quality. Adopting the EXCASS system, the optimum method of excavation (EPI_{opt}) in the Pelosika dam spillway tunnel are ripper easy and digger. While RMR values of 9 – 25 are classified as poor quality. Consequently, the appropriate method for the excavation sequence is using the top heading and bench and multiple drifts. And the recommended support system is a systematic rock bolt with wire mesh and shotcrete in the crown and walls with medium to light steel rib.

1. Introduction

The flood disaster in Sulawesi Island is a crucial issue to be tackled by dam construction. Pelosika Dam is one of the dam projects built in Southeast Sulawesi. Pelosika Dam is located in Asinua Jaya Village, Asinua District, Konawe Regency, Southeast Sulawesi, which can be seen in **Figure 1**. The Pelosika Dam is designed to have a spillway tunnel length of 597.8 m [1]. The tunnel design is in the shape of a horseshoe height of 10 m. A Spillway tunnel has the function of draining flood discharge and preventing overtopping when the dam and reservoir are already operated. The tunnel construction is a practical choice when the construction site passes through hilly and steep areas.

Geotechnical engineering, such as underground structures, is strongly influenced by the uncertainty of geological conditions and geotechnical parameters [2]. Geological and engineering geological investigations were carried out to determine surface and subsurface conditions. A recent investigation in 2019 conducted by BWS Sulawesi IV Kendari resulted in the spillway tunnel location being located in a metamorphic geological unit (phyllite and schist) [1]. Metamorphic rock has a unique foliation structure. Foliation affects the mechanical properties, which causes high anisotropy and heterogeneity conditions [3].

Rock mass classification has been carried out around the spillway tunnel by several methods, namely Q-system, CRIEPI, and RMR [4], [1]. The results of rock mass classification are obtained in the category of *exceptionally poor* to *extremely poor*, *soft rock* (CL) to *very soft* (D), and *poor* to *very poor*, respectively. Some rock mass classifications are only carried out around spillway tunnels. The rock mass classification has not been done specifically in the spillway tunnel.

The construction of the spillway tunnel requires planning the excavation method. Geotechnical material properties affect the rock excavability, work methods selection, and the type and size of excavation equipment used [5]. In addition, geotechnical material properties also minimize excavation costs and difficulties encountered during excavation [6]. Previous study has not yet conducted a study of the excavation method. One of the empirical excavation methods can use The EXCASS system developed by Dagdelenler [6]. The EXCASS system has a prediction performance of more than 74.2% and the first flexible excavation system in rock mechanics literature [7].

Based on these conditions, this study tries to provide an alternative in rock mass classification and the selection of excavation methods. Classification of surface rock mass using Geological Strength Index (GSI) [8] and subsurface classification using quantification Hoek et al.'s GSI [9] and Osgoui et al.'s GSI [10]. Selection of excavation method using EXCASS System [6]. At the same time, the recommendation for the support system uses the empirical method of the RMR classification [11]. Thus, to attain the above objectives, the research steps that were carried out include mapping of surface geology, mapping of engineering geology, and evaluation of core drilling results.

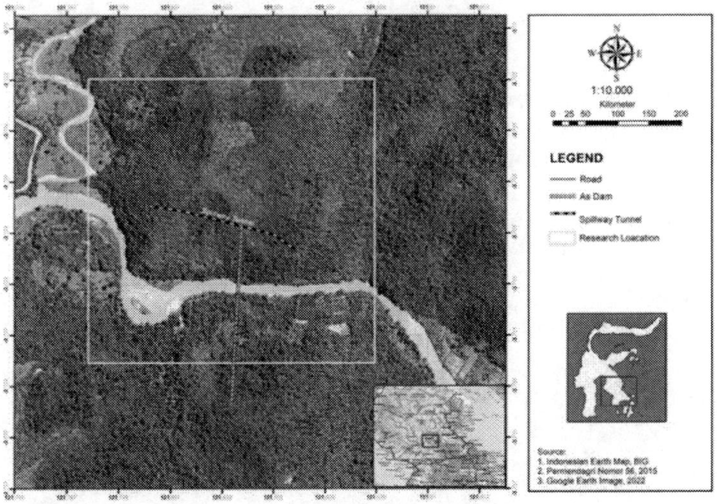

Figure 1 Location of the Pelosika dam spillway tunnel
(Google Earth Image, 2022; Permendagri, 2015; Indonesian Earth Map, BIG, 2019)

2. Methodology

Research activities include observations of surface and subsurface geological conditions. Identification of the study area lithology was carried out by geological mapping including megascopic observations of rock outcrops, microscopic observations of rock samples and core drill observation. Geological and engineering geological mapping area is 1 km x 1 km by observing 49 rock outcrop and 3 core drills (BHT-04, KSZK-01, and KSZK-44).

Analysis of the subsurface rock mass quality using the GSI method and the RMR method. The GSI method is divided into 2 which are influenced by the RQD value, if the RQD value is > 10% using the

GSI Hoek et al [9] and whereas if the RQD value is <10% using the GSI Osgui et al [10]. The Hoek's GSI was influenced by 2 parameters, *JCond* RMR89 [11] and *RQD* [12] by the equation (1).

$$GSI = 1{,}5JCond_{89} + \frac{RQD}{2} \tag{1}$$

Osgoui's GSI uses the parameters broken structure domain (BSTR), structure rating (SR), and joint condition index (I_{jc}) by the equation (2), *bs* is the value of Broken Structural Domain (BSTR), and W is the filling and weathering condition of the core material.

$$I_{jc} = (bs) + (W/2) + 4 \tag{2}$$

Another rock mass classification using Rock Mass Rating (RMR) by Bieniawski [11] has 6 parameters consisting of (1) UCS/Point Load, (2) RQD value, (3) discontinuity spacing, (4) discontinuity condition, (5) groundwater conditions, and (6) discontinuity orientation.

Analysis of the excavation method using the EXCASS system [6] based on point load index (Is_{50}) and GSI value. Point load values obtained from laboratory tests. The EXCASS system classified the level of excavation into 5 groups: digger, ripper easy, ripper hard, hammer, and blasting. In addition, selection of excavation sequence and support system of tunnel using RMR rock mass classification [11].

3. Result and discussion

3.1. Geological condition

Based on geological mapping, the lithology of the study area is divided into 2 lithological units, namely quartz muscovite phyllite and pebblely sand deposits. The quartz muscovite phyllite unit has an area of 80.39% of the study area. The quartz muscovite phyllite unit is characterized by brownish-gray color, fine-grained foliation (slaty-schistose) structure and lepidoblastic texture as shown in the **Figure 2**. Through microscopic observation with petrographic tests, the constituent minerals consist of muscovite (Ms), quartz (Qz), chlorite (Chl), and opaque minerals (Opq). The quartz muscovite phyllite unit is part of Carbon Paleozoic Metamorphic formation (Pzm) [13]. Pebblely Sand deposits are the youngest units found in the study area. Pebblely sand deposits are part of Holocene Alluvium (Qa) Formation. Pebblely sand deposits has an area of 19% which spread along the floodplain of Konaweha River. Pebblely Sand deposits are dominated by sand with some gravel-sized material with a thickness of 1 – 4 m. Generally, Pebblely sand deposits shown in **Figure 3** have brownish gray color with medium to coarse sand matrix and pebble-sized fragments.

Figure 2 Quartz muscovite phyllite unit (STA 43)

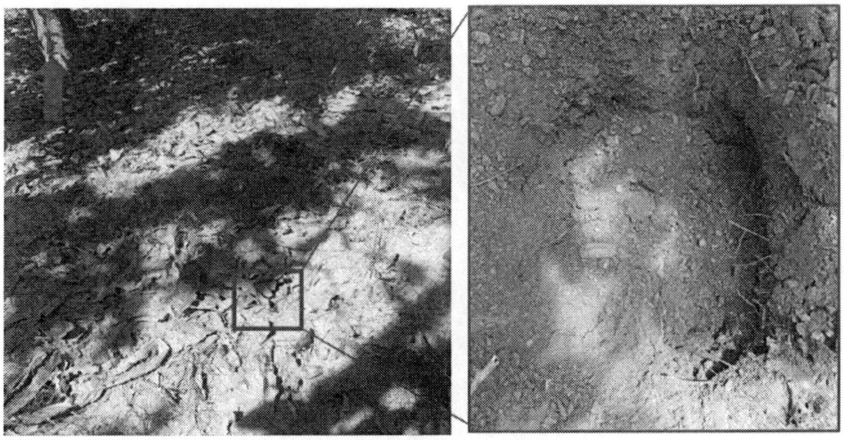

Figure 3 Pebblely Sand deposits (STA 49)

This research also conducted observations of geological structures. Determination of geological structure by strike dip measurements on outcrops and lineament analysis. The result showed that the study area has a northwest-southeast trending inferred fault. The faults and lineaments generally parallel to the active Lasolo fault. The fault is estimated to be located west of the main dam. Geological mapping results are presented in the form of geological map and geological section shown in **Figure 4** and **Figure 5.**

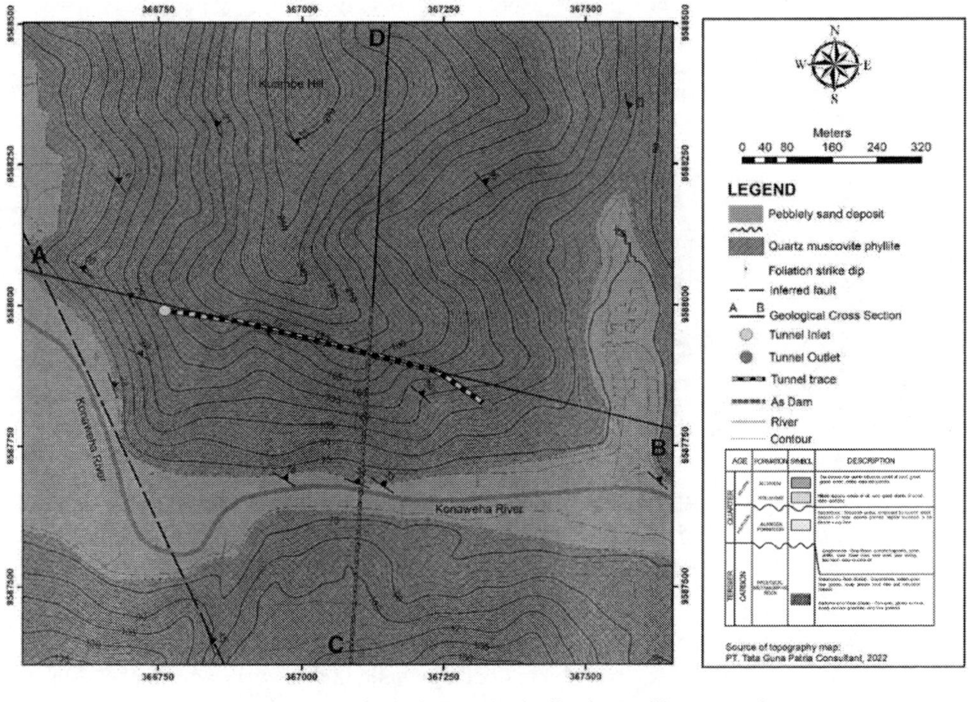

Figure 4 Geological map of Pelosika dam spillway tunnel

IOP Conf. Series: Earth and Environmental Science 1233 (2023) 012008 doi:10.1088/1755-1315/1233/1/012008

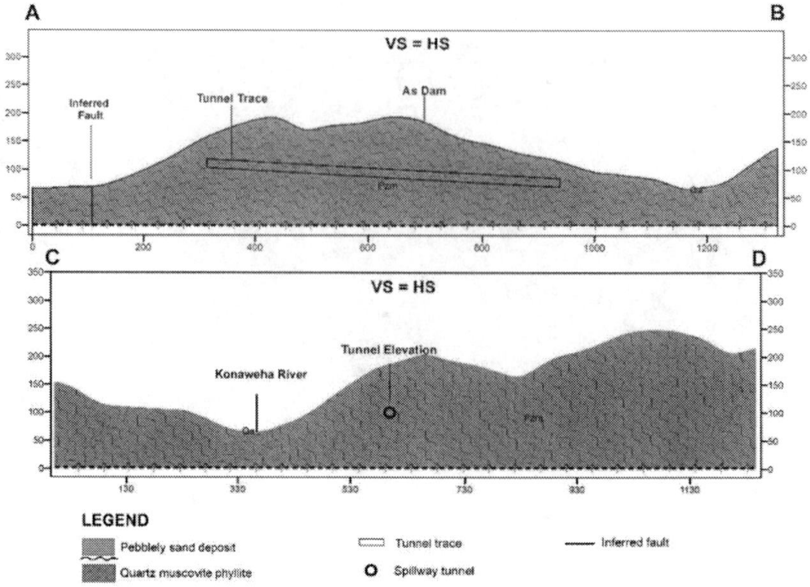

Figure 5 Geological profile section A-B and section C-D

3.2. *Engineering geological condition*

Information on subsurface rock conditions was obtained based on engineering geological investigation data conducted at 3 core drill, namely BHT-04, KSZK-01, and KSZK-44. Analysis of the subsurface rock mass quality is carried out per meter of depth and grouped based on lithology and rock mass quality class. The results of the analysis are made into subsurface engineering geological cross section. Core box samples are shown in **Figure 6**. Some calculation of subsurface rock mass quality using the GSI method and the RMR method is shown in the **Table 1** and **Table 2**.

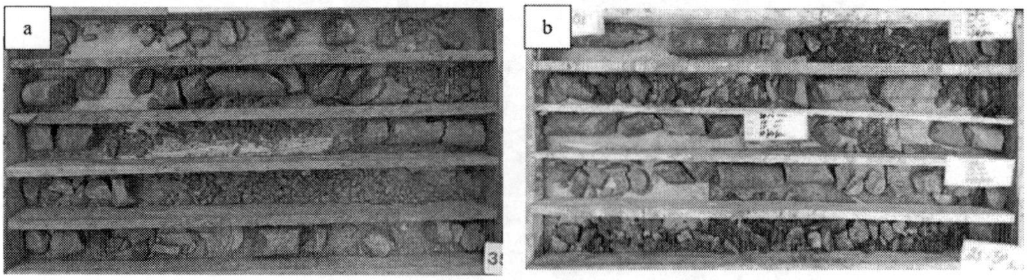

Figure 6 Core box sampel, a) BHT-04 depth 30 – 35 m, b) KSZK-01 depth 25 – 30 m

GSI analysis results at the BHT-04 drill point show that the quartz muscovite phyllite unit has very poor quality at a depth of 2 – 25 m and poor quality at a depth of 25 – 45 m. GSI analysis results at the KSZK-01 drill point show very poor quality in the muscovite quartz phyllite unit at a depth of 2 – 24 m and at a depth of 24 – 130 m dominated by poor quality. The results of the GSI analysis on the KSZK-44 drill point with quartz muscovite phyllite units have very poor quality at depths of 4 – 24 m and poor quality at depths of 24 – 60 m. Very poor rock masses are found at the KSZK-01 drill point at a depth of 83 m to 103 m from the surface. The rock mass quality condition is very poor and is estimated to extend towards the KSZK-44 drill point at a depth of between 40 m and 49 m with a thickness of between 4 and 20 m.

Table 1 Results of rock mass quality calculation using GSI method at depth 25 – 30 m of KSZK-01 core dri

eathering Degree	RQD (%)	JCond89	JCond89*					GSI**	BSTR	I_{jc}			
			a	b	c	d	e			Joint Condition	bs	W	Value
derately Weathered	13	25	x	x	x	x	x	44	x	no filling	x	x	x
derately Weathered	0	20	x	x	x	x	x		BSTR 5	no filling	8	6	15
derately Weathered	12	25	x	x	x	x	x	43.5	x	no filling	x	x	x
derately Weathered	12	25	x	x	x	x	x	43.5	x	no filling	x	x	x
derately Weathered	0	10	x	x	x	x	x		BSTR 4	no filling	6	6	13

**GSI (Hoek, et al., 2013) ***GSI (Osgui, et al., 2010)

m) d = Gouge
e = Weathering
x = N/A

Table 2 Results of rock mass quality calculation using RMR method at depth 30 – 35 m of BHT-04 core dri

eathering Degree	UCS	RQD	Joint Spacing	JCond89	Ground water	Joint Orientation	JCond89*				
							a	b	c	d	e
derately Weathered	4	3	5	10	0	-5	x	x	x	x	x
lighty Weathered	4	3	8	25	0	-5	x	x	x	x	x
lighty Weathered	4	3	8	25	0	-5	x	x	x	x	x
derately Weathered	2	3	8	10	0	-5	x	x	x	x	x
lighty Weathered	2	3	8	25	0	-5	x	x	x	x	x

m) d = Gouge
e = Weathering
x = N/A

Meanwhile, the results of the rock mass quality assessment using RMR method at the BHT-04 drill point show that the quartz muscovite phyllite unit at a depth of 2 – 25 m has an RMR value of <20 or very poor quality. The results of the KSZK-01 point drill analysis show that the quartz muscovite phyllite unit has an RMR value of <20 (very poor) at a depth of 2 – 24 m, an RMR value = 25.8 (poor) at a depth of 24 – 83 m, and an RMR value = 9 (very poor) at a depth of 83 – 103 m. Analysis at the KSZK-44 point drill showed that the muscovite quartz phyllite unit at a depth of 4 – 24 m had an RMR = 4 or very poor and an RMR = 25.5 (poor) at a depth of 24 – 40 m.

At the elevation of the tunnel alignment, the subsurface rock mass quality based on the GSI method and the RMR method showed similar results. The BHT-04 and KSZK-44 drill points have poor rock mass quality and are in class IV (RMR 21 – 40). Meanwhile, the KSZK-01 drill point has very poor rock mass quality with an RMR value of <20. **Table 3** shows the results of assessing the quality of the subsurface rock mass using the GSI method and the RMR method at each drill point. **Figure 7** shows the engineering geological section of surface rock mass quality based on the GSI method and the RMR method.

Table 3 Summary of rock mass quality analysis using the GSI method and the RMR method

Core Drill	Depth	Lithology	GSI Value	RMR Value	Quality Class
BHT-04	0 - 2	Residual Soil	0	0	-
	2 - 25	Quartz Muscovite Phyllite	10	4	Very Poor Rock
	25 - 39	Quartz Muscovite Phyllite	25.3	21.1	Poor Rock
	39 - 45	Quartz Muscovite Phyllite	26	21	Poor Rock
KSZK-01	0 - 2	Residual Soil	0	0	-
	2 - 10	Quartz Muscovite Phyllite	10	4	Very Poor Rock
	10 - 15	Quartz Muscovite Phyllite	28.3	26.2	Poor Rock
	15 - 24	Quartz Muscovite Phyllite	11.4	6.6	Very Poor Rock
	24 - 83	Quartz Muscovite Phyllite	28.7	25.8	Poor Rock
	83 - 103	Quartz Muscovite Phyllite	14.4	9	Very Poor Rock
	103 - 130	Quartz Muscovite Phyllite	32.5	29.5	Poor Rock
KSZK-44	0 - 4	Residual Soil	0	0	-
	4 - 24	Quartz Muscovite Phyllite	11.8	4	Very Poor Rock
	24 - 40	Quartz Muscovite Phyllite	24.7	25.5	Poor Rock
	40 - 49	Quartz Muscovite Phyllite	15.8	10.1	Very Poor Rock
	49 - 60	Quartz Muscovite Phyllite	38.1	30.5	Poor Rock

(shaded) Spillway tunnel elevation

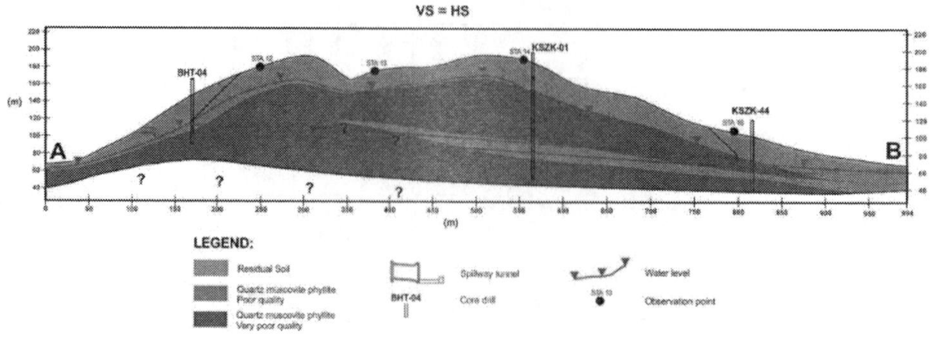

Figure 7 Subsurface rock mass quality cross section

3.3. Excavation methods and tunnel support system

The determination of excavation method considers rock mass structure condition and discontinuity condition. analysis of excavation method in this study was divided into 3 sections, namely Section 1 (S1), Section 2 (S2), and Section 3 (S3). Each section represents by geology and geological engineering condition resulting from the interpretation of each drill point. The analysis sections shown in **Figure 8**.

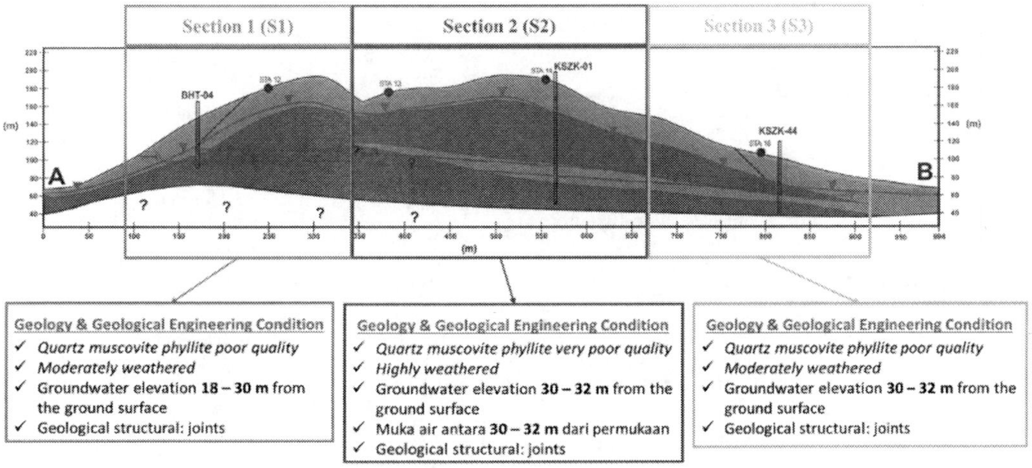

Figure 8 Section of excavation method analysis

The results of excavation method analysis using the EXCASS System show that S1 has an EPI_{opt} value of 24, so the optimum recommended excavation method is Ripper Easy. Meanwhile, S2 and S3 used the Digger excavation method with an EPI_{opt} value obtained <21 (showed in **Figure 9**). The recommended equipment based on the EXCASS System on the S1 using a hydraulic face shovel. While the S2 and S3 use digging with a hydraulic backhoe.

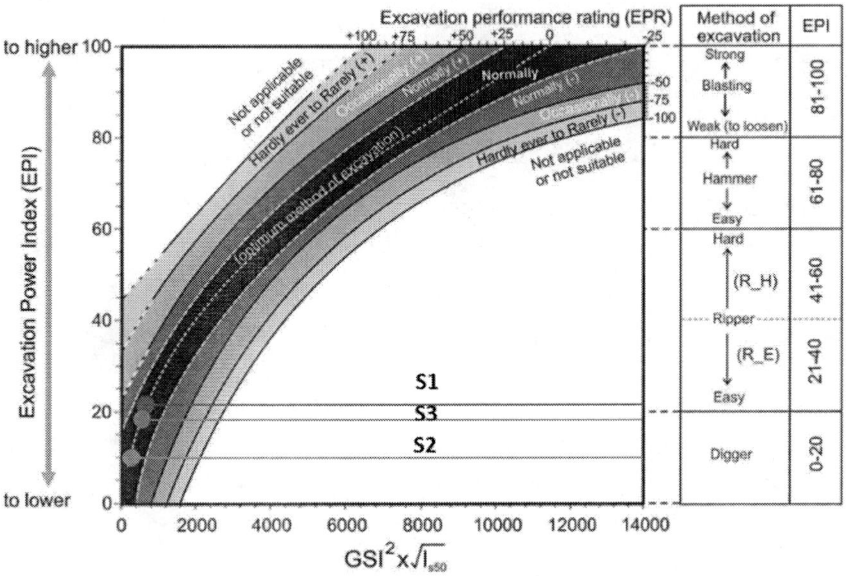

Figure 9 Tunnel excavation method based on EXCASS system chart [6]

The results of selecting excavation sequences using the RMR method are influenced by engineering geological conditions based on the RMR method. S2 has very poor rock mass quality (RMR < 21) so the excavation sequence uses the multiple drifts method, namely dividing the excavation area into several parts to avoid collapse. The multiple drifts method has an excavation length of 0.5 – 1.0 m at the top and a support system is installed immediately after excavation. S1 and S3 use the top heading and bench excavation methods which have poor rock mass quality (RMR 21 – 40). The stages of excavation using this method are to dig 1.0 – 1.5 m long at the top and then install a support system every 10 m from the face of the excavation. The summary of excavation sequence and tunnel support system using RMR are shown in **Table 4**. An overview of excavation methods, excavation sequence, and support systems are shown in the **Figure 10**.

Table 4 Excavation sequence and tunnel support system based on RMR classification [11]

Bore Hole	BHT-04	KSZK-01	KSZK-44
Rock Quality	Poor	Very Poor	Poor
Excavation Sequence	Top Heading and bench: 1.0 – 1.5 m advanced in top heading; Install support concurrently with excavation 10 m from face	Multiple drifts: 0.5 – 1.5 m advanced in top heading; Install support concurrently with excavation. Shotcrete as soon as possible after blasting	Top Heading and bench: 1.0 – 1.5 m advanced in top heading; Install support concurrently with excavation 10 m from face
Tunnel Support System		**Rockbolt** (20 mm diameter fully grouted)	
	Systematic bolts 4-5 m long, 1-1.5 m spacing in crown and walls with wire mesh	Systematic bolts 5 - 6 m long, 1-1.5 m spacing in crown and walls with wire mesh. Bolt invert	Systematic bolts 4-5 m long, 1-1.5 m spacing in crown and walls with wire mesh
		Shotcrete	
	100-150 mm in crown and 100 mm in sides	150-200 mm in crown, 150 mm in sides, and 50 mm on face	100-150 mm in crown and 100 mm in sides
		Steel Set	
	Light to medium ribs spaced 1.5 m where required	Medium to heavy ribs spaced 0.75 m where requiredwith steel lagging and forepoling if required. Close invert	Light to medium ribs spaced 1.5 m where required

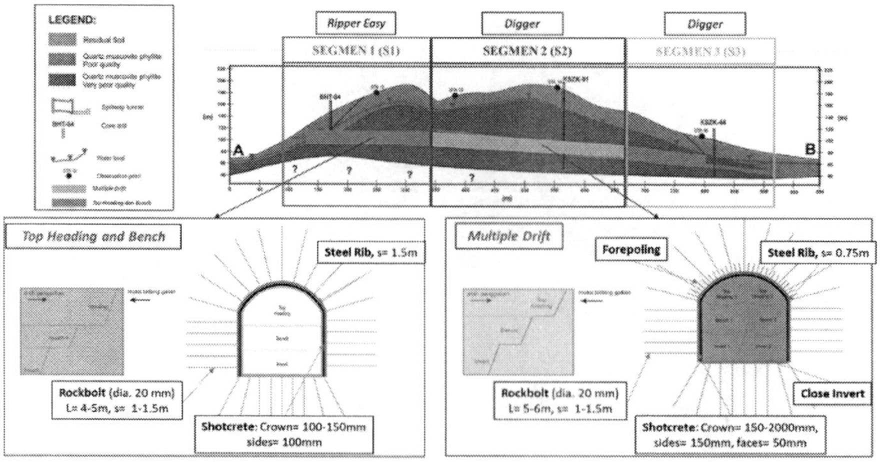

Figure 10 Overview of excavation methods, excavation sequence, and support systems

4. Conclusion

This study reviews the geological engineering conditions, including lithological conditions and rock mass quality in the study area tunnel of the Pelosika Dam. The excavation method and tunnel support system are determined according to the identification of geological conditions. Lithological conditions in the study area consist of 2 lithological units, namely quartz muscovite phyllite and pebblely sand deposits.

The subsurface rock mass quality at the tunnel elevation has a GSI value of 14–25 or a very poor to poor class. GSI and point load index (I_{S50}) parameters are used as input data to determine the excavation method. The optimum method of excavation based on EXCASS system uses ripper easy and digger. Recommended equipment are hydraulic face shovel and hydraulic backhoe.

Rock mass quality analysis using RMR at drill points BHT-04, KSZK-01, and KSZK-44 have RMR values of 9 – 25, classified into poor and very poor quality. While the RMR classification recommends excavation sequences with top heading and bench and multiple drifts. As a result, the tunnel support system uses (1) systematic rockbolt 4–6 m long, 1–1.5 m apart with wire mesh on the crown and walls, (2) 100–200 mm thick shotcrete on the crown and 100-150 mm on the sides, (3) light to heavy steel rib with 1.5 m spacing.

In further research, numerical modeling is needed to determine the tunnel's stability and select an effective and efficient support system. It is also necessary to review the slope stability of the tunnel portal.

Acknowledgment

The author would like to thank the Ministry of Public Works and Public Housing for the opportunity and scholarship to conduct this research. We acknowledge the Balai Wilayah Sungai Sulawesi IV Kendari for the permission to research and access the data provided.

References

[1] BWS Sulawesi IV Kendari 2019 *Sertifikasi Desain Bendungan Pelosika Tahap III Kab. Konawe* (Kendari)

[2] Lu H, Gutierrez M and Kim E 2022 Empirical approach for reliability evaluation of tunnel excavation stability using the Q rock mass classification system *Underground Space*

[3] Weng M-C, Li H-H, Fu Y-Y, Fang C-H, Chen H-R and Chang C-Y 2022 A failure criterion for foliation and its application for strength estimation of foliated metamorphic rock *International Journal of Rock Mechanics and Mining Sciences* **153** 105086

[4] Nurfarisa M N 2017 *Kelas massa batuan dan penyangga Terowongan Pengelak menggunakan metode Q-System di Desa Asinua Jaya, Kabupaten Konawe, Provinsi Sulawesi Tenggara* (Jatinangor: Universitas Padjajaran)

[5] Pettifer G S and Fookes P G 1994 A revision of the graphical method for assessing the excavatability of rock *Quarterly Journal of Engineering Geology* **27** 145–56

[6] Dagdelenler G, Sonmez H and Saroglou C 2020 A flexible system for selection of rock mass excavation method *Bulletin of Engineering Geology and the Environment* **79** 5355–69

[7] Dagdelenler G 2021 Comparison of the efficiency evaluations of selected excavatability classifications for rock masses *Arabian Journal of Geosciences* **14** 1281

[8] Hoek E, Marinos P and Benissi M 1998 Applicability of the geological strength index (GSI) classification for very weak and sheared rock masses. The case of the Athens Schist Formation *Bulletin of Engineering Geology and the Environment* **57** 151–60

[9] Hoek E, Carter T G and Diederichs M S 2013 Hoek, dkk., 2013, Quantification of the GSI Chart *47th US Rock Mechanics / Geomechanics Symposium*

[10] Osgoui R R, Ulusay R and Unal E 2010 An assistant tool for the Geological Strength Index to better characterize poor and very poor rock masses *International Journal of Rock Mechanics and Mining Sciences* **47** 690–7
[11] Bieniawski Z T 1989 *Engineering rock mass classifications: a complete manual for engineers and geologists in mining, civil, and petroleum engineering* (John Wiley & Sons)
[12] Deere D U 1963 Technical description of rock cores for engineering purposes, Felsmechanik und Ingenieurgeologie (Rock Mechanics and Engineering Geology), **1** 16–22
[13] Rusmana E, Sukido, Haryono E and Simandjuntak T O 1993 *Peta Geologi Lembar Lasusua-Kendari* (Bandung)

IOP Conf. Series: Earth and Environmental Science 1233 (2023) 012009 doi:10.1088/1755-1315/1233/1/012009

Greenhouse Gasses Emission from Coal Mining Operation in North Kalimantan, Indonesia

O W Lusantono[1,3], F Anggara[1], D H Amijaya[1], and D A Kusumaningsih[2]

[1] Geological Department, Faculty of Engineering, Gadjah Mada University, Indonesia
[2] Mining Engineering Department, Faculty of Science and Technology, Islamic State University Syarif Hidayatullah Jakarta, Indonesia
[3] PT Studio Mineral Batubara, Mining Consulting, Sleman, Yogyakarta, Indonesia

Corresponding author: oktarianwisnulusantono@mail.ugm.ac.id.

Abstract. Greenhouse gasses (GHG) are gases that could trap heat in the atmosphere and mainly made up of carbon dioxide (CO¬2¬) and methane (CH4). Coal is still one of the primary energy sources in several countries such as South Africa (71%), China (57%), Australia (30%), and Indonesia (44%). Therefore, coal mining operation is indicated as one of the biggest GHG contributor in the world. However, coal as commodity also indicated emitting GHG emission sequentially but its pathways are still debatable. GHG emission from coal mainly stated as an emission factor on both CO_2 and CH_4 because of coal combustion and/or existing gas in the coal bearing strata. The study of GHG emission or so-called Carbon Emission has been done based on several regions, coal rank, and condition. This study focuses on identify and projecting the GHG emission based on coal mining operation plan in North Kalimantan Province, Indonesia. This study aims to identify and adds more broader perspective about GHG emission from coal mining. The GHG emission estimation is based form Tier 1 IPCC Calculation method and estimated from a different scenario of production along the year. The GHG emission in this study estimated from fuel combustion and fugitive emission. The total emission estimated from this study was 1.31 x 10-02 kg CO_{2eq} / kg coal.

1. Introduction

Greenhouse gasses (GHG) are gases that could trap heat in the atmosphere and mainly made up of carbon dioxide (CO_2) and methane (CH_4) [1]. GHG typically result from fossil fuel combustion in many sector such as energy sector (25%), industrial process (21%), agriculture (24%), transportation (14%), building (6%) and other energy (10%) [2]. Coal is still one of the primary energy sources in several countries such as South Africa (71%), China (57%), Australia (30%), and Indonesia (44%). Therefore, coal mining operation is indicated as one of the biggest GHG contributor in the world [3].Coal mining conducted through surface mining and underground mining which has different threads and consequence in term of GHG emission. However, coal as commodity also indicated emitting GHG emission sequentially.

GHG emission from coal mainly stated as an emission factor on both CO_2 and CH_4 because of coal combustion and/or existing gas in the coal bearing strata [4]. The study of GHG emission or so-called Carbon Emission has been done based on several countries, coal rank, and coal characteristics. The emission study are generally using burned-coal sample to account the carbon emission factor. A study form US stated that the emission factor from various coal rank such as lignite, subbituminous, bituminous, and anthracite are within 2.93; 1.60; 2.4; and 6.42 kg CO_2 eq/GJ respectively [5], [6]. Study from Indonesia with the same coal rank the emission factor are 115; 100; 99.7; 101 kg CO2 eq/TJ respectively [7]. Some studies shown that an emission factor in certain region or countries of several coal rank are lower than stated in the guidelines of IPCC [8]. A study in Poland shown that a subbituminous coal getting (mining) generates 1.94 kg CO_2 eq/GJ while transportation of coal to the processing facilities generates 0.89 kg CO_2 eq/GJ [9]. However, those study only focused on in-situ GHG emission not focusing on its pathways.

This study aims to give a broader perspective about GHG emission from coal mining operation. Throughout this study GHG emission pathways are investigated and focuses on identify and projecting the GHG emission based on coal mining operation plan in North Kalimantan Province, Indonesia.

2. Methodology

The typical GHG emission in a surface mining operation categorized into two pathways: direct emission and indirect emissions source [4]. The direct emissions are emission from sources that are owned/controlled within the mining company. The indirect emissions are emission from source that resulting from activity that outside from the mining company. Thus, this study only focusing on direct emission sources that occurs in the main business process of surface mining operation. GHG emission in coal mining operation generally occurs from fuel combustion, explosive, electricity, and fugitive emission from coal bearing strata [10]. The illustration of GHG emission in a typical surface mining shown in Figure 1. This study only accounts GHG emission from the main operation process such as topsoil and overburden stripping, coal mining, and coal transporting due to the data availability. Thus the total emission is calculated based on the coal production (CO2eq/kg coal). The methodology

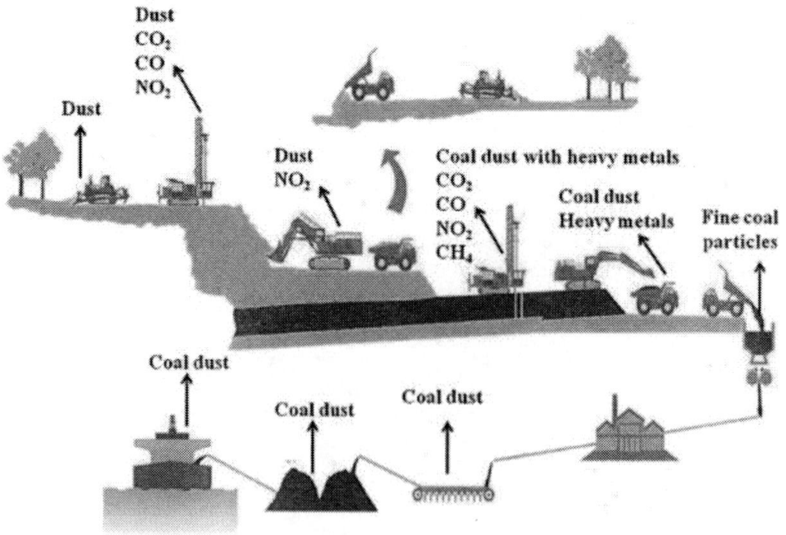

Figure 1. Illustration of Typical GHG Emission in a Surface Mining Operation [4]

The GHG emission estimation is based form Tier 1 IPCC Calculation method [11], [12]. The GHG emission from mining operation is typically quantified using the concept of Emission Factor (EF). EF is derived from available data by country, specific basin, and mine specific and so-called Tier 1, Tier 2, and Tier 3 respectively [8]. Thus, the calculation of GHG emission in this study is a summation of total emission from fuel combustion in the mining activities and fugitive emission. It can be expressed using mathematical equation in equation (1), (2), and (3). The graphical representation of methodology shown in Figure 2.

IOP Conf. Series: Earth and Environmental Science 1233 (2023) 012009 doi:10.1088/1755-1315/1233/1/012009

$$E_{total} = E_{fuel} + E_{fugitive} \tag{1}$$
$$E_{fuel} = \sum Fuel\ x\ EF \tag{2}$$
$$E_{fugitive} = Coal\ Production\ x\ EF_{CH_4} \tag{3}$$

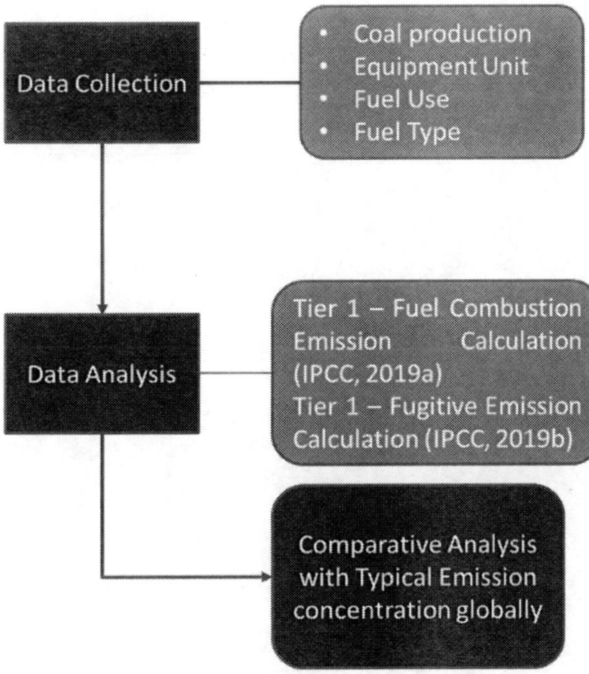

Figure 2. Research Methodology

3. Result and Discussion

3.1. Study Area and Baseline Data

Company PT X is mining company located in Tana Tidung Regency, North Kalimantan Province that has been operating since 2011. The geological map of study area shown in the Figure 3. There are three main lithology type in the area such as alluvial, sandstone, and claystone. The coal bearing strata is controlled by major structure such as fault and resulting folding on the coal seam. The main commodity is coal with an average calorific value of 4,627.3 kcal/kg. The typical coal characteristics in the study area is shown in the Table 1. The company planned the coal production for 17 years with maximum coal production of 20.000.000 metric tonnes (Figure 4). The mining process and overburden removal heavily rely on a combination of excavator and dump truck with a support of dozer. The main mining equipment needs for every year and stages listed in the Table 2. The typical operation of this mine is 20 EWH/day (effective working hour/day) and 300 EWD/year (effective working days/year).

Table 1. Coal Characteristics

Parameter	Value	Unit
Total Moisture	28.19	% (ar)
Inherent Moisture	21.5	% (ad)
Ash Yield	5.36	% (ar)
Total Sulphur	0.57	% (ar)
Calorific value	4627.3	kcal/kg (gar)
Density	1.3	gr/cc

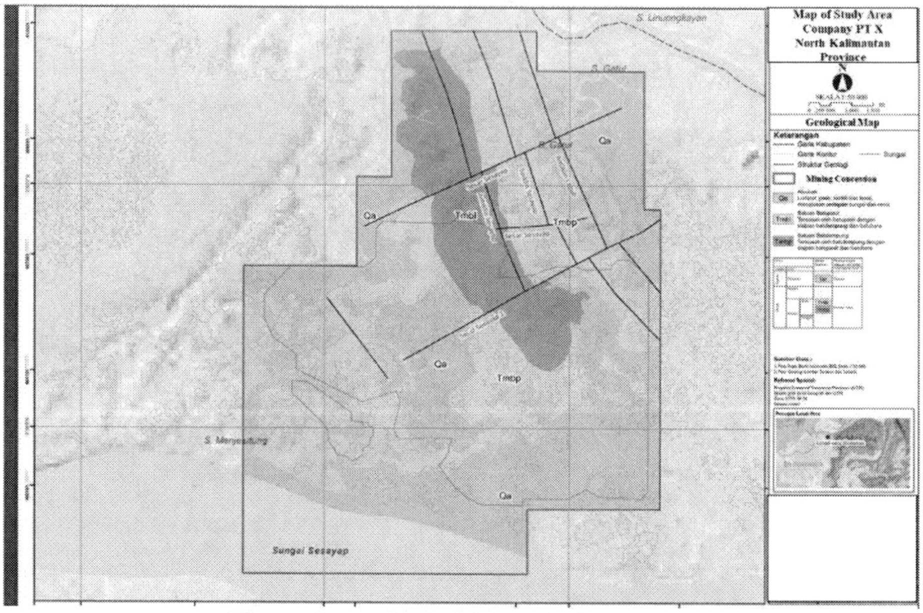

Figure 3. Geological Map of Study Area

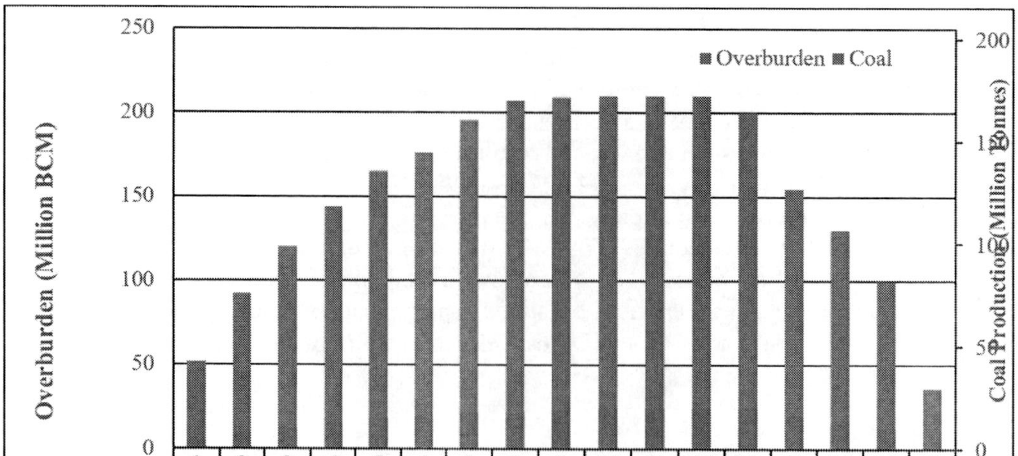

	1	2	3	4	5	6	7	8	9	10	11	12	13	14	15	16	17
Overburden	51,8	92	120	144	165	176	196	207	209	210	210	210	200	155	130	100	35,9
Coal	4,5	8	10	12,5	15	16	17	18	19	20	20	20	20	16,5	13	10	4,3

Figure 4. Production of Overburden and Coal on 17 Years of Mining Plan

Table 2. Mining Equipment Unit on Yearly Basis

Mining Year	Total Unit - Activity - Unit Type							
	Topsoil Removal			Overburden Removal		Coal Mining		Coal Transport
	Dozer D85SS	Excavator PC 400	ADT Volvo A30H	Excavator PC 2000	DT 785HD	Excavator PC 1250	DT 767HD	DT 767 HD
1	4	10	20	9	45	2	9	9
2	4	10	20	15	80	3	15	15
3	5	10	20	20	104	4	19	19
4	5	12	24	23	125	4	23	23
5	6	15	30	27	143	5	28	28
6	7	16	30	29	152	5	29	29
7	8	18	30	32	169	6	31	31
8	10	20	30	34	179	6	33	33
9	10	20	30	34	181	6	35	35
10	10	20	30	34	182	7	37	37
11	10	20	30	34	182	7	39	39
12	10	20	30	34	182	7	39	39
13	8	18	30	32	173	7	39	39
14	8	15	30	25	134	5	30	30
15	5	14	28	21	113	4	24	24
16	5	12	24	16	87	4	19	19
17	5	10	20	6	31	2	8	8

3.2. Emission from Fuel

The emission from fuel use calculated using equation (2). Total fuel obtained from a multiplication of an average fuel use of mining equipment with the unit of mining equipment. EF of fuel taken from IPCC [13] shown in the Table 3; the default value used for emission estimation. The fuel density is estimated around 0.85 kg/L [14]. The average fuel used (L/hour) for each mining equipment as listed in the table 2 shown in Table 4. The emission estimation is shown in Figure 5. Based on the result shown that the highest activity producing GHG Emission is overburden removal (15.32 – 89.39 ton CO_2). Meanwhile the lowest activity is the coal transport (2.3 – 11.41 ton CO_2).

Table 3. Emission Factor of Diesel Fuel Combustion

GHG	Default (kg GHG/kg Diesel)	Lower (kg GHG/kg Diesel)	Upper (kg GHG/kg Diesel)
CO_2	3.186	3.03	3.13
CH_4	1.73×10^{-04}	6.98×10^{-05}	4.73×10^{-04}

Table 4. Average Fuel Use

Unit Type	Average Fuel Use (L/Hour)
Dozer D85SS	15
Excavator PC 400	12
ADT Volvo A30H	13
Excavator PC 2000	28
DT 785HD	25
Excavator PC 1250	19
DT 767HD	18

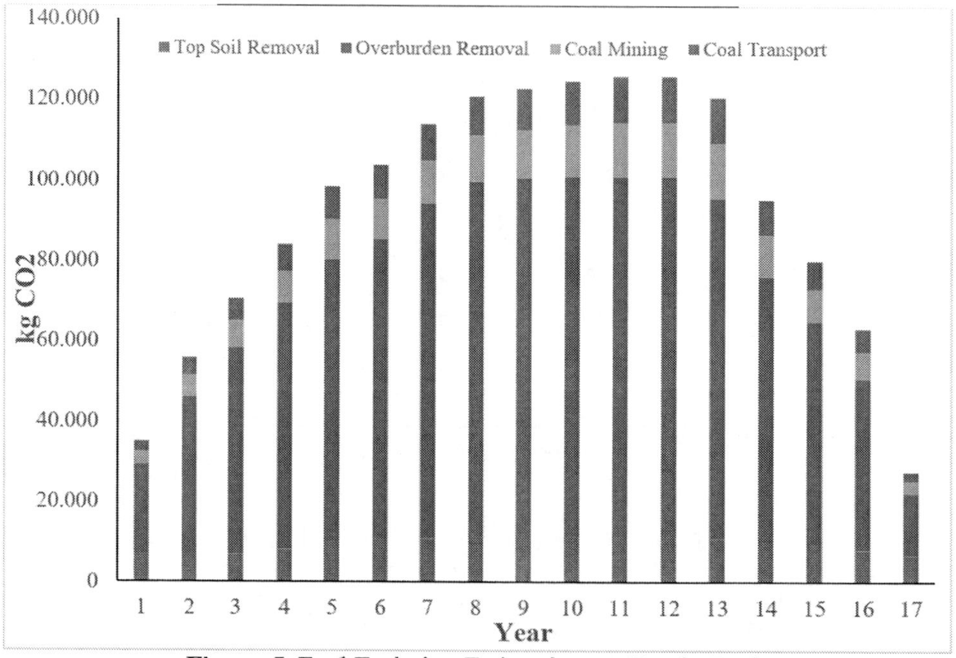

Figure 5. Fuel Emission Estimation on Yearly Basis

3.3. Fugitive Emission

Fugitive emission coming from the coal itself and it needs to be accounted for consideration of GHG emission in a coal mining operation. CH_4 is the most dominant of fugitive emission and account for around 8% of anthropogenic sources of methane [15]. Methane is commonly found in a coal bearing stratum as a result from coalification process and a function of organic matter composition and the deposit condition. Coal properties such as ash yield and moisture content have effect on methane content. In general, the ash yield is equally referred as mineral matter; a gangue without adsorbing capacity that may affect gas adsorption in a coal surface [16]–[18]. Thus, the higher ash yields may reduce the gas storage capacity of coals. Moisture content could be stated equally as water content in a coal. Water could inhibit the pores of coal and competing with methane gas from entering the coal pores [17]–[19]. Therefore, moisture content has negative correlation with methane gas content in coal. Inhibition of methane gas in a coal also affected from its maceral composition. Maceral composition affects coal structure and adsorption capacity. Usually vitrinite-rich coals have more adsorption capacity compared with inertinite-rich coal [20]–[22].

IOP Conf. Series: Earth and Environmental Science 1233 (2023) 012009 doi:10.1088/1755-1315/1233/1/012009

Mining activity having an impact to disturb pressure that contains methane gas in a coal bed and surrounding strata [23]. It caused methane gas released into atmospheric condition and could be referred as a lost gas [24]. IPCC [13] defines three different approaches to estimate fugitive emission from coal mine with different level of details. The first approach (Tier 1) is an estimation based on mining methods (surface or underground). The second approach (Tier 2) is based on specific EF of the coal basin and/or country. The third approach (Tier 3) is based on site-specific measurements. The estimation of fugitive emission in this study estimated using equation (3). Coal production accounted in kg. Due to data availability, the estimation of fugitive emission in this study is using Tier 1 approach with emission factor shown in Table 5; default value is used for estimation during mining operation and upper value is used for post-mining estimation. Data from [13] is converted using the average CH_4 standard density of 0.654 kg/m^3. The result of fugitive emission shown in Figure 6.

Table 5. Fugitive Emission Factor

Emission Source	Default (kg CH_4/kg Coal)	Lower (kg CH_4/kg Coal)	Upper (kg CH_4/kg Coal)
Mining Operation	2.0×10^{-4}	2.00×10^{-4}	1.33×10^{-3}
Post-Mining Operation	6.66×10^{-5}	0	1.33×10^{-5}

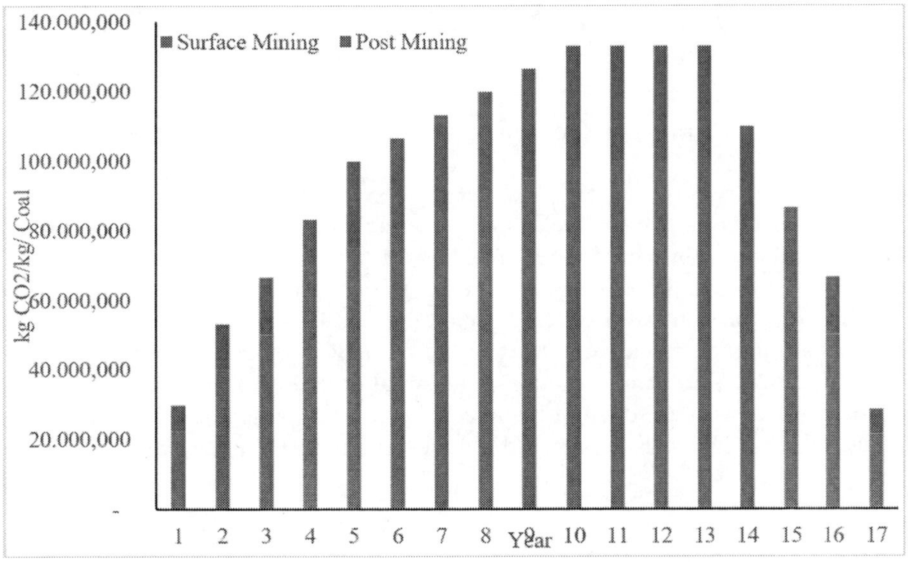

Figure 6. Fugitive Emission Estimation on Yearly Basis

3.4. *Total Emission*

The total emission is a summation of fuel combustion emission and fugitive emission as stated in the equation (1). The result is shown in the Figure 6. Based on the result, emission from fuel combustion and fugitive has shared a proportional value of total emission throughout production year. Thus, while the coal production is high during year 12– 14, the fugitive emission is overpassing the fuel combustion emission. The total emission value from this mining operation on average value is 1.31×10^{-2} kg CO_{2eq} /kg Coal.

The total emission on this mining project is lower than an average emission data from coal mining project around the world such as Australia (7.12×10^{-2} kg CO_{2eq}/kg Coal); USA (4.69×10^{-2} kg CO_{2eq}/kg coal); and Ukraine (3.00×10^{-2} kg CO_{2eq}/kg coal) [10]. This study needs to be addressed carefully due to two main reasons. First reason is the life cycle of mining operation, and the data inventory are portrayed partially. While to use a LCA (life-cycle assessment) approach needs more data especially from electricity and others activity within the mine. The second reason is the calculation approach. This study only using Tier 1 approach that typically taking conservative value for GHG emission estimation. In addition, this study also opted the mine reclamation activity that may reduce GHG emission.

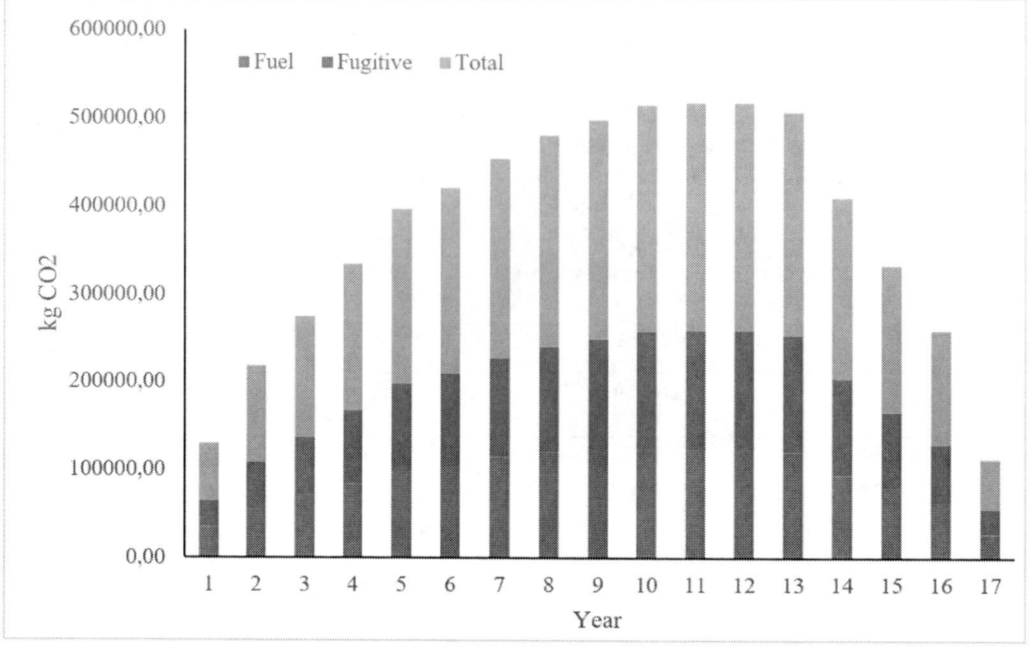

Figure 7. Estimation of Total Emission on Yearly Basis

4. Conclusion

The study of GHG emission from fuel combustion and fugitive emission in a surface coal mining operation was conducted. Overburden removal becomes the highest activity source of GHG emission form the fuel combustion. It has equally similar with fugitive emission from the surface mining operation. The total emission estimated from this study was 1.31 x 10^{-02} kg CO_{2eq}/kg coal. Since its only two pathways for emission calculation, the proportion between fuel combustion and fugitive emission was more likely equal. The estimated emission is smaller than emission identified in other countries. However, the GHG emission reduction plan such as reclamation is not included in this study. Further research must address GHG emission reduction plan and more adequate data for analysis.

Acknowledgement
The authors would like to thank you PT Studio Mineral Batubara for providing permission to use the data for this publication and Research Directorate Gadjah Mada University (Program Rekognisi Tugas Akhir Batch I Tahun Anggaran 2022) for its funding to publish this research.

References

[1] L. Jeffry, M. Y. Ong, S. Nomanbhay, M. Mofijur, M. Mubashir, and P. L. Show, "Greenhouse gases utilization: A review," Fuel, vol. 301, no. January, p. 121017, 2021, doi: 10.1016/j.fuel.2021.121017.

[2] IPCC, "2019 Refinement to the 2006 IPCC Guidelines for National Greenhouse Gas Inventories. Chapter 0. Overview. Glossary," pp. 1–19, 2019, [Online]. Available: https://www.ramsar.org/news/ramsar-topics-report-on-canal-estates

[3] N. Kholod et al., "EPA Public Access," 2021, doi: 10.1016/j.jclepro.2020.120489.Submit.

[4] B. Pandey, M. Gautam, and M. Agrawal, Greenhouse gas emissions from coal mining activities and their possible mitigation strategies. Elsevier Inc., 2017. doi: 10.1016/B978-0-12-812849-7.00010-6.

[5] J. C. Quick and T. Brill, "Provincial variation of carbon emissions from bituminous coal: Influence of inertinite and other factors," Int. J. Coal Geol., vol. 49, no. 4, pp. 263–275, 2002, doi: 10.1016/S0166-5162(01)00068-4.

[6] J. C. Quick, "Carbon dioxide emission factors for U.S. coal by origin and destination," Environ. Sci. Technol., vol. 44, no. 7, pp. 2709–2714, 2010, doi: 10.1021/es9027259.

[7] R. Damayanti and H. Khaerunissa, "Carbon dioxide emission factor estimation from Indonesian coal," Indones. Min. J., vol. 21, no. 1, pp. 45–58, 2018, doi: 10.30556/imj.vol21.no1.2018.687.

[8] IPCC, the Refinement To the 2006 Ipcc Guidelines for National Greenhouse Gas Inventories, vol. 2. 2019. doi: 10.21513/0207-2564-2019-2-05-13.

[9] P. Burmistrz, T. Chmielniak, L. Czepirski, and M. Gazda-Grzywacz, "Carbon footprint of the hydrogen production process utilizing subbituminous coal and lignite gasification," J. Clean. Prod., vol. 139, pp. 858–865, 2016, doi: 10.1016/j.jclepro.2016.08.112.

[10] M. Guimarães da Silva, A. R. Costa Muniz, R. Hoffmann, and A. C. Luz Lisbôa, "Impact of greenhouse gases on surface coal mining in Brazil," J. Clean. Prod., vol. 193, pp. 206–216, 2018, doi: 10.1016/j.jclepro.2018.05.076.

[11] B. Pandey, M. Agrawal, and S. Singh, "Assessment of air pollution around coal mining area: Emphasizing on spatial distributions, seasonal variations and heavy metals, using cluster and principal component analysis," Atmos. Pollut. Res., vol. 5, no. 1, pp. 79–86, 2014, doi: 10.5094/APR.2014.010.

[12] M. L. Gitarskiy, "the Refinement To the 2006 Ipcc Guidelines for National Greenhouse Gas Inventories," Fundam. Appl. Climatol., vol. 2, pp. 5–13, 2019, doi: 10.21513/0207-2564-2019-2-05-13.

[13] IPCC, "Volumen 2. Energy. Chapter 4. Fugitive Emissions. IPCC Guidelines for National Greenhouse Gas Inventories.," in Language, Society and Power. An Introduction, vol. 3, no. Chapter 4, 2006, pp. 56–74.

[14] J. A. de Carvalho, A. de Castro, G. H. Brasil, P. A. de Souza, and A. Z. Mendiburu, "CO2 Emission Factors and Carbon Losses for Off-Road Mining Trucks," Energies, vol. 15, no. 7, pp. 1–17, 2022, doi: 10.3390/en15072659.

[15] J. Gao, C. H. Guan, and B. Zhang, "China's CH4 emissions from coal mining: A review of current bottom-up inventories," Sci. Total Environ., vol. 725, p. 138295, 2020, doi: 10.1016/j.scitotenv.2020.138295.

[16] Y. Ju et al., "A new approach to estimate fugitive methane emissions from coal mining in China," Sci. Total Environ., vol. 543, pp. 514–523, 2016, doi: 10.1016/j.scitotenv.2015.11.024.
[17] P. J. Crosdale, B. B. Beamish, and M. Valix, "Coalbed methane sorption related to coal composition," Int. J. Coal Geol., vol. 35, no. 1–4, pp. 147–158, 1998, doi: 10.1016/S0166-5162(97)00015-3.
[18] Z. Guo, Y. Cao, Z. Zhang, and S. Dong, "Geological Controls on the Gas Content and Permeability of Coal Reservoirs in the Daning Block, Southern Qinshui Basin," ACS Omega, vol. 7, no. 20, pp. 17063–17074, 2022, doi: 10.1021/acsomega.2c00371.
[19] R. M. Bustin and C. R. Clarkson, "Geological controls on coalbed methane reservoir capacity and gas content," Int. J. Coal Geol., vol. 38, no. 1–2, pp. 3–26, 1998, doi: 10.1016/S0166-5162(98)00030-5.
[20] S. Hou, X. Wang, X. Wang, Y. Yuan, X. Zhuang, and X. Wang, "Geological controls on gas saturation in the Yanchuannan Coalbed Methane Field, Southeastern Ordos Basin, China," Mar. Pet. Geol., vol. 78, no. 2, pp. 254–270, 2016, doi: 10.1016/j.marpetgeo.2016.09.029.
[21] J. Shen, Y. Qin, and J. Zhao, "Maceral Contribution to Pore Size Distribution in Anthracite in the South Qinshui Basin," Energy and Fuels, vol. 33, no. 8, pp. 7234–7243, 2019, doi: 10.1021/acs.energyfuels.9b01514.
[22] L. Liu et al., "Coalbed methane adsorption capacity related to maceral compositions," Energy Explor. Exploit., vol. 38, no. 1, pp. 79–91, 2020, doi: 10.1177/0144598719870325.
[23] D. A. Kirchgessner, S. D. Piccot, and S. S. Masemore, "An improved inventory of methane emissions from coal mining in the united states," J. Air Waste Manag. Assoc., vol. 50, no. 11, pp. 1904–1919, 2000, doi: 10.1080/10473289.2000.10464227.
[24] N. Kholod et al., "Global methane emissions from coal mining to continue growing even with declining coal production," J. Clean. Prod., vol. 256, 2020, doi: 10.1016/j.jclepro.2020.120489.

IOP Conf. Series: Earth and Environmental Science 1233 (2023) 012010 doi:10.1088/1755-1315/1233/1/012010

Groundwater Facies in the South Parts of Tegal-Brebes Groundwater Basin

T R Puspita[1,2] and D A Kuncoro[2]

[1]Department of Geological Engineering, Faculty of Engineering, Gadjah Mada University

[2]Ministry of Public Works and Housing

tita.rpuspita@gmail.com

Abstract. The southern part of the Tegal Brebes Groundwater Basin has various rock units originating from the volcanic activity of the Slamet Volcano. The groundwater interacts with these rocks and influences the hydrogeochemical process that may alter its chemical components. The objective of this research is to assess the relationship between groundwater chemistry facies and the rock aquifer. The finding of this research will help us better understand how hydrogeochemical processes work, perform thorough research, and determine the water quality. The groundwater characteristics are determined based on the chemical composition of the water taken from the citizen's dug wells and springs. These samples were collected to represent a difference in rock's formation, namely Linggopodo Formation, Tapak, and Slamet Undifferentiated Volcanic Rock. The analysis of the Piper diagram shows that all the samples fall into one facies: the calcium bicarbonate type (Ca-HCO_3). The groundwater's physical and chemical properties in the study area indicate that groundwater quality has not interacted much with lithology and is still greatly influenced by meteoric water. The groundwater facies are dominated by the Ca-HCO_3, a characteristic of shallow groundwater and local water flow systems.

1. Introduction

Water is necessary to support domestic needs and industrial activities obtained from surface water or groundwater. At present, groundwater resources are more reliable than surface water due to their advantages. Groundwater has a wide dispersal below the surface, the water quality of groundwater is relatively better than surface water, the infrastructure required is simple, and the regulation of its utilization is easier[1]. In natural hydrological cycling, the groundwater interacts with the surrounding rocks, causing various hydrogeochemical processes that alter groundwater chemical components on a local or regional scale[2]. Groundwater cations and anions are used to estimate the characteristics and origin of groundwater. In this study, we attempt to assess groundwater quality by comparing the physiochemical characteristics and distribution of cations and anions in groundwater from the study area[3]. The groundwater chemistry is influenced by several variables, such as the bedrock's mineral composition, the weather, and the physical setting of the aquifer[4]. Groundwater aquifer systems are also dynamic and susceptible to human activity, particularly the use of chemicals in agriculture and industry. The current season greatly influences also their properties[5].

IOP Conf. Series: Earth and Environmental Science 1233 (2023) 012010 doi:10.1088/1755-1315/1233/1/012010

The groundwater conditions of the southern part of Tegal Brebes Groundwater Basin have a very diverse distribution of rock formations (see Figure 1). Its mean that the groundwater chemistry may varies significantly across the study area, even within small local sections, and other locations' chemical compositions are strikingly similar. To summarize, the chemical quality of groundwater varies from place to place, and the properties of the material through which the water moves are likely to be the most important. As a result, differences in water quality in the study area may be a major issue in the distribution of hydrochemical facies. A detailed, comprehensive study on the hydrogeochemistry of the Tegal Brebes groundwater basin is currently lacking. As a result of this issue, the study purposes to to determine the groundwater chemistry facies in the study area based on their chemical compositions and to look on the relation between groundwater chemistry and the rock's formation.

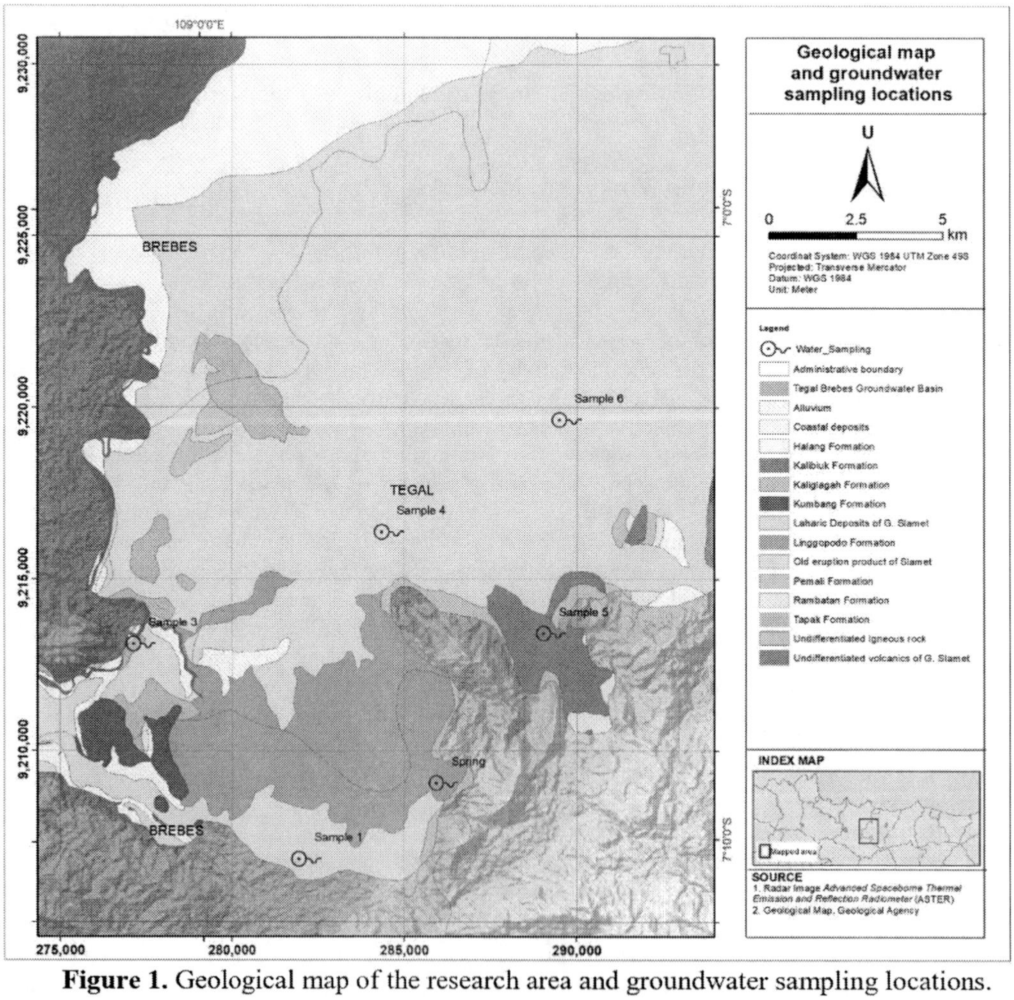

Figure 1. Geological map of the research area and groundwater sampling locations.

1.1. Theoretical Background

Groundwater facies identify groundwater types based on the differences and genesis of water related to the groundwater flow system. Groundwater facies can be found in the chemical properties of groundwater. The chemical nature of groundwater is one of the main properties of water that affects groundwater quality and is useful for determining groundwater quality. The physicochemical properties of groundwater include total dissolved solids (TDS), electrical conductance (EC), salinity, and ion content. The chemical properties of groundwater were carried out in research on groundwater facies in the Tegal Brebes area using the ion content method. The major ions that will be discussed are calcium (Ca), magnesium (Mg), sodium (Na), potassium (K), bicarbonate (HCO_3), sulfate (SO_4), chloride (Cl), and carbonate (CO_3).

The chemical composition of groundwater is affected by the aquifer where the groundwater is located. The contact between groundwater and the lithology of the aquifer itself causes a reaction that can change the chemical composition of groundwater. Dissolving rock-forming minerals into aquifers by passing groundwater will result in adding different types of ion composition (anions and cations) in groundwater.

1.2. Geological Setting
Groundwater sampling locations in the southern part of the Tegal-Brebes Groundwater Basin are in the Linggopodo formation (Qpl), Tapak (QTk) formations, and Undifferentiated Volcanic Rocks of Mount Slamet (Qvs) as listed on the Majenang Regional Geological map[6] and Purwokerto - Tegal Regional Geological map[7]. The rocks that make up the Linggopodo formation are volcanic breccia, tuff, and lava, which are thought to be the result of the activities of Mount Slamet Tua or Mount Copet[8]. The rocks that made up the Tapak formation are coarse sandstone, calcareous sandstone with marl containing mollusks of Early – Middle Pliocene age with a depositional environment in the form of a coastal area influenced by tides. Undifferentiated Volcanic Rocks of Mount Slamet consist of volcanic breccia, lava, and tuff. The distribution forms plains and hills.

1.3. Hydrogeology
A study of hydrogeological conditions is required to obtain an overview of the aquifer system and its physical properties, groundwater level conditions, recharge-discharge areas, and groundwater quality. Based on the hydrogeological map of the Pekalongan[9], the productivity of the aquifer of the southern part of the Tegal-Brebes Groundwater Basin is divided into three categories. The first category is aquifers (fissured or porous) of poor productivity, and regions without exploitable groundwater consist of two types: regions without exploitable groundwater and poorly productive aquifers. The second is aquifers, which flow in intergranular and consist of extensive-moderately aquifers. Third is aquifers, which flow through fissures and interstices, consisting of extensive-moderately aquifers and locally-productive aquifers.

2. Methodology
Field study activities include observing rock outcrops and measuring the water level of dug wells and groundwater physical and chemical parameters. Physical and chemical parameters were taken in the form of depth to the electrical conductivity (EC), the temperature, TDS, pH. Six (6) water samples were taken once during the rainy season and examined using standard analytical techniques. These water samples of groundwater analyzed in The Laboratory of Coal and Geothermal Mineral Research Center to get the value of anion and cation content, Ministry of Energy and Mineral Resources, Bandung, East Java. The coordinates of the sample point positions were recorded using the global positioning system (GPS). Mapping and measurement of physical and chemical parameters of groundwater were carried out at six new observation points in the form of dug wells and springs (Table 1 and Table 2).

Table 1. Physico-chemical properties of field data.

No	Name	Lithology	Water level (m)	EC (μS/cm)	Temperature (°C)	TDS (mg/l)	pH
1	Sample 1	Laharic deposit of M.Slamet	203.79	530	27.5	260	6.5
2	Spring-Sample 2	Linggopodo Formation	-	90	27.2	30	5.6
3	Sample 3	Tapak Formation	29.28	680	28.5	320	6.5

No	Name	Lithology	Water level (m)	EC (μS/cm)	Temperature (°C)	TDS (mg/l)	pH
4	Sample 4	Laharic deposit of M.Slamet	67.47	530	26.5	260	4.7
5	Sample 5	Undifferentiated volcanocs rock of M. Slamet	186.66	220	25.8	100	5
6	Sample 6	Laharic deposit of M.Slamet	105.12	460	28.4	220	6.5

In analyzing the existing groundwater facies from the result of ion content data from laboratory, various methods can be used, including the following three methods described in Piper Diagram, Fingerprint Diagram, and Composition Diagram.

2.1. Piper Diagram Method

Piper Diagram is the most important method for the groundwater genesis study. It is very effective in separating data analysis for studies on the sources of dissolved constituents in groundwater, changes or modifications in the properties of water passing through a certain area, and its relationship to geochemical problems[10][11].

This diagram consists of two equilateral triangles and one parallelogram. One triangle is for plotting cations, and the other is for plotting anions. The value of cations percentage and anions percentage are drawn up into the parallelogram. The type of groundwater quality can be interpreted from the position of these points on the parallelogram.

The Piper diagram is used using laboratory test results which are converted to meq/l units and then the percentage value is plotted onto the trilinear diagram. The data entered is data that has a Charge Balance Error (CBE) value of less than ±5%. Some of these parameters are used for hydro-geochemical facies analysis of groundwater in the study area.

The sum of positive ionic charges equals to the sum of the negative ionic charges where the quantities on left and the right side should be approximately equal as follows[12]

$$\sum zm_c = \sum zm_c \tag{1}$$

To determine the quality of the results of the analysis, a charge balance error (CBE) was calculated using the formula [11]:

$$\%\ CBE = \frac{\sum zm_c - \sum zm_a}{\sum zm_c + \sum zm_a} \times 100 \tag{2}$$

where:

z: ionic valence
m_c: cation molality
m_a: anion molality

2.2. Fingerprint Diagram Method

The Fingerprint diagram method provides a visual description of the relative abundance pattern of dissolved ions in the form of a line and the relative salinity seen from the position of the line either at the bottom or at the top of the diagram[13]. Each line represents a representation of a groundwater sample as the compositional imprint of a water sample and a similar pattern means a relatively similar water type and sources.

Several groundwater samples can be included in this diagram for comparison. In plotting the ions in this diagram, the cations and anions are separated. Cations are plotted on the left and anions are plotted on the right. All cations are connected by one line and all anions are also connected by one line so that the pattern of balance between cations and anions can be seen, all of these ions are called as major ions.

2.3. *Composition Diagram Method*

The pair of measured parameters can be plotted on an x-y diagram, or more channels on a composition diagram in which concentrations are given in meq/l. The composition diagram provides a handy way to visually express large amounts of data, complementing the fingerprint diagram. Plotted data, for example, between major ions and physical – chemistry parameters of groundwater. This diagram provides a quick way to see the relationship between groundwater samples.

The common patterns formed on these composition diagrams are: 1) One cluster, this indicates that the plotted water samples come from the same water type or in other words come from the same aquifer; 2) Two clusters this indicates that the water sample has two different types of water and no mixing occurs. In addition, three or more water groups may appear as shown in the composition diagram; 3) Line, line patterns can be formed as a result of mixing freshwater with saline water with various percentages; 4) Triangular distribution, often the plotted data forms a triangular pattern. This indicates that there is a mixture of three different types of water; 5) Random distribution, it can be caused by the water sample comes from an unrelated source with a different composition and/or poor quality of data analysis.

3. Result and Discussion

The ion content tested in groundwater and spring water samples were Calcium (Ca), Magnesium (Mg), Sodium (Na), Potassium (K), Bicarbonate (HCO_3), Sulfate (SO_4), Chloride (Cl), and Carbonate (CO_3). At the same time, the percentage of CBE was calculated using Equation (2) has a good value which less than 5%, as seen in Table 2 below.

Table 2. Ion content laboratory test results.

No	Name	Cation				Anion			
		Ca^{2+} (mg/l)	Mg^{2+} (mg/l)	Na^{+} (mg/l)	K^{+} (mg/l)	HCO_3^- (mg/l)	SO_4^{2-} (mg/l)	Cl^- (mg/l)	CBE (%)
1	Sample 1	78.70	25.71	23.14	7.29	354.55	24.78	10.66	4.31
2	Spring/Sample 2	7.29	1.34	3.64	1.15	34.35	1.00	5.00	4.56
3	Sample 3	67.59	18.56	30.71	1.98	285.30	40.00	20.00	1.71
4	Sample 4	57.08	15.59	26.96	17.95	252.83	25.91	20.00	3.34
5	Sample 5	28.45	10.06	14.20	2.71	140.00	1.00	20.00	0.93
6	Sample 6	39.25	11.30	15.41	13.56	201.13	5.00	30.00	4.20

There are 3 (three) types of lithology were samples taken. Sample 1, sample 4 and sample 6 located in Laharic Deposit of Mount Slamet, sample 2 in Linggopodo Formation, sample 3 in Tapak Formation and sampel 5 located in Undifferentiated volcanics of Mount Slamet. Based on the results of the laboratory analysis from the Table 2, it can be seen that in the study area, calcium (Ca^{2+}) is the dominant anion and bicarbonate is the dominant cation (HCO_3^-) while the sample 1 and sample 2 have the highest and lowest of calcium dan bicarbonate content.

This result is in accordance with the statement that the availability of Ca^{2+} ion in groundwater is derived from igneous rock, particularly minerals containing silicate, pyroxene, amphibole, and feldspar. Furthermore, calcium can be found on minerals in sedimentary rock along with sulphate, gypsum, and anhydrite. The anion HCO_3^- exists in groundwater derived from rainwater or meteoric water that has fallen to the ground. HCO_3 - CO_2 is dissolved in rainwater and falls to the ground[14].

All groundwater samples fall into one area with medium to low Magnesium and medium to high Calcium cation content of 40 mg/l until 70 mg/l in Laharic Deposit of Mount Slamet, around 60 mg/l in Tapak Formation, 28.45 mg/l in Undifferentiated volcanics of Mount Slamet and around 7 mg/l in Linggopodo Formation. At the same time, the anion content has very low Chloride in spring or sample 2 with Lithology of Linggopodo and very high Bicarbonate of Laharic deposit in sample 1. All of these cation and anion data were plotted on the Piper and Fingerprint diagram (see Figures 3 and 4).

Based on this Piper Diagram, the research area is categorized into one groundwater facies: Calcium Bicarbonate ($Ca-HCO_3$). Water with bicarbonate content ($Ca-HCO_3$) generally comes from shallow and young groundwater types. The main source of HCO_3 ions in groundwater is CO_2 ions which are dissolved in rainwater and enter the soil and dissolve more CO_2. In the type of groundwater facies $Ca-HCO_3$, the water changes by dissolving CO_2 ions with carbonate minerals through the soil and rocks, then releasing HCO_3. The $Ca-HCO_3$ facies is characteristic of fresh groundwater and is relatively shallow in depth[15]. From the piper diagram, it is not possible to directly see the relationship between lithology and the chemical facies of groundwater.

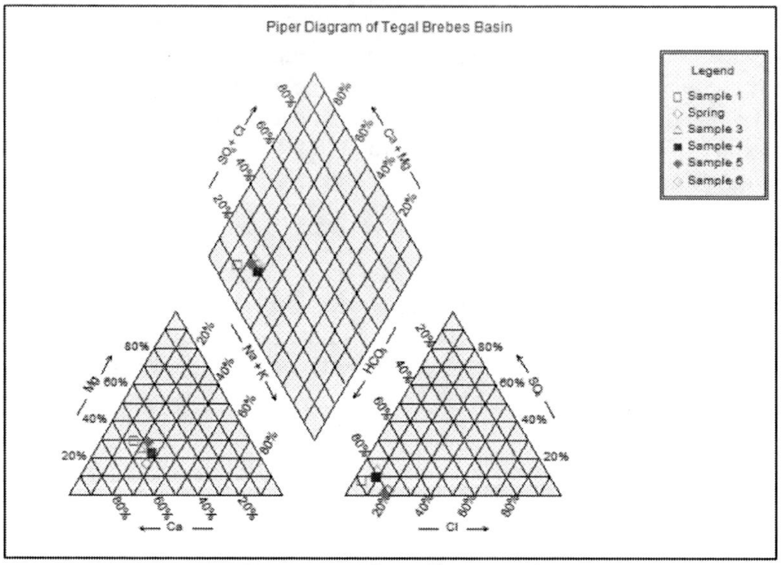

Figure 2. Piper diagram plot.

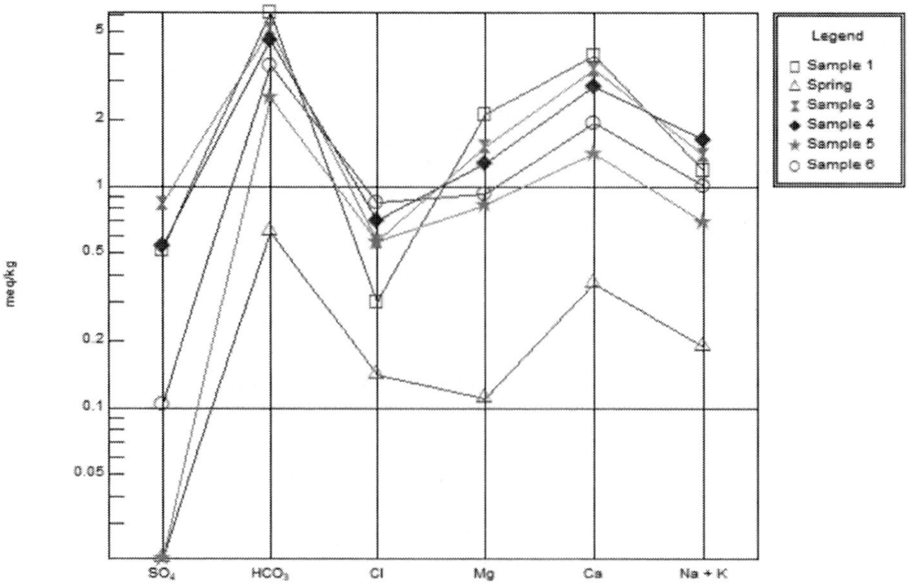

Figure 3. Fingerprint diagram plot.

Figure 3 is a fingerprint diagram of the groundwater chemical composition data, Major cations (Ca^{2+}, Mg^{2+}, Na^{+}, and K^{+}) were plotted on the left side of the diagram while the major anions (HCO_3^-, $SO4^-$, and Cl^-) plotted on the rightside. In this figure, each value of major ions is represented by one line. This line visually describes the relative abundance pattern of the dissolved ions (from left to right) and the concentration (the position of the line at the upper or lower part of the diagram). Figure 3 showed the water samples from sample 3, 4, 5, and 6 tend to have the simillar pattern among others. Sample 2 which located in Linggopodo formation has the low concentration of Mg^{2+} while other sample are relatively high. On the other hand, sample 1 has a lower chloride (Cl^-) value than samples from other dug wells and is directly adjacent to the Tegal-Brebes Basin boundary. According to Figure 4, a line patterns diagram can be formed as a result of mixing freshwater. Samples 1, 3 and sample 4 have a longer residence time than samples 5, 6 and spring.

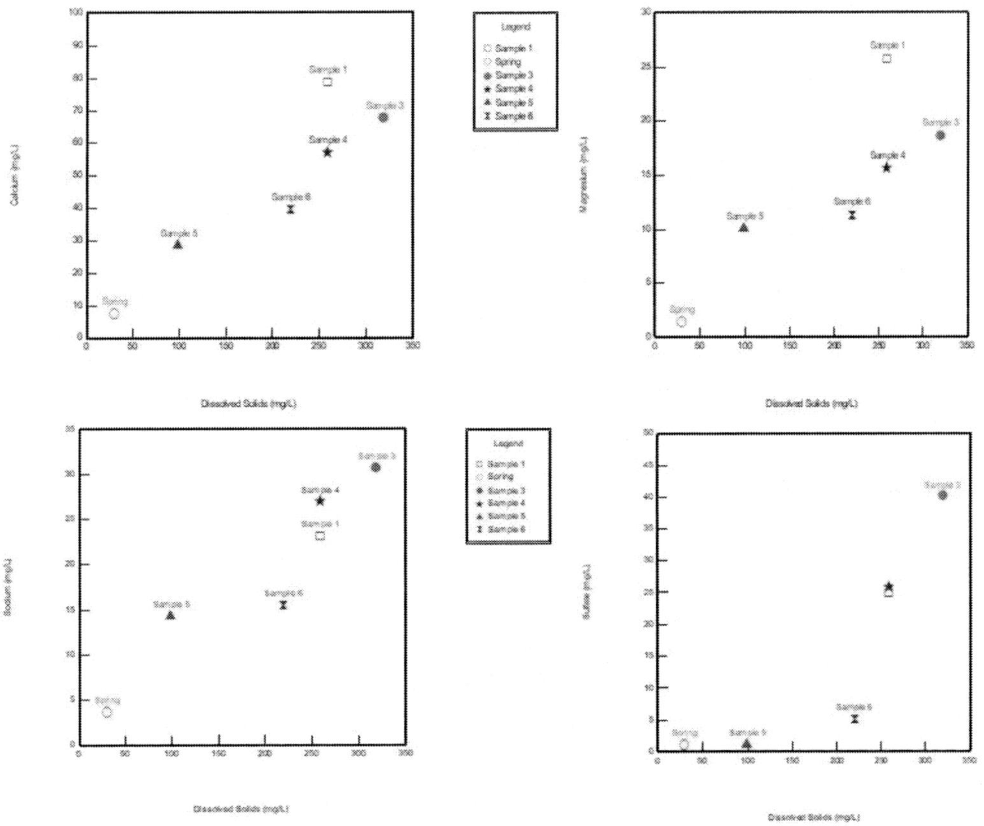

Figure 4. Composition diagram of TDS and major ion.

4. Conclusion

Based on temperature, pH, electrical conductivity, and total dissolved solids, the groundwater's physical and chemical properties in the research area indicate that groundwater quality has not interacted much with lithology and is still heavily influenced by meteoric water. The groundwater facies is dominated by the Ca-HCO_3 facies, which shows the character of shallow groundwater and local water flow systems. The fact of low electrical conductivity (EC) values less than 1000 μS/cm, as well as the TDS and pH. Therefore, the groundwater quality of the research area is good.

The research area is classified into one groundwater facies based on the Piper Diagram is Calcium Bicarbonate (Ca-HCO_3). Water with a bicarbonate content (Ca-HCO_3) is typically found in shallow and young groundwater. As a result, water on this area is mixing while the sample in laharic deposit area have a longer residence time than in other place in study area. The method that can best show the relationship between lithology and water chemistry is a fingerprint diagram with the results of 3 types of groundwater.

The geological condition of the research area correlates well with groundwater's hydrochemical properties. The findings of this study will help us better understand how hydrogeochemical processes work, perform more thorough research, and manage our water supplies.

5. References

[1] Carrard, N., Foster, T., and Willetts, J. (2019). Groundwater as a source of drinking water in southeast Asia and the Pacific: A multi-country review of current reliance and resource concerns. *Water (Switzerland)*, *11*(8). https://doi.org/10.3390/w11081605

[2] X. Li, H. Wu, H. Qian, and Y. Gao, "Groundwater chemistry regulated by hydrochemical processes and geological structures: A case study in Tongchuan, China," *Water (Switzerland)*, vol. 10, no. 3, 2018, doi: 10.3390/w10030338.

[3] Hwang, J. Y., Park, S., Kim, H.-K., Kim, M.-S., Jo, H.-J., Kim, J.-I., Lee, G.-M., Shin, I.-K., & Kim, T.-S. (2017). Hydrochemistry for the Assessment of Groundwater Quality in Korea. Journal of Agricultural Chemistry and Environment, 06(01), 1–29. https://doi.org/10.4236/jacen.2017.61001

[4] Thin, P.P., Hendrayana, H., Wilopo, W., and Kawasaki S. (2018). Assessment of groundwater facies in Wates Coastal Area, Kulon Progo, Yogyakarta, Indonesia. Journal Of Degraded And Mining Lands Management. https://doi:10.15243/jdmlm.2018.054.1389

[5] Putranto, T. T., Putra, B. R., and Marin, J. (2019). Groundwater Quality Analysis to Determine Groundwater Facies in Pati-Rembang Groundwater Basin. *IOP Conference Series: Earth and Environmental Science*, *246*(1), 0–8. https://doi.org/10.1088/1755-1315/246/1/012001

[6] Kastowo and Suwarna, N., 1996. Geological Regional Map of Indonesia 1: 100.000, Majenang Sheet. Geological Research dan Development Center.

[7] M. Djuri et al., 1996. Geological regional map of Indonesia 1: 100.000, Purwokerto – Tegal Sheet. Geological Research dan Development Centre.

[8] Van Bemmelen, R.W., 1949. The Geology of Indonesia Vol. IA. The Hague: Government Printing Office.

[9] Effendi, A.T., 1985. Hydrogeological map of Indonesia 1:250,000. Sheet VI, Pekalongan. Directorate of Environmental Geology, 5(3) 158-164.

[10] Suharyadi. 1984. Geohydrology Handbook. Yogyakarta: Department of Geological Engineering, Faculty of Engineering, Gadjah Mada University.

[11] Piper, A.M. 1944. A graphic procedure in the geochemical interpretation of water analyses, transition. Journal of American Geophysical Union 25:914-923.

[12] Freeze, R.A., and Cherry, J.A., 1979. Groundwater, Prentice-Hall.

[13] Mazor, E., 2004. Chemical and Isotopic Groundwater Hydrology, 3rd edition, 453 p, Marcel Dekker Inc.

[14] K Ramesh, P. J. B., 2013. Contamination of Groundwater due to Solid Waste Disposal and Textile Effluent in and Around Erode City, Tamil Nadu, International Journal of Research in Chemistry and Environment, 3, 262–271.

[15] Gemilang, W.A., Kusumah, G., and Rahmawan, G.A., 2018. Hydrogeochemical Of Unconfined Groundwater Of Coastal Area At Fisherman's Settlement In Southern Teupah, Simeulue District, Aceh Province, Riset Geologi dan Pertambangan, Vol. 28, No.1, June 2018 (25-35).

IOP Conf. Series: Earth and Environmental Science 1233 (2023) 012011 doi:10.1088/1755-1315/1233/1/012011

Rock Mass Classification for Support System Design and Stand-up Time of Asahan 3 Hydroelectric Power Plant Tunnel (STA 7+085 - STA 7+717)

A Rochmadi[1,2*] and D Karnawati[1]

[1]Department of Geological Engineering, Faculty of Engineering, Universitas Gadjah Mada, Yogyakarta, Indonesia
[2] PT. Adhi Karya (Persero) Tbk, Jakarta, Indonesia

*email: andirochmadi1990@mail.ugm.ac.id

Abstract. This Engineering geology research on rock mass classification to determine the support system's design and stand-up time at the Asahan 3 hydroelectric tunnel STA 7+085 – STA 7+717 has not been thoroughly carried out in detail. This study aims to estimate the support requirements. Rock mass classification is determined through core drill observations and then analyzed by Q-system and RMR methods. The results show that the rock mass at the tunnel depth elevation consists of sandstone with good quality based on RMR and Q-system. Based on the Q system classification, good quality sandstone is obtained. Based on RMR, the suggested support system is shotcrete in the crown and rockbolt. Meanwhile, a rock bolt is the appropriate support system based on the Q-system. The use of an RMR based support system is recommended because it provides a higher level of security with the addition of shotcrete. The tunnel stand-up time is 2,200 - 2,700 hours based on RMR. The result of this research is expected to be applied in other tunnel construction having similar engineering geological condition.

1. Introduction

Asahan 3 hydroelectric power plant is a planned power plant with a capacity of 2 x 87 megawatts built to accelerate the transition of electrical energy in North Sumatra. It was constructed by tunneling through a hill with a length of 7,794 m to drain water from the intake to the powerhouse. The research tunnel is located at STA 7+085 - STA 7 +717, with a tunnel length of 632 m. The research location is in Aek Songsongan village, Aek Songsongan sub-district, Asahan district, North Sumatra.

Rock types and geological structures affect the implementation of tunneling through a hill or mountain. The quality of rock mass in the tunnel area and its surroundings will affect what type of support system will be used safely for tunnel stability. Therefore, the objective of this study is to determine the appropriate design support system and to estimate the stand-up time based on rock mass classification. Tunneling media can be stable when used for construction, one of which depends on the stability of the rock mass.

The geological investigation and tunnel geometry design have been completed by PT PLN (Persero) with the assistance of Nippon Koei Co., Ltd as a planning consultant. The CRIEPPI Japan Standard was used to classify rock masses for design purposes. Because of the high uncertainty in a rock mass classification, a rock mass quality analysis using other methods is required for tunnel design. The rock mass classification is used in this study were determined empirically using RMR [3] and the Q-system [5].

Meanwhile, engineering geology study to determine support system and stand up time (time that rock mass of tunnel can stand unsupported itself before collapsing after excavation of tunnel openings) at research area has not been thoroughly carried out in detail. Determination of the support system and stand-up time is important because it is related to tunnel stability and excavation methods during tunnel construction period.

The results of the support system design and stand-up time in the Asahan 3 hydroelectric power plant tunnel (STA 7+085 - STA 7+717) are expected to be relevant based on the geological conditions found in the field during the tunnel face mapping, and generally are expected to be applied in other tunnel construction having similar engineering geological condition.

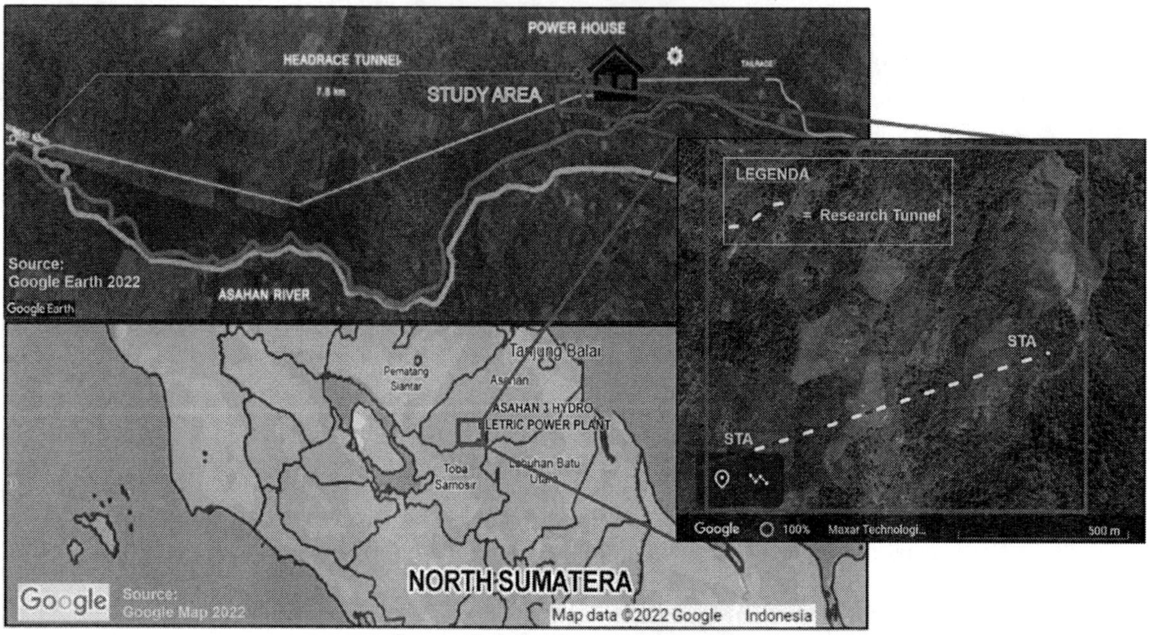

Figure 1. Location of Study Area

2. Methodology

2.1. Engineering Geological Mapping and Field Sample Parameters

The geological conditions at the research site were investigated at the surface and below the surface. The initial approach to surface conditions is carried out with regional geological maps to determine rock formations at the research site, and then megascopic observations are carried out on the outcrop to determine lithology and geological structure and the results are presented in the form of a geological map. Meanwhile, the subsurface investigation was carried out by observing core drills at 2 boreholes to determine the quality of the rock mass using an empirical method approach. The data will be used as a reference in determining support system and stand-up time of the tunnel based on empirical methods. Surface conditions can be used to correlate with subsurface conditions.

2.2. Rock Mass Classification Determination

Rock mass classification based on empirical approach methods for determining rock mass quality are widely used in rock engineering, with several examples including rock classification Rock Mass Rating (RMR) [3] and Q-system [5]. The main reason for using the two rock masses classifications in this study is that they are commonly used to determine rock mass quality in subsurface tunnels dominated by rock. Determination of rock mass classification in this study is based on observations of 2 borehole data, namely HT 8 and HT 9 at the tunnel elevation. The observation data then is analyzed to obtain RMR and Q-system values.

2.2.1. Rock Mass Rating (RMR). According to Bieniawski [3], the RMR system's rock mass classification can be determined in the field as well as interpretation of core drilling data using six parameters: 1) Rock quality designation (RQD) [8]; 2) Uniaxial compressive strength (UCS) of intact rock; 3) Discontinuity spacing; 4) Discontinuity condition; 5) Groundwater condition; and 6) Discontinuity orientation. These parameters determine the RMR value at each borehole in the Asahan 3 hydroelectric tunnel from STA 7+085 - STA 7+717. The RMR value is calculated by adding the six parameters obtained from testing rock samples and field observations from core drills or face mapping analysis [10].

2.2.2. Q-system. According to Barton et al. [5], the determination of rock mass quality and support system is based on the evaluation of several underground excavation cases. The value of the Q-system on a logarithmic scale from 0.001 to 1000. The Q-system classification of rock masses is based on six parameters, namely 1) RQD index [8]; 2) The number of joint sets (Jn); 3) The roughness of the weakest joints (Jr); 4) The degree of Alteration or filling along the weakest joints (Ja); 5) The joint water reduction factor (Jw) and 6) The stress reduction factor (SRF). The six parameters can be measured based on surface mapping and form core logging [6]. Determination of the value of Q to determine rock mass quality is obtained from equation (1) below [5]:

$$Q = \frac{RQD}{Jn} \times \frac{Jr}{Ja} \times \frac{Jw}{SRF} \quad (1)$$

The following ratios indicate the basic geotechnical parameters in determining the Q value: 1) RQD/Jn = Rock block size, 2) Jr/Ja = inter-block shear strength, and 3) Jw/SRF = active stress [6].

2.3. Support System Determination

The determination of support system in this study uses the results of rock mass classification at the tunnel level. The RMR value can determine the tunnel support system, which includes shotcrete with a specific thickness, rock bolt with or without occasional wire mesh with a specific length and diameter, and steel rib [3]. The RMR classification can also determine the stand-up time or the amount of time the rock mass can stand without collapsing before the support system is installed. According to Biewniaski [3], the stand-up time is determined using a graph of the relationship between the RMR value and unsupported roof span. An unsupported roof span is a distance between the excavated surface and the previously installed support system. In the case of other tunnels, the determination of the roof span can also be determined using the tunnel width based on Lauffer [4]. In this study, the tunnel width is equal to the tunnel diameter.

In determining the support system based on Q-system [5], additional geotechnical parameters are required, which will later be associated with the Q index value, namely the equivalent dimension of the excavation (De) [5]. The value of De is obtained by dividing the span, diameter or wall height of the excavation by the excavation support ratio as shown in equation (2). According to Barton et al., the use of the ESR value for tunnels based on the type of excavation is 1 [5].

$$De = \frac{Height\ of\ Tunnel}{Excavation\ Support\ Ratio} \quad (2)$$

3. Results and Discussion

3.1. Surface Geological Condition

According to Clarke et al. [7], the rock formation on the surface of the study site is the Toba Tuff Formation (Qvt), which mostly consists of volcanic products from the ancient Toba volcanic eruption, and the Bohorok Formation (Pub), which consists of massive sandstone units [9]. Surface rock outcrops observation shows that the research location consists of two rock units: lapilli tuff units and sandstone units. Figures 2 and 3 show a map of the distribution of surface rock units and geological cross-sections in the research area.

According to Indonesia's hazard map and earthquake source [2], the Sumatran active fault segment is divided into the northern and southern parts. The research location has the closest distance to active faults in the northern part of Sumatra, namely the Renun A segment, which is approximately 100 km away. However, based on geological mapping of surface conditions at the research location, no geological structures in the form of faults or folding were discovered, there is only the appereance of joints.

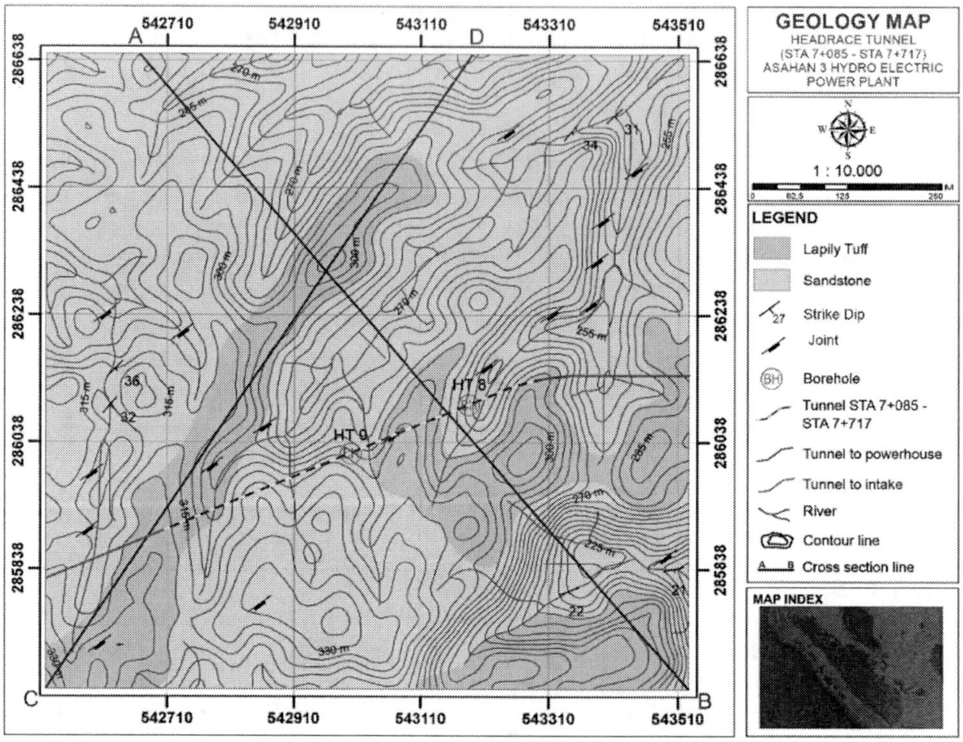

Figure 2. Geology map of the research location.

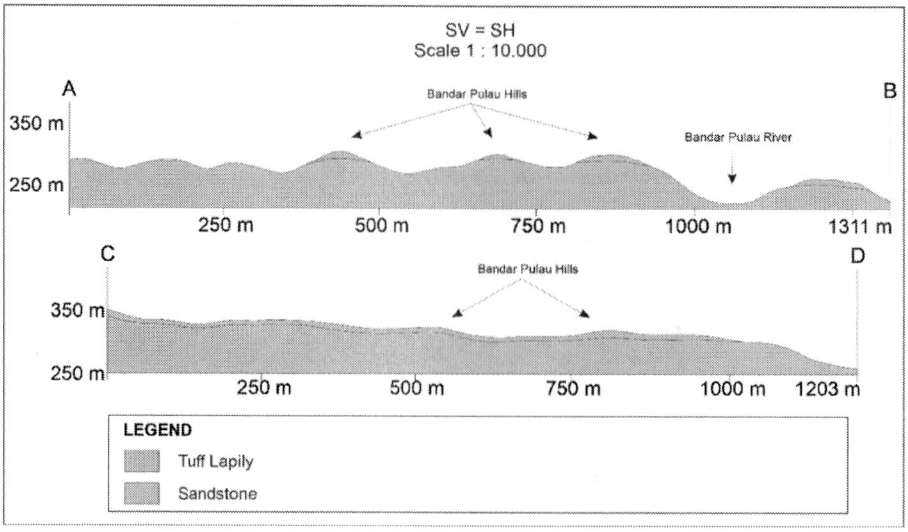

Figure 3. Geology cross section of the research location.

IOP Conf. Series: Earth and Environmental Science 1233 (2023) 012011 doi:10.1088/1755-1315/1233/1/012011

3.2. Sub Surface Geological Condition

Subsurface engineering geological mapping is carried out by observing core drill results from two boreholes to determine the quality of the rock mass at the subsurface location (tunnel elevation). Laboratory test results are used to determine one of the rock mass quality parameters. The tested samples are taken directly from the core drilling results based on the level of weathering [1] and its lithology. Furthermore, these data can determine the support system and tunnel stand-up time of Asahan 3 hydroelectric power plant tunnel STA 7+085 - STA 7+717.

Rock mass quality based on RMR [3] is obtained from the rating assessment of all parameters. RMR data was obtained from core drill observations of two boreholes, namely HT 8 and HT 9 (Figures 4 and 5).

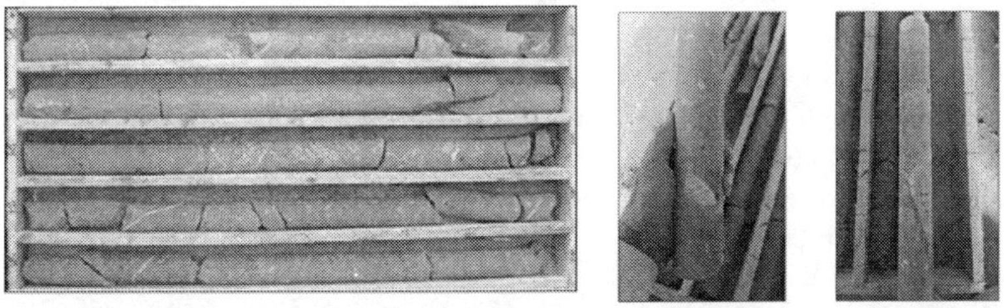

Figure 4. The typical core drill of HT 8 at elevation of tunnel consist of sandstone.

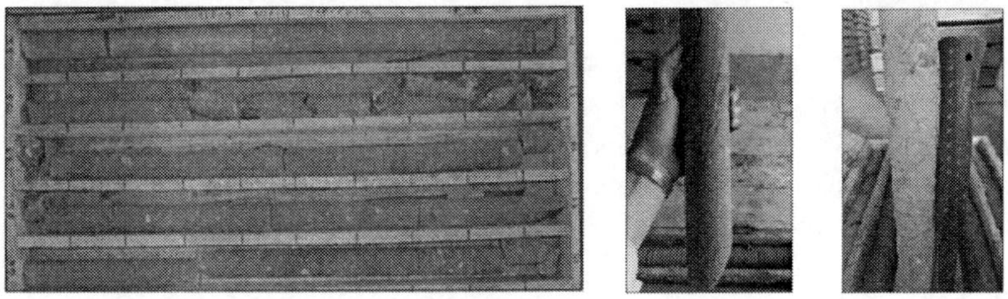

Figure 5. The typical core drill of HT 9 at elevation of tunnel consist of sandstone.

As a comparison of rock masses from core drills, Figure 6 shows the rock masses from excavation tunnel openings at STA 7+069 (tunnel openings before the research location), showing the lithology in the form of massive sandstone with slightly weathered weathering level and good rock mass quality.

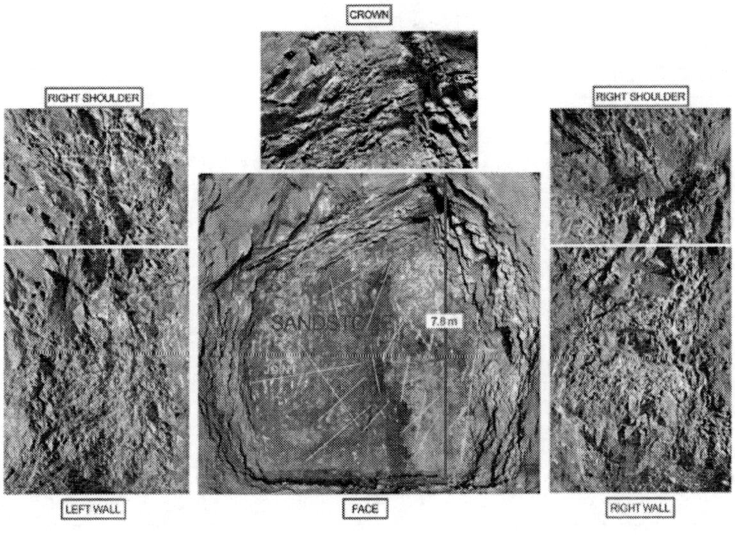

Figure 6. The appearance of sandstone on tunnel excavation face opening at STA 7+069.

IOP Conf. Series: Earth and Environmental Science 1233 (2023) 012011 doi:10.1088/1755-1315/1233/1/012011

The final result of the RMR value is classified into rock mass classes based on its strength and parameters. The core drill observations at the tunnel depth elevation according to the boreholes HT 8 and HT 9 consist of sandstone of good quality. Determination of assessment rock mass classification based on RMR [3] through core drill observations HT 8 and HT 9 can be seen in Tables 1 and 2. The resume of the RMR result is presented in Table 3.

Based on Tables 1 and 2, the UCS rating research was obtained from testing the intact rock samples associated with the UCS value based on the field [11]. Laboratory testing of UCS samples based on Indonesia National Standard number 2825 [12] was carried out on the intact rock of borehole HT 8 at tunnel elevation 36.6 m and borehole HT 9 at tunnel elevation 57.6 m, obtaining values of 51.7 Mpa and 62.1 MPa. While the UCS value is based on field estimates for each core length [11], the value ranges between 50-100 MPa (grade R4), which is characterized by several cores that can be broken by hitting using a geological hammer, with this value the UCS rating is obtained 10.

The RQD value rating is obtained from measurement-based solely on core pieces larger than 10 cm. The RQD has a value of 75-90 % to 90-100 %, so the RQD rating is 17-20. The rating of spacing discontinuity is obtained from the observed spacing exposed on core pieces, and the value is 60-200 mm until 0.6-2 m, so the rating is 8 - 15.

Discontinuity conditions (joint, fault or bedding) were observed in the core piece to assess persistence, aperture, roughness, infilling and weathering. Groundwater condition is obtained from observing the general condition of groundwater from the point of view: of whether the exposed surface is dry, damp, wet, or dripping based on the groundwater level, which is known in core data. Due to the groundwater level being above the tunnel, the groundwater condition assessment tunnel is assumed wet, so the rating is 7.

The rating adjustment is made based on the evaluation of strike and dip of discontinuity orientation against tunnel excavation. The value of discontinuity orientation is given as a qualitative description. Based on previous observations on tunnels that have been excavated, the value of discontinuity orientation is categorized as a favourable condition. It is indicated by the strike perpendicular to the tunnel axis and drive with dip (Dip 20-45°) so that the strike and dip orientation rating is -2.

Table 1. Assessment of rock mass quality based on RMR borehole HT 8.

Depth (m)	Lithology	Weathering Degree	RMR (Biewniaski 1989) [3]										
			UCS	RQD	Joint spac.	Joint cond.: Persistence	Aperture	Roughness	Infilling	Weathernig	GW cond.	Strike & dip orientation	Value
29 - 30	Sandstone	Slightly weathered	10	20	9	4	4	3	4	5	7	-2	64
30 - 31	Sandstone	Slightly weathered	10	17	9	4	4	3	4	5	7	-2	61
31 - 32	Sandstone	Slightly weathered	10	20	10	4	4	3	2	5	7	-2	63
32 - 33	Sandstone	Slightly weathered	10	20	10	4	4	3	2	5	7	-2	63
33 - 34	Sandstone	Slightly weathered	10	17	8	4	4	3	4	5	7	-2	60
34 - 35	Sandstone	Slightly weathered	10	20	9	4	4	3	4	5	7	-2	64
35 - 36	Sandstone	Slightly weathered	10	20	12	4	4	5	4	6	7	-2	70
36 - 37	Sandstone	Slightly weathered	10	20	14	4	4	5	4	6	7	-2	72
												Average RMR Value	**64.6**
												Quality	**Good**

IOP Conf. Series: Earth and Environmental Science 1233 (2023) 012011 doi:10.1088/1755-1315/1233/1/012011

Table 2. Assessment of rock mass quality based on RMR borehole HT 9.

Depth (m)	Lithology	Weathering Degree	RMR [2]											
			UCS	RQD	Joint spac.	Joint cond.					GW condition	Strike & dip orientation	Value
						Persistence	Aperture	Roughness	Infilling	Weathering			
29 - 30	Sandstone	Slightly weathered	10	17	8	4	4	5	4	5	7	-2	62
30 - 31	Sandstone	Slightly weathered	10	20	10	4	4	5	4	5	7	-2	67
31 - 32	Sandstone	Slightly weathered	10	20	9	4	4	5	4	5	7	-2	66
32 - 33	Sandstone	Slightly weathered	10	20	10	4	4	5	4	5	10	-2	70
33 - 34	Sandstone	Slightly weathered	10	20	9	4	4	5	4	5	10	-2	69
34 - 35	Sandstone	Slightly weathered	10	20	9	4	4	5	4	5	7	-2	66
35 - 36	Sandstone	Slightly weathered	10	20	9	4	4	5	4	5	7	-2	66
36 - 37	Sandstone	Slightly weathered	10	20	9	4	4	5	4	5	7	-2	66
											Average RMR Value		**66.5**
												Quality	**Good**

Table 3. The RMR result from core drill at the tunnel elevation

Borehole	Tunnel Elevation (m)	Lithology	RMR (Bieniawski, 1989)	Quality	Class
HT 8	29.44 - 37.24	Sandstone	64 - 72	Good	II
HT 9	56.52 - 64.32	Sandstone	62 - 70	Good	II

The RMR values on boreholes HT 8 and HT 9 at the tunnel elevation are classified as class II rock mass of good quality with sandstone lithology, and the RMR value range is 62-72.

Classification RMR [3] can also determine the stand-up time or how long a tunnel is safe without a support system and the tunnel support system using empirical methods. The value of stand-up time to determine the criteria for rock mass stability in tunnel openings without a support system can be expressed in a graph of the relationship between the RMR value and the unsupported span by correlating it to the stand-up time value (Figure 7).

Figure 7 shows that the recommended stand-up time value based on the RMR value and unsupported roof span (7.8 m) for boreholes of HT 8 without a support system is 2,200 hours (±3 months) and 2,700 hours (±3.5 months) for boreholes of HT 9 without a support system.

The recommended support system based on the RMR classification is shotcrete and rock bolt, shown in Table 4. More details, the tunnel support system for boreholes HT8 and HT 9 are shotcrete thick 50 mm in the crown, rock bolts 20 mm diameter and fully grouted, with a length of 3 m, spacing of 2.5 m with occasional wire mesh.

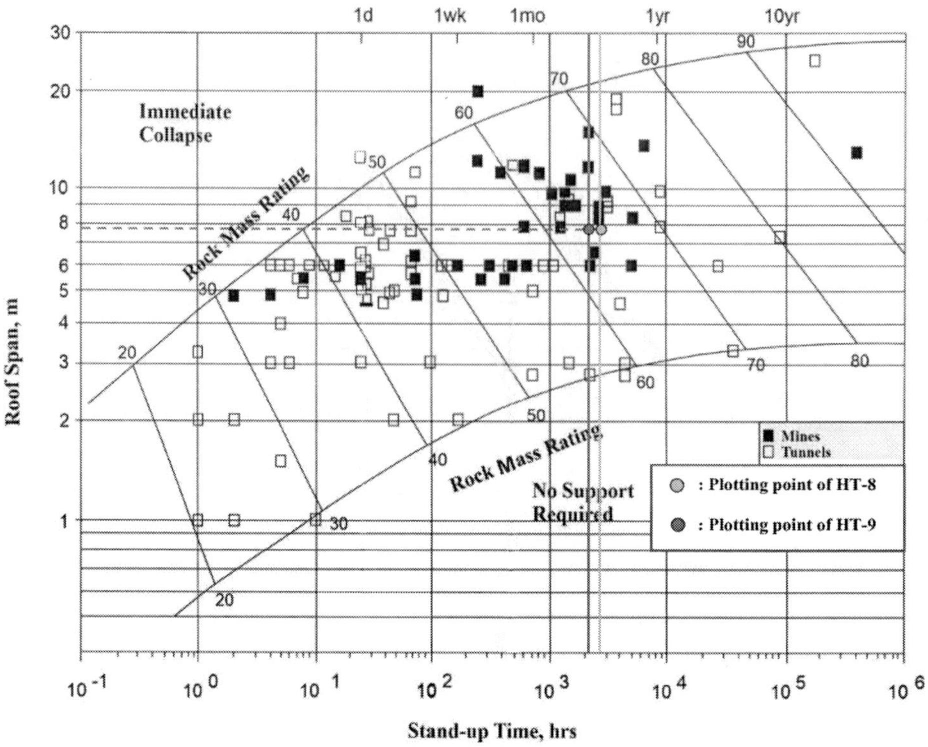

Figure 7. The relationship chart between unsupported span and stand-up time [3].

Table 4. The support system based on RMR.

Borehole	Tunnel Elevation (m)	Lithology	RMR (Bieniawski, 1989)		
			Rating (Quality)	Support System	
HT 8	29.44 - 37.24	Sandstone	64 - 72 Good (II)	- Shotcrete - Rock bolt	50 mm in crown where required 20 mm diameter, fully grouted bolts in crown 3 m long spaced 2.5 m with occasional wiremesh
HT 9	56.52 - 64.32	Sandstone	62 - 70 Good (II)	- Shotcrete - Rock bolt	50 mm in crown where required 20 mm diameter, fully grouted bolts in crown 3 m long spaced 2.5 m with with occasional wiremesh

Other empirical methods to determine rock mass classification need to be done to compare rock mass quality, such as the Q-system classification. According to Barton et al. [15], the Q-system classification was used to determine rock mass quality and a support system based on core drill analysis on boreholes

IOP Conf. Series: Earth and Environmental Science 1233 (2023) 012011 doi:10.1088/1755-1315/1233/1/012011

HT 8 and HT 9 located at the tunnel elevation. Determination of assessment rock mass classification based on Q-system [5] through core drill observations HT 8 and HT 9 can be seen in Tables 5 and 6. The resume of the Q-system result is presented in Table 7.

Table 5. Assessment of rock mass quality based on Q-system borehole HT 8.

Depth (m)	Lithology	Weathering Degree	Q-SYSTEM (Barton et al.) [5]						
			RQD Value	Join set	Joint roughness	Joint alteration	Joint water	SRF	Value
29 - 30	Sandstone	Slightly weathered	95	3	3	2	0.7	2.5	12.5
30 - 31	Sandstone	Slightly weathered	80	3	3	1	0.7	2.5	21.1
31 - 32	Sandstone	Slightly weathered	90	3	3	2	0.7	2.5	11.8
32 - 33	Sandstone	Slightly weathered	90	3	3	1	0.7	2.5	23.7
33 - 34	Sandstone	Slightly weathered	80	4	3	1	0.7	2.5	15.8
34 - 35	Sandstone	Slightly weathered	100	4	3	1	0.7	2.5	19.8
35 - 36	Sandstone	Slightly weathered	100	2	3	1	0.7	2.5	39.6
36 - 37	Sandstone	Slightly weathered	100	2	3	1	0.7	2.5	39.6
							Average RMR Value		**23.0**
								Quality	**Good**

Table 6. Assessment of rock mass quality based on Q-system borehole HT 9.

Depth (m)	Lithology	Weathering Degree	Q-SYSTEM (Barton et al.) [5]						
			RQD Value	Join set	Joint roughness	Joint alteration	Joint water	SRF	Value
29 - 30	Sandstone	Slightly weathered	79	6	3	1	0.7	2.5	10.4
30 - 31	Sandstone	Slightly weathered	95	3	3	1	0.7	2.5	25.0
31 - 32	Sandstone	Slightly weathered	95	3	3	1	0.7	2.5	25.0
32 - 33	Sandstone	Slightly weathered	100	3	3	1	0.7	2.5	26.4
33 - 34	Sandstone	Slightly weathered	100	3	3	1	0.7	2.5	26.4
34 - 35	Sandstone	Slightly weathered	90	3	3	1	0.7	2.5	23.7
35 - 36	Sandstone	Slightly weathered	90	4	3	1	0.7	2.5	17.8
36 - 37	Sandstone	Slightly weathered	90	4	3	1	0.7	2.5	17.8
							Average RMR Value		**21.6**
								Quality	**Good**

Based on Tables 4 and 5, the RQD value is determined the same way as the RQD assessment on RMR, based solely on core pieces larger than 10 cm. The RQD has a value of 80-100%. A joint set number (Jn) is a collection of joints that have orientations parallel to each other and the same repetition [5]. Boreholes HT 8 and HT 9 found 1 - 2 joint sets, and there are also random joints so that the value of Jn obtained varies for each core piece between 2-6. Joint Roughness (Jr) shows the surface roughness of the discontinuities in each joint set [5]. This value assessment is obtained from observing each core broken by the joint and its surface. Figure 8 shows a typical appearance of surface roughness conditions. On the observation of boreholes HT 8 and HT 9, the joint surface roughness level is rough or irregular, undulating so that the value of Jr is 3.

Figure 8. The typical of joint surface from core pieces.

Joint alteration is an alteration condition between joints [5]. The alteration condition is assessed based on the observed infilling exposed on core pieces. The joint wall looks clean, but there are stains due to oxidation, and some joint walls are slightly altered by non-softening mineral coating minerals so that the value of Ja is 1-2. The joint water reduction factor (Jr) is a groundwater factor that affects joint filling minerals [5], determining the water reduction factor based on the groundwater level known from the core data. Due to the groundwater level being above the tunnel, it is assumed that seepage with moderate water pressure will occur at the time of excavation of tunnel openings and occasional outwash of joint filling so that the value of Jw is 0.66.

The stress reduction factor is the relationship between rock strength around the tunnel opening and stress [5]. The determination of SRF is determined based on the category of stress conditions. The research location is categorized as competent rock, mainly massive with stability problems. The tunnel elevation is still categorized as shallow. The stress that occurs is still within the category of low stress and near surfaces, so the value of SRF is 2.5.

Based on observations and analysis of core drill results, the Q-system value ranges from 10.43 to 39.6, as shown in Table 7, indicating that the rock mass can be classified as good quality.

Table 7. The Q-system result from core drill at the tunnel elevation.

Borehole	Tunnel Elevation (m)	Lithology	Q-system (Barton et al, 1974)	Quality
HT 8	29.44 - 37.24	Sandstone	11.8 - 39.6	Good
HT 9	56.52 - 64.32	Sandstone	10.43 - 26.4	Good

The value of the Q-system [5] is also carried out to determine the tunnel support system using empirical methods. To find out the tunnel support system, first, it is necessary to know the value of the Equivalent Dimension (De). The Equivalent Dimension (De) value is obtained by dividing the tunnel diameter by the ESR value. The tunnel diameter is 7.8 m and ESR 1, so the Equivalent Dimension (De) value is 7.8.

The borehole of HT 8 (elevation of the tunnel) has an average Q value of 23, while a value of equivalent dimension (De) from the previous results is 7.8. After both values are plotted on a support system chart based on the classification Q index [6], as shown in Figure 9, the recommended support system is a rock bolt length of 2.6 m with a spacing of 2.5 m. Meanwhile, in HT 9, with an average Q value of 21.6, the same reinforcement was obtained as in HT 8. The appropriate support system based on the Q-system classification is shown in Table 8. In addition, Figure 10 show summary engineering geological condition in research area.

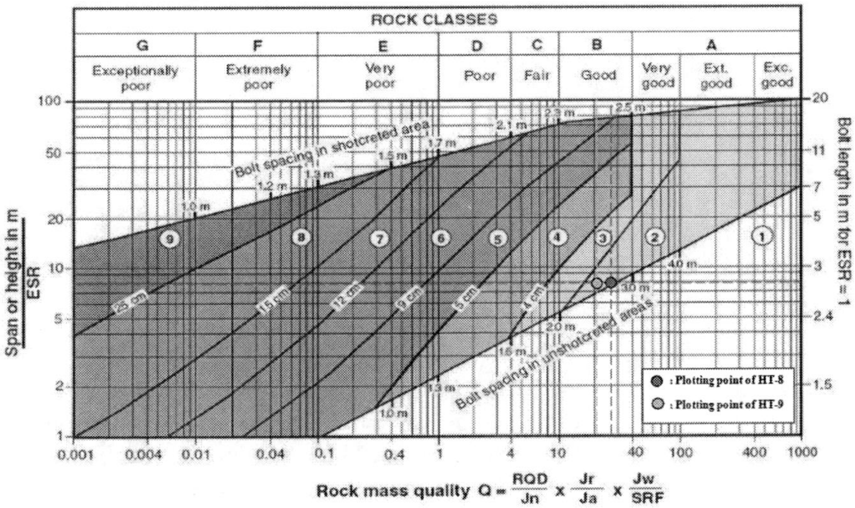

Figure 9. Support system based on the chart of classification Q index [6]

Table 8. The support system based on RMR

Borehole	Tunnel Elevation (m)	Lithology	Q-System (Grimstad & Barton, 1993)		
			Rating (Quality)	Support System	
HT 8	29.44 - 37.24	Sandstone	11.8 - 39.6 Good (II)	- Rock bolt	20 mm diameter bolts in crown 2.6 m long spaced 2.5 m with
HT 9	56.52 - 64.32	Sandstone	10.43 - 26.4 Good (II)	- Rock bolt	20 mm diameter bolts in crown 2.6 m long spaced 2.5 m with

Meanwhile, the results from in light of previous work, which have been carried out by previous consultants [9] show that the rock quality at the study site is slightly weathered and relatively solid rock grade CH and the reinforcement is only shotcrete with a thickness of 50 mm based on CRIEPPI Japan Standard.

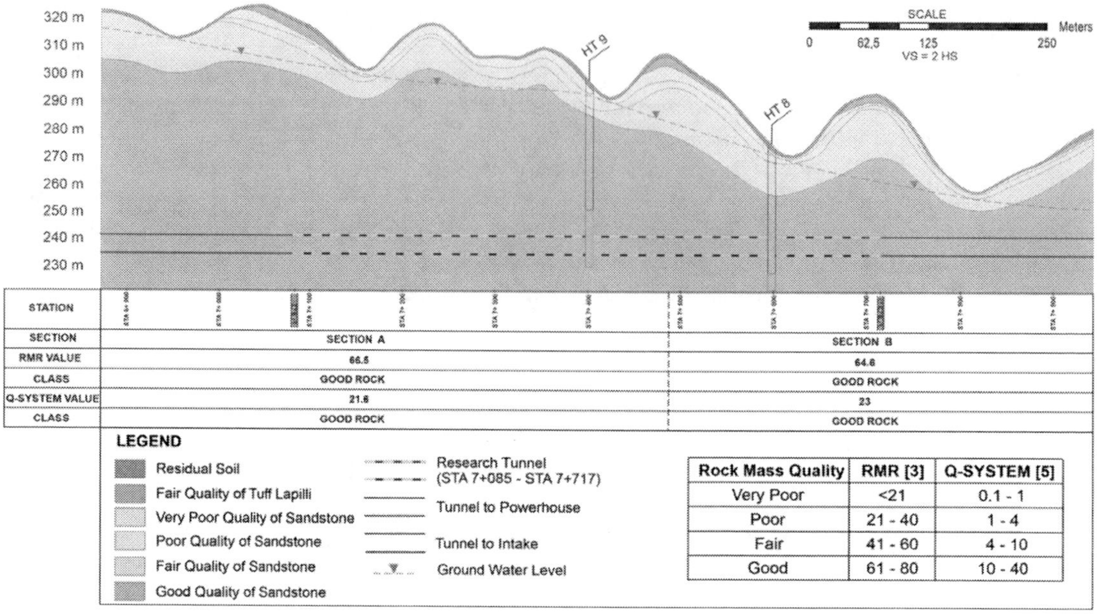

Figure 10. Engineering geologiccal condition along tunnel

Based on Figure 10, the rock type on the surface consists of sandstone and lapilli tuff, but after entering the subsurface the rock is dominated by sandstone. On the surface a geological structure in the form of joint is found which is shown on the geological map Figure 2. On the subsurface, sandstone with poor to good quality is found, but the tunnel elevation is dominated by good quality sandstone with slightly weathered. The geological structure in the form of joint on the subsurface is shown on the face of the excavated plane as seen on the face mapping of the tunnel openings before the research location (Figure 6). Meanwhile, based on coredrill observations, the geological structure in the form of a joint is also shown in the coredrill results which are cut into several parts. The existence of geological structures both on the surface and below the surface is one of the controlling factors in determining the quality of rock mass, especially at tunnel elevation. Determination of rock mass quality in Figure 10 is carried out by dividing 2 sections, namely A and B. Section A is based on the analysis of the HT 9 core drill while section B is based on the HT 8 core drill.

4. Conclusions and Recommendations

The RMR values on boreholes HT 8 and HT 9 at the tunnel elevation are classified as class II rock mass of good quality with sandstone lithology, and the RMR value range is 60-72 (Table 1). The analysis of rock mass quality is carried out using the Q-system method [5] (Table 2). On the borehole HT 8, sandstone with good quality is found with a Q value range of 11.8 – 39.6, while in the borehole HT 9, sandstone with good quality is obtained with a range of the value of Q is 10.4 – 26.4.

Based on the value of rock mass classification using RMR [3], the recommended tunnel support system for boreholes HT 8 and HT 9 is shotcrete with 50 mm thickness in the crown and rock bolt 20 mm diameter with length 3.0 m installed in the crown position with spacing 2.5 m and occasional wire mesh. Meanwhile, based on the Q-system method [5], the recommended tunnel support system for boreholes HT 8 and HT 9 are rock bolts with a length of 2.6 m and installed with a spacing of 2.5 m.

Meanwhile, based on RMR [3], the stand-up time value on a tunnel with a diameter of 7.8 is predicted to be unstable without a support system for 2,200 hours (3 months) for HT 8 and 2,700 hours (3.5

months) for HT 9. The value of stand-up time is relatively high so that the support system can be installed not immediately after the tunnel excavation.

In this study, the analysis of determining the support system using an empirical method based on RMR [3], which is applied to the boreholes HT 8 and HT 9 according to the depth of the tunnel elevation, shows that the support system is safer than using the Q-system method [5] and previous work [9]. However, further research such as numerical analysis is required to evaluate which support system while provide safer and better performance.

Acknowledgments
The author expresses his deepest gratitude to PT PLN (Persero) and SHIMIZU - ADHI JO for a research permit and the opportunity to conduct research and collect data at the location of the Asahan 3 Hydroelectric Power Plant Project. Thanks also to PT. Adhi Karya (Persero) Tbk has financed the superspecialist education program in collaboration with Indonesia's public work and housing ministry.

References

[1] ISRM 1978 Suggested methods for the quantitative description of discontinuities in rock masses *Int. J. Rock Mech. Min. Sci. Geomech. Abstr.* **15** 319-368
[2] PUSGEN 2017 *Indonesia Earthquake Source and Hazard Map* (Bandung: Minister of Public Works and Public Housing) p 34
[3] Bieniawski Z T 1989 *Engineering rock mass classifications: a complete manual for engineers and geologists in mining, civil, and petroleum engineering* (New York: John Wiley & Sons)
[4] Lauffer H 1958 *Gebirgsklassifizierung fur den stol lenbau Geo. Bau.* **24** 46
[5] Barton N, Lien R and Lunde J 1974 Engineering classification of rock masses for the design of tunnel support *Rock Mech.* **6** pp 189-236
[6] Grimstad E and Barton N 1993 Updating of the Q-system for NMT *Proc. Int. Symp. on Sprayed Concrete for Underground Support* (Fagernes) pp 46-66
[7] Clarke M, Ghazali S, Harahap H, Kusyono and Stephenson B 1982 Geologic Map of The Pematangsiantar Quadrangle, Sumatra p 0718
[8] Deere D U, Hendron A J, Patton F D and Cording E J 1966 Design of surface and near-surface construction in rock *the 8th US symp. on rock mechanics.* (American Rock Mechanics Association)
[9] PT. PLN 2007 *Geological Investigation Report of Asahan 3 Hydroelectric Power Plant Construction Project* (Indonesia: Nippon Koe with association)
[10] Putra D, Hendrayana H and Setiawan H 2021 Analysis of Support Systems and Excavation Methods Based on Rock Mass Quality in the Intake Tunnel of the Jlantah Dam *E3S web of conf.* ***325*** p 2
[11] Hoek E, Marinos P and Benissi, M 1998 Applicability of the geological strength index (GSI) classification for very weak and sheared rock masses. The case of the athens schist formation *Bull. of Eng. Geol. Env.* **57** pp 151-160
[12] SNI 2825 2008 The instruction of uniaxial compressive strength test for rock (Indonesia: Association of National Standardization) p 8

IOP Conf. Series: Earth and Environmental Science 1233 (2023) 012012 doi:10.1088/1755-1315/1233/1/012012

Seepage Analysis in Construction of Ciuyah Tunnel, Lebak, Banten, Indonesia

Dendi Darmawan[1,2] **and Wawan Budianta**[1]

[1]Department of Geological Engineering, Faculty of Engineering, Universitas Gadjah Mada, Yogyakarta, Indonesia
[2]Ministry of Public Works and Housing of Republic of Indonesia

Abstract. The Ciuyah Tunnel is 1,329 meters long, with excavation progress reaching 720 meters from the inlet and 120 meters from the outlet. However, the excavation process from the inlet had to be stopped because of groundwater leakage. Groundwater management has become essential in reducing this tunnel project's environmental disasters. Seepage analysis is needed to determine the potential leaks in the following excavation process. So the determination of handling methods for leaks and the impact of leaks can be anticipated. The objective of this study is to identify how groundwater leakage occurs during excavation based on the characteristics of the rock masses on the tunnel excavation faces and surface outcrops. This study also predicts the potential seepage point in unexcavated sections. The results show that the tunnel construction area consists of slightly to moderately weathered intercalated tuffaceous sandstone and claystone with many discontinuities. The groundwater leakage on the inlet side occurred due to excavation cutting the discontinuities on the roof. While at the outlet side, it is influenced by syncline geometry and joint planes. The potential for seepage points can occur at excavation points that affect the geometry of the fold where the water goes to a lower place.

1. Introduction

Groundwater seepage is typical in tunnel construction, particularly in tunnels below the groundwater level. Due to the difference in water head, groundwater can seep while tunneling in a water-rich stratum. The groundwater seepage can readily cause water inrush and mud surges to the tunnel face, hastening the consolidation and settlement of the stratum. Tunnel collapse accidents may occur as a result of excessive stratum settlement deformation [1]. Therefore, constructing a tunnel in a water-rich area has become an important factor restricting the safety, progress, and environment.

Based on previous studies, Wei & Zhu [1] studied the reason for seepage collapse in loess lithology. Li et al. [2] used the step and CRD methods to control tunnel water seepage and pore water pressure distribution. Zhang et al. [3] studied how to predict the water inflow by the superposition principle. Liu et al. [4], Niu et al. [5], and Rana & Nair [6] studied the water inrush disaster in the tunnel and how to handle it. This study aims to identify how groundwater leakage occurs during excavation based on the characteristics of the rock masses on the tunnel excavation faces and surface outcrops. This study also predicts the potential seepage point by two-dimensional models in unexcavated sections.

The groundwater leaked event in Ciuyah Tunnel serves as the backdrop for this study. Ciuyah Tunnel is an intake tunnel that is part of the Karian Dam project in Lebak Regency, Banten Province, Indonesia. As illustrated in Figure 1, The tunnel has 1,329 meters long and is designed in the shape of a horseshoe. The average thickness of the overburden is roughly 50 meters.

IOP Conf. Series: Earth and Environmental Science 1233 (2023) 012012 doi:10.1088/1755-1315/1233/1/012012

(a) (b)

Figure 1. Location and portal appearance of Ciuyah Tunnel

2. The Research Method

This project involved geological mapping in identifying geological structure, lithology, and rock geometry. Then surface and subsurface geotechnical conditions include groundwater level, rockmass classification, and weathering grade determination. Also, permeability testing to get hydraulic conductivity value. Seep/w would model groundwater flow using rock geometry, tunnel geometry, hydraulic conductivity values, and groundwater level conditions.

A seepage study is required to determine the possibility of leaks during the next excavation operation. The basic situation of the project is detailed first, followed by a systematic description of the groundwater leak and an analysis of the cause of seepage. Based on geological conditions and geotechnical parameters this study tries to determine how groundwater leakage occurs. At the same time, the possible seepage point in the unexcavated area would be depicted from the two-dimensional Seep/w model.

3. Result and Discussion

3.1. Geological Condition

Geological conditions in the Ciuyah Tunnel site get affected by anticlinorium from The Bogor Zone [7]. Along the tunnel path, there are synclines and anticlines fold geometries. The rock offsets in normal faults also can be found near the Ciuyah tunnel's outlet portal. This geological structure creates many discontinuities in the rock body. For the lithology, the tunnel site is composed of Pliocene Genteng Formation [8]. It is dominated by the debris of vulcanic matter, including tuffaceous sandstone and tuffaceous claystone. The rock material consists of quartz, feldspar, and glass mineral compositions. Charcoal and wood fossils have also been founded in several locations. The rock's striking direction is northwest-southeast to north-south, with a dip of 12-20°.

3.2. Geotechnical Condition

The subsurface geotechnical conditions along the tunnel from inlet to outlet have slightly weathered to moderately weathered rock. The rock mass quality is poor based on RMR [9] or extremely poor based on Q-system [10], as shown in Figure 2.

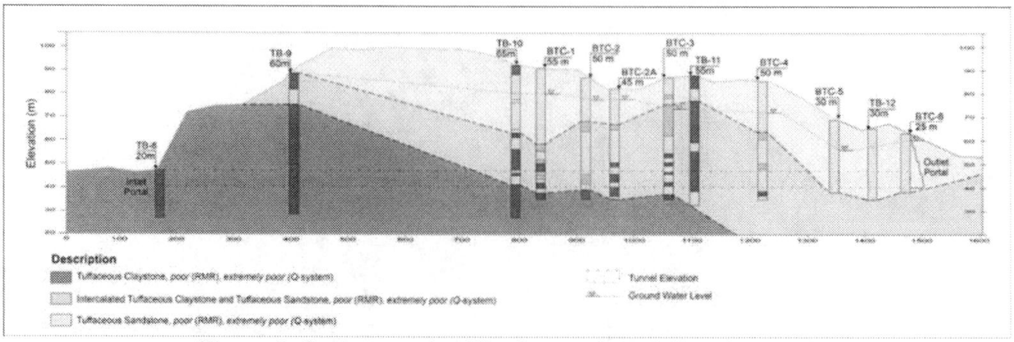

Figure 2. Rock mass classification section of Ciuyah Tunnel

3.3. Hydrogeological Condition

Seven boreholes location were observed to check groundwater conditions. These data show that the groundwater level is at an elevation of 55-82 m. This condition is above the tunnel construction elevation. From the permeability test results, the tunnel excavation area from STA 0+677 to 1+320 has a lugeon value in the range of 1-15 and a hydraulic conductivity value as shown in Table 1.

Table 1. Lugeon and permeability index along outlet portal route (STA 0+677 to 1+320) [11]

Lithology	Lugeon (Lu)	k (cm/sec)
Tuffaceous Claystone	1-5	10^{-5}
Tuffaceous Sandstone	5-15	1.3×10^{-6} - 2.4×10^{-4}
Intercalated Tuffaceous Sandstone and Tuffaceous Claystone	1-5	8.8×10^{-6} - 2.4×10^{-3}

Based on the surface water flow pattern (Figure 3), the location of the tunnel route receives surface runoff from the north, which is divided into two directions towards the Cibereum River on the east side and the Cibeurang River on the west side of the study area.

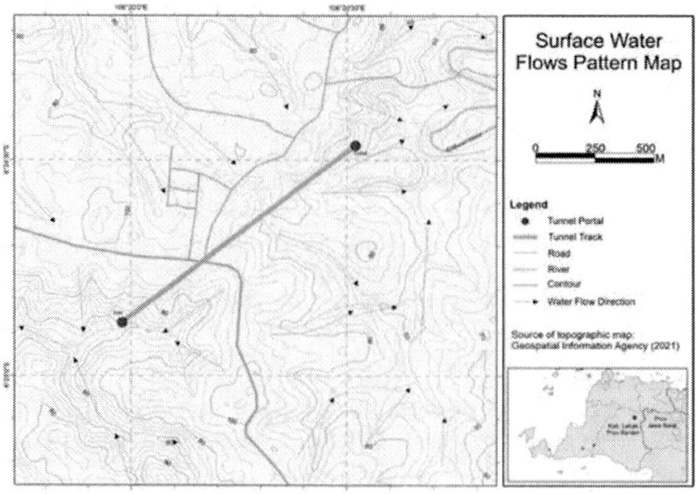

Figure 3. Surface water flows pattern map.

IOP Conf. Series: Earth and Environmental Science 1233 (2023) 012012 doi:10.1088/1755-1315/1233/1/012012

3.4. Water Inrush Incident

At STA 0+719.8, there was a groundwater leak while installing the roof supports, as shown in Figure 4. it suddenly made a puddle around the excavation face (Figure 5). The flow rate measurement results show that water flow is 1.0-1.25 L/s. Groundwater flow can erode the fine soil particles and make pore space and seepage channels. It also softened the surrounding rock and reduced the bearing capacity [4].

Figure 4. Groundwater leak at STA 0+719.8

Figure 5. Puddles of water leak at STA 0+719.8

The last condition before the leak can be known based on the face mapping report. The lithology found on the face of the excavation is tuffaceous sandstone and tuffaceous claystone with moderate-slightly weathered. Joint conditions varied open-closed, there was mineral filling in the form of soft clay sediments, and the joint axis was parallel to the direction of the tunnel. There is a slickenside on the excavation's top. The water pours in from the top, seeping through exposed fractures. The discontinuity at the top becomes weak, triggering rock falls to create a void with an area of about 1.5m^2 and becoming a path for groundwater to exit (Figure 6).

(a) (b)

Figure 6. (a) Slickenside at the top of the excavation; (b) Water flows through rockfall voids

After installing the steel ribs, the fall of rocks began to stop, and then construction activities continued with installing the following support system: a pipe roof with steel pipe support. At the first pipe roof installation at the crown apex and the depth reaches 5 meters, groundwater begins to discharge 1.67 L/s. Pipe roof installations were also carried out at STA 0+714.6, with almost the same incident at hole number 1 and a depth of 6 meters, water starting to flow out at 1.0 L/s. Monitoring water flow is carried out for 9 (nine) days until the water flow rate is stable (Table 2).

Table 2. Water rate monitoring [12]

No	Day-	Time	Flow Rate (L/s)		Remark
			STA 0+719.8	STA 0+714.6	
1	1	08:30	1.67	-	
2		12:30	1.67	-	
3		16:30	1.25	-	Water is channeled to the portal through pipes.
4		23:00	1.25	-	
5	2	09:20	1.25	-	
6		13:00	1.25	-	
7		21:00	1.25	-	
8	3	00:00	1.25	-	
9		07:00	1.25	-	
10		21:00	1.25	-	
11	4	00:00	1.25	-	
12		09:00	1.25	-	Water flow appears at STA 0+714.6 after pipe roofing
13		15:45	1.25	0.77	
14	5	09:00	1.11	0.71	
15		15:00	1.00	1.00	Water is channeled to the portal through pipes.
16		23:00	1.25	1.25	
17	6	03:00	1.00	1.25	
18		16:00	1.25	0.71	
19		21:00	1.11	0.63	
20	7	03:00	1.25	1.00	
21		16:00	1.25	1.00	
22		21:00	1.25	1.00	
23	8	16:00	1.25	1.25	The water flow from the two pipes combined flows into the portal
24		21:00	1,67	1,67	
25	9	08:00	1.00	0.71	
26		13:00	1.00	0.67	
27		24:00	1.00	0.67	

The water conditions are not carrying fine material (sediment). There is no difference in physical water conditions (color, sediment, discharge) during rainy or sunny conditions. The survey results on the surrounding settlements also did not experience a shortage of water supply. The monitoring land subsidence above the STA 0+719.8 with a radius of 30 meters using the 4-point polygon observation method and 1 benchmark (BM) for 7 (seven) days, no significant signs of subsidence were found on the surface (Table 3).

Table 3. Monitoring land subsidence [12]

Day-		Point					Difference				
		CL	CP1	CP2	CP3	CP4	CL	CP1	CP2	CP3	CP4
	x	647931.459	647954.598	647950.565	647912.353	647908.322					
1	y	9291234.983	9291254.102	9291211.812	9291258.124	9291215.899					
	z	91.219	87.844	89.670	92.466	91.597					
	x	647931.455	647954.609	647950.570	647912.341	647908.337	0.004	-0.011	-0.005	0.012	-0.015
2	y	9291234.979	9291254.090	9291211.813	9291258.135	9291215.879	0.004	0.120	-0.001	-0.011	0.020
	z	91.223	87.847	89.679	92.467	91.603	-0.004	-0.003	-0.009	-0.001	-0.006
	x	649731.450	647954.615	647950.560	647912.345	647908.315	0.009	-0.017	0.005	0.008	0.007
3	y	9291234.985	9291254.091	9291211.806	9291258.134	9291215.878	-0.002	0.011	0.006	-0.010	0.021
	z	91.226	87.845	89.679	92.467	91.598	-0.007	-0.001	-0.009	-0.001	-0.001
	x	649731.448	647954.613	647950.558	647912.343	647908.317	0.011	-0.015	0.007	0.010	0.005
4	y	9291234.981	9291254.088	9291211.807	9291258.130	9291215.876	0.002	0.014	0.005	-0.006	0.023
	z	91.224	87.844	89.677	92.465	91.597	-0.005	0.000	-0.007	0.001	0.000
	x	649731.450	647954.615	647950.560	647912.345	647908.315	0.009	-0.017	0.005	0.008	0.007
5	y	9291234.979	9291254.102	9291211.806	9291258.131	9291215.878	0.004	0.000	0.006	-0.007	0.021
	z	91.224	87.843	89.679	92.467	91.598	-0.005	0.001	-0.009	-0.001	-0.001
	x	649731.458	647954.619	647950.562	647912.340	647908.312	0.001	-0.021	0.003	0.013	0.010
8	y	9291234.983	9291254.092	9291211.809	9291258.136	9291215.877	0.000	0.010	0.003	-0.012	0.022
	z	91.228	87.847	89.681	92.474	91.605	-0.009	-0.003	-0.011	-0.008	-0.008

Modeling using Seep/W software shows groundwater flow towards the tunnel's outlet side. accumulating around point BTC-5 (Figure 7). This area has a syncline fold geometry that can be observed through geological cross-section reconstruction and groundwater correlation.

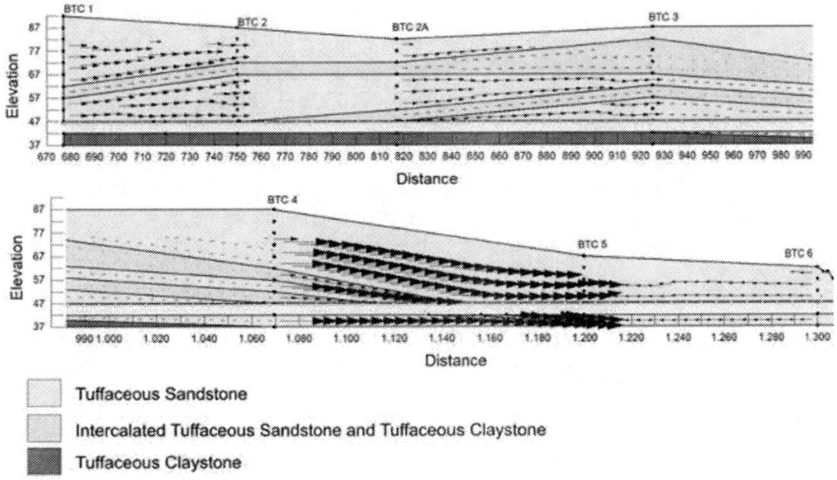

Figure 7. Direction modeling of groundwater flow

3.5. *Seepage at STA 0+719.8*

Based on seepage modeling at the STA 0+719.8, the opening of the tunnel excavation causes groundwater to move towards the excavation, as shown in Figure 8. The excavation face has tuffaceous claystone, while the top has tuffaceous sandstone, which has a permeable surface. The rock joints and slickenside cannot withstand water's hydrostatic pressure. The voids of 1.5 m^2 are becoming a groundwater way. The direction of rock dip also leads to excavation so that groundwater can flow through sedimentary rock layers. Bedding planes and rock joints control groundwater movement as secondary porosities. At the pipe roof installation, the discontinuity areas that became the groundwater path were cut off and caused groundwater to be diverted through the pipe with a flow rate of 1.67 L/s. Pipe roof installation at STA 0+714.6 also causes the previously concentrated water at STA 0+719.8 to flow to the rear, dividing the discharge in both holes into 1.0 – 1.25 L/s.

The groundwater comes from a deeper semi-confined aquifer layer and is covered by an aquitard layer. The semi-impermeable aquitard layer can filter water flow from the aquifer above to the aquifer below. It makes deposits, colors, and discharge that should be different in rainy conditions do not appear at the seepage point. The claystone lithology at the excavation face also acts as an impermeable cover layer so that the water flow does not appear through the face and tunnel walls but only flows through the tunnel roof.

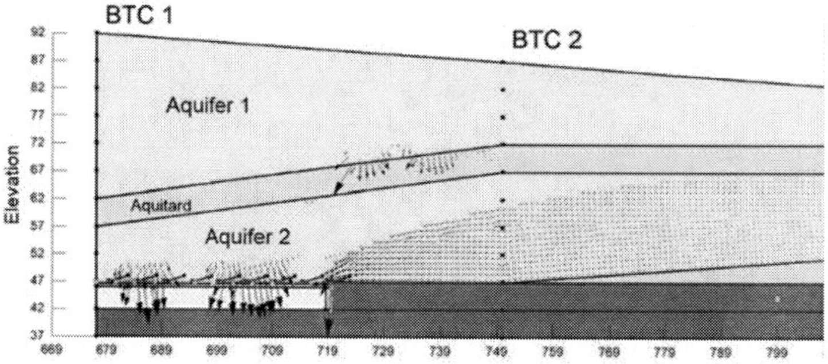

Figure 8. Groundwater movement conditions model around STA 0+719.8

3.6. *Seepage on Outlet Portal*

An aquitard layer no longer covers the aquifer character at the outlet side. The tuffaceous sandstone layer on the outlet side continues to the tunnel elevation, so the aquifer is unconfined. The model in Figure 9 shows groundwater movement concentrated around STA 1+224. The seepage flows through the tunnel's roof and passes through the excavation's face (Figure 10). The rock geometry at that location is syncline folds that make groundwater flows through the fold geometry. Based on the face mapping condition, the rocks have moderately weathered, open-closed joint conditions with soft clay material filling, rough-smooth discontinuity, and groundwater flowing through the excavation face. Under these conditions, the tunnel excavated field can withstand the overburdened load and water pressure. When installing the pipe roof, there is a flow of groundwater with a discharge that is not too large and can still be controlled through drainage installation.

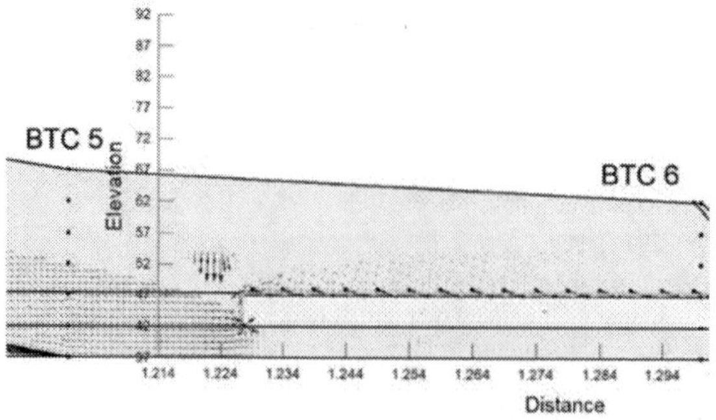

Figure 9. Groundwater movement conditions model around STA 1+224

Figure 10. (a) Puddles of water leaks at STA 1+224; (b) Water flow after pipe roofing at STA 1+224

3.7. Seepage Prediction

Based on the modeling of the tunnel alignment that has not been excavated. the potential for seepage can occur at points with a folded geometry (Figure 11). The water inflow can increase with the permeability coefficients, tunnel depth increasing, and water depth [3]. The pore water pressure would change with excavation. The pore water pressure on the face is set to zero, while the pore water in the soil would seep into the tunnel under the pressure of the pore. After that, the seepage field would continuously form a new seepage field [2].

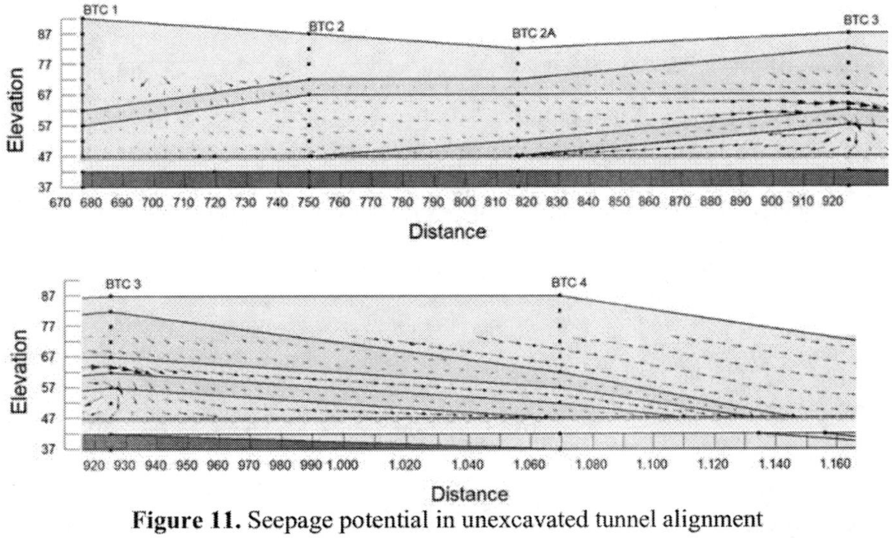

Figure 11. Seepage potential in unexcavated tunnel alignment

Based on the typical rock conditions outside the tunnel excavation, there are many points of groundwater seepage on the slope walls (Figure 12). The seepage point is not only controlled by the rock layers but also by the joint structure with openings of 0.3 – 5 mm. This condition can reflect the lithological conditions in the tunnel excavation where groundwater flows through the rock geometry and is controlled by a joint structure that becomes secondary porosity for groundwater flow.

Figure 12. Seepage locations in typical rock outside tunnel excavation

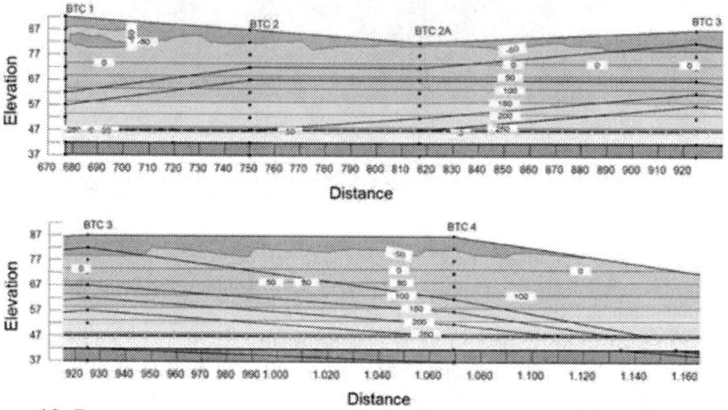

Figure 13. Pore water pressure contours in the unexcavated tunnel alignment

The pore water pressure value based on the modeling is 250 KPa (Figure 13). Seepage's impact depends on the rock mass's strength and quality. The more dominant discontinuities, especially the joint and slickenside areas, can be a weak point for collapse and groundwater entry into the tunnel.

4. Conclusion and Recommendation

The seepage analysis in this study was based on geological conditions, geotechnical parameters, and two-dimensional possible seepage point models. It revealed that the leak at STA 0+719.8 was due to the tunnel excavation cutting slickenside and joints on the tunnel roof. The weak area of rock became a pathway for groundwater to enter the tunnel excavation site. The semi-confined aquifer influences the water's character, not affected by surface conditions. Meanwhile, the leak at STA 1+224 was influenced by the syncline geometry and tuffaceous sandstone lithology. It makes groundwater can seep through the face of the excavation and the tunnel's foot. Potential seepage points can occur at excavation points that affect the geometry of the fold where the water goes to a lower place. However, the impact of leakage is influenced by the intensity and strength of the discontinuity plane.

Further research is needed to examine methods of handling Ciuyah Tunnel groundwater leaks, including grouting, dewatering, and drainage methods.

Acknowledgment

The authors would like to thank BBWS Cidanau-Ciujung-Cidurian, Korean Rural Community Corporation, Consultant Supervision (Korea Engineering Consultants Corp., PT. Indra Karya (persero), PT. Wiratman, PT. Mettana), and contractors (Daelim-Wika-Waskita, JO) for the permission to conduct this research and supplying the data. The first author would also like to thank The Ministry of Public Works and Housing of Indonesia for providing financial support.

References

[1] Wei Z and Yanpeng Z 2021 Seepage in Water-Rich Loess Tunnel Excavating Process and Grouting Control Effect *Geofluids* 2021: 1-13.

[2] Li X Yongquan Z and Jia Z 2019 Seepage Analysis and Optimization of Engineering Method for Municipal Underwater Tunnel Construction in Urban Coastal Stratum *Journal of Coastal Research* 94 (sp1): 260-263.

[3] Zhang L Dukun Z Jing W Weimin Y Wen W and Dongdong X 2020 Prediction of Water Inflow in Tsingtao Subsea Tunnel Based on the Superposition Principle *Tunnelling and Underground Space Technology* **97**: 103243.

[4] Liu JQ Ka VY Wei ZC Xiao SZ and Wei W 2020 Grouting for Water and Mud Inrush Control in Weathered Granite Tunnel: A Case Study *Engineering Geology* **279**: 105896.

[5] Niu J Yong S Bin W Keneng Z Youhan H Shiwu H Jiaqin Y and Liangliang Q 2020 Grouting Treatment of Water and Mud Inrush in Fully Weathered Granite Tunnel: A Case Study *Geofluids* 2020: 1-18.

[6] Rana N and Nair R 2016 Performance of polyurethane grouting to handle heavy seepage in tunnels *Proceedings of the conference on Recent Advances in Rock Engineering*. 241-248

[7] Bemmelen R W van 1970 *The Geology of Indonesia Vol.IA : General Geology of Indonesia and Adjacent Archipelagoes* Martinus Nijhoff.

[8] Rusmana E Suwitodirdjo K and Suharsono 1991 *Geological Map of Serang Quadrangle, Jawa* Geological Research and Development Centre.

[9] Bieniawski Z T 1989 *Engineering rock mass classifications: A complete manual for engineers and geologists in mining, civil, and petroleum engineering*: New York, Wiley, 272p.

[10] Barton NR Lien R and Lunde, J 1974 *Engineering classification of rock masses for the design of tunnel support: Rock Mechanics*, v. **6**, p. 189-239.

[11] Korea Rural Community Corporation in consortium with Korea Engineering Consultants Corp, PT. Indra Karya (Persero), PT. Wiratman, PT. Mettana. 2021. A Review of The Excavation and Supporting System Ciuyah Tunnel. *Technical Advice Note*.

[12] Daelim Industrial Co., LTD - PT. Wijaya Karya (Persero) – PT. Waskita Karya (Persero) Joint Operation. 2021. *Method Statement* for Treatment Water Flow STA 0+719.8 & Additional Ribs STA 0+719.8 – 0+714.6.

Stability analyses of diversion tunnel portal slopes at the Cidanau Dam, Banten, Indonesia

B A R Amalia[1,2] **and W Wilopo**[1]

[1]Department of Geological Engineering, Gadjah Mada University, Indonesia
[2]Ministry of Public Works and Housing, Indonesia
[*]Corresponding author: wilopo_w@ugm.ac.id

Abstract. The inlet and outlet portal slopes of the Cidanau Dam diversion tunnel were designed with 55° and 46° overall slope angles, respectively, based on stability chart method static analyses. However, the performance of tunnel portal slopes under earthquake loads has not been verified whereas the research area is located in a high earthquake hazard zone. This research was conducted to verify the slope performance under static and earthquake load using limit equilibrium (LE) and finite element (FE) methods. The engineering geological mapping and rock cores evaluation showed that both portal slopes comprised four materials: residual soil, very poor-lapilli tuff, poor-lapilli tuff, and fair-tuff breccia. The stability analyses results showed that the designed tunnel portal slopes were in a stable state under static and earthquake loads. The lowest safety factor (FS) values at the inlet section under static and seismic load are 1.9 and 1.24, and at the outlet section are 2.29 and 1.69, respectively. However, the weathered pyroclastic volcanic rocks comprising the slope is required to be protected from severe degradation upon exposure during and after construction. Results of this study are expected to contribute to the safe design of rock slopes in tropical areas, particularly in Indonesia.

1. Introduction

This research was conducted at the Cidanau Dam site, Cinangka District, Serang Regency, Banten, Indonesia, as shown in figure 1. The Cidanau dam has been designed to have a diversion tunnel on the dam's left abutment to divert the Cidanau river water during the earth-fill dam construction, and then the tunnel will be used as a waterway during dam operation. BBWS Cidanau Ciujung Cidurian in 2018 designed the geometries of tunnel's inlet and outlet portal slopes with a 17 m and a 25 m high cut with 55° and 46° overall face angles, respectively, based on static analyses using the stability chart method [1]. However, the performance of the designed tunnel portal slopes under seismic loads has not been assessed in previous study. Stability analysis of the tunnel portal slopes under seismic load is necessary because the research area is in a high earthquake hazard zone [2].

The widely used stability analyses approaches are the limit equilibrium (LE) and finite element (FE) methods. LE method is commonly used to perform stability analysis because it is user-friendly and straightforward, with a reasonably good factor of safety (FS). But it requires dividing the soil/rock mass into slices, and the side forces directions between each slice in the slope must be assumed. On the other hand, in the FE analysis, no assumptions about the location or shape of the failure surface, side forces of the slices, or orientations must be made in advance [3] – [5]. However, previous studies showed differences in the FS determined by FE and numerous LE methods, so the engineer is recommended to analyze slopes stability using both FE and LE approaches [6]. Therefore, this research was conducted using limit equilibrium (LE) and finite element (FE) methods to verify the slope performance under static load and to assess the slope performance under earthquake load. In addition, the engineering geological condition from field mapping, rock cores evaluation, and laboratory testing are presented in this paper as the factors that influence slope stability.

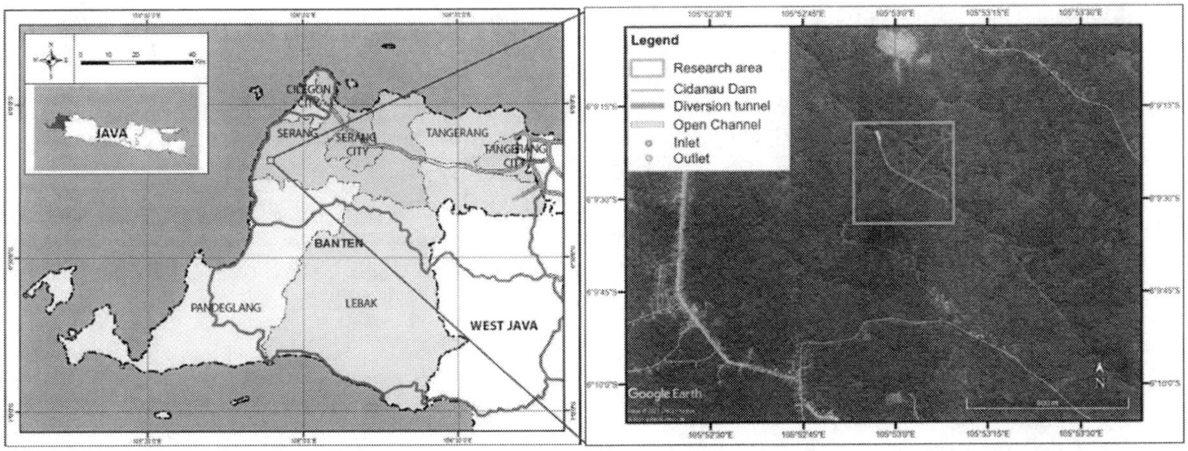

Figure 1. Research area. Image source: Spatial plan (RTRW) map of Banten Province 2010-2030 (left); Google Earth (right)

2. Method

In a slope design, it is necessary first to understand the geology of the area. Each slope performance is influenced significantly by local geological characteristics such as subsurface geometry, soil/rock properties, and groundwater [7]. This research obtained the subsurface geometry from an engineering geological mapping conducted around the tunnel portal and rock cores evaluation. The data of rock mass layers comprising the portal slope were determined using the Geological Strength Index (GSI) classification proposed by Hoek et al. (2013) [8]. Soil/rock properties were determined from laboratory testing of soil and rock samples, e.g., index properties, direct shear, and uniaxial compressive strength (UCS) tests. The groundwater table used in the model was inferred from the Cidanau Ciujung Cidurian river basin organization's borehole logs [1]. The slope models were created based on rock mass layers data, and material properties were defined for each layer of materials.

The stability analysis of the tunnel portal slope under static and seismic loads was then conducted using the LE and FE methods. The LE analysis measures stability by the FS, which is the ratio of resisting and driving forces working on the slope. Slope stability calculations must ensure that the driving forces are not more significant than the resisting forces that cause a slope to fail [9]. The LE analyses in this research were performed using the Bishop Simplified, Spencer, and GLE methods in the SLIDE slope stability package developed by Rocscience, Inc. The failure criterion used in the analysis was Mohr-Coulomb for the soil and Generalized Hoek Brown for the rock mass. Considering the weathering degree of rock and assuming that structure does not influence slope stability, the failure mode used in the analysis is circular (rotational) [10], [11]. The interslice forces were computed using a half-sine function on an assumption of 25 slices of the slip surface. The model's basic input settings included a grid search option for a circular shear surface.

On the other hand, the numerical method determines slope stability by the ratio of slope's stresses to the rock strength. In a FE approach, the FS is computed by reducing the shear strength until it collapses. Thus, the FS is the rock's actual shear strength in proportion to its reduced shear strength upon failure. Zienkiewicz et al. (1975) [12] was first used this shear-strength reduction (SSR) approach using finite elements to calculate the FS of a slope made of various materials [11]. The FE analyses in this work were performed using the RS2 software program (Rocscience, Inc.). The slope model and the failure criterion used are the same as the LE's. The finite element mesh was generated using a plane strain model, considering that the out-of-plane excavations are infinitely long and that consequently there is no strain in that direction, with 6-noded triangular elements. The field stress is considered under gravity load, to define an in-situ stress field which varies linearly with depth. FS of the slope was determined by applying the SSR technique, and the FS value in this FE method is represented by the value of critical Strength Reduction Factor (SRF) [13].

The stability of the slopes during a seismic condition was analyzed using the pseudo-static analysis method. The pseudo-static approach includes modeling ground motions as a static horizontal force (FH) acting out of the slope face for slope design under seismic load. FH has a magnitude of = W*k, where W is the block's weight, and k is the seismic coefficient (units: g) [11]. The seismic coefficient was calculated following the procedures outlined in SNI 8460 [14]. It is calculated as half of the peak ground acceleration (PGA) and multiplied by a factor of amplification (FPGA) based on the site class. The PGA used in this study was determined based on the Indonesian earthquake source and hazard map for ground motions with a 10% probability of exceedance in 50 years (≈ 500 years return period) [15]. Based on SNI 8460 [14], the required FS values for permanent slopes under static loads is 1.5, and under seismic loads is 1.1. Results of this study are expected to contribute to the safe design of rock slopes in tropical areas, particularly in Indonesia.

3. Result and Discussion

3.1. Geological Condition

The surface geological mapping and rock cores evaluation show that the research area comprises two lithological units: the lapilli tuff and the tuff breccia, as shown in figure 2 and figure 3. The diversion tunnel inlet and outlet portal slopes surface were located at the lapilli tuff unit. Based on the Regional Geological Map of Anyer [16], the lapilli tuff and tuff breccia lithological unit at the research area was likely the member of the Upper Banten Tuff Formation, which is presumed to be Late Pleistocene age, and was deposited in a terrestrial environment. The lapilli tuff and tuff breccia units are pyroclastic volcanic rocks, in which the central problem of these rocks is the rapid disintegration upon exposure to drying and wetting. It is, therefore, essential to protect these materials from severe deterioration upon exposure during construction [7].

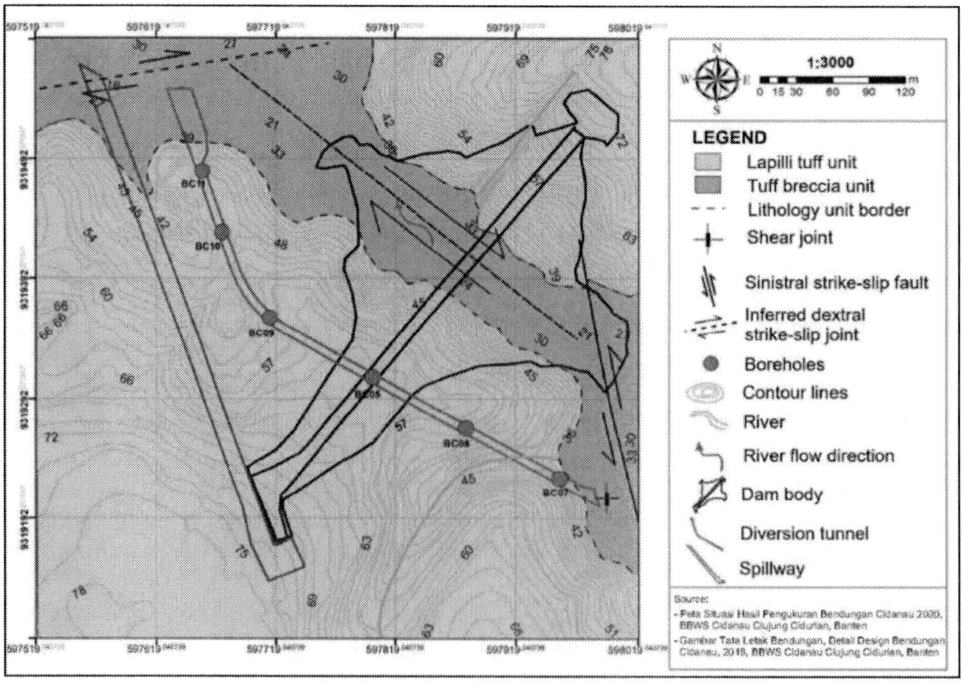

Figure 2. Geological map of the research area

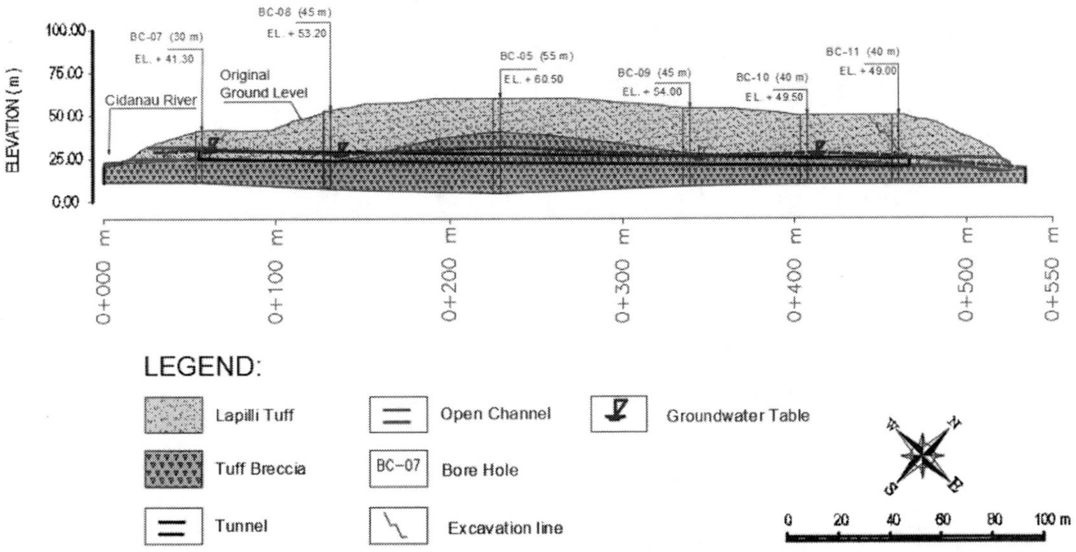

Figure 3. Geological section along the tunnel alignment

3.2. Rock Mass Quality

As mentioned in the previous chapter, the lapilli tuff and tuff breccia units are pyroclastic volcanic rocks, in which the central problem of these rocks is the rapid disintegration upon exposure to drying and wetting [7]. This condition, in addition to the fact that the tunnel is located at a high earthquake hazard zone [2], affects the weathering grade and quality of rockmass at the tunnel alignment.

The weathering grade of rock mass at the tunnel alignment based on geological mapping and rock core evaluation, referring to the weathering level by ISRM (1978) [17], ranges from slightly weathered rock at depth to residual soil at the surface. While based on the GSI classification results, rock masses at the inlet and outlet tunnel portal slopes were divided into four classes, namely residual soil, very poor quality of lapilli tuff, poor quality of lapilli tuff, and fair quality of tuff breccia, as shown in table 1 and figure 4. Figure 4 shows that the typical models used in the slope stability analyses at the inlet and outlet tunnel portal slopes have the same rock mass quality configuration. In addition, figure 5 shows the representative cores of each rock mass class, which consist of residual soil – very poor lapilli tuff, poor lapilli tuff, and fair tuff breccia, respectively. From the UCS laboratory testing, the UCS values of the intact rocks range from 3.11 – 12.43 MPa, which is classified as weak rock [17], verifying the extensive weathering degree of the rocks. However, the rock cores samples used in this study were drilled in 2018 so that it might be has been deteriorated and resulting in lower strength and quality than it used to be.

3.3. Material Properties

The properties of each rock mass layer inferred from the laboratory testing were the cohesion (c), internal friction (ϕ), uniaxial compressive strength of intact rock (σci), Young's modulus (Ei), unit weight (γ), and Poisson's ratio (v). Those properties were employed as input parameters in the analysis, as shown in table 2. The intact rock constant (mi) was estimated using the RocLab database's range mi value based on rock type, while the rock mass strength parameters for the Hoek-Brown failure criterion were calculated using the RocLab software application. On the other hand, the disturbance factor values (D) were set to zero by presuming that the slope excavation method created a low disturbance.

Table 1. Weathering degree and rock mass quality at inlet and outlet slope

Section	Depth (m)	Lithological Unit	Weathering degree	GSI	Rock mass quality
Inlet	0 – 2	Lapilli tuff	Residual soil	-	-
	2 – 6	Lapilli tuff	Highly weathered	15	Very poor
	6 – 15	Lapilli tuff	Moderately weathered	23	Poor
	15 – 30	Tuff breccia	Slightly weathered	55	Fair
Outlet	0 – 2	Lapilli tuff	Residual soil	-	-
	2 – 7	Lapilli tuff	Highly weathered	15	Very poor
	7 – 30	Lapilli tuff	Moderately – slightly weathered	28	Poor
	30 – 40	Tuff breccia	Slightly weathered	55	Fair

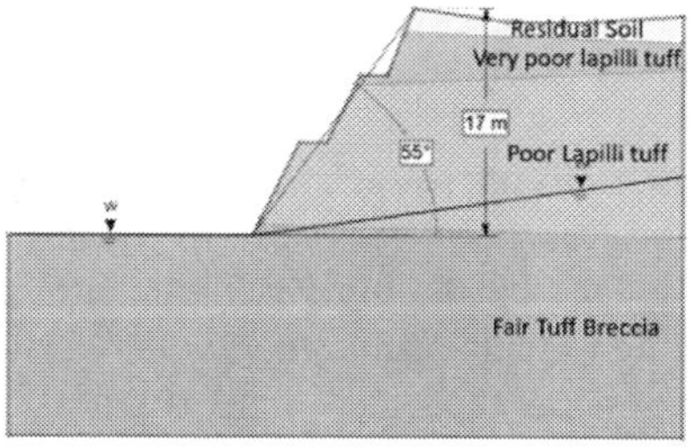

(a)

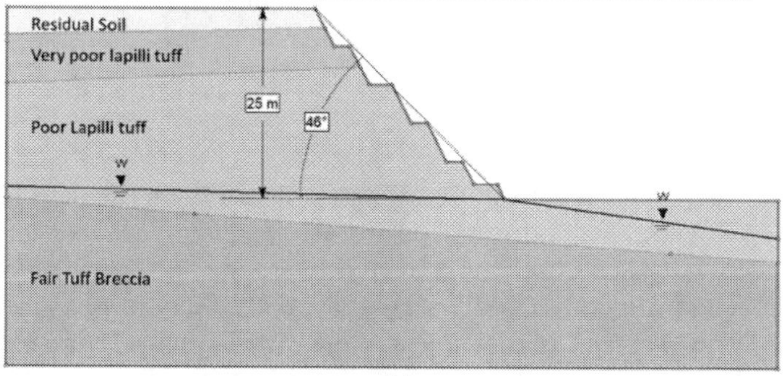

(b)

Figure 4. A typical model for slope stability analysis: (a) inlet slope, (b) outlet slope

Table 2. Material Properties

Section	Depth (m)	Lithological Unit	Rock mass GSI	Rock mass Quality	σ_{ci} (kPa)	m_i	D	E_i (kPa)	γ (kN/m^3)	v	c (kPa)	ϕ (°)
Inlet	0-2	Residual soil	-	-	-	-	-	27600	13.15	0.3	30.5	52.09
	2-6	Lapilli tuff	15	Very poor	-	-	-	27600	13.15	0.3	30.5	52.09
	6-15	Lapilli tuff	23	Poor	3110	15	0	2271000	11.28	0.42	-	-
	15-30	Tuff breccia	55	Fair	12430	18	0	1763680	17.56	0.3	-	-
Outlet	0-2	Residual soil	-	-	-	-	-	27600	9.51	0.3	32.99	58.07
	2-7	Lapilli tuff	15	Very poor	-	-	-	27600	9.51	0.3	32.99	58.07
	7-30	Lapilli tuff	28	Poor	3110	15	0	2271000	11.28	0.42	-	-
	30-40	Tuff breccia	55	Fair	12430	18	0	1763680	17.56	0.3	-	-

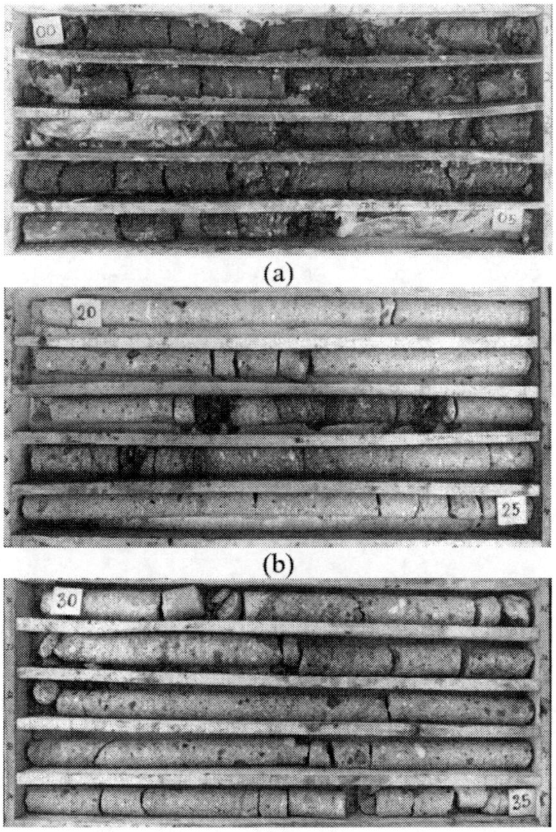

(a) (b) (c)

Figure 5. Examples of drill core photographs: (a) residual soil – very poor lapilli tuff at BC07, (b) poor lapilli tuff at BC11, and (c) fair tuff breccia at BC11. Image source: BBWS Cidanau Ciujung Cidurian (2018)

3.4. Seismic Load

With a 10% probability of exceedance in 50 years, the value of PGA in the research area is 0.25 – 0.3 according to the Indonesian earthquake source and hazard map published by the National Center for Earthquake Studies (2017) [15]. On the other hand, the SPT N values based on the previous study's data vary from 14 – 25 [1]. Therefore, based on SNI 8460, the seismic site classification is stiff soil (SD), the corresponding amplification factor (FPGA) is 1.1, and the calculated seismic load coefficient for the pseudo-static analysis was 0.165.

3.5. Slope Stability Analysis

The stability analyses results show that in a static condition, the tunnel's inlet slope has a 1.90 FS based on the FE analysis method and 2.03 FS based on the LE analysis method. At the same time, in a static condition the tunnel's outlet slope has 2.29 FS based on the FE analysis method and 2.35 – 2.36 FS based on the LE analysis method. These results were found to be in agreement with that of the previous study, in which the inlet and outlet portal slopes of the diversion tunnel at the Cidanau dam are in a stable condition under a static load. On the other hand, the results of stability analyses in a seismic condition show that the tunnel's inlet slope has a 1.24 FS based on the FE analysis method and 1.6 – 1.63 FS based on the LE analysis method. At the same time, the tunnel's outlet slope in a seismic condition has 1.69 FS based on the FE analysis method and 1.83 – 1.85 FS based on the LE analysis method. These results show that seismic load reduces the FS values of the slopes, however the inlet and outlet portal slopes of the diversion tunnel at the Cidanau dam are in a stable condition under a seismic load.

As shown in table 3, the FS values obtained from the LE method, namely Bishop Simplified, GLE, and Spencer methods, were similar, while the FS values obtained from the FE method were slightly lower than those obtained from the LE methods. These results are consistent with previous studies in which there are differences between the safety factors obtained using LE and FE methods [3], [6]. However, in general, the designed cut slopes at the tunnel's inlet and outlet portal complied with SNI 8460 stability standards [14], in which the slopes must have more than 1,5 safety factors under static load and more than 1,1 safety factors under seismic load.

In addition, since the rock masses comprising the slopes are pyroclastic volcanic rocks with varying degrees of weathering, it is susceptible to rapid disintegration when exposed to drying and wetting [7]. Therefore, the slope's face was designed to be protected with shotcrete to prevent slope degradation upon exposure during and after construction. In this condition, drain holes should be drilled through the shotcrete to prevent water pressure from building up behind the face [11], and the slopes benches should be equipped with a lined drainage ditch to collect and dispose of the surface water.

Table 3. Slope stability analysis results

METHOD		FS				FS_{min}[a]	
		INLET		OUTLET			
		STATIC	SEISMIC	STATIC	SEISMIC	STATIC	SEISMIC
Limit Equilibrium	**BISHOP**	2.03	1.60	2.35	1.83	1.5	1.1
	GLE	2.03	1.60	2.35	1.83		
	SPENCER	2.00	1.63	2.36	1.85		
Finite Element		1.90	1.24	2.29	1.69		

[a]According to SNI 8460:2017

4. Conclusions

From the geological engineering investigation, the Cidanau dam's portal slopes comprise lapilli tuff and tuff breccia lithological units (pyroclastic volcanic rocks), in which the central problem of these rocks is the rapid disintegration upon exposure to drying and wetting. The weathering grade of the rock mass ranges from slightly weathered rock at depth to residual soil at the surface, while the quality of the rock masses comprising the inlet and outlet portal slopes were divided into four classes, namely: residual soil, very poor quality of lapilli tuff, poor quality of lapilli tuff, and fair quality of tuff breccia. The UCS values of the intact rocks range from 3.11 – 12.43 MPa, which is classified as weak rocks, verifying the extensive weathering degree of the rocks.

The calculated seismic load coefficient used in the pseudo-static analysis was 0.165. The analyses using LE method, namely Bishop Simplified, GLE, and Spencer methods result in a similar FS value, while the FE method analysis results were slightly lower than those obtained from the LE methods. The results show that in a static condition the tunnel's inlet slope geometries have a 1.90 FS value based on the FE analysis method and a 2.03 FS value based on the LE analysis method. At the same time, the tunnel's outlet slope in a static condition has a 2.29 FS value based on the FE analysis method and 2.35 – 2.36 FS based on the LE analysis method. These results were found to be in agreement with that of the previous study, in which the inlet and outlet portal slopes of the diversion tunnel at the Cidanau dam are in a stable condition under a static load. On the other hand, the results of stability analyses under seismic load show that the tunnel's inlet slope has a 1.24 FS value based on the FE analysis method and a 1.6 – 1.63 FS value based on the LE analysis method. At the same time, the tunnel's outlet slope under seismic load has a 1.69 FS value based on the FE analysis method and 1.83 – 1.85 FS value based on the LE analysis method. The analyses results show that the seismic load reduces the FS values of the slopes. In conclusion, the previously designed inlet tunnel portal slope with 17 m high cut and 55° overall slope angle, and outlet tunnel portal slope with 25 m high cut and 46° overall slope angle are in a stable condition under static and seismic loads, fulfilled the SNI 8460 requirements [14].

In addition, to protect the weathered pyroclastic volcanic rocks comprising the slope from severe degradation upon exposure during and after construction, the slope face is required to be covered with shotcrete and equipped with a lined drainage ditch at the slope's bench to collect and dispose of the surface water.

Acknowledgment

The authors would like to thank BBWS Cidanau Ciujung Cidurian for the research permit and the Ministry of Public Works and Housing Indonesia for the research funding.

References

[1] PT. WECON 2018 *Laporan Akhir Pekerjaan Detail Desain Bendungan Cidanau Tahap I* (Banten: BBWS Cidanau Ciujung Cidurian).
[2] Syahbana D K, Suantika G and Solikhin A 2008 *Earthquake Hazard Map Western Part Of Java Scale 1:500.000* (Bandung: Center for Volcanology and Geological Hazard Mitigation Ministry of Energy and Mineral Resources).
[3] Aryal K P 2006 *Slope Stability Evaluation By Limit Equilibrium And Finite Element Methods* Norwegian University Science and Technology Ph.D. thesis.
[4] Griffiths D V and Lane P A 1999 *Géotechnique* **49** No. 3 387–403.
[5] Rocscience Inc. 2004 *Application of the Finite Element Method to Slope Stability* (Toronto: Rocscience Inc) pp 1–2.
[6] Hammouri N A 2008, Husein Malkawi A I and Yamin M M A *Bull. Eng. Geol. Environ.* **67** 471–478.
[7] Abramson L W, Lee T S, Sharma S and Boyce G M 2002 *Slope Stability and Stabilization Methods* 2nd Edition (New York: John Wiley & Sons, Inc.) pp 76.
[8] Hoek E, Carter T G and Diederichs M S 2013 Quantification of the geological strength index chart *47th US Rock Mechanics / Geomechanics Symposium* 1757–1764.
[9] Duncan J M, Wright S G and Brandon T L 2014 *Soil Strength and Slope Stability* (New Jersey: John Wiley & Sons, Inc.) pp 81.
[10] Singh B and Goel R K 2011 *Engineering Rock Mass Classification: Tunneling, Foundation, and Landslides* (Waltham: Butterworth-Heinemann) pp 211-230.
[11] Wyllie D C 2018 *Rock Slope Engineering: Civil Applications, Fifth Edit.* (Boca Raton: CRC Press).
[12] Zienkiewicz O C, Humpheson C, and Lewis R W 1975 *Géotechnique* **25** No. 4 671–689.
[13] Rocscience Inc. *RS2 User Guide* see http://https://www.rocscience.com/help/rs2/tutorials
[14] Badan Standardisasi Nasional 2017 *SNI 8460:2017 Persyaratan Perancangan Geoteknik* (Jakarta: Badan Standardisasi Nasional).
[15] Pusat Studi Gempa Nasional 2017 *Peta Sumber Bahaya dan Gempa Indonesia* (Bandung: Pusat Penelitian dan Pengembangan Permukiman Kementerian Pekerjaan Umum dan Perumahan Rakyat).
[16] Santosa S, Sutrisno, Turkandi T and Sukanta U 1991 *Geological Map of the Anyer Quadrangle, West Java* (Bandung: Pusat Penelitian dan Pengembangan Geologi).
[17] International Society for Rock Mechanics 1978 *Int. J. Rock Mech. Min. Sci. Geomech.* Abstr. **15** 319-368.

IOP Conf. Series: Earth and Environmental Science 1233 (2023) 012014 doi:10.1088/1755-1315/1233/1/012014

Static and Pseudo Static Analysis of Embankment Slope Stability of Bener Dam, Purworejo Regency, Central Java

Thuon Ranysakol[1]

[1]Department of Geological Engineering, Universitas Gadjah Mada; Jl. Grafika Bulaksumur No.2, Kampus UGM, Yogyakarta, Indonesia

Abstract. The Bener Dam is considered as one of the largest rock-fill dams in Indonesia. This dam will impact the inundation area of three main villages in the Wonosobo and Purworejo regency. Since this zoned embankment dam consists of rock-fill and the core material, the stability of the dam slope has to be done to ensure that the dam is stable against any failures including the seismic effects. Moreover, the seismic coefficient in the study area in the seismic analysis of the large dam slope is needed to be used in the analysis which is not yet done from the previous studies. This paper presents the analysis of the embankment slope stability of the Bener dam under static condition and pseudo static condition. The analysis takes into account the end stage of construction, the steady-state stage, and rapid drawdown stage. The results of utilizing the Finite Element Method to analyse the slope stability demonstrates that the dam embankment slopes are stable under static and seismic stress conditions in all the reservoir condition stages. After applying the seismic loading, the result of safety factor of slope stability is reduced while the rapid drawdown stage has the lowest safety factor.

1. Introduction

This research is carried out in Bener Dam, which is built on the Bogowonto river, in Guntur village, Bener district, Purworejo regency, Central Java province, Indonesia. Bener dam is planned to be a zonal dam type as embankment dam with a layer of concrete as waterproof layer in the upper part called rock fill concrete membrane or concrete face rockfill dam with the 156 meters height of the dam [1]. Since the Bener dam is set to be one the largest rock-fill dam in Indonesia, and it will cover the inundation area of three district at the downstream. Administratively, the inundation reservoir is located at the site of three different villages. There are Limbangan Village, Guntur, Nglaris District Bener Regency Purworejo, Kemiri Village, Gebang District Purworejo Regency Burat Village, Bener, Gadingrejo Kepil, Wonosobo district. The embankment of main dam body has to be stable against any failure factors. Moreover, the seismic force has to be included to the analysis of slope stability in the earthquake prone area since the failure of dam slope can cause the catastrophic flood which leads to the loss of life, existing property damages, infrastructure damages, and dam function damages. The slope is susceptible to fail due to the seismic event that can be determine by using the critical seismic acceleration coefficient [2]. The seismic coefficient is an important parameter to be used in the soil or rock mass measurement which is resisted to earthquake induced sliding. Pseudo static slope stability technique is widely used for evaluating the seismic performance of slope stability analysis since the critical seismic acceleration coefficient is used from the peak ground acceleration (PGA) value in the research area [3].

The embankment slope of the proposed dam has to be properly studied in both static and seismic condition. since the seismic coefficient required, and the slope. Moreover, the engineering geological characteristics need to be focused for estimating the slope safety and prevent failure. In the study of [4] the engineering geological characteristics and the rock stability in the diversion tunnel in the Bener dam

IOP Conf. Series: Earth and Environmental Science 1233 (2023) 012014 doi:10.1088/1755-1315/1233/1/012014

were presented, yet the stability of dam slope is not discussed. The study of the embankment slope of the Bener dam is necessary to analyse since the main dam body is the most important to ensure that the whole project is in the safe condition. In addition, the seismic slope stability is needed to considered in the study.

This paper aims to analyse the slope stability of the main dam body on the Bener dam project in the static and pseudo static methods in terms of end construction stage, maximum reservoir level stage, and drawdown stage. This study is intended to complete the information of slope stability of dam slope which have not yet been done by previous research. Since the Bener dam is the largest rock-fill dam in Indonesia including its zoned type and concrete layer on the top of the upstream, the analysis in both static and seismic condition of the slope will give the important result towards the slope stability analysis problems in the future dam slope analysis. In addition, the stability analysis is conducted to estimate the factor of safety as the output from static and pseudo static analyses based on Finite Element Method (FEM) of the embankment slope of the main dam body. However, the limitation in this study is that the material properties are set to be in the isotropic condition.

1.1. Research Location

This research is conducted in the Bener Dam Development Project, which is situated on the Bogowonto river in Guntur Village, Bener District, Regency of Purworejo, Central Java Province. The dam is located in the coordinate area (UTM) X = 391256-391782 E and Y =9160215-9160419 S. For more details, the location can be seen on the project location map in the Figure 1.

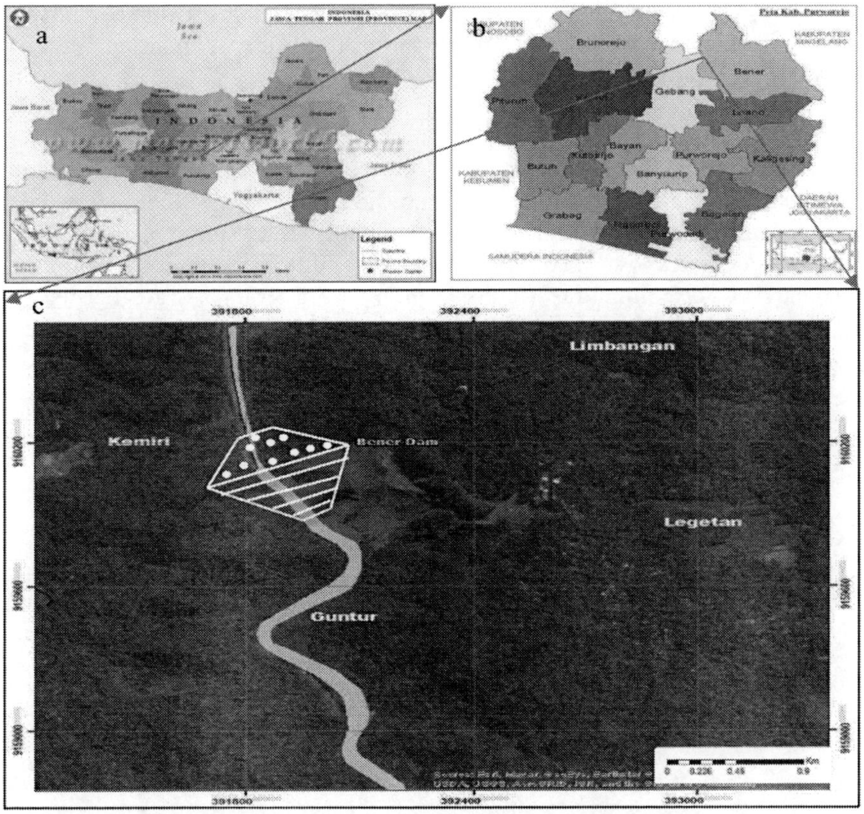

Figure 1. a. Map of Indonesia showing the location of purworejo regency, b. Purworejo regency map showing the Bener District, c. The Location of Bener Dam, Purworejo Regency, Central Java Province (Source: https://www.google.com/maps).

1.2. Regional Geology

The stratigraphy regional rock units of the research area consist of Alluvium Unit, the Sentolo Formation Unit and the Kebobutak Formation [5]. In addition, the investigation location of the research consists of the Kebobutak Formation. This formation unit consists of andesite, tuff, tuff lapilli, agglomerates, and inserts of andesite lava. Andesite breccia is a volcanic rock, and its composition ranges from hypersthene andesite to hornblende-augite andesite and trachyandesite. Andesite Igneous intrusions have varied compositions from hypersthene andesite to hornblende-augite andesite and trachyandesite, most of which have experienced propylitization. According to the study of [4], the andesitic breccia is the dominant rock type in this research location. Moreover, the report of [6] indicates that the subsurface and surface rock are also dominated by andesitic breccia rock type. Then the andesitic breccia rock type is perfectly used as the rockfill material for the embankment of dam construction.

2. Methodology

2.1. Slope Stability Analyses

According to [7], the main indicator of the stability of slope is the safety factor criteria that is the ratio of the resisting strength of the slope needed to retain or equalise the driving stress on the potential failure surface of the slope. The aspect of safety factor that represents stability have to be greater than 1.0. Moreover, the factor of safety is the indication of the stable or unstable condition of the analysed slope. According to [8], the minimum factor of safety of upstream and downstream slope at the end of construction stage, maximum flood level, and drawdown stage are 1.3, 1.5, and 1.1, respectively.

There are various methods of slope stability that are broadly used, namely, Numerical Method Ex. finite element method (FEM), and limit equilibrium method (LEM) [7].

Furthermore, Rocscience., Inc developed the two-dimensional software to analyse the slope stability by using the finite element method. In this research, the finite element analysis is applied by using Phase2 software.

2.2. The Input Parameters

The influenced parameters that determine the stability of slopes are the geometry of the slope (the height and slope inclination), the material properties (cohesion, angle of friction, unit weight, and rock quality parameters), and the seismic coefficient in the research area. Moreover, the engineering geological aspects such as geomorphology (geological surface conditions), lithology (soil and rock), geological structures (rock layer discontinuities, faults, joints), hydrogeological are needed to study geological parameters in the research location [9]. The geometry of main dam body has the height of 156 m with the crest of 13.40 m width. The downstream slope inclination is 1:1.4 while the upstream slope inclination is 1:1.5 as shown in the Figure 2 [1].

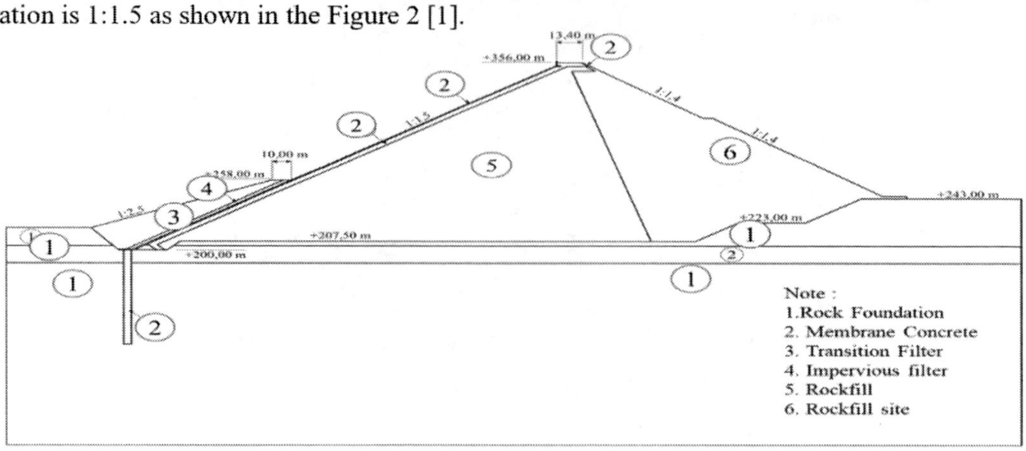

Figure 2. The Bener dam geometry and its functions, after [1]

This rockfill dam is divided into 8 functions such as rockfill, rockfill site, filter transition, impervious layer, membrane concrete, riprap, and foundation rock. The material properties of each dam function are studied by [6] and it is summarized in the table.

The Geological Strength Index classification (GSI) is a way to facilitate the rock mass properties determination for engineering projects for both hard and weak rock masses [7]. Furthermore, this classification system formed between four basic rock mass fracture intensities and the respective quality of the discontinuity surfaces. The engineering rock mass quality is classified into four rock mass quality classes, namely very poor-quality rock mass (0 – 20), poor quality (21 – 40), medium quality (41 – 55), and good quality (56 – 75) based on the geological strength index classification system.

[10] proposed a chart to quantify the GSI using the rock quality designation value (RQD) and the joint condition rating (JCond89) as shown in Table 1 below. By using the value of rock quality designation value (RQD) from borehole in the research site, we can make the calculation to get GSI (Geological strength index) afterward by using the formulas as shown below which is taken from [11].

$$GSI = 1{,}5JCond_{89} + {}^{RQD}/_{2} \quad (1)$$

Table 1. The Description of Joint Condition of Rock and its rating, after [11]

Description	Joint Separation	Rating
Rock with the rough surfaces with no any separation or joints	0	30
Rock walls with slight weathering and less than 1mm the separation slightly rough surfaces	<1	25
Rock walls with high weathering and less than 1mm the separation slightly rough surfaces	<1	25
Rock with less than 5mm or 1-5 mm separation of slickensided surfaces or Gouge.	1-5	10
Rock with greater than 5 mm thick or greater than 5 mm of soft gouge.	>5	0

Based on the observation of rock condition in the five boreholes rock sampling, the condition of rock which is slightly weathered andesite breccia rock. According to the table and the RQD results from the borehole, we calculate the GSI value in each andesitic breccia rock layer of all boreholes. The results indicate that there are two types of andesitic breccia rock which are the good quality andesitic breccia rock with 72 of GSI value, and the fair quality andesitic breccia rock with 51 of GSI value as shown in Table 2. These two types of andesitic breccia are used as rockfill material of the embankment body of the dam. The mb, s, and a are the constant of rock mass material can be calculated by using formulas as shown below by [12].

$$m_b = m_i \exp[\tfrac{GSI-100}{28-14D}] \quad (2)$$

$$s = \exp[\frac{GSI-100}{9-3D}] \quad (3)$$

$$a = \frac{1}{2} + \frac{1}{6}(e^{\frac{-GSI}{15}} - e^{\frac{-20}{3}}) \quad (4)$$

The disturbance factor (D) is a factor that indicate the disturbance degree of rock mass from the excavation or blasting degree. Additionally, the strength of the rock can be obtained through the description at the time of taking field data or obtained through laboratory testing. Identification can be made by several methods, either directly in the area or by laboratory test. The following is an identification method to determine strength rock press, namely uniaxial compressive strength test is the test of value possessed by rocks for withstanding the maximum vertical stress before experiencing collapse. The point load index (PLI) test is an index used to know the value of uniaxial compressive strength (UCS) in a rock indirect [13]. According to [4], the andesitic breccia samples were taken to do compression test in laboratory which gives the results of rock strength and stiffness parameters of rockfill site in this project as shown in the Table 2 below. Based on [5], we consider the andesite breccia with good quality of rock from the table below is used as the rockfill material of the embankment dam.

Table 2. The input parameters of composed materials of the dam for modelling, after [5], [6]

Material Modelling	Unit Weight kN/m³	Young Modulus	UCS (MPa)	GSI	Poisson's ratio	mb	s	a	C (kPa)	φ′ (°)
Good Andesitic Breccia (RockFill)	23.978	5458	31.05	72	0.35	8.84	0.04	0.50	-	-
Fair Andesitic Breccia (RockFill Site)	20.601	5458	31.05	51	0.35	3.301	0.004	0.51	-	-
Membrane Concrete	23	50000	-	-	0.4	-	-	-	10000	40
Filter Transition (Random Material)	18.73	50000	-	-	0.4	-	-	-	15840	35.82
Impervious Filter	11.125	50000	-	-	0.4	-	-	-	30.519	15.67
Foundation Rock	23.29	11030	26.12	70	0.35	6.507	0.03	0.50	-	-

The pseudo-static method is one of the most common analysis slope stability analyses. It is mostly applied for earth dams with materials which not undergo severe strength loss that has crest. Pseudo-static approach is an approximate method that expresses the seismic accelerations coefficients in both horizontal and vertical directions in terms of equivalent static forces, called pseudo-static accelerations (k_h) and (k_v). Based on [6], the Bener dam is the large dam which is designed of earthquake condition in Operating Basis Earthquake (OBE) with the acceleration of 0.120g. The Operating Basis Earthquake (OBE) has the return period of 145 years with 50% of probability of occurrence. According to [14], the seismic coefficient is set to be equal to the half of PGA value, then the seismic coefficient to be used is 0.06. Nevertheless, the traffic loading on the dam crest has to be included since the road is classified as road type 1 with 15 kPa traffic loading based on [15]. In this study, we analyse our embankment slope in static and pseudo static condition based on Finite Element Method with the computational software Phase2 by Rocscience.

3. Results and Discussion

The analysis is conducted by using the Phase2 software. Moreover, the slope is divided into two slope analysis conditions with dry condition, maximum reservoir condition, and drawdown condition as the end construction stage, steady-state stage, and rapid drawdown stage. In the analysis, the static and pseudo static analysis method are based on the Finite Element Method. In addition, the seismic coefficient 0.06 are input to pseudo static analysis. After running the analysis on Phase2 software with the input parameters, we got the result of slope stability analysis which is expressed as Strength Reduction Factor (SRF) value as shown in Table 3.

Table 3. The results of the strength reduction factor and the maximum displacement of the dam slope in different loading stage.

Reservoir Condition	Loading Stage	SRF	Maximum Displacement (m)
End Construction	Dry condition with Traffic	3.13	0.011
	Dry condition with Traffic and Seismic Loading	2.84	0.0072
Steady-State	Maximum Reservoir with Traffic Loading	3.12	0.011
	Maximum Reservoir with Traffic and Seismic Loading	2.88	0.0073
Rapid Drawdown	Drawdown condition with Traffic Loading	3.11	0.04
	Drawdown condition with Traffic and Seismic Loading	2.86	0.0075

The results of the modelling and analyses were obtained from the Phase2 software analysis. The results are the critical strength reduction factor and total displacement of the slope as shown in the Figure 3 Figure 3 & Figure 4.

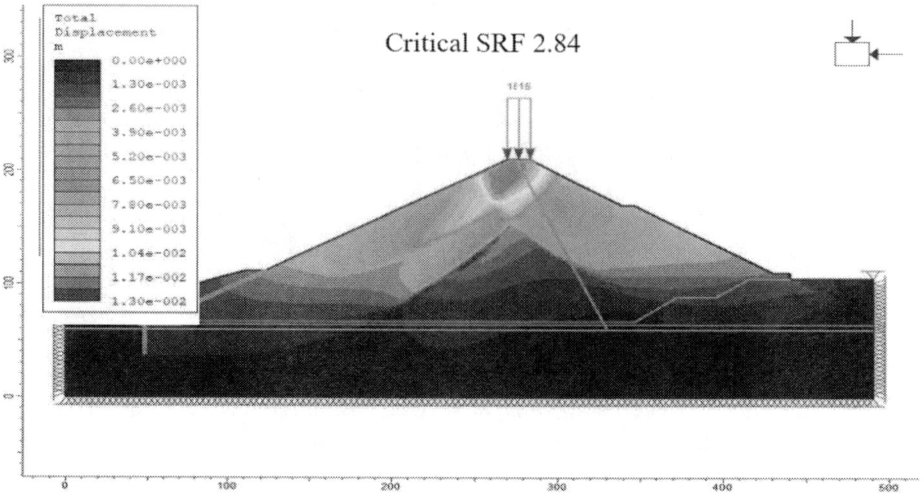

Figure 3. The Interpretation of total displacement static slope stability analysis in maximum reservoir level stage, critical SRF of 2.84, max displacement of 0.0072.

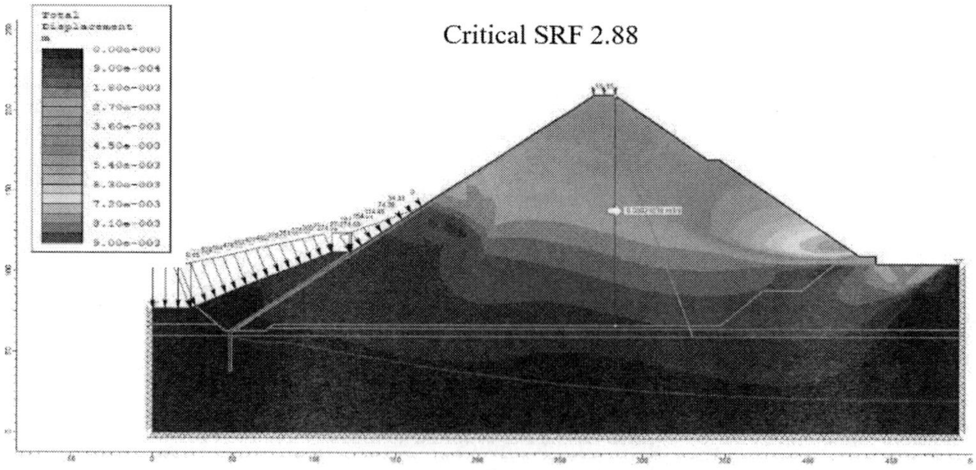

Figure 4. The Interpretation of total displacement of the pseudo static slope stability in maximum reservoir level stage, critical SRF of 2.88, max displacement of 0.0073.

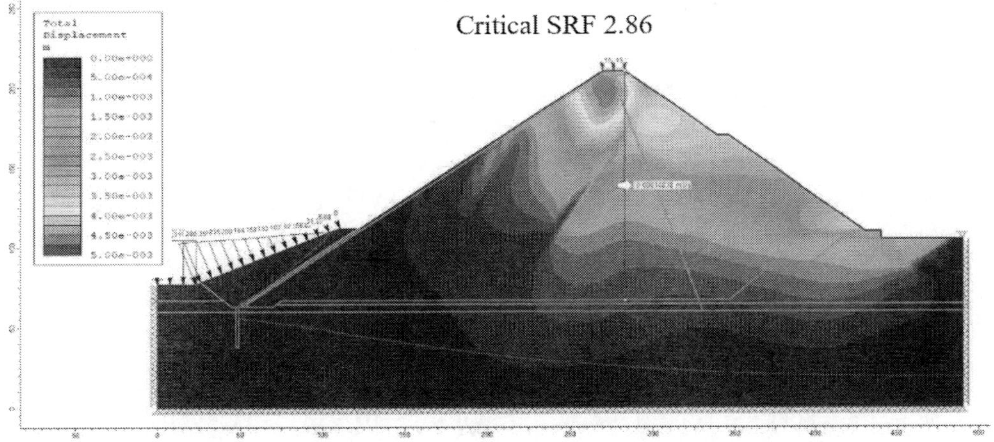

Figure 5. The Interpretation of total displacement of the pseudo static slope stability in maximum reservoir level stage critical SRF of 2.86, max displacement of 0.0075.

This result indicate that the safety factor of the slope is reduced in the order of the stage causing by the effects of water at the reservoir, since the slope is more stable when the upstream reservoir is dry without the effect of water. For the modelling the steady state seepage condition, all materials are fully saturated. The analyses using drained strengths are expressed in terms of effective stress parameters such as cohesion and frictional angle as the result the steady-stage seepage condition have the higher factor of safety value than the rapid drawdown stage since it starts reducing the water level at the upstream. The values of safety factor for the right after construction, during steady state seepage condition and the rapid drawdown stage were greater than the acceptable safety factor 1.3, 1.5 and 1.1 which indicated that the slope was stable against earthquake effects.

4. Conclusion

The embankment of main dam body slope consists of the good to fair rockfill material quality of andesite breccia. The analysis is conducted by using the Phase2 software. The slope is divided into two slope analysis conditions with dry and full reservoir level condition. Moreover, the finite element method (FEM) is applied in the analysis. The static Strength Reduction Factor (SRF) of our embankment slope is 3.13 in dry condition, 3.12 in full reservoir condition, and 3.11 in drawdown condition with 0.011 m, 0.011, and 0.04 maximum total displacement for these three phases, respectively. The pseudo static Strength Reduction Factor (SRF) in the dry condition are 2.84, 2.88 and 2.86 in end construction stage, steady-state stage, and rapid drawdown stage for seismic coefficient of 0.06. Whilst, the pseudo static analysis outcome shows the maximum of the displacement of each stage as 0.0072, 0.0073, and 0.0075, respectively. The results from both analysis method show that the slopes of all the sections are stable in both before and after reservoir filling and with seismic loading applied conditions. Additionally, the material in the anisotropic condition and sample testing are recommended to be considered in the future research to enhance the accuracy and reliable outcome towards the site condition.

References

[1] B. Masfianto, I. Ngurah, and A. Parbawa, "Metode kerja struktur concrete face slab bendungan tipe cfrd, pada pembangunan bendungan bener, kabupaten purworejo – jawa tengah," 2019, (unpublised).

[2] M. Ali, "Static Analysis and Pseudostatic Slope Stability," vol. **9**, no. 6, pp. 2013–2016, 2015.

[3] J. D. Bray and T. Travasarou, "Pseudostatic Slope Stability Procedure," 5th Int. Conf. Earthq. Geotech. Eng., pp. 1–12, 2011, [Online]. Available: papers3://publication/uuid/221A655B-DB42-4E5C-8293-541FEF166DCC.

[4] D. J. Sasangka, R. Taufik, and I. G. B. Indrawan, "Engineering Geology Investigation for Slope Geometry Design of Diversion Tunnel Portal in Bener Dam Purworejo," Proceeding of SLOPE,pp.1–8,2019,[Online].Available: https://www.researchgate.net/profile/Daru_Sasangka/publication/343788725_ENGINEERING_GEOLOGY_INVESTIGATION_FOR_SLOPE_GEOMETRY_DESIGN_OF_DIVERSION_TUNNEL_PORTAL_IN_BENER_DAM_PURWOREJO/links/5f3f9e13299bf13404da6adb/ENGINEERING-GEOLOGY-INVESTIGATION-FOR-SLO.

[5] W. Rahardjo and R. H. M. D. Sukandarrumidi, "Peta Geologi Lembar Yogyakarta, Jawa, skala 1: 100.000," Pus. Penelit. dan Pengemb. Geol. Bandung, 1995.

[6] PT. Virama Karya, "Geological and Hydrological Data Bore Hole Bendungan Bener.pdf." 2015, (unpublished).

[7] D. C. Wyllie and C. W. Mah, "Rock slope engineering: Civil and mining, 4th edition," Rock Slope Eng. Fourth Ed., pp. 1–432, 2017, doi: 10.1201/9781315274980.

[8] R. Fell, P. MacGregor, D. Stapledon, and G. Bell, Geotechnical Engineering of Dams. 2005.

[9] J. S. Griffiths, "Engineering geological mapping," Geol. Soc. London, Eng. Geol. Spec. Publ., vol. **18**, no. 1, pp. 39–42, 2001.

[10] E. Hoek, C. Carranza-Torres, and B. Corkum, "Hoek-Brown failure criterion-2002 edition," Proc. NARMS-Tac, vol. **1**, no. 1, pp. 267–273, 2002.

[11] Z. T. Bieniawski, Engineering rock mass classifications: a complete manual for engineers and geologists in mining, civil, and petroleum engineering. John Wiley & Sons, 1989.

[12] E. Hoek and E. T. Brown, "The Hoek–Brown failure criterion and GSI – 2018 edition," J. Rock Mech. Geotech. Eng., vol. 11, no. 3, pp. 445–463, 2019, doi: 10.1016/j.jrmge.2018.08.001.

[13] R. Chowdhury, P. Flentje, and G. Bhattacharya, Geotechnical slope analysis. 2009.

[14] M. E. Hynes Griffin and A. G. Franklin, "Rationalizing the seismic," p. WES MP GL-84-13, 1984, doi: 10.1002/oby.20709.

[15] Badan Standarisasi Nasional, "Sni 8460-2017," Persyaratan Peranc. Geotek., vol. 8460, p. 2017, 2017.

IOP Conf. Series: Earth and Environmental Science 1233 (2023) 012015 doi:10.1088/1755-1315/1233/1/012015

The Effects of Palu-Koro Fault on Predicting Ground Motion for Estimating Liquefaction-Induced Settlement of the Educational Infrastructure Reconstruction Project at the University of Tadulako

H R Pasin[1,2], S Ismanti[1*] and H C Hardiyatmo[1]

[1] Department of Civil and Enviromental Engineering, Faculty of Engineering, Universitas Gadjah Mada Yogyakarta, Indonesia
[2] Directorate General of Human Settlement, Ministry of Public Works and Housing, Republic of Indonesia.

*Corresponding author: sito.ismanti@ugm.ac.id

Abstract. The University of Tadulako suffered significant damage after a tectonic earthquake with a magnitude of M 7.5 with a depth of 20 km that occurred on September 28, 2018. This incident happened because of its location very close to the Palu-Koro fault. The Ministry of Public Works and Public Housing was assigned to reconstruct damaged educational infrastructure there. Seismic hazard analysis was needed to prepare reconstruction development projects and mitigation efforts. Many studies are related to predicting ground motion due to seismic fault activity, one of which is the attenuation NGA West 2 models. This study aims to determine the relationship between fault effects and ground motion to estimate liquefaction-induced settlement at this project. The results show that the relationship between the fault effects and ground motion affects various PGA values. PGA value resulting from the CY14 model was used to estimate settlement because it considers the worst condition aspect, where the most significant settlement occurred at the drill point DP-05 of 10.234 cm. This settlement can happen because the liquefied soil layer at the drill point was quite deep compared to other drill points.

1. Introduction

Central Sulawesi is located between the three earth's tectonic plates, such as the Indo-Australian Plate, the Pacific Plate, and the Eurasian Plate, which converge [1] [2]. Therefore, this region is very vulnerable to natural disaster events, especially those caused by tectonic movement, which causes a shift in the Palu-Koro fault and other identified faults in this region [1]. The Palu-Koro Fault is the main structure of the island of Sulawesi, which extends from Central Sulawesi to the Karimata Strait [3] with a length of more than 300 km [2]. These faults are always moving and colliding with each other, which resulted in such a large fault, and was followed by movements on the earth's surface [4]. The latest research estimates the Palu-Koro Fault is actively moving at 41-45 mm/year [5].

One of the consequences of the movement of the Palu-Koro fault is an earthquake. Earthquakes caused by fault movements are also known as tectonic earthquakes [6] [7]. Tectonic earthquakes are the most common natural disaster phenomena in Palu City [1]. Tectonic earthquakes are generally the largest compared to other types of earthquakes in terms of their causes [8]. One of the recorded tectonic earthquakes originating came from the Palu-Koro fault was a Palu earthquake event on September 28, 2018, measuring magnitude M 7.5 with a depth of 20 meters, which devastated the region of Palu, Sigi, and Donggala [9]. Pawirodikromo (2012) defines earthquakes as shaking the ground surface due to the sudden release of energy as a result of the shifting of rock masses in the earth's crust so that the ground surface vibrates due to earthquake energy that propagates from the epicenter of the earthquake in all directions [8]. In seismic hazard mitigation, the magnitude and distance of the earthquake greatly affect the amplitude of ground motion [10]. Earthquake energy waves that have passed through the medium far enough (distant epicenter distance) will have a relatively low frequency and relatively long duration of vibration, so the effect of the epicenter distance (from the earthquake source to the site) will affect the damage to buildings that occur [8]. Therefore, seismic hazard analysis needs to understand the characteristics of earthquakes and their possible impacts [11]. There are many studies on seismic hazard assessment related to ground motion prediction equation, one of which is the attenuation models developed by the Pacific Earthquake Engineering Research (PEER) Institute, namely the Next Generation Attenuation models project, which is used to predict ground motion due to seismic fault activity in the western region The United States of America [12].

This earthquake also resulted in the hazard of liquefaction following in several locations such as Balaroa, Petobo, Jono Oge, Lolu Village, and Sibalaya [13]. Liquefaction is an event of changes in soil conditions from solid-state to liquid [14]. The liquefaction hazard is a phenomenon triggered by an earthquake [15], shallow groundwater depth [16], and loose sand and silt grains [17]. The effects of liquefaction are sand boiling [11] [18] [19], lateral spreading, ground oscillation [20], loss of soil bearing capacity [20] [21], ground settlement [18] [21] [22], flow liquefaction and cyclic mobility [11] [23]. The effects of the liquefaction need to be taken into account because liquefaction can have many effects on the soil, especially if buildings are standing on it that will collapse, causing soil liquefaction-induced ground settlement [18] [21] [22]. Besides, areas experiencing earthquake-induced soil liquefaction events can be repeated if the parameter of the earthquake that occurs is equal to or higher than the Palu earthquake event in 2018 [1]. The case study of the accident at the Fukushima Daiichi reactor was caused by the massive Japanese earthquake on March 11, 2011, with a magnitude M 9.0 earthquake that was centered in the Pacific Ocean about 80 kilometers east of the city of Sendai. This reactor was built to withstand the worst earthquake that has ever occurred in history in that region. The 1995 Kobe earthquake with a magnitude M 7.5 and its construction did not envisage a worse situation as it had never happened before. This is a serious warning against overconfidence and underestimating seismic hazard predictions, despite Japan's investment in seismology [24]. If the disaster risk is not calculated properly, it will damage public facilities a lot and even cause many fatalities and massive infrastructure damage [25] [26] [27].

The University of Tadulako was chosen as the study area because its location is very close to the Palu-Koro fault. This area is very vulnerable to geological disasters such as earthquakes and liquefaction. During the 2018 earthquake, the condition of the University of Tadulako suffered significant damage. Based on the triggering parameters in the soil investigation report by the geotechnical consultant, soil liquefaction was highly possible in the study area [28]. The Ministry of Public Works and Public Housing was assigned to accelerate the reconstruction as a restoration of the function of education in Central Sulawesi Province, especially in Palu City [29]. This study aims to determine the performance of ground motion if the study area is close to the fault site as mitigation efforts in designing foundations for earthquake-resistant structures in the educational infrastructure reconstruction project at the University of Tadulako, Palu City, Central Sulawesi. Besides, mapping which areas have the potential for settlement due to liquefaction helps planning for the preparation of reconstruction development projects.

2. Research Methodology

2.1. Data Collection and Description of Study Area

The study area was at the University of Tadulako, Tondo Village, Mantikulore District, Palu City, Central Sulawesi, Indonesia, as seen in Figure 1. Soil investigations were carried out using 14 drill points as a result of the Standard Penetration Test (SPT) test belonging to a geotechnical planning consultant scattered in the study area, as seen in Figure 2. Based on the triggering factors, soil liquefaction is possible to happen at some drill points in this area, as presented in Table 1 [28].

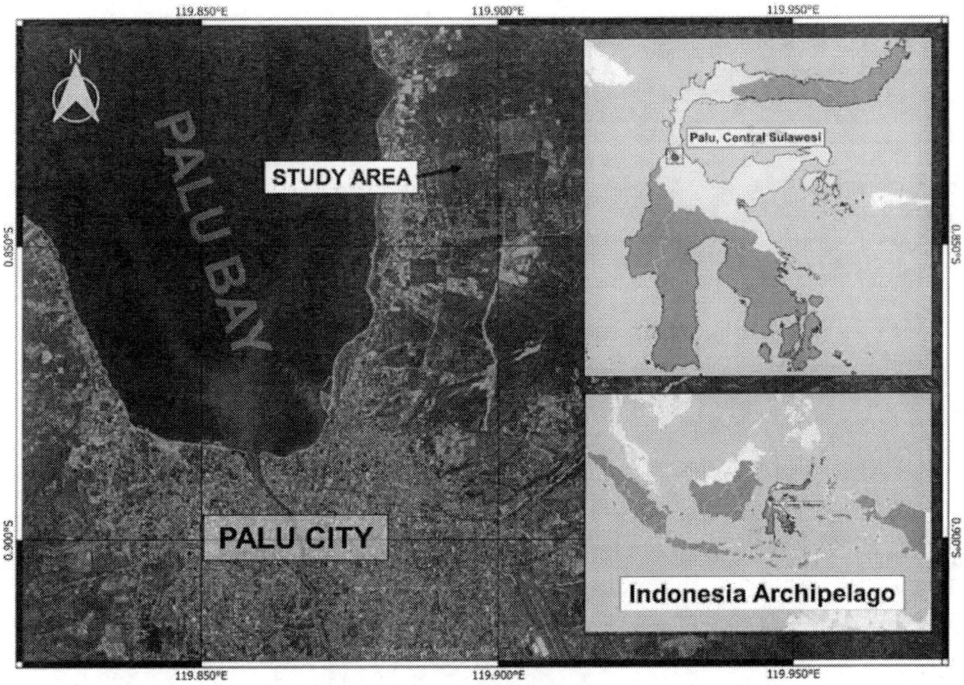

Figure 1. Map of the Study Area

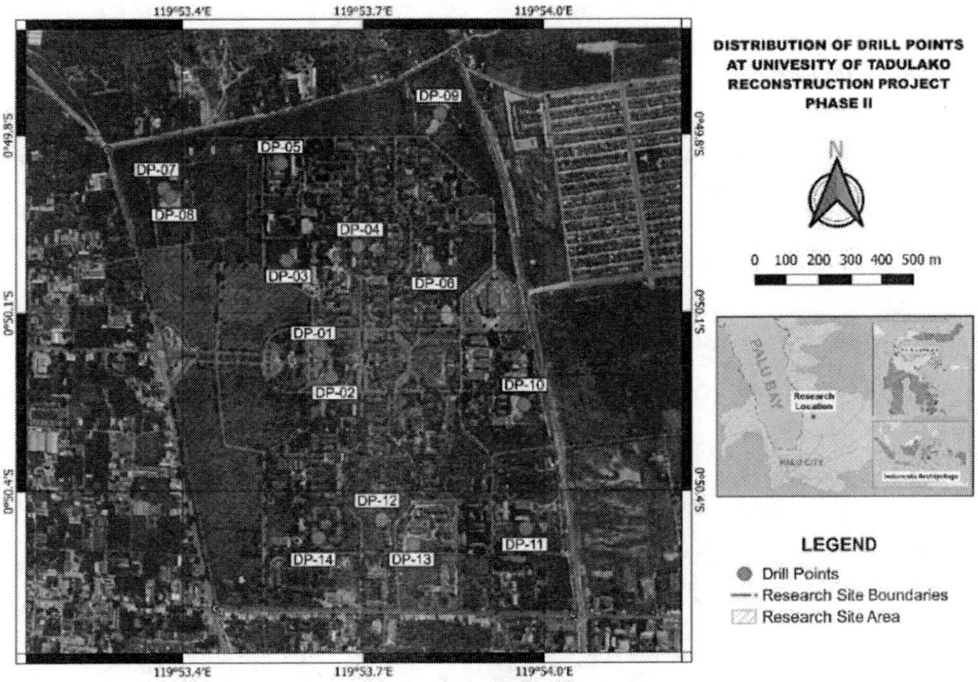

Figure 2. Distribution of Drill Points in Reasearch Location

Table 1. Data Collection and Description of Study Area Based on Soil Investigation Report [28]

Drill Points	Reconstruction Building	The trigger soil liquefaction factors				Preliminary Investigation for Liquefaction Potential
		Earthquake Magnitude	GWL (meter)	Soil Type (USCS)	Dominant Soils	
DP-01	Rectorate	M 7.5	3 meters	SC.	Clayey Sand	No Potential
DP-02	Rectorate		3 meters	SM.	Silty Sand	Possible
DP-03	Dean of Social and Political Science		3 meters	SC.	Clayey Sand	No Potential
DP-04	College of Teacher Training and Education A51		3 meters	SC.	Clayey Sand	No Potential
DP-05	Dean of Teacher Training and Education		3 meters	SM.	Silty Sand	Possible
DP-06	Dean & College of Law		3 meters	SM.	Silty Sand	Possible
DP-07	Physical Education, Health and Recreation Study		3 meters	CL	Sandy Silt	No Potential
DP-08	Physical Education, Health and Recreation Study		3 meters	CL	Sandy Silt	No Potential
DP-09	Dormitory		3 meters	CL	Sandy Silt	No Potential
DP-10	Forestry		3 meters	SM.	Silty Sand	Possible
DP-11	Animal Husbandry and Fisheries		3 meters	SM.	Silty Sand	Possible
DP-12	Public Health		3 meters	SM.	Silty Sand	Possible
DP-13	Medical		3 meters	SM.	Silty Sand	Possible
DP-14	Architectural Design and Theory Laboratory		3 meters	SM.	Silty Sand	Possible

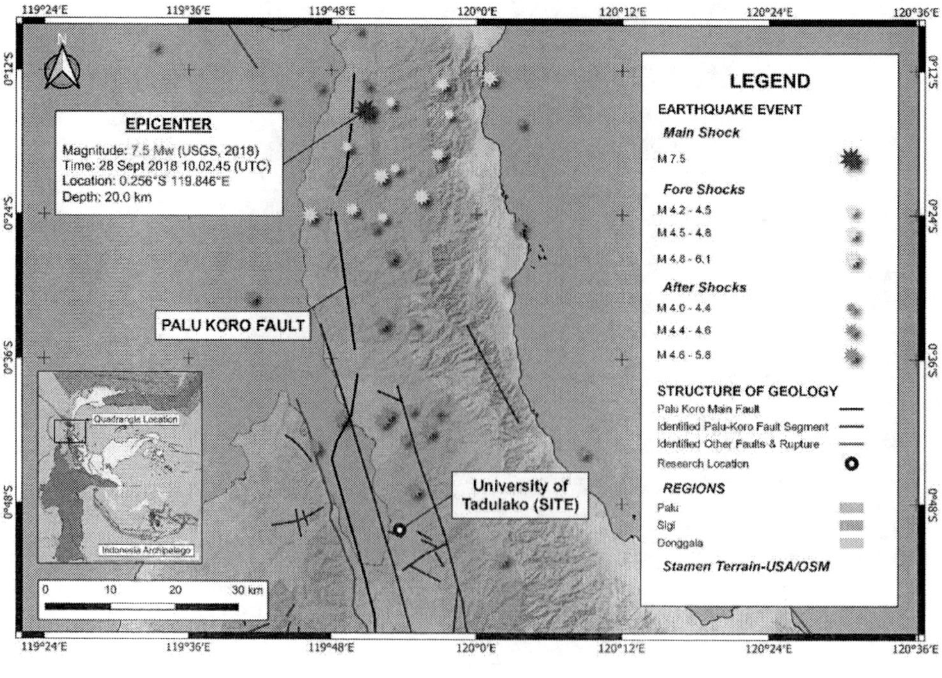

Figure 3. An Earthquake Event Occurred in the Region of Palu, Sigi, and Donggala in 2018

IOP Conf. Series: Earth and Environmental Science 1233 (2023) 012015 doi:10.1088/1755-1315/1233/1/012015

The University of Tadulako is located close to the Palu-Koro fault and several other identified faults. Besides, the site is close to the earthquake epicenter (Latitude: -0.256 and Longitude: 119.846), with a depth of 20 km and an average distance of 64,681 km to the site, and the hypocenter is 67.703 km. The earthquake that occurred on September 28, 2018, devastated the region of Palu, Sigi, and Donggala. This earthquake was caused by Palu-Koro fault movement with the main shock of M 7.5. It was followed by several earthquake events, with foreshocks ranging from M 4.2 to M 6.1 and aftershocks ranging from M 4.0 to M 5.8. [9]. The earthquake event occurs in the region of Palu, Sigi, and Donggala in 2018 can be seen in **Figure 3**.

Based on interviews with earthquake observers at Balaroa geophysical station (BMKG) in Palu city, the depth of fracture (Z_{TOR}) is 15 km. The average close distance between the site to the surface projection of the Palu-Koro fault (R_{JB}) is 7.128 km. The average distance to the rupture (R_{RUP}) is 16.61 km. The difference between the minimum to maximum distance is 0.885 km for R_{JB} and 0.380 km for R_{RUP}. Distance from the site to the Palu-Koro fault and earthquake source are presented in **Table 5**, and the distribution of drill points in the research location can be seen in **Figure 2**.

2.2. *Ground Motion Prediction Equation using Attenuation NGA West 2 Models*

The Next Generation Attenuation for the Western part of the United States (The NGA West) was a coordinated project by the Pacific Earthquake Engineering Research Center (PEER) and collaboration with the US Geological Survey (USGS) and Southern California Earthquake Center for the development of the ground motion prediction equation (GMPE) of the new attenuation models in the active tectonic region for shallow crustal seismic sources or earthquake sources originating from faults and shallow backgrounds obtained from statistical regression analysis which was developed in 2008 [3] [8] [12]. This project was later updated in 2014, known as The NGA West 2. There are three GMPE PEER NGA West 2 used in the development of 2017 Indonesia's national earthquake hazard map, especially for the shallow crustal seismic sources model, namely, Boore-Atkinson (2014) or BSSA14, Campbell-Bozorgnia (2014) or CB14, and Chiou-Youngs (2014) or CY14 [3]. Those attenuation models have been applied in other regions of the world, such as a study conducted by Mase (2019), to predict ground motion during the 1995 Kobe strong earthquake with magnitude M 6.9, which was caused by the tectonic activity of the Awaji Fault in Japan region, which this fault has the same category of sense mechanism as the Palu-Koro fault, namely the strike-slip fault [12].

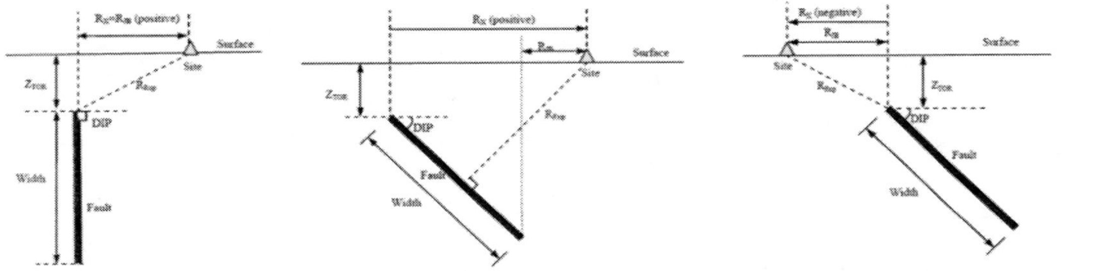

(a) Strike slip faulting (b) Reverse or normal faulting, hanging-wall site (c) Reverse or normal faulting, foot-wall site

Figure 4. Type of Fault Based on the Relative Displacement of the Moving Block Along with Dimensional Features [30]

In general, the NGA model is used to estimate the level of ground motion due to uncertainty in an area where a strong earthquake occurs by linking the diversity of parameters and factors into the NGA's equation. Those have a significant influence on the results of the analysis, in which the data required is quite complex, and accuracy is required in the analysis, to provide more accurate and realistic calculation results. The simpler the parameters used in the GMPE equation, the easier it is to use, but the analysis results obtained are less accurate and less realistic on the factual values in the field [3] [12]. The parameters calculated in the NGA attenuation include medium-high earthquakes at close range, earthquake distance, rupture directivity, type of fault based on the relative displacement of the moving block along with dimensional features (hanging wall, footing wall, and dip angle) as shown in **Figure 4**, style of faulting (strike-slip, reverse slip, normal), depth to faulting (surface rupture, buried rupture), static stress drop/rupture area, site amplification relative to rock condition, 3-D amplification basin and depth [8].

2.2.1. *Boore and Atkinson (2014)*

Boore and Atkinson (2014) is a ground motion prediction equation (GMPE) developed as part of the NGA West 2 project (Bozorgnia *et al.*, 2014), with the main variable controlling ground motion at the site being the moment of earthquake magnitude M which can be applied to M 3.0 to M 8.5 earthquakes, for fault types classified as strike-slip (SS), and reverse slip (RS), and M 3.0 to M 7.0 earthquakes for fault types with a normal slip classification. (NS), with the distance to the fault, R_{JB} from 0 to 400 km, the mean time shear wave velocity is above 30 meters from the V_{S30} soil profile which ranges from 150 m/s to 1500 m/s, and the spectral period T from 0.01 to 10 seconds. The depth of the groundwater basin, z_1 ranges from 0 km to 3.0 km. The type of earthquake is class 1 (CL1: main shocks) and class II earthquake (CL2: aftershocks) [31]. Boore and Atkinson's (2014) attenuation equation can be seen in **Equation (1):**

$$\ln Y = F_E(\mathbf{M}, \text{mech}) + F_P(R_{JB.}, \mathbf{M}, \text{region}) + F_S(V_{S30}, R_{JB.}, \mathbf{M}, \text{mech}, z_1) + \varepsilon_n \sigma(\mathbf{M}, R_{JB.}, V_{S30}) \quad (1)$$

where $ln\, Y$ represents the natural logarithm of the IM ground motion (PGA, PGV, or PSA), F_E, F_P, and F_S represent functions for the effect source ("E" for "event"), path ("P"), and site ("S"), respectively; $\varepsilon_n$is the fractional standard deviation of the single predicted value $ln\, Y$ away from the mean (for example, ε_n= -1.5 is 1.5 standard deviations less than the mean), and σ is the total standard deviation of the model. The predictor variables were M, mech, R_{JB} (in km), region, V_{S30} (in m/s), and z_1 (in km). Mech parameters = 0, 1, 2, and 3 for unspecified, SS, NS, and RS, respectively. The region parameter is 0 if no regional correction is desired. The default value is set, 1 for California, New Zealand, and Taiwan (this also provides no correction); 2 for China and Turkey; and 3 for Italy and Japan. The PGA and PSA units are g, and the PGV units are cm/s.

2.2.2. *Campbell and Bozorgnia (2014)*

Campbell and Bozorgnia (2014) developed the PEER NGA West 2 ground motion prediction equation (GMPEs) to estimate the PGA, PGV, and PSA values in the period T = 0.001 – 10 seconds from earthquakes in shallow crustal tectonic areas active with a magnitude ranging from M 3.0 – M 7.9 and the closest distance to the R_{RUP} fault plane ranges from 0 to 500 km [32]. Campbell and Bozorgia's (2014) attenuation equation can be seen in **Equation (2)**;

$$\ln Y = f_{mag} + f_{dis} + f_{flt} + f_{hng} + f_{site} + f_{sed} + f_{hyp} + f_{dip} + f_{atn} + \eta_i + \varepsilon_{ij} \quad (2)$$

where Y is a measure of the intensity of the ground motion under consideration, and the function (sq-f) is a fixed model of the effect that represents the scale of ground motion on earthquake magnitude, attenuation geometry, fault force, hanging-wall geometry, shallow location response, shallow and deep basin response, depth hypocentre fault slope, and elastic damping. Random effects and random error terms, η_i and ε_{ij}, representing the distribution of the residuals between events for the event i and the in-event residuals for site j of event i, respectively. The between-event and in-event residues are normally distributed with a mean of zero and standard deviations τ and $\emptyset$, respectively. The total number of standard deviations σ is the square root of the sum of the squares of τ and $\emptyset$. Due to the limitation of the number of paper pages, the functional form of these terms is not presented. Instead, the reader is referred to the detailed documentation of functional forms given in Campbell and Bozorgnia (2014). CB14 may be used to estimate ground motion for M = 3.3–7.5 or 8.5, depending on the shear force,

$R_{RUP} < 300$ km, and V_{S30} = 150 - 1500 m/s, where V_{S30} is the average shear wave velocity average time in the top 30 m of a site. Companion GMPE for the vertical component and the vertical-to-horizontal (V/H) ratio is given in Bozorgnia and Campbell, 2015 a, and 2015 b).

2.2.3. *Chiou and Youngs (2014)*

Chiou and Youngs (2014) is an updated model of the NGA 2008 horizontal ground motion prediction equation (GMPE) caused by shallow crustal earthquakes that occur in active plate areas. This equation can be applied to earthquake conditions of M 3.5 to M 8.5 for strike-slip earthquakes and M 3.5 to 8.0 for earthquakes on the reverse and normal faults. The depth to the top of the rupture or Z_{TOR} is less than equal to 20 km. The collapse distance of R_{RUP} 0 to 300 km, and shear wave velocity above 30 meters, V_{S30} ranges from 180 m/s to 1,500 m/s [33]. **Equation (3)** is an equation for predicting ground motion developed by Chiou and Youngs (2014), which is as follows:

$$\ln\left(y_{ij}\right) = \ln(y_{ref_{ij}}) + \eta_i + \emptyset_1 \min\left(\ln\left(\frac{V_{S30j}}{1130}\right), 0\right) + \emptyset_2\left(e^{\emptyset_3(\min(V_{S30j},1130)-360)} - e^{\emptyset_3(1130-360)}\right) \ln\left(\frac{y_{ref_{ij}} e^{\eta_i} + \emptyset_4}{\emptyset_4}\right) + \emptyset_5\left(1 - e^{-\frac{\Delta Z_{1.0j}}{\emptyset_6}}\right) + \varepsilon_{ij} \quad (3)$$

where the variable y_{ij} in the equation is the amplification of earthquake i at station j. The variable $y_{ref_{ij}}$ is the median population for the reference condition V_{S30}= 1130 m/s. The random variables η_i (inter-event residual, or event term) and ε_{ij} (the remainder in occurrence) the equation represents two modeling errors contributing to the aleatory variability of the predicted motion.

2.3. *Estimating Earthquake-Induced Soil Liquefaction*

Surface waves have large energy which affects the peak ground acceleration due to earthquakes. Surface waves have a shear effect that causes the ground surface to move horizontally, both parallel and perpendicular to the direction of propagation. This wave causes damage. The damage caused by the earthquake can be in the form of damage to building structures that stand on the ground and damage to the surface or within the soil itself. One of the effects of soil damage that occurs such as soil liquefaction. If liquefaction occurs, the soil structure will be damaged so that the soil layer does not have the bearing capacity of the soil. As a result, ground settlement, soil surface cracks, the release of fine slurry to the soil surface, and loss of soil frictional force against deep foundations can cause the foundation of the building structure to overturn [8]. For this reason, it is necessary to estimate the earthquake-induced soil liquefaction.

2.3.1. *The Factor of Safety Against Soil Liquefaction*

Many methods are used in estimating earthquakes-induced soil liquefaction, one of which is using the SPT data procedure method developed by Idriss and Boulanger [18]. To estimate the factor of safety against soil liquefaction, it can be formulated as in the **Equation (4)**, which is as follows:

$$FS = \frac{CRR_{M, \sigma'_{vc}}}{CSR} \quad (4)$$

where FS. is a factor of safety against liquefaction, $CRR_{M, \sigma'_{vc}}$ is Cyclic Resistance Ration (CRR) value with Overburden Effective Stress and Magnitude Scaling Factor (MSF) correction factor, CSR is Cyclic Stress Ratio value.

2.3.2. *Soil Liquefaction-Induced Ground Settlement*

Ground settlement is the effect of soil liquefaction. One of the methods used to estimate the soil liquefaction-induced ground settlement is the method developed by Zhang *et al.* (2002) [22] by using

Cone Penetration Test (CPT) data which is corrected for fines content $\left((qc_{1N})_{cs}\right)$. In this study, the data used were Standard Penetration Test (N-SPT) data which corrected for fine grain content ($(N_1)_{60cs}$). Andrus *et al.* (2003) [34] proposed the conversion of the value $(N_1)_{60cs}$ to the value $(qc_{1N})_{cs}$ as defined in **Equations (5)** and **(6)** as follows:

$$(N_1)_{60cs}=0,488[(qc_{1N})_{cs}]^{0,779} \tag{5}$$

$$(qc_{1N})_{cs}=\left(\frac{(N_1)_{60cs}}{0,488}\right)^{\frac{1}{0,779}} \tag{6}$$

where $(N_1)_{60cs}$ is the fines content corrected N-SPT, and $(qc_{1N})_{cs}$ is the end resistance value of the fines content corrected CPT. The empirical equation for volumetric strain (ε_V) due to liquefaction in percent (%) can be seen in **Table 2**. for various factors of safety.

Table 2. Empirical Equation of Volumetric Strain due to Liquefaction [22].

Factor of Safety (FS.)	Corrected CPT end-resistance fines content $(qc_{1N})_{cs}$	Volumetric Strain ε_V (%)
≤ 0,5	$33 \le (qc_{1N})_{cs} \le 200$	$102(qc_{1N})_{cs}^{-0,82}$
0,6	$33 \le (qc_{1N})_{cs} \le 147$	$102(qc_{1N})_{cs}^{-0,82}$
	$147 \le (qc_{1N})_{cs} \le 200$	$2411(qc_{1N})_{cs}^{-1,45}$
0,7	$33 \le (qc_{1N})_{cs} \le 110$	$102(qc_{1N})_{cs}^{-0,82}$
	$110 \le (qc_{1N})_{cs} \le 200$	$1701(qc_{1N})_{cs}^{-1,42}$
0,8	$33 \le (qc_{1N})_{cs} \le 80$	$102(qc_{1N})_{cs}^{-0,82}$
	$80 \le (qc_{1N})_{cs} \le 200$	$1690(qc_{1N})_{cs}^{-1,46}$
0,9	$33 \le (qc_{1N})_{cs} \le 60$	$102(qc_{1N})_{cs}^{-0,82}$
	$33 \le (qc_{1N})_{cs} \le 200$	$1430(qc_{1N})_{cs}^{-1,48}$
1,0	$33 \le (qc_{1N})_{cs} \le 200$	$64(qc_{1N})_{cs}^{-0,93}$
1,1	$33 \le (qc_{1N})_{cs} \le 200$	$11(qc_{1N})_{cs}^{-0,65}$
1,2	$33 \le (qc_{1N})_{cs} \le 200$	$9,7(qc_{1N})_{cs}^{-0,69}$
1,3	$33 \le (qc_{1N})_{cs} \le 200$	$7,6(qc_{1N})_{cs}^{-0,71}$
2,0	$33 \le (qc_{1N})_{cs} \le 200$	0

Determination of the estimation of soil liquefaction-induced ground settlement is obtained from the sum of the total volumetric strain of each soil layer as defined in **Equation (7)**, which is as follows:

$$S=\sum_{i=1}^{n} \varepsilon_{Vi}\, \Delta z_i \tag{7}$$

where S is soil liquefaction-induced ground settlement (cm), ε_V is a volumetric strain (%), Δz is soil thickness per layer (m). i is the soil layer. Categories of soil liquefaction-induced ground settlement composed by Ishihara *et al.* (1992) [35] are shown in **Table 3**.

Table 3. Categories of Soil Liquefaction-Induced Ground Settlement [35].

Settlement Thickness (cm)	Settlement Categories	Phenomena at Ground Surface Level
0 - 10	Low to No Damage	Minor crack
10 – 30	Medium	Small cracks, streams of fine sand come to the surface
30 - 70	Very High	Massive cracks, sands spouting, large offsets, and lateral displacement

3. Results and Discussion

3.1. Predicting Ground Motion using Attenuation NGA Models

This study used several analytical approaches, consisting of a qualitative approach and a quantitative approach. The following are some approaches to predicting ground motion using attenuation NGA models; in this case, it is NGA West 2.

3.1.1. Qualitative Approach for Predicting Ground Motion using Attenuation NGA Models

The qualitative approach used for ground motion prediction analysis with the attenuation NGA West 2 models, which aims to provide an overview term with site condition with several threshold parameters such as site condition, the magnitude, the distance, and the fault effect are considered in Boore and Atkinson (2014) model, Cambell and Bozorgnia (2014) model, and Chiou and Youngs (2014) model [12], which presented in **Table 4**.

Table 4. Qualitative Approach for Predicting Ground Motion

Parameters	Models Applicability	The GMPE PEER NGA West 2 Model for Shallow Crustal		
		Boore and Atkinson (2014)	**Campbell and Bozorgnia (2014)**	**Chiou and Youngs (2014)**
Earthquake Source	Term / Threshold	Strike-Slip, Reverse, Normal	Strike-Slip, Reverse, Normal	Strike-Slip, Reverse, Normal
	Site Condition		Strike-Slip	
	Conclusion	OK	OK	OK
Moment Magnitude (M)	Term / Threshold	M 3 - 8.5 for Strike-slip and Reverse-slip; M 3 - 7 for Normal-slip	M 3.5 - 8.5 for strike-slip; 3.5-8.0 for Reverse dan Normal Slip	M 3.5 - 8.5 for Strike-slip; 3.5 - 8.0 for Reverse and Normal faulting
	Site Condition		M 7.5	
	Conclusion	OK	OK	OK
Distance to Surface Projection, R_{JB} (km)	Term / Threshold	0 - 400 km	0 - 300 km	0 - 300 km
	Site Condition		6.689 to 7.574 km	
	Conclusion	OK	OK	OK
The Average Time of Shear Wave Velocity to a depth of 30 meters, V_{S30} (m/s)	Term / Threshold	150 - 1500 m/s	150 - 1500 m/s	180 - 1500 m/s
	Site Condition	300 m/s	300 m/s	300 m/s
	Conclusion	OK	OK	OK
Distance to the Rupture, R_{RUP} (km)	Term / Threshold	0 - 500 km	< 300 km	0 - 300 km
	Site Condition		16.424 to 16.804 km	
	Conclusion	OK	OK	OK
Depth to fracture, Z_{TOR} (km)	Term / Threshold	Not Considered	< 300 km	0 - 300 km
	Site Condition		15 km	
	Conclusion	OK	OK	OK
Terms with Site Conditions		**OK**	**OK**	**OK**

Based on the results of the qualitative analysis shown in **Table 4**, it was found that the Palu-Koro fault conditions met the requirements specified by the NGA West 2 model, so these three models were suitable for site conditions, and all of them could be used for further analysis.

3.1.2. Quantitative Analysis for Predicting Ground Motion using Attenuation NGA Models
In this study, Microsoft Excel Program by PEER [30] was used for the NGA West 2 analysis with input parameters, the type of fault is M 7.5 [9]. The average time of shear wave velocity up to a depth of 30 meters at the site location V_{S30} is 300 m/s [36]. **Figure 4.** shows the type of fault based on the relative displacement of the moving block along with dimensional features, of which the Palu-Koro fault's sense mechanism is the strike-slip fault (a) with inclined fractures and dip-slip fault 90° [3].

Table 5. Quantitative Approach for Predicting Ground Motion

Distance Effects			Site Condition				PGA Value Resulting					
Drill Points	Distance site to fault (km)		Site Class				PGA Bedrock			PGAM Surface		
	RJB	RRUP	Cate-gories	FPGA BSSA14	FPGA CB14	FPGA CY14	BSSA 14	CB14	CY14	BSSA 14	CB14	CY14
DP-01	7.015	16.559	SD	1.175	1.359	1.138	0.425	0.241	0.462	0.499	0.327	0.526
DP-02	7.026	16.564	SD	1.176	1.359	1.138	0.424	0.241	0.462	0.499	0.327	0.526
DP-03	6.996	16.551	SD	1.175	1.359	1.138	0.425	0.241	0.462	0.499	0.327	0.526
DP-04	7.216	16.645	SD	1.180	1.360	1.140	0.420	0.240	0.460	0.496	0.326	0.525
DP-05	7.059	16.578	SD	1.176	1.359	1.139	0.424	0.241	0.461	0.498	0.327	0.525
DP-06	7.423	16.736	SD	1.184	1.361	1.141	0.416	0.239	0.459	0.492	0.326	0.524
DP-07	6.689	16.424	SD	1.168	1.358	1.136	0.432	0.242	0.464	0.505	0.329	0.527
DP-08	6.702	16.429	SD	1.168	1.358	1.136	0.432	0.242	0.464	0.504	0.329	0.527
DP-09	7.572	16.803	SD	1.187	1.361	1.142	0.413	0.239	0.458	0.490	0.325	0.523
DP-10	7.574	16.804	SD	1.187	1.361	1.142	0.413	0.239	0.458	0.490	0.325	0.523
DP-11	7.480	16.762	SD	1.185	1.361	1.142	0.415	0.239	0.458	0.491	0.325	0.523
DP-12	7.056	16.577	SD	1.176	1.359	1.138	0.424	0.241	0.462	0.498	0.327	0.525
DP-13	7.132	16.609	SD	1.178	1.360	1.139	0.422	0.240	0.461	0.497	0.327	0.525
DP-14	6.860	16.494	SC	1.200	1.200	1.200	0.428	0.241	0.463	0.514	0.290	0.556
Min.	**6.689**	**16.424**	**Min.**	**1.168**	**1.200**	**1.136**	**0.413**	**0.239**	**0.458**	**0.490**	**0.290**	**0.523**
Max.	**7.574**	**16.804**	**Max.**	**1.200**	**1.361**	**1.200**	**0.432**	**0.242**	**0.464**	**0.514**	**0.329**	**0.556**
Avg.	**7.128**	**16.610**	**Avg.**	**1.180**	**1.348**	**1.144**	**0.422**	**0.240**	**0.461**	**0.498**	**0.324**	**0.527**
Dif.	**0.885**	**0.380**	**Difference Min. to Max**				**0.020**	**0.003**	**0.007**	**0.024**	**0.039**	**0.033**

Table 5. was the result of quantitative analysis for predicting ground motion using attenuation NGA West 2 models, where for the site class dominated by medium soil (SD), there was one drill point with hard soil site class (SC) [37]. Peak Ground Acceleration (PGA) calculation results at bedrock with a deterministic method. The attenuation NGA West 2 models are obtained from 0.239 to 0.464. According to Idriss and Boulanger (2008) [18], PGA required in potential liquefaction analysis is PGA at the surface. PGA at bedrock is then modified to PGA at the surface (PGA_M) by multiplying the site class factor (F_{PGA}) as given in the Indonesian Seismic Code, SNI 1726:2019, for each drill point [37]. Hence, the PGA at surface values ranges from 0.290 to 0.556.

To accommodate the worst situation that never happened before, such as the case study of the Fukushima Daiichi nuclear reactor disaster [24], the highest value of PGA at the surface was chosen, namely Chiou and Young's NGA West 2 model (CY14) for soil liquefaction-induced ground settlement analysis, which can be seen in **Table 5**.

3.1.3. Relationship Between Fault Effects and Ground Motion

To determine fault effects on ground motion performance, it can be done by comparing several condition parameters, such as the fault type classification (strike-slip, reverse-slip, normal slip), distance effects (distance to surface projection (R_{JB}), distance to the rupture (R_{RUP})), site condition effects (soil type, site class) on the resulting PGA value.

PGA at bedrock was strongly influenced by distance from the site to fault, because waves that have passed through a distant medium will have a relatively low-frequency content and a relatively long vibration duration. This is related to the amplification and filtering effect. The closer the distance, the greater the PGA value [8], whereas at the drill point DP-07, the distance between the site and the fault was very close. **Table 5** shows the minimum value for the distance site to the fault found in DP-07, which was 6.689 km for R_{JB}, and 16.424 km for R_{RUP} and resulting in the maximum value of the PGA at bedrock, which was 0.432 g for BSSA14, 0.242 g for CB14, and 0.464 g for CY14. If the difference between the minimum and maximum distances was taken into account, the resulting PGA at bedrock was shown in **Table 5** that the difference in the R_{JB} distance of 0.885 km produces 0.02 g for BSSA14, 0.003 g for CB14, and 0.007 g for CY14.

Meanwhile, PGA at the surface is strongly influenced by site effect conditions. Site effect condition is the effect of soil properties on ground motion due to an earthquake. It is related to ground motion amplification. Amplification is the increase in the response of the soil (acceleration, velocity, or deviation) to soil and sediment properties. The looser the material, the greater the amplification, the greater the soil and structure damage that occurs, and the larger the amplification factor or site class coefficient (F_{PGA}) because the soil can absorb energy and vibrations when an earthquake occurs. This is related to the duration of the earthquake (surface waves), which becomes longer duration at longer distances [8].

The Indonesian Seismic Code (SNI 1726:2019) classifies five types of site classes, starting from the hardest to the loosest material, which include hard rock (SA), rock (SB), hard soil (SC), medium soil (SD), soft soil (SE), and special soil (SF). However, in this code, the F_{PGA} gets smaller as the PGA gets bigger, especially in on-site class SC and SD, where the F_{PGA} value was influenced by PGA at bedrock, and also PGA at bedrock value was influenced by distance. This was what caused the varying the PGA at surface values because of various distances and site effect conditions [37].

3.2. Estimating Soil Liquefaction-Induced Ground Settlement

After predicting ground motion using attenuation NGA West 2 models, the PGA value was used for further analysis, namely estimating soil liquefaction-induced ground settlement at each drill point. To determine the soil liquefaction-induced ground settlement, the factor of safety against soil liquefaction (FS) parameter was needed to find the total volumetric strain. The FS parameter was obtained from the SPT data procedure method developed by Idriss and Boulanger (2008) [18], where this analysis can determine the depth prone to liquefaction, as shown in **Table 6**.

Based on the results of the soil liquefaction-induced ground settlements analysis developed by Zhang *et al.* (2002) [22], as shown in **Table 6**, the largest settlement value was obtained at the drill point DP-05 of 10.234 cm, where according to Ishihara *et al.* (1992) [35], the settlement ranging from 10 cm to 30 cm can be categorized as medium category, with the phenomenon on the surface are small cracks, there are manifestations of sand flow that occur above the ground surface [35]. The low category (S = 0 cm - 10 cm) was found at drill points DP-01, DP-02, DP-04, DP-06, DP-10, and DP-11, with settlement values ranging from 0.590 cm to 2.907 cm, which is one of the signs of the phenomenon found above the ground surface, namely minor cracks [35]. In this study, settlement value equal to 0 centimeters (S = 0 cm) was considered to have no impact on the settlement found at drill points DP-03, DP-07, DP-08, DP-09, DP-12, DP-13, and DP-14.

In this study, the estimated settlement was only due to liquefaction during an earthquake like what happened in event 2018 ago. This was done because the soil with the liquefaction potential must consider the effect of liquefaction on the structure to be built on it [38]. Overall, the settlement that occurred in this area did not have much impact, as presented in **Table 6.** It was dominated by low to no damage to the settlement when liquefaction occurred due to the earthquake. Only at the drill point, DP-05 has a medium category than other drill points, which means that the payment in this drill point was greater than at other points.

Table 6. Estimation of Soil Liquefaction-Induced Ground Settlement

Drill Points	Depth Prone to Soil Liquefaction [18]	Settlement (cm) [22]	Settlement Categories [35]	Phenomenon on the ground surface level [35]
DP-01	No Potential	0.955	Low	Minor crack
DP-02	5 meters	2.907	Low	Minor crack
DP-03	No Potential	0.000	No Damage	-
DP-04	No Potential	0.590	Low	Minor crack
DP-05	5 to 9 meters	10.234	Medium	Small crack, and streams of fine sand come to the surface
DP-06	7 to 9 meters	2.212	Low	Minor crack
DP-07	No Potential	0.000	No Damage	-
DP-08	No Potential	0.000	No Damage	-
DP-09	No Potential	0.000	No Damage	-
DP-10	5 meters	2.274	Low	Minor crack
DP-11	5 meters	2.154	Low	Minor crack
DP-12	No Potential	0.000	No Damage	-
DP-13	No Potential	0.000	No Damage	-
DP-14	No Potential	0.000	No Damage	-

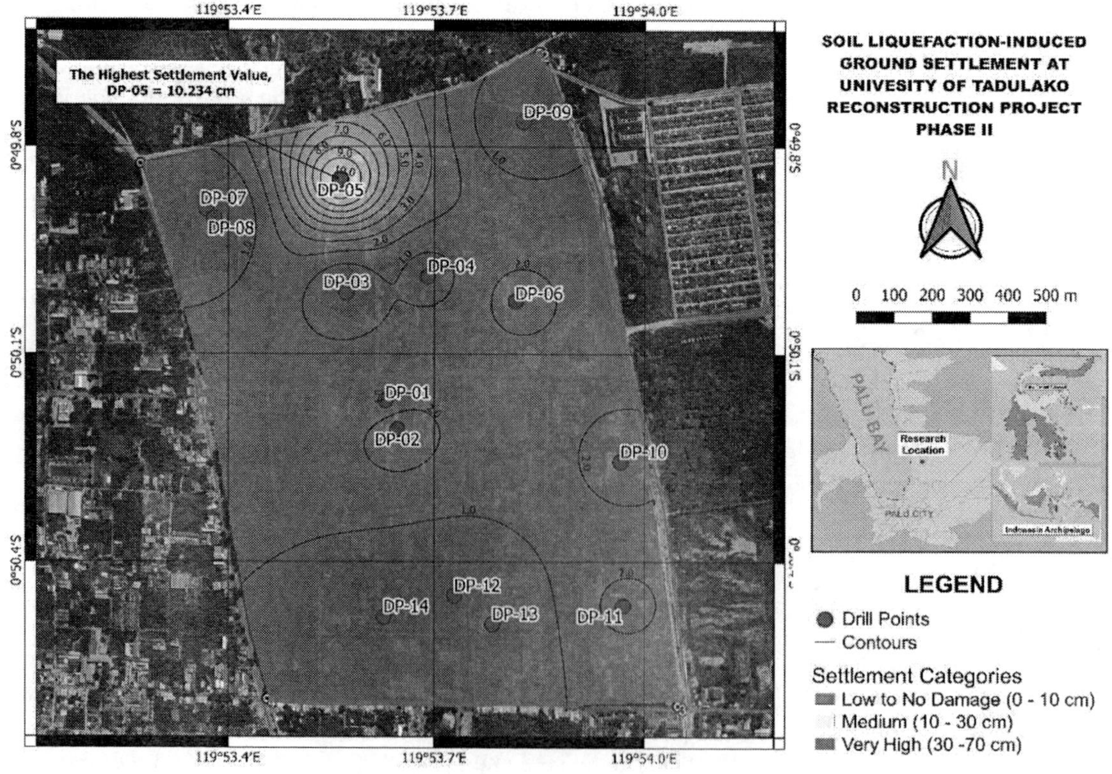

Figure 5. Soil Liquefaction-Induced Ground Settlement Map

Figure 5 shows a liquefaction-induced ground settlement map. This was related to the concept of disaster risk management which was presented in the spatial form in the form of maps. With maps, information related to disaster locations can be easily accessed. In this way, it was easy to know which areas were affected and which were not, where the settlement due to liquefaction had no overall impact, except for the drill point DP-05, which is slightly lighter in color than the other drill points [40]. Why does a large settlement anomaly occur only at the drill point DP-05? This can happen because the liquefied soil layer at the drill point was quite deep compared to other drill points, as shown in **Table 6**, where the depth was prone to soil liquefaction at a depth of 5 to 9 meters.

4. Conclusion

The effects of the Palu-Koro fault on predicting the ground motion using attenuation NGA West 2 models for estimating soil liquefaction-induced ground settlement based on SPT data of educational infrastructure reconstruction project at the University of Tadulako, Palu City, Central Sulawesi has been studied. This study aims to determine the performance of ground motion if the study area is located in an area close to the fault site. In this study, the project was located close to the fault site, namely the Palu-koro fault. The seismic hazards assessment was needed as one of the mitigation efforts against geological disasters caused by earthquakes. One of the methods to analyze seismic hazards in this study was the attenuation method of the NGA West 2 models, which with this method, can determine how the fault site performs against the ground motion that occurs during an earthquake.

Based on the results and discussion of the prediction analysis of ground motion using the attenuation NGA West 2 models, the overall threshold parameters compared have met the requirements in the qualitative approach. The Palu-Koro fault has a significant influence on the prediction of ground motion. The relationship between the fault effects and ground motion affects various PGA values. This was due to variations in input parameters such as distance for PGA at bedrock and site class condition for PGA at the ground surface. The closer the distance of fault to the site, the greater the PGA at bedrock value obtained. Whereas for PGA at the surface, the looser the material, the greater the amplification, the greater the soil and structure damage that occurs, and the larger the site class coefficient (F_{PGA}). However, in Indonesian Seismic Code SNI 1726:2019, the F_{PGA} gets smaller as the PGA gets bigger. The F_{PGA} value was influenced by PGA at bedrock, where this value was influenced by distance. This was what caused the varying the PGA at surface values because of various distances and site effect conditions.

The PGA at surface value was then used to estimate the soil liquefaction-induced ground settlement. The PGA used was the largest PGA obtained from the NGA West 2 Chiou and Young (2014) model because of the uncertainty about the occurrence of earthquakes in the future. Estimation of the soil liquefaction-induced ground settlement was necessary for this study because it will affect the design of the building foundation in this project. This is related to the aspect of life safety. If there is a building structure standing on soil that is experiencing liquefaction, then the building will collapse and will result in a lot of economic losses, loss of life, and loss of function.

Based on the results, the settlement due to liquefaction has no overall impact in this study area, except for the drill point DP-05, where this point had the largest settlement value compared to other drill points, which was 10.234 cm, with the category of medium settlement, with phenomena that occur above the surface, namely small cracks, and the flow of sand grains to the surface. This can happen because the liquefied soil layer at the drill point was quite deep compared to other drill points, where the depth was prone to soil liquefaction at a depth of 5 to 9 meters. Further analysis is needed to estimate the total and differential settlement due to the load of the building structure itself, and mitigation effort for liquefaction disaster is suitable to be applied in this location, especially for ground settlement due to liquefaction such as ground improvement techniques, build liquefaction resistant structures, or avoid liquefaction susceptible soils. In addition, the concept of mapping was used as spatial disaster risk management so that it is more informative and easily accessible.

5. Acknowledgments
The authors would like to express their sincere gratitude for the support given by the Settlement Infrastructure Agency for Central Sulawesi Region, Directorate General of Human Settlement, Ministry of Public Works, and Public Housing, Republic of Indonesia.

6. References
[1] Badan Geologi (2018). "Di Balik Pesona Palu Bencana Melanda Geologi Menata." Bandung, Indonesia.
[2] Masyhur I, Hanifah R, Djarwadi D, and Sarsito A (2018). "Kajian Gempa Palu Provinsi Sulawesi Tengah 28 September 2018 (M7.4)." *National Center for Earthquake Study (PUSGEN)*, Bandung, Indonesia.
[3] Masyhur I, et al. (2017). "Peta Sumber dan Bahaya Gempa Indonesia Tahun 2017." National Center for Earthquake Study (PUSGEN), Bandung, Indonesia.
[4] Sukandarrumidi, et al. (2021). "Geologi Umum Bagian Kedua." Gadjah Mada University Press, Yogyakarta, Indonesia.
[5] Socquet A, Simons W, Vigny C, McCaffrey R, Subarya C, Sarsito D, Ambrosius B and Spakman W (2006). "Microblock Rotations and Fault Coupling in South East Asia Triple Junction (Sulawesi, Indonesia) from GPS and Earthquake Slip Vector Data." Journal of Geophysical Research, Vol.111.
[6] Bolt, B. A (1978). "Earthquake A Primer." W.H. Freeman and Company, San Fransisco, United States of America.
[7] Bolt, B. A (1995). "Intraplate Seismicity and Zonation." Proceeding of The Pacific Conference on Earthquake Engineering. University of Melbourne, Australia.
[8] Pawirodikromo, Widodo (2012). "Seismologi Teknik dan Rekayasa Kegempaan." Yogyakarta, Indonesia.
[9] The United States Geological Survey, USGS (2018). "M7.5 Palu, Indonesia Earthquake of September 28, 2018. Poster."
[10] Atkinson, G. M, and Boore, D.M. (2006). "Earthquake Ground-Motion Prediction Equation for Eastern North America." Bulletin of the Seismological Society of America, Vol. 96, No. 6. Pp. 2181 - 2205.
[11] Kramer, S. L. (1996). "Geotechnical Earthquake Engineering". Prentice-Hall International Series, New Jersey, USA.
[12] Mase, L. Z., (2019). "Performance of NGA Models in Predicting Ground Motion Parameter of the Strong Earthquake." Journal of the Civil Engineering Forum, Vol.5 No.3.
[13] Montgomery J, Wartman J, Reed N, Gallant P, Hutabarat D, and Manson B (2021) "Field reconnaissance data from GEER investigation of the 2018 M 7.5 Palu-Donggala earthquake."
[14] Hakam, Abdul, (2020). "Analisis Praktis Potensi Likuifaksi." Andalas Press, Padang, Indonesia.
[15] Green, et al (2019). "What is the Smallest Earthquake Magnitude that Needs to be Considered in Assessing Liquefaction Hazard?" Earthquake Spectra, Volume 35, No. 3, pages 1441-1464.
[16] Youd, T. L (1979). "Liquefaction Potential Map of San Fernando." Proceeding of The Second International Conference on Microzonation for Safer Construction, Research and Application. San Francisco, California, USA.
[17] Tsuchida, H (1970). "Prediction and countermeasure against the liquefaction in sand deposits." Abstract of the Seminar in the Port and Harbor Research Institute. pages 31-333.
[18] Idriss, I. M. and Boulanger, R.W. (2008). "Soil Liquefaction During Earthquakes." Earthquake Engineering Research Institute, the University of California at Davis, USA.
[19] Day, R. W. (2012). "Geotechnical Earthquake Engineering Handbook Second Edition." The McGraw-Hill Companies Inc, USA.
[20] Youd, T. L, and Wieczorek, G.F (1984). "Liquefaction During the 1981 and Previous Earthquakes Near Westmorland, California." The United States Geological Survey. Menlo Park, California, USA.

[21] Ishihara, K., and Koga, Y (1981). "Case Studies of Liquefaction in the 1964 Niigata Earthquake." Japanese Society of Soil Mechanics and Foundation Engineering. Vol.21.
[22] Zhang, G., Robertson, P.K., and Brachman, R., W., I. (2002). "Estimating liquefaction-induced ground settlements from CPT for level ground." Can, Geotech. J., 39: 1168-1180.
[23] Faris, F. and Wang, F. (2014). "Investigation of the initiation mechanism of an earthquake-induced landslide during rainfall: a case study of the Tandikat landslide, West Sumatera, Indonesia.
[24] Acton, J., M., and Hibbs, M. (2012). "Why Fukushima Was Preventable." Carnegie Endowment for International Peace.
[25] Robertson I, et al (2019). "StEER: Structural Extreme Event Reconnaissance Network Palu Earthquake and Tsunami, Sulawesi, Indonesia." Field Assessment Team 1 (FAT-1).
[26] Asian Development Bank and Ministry of Public Works and Housing Republic of Indonesia (2019). "The Initial Environmental Examination (IEE) Reconstruction and Rehabilitation Palu Coastal Protection"
[27] Masyhur I, Hanifah R, Djarwadi D, and Sarsito A (2018). "Kajian Gempa Palu Provinsi Sulawesi Tengah 28 September 2018 (M7.4)." National Center for Earthquake Study (PUSGEN), Bandung, Indonesia.
[28] PT. Geocipta Bangun Optima (2020). "Laporan Penyelidikan Tanah Universitas Tadulako Zona 1." Bandung, Indonesia.
[29] Indonesia (2018), "Instruksi Presiden Republik Indonesia No.10 Tahun 2018 tentang Percepatan Rehabilitasi dan Rekonstruksi Pascabencana Gempa Bumi dan Tsunami di Provinsi Sulawesi Tengah dan Wilayah Terdampak Lainnya." Jakarta, Indonesia.
[30] Seyhan, E. (2014), "Weighted Average of 2014 NGA WEST-2 GMPEs Spreadsheets_V5.6_Last updated: 070514". Pacific Earthquake Engineering Research Center.
[31] Boore, D. M., Stewart, J. P., Seyhan, E., and Atkinson, G., M. (2014). "NGA-West2 Equations for Predicting PGA, PGV, and 5% Damped PSA for Shallow Crustal Earthquakes." Earthquake Spectra, Vol. 30, No. 3, pages 1057 – 1085.
[32] Campbell, K.W., and Bozorgnia, Y. (2014). "NGA-West2 Ground Motion Model for the Average Horizontal Components of PGA, PGV, and 5% Damped Linear Acceleration Response Spectra." Earthquake Spectra, Vol.30, No3, pages 1087-1115.
[33] Chiou, B. S., and Youngs, R., R. (2014). "Update of the Chiou and Youngs NGA Model for the Average Horizontal Component of Peak Ground Motion and Response Spectra." Earthquake Spectra, Vol.30, No3, pages 1117-1153.
[34] Andrus, R., D., Piratheepan, P., Ellis, B., S., Zhang, J., and Juang, C., H. (2003). "Comparing liquefaction evaluation methods using penetration-Vs relationships."
[35] Ishihara, K., and Yoshimine, M. (1992). "Evaluation of Settlements in Sand Deposits Following Liquefaction During Earthquakes." Japanese Society of Soil Mechanics and Foundation Engineering, Vol. 32, No. 1, 173 - 188.
[36] Nugroho, Cahyo (2018). "Laporan Akhir Analisis Indeks Kerentanan Seismik Kota Palu". Badan Penanggulangan Bencana Daerah Kota Palu.
[37] BSN (2019), "SNI 1726: 2019-Tata Cara Perencanaan Ketahanan Gempa Untuk Struktur Bangunan Gedung Dan Nongedung."
[38] Maithili, KL (2017). "A Discussion of Liquefaction Mitigation Methods." International Research Journal of Engineering and Technology (IRJET), Volume: 04. Issue: 12.
[39] Harijoko, A., Puspitasari, D., Prabaningrum, I., Prastika, K.P., and Wijayanti, N. (2021), "Manajemen Penanggulangan Bencana dan Pengurangan Resiko Bencana di Indonesia." Gadjah Mada University Press, Yogyakarta, Indonesia.

IOP Conf. Series: Earth and Environmental Science 1233 (2023) 012016 doi:10.1088/1755-1315/1233/1/012016

The handling of tunnel collapse and settlement (case study: Jragung Dam diversion tunnel, Indonesia)

Laela Fitriyantina[1*]

[1]Directorate General of Water Resources, Ministry of Public Works and Housing of Indonesia, Jakarta

*Corresponding author: lala.fitriya@pu.go.id

Abstract. A diversion tunnel is a crucial part of dam construction because of its function to derivate the river during earth dam filling works. The Jragung Dam diversion tunnel is located in the Kalibeng Formation, which is dominated by claystone and sandstone lithology, and passed by an anticline fold structure. The geological conditions provide challenges in tunnel construction. The collapse of rock material in the tunnel overburden and settlement on the slopes hill requires a comprehensive approach and handling by prioritizing factors of safety, cost efficiency, and speed of time. Field observation including rock mass quality analysis, reviewing the design, also monitoring of discharge and deformation was conducted to find the source of collapse. The groundwater flow and deformed soft rock were accused of being the tunnel collapse determinant. The rock mass improvements used pore-filling of sodium silicate mixture and fore-polling. Then dewatering was carried out using drainage pipes and deep wells. Because of the steep rock's dip in the tunnel, the release wells and observation wells location need to design effectively. Geoelectric investigations correlated with bore log analysis aimed to find effective locations for groundwater control. The settlement was refilled with plastic concrete because of its lightweight and durability.

1. Introduction

Tunnels are genuinely liable to geological conditions in the field. A diversion tunnel is used to divert a river during dam construction. Tunnels in Indonesia are generally built using the NATM (New Austrian Tunneling Method) system or commonly called Mountain Tunneling, where tunnel excavation is driven by digging, ripping, and blasting methods [1]. The difference with the excavation using a drilling machine is there is no support in the excavation face, affecting difficulties when applying to highly deformed, highly weathered, high swelling rate, or low permeability rock [2]. The earth-filling dam is impossible to be done without the derivation works of the river, so the tunnel diversion work needs to be accomplished within a limited time. These things are the challenges in tunnel diversion works.

The collapse happened in the 5 m of excavation from the outlet portal. The accident was indicated by the presence of groundwater flowing through the cracks, mostly from the right side of the crown. When heavy rains occur, the groundwater brings brittle materials that have been deformed by the geological structure. The slope settlement above the tunnel occurs due to the loss of rock mass. The collapse in the tunnel can be seen in Figure 1. From the picture, the rock material has turned into a porridge form that indicates the claystone at the tunnel site has undergone high deformation. Based on the regional geological map of Salatiga [3], the research area is located in the multiple members of the Kalibeng Formation (Tmkb) the age of late Miocene. The lithology was formed by alternating tuffaceous sandstones, limestone siltstones, sandstones, and gravel sandstones. The geological structure is found in the form of syncline and anticline folds, which are seen from the direction of rock subduction in the opposite direction, stretching from northeast to southwest. The anticline folds created the weak zone which spreads around the tunnel lines and forms many cracks. Those cracks become the way for water or other fluids and minerals to flow [4].

(a) (b)

Figure 1. The tunnel collapse (a), settlement above the tunnel (b) (documentation of 2022)

Based on the tunnel's design, the underground investigation in 380 m of tunnel length is just conducted on three boreholes, whereas the steep rock's dip makes the rock's conditions between boreholes almost impossible to analyze. The limitation of the study in the tunnel's design provided the need for evaluation. In Indonesia, dam construction is very popular these days, but the study about this subject does not have much to be approached, so this study is aimed to give the recommendation of options in handling the tunnel collapse and ground settlement.

2. Methods

This study aimed to find the most suitable, safe, and efficient method to handle tunnel collapse and ground settlement, include of drainage management. In this case, the steps of the study are represented below.

2.1. Review the design

The Jragung Dam diversion tunnel has a design length of 380m with an outer diameter of 6 m. Design data shows that the subsurface investigations were carried out on 3 drilling points along the tunnel path, with the depth of each drilling point from upstream to downstream 40 m, 95 m, and 35 m. The excavation stages are divided into two sections: "upper half section", and "lower half section". Excavation of the upper half section using a rock breaker was excavated with a height of 4.6 m every 0,5 m. Furthermore, a tunnel support system is provided in form of safety shotcrete after excavation with a thickness of 50 mm, H beam at the top, wire mesh, shotcrete with a thickness of 100 mm, and rock bolts with a length of 3 m. After the excavation of the upper half section reaches 25 m, it is continued with the excavation of the lower half section. The support system used in the lower half section is the same as in the upper half section. Meanwhile, the groundwater drips were maintained by collecting into a pond and then drained out using hoses [5].

The RMR (Rock Mass Rating) [6];[7] face mapping was conducted to ensure that the rock condition is still in the same condition as the design. The RMR was a common method to determine the rock mass quality that have been proposed as earliest as the 1970s, but the 1989 version of RMR has been widely accepted and applied [8]. This method follows six parameters for the rock mass classification; i) uniaxial compressive strength [9]; ii) Rock Quality Designation (RQD); iii) spacing of discontinuities; iv) condition of discontinuities; v) groundwater condition; and vi) orientation of discontinuities. Figure 2 shows the rock condition inside the tunnel that needs to be determined their quality. The right side shows that the alternate claystone and sandstone have a shaly structure, while the left side shows a blocky structure of claystone.

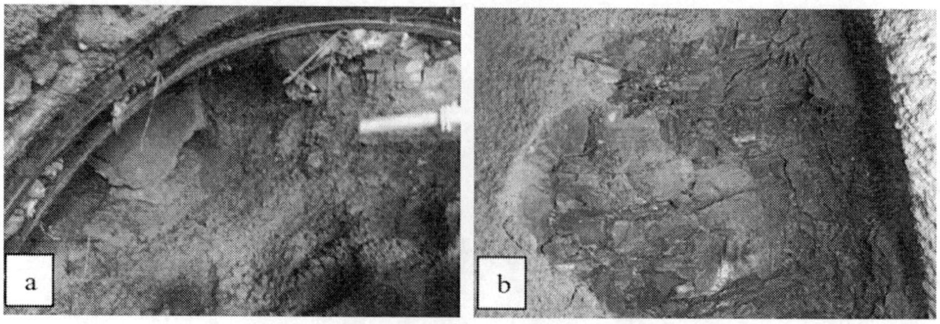

Figure 2. Crown area on the tunnel right side with shaly structure (a) crown area on the tunnel left side with blocky structure (b) (documentation of 2022)

2.2. *Deformation and discharge monitoring*

Monitoring of the deformation that occurs in the tunnel is necessary throughout the construction and post-construction period to ensure that the tunnel building is safe. The deformation monitoring was using a theodolite every 12 hours. For the discharge, monitoring was carried out by counting the water level at a measurable glass and the time of filling up was checked by a stopwatch [10].

The deformation and discharge monitoring data are needed to achieve the perception of the rainfall effect on deformation and discharge in the tunnel. Also, to secure the workers when excavation work continues. The observations serve the information to conclude the origin of collapse and settlement incidents [11].

2.3. *Additional investigation*

From the geological mapping, it is known that the dip of rock at the tunnel area was very steep (± 68°) so the three boreholes can't precisely describe the geological condition along the tunnel. The need for additional investigation is a requirement for geological evaluation. Also, the requirement of water concentration data to draw the location of deep wells. For these geological and hydrogeological purposes, the additional investigation that is suitable is geoelectrical methods [12].

Geoelectrical is applied to map the resistivity structure of the underground. The measurement of resistivity aims to determine the variation in the resistivity of subsurface rock which reflects the different types of soil and rock layers. The geoelectrical method that was used is 2-D. The 2-D model is a subsurface model consisting of blocks of different sizes. In this case, the parameter of the 2-D model is the resistivity value of each block which has lateral (x) and vertical (z) dimensions. The basic principle of the geoelectrical investigations is by injecting currents and the resulting potential differences at the surface, then the apparent resistivity can be calculated using Equation (1):

$$\rho_a = K\frac{\Delta V}{I} \tag{1}$$

Where: K : geometry factor (m)
V : potential (Ω/ Ohm)
I : current (Ω/ Ohm)
ρ_a : apparent-resistivity (Ωm/ Ohm.m)

The geoelectrical configuration is using dipole-dipole, which requires two pairs of electrodes: a pair of electrodes for current injections, and a pair of electrodes for potential difference measurements. The advantage of this configuration is that the variation of the lateral and vertical resistivity of the sounding can be measured properly. The geometry factor of the dipole-dipole electrode configuration is formulated as follows in Equation (2) [12];[13];[14]:

$$K = \pi . n(n + 1)(n + 2)a \tag{2}$$

Where: n : number of measurements
a : space (m)

The list of resistivity values from various types of minerals is presented in Table 1. The sketch of the dipole-dipole method and the documentation of geoelectrical investigations of this study are represented in Figure 3. The machine that was used is ARES, with the length of each line across the tunnel being 280 m, and the space for each line is 50 m. The number of electrodes at one line is 48, so the space between electrodes is 6 m [14].

Table 1. The Resistivity values from various types of minerals (Telford,1990)

Material	Resistivity (Ωm)	Material	Resistivity (Ωm)
Air	~	Shales	20 - 2,000
Pyrite	0.01 - 100	Sand	1 - 1,000
Quartz	500 - $8x10^5$	Clay	1 - 100
Calcite	$1x10^{12}$ - $1x10^{13}$	Groundwater	0.5 - 300
Rock salt	30 - $1x10^{13}$	Sea water	0.2
Granite	200 - 10,000	Magnetite	0.01 - 1,000
Andesite	1,700 - $4.5x10^5$	Dry gravel	600 - 10,000
Basalt	200 - $1x10^5$	Alluvium	10 - 800
Limestone	500 - 10,000	Gravel	100 - 600
Sandstone	200 - 8,000		

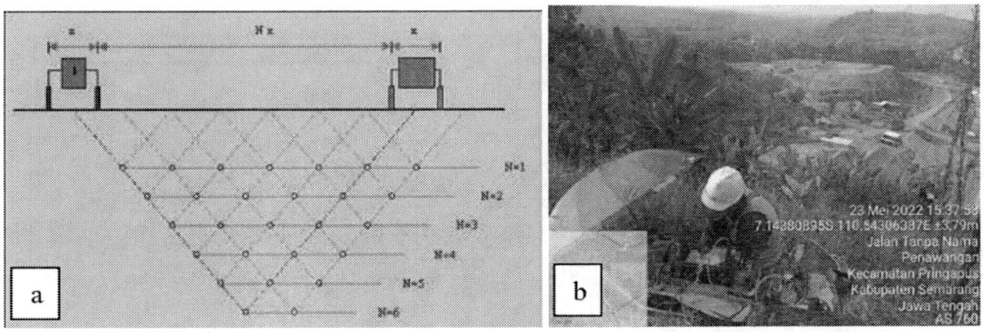

Figure 3. Dipole-dipole method of geoelectrical investigation (a) documentation of geoelectrical (b)

3. Results

3.1. Geological Conditions

The physiographic conditions are based on Van Bemmelen classification, the study area located in Kendeng Zone, consists of a row of mountains elongated west-east direction, located directly to the north of the Ngawi sub-zone. These mountains are composed of deep-sea sedimentary rocks that have undergone intensive deformation to form an anticlinorium. The morphological feature of the Kendeng Zone is a range of low hills with a wavy morphology, with a height of between 50 and 200 meters. This west-east trending line reflects the presence of folds and faults that are trending west-east and gradually weakening in the east. [15];[16].

Based on detailed geological mapping as seen in the geological map of the Jragung Dam diversion tunnel (Figure 4), it is suspected that the research area is influenced by the Kerek Formation (Tmk) by the presence of flysch minerals. Meanwhile, the tunnel location is dominated by claystone and sandstone that contain a few carbonate minerals. The geological structures that were found are the syncline and anticline fold. The joints indicated that the tunnel location is included in the fracture zone.

The rock mass quality analysis results in Table 2 are suitable for the geological condition. Then it is proofing that the rock mass has undergone high deformation. The rock mass will immediately collapse when the tunnel is excavated.

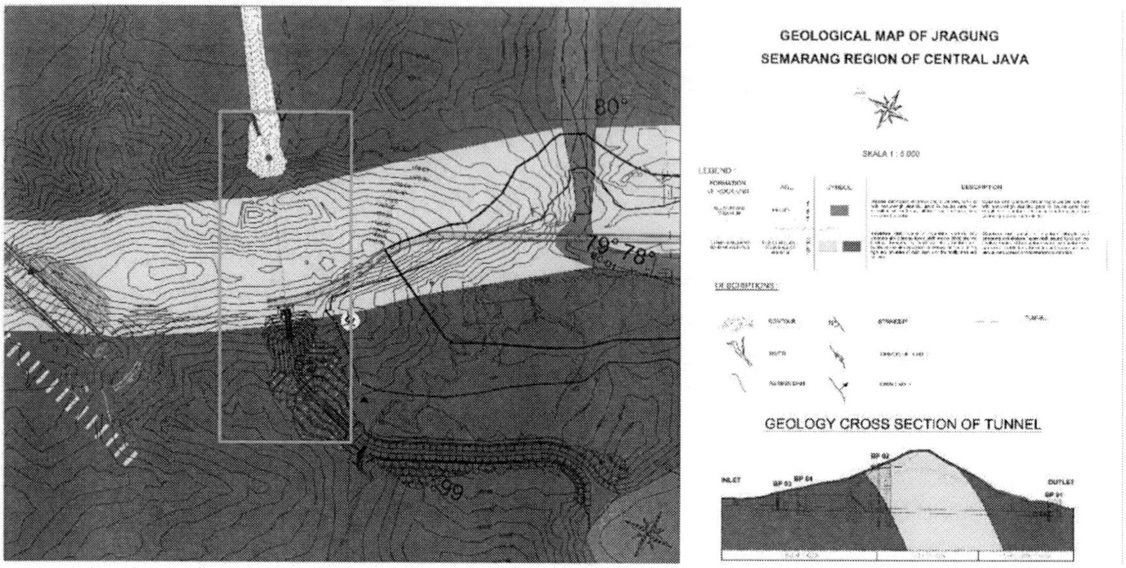

Figure 4. Geological Map of Jragung scale of 1:5.000

Table 2. The RMR analysis at the collapse location

Location	UCS (MPa)	RQD (%)	Joint spacing (mm)	Condition of joints	Ground water	Orientation of discontinuity	Total rating	Rock mass class
Right	< 1	< 25	< 60	Gouge, separation 1-5 mm	Flowing	Fair	13	Very poor rock
rating	0	3	5	10	0	-5		
Left	< 1	< 25-50	< 60	Slickenside, separation 1-5 mm	Flowing	Fair	18	Very poor rock
rating	0	8	5	10	0	-5		

3.2. *Reinforcement for Tunnel Support*

After checking geological conditions, including the lithology, structure geology, weathering conditions, and RMR face mapping, the design and construction methods both had good work. In the RMR analysis (Table 1), both sides of the tunnel are in the range of very poor rock mass class, with a total rating is 13 for the right side, and 18 for the left side. The space between steel support, length of the rock bolts, and excavation sequence were unharmed. Then the external factor such as the heavy rain days before must be considered. The interpretation of tunnel conditions by the field observation is illustrated in Figure 5. Then the treatment method is chosen using pore-filling and fore-polling [17] also illustrated in Figure 5.

a. Fore-polling

Fore-polling is one of the support methods to enhance rock shear strength. The fore-polling is using a bolt, steel bar/ steel pipe with a length of 3 m and angle of 6° that is installed in the tunnel's crown. Fore-polling is a pre-support that is installed immediately after excavation. The installation of fore-polling is illustrated in Figure 5.

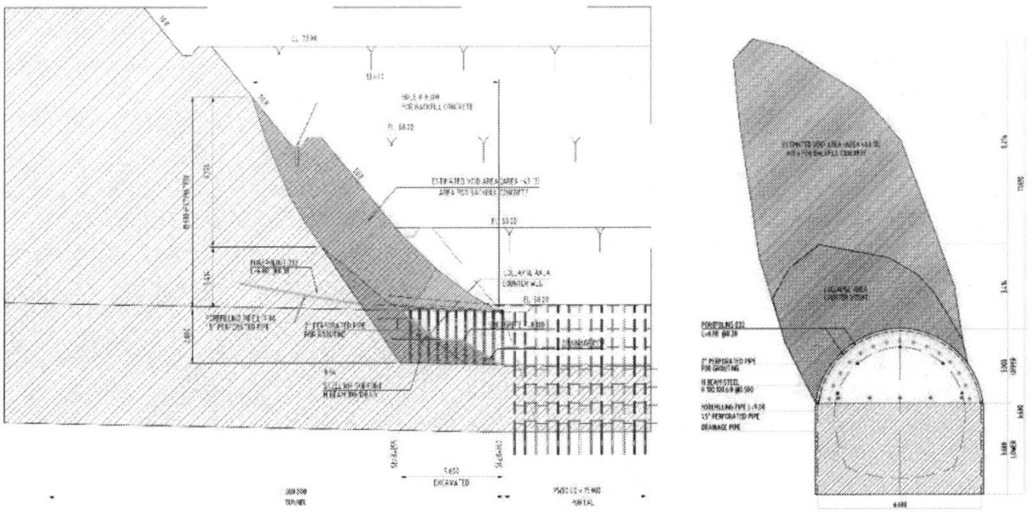

Figure 5. The Interpretation of tunnel collapse condition and the handling method

b. Pore-filling with sodium silica

Consolidation grouting with the pore-filling pipe as the casing was used to strengthen rock mass on the overburden area. The consolidation grouting is a mixture containing the material which can fill the void, cracks, and rock discontinuity. In this case, the injection is divided into two steps; first is the solutions of sodium silica (Na_2SiO_3) and water, and then the solutions of cement and water. The sodium silica solutions are intended to link the groundwater and then transform it into a hard plastic gel. The setting time of this solution is quite fast compared with cement. The composition of sodium silica - water can see in Table 3, and cement - water can see in Table 4.

Table 3. The composition of water – sodium silica

Water : Sodium Silica Ratio	Water (lt)	Sodium Silica (kg)	Mixture volume (lt)
10:1	200	20	208.23

Table 4. The composition of water–cement

Water : Cement Ratio	Water (lt)	Cement (kg)	Mixture volume (lt)
4:1	80	20	86.35
2:1	80	40	92.70
1:1	80	80	105.40

The increase in thickness of water – cement mixture aims to gain the allowance pressure (P), which depends on the volume mixture in 30 minutes. The grouting is called complete if the allowable pressure is achieved. If it is not the injection process must be stopped for 2 hours, then continue with a 1:1 water-cement ratio mixture at the same hole. If there is a leakage, the grouting must be temporarily suspended and clogged. After that, the injection continues with 50% of maximum pressure and waits for clogging to be set.

The installation of the 6 m pore-filling pipes with the angle of 6° is using the pneumatic bore machine CRD (Crawler Rock Drilling). The drilling depth is 6.5 m with a diameter of 57 mm. The total hole in each segment is 15 holes, of which the odd hole is the primary and the even hole is secondary. All the primary holes are injected first before the injection of the secondary holes are begin.

3.3. Geoelectrical investigation results

The geoelectrical investigation was carried out in a long section and crosslines of the tunnel. Figure 6 shows the geoelectrical results near the collapse location. The interpretation of geoelectrical investigations was correlated with the bore log data, resulting in the description: for the very low resistivity group, which has a value of < 6 Ohm.m with dark blue colour, it is a water-saturated clay. For the low resistivity group, around 6 – 30 Ohm.m, the water-saturated sandstone area is represented by light blue - green. For medium resistivity with a value of 30 – 80 Ohm.m, it is assumed that it is relatively unsaturated clay represented by a dark green colour. For high resistivity, 80 – 250 Ohm.m, it is assumed that the sandstone is not saturated with water, represented by the yellow colour. The high resistivity group that is > 250 Ohm.m is thought to be a boulder/limestone/void area represented in brown - red - purple [13].

From the results, we can describe that the wet sandstone area which is usually suspected as the water bag is located on the right side of the tunnel. This result is matching with the location of the water sources and the collapsed hole is dominated by the right side. The wet underground claystone indicated that the claystone has been deformed and had many cracks. Thus, the treatment should focus on the right side of tunnel, especially for dewatering work. Dewatering work is carried out using drainage pipes installed inside of tunnel wall and crown. These pipes are expected to reduce the groundwater pressure that causes the collapse.

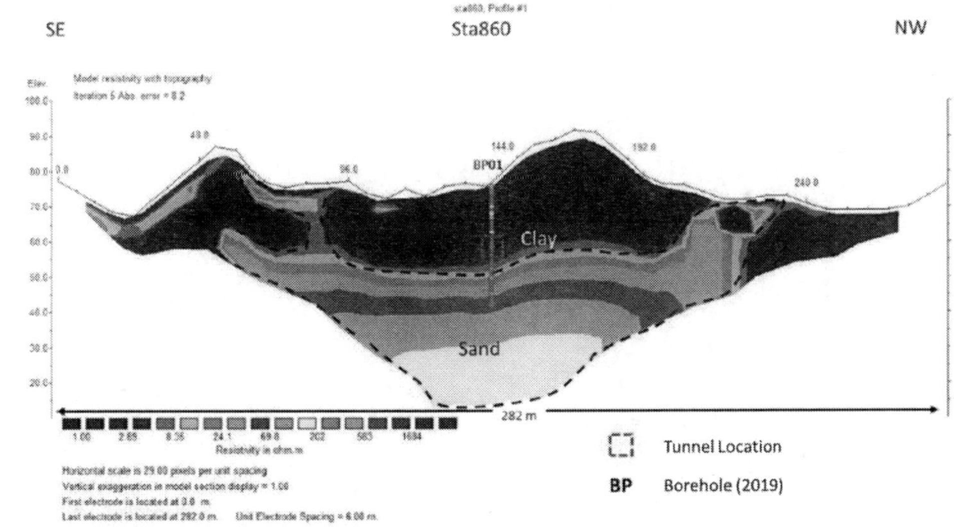

Figure 6. Geoelectrical results at the cross of portal outlet

In the meantime, the monitoring of tunnel deformation was executed. The data shows that the steel rib has deformed around 10 cm into the tunnel. The monitoring data of discharge also show a very high groundwater flow at a maximum of 1 litter/s. The groundwater pressure is accused to be the origin of deformations. Therefore, the release well is considered. The main purpose of this well is to decrease the groundwater level. The release well is planned to be in the wet sandstone area which is suspected as the waterbag. Located on the tunnel's right side, which parallels the collapse location. The deep of the well must be lower than the tunnel's invert, which is 21 m. The observation well is added to control the effectiveness of sinking in groundwater. The observation well location is the same as the release well. The automatic deep meter was installed to check the water level in the well. The checking is conducted 5 minutes after pumping has stopped. The design of the release well and observation well is illustrated in Figure 7.

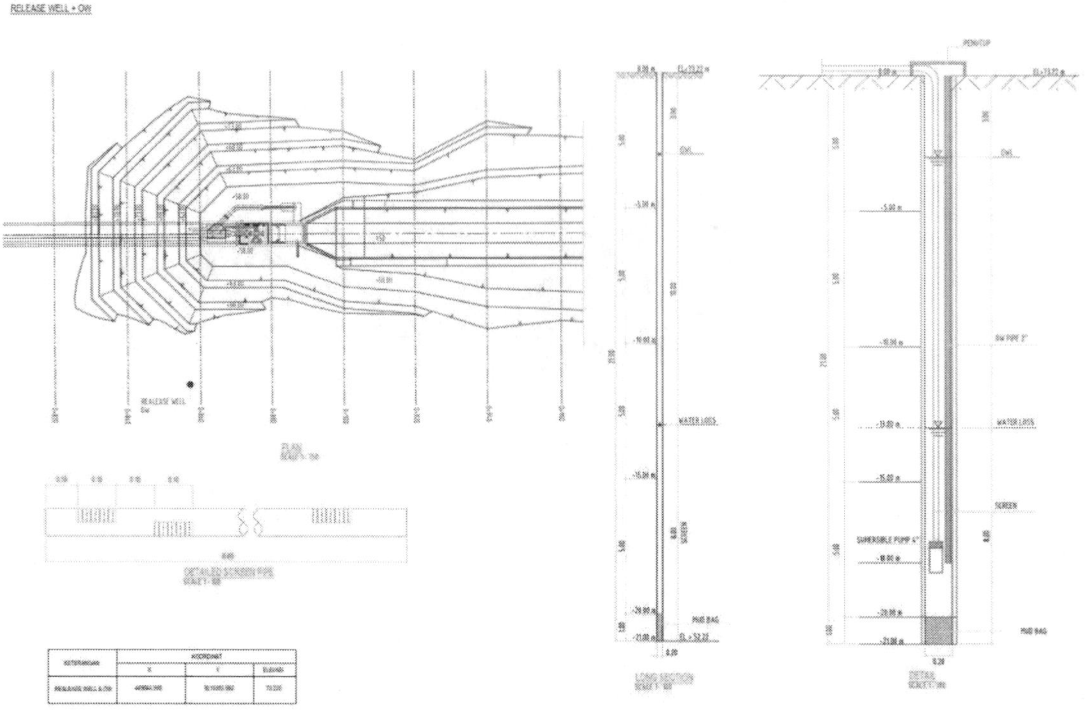

Figure 7. Design of release well and observation well

3.4. *Backfilling the settlement*

Backfilling the settlement intended to hold the shotcrete up to keep that function as the surface protection, also to keep the overburden above the tunnel that function as the counterweight. The slope cutting is not recommended in this case because the hill has a very steep topography, which will decrease the slope stability in the portal. The filler material should lightweight, unyielding to vibration, and have lifetime durability (not an organic material).

The option of filler material is foam mortar and plastic concrete. The foam mortar is a mortar with an additional foam material, while the plastic concrete has additional bentonite. The foam mortar has a unit weight of around 800-900 kg/m^3, strength of 8 – 19 MPa, slump is around 16 – 20 cm, high permeability, and stiff deformability. Meanwhile, the plastic concrete has a unit weight of around 1.200 kg/m^3, strength of 2,51 MPa, slump is around 18 ± 2 cm, permeability at 10^{-9}, and high deformability. [18];[19]. The cycle time of foam mortar is calculated at about 16 days, and for plastic concrete is about 5 days.

Based on the unit weight number of materials, the foam mortar is more effective than the plastic concrete. But the deformability is unsure to bear the vibration of the excavation and drilling process on the tunnel's lower section. Moreover, the plastic concrete is very permeable against the groundwater flow. When plastic concrete is applied, the repairing work of drainage pipes is not needed. Based on those criteria, then plastic concrete was chosen as the backfilling material. For the operational methods, the plastic concrete was injected for each 1 m layer. After the plastic concrete was settled, then it has monitored in tunnel support deformation. The methods are illustrated in Figure 8.

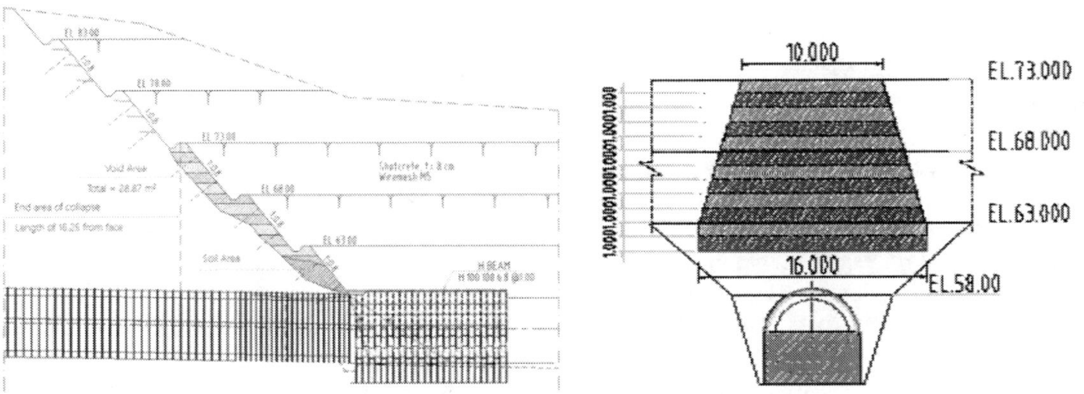

Figure 8. Illustration of plastic concrete injection

4. Conclusion

According to the study above, the derivation of tunnel collapse and settlement is the groundwater pressure triggered by the highly intense rain. The very poor rock mass quality also contributes, while the deformed soft rock met the high groundwater flow causing piping. This cause is identified after a few days of observation on rainy and sunny days, evaluation of the geological condition and the design, and also the geoelectrical investigation results, all of which indicate the same point. The decrease in rock mass quality caused the stand-up time to be not available [7]. These problems are demanded a comprehensive solution to enhance the rock mass quality and decrease the groundwater flow.

The additional tunnel support is fore-polling with 3 m of length and 6° of angle, and perforated pore-filling grouting of sodium silica as long as 6 m with an angle of 6°. The drainage pipes were also installed inside the tunnel to decrease the effect of groundwater pressure. But after observing groundwater discharge, installing a release well is considered to achieve an effective loss in groundwater level. The location of the water bag on the hill is identified by additional geoelectrical investigations. The results of the geoelectrical investigation presented the wet sand lithology that is ideal for being released well location, is on the tunnel's right side. With the steep dip in the tunnel location, the release well location must be parallel with the collapse location. The depth of the tunnel is about 21 m which is 1 m deeper than the tunnel's invert to prevent sedimentation in the bottom of wells.

To refill the void in the settlement area, plastic concrete material was chosen. Plastic concrete is inorganic light material (unit weight of around 1.200 kg/m^3), impermeable (permeability at 10^{-9}), vibration resistant, and easy installation with a total time of 5 days. The installation of plastic concrete needs to be monitored for each 1 m layer for safety reasons.

The suggestion for safety reasons in the next excavation is the additional investigation to identify the structures, weak zones, and deformed rocks by Multichannel Analysis of Surface Waves (MASW). This method is suitable for medium-depth underground exploration.

Acknowledgment

The writer is pleased to acknowledge the Pemali Juana River Basin Region, Directorate General of Water Resource for the data publication permission. Thanks to the Jragung Dam team for helping collect the data.

References

[1] Wern-ping Chen and Hugh Caspe 2004 NATM and Its Practice in The US *Proceedings of The North American Tunneling Conference, Georgia* 473-476 ISBN 90 5809 669 6.

[2] F. Tonon 2010 Sequential excavation, NATM, and ADECO: What They Have in Common and How They Differ *Journal of Tunneling and Underground Space Technology* 25(2) 245-265.

[3] Sukardi and T Budhitrisna 1992 Systematic Geological Map of Indonesia Salatiga Sheet Scale of

1:100.000 Pusat Penelitian dan Pengembangan Geologi: Bandung.
[4] T.H. Yang, et.al 2014 Seepage-stress coupled analysis on anisotropic characteristics of the fractured rock mass around roadway *Tunneling and Underground Space Technology* 34, 11-19.
[5] PT. Indra Karya (Persero) Consulting Engineers 2019 Preparation of Jragung Multipurpose Dam Project Jakarta: Government of The Republic of Indonesia, Ministry of Public Works and Housing, The Directorate General of Water Resources
[6] Z. T. Bieniawski 1989 Engineering Rock Mass Classifications: A Complete Manual for Engineers and Geologists in Mining, Civil, and Petroleum Engineering Canada: John Wiley & Sons.
[7] S. M. Abbas and H. Konietzky 2017 Rock Mass Classification Systems Peshawar: TU Bergakademie Freiberg, Geotechnical Institute.
[8] Qi Zhang, Xianbin Huang, Hehua Zhu, Jianchun Li 2019 Quantitative assessments of the correlations between rock mass rating (RMR) and geological strength index (GSI). *Journal of Tunnelling and Underground Space Technology* 83(10) 73-81
[9] Badan Standardisasi Nasional 2008 SNI 2825:2008 Cara uji kuat tekan batu uniaksial Jakarta: Badan Standardisasi Nasional.
[10] E.T. Brown 1981 The ISRM Suggested Methods for Rock Characterization, Testing, and Monitoring.
[11] H. Mirmehrabi, J. Hassanpour, M. Morsali, and S. T. Azali 2008 Experiences Gained From Gas and Water Inflow Toward The Tunnel, Case Study: Aspar Anticline, Kermanshah, Iran. *ISRM Int. Symp. - 5th Asian Rock Mech. Symp.* 5(11) 1469–1476.
[12] K. Reinhard 2009 Groundwater Geophysics A Tool for Hydrogeology 2nd edition Berlin: Springer.
[13] M.H. Loke, J.E. Chambers, D.F. Rucker, O. Kuras, P.B. Wilkinson 2013 Recent Developments in The Direct-current Geoelectrical Imaging Method *Journal of Applied Geophysics* 95(3) 135-156.
[14] PT. Geochem Survey Internasional 2022 Laporan akhir penyelidikan geolistrik pada lokasi bangunan pengelak Semarang: Abipraya-Pelita, KSO
[15] R.W. Van Bemmelen 1949 The Geology of Indonesia Vol. 1A - General Geology of Indonesia and Adjacent Archipelagoes Batavia: The Hague.
[16] N.W. Purasongka, I. Syafri, L. Jurnaliah 2015 Karakteristik batuan sedimen berdasarkan analisis petrografi pada formasi kalibeng anggota banyak *Bulletin of Scientific Contribution* 13(1) 1-15.
[17] Ziyuan Li, Liqiong Wang, Bo Feng, Jie Xiao, Qi Zhang, Liying Li, Jie Liang 2020 Comprehensive Collapse Investigation and Treatment: An Engineering Case from Qingdao Expressway Tunnel. *Journal of Cleaner Production* 270(6) 1-15.
[18] Abipraya-Pelita, KSO 2020 Laporan Trial Mix Beton Plastis pada Proyek Pembangunan Bendungan Semantok Jakarta: Kementrian Pekerjaan Umum dan Perumahan Rakyat, Direktorat Jendral Sumber Daya Air.
[19] ICOLD 2005 Dam Foundations, Geologic Considerations, Investigation Methods, Treatment, Monitoring *Bulletin 129*.

Fuzzy Logic Method To Manage The Risks Of Oil And Gas Pipeline Projects To The Environment

N N Rodhi[1*], I P A Wiguna[2] and N Anwar[2]

[1]Fakultas Sains dan Teknik, Universitas Bojonegoro, Bojonegoro
[2]Fakultas Teknik Sipil, Perencanaan dan Kebumian, Institut Teknologi Sepuluh Nopember Surabaya
*Corresponding author: nova.nevila@gmail.com

Abstract. Pipes are one of the supporting infrastructures that are needed in oil and gas projects. However, oil and gas pipeline construction projects have very complex risks and have a major impact on the environment. On the other hand, environmental factors are one of the risk factors for oil and gas pipeline projects with a value of 14% [2]. The existence of risk factors that are not managed properly will have a negative impact that accumulates and causes a hazard. Therefore, an affective risk analysis method is needed. The purpose of this study is to analyze the risks of oil and gas pipeline construction projects to the environment using the Fuzzy Logic method based on data obtained from questionnaires distributed to Oil and Gas Cooperation Contract Contractors operating in East Java – Indonesia. The results of the analysis show that of the 20 (twenty) risk factors, there are 2 risk factors that are at a high level, namely financial performance risk with a value of 14.9 and contractor experience risk with a risk value of 14.9 and the fuzzy logic method are considered to be used as an effective reference in decision making, this can be seen from the results of the calculation of the accuracy value which shows that the risk analysis system for the environment is accurate with an accuracy value of 0.90 which means the accuracy level is very high.

Keywords: Oil and Gas Pipeline, Risk Management, Environmental Risk, Risk Analysis, Fuzzy Logic

1. Introduction

One of the supporting infrastructure that is needed in the project is a pipeline network that functions to support the transmission and distribution of oil and natural gas both in onshore and offshore areas. The use of pipelines for the transmission process is considered effective and efficient when compared to using other modes of transportation, but there are also consequences caused by oil and gas pipeline projects, such as negative impacts caused by the use of hazardous materials, inappropriate methods, human resources. incompetence and the existence of equipment and material transportation, especially if the project is located in a densely populated area [1].

According to Rodhi et al (2019), oil and gas pipeline construction projects have very complex risks and have a major impact on the environment [2]. The risks contained in the oil and gas pipeline project can certainly be anticipated with risk management applications, but if risk management has been applied and still causes negative impacts, then this is one of the reasons for the application of

IOP Conf. Series: Earth and Environmental Science 1233 (2023) 012017 doi:10.1088/1755-1315/1233/1/012017

unsystematic and ineffective risk management. Therefore, in this case it is necessary to develop a risk management concept that can build an effective risk management system [3].

Risk management is one of the keys to project success, but in its application, risk management standards are not widely understood and their application is still poor, including in the piping sector [4]. One of the reasons for this is that the risk analysis process often encounters obstacles related to limited data and ambiguous data, such as incomplete or unreliable data, and subjective information that depends on the respondent's expert judgment. To anticipate such data, previous researchers proposed fuzzy logic as a tool that can be used for risk analysis [5] [6]

The most important stage in fuzzy is forming membership functions [7]. The degree of membership is a thing that controls the fuzzy, the control is a curve that maps the data input points. There are several forms of membership degrees in fuzzy, triangular membership degrees are considered to have better behavior with fuzzy sets that are evenly distributed and can represent fuzzy sets [8].

2. Research methods

The identification of risk factors was obtained based on the results of the literature review which was then verified with real conditions by distributing a preliminary questionnaire. Then the risk factor assessment was obtained based on the results of distributing questionnaires to respondents consisting of Health, Safety, Environment (HSE) managers and Engineering Project Risk Managers in 4 (four) oil and gas Cooperation Contract Contractors (KKKS) companies in East Java. Furthermore, the data were analyzed using the Fuzzy Logic method. The study location is an oil and gas pipeline network on the island of East Java, as can be seen in the following map

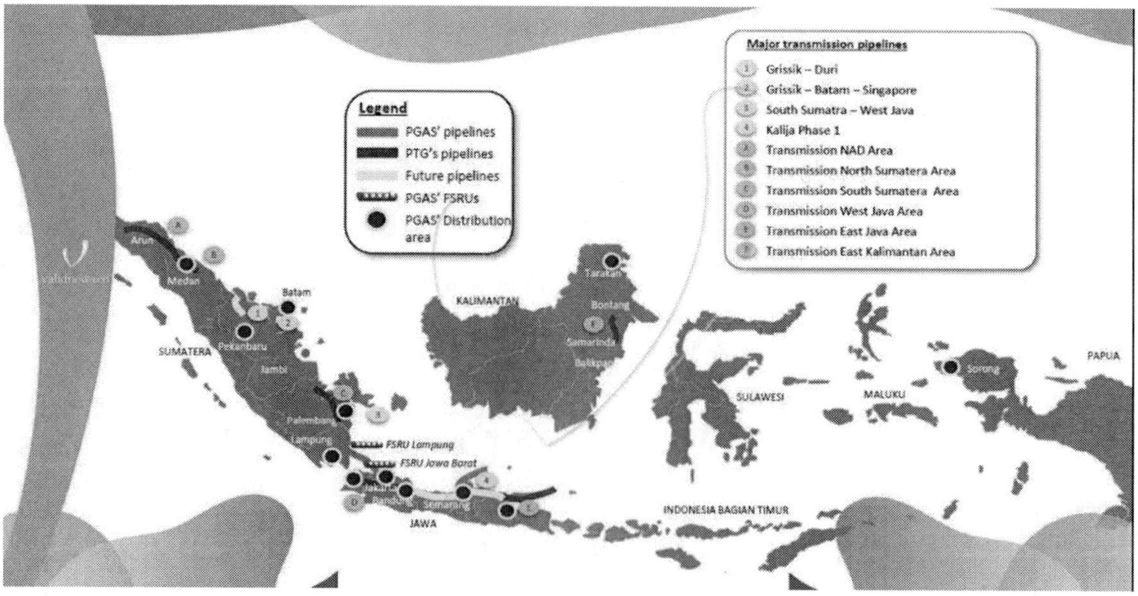

(Source: PGN, 2019)
Figure 1. Oil and Gas Pipelines in Indonesia

3. Results and Discussion

From the results of the identification of risk factors obtained from the literature study and adapted to field conditions, it can be seen that the risk factors that occur in the oil and gas pipeline project are in table 1 below.

Table 1. Risk factors for oil and gas pipelines

No	Variable	Description
1	Policy	There is a change in government policy (eg local content regulations)
2	Construction technology	Technological failure
3	Health and safety	Compensation in the form of public services
4	Human Resources	Incompetent human resources
5	Employment Opportunity	The emergence of job opportunities for local communities
6	Communication to the public	Poor method of communicating to the public
7	Land use	There is a change in population
8	Conservation of cultural and natural heritage	Does not minimize damage to sensitive landscapes, including areas valuable from a scenic, cultural, historical, or architectural point of view
9	Financial performance	Uncontrolled company finances (Start-up costs and long-term expenses)
10	Energy efficiency	There is no instrument that can measure the efficiency of energy use
11	Design	Environmentally conscious design
12	Material	Inappropriate material selection and use (supplier type and delivery)
13	Metode	Choosing the wrong construction method
14	Standard Operating Procedure (SOP)	SOPs that do not apply the concept of Sustainable development
15	Contractor experience	Inappropriate Contractor Safety Management System in Selecting environmentally responsible suppliers and contractors who can demonstrate environmental performance
16	Reduce	(Design) does not reduce the use of the four common sources used in construction; energy, water, materials, and land, at every stage in the project life cycle
17	Reuse	(Design) does not maximize resource reuse, and/or recycling to reduce waste
18	Recycle	(Design) not using recycled products (reduction of raw materials used in new products)
19	Renewable	(Design) do not use renewable resources to choose non-renewable resources
20	Natural disasters	The occurrence of disasters caused by matters related to climate and geographical factors.

(Source: Identification Results, 2016)

From table 1, it can be seen that there are 20 risk factors identified and validated in the oil and gas pipeline project. The identification results are then analyzed using Fuzzy Logic. In this method the selected function is a triangular function. Meanwhile, the input and output membership degrees are built based on the risk categories that have been used in oil and gas industrial companies, namely low, medium and high as shown in Figure 1 below.

IOP Conf. Series: Earth and Environmental Science 1233 (2023) 012017 doi:10.1088/1755-1315/1233/1/012017

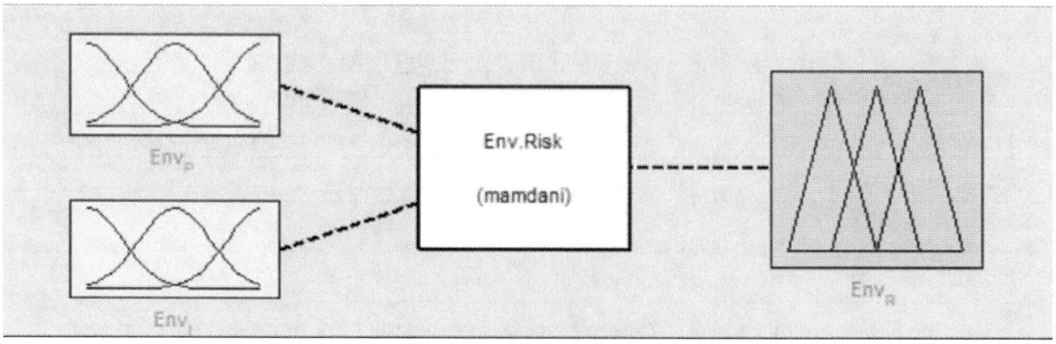

(Source: Analysis Results, 2020)
Figure 2. Membership function of oil and gas pipeline network risk

The next step is to develop system rules that refer to the membership degrees that have been prepared previously. In this study, the rules used are connecting and because it is a form of multiplication, namely

R = P x I
Where:
R : Risk
P : Probability
I : Impact

The results of the fuzzy analysis can be seen in Figure 2

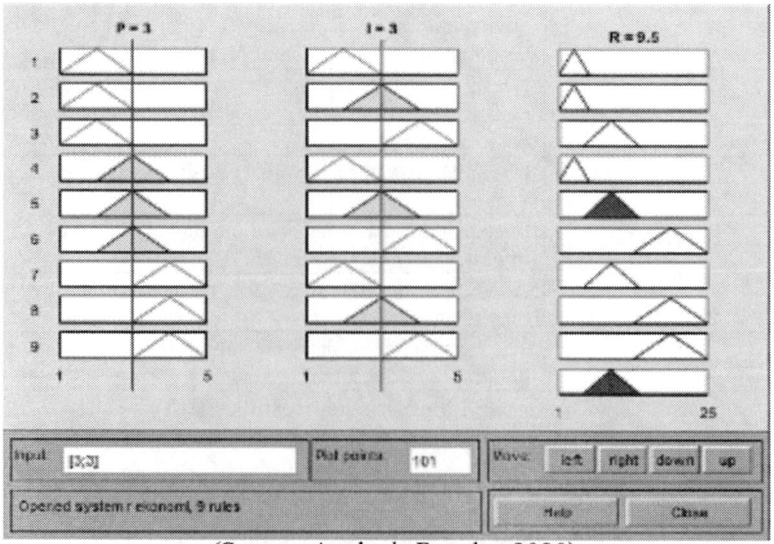

(Source: Analysis Results, 2020)
Figure 3. Graph of the simulation results of one of the risk factors with fuzzy

The graph shows that, if the probability (P) has a value of 3 and the impact (I) is also 3 then the resulting risk value is 9.5. All 20 existing risk factors were simulated as shown in Figure 2. Overall, the results of risk analysis using fuzzy are presented in table 2.

Table 2. The results of the risk analysis of the oil and gas pipeline

No (Code)	Risk Factor	Konvensional				*Fuzzy*	
		P	I	R	Level	R	Level
1	Policy	2	3	6	M	7.4	M
2	Construction technology	3	3	9	M	9.5	M
3	Health and safety	3	2	6	M	7.4	M
4	Human Resources	3	3	9	M	9.5	M
5	Employment Opportunity	3	3	9	M	9.5	M
6	Communication to the public	3	3	9	M	9.5	M
7	Land use	3	3	9	M	9.5	M
8	Conservation of cultural and natural heritage	3	3	9	M	9.5	M
9	Financial performance	2	4	8	M	14.9	H
10	Energy efficiency	3	3	9	M	9.5	M
11	Design	3	3	9	M	9.5	M
12	Material	3	3	9	M	9.5	M
13	Metode	3	3	9	M	9.5	M
14	Standard Operating Procedure (SOP)	3	3	9	M	9.5	M
15	Contractor experience	3	4	12	M	14.9	H
16	Reduce	3	3	9	M	9.5	M
17	Reuse	3	3	9	M	9.5	M
18	Recycle	2	3	6	M	7.4	M
19	Renewable	2	3	6	M	7.4	M
20	Natural disasters	2	3	6	M	7.4	M

(Source: Analysis Results, 2020)

From table 2 it can be explained that the overall risk value obtained conventionally is at level S (medium), while with the fuzzy method there are 2 risks with codes 9 (financial performance) and 15 (contractor experience) having a higher risk value and being in the middle. T level (high).

In detail, the results in table 2 can be explained that for risk factors code 1 is for policy risk factors, code 2 is for construction technology risk factors, code 3 is for health and safety risk factors, code 4 is for human resources risk factors, code 5 is for human resources risk factors. employment risk factors.

Code 6 for public communication risk factors, code 7 for land use risk factors, code 8 for cultural and natural heritage conservation risk factors, code 9 for financial performance risk factors, code 10 for energy efficiency risk factors, Code 11 for design risk factors , code 12 for material risk factors, code 13 for method risk factors, code 14 for standard Operating Procedures (SOP) risk factors.

Code 15 is for contractor experience risk factor, code 16 is for reduce risk factor, code 17 is for reuse risk factor, code 18 is for recycle risk factor, code 19 is for renewable risk factor and code 20 is for natural disaster risk factor.

The results of conventional analysis for code 1 produce a risk value of 9 which is at the moderate level (S), while the fuzzy method produces a risk value of 9.5 which is at the moderate level (S). Code 2, conventionally produces a risk value of 6 which is at the medium level (S), while the fuzzy method produces a risk value of 7.4 which is at the medium level (S).

Code 3, conventionally produces a risk value of 9 which is at the medium level (S), while the fuzzy method produces a risk value of 9.5 which is at the medium level (S). Code 4, conventionally produces a risk value of 9 which is at the medium level (S), while the fuzzy method produces a risk value of 9.5 which is at the medium level (S).

Code 5, conventionally produces a risk value of 9 which is at the medium level (S), while the fuzzy method produces a risk value of 9.5 which is at the moderate level (S). Code 6, conventionally produces a risk value of 9 which is at the medium level (S), while the fuzzy method produces a risk value of 9.5 which is at the medium level (S).

Code 7, conventionally produces a risk value of 9 which is at the medium level (S), while the fuzzy method produces a risk value of 9.5 which is at the medium level (S). Code 8, conventionally produces a risk value of 9 which is at the medium level (S), while the fuzzy method produces a risk value of 9.5 which is at the moderate level (S).

Code 9, conventionally produces a risk value of 8 which is at the moderate risk level (S), while the fuzzy method produces a risk value of 14.9 which is at the high level (T). Code 10, conventionally produces a risk value of 9 which is at the moderate level (S), while the fuzzy method produces a risk value of 9.5 which is at the moderate level (S).

Code 11, conventionally produces a risk value of 9 which is at the moderate level (S), while the fuzzy method produces a risk value of 9.5 which is at the moderate level (S). Code 12, conventionally produces a risk value of 9 which is at the moderate level (S), while the fuzzy method produces a risk value of 9.5 which is at the moderate level (S).

Code 13, conventionally produces a risk value of 9 which is at the moderate level (S), while the fuzzy method produces a risk value of 9.5 which is at the moderate level (S). Code 14, conventionally produces a risk value of 9 which is at the moderate level (S), while the fuzzy method produces a risk value of 9.5 which is at the moderate level (S).

Code 15, conventionally produces a risk value of 12 which is at the moderate risk level (S), while the fuzzy method produces a risk value of 14.9 which is at the high level (T). Code 16, conventionally produces a risk value of 9 which is at the medium level (S), while the fuzzy method produces a risk value of 9.5 which is at the moderate level (S).

Code 17, conventionally produces a risk value of 9 which is at the moderate level (S), while the fuzzy method produces a risk value of 9.5 which is at the moderate level (S). Code 18, conventionally produces a risk value of 6 which is at the medium level (S), while the fuzzy method produces a risk value of 7.4 which is at the moderate level (S).

Code 19, conventionally produces a risk value of 6 which is at the medium level (S), while the fuzzy method produces a risk value of 7.4 which is at the medium level (S). Code 20, conventionally produces a risk value of 6 which is at the medium level (S), while the fuzzy method produces a risk value of 7.4 which is at the medium level (S).

With the results obtained, further control is carried out on the accuracy value, which serves to determine whether the fuzzy analysis is accurate. The accuracy assessment is carried out by calculating the accuracy value of the expert domain scenario with the following formula

$$V = \frac{\Sigma \text{case} - \Sigma \text{ inequality}}{\Sigma \text{case}}$$

$$= \frac{20 - 2}{20} = 0.90$$

The results of the accuracy calculation show that the risk analysis system for the environment is accurate with an accuracy value of 0.90 which means the accuracy level is very high.

4. Conclusion

The results of the analysis show that of the 20 (twenty) risk factors, there are 2 risk factors that are at a high level, namely financial performance risk with a value of 14.9 and contractor experience risk with a risk value of 14.9.

In this study, the results of the analysis using the fuzzy logic method can be used as a decision to make decisions, it can be seen from the results of the calculation of the accuracy value which shows that the risk analysis system for the environment is accurate with an accuracy value of 0.90 which means the accuracy is very high.

References

[1] A. Hasan, "Security of Cross-Country Oil and Gas Pipelines:A Risk-Based Model," *Journal of Pipeline Systems,* no. ASCE, ISSN 1949-1190., pp. 04016006-1 to 04016006-8, 2016.

[2] N. N. Rodhi, I. P. A. Wiguna and N. Anwar, "DISASTER RISK MITIGATION OF OIL AND GAS PIPELINES PROJECT IN JAVA ISLAND INDONESIA," *ARPN Journal of Engineering and Applied Sciences,* vol. Vol 14 No 24, p. 4160 to 4163, December, 2019.

[3] I. P. A. Wiguna, N. Anwar, A. Widodo and N. N. Rodhi, "Risk Management Effectiveness of Oil And Gas Pipeline Cunstruction in Java In Java Island - Indonesia," in *International Conference Civil Engineering Research (ICCER)*, Surabaya, 2017.

[4] K. Nielsen, "Risk Management: Lessons From Six Continennts," *Journal of Management in Engineering,* vol. Vol 22 No 2, no. 2, 2006.

[5] A. Shapiro and M. Koissi, "Risk Assessment Aplications of Fuzzy Logic," Canadian Institute of Actuaries, Canada, 2015.

[6] I. Orji and S. Wei, "An Innovative Integration of Fuzzy-Logic and Systems Dynamics in Sustainable Supplier Selection: A Case on Manufacturing Industry," *Computers and Industrial Engineering,* p. 1 to 12, 2015.

[7] G. Dag and M. Bagriyanik, "The Effect of Different Fuzzy Membership Function Forms on Controlling Loop Flows," in *IEEE Xplore*, 2009.

[8] N. N. Rodhi, "Model Analisis Risiko Proyek Jaringan Pipa Migas Onshore Untuk Sustainable Construction di Pulau Jawa," Institut Teknologi Sepuluh Nopember, Surabaya, 2021.

IOP Conf. Series: Earth and Environmental Science 1233 (2023) 012018 doi:10.1088/1755-1315/1233/1/012018

Lithium Extraction from Geothermal Brine Using Electrodialysis Method

A D Jati[1], V S H Sujoto[1], Sutijan[1], H T B M Petrus[1] and P Mulyono[1,*]

[1]Department of Chemical Engineering (Sustainable Mineral Processing Research Group), Faculty of Engineering, Universitas Gadjah Mada, Jl. Grafika 2, D.I. Yogyakarta 55281, Indonesia
*Corresponding author: pmulyono@ugm.ac.id

Abstract. The global market share of electric vehicles is expected to increase significantly. The lithium-ion battery is one type of battery that can power electric vehicles because light, inexpensive, has a high energy density, and has a long service life. Lithium may be obtained economically from hard rock ores and brines. Indonesia has a huge geothermal potential and lithium concentration in geothermal brine can reach 70 mg L^{-1}. With the high concentration of lithium in brine, this is a great opportunity for Indonesia to produce lithium independently. The separation of lithium from brine can be separated by various methods. One method that has high effectiveness is by using the electrodialysis method. This study will extract lithium from geothermal brine with the electrodialysis method using monovalent membranes. Anion exchange membrane (AEM) and cation exchange membrane (CEM) are used in this system. The recovery solution to collect lithium from brine is HCl 0,1 M. We variate the voltages (2 volt and 3 volts) to see the effect on flux and conductivity. From the experiment, the 3 volts give more flux and conductivity than the 2 volts. The 3 volts has 10 mL cm^{-2} h^{-1} of flux and 2110 µS cm^{-1} of conductivity while the 2 volts has 4 mL cm^{-2} h^{-1} of flux and 1770 µS cm^{-1} of conductivity. The higher voltages given to the system, the easier separation can be done.

1. Introduction

Technological developments nowadays have created breakthroughs like electric vehicles. The focus on renewable energy has also increased in the 21st century. To increase the adoption of electric vehicles, the government gives incentives to the public around the world [1]. The global share of electric vehicles is expected to increase from 2% in 2016 to 22% in 2030. For example, new electric vehicles global sales passed a million units in 2017. In 2020 EVs had sold 4.5 million units or equal to 5% of the overall global light-vehicle market. From that point, sales of EVs are expected to be 10 million in 2025, 28 million in 2030, and 56 million in 2050 [2].

Electric vehicles have many advantages; they can reduce fossil fuel dependence and have high energy efficiencies compared to conventional cars. An electric vehicle is powered by electricity from a battery. There are types of batteries that are rechargeable, like lithium batteries, nickel batteries, lead batteries, etc [3]. Compared to other battery types, lithium-ion batteries have several benefits, such as being light, inexpensive, having high energy density, low charge loss, long service life, and large number of charge/discharge cycles [4-5].

Lithium may be obtained economically from hard rock ores and brines [6]. Approximately 62% of commercial lithium comes from underground brine, while 38% comes from rock minerals like spodumene that are mined [7]. Brine is the primary source of lithium production in Chile and Argentina. Whereas Australia, Canada, Zimbabwe, Portugal, and Brazil are the top exporters of lithium concentrates from hard rock ores [8]. Gravity separation (DMS), magnetic separation, and froth flotation are the main methods utilized in the beneficiation of lithium materials. In recent years, sorting has also been acknowledged as a promising technique [8]. Lithium extraction from brine has been accomplished using a variety of desalination techniques. The most common process is solar evaporation, which has many steps for precipitation and crystallization. However, this process takes a long time and depended on the weather [9].

Indonesia has a huge geothermal potential since the country is in the ring of fire that is equivalent to about 30 GW [10]. Lithium content in geothermal fluids that emerge on the surface as geothermal representations are mainly less than 20 mg L^{-1} [11]. Brine water from geothermal systems, including Dieng and Wayang Windu, appears to have more lithium. Lithium concentrations of up to 70 mg L^{-1} can be found in the chloride brine water from deep reservoirs [12–14]. However, there are just a few milligrams per liter of lithium in steam-dominated systems, as those in Kamojang and Darajat. The high lithium anomaly in the brine water is probably the result of intensive water circulation in a closed system during geothermal exploitation [14].

It has been investigated how to extract lithium from brine using lithium adsorbents [15-16]. However, to achieve cost-effective manufacturing, the efficiency of lithium absorption must be enhanced because it has low selectivity. Other separation technologies such as solvent extraction, ion exchange, and nanofiltration (NF) have been developed to extract lithium from brine [17]. Electrodialysis is one technique of separation with ion exchange membranes. The process is transferring ions through a membrane because of the difference in potential electrical charge [18]. Electrodialysis is alternately arranged between an anion exchange membrane (AEM) and a cation exchange membrane (CEM), an anode electrode, and a cathode electrode [9]. The majority of the electrical systems in electrodialysis used direct current (DC) [14–16]. Compared to other methods, electrodialysis has high efficiency in lithium extraction.

Electrodialysis has significant features in seawater desalination, metal resource extraction from salt lake brines, and industrial effluent treatment [19]. Electrodialysis is used as a potentially effective approach in concentrated brines due to its excellent desalination performance and high-efficiency enrichment of high concentrations of salt solution [20]. Selectrodialysis (SED), a new type of electrodialysis, is created when monovalent selective ones partially replace the conventional ion exchange membranes [21]. Nie et al [22]. researched the electrodialysis separation of lithium ions from magnesium ions, and the findings demonstrated that electrodialysis performs better NF both technically and economically for the fractionation of Li^{+}/Mg^{2+} [23].

This study will extract lithium from geothermal brine with the electrodialysis method using a monovalent membrane. Flux is observed to know how fast electrodialysis membrane can separate lithium from brine. The quantity of lithium can be represented by the conductivity changes in permeate solution. The higher conductivity means ions in the solution are highly concentrated. Selectivity of lithium separation with other ions can be calculated after specific ions are measured. In this study, we will observe the effect of potential voltages on flux and conductivity in permeate.

2. Materials and method

2.1. Geothermal Brine and Recovery Solution

Geothermal brine samples were obtained from PT Geo Dipa Energi Dieng, Central Java, Indonesia. Brine was added by acid to modify in acid conditions so precipitation of salt could be reduced. The water type of brine in Dieng is chloride type. Hydrochloric acid (HCl) 0,1 M was used for recovery solutions that were suitable for this brine. Recovery solution made from high concentration HCl (37%) and diluted with water.

IOP Conf. Series: Earth and Environmental Science 1233 (2023) 012018 doi:10.1088/1755-1315/1233/1/012018

2.2. Electrodialysis Set Up

This research uses two types of membranes: anion exchange and cation exchange. The monovalent ion exchange membrane used in this research was collected from Hangzhou Huamo Technology Co., Ltd. Material of the membrane is made from polyethylene. The membranes must to soaked in NaCl 5% solutions for one day. Configuration for the electrodialysis stack showed in Figure 1.

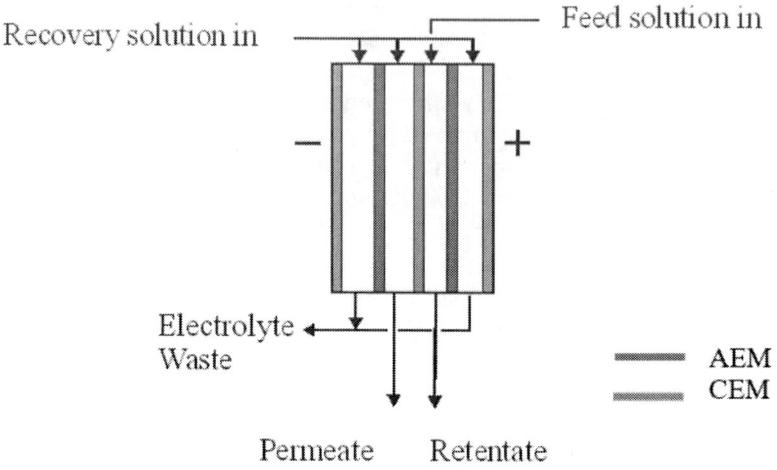

Figure 1. Configuration of Electrodialysis Stack

Product from the electrodialysis system is collected in permeate. Ions that cannot be transferred from feed will be collected in the retentate. In this study, we analyzed the increased volume flow rate and conductivity in permeate. Volume flow rate is recorded every 5 minutes, and conductivity is done every 15 minutes. During the experiment, voltages and electric current were recorded every 5 minutes.

2.3. Flux Calculation

Mass transfer in electrodialysis system based on Nernst-Planck equation divided into three terms [24]:

1. Diffusion that affected by the difference in concentration
2. Electro-potential
3. Convection that affected by the osmotic pressure

The flux equation for the electrodialysis system can be simplified with equation 1

$$Jv = \frac{V}{A \times t} \quad (1)$$

Where Jv is the flux of solutions, V is the volume of solutions, A is the effective membrane surface area, and t is the time.

3. Result and discussion

3.1. Flux

Flux is calculated every 5 minutes from t=0 until 120 minutes with a variety of potential voltage. This study variates voltages to 2 and 3 volts. Electrodialysis separation is done with atmospheric pressure and temperature. Figure 2 shows that the higher voltages the higher flux that can be generated.

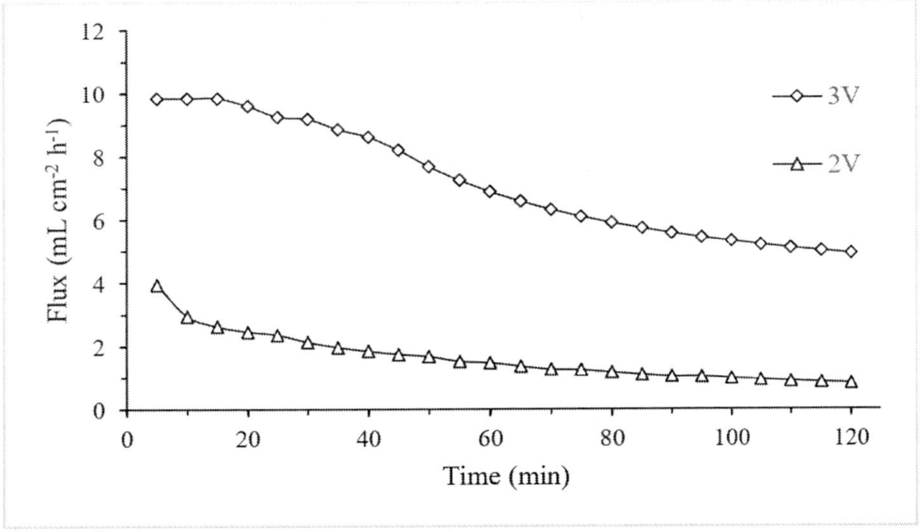

Figure 2. Effect of Voltages on Flux

After 5 minutes initial flux for 2 volts and 3 volts can be observed. From Figure 2 initially, 3 volts has more flux (10 mL cm^{-2} h^{-1}) than the 2 volts (4 mL cm^{-2} h^{-1}). During the observation, both fluxes are decreasing. The decrease is caused by blocking from precipitated salt that maybe occur and the resistance that comes from the layer of high-valent ions. Until the end of observation, the 3 volts still has more flux (5 mL cm^{-2} h^{-1}) than the 2 volts (1 mL cm^{-2} h^{-1}). Theoretically, the more voltages are given to the system, the easier separation can be done. High voltages carry more energy than low voltages and the ions in the system can move rapidly due to high attraction by potential charges.

3.2. Conductivity

Conductivity is a representation of how many ions are in the solutions. In this study, we observe the effect of voltages on conductivity in permeate solution. The result of conductivity in the experiment of electrodialysis separation with selective monovalent ion membrane showed in Figure 3.

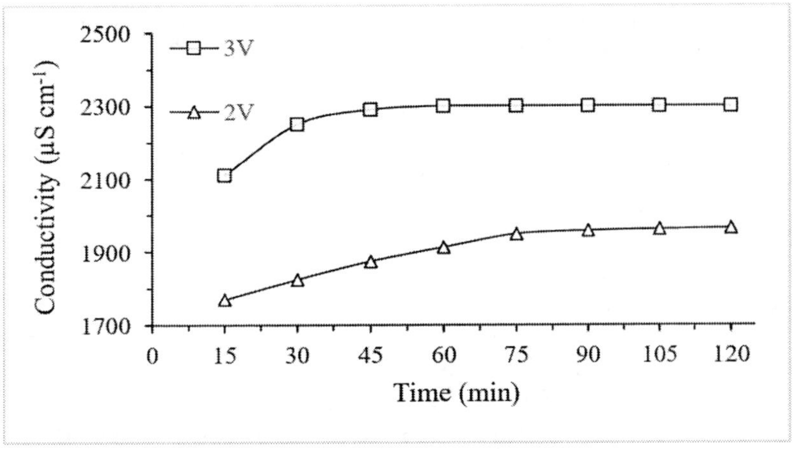

Figure 3. Effect of Voltages on Conductivity in Permeate

Permeate solution conductivity was measured every 15 minutes. Initially, the 2 volts has conductivity at 1770 μS cm^{-1} and the 3 volts has 2110 μS cm^{-1}. The 3 volts has a high conductivity because high voltages have more energy to attract the ions so ions can be more collected than low voltages. As time increases, the conductivity will also increase. It means that ions in feed solutions transfer to permeate solution through membranes. From the graphic in Figure 3, there is a flat line which means there is no ion that is transferred through the membrane. Those phenomena are caused by blocking from precipitated salt and high valent ions or the system has no more energy to transfer the ions.

4. Conclusion

This research investigated how electrodialysis system can separate lithium from geothermal brine. The difference in voltages give effect on the flux in membrane and conductivity in permeate. Increasing voltages can increase the flux and the conductivity. High voltages give more energy to attract the ions in brine solutions. During the experiment, the membrane will be blocked by precipitation salt or layer of high-valent ions as time increases. The optimum condition for separating lithium from brine solutions in this research obtained at 3 volts.

References

[1] Journal I 2020 IJERT-Global Perspective on Electric Vehicle 2020 Global Perspective o n Electric Vehicle 2020

[2] Lu, Sophie. "Will the Real Lithium Demand Please Stand Up? Challenging the 1Mt-by-2025 Orthodoxy". https://about.bnef.com/blog/will-the-real-lithium-demand-please-stand-up-challenging-the-1mt-by-2025-orthodoxy/.

[3] Costa C M, Lee Y H, Kim J H, Lee S Y and Lanceros-Méndez S 2019 Recent advances on separator membranes for lithium-ion battery applications: From porous membranes to solid electrolytes *Energy Storage Mater.* **22** 346–75

[4] Scrosati B and Garche J 2010 Lithium batteries: Status, prospects and future *J. Power Sources* **195** 2419–30

[5] Izzati A U N, Petrus H T B M and Prasetya A 2021 Nickel recovery from precipitate of NCA lithium-ion battery leach liquor by using disodium ethylene diamine tetraacetate *IOP Conf. Ser. Earth Environ. Sci.* **882**

[6] Ebensperger A, Maxwell P and Moscoso C 2005 The lithium industry: Its recent evolution and future prospects *Resour. Policy* **30** 218–31

[7] Grosjean C, Herrera Miranda P, Perrin M and Poggi P 2012 Assessment of world lithium resources and consequences of their geographic distribution on the expected development of the electric vehicle industry *Renew. Sustain. Energy Rev.* **16** 1735–44

[8] Tadesse B, Makuei F, Albijanic B and Dyer L 2019 The beneficiation of lithium minerals from hard rock ores : A review *Miner. Eng.* **131** 170–84

[9] Sutresno V, Sujoto H, Widi S, Slamet A, Isana S and Yosephine S 2022 Effect of Operating Conditions on Lithium Recovery from Synthetic Geothermal Brine Using Electrodialysis Method *J. Sustain. Metall.* **8** 274–87

[10] Pambudi N A, Itoi R, Yamashiro R, CSS Syah Alam B Y, Tusara L, Jalilinasrabady S and Khasani J 2015 The behavior of silica in geothermal brine from Dieng geothermal power plant, Indonesia *Geothermics* **54** 109–14

[11] Nicholson K 1993 Geothermal Systems BT - Geothermal Fluids: Chemistry and Exploration Techniques ed K Nicholson (Berlin, Heidelberg: Springer Berlin Heidelberg) pp 1–18

[12] Sutijan, Wahyudi S, Ismail M F, Mustika P C B, Astuti W, Prasetya A and Petrus H T B M 2022 Forward Osmosis to Concentrate Lithium from Brine: The Effect of Operating Conditions (pH and Temperature) *Int. J. Technol.* **13** 136–46

[13] Mustika P C B W, Sutijan S and Petrus H T B M 2021 Effect of operating conditions on the flux recovery of lithium from geothermal brine using forward osmosis *AIP Conf. Proc.* **2349**

[14] Herdianita N R, Sucipta I G B E and Kencana A Y 2019 Lithium in Brine Waters from The Indonesian Geothermal Systems : Could it Meet the National Needs of Making Lithium Batteries ? *7th Indones. Int. Geotherm. Conv. Exhib.*

[15] Ooi K, Miyai Y and Katoh S 1986 Recovery of Lithium from Seawater by Manganese Oxide Adsorbent *Sep. Sci. Technol.* **21** 755–66

[16] H Tangkas I W C W, Astuti W, Sutijan, Sumardi S and Petrus H T B M 2021 Lithium titanium oxide synthesis by solid-state reaction for lithium adsorption from artificial brine source *IOP Conf. Ser. Earth Environ. Sci.* **882**

[17] Hoshino T 2013 Preliminary studies of lithium recovery technology from seawater by electrodialysis using ionic liquid membrane *Desalination* **317** 11–6

[18] Sujoto V S H, Sutijan, Astuti W, Sumardi S, Louis I S Y and Petrus H T B M 2022 Effect of Operating Conditions on Lithium Recovery from Synthetic Geothermal Brine Using Electrodialysis Method *J. Sustain. Metall.* **8** 274–87

[19] Chen Q, Ji Z, Liu J, Zhao Y and Wang S 2018 Development of recovering lithium from brines by selective-electrodialysis : E ff ect of coexisting cations on the migration of lithium *J. Memb. Sci.* **548** 408–20

[20] Zhang Y-F, Liu L, Du J, Fu R, Van der Bruggen B and Zhang Y 2017 Fracsis: Ion fractionation and metathesis by a NF-ED integrated system to improve water recovery *J. Memb. Sci.* **523** 385–93

[21] Zhang Y, Paepen S, Pinoy L, Meesschaert B and Van der Bruggen B 2012 Selectrodialysis: Fractionation of divalent ions from monovalent ions in a novel electrodialysis stack *Sep. Purif. Technol.* **88** 191–201

[22] Nie X-Y, Sun S-Y, Sun Z, Song X and Yu J-G 2017 Ion-fractionation of lithium ions from magnesium ions by electrodialysis using monovalent selective ion-exchange membranes *Desalination* **403** 128–35

[23] Ji Z yong, Chen Q bai, Yuan J sheng, Liu J, Zhao Y ying and Feng W xian 2017 Preliminary study on recovering lithium from high Mg2+/Li+ ratio brines by electrodialysis *Sep. Purif. Technol.* **172** 168–77

[24] Moshtarikhah S, Oppers N A W, de Groot M T, Keurentjes J T F, Schouten J C and van der Schaaf J 2017 Nernst–Planck modeling of multicomponent ion transport in a Nafion membrane at high current density *J. Appl. Electrochem.* **47** 51–62

Springs and Tourism Facilities Development in Purwosari, Gunungkidul, Special Region of Yogyakarta, Indonesia

N N Soviana[1*], D P E Putra[2] and G I Marliyani[3]

[1]Master student in the Master Program of Geological Engineering Department, Faculty of Engineering, Universitas Gadjah Mada (UGM)
[2,3]Lecturer in Geological Engineering Department, Faculty of Engineering, Universitas Gadjah Mada (UGM)
*Corresponding author: nabila.novasari@mail.ugm.ac.id

Abstract. There have been progressive development and construction of tourism facilities such as hotels and restaurants in Purwosari. Besides the tourism development potentials, Purwosari, also has problems related to surface water sources. The research on the existence of springs in this place is urgent to support water needs for tourism facility development near tourism sites. This research consists of two parts: (1) to observe springs' existence, water quality, relative water quantity, existence throughout the year, and spring's relative distance to tourism sites, and (2) to determine suitable springs near tourism sites, related to the development of tourism facilities along the southern coast of Purwosari area. Based on field observation and interviews with five key persons in each village, the results show that 27 springs spread over the research area are unevenly distributed. The quality of spring water is good according to its physical-chemical properties. There are 13 springs with relatively abundant discharge, nine springs are relatively near to tourism sites, and 23 flows the water throughout the year. Springs evaluation with the MCDA method for developing tourism facilities along the waterfront of the Purwosari shows that only 11 springs have the potential to be economical and efficient sources of clean water.

Keywords: Springs; Springs evaluation; Tourism facilities development; Purwosari; Gunungkidul

1. Introduction

Purwosari is one of the subdistricts located on the west side of Gunungkidul Regency, Special Region of Yogyakarta. It faces the Indian Ocean directly on the southwest-south side (**Figure 1**). Purwosari subdistrict is also near one of the Yogyakarta tourism centers, Parangtritis beach. Purwosari also has tourism potential since tourism developers are also expanding their development area from Parangtritis beach to Purwosari. The southern coastal areas of the province are designed to be the main tourism sites oriented toward the strategic values of the Indian Ocean [1]. The tourism sites include Parangndog Cliff, Watu Gupit Paragliding Hill, Langse Cave, and Bekah Beach, located near the southern coastal area, Purwosari. The tourism potential in the southern part of the region has made the Purwosari prepare

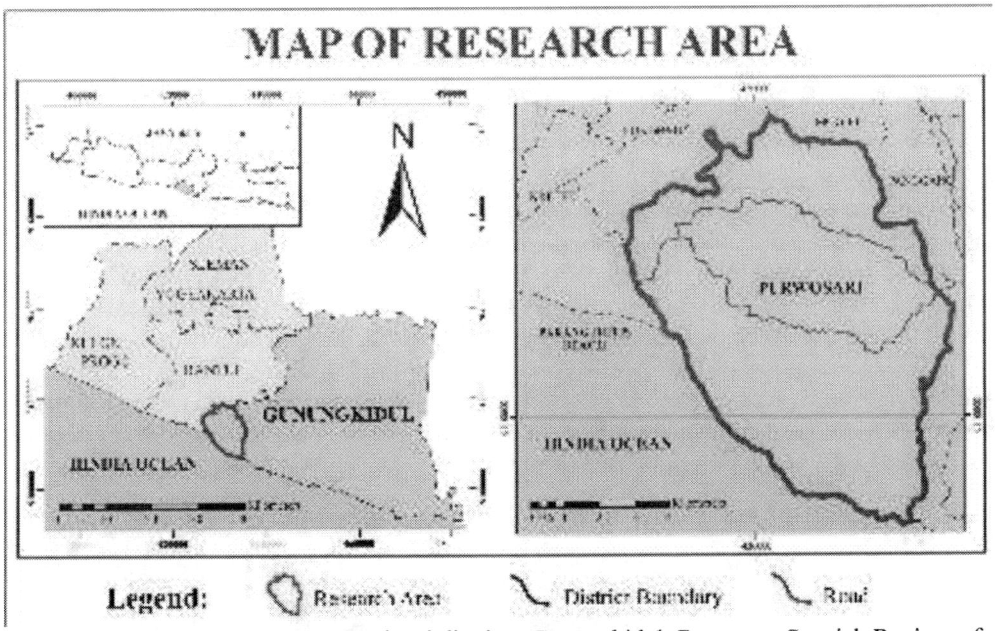

Figure 1. Research area, Purwosari subdistrict, Gunungkidul Regency, Special Region of Yogyakarta [2]

anything needed to attract tourists and tourism developers, such as tourism facilities development located near the tourism sites. Besides the tourism development potential, Purwosari, as a part of Gunungkidul regency that located in the Gunung Sewu karst area [3], also has a problem related to surface water sources [4]. Water source development aims to fulfill water needs for households, agriculture, industry, and tourism [5]. Water source is one of the fundamental parameters that must be noticed for regional development [6] [7] [8] [9]. As water sources are one of the crucial factors when developing tourism facilities and tourism sites, it is essential to look carefully at water sources in Purwosari to ensure that tourism facilities located near tourism sites have a suitable water source.

There have been progressive development and construction of tourism facilities that develop near the tourism sites, such as hotels and restaurants, to support the development of tourism sites in that area, which are expected to increase over time. The tourism facilities development must be in line with the availability of water. The water pipeline network from PDAM Tirta Handayani, a local state-owned water utility company in Gunungkidul, covers only a tiny part of the Purwosari area [4]. Deep wells by a tourism developer drilled near one of the tourism sites in Purwosari produced saline water [10]. It is essential to develop other water sources in Purwosari to ensure tourism facilities located near tourism sites have a suitable water source. According to an interview with key persons in Purwosari in 2022, springs as water source potential have an essential role in Purwosari, supporting tourist and communities' water needs.

Due to these reasons, a study on the existence of springs in the Purwosari, Gunungkidul Regency of Yogyakarta is necessary. The research objective is to (1) to observe the existence of springs' water quality, water relative quantity, existence throughout the year, and spring relative distance to tourism sites, and (2) to determine suitable springs near tourism sites related to the development of tourism facilities along the southern coast of Purwosari area. This research can be a recommendation for the government, tourism developers, and communities to maximize the most suitable springs as water sources wisely to support tourism facilities development located near tourism sites in Purwosari.

2. Hydrogeology of Research Area

Purwosari subdistrict is located within the karst area of Gunung Sewu, Gunungkidul [3]. It has a complex hydrogeological system [10]. Gunung Sewu hydrogeological system can be divided into five subsystems, namely (1) Panggang subsystem; (2) Bribin-Baron-Seropan subsystem, (3) Ponjong subsystem, (4) Pracimantoro and Giritontro subsystem, and (5) Donorojo-Pringkuku subsystem [11]. The research area is in the Panggang subsystem. Springs water is a part of groundwater that comes out naturally at the earth's surface [12]. The springs in this subsystem are unique because they lie in the middle of the Gunung Sewu karst. Meanwhile, the springs in other subsystems emerge in the formation contact zone or Graben Bantul [11]. The springs in Panggang subsystem have some typical characteristics, such as small underground river flows that generally appear in the form of springs with small discharges; even some springs run out of water during the dry season [13].

Lithologically, the research site is dominant with limestone [14] with karst topography [15]. Rock composition significantly influences water quality [16]. The limestone has a low salinity level, resulting in good water quality [16]. The springs generally emerge at specific points [17]. The springs in Gunungkidul are inversely proportional to the success of drilling production wells [17]. The springs appear in areas where production wells have not been drilled and do not exist in places where the production wells have been drilled [17]. Based on the Yogyakarta Regional Hydrogeological Map Sheet IX [18] (**Figure 2**), the research area covers two types of groundwater groups: (a) aquifers with medium productivity and (b) areas with limited groundwater.

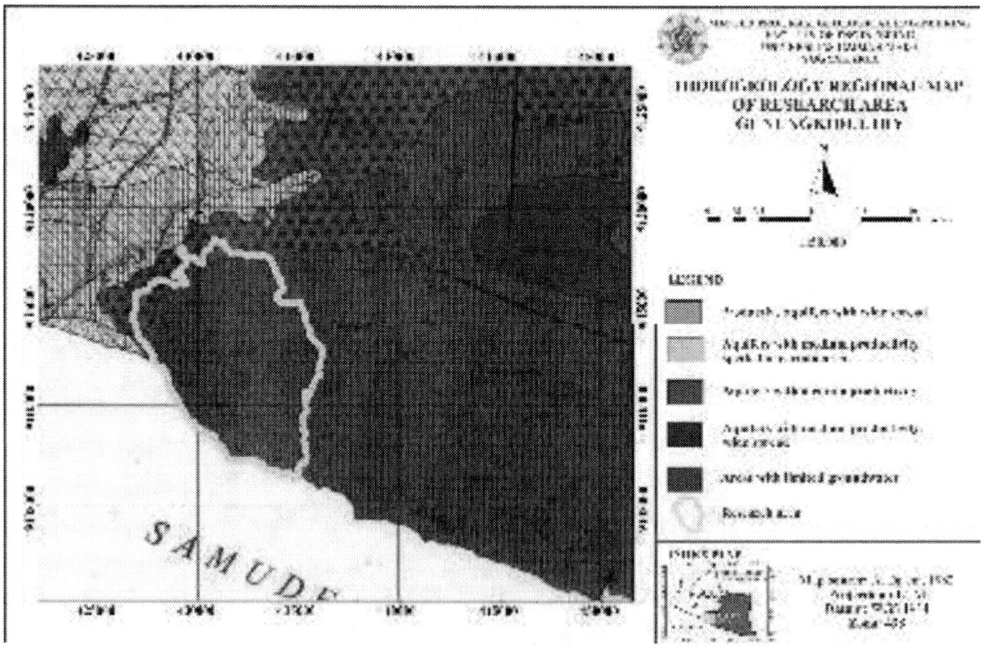

Figure 2. Yogyakarta Regional Hydrogeological Map Sheet IX [18]

3. Methods

3.1. Data collection

This research used primary and secondary data. Secondary data explain preliminary data about springs location and springs condition in Purwosari. Primary data consisted of springs' water quality and springs' distance to tourism sites. Preliminary data about the location and condition of spring in Purwosari were collected with stratified sampling [19] by interviewing five key persons in each village in Purwosari, Girijati Village, Giriasih Village, Giritirto Village, Giripurwo Village, and Giricahyo Village. The information collected for springs condition consists of relative quantity and spring flow throughout the year. Primary data, which is spring water quality and the relative distance from springs to tourism sites, were collected through field observation. Field observation measured physical and chemical parameters: water color, taste, odor, temperature, electrical conductivity, pH, and total dissolved solids (TDS) by using Hannameter water test kit during the rainy season (February-March 2022).

3.2. Data analysis technique
Multicriteria Decision Analysis (MCDA) is a common method in regional development [20] [21] [6] [7] [8]. MCDA is used to determine springs' suitability for tourism facilities development in this research [18]. This research has a few steps of MCDA: determine the springs suitability factors criteria, standardization of factor measurements, development of weight, and springs suitability determination.

3.2.1. Springs suitability criteria factors
Springs' suitability criteria factors in Purwosari were water quality, relative water quantity, distance to tourism sites, and existence throughout the year. It is essential to determine water quality in each spring to ensure spring sanitation. Water relative quantity is related to the spring's capability to fulfill tourist water needs, and it is also related to whether springs have an economical and efficient value or not. Distance to tourism site is related to whether springs have an economical and efficient value or not. Existence throughout the year is related to how annual spring flow is.

3.2.2. Standardization of factor measurements
Standardization of factor measurements can refer to literature, preliminary study, and field observation. In this research, it was determined a potential rating scale of 1 up to 3. Low, medium, and high potential rating range for 1-3. Physical and chemical data are examined based on environmental health standards for water media for sanitation hygiene purposes according to Minister of Health Regulation number 32 of 2017 [22] concerning Environmental Health Quality Standards and Water Health Requirements for Sanitary Hygiene, Swimming Pools, Solus per Aqua, and Public Baths (**Table 1**) to determine the springs water quality.

Table 1. The environmental health standards for water media for sanitation hygiene purposes according to the Regulation of the Minister of Health number 32 of 2017 [22]

No.	Parameter	Unit	Quality Standards (maximum grade)
1	Turbidity	NTU	25
2	Color	TCU	50
3	*Total Dissolved Solids*	mg/l	1000
4	Temperature	Celsius degree	temperature +/- 3
5	Taste		plain
6	Odor		Plain
7	pH	mg/l	6,5-8,5

Good water quality can be concluded if the parameter value of data is appropriate to the parameter value in **Table 1,** so water quality was classified into two categories: good and poor. Springs with good water quality have a potential rating of 3, while springs with poor quality have a potential rating of 1. The relative distance from springs to tourism sites was classified into three categories: near, medium, and far. Springs relatively near to tourism sites have a potential rating of 3, while springs relatively far from tourism sites have a potential rating of 1. Relative water quantities were classified into limited, enough, and abundant. Springs with relatively abundant water have a potential rating of 3, while springs with relatively limited water have a potential rating of 1. Existence throughout the year was classified into annual springs (yes) and intermittent springs. Annual springs have a potential rating of 3, while intermittent springs have a potential rating of 1. The standardization of factor measurements in this research can be seen in **Table 2.**

Table 2. Parameter criteria rating for evaluating the springs' quantity and quality

Rating of Water Quality		
Poor		Good
1		3
Rating of Relative Water Quantity		
Limited	Enough	Abundant
1	2	3
Rating of Existence Throughout the Year		
No		Yes
1		3
Rating of Distance to Tourism sites		
Far	Medium	Near
1	2	3

3.2.3. *Development of Weight*
The weight development in this research used a semi-quantitative method, assigning a weight to each factor [21] of springs suitability. The weights of one factor are normalized to another factor. The higher the weight, the higher the level of importance is. The weights used in this research can be seen in **Table 3.**

Table 3. The weighting evaluation of the springs' quantity and quality.

Weighting	
Water quality	0,2
Relative Water quantity	0,25
Distance with tourism sites	0,25
Existence throughout the year	0,3

3.2.4. *Springs suitability determination*
MCDA, as the spring suitability determination, is a combination of all spring suitability factors in a weighted linear combination [20]. From the formula below (1), it can be concluded that a higher S value means the springs are more suitable than another.

$$S = \sum wi \,.\, xi \tag{1}$$

S = suitability value of springs
wi = weight of factor i
xi = potential rating of factor i

4. Result and Discussion
Purwosari consists of five villages, namely Girijati, Giriasih, Giricahyo, Giritirto, and Giripurwo Village. The research area is done throughout the Purwosari area. The results show that there are 27 springs spread unevenly. The existence throughout the year, water quality, relative water quantity, and distance to tourism site data for each spring can be seen in **Table 4.**

Table 4. Table of Springs' existence throughout the year, water quality, relative water quantity, and distance to the tourism site

Location	Spring	Observation Code	Coordinate X	Coordinate Y	Color	Odor	Taste	pH	TDS	Electrical Conductivity	Temperature	Water Quality	Relative Water Quantity	Distance to Tourism Site	Existence Throughout the Year
Girijati Village	Ndimoro 1	STA-H1	426551	9114470	Clear	No smell	Plain	6,7	660	320	27	Good	Limited	Near	No
	Ndimoro 2	STA-H2	426535	9114611	Clear	No smell	Plain	6,8	670	330	27	Good	Enough	Near	Yes
	Mudal	STA-H3	426832	9114896	Clear	No smell	Plain	6,8	670	330	27,2	Good	Abundant	Near	Yes
	Guntur	STA-H4	427980	9116400	Clear	No smell	Plain	7	650	320	26,9	Good	Abundant	Medium	Yes
	Kalijambu 1	STA-H5	427986	9116156	Clear	No smell	Plain	6,8	640	310	26,9	Good	Abundant	Medium	Yes
	Kalijambu 2	STA-H6	428088	9116210	Clear	No smell	Plain	6,7	590	300	27,5	Good	Limited	Medium	No
	Kalikulon	STA-H7	427593	9116090	Clear	No smell	Plain	6,9	630	310	26,2	Good	Abundant	Medium	Yes
	Sendang Beji	STA-H8	427671	9113600	Clear	No smell	Plain	6,7	780	360	27,5	Good	Abundant	Near	Yes
	Kali Anyar 1	STA-H9	427766	9113747	Clear	No smell	Plain	6,7	690	380	28,2	Good	Enough	Near	Yes
	Gembirowati Site	STA-H10	427278	9113862	Clear	No smell	Plain	6,9	580	260	27,2	Good	Limited	Near	No
	Kali Anyar 2	STA-H11	429155	9117701	Clear	No smell	Plain	6,9	630	310	28,7	Good	Limited	Far	Yes
	Kali Gayam	STA-H12	429477	9117526	Clear	No smell	Plain	7	670	330	28,2	Good	Enough	Far	Yes
	Kali Lo	STA-H13	429030	9117360	Clear	No smell	Plain	6,8	630	300	27,6	Good	Enough	Far	Yes
	Kali Setro	STA-H14	428473	9116996	Clear	No smell	Plain	7,2	630	310	27,6	Good	Abundant	Far	Yes
	Kali Adoh	STA-H15	430112	911795	Clear	No smell	Plain	7,2	650	310	27,6	Good	Abundant	Far	Yes
Giriasih Village	Manggung Cave	STA-H16	428752	9115843	Clear	No smell	Plain	6,9	620	300	27,4	Good	Abundant	Medium	Yes
	Kali Kidul Cave	STA-H17	428924	9115205	Clear	No smell	Plain	6,7	630	310	26,9	Good	Limited	Near	Yes
	Kali Peguh Cave	STA-H18	430622	9115759	Clear	No smell	Plain	7	670	330	27,1	Good	Enough	Far	Yes
	Sendang Arum	STA-H19	428504	9114581	Clear	No smell	Plain	7,2	720	330	27,6	Good	Abundant	Near	Yes
Giricahyo Village	Plawan Cave	STA-H20	428556	9112934	Clear	No smell	Plain	7,2	670	310	29	Good	Abundant	Near	Yes
Giritirto Village	Ngeleng	STA-H21	431696	9115899	Clear	No smell	Plain	6,9	670	330	27,6	Good	Abundant	Far	Yes
	Telaga Tengah	STA-H22	432441	9116716	Clear	No smell	Plain	6,9	470	200	29,9	Good	Limited	Far	No
	Kali Cabe	STA-H23	429959	9118129	Clear	No smell	Plain	7,5	640	310	26,5	Good	Abundant	Far	Yes
	Cerme Cave	STA-H24	431740	9117860	Clear	No smell	Plain	7,2	630	300	26,8	Good	Abundant	Far	Yes
Giripurwo Village	Cacahan	STA-H25	435118	9113190	Clear	No smell	Plain	7	640	310	24,4	Good	Enough	Far	Yes
	Tangis	STA-H26	435587	9112939	Clear	No smell	Plain	6,9	690	340	27,5	Good	Enough	Far	Yes
	Pego Cave	STA-H27	431814	9112583	Clear	No smell	Plain	6,9	650	300	27,4	Good	Abundant	Medium	Yes

Based on physical and chemical properties, the water quality in all springs is good and already meets the requirements of clean water based on [22]. Five springs have relatively limited discharge, nine springs with relatively enough discharge, and 13 springs with relatively abundant discharge, according to 5 key persons in each village. The springs that run out of water during the dry season are Ndimoro spring 1, Kalijambu spring 2, Gembirowati spring located in Girijati Village, and Telaga Tengah spring at Giritirto Village. Nine springs are relatively near to tourism sites, six are relatively medium in distance to tourism sites, and 12 are relatively far from tourism sites. **Figure 4** shows the location of the springs and tourism sites.

The data are analyzed with MCDA [20] [21] [6] [7] [8] [9] with a semi-quantitative method [23] to evaluate the economic and efficient water supply that can support the development of tourism facilities near tourism sites along the waterfront of the Purwosari area. Springs with good water quality, abundance in relative quantity, and are located near tourism sites are expected to fulfill the water needs economically and efficiently for the ongoing development of tourism facilities along the waterfront of Purwosari based on the rating and weighting results seen in **Table 2** and **Table 3**. **Table 5** displays the results of the evaluation.

The springs with a suitability value of ≥ 2.75 are set to be the economic and efficient water sources to support the development of tourism facilities in the coastal areas of the Purwosari. There are 11 springs in Purwosari that are suitable to support the development of tourism facilities near tourism sites along the waterfront of the Purwosari area: Ndimoro Spring 2, Mudal Spring, Guntur Spring, Kalijambu Spring 1, Kalikulon Spring, Sendang Beji Spring, Kali Anyar Spring 1 in Girijati Village, Manggung cave Spring and Sendang Arum Spring in Giriasih village, Plawan Cave Spring in Giricahyo Village, and Pego Cave Spring in Giripurwo Village. This research shows a corresponding result with another regional development research that clean water is one of the factor criteria that should be noted related to regional development [6] [7] [8] [9].

5. Conclusion

Clean water is one of the basic needs in daily human activities. Springs have an important role in Purwosari in fulfilling communities' demand for clean water. The water quality in all springs is good and already meets the requirements of clean water. Five springs have a relatively limited quantity, nine springs with relatively enough quantity, and 13 springs with relatively abundant quantity, according to 5 key persons in each village. The springs that run out of water during the dry season are Ndimoro spring 1, Kalijambu spring 2, Gembirowati spring located in Girijati Village, and Telaga Tengah spring at Giritirto Village, according to 5 key persons in each village. Nine springs are relatively near to tourism sites, six are relatively medium in distance to tourism sites, and 12 are relatively far from tourism sites.

The evaluation of the springs along the coastal areas of the Purwosari shows that only 11 springs have the potential as economical and efficient sources of clean water: Ndimoro Spring 2, Mudal Spring, Guntur Spring, Kalijambu Spring 1, Kalikulon Spring, Sendang Beji Spring, Kali Anyar Spring 1 in Girijati Village, Manggung cave Spring and Sendang Arum Spring in Giriasih village, Plawan Cave Spring in Giricahyo Village, and Pego Cave Spring in Giripurwo Village. These Springs are suitable for development related to tourism facility development near tourism sites. The results of this study are expected to be one of the considerations in decision-making by the government so that it can provide the water needs economically and efficiently for the ongoing development of tourism facilities along the waterfront of Purwosari. Since clean water is a part of the factors in evaluating regional development, further study related to other regional development criteria is needed.

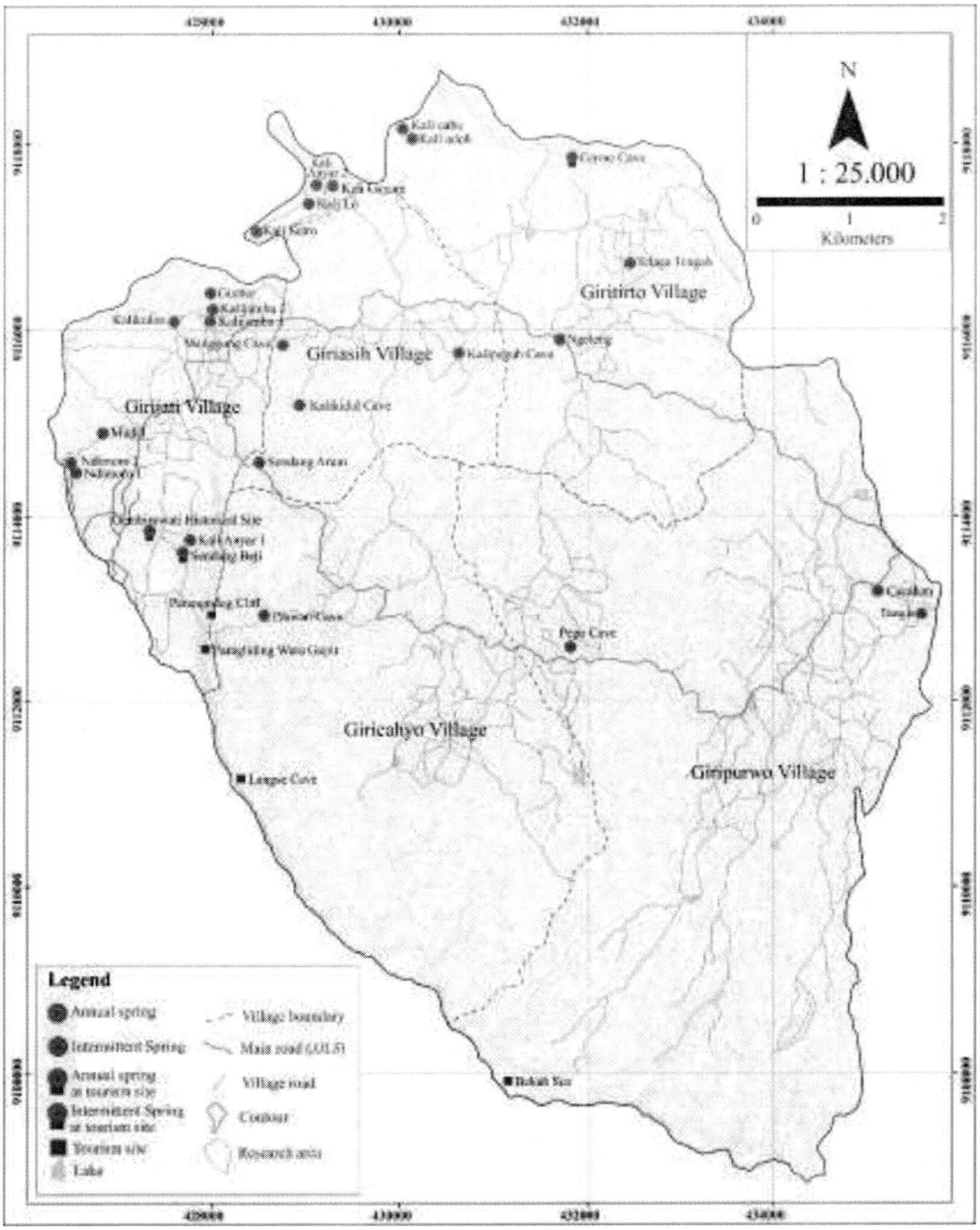

Figure 3. Springs and tourism sites distribution in Purwosari, Gunungkidul, DIY, Indonesia

Table 5. Weighting result of existence throughout the year, water quality, relative water quantity, and distance to the tourism site

Location	Spring	Water Quality	Relative Water Quantitiy	Existence Throughout the Year	Distance to Tourism Site	Suitability value
		Weight				
		0,2	0,25	0,3	0,25	
Girijati Village	Ndimoro 1	Good	Limited	No	Near	1,9
	Rating	3	1	1	3	
	Ndimoro 2	Good	Enough	Yes	Near	2,75
	Rating	3	2	3	3	
	Mudal	Good	Abundant	Yes	Near	3
	Rating	3	3	3	3	
	Guntur	Good	Abundant	Yes	Medium	2,75
	Rating	3	3	3	2	
	Kalijambu 1	Good	Abundant	Yes	Medium	2,75
	Rating	3	3	3	2	
	Kalijambu 2	Good	Limited	No	Medium	1,65
	Rating	3	1	1	2	
	Kalikulon	Good	Abundant	Yes	Medium	2,75
	Rating	3	3	3	2	
	Sendang Beji	Good	Abundant	Yes	Near	3
	Rating	3	3	3	3	
	Kali Anyar 1	Good	Enough	Yes	Near	2,75
	Rating	3	2	3	3	
	Gembirowati Sites	Good	Limited	No	Near	1,9
	Rating	3	1	1	3	
	Kali Anyar 2	Good	Limited	Yes	Far	2
	Rating	3	1	3	1	
	Kali Gayem	Good	Enough	Yes	Far	2,25
	Rating	3	2	3	1	
	Kali Lo	Good	Enough	Yes	Far	2,25
	Rating	3	2	3	1	
	Kali Setro	Good	Abundant	Yes	Far	2,5
	Rating	3	3	3	1	
	Kali Adoh	Good	Abundant	Yes	Far	2,5
	Rating	3	3	3	1	
Giriasih Village	Manggung Cave	Good	Abundant	Yes	Medium	2,75
	Rating	3	3	3	2	
	Kali Kidul Cave	Good	Limited	Yes	Near	2,5
	Rating	3	1	3	3	
	Kali Peguh	Good	Enough	Yes	Far	2,25
	Rating	3	2	3	1	
	Sendang Arum	Good	Abundant	Yes	Near	3
	Rating	3	3	3	3	
Giricahyo Village	Plawan Cave	Good	Abundant	Yes	Near	3
	Rating	3	3	3	3	
Giritirto Village	Ngeleng	Good	Abundant	Yes	Far	2,5
	Rating	3	3	3	1	
	Telaga Tengah	Good	Limited	No	Far	1,4
	Rating	3	1	1	1	
	Kali Cabe	Good	Abundant	Yes	Far	2,5
	Rating	3	3	3	1	
	Cerme Cave	Good	Abundant	Yes	Far	2,5
	Rating	3	3	3	1	
Giripurwo Village	Cacahan	Good	Enough	Yes	Far	2,25
	Rating	3	2	3	1	
	Tangis	Good	Enough	Yes	Far	2,25
	Rating	3	2	3	1	
	Pego Cave	Good	Abundant	Yes	Medium	2,75
	Rating	3	3	3	2	

Acknowledgements
The first author would like to thank LPDP, a scholarship by the Indonesian Goverments, for fully supporting the financial of this research.

References

[1] Government of Yogyakarta. (2011). Induk Pembangunan Kepariwisataan Daerah Provinsi Daerah Istimewa Yogyakarta Tahun 2012 – 2025
[2] Geospasial untuk Negeri. (2017). Topography map of Indonesia.
[3] Husein, S. and Srijono. (2007). Tinjauan Geomorfologi Pegunungan Selatan DIY/Jawa Tengah:endogenik dan eksogenik dalam proses pembentukan pegunungan. A conference paper
[4] Irshabdillah, M. R., & Widyastuti, M. (2020). Water quality analysis of the PDAM drinking water distribution network at the Baron-Ngobaran management unit, Gunungkidul Regency - Indonesia. E3S Web of Conferences, 200. https://doi.org/10.1051/e3sconf/202020002027
[5] Intining. (2007). *Upaya Mengatasi Kekeringan Di daerah Pegunungan Sewu Bagian Barat,* Undergraduate thesis, Teknik geologi, UGM (not published)
[6] Saputro, Rifqi P. W. (2020). *Geologi Pengembangan Wilayah Kecamatan Girisubo, Kabupaten Gunung Kidul, Daerah Istimewa Yogyakarta.* Undergraduate thesis. Teknik geologi, Universitas Gadjah Mada (not published)
[7] Indriyani, D. (2018). *Geologi Pengembangan Wilayah Pantai Selatan Kabupaten Kulon Progo, Daerah Istimewa Yogyakarta,* master thesis, Teknik Geologi, Universitas Gadjah Mada (not published).
[8] Alpina. (2017). *Evaluasi Pengembangan Wilayah Berdasarkan Aspek Geologi Lingkungan di Kecamatan Pemenang dan Tanjung Kabupaten Lombok Utara Nusa Tenggara Barat.* master thesis. Teknik Geologi, Universitas Gadjah Mada (not published)
[9] Khasanah, Uswatun. (2015). *Geologi Pengembangan Wilayah Kecamatan Kretek, Kabupaten Bantul, Daerah Istimewa Yogyakarta.* Undergraduate thesis. Teknik geologi, Universitas Gadjah Mada (not published)
[10] Department of Environment and Forestry of Yogyakarta, 2021, *Analisis Dampak Lingkungan Hidup (ANDAL)Rencana Pembangunan Hotel Grand Queen of The South Di Padukuhan Parangrejo, Kalurahan Girijati, Kapanewon Purwosari, Kabupaten Gunungkidul*
[11] Haryono, E. Barianto, D. H., and Cahyadi., A. (2017). *Hidrogeologi Kawasan Karst Gunung Sewu, PIT PAAI Yogyakarta.* https://doi.org/10.31227/osf.io/t5dgp
[12] Ansari, F., Syahidan, Dewantoro, A., & Putri, I. A. S. L. P. (2021). Assessment of communities' water springs in Gowa Regency, South Sulawesi Province, Indonesia. *IOP*
[13] Cahyadi, A., Riyanto, I. A., Adji, T. N., Tivianton, T. A., Agniy, Romza Fauzan, Ramadhan, F., Naufalg, M., & Saputro, T. C., (2018*), Hidrostratigrafi dan Dampaknya Pada Kemunculan Mataair di Sub-Sistem Panggang, Kawasan Karst Gunungsewu, Kabupaten Gunungkidul,* Seminar Nasional Geografi, 2nd, Yogyakarta, Himpunan Mahasiswa Pascasarjana Geografi (HMPG), Fakultas Geografi UGM, https://doi.org/10.31227/osf.io/pc9fv .
[14] Rahardjo, W., Sukandarrumidi, dan Rosidi, H.M.D. (1995). *Peta Geologi Lembar Yogyakarta, Jawa,* Pusat Penelitian dan Pengembangan Geologi, Bandung, Skala 1:100.000.
[15] Van Bemmelen, R. W. (1949). The Geology of Indonesia. General Geology of Indonesia and Adjacent Archipelagoes, In Government Printing Office, The Hague, p. 732.
[16] Mazor, Emanuel. (1997). Chemical and Isotopic Groundwater Hydrology Third Edition, New York, Marcel Dekker, 51p.
[17] MacDonald & Partners. (1984). Greater Yogyakarta Groundwater Resources Study, Volume 3: Groundwater. Directorate General of Water Re-sources Development Project (P2AT), Ministry of Public Works, Government of the Republic Indonesia

[18] Djaeni, A. (1982). *Peta Hidrogeologi Lembar* IX (Yogyakarta): Direktorat Geologi Tata Lingkungan, Bandung, Skala 1:250.000, 1 lembar.

[19] The, I., Singh, R., Science, S., & Dordrecht, B. M. (1996). Stratified Sampling *5.1.* 5–6.

[20] Dai, F. C., Lee, C. F., & Zhang, X. H. (2001). GIS-based geo-environmental evaluation for urban land-use planning: A case study: Engineering Geology, v. 61, p. 257–271. doi: 10.1016/S0013-7952(01)00028-X.

[21] Huang, I. B., Keisler, J., & Linkov, I. (2011), Multi-criteria decision analysis in environmental sciences: Ten years of applications and trends. *Science of the Total Environment,* **v. 409**, p. 3578–3594. doi: 10.1016/j.scitotenv.2011.06.022.

[22] The ministry of health Indonesia (2017). *Nomor 32 Tahun 2017 tentang Standar Baku Mutu Kesehatan Lingkungan dan persyaratan Kesehatan Air untuk Keperluan Higiene Sanitasi Kolam Renang, Solus Per Aqua, dan Pemandian Umum*

[23] Andiani, Darmawan, A., Badri, I., & Kurniawan, A. (2011). Peranan geologi tata lingkungan dalam penataan ruang Kota Padang pasca Gempa Bumi 30 September 2009. *Jurnal Lingkungan Dan Bencana Geologi*, **v. 2(2),** 95–112.

IOP Conf. Series: Earth and Environmental Science 1233 (2023) 012020 doi:10.1088/1755-1315/1233/1/012020

Surface and Subsurface Engineering Geological Characterization for Empirical Design Excavation Method and Support System of Jragung Dam Diversion Tunnel, Indonesia

Anggoro Bagyo Mulyo[1,2] and Ferian Anggara[1]

[1] Geological Engineering Department, Engineering Faculty, Universitas Gadjah Mada, Indonesia
[2] PT. Hutama Karya (Persero), Indonesia

Email: Anggoro.bagyo.mulyo@mail.ugm.ac.id

Abstract. The diversion tunnel of Jragung Dam was initially designed using the New Austrian Tunnelling Method (NATM) based on assessment of limited drill cores. Due to the uncertainty in the use of single empirical tunneling method and limited data for the tunnel design, this research was conducted to better characterize the rock masses and provide design alternatives for the tunnel construction. In addition to drill core evaluation, surface engineering geological mapping was carried out to obtain complete parameters of Rock Mass Rating (RMR) and Q rock mass classifications for designs of the tunnel excavation method and support system. The results show that the tunnel area consists of highly weathered sandstone and claystone layers with extremely poor to fair rock mass quality. The rock masses have very favorable and fair discontinuity strike and dip orientations with respect to the tunneling direction. The recommended excavation methods are top heading and bench and multiple drifts. The tunnel mostly requires rock bolt and shotcrete support system and the tunnel sections with poor rock mass quality require additional installation of steel sets. This research highlights potential collapse of the extremely poor quality rock masses near the tunnel portals during excavation of the diversion tunnel.

1. Introduction

The Jragung dam development project is one of the largest water resources projects in Indonesia that began construction in 2020. This large dam planned to have a capacity of 113.03 million m^3 and is expected to provide benefits to meet raw water needs of 1,000 liters/second for the people of Semarang, Demak, and Grobogan, irrigate agricultural land irrigation covering an area of 4,528 hectares, control floods downstream of the Jragung river and potentially supply electricity from micro hydropower plant with a capacity of 1.4 MW (PT. Indra Karya, 2019)[12].

The research location was conducted on the Jragung dam diversion tunnel, administratively located in Candirejo village, Pringapus sub-district, Semarang, Central Java Province (Figure.1). The Jragung dam is planned to have single diversion tunnel of 380 m lenght.

The geological conditions at the project location and. the surrounding area will significantly affect the stability of the. portal slope, excavation methods, and preliminary support systems before the permanent lining concrete. Classification of subsurface rock mass along tunnel path through engineering geological approaches and evaluation of observations of discontinuity conditions of core drill results can be used as a guide for enginneers to predict the exact and safe methods of excavation and installation of tunnel wall reinforcement from the risk of collapse according to the parameters of the quality of the rock masses at tunnel elevation.

Various researchers had published their work using similar methods in the sedimentary rocks. Putera et al [10], (2021) publish their work on the Jlantah Dam intake tunnels were designed utilizing an analysis of excavation techniques and support systems based on the assessment of rock mass quality using RMR and Q. Zekrinaldi et al [14]. (2021) also analyze the excavation method and support system in the Tiga Dihaji dam diversion tunnel, South Sumatera, Indonesia, based on the quality of the rock masses. According to Hamdhani [6] (2020), the rocks at the research site are sedimentary rocks of a heterogeneous nature consisting of a layer of Claystone and Sandstone.

During the pre construction stage, BBWS Pemali Juana as the owner appoint PT. Indra Karya (Persero) [12] to conduct surveys and geological investigations, including tunnel method and preliminary support but based only from New Austrian Tunnelling Method (NATM) Japanese standard for water tunnel guideline. Based on previous research with insufficient data for the classification of rock masses for tunnel support system design, it is necessary to Figure out tunnel preliminary support systems and apply other rock mass classification, especially applying the RMR [3] and Q-system [1] empirical approach methods. Furthermore, it is critical to evaluate the quality of the rock mass using the RMR based on the findings of the geological condition evaluation, since the results will be utilized to decide the excavation procedure.

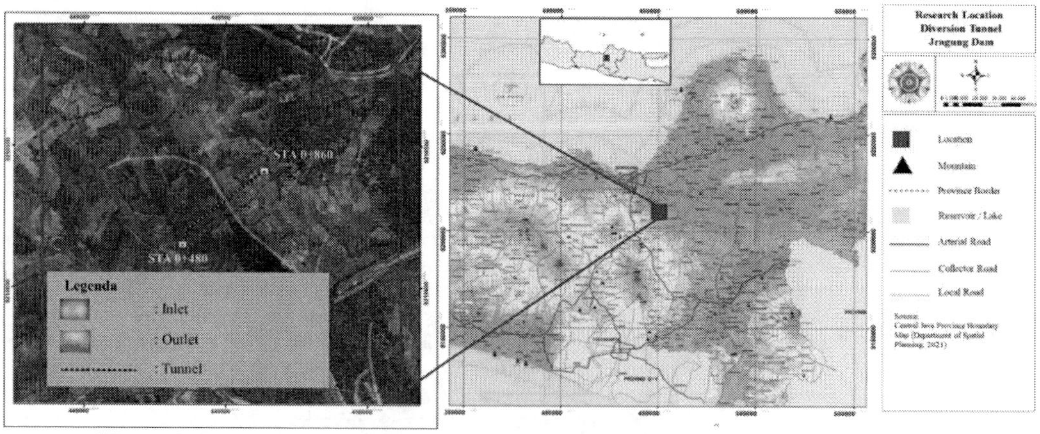

Figure 1. Research location (source : Google Earth & Map of Central Java Province)

2. Geological Conditions

Referring to the regional geological map of salatiga sheets (Sukardi and Budhitrisna, 1992) [13] on a scale of 1: 100.000, The research is situated on the Late Miocene Banyak beds formation (Tmkb) (as part of the Kalibeng Formations (Tmpk) and this formation consist of Alternation tuffaceous sandstone, calcareous siltstone and gravel sandstone - see Figure. 2.

This area have geological structures such as folds, the existence of these geological structures can have an impact on dam infrastructure, both during the construction process and operational period. The fold structure at the research site is in the form of a syncline located on the north and southwest sides of the dam body site, while the axis of the syncline structure is directed northwest-southeast. The research site is also between two trust faults, these two faults are to the north – northeast of the project site and to the south-southwest. The direction of these fault relatively same to the direction of the syncline axis, namely northwest-southeast.

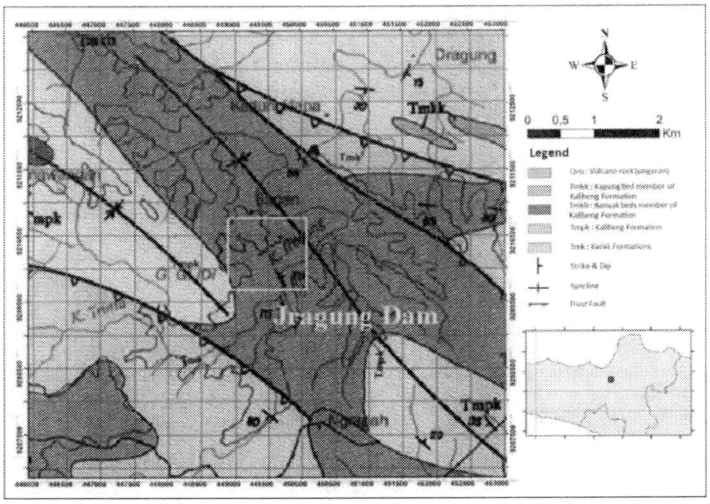

Figure 2. Regional geological map of Salatiga Sheet (Sukardi and Budhitrisna, 1992)

3. Method of Research

3.1. Field and sample parameters

The paper's parameter data consisted engineering geological mapping in an area of 1 km x 1 km and a core drill from a subsurface sample. Quality assessment of rock masses at the construction site was carried out utilizing subsurface data from the sample core drilled in 2021 by PT. Brantas Abipraya (Persero) [11], namely BP-01 with a depth of 35 m, BP-02 with a depth of 95 m, and BP-03 with a depth of 40 m. (Figure. 3). Observational analysis on the core drill and discontinuity conditions are used to determine the class of rock mass quality at the tunnel level. The data will be used as a reference in determining excavation methods and installation of tunnel wall reinforcement methods based on empirical classifications

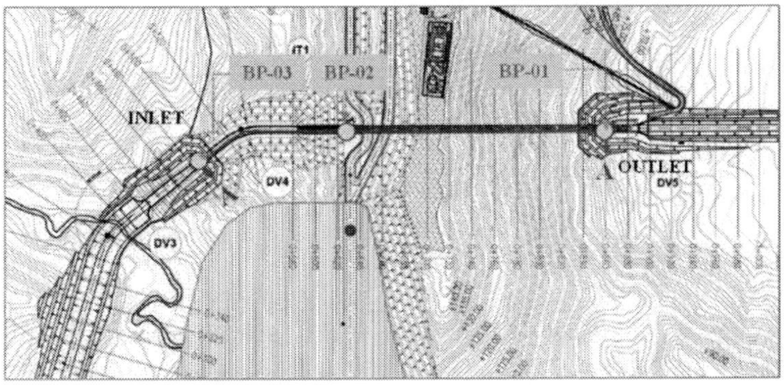

Figure 3. Diversion tunnel layout plan
(Source : PT. Indra Karya, Preparation of Jragung Multipurpose Dam Project)

3.2. Rock Mass Classification

Generally, a condition of the excavation surface are converted into scores and quantitative evaluation is performed using the evaluation points, and design the tunnel support system and excavation method using the evaluation result. Currently, many evaluation methods of tunneling are applied, such as the Geologic Strength Index (GSI), Rock Mass Rating (RMR) [3], Japanese Society of Civil Engineers (JSCE), Rock Quality Designation (RQD) [4], Rock Structure Classification (RSR), and Q-system [1]. In this research, the Rock Mass Rating (RMR) and Q-system were implemented to characterize rock masses.

IOP Conf. Series: Earth and Environmental Science 1233 (2023) 012020 doi:10.1088/1755-1315/1233/1/012020

3.2.1. Rock Mass Rating

According to Bieniawski (1989), this RMR [3] categorization contains six characteristics, five of which are primary parameters, rock strength / UCS (strength of intact rock)-F_1, length of core drill quality or fracture intensity or Rock Quality Design (RQD) [4]-F_2, spacing Discontinuity-F_3, conditions of Discontinuity-F_4, groundwater conditions-F_5 and one control parameter, especially orientation of discontinuities. To determine RMR using the following equation:

$$RMR = F_1+F_2+F_3+F_4+F_5+F_6 \quad (1)$$

3.2.2. Q-System

Rock quality classification using the Rock Tunnel Quality Index approach, often known as the Q-system [1], created by (Barton et al., 1974) based on tunneling experience in Norway. With this Q-system method, rock properties and support systems in tunnel construction are generally determined in the form of rock media. According to Barton et al. (1974) to determine Q value in this system using the following equation:

$$Q = \frac{RQD}{Jn} \times \frac{Jr}{Ja} \times \frac{Jw}{SRF} \quad (2)$$

Assign the value to determine the Q-system value based on the following 6 parameters: Rock Quality Design (RQD)[4], Joint Set Number (Jn), Hardness/Joint. Roughness (Jr), in relation to type of mineral in a joint filling (Ja), Joint Water Reduction Factor (Jw) and Stress Reduction Factor (SRF). Q's value might range between 0.001 and 1.000. In addition to the rock mass quality (the Q-value) two other factors are decisive for the support design in underground openings and caverns. These factors are the safety requirements and the dimensions, i.e., the span or height of the underground opening. height. Safety requirements will depend on the use (purpose) of the excavation. A road tunnel or an underground power house will need a higher level of safety than a water tunnel or a temporary excavation in a mine. To express safety requirements, a factor called ESR (Excavation Support Ratio) In addition to the span (or wall height) ESR gives the "Equivalent dimension" in the following way:

$$De = \frac{\text{Height of Tunnel}}{\text{Excavation Support Ration (ESR)}} \quad (3)$$

To meet safety requirements, a factor called ESR (excavation Support Ration) is used. In this study, the height of the tunnel was 6,6 meter and ESR value by Barton for water tunnel for hydro power category has been determined at 1.6.

4. Results and Discussions

4.1. Surface Geological Condition

According to geological mapping, rock outcrops in this location consist of three rock unit ranging in age from old to young, namely Claystone intercalated with Siltstone unit, above it alternation of Sandstone, Siltstone and Claystone and last is Alluvial unit. The geological structure from the results of field observations shows that there is a syncline structure oriented northwest - southeast located in southwest inlet tunnel and the rocks at the tunnel location are in sedimentary rocks which have a northwest – southeast striking direction with a dip of 56°-70°. Figure 4 shows geological maps from the research area. All rock units in the study area were included in the Tertiary Era Late Miocene Period rock formation.

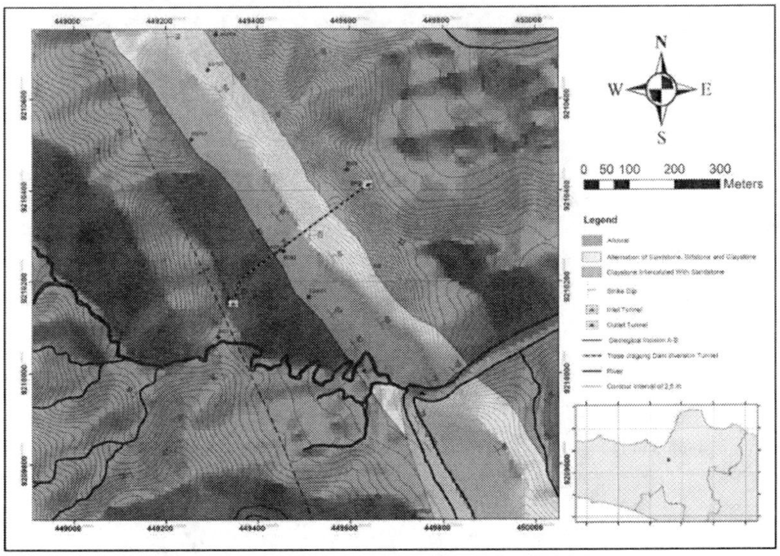

Figure 4 Geological map of research location

4.2. Subsurface Engineering Geological Condition

Based on the core observation, drilling point position and the tunnel geometry, lithology along tunnel elevation consist of 2 rock unit namely Claystone intercalated with Siltstone and Alternation of Sandstone, Siltstone and Claystone with have highly weathered sandstone and claystone layers with extremely poor to fair rock mass quality. The rock mass quality assessment along the tunnel was grouped into five sections (Figure. 5). Furthermore, each section's rock mass was assessed using a sample rock core. (Figure. 6) is presented in Table 1 for the RMR classification and Table 2 for the Q-system classification.

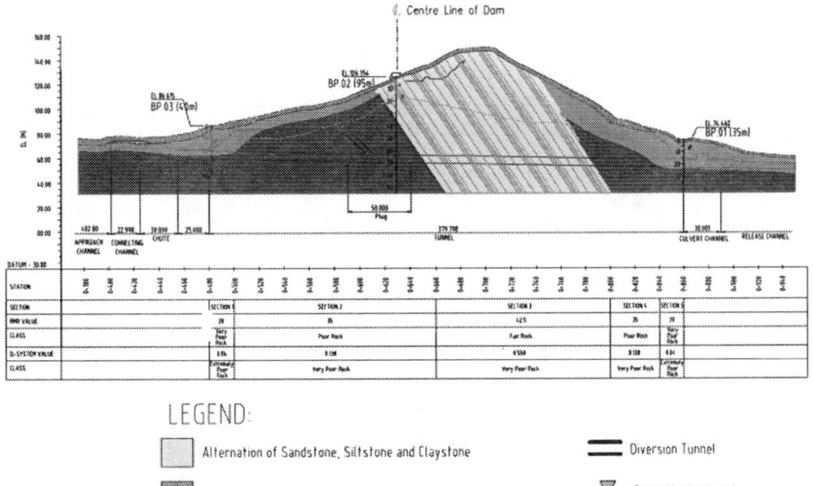

Figure 5. Engineering geological cross-section along tunnel

Figure 6. Drill core of sandstone at 35 m depth in BP-02 having moderate weathering degree RMR = 42 and Q = 0,550

Table 1. RMR assessment along the tunnel

Parameter	Section 1	Section 2 & 4	Section 3	Section 5
	Rating			
UCS	0	1	1	0
RQD	3	8	8	3
Discontinuity Spacing	5	8	8	5
Discontinuity Condition	10	16	20	10
Groundwater	7	7	10	7
Adjustment for Discontinuities Orientation	-5	-5	-5	-5
Total Rating	**20**	**35**	**42**	**20**
Rock Mass Classes	Very Poor Rock (V)	Poor Rock (IV)	Fair Rock (III)	Very Poor Rock (V)

Table 2. Q-System assessment along the tunnel

Parameter	Section 1	Section 2 & 4	Section 3	Section 5
	Rating			
RQD	10	25	25	10
Jn	12	9	9	12
Jr	1.5	1.5	3	1.5
Ja	2	2	2	2
Jw	0.33	0.33	0.66	0.33
SRF	5	5	5	5
Q Value	**0.04**	**0.138**	**0.550**	**0.04**
Rock Mass Quality	Extremely Poor (F)	Very Poor (E)	Very Poor (E)	Extremely Poor (F)

4.3. Excavation Method

From Table 2, RMR [3] value results from drill points BP-01 & BP-02 at elevation tunnel in the section 1 and section 5 located in outlet and inlet portal tunnel classified as class V (very poor rocks) with a total rating of 20. While at the drill point, BP-02 section 2 and 4 is classified in class IV (poor rocks) and section 3 class III (fair rocks) with a total. rating of 35 and 42. Based on RMR [3] values (Table 1), the preferred excavation methods at the project site can be seen in Table 3.

Table 3. Recommended excavation methods based on RMR classification

Bore Hole	Section	STA	RMR		
			Rating	Rock Mass Class	Excavation
BP-01	Section 1	0+480 – 0+500	20	Very Poor Rock (Class V)	Multiple drifts distance 0.5 m - 1.5 m top heading advance, simultaneously with excavation, install reinforcement; shotcrete as soon as possible following excavation.
BP-02	Section 3	0+660 – 0+800	42	Fair Rock (Class III)	Top heading and bench distance 1.5 m – 3.0 m advance in top heading; Install support simultaneously with excavation - 10 m from face
BP-02	Section 2 & Section 4	0+500 – 0+660 0+800 – 0+840	35	Poor Rock (Class IV)	Top heading and bench distance 1.0 m - 1.5 m advance in top heading; Install support simultaneously with excavation - 10 m from face
BP-03	Section 5	0+840 – 0+860	20	Very Poor Rock (Class V)	Multiple drifts distance 0.5 m - 1.5 m top heading advance; Simultaneously with excavation, install reinforcement; shotcrete as soon as possible following excavation.

4.4. *Support System*

RMR [3] class is also recommended to decide on stand-up time while tunneling without assistance the use of a support system. Figure 7 illustrates that the suggested stand-up time supported RMR for section 1 and section 5 is 1 m long while not given tunnel wall reinforcement for 30 minutes. In comparison, the recommended stand-up. time for section 2 & 4 is 2.5 m long without reinforcement for 3.5 hours for poor rock mass quality and 11 hours for fair rock mass quality.

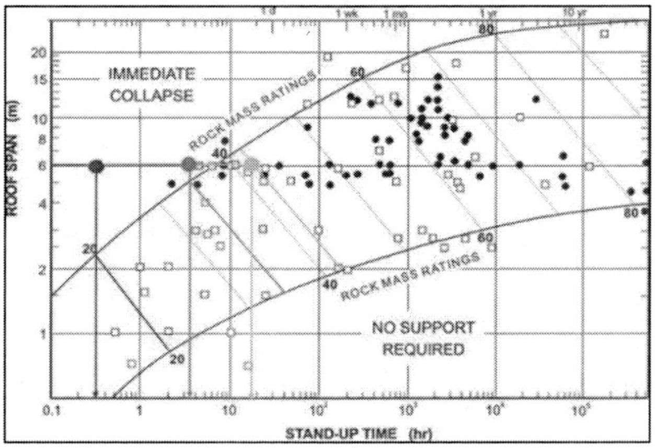

Figure 7. The relationship between stand-up time and unsupported span for rock mass classes

After knowing the method of excavation and being strong enough to stand during the tunnel's excavation without reinforcement, After that, the tunnel reinforcement method has been selected. RMR classification recommends rockbolts, steel sets, and shotcrete tunnel support systems (Table 4).

Table 4. Support System Recommendations RMR classification

Bore Hole	Section	RMR			
		Rating	Rock Mass Class	Support System	
BP-01 & BP-03	Section 1 & Section 5	20	Very Poor Rock (Class V)	Rockbolt	Systemic bolts 5-6 m long, c/c 1-1.5 m in crown and walls with wire mesh. Bolt invert
				Steel Sets	Medium to heavy ribs c/c 0.75 m if needed, with steel lagging and forepoling. Close invert
				Shotcrete	150-200 mm on the roof, 150 mm on the wall, and 50 mm on face
BP-02	Section 3	42	Fair Rock (Class IIII)	Rockbolt	Systemic bolts 4 m long, c/c 1.5-2.0 m in crown and walls with wire mesh in crown
				Steel Sets	None
				Shotcrete	50-100 mm on the roof, 30 mm on the wall
BP-02	Section 2 & Section 4	35	Poor Rock (Class IV)	Rockbolt	Systemic bolts 4-5 m long, c/c 1-1.5 m in crown and walls with wire mesh
				Steel Sets	Light ribs c/c 1.5 m
				Shotcrete	100-150 mm on the roof and 100 mm on the wall

c/c = steel rib and systemic rockbolt spacing, center to center

4.5. Q-System

From data Table 3, the Q value [1] at the section 1 and section 5 from drill point BP-01 and BP-02 located in the outlet and tunnel inlet is in the category of F rock (extremely poor) with a total value of Q 0.04. While at section 2, section 3 and section 4 is classified in E rock (very poor rock) with value 0.138 and 0.550. The Equivalent Dimension value (De) is obtained a calculation of a value of 4.125. It is also recommended that the mass of Q-system rocks be classified in order to determine the tunnel reinforcement support system using empirical approaches. The tunnel support system proposed by the Q-system rock mass categorization based on Q values (Table 2) can be seen in Table 5.

Table 5. Support System Recommendations Q-system classification

Bore Hole	Section	Q-System			
		Q-Value	Rock Mass Quality	Rock Support	
BP-01 & BP-03	Section 1 & Section 5	0.041	Extremely Poor Rock (F)	Rockbolt	Rockbolt 2.4 m long, c/c 1.2 m in crown and walls using wiremesh
				Shotcrete	Fibre reinforced sprayed concrete 12 cm in crown and sides with energy absorption E700
				RRS	Single layer of 6 rebars ø16-ø20 mm + 35 cm thickness of sprayed concrete (RRS II) c/c 2.9 m

Bore Hole	Section	Q-System			
		Q-Value	Rock Mass Quality	Rock Support	
BP-02	Section 3	0.550	Very Poor Rock (E)	Rockbolt	Rockbolt 2.4 m long, c/c 1.5 m in crown and walls using wiremesh
				Shotcrete	Fibre reinforced sprayed concrete 6 cm in crown and sides with energy absorption E500
				RRS	Single layer of 6 rebars ⌀16-⌀20 mm + 30 cm thickness of sprayed concrete (RRS I) c/c 4.0 m
BP-02	Section 2 & Section 4	0.138	Very Poor Rock (E)	Rockbolt	Rockbolt 2.4 m long, c/c 1.3 m in crown and walls using wiremesh
				Shotcrete	Fibre reinforced sprayed concrete 9 cm in crown and sides with energy absorption E700
				RRS	Single layer of 6 rebars ⌀16-⌀20 mm + 30 cm thickness of sprayed concrete (RRS I) c/c 4.0 m

c/c = Reinforced ribs of sprayed concrete (RRS) and systemic rockbolt spacing, center to center

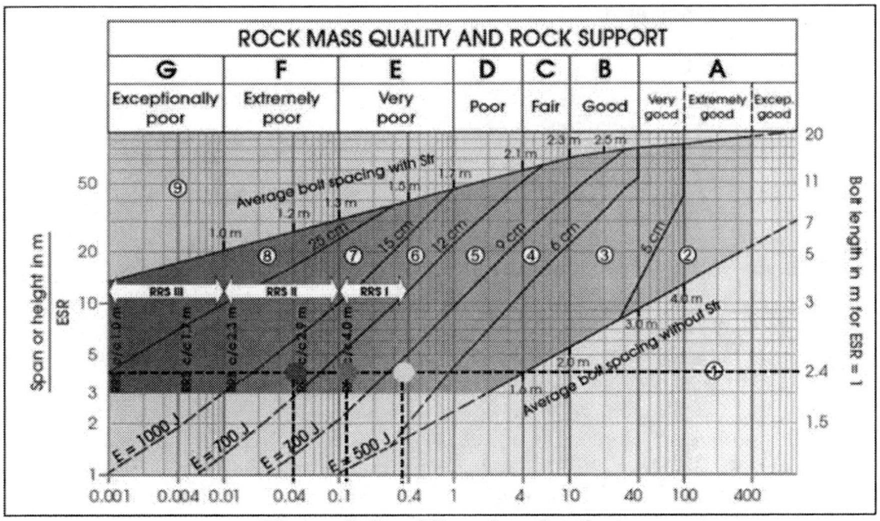

Figure 8. Q and De value plot chart

Based on the results of this study through an empirical approach to the tunnel support reinforcement system, there were differences where the NATM initial design had been determined for the entire 380 meter section of the tunnel using a light steel rib, 10 cm thickness of shotcrete plus wiremesh and 3.0 m long of rockbolt reinforcement system with the same installation distance of 1.2 meters in each section. In comparison, our research using the RMR and Q classifications resulted in a tunnel reinforcing system that dynamically adapts the installation distance of steel reinforcement, rock bolts, and shotcrete thickness based on the level of rock mass quality in each section, but more research, particularly numerical analyses, are required to compare and evaluate the performance of the initial design and recommended support systems.

5. Conclusions and Recommendations

The following conclusions can be drawn from the overall data analysis results:

1. Lithology along tunnel elevation consist of 2 rock unit Claystone dan Sandstone with very poor rock mass until fair rock mass. According to the data, as a result of the abovementioned assessment, rock discontinuities from weathering degrees and joints have a considerable influence on engineering geological properties.
2. Based on the RMR class, the suggested tunnel excavation method to use top heading and bench excavation with the distance of excavation 1.0 m - 1.5 m location of fair - poor rock mass quality, as well as excavation of multiple drift method with a distance of 0.5 m - 1.0 m in poor rock - very poor mass quality at location around outlet and inlet portals.
3. Based on the two methods of classification of rock masses above, a suitable support system is the installation of a medium-sized steel h-beam with a distance of 1.5 m at the location of fair rock - poor rock. Mass quality, and as soon as possible be installed support with a tighter space at the site of the quality of the rock mass is poor rock - very poor rock at the location around the outlet and inlet and reinforcement by the installation of systemic rock bolts with a length of 2.4 m, distance centre to centre 1.2 m on the roof and walls using wire mesh on the roof and walls and covered by a thickness 120 mm of shotcrete.
4. In practical terms, It is recommended that these two systems be used following the needs of the field, with each other complementing or correcting each other to produce appropriate decisions. It is crucial to conduct further research, among others, through numerical analysis to evaluate the performance of the two tunnel excavation reinforcement systems presented in this work.

Acknowledgments

We would like to thank BBWS Pemali Juana especially Mr. Rino Ari Wibowo as mentor and Mr. Chandra Adriawan as team leader for given permission and provided access and also important data to conduct research. Not forgetting, We also thanked to PT. Hutama Karya (Persero) for its financial support for education and research through Master of Engineering program organized by the Ministry of Public Works and Housing of Indonesia in 2021.

References

[1] Barton N, Lien R and Lunde J 1974 *Engineering Classification of Rock Masses for the Design of Tunnel Support Rock Mechanics.*
[2] Bieniawski Z T 1989 *Engineering rock mass classifications a complete manual for engineers and geologists in mining*, civil, and petroleum engineering p 251.
[3] Bieniawski Z T 1993 *Classification of rock masses for engineering*: The RMR system and future trends Comprehensive Rock Engineering, Vol. **3**: Rock Testing and Site Characterization-Principles, Practice & Projects ed Hudson J A (New York: Pergamon Press) pp 553-573).
[4] D.U. Deere and D. W. Deere 1989 *Rock quality designation (RQD) after twenty years*, 11-19 (US Army Corps of Engineers, Washington).
[5] E. Hoek, T. G. Carter, M. S. Diederichs 2013 *Quantification of the geological strength index chart*, in US Rock Mechanics /Geomechanics Symposium, San Francisco.
[6] Hamdani 2020 *Evaluation of geological conditions of engineering and analysis of the stability of diversion tunnel Jragung Dam, Semarang – Central Java* ,Unpublished.
[7] H. Rehman, W. Ali, A. Naji 2018 *Review of rock-mass rating and tunneling quality index system for tunnel design*: development, refinement, application and limitation.
[8] ISRM 1978, *Suggested methods for the quantitative description of discontinuities in rock masses* In International Journal of Rock Mechanics and Mining Sciences & Geomechanics Abstracts **15** 319.
[9] Ministry of PUPR 2015 *Circular Letter Of The Minister Of Public Works And Public Housing No. 30/SE/M/2015 guidelines for excavation planning methods and road tunnel support system on mixed soil-rock media*, Jakarta, Indonesia.
[10] Putera et al 2021 *Analysis of support systems and excavation methods based on rock mass quality in the Intake Tunnel of the Jlantah Dam*, ICST 2021 – The 2nd Geoscience and Environmental Management Symposium Volume **325**.
[11] PT. Brantas Abipraya (Persero) – PT. Pelita Nusa KSO 2021 *Final Report Pekerjaan Penyelidikan Geologi Tambahan Proyek Pembangunan Bendungan Jragung Paket 3*, Unpublished.
[12] PT. Indra Karya (Persero) 2021, *Engineering Design Report Preparation Of Jragung Multipurpose Dam Project,* Unpublished, Jakarta.

[13] Sukardi and Budhitrisna 1992 *Regional geological map of salatiga sheet.*
[14] Zekrinaldi et al 2021 *Excavation method and support system in the diversion tunnel of Tiga Dihaji Dam, South Sumatera, Indonesia based on the rock mass quality*, IOP Conference Series: Earth and Environmental Science, Volume **871**, The 2nd International Seminar on Civil and Environmental Engineering (ISCEE 2021) 6th-8th September, Bogor, Indonesia.

The Effect of Kacang Ratu BW on Ration to Production and Business Income of Layer Chickens

S. K. Rahmandriani[1a], H. Sasongko[1b], M. F. Habibi[1c] and B. Suwignyo[1d*]

[1]Faculty of Animal Science, Universitas Gadjah Mada, Indonesia

[a)]salsabila00@mail.ugm.ac.id, [b)]herusasongko@ugm.ac.id,
[c)] mochammad.fahmi.h@ugm.ac.id, [d)] bsuwignyo@ugm.ac.id
* Corresponding author

Abstract. The research was conducted to determine the effect of Alfalfa (*Medicago sativa* L.) substitution for laying hen performance. 108 *Hy-Line* strains on 26 weeks were put in battery cage and divided randomly into three treatments, (with six replications and every replication consisted six hens). The treatments were used 100% basal diet for control (A0), 99% basal diet + 1% Alfalfa (A1), and 97% basal diet + 3% Alfalfa (A2). Data recorded every day on three periods, with twenty-eight days per period. Feed and drink were given by *ad libitum*. The data collected every day consist of egg production, egg weight, while laying hen performance counted every week consist of *Feed Conversion Ratio* (FCR), *Hen Day Average* (HDA). In addition, the estimation of *Income Over Feed Cost* (IOFC) counted at the end of this period. Data were analyzed by factorial of classification of variance analysis (ANOVA) between Alfalfa and period. of treatment followed by testing significant means Duncan's Multiple Range Test (DMRT). According to this research, it can be concluded that Alfalfa can be used for substitution on laying hen's diet due to decrease the feed cost (increase feed efficiency) moreover without declined its production.
Keywords : Alfalfa, Productivity, IOFC

1. Introduction

Eggs are a favorite food for people in Indonesia. Besides being easy to get, chicken eggs have a relatively affordable price so that they can be purchased by various groups. The demand for egg consumption in Indonesia is increasing in line with public understanding and awareness of the importance of the animal protein derived from eggs [1]. Egg production increased from 2019 which was 4.7 million tons to 5.1 million tons (temporary data) in 2020 [2]. The increasing demand for eggs in the community requires higher production results from producers. Many factors affect production in livestock management. Feed is critical aspect in livestock management. Feed accounts for the largest production costs and can reach 70% of all maintenance costs (layer). It is necessary to use alternative feed ingredients that are cheap, easy to obtain, and contain nutrients needed by livestock.

An alternative feed ingredient that can be used as a protein source is Tropical Alfalfa (Medicago sativa cv Kacang Ratu BW). Kacang Ratu BW is a legume that has been developed in Indonesia [3, 4, 19], has high nutrition [5] and complete amino acids [6] so that it has potential as poultry feed [7, 8, 9]. However, the presence of saponins and high crude fiber content in Alfalfa makes its use to be limited to poultry. Therefore, further research is needed to obtain the appropriate level of use of

Alfalfa as a substitute for poultry feed ingredients so as to save on feed costs and increase the productivity of laying hens.

2. Material and Method

This research was allowed by Institutional Animal Care And Use Committee (IACUC)Veterinary Faculty of Gadjah Mada University (by the Ethical Clearance Letter 0002/EC-FKH/Eks./2021. The research was conducted in the research enclosure of the Integrated Technical Implementation Unit (UPT) of the Faculty of Animal Science, Universitas Gadjah Mada. The study was conducted for 3 periods (each period consisted of 4 weeks), so that the total data collection was 12 weeks, with 1 week of adaptation period before data collection. The research was conducted from January 2022 to May 2021.

The feed used was self-mixed feed for laying hens in the production phase which was added with Kacang Ratu BW according to the treatment. The equipment used in this research is a digital scale with a capacity of 50 kg Furano® with an accuracy of 0.5 g to weigh chicken feed and eggs as well as cleaning tools such as trash cans, brooms, and buckets.

108 laying hens were placed randomly in individual battery cages made of wire and divided into 3 treatment groups treatments. The cage is equipped with a feed container in the form of a pvc pipe and a drinking place in the form of a nipple drinker. The enclosure building has a fan, electric light and thermometer.

Each treatment was repeated 6 times, and each replication consisted of 6 laying hens. Feeding and drinking was carried out three times a day, namely at 6.00, 12.00, 17.00 WIB. Feeding and drinking was provided ad libitum. The nutrient content of feed in each treatment is presented in Table 1 below:

Table 1. Nutrient content of self-mixed for laying hens

Nutrient	Treatment		
	A0	A1	A2
CP (%)	16.92	16.65	16.54
EE (%)	4.50	4.49	4.96
CF (%)	2.56	2.85	3.38
ME (Kcal/Kg)	2.798.73	2,778.19	2,778.27
Ca (%)	4.52	4.5	4.53
P total (%)	0.79	0.75	0.74
Methionin *(%)*	0.58	0.57	0.56
Lysin *(%)*	1.32	1.20	1.17
Threonin *(%)*	0.73	0.72	0.70
Triptofan *(%)*	0.22	0.22	0.21
Fenilalanin *(%)*	0.94	0.92	0.9
P *Available* (%)	0.47	0.58	0.59
Price	IDR 5,869	IDR 5,761	IDR 5,744

A0: Basal Feed 100%, A1: Basal Feed 99% + 1% Kacang Ratu BW, A2: Basal Feed 97% + 3% Kacang Ratu BW

Production records were carried out every afternoon. The data recorded daily were the number of eggs, egg weight and the amount of feed consumption. Calculation of productivity as an observed parameter is carried out once a week including feed conversion or Feed Conversion Ratio (FCR) and egg production (HDA%). Income Over Feed Cost (IOFC) analysis was performed at the end of maintenance.

The percentage of egg production or Hen Day Average (HDA by %) was obtained by counting the number of eggs each week divided by the number of live chicken populations in each replication multiplied by 100%. Egg weight (g/egg) was obtained by weighing the total weight of egg production divided by the number of eggs. Feed conversion ratio (FCR) was calculated from the quotient between the amount of feed consumption and egg mass. Egg mass was obtained by multiplying HDA by egg weight. The value of Income Over Feed Cost (IOFC by IDR) was the result of income obtained by reducing the selling price of eggs with the cost of feed. The feed cost was obtained by multiplying the FCR by the price per kg of feed. IOFC = egg sales value – feed cost [10].

The data obtained were analyzed using analysis of variance based on a completely randomized design (Completely Randomized Design) factorial between Alfalfa treatments with maintenance period. If there was a significant difference between treatments, it was continued with a different mean test using Duncan's Multiple Range Test (DMRT). Data analysis using computer calculations with SPSS Statistics version 20 software.

3. Result and Discussion

3.1. Egg production

The number of egg production or Hen Day Average (HDA%) of laying hens given the addition of Tropical Alfalfa or Kacang Ratu BW to the ration is presented in Table 2 below.

Table 2. Egg production or Hen Day Average (HDA%)

Treatments	Period 1	Period 2	Period 3	Average ^{ns}
A0	87.2±6.2	94.7±4.2	94.4±4.4	92.1±3.5
A1	92.4±3.9	90.5±6.1	87.9±8.5	90.3±5.4
A2	90.3±7.9	89.4±8.6	91.8±5.4	90.5±6.6
Average ^{ns}	88.4±5.9	91.4±6.7	91.0±6.0	90.6±4.7

ns : *nonsignificant*

The results of statistical analysis showed [19] that there was no significant difference between the three feed treatments and age (maintenance period) on the daily average production. The results of this study are in accordance with the research of [8] who found that giving Alfalfa did not affect the percentage of egg production. Research by [11] found that giving Alfalfa had no significant effect on egg production.

The daily production standard for Hy-Line laying hens at week 26 to week 37 is 94 to 96% [12]. The average daily egg production for the three treatments was lower than the standard. Rahayu and Widjastuti [13] stated that the factors that affect the percentage of daily egg production are genetic factors and environmental factors including nutrient content of feed, maintenance management and environmental temperature. The results of statistical analysis showed that there was no interaction between the effects of the addition of Tropical Alfalfa treatment with the maintenance period on the daily average egg production.

3.2. Egg weight

The weight of chicken eggs from laying hens given the addition of Tropical Alfalfa or Peanut Ratu BW to the ration is presented in Table 3 below.

Table 3. Egg weight for each treatment (g/egg)

Treatments	Period 1	Period 2	Period 3	Average ^{ns}
A0	55.9±1.7	58.5±0.9	59.4±1.6	57.9±1.0
A1	56.6±1.8	57.6±0.51	60.1±3.8	58.1±1.3
A2	57.3±1.5	58.8±2.3	59.0±1.9	58.4±1.9
Average **	56.6±1.7^{p}	58.3±1.5^{q}	59.5±2.5^{q}	58.1±1.4

**: *significant* (P<0,01), p,q : different superscript in the same row show *significant* differences ns : *nonsignificant*

The results of statistical analysis showed [19] that there was no significant difference for the three feeding treatments on egg weight. The results of this study are in accordance with the research of [8] who found that giving Alfalfa did not affect egg weight. Research by [11] found that giving Alfalfa had no significant effect on egg weight.

The results of statistical analysis showed that there was a very significant difference (P<0.01) in egg weight in the three rearing periods. This is in accordance with the opinion of [14] which states that egg weight will increase with increasing age of chickens as a result of enlargement of the liver which plays a role in increasing egg yolk production and deposition. The standard egg weight for laying hens of Hy-Line strain at week 26 to week 37 is 57.3 to 63.2 g per egg [12]. The egg weight for the three treatments corresponded to the standard egg weight range. The results of statistical analysis showed that there was no interaction between the effect of the addition of Tropical Alfalfa treatment with the maintenance period on egg weight.

3.3. Feed conversion ratio

The feed conversion or Feed Conversion Ratio (FCR) of laying hens given the addition of Tropical Alfalfa or Kacang Ratu BW to the ration is presented in Table 4 below.

Table 4. *Feed conversion ratio*

Treatments	Period 1	Period 2	Period 3	Average ^{ns}
A0	2.5±0.2	2.2±0.1	2.2±0.1	2.3±0.1
A1	2.2±0.1	2.3±0.1	2.3±0.1	2.3±0.1
A2	2.2±0.2	2.2±0.2	2.2±0.1	2.2±0.1
Average ^{ns}	2.3±0.2	2.2±0.1	2.3±0.1	2.3±0.1

ns : *nonsignificant*

The results of statistical analysis showed [19] that there was no significant difference between the three feeding treatments and the age of the chickens on feed conversion. Feed conversion is the ratio of the amount of feed consumed to the mass of eggs. The smaller the feed conversion value, the less amount of feed needed to produce eggs. Dozier et al. [15] stated that a small feed conversion value will be more efficient for spending on feed costs.

The results of this study were in accordance with the research of [8] who found Alfalfa administration did not affect feed conversion. Regarding the maintenance period from 26 to 37 weeks, [16] stated that chicken age did not significantly affect feed conversion. The feed conversion standard for Hy-Line laying hens at week 26 to week 37 is 1.92 to 1.97 [12]. The results of the calculation of feed conversion for the three treatments were higher than the standard. Anne et al. [17] stated that the factors that influence feed conversion are genetic, physiological, environmental factors and rearing management. The results of statistical analysis showed that there was no interaction between the effects of the addition of Tropical Alfalfa treatment with the maintenance period on feed conversion.

Income Over Feed Cost (IOFC) is a simple calculation for analyzing livestock business. This was in accordance with the statement of [10] which states that IOFC is an analysis of livestock business by calculating the difference between income and costs for feed. Income is the multiplication of production with the selling value of eggs and the cost of feed is the cost incurred to produce one kg of eggs

3.4. Income over feed cost

The calculation of simple income analysis or IOFC on the laying hen production system given the addition of Tropical Alfalfa or Kacang Ratu BW to the ration is presented in Table 5 below.

Table 5. *Income over feed cost* (IOFC)

Variables	Treatments		
	A0	A1	A3
Egg price (IDR/Kg)	20,000	20,000	20,000
FCR	2.3	2.3	2.2
Cost of feed (IDR/Kg)	5,869	5,761	5,744
Feed cost (IDR)	13,498	13,250	12,636
IOFC (IDR)	6,501	6,749	7,363

The selling price of eggs during the research was IDR 20,000 per kg. The cost of feed is the cost incurred for the feed used to produce 1 kg of eggs. The cost of feed is obtained by multiplying the average FCR by the price of feed per kg. The IOFC value is obtained by subtracting the value of egg sales during the research with the cost of feed. The highest IOFC value was in treatment A2. These results indicate that [19] 3% substitution of Kacang Ratu BW in the diet is more effective in reducing feed costs compared to 1% substitution or without substitution with Kacang Ratu BW. Sjofjan [18] states that the greater the value of IOFC is due to the greater the difference between the value of selling eggs and the cost of feed spent.

4. Conclusion

Based on the research that has been done, it can be concluded that the addition of Tropical Alfalfa or Kacang Ratu BW to the ration of laying hens has no effect on productivity (egg weight HDA and FCR). The highest IOFC value is the addition of Kacang Ratu BW in the ration as much as 3%. Kacang Ratu BW can be used as a substitute feed ingredient for laying hens to provide higher profits.

Acknowledgments

The authors want to express an appreciation to research team: Mr. Bambang Suwignyo, Mr. Mochammad Fahmi Habibi, Yoga Michola, Mia Agusthia, Kanya, Yanuar and Ibnu. Additionally, to Laboratory of Poultry Science, Faculty of Animal Science and Industry, Universitas Gadjah Mada that supported this project.

References

[1] Nizhamuddien M Isdiantoni and Purwati R W 2019 Analisis permintaan telur konsumsi di Pasar Anom Baru Kabupaten Sumenep *Seminar Nasional Optimalisasi Sumberdaya Lokal di Era Revolusi Industri 4.0* 314-323.

[2] Information on https://www.bps.go.id/indicator/24/491/1/produksi-telur-ayam-petelur-menurut-provinsi.html

[3] Suwignyo B B Suhartanto C T Noviandi N Umami and Suseno N 2017 Generative plant characteristics alfalfa (*Medicago sativa* L.) on different levels of dolomite and lighting duration *Proceeding of the 1st International Conference on Tropical Agriculture.*

[4] Suwignyo B L Arifin N Umami Muhlisin and Suhartanto B 2021 The performance and genetic variation of first and second generations tropical Alfalfa *(Medicago sativa* L.*)* Biodiversitas *Journal of Biological Diversity* **22** 3265-3270.

[5] Suwignyo B A Mustika Kustantinah L M Yusiati and Suhartanto B 2020 Effect of drying method on physical-chemical characteristic and amino acid content of tropical Alfalfa (*Medicago sativa* L.) hay for poultry feed *American Journal of Animal and Veterinary Science* **5** 118-122.

[6] Suwignyo B F Izzati A Astuti and Rini A 2020 Nutrient content of Alfalfa (*Medicago sativa* L.) regrowth I in different fertilizers and lighting *IOP Conference: Improving Tropical Animal Production for Food Security* 465 12035.

[7] Xie Z J Huang X Xu and Jin Z. Jin 2008 Antioxidant activity of peptides isolated from Alfalfa leaf protein hydrolystat *Journal of Food Chemistry* **15** 474-487.

[8] Laudadio V E Ceci N M B Lastella M Introna and Tufarelli V 2014 Low-fiber Alfalfa *(Medicago sativa* L.*)* meal in the laying hen diet: effects on productive traits and egg quality *Journal of Poultry Science* **93** 1868–1874.

[9] Suwignyo B E Suryanto H Sasongko Y Erwanto amd Rini E A 2020 The effect of fresh and hay Alfalfa (*Medicago sativa* L.) supplementation on carcass quality of hybrid duck *IOP Conference Series: Earth and Environmental Science* 478 12024.

[10] Mayulu H B Suryanto Sunarso M Christiyanto F I Ballo and Refa'i 2009 Feasibility of complete feed based on ammoniated fermented rice straw utilization on the beef cattle farming *Journal of The Indonesian Tropical Animal Agriculture* **34** 74-78.

[11] Najm E K and Cufadar Y 2020 Effect of enzyme addition to diets containing different levels of Alfalfa meal on performance and egg quality parameters of laying hens *Selcuk Journal of Agriculture and Food Sciences* **34** 8-14.

[12] Anonymous 2018 Hy-Line Brown Commercial Layer Management Guide *Hy-Line International*.

[13] Rahayu N and Widjastuti T 2019 Egg weight and hen day production (HDP) layer commercial at high and low altitude *Journal of Physics: Conference Series* 1179 1-5.

[14] Stadelman W J and Cotterill O J 1995 *Egg Science and Technology fourth ed., Binghamton* New York Food Product Press Haworth Press.

[15] Dozier III W A K C Behnke C K Gehring and Branton S L 2010 Effects of feed form on growth performance and processing yields of broiler chickens during a 42-day production period *Journal of Applied Poultry Research* **19** 219-226.

[16] Okoro V M O K E Ravhuhali T H Mapholi E F Mbajiorgu and Mbajiorgu A 2017 Effect of age on production characteristics of Boschveld indigenous chickens of South Africa reared intensively *South African Journal of Animal Science* **47** 157-167.

[17] Anene D O Y Akter P C Thomson P Groves S Liu and O'Shea C J 2021 Hens that exhibit poorer feed efficiency produce eggs with lower albumen quality and are prone to being overweight *Animals Journal* **11** 2986.

[18] Sjofjan O 2008 Efek penggunaan tepung daun kelor (*Moringa oleifera*) dalam pakan terhadap penampilan produksi ayam pedaging Prosiding Seminar Nasional Teknologi Peternakan dan Veteriner, Fakultas Peternakan Universitas Brawijaya. Malang.

[19] Suwignyo, B., E. Aristia Rini, S. Helmiyati. 2023. The profile of tropical alfalfa in Indonesia: A review. Saudi Journal of Biological Sciences, 2023, 30(1), 103504

Mineralogical Distribution and Characteristics of Fe-, Ti-, and V-Bearing Beach Ironsand Deposit in Adikarto Bay Kulonprogo, Yogyakarta, Indonesia

L M L Harahap[1], A Idrus[1], Ernowo[2], I G Sukadana[2] ,Suwahyadi[3] and T Handayani[3]

[1] Universitas Gadjah Mada, Yogyakarta, Indonesia
[2] National Research and Innovation Agency (BRIN), Jakarta, Indonesia
[3] Geological Agency of Indonesia, Bandung, Indonesia

Abstract. Identification of beach sand deposits was carried out in the south coastal area of Adikarto Bay, Kulonprogo, Yogyakarta in Indonesia. A total of 25 points were observed to identify the characteristics and mineralogy of beach sand related to the iron (Fe), titanium (Ti), and vanadium (Fe) bearing minerals that have an important role in the transition of fossil energy into clean energy as an energy storage technology known as Vanadium Redox flow batteries (VRBs) and lithium-vanadium-phosphate batteries [1]. Laboratory analysis was performed by mineral grain analysis, ore microscopy, and micro-XRF. Based on the mineralogical analysis, the beach sand deposit is composed of heavy minerals (magnetite, titanomagnetite, and hematite) and light minerals (quartz, clinopyroxene: hedenbergite and augite), biotite, hornblende, and ilvaite. The results of the analysis show that the distribution of heavy minerals is dominantly in the fraction size <0.106 mm (70-80%,) while light minerals are distributed in the fraction size 0.212-2 mm (80-90%). Mico-XRF for elemental analysis identification shows that magnetite is the mineral bearing of Fe, Ti, and V elements with a percentage of (79.02 - 92.43 wt%), Ti (6.08-7.89 wt%), and V (0.56-1.15 wt%). Therefore, it is important to characterize the mineralogy and distribution of magnetite, since it can be used as a "mineral pathfinder" for advanced exploration of Fe, Ti, and V elements. So, the ironsand is for the cement indsutry and has advanced utilization.

1. Introduction

Beach ironsand in the Kulonprogo area has been identified as containing economic minerals of iron and is known as black ironsand deposits [2]–[4]. There are two types of beach ironsand deposits based on their setting deposition well known as the ironsand dunes and the coastline [2]. The beach ironsand material is interpreted as the result of weathering from the Old Andesite Formation (OAF) which was transported by river flow and accumulated in the coastal environment [5]. Several previous study explained that beach ironsand deposits in the Kulonprogo area contain of magnetite (Fe_3O_4), maghemite (Fe_2O_3), ilmenite ($FeTiO_3$) and titanomagnetite (Fe_2TiO_4) ([3], [5]). These minerals are the main mineral bearing of iron (Fe), titanium (Ti) and vanadium (V) [1], [6]. This study is determine the mineral textural and elemental mapping of Fe-,Ti-, and V- to know the distribution in ironsand deposits.

Iron application is dominantly used in industrial aspects as the raw material of building construction [2]. Titanium is a very hard metal used as a metal alloy in the plastics and paper industry [7]. Titanium and vanadium are also used in the aircraft industry [8]. Vanadium has an important role in the transition of fossil energy into clean energy as an energy storage technology known as

Vanadium Redox flow batteries (VRBs) and lithium-vanadium-phosphate batteries [1]. Based on these conditions, beach ironsand deposits have the potential for the mineral-bearing of vanadium and titanium, so further identification is needed regarding mineralogy characterization and distribution of beach ironsands in the Kulonprogo area. The importance of research is used to develop advanced exploration activities for Fe, Ti, and V target elements by determine the mineralogy in coastal ironsand deposits in the study area and in the region. So, the ironsand is not only for cement industrial but also have advance utilization.

1.1 Study Area

The administrative location of the research area is in the southern part of the Kulonprogo, Karangwuni Sub district (Figure 1). The regional stratigraphy of this area is the Quaternary Alluvium Formation (Qa) compose of ironsand, silt, and clay rivers and coastal plains [9]. The discovery of beach ironsand deposits in the Kulon Progo area has been recognized since 1979 in the types of dunes and beach ironsand. It was also mapped as the location of the local community's iron mining in the northwest part of the research area [9] (Figure 1).The straight coastline show the flat topography (Figure 1) consequently, the beach in the Kulonprogo area is classified as a Type 2 [10].

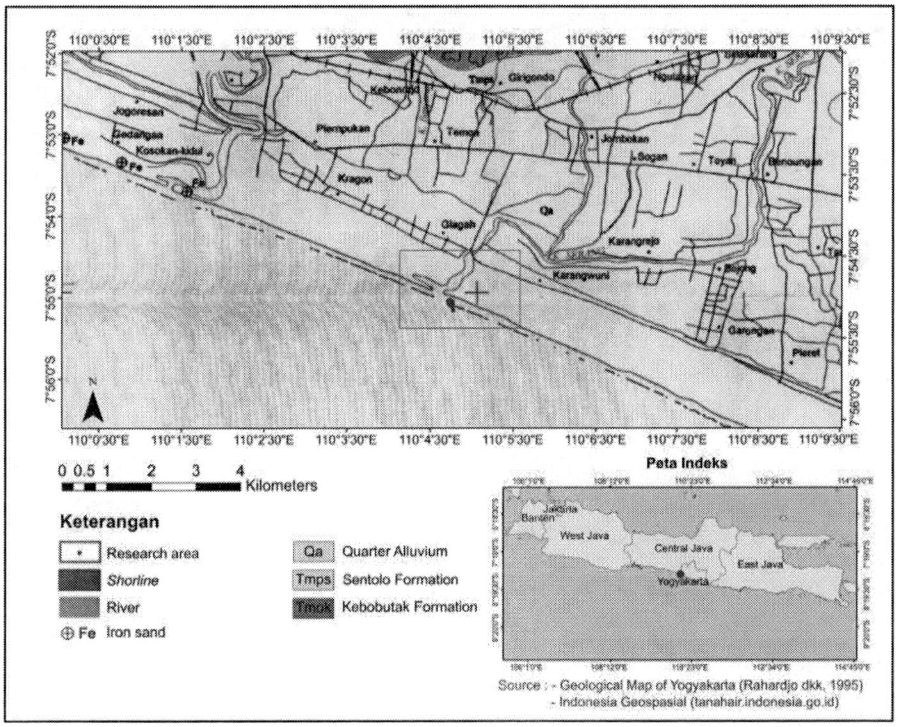

Figure 1. The southern part of Yogyakarta geological map scale 1:100.000 of [9] and the study area is marked by red square

2. Material and Methods

Field observations were followed by beach ironsand sampling of 25 drilling points (Figure 2A). The sample preparation was carried out in the field by hand auger drilling (Figure 2B) (depth : 2 meter) and manually separating the magnetite concentrate using a 300 gauss magnet. Thus, from one drilling hole, two types of samples were obtained : crude and concentrate samples. There are 3 (three) samples for grain mineralogy analysis using the crude sample (sample codes: AK3, AK7, and AK 12). The analysis aims to identify the mineralogy and the distribution of grain fractions in the beach ironsand sample. The sieving process uses the ASTM-E11 mesh size standart 2 mm (+10); <2-1 mm (-10+18); <1-0.5 mm (-18+35); <0.5-0212 mm (-35+70); <0.212-0.106 mm (-70+140); and <0.106 mm (-140). Ore Microscopic analysis uses three concentrate samples polished sections (AK3, AK7, and AK12). This analysis is intended to identify ore mineralogy and ore texture which aims to interpret the paragenesis of ore minerals.

The micro-XRF analysis uses 1 (one) concentrate sample (AK3) to determine the composition and distribution of the highest Fe, Ti, and V content in mineral-bearing beach ironsand deposit. This method uses the M4 Tornado devices and AMICS software (Advanced Mineral Identification and Characterization System). Advance analysis using The AMICS software can identify mineral groups (mixtures) and a single mineral in each sample and also give results for mineral identification with the characterization of elements set spectrum in every single pixel [11]. The sample scanning process uses measurement resolution (pixel size: 20 μm) on glass preparations (Figure 2C) surface with the mapping area of 15x25 mm and the relative error percentage range between 0-0.01%.

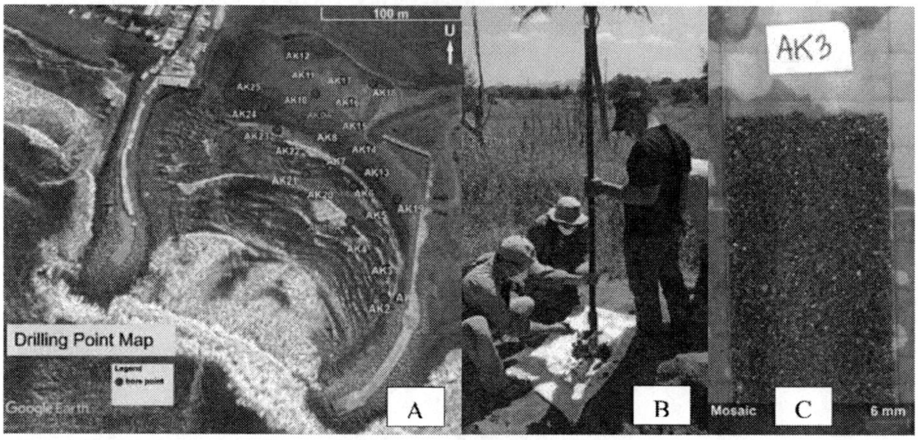

Figure 2. (A) Hand auger sampling location (source:https://earth.google.com), (B) hand auger drilling and (C) Glass preparation

3. Results

3.1 Mineral Grain Analysis

Based of identification from mineral fraction analysis, beach sediment in study area composed of heavy mineral ironsands (magnetite, hematite, ilmenite, and garnet) and light mineral ironsand (quartz, pyroxene, and plagioclase). Magnetite has a black color with a metallic luster and sub-angular. Ilmenite has a black color and metallic luster, grains are sub angular to sub rounded. Hematite, reddish-brown color, dull luster. Pyroxene, green, transparent, prismatic habit. Quartz, colorless, transparent, and glassy. Plagioclase is gray to dull white and slightly transparent. Garnet has a reddish color, transparent rounded habit (Figure 3).

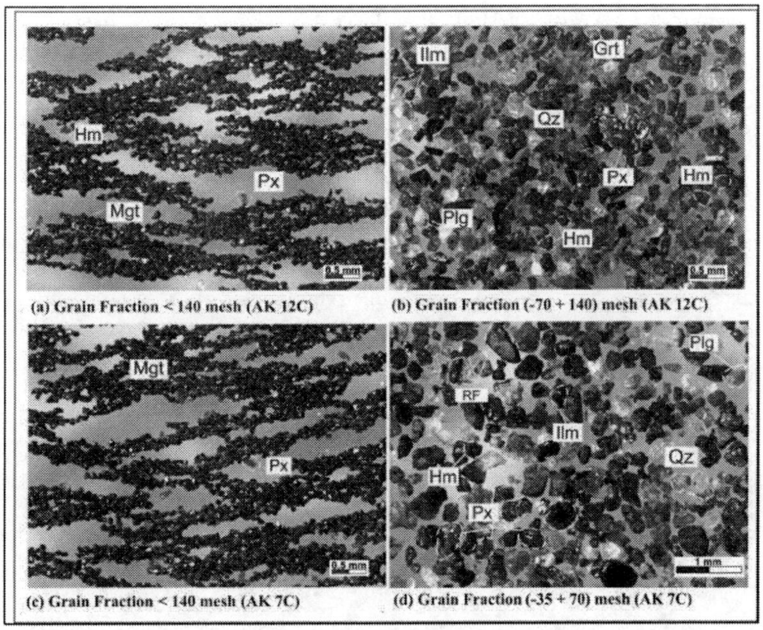

Figure 3. Grain fraction images : (a) and (b) Sampel AK 12C, (c) and (d) Sampel AK 7C (Mgt: magnetite, Px: Pyroxene, Hm: Hematite, Qz: Quartz, Ilm: Ilmenite, Plg: Plagioclase, RF: Rock fragment, and Grt: Garnet)

The grain mineralogy observations showed that heavy minerals such as magnetite were dominantly distributed in grain sizes ranging from 0.212 (-70+140 mesh) to <0.106 mm (<140 mesh) and associations with pyroxene and a small amount of ilmenite. The presence of light minerals such as quartz, plagioclase, and pyroxene was distributed in abundance in the grain size of 0.212 – 2 mm (-18+35 mesh to – 35+70 mesh) (Table 1).

Table 1. Heavy mineral and light mineral average distribution based on grain size

	Fraction size (mm)	2	-2+1	-1+0.5	-0.5+0.212	-0.212+0.106	<0.106
	(mesh)	+10	-10+18	-18+35	-35+70	-70+140	<140
				Heavy mineral (%)			
1	Magnetite	-	-	0.87	9.46	88.36	74.76
2	Ilmenite	-	-	1.84	1.56	1.8	2.11
3	Hematite	-	-	*trace*	*trace*	*trace*	*trace*
4	Garnet	-	-	*trace*	*trace*	*trace*	*trace*
				Light mineral (%)			
5	Piroxene	-	-	73.31	63.96	50.69	18.07
6	Quartz	-	-	18.14	26.09	17.41	5.07
7	Plagioclase	-	-	2.93	2.63	0.95	0.00
8	Muscovite	-	-	*trace*	*trace*	*trace*	*trace*
9	Apatite	-	-	*trace*	*trace*	*trace*	*trace*
10	Corundum	-	-	*trace*	*trace*	*trace*	*trace*
				Other components (%)			

Fraction size (mm)	2	-2+1	-1+0.5	-0.5+0.212	-0.212+0.106	<0.106
(mesh)	+10	-10+18	-18+35	-35+70	-70+140	<140
11 Rock fragmen	-	-	2.91	1.46	0.14	0
12 Fossil/shell	-	-	*trace*	0.02	*trace*	*trace*

3.2 Ore Microscopy

Ore microscopic identification of 3 (three) polish sections identified that the average mineral composition in the study area, compose of magnetite (55%), ulvospinel (5%), titanomagnetite (10%), ilmenite (25%), and hematite (10%). Ilmenite is present and grows together with magnetite with a thin lamella texture, reflection color gray to light brown. Myrmekitic texture as exsolution texture of ilmenite and magnetite as a host mineral (Figure 4e). The texture of the ilmenite trellis has the angular lamella in magnetite as the host mineral. The texture of the ilmenite ironsandwich on the polished shows thick lamellae of ilmenite. Furthermore, replacement textures are also found in magnetite and hematite (Figure 4h)

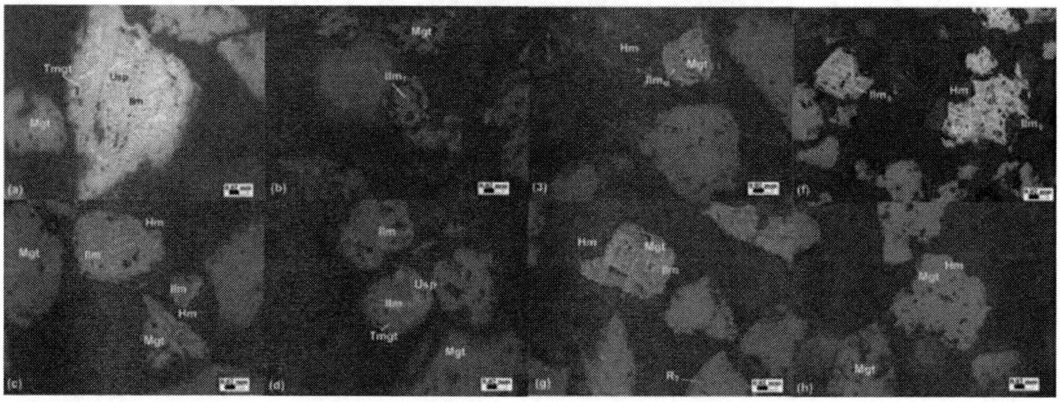

Figure 4. Ore microscopic images of, (a) Titanomagnetite as host-mineral (light white), (b) ilmenite trellis type texture, (c) Hematite (dark color), (d) exsolution lamella ilmenite, (e) myrmekitic texture in ilmenite, (f) ilmenite ironsandwich-type texture, (g) magnetite replace by hematite (red color), and (h) hematite replacement magnetite by the rim. (Ilm_S: Ilmenite Ironsandwich, R_T: Rudle texture, Ilm_M: Myrmekitic ilmenite, Mgt: Magnetite, Usp: Ulvospinel, Hm: Hematite)

3.3 Mineral Chemistry (Micro-XRF)

Based on the components of chemical elements, the mineralogical composition of the study area is composed of 43.89 wt% magnetites; clinopyroxene (hedenbergite 15.97 wt%, augite 2.56%); biotite (15.29 wt%); hornblende 12.89 wt%; and ilvaite 7.24% (

Table 2). Meanwhile, olivine and ilmenite were not identified as mineral composition of beach ironsand deposits because the relative error percentage values exceeded 0.01 (0.03 and 0.07).

Table 2. Mineral percentage (%) of AMIC analysis

Name	Wt %	Area %	Area (μ2)	Segment Number	Relative Error
Magnetite	43.86	33.43	125657036	313392	0
Hedenbergite	15.97	17.86	67127991	167419	0
Biotite	15.29	19.27	72424646	180629	0
Hornblende	12.89	15.65	58828962	146721	0

Name	Wt %	Area %	Area (μ2)	Segment Number	Relative Error
Ilvaite	7.24	7.06	26529387	66165	0.01
Augite	2.56	2.94	11057620	27578	0.01
Olivine	0.43	0.5	1894125.7	4724	0.03
Ilmenite	0.11	0.09	327983.66	818	0.07
Total	98.24				

The results of the mapping are presented in (Figure 4) where each mineral shows a different distribution of the elements contained especially target elements such as iron (Fe), titanium (Ti) and vanadium (V). There are 12 mineral observation points, these point are scaning based on a low relative error value of "0" such a magnetite, hedenbergite, biotite, and hornblende.

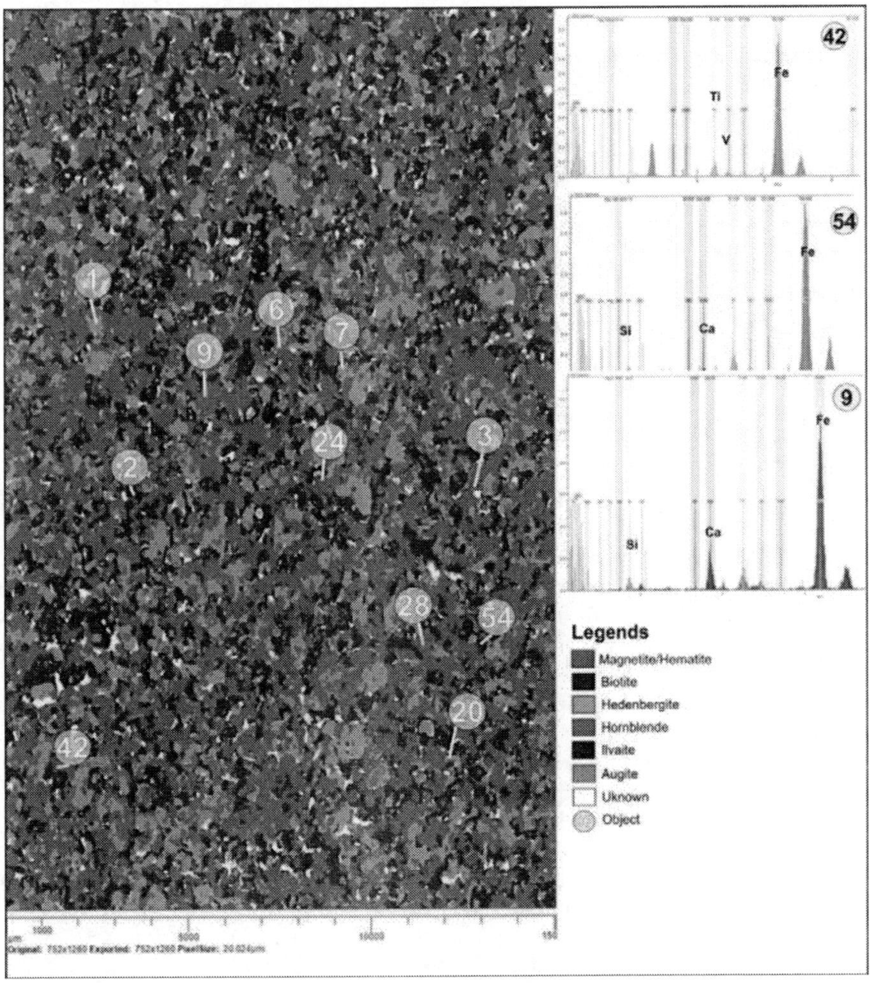

Figure 5. Mineral mapping image by AMICS, 42 and 97 are mineral spectrum of magnetite and 54 is mineral spektrum of biotite

Table 3.AMICS identification of elemental distribution object mapping

Nama Mineral	Presentasi Element (wt%)											Total
	Fe	Ti	V	Si	Al	Zn	Ca	Mg	K	Zr	Cr	
Hedenbergite	70.63	6.89	0.41	12.58	6.18	0.11	2.53	0	0.67	0	0	100.0
Hedenbergite	81	5.57	0.34	5.79	4.69	0.28	2.17	0	0.11	0	0	100.0
Hedenbergite	79.95	4.79	0.22	7.91	4.71	0.06	1.56	0	0.64	0.15	0	100.0
Biotite	92.54	6.27	0.16	0	0	0.25	0.36	0	0.37	0	0	100.0
Biotite	73.26	8.72	0.3	11.88	2.31	0.32	2.37	0	0.67	0.13	0	100.0
Biotite	82.03	7.39	0.29	6.26	2.11	0.16	0.82	0	0.81	0.09	0	100.0
Hornblende	78.69	3.94	0.22	6.15	2.74	0.14	0.66	7.31	0.15	0	0	100.0
Hornblende	72.01	6.54	0.35	11.68	6.7	0	0.46	2.16	0.1	0	0	100.0
Hornblende	83.55	6.63	0.24	4.08	2.02	0.22	0.42	2.73	0.11	0	0	100.0
Magnetite	90.99	7.89	0.81	0	0	0.14	0	0	0.1	0	0.07	100.0
Magnetite	79.02	6.08	1.15	3.78	6.84	0.35	2.71	0	0	0	0.07	100.0
Magnetite	92.43	6.65	0.56	0	0	0	0	0	0.37	0	0	100.0

Based on the distribution and composition of elements in each object, which is presented in detail in (Table 3). Hedenbergite, biotite, and hornblende as the light minerals have a high relative value of Si (7.37 wt%) compared to heavy minerals such as magnetite with an average Si value of 1.26 wt%. Hedenbergite has an average Ti value of 5.75 wt%; V 0.32 wt% . Biotite has an average Ti of 7.46 wt%; V 0.25 wt%. Hornblende has an average Ti of 5.70 wt%; V 0.27 wt%. Magnetite has an average Ti value of 6.97 wt%; V 0.84 wt%. So that the value of Ti content is from high to low are minerals biotite, magnetite, hornblende, and hedenbergite. The value of V content from high to low is the minerals magnetite, hedenbergite, hornblende, and biotite.

4. Discussion

4.1 Mineralogy Composition and Distribution

Mineralogy composition which classified as light mineral plagioclase, quartz, and pyroxene. The presence of pyroxene minerals detailed as clinopyroxene (hedenbergite and augite) using advance elemental maping (mikro-XRF) and also the presence of hornblende and biotite. The heavy mineral compositions identified are magnetite, ilmenite, and hematite. Based on the textural and optic properties identified in more detail the presence of titanomagnetite and ulvospinel in the sample. This is also conformable to the research done by [3] on the east coast of the Opak River estuary. The fraction distribution of 0.212-2 mm has a total average of light mineral (80-90%) and heavy mineral (10-20%), in contrast to the fraction size of <0.106 mm has a total light mineral range of (20-30%) and heavy minerals range in value (70-80%) (Figure 6a). This impact on the interpretation of the deposition process in the research area is influenced by specific gravity sorting [12]. The propensity of magnetite with relatively homogeneous sorting indicates that the coastal wave energy is relatively the same and the currents are relatively stable during the deposition process [13].

4.2 Fe-; Ti-; and V-Bearing Minerals

[1], [6] explained that the mineral-bearing for Fe, Ti, and V are magnetite and hematite (high Fe element composition) as well as ilmenite, perovskite, and rutile (high Ti element composition). In the research area, magnetite, hematite, and ilmenite are identified by ore microscopy as mineral-bearing of Fe, Ti, and V. The presence of ilmenite based on the identification of micro-XRF were not found because spectrally ilmenite standard contains ±50% Fe and 50% Ti. In contrast, the spectral samples of Fe minerals ranged between 70-95 wt% and Ti 3-8 wt% (Table **3**) present as magnetite spectral.

Magnetite has a high percentage of element V content compared to other minerals in the range of 0.5-1.15 wt% (Figure 6b). Based on chemical conditions, magnetite mineral is stable at low metamorphic conditions, and when in high metamorphic conditions magnetite will associate with Ti, V, Cr, and Zn [14]. [15] explained that magnetite's relatively low Ti content is found in andesitic rocks. So it can be interpreted that magnetite is a mineral-bearing of element V in the condition of low temperature and pressure and comes from andesitic rocks (relatively low Ti content). This is also explained by [16] who describe that magnetite is the main "host mineral" for vanadium-bearing minerals.

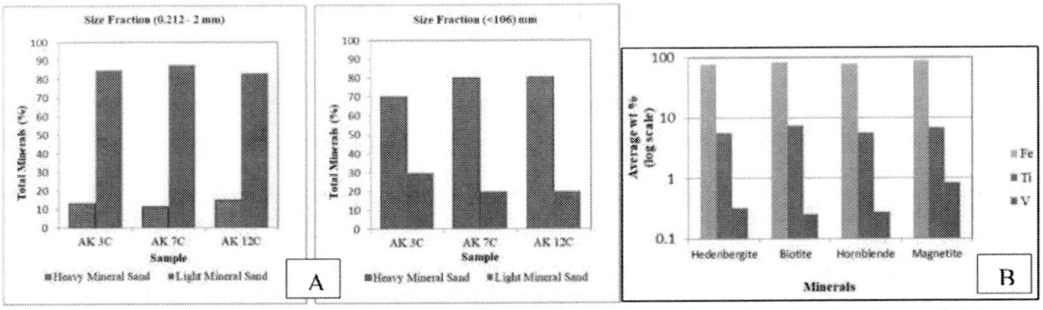

Figure 6. (a) Comparison of heavy mineral and light mineral based on size fraction and (b) Compositition Fe, Ti, and V by AMICS Identification

4.3 Mineralogy Paragenesis

Microscopy ore identification shows that the research area has 2 (two) different mineral hosts magnetite and titanomagnetite. The presence of titanomagnetite in this sample indicates an oxidation process at low temperature (600°C) known as "oxidation-exsolution" or "oxyexsolution" [17]. Exsolution process of ulvospinel in titanomagnetite becomes the limit of miscibility in the "solid solution" series. [18]. [17], [19] describe the stages of paragenesis of minerals under an "oxidation-exsolution" process known as the C1-C4 stages : C1 stage : The presence of ulvospinel-rich homogeneous magnetite is also known as Ti-rich magnetite [20]. C2 stage : Magnetite$_{SS}$ solid solution with lamellae ilmenite under exsolution process. C3 stage : Magnetite$_{SS}$ decrease Ti content and the presence of ilmenite trellis type. C4 stage : ilmeno-magnetite, this indicates the formation of ilmenite is more dominant than magnetite. The beach ironsand at the research area has another anomaly after the C4 stage, which is with the presence of hematite as the last phase of the oxidation process in mineral paragenesis. In addition, there is a rudle texture which is indicated by the presence of water spots due to the "progressive high-temperature oxidation" condition in the ilmenite mineral grains [21]. The following is a schematic ore mineral paragenesis (Figure 7).

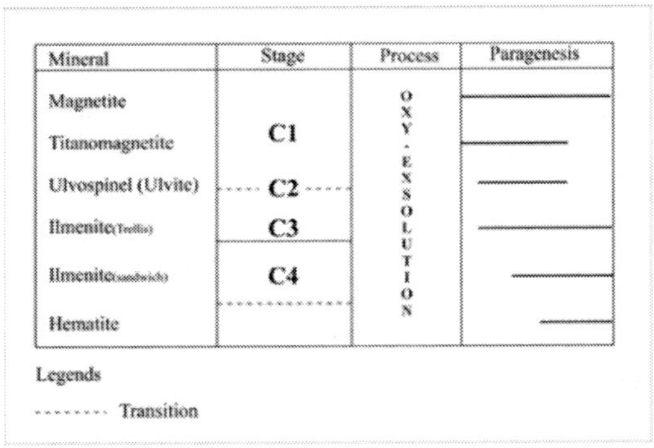

Figure 7. Mineral paragenesis of the samples

5. Conclusion

The mineralogy distribution of light minerals is more dominant in the grain size of 0.212-2 mm and the distribution of heavy minerals is abunndant in the size fraction of <0.106 mm. The mineralogy of beach ironsand deposits in the study area is composed of two classifications, heavy minerals (magnetite, titanomagnetite, and hematite) and light minerals (quartz, clinopyroxene (hedenbergite and augite), biotite, hornblende, and ilvaite. The Fe, Ti, and V-bearing mineral is magnetite with an average of Fe content (79.02 - 92.43 wt%), Ti (6.08-7.89 wt%), and V (0.56-1.15 wt%). The Ti element distribution is not correlated with the presence of V, while the content of the V correlates with the Fe. Thus it indicate that the magnetite mineral can be used as the main target or "mineral pathfinder" for advanced exploration of Fe and critical raw minerals benefication such as, Ti-, and V at the Adikarto Bay, Kulonprogo Yogyakarta particularly and in general along southern Java beach.

Acknowledgements

This paper is a part of the first author thesis research. PPSDMBP Bandung provides samples, We thank to all parties who support this work.

References

[1] K. D. Kelley, C. T. Scott, D. E. Polyak, and B. E. Kimball, "Vanadium," Professional Paper, 2017.

[2] H. Elsner, *Heavy Minerals of Economic Importance.* Germany: Bundesanstalt für Geowissenschaften und Rohstoffe (BGR), 2010.

[3] P. A. Nugraha, S. P. Sari, W. N. Hidayati, C. R. Dewi, and D. Y. Kusuma, "The origin and composition of iron sand deposit in the southern coast of Yogyakarta," Yogyakarta, Indonesia, 2016, p. 020028. doi: 10.1063/1.4953953.

[4] M. Nurcholis and D. Mulyanto, "Study on the Residue Resulted from the Metallic Minerals separations to the Coastal Iron-sand of Yogyakarta Indonesia," vol. 12, no. 15, p. 9, 2017.

[5] H. Kurnio, "Coastal Characteristics Of Iron Sand Deposits In Indonesia," vol. 10, no. 09, p. 12, 2007.

[6] S. Lee, "A Review on Types of Vanadium Deposits and Process Mineralogical Characteristics," *ksmer*, vol. 57, no. 6, pp. 640–651, Dec. 2020, doi: 10.32390/ksmer.2020.57.6.640.

[7] M. B. Verma, "Atomic Minerals Directorate For Exploration And Research," *PINSA*, vol. 86, Feb. 2020, doi: 10.16943/ptinsa/2020/49824.

[8] A. Kerr, J. A. Walsh, G. W. Sparkes, and J. G. Hinchey, "Vanadium Potential In Newfoundland And Labrador: A Review And Assessment," p. 30, 2013.

[9] W. Rahardjo, Sukandarrumidi, and H. M. D. Rosidi, "Peta Geologi Lembar Yogyakarta, Jawa," Pusat Penelitian dan Pengembangan Geologi, 1995.

[10] Muswerry, M. H. Azkab, Fahmi, D. E. D. Setiono, H. Thoha, and S. Natsir, *Sumber Daya Laut di Perairan Pesisir Gunung Kidul, Yogyakarta,* 1st ed., vol. xvi. Jakarta: LIPI Press, 2015.

[11] I. G. Sukadana, I. W. Warmada, F. Pratiwi, A. Harijoko, T. B. Adimedha, and A. W. Yogatama, "Elemental Mapping for Characterizing of Thorium and Rare Earth Elements (REE) Bearing Minerals Using μXRF," *Atom Indo.*, vol. 48, no. 2, p. 87, Jul. 2022, doi: 10.17146/aij.2022.1215.

[12] B. S. V. Gosen, D. L. Fey, A. K. Shah, P. L. Verplanck, and T. M. Hoefen, "Deposit Model for Heavy Mineral Sand in Coastal Environments," p. 51, 2014, doi: http://dx.doi.org/10.3133/sir20105070L.

[13] V. Hernández-Hinojosa, P. C. Montiel-García, J. S. Armstrong-Altrin, R. Nagarajan, and J. J. Kasper-Zubillaga, "TEXTURAL AND GEOCHEMICAL CHARACTERISTICS OF BEACH SANDS ALONG THE WESTERN GULF OF MEXICO, MEXICO," *Carpath. J. Earth Environ. Sci.*, vol. 13, no. 1, pp. 161–174, Feb. 2018, doi: 10.26471/cjees/2018/013/015.

[14] E. M. A. Mitwally and B.-S. Yu, "Geochemistry of magnetite in beach sands, stream sediments, and in situ magnetites in surrounding rocks at north Taiwan island," *Acta Geochim*, vol. 41, no. 3, pp. 434–452, Jun. 2022, doi: 10.1007/s11631-021-00521-y.

[15] Q. Duparc, S. A. S. Dare, P. A. Cousineau, and J. Goutier, "Magnetite Chemistry As A Provenance Indicator In Archean Metamorphosed Sedimentary Rocks," *Journal of Sedimentary Research*, vol. 86, no. 5, pp. 542–563, May 2016, doi: 10.2110/jsr.2016.36.

[16] T. Grammatikopoulos, A. McKen, and C. Hamilton, "Vanadium-bearing magnetite and ilmenite mineralization and beneficiation from the Sinarsuk V-Ti project, West Greenland," *CIM Bulletin*, vol. 95, p. 9, 2002.

[17] S. E. Haggerty, "Oxide Texture-A Mini-Atlas," in *OXIDE MINERALS: THEIR PETROLOGIC AND MAGNETIC SIGNIFICANCE*, vol. 25, 1991, pp. 129–137.

[18] W. Tan, P. Liu, H. He, C. Y. Wang, and X. Liang, "Mineralogy and Origin of Exsolution In Ti-Rich Magnetite From Different Magmatic Fe-Ti Oxide-Bearing Intrusions," *Can Mineral*, vol. 54, no. 3, pp. 539–553, May 2016, doi: 10.3749/canmin.1400069.

[19] A. Mucke, "Magnetite, ilmenite and ulvite in rocks and ore deposits: petrography, microprobe analyses and genetic implications," vol. 77, no. Mineralogy and Petrology, pp. 215–234, 2003, doi: 10.1007/s00710-002-0216-1.

[20] G. Ivanyuk *et al.*, "Subsolidus Evolution of the Magnetite-Spinel-UlvöSpinel Solid Solutions in the Kovdor Phoscorite-Carbonatite Complex, NW Russia," *Minerals*, vol. 7, no. 11, p. 215, Nov. 2017, doi: 10.3390/min7110215.

[21] M. W. McElhinny and P. L. McFadden, "Chapter 2-Rock Magnetism," in *International Geophysics*, vol. 73, Elsevier, 2000, pp. 31–77. doi: 10.1016/S0074-6142(00)80095-9.

IOP Conf. Series: Earth and Environmental Science 1233 (2023) 012023 doi:10.1088/1755-1315/1233/1/012023

Microzonation study around Lake Toba, North Sumatra, Indonesia based on horizontal – vertical spectral ratio

T Anggono[1*], D Anjali[2], E Marinda[2], S Haryanti[2], L Handayani[1], MM Mukti[3], AM Lubis[4], KL Gaol[1], Y Sudrajat[1], S Syuhada[1], FRT Saputra[1], AD Prasetio[1], M Hasib[1], F Febriani[1], CN Dewi[1], W Setyonegoro[1]

[1]Reserch Center for Geological Disaster – National Research and Innovation Agency
[2]Physics Department - Syarif Hidayatullah State Islamic University
[3]Research Center for Geological Resources – National Research and Innovation Agency
[4]Physics Department – Bengkulu University

*Corresponding author: titi.anggono@brin.go.id

Abstract. Lake Toba, North Sumatra, Indonesia, is one of the biggest calderas in the world. The area is also prone to earthquake damage because of the tectonic activity due to its location in the subduction zone of Indo-Australian and Eurasian Plates. In recent years, Lake Toba has become one of the most attractive tourist destinations in Indonesia. Hence, the Indonesian government has been considering infrastructure development in this region. The region vulnerability cause by soil amplification due to the earthquake activity can be quantitatively investigated. We carried out a microzonation analysis around Lake Toba using Horizontal-Vertical (H/V) Spectral Ratio method. A total of about 30 short-period seismometers around Lake Toba were used for the analysis. We estimated several parameters that may define some soil characteristics, such as H/V peak amplitudes, peak frequencies, average shear wave velocity of 30 m depth (Vs30), and vulnerability index (Kg). Spatial distribution of peak frequencies and Vs30 suggest that the region may fall into sites SC and SD showing characteristics of dense to stiff soil. We also observed areas with a high vulnerability index (Kg > 20), which may indicate the high possibility of soil vulnerability due to strong earthquakes. The findings may provide preliminary results or sources for future geophysical, geological, and geotechnical analysis for disaster risk reduction in this high-risk region.

1. Introduction

Site amplification sometimes is defined as the probability of how the ground shakes with an incoming earthquake. In risk reduction, it is an important factor for earthquake hazard mitigation besides the earthquake mechanism. Several studies suggested that the ground shaking depends not only on the earthquake mechanism but also on the site conditions such as soil thickness, surface condition, and nonlinearity [1, 2]. The spectral ratio method has been used to measure the site effect of a location due to an earthquake effect [3, 4]. Nakamura [5] proposed the use of microtremor records to measure H/V spectral ratio. The proposed method uses continuous seismic records and can be easily applied for the acquisition. The method can extract information on the peak/resonance or predominant soil frequency, especially with soil having strong impedance contrast [6, 7]. Horike et al. [7] suggested the peak frequency from the microtremor measurement seems in agreement with the peak frequency from the seismic shear-wave spectra. However, peak amplitude is debatable if it may represent the true site amplification. Hagshenas et al. [4] suggested that the H/V spectral ratio peak amplitude can be considered a lower bound of site amplification. The resonance frequency may correlate to the soil dynamic properties. Microtremor (weak ground motion) sources always exist in the form of natural or artificial processes such as wind, sea waves, or human activities. They can be composed of body and surface wave components [8]. Due to the always existence of microtremor sources, microtremor records have some advantages in the microzonation studies compared to the analysis based on the earthquake records [9, 10]. Inversion of H/V spectral ratio to estimate the shallow velocity structure considers that the curve/ellipticity consists only the Rayleigh wave. However, it is suggested that that the H/V spectral ratio curve may also contain Love wave and vary with frequency and time [8, 11]. The inversion from the H/V spectral ratio curve is nonlinear and there is a trade-off between medium velocity and thickness. To reduce the trade-off, borehole data can be used to constraint either the medium thickness or velocity [9]. Ghofrani and Atkinson [12] developed a relationship between H/V spectral ratio parameters (peak amplitude, peak/resonance frequency) and soil's shallow shear wave velocity (Vs30). Indonesia National Standardization Agency classified site location based on Vs30 in the SNI 1726-2019 (Table 1) [13].

Table 1. Site classification from SNI 1726 – 2019 [13]

Site class	Site description	Vs (m/s)
SA	Hard rock	>1500
SB	Rock	750 – 1500
SC	Very dense soil and soft rock	350 – 750
SD	Stiff soil	175 – 350
SE	Soft soil	<175
SF	Site specific	

Lake Toba, North Sumatra, Indonesia has become one of the most visited tourist destinations in Indonesia with a large population living in the surrounding region. It is located in the northern part of Sumatra Island, which is prone to earthquakes and volcanoes. The Island is in the boundary of Eurasian-Indo Australian Plates. Along the subduction zone, the Indo-Australian plate subducts beneath the Eurasian plate in the oblique direction. This motion accommodates the dextral slip of the Sumatra fault zone [14, 15]. The subduction zone and the Sumatra fault zone cause high earthquake seismicity in this region. Large and devastating earthquakes have struck this region, such as the 2004 Mw 9.2 Sumatra and the 2005 Mw 9.7 Nias earthquakes. Those large earthquakes occurred in the subduction zone near the North Sumatra region. Due to the proximity to the earthquake source, it is very important to have information about the vulnerability of this region. Figure 1 shows the geologic map of Lake Toba and its surrounding area. The geological processes cause a complex geological setting in the region. Hence, it suggests the possibility of spatial differences in site characteristics. We investigate the site information or characteristics in the area using the H/V spectral ratio method.

2. Data and method

A total of 30 seismometers were used to calculate the H/V spectral ratio (Figure 1). The seismometers were part of 42 seismometers of a temporary network installed around Lake Toba from May to October 2008. The temporary network was equipped with three components short-period seismometers (L4-3D) with a 1 Hz natural frequency. Initially, the temporary network was installed to study the Lake Toba structure. The data were digitized at 100 samples per second. Between 30-minutes and 1-hour of data were selected for the continuous ambient noise recording. We chose these periods to exclude data contaminated with large amplitude due to earthquakes. The selected records were then manually inspected to avoid instrumental glitches, which may cause signal saturation in the selected period.

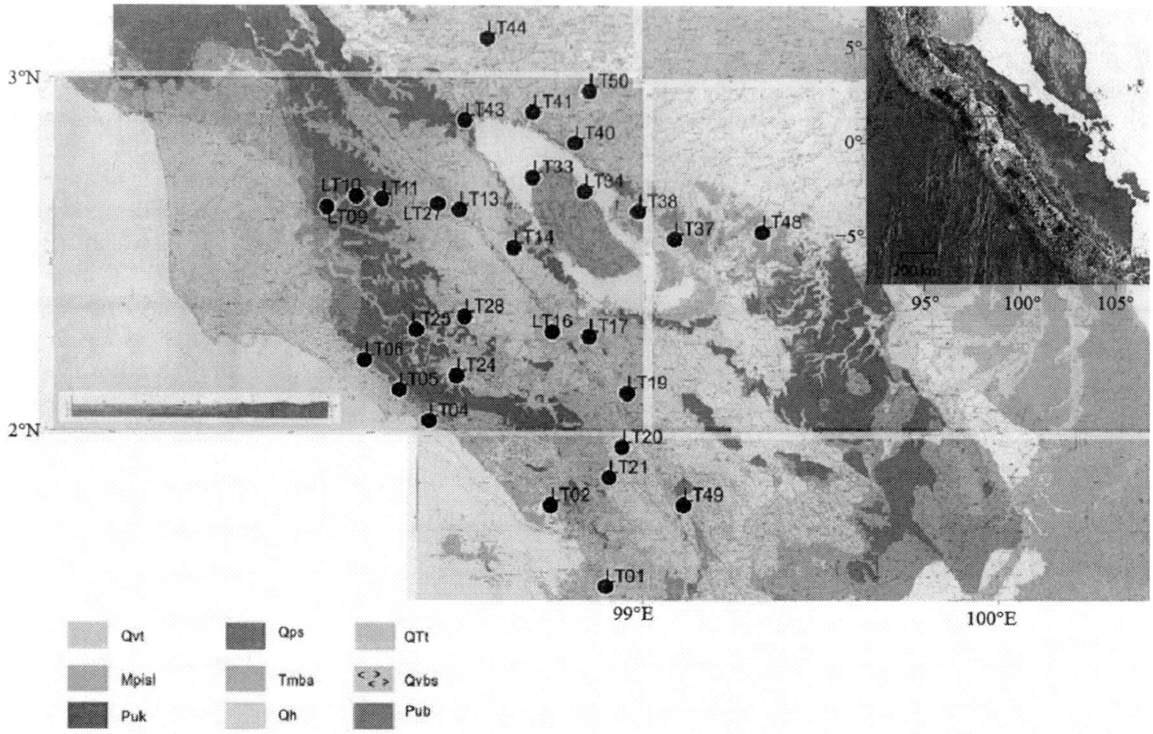

Figure 1. Geological map around Lake Toba, North Sumatra, Indonesia from Cameron et al. [16], Cameron et al. [17], Aldiss et al. [18], Aspden et al. [19], and Clarke et al. [20]. Solid black circles represent short-period seismometers used in this study. Qvt: Toba tuffs, MPisl: Sibolga complex, Puk: Kluet formation, Qps: Samosir formation, Tmba: Barus formation, Qh: Alluvium, QTt: Tutut formation, Qvbs: Singkut unit, Pub: Bohorok formation. Only geological units with the analyzed short-period seismometers are listed. Inset map shows earthquake seismicity in Sumatera, Indonesia in the period of 2000 – 2020, red square is the location of the study area.

We processed the HVSR using an open source Geopsy software. Fourier amplitude spectra of each component were calculated with a 40 s window length after applying 5% cosine taper. The amplitude spectra were then smoothed using the Konno-Ohmachi algorithm [21]. The amplitude spectra of two horizontal components were then averaged. H/V spectral ratio was computed for each window and then averaged. H/V spectral ratio was computed in in the frequency range of 0.1 – 15 Hz. We applied the following criteria for reliable measurement of H/V spectral ratio: the H/V peak frequency f_0 at a given window length lw should be larger than $10/lw$, the number of significant cycles for the selected noise window should be larger than 200, and the standard deviation of the H/V amplitude at peak frequency f_0 should be less than the threshold values of 2.0 ($f_0 > 0.5$ Hz) and 3.0 ($f_0 < 0.5$ Hz).

3. Results and discussion

The obtained H/V spectral ratios are shown in Figure 2. Figure 2a shows examples of observed ellipticity of H/V spectral ratio. We obtained that the amplitudes of H/V spectral ratios are above 2.0, which can be considered a clear amplitude [9, 22]. We also observed not only single peaks, but also multiple peaks such as LT05 and LT09. A single clear peak was suggested to be related to the depth of a sharp velocity increase or soil deposit [23]. In these cases, the peak/resonance frequencies were estimated using information from the adjacent stations with single peaks (LT04 for LT05 and LT10 for LT09). It was suggested that the multiple peaks in the observed H/V spectral ratio might indicate the existence of complex subsurface structures vertically or horizontally [24]. In contrast, low impedance contrast may cause a broad or low H/V spectral ratio peak that is typically observed at hard rock or bedrock sites. The peak frequencies, which represent the frequency resonance (f_0), are observed at a range of 0.27 – 6.5 Hz (Figure 2b). It is typically assumed that thin soil has a range of f_0 at 1.7 – 9.0 Hz and thick sediment thickness for f_0 less than 1 Hz. The observed H/V peak amplitude, which may represent the amplification factor (A_0), ranges between 2.10 – 8.99. It was suggested that the peak amplitudes obtained from HVSR are generally smaller than the actual amplification and can be the lower bound of the site amplification [4, 25].

The spatial distribution of A_0 and f_0 are shown in Figures 3a and 3b. The spatial distribution was plotted by interpolating each station's A_0 and f_0 values. The high H/V peak amplitude (> 6) region, which may represent the high probability of soil amplification, is shown in the west down to the south region area from Lake Toba. Several studies have proposed soil type classification based on resonance frequency from microtremor measurement [26-28]. Kanai and Tanaka [26] suggested that the changes in resonance frequency can be related to the soil type. They classified the soil type into four types based on the resonance frequency, which are type 1 (5.0 – 10.0 Hz), type 2 (2.5 – 5.0 Hz), type 3 (1.25 – 2.5 Hz), and type 4 (0.5 – 1.0 Hz). The interpolated resonance frequency (Figure 3b) from our measurement shows that around Lake Toba the resonance frequency generally falls around 2.0 – 5.0 Hz. The results suggest that it may fall into type 2 – 3 from Kanai and Tanaka soil classification, which suggests the presence relatively hard soil.

We also calculated *Vs30*, which is shear wave velocity down to a depth of 30 m, from the obtained H/V spectral ratio. The *Vs30* is used as a site condition in the ground motion model. Even though Vs30 represents the shallow velocity structure, it may correlate to the deeper soil depth if no significant changes occur in the soil structure. In this study, we calculate the *Vs30* using the equation proposed by Ghofrani and Atkinson [12]. Their relationship was developed using the Japanese database of K-Net dan KiK-Net of NIED-Japan. It is also shown to be applicable to the NGA-West 2 database that has global earthquake records [29]. From our calculation, site classification around Lake Toba based on Indonesia National Standard can be classified into two sites, SC (very dense soil and soft rock, 350 < *Vs30* < 750 m/s) and SD (stiff soil, 175 < *Vs30* < 350 m/s) (Figure 3c). The vulnerability index (*Kg*), which was calculated from H/V peak amplitude and the resonance frequency, is shown in Figure 3d. *Kg* value depends on the soil's dynamic properties. The vulnerability index may define the vulnerability damage possibility of soil during a strong or large earthquake [30, 31]. It has been suggested that if *Kg* > 20, then the soil is susceptible to deformation, *Kg* > 10 is relatively at risk, while low *Kg* may reflect the low-risk region [30-32]. From H/V spectral measurement, we observed several areas or sites that might have a high vulnerability index (> 20).

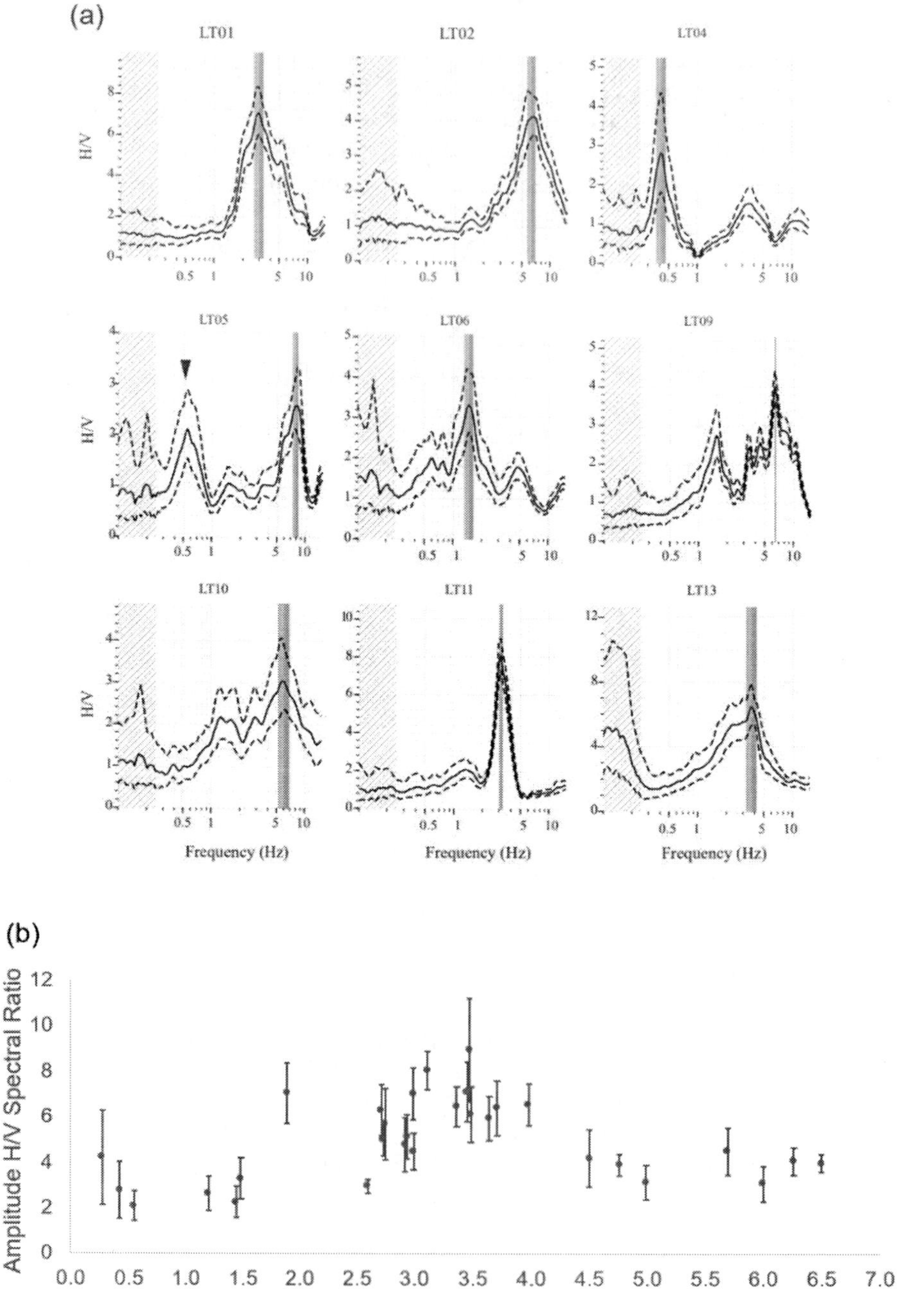

Figure 2. (a) Calculated H/V spectral ratio from the seismic network around Lake Toba. Solid black lines represent the mean H/V spectral ratio, and the dashed lines represent its standard deviation. (b) Observed H/V peak amplitudes as a function of peak frequency. The vertical lines represent the standard deviation of peak amplitudes.

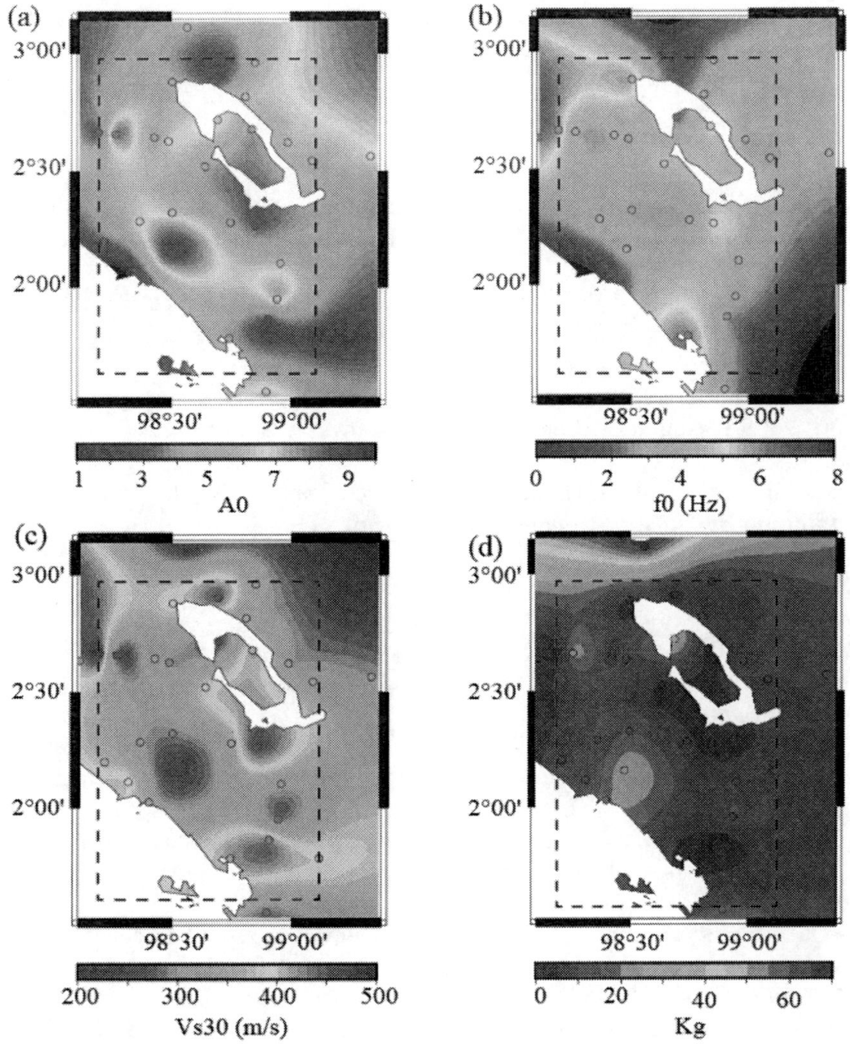

Figure 3. Observed (a) H/V peak amplitude (A_0), (b) resonance frequency (f_0), (c) Vs30 and (d) vulnerability index (*Kg*) from the measurement of H/V spectral ratio around Lake Toba, Indonesia. Dashed square is the area that we interpreted is quite reliable interpolation area, which beyond the square may consist of artifact due to no or a small number of data.

4. Conclusions

In this study, we applied the horizontal to vertical spectral ratio method for seismic microzonation study and estimation of the soil type. We determined the H/V peak amplitude and resonance frequency. Average shear velocity down to 30 m depth (Vs30) was also investigated. The site classification is estimated to be SC and SD based on Indonesian National Standard SNI 1726-2019. From the calculated *Kg*, we found areas with high vulnerability index *Kg* (> 10). High vulnerability index may represent the high susceptibility of soil deformation when a strong earthquake strikes. We hope our findings might serve as preliminary data sources or results for other geophysical, geological, and geotechnical analysis to investigate hazard mitigation in this region.

Acknowledgement
Seismic data were obtained from the GEOFON data center of the GFZ German Research Center for Geoscience. Some figures were prepared using Generic Mapping Tools (GMT) by Wessel and Smith [33]. This study was supported by National Research and Innovation Agency Rumah Program Kebencanaan.

References

[1] Aguirre J and Irikura K 1987 Nonlinearity, Liquefaction, and Velocity Variation of Soft Soil Layers in Port Island, Kobe, during the Hyogo-ken Nanbu Earthquake *Bulletin of Seismological Society of America* 87(5) 1244-58.

[2] Fukushima Y Irikura K Uetake T and Matsumoto H 2000 Characteristics of Observed Peak Amplitude for Strong Ground Motion from the 1995 Hyogoken Nanbu (Kobe) Earthquake *Bulletin of Seismological Society of America* 90(3) 545-65.

[3] Akinci A Malagnini L and Sabetta F 2008 Characteristics of the strong ground motions from the 6 April 2009 L'Aquila earthquake, Italy *Soil Dynamics and Earthquake Engineering* 30(5) 320- 35.

[4] Haghshenas E Bard PY and Theodulidis N 2008 Empirical evaluation of microtremor H/V spectral ratio *Bulletin of Earthquake Engineering* 6(1) 75-108.

[5] Nakamura Y 1989 A Method for Dynamic Characteristics Estimation of Subsurface using Microtremor on the Ground Surface *Quarterly Report of Railway Technical Research Institute* 30(1) 25-33.

[6] Duval A-M Bard PY Lebrun B Lacave-Lachet C Riepl J and Hatzfeld D 2001 H/V technique for site response analysis. Synthesis of data from various surveys *Bollettino di Geofisica Teorica ed Applicata* 42(3-4) 267-80.

[7] Horike M Zhao B and Kawase H 2001 Comparison of site response characteristics inferred from microtremors and earthquake shear waves *Bulletin of Seismological Society of America* 91(6) 1526-36.

[8] Bonnefoy-Claudet S Cornou C Bard P-Y Cotton F Moczo P Kristek J et al. 2006 H/V ratio: a tool for site effects evaluation. Results from 1-D noise simulations *Geophysical Journal International* 167(2) 827-37.

[9] Vella A Galea P and D'Amico S 2013 Site frequency response characterisation of the Maltese islands based on ambient noise H/V ratios *Engineering Geology* 163 89-100.

[10] Flores H Malischewsky P and Jentzsch G 2013 H/V spectral ratio analysis and Rayleigh modelization in Eastern Thuringia, Germany *Geofisica Internacional* 52(4) 355-64.

[11] Knapmeyer-Endrun B Golombek MP and Ohrnberger M 2016 Rayleigh Wave Ellipticity Modeling and Inversion for Shallow Structure at the Proposed InSight Landing Site in Elysium Planitia *Mars Space Science Reviews* 211(1-4) 339-82.

[12] Ghofrani H and Atkinson GM 2014 Site condition evaluation using horizontal-to-vertical response spectral ratios of earthquakes in the NGA-West 2 and Japanese databases *Soil Dynamics and Earthquake Engineering* 67 30-43.

[13] Badan Standardisasi Nasional (BSN) 2019 Tata cara perencanaan ketahanan gempa untuk struktur bangunan gedung dan non gedung SNI 1726-2019.

[14] Tilmann FJ Craig TJ Grevemeyer I Suwargadi B Kopp H and Flueh E 2010 The updip seismic/aseismic transition of the Sumatra megathrust illuminated by aftershocks of the 2004 Aceh-Andaman and 2005 Nias events *Geophysical Journal International* 181 1261-74.

[15] Collings R Lange D Rietbrock A Tilmann F Natawidjaja D Suwargadi B et al 2012 Structure and seismogenic properties of the Mentawai segment of the Sumatra subduction zone revealed by local earthquake traveltime tomography *Journal of Geophysical Research: Solid Earth* 117(B1) B01312.

[16] Cameron NR Aspden JA Miswar M and Syah HH 1981 Geologic map of the Tebingtinggi Quadrangle, Sumatra *Geological Research and Development Centre*.

[17] Cameron NR Aspden JA Bridge DMC Djunuddin A Ghazali SA Harahap H et al 1982 Geological Map of the Medan Quadrangle, Sumatra *Geological Research and Development Centre*.

[18] Aldiss DT Whandoyo R Ghazali SA and Kusyono 1983 Geological Map of the Sidikalang and (part of) Sinabang Quadrangles, Sumatra *Geological Research and Development Centre*.

[19] Aspden JA Kartawa W Aldiss DT Djunuddin A Diatma D Clarke MCG et al 1982 Geologic Map of the Padangsidempuan and Sibolga Quadrangles, Sumatra *Geological Research and Development Centre*.

[20] Clarke MCG Ghazali SA Harahap H Kusyono and Stephenson B 1982 Geologic Map of the Pematangsiantar Quadrangle, Sumatra *Geological Research and Development Centre*.

[21] Konno K and Ohmachi T 1998 Ground-Motion Characteristics Estimated from Spectral Ratio between Horizontal and Vertical Components of Microtremor *Bulletin of the Seismological Society of America* 88(1) 228-41.

[22] Macau A Figueras S Roulle A Kingne L Bellmunt F and Meneses C 2012 Seismic microzonation in two Pyrenean Valleys: Val d'Aran and Luchonnais *Proceedings of the Fifteenth World Conference of Earthquake Engineering* Lisbon Portugal.

[23] Bignardi S Yezzi AJ Fiussello S and Comelli A 2018 OpenHVSR - Processing toolkit: Enhanced HVSR processing of distributed microtremor measurements and spatial variation of their informative content *Computers & Geosciences* 120 10-20.

[24] Castellaro S and Mulargia F 2009 Vs30 estimates using constrained H/V measurements *Bulletin of the Seismological Society of America* 99(2A) 761-73.

[25] Parolai S Richwalski SM Milkereit C and Bormann P 2004 Assessment of the stability of H/V spectral ratios from ambient noise and comparison with earthquake data in the Cologne area (Germany) *Tectonophysics* 390(1-4) 57-73.

[26] Kanai K and Tanaka T 1961 On Microtremors *Bulletin of the Earthquake Research Institute* 39 97-114.

[27] Nogoshi M and Igarashi T 1970 On the propagation characteristics of microtremors *Journal of the Seismological Society of Japan* 23 264-80.

[28] Panah AK Moghaddas NH Ghayamghamian MR Motosaka M Jafari MK and Uromieh A 2002 Site effect classification in east-central of Iran *Journal of Seismology and Earthquake* Engineering 4 37-46.

[29] Anbazhagan P Srilakshmi KN Bajaj K Moustafa SSR and Al-Arifi NSN 2019 Determination of seismic site classification of seismic recording stations in the Himalayan region using HVSR method *Soil Dynamics and Earthquake Engineering* 116 304-16.

[30] Akkaya İ 2020 Availability of seismic vulnerability index (Kg) in the assessment of building damage in Van, Eastern Turkey *Earthquake Engineering and Engineering Vibration* 19(1) 189- 204.

[31] Öztürk S Beker Y Sari M Pehlivan L 2021 Estimation of ground types in different districts of Gümüşhane province based on the ambient vibrations H/V measurements *Sigma Journal of Engineering and Natural Sciences* 39(4) 374-91.

[32] Babacan AE Ceylan S 2021 Evaluation of soil liquefaction potential with a holistic approach: a case study from Araklı (Trabzon, Turkey) *Bollettino di Geofisica Teorica ed Applicata* 62(1) 173- 98.

[33] Wessel P Luis JF Uieda L Scharroo R Wobbe F Smith WHF and Tian D 2019 The Generic Mapping Tools Version 6 *Geochemistry, Geophysics, Geosystems* 20(11) 5556-64.

IOP Conf. Series: Earth and Environmental Science 1233 (2023) 012024 doi:10.1088/1755-1315/1233/1/012024

Landslide Susceptibility Mapping in Jurangjero and Kampung Villages, Ngawen Sub-District, Gunung Kidul Regency, Special Region of Yogyakarta, Indonesia

B Manyvanh[1], H Setiawan[1,2]

[1]Department of Geological Engineering, Universitas Gadjah Mada, Yogyakarta, Indonesia
[2]Center for Disaster Mitigation and Technological Innovation (GAMA-InaTEK), Universitas Gadjah Mada, Indonesia

Abstract. Jurangjero and Kampung villages are one of the areas in the Ngawen sub-district, Gunung Kidul Regency, Special Region of Yogyakarta, Indonesia. The research area is located in the Southern Mountain regionally, especially the Baturagung Sub-Zone. This area is part of The Baturagung Hills, dominated by weathered andesite breccia, tuff, sandstone, and limestone. The topography of these hills can be divided into three slopes, 15-30%, 30-45%, and more than 45%. A soil movement event happened in Jurangjero Village, the biggest, on January 15, 2020, which caused heavy damage to people's homes, road damage, and breaking bridges. Geographic Information System (GIS) and Analytic Hierarchy Process (AHP) model were applied to identify landslide conditioning factors and produce the landslide susceptibility map. Four conditioning factors including slope, lithology, faults to distance, and land use, were combined using ArcGIS 10.8 to create the landslide susceptibility map of the research area. The research area is divided into four landslide susceptibility zone very low, low, moderate, and high, with each zone of 36%, 20%, 41%, and 3%, respectively. A significant percentage (3%) of the research area with high landslide susceptibility is located along the steep slope, showing the landslides' locations.

1. Introduction

A landslide is an occurrence or phenomenon of geological disasters where rock or soil mass or debris moves down a slope. Landslide mechanisms include sliding, falling, or flowing of material down a slope due to gravitational force [1]. Indonesia is situated at the intersection of three global plates: the Australian, Eurasian, and Pacific plates, constantly moving and colliding [2]. In Indonesia, the intersection of the collision has resulted in the formation of volcanic paths. Because of these volcanic paths, some zones of Indonesia's mountains and hills have various slopes to steep slopes [3]. As a result, Indonesia is at risk of landslides, leading to fatalities and property damage, infrastructure, and environmental damages[4]. Landslide disasters are often localized; however, numerous occur across Indonesia. Landslides inflict more losses in the long run than other natural disasters. From January to August 2014, Indonesia occurred 972 times of landslides, with 374 dead and missing people, 1,764,227 victims, and 39,823 damaged houses [5].

Jurangjero and Kampung villages are one of the Ngawen sub-district areas with high elevations and steep slopes to the west, which cause landslide disasters. Ngawen sub-district has much hilly topography and mountainous areas. These conditions can contribute to the high landslide susceptibility. Significant slope factors cause landslides that occur in the research area and natural process factors that have occurred in landslides since the beginning of the formation of the region and still last today. Soil movement event happened in Jurangjero Village on January 15, 2020, which caused heavy damage to people's homes, road damage, and breaking bridges [2].

Analytic Hierarchy Process (AHP) and Geographic Information System (GIS) rate two each other assist the development of convergence and the long-term viability of decision alternatives [6]. The current research aims to conduct a landslide susceptibility mapping using AHP and GIS to identify high susceptibility areas. This research will help the designing, planning, and resettlement, particularly in terms of social and economic development of Jurangjero and Kampung villages (in particular) and for Indonesia (in general).

2. Study Area

The research area is Jurangjero and Kampung Villages, Ngawen sub-district, Gunung Kidul regency, Special Region of Yogyakarta, Indonesia. The Semin Sub-district borders the villages to the east; Nglipar Sub-districts to the west; Tancep Village to the north; and Beji Village to the south. According to the coordinate system of WGS 1984 UTM Zone 49S: E= 461781.851 – 468231.147, and N = 9131884.505 – 9136755.647.

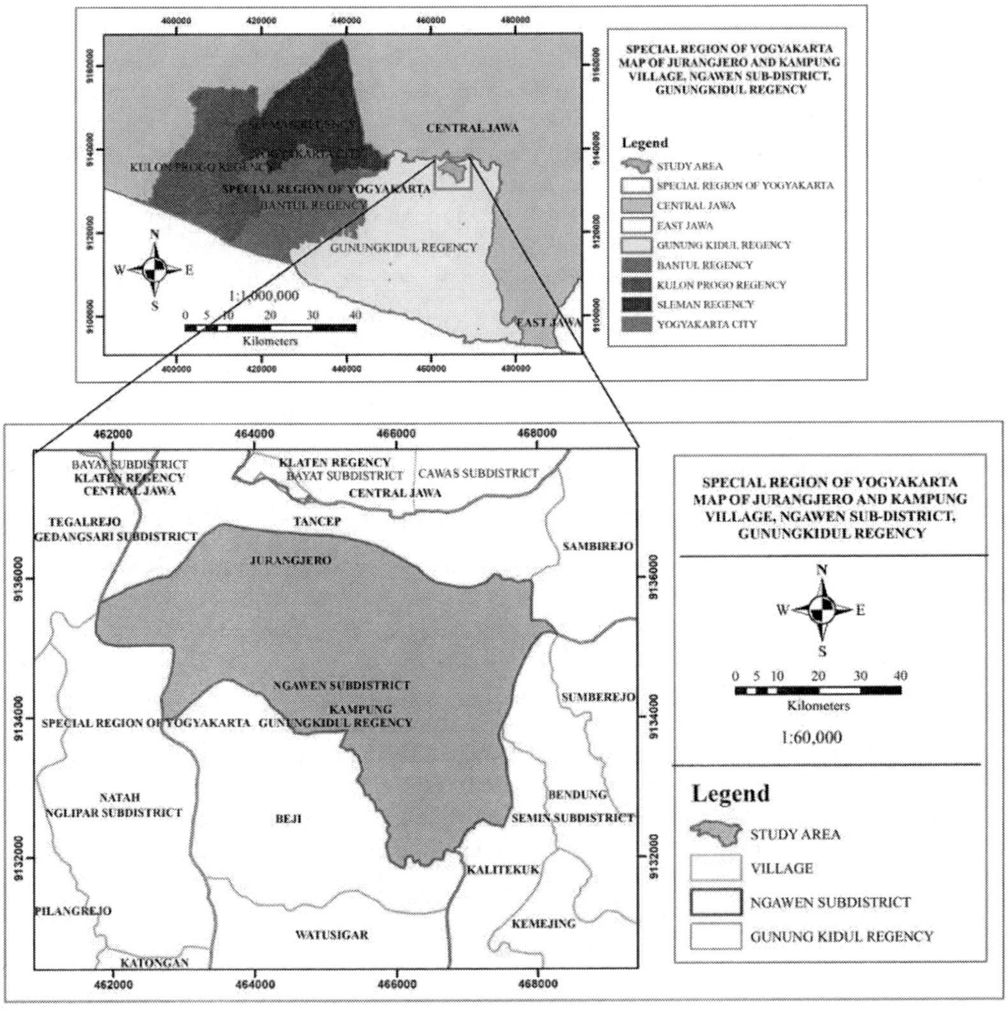

Figure 1. Study Area Map in Jurangjero and Kampung Villages, Ngawen Sub-District, Gunung Kidul Regency, Special Region of Yogyakarta, Indonesia (source: https://tanahair.indonesia.go.id/portal-web)

IOP Conf. Series: Earth and Environmental Science 1233 (2023) 012024 doi:10.1088/1755-1315/1233/1/012024

3. Geology

3.1. Geomorphology

The research area is part of the Southern Mountain Zone, the Baturagung Sub-Zone. The Baturagung hills are in the Baturagung Sub-Zone, the structural origin of the southern part of Java Island, which has experienced uplift and fracture. These structural hills are formed by a diastrophism process in the form of a multilevel fault [7]. The northern part has a mountainous topography with sharp ridges, and the highest peak is Baturagung Mountain with an altitude of approximately 700 meters above sea level, while the lowest place is around the terraces of the Oyo River with an altitude of approximately 150 meters above sea level, with a topography of low hills and rounded edges. The topography of these hills has slopes at the bottom (15-30%) to steep at the top (30-45%), and ridges extend from west to east in the north with very steep slopes (>45%), which is the fault plane (escarpment). The constituent rocks are old volcanic materials that have undergone much-advanced weathering, lots of cracks and fractures, relatively thin soil layer, lots of rocks, with fairly high rainfall, causing the process of erosion and landslides to be quite intensive and very frequent. This area, as happened in the Ngawen area and its surroundings [8].

3.2. Regional stratigraphy

The stratigraphy of the research area is made up of volcanic rock and carbonate, and it is divided into several rock formations, including the Semilir Formation, Nglanggeran Formation, Oyo Formation, and Wonosari Formation. The Semilir Formation is scattered in the northern areas of research, forming the morphology of slopes and hills in the Ngawen Region at elevations ranging from 200 to 365 meters above sea level [9]. The formation is made up of tuff, lapilli, and volcanic rocks. Semilir Formation started from Oligocene Early Miocene and interfingering with Nglanggran [10]. The formation was deposited during the period of Volcanism Southern Mountains in the Early Oligocene-Middle Miocene [11]. The Nglanggeran Formation is dispersed in the northwest of a locally dispersed research area that forms the morphology of the hill in the Ngawen Region at elevations ranging from 225 to 290 meters above sea level. This unit comprises polymic breccia lithology with andesite and basalt fragments [9]. Nglanggeran Formation was deposited with the Semilir Formation in the Miocene Early [11], [12]. The Oyo Formation is found in the valleys of Baturagung Mountain in the northwest of the Ngawen Region, at elevations of less than 250 meters above sea level. This formation comprises the lithology of sandstone, which has a carbonate content of more than 70% and is composed of shells of fossil shells, small lithic fractions, and feldspar [9]. This formation was deposited in the period of the Middle Miocene [11]. The thickness of this formation reaches 140m [11]. The Wonosari Formation depicts the syncline valleys that form the plains spread throughout Ngawen and Semin Regions. This formation is constituted of layered limestone lithologies that are mostly made up of calcarenite and calcilutite. Chemically, this formation is a limestone with a carbonate content greater than 90%, compact physical properties, and a growing pore of secondary pores caused by dissolution. This is the most recent formation [9].

3.3. Geological Structure

The geological structure of the research area has one fold structure, two faults, and several joints. The research area comprised two folds, namely, syncline fold. The rock layers found in the research area have the opposite slope due to the structure of syncline fold. Several joints can be found in every rock that composes Baturagung Mountain in the Ngawen Region, generally one set of paired joints parallel to the fault [13]. In the research area, there is a pattern of the direction of the fault, which is the northeast fault pattern, which is identified from the appearance of the dispersion pattern of the rock in the area. A dextral horizontal fault characterizes the northeast fault pattern.

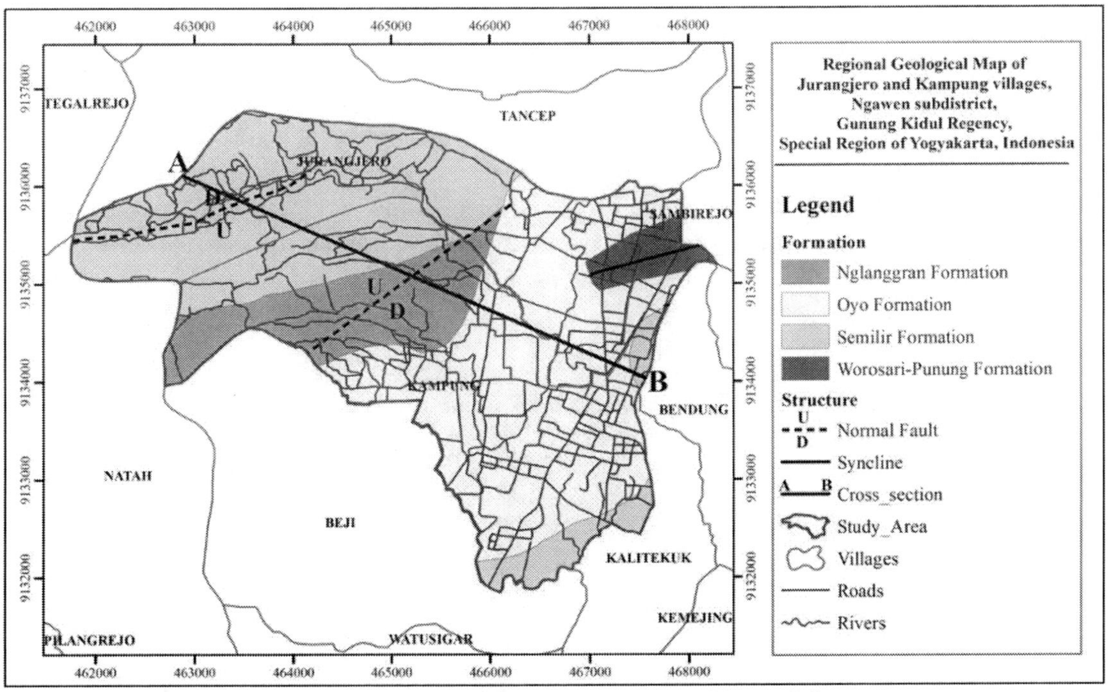

Figure 2. Geological Map of the Study area [7]

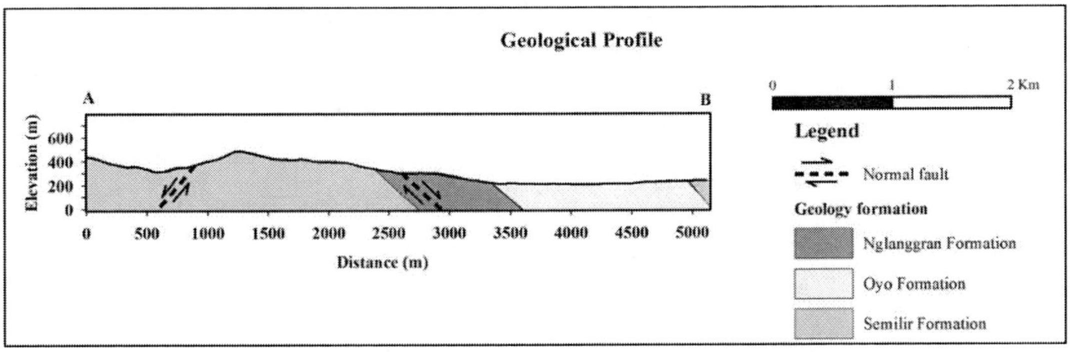

Figure 3. Geological Profile of the Study Area [7]

4. Method

Data sets are secondary data collected from different websites, as shown in Table 1. Data sets are applied to analyze thematic maps: slope angle map, lithological map, distance to faults map, and land-use map. These maps are applied to analyze a landslide susceptibility map.

Table 1. Data Sets for Landslide Susceptibility Mapping [7], [14], [15]

No.	Data	Sources
1	Geology (Shapefile)	Geological Research and Development Center, Geological Agency
2	Digital Elevation Model, DEM (12.5m)	Alos Palsar Rtc
3	Land use Land Cover (Shapefile)	ESRI

In this research, landslide susceptibility mapping is dependent on the determined parameters or factors: slope angle, lithology, distance to faults, and land use. Each factor is classified to obtain each class of the factors. In the next step, Analytic Hierarchy Process(AHP) was used to determine the important score of the factors by comparing each factor with others from 1 to 9 scores depending on the importance level, which is referred to in Table 2, calculating the weight percentage of the factors and then verifying the calculated CR value in Equation 1. If the CR value is less than 0.1, determining the importance scores is suitable. If the value of CR is 0.1 over, that indicates the determination of the importance scores is unsuitable needed to revise. In this step, the secondary data in Table 1 is used to analyze the thematic maps: slope angle map, lithological map, distance to faults map, and land use map. Finally, all of the thematic maps are overlayed to obtain the landslide susceptibility map by using ArcMap 10.8.

Table 2. Scale of relative importance [16]

Intensity of importance	Definition	Explanation
1	Equal importance	Two factors contribute equally to the objective.
3	Somewhat more importance	Experience and judgment slightly favor one ever the other.
5	Much more importance	Experience and judgment strongly favor one over the other.
7	Very much more importance	Experience and judgment strongly favor one over the other. Its importance is demonstrated in practice.
9	Absolutely more importance	The evidence favoring one over the other is of the highest possible validity.
2, 4, 6, 8	Intermediate values	When compromise is needed.

Saaty's eigenvector method is used to evaluate the consistency ratio. The following formula can solute the consistency ratio[10]:

$$CR=\frac{CI}{RI} \qquad (1)$$

The consistency index (CI) can be calculated using equation 2. The judgment is acceptable if the calculated CR is less than or equal to 0.1. In opposition, revise the judgment if the CR exceeds 0.1 [11].

$$CI = \frac{\lambda_{max} - n}{n - 1} \qquad (2)$$

Where: λ_{max} : Eigenvalue Maximum (Perron Root)

n : Column and Rows of Number

Random index (RI) is determined based on the number of parameters (n) according to Table 3.

Table 3. Determination of Random Index depending on n [16]

n	1	2	3	4	5	6	7	8	9	10
Random index	0	0	0.52	0.89	1.11	1.25	1.35	1.4	1.45	1.49

5. Results and Discussion

Various thematic data layers represent landslide conditioning factors: lithology, slope angle, distance to faults, and land use.

5.1 *Slope Angle*

The slope angle determines the sharpest downhill slope for a place on a surface. The slope angle is determined for each triangle in a TIN and each cell in a raster. This is the maximum rate of elevation change across each triangle for a TIN. It is the maximum rate of change in elevation for the raster over each cell. The slope angle command accepts an input surface raster and calculates the slope angle at each cell in an output raster. The flatter the terrain, the lower the slope angle value; the steeper the terrain, the greater the slope angle value. The output slope angle raster can be calculated as the percent slope angle or degree of slope angle, and Figure 4 shows the slope map (in degree) in the study area. The classes of slope represent the terrain morphology, namely: flat (<10°), gently slope (10°-20°), moderately steep (20°-40°), and steep (>40°). The slope gradient map is illustrated in Figure 4.

5.2 Lithology

The lithology map is prepared by comparing the secondary data. Different rock types behave differently concerning landslide occurrences because their mineral composition determines their strength. The stronger rocks give more resistance and are less prone to landslide occurrences. Based on the secondary data, the rock types found in the area are andesitic breccia, limestone, sandstone, and tuff. The most landslides susceptibility in the study area which found on the tuff and andesite breccia more than others. Because the engineering properties of this lithology are very poor, it also has the highest weathering or erosion in this lithology. The lithology map is generated from the lithology data that is obtained in the digital dataset. Then, the lithology map is converted from feature into raster grid format by using a spatial analysis tool. The cell size of this thematic layer is the same as one of slope (5mx5m resolution) and then setting the environmental processing extent of the area as shown in Figure 4.

5.3 Distance to faults

Distance from tectonic structural features as landslide susceptibility. Fault form a line or weak zone characterized by heavily fractured rocks. Selective erosion and movement of water along the fault plane promote such phenomena. The distance to fault map is created from a geological structure dataset and buffered surrounding the fault feature based on the classification of the geological structure [17]. The fault zone is buffered using a multiple ring buffer in ArcGIS and ranked from features as the probability of landslide occurrences is less than 500m, 500m-1000m, and larger than 1000m. Then, the distance to fault map also converts the feature line into raster grid format in the same cell size as lithology. The result of the distance to fault map is illustrated in Figure 4.

5.4 Land use

The stability of a slope is also influenced by land use. The area covered by forest maintains continuous water flow and ensures that water regularly infiltrates, whereas cultivated land impacts slope stability due to saturated soil. In this study, the Land-use map is derived from the land-use dataset. Land use distribution in the study area consists of barren land, forest, settlement, and rice field or farmland. Then, the land use map converts into gird format and sets the environmental processing extent as big as the cell size of lithology resolution. The resulting land-use map is given in Figure 4.

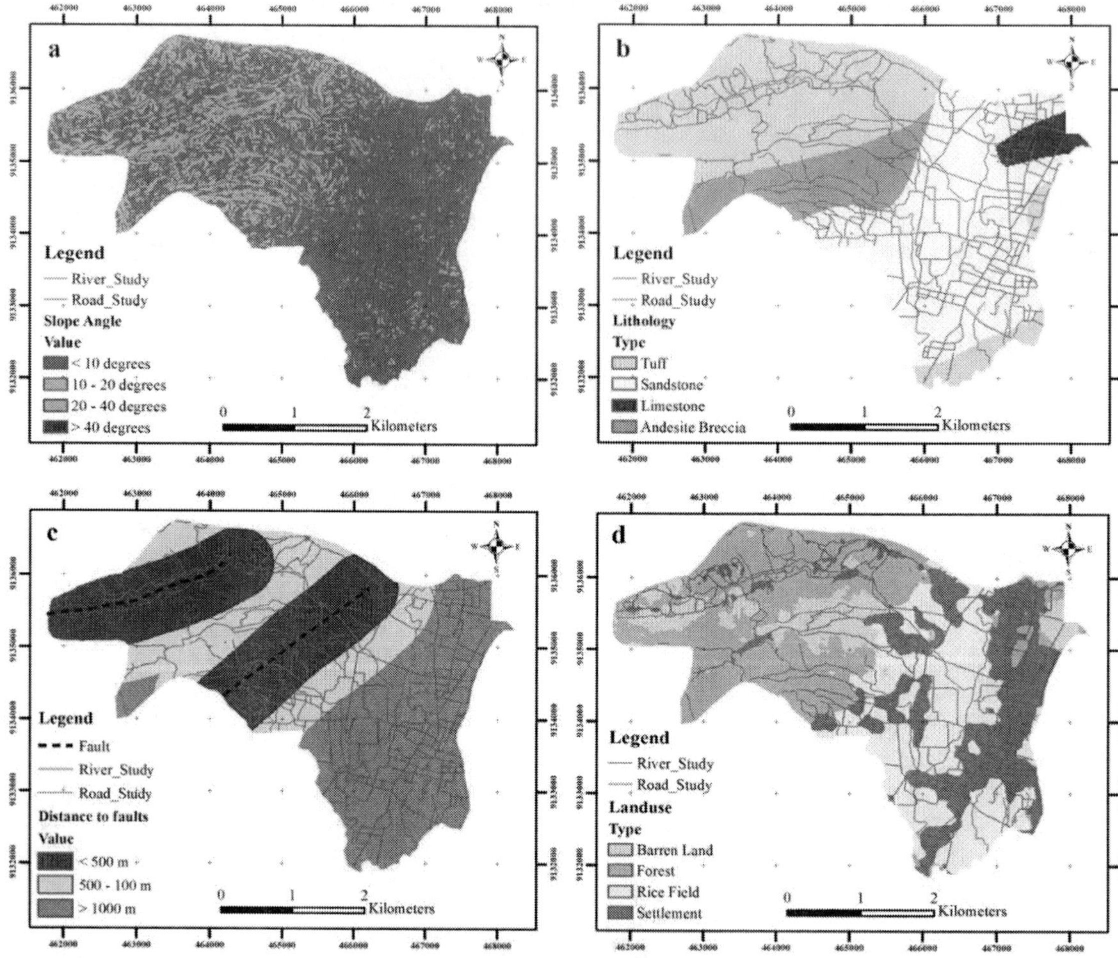

Figure 4. Landslide conditioning maps **(a)** Slope, **(b)** Lithology, **(c)** Distance to, and **(d)** Land use

Analytic Hierarchy Process (AHP) that is the approach used to provide priority scores (in Table 2) and find the weighting of the factors. The provision of the priority score is based on consideration of what factor influences landslide susceptibility in the study area. From the consideration, the slope angle factor mostly influences landslide susceptibility, hence, the slope angle factor is given the highest score more than other factors. Then the weight of the factors is calculated from Equation 1. CR value is used to certify the provision of the priority score and the weight of the factors, the CR value of 0.02 is acceptable. As the result, the weight of slope angle, lithology, distance to fault, and land use are 47%, 32%, 12%, and 9%, respectively. The details are shown in Table 4.

Table 4. Pairwise Comparison matrix and weight of each parameter by AHP

Factors	Slope Angle	Lithology	Faults to Distance	Land use	Weights	CR
Slope Angle	1	2	3	5	47%	0.02
Lithology	1/2	1	3	4	32%	
Distance to Fault	1/3	1/3	1	1	12%	
Land use	1/5	1/4	1	1	9%	

Each class of factors differently affects landslide susceptibility. Steep slope, weathered tuff is low strength, areas nearby fault zone, and barren land usually have landslide occurrences if likening other classes, that why they are provided the highest score seen in Table 5.

Table 5. Score for each class and the weight corresponding to the parameter

Data layer (Factor)	Class	Scoring	Weighting
Slope Angle	<10°	1	47%
	10° - 20°	2	
	20° – 40°	4	
	>40°	5	
Lithology	Limestone	1	32%
	Sandstone	2	
	Andesite Breccia	4	
	Tuff	5	
Distance to Fault	<500 m	4	12%
	500 – 1000 m	3	
	>1000 m	1	
Land use	Rice Field	1	9%
	Settlement	2	
	Forest	3	
	Barren Land	4	

The result analysis in ArcMap based on the classification of Natural Breaks (Junks) method found the landslide susceptibility index (LSI) had a minimum value of 1 and a maximum value of 5, with an average value of 2.17 and a standard deviation of 0.95. The LSI represents the relative susceptibility of a landslide occurrence. There is a higher sensitivity to landslides if the LSI value is high; a lower value implies a lesser susceptibility to landslides. As a result, the higher the rating, the more prone to landslides the area is. The zone of landslide susceptibility is based on the LSI value: very low the LSI value of 0 to 1, 1 to 2 low, 2 to 3 moderate, and 3 to 5 high, as given in Figure 5.

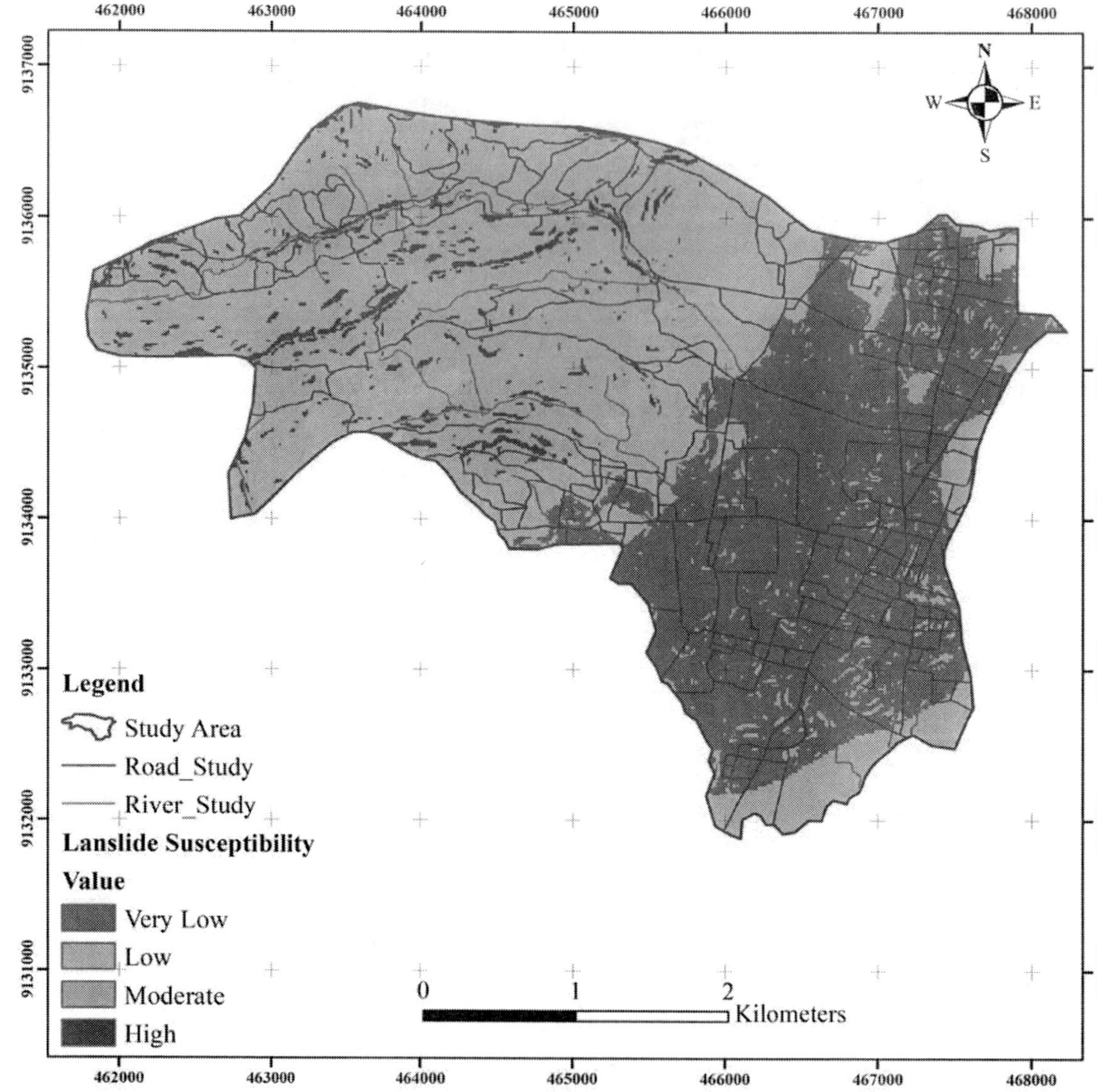

Figure 5. Landslide Susceptibility Map in Jurangjero and Kampung Villages, Ngawen Sub-District, Gunung Kidul Regency, Special Region of Yogyakarta, Indonesia

The preparation of landslide susceptibility is the great interest to planning agencies for preliminary hazard studies, especially when a regulatory planning policy is to be implemented as expressed in the form of visualization maps which is quite convenient for interpretation and explanation. These maps are easy to read as they show the legend that contains the information on the map. Moreover, a landslide susceptibility map can be used for landslide mitigation and land use development as it shows the level of susceptibility.

The landslide susceptibility map shows that the highest susceptible zones lie on this slope inclination range. This fact can be proved by comparing the landslide susceptibility map and slope inclination map by conducting a pixel-to-pixels comparison. Moreover, the closer to the lithology area are more susceptible to landslides it would be. The result of the landslide susceptibility map is very acceptable in this case as it can be found that most of the highest susceptibility zones are located near the lithology (tuff and andesite breccia). This study found that most landslide susceptibility is close to the faults (0 m

– 500 m), and land use plays the least important role as the landslide susceptibility can be higher on barren land areas. Therefore, the land cover has a minor influence on landslide susceptibility in this case. By overlaying the analysis result with the location of the landslide, the coincidence of the location of the landslide that can be calculated the landslide area per square kilometers and susceptibility classes show that the high susceptibility class 0.53 km^2, moderate susceptibility class 6.34 km^2, low susceptibility class 3.05 km^2, and very low susceptibility class 5.7 km^2 shown in Table 6.

Table 6. landslide susceptibility area by comparing each class

Landslide classes	Area (km^2)	% of Area
Very low susceptibility	5.70	36
Low susceptibility	3.05	20
Moderate susceptibility	6.34	41
High susceptibility	0.53	3
Total	15.61	100.00

6. Conclusion

There are several types of research on land susceptibility mapping by using Analytic Hierarchy Process (AHP) and Geographic Information System (GIS). In this research, AHP and GIS approaches were applied to identify and analyze landslide susceptible areas. Based on the result of the landslide susceptibility map. Most of the landslide susceptibility occurred on the slope inclination (steep-slope) and lithology class of tuff and andesitic breccias unit. It is found that most of the landslide susceptibility is close to the faults (0 m – 500 m next to faults), and land use is located in the barren land. The landslide susceptibility map development in Jurangjero and Kampung areas is based on computer software (ArcMap 10.8). The landslide susceptibility zonation map of the research area shows that four classes are very low, low, moderate, and high landslide susceptibility 36 %, 20 %, 41 %, and 3 %, respectively. The very low and low susceptibility is scattered in the east and central, and the moderate and high susceptibility is scattered in the west. This landslide susceptibility map is useful for planning settlements, mitigating, and avoiding landslide risks that will happen in the study area in the future.

References

[1] F. E. S. Silalahi, Pamela, Y. Arifianti, and F. Hidayat, "Landslide susceptibility assessment using frequency ratio model in Bogor, West Java, Indonesia," *Geosci. Lett.*, vol. 6, no. 1, 2019, doi: 10.1186/s40562-019-0140-4.

[2] T. R. Charlton, "Tertiary evolution of the Eastern Indonesia collision complex," *J. Asian Earth Sci.*, vol. 18, no. 5, pp. 603–631, 2000, doi: 10.1016/S1367-9120(99)00049-8.

[3] E. A. Silver, R. McCaffrey, and R. B. Smith, "Collision, rotation, and the initiation of subduction in the evolution of Sulawesi, Indonesia.," *J. Geophys. Res.*, vol. 88, no. B11, pp. 9407–9418, 1983, doi: 10.1029/JB088iB11p09407.

[4] M. Case, F. Ardiansyah, and E. Spector, "Climate Change in Indonesia," *Clim. Energy WWF-Indonesia*, pp. 1–13, 2007.

[5] O. Oktorie, "A Study of Landslide Areas Mitigation and Adaptation in Palupuah Subdistrict, Agam Regency, West Sumatra Province, Indonesia," *Sumatra J. Disaster, Geogr. Geogr. Educ.*, vol. 1, no. 1, p. 43, 2017, doi: 10.24036/sjdgge.v1i1.34.

[6] M. R. D. Mansouri, "Landslide susceptibility zonation using analytical hierarchy process and GIS for the Bojnurd region, northeast of Iran," *Landslides*, vol. 11, no. 6, pp. 1079–1091, 2014, doi: 10.1007/s10346-013-0458-5.

[7] Surono, B. Toha, and I. Sudarno, "Geological Map of The Surakarta - Giritontro Quadrangles, Jawa, Scale 1 : 100.000." Geological Research and Development Centre, Bandung, 1992.

[8] L. W. Santosa, "Identification of Land Degradation and Method of Solution in Zone of Baturagung Hill at Gunung Kidul Regency," *Forum Geogr.*, vol. 19, p. 30, 2017, doi: 10.23917/forgeo.v19i1.4571.

[9] B. Prastistho, A. Rodhi, C. Prasetyadi, P. Pratiknyo, M. R. Massora, and Y. K. M, "Potency of Groundwater Resources in Ngawen Region , The Non Groundwater Basin Area , North-Eastern Part of Gunungkidul," *3rd Int. Conf. Green Agro-Industry*, vol. 5, no. 1983, pp. 143–153, 2008.

[10] V. R. W. Bemmelen, "The Geology of Indonesia. General Geology of Indonesia and Adjacent Archipelagoes," *Government Printing Office, The Hague*. pp. 1–766, 1949.

[11] Surono, "Litostratigrafi Pegunungan Selatan Bagian Timur Daerah Istimewa Yogyakarta dan Jawa Tengah," *J. Geol. dan Sumberd. Miner.*, vol. 19, no. 3, pp. 209–221, 2009.

[12] B. Toha, R. D. Purtyasti, Srijono, Soetoto, and S. Rahardjo, W. Pramumijoyo, *"Geologi Daerah Pegunungan Selatan: Suatu Kontribusi", in Srijono, Hendrayana, H., Rahardjo, W. and Wijono, S., eds., Geologi dan Geotektonik P. Jawa, Sejak Akhir Mesozoik Hingga Kuarter*. Yogyakarta: Jurusan Teknik Geologi FT UGM, 1994.

[13] Sudarno, "Kendali Tektonik Terhadap Pembentukan Struktur Pada Batuan Paleogen Dan Neogen Di Pegunungan Selatan, Daerah Istimewa Yogyakarta Dan Sekitarnya," Institur Teknologi Bandung, 1997.

[14] A. S. F. A. PALSAR_Radiometric_Terrain_Corrected_low_res, "Includes Material© JAXA/METI 2007."

[15] Karra, Kontgis, and Et Al, "Global land use/land cover with Sentinel-2 and deep learning.," *IGARSS 2021-2021 IEEE International Geoscience and Remote Sensing Symposium. IEEE*, 2021.

[16] R. W. Saaty, "The analytic hierarchy process-what it is and how it is used," *Math. Model.*, vol. 9, no. 3–5, pp. 161–176, 1987, doi: 10.1016/0270-0255(87)90473-8.

[17] D. Varnes, *In International Association of Engineering Geology Commission on Landslides and Other Mass Movements on Slopes: Landslide hazard zonation: a review of principles and practice*. UNESCO, 1984.

IOP Conf. Series: Earth and Environmental Science 1233 (2023) 012025 doi:10.1088/1755-1315/1233/1/012025

Real Time Monitoring Using Robotic Total Station on the Outlet Slope of the Cisumdawu Tunnel, West Java

R B Prasetya[1,2] and H Hendarto[2]

[1]Department of Geological Engineering, Faculty of Engineering, Universitas Gadjah Mada, Yogyakarta, Indonesia
[2]The Ministry of Public Works and Housing, Directorate General of Highways, Jakarta, Indonesia

*email: ranggabayuprasetya@mail.ugm.ac.id, hendarto@pu.go.id

Abstract. Slope failure is still possible even though the slope has been designed and reinforcement has been carried out, as an example the slope of the Cisumdawu Tunnel Outlet which collapsed in October 2021. Therefore, monitoring technology is needed to anticipate slope collapse. In this study, the repaired slopes of the Cisumdawu Tunnel Outlet was monitored using a Robotic Total Station (RTS) to measure surface displacement in real time. Monitoring using RTS was carried out for approximately 40 days. The purpose of this study is to determine value and characteristic of slope displacement using an empirical approach from previous researchers. The displacement starts April 23rd at night until it reaches stability on April 24th 2022. The most significant displacement occurred in the prism target 08 which is located in the foot of slope and between the twin tunnels. It has 8 mm longitudinal displacement, the transverse displacement is 0.8 mm and the height has dropped by 13 mm. Almost the rest of the prism is in a stable state. From this study, the slope movement class is found to be regressive and categorized as velocity < 15 mm/day which is no failure expected within 48 hours. From this study, it was concluded that during the monitoring period using the RTS the outlet slope of the Cisumdawu tunnel was safe

1. Introduction

The Cisumdawu Tunnel is the first highway tunnel built in Indonesia which is part of the Cileunyi-Sumedang-Dawuan Highway (Cisumdawu) with a tunnel length of 472 m [1]. The Cisumdawu Tunnel is located at Sta 12+628 – Sta 13+100 in the Pamulihan area, Rancakalong District, Sumedang Regency, West Java Province with coordinates in UTM system are easting 812949.48, northing 9239664.95 for tunnel inlet and easting 813164.89, northing 9240082.72 for tunnel outlet (Figure I).

The slope of the Cisumdawu Tunnel has several times experienced failure, as happened on the outlet slopes of the tunnel in October 2021 (Figure II). It is estimated that one of the causes of the slide is due to the influence of rising and widespread groundwater levels. According to [2], the groundwater level at the location is above of the tunnels. Hence, if the rainfall occurs heavily and lasts a long time it will result in saturated soil that can trigger a landslide on the slopes. At the time of writing this paper, the slope has just been repaired

Even if the slope has been designed and reinforced, it is still possible for the slope to fail. However, there are typically signs of the slope before it entirely collapses [3]. Therefore, these signs should be considered carefully through monitoring activities so that appropriate treatment can be carried out before slope failure occurs. Based on [4] that slope behavior is not easy to predict, therefore slope monitoring programs are needed that can be useful in managing landslide hazards, and they provide useful information for the design of repair work. The benefits of monitoring activities according to [5] especially at the post-construction stage, monitoring slope can be a solution to ensure long-term slope safety.

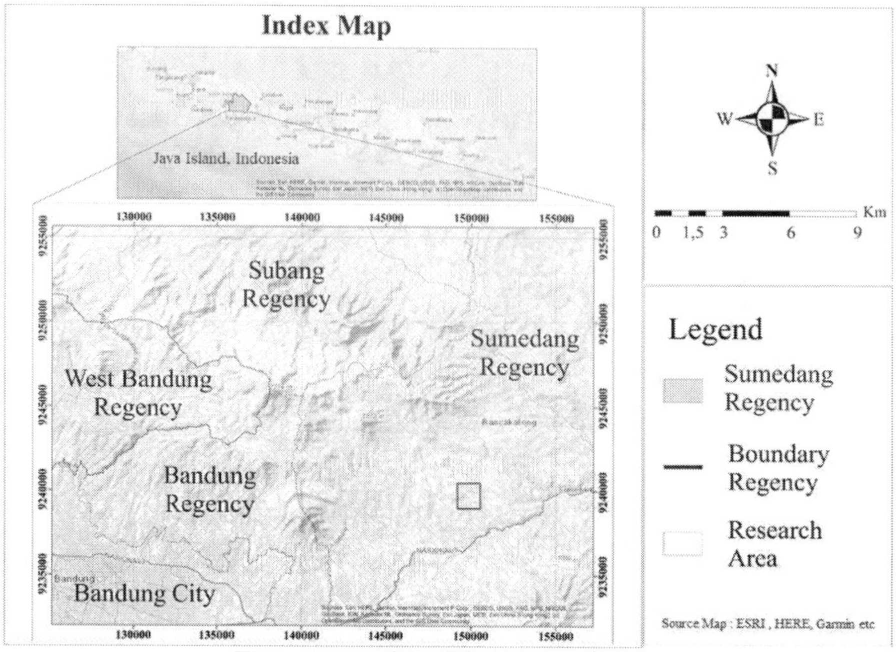

Figure I. Location of the study on the map

Figure II. Landslide on west side of Outlet Slope Cisumdawu Tunnel

According to [6] modern Robotic Total Stations (RTS) can automatically find and track prism targets. RTS can capture displacement data in 3 dimensions (X,Y, Z coordinates). RTS are frequently used for tunnel construction monitoring as well as assessing the stability of water dams, erosions, and rock faces. Conforming to [7] by using a case research of open pit mining, the use of survey instruments in the form of total stations to monitor ground surface movements was commonly used. Therefore, the Cisumdawu tunnel outlet slope case was monitored using RTS to determine characteristic and 3D value of slope displacement.

2. Research Method

2.1 Tools and Materials

The tools and materials used in this research can be seen in Table I. as follows :

Table I. Tools and materials

No.	Item	Qty (unit)	Description
1	RTS; Leica TS15, Horizontal Accuracy 2" (± 0.5 mm within 50 m)	1 unit	
2	Prism; GMP104	20 units	- 3 as Reference - 17 as Target
3	Mini PC; 8Gb RAM	1 unit	
4	Solar Panel, 120WP	2 units	
5	AccuBattery 12V, 65MaH	1 unit	
6	Mounting/ Bracket for RTS	1 unit	
7	Leica GeoMoS	1 Bundle	As management RTS data
8	Receiver GNSS; Leica GS18	1 Unit	To define control point coordinate
9	Modem 4G	1 Unit	

2.2 Research Stages

a. The acquisition of sufficiently accurate and consistent displacement values for the inspected item requires the stability of the RTS setup and the points constituting the reference system. The station setup is normally positioned on a massive concrete suitable pillar in the case of structures exposed to constant monitoring, ensuring stability. In this research, the RTS was installed on the bridge piers near tunnels using mounting brackets made of steel Figure III (a). Three prism as a reference mounted on the bridge abutment Figure III (b). The installation was carried out on March 28, 2022, followed by the installation of supports instruments such as MiniPC, Modem 4G, Solar panels and batteries in a box.

b. In accord with [8] that control points from the United States Geological Survey (USGS) are used as reference points for monitoring using geodetic sensors such as RTS and GNSS receivers. In Indonesia, control points such as USGS points are provided by InaCORS. InaCORS stands for Indonesia Continuously Operating Reference Station (CORS) which is managed by the Indonesia Geospatial Information Agency as a fixed/continuous geodetic observation station that is used as a reference in surveying, mapping, and organizing other Geospatial Information [9]. Data from InaCORS will be used as a reference for control points in the GNSS measurement on site (Figure III (c), then the site control points will be used to define 3 prism reference. InaCORS can be accessed via the link. http://inacors.big.go.id/.

c. The installation of target prisms was carried out on March 29, 2022. There were 17 target prisms installed spread across the outlet slopes of the Cisumdawu Tunnel. The prism distribution considers locations that have the potential for landslides based on models obtained from previous researchers [1] (Figure IV). The target prism is planted in the ground to a depth of 30-45 cm. (Figure V).

Figure III. (a) Mounting of RTS Installed on Bridge Pier; (b) Reference Prism Installed on Bridge Abutment, (c) GNSS measurement for defining site control points

d. Measurements started April 17, 2022 due to considering the readiness of software, electricity and soil stability on the prism due to the action of prism planting. Measurements were carried out for approximately 40 days until May 28, 2022. Displacement that obtained in this research are longitudinal displacement, transverse displacement and height displacement.

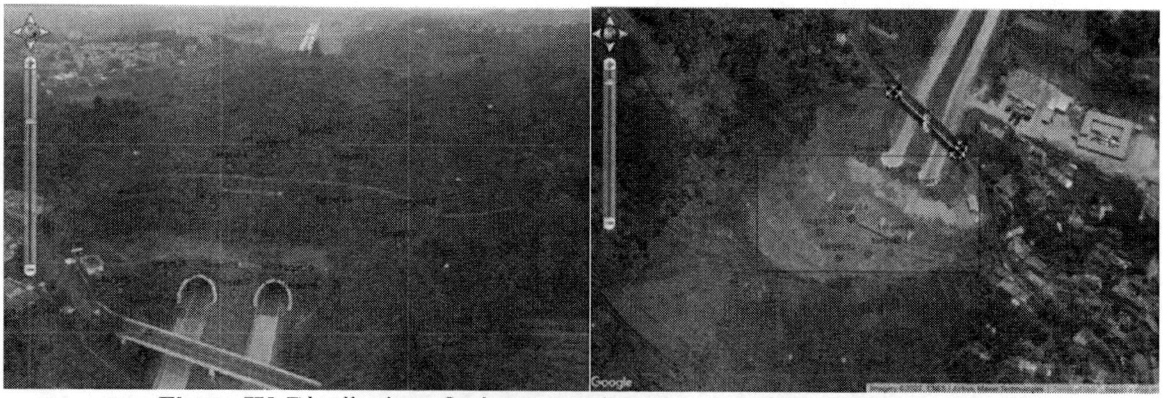

Figure IV. Distribution of prism target (Source : Google Earth and GeoMoS)

Figure V. Planting prism as target

e. Position measurement using Robotic Total Station is done automatically and continuously 24/7. After getting position data for a certain period of time, the displacement data is obtained. Data is stored in a miniPC and calculated by GeoMoS software. The basic method of determining position using Total Station [10] is explained using Figure VI and equation 2.1, 2.2, 2.3.

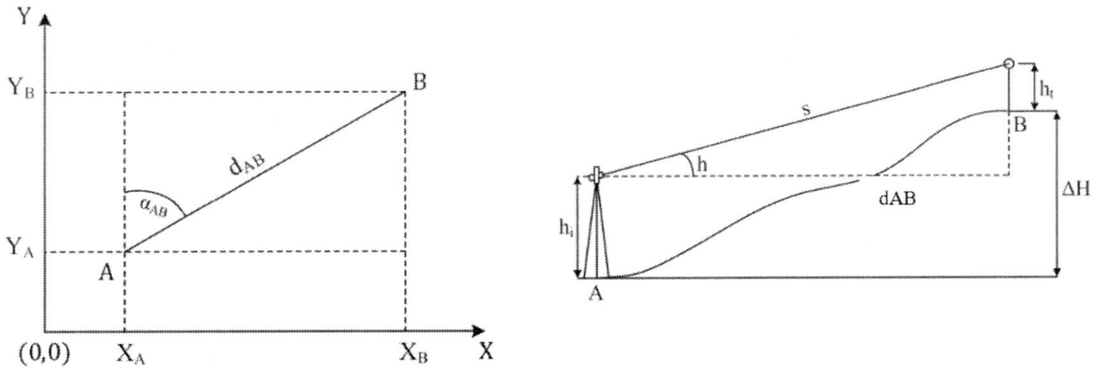

(a) (b)

Figure VI Position measurement method using Total Station [10] horizontal (a) vertical (b)

$$XB = XA + dAB\ sin\ sin\ \alpha\ AB \quad 2.1$$

$$YB = YA + dAB\ cos\ cos\ \alpha\ AB \quad 2.2$$

$$\Delta H = hi + dAB\ \tan(h) - ht \quad 2.3$$

Description :

dAB : Horizontal distance from instrument to object
αAB : Azimuth from instrument to object
A : Instrument coordinate
B : Object coordinate
hi : Height of instrument
ht : Height of target
h : Inclination angle
s : Slope distance between A-B
ΔH : Height different between A-B

2.3 Slope Movement Classification

Slopes rarely collapse without giving a sign. Slope failure is usually initiated by movement and other indications of instability, such as fracture formation. Empirical studies by [11] [12] show that slope movements can be divided into four types (Figure VII), such as :

a. Regressive
Characterized by the occurrence of slope movement that gets slower until it finally stops. This movement lasts for a short period of time.
b. Progressive
Characterized by a thorough and slowly accelerating collapse movement.
c. Transition
It starts with a regressive type and ends like a progressive type. This usually occurs as a result of changes in external conditions, such as pore pressure or continuous activity around the slope so that the rate of movement of the slope exceeds the normal limit.
d. Stickslip
This type, introduced by [12] is characterized by sudden movements followed by little or no movement and then there will be another sudden movement.

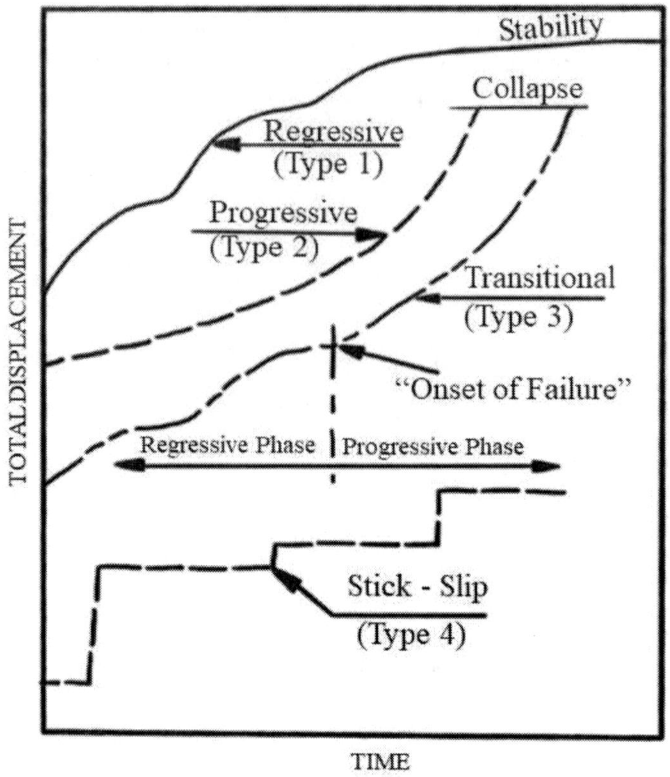

Figure VII. Slope movement Classification by comparing total displacement and time [11][12]

2.4 Displacement Criteria

Knowing the rate of slope displacement is one of the keys in predicting the behavior of the slope in the future. As stated in [13] made several generalizations from empirical data obtained from observations at several mines to help geotechnical experts in the field Table II.

Table II. Displacement Criteria [12]

Velocity Measurement (Displacement/ Time)	Description
0.1mm/day (0.004 mm/hr)	Initial Response
< 17mm/day (0.71mm/hr)	No failure expected within 24 hours
< 15 mm/day (0.61 mm/hr)	No failure expected within 48 hours
> 50 mm/day (2.1 mm/hr)	Indicates progressive failure (total collapse expected within 48 days)
> 100 mm/day (4.2 mm/hr)	Clear mining area (Progressive geometry and progressive velocity)
150 mm/day (5.25 mm/hr)	Clear mining area (regressive geometry)

2.5 RTS External Obstacles

In monitoring activities using Robotic Total Stations, there are external influences including atmospheric influences. According to [7], atmospheric influences that need to be considered in monitoring measurements are heavy rain and thick fog. In this research, atmospheric effects can be reduced by paying attention to the consistency of the data taken continuously 24/7 and smoothing feature of GeoMoS.

3. Result

Monitoring results recorded and managed by GeoMoS. GeoMos displays longitudinal displacement (Figure VIII), transverse displacement (Figure IX), height displacement (Figure X) vector of horizontal displacement (Figure XI) and vector of height displacement (Figure XII). In this research, the longitudinal direction is in same the direction of the highway or +26 degree from north pole (clockwise). The following are the results of monitoring for approximately 40 days on the outlet slope (Table III).

Table III. Monitoring surface displacement result

Target	Distance to Prism (m)	Longitudinal Displacement (m)	Transverse Displacement (m)	Height Displacement (m)
target01	127.763	0.0023	0.0005	0.0026
target02	135.457	0.0022	-0.0023	0.0038
target03	119.908	0.0024	0.0037	0.0028
target04	111.639	0.0045	-0.0002	0.0030
target05	98.765	0.0019	-0.0016	0.0019
target06	50.295	0.0028	-0.0007	-0.0047
target07	46.552	0.0054	0.0003	-0.0078
target08	50.103	0.0080	-0.0008	-0.0132
target09	52.939	0.0005	0.0008	-0.0010
target10	55.728	0.0001	-0.0002	-0.0003
target11	59.907	0.0003	0.0006	-0.0002
target12	128.229	0.0025	-0.0009	0.0030
target13	109.365	-0.0001	0.0010	0.0013
target14	100.678	0.0012	0.0013	0.0008
target15	56.631	0.0012	0.0007	-0.0002
target16	60.134	-0.0006	-0.0014	0.0007
target17	57.061	-0.0004	0.0005	0.0004

The biggest displacement happens on the prism of the target 08, which is located in the foot of slope and between the twin tunnels. The displacement values that occur are as follows, a longitudinal displacement is 0.0080 m, a transverse displacement is 0.0008 m, and a height displacement of 0.0132 m, as indicated in the Table III. It is estimated that the displacement was caused by heavy rain, loose and saturated embankment soil conditions at the foot of the slope. The smallest displacement is in target 13, which is a longitudinal displacement of -0.0001 m, a transverse displacement of 0.0010 m, and a height reduction of 0.0013 m. Target 13 is located in west side of outlet slope that facing east. The displacement starts from April 23rd 2022 at night until it reaches stability again on April 24th 2022.

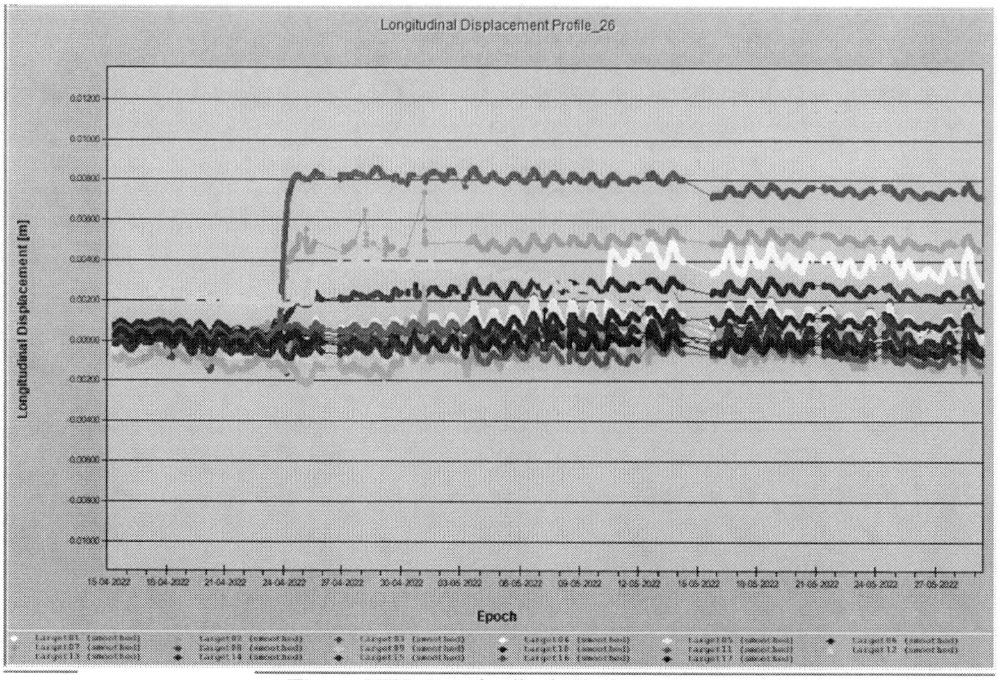

Figure VIII. Longitudinal Displacement

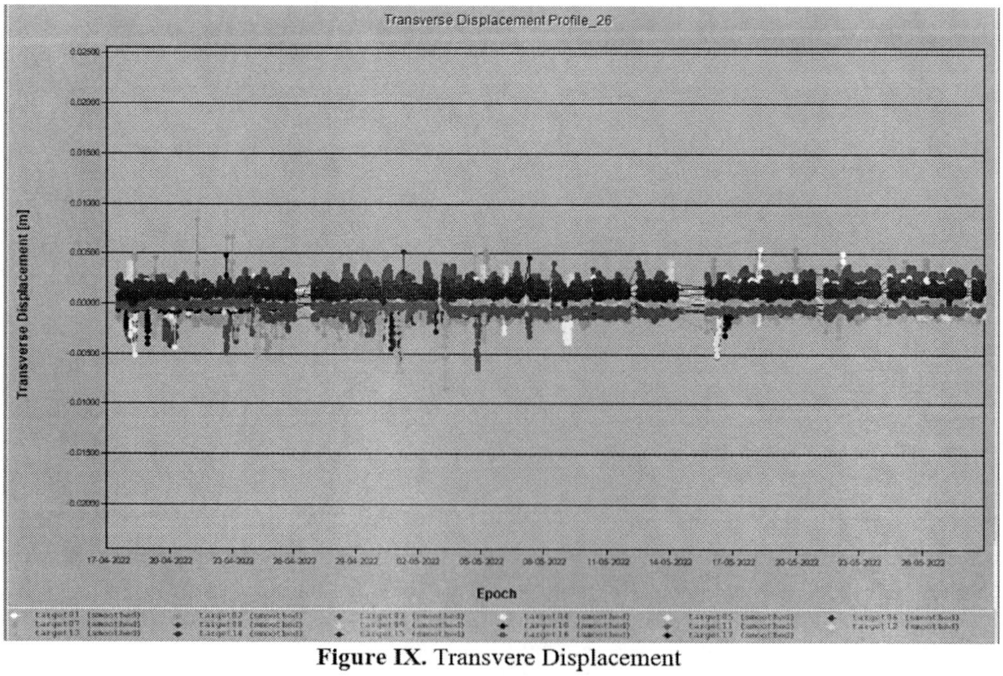

Figure IX. Transvere Displacement

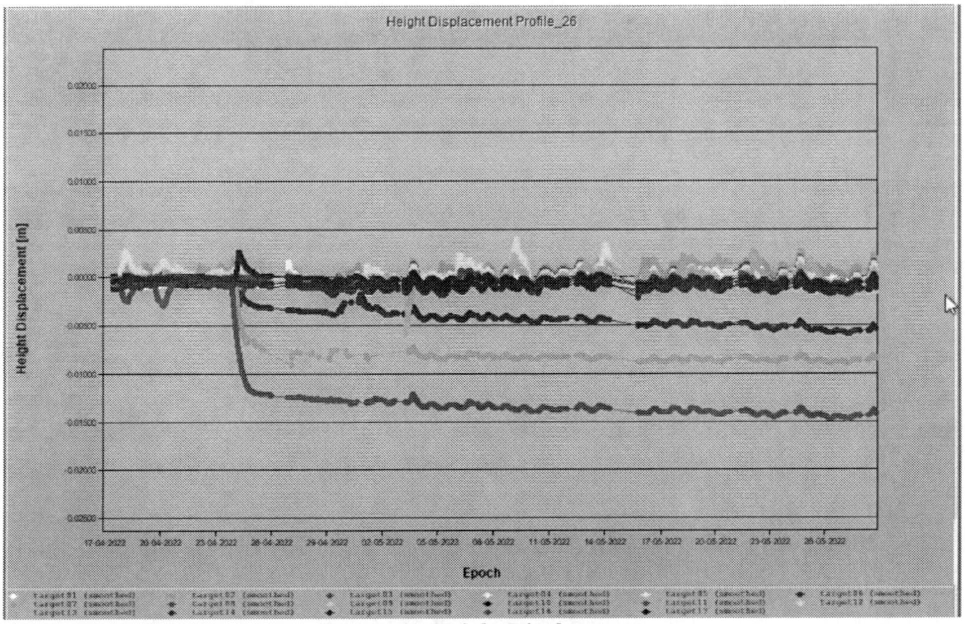

Figure X. Height Displacement

The availability of data based on the **Figure VIII, IX, X,** there is a data gap on 14 May 2022 around 06:00 WIB to 15 May 2022. However, the data is reconnected again. The gap happens because the internet data runs out. Actually, this activity does not require a lot of internet data. However, due to online remote activities, it seems quite large. So it is necessary to pay attention to the internet data package. In addition, there is a gap in target 07, presumably because the prism is covered by leaves or plant on the slopes.

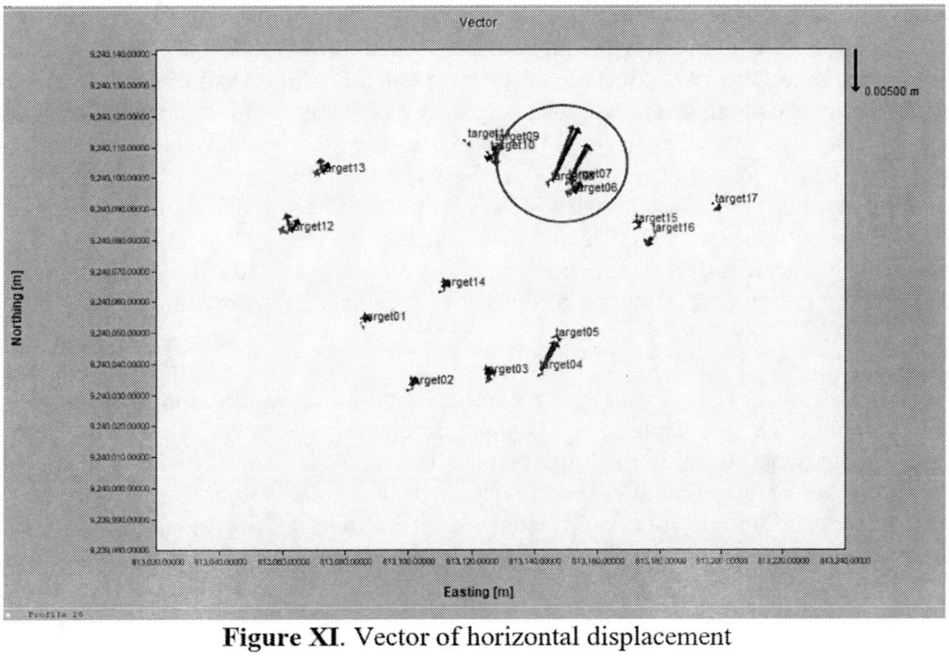

Figure XI. Vector of horizontal displacement

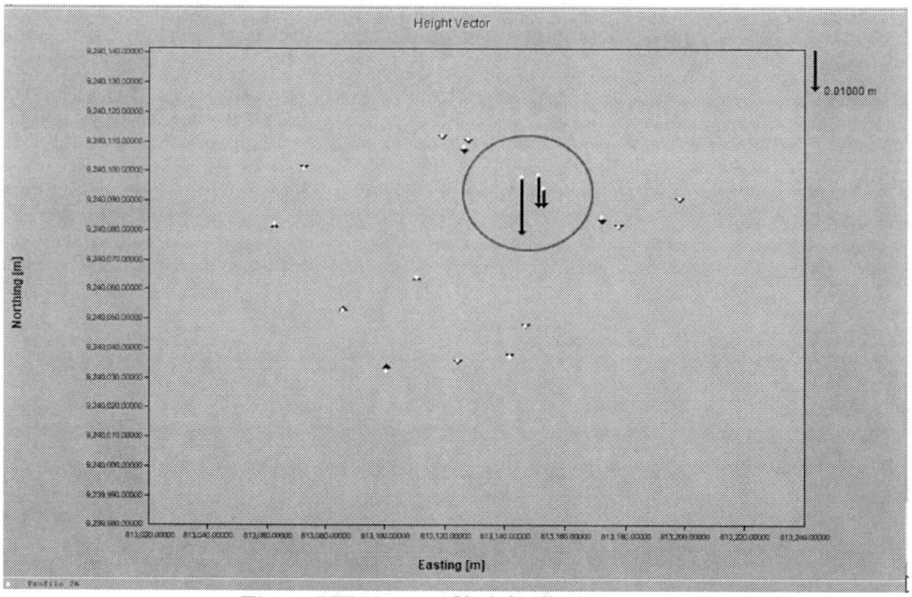

Figure XII Vector of height displacement

From the Figure XI and XII indicates that that the target group 06,07,08 experienced a horizontal displacement (longitudinal & transverse) as well as the largest decrease. The rest of the other prisms were in stable condition during 40 days of monitoring.

Based on slope movement classification by [11][12], this slope movement categorized as regresive, because the slope movement has short period and gets slower until it finally stops. Based on displacement criteria by [13] this slope movement categorized as velocity < 15 mm/day which is no failure expected within 48 hours, and it was proven that the next day after the displacement occurred there was no collapse.

4. Conclusion and Recommendation

This paper describes how to monitor the movement of the slope surface using Robotic Total Station in the Cisumdawu Tunnel, West Java. From monitoring using RTS, the coordinates of X, Y, Z are obtained which show the longitudinal, transverse and height displacements. Throughout the monitoring for 40 days, the outlet slope of Cisumdawu Tunnel have experienced of regressive slope movement. Group of target 06, 07, 08 which is located in foot of slope between twin tunnels have deformation < 15 mm and reaches stability within 1 day. Based on that information, it can be said that during the monitoring period using the RTS the outlet slope of the Cisumdawu tunnel was safe.

The obstacle experienced in the implementation of this monitoring is that the data is cut off for about 1 day due to running out of internet data. Therefore it is necessary to pay attention to the internet data.

5. Acknowledgement

This research is funded and supported by Ministry of Public Works and Housing of Indonesia. The author would like to thank Mr. Hendy Setiawan as supervisor from Department of Geological Engineering, Faculty of Engineering, Universitas Gadjah Mada for his guidance in internship and paper writing activities. And thank to Mr. Fahmi Aldiamar from Directorate General of Highway, Ministry of Public Works and Housing of Indonesia, for his guidance as mentor at the research location.

References

[1] Halomoan RP 2018 Analysis of Excavation Methods and Stability of the Cisumdawu Toll Road Tunnel (Cileunyi – Sumedang – Dawuan) West Java Province Faculty of Engineering Department of Geological Engineering Universitas Gadjah Mada Yogjakarta Indonesia.
[2] Febriyanto WD 2018 Characteristics of Geological Engineering at Cisumdawu (Cileunyi-Sumedang-Dawuan) Toll Road Tunnel Construction Site, Faculty of Engineering Department of Geological Engineering Universitas Gadjah Mada Yogjakarta Indonesia.
[3] Arif I 2016 Mining Geotechnical: Realizing Sustainable Mining Production by Maintaining Slope Stability PT Gramedia Pustaka Utama Jakarta Indonesia.
[4] Wyllie DC and Mah CW 2004 Rock slope engineering: Civil and mining 4th edition Rock Slope Engineering.
[5] Dunnicliff J 1993 Geotechnical Instrumentation for Monitoring Field Performance J Wiley and Sons Inc New York.
[6] Lienhart W 2017 Geotechnical monitoring using total stations and laser scanners: critical aspects and solutions. Journal of Civil Structural Health Monitoring *7*(3) 315–324. https://doi.org/10.1007/s13349-017-0228-5.
[7] Afeni TB and Cawood FT 2013 Slope monitoring using Total Station: What are the challenges and how should these be mitigated? *South African Journal of Geomatics 2*(1) 41–53.
[8] FHWA (Federal Highway Administration) 2009 Technical Manual for Design and Construction of Road Tunnels-Civil Elements FHWA-NHI-10-034 Washington DC.
[9] Bidang Geodinamika 2018 InaCORS BIG One Indonesia Mapping Reference. Cibinong: Department of Geodesy and Geodynamic Reference Network of Indonesia Geospatial Agency. (Cibinong, West Java: Indonesia Geospatial Information Agency)
[10] Basuki S 2011 Ilmu Ukur Tanah (Edisi Revisi) Universitas Gadjah Mada Press Yogyakarta.
[11] Broadbent CD. and Zadvoni ZM 1982 Influence of Rock Structure on Stability Stability in Surface Mining Vol 3 Ch 2 Society of Mining Engineers Denver CO.
[12] Sullivan TD 1986 Mine and Monitor an Approach to Coping with Unstable Pit Walls 13th CMMI Conf Singapore.
[13] Zavodni ZM 2001Time dependant Movements of Open Pit Slopes' Proceedings Slope Stability in Surface Mining (Hustrelid, McCarthur and vanZyl eds) SME. pp 81 – 87.

Estimation of Rock Mechanical Parameters Using Well Log Data in the Well Poseidon 1, Lower Cretaceous, Browse Basin, Northwest Shelf, Australia

K L Peña Cerón[1], S S Surjono[1] and I G B Indrawan[1]

[1]Geological Engineering Department, Faculty of Engineering, Universitas Gadjah Mada. Jl. Grafika No.2., UGM Campus, Yogyakarta, Indonesia. 55283

Corresponding author: sugengssurjono@ugm.ac.id

Abstract. This work implements a well-logging approach to evaluate the rock Elastic Parameters, basic Rock Mechanics, and in-situ Stresses in an interval of the Well Poseidon 1 located in the offshore Browse Basin in the Timur Sea of Australia. For this aim, there were used several correlations for the Young's Modulus, Uniaxial Compressive Strength; one formula for Poisson's Ratio and one for Angle of Internal Friction, and the Poroelastic Model with tectonic strains for the horizontal stresses with calibrations from logs. All these estimations were compared to a homologous section with triaxial test laboratory data and with previous studies. The results reached indicates that the chosen correlations can replicate under the adequate considerations the values of the rock mechanics and they give also good estimations for in-situ stresses in the interval studied. Moreover, these can be used to model the Young's Modulus, Uniaxial Compressive Strength, Internal Friction Angle with fair precision. This methodology can be used in order to create geomechanical models by estimating the basic parameters needed, these models can be applied to Improved Oil Recovery, Enhanced Oil Recovery, Carbon Capture and Storage, and Carbon Capture Utilization and Storage projects.

1. Introduction

The Reservoir Geomechanics is an interdisciplinary study of the rock mechanics, geology, and petroleum engineering areas in order to dealt with several issues in the whole chain of the hydrocarbon industry, from the exploration phase as far as the decommission and reconversion; it is also applicable in other industries, for example geothermal [1], CCS (Carbon Capture and Storage), and CCUS (Carbon Capture, Use, and Storage) [2]. The application of the Reservoir Geomechanics principles may support the estimation of parameters which relies on surveys and laboratory tests not available afoot during particular work or study; amongst them: the Elastic Rock Modules, in situ stresses, Rock Mechanics, e.g., UCS (Uniaxial/Unconfined Compressive Strength) [1].

The well Poseidon 1 was perforated in the Caswell Sub-basin of the Browse Basin, currently plugged and abandoned, with 5112.1 m MDRT in depth, 152 m of net conventional pay zones identified, and with no coring on the strata [3]. The Browse Basin NW (North West) Shelf Australia have been on focus and interest for hydrocarbon exploitation and geological applications per decades, but the only projects carried out and currently producing oil and gas are the Ichthys and Prelude gas accumulations [4], [5]. The future projects over the Browse basin and on the wells perforated within may require information related to the reservoir geomechanics [1], [2]: in the case of exploration (drilling), wellbore stability models; hydrocarbon production, sand production and compaction related geomechanics; Improved

IOP Conf. Series: Earth and Environmental Science 1233 (2023) 012026 doi:10.1088/1755-1315/1233/1/012026

Oil/gas Recovery (fluids injection) also CCS and CCUS projects, geomechanical, e.g., fault stability, seals and rock integrity. Particular interest in the well Poseidon 1 and other wells in the Basin, it is the Lower Cretaceous age stratum as hydrocarbon and storage potential [3 – 6]. The reservoir geomechanics studies mentioned beforehand use as inputs rock mechanic properties that in the case of the Browse Basin has not been publicly disclosed and in the Poseidon 1 well has not been fully carried out [3], [6]. At basin scale the studies done in the area uses valid, but apparently are limited approaches to determine some of the parameters and studies at well context only have been focused on the Prelude 1 and 1A wells [7].

To overcome this condition in the Poseidon 1 well and may be the Basin; the use of well logs come advantageous, for example, implementing formulations that relate directly surveys as Sonic Compressional (DTC), Sonic Slowness (DTS), and Bulk Density (BD) to the Young's Modulus and Poisson's Ratio [8]. Similarly, to predict basic rock mechanical properties, well log approaches are convenient, at discretion, where the previously estimated parameters work as inputs into the correlations for UCS and Angle of Internal Friction (FANG) [1], [9]. Finally, all these, along with pore pressure, are used to determine In Situ Stresses estimated with a Poroelastic model [10].

2. Geological Setting

The Browse Basin is located in the offshore Timor Sea of Australia, figure 1 (a), with an approximated area of 140 000 km^2, sediments from Paleozoic to Cenozoic with an excess of 15 km in its depocenter [11], and with the water depth averaging 200 m as minimal in the whole Basin [4]. The Basin sits enclosed by: to the South, the Leveque Shelf; to the East, the Kimberly Block; to the North, the Ashmore Platform and Scott Plateau; and to the southwest, grading to Offshore Canning Basin [4].

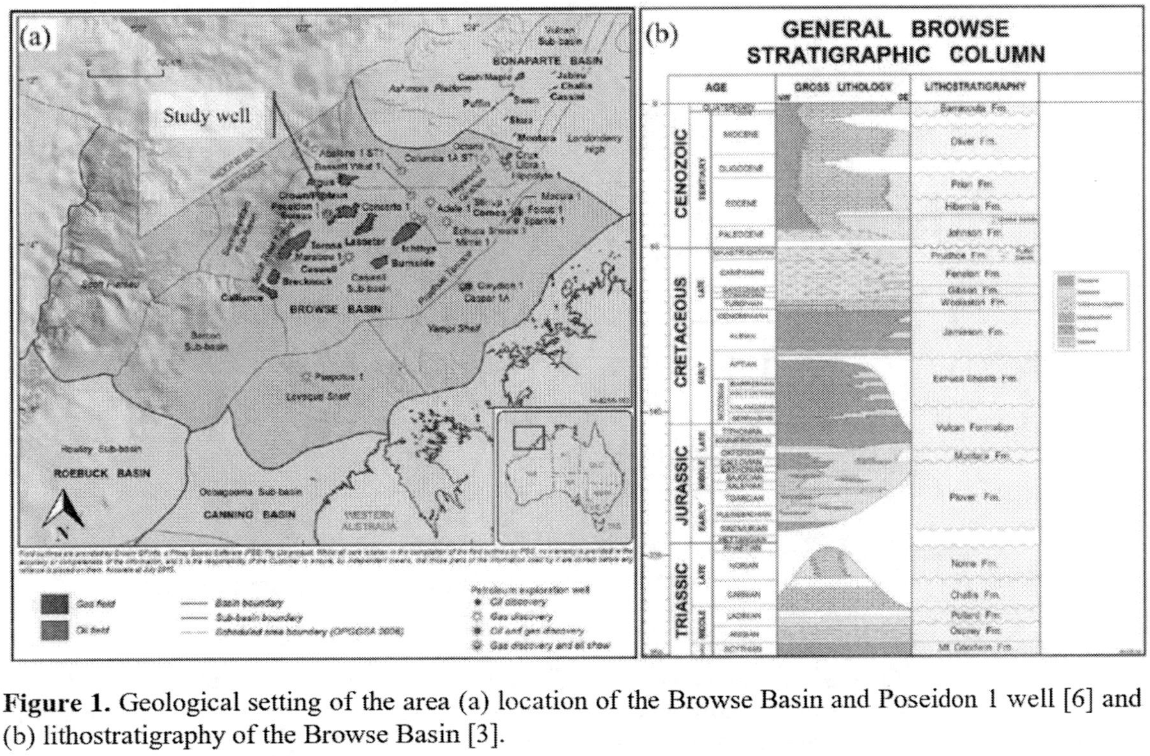

Figure 1. Geological setting of the area (a) location of the Browse Basin and Poseidon 1 well [6] and (b) lithostratigraphy of the Browse Basin [3].

The tectonic evolution of the basins is divided in six major events: extension (Late Carboniferous to Early Permian), thermal subsidence (Late Permian to Triassic), inversion (Late Triassic to Early Jurassic), extension (Early to Middle Jurassic), thermal subsidence (Late Jurassic to Cenozoic), and inversion (Middle to late Miocene) [3]. Sediment deposition contains, figure 1 (b), fluviodeltaics, silts, shales, marine shales, sands and carbonates; with small amounts of volcanics [3]. A regional unconformity is present at the Base of the Plover Formation [11], figure 1 (b).

The Poseidon 1 well is located in the NW Shelf of Western Australia to 480 km in distance to Broome, Australia, with a total depth of 5 112.1 m Measure Depth from Rotary Table (MDRT) and a sea water depth of 494 m to Datum Mean Sea Level (MSL) [3]. The well was perforated through thirteen formations, starting with the Barracouta Formation at the seafloor and ending at the Nome Formation. The primary objective was the Plover Formation sandstones with top encountered at 4 798.4 m MDRT [3]. The section of the early Cretaceous in this study comprehends the intervals with top in 4 592 m MDRT – 4620 m MDRT, 28 m in thickness, in the base of the Jamieson Formation [3].

3. Data and Methods

All the data utilized in this work (from raw datasets to images), are available in the National Offshore Petroleum Information Management System (NOPIMS) from Geoscience Australia which is © Commonwealth of Australia and is provided under a Creative Commons Attribution 4.0 International License [12]. The figure 2, shows the detailed workflow in this study, it is divided in three main groups: 1. Elastic Parameters, 2. UCS and FANG, and 3. Stresses; moreover, it includes a preparation stage, preliminary, and validation parts.

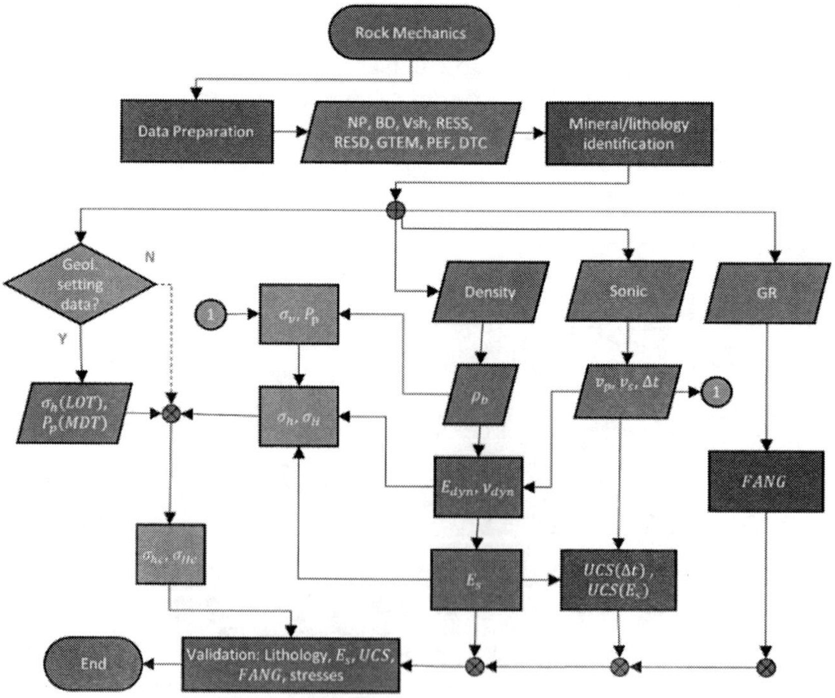

Figure 2. Flowchart of the Methodology indicating processes and inputs. (Green) Elastic Parameters. (Purple) UCS and FANG. (Orange) Stresses.

3.1. *Data Preparation and Preliminary Stage*

A quality control of the logs used was done as suggested by [13] for checking the reliability of the data, also in this stage the necessary minimal interventions were implemented, i.e., splicing, depth shifting; these were done so as to create variable datasets ready to use.

The mineral/lithological identification was intended to stablish a baseline to contrast the different criteria on the correlations used as this parameter can be linked to a particular lithology [8,9]. The Neutron Porosity, Bulk Density, Shallow Resistivity (RESS), Deep Resistivity (RESD), Generalized Borehole Temperature (GTEM), Photoelectric Factor (PEF), Compressional Slowness, along with estimated Shale Volume (Vsh) from Neutron-Density calculation, were use in a model of 3 lithologies/minerals volumes [13]: Volume of Shale, Quartz, and Limestone.

3.1.1. *Shale Volume.* The volume of shale was estimated using the parameters and formula (Neutron-Density) defined in [3]. The Parameters are summarized in the table 1.

Table 1. Parameters for the neutron-density shale volume calculation [3].

Log type	Matrix	Fluid	Shale
BD (g/cm^3)	2.65	0.7	2.71
NP (v/v)	0.34	1	-0.1

3.1.2. *Mineralogy/Lithology.* The mineralogy/Lithology identification was done calculating a volume model: Shale Volume, Quartz, and Limestone implementing Effective Porosity (corrected by shale and hydrocarbons) [3]. The criteria use to estimate mineralogy/lithology based upon well logs are listed in the table 2, considering ranges and formulae listed in [3] and the others defined in [13].

Table 2. Mineralogy and fluids cut-off for mineralogy/lithology identification.

		Invaded Zone	**Saturation**	
	HCs density (g/cm^3)	**0.23**	**n**	**2**
Fluids[a]		**Virgin Zone**	**m**	**2**
	Sxo	**Exponent factor of Sw**	**a**	**1**
	Exp Factor	**0.2**	**Rw (ohm m)**	**0.09**
	Parameter	**Shale**	**Quartz**	**Limestone**
	Res (ohm.m)	**0.7[c]**	**-**	**-**
Minerals[b]	**Den (g/cm^3)**	**2.5 (wet shale)[c], 2.71 (dry shale)[a]**	**2.65**	**2.71**
	DTC (us/m)	**328[c]**	**182**	**150.6**
	Pef (b/e)	**4[c]**	**1.8**	**5.1**

[a] From the petrophysical report in [3].

[b] From common ranges stablished in [13].

[c] From the well strata studied in 100% shale.

3.2. Elastic Parameters

The Elastic parameters estimated were the dynamic Young's Modulus and Poisson's Ratio [8], and several models of dynamic-static Young's Moduli were calculated defined by [14], [15], and [16].

3.2.1. *Dynamic Estimations.* For the dynamic parameters, seen in equation (1) and equation (2) [8], the units consistently coincide with the International System (SI) to produce Pascals in the Young's Modulus and nondimensional in the Poisson Ratio; however, it is convenient to have the Young's Modulus in Giga Pascals, which need density in grams per cubic centimeter, primary and shear velocities in Kilometers per second.

$$E_{dyn} = \frac{\rho v_s^2 (3v_p^2 - 4v_s^2)}{(v_p^2 - v_s^2)} \tag{1}$$

$$\nu_{dyn} = \frac{v_p^2 - 2v_s^2}{2(v_p^2 - v_s^2)} \tag{2}$$

3.2.2. *Static Correlations of Youngs Modulus.* In order to estimate the static Young's Modulus in Giga Pascals, the correlations in equation (3) [14], equation (4) [15], and equation (5) [16] have as inputs the dynamic Young's Modulus (Edyn) in Giga Pascals and the density in grams/cubic centimeter.

$$E_s = 0.018 E_{dyn}^{\ 2} + 0.422 E_{dyn} \tag{3}$$

$$E_s = \left(\frac{E_{dyn}}{3.3674}\right)^{2.042} \tag{4}$$

$$E_s = 0.3361 E_{dyn} (\rho_b^{0.8}) - 2.4603 \tag{5}$$

3.3. *Unconfined Compressive Strength and Internal Friction Angle*

In order to link the correlation that best reflects the UCS and FANG on the section evaluated, there were calculated five UCS based upon the equation given in [9], [17], [18], [10], and equation (10) correlated based on data from triaxial rock test [19], as well as one FANG [1] formula.

3.3.1. *UCS.* The units of equation (6) – (10) for the UCS are in Mega Pascals and their inputs: for the Young's Modulus in Giga Pascals [9],[10], [17], and for Compressional Slowness in micro seconds/foot in equation (8) [18]. For this work, the Young's Modulus inputs were the static one estimated from equation (5). The values of UCS of reference from the well Prelude 1A were estimated using the linear Mohr - Coulomb criteria formula [1] with data from the reports in [19], resulting in equation (10).

$$UCS = 7.22 E_s^{\ 0.712} \tag{6}$$

$$UCS = 2.28 + 4.1089 E_s \tag{7}$$

$$UCS = 1200 exp(-0.036 \Delta t) \tag{8}$$

$$UCS = 14.86 E_s^{\ 0.464} \tag{9}$$

$$UCS = -125.97 + 9.2769 E_s \tag{10}$$

3.3.2. *FANG.* The Angle of internal friction in equation (11) has for unit degree and the Gamma Ray in gamma ray units of the American Petroleum Institute (gAPI) [1]. The Angle of Internal Friction of reference from the well Prelude 1A is set as in the original report in [19].

$$FANG = 70 - 0.417 GR \tag{11}$$

3.4. *In-situ stresses*

In order to estimate the in-situ stresses, several calculations were carried out with two types of calibrations on them. For the overburden, the density all over the overlying strata (shallow and not available in the logs) was extrapolated using a tie in the mudline [20] and the vertical stress integrates the effects of the water column with a fixed water density [1]. The pore pressure used was calculated from the Sonic Eaton model [21] and the normal pore pressure method [1], these calibrated with Modular Formation Dynamic Tester (MDT) data [22]. The horizontal stresses, maximum and minimum, were estimated with the Poroelastic theory with tectonic strains [10] and calibrated with minimal horizontal stress data from Leak-off Test (LOT) from other wells in the Browse Basin [6].

3.4.1. *Overburden.* The vertical stress in equation (12) was estimated using a seawater constant density of 1.03 (g/cm³) [1].

$$\sigma_v = \rho_w g z_w + \int_0^z \rho(z) g dz \tag{12}$$

3.4.2. *Pore Pressure.* The normal pore pressure gradient in (13) is referred as the hydrostatic pore pressure, with a gradient of 10.18 (kPa · m) [1], and in equation (14) the Eaton Pore Pressure with its fitting parameters a = 1 and n = 2 [21]. The delta t norm and delta t in equation (14), makes reference to the compressional slowness value in the normal pressure conditions and the compressional slowness readings correspondingly [21]. Calibration of pore pressure was done using MDT data from the Poseidon 1 well available in [22].

$$P_{pnorm} = \rho_w g z_w \tag{13}$$

$$P_p = \sigma_v - a(\sigma_v - P_{pnorm}) \left(\frac{\Delta t_{norm}}{\Delta t}\right)^n \tag{14}$$

3.4.3. *Minimum and Maximum Horizontal Stresses.* The minimum and maximum horizontal stresses, equation (15) and equation (16), in an isotropic Poroelastic model with horizontal tectonic strains [10], was estimated iteratively with the inclusion of the tectonic horizontal strains ε_h and ε_H , these were being calibrated to the Minimum Horizontal Stresses available of the Browse Basin from LOT [6]. The Biot's Coefficient (alpha) was assumed to be 1.

$$\sigma_h = \left(\frac{v}{1-v}\right)(\sigma_v - \alpha p_p) + \alpha p_p + \left(\frac{E}{1-v^2}\right)(\varepsilon_h + v\varepsilon_H) \tag{15}$$

$$\sigma_H = \left(\frac{v}{1-v}\right)(\sigma_v - \alpha p_p) + \alpha p_p + \left(\frac{E}{1-v^2}\right)(v\varepsilon_h + \varepsilon_H) \tag{16}$$

4. Results and Discussions

In the table 3 are summarized the results of the calculations done in the interval 4 592 m MDRT – 4 620 m MDRT of the Poseidon 1 well (P1 Zone) and the ones derived from the triaxial test from the well Prelude 1A (P1A). The parameters shown for the Poseidon 1 well with their formulation are as listed: Static Young's Modulus (Es) from equation (5), Poisson's Ratio (v) from equation (2), UCS from equation (8), and FANG from equation (11). The mean, median, and standard deviation were estimated [23] in order to analyze the differences amongst the estimated data from the study well P1 to those calculated from the reference well P1A.

Table 3. Results for the rock mechanics estimations.

Property	Well	Min	Max	Mean	SD	Median
Es (GPa)	P1A	11.24	50.60	24.40	6.68	22.87
	P1_Zone	18.40	34.18	24.97	3.18	24.83
v (dimensionless)	P1A	0.26	0.67	0.37	0.09	0.34
	P1_Zone	0.13	0.25	0.19	0.02	0.19
UCS (MPa)	P1A	21.75	210.86	98.60	54.64	91.24
	P1_Zone	60.46	114.82	83.14	12.04	82.40
FANG (°)	P1A	46.60	57.20	50.93	3.31	50.90
	P1_Zone	33.87	59.84	49.99	5.03	49.98

IOP Conf. Series: Earth and Environmental Science 1233 (2023) 012026 doi:10.1088/1755-1315/1233/1/012026

The figure 3, shows the composite well log of the section evaluated in the Poseidon 1 well. The log tracks are defined clearly by variable, in the case of the track 3 (left-right) TVDSS makes reference to True Vertical Depth Sub-Sea, distance measure from the MSL; in the track 7, the 3 minerals/lithology are grouped along with the PHIE (effective porosity, red color) calculated from the Neutron-Density approach.

4.1. Elastic Parameters

4.1.1. Static Young Modulus. The three equations tested are represented in box-plots in figure 4. Equation (4) in figure 4, gray color, named Es_Barree with a depicted range of 80 – 280 GPa, has the biggest dispersion and with an arithmetic mean of 149.96 (GPa). Equation (3), light blue color, named Es_Lacy with range depicted 0 – 100 (GPa), has an arithmetic mean of 44.01 (GPa). Finally, in equation (5), green color, named Es_Onalo with a range shown from 0 – 50 (GPa), has a mean of 24.97 (GPa).

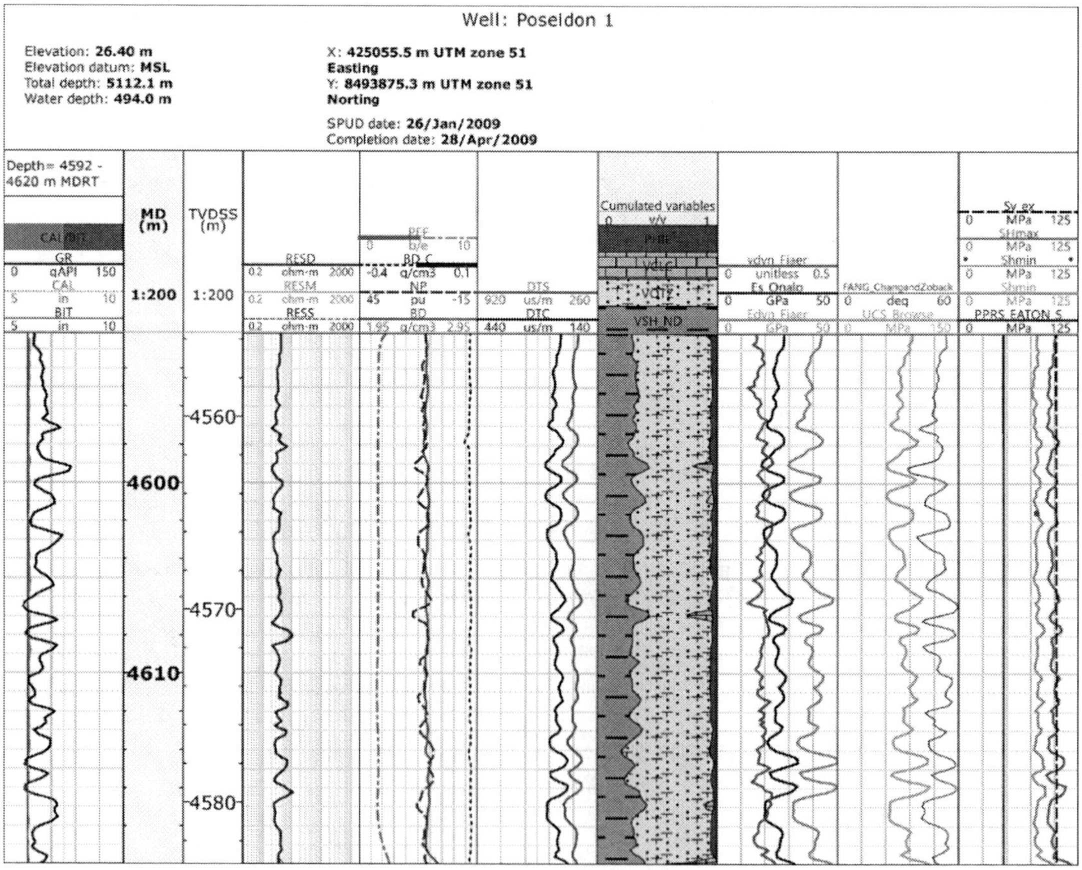

Figure 3. Composite well log of the section 4 592 m MDRT – 4 620 m MDRT, Poseidon 1 well.

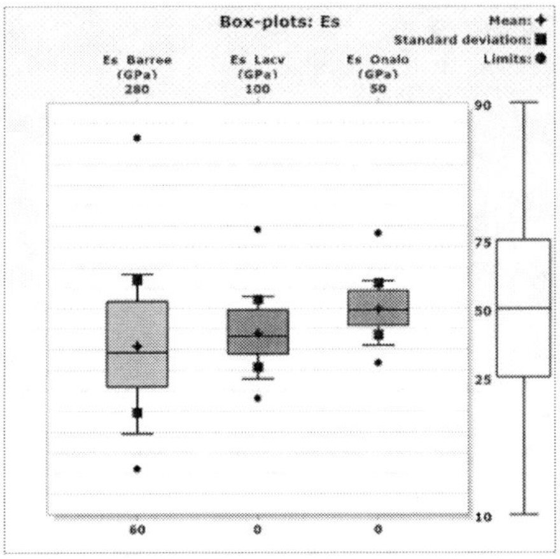

Figure 4 Box-Plots of the Static Young's Modulus results from the equation (3) – (5) calculated.

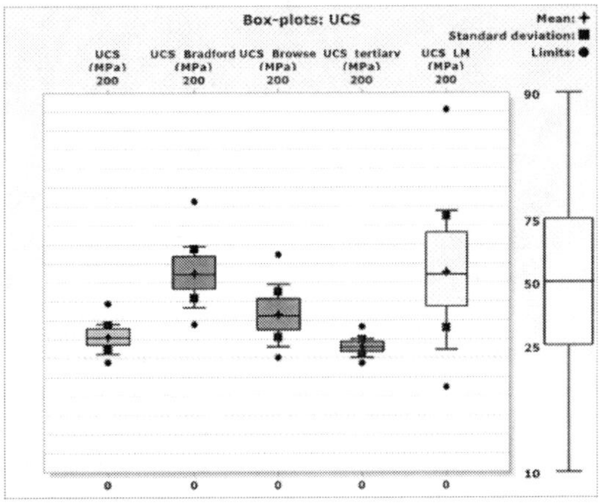

Figure 5. Box plots of the UCS models from equation (6) – (10) into the Poseidon 1 well.

Equation (4) was designed to perform over all the lithologies and especially tight formations [15], for the same reason, its dispersion is wide and in this particular section of the Poseidon 1 well the values are overpredicted by more than 120 (GPa) compared to the mean of P1A. In the case of equation (3), it was also designed to cover all the major lithologies [14], in this study its performance lacks of validity: 80% of error in the means compared to the reference Prelude 1A. Equation (5) [16], represents the best performance when dealing with the estimation of the Static Young's Modulus in the section evaluated of the Poseidon 1 well, as can be seen in the table 3, the values of the basic statistics agree with the ones calculated from the well P1A.

4.1.2. Poisson Ratio. The static Poisson's Ratio was not determined taking into account that there is not well a recognized trend between the dynamic and the static Poisson's Ratio [24] under confining stresses. Despite this, the values obtained from equation (2), were run into a Welch Two Sample t-test with 95% confidence interval compared to the ones given in the data sets included in [22], resulting the two sets being equal.

4.2. Unconfined Compressive Strength and Internal Friction Angle

4.2.1. UCS. The UCS calculated from equation (6) – (10), appears in the figure 5 keeping the same order from left to right as numbered in ascending number. Consecutively: the gray UCS boxplot was designed for strong compact shales [9], its mean is 71.24 (MPa); light blue UCS was modelled for sandstones considered weak [17], its mean is 104.86 (MPa); the green was conceived specifically for the reservoir sandstones from the Browse Basin [18]; the light brown was correlated from tertiary sandstones [10], its mean is 66 (MPa); the far right UCS, correlated from the triaxial test data of the well Prelude 1A has a mean of 105.64 (MPa).

As the values of equation (8) can be accountable for the whole Basin as was conceived, this formula compared to equation (10), it is more reasonable for assessing the section studied in the Poseidon 1 well (early cretaceous). Furthermore, in equation (10), the modelling can lead to overfitting due to the small sample size and spatial coverage (only one well); in spite this, it may be used as reference for validation of equation (8). Equation (7) demonstrated to have a very good performance to the values of equation (10); nevertheless, this was designed to reflect global weak reservoir sandstones [17] and for this study it was decide to proceed with equation (8) in the evaluation of UCS as the precision with equation (10) can be compromised.

4.2.2. FANG. The Angle of Internal Friction in equation (11), designed for silty sandstones [1], and the FANG calculated from the triaxial data presented in table 3, represents a good match. The means and the medians errors are below the 2%, and the minimum and maximum values are also close in magnitude. The difference errors in percentage can be seen in table 4, for this and the previous parameters between the estimations and models chosen for the well P1 and the values of P1A.

In the table 4, it can be seen more clearly the congruity of the parameters estimated. Equation (5) performs in this study inside the ranges of the mean percentage error designed: 13% [16] for the Young's Modulus. This can be due to the good quality of the well logging in the section for the Bulk Density, Compressional Slowness, and Shear Slowness. The good quality of the borehole as can be seen by the difference between the Bit Size and the Caliper reading, figure 1, track 1, may allowed good measurements of BD. Conversely, the Poisson Ratio exhibits high differences, it may be due to the nature of the variable, comparing a dynamic estimation to static values; however, this was already previously discussed in the Poisson Ratio sub-section.

In order to analyze the correctness of the mineralogical/lithological interpretation, figure 6 shows the results (track 4) compare to the mineralogy from the Elemental Capture Spectroscopy (ECS) obtained from the same interval (track 5), raw data from [22], also sided by the official interpretation from [3] (track 6). On average the section has 30 % in shale volume and 63 % in quartz volume and along with the other log curves such as Gamma Ray [25], including the mud logs in [22] can be classified as silty sandstone or siltstone. It is important to know that the estimations done here are mineral volumetric based and the ones from the ECS are on mineral weight; despite this, the percentual values are in general accordance through the section. In the same figure 6, the curves for the UCS from equation (8) in green color and equation (10) in light blue color are displayed, showing their shared behavior, with equation (10) curve exhibiting bigger magnitudes in the peaks of UCS, which are at the same depth to the lowest Gamma Ray readings, track 1, and to the more relative quartz volume, track 4.

Table 4. Error in percentage between P1 estimations and P1A calculated data.

Parameter	Min	Max	Mean	SD	Median
E (%)	63.73	32.45	2.33	52.36	8.57
ν (%)	50.90	63.19	49.25	76.17	45.17
UCS (%)	178.02	45.55	15.68	77.97	9.69
FANG (%)	27.32	4.62	1.83	52.03	1.81

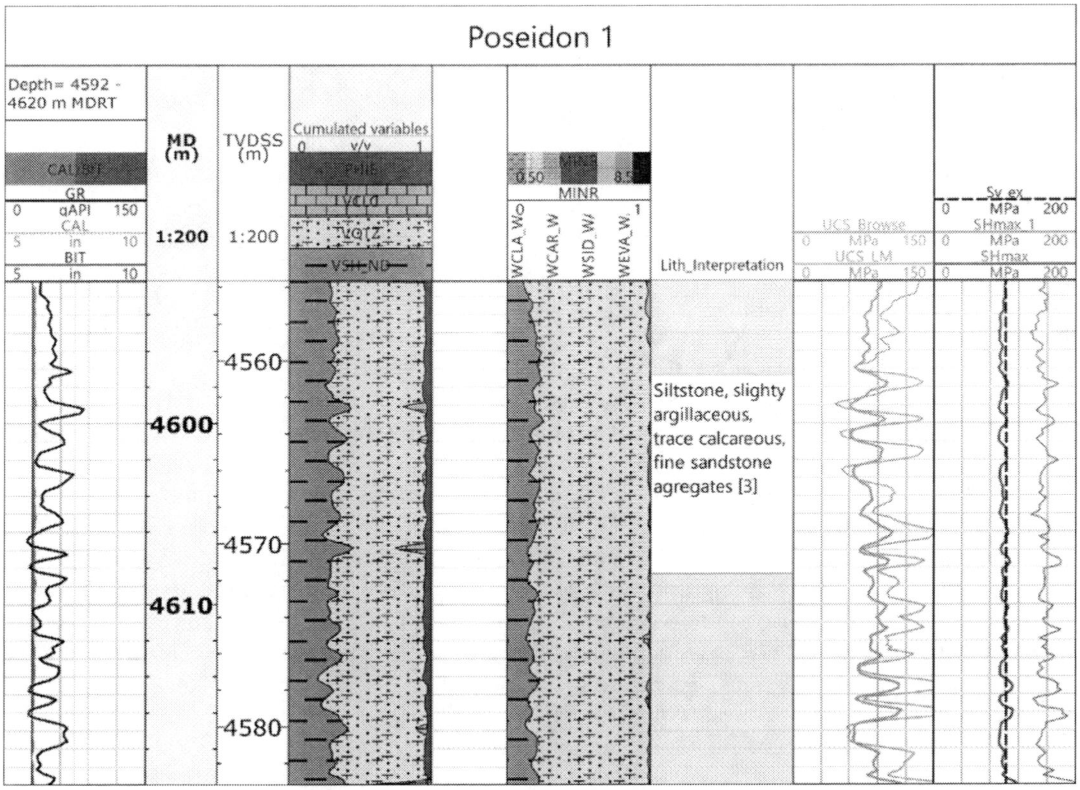

Figure 6. Composite well log of selected estimations for comparing in the Poseidon 1 well.

4.3. *In-situ stresses.*

The pore pressure calculated in the section shown in the figure 3 has a normal pore pressure gradient, mainly because the definition of clean shales for the Eaton Method trend in the Sonic Curve that stablished this section as "no shale" section, at the top of the section evaluated the pore pressure was found 46.33 (MPa) and at the bottom 46.61 (MPa). The minimum horizontal stress calibrated with LOT data, as seen in figure 3, track 10, has values ranging in between 75.12- 97.78 (MPa), the relative high values in the section are in synch with the relative high values in UCS and FANG; the overburden goes from top to bottom from 101.87 – 102.55 (MPa) with gradual increments.

The maximum horizontal stress, figure 6, track 7, SHmax, has values ranging from 91.34 (MPa) to 118.25 (MPa) and these values follow the same trend as the minimum horizontal stress estimated beforehand. The SHmax from equation (16) differs significantly from the SHmax_1 calculated using the method utilized by [6] with a value of 58.09 (MPa) in the absolute mean difference.

This difference between the frictional equilibria method [1] and the use of tectonic strain equilibria in equation (15) and equation (16) also results in a different fault regime; the first it is characterized by a Strike-slip faulting, where the principal stresses are configurated as $\sigma_H > \sigma_v > \sigma_h$, in the second, produces a Normal/Strike-slip faulting regime with $\sigma_H \sim \sigma_v > \sigma_h$ [26]. A clear advantage of the method defined by the equation (15) and (16) is the possibility of integrate calibration points directly, in this case the Shmin from LOT, iterating the tectonic stresses until reaching a fitting in the trends of the Shmin from the LOT and the Shmin estimated from the equation (15), which simultaneously will estimate the SHmax with equation (16). In [6] the definition of the Shmin by depth was with a linear regression from the LOTs that may led to reduce the resolution of the values calculated due to oversimplification.

5. Conclusions and Recommendations

5.1. Conclusions

The conclusions that can be drawn from the application of this methodology, the analysis of the results and discussion in the fashion done beforehand are:

The evaluation of the basic elastic rock parameters: Young's Modulus and Poisson's Ratio can be done using well logs in the Poseidon 1 well; in the case of the static Young's Modulus estimation, the equation (5) demonstrated to have an ample and precise range of application for the section of the Early Cretaceous in the depths 4 592 - 4 620 m MDRT. The percentage error in the mean to the homologous rocks used to reference the Static Young's Modulus is less than 3%, this confirms also that equation (5) is highly versatile in any geological setting and independent of lithology. Moreover, the application of this equation may be spread over the Browse Basin for the Estimation of Young's Modulus with fair precision. The dynamic Poisson's Ratio is properly calculated agreeing with the previous work done, validating equation (2) as standard for this calculation; there is a marked difference in the static Poisson's Ratio to the dynamic calculated, more works should be done in order to help to specify the correlation existent, especially in the Browse Basin.

The evaluation of the Uniaxial Compressive Strength and the Angle of Internal Friction can be performed by equation (8) and equation (9) in the interval of the Early Cretaceous of the Poseidon 1 well. The percentage of error for the correlations chosen are neighboring 15% and below 2% correspondingly. The relatively high percentage of quartz and relative low porosity are comparable to the homologous sandstone use as reference, enabling the use of (8) and there is a direct relation between the percentage of volume quartz calculated to the UCS, rocks with similar characteristics may be modeled with these two equations. Equations (8) and equation (9) that were designed for sandstone and silty sandstones are valid for the lithologies present in the Early Cretaceous of the Poseidon 1 well, which is composed by a mix of siltstones and sandstones.

The stress evaluation in the Poseidon 1 well can be worked with the use of the Poroelastic Model, the use of tectonic strains produces a Normal/Strike-slip faulting regime with the vertical and maximum horizontal stresses near in magnitudes, in the Early Cretaceous in the depths 4 592 - 4 620 m MDRT. The equation (15) and equation (16) for modelling the horizontal in-situ stresses offer an alternative for estimating their values allowing as input direct calibrations with the Shmin determined from LOT and the pore pressures with the MDT, these enables to alleviate the loss of resolution or the lack of precision due to excessively simplified workflows. The pore pressure in the section studied resulted in a normally pressured interval and the minimum horizontal has less than 100 (MPa) (75 – 98).

The general methodology presented in this study also provide with an alternative workflow that favors the estimation of reservoir geomechanical parameters avoiding both oversimplification and excessive processing. In the Poseidon 1 well, the correlations for these parameters on the early cretaceous were defined and such equations may work over the Browse Basin under certain conditions also indicated in this work.

5.2. Recommendations

The highlighted recommendations derived from this work are:

To evaluate different Lithologies on intervals of interest and compare to other or additional available laboratory data of Young's Modulus, preferable from the same Browse Basin or age, in order to have more certainty on the use of the correlations implemented in this work.

To test the UCS and FANG correlations that were chosen here in other reservoir rocks of the Timor Sea and verify its applicability by using the workflow defined in this study.

To apply anisotropy on the Elastic Parameters in sections with different lithology beddings or with high stress contrast amongst the vertical and the horizontal stresses.

To integrate a Biot's Coefficient according to the interval rocks and fluid properties evaluated.

To create a fault and stability model based on the parameters estimated with this methodology. These can be integrated in geomechanical studies in injection and characterization in Enhanced Oil Recovery

(EOR), Improved Oil Recovery (IOR), Carbon Capture and Storage (CCS), and Carbon Capture Utilization and Storage (CCUS) processes.

Acknowledgment
This work is one of the products from the Gadjah Mada International Fellowship 2020 (GMIF 2020), the authors express their gratitude for the funding of the international student involved.
The authors extend thanks to the Occam Technology Pty. Ltd., for supplying part of the well log data already preprocessed for the completion of this work .

References
[1] Zoback M D 2007 *Reservoir Geomechanics* (Cambridge: Cambridge University Press) p 449.
[2] Zoback M D 2021 *Geomechanical Challenges Associated with Massive Storage of CO2* (Online: SPE Education Online) viewed 8 January 2022 https://webevents.spe.org/products/geomechanical-challenges-associated-with-massive-storage-of-co2.
[3] ConocoPhillips (Brose Basin) Pty Ltd 2011 *Poseidon 1 Well Completion Report (Interpretative Data* vol 2*)* National Offshore Petroleum Information Management System viewed December 2021 https://dnxxuwuw8tglo.cloudfront.net/Wells/P00685603.zip.
[4] Government of Western Australia 2014 *Western Australia's Petroleum and Geothermal Explorer's Guide* 2014 Edition (Perth: Resource and Environmental Regulation Group) viewed November 2021 http://www.dmp.wa.gov.au/Petroleum/Browse-Basin-10988.aspx.
[5] Geoscience Australia 2021 *Regional Geology of the Browse Basin* 2021 update (Canberra: Geoscience Australia) viewed June 2022 https://www.ga.gov.au/scientific-topics/energy/province-sedimentary-basin-geology/petroleum/acreagerelease/browse.
[6] Rollet N, Abbott S, Lech M, Romeyn R, Grosjean E, Edwards D, Totterdell J, Nicholson C, Khider K, Nguyen D, Bernardel G, Tenthorey E, Orlov C and Wang L 2016 *A regional assessment of CO2 storage potential in the Browse Basin: Results of a study undertaken as part of the National CO2 Infrastructure* Plan Record 2016/17 eCat 89867 (Canberra: Geoscience Australia) http://dx.doi.org/10.11636/Record.2016.017.
[7] Hansen K S, Wang G, Adeleye O, McNeil, K V, Couzens-Schultz B A, Azbel K, Sarfare M and Tare U 2014 Integrated Predrill Pore-Pressure and Borehole-Stability Prediction for Prelude Development *SPE Drill & Compl* **29**(04) 418–430 https://doi.org/10.2118/173179-PA.
[8] Fjær E, Holt R, Horsrud P, Raaen A M and Risnes R 2008 *Petroleum Related Rock Mechanics (Developments in Petroleum Science* vol 53) 2nd edn (Amsterdam: Elsevier Science) p514.
[9] Chang C, Zoback M D and Khaksar A 2006 Empirical relations between rock strength and physical properties in sedimentary rocks *J. Pet. Sci. Eng.* **51**(3 – 4) 223 – 237 https://doi.org/10.1016/j.petrol.2006.01.003.
[10] Zhang J 2019 *Applied Petroleum Geomechanics* 1st edn (Cambridge: Gulf Professional Publishing) p 534.
[11] ConocoPhillips (Brose Basin) Pty Ltd 2012 *WA-315-P & WA-398-P Browse Basin Western Australia: 2009 Poseidon 3D Marine Surface Seismic Survey (Interpretation Report)* National Offshore Petroleum Information Management System viewed December 2021 https://dnxxuwuw8tglo.cloudfront.net/Surveys/P00696639.zip.
[12] Geoscience Australia 2021 *Copyright and Disclaimer Notice* (Canberra: Geoscience Australia) viewed November 2021 https://www.ga.gov.au/copyright.
[13] Serra O and Serra L 2004 *Well Logging Data Acquisition and Applications* (Méry Corbon: Editions Serralog) p 696.
[14] Lacy L 1997 Dynamic Rock Mechanics Testing for Optimized Fracture Designs *SPE Annual Technical Conference and Exhibition (San Antonio, Texas, USA, October 1997*) SPE-38716-MS https://doi.org/10.2118/38716-MS.
[15] Mullen M, Roundtree R and Barree B 2007 A composite Determination of Mechanical Rock

Properties for Stimulation Design (What To Do When You Don't Have a Sonic Log) *Rocky Mountain Oil & Gas Technology Symposium (Denver, Colorado, USA, April 16-18, 2007)* SPE-108139-MS p 13 https://doi.org/10.2118/108139-MS.

[16] Onalo D, Oloruntobi O, Adedigba S, Khan F, James L and Butt S 2018 Static Young's modulus model prediction for formation evaluation *J. Pet. Sci. Eng.* **171** 394 – 402 https://doi.org/10.1016/j.petrol.2018.07.020.

[17] Bradford I D , Fuller J, Thompson P J and Walsgrove T R 1998 Benefits of Assessing the Solids Production Risk in a North Sea Reservoir using Elastoplastic Modelling *SPE/ISRM Rock Mechanics in Petroleum Engineering (Trondheim, Norway, July 1998)* SPE-47360-MS https://doi.org/10.2118/47360-MS.

[18] McNally G H 1987 Estimation of coal measures rock strength using sonic and neutron logs *Geoexploration* **24**(4–5) 381-395 https://doi.org/10.1016/0016-7142(87)90008-1.

[19] Shell Development (Australia) Pty Ltd 2008 *PRELUDE-1, 1-A, Well completion Report (Basic Data)* National Offshore Petroleum Information Management System viewed June 2022 https://dnxxuwuw8tglo.cloudfront.net/Wells/P00596371.zip.

[20] Elmahdy M, Farag A E, Tarabees E and Bakr A 2018 Pore Pressure Prediction in Unconventional Carbonate Reservoir *SPESATS* p 11 https://doi.org/10.2118/194224-MS.

[21] Eaton B A 1975 The Equation for Geopressure Prediction from Well Logs *Fall Meeting of the Society of Petroleum Engineers of AIME* (Dallas, Texas, September 1975) SPE-5544-MS https://doi.org/10.2118/5544-MS.

[22] ConocoPhillips (Brose Basin) Pty Ltd 2010 *Poseidon 1 Well Completion Report (Basic Data* vol 1) National Offshore Petroleum Information Management System viewed December 2021 https://dnxxuwuw8tglo.cloudfront.net/Wells/P00685602.zip.

[23] Bonamente M 2017 *Statistics and Analysis of Scientific Data* 2nd edn (New York: Springer) p 335.

[24] McPhee C, Reed J and Zubizarreta I, 2015 *Core Analysis: A Best Practice Guide* (*Developments in Petroleum Science* vol 64) 1st edn (Amsterdam: Elsevier) p 852.

[25] Serra O 2007 *Well Logging and Reservoir Evaluation* (Paris: Editions Technip) p 250.

[26] Zoback M D and Kohli A 2019 *Unconventional Resources Geomechanics: Shale Gas, Tight gas, and induced seismicity* (Cambridge: Cambridge University Press) p 492.

IOP Conf. Series: Earth and Environmental Science 1233 (2023) 012027 doi:10.1088/1755-1315/1233/1/012027

Characterization of Metapelite and Metasiltstone as Uranium-REE hosted rocks in Rirang Area, West Kalimantan

R A Farrenzo[1], R D Nugraheni[1*], I G Sukadana[2], F D Indrastomo[2] and T B Adimedha[2]

[1]Depatement of Geological Engineering, Faculty of Earth Technology and Energy, Universitas Trisakti, Kyai Tapa St. No. 1 Grogol, Jakarta 11440, Indonesia
[2]Research Center for Nuclear Fuel Cycle and Radioactive Waste Technology, The National Research and Innovation Agency, Indonesia, Lebak Bulus Raya St. No. 9 Ps. Jumat, Jakarta 12440, Indonesia
*Corresponding author: rosmalia.dn@trisakti.ac.id

Abstract. The Rirang River, Kalan, West Kalimantan has U and REE resources in the form of mineralization found at the intercalation of metapelite and metasiltstone. However, there are only a few studies about the characterization of U and REE-bearing host rocks in Indonesia. Determining the geological influence of the metasedimentary rock on the mineralization of U and REE is crucial. Thus, the characterization of metapelite and metasiltstone has been substantiated in this study to reveal the potency of the host rock in hosting the U and REE. Petrographic, micro-XRF, and AMICS methods were used to conduct mineralogical analysis. These two rocks share a mineral composition that includes clay minerals, hydrothermal alteration products, and several minerals that indicate metamorphism processes. The foliation structure with a spotted shale texture further indicates the metamorphism process in metapelite and metasiltstone. REE and trace elements associated with U are constituents of clay minerals such as wonesite, staurolite, montmorillonite, chlorite, illite, and kaolinite. The geochemical properties of metapelite and metasiltstone suggest that REE mineralization and trace elements associated with U come from hydrothermal alteration processes.

1. Introduction

The energy transition and the fulfilment of carbon emission free by 2060 agendas, as proclaimed by the Government of Indonesia, can be realized using renewable energy resources. One is using nuclear energy, which uses uranium as fuel in nuclear power plants. Developing nuclear power plants also requires rare earth elements (REE) as instrumentation in nuclear power plants. For example, generators use neodymium (Nd) as an alloy with iron and boron to form permanent solid magnets. In addition, there are also europium (Eu) and gadolinium (Gd), which have good neutron absorption capabilities, so they are used as control rods in nuclear reactors. The development of nuclear power plants in Indonesia has not yet taken place. However, with the abundant potential, it is necessary to conduct detailed research to support energy transition policies. Especially in the Rirang River, Kalan, West Kalimantan has abundant potential for U and REE resources [1], with an estimated resource of 20.2017 tons of U_3O_8 and 855,2639 tons of REE [2].

Research on the U and REE resources in the Rirang River conducted by Suhardji et al. [2] was calculated using mineralized vein bodies at the intercalation of metapelite and metasiltstone, as well as ores containing U and REE. It is known that the mineralization characterization of U and REE has been extensively studied by several previous researchers, among others, including Tjokrokardono [3]–[5], Ngadenin [6], and Sartapa [7]. Likewise, the characterization of ores containing U and REE has been previously discussed by Tjokrokardono [8] and summarized by Dahlkamp [9]. The abundance of U and REE in metasedimentary rocks in the Rirang area has similarities with U deposits in other areas such as the Valhalla uranium deposit, Australia [10] and the Olympic Dam IOCG deposit [11]. Those deposits have also been previously investigated by Alexandre et al. [12] and Cuney and Kyser [13], but it focuses on identifying the deposit type and the process of mineralization, not on the characteristics of the host rock. The Rirang area has a high abundance of U mineralization associated with REE [1], [2]. Meanwhile, there is a minimal discussion on the characterization of U and REE-bearing host rock, especially in Indonesia. It is crucial to know the characterization of the metasedimentary rock carrying U and REE as a basis for determining the geological influence of the host rock on the mineralization of U and REE. Thus, it is essential to know the potency of the host rock in hosting the U and REE. This research focused on identifying metapelite and metasiltstone characteristics found in the Rirang River based on mineralogy and geochemical features.

2. Regional Geology

The Kalan area is a sedimentary basin known as the Kalan Basin, which consists of fine-grained Permo-Carbon sediments with intercalation of volcanic materials. The sedimentary rocks underwent regional metamorphism at a pressure of 2,000 bar and a temperature of 540 °C [14]. The metamorphic rock is called Pinoh Metamorphic Rock (PzM). This rock was intruded by Early Cretaceous tonalite/granitoid, which is equal to the magamtism period of Sepauk Tonalite Group. The metamorphism process produces several lithologies in the Kalan Basin, as shown in Fig. 1.

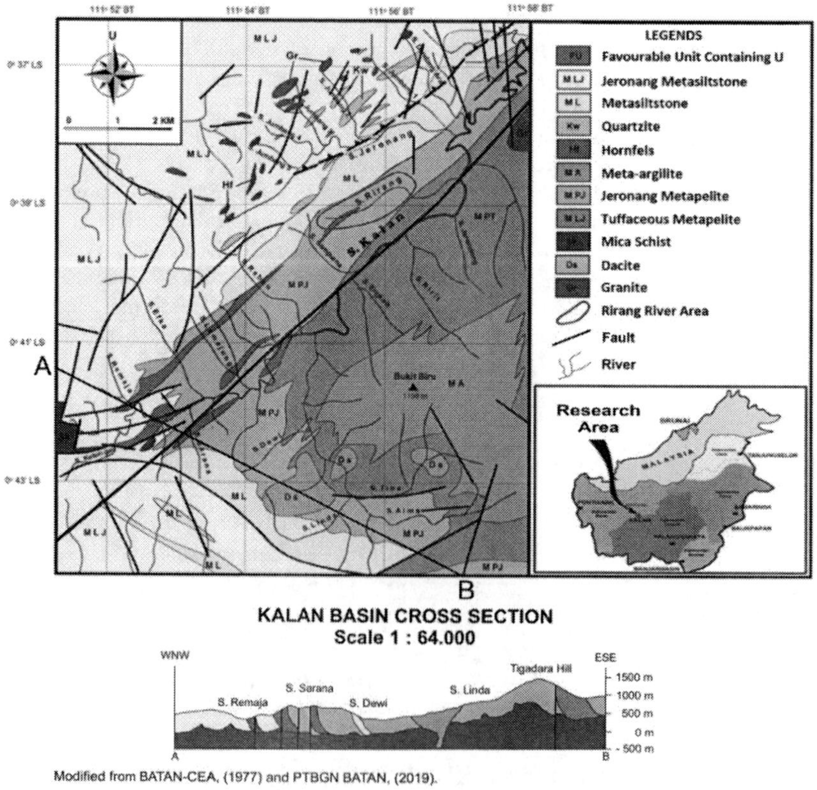

Figure 1. Geological Map of the Kalan Basin, West Kalimantan. Modified from BATAN-CEA and PTBGN BATAN [14], [15].

The stratigraphy of the Kalan Basin is divided into three series, namely the lower, middle, and upper series (Fig. 2) [14]. The Rirang area is included in the upper series, characterized by intercalation of metapelite and metasiltstone with fracture schistosity, and formed in greenschist metamorphism facies [6]. The intercalation of metapelite and metasiltstone is a favorable lithological unit containing U mineralization [4], [7]. Metapelite is characterized by greenish-grey color in fresh conditions and grey-brown color in weathered conditions has a grain size of clay, showing a spotted shale texture that has an orientation in the direction of foliation, and is composed of feldspar, sericite, quartz, biotite, chlorite, and opaque minerals. Spotted shale texture is a collection of clay minerals that are grouped and oriented due to an increasing level of metamorphism. The overall thickness of the metapelite is more than 500 m [4]. Metasiltstone is gray when fresh and brownish gray when weathered, has a grain size of silt to fine sand, and consists of quartz, feldspar, biotite, tourmaline, iron oxide, pyrite, and glass material. Metasiltstone still shows sedimentary structures like parallel lamination to layering. The thickness of metasiltstone is identified as 25 – 100 m [4].

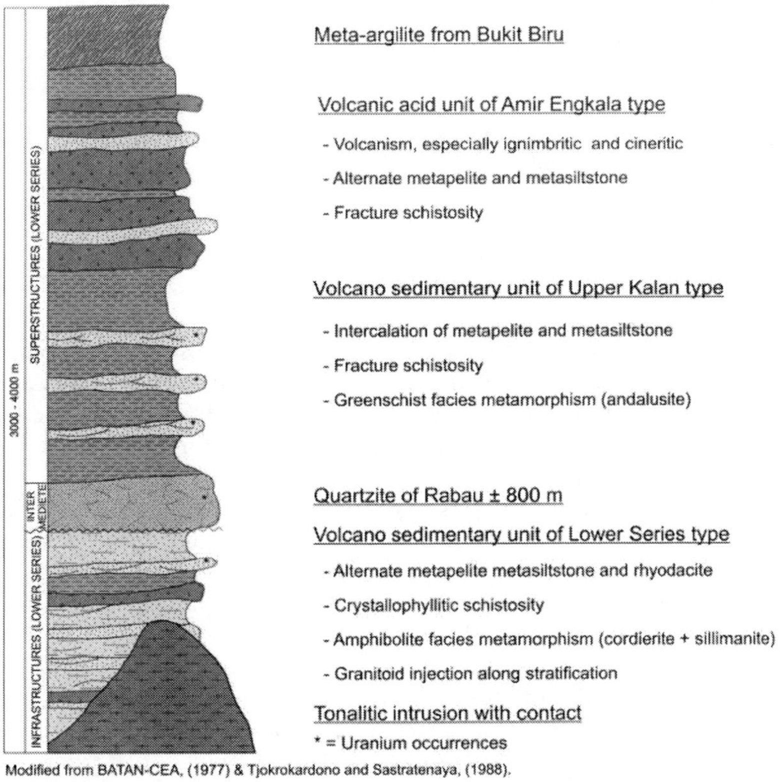

Figure 2. Kalan Basin Stratigraphy. Modified from Tjokrokardono and Sastratenaya and BATAN-CEA [8], [14].

2.1. *Geology of Rirang Area*

The geology of the Rirang area is divided into two rock units, namely the metapelite unit and the metasiltstone unit [15]. The metapelite unit megascopically has lithological characteristics of dark grey – reddish brown, clay-sized grain size, and foliation texture. Meanwhile, the metasiltstone unit has megascopic features of fresh grey – red-brown color, fractured, clay–silt-sized grain size, and still has lamination structure. The altered rocks unit is composed of an altered metasiltstone unit with the mineralized boulders that spread along the valley. Outcrop found in this area is heavily weathered. The distribution of rock units in the Rirang area is shown in Fig. 3.

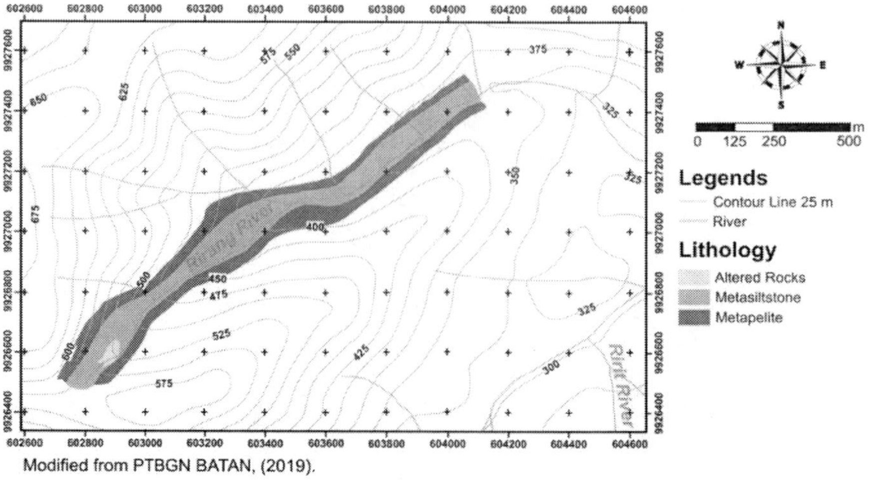

Figure 3. Geological Map of the Rirang Area. Modified from PTBGN BATAN [15].

3. Methods

The identification of mineralogy characteristics is carried out using several analyses. Petrographic analysis was used to determine the mineral texture and composition. The petrographic analysis was carried out using a Nikon Eclipse E200 polarizing microscope and photographed using a Nikon D5200 camera at the Petrology and Mineralogy Laboratory, Geological Engineering Departement, Universitas Trisakti, Indonesia. Sample preparation into thin sections was carried out at Obsidian Laboratory, Bandung, Indonesia. Then Micro-XRF (μXRF) analysis was conducted to determine the distribution and abundance of elements in the sample both qualitatively and quantitatively [16]. μXRF is an elemental analysis technique that collects the X-ray spectrum of each pixel representing each element's content, with a spatial resolution of 10 μm. μXRF analysis was performed using the Bruker M4 Tornado Plus in a vacuum state at 50 kV 600 μA. Scanning was performed in area mode with an X-ray spot size of 20 μm and an acquisition time of 12 ms/pixel, with a pixel size of 35 μm. The μXRF result was then used for Advanced Mineral Identification and Characterization Software (AMICS) analysis provided by Bruker Nano Analytics. AMICS is an application to identify the mineralogy of a sample based on the spectrum generated by μXRF analysis. The spectrum is then compared with the spectrum database on AMICS, which can produce a mineralogical map and mineral abundance of the sample. This method could identify mineral groups and a single mineral in each sample because each specific mineral has a composition characteristic represented by the backscattered electron images generated from μXRF analysis. Mineralization and genesis of some minerals could be identified using AMICS due to its characteristics to quickly identify minerals carrying U, REE, and trace elements [17]. μXRF and AMICS analyses were carried out at the Research Center for Nuclear Fuel Cycle and Radioactive Waste Technology, The National Research and Innovation Agency, Indonesia.

4. Results and Discussion

4.1. Mineralogy

Metapelite is a metamorphic rock with a phyllitic foliation structure, a nematooblastic texture, and a blastopelitic palimpsest. Then, some minerals are clustered and have a mineral orientation called spotted shale texture (Fig. 4). The clustered minerals are composed of illite, kaolinite, and chlorite. This mineral assemblage is the main mineral, while the other part that surrounds those minerals is formed by wonesite. The main constituent minerals of metapelite are included in the phyllosilicate mineral group, which were identified as coming from the mica minerals group, such as illite and chlorite.

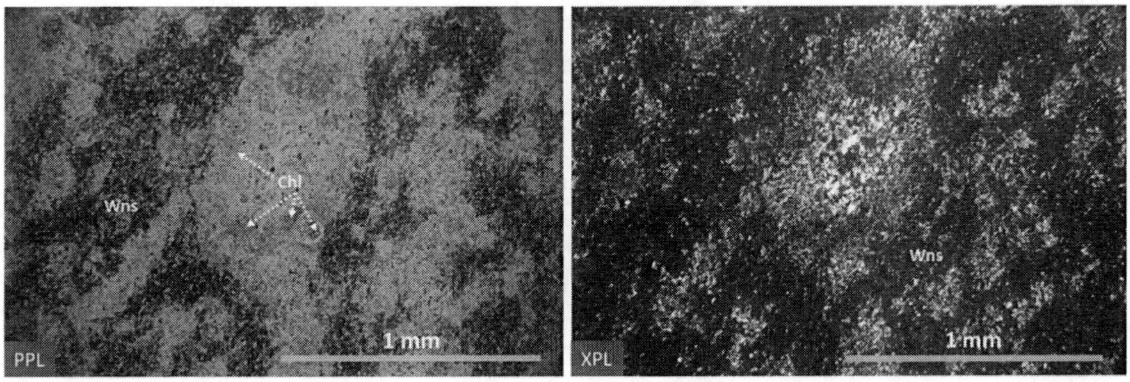

Figure 4. Photomicrograph of metapelite on petrographic analysis, showing the presence of illite, kaolinite, chlorite, and wonesite. Those minerals constitute fragments of spotted shale texture. This texture is a manifestation of the metamorphic process that occurs in metapelite. There is a wonesite layer wrapping around these fragments. (Chl = Chlorite, Wns = Wonesite, Kln = Kaolinite. Mineral abbreviation based on IMA-CNMNC [18]).

Illite is colorless in plane-polarized light (PPL) and white–grey in cross-polarized light (XPL), has a subhedral–anhedral shape measuring 0.3 – 0.5 mm, and has low relief. Kaolinite has a characteristic subhedral-anhedral shape, measuring < 0.3 mm. Kaolinite is colorless in PPL and grey in XPL, with low relief. In smaller amounts, there is chlorite between the fragments. Chlorite is colorless in PPL and grey-dark green in XPL, with a fragmented anhedral plate of < 0.1 mm. Chlorite has low pleochroism and moderate relief. Wonesite is a clay mineral with a characteristic muddy color in PPL and grey-brown in XPL. Wonesite has an anhedral shape of < 0.3 mm and a fine texture with moderate relief.

Metasiltstone is a metamorphic rock that megascopically has a grey-black color with clay-silt grain size. Based on petrographic observations, metasiltstone has a slaty foliated structure with a lepidoblastic texture with a blastopelitic palimpsest (Fig. 5). Metasiltstone can be distinguished from metapelite based on the texture and structure of the rock. Metasiltstone has mineral clusters composed of muscovite, oligoclase, and illite, surrounded by chlorite and hematite. These minerals are petrographically identified as the main constituents of metasiltstone.

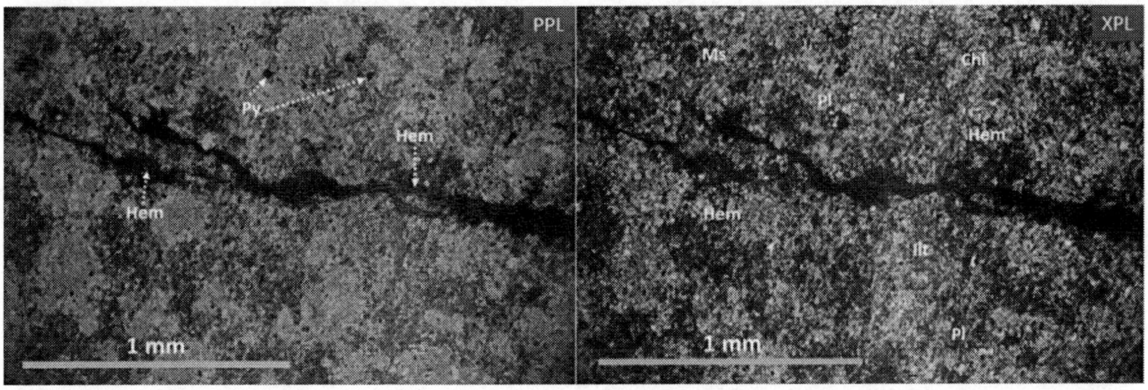

Figure 5. Photomicrograph of metasiltstone on petrographic analysis, showing the presence of muscovite, illite, chlorite, and plagioclase. Muscovite, as the predominant constituent mineral, shows sericitization texture. There are also hematite veins, with pyrite present, among other minerals. (Ms = Muscovite, Ilt = Illite, Chl = Chlorite, Pl = Plagioclase, Hem = Hematite, Py = Pyrite. Mineral abbreviation based on IMA-CNMNC [18]).

Muscovites have muddy color in PPL and a grey-orange in XPL. It has a fine sericitization texture measuring < 0.3 mm with medium relief. Muscovite has shown changes to clay minerals. Oligoclase has a characteristic subhedral-anhedral shape measuring < 0.3 mm, colorless in PPL and grey to white in XPL. Illite has a characteristic of subhedral - anhedral shape measuring < 0.3 mm, colorless on PPL, white-grey on XPL, and has low relief. Illite is present as one of the minerals that make up oriented fragments. Like chlorite, illite is a clay mineral resulting from hydrothermal alteration processes [19]. Chlorite has a colorless characteristic in PPL and grey-dark green in XPL. Chlorite has the form of fragmented anhedral plates measuring < 0.1 mm and has low pleochroism and moderate relief. Hematite veins cut through the body of the metasiltstone, characterized by reddish black in both PPL and XPL, measuring < 0.3 mm. Pyrite is an opaque mineral that can be identified in metasiltstone, has black color in PPL and XPL, with an isometric shape. Pyrite is a product of hydrothermal alteration mineralization [19].

4.2. Elemental Distribution

Metapelite is identified as having a high abundance of the major oxides, especially the higher Al_2O_3 and SiO_2 content than Fe_2O_3, MgO, CaO, and TiO_2 so that it can be classified as high-Al metapelite [20]. In addition, the high abundance and similar position of Al_2O_3 with SiO_2 and Fe_2O_3 indicate that Al_2O_3 is concentrated in phyllosilicate groups such as wonesite, sudoite, chlorite, illite, staurolite, montmorillonite, and mica minerals [21]. These elements are also classified as incompatible elements, so their presence is interpreted as compiling the composition of the main rock-forming minerals such as muscovite, albite, and oligoclase. Based on the total mineral content and also the relatively low content of Fe_2O_3 (6 – 12 Wt%), CaO (1 – 2 Wt%), and TiO_2 (1 – 1.2 Wt%) in metapelite, accompanied by a relatively high abundance of SiO_2 and Al_2O_3, indicates a low abundance of mafic minerals [22], therefore shows metapelite derived from felsic rock. On the other hand, trace elements have an abundance of up to 300 ppm. The trace elements identified were Co, Zr, Pm, Mn, Y, Rb, Sr, W, Zn, Cu, V, Mo, and Pb. These trace elements are known to originate from hydrothermal fluids originating from the Sukadana Granite Intrusion, which intrudes metapelite and metasiltstone [3]. Zr concentrations around 230 – 300 ppm are commonly found as metapelite materials [23]. In smaller quantities, identified REEs such as Y only have concentrations that do not exceed 30 ppm. Meanwhile, it is identified that the concentration of F is quite abundant (1,000 – 10,000 ppm). It is one of the manifestations of the hydrothermal alteration process [4], [24]. A high fluorine concentration indicates that a metasedimentary rock has undergone advanced alteration or is in a proximal alteration zone [25].

Based on the identification of the mapped elements, several groups of elements can be grouped based on their position and abundance (Fig. 6). The major element groups are mainly positioned in the oriented fragments, with trace elements dispersed in all parts. The dark bands are identified as composed of Fe, Ti, and V. Chemically, these elements can form minerals mainly composed of iron oxide or titanium oxide. Wonesite ($(Na;K)_{0.5}(Mg;Fe;Al)_3(Si;Al)_4O_{10}(OH;F)_2$) was composed mainly of Fe, thus confirming wonesite as a geochemical constituent of metapelite. The presence of dispersed Al and Mg in the dark bands indicates the occurrence of elemental substitution with Fe or Ti. A low concentration of V can be considered a trace element that is adsorbed by other elements that make up wonesite.

IOP Conf. Series: Earth and Environmental Science 1233 (2023) 012027 doi:10.1088/1755-1315/1233/1/012027

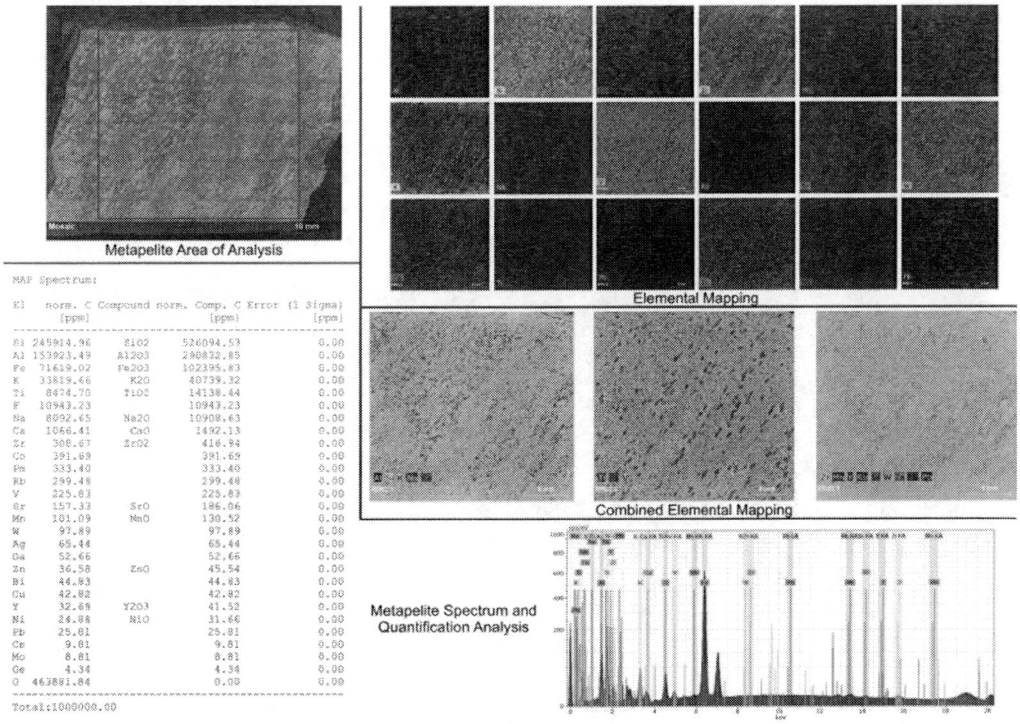

MAP Spectrum:

El	norm. C (ppm)	Compound	norm. Comp. C (ppm)	Error (1 Sigma) (ppm)
Si	245914.96	SiO2	526094.53	0.00
Al	153923.49	Al2O3	290832.85	0.00
Fe	71619.02	Fe2O3	102395.83	0.00
K	33819.66	K2O	40739.32	0.00
Ti	8474.70	TiO2	14138.44	0.00
F	10943.23		10943.23	0.00
Na	8092.65	Na2O	10908.63	0.00
Ca	1066.41	CaO	1492.13	0.00
Zr	308.67	ZrO2	416.94	0.00
Co	391.69		391.69	0.00
Pm	333.40		333.40	0.00
Rb	299.48		299.48	0.00
V	225.83		225.83	0.00
Sr	157.33	SrO	186.06	0.00
Mn	101.09	MnO	130.52	0.00
W	97.89		97.89	0.00
Ag	65.44		65.44	0.00
Ga	52.66		52.66	0.00
Zn	36.58	ZnO	45.54	0.00
Bi	44.83		44.83	0.00
Cu	42.82		42.82	0.00
Y	32.69	Y2O3	41.52	0.00
Ni	24.88	NiO	31.66	0.00
Pb	25.81		25.81	0.00
Ce	9.81		9.81	0.00
Mo	8.81		8.81	0.00
Ge	4.34		4.34	0.00
O	463881.84		0.00	0.00

Total:1000000.00

Figure 6. Schematic distribution of major elements and trace elements mapped on metapelite using μXRF analysis.

The abundance of other trace elements was identified as a form of elemental mobility during the leaching process [16]. It is also characterized by the distribution of trace elements dispersed in almost the entire rock body — especially in oriented fragments — indicating associations with SiO_2 and Na_2O, which also have a significant position in that section [20]. On the other hand, several trace elements also have a major position in the dark bands, like V. Metapelite spectral analysis showed the highest peak of the Fe spectrum, followed by Ti, K, Si, and Al spectra. The trace elements show peak morphology that distinguishes them from other elements. The resulting quantification of the mapped spectrum also illustrates the spectral analysis, showing a high concentration of Fe, followed by a lower concentration of Ti, and V which has a concentration of up to 225 ppm. The presence of trace elements in metapelite is produced through the interaction process between hydrothermal fluid and meteoric fluid in fractures in the parent rock in the form of metapelite. This interaction resulted in the mineralization of REE and other trace elements, which were adsorbed by some minerals in metapelite.

Metasiltstone is identified as having a high abundance of major elements. Accompanied by the presence of identified trace elements, which consist of Co, Rb, Zr, Zn, Sr, Y, Cu, Nd, Mo, and V. In addition, the identified concentration of F which is relatively abundant (1,000 – 10,000 ppm) and it is one of the manifestations of the hydrothermal alteration process [4], [24]. Fluorine is an essential ligand in hydrothermal fluids. F has a significant role in the transport and precipitation of U and other elements, such as Si, Al, Cu, Zr, and W [26], therefore precipitating new minerals. Metasiltstone is indicated to be passed by hydrothermal fluid, characterized by an increase in the concentration of F due to the anion substitution between fluorine (F-) and oxyhydroxide (OH-) [25].

Major element groups have clustered positions in the fragment, along with some trace elements. Meanwhile, Fe, Ti, Mn, Nd, Co, and V are wrapping those fragments (Fig. 7). Spectral analysis of metasiltstone produced a spectrum with high peaks of Si, Al, Fe, Ti, K, Ca, and S. The trace elements spectrum showed a distinct morphology compared to other elements. The quantification results of the metasiltstone spectrum showed that there was sulfur with a concentration of more than 30,000 ppm, followed by trace elements such as Co and V with the highest concentrations reaching 500 ppm and 200 ppm, respectively. In smaller quantities, there are REE in the form of Nd (220 ppm) and Y (59 ppm). The presence of Y and Nd are interpreted as being present as trace elements that bind to clay mineral through ion adsorption process that causes these elements to be enriched in clay minerals because of their ability to bind trace elements [27]. The increased mobility of REE is influenced by hydrothermal fluids rich in F [28], so clay minerals can easily adsorb it. The process of releasing HF gas at the time of volcanism can produce a fluid rich in F, thereby increasing the mobility of REE [29]. Trace element concentrations are also controlled by the number of clay minerals and accessory minerals in sedimentary protoliths [30].

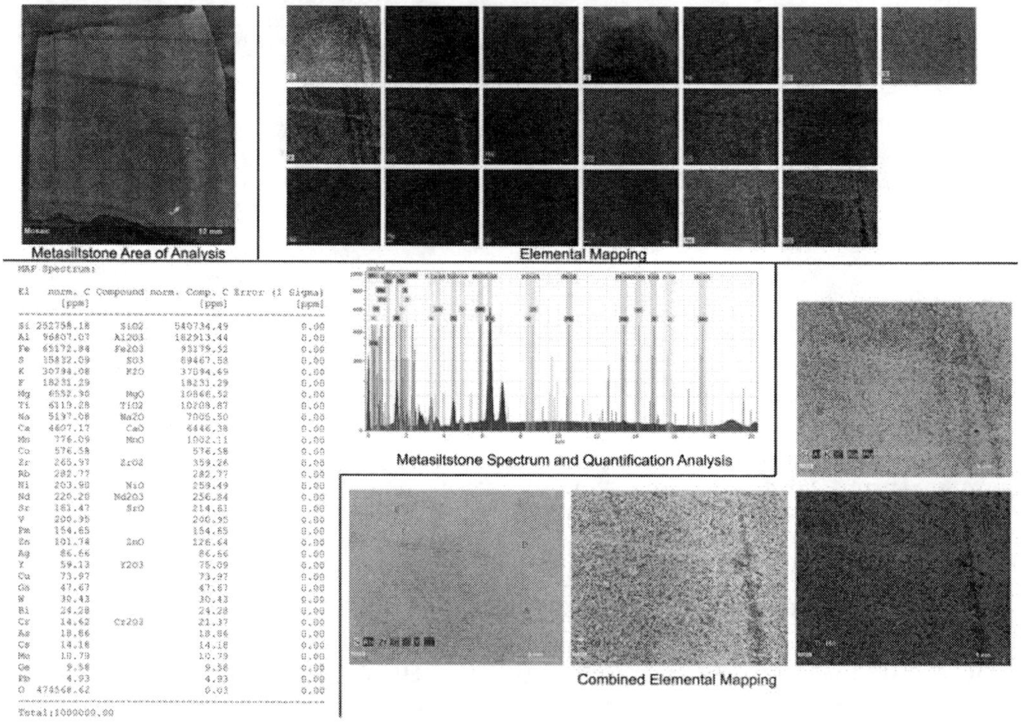

El	norm. C (ppm)	Compound	norm. Comp. C (ppm)	Error (1 Sigma) (ppm)
Si	252758.18	SiO2	540734.49	0.00
Al	96807.07	Al2O3	182913.44	0.00
Fe	65172.84	Fe2O3	93179.52	0.00
S	35832.09	SO3	89467.58	0.00
K	30794.08	K2O	37094.69	0.00
F	18231.29		18231.29	0.00
Mg	6552.90	MgO	10866.52	0.00
Ti	6119.28	TiO2	10208.87	0.00
Na	5197.08	Na2O	7006.50	0.00
Ca	4607.17	CaO	6446.38	0.00
Mn	776.09	MnO	1002.11	0.00
Co	576.58		576.58	0.00
Zr	265.97	ZrO2	359.26	0.00
Rb	282.77		282.77	0.00
Ni	203.90	NiO	259.49	0.00
Nd	220.28	Nd2O3	256.84	0.00
Sr	181.47	SrO	214.61	0.00
V	200.95		200.95	0.00
Pm	154.65		154.65	0.00
Zn	101.74	ZnO	126.64	0.00
Ag	86.66		86.66	0.00
Y	59.13	Y2O3	75.09	0.00
Cu	73.97		73.97	0.00
Ga	47.67		47.67	0.00
W	30.43		30.43	0.00
Bi	24.28		24.28	0.00
Cr	14.62	Cr2O3	21.37	0.00
As	18.86		18.86	0.00
Cs	14.18		14.18	0.00
Mo	10.79		10.79	0.00
Ge	9.58		9.58	0.00
Pb	4.93		4.93	0.00
O	474568.62		0.03	0.00

Total:1000000.00

Figure 7. Schematic distribution of major elements and trace elements mapped on metasiltstone using μXRF analysis.

The presence of these elements other than U indicates several possibilities. Among others, U is not detected in the analyzed section. Another possibility is that the U concentration is deficient, so a bulk sample geochemical analysis with a lower detection limit, such as an ICP-MS analysis, is needed. The identified trace elements are associated with the U and REE, indicating that the mineralization is related to the hydrothermal alteration process in metapelite and metasiltstone. The association occurs because the hydrothermal fluid that transports U is rich in trace elements (Mo, F, Pb, Zn, V, Y, Rb, and REE) and behaves as a strong acid that binds U [11], [26], [31]. It is also known that hydrothermal alteration processes cause U and trace elements mobilization [32]. Under other conditions, enrichment of trace elements associated with U can be a criterion for determining the type of mineralization [33]. The abundance of F can also indicate the presence of hydrothermal U deposits [34], fluorine in metapelite and metasiltstone was identified as being present in muscovite, biotite, and other mica minerals. At low-medium temperature hydrothermal U deposits in metasedimentary rocks, it is common to find associations between U and Cu, Co, Mo, and V forming sulfide and aluminosilicate bonds [35]. An intense hydrothermal alteration process might be the reason for uranium oxides absence [31].

4.3. Mineral Characterization

The results of the metapelite mineralogical map confirm the interpretation of the petrographic analysis, which shows that the oriented fragments are composed of illite, kaolinite, and chlorite minerals (Fig. 8). Kaolinite, one of the minerals that make up oriented fragments, can act as an element exchanger. Kaolinite has a single-layer structure; the layers are composed of $[SiO_4]^{-4}$ tetrahedra connected by three corners to form a sheet. Between these layers are Al ions bonded to the fourth corner of the $[SiO_4]^{-4}$ tetrahedra, providing, on average, three hydroxyls from one layer and one hydroxyl plus two O_2 from the next layer. In clay minerals, ion exchange can occur in imperfect structures (broken bonds) or exposed hydroxyl edges [36]. So that Al ions can be replaced by other ions such as Y, Mn, Rb, Sr, Zr, V, and Cu. Yttrium suspected occurs through ion-adsorption and mainly adsorbed on the clay crystal surface, particularly on illite and chlorite which exhibits high CEC and specific surface area. Apart from silicates, it is known that iron and manganese oxides also have an essential role in the distribution of trace elements. Both can bind trace elements through coprecipitation, adsorption, and ion exchange, but not by isomorphic substitution [37].

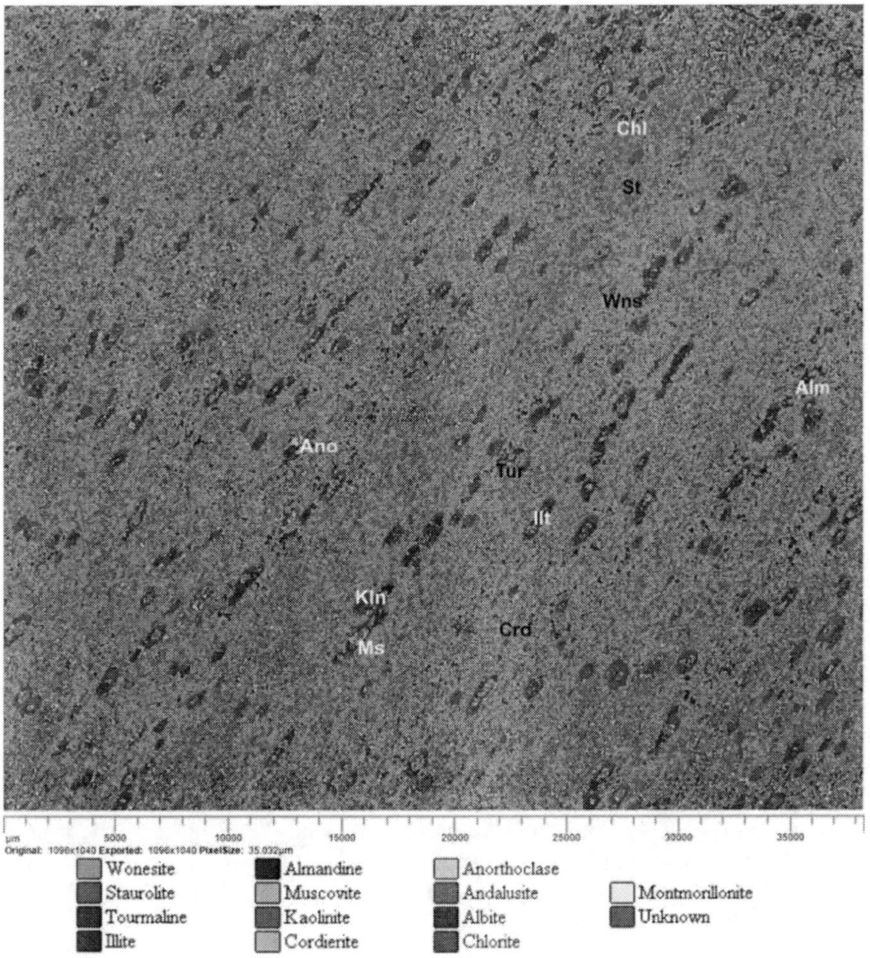

Figure 8. Mineral mapping of metapelite based on AMICS analysis. (Tur = Tourmaline, Wns = Wonesite, St = Staurolite, Chl = Chlorite, Alm = Almandine, Ano = Anorthoclase, Ms = Muscovite, Kln = Kaolinite, Crd = Cordierite, Ilt = Illite. Mineral abbreviation based on IMA-CNMNC [18]).

The precursor minerals in metapelite are illite, kaolinite, and chlorite, which occasionally recrystallize into the well-oriented nematoblastic fabric of muscovite, anorthoclase, and albite. The nematobastic texture in metapelite evidence that the protolith is heterogeneous. Silicate minerals such as muscovite, anorthoclase, tourmaline, and albite are found in an oriented fragment core, indicating that those minerals are the product of the metamorphism process. One sign of a metamorphic process is the foliation of mica minerals. Almandine, a garnet group, together with cordierite, andalusite, and staurolite, present as index minerals for the metamorphism process [19]. These minerals also have similarities with wonesite, that composed of Fe as one of the primary constituent elements. So, it is known that Fe composes not only wonesite but also part of other minerals.

Metapelite is composed of 13 minerals, all of which are dominated by clay minerals, like wonesite, with an abundance of 33.99 Wt%. Other clay minerals include staurolite, illite, kaolinite, chlorite, and montmorillonite (21.25 Wt%, 8.21 Wt%, 6.26 Wt%, 0.35 Wt%, 0.21 Wt%, respectively). Also, some minerals that characterize metamorphism processes, such as cordierite, tourmaline, muscovite, anorthoclase, almandine, albite and andalusite (14.59 Wt%, 5.11 Wt%, 4.37 Wt%, 2.72 Wt%, 2.54 Wt%, 0.16 Wt%, and 0.13 Wt%, respectively). The abundance of metapelite constituent minerals can be seen in Table 1.

Table 1. Metapelite modal mineralogy based on AMICS analysis.

No.	Name	Wt%	Area%	Area (μ^2)	Segment Number	Relative Error
1	Wonesite	33.99	33	458079493.8	373261	0
2	Staurolite	21.25	22.33	310052691.2	252643	0
3	Cordierite	14.59	14.89	206691158.1	168420	0
4	Illite	8.21	8.26	114615291.1	93393	0.01
5	Kaolinite	6.26	6.55	90879311.47	74052	0.01
6	Tourmaline	5.11	4.67	64883989.59	52870	0.01
7	Muscovite	4.37	4.49	62365700.46	50818	0.01
8	Anorthoclase	2.72	2.93	40627661.73	33105	0.01
9	Almandine	2.54	1.89	26270222.83	21406	0.01
10	Chlorite	0.35	0.35	4803403.35	3914	0.03
11	Montmorillonite	0.21	0.23	3219041.13	2623	0.04
12	Albite	0.16	0.17	2331749.2	1900	0.05
13	Andalusite	0.13	0.12	1705858.63	1390	0.05
14	Unknown	0.09	0.08	1087331.47	886	0.07

The mineralogical map of the metasiltstone shows a good correlation with the interpretation of the petrographic analysis (Fig. 9). Almost the entire dark layer is composed of iron oxide. Hematite and pyrite comprise the metasiltstone and mainly occupy the nearby vein and veinlet, indicating that this mineral is associated with the hydrothermal alteration process. Abundant pyrite manifests that this sulfide mineral can be bounded by Mo, Cu, and Pb [19]. Apart from pyrite, the presence of tourmaline and beryl can also indicate a hydrothermal alteration process. Beryl, part of the tourmaline group, is commonly found as a constituent in granite, metamorphic rocks, and low-high-temperature hydrothermal veins [19]. The abundance of montmorillonite, along with other clay minerals such as chlorite, illite, and kaolinite, are identified as the precursor minerals.

IOP Conf. Series: Earth and Environmental Science 1233 (2023) 012027 doi:10.1088/1755-1315/1233/1/012027

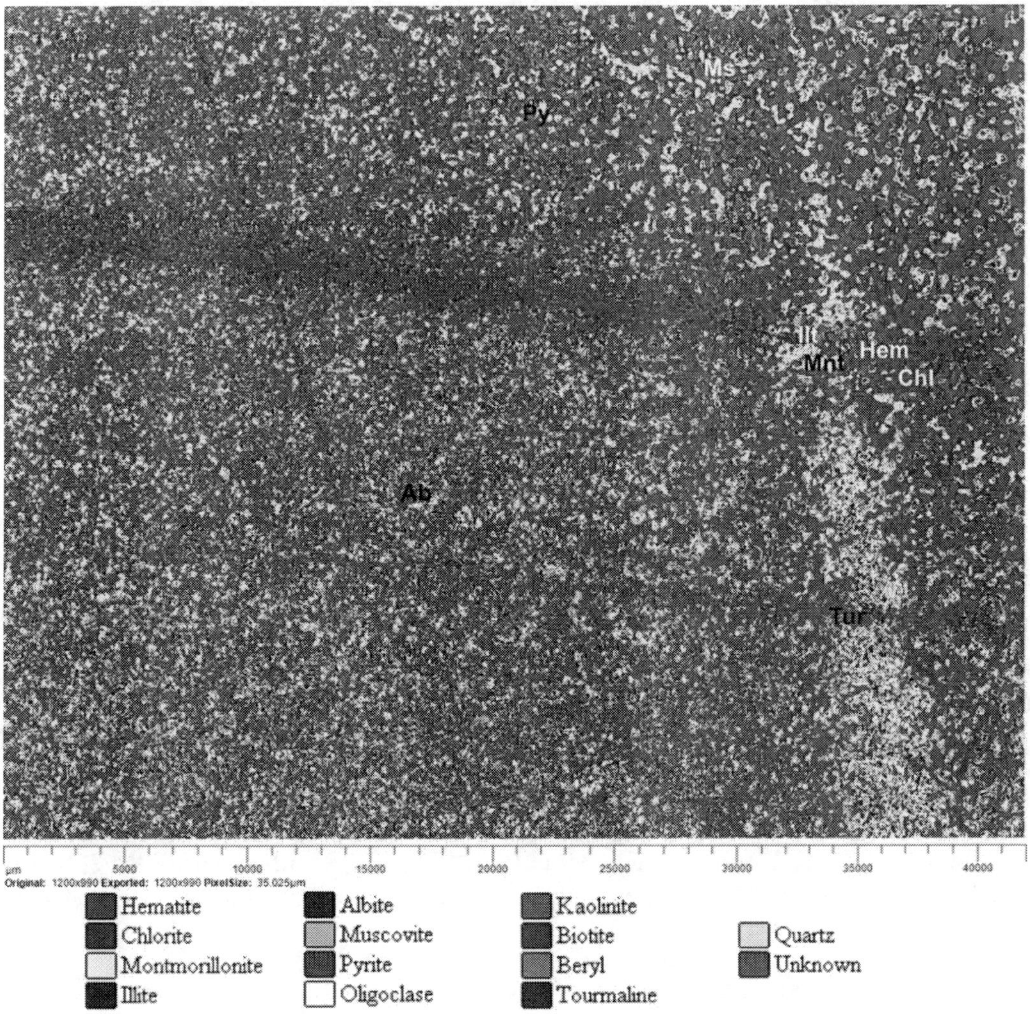

Figure 9. Mineral mapping of metasiltstone based on AMICS analysis. (Tur = Tourmaline, Mnt = Montmorillonite, Hem = Hematite, Ilt = Illite, Chl = Chlorite, Py = Pyrite, Ab = Albite, Ms = Muscovite. Mineral abbreviation based on IMA-CNMNC [18]).

Metasiltstone comprises 13 minerals, all of which are dominated by clay minerals, like montmorillonite, with an abundance of 23.44 Wt%. Other clay minerals include chlorite, illite, and kaolinite (16 Wt%, 14.23 Wt%, and 0.08 Wt%, respectively). However, iron oxides were identified as having a more abundant amount, represented by an abundance of hematite as much as 31.28 Wt%, caused by the iron oxides were the cement material that made up some minerals, also supported by the fact that metasiltstone has many hematite veins. The presence of hematite can also indicate a hydrothermal alteration process, accompanied by pyrite, tourmaline, and beryl (2.2 Wt%, 0.05 Wt%, 0.03 Wt%, respectively). To a lesser extent, there are minerals that indicate metamorphism processes, such as albite, muscovite, oligoclase, biotite, and quartz (9.2 Wt%, 1.79 Wt%, 0.94 Wt%, 0.64 Wt%, 0.01 Wt%, respectively). The abundance of minerals that make up metasiltstone can be seen in Table 2.

Table 2. Metasiltstone modal mineralogy based on AMICS analysis.

No.	Name	Wt%	Area%	Area (μ^2)	Segment Number	Relative Error
1	Hematite	31.28	26.28	379275717.5	309171	0
2	Montmorillonite	23.44	24.38	351975609.1	286917	0
3	Chlorite	16	20.24	292107725.1	238115	0
4	Illite	14.23	14.57	210268737.4	171403	0
5	Albite	9.2	8.98	129661407.3	105695	0.01
6	Pyrite	2.2	1.88	27171299.59	22149	0.01
7	Muscovite	1.79	1.71	24728839.1	20158	0.01
8	Oligoclase	0.94	1.03	14809333.54	12072	0.02
9	Biotite	0.64	0.58	8351718.25	6808	0.02
10	Unknown	0.09	0.16	2363948.45	1927	0.05
11	Kaolinite	0.08	0.08	1202215.61	980	0.06
12	Beryl	0.05	0.06	873446.44	712	0.07
13	Tourmaline	0.03	0.04	571665.79	466	0.09
14	Quartz	0.01	0.01	95686.55	78	0.23

Metapelite and metasiltstone texture and mineralogical characteristics define the protoliths of the two lithologies. The texture of these two rocks has a fine grain size and deformed mineral forms. Metasiltstone also still shows a laminated sedimentary structure. Indicating a fine-grained deformed protolith that originates from sedimentary rock [38]. The abundance of cordierite, muscovite, biotite, quartz, plagioclase, almandine, wonesite, staurolite, and andalusite in metapelite and metasiltstone indicates these minerals were formed in the greenschist – amphibolites facies [38]. The presence of quartz, together with oligoclase, albite, and anorthoclase, is also commonly found as a constituent of metapelite [39]. The presence of andalusite and muscovite in metapelite indicates that the metamorphism process probably occurred at 450 °C [39]. The main minerals of hydrothermal alteration products in metapelite and metasiltstone are chlorite, montmorillonite, illite, hematite, pyrite, kaolinite, tourmaline, and beryl.

The parallel orientation of mica minerals covered by clay minerals defines the textural characteristics of metapelite and metasiltstone, showing the earliest form of metamorphism processes. Meanwhile, the abundance of clay minerals oriented toward foliation indicates that the precursor minerals are affected by metamorphism processes in metapelite and metasiltstone. The Sepauk Tonalite Intrusion was identified as causing the metamorphism process, while the hydrothermal alteration process and the abundance of trace elements were known to originate from the Sukadana Granite Intrusion [3]. Detailed mineralogical and geochemical analysis of the Sukadana granite intrusion is needed to identify the hydrothermal alteration process and also the characteristics of U and REE-bearing magmas.

5. Conclusion

The mineralogical characteristics of metapelite and metasiltstone are characterized by the dominance of clay minerals such as wonesite, staurolite, montmorillonite, chlorite, illite, and kaolinite. The presence of pyrite, hematite, tourmaline and beryl are identified as hydrothermal alteration products. At the same time, the metamorphism process formed in the greenschist – amphibolites facies is shown by the abundance of cordierite, muscovite, biotite, quartz, plagioclase, almandine, wonesite, staurolite, and andalusite. The process of metamorphism is also manifested by the presence of foliation structure with a spotted shale texture. The Sepauk Tonalite Intrusion was identified as causing the metamorphism process. The presence of REE and trace elements associated with U indicates the occurrence of mineralization related to the hydrothermal alteration process coming from Sukadana

Granite Intrusion. REE, along with other trace elements suspected, occurs through ion-adsorption and is mainly adsorbed on the clay crystal surface.

6. Acknowledgment

The author would like to thank the Research Center for Nuclear Fuel Cycle and Radioactive Waste Technology, The National Research and Innovation Agency, which has provided the opportunity to collect and finance data processing. The authors also thank Universitas Trisakti, which provided the tools to prepare this paper.

References

[1] H. Syaeful, K. S. Widana, I. G. Sukadana, and A. G. Muhammad, "Rare Earth Element Exploration in Indonesia," in *Proceedings Of Sundaland Resources 2014 MGEI Annual Convention*, 2014, no. November, pp. 205–217.

[2] Suharji, Ngadenin, Wagiyanto, and Sumarno, "Peningkatan Kwalitas Estimasi Cadangan Uranium dan Unsur Tanah Jarang Sebagai Asosiasinya di Sektor Rirang Hulu, Kalimantan Barat," *Pros. Semin. IPTEK Nukl. dan Pengelolaan Sumber Daya Tambang*, pp. 51–65, 2002.

[3] S. Tjokrokardono, D. Soetarno, S. MS, S. Lilik, and R. Witjahyati, "Studi Geologi Regional dan Mineralisasi Uranium di Pegunungan Schwaner Kalimantan Barat dan tengah," *Semin. Geol. Nukl. dan Sumberd. Tambang Tahun 2004*, pp. 64–84, 2004, [Online]. Available: http://www.iaea.org/inis/collection/NCLCollectionStore/_Public/39/123/39123070.pdf.

[4] S. Tjokrokardono, S. Lilik, and M. Widodo, "Sintesis Geologi dan Mineralisasi Uranium Kalan dan Sekitarnya, Kalimantan Barat," Jakarta, 2006.

[5] S. Tjokrokardono, "Studi Provinsi Uranium Kalimantan: Kajian Mineralisasi Uranium pada batuan metamorf dan granit Di Pegunungan Schwaner," *Semin. Geol. Nukl. dan Sumberd. Tambang Tahun*, pp. 66–77, 2002.

[6] Ngadenin, A. Sumaryanto, H. Syaeful, and I. G. Sukadana, "Geologi dan Mineralisasi Uranium di Daerah Kalan, Kabupaten Melawi, Kalimantan Barat," in *Seminar Nasional Kebumian XII*, 2017, pp. 108–114.

[7] Sartapa and I. G. Sukadana, "Geology and uranium mineralization in Sarana Sector, Kalan, West Kalimantan based on drilling data," *Eksplorium*, vol. 32, no. 2, pp. 91–102, 2011.

[8] S. Tjokrokardono and A. S. Sastratenaya, "Rich mineralized boulders of the Rirang River, West Kalimantan," *Uranium Depos. Asia Pacific; Geol. Explor. Panel Proc. Ser. Int. At. Energy Agency. Uranium Depos. Asia Pacific; Geol. Explor. Jakarta, Dec. 16-19, 1985*, pp. 79–95, 1988.

[9] F. J. Dahlkamp, *Uranium Deposits of the World: Asia*. 53343 Wachtberg b. Bonn: Springer-Verlag Berlin Heidelberg, 2009.

[10] P. A. Polito, T. K. Kyser, and C. Stanley, "The proterozoic, albitite-hosted, Valhalla uranium deposit, Queensland, Australia: A description of the alteration assemblage associated with uranium mineralisation in diamond drill hole V39," *Miner. Depos.*, vol. 44, no. 1, pp. 11–40, 2009, doi: 10.1007/s00126-007-0162-2.

[11] R. G. Skirrow *et al.*, *Uranium Mineral Systems: Processes , exploration criteria and a new deposit framework*, no. January. 2009.

[12] P. Alexandre, K. Kyser, D. Layton-Matthews, B. Joy, and Y. Uvarova, "Chemical compositions of natural uraninite," *Can. Mineral.*, vol. 53, no. 4, pp. 595–622, 2015, doi: 10.3749/canmin.1500017.

[13] M. Cuney and K. Kyser, "Deposits related to Na metasomatism and high-grade metamorphism. Pada: Cuney M, Kyser TK (eds) Recent and not-so-recent developments in uranium deposits and impli- cations for exploration.," in *Mineralogical Association of Canada Short Course Series Volume 39*, vol. 39, no. May, Quebec: Mineralogical Association of Canada (MAC) and the Society for Geology Applied to Mineral Deposits (SGA), 2008, pp. 97–117.

[14] BATAN-CEA, "Prospect to Develop Uranium Deposits in Kalimantan Volume I and II, Introduction General Reconnaissance," Jakarta, 1977.
[15] PTBGN Badan Tenaga Nuklir Nasional, "Geologi & Potensi LTJ Rirang," 2019.
[16] D. Savira, M. A. Gunawan, W. A. Draniswari, I. G. Sukadana, and F. Sihombing, "µxRF Application for Uranium Exploration (Case Study: Mamuju Deposit, Indonesia)," in *IOP Conference Series: Earth and Environmental Science*, 2021, vol. 830, no. 1, doi: 10.1088/1755-1315/830/1/012076.
[17] I. G. Sukadana, I. W. Warmada, F. Pratiwi, A. Harijoko, and T. B. Adimedha, "Elemental Mapping for Characterizing of Thorium and Rare Earth Elements (REE) Bearing Minerals Using µXRF," *Atom Indones.*, vol. 48, 2022, doi: https://doi.org/10.17146/aij.2022.1215.
[18] L. N. Warr, "IMA–CNMNC approved mineral symbols," *Mineral. Mag.*, vol. 85, no. 3, pp. 291–320, 2021, doi: 10.1180/mgm.2021.43.
[19] J. W. Anthony, R. A. Bideaux, K. W. Bladh, and M. C. Nichols, *Handbook of Mineralogy*. Chantilly, VA 20151-1110, USA: Mineralogical Society of America, 2001.
[20] C. Wang, L. Zhang, Y. Dai, and C. Lan, "Geochronological and geochemical constraints on the origin of clastic meta-sedimentary rocks associated with the Yuanjiacun BIF from the Lüliang Complex, North China," *Lithos*, vol. 212–215, pp. 231–246, 2015, doi: 10.1016/j.lithos.2014.11.015.
[21] K. C. Condie, P. D. Jr. Noll, and C. M. Conway, "Geochemical and detrital mode evidence for two sources of Early Proterozoic sedimentary rocks from the Tonto Basin Supergroup, central Arizona-reply," *Sediment. Geol.*, vol. 77, pp. 51–76, 1992, doi: 10.1016/0037-0738(93)90007-R.
[22] A. Šoster *et al.*, "Geochemistry of Bashibos-Bajrambos metasedimentary unit, Serbo-Macedonian massif, North Macedonia: Implications for age, provenance and tectonic setting," *Geochemistry*, vol. 80, no. 4, 2020, doi: 10.1016/j.chemer.2020.125664.
[23] L. P. Gromet, R. F. Dymek, L. A. Haskin, and R. L. Korotev, "The 'North American shale composite': Its compilation, major and trace element characteristics," *Geochim. Cosmochim. Acta*, vol. 48, no. 12, pp. 2469–2482, 1984, doi: 10.1016/0016-7037(84)90298-9.
[24] Y. Xing *et al.*, "The role of fluorine in hydrothermal mobilization and transportation of Fe, U and REE and the formation of IOCG deposits," *Chem. Geol.*, vol. 504, 2018, doi: 10.1016/j.chemgeo.2018.11.008.
[25] N. Gaillard *et al.*, "Mica composition as a vector to gold mineralization: Deciphering hydrothermal and metamorphic effects in the Malartic district, Quebec," *Ore Geol. Rev.*, vol. 95, pp. 789–820, 2018, doi: 10.1016/j.oregeorev.2018.02.009.
[26] J. A. Plant, P. R. Simpson, B. Smith, and B. F. Windley, "Uranium ore deposits-products of the radioactive earth," *Uranium Mineral. Geochemistry, Environ.*, pp. 255–319, 2019, doi: 10.1515/9781501509193-011.
[27] E. M. Elias and Z. J. Al-Jubory, "Provenance and tectonic setting of the metapelites deposits in the Bulfat complex, NE-Iraq," *Arab. J. Geosci.*, vol. 7, no. 9, pp. 3589–3598, Sep. 2014, doi: 10.1007/s12517-013-1031-1.
[28] C. S. Cole, R. H. James, D. P. Connelly, and E. C. Hathorne, "Rare earth elements as indicators of hydrothermal processes within the East Scotia subduction zone system," *Geochim. Cosmochim. Acta*, vol. 140, no. 2014, pp. 20–38, 2014, doi: 10.1016/j.gca.2014.05.018.
[29] A. A. Migdisov and A. E. Williams-Jones, "Hydrothermal transport and deposition of the rare earth elements by fluorine-bearing aqueous liquids," *Miner. Depos.*, vol. 49, no. 8, pp. 987–997, 2014, doi: 10.1007/s00126-014-0554-z.
[30] M. M. Abu El-Enen, "Geochemistry, provenance, and metamorphic evolution of Gabal Samra Neoproterozoic metapelites, Sinai, Egypt," *J. African Earth Sci.*, vol. 59, no. 2–3, pp. 269–282, Feb. 2011, doi: 10.1016/j.jafrearsci.2010.11.002.
[31] I. G. Sukadana and H. Syaeful, "Implication of Hydrothermal Process on Radioactive and Rare Earth Element Enrichment in Mamuju, West Sulawesi.," 2017.

[32] I. G. Sukadana and H. Syaeful, "Evaluasi Sistem Pengendapan Uranium Pada Batuan Sedimen Formasi Sibolga, Tapanuli Tengah," *Eksplorium*, vol. 37, no. 2, pp. 125–138, 2016.
[33] S. Akhtar, X. Yang, and F. Pirajno, "Sandstone type uranium deposits in the Ordos Basin, Northwest China: A case study and an overview," *J. Asian Earth Sci.*, vol. 146, no. May, pp. 367–382, 2017, doi: 10.1016/j.jseaes.2017.05.028.
[34] Y. Xing, Y. Mei, B. Etschmann, W. Liu, and J. Brugger, "Uranium Transport in F-Cl-Bearing Fluids and Hydrothermal Upgrading of U-Cu Ores in IOCG Deposits," *Geofluids*, vol. 2018, 2018, doi: 10.1155/2018/6835346.
[35] Y. D. Gotman and I. N. Zubrev, "Genetic classification of uranium deposits," *Int. Geol. Rev.*, vol. 7, no. 3, pp. 518–525, 1965, doi: 10.1080/00206816509474136.
[36] A. Dyer, *Ion exchange*. Salford: Elsevier, 2013.
[37] Z. M. Yusoff, B. T. Ngwenya, and I. Parsons, "Mobility and fractionation of REEs during deep weathering of geochemically contrasting granites in a tropical setting , Malaysia," *Chem. Geol.*, vol. 349–350, pp. 71–86, 2013, doi: 10.1016/j.chemgeo.2013.04.016.
[38] I. I. Likhanov, V. V. Reverdatto, and A. E. Vershinin, "Fe- and Al-Rich metapelites of the Teiskaya Group, Yenisei Range: Geochemistry, protoliths, and the behavior of their material during metamorphism," *Geochemistry Int.*, vol. 46, no. 1, pp. 17–36, 2008, doi: 10.1134/s0016702908010023.
[39] L. Zeng, J. B. Saleeby, and M. Ducea, "Geochemical characteristics of crustal anatexis during the formation of migmatite at the Southern Sierra Nevada, California," *Contrib. to Mineral. Petrol.*, vol. 150, no. 4, pp. 386–402, 2005, doi: 10.1007/s00410-005-0010-2.

IOP Conf. Series: Earth and Environmental Science 1233 (2023) 012028 doi:10.1088/1755-1315/1233/1/012028

Social Vulnerability Analysis to Disaster COVID-19 in Central Java

N Qomariyah[1], E T W Mei[1]

[1]Regional Development Study Program, Faculty of Geography, Universitas Gadjah Mada, Yogyakarta 55281, Indonesia

Abstract. The COVID-19 disaster has socio-economic impacts in various regions, including Central Java. COVID-19 has the potential to exacerbate social vulnerability due to the limited internal conditions of individuals to survive. This study tries to analyze social vulnerability based on indicators of individual conditions demographically, socio-economically, and the ability to access service facilities which can be exacerbated by the potential danger of COVID-19. This study aims to analyze the level of social vulnerability and the spatial and temporal distribution of the COVID-19 disaster in Central Java. This study uses a quantitative analysis method by calculating Principle Component Analysis (PCA) and Social Vulnerability Index (SoVI) to identify the level of social vulnerability in Central Java. Analysis of the level of the COVID-19 disaster uses cumulative positive case data which is analyzed spatially and temporally using ArcGIS. This study shows that the higher the SoVI value in a region, the higher the level of individual social vulnerability in that region. The results showed that the highest level of social vulnerability to the COVID-19 disaster in Central Java was the cities of Semarang, Cilacap and Banyumas which were influenced by the conditions of individual vulnerability and the number of positive cases of COVID-19.

1. Introduction

Coronavirus Disease 2019 (COVID-19) is an acute respiratory infection disease caused by the Corona virus or Severe Acute Respiratory Syndrome Corona Virus 2 (SARS-CoV-2) [1]. COVID-19 is a disaster that threatens people's lives and livelihoods [2] which causes fatalities and socio-economic impacts for the community. Indonesia is one of the countries affected by COVID-19. Based on data released by the Ministry of Health of the Republic of Indonesia on July 31, 2021, the number of positive cases of COVID-19 in Indonesia reached 3.409.658 people, which is the peak of the second wave [3]. Central Java is one of the provinces in Indonesia that has the 3rd highest number of positive COVID-19 cases in July 2021 reaching 522.589 people [4]. The COVID-19 disaster had an impact on the socio-economic sector in Central Java, it increase in the percentage of the poor by 1.26% to 11.84% compared to 2019 [5]. The COVID-19 disaster has the potential to exacerbate social vulnerabilities.

The condition of individual social vulnerability can be exacerbated by the potential danger due to COVID-19. COVID-19 is categorized as a disaster because it causes casualties and socio-economic impacts on society [2][6]. COVID-19 disaster events can affect and potentially exacerbate individual social vulnerability conditions in an area [2] including those that occurred in Central Java. This research attempts to analyze social vulnerability and the COVID-19 disaster in Central Java to describe the conditions of individual vulnerability in each region in order to see the ability to survive against the impact that can occur due to the severity of the COVID-19 disaster [6]. Social vulnerability analysis in this study focused on quantitatively assessing social vulnerability based on PCA (Principle Component Analysis) and SoVI (Social Vulnerability Index) calculations. The results of the calculation of the social vulnerability index are used as a basis for spatial analysis to describe the condition of individual social vulnerability in the regional context of each district/city in Central Java.

Social vulnerability is a system that cannot be separated from the internal social characteristics of the community [7] related to individual conditions that affect access limitations, causing potential vulnerability and severity to the impact of hazards and disaster events [8]. Social vulnerability analysis includes a study of vulnerability as a potential impact, damage or loss with a scale depending on the social, economic and environmental conditions of a population during a particular hazard [9].

Vulnerability analysis is important to determine the ability of individuals or communities in an area to change pressures that occur associated with social conditions and location [10]. The condition of social vulnerability can be influenced by several factors, for example: community characteristics, economic conditions, social status, gender, access to health facilities, housing quality and community response to disasters [6][7][10][11][12]. Social vulnerability analysis is not only limited to assessing individual vulnerability conditions but is also able to represent regionally related social vulnerability conditions based on individual community vulnerability conditions in each region [2]. Analysis of social vulnerability based on individual objects in a regional context becomes interesting and relevant to be studied in this study to show the spatial distribution of social vulnerability conditions in each region is different.

This research aims to; 1) analyzing the level of regional social vulnerability, 2) analyzing the spatial and temporal distribution of the COVID-19 disaster and 3) analyzing the spatial distribution of the regional social vulnerability level to the COVID-19 disaster in Central Java. This study is useful for analyzing regional social vulnerability which is seen based on individual conditions in the spatial aggregate, namely the district administration area. The analysis of the spatial level of social vulnerability used in this study is new and interesting to identify the vulnerability conditions of each region that are associated with the occurrence of the COVID-19 disaster in Central Java.

Social vulnerability analysis becomes relevant to be studied to identify the spatial distribution of social vulnerability levels in areas affected by the COVID-19 disaster. This research topic is quite relevant and interesting to study at this time because it tries to combine analysis related to social vulnerability and COVID-19 disaster analysis. The condition of social vulnerability and the occurrence of the COVID-19 disaster may exacerbate and affect the ability to survive the impact of the COVID-19 disaster that occurred in Central Java.

So this study tries to identify and analyze the condition of individual social vulnerability in the regional or regional context in Central Java which may be exacerbated by the occurrence of the COVID-19 disaster. The spatial analysis used in this study is to show the distribution of social vulnerability and COVID-19 disasters in each district/city in Central Java. The spatial distribution analysis is to describe the condition of vulnerability seen from the social vulnerability of individuals and the impact of the COVID-19 disaster. It is hoped that the results of this study can be used as a basis for research development and considerations in the preparation of regional development strategies based on disaster reduction based on aspects of social vulnerability and the COVID-19 disaster in Central Java.

2. Method

The scope of this research location covers 35 districts in Central Java. This study uses quantitative analysis methods with the calculation of Principle Component Analysis (PCA) and Social Vulnerability Index (SoVI) to identify the level of regional social vulnerability in Central Java Province. The choice of PCA and SoVI methods in this study refers to the previous research [2]. PCA method in this study serves to significantly reduce variables including social vulnerability factors in Central Java Province. While the use of the SoVI method serves to calculate the value of social vulnerability quantitatively

IOP Conf. Series: Earth and Environmental Science 1233 (2023) 012028 doi:10.1088/1755-1315/1233/1/012028

which will later be used as the basis for classifying the level of individual social vulnerability in each district/city in Central Java Province.

The social vulnerability analysis in this study is based on the concept of social vulnerability by considering the individual's social condition including demographic, socio-economic conditions, and the ability to access service facilities. The analysis of social vulnerability in this study uses variables that refer to previous research so that it is expected to be able to describe the condition of individual social vulnerability. Identification of social vulnerability factors in Central Java using 27 variables (Table 1) that are expected to be able to represent to explain the condition of social vulnerability.

The use of these 27 variables is based on several previous studies related to social vulnerability so that it is expected to be able to show the conditions of social vulnerability in Central Java. The data used in this social vulnerability analysis is statistics data Central Java in 2021. The analysis of social vulnerability in this study uses weighting based on the results of the calculation of factor components using the PCA method [13]. Social Vulnerability Index (SoVI) used the total score component factors (f_i), weight (w_i) which is the percent (%) of the total variance of each factor component from the PCA calculation results and the number of component factors formed (n) [13].
The formula of Social Vulnerability Index (SoVI) :

$$SoVI = \sum_{1}^{n} fiwi$$

The results of the calculation of the value of the Social Vulnerability Index (SoVI) in each district/city are classified and mapped using GIS method. The results of the analysis of the social vulnerability index are classified into 5 (five) categories, namely very low social vulnerability, low social vulnerability, moderate social vulnerability, high social vulnerability, and very high social vulnerability. Meanwhile, the disaster analysis in this study focused on the level of the COVID-19 disaster as seen from the number of positive cases. The COVID-19 data used in this study is the number of positive cases in March 2020 - July 2021 for each district/city in Central Java. The level of social vulnerability and COVID-19 disaster analyzed with overlapping maps to identify the typology of vulnerability in each region.

Table 1. Details of Social Vulnerability Variables

Variables
Total population
Number of female population
Number of households
Number of health centers
Number of population receiving health insurance
Unemployment number
Number of poor families
Percentage of home ownership
Percentage of population receiving health insurance
Number of auxiliary health centers
Percentage of poor people
Percentage of households with easy access to drinking water sources
Unemployment Rate
Percentage of National Health Insurance
Population density
Number of medical personnel

Variables
Number of polyclinics / treatment centers
Number of hospitals
Sex Ratio
Regency Minimum Salary
Percentage of households having access to private toilets
Poverty depth index
Poverty severity index
Percentage of households with easy access to sanitation
Average number of households
Average income of informal workers
Total population

3. Result and Discussion

3.1. Analysis of Social Vulnerability Levels

The condition of regional social vulnerability referred to in this study is the condition of individual social vulnerability which is represented in the aggregate of regional space [2][13]. The analysis of social vulnerability in this study tries to describe the condition of social vulnerability with individual objects in the area that is analyzed and compared based on the administrative area so that it is expected to be able to represent the condition of social vulnerability of each district / city in Central Java. The social vulnerability analysis in this study is focused on describing the individual social vulnerability conditions of each district/city in Central Java spatially and displayed in the form of a map. Social vulnerability analysis in this study uses a quantitative calculation method based on the social vulnerability index or SoVI (Social Vulnerability Index) which is classified into 5 categories of social vulnerability, namely very low, low, medium, high, and very high.

The factors that influence the condition of individual social vulnerability are quite varied and differ from region to region [2][13][14]. This condition is due to each region having different characteristics as well as diverse communities. The determinants of social vulnerability are closely related to the social conditions of the community itself, for example related to demographics and socio-economics. Although on the other hand social vulnerability is also influenced by environmental and disaster aspects. Vulnerable individuals have the potential for more severe disaster impacts [11].

The results of Principle Component Analysis based on Table 2 explained the social vulnerability of Central Java by 82.301% total cumulative variance. The results of PCA calculations in this study were used to identify the components of social vulnerability factors in Central Java. These factor components include: Social and Infrastructure Access (factor component 1), Population and Health Access (factor component 2), Welfare (factor component 3), House Conditions (factor component 4), and Economy (factor component 5). The naming of each component of these factors is adjusted based on the details of the constituent factors of each variable resulting from the PCA analysis.

Table 2. PCA Calculation Results

Component	Initial Eigenvalues			Extraction Sums of Squared Loadings			Rotation Sums of Squared Loadings		
	Total	% of Variance	Cumulative %	Total	% of Variance	Cumulative %	Total	% of Variance	Cumulative %
1	8.864	32.829	32.829	8.864	32.829	32.829	7.788	28.846	28.846
2	5.595	20.724	53.553	5.595	20.724	53.553	5.375	19.908	48.755
3	3.974	14.718	68.271	3.974	14.718	68.271	3.826	14.171	62.925

Component	Initial Eigenvalues			Extraction Sums of Squared Loadings			Rotation Sums of Squared Loadings		
	Total	% of Variance	Cumulative %	Total	% of Variance	Cumulative %	Total	% of Variance	Cumulative %
4	2.304	8.533	76.804	2.304	8.533	76.804	2.922	10.824	73.749
5	1.484	5.497	82.301	1.484	5.497	82.301	2.309	8.551	82.301

The percentage value of each factor component in Table 3 is used as the basis for weighting in the calculation of the social vulnerability index. The basis for selecting the weighting method used in this study refers to the research of Sarkar and Chouhan (2020) which also analyzes social vulnerability to the COVID-19 disaster based on PCA calculations. The weight value used in this study is purely determined based on the percent (%) of the total variance of each factor component from the results of the PCA calculation.

Table 3. Central Java Social Vulnerability Factor Analysis Results

Factor Analysis Results		**Percentage**
Component Factor 1 (Social and Infrastructure Access)	Total population Number of female population Number of households Number of health centers Number of population receiving health insurance Unemployment number Number of poor families Percentage of home ownership Percentage of population receiving health insurance Number of auxiliary health centers Percentage of poor people Percentage of households with easy access to drinking water sources	32.83%
Component Factor 2 (Population and Health Access)	Unemployment Rate Percentage of National Health Insurance Population density Number of medical personnel Number of polyclinics / treatment centers Number of hospitals *Sex Ratio* Regency Minimum Salary	20.72%
Component Factor 3 (Welfare)	Percentage of households having access to private toilets Poverty depth index Poverty severity index	14.72%
Component Factor 4 (House Condition)	Percentage of households with easy access to sanitation Average number of households	8.53%
Component Factor 5 (Economy)	Average income of informal workers	5.50%

The results of the calculation of the social vulnerability index show that each district in Central Java has varying social vulnerability values, for example: Brebes Regency is the area with the highest social vulnerability index value in Central Java at 274.84 while the area with the lowest value is Salatiga City at -238, 91. The higher the SoVI value, the higher the social vulnerability, while the lower the SoVI value, the lower the social vulnerability [2].

Table 4. Classification of Social Vulnerability Level in Central Java

Districts / City	Score SoVI	Social Vulnerability Level
Cilacap	194.27	Very high
Banyumas	182.16	Very high
Purbalingga	11.74	Moderate
Banjarnegara	7.61	Moderate
Kebumen	90.47	High
Purworejo	-112.45	Low
Wonosobo	-48.73	Moderate
Magelang	8.63	Moderate
Boyolali	-17.27	Moderate
Klaten	64.97	High
Sukoharjo	-36.35	Moderate
Wonogiri	-35.07	Moderate
Karanganyar	-47.50	Moderate
Sragen	-17.48	Moderate
Grobogan	57.84	High
Blora	-71.74	Low
Rembang	-88.36	Low
Pati	27.01	Moderate
Kudus	-25.28	Moderate
Jepara	28.92	Moderate
Demak	86.50	High
Semarang	-40.76	Moderate
Temanggung	-110.45	Low
Kendal	12.70	Moderate
Batang	-45.21	Moderate
Pekalongan	7.77	Moderate
Pemalang	142.06	High
Tegal	130.08	High
Brebes	274.84	Very high
Magelang City	-217.80	Very low
Surakarta City	-70.05	Low
Salatiga City	-238.91	Very low
Semarang City	202.16	Very high
Pekalongan City	-148.54	Low
Tegal City	-157.81	Low

The results of the calculation of the SoVI value in table 4 were used as the basis for analyzing the social vulnerability of each district/city in Central Java. The use of the SoVI index is able to facilitate the analysis and identification of the level of regional social vulnerability quantitatively. The SoVI value is proportional to the condition of social vulnerability, the higher the SoVI value, the higher the level of social vulnerability in an area [13]. Based on the results of the calculation of the social vulnerability index (Table 4), it shows that each district/city in Central Java has varying values of social vulnerability. Brebes Regency is an area that has the highest social vulnerability index value in Central Java, which is 274.84 while the area with the lowest value is Salatiga City at -238.91. The higher the SoVI value, the higher the social vulnerability, while the lower the SoVI value, the lower the social vulnerability in each district/city in Central Java.

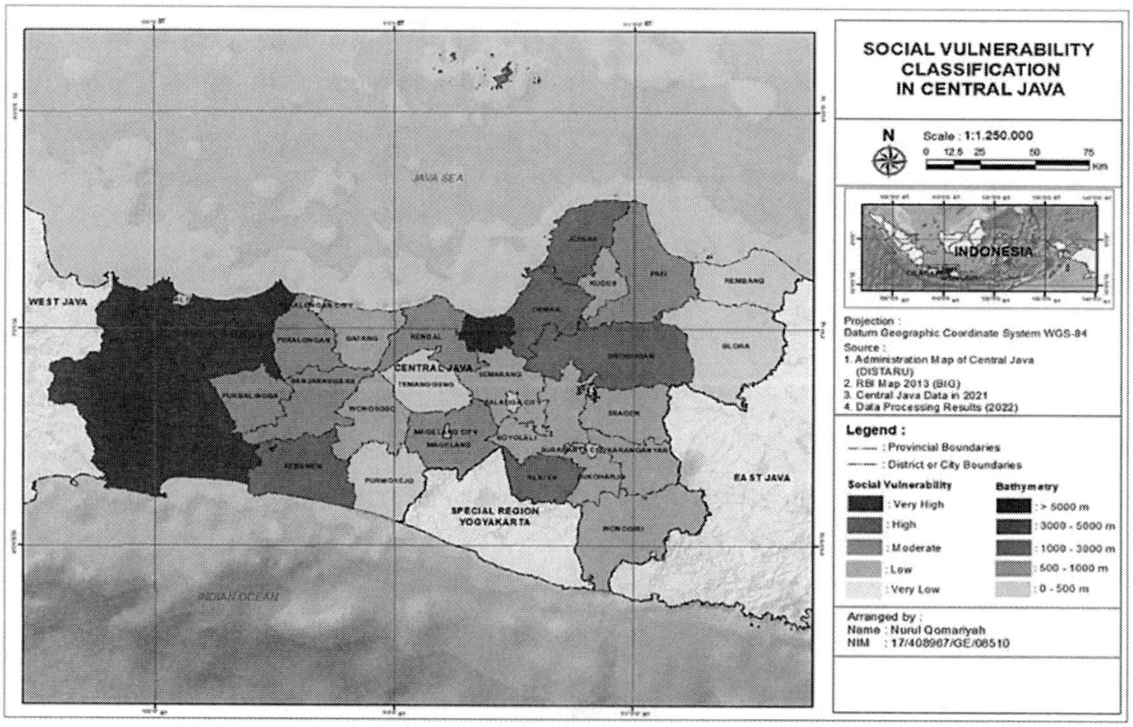

Figure 1. Social Vulnerability Classification in Central Java

Central Java has varying conditions of social vulnerability (Figure 1). Based on the results of the classification of social vulnerability in Table 3, Central Java is dominated by moderate social vulnerability conditions of around 46% or as many as 16 districts. Variations in social vulnerability in Central Java are influenced by social conditions, access to infrastructure, population, access to health, welfare, housing conditions, and the economy. Areas with very high social vulnerability conditions in Central Java are dominated by suburban areas with a high female population, namely Brebes Regency, Cilacap Regency, and Banyumas Regency. In addition, there is one urban area that has a very high social vulnerability, namely the city of Semarang. The high social vulnerability in the four areas is influenced by the population, the number of female residents, the number of households, and the highest number of unemployed in Central Java.

3.2. Analysis of the Spatial and Temporal Distribution of COVID-19 Disaster

The analysis of the COVID-19 disaster in this study used data on the number of positive cases in every 35 districts in Central Java. The higher the number of positive cases, the higher the potential danger and

IOP Conf. Series: Earth and Environmental Science 1233 (2023) 012028 doi:10.1088/1755-1315/1233/1/012028

disaster level of COVID-19 [6]. Data on the number of positive cases were classified into 5 categories (very low, low, moderate, high, very high). Based on the results of processing existing COVID-19 data, it shows that areas with high levels of COVID-19 disasters in Central Java tend to be concentrated in urban areas with high populations, such as Semarang City. The city of Semarang is an area with a high COVID-19 disaster rate when viewed from the number of positive cases and deaths. The city of Semarang is the area with the highest total positive cases and deaths from COVID-19 in Central Java.

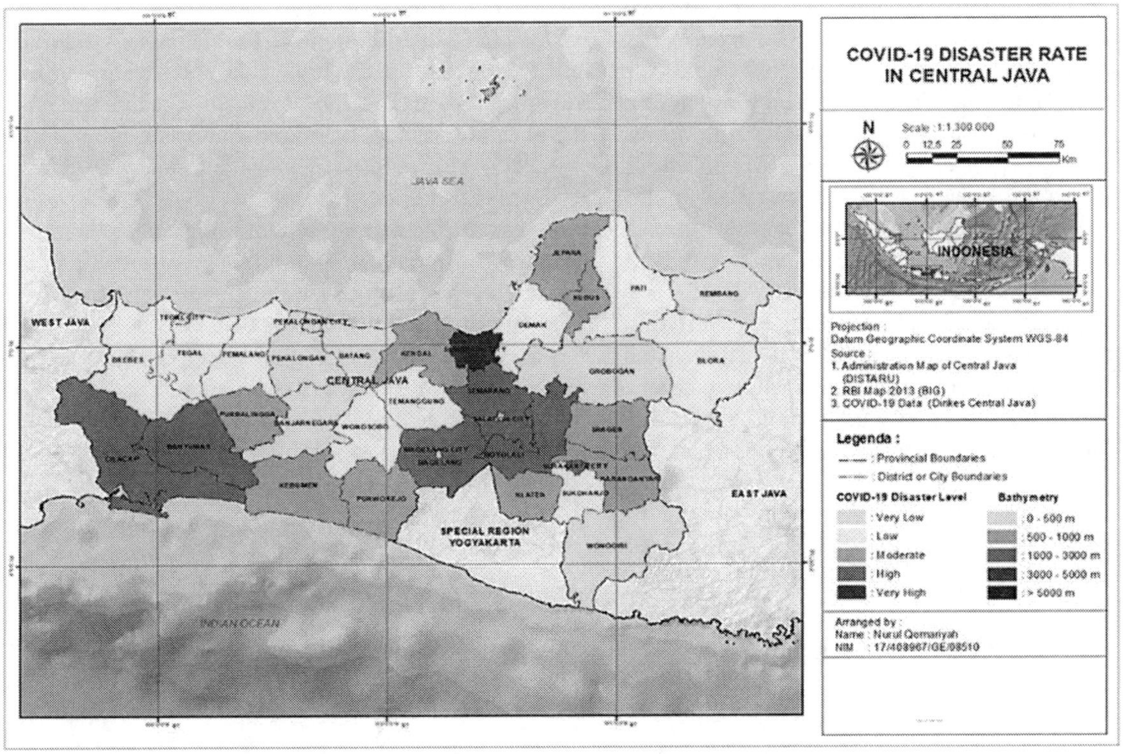

Figure 2. COVID-19 Disaster Rate in Central Java

The high COVID-19 disaster rate in Central Java tends to be concentrated in urban and suburban areas with a high population and population density. Based on the results of data processing, there are 6 regions with high COVID-19 disaster levels, including: Cilacap Regency, Banyumas Regency, Magelang Regency, Boyolali Regency, Semarang Regency, and Surakarta City. The spatial distribution of the highest level of COVID-19 disaster in Central Java is concentrated in urban areas with a high population, namely Semarang City (Figure 2). Areas with a high population and population density cause the potential danger of transmission and the severity of the COVID-19 disaster is high [13]. The city of Semarang is included in the category of very high COVID-19 disaster level because the region has the highest total positive cases reaching 80,646 people or around 15.4% of the total positive cases of Central Java in July 2021. This figure shows the level of COVID-19 disaster in Indonesia. The city of Semarang is very high.

3.3. Analysis of Social Vulnerability Levels to COVID-19 Disasters

The condition of social vulnerability in Central Java is different. The variation in the level of social vulnerability of each district/city in Central Java is influenced by the social characteristics of the community. On the other hand, social vulnerability is not only influenced by the social conditions of the community but can be exacerbated by the occurrence of disasters so that it can increase regional vulnerability [15]. Disaster events that occur in an area can worsen the existing conditions of social vulnerability. So the analysis of social vulnerability is often associated with disaster events and potential hazards [2][6][15]. Based on the results of the calculation of the social vulnerability index, it shows that each district/city in Central Java has varying values of social vulnerability. The condition of social vulnerability in Central Java is not only exacerbated by the socio-economic conditions of the community but also the COVID-19 disaster. COVID-19 is an acute respiratory infection disease caused by the Corona virus [1] which has the potential to threaten people's lives and livelihoods [2].

Conditions of social vulnerability have a significant relationship with the number of positive cases of COVID-19 in Central Java (Table 5). These conditions indicate that the higher the number of positive cases of COVID-19, the greater the social vulnerability in Central Java. Regencies/cities in Central Java that have a high number of positive cases of COVID-19 have the potential for higher social vulnerability when compared to areas with low social vulnerability. Conversely, the higher the social vulnerability of an area, the higher the potential for danger and transmission of COVID-19, thus increasing the number of positive cases of COVID-19. Conditions of social vulnerability in an area can affect the number of positive cases of COVID-19 [16]. Based on the results of this study, it shows that conditions of social vulnerability have a significant effect on the COVID-19 disaster.

Based on the calculation results of the correlation test with Pearson Correlation in Table 5 shows that the significance value is 0.005 or sig. <0.05 so that there is a statistically significant relationship between social vulnerability and the COVID-19 disaster. Based on the results of the interpretation, it shows that the higher the level of regional social vulnerability, the higher the number of COVID-19 cases. The regional severity of the COVID-19 disaster in Central Java is shown by the large number of positive cases. The high number of positive cases indicates a high rate of transmission and spread of COVID-19. So that areas with high social vulnerability conditions have a high potential for COVID-19 danger, for example, Semarang City is an area with social vulnerability conditions and a very high level of COVID-19 disaster.

Table 5. Classification of Social Vulnerability Level in Central Java

		SoVI	COVID_July2021
SoVI	Pearson Correlation	1	.465**
	Sig. (2-tailed)		.005
	Sum of Squares and Cross-products	456040.637	456.579
	Covariance	13412.960	13.429
	N	35	35
COVID_July2021	Pearson Correlation	.465**	1
	Sig. (2-tailed)	.005	
	Sum of Squares and Cross-products	456.579	2.114
	Covariance	13.429	.062
	N	35	35

***. Correlation is significant at the 0.01 level (2-tailed).*

Based on the results of the relationship analysis in this study, it shows that the higher the level of social vulnerability in a region, the higher the number of COVID-19 cases. A high number of positive cases indicates a high rate of transmission and spread of COVID-19. So that areas with conditions of high social vulnerability have a high number of positive cases of COVID-19, for example Semarang City. Semarang City is a region with conditions of social vulnerability and a very high level of COVID-19 disaster in Central Java (Table 4 and Figure 2). Based on the results of this relationship analysis, it shows that COVID-19 can be included as one of the variables that has the potential to exacerbate the condition of individual social vulnerability in a region. However, COVID-19 cannot be permanent to be used as a basic indicator for analyzing conditions of social vulnerability because COVID-19 is only a temporary disaster event and has the potential to exacerbate individual social vulnerability, especially in facing and surviving the impacts.

Typological analysis of regional vulnerability in this study is divided into four quadrants which function to group districts/cities in Central Java based on the level of regional vulnerability and the condition of the COVID-19 disaster [16]. The grouping aims to identify regencies/cities based on the condition of the community's social vulnerability as well as the severity of the COVID-19 disaster. Regions with a high number of positive cases have a higher COVID-19 disaster rate. Determination of the limits of each quadrant in this study is based on the median value between the social vulnerability score and the COVID-19 disaster score which is a value of 50. The limits of each quadrant in the typology analysis can be determined based on the average value of two variables, namely social vulnerability and the number of positive cases COVID-19 [16]. Typological analysis in this study is displayed on a chart which is divided into four quadrants as shown in Figure 3.

Based on the results of the typological analysis, it shows that the conditions of social vulnerability and COVID-19 disasters in Central Java (Table 6). Based on the typology chart of regional vulnerabilities (Figure 3), it shows that the conditions of social vulnerability and COVID-19 disasters from 35 districts in Central Java have an even distribution and vary in each quadrant. Regions with high social vulnerability scores tend to fall into quadrant I and III categories. Meanwhile, areas with low vulnerability tend to fall into quadrant II and IV categories. Based on the results of the typological analysis in Table 5. The area that belongs to quadrant I has social vulnerability conditions and the number of positive COVID-19 cases is high, namely the City of Semarang. This condition shows that Semarang City is not only developing as an urban area but on the other hand has a high social vulnerability condition as well as a high risk of COVID-19 transmission when viewed from the large number of positive cases.

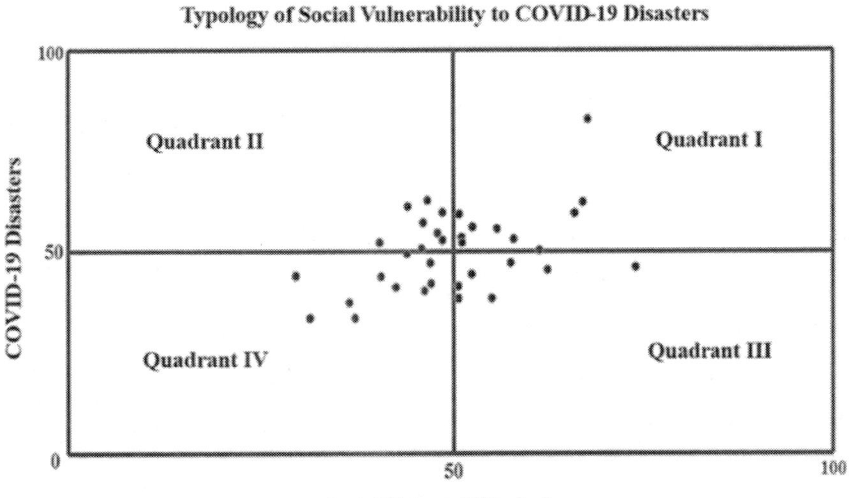

Figure 3. Typology of Social Vulnerability to COVID-19 Disasters in Central Java

Analysis of the level of social vulnerability and the level of the COVID-19 disaster using calculations based on the scores in Table 7 and Table 8. The higher the level of social vulnerability and COVID-19 disaster, the higher the score for the region [17]. Regions with the highest conditions of social vulnerability and COVID-19 disaster had the highest score of 50. The classification of regional vulnerability in this study was determined based on the total score classified into 5 (five) regional vulnerability classes, including: very low, low, medium, high, and very high vulnerability (Table 9). Regions with the highest social vulnerability conditions and the level of COVID-19 disaster have the highest score of 100. The higher the social vulnerability and the level of the COVID-19 disaster, the higher the regional vulnerability score. The total score of social vulnerability to the COVID-19 disaster in Central Java is shown in Table 9.

Table 6. Regional Typology Based on Social Vulnerability to COVID-19 Disasters in Central Java

Category	Districts / City
Quadrant I (Social vulnerability and severity of the COVID-19 disaster are high)	Semarang City, Cilacap, Banyumas, Purbalingga, Kebumen, Magelang, Klaten, Jepara, Kendal, Tegal
Quadrant II (Low social vulnerability and high severity of the COVID-19 disaster)	Purworejo, Wonosobo, Boyolali, Karanganyar, Sragen, Kudus, Semarang, Surakarta City
Quadrant III (High social vulnerability and low COVID-19 disaster severity)	Banjarnegara, Grobogan, Pati, Demak, Pekalongan, Pemalang, Brebes,
Quadrant IV (Social vulnerability and severity of the COVID-19 disaster are low)	Sukoharjo, Wonogiri, Blora, Kabupaten Rembang, Temanggung, Batang, Magelang City, Salatiga City, Pekalongan City, Tegal City

Analysis of the spatial distribution of the level of social vulnerability and the COVID-19 disaster in this study is to describe the condition of regional vulnerability in terms of individual social vulnerability and the impact of the COVID-19 disaster. Spatial analysis of the level of social and disaster vulnerability of COVID-19 uses scoring methods and overlay analysis to group and classify areas based on individual vulnerability conditions as well as the severity of the COVID-19 disaster from data on the number of positive cases. The results of the spatial analysis of the level of social vulnerability and the COVID-19 disaster in this study are displayed in the form of a map so that they can show the distribution of conditions for each district/city in Central Java.

The process of analyzing the vulnerability of the Central Java region uses the results of the classification of social vulnerability and the classification of the level of the COVID-19 disaster combined with the scoring method. The scoring method was chosen to make it easier to identify regional vulnerabilities according to social vulnerability and the level of the COVID-19 disaster. The use of the scoring method in the analysis of the level of regional vulnerability associated with aspects of hazard and disaster events has been carried out by previous research [17]. The scoring in this study is based on the level of social vulnerability and the level of the COVID-19 disaster. The higher the level of social and disaster vulnerability of COVID-19, the higher the score of the region [17]. Regions with the highest conditions of social vulnerability and disasters of COVID-19 have the highest score of 50. Details of social vulnerability and disasters of COVID-19 are shown in Table 7 and Table 8.

Table 7. Social Vulnerability Level Score in Central Java

Social Vulnerability Level	Score
Very low	10
Low	20
Moderate	30
High	40
Very High	50

IOP Conf. Series: Earth and Environmental Science 1233 (2023) 012028 doi:10.1088/1755-1315/1233/1/012028

Table 8. COVID-19 Disaster Level Score in Central Java

COVID-19 Disaster Level	Score
Very low	10
Low	20
Moderate	30
High	40
Very High	50

Table 9. Regional Vulnerability Classification based on the Level of Social Vulnerability and COVID-19 Disaster

Score	Category
0 - 20	Very low
20 - 40	Low
40 - 60	Moderate
60 - 80	High
80 - 100	Very High

Each region in Central Java has a varying level of COVID-19 disaster. The difference in the level of the COVID-19 disaster for each district/city in Central Java is affected by the large number of positive cases of COVID-19. The higher the number of positive cases, the higher the potential danger of transmission and the severity of the COVID-19 disaster in Central Java. Based on the scoring results, it shows that each district/city in Central Java has a different score for the level of the COVID-19 disaster. The following is the result of scoring the level of the COVID-19 disaster for each district/city in Central Java which is shown in Table 10.

The areas with the highest social vulnerability conditions to the COVID-19 disaster in Central Java are Cilacap Regency, Banyumas Regency, and Semarang City. These four areas have the highest regional vulnerability due to social vulnerability conditions and a higher level of COVID disaster compared to other areas in Central Java. These four areas have a high potential for vulnerability and severity to the COVID-19 disaster. Meanwhile, the area in Central Java with the lowest level of vulnerability is Magelang City because it has low social vulnerability conditions and the COVID-19 disaster.

Table 10. Classification Regional Vulnerability in Central Java

Districts / City	Score	Category
Cilacap	90	Very High
Banyumas	90	Very High
Purbalingga	60	Moderate
Banjarnegara	40	Low
Kebumen	70	High
Purworejo	50	Moderate
Wonosobo	50	Moderate
Magelang	70	High
Boyolali	70	High
Klaten	70	High
Sukoharjo	50	Moderate
Wonogiri	40	Low

Districts / City	Score	Category
Karanganyar	60	Moderate
Sragen	60	Moderate
Grobogan	50	Moderate
Blora	40	Low
Rembang	30	Low
Pati	50	Moderate
Kudus	60	Moderate
Jepara	60	Moderate
Demak	60	Moderate
Semarang	70	High
Temanggung	40	Low
Kendal	60	Moderate
Batang	40	Low
Pekalongan	40	Low
Pemalang	60	Moderate
Tegal	60	Moderate
Brebes	70	High
Magelang City	20	Very low
Surakarta City	60	Moderate
Salatiga City	30	Low
Semarang City	100	Very High
Pekalongan City	30	Low
Tegal City	30	Low

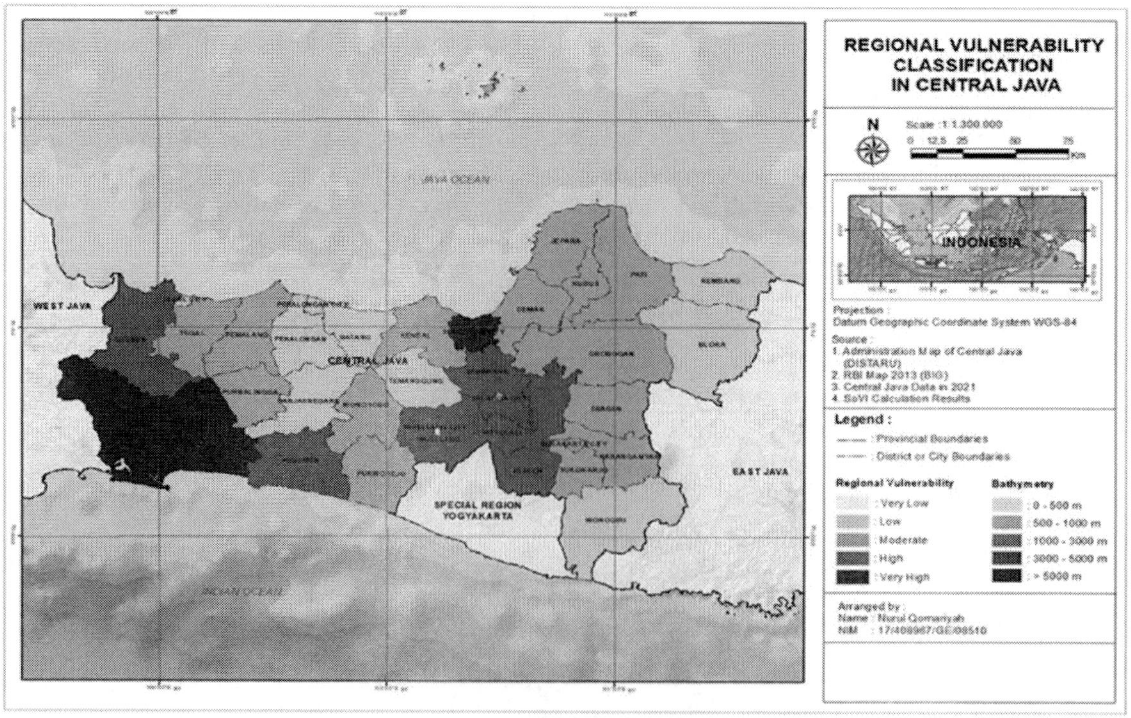

Figure 4. Regional Vulnerability Classification in Central Java

Central Java has various regional vulnerability conditions. Based on the results of the classification of the level of social vulnerability to the COVID-19 disaster shown in Table 10, it shows that Central Java is dominated by areas with moderate vulnerability conditions. As many as 43% of the area in Central Java or around 15 districts are included in the category medium level of vulnerability. Areas that are adjacent to each other and directly adjacent to each other have the same vulnerability conditions so that spatially they form a cluster pattern based on the condition of social vulnerability to the COVID-19 disaster in Central Java (Figure 4).

4. Conclusion

Central Java has various social vulnerabilities that are influenced by social conditions, access to infrastructure, population, access to health, welfare, housing conditions, and economic factors. Based on the results of the SoVI calculation, the social vulnerability conditions in Central Java have 4 (four) areas with very high levels of social vulnerability, namely Cilacap Regency, Banyumas Regency, Brebes Regency, and Semarang City. In addition, about 46% or areas in Central Java have a moderate level of social vulnerability. The spatial and temporal distribution of the COVID-19 disaster is concentrated in areas with high population, including: Semarang City, Surakarta City, Semarang Regency, Banyumas Regency, Cilacap Regency, Boyolali Regency, and Magelang Regency. The number of COVID-19 cases has exacerbated the condition of social vulnerability in Central Java. Based on the results of the analysis, Central Java has 3 regions with a very high level of social vulnerability to the COVID-19 disaster, namely Semarang City, Cilacap Regency, and Banyumas Regency. Meanwhile, around 43% areas in Central Java have a moderate level of social vulnerability to the COVID-19 disaster.

Acknowledgements

This research is part of the 2021 Universitas Gadjah Mada final project led by Dr. Estuning Tyas Wulan Mei, M.Sc. The authors would like to thank Universitas Gadjah Mada for supporting this research based on the funds.

References

[1] Qu J M Cao B and Chen R C 2020 *COVID-19: The Essentials of Prevention and Treatment* (China : Shanghai Jiao Tong University Press)

[2] Sarkar A and Chouhan P 2020 COVID-19: District level vulnerability assessment in India *J. Clin.l Epdm. and Glob. Health* doi:10.1016/j.cegh.2020.08.017

[3] Puspa A 2021 Indonesia Masuk Gelombang Kedua COVID-19, Daerah Harus Susun Strategi. Media Indonesia *Media Indonesia* https://m.mediaindonesia.com/indonesia-masuk-gelombang-kedua-covid-19-daerah-harus-susun-strategi/

[4] Kementrian Kesehatan Republik Indonesia 2021 *Update Kolibri Situasi COVID-19* (Jakarta : Kementrian Kesehatan Republik Indonesia)

[5] Badan Pusat Statistik 2021 *Provinsi Jawa Tengah dalam Angka 2020* (Jawa Tengah : Badan Pusat Statistik)

[6] Acharya R and Porwal A 2020 A vulnerability index for the management of and response to the COVID-19 epidemic in India: an ecological study *J. Lancet Glob Health* **8** 42–51

[7] Adger W N 2006 Vulnerability *J. Glob. Env. Chg.* **16** 268–281

[8] Adger W N and Kelly P M 1999 Social vulnerability to climate change and the architecture of entitlements *J. Mitgn. and Adpt. Strg. for Glob. Chg.* **4** 253-266

[9] Hizbaron D R Hadmoko D S Mei E T W Murti S H Laksani M R T Tiyansyah A F Siswanti E and Tampubolon I E 2018 Towards measurable resilince: mapping the vulnerability of at risk community at Kelud Volcano, Indonesia J. App. Geog. **97** 212-227

[10] Cutter S L Boruff B J and Shirley W L 2003 Social vulnerability to environmental hazards *J. Soc. Sci. Quarterly* **84** 242-261

[11] Hizbaron D R Rahmat P R Setyaningrum A and Malawani M N 2015 Kajian pola spasial kerentanan sosial, ekonomi, dan fisik di wilayah rawan erupsi Gunungapi Merapi *J. Riset*

Kebencanaan Indonesia **1** 16-24
[12] Mei E T W Sari I M Fajarwati A and Safitri D 2016 Assessing the social economic and physical vulnerabilities to Gamalama Volcano. *J. Advn. in Soc. Sci. Edu. and Hum.* **79** 33-40
[13] Mavhura E Manyena B and Collins A E 2017 An approach for measuring social vulnerability in context : the case of flood hazards in Muzarabani District, Zimbabwe *J. Geoforum* **86** 103-117
[14] Armas I and Gavris A 2013 Social vulnerability assessment using spatial multi-criteria analysis (SEVI Model) and the social vulnerability index (SoVI Model) : a case study for Bucharest, Romania *J. Nat. Haz. and Earth Sys. Sci.* **13** 1481-1499
[15] Eryando T Sipahutar T and Rahardiantoro S 2020 The risk distribution of COVID-19 in Indonesia : a spatial analysis *J. Asia Pacific of Pub. Health* **32** 450 – 452
[16] Sari L K 2021 Pengaruh Kerentanan Sosail terhadap Penyebaran COVID-19 di Tingkat Wilayah Studi Kasus : Kabupaten Pubalingga. *Tesis*. ITB : SAPPK.
[17] Yesiana R Yuniartanti R K and Artiningsih 2015 Tipologi kerentanan masyarakat pesisir terhadap perubahan iklim di Kota Semarang *J. Riptek* **9** 61 – 70

IOP Conf. Series: Earth and Environmental Science 1233 (2023) 012029 doi:10.1088/1755-1315/1233/1/012029

Study of The Impact of Convectively Coupled Equatorial Kelvin and Rossby Waves on Rainfall Extremes in Sulawesi, Indonesia

Muhammad Subhan Al Zibrah[1] and Novvria Sagita[1,2*]
[1]State School of Meteorology, Climatology, and Geophysics (STMKG)
[2]Indonesian Agency for Meteorology, Climatology and Geophysics

*Corresponding author: novvria.sagita@stmkg.ac.id

Abstract. This study examined the impact of Equatorial Kelvin and Equatorial Rossby (ER) waves on rainfall extremes over Sulawesi Island, Indonesia. Impacts are assessed by changes in the probability of rainfall events that exceeded extreme value and changes in the fraction of wet days. Using Asian Precipitation – Highly Resolved Observational Data Integration Towards Evaluation (APHRODITE) rainfall data and Outgoing Longwave Radiation (OLR) from NOAA, impacts are quantified with the respect to wave's local phase grouping. The local wave phase is based on the standardized OLR anomalies and their temporal tendency. Results showed that Kelvin and ER wave can modulate daily rainfall extremes over Sulawesi Island. Kelvin wave has the strongest impact with changes in the probability of extreme rainfall by up to +100% (-70%) in phases 5 and 6 (1 and 2) and increases the value of wet day fraction by up to > 0.04. Followed by the ER wave with changes in the probability of extreme rainfall by up to +50% (-50%) in phase 4 (1 and 2) and increases the value of the wet day fraction by up to 0.05. The superposition between ER and Kelvin wave occurrence increases the possibility of extreme rainfall occurrence by up to 100 % in almost every part of Sulawesi island in phases 4,5 and 6. The results suggest that Kelvin and ER waves can be another important source of predictability for extreme rainfall on Sulawesi Island.

1. Introdcution

Flood is one of the most frequent hydrometeorological disasters in Indonesia, including Sulawesi Island. The seasonal pattern on Sulawesi Island is a mixture of monsoon and equatorial climates with the normal value of rainfall exceeding 3500 mm [1]. BNPB reported that there were 1445 flood events associated with heavy rains that occurred in Sulawesi in the 1996 – 2020 period, with the most affected area is South Sulawesi [2]. The impact is not only in the form of physical losses/buildings owned by the community and public facilities, but also causes economic losses in the form of land damage, and even fatalities (death, injured, and missing). Based on these data, it is very important to understand the factors controlling heavy rain to improve the ability to predict high impact extreme events in the Sulawesi region.

Located close to the equator, rainfall in Sulawesi is influenced by various modes of tropical climate variability. One of the modes of variability that affects tropical rainfall is the Madden-Julian Oscillation (MJO) which belongs to the sub-seasonal scale with a period of 30-90 days [3]. The Sulawesi region is also within the Asian and Australian monsoon region. Then for the inter-annual scale, there is the El Nino-Southern Oscillation (ENSO) [4] [5]. In addition, there is also another mode of climate variability with a shorter timescale that can also control tropical rainfall but has not been studied much for the Sulawesi region, namely Convectively Coupled Equatorial Waves (CCEWs).

CCEWs are the type of weather disturbance in the form of waves that propagate in the tropics in a zonal direction and interact with convective activity [6]. CCEWs rise as a solution to the shallow water equation which was solved by Matsuno [7]. There are two dominant types of CCEW with a symmetrical characteristic, namely the equatorial Kelvin wave and the equatorial Rossby Wave (ER). ER is a long wave that propagates westward along the equator, while the Kelvin wave is a wave with a higher frequency and propagates eastward.

Due to its interaction with convective activity, CCEWs have an influence on rainfall in the tropics, especially in Indonesia [8][9][10]. Lubis and Jacobi [8] found that CCEWs contribute up to 16-20% of the total intraseasonal (2.5-72 days) rainfall variance, with Kelvin and ER waves showing values of 6.99 ± 3.30% and 3.77 ± 1.79%, respectively. A study by Ferret et al. [9] has shown that by using the Tropical Rainfall Measuring Mission (TRMM), CCEWs affected the frequency and intensity of extreme rainfall in South East Asia. However, their research didn't assess the more localized impact of CCEWs on extreme rainfall, and also satellite based rainfall data is subject to biases for rainfall concentration over the ocean. Therefore, there is a need for a more localized study of the impact of CCEWs on rainfall extremes on a certain region using a rainfall dataset from land based rain gauge network. Lubis and Respati [10] conducted research that covered the previous needs. It is shown that CCEW significantly modulates daily extreme rainfall on Java Island during the rainy season (November to April). Their results open the possibility for research in other parts of Indonesia related to equatorial wave activity on rainfall extremes.

In this study, using the same principle as Lubis and Respati [10], the impact of Equatorial Kelvin and ER waves on daily rainfall extremes over Sulawesi Island during the rainy season (November to April) is investigated and quantified. The impact is quantified by the value of changes in extreme-event probability and changes in wet day fraction with respect to the different phases of the waves.

2. Methods

2.1. Study Area

The research location covers the Sulawesi region (Fig. 1). The geographical location of Sulawesi itself is 02°N – 06° South and 118°45' – 125°15' East. The research time chosen was based on the availability of the APHRODITE rainfall dataset, which was 1980 – 2015, by focusing on the days of the rainy season (November – April).

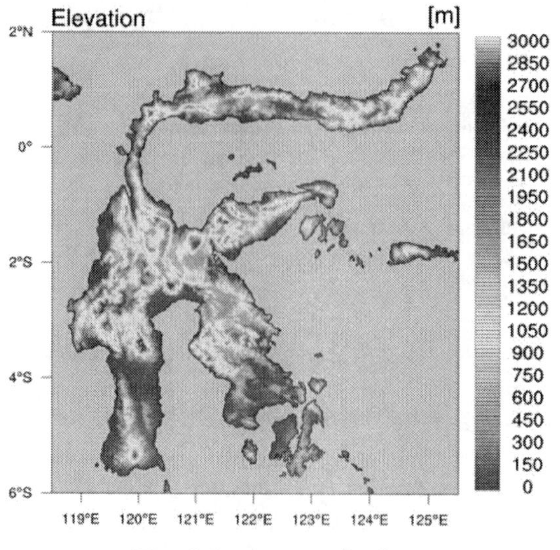

Fig. 1 Study Area [11]

2.2. Data

To investigate the impact of Kelvin and ER waves on rainfall extremes, daily gridded rainfall data of a continental-scale Asian Precipitation - High-Resolved Observational Data Integration Towards Evaluation (APHRODITE) [12]. APHRODITE consists of a rain gauges data network spread across the ASIA region. The rainfall data of the APHRODITE grid has a spatial resolution of 0.25° x 0.25°, Asian Monsoon V1101 for the period January 1, 1980-December 31, 2007, and Monsoon Asia V1101EX_R1 for the period January 1, 2008 - December 31, 2015.

To isolate the individual wave event and to determine the respective phases, Outgoing Longwave Radiation data from NOAA is used. The data has a spatial resolution of 1° x 1° and extended from 1 January 1979 to 31 December 2020.

2.3. Wave Filtering

To isolate the individual wave event, a wavenumber frequency filtering technique by WK99 has been used. First, the OLR is transformed into wavenumber and frequency using a Fast Fourier Transform (FFT) method. All elements that have a wavenumber and frequency outside of the wavenumber and wave period values according to the ER and Kelvin characteristics based on table 1 are not used. Then the data is transformed back into its normal space and time dimension, resulting in a filtered dataset. NCAR Command Language (NCL) is applied to run all the calculations.

Table 1 Wave characteristics [6]

Wave	Direction	Periods	Wavenumber	Depth (m)
ER	West	9 – 72 days	1 - 10	1 - 90
Kelvin	East	2.5-17 days	1 - 14	1 - 90

2.4. Wave Phase Determination

After obtaining a dataset of the occurrence of ER waves and active Kelvin waves, the phases of the waves were then carried out. Each phase is carried out to quantify the impact of active ER and Kelvin waves on rainfall in the research area. The filtered OLR is averaged in the latitude reference range, 2° N - 7° S. The tendency (temporal change) of the wave is calculated based on the reference longitude, 122° E. Then, we standardize the OLR filter results and the OLR temporal tendency of the filter results. After standardization, the standardized OLR filter data and its tendencies are transformed from standardized values into harmonic values. Furthermore, this study selected events are determined based on wave activity that has an amplitude value above 1. Then the occurrence of ER waves and Kelvin waves can be grouped into 8 phases of propagation and arranged in the form of a time series. The daily time series of wave phases were then used for composite analysis and calculation of the effect of ER and Kelvin wave propagation on rainfall in Sulawesi.

2.5. Changes in Probability of Extreme Rainfall

The effect of ER and Kelvin wave propagation on extreme rainfall was quantified by calculating the percentage change in cumulative probability using calculations based on the description of Lubis and Respati [10]. The change in the percentage of rainfall events that exceed the threshold values for each phase of the ER wave and Kelvin wave relative to the overall probability of the day is calculated at each grid point. Lubis and Respati [10] used the 95th percentile threshold value for the rainy season on the island of Java (November – April). But here we used the 90th percentile as the threshold. The calculation of changes in the probability of extreme rain events are:

$$\Delta P_n = \frac{P_n(x \geq x_c) - P_{all}(x \geq x_c)}{P_{all}(x \geq x_c)} \quad (1)$$

where P_n is the change in the probability of extreme rain events due to wave activity during n-phase, P_n is the cumulative probability of rainfall exceeding the threshold value during n-phase wave activity, and Pall for all days.

2.6. *Wet Day Analysis*
Wet days are defined as days where there is a high rainfall of 1 mm or more. The change in wet day fraction was calculated using:

$$\Delta f_n = \frac{n_n}{N_n} - \frac{n}{N} \qquad (2)$$

where f_n is the change in the fraction of wet days in the nth phase, n_n is the number of wet days in the nth phase, N_n is the number of days in the nth phase, n is the total number of wet days in the rainy season, and N is the number of days overall in the rainy season. Wet day analysis was carried out to determine whether the increase in rainfall was due to an increase in frequency or due to incidental events (high intensity) [9].

3. Results and Discussion
3.1. *Intensity of Wave Activity*
The intensity of wave activity can be seen from the OLR variance value. The value of variance can show how fluctuating a disturbance is. Kelvin waves have strong activity in the ITCZ region. While the ER wave has a maximum variance value in the off-equator region. In general, the activity of all wave modes follows the apparent movement of the sun where the peak of activity shifts slightly to the south in November-April and becomes slightly weaker. In the Sulawesi region itself, the wave mode that has the strongest activity is the Kelvin wave with a variance value of 84 – 104 $(W/m^2)^2$ (Fig. 2), while the ER wave has a value in the range of 60 – 80 $(W/m^2)^2$ (Fig. 2).

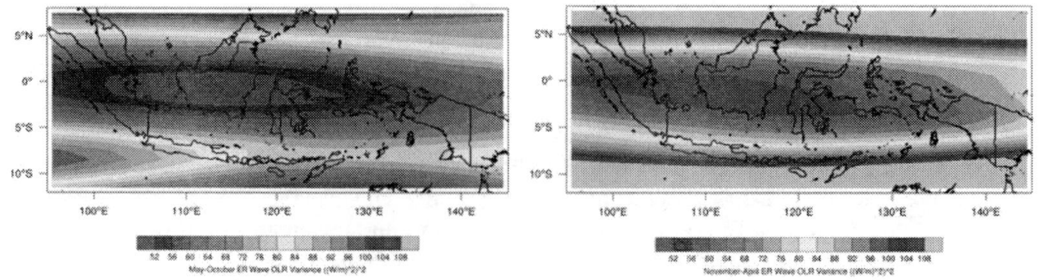

Fig. 2 Fitered OLR Variance

3.2. *Daily Rainfall Modulation*
Figures 3 to 4 show how the equatorial Kelvin and ER waves influence the daily rainfall during the rainy season in Sulawesi. There is an increase (suppressed) in the value of daily rainfall during the wet (dry) phase.

For the ER wave mode, the increase in rainfall values occurred in the 4th phase for several areas of Sulawesi (Fig. 3), with the highest positive anomaly in the South and Southeast Sulawesi regions, with a positive anomaly value of > 1 mm. The positive anomaly persisted until the 6th phase except for Central and West Sulawesi, which in the 5th phase showed a negative anomaly. In the 7th phase, rainfall suppression has begun, except for the Central and North Sulawesi regions. The suppression of rainfall is indicated by a negative anomaly value, and this rainfall suppression continues and peaks in phases 1 and 2. In phases 1 and 2 the negative anomaly peaks to show a value that exceeds 1 mm (negative) in the southern part of Sulawesi, but North Sulawesi still shows a positive anomaly. Then the anomalous value begins to weaken and approaches a positive value in the 3rd phase indicating the transition phase to the wet phase.

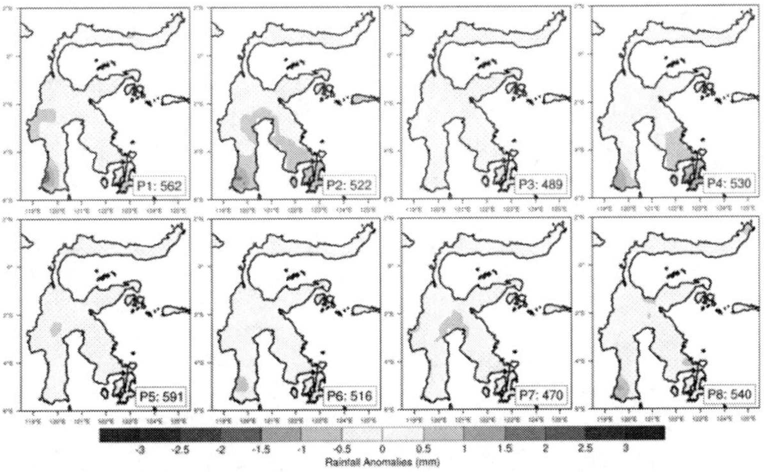

Fig. 3 Daily rainfall anomaly at each phase of ER wave

For the equatorial Kelvin wave mode, the increase and suppression of rainfall reach the value of > 3 mm (Fig. 4). The wet phase was recorded starting from the 4th phase, and then peaked at the 5th and 6th phases. The largest positive anomaly was in the southern part of Sulawesi, with a value of > 3 mm. In the 6th phase, the entire Sulawesi region shows a positive rainfall anomaly value of > 1 mm. Entering the 7th phase, the northern part of Sulawesi shows a negative anomaly, and the positive anomaly value has started to weaken, except for the southeastern part of Sulawesi. The 8th to 2nd phases show a dry phase, namely the rainfall suppression phase, marked by an increase in negative anomaly values. The largest negative anomaly value is in the southern part of Sulawesi with a negative value reaching > 3 mm (negative). The third phase is the transition phase to the wet phase, marked by a weakening of the negative anomaly values for the entire Sulawesi region.

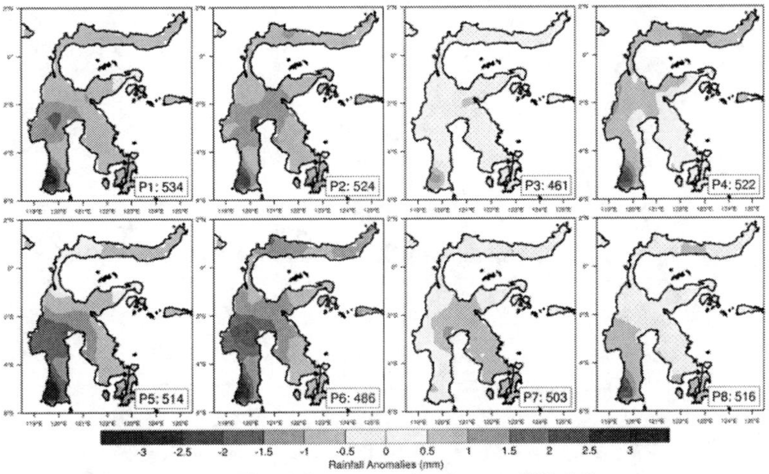

Fig. 4 Same as Fig. 3 but at each phase of Kelvin wave

3.3. *Rainfall Extremes Modulation*

Equatorial wave modes can also modulate extreme rainfall by increasing or decreasing the chance of extreme rainfall events [10]. Here, the 90th percentile rainfall is used as the threshold for extreme rainfall.

For the ER wave (Fig. 5), the peak of the probability of extreme rainfall occurs in phase 4, with the largest increase being in the Southeast Sulawesi region with an increased probability of > 50% or 1.5 times more likely. For the suppression of rainfall, the ER wave mode reduces the probability of extreme rain events by 40 – 50%. However, the phase and location of the decrease in the chance of extreme rain are quite volatile. In phase 1, the peak of suppression is in parts of Southern Sulawesi. However, in phase 1, the northern part of Sulawesi also experienced an increase in the probability of extreme events. In phase 2, the peak of suppression is in the Southeast Sulawesi region. In the 7th phase, the peak of suppression is in the western part of Sulawesi, and in the 8th phase, the peak of suppression is in the Central Sulawesi region. However, in the 8th phase, parts of the northern part of Sulawesi experienced an increase in the value of the changes in the probability of extreme rainfall.

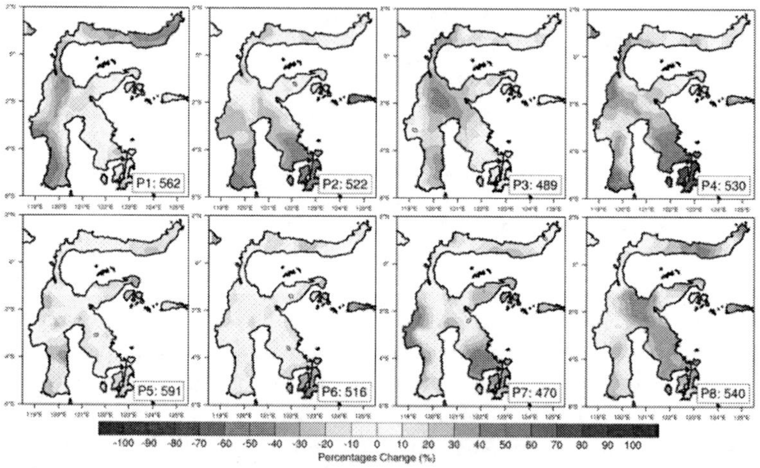

Fig. 5 Changes of Probability of extreme rainfall for each phase of ER wave

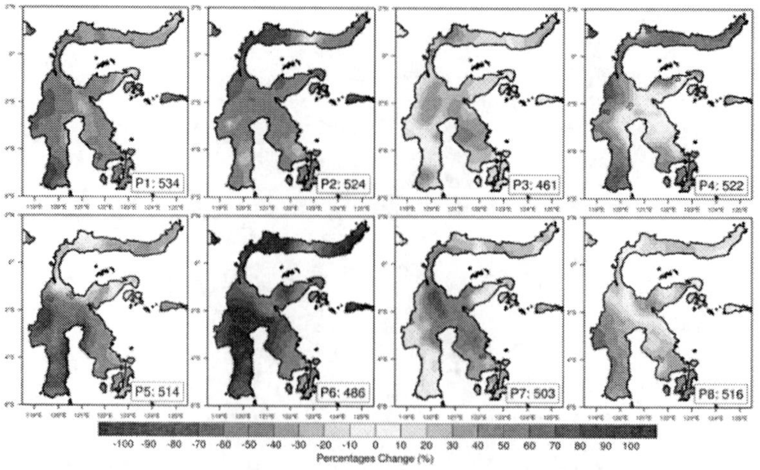

Fig. 6 Same as Fig. 5 but for each phase of Kelvin wave

For the equatorial Kelvin wave mode (Fig. 6), the changes in the probability of extreme rain events is consistent with the wet and dry phases as shown in the daily rainfall modulation section. The probability of extreme events in phase 6 increases to 100% or becomes twice as likely to occur. The

IOP Conf. Series: Earth and Environmental Science 1233 (2023) 012029 doi:10.1088/1755-1315/1233/1/012029

largest increase occurred in the Southern part of Sulawesi, and parts of North Sulawesi. Meanwhile, the decrease in the probability of extreme rain events reaches the range of 60-70%, with the center of suppression in South Sulawesi in the 1st phase and parts of North Sulawesi in the 2nd phase.

3.4. Linear Superpostion

The Kelvin wave is a high-frequency wave, while the ER wave is a longer wave. This causes the possibility of the two waves to experience superposition [10]. In the event of wave superposition (Fig. 7), there was a strong increase and suppression of rainfall, with positive and negative anomaly values reaching values > 6 mm. The strongest increase occurred in the Central and Southern Sulawesi region, while the strongest emphasis occurred in the Southern Sulawesi region.

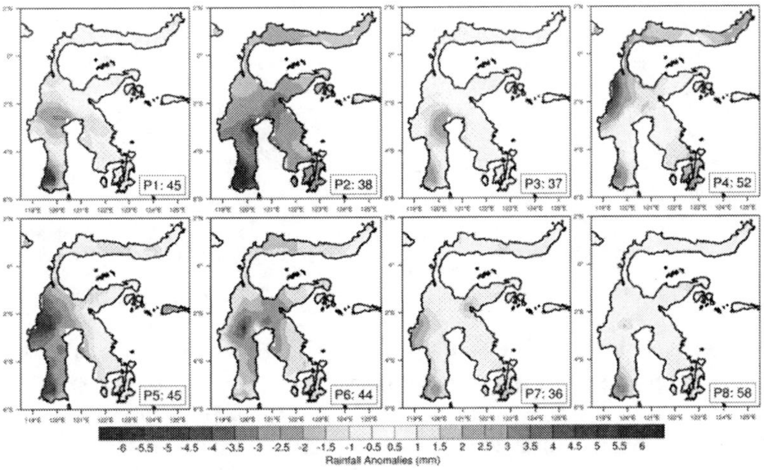

Fig. 7 Rainfall anomaly in the superposition events

Besides that, in the driest and wettest phases of superposition events, the increase and decrease in the probability of extreme rain events can reach 100% or double, and occur in almost all areas of Sulawesi (Fig. 8). This indicates that when the wave superposition occurs, extreme rain tends to occur.

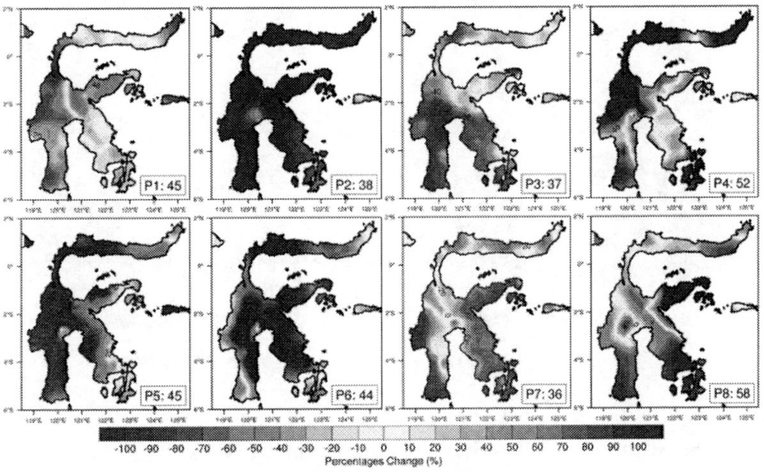

Fig. 8 Same as Fig. 7 but for changes of probability in extreme rainfall

3.5. Wet Day Analysis

Wet day analysis was carried out to determine whether the increase in rainfall was due to an increase in frequency or due to incidental events (high intensity) [9]. Fig. 9 shows the change in the wet day fraction for each phase of the ER wave. In the wet phase of the ER wave (phase 4), the Southeast Sulawesi region shows changes in the wet day fraction with values in the range of 0.01 – 0.05. In phase 1, the North Sulawesi region also shows values in the range of 0.01 – 0.05. Then in phases 6 and 7, almost the entire Sulawesi region shows values in the range of 0.01 – 0.02. In phases 2 and 5, almost the entire Sulawesi region shows negative values.

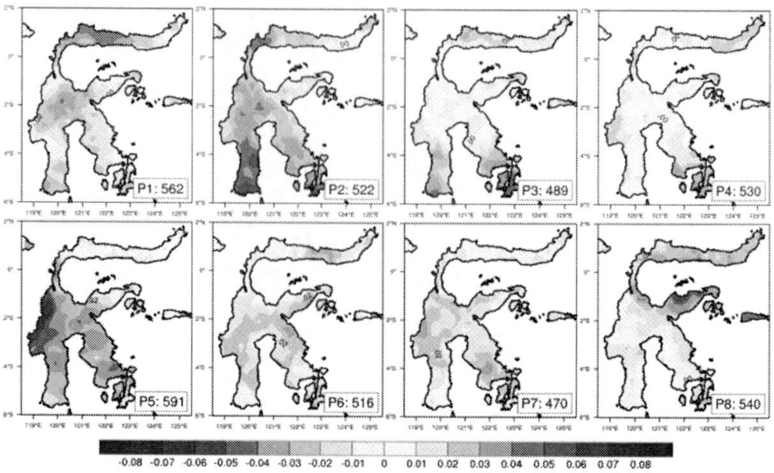

Fig. 9 Changes in fraction of wet days for each ER phase

For the Kelvin wave (Fig. 10), the wet phase (5 and 6) shows the changes in the wet day fraction > 0.04, with the highest changes in the North Sulawesi region. In general, phases 4, 5, and 6 for the equatorial Kelvin wave activity show positive changes in the wet day fraction for the Sulawesi region as a whole. Meanwhile, the negative value for the entire Sulawesi region started from phase 8 and continued until the second phase which showed the largest negative value.

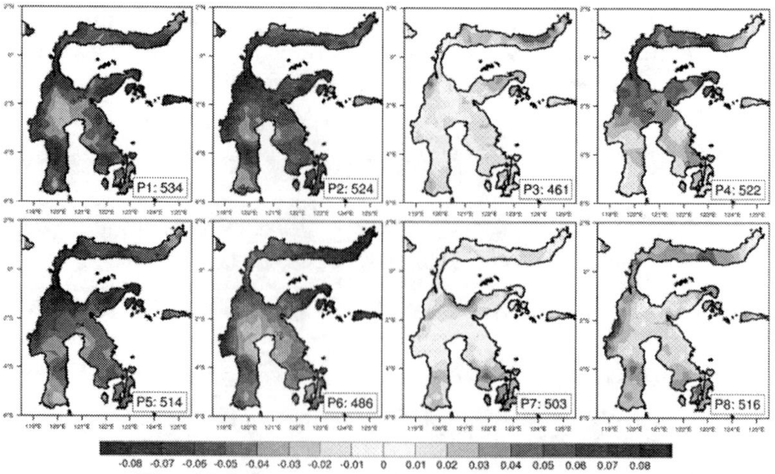

Fig. 10 Same as Fig. 9 but for Kelvin wave

4. Conclusions

It was concluded that Equatorial Kelvin and Equatorial Rossby (ER) waves can modulate extreme rainfall events. Kelvin wave shows the stronger impacts, by increasing the rainfall anomalies by up to 3 mm/day and causing extreme rain events to be twice as likely to occur (100%) in phases 5 and 6. The ER wave shows smaller impacts in comparison to the Kelvin wave, by only increasing the rainfall anomalies to a value of > 1 mm/day and the probability of extreme rainfall to a value of 40 - 50% in phase 4. The results also indicate that the impacts are not equally distributed across the Island and tend to be localized over a certain region. When the ER wave is active, the Southern, South Eastern, and Northern parts of the Island have a higher impact compared to the Center part of the Island. It is likely due to the off-equatorial characteristic of the ER wave [6][8]. When the Kelvin wave is active, The Center, Western to Southern parts of Sulawesi have a higher impact in comparison to the other parts of the Island.

In addition, the two waves can experience linear superposition which can increase (reduce) the daily rainfall value by up to 6 mm/day while increasing (reducing) the probability of extreme rainfall events by up to 100%. The activities of the ER and Kelvin waves also increase the frequency of rainfall on wet days, indicated by an increase in the value of the wet day fraction. The findings suggest that both the Equatorial Kelvin and ER waves can be added as another important source of predictability for extreme rainfall prediction in Sulawesi Island.

References

[1] Badan Meteorologi Klimatologi dan Geofisika (BMKG) 2021 *Prakiraan Musim Hujan 2021/2022 di Indonesia* (Badan Meteorologi Klimatologi dan Geofisika).
[2] Badan Nasional Penanggulangan Bencana (BNPB) (2021, 13 Dec). Accessed from https://dibi.bnpb.go.id/.
[3] Purwaningsih A Harjana T Hermawan E dan Andarini D 2020 *J. Sains & Tek. Mod. Cuaca* **21** 85-94.
[4] As-syakur A R Osawa T Miura F Nuarsa I W Ekayanti N Dharma I G B Adnyana I W Arthana I W and Tanaka T 2016 *Dy. Atmo. Oce* **75** 58-77.
[5] Supari T F Salimun E Aldrian E Sopaheluwakan A Juneng L 2017 *Clim. Dy.* **51** 2559-2580.
[6] Kiladis G N Wheeler M C Haertel P T Straub K H Roundy P E 2009 *Rev. Geophys.* **47** RG2003.
[7] Matsuno T 1966 *J. Meteor. Soc. Japan* **44(1)** 25-43.
[8] Lubis S W and Jacobi C 2015 *Int. J. Climatol.* **35** 1465-1483.
[9] Ferrett S Yang G-Y Woolnough S Methven J Hodges K and Holloway C 2020 *Q. J. R. Meteorol. Soc.* **146** 665–84.
[10] Lubis S W and Respati M R 2021 *Int. J. Climatol.* **41** 2418–40.
[11] GLOBE, Task "Team and others (Hastings, David A., Paula K. Dunbar, Gerald M. Elphingstone, Mark Bootz, Hiroshi Murakami, Hiroshi Maruyama, Hiroshi Masaharu, Peter Holland, John Payne, Nevin A. Bryant, Thomas L. Logan, J.-P. Muller, Gunter Schreier, and John S. MacDonald)" 1999 The Global Land One-Kilometer Base Elevation (GLOBE) Digital Elevation Model, Version 1: 80305-3328.
[12] Yatagai A Kamiguchi K Arakawa O Hamada A Yasutomi N and Kitoh A 2012 *Bulletin of the American Meteorol. Soc.* **93** 1401-1415.

IOP Conf. Series: Earth and Environmental Science 1233 (2023) 012030 doi:10.1088/1755-1315/1233/1/012030

Subsurface Structure Evaluation Beneath the Sileri Crater's East Flank Using the Misfit Function Gradient of the Full Waveform Modeling

D P D Ramadhan, P A Paninggar, H Darmawan and T M Irnaka*
Geophysics Laboratory, Faculty of Mathematics and Natural Sciences, Universitas Gadjah Mada, Sekip Utara BLS 21 Yogyakarta 55281

*corresponding author: marwan.irnaka@ugm.ac.id

Abstract. Sileri crater is one of the prominent craters in Dieng Volcanic Complex. It is one of the craters in the area with the highest activity of phreatic eruption. This type of eruption is characterized by a minimal to no precursor beforehand, making mitigation more challenging. Previously, there is minimum number of research which tries to understand the subsurface properties of the crater. Therefore, in this research, we integrate two methodologies to understand Sileri crater's characteristics better. An Unmanned Aerial Vehicle (UAV) survey was conducted in parallel with the surface seismic survey. The UAV data was then processed, yielding a Digital Elevation Model (DEM) and orthophoto for surface analysis. The ground seismic survey on the east flank of the crater was processed to obtain the gradient with respect to the misfit function. The obtained gradient information can be used to interpret the tendency of the subsurface velocity model. Through this process, a vertical shape positive gradient anomaly is detected at the crater's northeast flank, which may correspond to the hot spring activity. The constant positive gradient anomaly layer throughout the model, which corresponds to the alteration zone, is also detected. Another 10 m depth positive anomaly is also found in the model's southeast part, which corresponds with the new potential geothermal activity. The location of the anomalies in our investigation indicating a continuous tendency in the east which correspond to the possible eastward expansion of the crater. This finding is important to better monitor and mitigate the future activities and eruption of the Sileri crater.

1. Introduction

Dieng Volcanic Complex is one of the areas with high community mobility. Besides its function as a residential area, the Dieng Volcano Complex is used as a plantation, tourism, and geothermal area. Located in the province of Central Java and part of the Sumbing-Sindoro-Dieng Quaternary mountain chain [1], the Dieng Volcanic Complex, which is located at an altitude of about 2000 meters above sea level, has several active craters with recorded eruptions, including Pakuwaja Crater, Sikidang Crater, Sinila Crater, Timbang Crater, Candradimuka Crater, and Sileri Crater. One of the craters with several eruption records is Sileri Crater. It is one of the most active craters in the area.

The last recorded eruption of Sileri Crater was in April 2021. Previously, Sileri Crater erupted in 1944, 1964, 2003, 2017, and 2018. Most of the eruptions were of the phreatic type, with the eruption product in the form of mud. The lack of eruptive precursors characterizes the type of phreatic eruption. This type of eruption makes the mitigation process more challenging and is also one of the causes of the Sileri Crater eruption that once claimed hundreds of casualties [2]. As one of the most active craters in the Dieng Volcano Complex, more research is needed to characterize the crater better, hoping to help the mitigation process and reduce losses caused by the eruption.

Sileri crater characterization can be done by analyzing the surface and the subsurface structure of the area around the crater. The surface observation can be carried out using photogrammetry through Unmanned Aerial Vehicle (UAV). UAV technology is widely used to reconstruct topography and to map structures of an active volcano such as at Merapi [3], Kelud [4] and Anak Krakatau [5]. The condition of Sileri Crater that continues to change along with the volcanic activity that occurs makes the use of UAV for surface mapping important. The main purpose of using UAV is to get the latest data with high resolution.

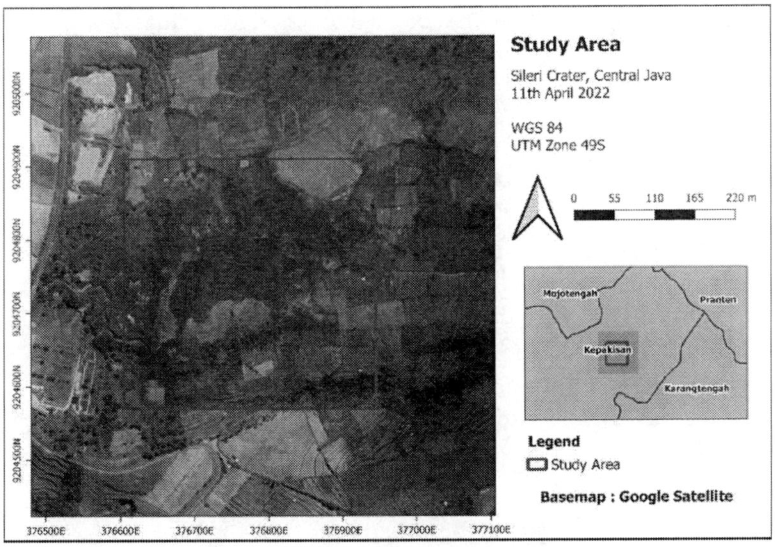

Figure 1. Study Area of Sileri Crater

On the other hand, the subsurface investigation can be performed using the geophysical method. One of the geophysical methods that are able to provide an overview of the subsurface structure is the active seismic method. However, appropriate processing methods are needed to provide a clearer picture of the subsurface structure. The processing method that can model the subsurface structure is the full waveform inversion (FWI) method, which has been assessed for a shallow target [6], [7], and [8]. In this study, the analysis of the subsurface structure of the Sileri crater was analyzed based on the gradient of the misfit function as a part of the standard FWI workflow. With the integration between surface and subsurface observation, we are able to study the potential subsurface geothermal activity at Sileri crater.

The advantage of integrating these two methods is to produce high-resolution surface and subsurface images. The integration of these two methods is also able to provide surface and subsurface images of a more localized area in the east of the Sileri Crater. Previous research, [9] using the Very Low Frequency (VLF) method to describe the distribution of the conductivity zone in the area around Mount Pagerkandang which is located east of Sileri Crater. These results have not been able to characterize Sileri Crater because it is not located right in Sileri Crater and the resolution of the resulting data is not too high. For this reason, it is hoped that this research will be able to provide a clearer picture of the condition of the Sileri crater.

The integration of geophysical data with UAV data has previously been carried out by [10]. The geophysical data used is Electrical Resistivity Tomography data and is integrated with the Digital Elevation Model from UAV data. The focus of this research is to monitor the active landslide area. Accurate electrical potential field modeling requires good information about the geomorphology of the research area obtained from the UAV survey. Changes in the geomorphology of the research area are also able to provide an understanding of the slope failure mechanism of the research area. The importance of UAV data in this study is equal to the importance of UAV data in our study, where UAV data in our study is used to model the research area and to help interpret the processing results.

2. Viscoelastic waveform modeling and gradient estimation

Accurate modeling of seismic waves is required to approximate better our experiment's synthetic data and the gradient of the misfit function. The gradient of the misfit function will reflect the tendency of the optimum velocity model update. In our data, aside from having a strong elastic effect due to the presence of the S wave and the surface waves. We also observe a strong attenuation due to the unconsolidated soil and rock alteration from geothermal activities. Moreover, in the geothermal area around the craters, we have a complex topography that needs to be tackled carefully. Our study uses

IOP Conf. Series: Earth and Environmental Science 1233 (2023) 012030 doi:10.1088/1755-1315/1233/1/012030

SEM46, a time domain visco-elastic waveform modeling and inversion developed by SEISCOPE consortium. It relies on the standard spectral element method for spatial discretization and the Newmark scheme as the time discretization [11,12]. The gradient of misfit function is calculated using the adjoint method [13,14].

2.1. *Viscoelastic waveform modeling*

In visco-elastic medium, the wave equation can be written as a set of equations:

$$\rho(\mathbf{x})\partial_{tt}u_i(\mathbf{x},t) = \partial_j\sigma_{ij}(\mathbf{x},t) + f_i(\mathbf{x},t), \tag{1}$$

$$\sigma_{ij}(\mathbf{x},t) = M_{ijkl}(\mathbf{x},t) * \varepsilon_{kl}(\mathbf{x},t) + \mathrm{T_{ij}}(\mathbf{x},t), \tag{2}$$

where ρ is the density of the medium, u_i id the displacement field in i direction, σ and ε are the second order stress and strain tensors, f_i is the external force in i direction, and T_{ij} is the stress failure in ij component. The effect of attenuation is written as a relaxation rate M_{ijkl}. Symbol $*$ represents convolution in time domain.

The inverse of the quality factor tensor $Q_{ijkl}(\mathbf{x})$, which represents the energy loss per cycle of the wave, is a ratio between the imaginary part and the real part of the relaxation rate $M_{ijkl}(\mathbf{x},\omega)$ in frequency domain:

$$\bar{Q}^{-1}_{ijkl}(\mathbf{x},\omega) = \frac{\mathrm{Imag}\left(\bar{M}_{ijkl}(\mathbf{x},\omega)\right)}{\mathrm{Real}\left(\bar{M}_{ijkl}(\mathbf{x},\omega)\right)}, \tag{3}$$

where the bars above the symbol represents frequency domain data.

Spectral element method discretizes the seismic wavefield based on Lagrange interpolants. The numerical integration is based on Gauss-Lobatto-Legendre (GLL) points. The discretization is performed on quadrilateral and hexahedral element in 2D and 3D, respectively. The visco-elastic wave equation is formulated in the weak formulation, resulting a discretized wave equation as

$$\breve{M}\frac{\partial^2 \boldsymbol{u}}{\partial t^2} = -\mathbf{Ku} + \mathbf{D}^w C^R \textstyle\sum_{\mathrm{A}=1}^{L} \psi_{\mathrm{A}} + \mathbf{F} + \mathbf{B}, \tag{4}$$

$$\frac{\partial \psi_{\mathrm{A}}}{\partial t} + w_{\mathrm{A}}\psi_{\mathrm{A}} = w_{\mathrm{A}} y_{\mathrm{A}}\mathbf{Du}, \tag{5}$$

where $\breve{M}$ represents the global mass matrix, $\mathbf{K}$ represents stiffness matrix, $\mathbf{D}^w$ represents the weighted spatial derivative matrix with the weight w, C^R represents the relaxed discrete stiffness tensor, $\mathbf{F}$ represents the source term, and $\mathbf{B}$ is the boundary term. A more detailed derivation of the method is written in [12]. SEM method has the advantage of having a flexible mesh design following complex topography an implicit free surface condition, ensuring an accurate surface wave modeling.

2.2. *Gradient of misfit function estimation*

Calculating the gradient of misfit function implying a first stage of inverse problem in Full Waveform Inversion (FWI). The misfit function $f(m)$ which is used in our experiment is L_2-norm:

$$f(m) = \frac{1}{2}\|\mathbf{d}_{cal}(m) - \mathbf{d}_{obs}\|^2, \tag{6}$$

where $d_{cal}(m)$ is the calculated data based on model parameter m and d_{obs} is the observed data.

The gradient of the misfit function with respect to the parameter is calculated following the adjoint-state method [14]. It requires calculation of zero-lag cross correlation between the adjoint displacement field $\mathbf{u}^\dagger(m)$ and the incident acceleration field $\ddot{\mathbf{u}}(m)$ with the additional right hand-side term related to memory variable on the viscoelastic case:

$$\nabla f(m) = \langle \mathbf{u}^\dagger(m), \frac{\partial C}{\partial C_{ij}}\ddot{\mathbf{u}}(m)\rangle - \langle \mathbf{u}^\dagger(m), \textstyle\sum_{s=1}^{L}\frac{\partial C^R}{\partial C_{ij}}\psi_s\rangle, \tag{7}$$

Our experiment mainly calculated the gradient of the misfit function with respect to the parameter V_S. It is given as:

$$\frac{\partial f(m)}{\partial V_S} = \textstyle\sum_{i=1}^{6}\sum_{j=1}^{6}\frac{\partial f}{\partial C_{ij}}\frac{\partial C_{ij}}{\partial V_S} + \frac{\partial f}{\partial \rho}\frac{\partial \rho}{\partial V_S}, \tag{8}$$

where ρ is the density of the medium and V_S is the shear wave velocity.

IOP Conf. Series: Earth and Environmental Science 1233 (2023) 012030 doi:10.1088/1755-1315/1233/1/012030

3. UAV and Seismic acquisition setup

3.1. UAV data acquisition

The UAV acquisition is focused on Sileri Crater and the area around Sileri Crater. The UAV flies at an altitude of 80m to 90m from the ground. The UAV is flown manually, considering the field conditions, which have extreme altitude changes. The UAV flies to form a grid in a north-south direction. A flight path is shown in Figure 2, starting from the red line and then the blue line. A total of 216 photos were obtained during the acquisition; four photos are less clear due to being covered in water vapor from the crater. So that these four photos were not included in the processing, while the other 212 photos were processed using Agisoft software. The processed images yield Digital Elevation Model (DEM) maps and orthophoto of Sileri Crater. The parameters used can be seen in Table 1. This parameter was chosen to get a high-resolution map and fast computation time. The DEM map is used to model the surface of the block model in the seismic method. Then, the orthophoto map is used to identify geological features. Based on field observations and orthophoto maps, geothermal manifestations and fractures were found, as shown in Figure 3. The distribution of these geological features is the basis for determining the seismic line survey.

Table 1. Agisoft Parameter

Parameters	Value
Accuracy	High
Generic Preselection	Yes
Reference Preselection	Yes
Adaptive Camera Model Fitting	Yes
Key Point Limit	40000
Tie Point Limit	4000
Quality	Ultra High
Depth Filtering	Mild
Calculate Point Color	Yes
Calculate Point Confidence	Yes

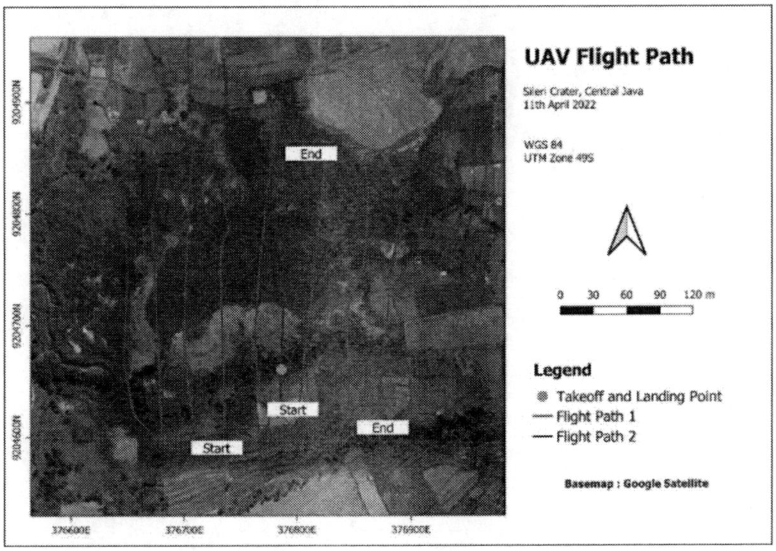

Figure 2. UAV Flight Path

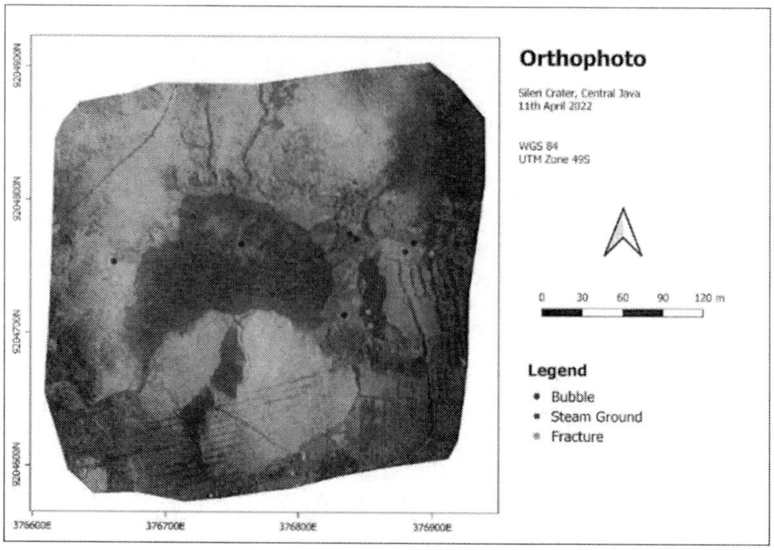

Figure 3. Orthophoto map with the observed geothermal activities and geology

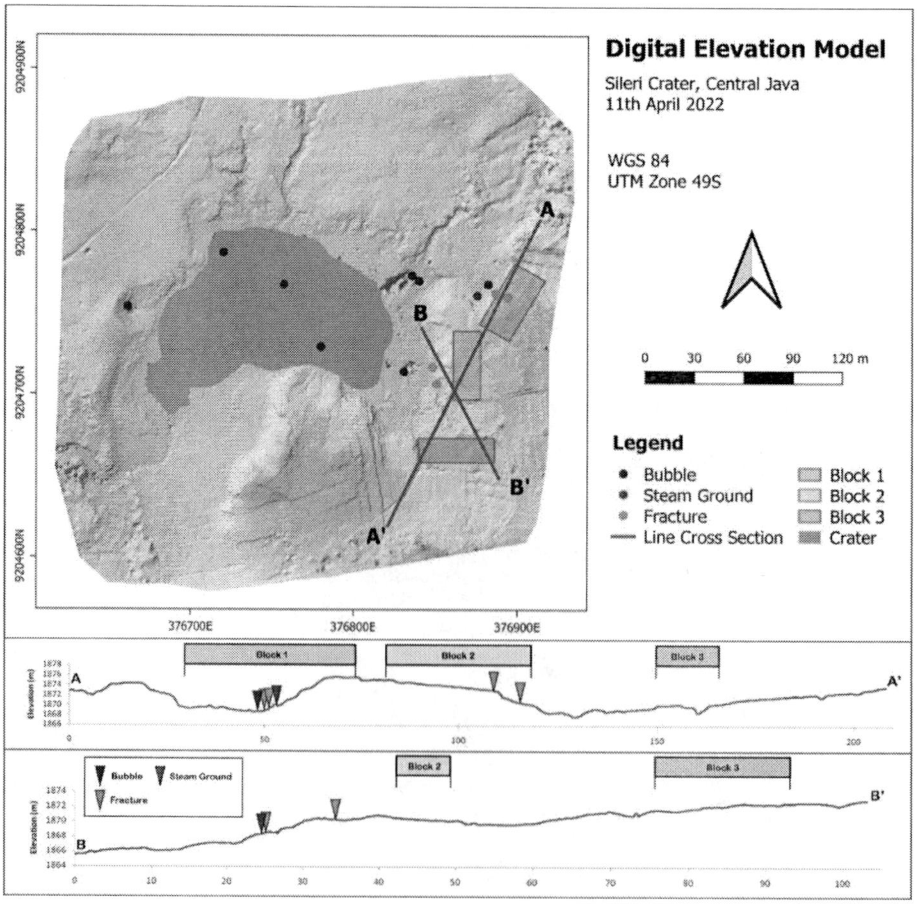

Figure 4. Digital Elevation Model and Cross Section

3.2. Seismic data acquisition

The research focused on the eastern part of Sileri Crater. The data used in this research was obtained during the acquisition on 10-13 April 2022. The selection of the acquisition location was based on the presence of geological features such as fractures and geothermal manifestations which were visible from the results of photogrammetry (Figure 3) and field observations. The total number of acquisition tracks is seven lines with each line having a length of 34.5 m. One line consists of 24 geophones arranged with a distance between geophones as far as 1.5 m. The seven lines are divided into 3 blocks, with block 1 consisting of 1 line and 12 sources and block 2 and block 3 consisting of three lines with a total of 4 sources on each line (Figure 5).

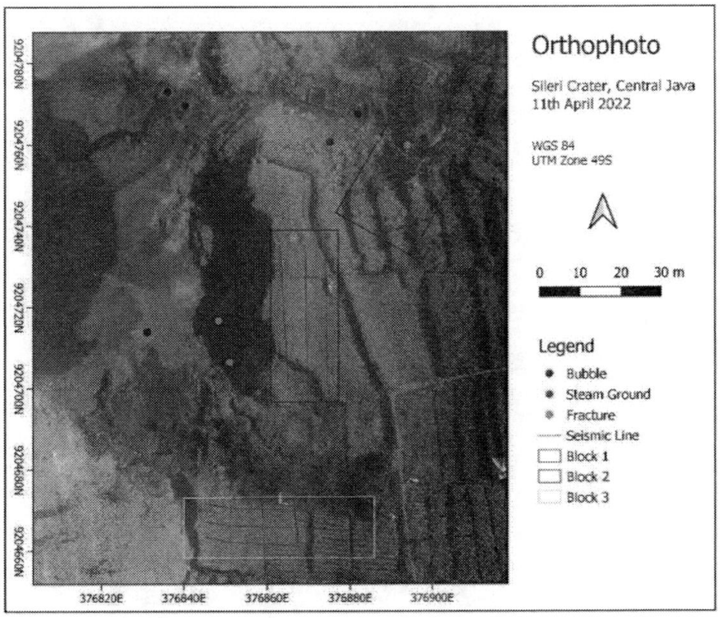

Figure 5. Aerial map of eastern Sileri Crater

4. Gradient of the misfit function with respect to V_S

There are several processing steps in estimating the gradient of the misfit function of full waveform inversion. These stages are shown in the following flowchart (Figure 6).

4.1. Initial model building

The initial model used in the gradient estimation process is a homogeneous model with parameters obtained from the observed data. The initial model for V_P is obtained from the arrival time of the P wave, and the result for block 1 is 314.5 m/s, block 2 is 332.35 m/s, and block 3 is 324.25 m/s. The initial V_S model is obtained from the relation $V_R = 0.9\ V_S$, where V_R is the velocity of the Rayleigh wave obtained from the arrival time of the Rayleigh wave. The V_S value for block 1 is 185.38 m/s, block 2 is 208.28 m/s, and block 3 is 181.3 m/s. The density value is calculated based on the Gardner relation and produces a density value for block 1 of 1305.48 kg/m^3, block 2 of 1323.6 kg/m^3, and block 3 of 1315.47 kg/m^3.

4.2. Source estimation

The source used at the time of acquisition is a vertical direction source. The sources from each seismic block are assumed to have the same energy. The source time function estimation is carried out using the deconvolution approach proposed by [15]. Each block's estimated source time function is then used for all source excitations on the respective acquisition block.

4.3. Gradient estimation
The estimation of the gradient of the misfit function is carried out using observation data that has been filtered using a bandpass filter in the frequency range of 6-25 Hz. It aims to reduce the complexity of the calculation because the initial model uses a homogeneous model. The consequence of gradient estimation with low-frequency data is reduced resolution at shallow depths. In this gradient estimation process, a smoothing process is carried out using the Bessel Filter proposed by [16] to reduce gradient artifacts.

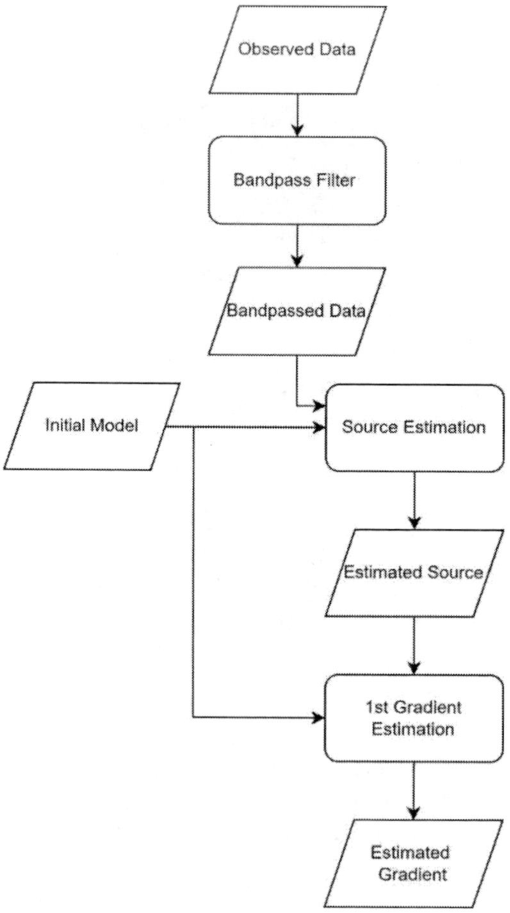

Figure 6. Flow chart of the gradient estimation process

5. Results
The gradient of the misfit function can give a hint regarding the subsurface structure. For example, the misfit gradient function can show an area in the model with a relatively lower or higher velocity than its surrounding. A negative gradient means that the initial velocity model has underestimated velocity and needs to be increased; on the other hand, a positive gradient means that the initial velocity has overestimated velocity and needs to be reduced.

As previously mentioned, the research area is divided into three blocks. The position of each block on the map can be seen in Figure 5 and the gradient estimation results can be seen in Figure 7. In the model shown, the red color on the color scale indicate a low-velocity area, and the blue color indicates a high-velocity area. In block 1, a low-velocity area is found in the center of the 2D profile at the receiver location (Figure 8). The low-velocity zone may be related to volcanic activity, namely the alteration

IOP Conf. Series: Earth and Environmental Science 1233 (2023) 012030 doi:10.1088/1755-1315/1233/1/012030

process as mentioned by [17]. This finding is supported by visual observations and DEM map from photogrammetry results (Figure 4 and Figure 5) that the zone corresponds to geological features in the form of fractures, active bubbles, and the hot springs found at an altitude of 1869 – 1871 meter above sea level (masl). It can be seen in Figure 4, the cross section A - A' is at a position of 50 - 52 meters. The low-velocity zone extends continuously downwards, which is probably the path of volcanic activity from below. In addition, the low-velocity zones also spread horizontally at an altitude of 1870-1875 masl. It is deducted that these low-velocity zones may correlate with the alteration zone.

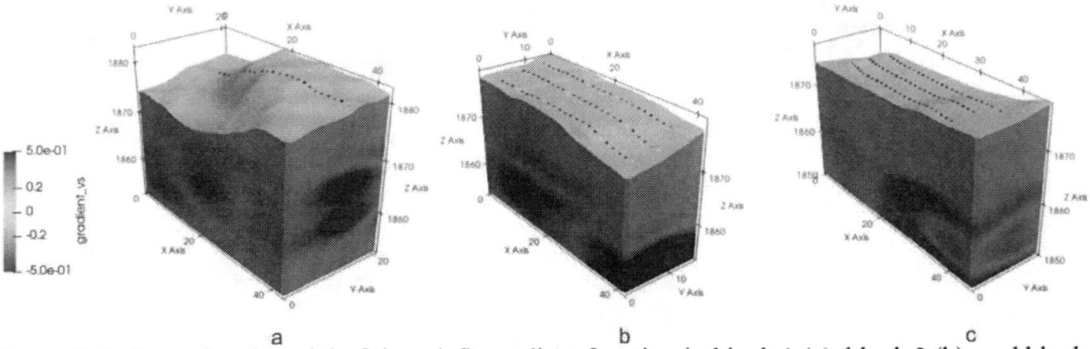

a b c

Figure 7. 3-dimensional model of the misfit gradient function in block 1 (a), block 2 (b), and block 3 (c). The green dot is the receiver position and the yellow dot is the source position.

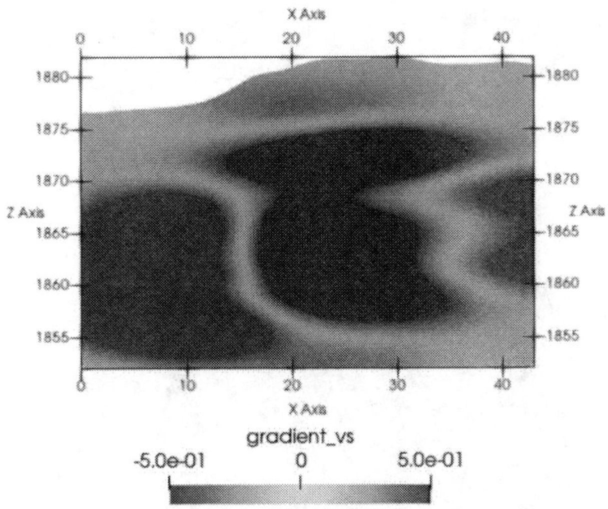

Figure 8. 2-dimensional profile on the seismic line of block 1.

In block 2 (Figure 9), the top of the model tends to have no change in the velocity value, marked in a gray color scale. There is also a low-velocity zone that continues horizontally at the elevation level of 1865-1870 masl. Those observations are supported by photogrammetric data, which shows that in the western area of the block, there are fracture features, which indicated by some active bubbles at the crater and geothermal manifestations (blue and orange dots in Figure 4 and Figure 11) found at an altitude of 1868 - 1870 masl. It can be seen in Figure 4, the cross section B - B' is at a position of 24 - 34 meters; the low-velocity zone is thought to be related to the activity that occurs or is an alteration zone. In the southern model, there is a zone that has a low velocity, which is probably a zone related to volcanic processes. However, the reliability of the interpretation is low because the zone's location is situated in the corner of the gradient model.

IOP Conf. Series: Earth and Environmental Science 1233 (2023) 012030 doi:10.1088/1755-1315/1233/1/012030

Although the low-velocity zone at the corner of the block two models is uncertain, it is interesting that it corresponds to the low-velocity zone also detected in block 3 (Figure 10). There are two low-velocity zones in block 3, a low-velocity zone at an altitude of 1850-1860 masl and an altitude of 1860-1865 masl. The zone that is meant to correlate with the low-velocity zone in block 2 is the low-velocity zone at an altitude of 1850-1860 masl. This suitability can be seen more clearly in Figure 11.

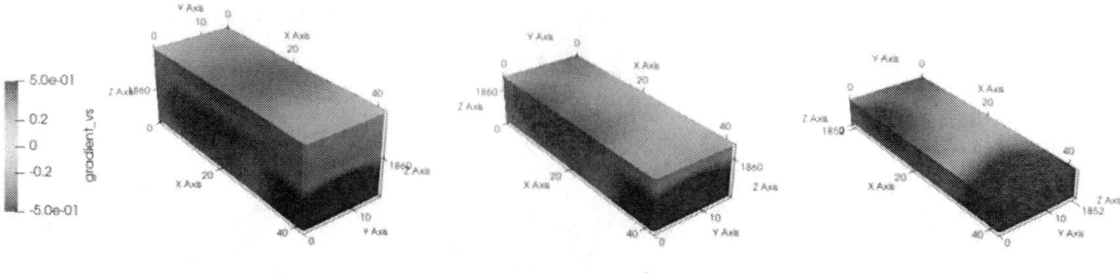

a b c

Figure 9. 3-dimensional model of horizontal cross-section of block 2. The top of the model is at an altitude of 1868 (a), 1863 (b), 1858 (c) meters above sea level.

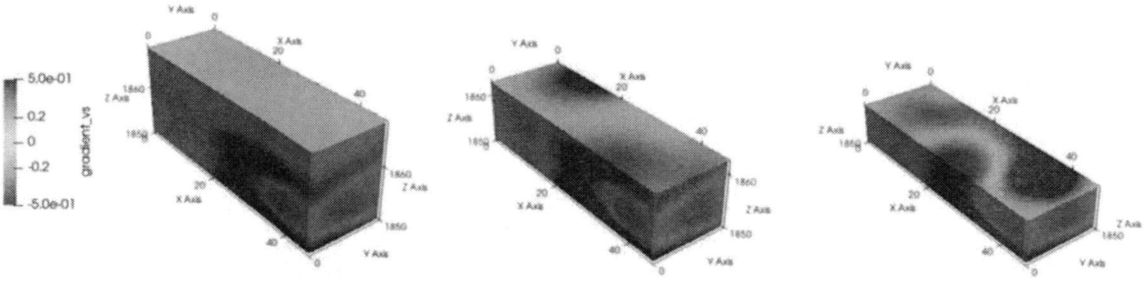

a b c

Figure 10. 3-dimensional model of horizontal cross-section of block 3. The top of the model is at an altitude of 1868 (a), 1863 (b), 1858 (c) meters above sea level.

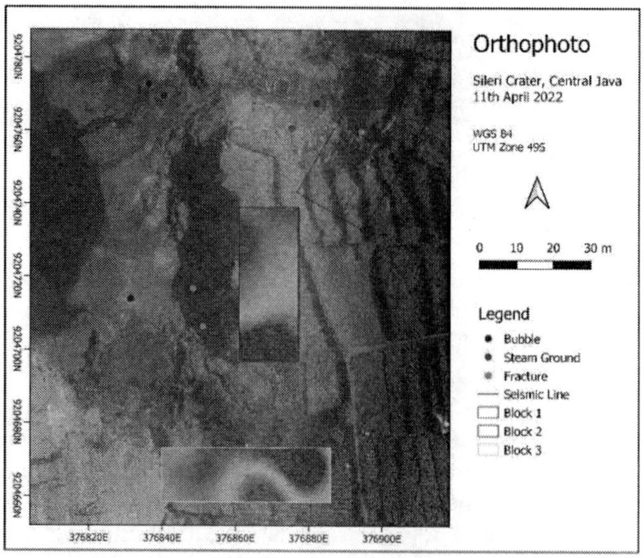

Figure 11. Horizontal cross-section of blocks 2 and 3 at an altitude of 1858 meters above sea level, or a depth of about 20 m, shown on an orthophoto map.

6. Conclusions and outlooks

In this experiment, an integrated surface and subsurface analysis were performed through photogrammetry and seismic method at Sileri crater. The observation from photogrammetry reveals several areas with different geothermal activities such as bubbles, steam grounds, and fractures. The seismic method, through the gradient with respect to the misfit function, reveals subsurface zonation, in which the initial velocity model is supposed to be lower or higher. From these results, we observe several subsurface features, which have been confirmed with the photogrammetry result. Gradient analysis on block 1 reveals a deep vertical positive gradient zone which is interpreted as volcanic activity. Gradient analysis in block 2 reveals a positive gradient zone layer which is interpreted as a zone of alteration due to the geothermal activity. Between block 2 and block 3, a possible layer of positive gradient zone might be related to the new geothermal activity between those two blocks. With these results, it can be seen that volcanic and geothermal activity in Sileri crater has a continuous tendency in the east. This makes it possible that the crater will expand eastward from the current main crater. This finding is important to better monitor and mitigate the future activities and eruption of the Sileri crater.

In order to better interpret the subsurface condition, a more comprehensive and complete full waveform inversion will be performed. The extension from a low-frequency band to a higher frequency band will also be performed to obtain higher resolution images.

Acknowledgements

This study is funded by Hibah Penelitian Dosen Dana Masyarakat Alokasi FMIPA UGM No 43/J01.1.28/PL.06.02/2022. The author would like to thank our friends Dipta Eka Satria, Dyah Retno Budiarti, Rafi Maulana Arya Dhia, and Ryzki Guruh Sanjaya Putra for their assistance with the field data acquisition throughout our research. We also acknowledge Center of Volcanology and Geological Hazard Mitigation for the permission to conduct this study at Sileri Crater.

References

[1] Harijoko A, Uruma R, Wibowo H E, Setijadji L D, Imai A, Yonezu K and Watanabe K 2016 Geochronology and magmatic evolution of the Dieng Volcanic Complex, Central Java, Indonesia and their relationships to geothermal resources *Journal of Volcanology and Geothermal Research* 310 209-224.

[2] Loeqman A *et al* 2020 *Gunung Api Indonesia dan Karakteristik Bahayanya: Bagian I, Wilayah Barat* Bandung: Pusat Vulkanologi dan Mitigasi Bencana Geologi Badan Geologi Kementerian Energi dan Sumber Daya Mineral.

[3] Darmawan H, Walter T R, Richter N and Nikkhoo M 2017 High resolution Digital Elevation Model of Merapi summit in 2015 generated by UAVs and TLS https://doi.org/10.5880/GFZ.2.1.2017.003.

[4] Wahyudi, et al 2020 Topography changes and thermal distribution at the Kelud crater after the 2014 Plinian eruption *Indonesian journal of Geography* 52(3) 411-417 https://doi.org/10.22146/ijg.51986.

[5] Darmawan H *et al* 2020 Topography and structural changes of Anak Krakatau due to the December 2018 catastrophic events *Indonesian journal of Geography* 52(3) 402-410 https://doi.org/10.22146/ijg.53740.

[6] Irnaka T M, Brossier R, Métivier L, Bohlen T and Pan Y 2022 3-D multicomponent full waveform inversion for shallow-seismic target: Ettlingen Line case study *Geophysical Journal International* 229(2) 1017-40.

[7] Köhn D, Wilken D, Wunderlich T, De Nil D, Rabbel W, Werther L, Schmidt J, Zielhofer, C Linzen S 2018 Seismic SH Full Waveform Inversion as New Prospection Method in Archeogeophysics 10.3997/2214-4609.201802514.

[8] Wittkamp F, Athanasopoulos N, Bohlen T 2018 Individual and joint 2-D elastic full-waveform inversion of Rayleigh and Love waves *Geophysical Journal International* 216 10.1093/gji/ggy432.

[9] Wulandari A, Asti E, Dwiasih N, Suyanto I 2018 Estimation of Existence Geothermal Manifestation Using Very Low Frequency (VLF) Method in the PagerkandangVulcanic, Dieng, Central Java *IOP Conference Series: Earth and Environmental Science* 132. 012023. 10.1088/1755-1315/132/1/012023.

[10] Boyd J *et al* 2021 A linked geomorphological and geophysical modelling methodology applied to an active landslide Landslides 18 1-16 https://doi.org/10.1007/s10346-021-01666-w

[11] Komatitsch D and Tromp J 1999 Introduction to the spectral element method for three-dimensional seismic wave propagation *Geophysical journal international* 139(3) 806-822.

[12] Trinh P T, Brossier R, M´etivier L, Tavard L and Virieux J 2019 Efficient 3D time-domain elastic and viscoelastic Full Waveform Inversion using a spectral-element method on flexible Cartesian-based mesh Geophysics 84(1) R75–R97.

[13] Tromp J, Tape C, and Liu Q 2005 Seismic tomography, adjoint methods, time reversal and banana-doughnut kernels *Geophysical Journal International* 160(1) 195-216.

[14] Plessix R E 2006 A review of the adjoint-state method for computing the gradient of a functional with geophysical applications *Geophysical Journal International* 167(2) 495-503.

[15] Pratt R G 1999 Seismic waveform inversion in the frequency domain, part I: theory and verification in a physical scale model Geophysics 64:888–901.

[16] Trinh P T, Brossier R, M´etivier L, Virieux J and Wellington P 2017 Bessel smoothing filter for spectral element mesh *Geophysical Journal International* 209(3):1489–1512.

[17] Shalev E *et al* 2010 Three-dimensional seismic velocity tomography of Montserrat from the SEA-CALIPSO offshore/onshore experiment *Geophysical Research Letters* - GEOPHYS RES LETT. 37. 10.1029/2010GL042498.

IOP Conf. Series: Earth and Environmental Science 1233 (2023) 012031 doi:10.1088/1755-1315/1233/1/012031

Performance of Regional Economic in East Kalimantan: Leading Sector of Export-Import Commodities in 2019-2020

R W Kusuma[1], R Farhan[1], S Budiman[1], S Kusumawati[1], A S Hayyati[1], and R F Putri[1]*

[1] Department of Environmental Geography, Faculty of Geography, Universitas Gadjah Mada, Indonesia

*Corresponding author: ratihfitria.putri@ugm.ac.id

Abstract. Industrial sector has an important role in encouraging the activities of economic sectors, such as trade, transportation, and the service sector. Industrial sector can increase the foreign exchange of a country in large amounts. The existence of industry is related to population distribution and economic growth. Commodities of a regional area are utilized in international trade to improve welfare of the society. International trade is required in economic importance because the whole countries need each other. This research aims to analyze the performance of regional economics in East Kalimantan Province. The used methods were LQ calculation and the amount of export-import in 2019 − 2020. The calculation is based on pre- and post- Covid-19 pandemic. The LQ of the industrial sector ranged from 0,32 − 6,92 with the highest LQ was coal and oil-gas refining industry. It was the leading sector in East Kalimantan Provinsi because their LQ was above 1. The advantages of East Kalimantan Province are supported by the prime commodity in the form of "other coal, whether or not pulverized but not agglomerated", even though it experienced a decline in 2020 due to the Covid-19 pandemic.

1. Introduction

One of the mainstay sectors that drive economic growth in Indonesia is the manufacturing industry sector. This sector also has a role as a driver of activity in other economic sectors, such as trade, transportation, services, and so on. The industrial sector is also considered to be able to increase the country's economic income by strengthening the balance of payments or foreign exchange reserves. However, the condition of the industrial sector cannot be separated from its ups and downs. One of the efforts to encourage economic growth in the industrial sector is to strengthen the management of existing resources and existing industrial facilities and infrastructure. The existence of inter-sector linkages can also affect the level of economic growth in the industrial sector in the region [1].

The economic condition of East Kalimantan Province from 1970 to 2010 was dominated by the forestry, oil and gas, and coal mining sectors. This is caused by the condition of East Kalimantan which has natural resources in the form of forests, oil and gas, and coal [2]. The state of natural resources has a major influence on the state of the economy of a region. As a province that still relies on the potential of natural resources, East Kalimantan has a high tendency in these three areas. Especially when the natural resources that are owned have high usability, not only in that area, but also in other areas. Utilization of natural and human resources should be maximized with the aim of maximum economic growth for each region, in this study is economic growth in East Kalimantan Province. Main commodities in East Kalimantan are in the forestry sector, oil and gas, and coal mining. The same thing is found in the manufacturing industry sector, where the processing of the three natural resources is the main sector in the manufacturing industry. This is in line with the theory that superior commodities can be seen from the main resources in the region. Analyzing leading commodities can be used with the *Location Quotient* (LQ) analysis method, where the results will describe the commodity as a basis or non-base. From these results, it can be seen which commodities can be favored in East Kalimantan in the effort of economic growth in this area.

The economic growth of East Kalimantan Province is supported by the existence of import-export activities as one of the biggest potentials in international trade. International trade is defined as global economic relations between countries with the maximum contribution of commodities owned to improve people's welfare [3]. One of the largest exporting regions in Indonesia is East Kalimantan. The largest export activity is in mineral natural resource commodities. This is in line with the existence of the main resources in East Kalimantan, namely coal and oil and gas. In addition to the oil and gas sector, East Kalimantan also exports a lot of non-oil and gas commodities, especially CPO or *Crude Palm Oil*. Export value in East Kalimantan can be said to fluctuate every year. This happened because of the policy of the Regional Government of East Kalimantan Province which began to boost non-oil and gas commodities such as CPO [4]. This increase in the value of non-oil and gas exports shows the government's seriousness in maximizing exports of other natural resources besides oil and gas in the region. This effort is carried out so that there is an even distribution of income between regions because the oil and gas industry only exists in certain areas.

The export destinations of East Kalimantan vary, but there are a total of 11 countries that always import goods from East Kalimantan. These countries are China, Japan, India, South Korea, Malaysia, Taiwan, Philippines, Thailand, Singapore, Bangladesh, and Australia. These eleven countries not only import, but also export to East Kalimantan, such as machinery and industrial supporting equipment. Seeing the relationship between East Kalimantan and export destination countries is to analyze it. These relationships can be reflected in the distance, total population, and GDP or *Gross Domestic Product* of export destination countries with the existing export value. By looking at the data pattern, it can be seen the tendency of the relationship between variables, where all three show a linear tendency. If the line shows an increase, it can be said that the two variables have a straight comparison. Conversely, if the line shows a decrease, then the two variables show an inverse comparison between the variables. This analysis can be used as a reference regarding the factors that most influence the occurrence of exports between these countries and East Kalimantan.

2. Methods

Data collection from this study came from several sources obtained secondarily. The data used to look at several industrial bases, main commodities, and import exports in East Kalimantan Province with a time span of 2019 and 2020. The year used is based on the year before and after the Covid-19 pandemic to compare whether there are differences in economic growth and differences in distribution patterns and international trade that occurred in that year span. The resulting data is in the form of an industrial base table of East Kalimantan Province which contains the Location Quotient (LQ) value and information about the LQ value results, a table of export values of major industrial commodities, export and import values in several destination countries (Japan, China, India, South Korea, Taiwan, Thailand, the Philippines, Malaysia, Singapore, Australia, and Bangladesh), and a map of export destination countries of East Kalimantan Province. Some of the calculations and data processing carried out in this analysis use the following formulas:

a. LQ Value Calculation

The LQ value is used to determine the economic growth of the industrial sector of East Kalimantan Province with national industry.

$$LQ = \frac{xi/_{xt}}{Xi/_{Xt}}$$

Information:
xi: East Kalimantan Province's industrial GRDP value
xt: Total value of industrial GRDP of East Kalimantan Province
Xi: National industrial GDP value
Xt: Total value of national industrial GDP

b. Determining the Industrial base is carried out by determining the value of LQ. If the LQ value is >1, it will be said that the sector is an industrial base, but if the LQ value<1 then the sector is not an industrial base.

3. Results and Discussions

3.1. Basis Industri

Analysis of the base sector in East Kalimantan in 2019-2020 uses the LQ method so that the base and non-base sectors in East Kalimantan can be known. LQ or Location Quotient is used to find out how much the level of specialization of the base or superior sector is by comparing its role in the regional economy with the role of similar sectors in the regional or larger regional economy [5]. The sector that has an LQ>1 value can be said to be a leading sector or a base sector, while LQ<1 can be said to be a non-base sector. The base sector is a sector that can serve the market within the region and outside the region. Indirectly, the regions have the ability to export goods and services to other regions so that it can be interpreted that the role of the sector in East Kalimantan is greater or contributes significantly in providing benefits to the East Kalimantan region. The non-base sector is the sector that provides goods and services to the people within the boundaries of the economic area. The non-base sector can be interpreted to mean that the role of the sector in East Kalimantan is smaller than the role of the sector nationally in research is Indonesia.

Table 1. East Kalimantan Industrial Base in 2019 and 2020

2019						
Industrial Sector	**National Sector**	**Industrial GDP**	**Provincial Sector**	**Industrial GDP**	**LQ**	**Information**
Coal Industry and Oil and Gas Refinery	217,402.2		63,971,424.08		6.62	Basis
Wood Industry, Wood Products and Cork and Woven Products from Bamboo, Rattan	59,498.3		2,754,582.99		1.04	Basis
Chemical Industry, Pharmacy and Traditional Medicine	195,040.9	2,276,667.8	13,113,506.79	96,802,393.71	1.51	Basis
Paper Industry and Paper Products; Printing and Media Reproduction	80,211		2,959,034.93		0.83	Non-Basis
Food and Beverage Industry	744,170.5		11,003,441.75		0.33	Non-Basis
2020						
Industrial Sector	**National Sector**	**Industrial GDP**	**Provincial Sector**	**Industrial GDP**	**LQ**	**Information**
Coal Industry and Oil and Gas Refinery	202,603.7	2,209,920.3	61,742,744.14	94,031,227.95	6.92	Basis
Wood Industry, Wood Products and Cork and Woven Products from Bamboo, Rattan	58,212.1		2,414,915.6		0.94	Non-basis

Chemical Industry, Pharmacy and Traditional Medicine	213,360.5		13,442,527.27		1.43	Basis
Paper Industry and Paper Products; Printing and Media Reproduction	80,385.3		2,885,507.39		0.81	Non-basis
Food and Beverage Industry	755,914.9		10,528,874.17		0.32	Non-basis

(Source: Statistic Data Processing East Kalimantan, 2019-2020)

Based on table 1, East Kalimantan Province in 2019 has three industrial base sectors, namely the Coal Industry and Oil and Gas Refining; Wood Industry, Goods from Wood and Cork and Woven Goods from Bamboo, Rattan and the Like; and the Chemical, Pharmaceutical and Traditional Medicine Industries. Meanwhile, in the 2020 results, Kalimantan Province has two industrial base sectors, namely the Coal Industry and Oil and Gas Refining and the Chemical, Pharmaceutical and Traditional Medicine Industries. Based on the results of the LQ analysis produced from the industrial base sectors in East Kalimantan, the Coal and Oil and Gas Refining Industry has a very high value with a fairly far interval comparison with the LQ value of other industrial base sectors. The LQ value of the Coal and Oil and Gas Refining Industry is between 6-7, while the Wood Industry, Goods from Wood and Cork and Woven Goods from Bamboo, Rattan and the Like and also the Chemical, Pharmaceutical and Traditional Medicine Industries only have an LQ value with a value of approximately 1.

This shows that the Mining and Oil Refining Industry sector still plays an important role for the Economy of East Kalimantan so that the role of the mining and oil refining sector in East Kalimantan can be said to be quite dominant. The high potential of mining and oil resources in East Kalimantan has caused the development of this industrial sector to be quite high. The high mining potential of East Kalimantan Province is coal. According to data from the Geological Resources Center (2014), the potential of East Kalimantan coal resources is 46,350.83 million tons or 37.14% of Kalimantan's coal resources. The huge potential of coal makes Kaltim one of the coal production centers in Indonesia. According to the "East Kalimantan Province Economic Report" (2021) [6], the mining industry is expected to grow higher in line with the potential for increasing demand from outside the region to foreign countries. Based on this potential in East Kalimantan, it is natural that the mining and oil refining industry sector plays an important role in the economy of East Kalimantan. This dependence on the mining and oil refining sector, which is a non-renewable natural resource, must certainly be balanced with wise and planned use as the basic capital to develop new sources of economic growth that are sustainable and equitable in the future.

The results of industrial GRDP in East Kalimantan from 2019 to 2020 have decreased. This decline was caused by several factors, one of which had the most impact was the occurrence of the Covid-19 pandemic in 2020 which caused new adaptations to human activities such as restrictions on human activities through direct contact. The government's policy of limiting community activities is carried out to overcome the spread of the virus due to human contact. These restrictions resulted in constraints on distribution activities that caused a decrease in industrial productivity so that the growth and economic development of the industrial sector were disrupted. This has caused the East Kalimantan industrial sector to only be able to produce for its own territory and cannot supply its goods to other regions.

3.2. Main Commodity

Commodity is raw materials in the form of crops, commercial goods, goods, merchandises, and crafts that can be used as goods or commodities that can be exported such as nickel, coal, wheat, coffee, and others. Main commodities are leading commodities that are profitable to be worked on or developed in an area that have competitiveness, either in national market or international market. A commodity is said to be a leading commodity if it has competitiveness so that it can ward off competing products in the domestic market and engage in the export market.

Table 2. East Kalimantan Leading Sector Export Import Values in 2019–2020

Code	Commodities	2019	2020
2701121000	Bituminous coal: coking coal	1,811,913,564.00	1,467,249,768.00
2701190000	Other coal, whether or not pulverised but not agglomerated	8,001,013,184.00	6,476,585,725.00
2711110000	Liquid natural gas	1,831,524,807.00	1,032,384,593.00
2702100000	Lignite, whether or not agglomerated	812,079,216.00	571,080,281.00
	Other	3,726,676,132.00	3,432,979,334.00
	Total	16,183,206,884.00	12,980,279,703.00

(Source: Statistic Data Processing East Kalimantan, 2019-2020)

East Kalimantan Province main commodities are minerals, such as coals and natural gasses. In general, the export trend of mining commodities in East Kalimantan Province experienced a declining phase in 2019 to 2020. One of the factors was COVID-19 pandemic in that year where there were many restrictions on outdoor activities due to the virus.

3.3. Exports-Imports

In general, the distribution of East Kalimantan mineral commodities exports has been well distributed. This is indicated by a number of export country destinations. Those countries are China, Japan, India, South Korea, Malaysia, Taiwan, Philippines, Thailand, Singapore, Bangladesh, and Australia (Figure 1).

According to Krugman [7], in international trade theory, factors that affect export can be seen from two sides, namely supply and demand. The volume of exports to a country is influenced by demand of the population in that country. The more people in an export destination country, the more demand will be needed. The number of demand will increase the volume of exports to a country so that the value of exports also increases. Therefore, the population of the export destination country had a significant effect on the East Kalimantan export value in 2020. Based on Table 3.3, the trend of East Kalimantan exports and imports value experienced a decline from 2019 to 2020 in all of the destination countries.

Declining in 2020 could occur due to the declining number of exports and imports in every country that is affected by COVID-19. COVID-19 pandemic brought huge influence to international trade. Exports and imports exist because of mobility. When COVID-19 occurred, mobility was reduced. There were even countries that had locked down, which means they did not receive any goods into the country so that it could hinder the export and import process.

Table 3. East Kalimantan Exports and Imports to Several Countries Value in 2019–2020

Country	Export		Import	
	2019	**2020**	**2019**	**2020**
Jepang	2,380,064,379	1,380,761,046	79,688,890	38,590,116
China	4,316,448,830	3,645,617,020	260,209,821	133,785,813
India	324,084,2751	2,361,992,039	19,042,740	18,511,113
Korea Selatan	83,298,5273	706,412,101	329,051,428	210,330,326
Taiwan	1,046,511,199	746,614,330	1,454,927	1,270,646
Thailand	586,409,719	444,432,439	3,765,398	2,624,558
Filipina	913,459,118	887,391,121	2,1762,009	22,502,630
Malaysia	1,177,037,399	1,015,097,744	334,114,547	387,434,438
Singapura	121,415,129	56,240,983	200,871,730	209,668,532
Australia	98,954,472	34,866,528	45,231,237	23,770,751
Bangladesh	315,221,194	396,352,970	142,333	858

(Source: Statistic Data Processing East Kalimantan, 2019-2020)

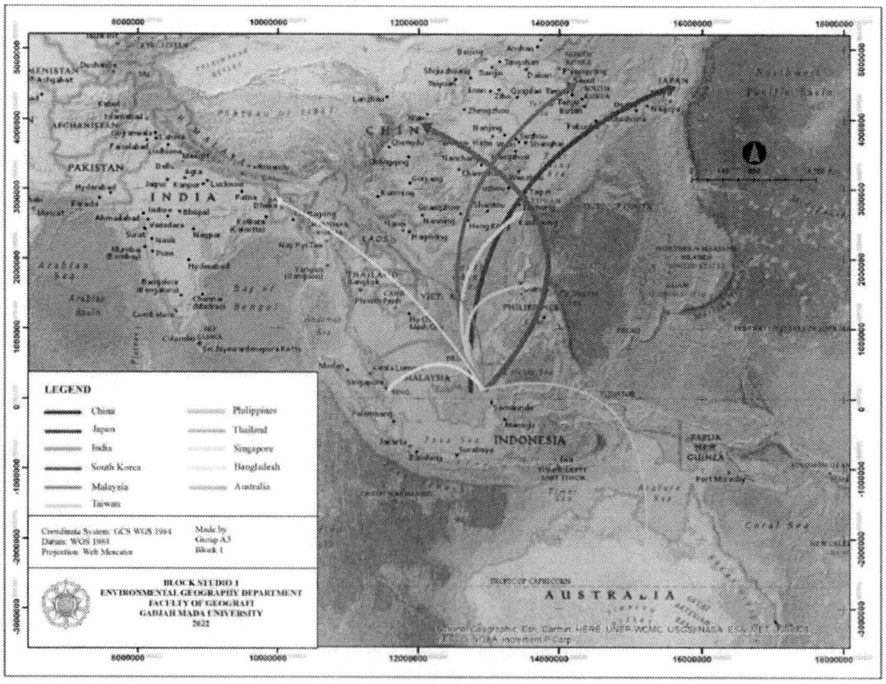

Figure 1. East Kalimantan Industrial Commodity Export Destination Countries Map. *(Source: Statistic Data Processing East Kalimantan, 2019-2020)*

International trade has an important role for the economy of a country, especially exports. Exports have many important roles to drive the international economy. A country that carries out export activities will increase its foreign exchange reserves. That foreign exchange reserves will later be used to finance imports and development of domestic economic sectors.

4. Conclusion

The industrial sectors that became the base in East Kalimantan Province in 2019 and 2020 were the Coal Industry and Oil and Gas Refinery sector as well as the Chemical Industry, Pharmacy and Traditional Medicine, shown through a fairly large LQ, namely 6.62 and 1.51 in 2019 and 6.92 and 1.43 in 2020. Meanwhile, the Wood Industry, Wood Products and Cork and Woven Products from Bamboo, Rattan sector was only the base sector in 2019. The Food and Beverage Industry sector was the sector with the lowest LQ value in the year period and was not a base sector.

The high potential of mining and oil resources in East Kalimantan has caused the development of this industrial sector to be quite high and has become an important contribution to the economy of East Kalimantan Province. The industry is expected to grow higher in line with the potential for increasing demand from outside the region to foreign countries. However, the coal and natural gas industries experienced a decline in the value of exports in 2020 due to the Covid-19 pandemic. In addition, the export and import value of East Kalimantan Province for all commodities in the above destination countries almost all also decreased in 2020. The Covid-19 pandemic has had a major impact on Indonesia's international trade and the global economy because it affects mobility. Some countries are in lockdown which means that they do not accept goods or anything entering the country so that it can hinder the export and import process.

References

[1] Winardi., Priyarsono, D. S., Siregar, H., Kustanto, H. 2017. Kinerja Sektor Industri Manufaktur Provinsi Jawa Barat berdasarkan Lokasi di Dalam dan di Luar Kawasan Industri. *Jurnal Manajemen Teknologi*. Vol. 16 (3). Hal. 241—257.

[2] Zaini, A. 2017. Pengaruh Kekayaan Sumberdaya Alam Batubara terhadap Ketimpangan Pendapatan di Provinsi Kalimantan Timur. *Jurnal Borneo Administrator*. Vol. 3 (2). Hal. 111—130.

[3] Baene, C. W. 2020. Analisis Pengaruh Ekspor Sektor Industri, PMA Sektor Industri dan Inflasiterhadap Produk Domestik Bruto Indonesia. *Tesis*. Fakultas Ekonomi dan Bisnis, Universitas Sumatera Utara.

[4] Nurmawaddah, S. 2019. Pengaruh Ekspor Non-Migas terhadap Pertumbuhan Ekonomi Provinsi Kalimantan Timur. *Jurnal Akuntabel*. Vol. 16 (1). Hal. 128—133.

[5] Imelia, E. 2006. *Modul Ekonomi Regional*. http://id.pdfsb.com/modul+ekonomi+regional+manual. [30 Maret 2022].

[6] Bank Indonesia. 2021. *Laporan Perekonomian Provinsi Kalimantan Timur*. Samarinda.

[7] Ekananda, Mahyus. 2014. *Ekonomi Internasional*. Jakarta: Erlangga.

IOP Conf. Series: Earth and Environmental Science 1233 (2023) 012032 doi:10.1088/1755-1315/1233/1/012032

Menoreh Mountain sustainable tourism development: a spatial outlook

H Reinhart[1], M R Rafida[1] and T Sofie[1]

[1]Department of Development Geography, Faculty of Geography, Universitas Gadjah Mada
*Corresponding author: hilary.reinhart@ugm.ac.id

Abstract. Increasing development of geotourism and Geopark raises awareness of more environmental-friendly and sustainable tourism which can be approached using spatial analysis. This paper aims to connect the spatial element with sustainability tourism in Menoreh Mountain and formulate sustainable tourism development. Using Moran's Index, this study examines the Geo-tourism site's spatial characteristics and how the development should be done based on that analysis. For the result, 120 sites have been collected through Google Maps data and a field survey for validation. The sites comprise caves, waterfalls, pine forests, summits/peaks, and religious sites. By plotting the density of the tourism destination to a certain village, the Global Moran's Index for it is 0.318. The Univariate Local Moran's I show 10 villages in the High-High quadrant, 4 villages in the Low-High quadrant, 3 villages in the High-Low quadrant, and 2 villages in the Low-Low quadrant or cluster. Based on the spatial pattern, it is crucial to focus on the characteristic of each quadrant to establish sustainable tourism comprising economic and geological, and ecological sustainability and strengthen the spatial policy. The characteristics of each cluster also become a basis for developing strategies regarding the administrative division of Menoreh Mountain.

1. Introduction

The development of the Geopark in Indonesia is steadily increasing. New Geopark is established annually with more concise regulations issued (e.g., Presidential Decree 9/2019 On Geopark). This development deserves quite an appreciation since the tourism potency, scientifically and economically, in Indonesia is very promising. The geological richness with a huge market and the commitment from the government become attracting factors for the visitors. Moreover, the establishment of National Geopark with more and more Geoparks in Indonesia joining the Global Geopark Network also benefits the tourism sector by promoting them to a wider market [1, 2]. The constant new Geopark has to have remained because not only contributes to the local and regional economic development, it also promises to conserve the geological heritage within the Geopark area [3, 4] and also the sustainable development of the local and regional level.

Menoreh Mountain is a new promising geo-tourism destination area. It is located near a tourism epicenter of Yogyakarta Special Region and adjacent to the super-priority destination of Borobudur Heritage Area. Menoreh Mountain has many remarkable tourist attractions and varies from waterfalls, cliffs, and caves, with high biodiversity and cultures. In terms of geological features, it has a fascinating geological history with unique geological and geomorphological features. Nevertheless, the planning and tourism concept in this region is still lagging due to administrative issues as it is located in three different regencies. Without firm and proper sustainable and spatial tourism planning, this area could be left unmonitored and gradually deteriorated along with the irresponsible tourism practice. Also, with an abundant mineral resource of andesite rocks, this area faces imminent threats from quarrying and other mineral exploitation.

The sites and the attraction factor are *there*, whether they are on the earth's surface or below the ground, in some cases. Spatial aspects take account of the quality of tourism and the flow of capital in the economic dimension of tourism. It ensures the visitors gain their motivation of travelling within their movement from one site to the others. Therefore, spatial planning in tourism planning and development needs to be satisfyingly addressed particularly to ensure its sustainability [5]. At the operational level, the tourism spatial aspect comprises the site's network, the trail, and the spatial distribution of the sites.

Sustainable tourism is inherently embedded within the concept of Geopark and intertwined with spatial planning. The effort to grasp the relationships between spatial planning, including all of its elements and sustainable development has successfully signified the linkage between those two. Spatial analysis as the technical component of spatial planning could be used to identify the tourism development area of focus as concluded by Deng [6] and Rangel and Rivero [7] and then established into spatial planning. It is also proven to be crucial as the political basis for sustainable tourism development [8, 9]. However, the findings from this previous study need to be examined in the Indonesian context and taking Menoreh Mountain as an example and advanced tourism development strategies covering both economic and geological sustainability.

To respond to those issues, an integrated, strong tourism policy and planning among stakeholders are desperately required. This paper aims to open the narration of the tourism development in the Menoreh Mountain departing from the spatial ground by examining several spatial aspects of the tourism sites including their location, patterns, and aggregates. From that departing point, it is expected that sustainable tourism planning could be initiated for the preservation and conservation of the geological features at Menoreh Mountain and to strengthen the sustainability aspects. In the long-term, this paper also aspires to the possibility of establishing the Menoreh Mountain as, at least, a national Geopark or integrated tourism area.

2. Method

This paper mainly consists of spatial analysis as the method to obtain the research objectives. Spatial analysis or perspective is one of the foundation methods in regional development and geography. The analysis works with quantitative data hence the quantitative analysis. It helps the policy and sharpens the findings and the discussion entails. A more detailed elaboration of the research method could be seen in the following sections.

2.1. Research location

Menoreh Mountain is a mountain ranges in the central-southern part of Java Island. Administratively, this mountain range is located within 3 regencies of Kulon Progo (Yogyakarta Special Region), Magelang, and Purworejo (Jawa Tengah Province). Briefly, the main geological formation that builds this area is the Kebobutak Formation, a Miocene igneous rock that acts as the base of the Menoreh Mountain. On top of it, several other geological formations are laid out. Jonggrangan Formation is a Miocene limestone that has been karstified and hosts some of the unique karstic features including caves, karst depression, and an underground river. Numerous intrusions hills are also scattered in this area. In the northern part, a geological contact between the limestone and the Sumbing volcanic activity alters the limestone and forms a small portion of marbles [10]. The research location can be seen in Figure 1 below.

IOP Conf. Series: Earth and Environmental Science 1233 (2023) 012032 doi:10.1088/1755-1315/1233/1/012032

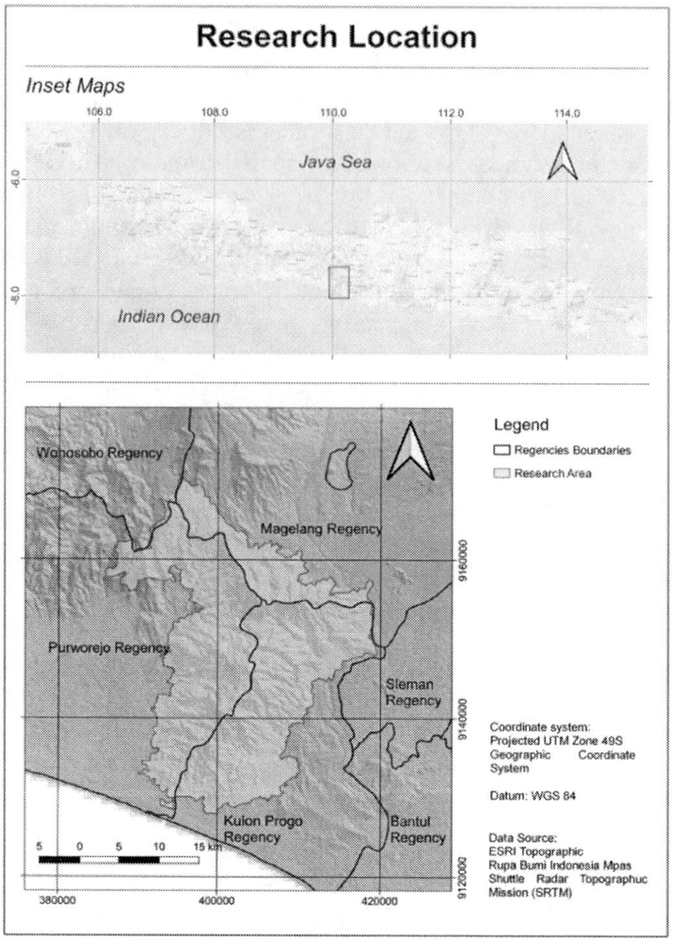

Figure 1. Research location takes place in three regencies sharing a border with each other's: Magelang, Purworejo, and Kulon Progo Regency.

2.2. Data collection

This research uses two types of data, namely primary and secondary data. The primary data are obtained from field surveys and direct observation while the secondary data are derived from previous studies or datasets, the data used are mostly spatial. Secondary data consist of the location of the tourism destination and the type of the destination. The location of the tourism and type of the destination is scrapped from Google Maps data [11]. We only select the data that have reviews, ratings, and pictures to ensure the data in Google Maps are valid. Data from the internet are then verified by field survey and direct observation. In the field survey, the point and location of the tourism destination are checked and documented using a camera. The location of the tourism destination is then combined with the administration and geological data. Administration data is downloaded from Ina-Geoportal (https://tanahair.indonesia.go.id/portal-web) and the geological data of geological formations and their properties are taken from a geological map. All of the data, either the secondary or the primary, is collected on 2022.

2.3. Data analysis

Spatial analysis as the primary analysis in this paper could take various forms. In this paper, we employ the spatial correlation of the tourism destinations and the village where they are located. The variable used in this spatial correlation analysis is the number of tourism destinations within a village. The spatial correlation is determined using the Moran's I, globally (Global Moran's I), and locally (Local Moran's I). Several previous studies use this instrument in the tourism sector including tourism revenue [6], tourism industry [12], and supply-demand [7]. Global Moran's I is calculated using this equation 1 below:

$$I = \frac{n}{W \sum_{i=1}^{n} z_i^2} \sum_{i=1}^{n} \sum_{j=1}^{n} w_{ij} z_i z_j \quad (1)$$

With Wij serves as the elements of the spatial weights matrix and n is the number of observations. Global Moran's I value varies from -1 (absolute dispersed) to 1 (absolute clustered). Between those two, zero value refers to no spatial correlation at all. Once the spatial correlation is obtained, the Univariate Local Moran's I could be deployed to examine the neighboring value of the spatial cluster. Univariate Local Moran's I yield 4 quadrants of the surrounding neighborhood. Both Global Moran's I and Local Moran's I require the spatial weight to assume the spatial relation. We use the Queen Contiguity where each neighbor or adjacent feature that shares a border is taken into account. Queen Contiguity is the most suitable due to the nature of the dispersion of tourism and the phenomenon of tourism spatial distribution occurs evenly in all the neighborhoods.

Univariate Local Moran's I result in 4 quadrants that explain the context of a particular feature toward its neighbor. Those 4 quadrants are High-High where a high-valued feature is surrounded by another high-valued, High-Low where a high-valued is surrounded by the low-valued, and Low-High where a low-valued feature is circled by the high-valued and Low-Low as the case where a low valued feature shares boundary with other low-valued features. This context would be the basis for the development of tourism in a village regarding the number of tourism destinations.

The mapping unit in this research is the village administration. Tourism destinations in each village are normalized to proportionate the value from the number (counts) to density (ranges) by dividing the number of tourism destinations in a particular village by its area. Data processing and compilation are done using QGIS software and the calculation of Moran's both Global and Local uses GeoDa. The spatial result from the Moran's I is then linked with sustainable tourism development by antagonizing the spatial cluster with the sustainable tourism principles, derived from the literature review and contextualized to the Indonesian context. This final step would be done using descriptive analysis.

3. Results and discussions

3.1. Tourism destinations of Menoreh

We collect 120 tourist spots in the whole of Menoreh Mountain and they are scattered in three regencies of Purworejo, Magelang, and Kulon Progo. These 120 tourist spots could be divided into five distinct features based on their main properties which are cave, pine forest, summit/peak, waterfall/cascade, and man-made. Caves, summit, and waterfall were formed and located due to the geological process and the hydrological characteristics of Menoreh Mountain hence it could be used and developed as a tourism destination.

Several tourism destinations have been already established and invite a significant number of visitors. Caves like Seplawan at Purworejo and Kiskendo Cave at Kulon Progo were developed in the 90s by the Regency authorities and initially created as cultural attractions. Kiskendo Cave is told to be the base of monkey troops in Ramayana folklore and a gold statue artifact was found in the Seplawan Cave. most of the summits or peaks are developed lately entailing the emergence of social media-selfie tourism. The most notable summit or peak tourism site and the most successful one is the Kalibiru photo spot in Kulon Progo which it experienced a sharp increasing visitor in the last five years. The Kalibiru phenomenon encourages many communities in Menoreh to open a similar tourism destination and it becomes typical: coniferous forest with a summit or a peak as the selfie spot. Another famous summit spot is the Suroloyo Peak which also had cultural value and Punthuk Setumbu at Magelang Regency where tourists could see the Borobudur Temple from above. Several tourism attractions in Menoreh Mountain can be seen in Figure 2.

IOP Conf. Series: Earth and Environmental Science 1233 (2023) 012032 doi:10.1088/1755-1315/1233/1/012032

Figure 2. Tourism Attractions in Menoreh Mountain (a) Ayunan Langit (b) Moyeng Peak (c) Kiskendo Cave (d) Gunung Gajah

Although they are fewer in number than the photo spot, the waterfalls are still favored by the tourist. Mudal Spring and Sidoharjo are the main stars. The community in Mudal Spring develops this site into a park to make it more attractive. Another hydrological attraction is the reservoir and the main and the largest example is the Sermo Reservoir, Kulon Progo. Water Reservoirs are mainly built to ensure the water availability for agriculture but then the community takes initiative to develop them into tourism destinations. Man-made features that serve as the attraction are religious spots such as Sendangsono and Lawaningsih Maria Cave, both at Kulon Progo. Sendangsono is a sacred spring for catholic people as it was a place where the early catholic people were baptized. The architecture of this site was designed by Mangunwijaya and bestowed by the award as it combined nature and man-made features. The summary of the geo-tourism destination could be seen in Table 1 while the spatial distribution of the tourism destination is shown in Figure 3.

Table 1. Several Existing Geo-Tourism Destinations at Menoreh Mountain

Types	Sub-Type	Examples
Caves		Seplawan, Kiskendo, Anjani, Sendangsri
Summits/Peaks		Suroloyo, Moyeng, Kunir, Gajah
Pine Forests		
Waterfall	Waterfall	Sidoharjo, Benowo

Types	Sub-Type	Examples
	Cascade (*curug*)	Curug Tawing
Man-Made	Water Reservoir	Sermo, Banjaroyo
	Historical Place	
	Religious Sites	Sendangsono, Lawaningsih

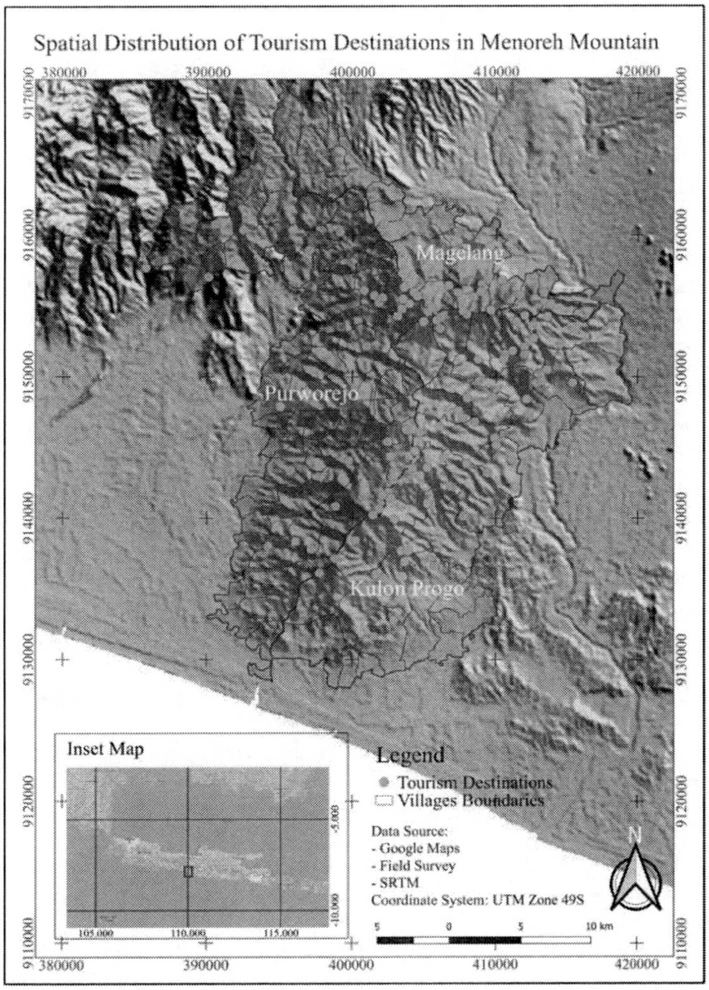

Figure 3. Spatial Distribution of Tourism Destination in Menoreh Mountain. Most of the destination is located in Kulon Progo Area

3.2. Spatial pattern

The number of tourism destination and the density of each village shows the clustered pattern with the Moran's I Global Value being 0.318. This positive Global Moran's Index indicates that the village with a tourism destination is aggregated into a certain location and holds a spatial significance context. The Moran's Scatter Plot from each village could be seen in Figure 4 below.

IOP Conf. Series: Earth and Environmental Science 1233 (2023) 012032 doi:10.1088/1755-1315/1233/1/012032

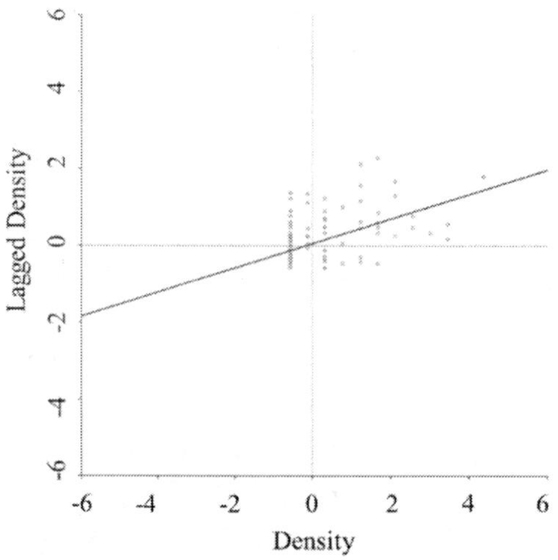

Figure 4. Global Moran's I Scatter Plot of Village with Tourism Density Distribution

After identifying the spatial relation and the Global Moran's I, the result could be derived to fathom how the spatial relation is in the local context using Local's I. The result from Local's I analysis yields several aggregates of the village with a certain number of densities in a couple of levels of significance as can be seen in Figure 5.

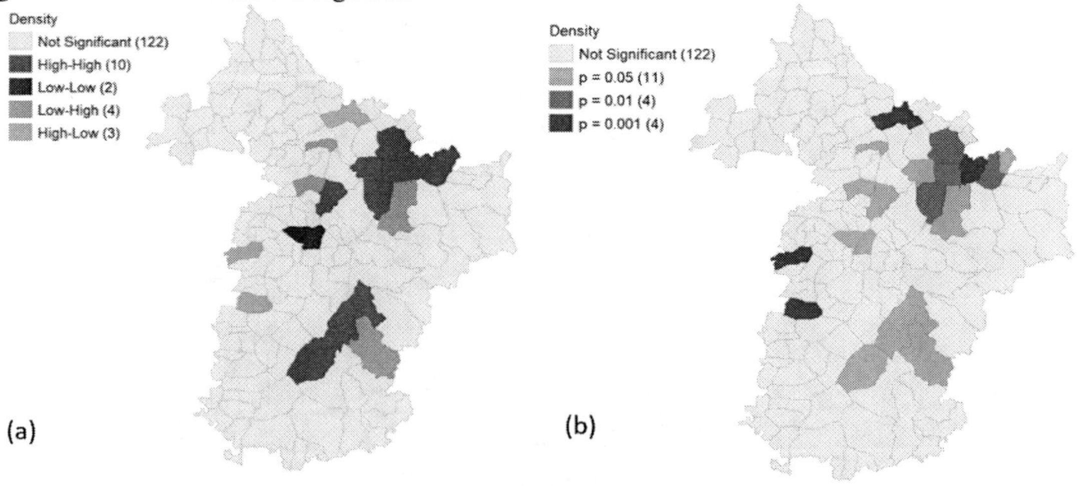

Figure 5. (a) LISA Maps for Tourism Destination Density at Each Village (b) Significance Value

Local Moran's I map visualizes several villages that are located at the four quadrants at the significant level of p = 0.05 (95%). 10 villages are located at the High-High quadrant (Jatimulyo, Hargowilis (Kulon Progo), Sedayu (Purworejo), Ngargosari, Ngargoretno, Giripurno, Ngadiharjo, Giritengah, Majaksingi, Ngargogondo (Magelang)), 2 villages at the Low-Low (Hardimulyo & Sudorogo (Purworejo), 4 villages at Low-High (Cacaban Lor, Ngargosari, Gerbosari, & Sidomulyo), and 3 villages at High-Low.

3.3. Enacting sustainable tourism development through spatial planning

Spatial planning is a crucial component in developing sustainable tourism and its roots in the spatial analysis of a particular region. As Aleksandar, et al pointed out [8] in the case of Bosnia and Herzegovina (B & H), spatial planning, from a political perspective of development, facilitates the zone of tourism. Spatial planning is also fruitful to find marketing spots for sustainable tourism development. In the case of Golija Mountain, Serbia, the spatial planning process yields five main areas of sustainable tourism development consists of biodiversity, tourism activity, rural development, research, and education as well [9]. A firm technical method will strengthen the policy and political process, and implementing the legal document will be more acceptable. GIS and Spatial clustering, including Moran's I, help the policy maker to decide on the program and the planning as part of the Decision Support System [13, 14] and also can be deployed in tourism planning [15].

Referring to the results of those studies, to establish sustainable tourism, the development of Menoreh Mountain should incorporate the spatial analysis at Spatial and Regional Planning (*Rencana Tata Ruang Wilayah* (RTRW)) of Detailed Spatial Planning (*Rencana Detail Tata Ruang* (RDTR)). The planning process also needs to be done by multi-level stakeholders and using a bottom-up process. GIS and spatial planning could also assist this process by illustrating spatial elements to the stakeholder. Spatial analysis in Menoreh Mountain also yields the area focus of sustainability, including the economical and geological sustainability which are the components of the Geopark framework.

Moran's I have been a handful of the instrument to display the economic sustainability of tourism development. Examples from Guangdong, China [6] and Extremadura Spain [7] demonstrate how the spatial cluster pictures economic factors and supply-demand as the basis to develop tourism sustainability. The spatial cluster has similar characteristics and properties due to the distance proximity, following the famous Tobler's first law of geography: everything is related but the nearer one is more related than the further [16]. Spatial planning for sustainable development in Menoreh Mountain also behaves under this law. The spatial cluster of High-High encompassing Jatimulyo, Hargowilis, Sedayu, Ngargosari, Ngargoretno, Giripurno, Ngadiharjo, Giritengah, Majaksingi, and Ngargogondo has more tourism sites than others and are more likely to develop a sustainable economic development by providing an integrated tourism pack or settling a travel pattern with their neighboring villages.

To complement the economical sustainability, spatial clustering and Moran's I are also could be used to obtain a basis for geological or ecological sustainability. It can be done through the characterization and defining the geological or ecological significance value and importance within the cluster. More than an economical aspect of development through tourism package in a particular cluster, the similarity of geological and ecological properties can also be expanded to formulate the sustainability of tourism.

3.4. Addressing the administrative authorization in the context of Menoreh Mountain

Menoreh Mountain is situated at a highly strategic location. It is adjacent to the Borobudur Temple Area which is designed to be a super-premium tourism destination. The Indonesian Government constantly provides the necessary infrastructure to execute the program and achieve the purposes of tourism development. Moreover, this area is located between Yogyakarta International Airport as the main gate for tourists to visit Borobudur Temple Area. By connecting those two through Menoreh Mountain, the government will be able to create numerous tourism destinations and trails which will impact the community within. That spatial context must be addressed to enforce sustainable development at Menoreh Mountain. From the result of this study, two fruitful tools for sustainable tourism development are: (i) spatial planning from the spatial analysis and (ii) the further Menoreh Mountain status to determine the most appropriate sustainable development framework.

From the Moran's I, it could be seen that the location of the tourism destination at Menoreh Mountain is patterned. It could be due to numerous factors but at a glimpse, based on the region, the administrative region holds a critical role. Kulon Progo and Magelang Area have spatially significant villages where they are related to other neighborhoods. In a nutshell, villages in the High-High quadrant can be integrated to create an integrated tourism area and employ the village authorities as the main drivers. Cooperation among village governments is relevant due to the financial power they have through village budgeting (*Dana Desa*). In different cases, the villages in the Low-High quadrant should be assisted by the surrounding and neighboring villages. Once the Low-High village is strengthened enough (in terms of having enough tourism destinations), it could be developed into an integrated area. A brief development strategy for each cluster can be seen at Table 2.

Table 2. Development Strategies for Each Cluster

Cluster	Villages	Sustainable Tourism Development Strategies
High-High	(Jatimulyo, Hargowilis (Kulon Progo), Sedayu (Purworejo), Ngargosari, Ngargoretno, Giripurno, Ngadiharjo, Giritengah, Majaksingi, Ngargogondo (Magelang))	Formulating village inter-cluster integrated tourism for economic sustainability, creating thematic geo-tourism/eco-tourism within the cluster
High-Low	Menoreh (Magelang), Pacekelan, Sidomulyo (Purworejo)	Dispersing tourists, strengthening low-density tourism village local institution from the high density.
Low-High	Cacaban Lor (Purworejo), Ngargosari, Gerbosari (Magelang), & Sidomulyo (Kulon Progo)	Focusing and allocating resources from the high-density village.
Low-Low	Hardimulyo & Sudorogo (Purworejo)	Establishing basic tourism elements (institution, destination)

A combination of the strength and the authority from adjacent villages could support each single destination spot into an area where it supports sustainability more satisfyingly by: (i) alleviating a concentrated environmental pressure from tourism activity by providing options, (ii) sharing and increasing profit and income from the tourism sector, (iii) decreasing the risk of social friction and conflict between people. Spatial planning is one of the foundations on which sustainable development is built [17-19]. Configuring the space and its utilization of it may strengthen the relationship among the social, ecological, and economical aspects as the main pillars of sustainable development.

Spatial planning needs to be boosted by the enactment of a status of an area for monitoring, regulation, and law enforcement. Evaluating the spatial distribution and the characteristic of the spatial aggregate where there is still a disparity in the tourism destination number between the three regencies, at a regional level, Kulon Progo, Purworejo, and Magelang have to establish cooperation to develop tourism in the Menoreh Mountain. One of the most appropriate statuses is the Geopark where it could integrate the three regencies into a collaboration based on geological and geographical integrity. Geopark also inherently contains sustainable development aspects through its three pillars of education, conservation, and local economic development. Furthermore, the geological richness and uniqueness are crucial assets for the development of geopark as the basis of geo-tourism.

4. Conclusion

Menoreh Mountain hosts numerous geo-tourism sites with various types and characteristics. Some of those sites have been developed into tourism destinations. A spatial analysis using Global and Local Moran's I show the spatial distribution of the destinations is aggregated regarding the adjacent neighbor. The Global Moran's Index is 0.318 which indicates a spatial relationship between the number of tourism destinations in a village and the Local Moran's I exhibit the context of the villages in 4 quadrants. This spatial pattern could be a basis for spatial development to help to achieve sustainable development. By further developing the High-High and Low-High quadrant, the government could create an integrated tourism area at the local or village level.

In Menoreh Mountain, spatial planning is required to enact tourism sustainable development and address the policy utilizing the technical method. Spatial cluster analysis can be used to develop economic and geological or ecological sustainability. Economic sustainability can be achieved by attracting and formulating an integrated tourism package among villages within a cluster whereas geological and ecological are used in the narration. Nevertheless, further study is required to formulate and link spatial analysis with geological and ecological sustainability.

Acknowledgments
This paper is funded from Young Lecturer Capacity Building Grant 2022 (Hibah Peningkatan Kapasitas Dosen Muda), based on the Letter 1312/UN1/DITLIT/Dit-Lit/PT.01.07/2022 and funded from the Directorate of Research, Universitas Gadjah Mada. The authors would like to thank the fund provider and all of the supporting parties.

References

[1] Rahmawati Y D, Agus R, Dirgantari P D and Nandi 2021 *Jurnal Geografi Gea* **21** 26-37 DOI: 10.17509/gea.v21i1.30618
[2] Yuliawati A and Budiman A 2017 *Proceedings of the 1st International Conference on Islamic Economics, Business, and Philanthropy* (ICIEBP 2017) 538-541
[3] Gonzalez-Tejada C, Du Y, Read M and Girault Y 2017 *Int. J. of Geoheritage* **5** 1-20. 10.17149/ijg.j.issn.2210.3382.2017.02.001. halshs-02087621
[4] Štrba L, Kolackovská J, Kudelas D, Kršák B and Sidor C 2020 *Sustainability* **12** 2979 DOI: doi:10.3390/su12072979w
[5] Risteski M, Kocevskia J and Arnaudov K 2012 *Procedia - Social and Behavioral Sciences* **44** 375–386 DOI: 10.1016/j.sbspro.2012.05.042
[6] Deng G 2020 Advances in Economics, Business and Management Research **153** 298-302
[7] Rangel M C R and Rivero M S 2020 *Sustainability* **12** 1651 doi:10.3390/su12041651
[8] Aleksandar A, Marija K, Rosalina G 2017 *Bus. Studies* **9** 271-284
[9] Tomić N and Stojsavljević R 2013 *European Researcher* **65** 2918-2929
[10] Rahardjo W, Sukandarrumidi and Rosidi H M D 1977 *Peta Geologi Lembar Yogyakarta, Jawa* (Bandung: Departemen Geologi)
[11] Iswandhani N and Muhajir M 2018 *J. Phys.: Conf. Ser.* **974** 012033
[12] Adabi G, Hajiha A, Lotfi F H 2021 *J. of Industrial Engineering and Management Studies* **8** 261-276 DOI: 10.22116/JIEMS.2022.138866
[13] Fedra K and Reitsma R F 1990 *Decision support and geographical information systems* ed Scholten H J, Stillwell J C H (Dordecht: Springer) https://doi.org/10.1007/978-94-017-1677-2_16
[14] Halbich C and Vostrovský V 2011 *Agris on-line Papers in Economics and Informatics* **3** 67-73
[15] Masron T, Mohamed B and Marzuki A 2015 *Theoretical and Empirical Res. in Urb. M.* 10 21-35
[16] Miller HJ 2004 *An. of the Ass. of American Geogr* **94** 284–289
[17] Shen Z and Kawakami M 2013 *Overview: Spatial Planning for Achieving Sustainable Urban Forms* ed Kawakami, M et al. (Dordrecht: Springer Science+Business Media) DOI 10.1007/978-94-007-5922-0_1,
[18] Gorzym-Wilkowski, W A 2017 *Barometr regionalny* **15** 75-85
[19] Silver C 2014 *Jurnal Perencanaan Wilayah dan Kota* **25** 115-125

Understanding Kulon Progo through Historical Maps

A Cahyono[1,2,*], B Susilo[2], N Wirasanti[3], N Rahardjo[2], Y O S Albadara[4], M C Berlyanti[4]

[1] Doctoral Program in Geography, Faculty of Geography, Universitas Gadjah Mada
[2] Geographic Information Science Department, Faculty of Geography, Universitas Gadjah Mada, Indonesia
[3] Archeology Department, Faculty of Cultural Sciences, Universitas Gadjah Mada, Indonesia
[4] Undergraduate Program in Cartography and Remote Sensing, Faculty of Geography, Universitas Gadjah Mada, Indonesia
*Corresponding author: aricahyono@ugm.ac.id

Abstract. Kulon Progo Regency, part of the Special Region of Yogyakarta, has undergone political and administrative divisions, thereby leading to a dynamic geographical name change. Presently, this regency does not yet have a database for geographical names, which are part of cultural heritage to reveal historical events. Therefore, this study aims to conduct an inventory of geographical names in the Kulon Progo Regency through multitemporal historical maps and interpret their meanings. The inventory was carried out by digitizing scanned topographic maps stored on the Leiden University repository website and interpreting the meaning of geographical names through dictionaries and history books. Inventory using historical maps is an effective method of revealing past events. The result found that physical geographical names dominated 78 or 54% of villages, with 45%, 35%, and 20% consisting of biotic, non-biotic and topological aspects. Meanwhile, non-physical geographical names comprised 61 villages, incorporating 26%, 8% and 10% social, economic and cultural aspects.

Keywords: Kulon Progo, geographical names, historical maps, Yogyakarta, administrative

1 Introduction

Toponymy assists scholars who study cultural geography by revealing place names' location, usage, and origin. Its interpretation tends to be helpful in describing cultural diversity and determining environmental changes in the past and present [1]. Geospatial Information Law No. 4 2011 was updated to government regulation no. 45 2021, stated that geographical names (toponyms) are part of the Base Map Layer [2]. These are attached to relevant human activities such as objects, trade, flora, fauna names, etc.

Changes in geographical names tend to occur occasionally, irrespective of their dynamic properties. The natural and artificial phenomena influence these in an area [3]. Kulon Progo Regency is typical proof of a change in the geographic name of the district. It is presently experiencing certain dynamics due to regional autonomous policies. Moreover, regional autonomy boosts societal welfare through improved services, empowerment, community participation, and increased competitiveness. Since its emergence in the 1950s, Kulon Progo Regency has resulted from the merger of Adikarto Regency and Kulon Progo Regency. This also led to the eradication of several administrative areas, which in turn had an impact on geographical names in the past. Presently, Kulon Progo Regency does not yet have a database for geographical names in its area. This aspect of cultural heritage is used to reveal events that occurred in the past or during the formation of the region.

Spatial analysis converts raw data into valuable information related to the set goals [4]. It is usually displayed on a map to provide information about a place or its geography. Historical maps are one of the sources used to develop a geographic name database in the Kulon Progo Regency. It is defined as a nation's physical and cultural portrait at a certain time. A map shows how a particular area looked before it was developed as well as provides a detailed view of its changes over time. Indonesia was formerly a Dutch colony, where several colonial countries had a database of old maps of their colonies. Therefore, historical or Dutch maps showing the territory of Indonesia were published during the Dutch colonial period in the nineteenth and twentieth centuries. It showed the areas equipped with geographical names that matched their pronunciation and were written in Dutch [5]. The maps produced by the Dutch were written in their language [6]. The rapid development of science and technology also impacted the development of cartography. This is influenced by the peoples' perspectives and way of life in every era, civilization, culture, science, and technology in describing the environment [7]. These physical and cultural portraits are currently developed in printed or virtual forms that digitally store geographic information or old maps [8]. Technological developments in the current era are used for digitizing Dutch maps.

Based on the description above, this study aims to carry out an inventory and interpretation of geographical names through multitemporal historical maps and develop a database for Kulon Progo Regency. This is intended to enrich the geographic name information because it is an important element of the base map layer. In addition, the information contained in the database is perceived as a cultural heritage of the Kulon Progo Regency.

2 Study Area

This research was carried out in Kulon Progo Regency, geographically located at 7°38'42" - 7°59'3" S and 110 1'37" - 110 16'26" E. It is part of the Special Region of Yogyakarta, which is situated in the country's westernmost region. Kulon Progo Regency is bounded in the west and north by the Province of Central Java and in the south by the Indian Ocean. Physiographically, it is a flat area surrounded by mountains, mostly located in the northern region, as shown in Figure 1. Interestingly, 17.58%, 15.20% and 22.85% of its area are situated at altitudes of <7 m, 8 to 25, and 26 to 100 m above sea level, respectively. Furthermore, 33% and 11.37% of its area are at an altitude of 101 to 500 m and > 500 m above sea level, respectively [9]. Historically, the current territory of Kulon Progo Regency is an amalgamation of two regencies. These are Kulon Progo and Adikarto Regencies situated in the territories of Kasultanan Ngayogyakarta Hadiningrat and Kadipaten Pakualaman, respectively [10].

Figure 1: Study area map in Kulon Progo Regency with terrain information represented by hillshade followed by the river stream
Source: Shuttle Radar Topography Mission (SRTM-30 M), 2021 Google Earth Engine, BaseMaps from Geospatial Information Agency

The main data used in this study is a historical map collection from the Royal Tropical Institute and uploaded on the Leiden University Library website (http://maps.library.leiden.edu/apps/s7). This was because it showed the initial geographical names in the Kulon Progo Regency before they were changed as well as the relationship of the geographical names with the historical or original physical condition of the region. According to Apriadsa et al. [11], the most recent maps of the Dutch colonial era contain more complete geographical names than the latest paper version of the RBI physical and cultural portrait in Java. Irrespective of the fact that there were slight alterations in some places due to the scale of the map, yet valid information was provided. The Archived Map consists of 1938 46-XLIId 25k Midden-Java en Gouvt Jogjakarta Kaligesing, 1934 46-XLIIh 25k Midden-Java en Gouvt Jogjakarta Sermo, 1936 46-XLIIl 25k Midden-Java en Gouvt Jogjakarta Bandjarsari, 1935 46-XLIIm 25k Gouvt Jogjakarta en Midden-Java Krijan, 1935 46-XLIIq 25k Java Gouvt Jogjakarta Garongan, 1938 46-XLIq 25k Midden-Java en Gouvt Jogjakarta Goenoeng Koenir, 1927 47-XLIIa 25k Java Res Jogjakarta Balong, 1927 47-XLIIb 25k Java Resn Jogjakarta en Kedoe Gagan, 1935 47-XLIIe 25k Java Gouvt Jogjakarta Nanggoelan, 1935 47-XLIIf 25k Java Gouvt Jogjakarta Godean, 1934 47-XLIIn 25k Java Gouvt Jogjakarta Sewoegaloer, 1934 47-XLIIo Gouvt Jogjakarta Kabonongan, 1935 48-XLIIk Jogjakarta 25k Pandak, and 1934 47-XLIIi 25k Java Gouvt Jogjakarta Wates including the first edition of AMS (1965 5020-III 50k AMS Series T725 Salaman). The entire map was published in the 19th and 20th centuries, during the Dutch colonial period in the East Indies.

2.1 Georeference
Before usage, the old map needs to be integrated into the same coordinate system through georeferencing. This is important because it shows the actual coordinates of the earth. Georeference is carried out at four points, including the map's right, left, top, and bottom corners, by entering the coordinates of the intersection between the latitude and longitude. The datum used is WGS 84 because it is commonly utilized in Indonesia, and the process involves adopting mapping software.

2.2 Map Digitalization
The geospatial data on the archived map is an analogue physical and cultural portrait. It is transformed into a digital map by converting the raster to a vector format through a digitization process, which was initially georeferenced to show the actual coordinates of the earth. The only digitized map information is geographic names, which was converted into vector format (shapefile) and point form.

2.3 Build Database
Geographical names that have been digitized are further grouped based on elements such as Administrative, Water, and Relief. The focus of this research is on Administrative features. Afterwards, a database of geographical names was generated for Kulon Progo Regency. This is a collection of data stored systematically, which is further processed to produce information. The geographic name database is in the form of a digitized geographical names point shapefile table containing information concerning the element. These include specific and general names on the map, the meaning of the geographic name, and its history. Incidentally, the meaning of geographic name was obtained from the Javanese-Indonesian, the Old Javanese Language Dictionaries, and the Kulon Progo Village History Book. After filling in the attributes of the table, then grouping based on the category was discussed in detail.

3 Result and Discussion
The digitization of geographical names is carried out on maps that were appropriately georeferenced to determine the appropriate coordinates of the earth. The present research focused on data generated from geographical names of villages in the Kulon Progo Regency. Through Governor Regulation No. 25 of 2019, the sub-district government unit in the region is called Kapanewon, while that of the city is referred to as Kemantren. In addition, the village government unit is called the Kalurahan. The inventory obtained from the digitized results shows dissimilarities in the number of village listed on the map and those registered in Declaration Number 5 of 1948 concerning changes in their names. The village's name may be changed, included, or occasionally removed due to human activities. The digitization results produced geospatial data in digital form, which was easily used in this study. Therefore, it is a known fact that the administration of Kulon Progo Regency consists of Kapanewon (districts) and Kalurahan (villages), besides the need to possess a similar geographic name.

The meaning of geographical names is deciphered by studying hermeneutical, which is how to interpret these terms. Its analysis was carried out descriptively based on the meaning of the names found in the Javanese-Indonesian, and the Old Javanese Language Dictionaries, as well as history books. The results are used as a symbol of the original formation of geographical names and serve as a reference during grouping. The geographical names studied are the specific names of villages in Kulon Progo Regency. It also consists of a single name, such as Garongan, with more than one syllable, including Temon Wetan.

Table 1. Category of geographical names in Kulon Progo

Category	Aspect	Sub-Aspect	Geographical Names
Physical	Biotic	Plant	Klepoe, Karangwoeni, Bandjaran, Sogan, Kembang, Penggoeng, Ploembon, Sambiroto, Pendjalin, Kemiri, Kamal, Kepoendoeng, Salamrejo, Tandjoeng, Glagah, Barongan, Degan, Kalinongko, Kedondong, Nogosari, Petet, Sambiroto, Soeloer, Soeren, Kalibawang, Karangardjo, Tegalardjo, Gadingan, Kalimenur, Kaliwinong, Tjerme, Sentolo, Worawari, Kemiriombo, Djatiroto
		Animal	Poetjoeng, Tjabean
	Non-Biotic	Water	Kaliredjo, Kaliandong, Djoeroeg, Banjoemeneng, Kalikoetoek, Kalisono, Kaliwiroe, Kaliboeka, Bodjong, Plered, Soemberredjo, Bantar, Kedoengtangkil, Kalibondol, Serang
		Land	Tegalsari, Banaran, Bonoredjo, Kenteng, Boto, Toekardjo, Watoemoerah, Sindoetan, Paras, Grahoelan, Wadas
	Topology	Shape	Balong, Djoemblangan
		Location	Tambak, Wates, Pereng, Tandjoenggoenoeng, Koetogiri, Djoenggrangan, Menggermalang, Ngargowahono, Poendak, Djangkaran, Ledok, Milir, Pandjatan
Non-physical	Social	Character Name	Djombokan, Koeloer, Kradenan, Boegel, Kranggan, Lendah, Trajoe, Sermo, Dekso, Samigaluh, Karangwoeloeh, Garongan, Demangan, Gotakan
	Economy	Trade	Plano, Bangeran
		Agriculture	Kebonredjo, Wonopeti, Krembangan, Wanasidi
	Culture	Building	Bendoengan, Nanggoelan, Prigiwatoe, Maesan, Pendem, Sewoegaloer
		Object	Pripih, Potrogaten, Dalangan
		Story	Gegerbadjing, Kalidengen, Temon Wetan, Djanten, Temon Koelon, Ngentak, Doekoeh, Kedoendang, Kaligintoeng, Demen, Kokap, Djatingarang, Pantaran, Depok, Kanoman, Soengapan, Brosot, Kempleng, Boro, Sanggrahan, Palihan, Sawon, Widjilan, Boetoeh, Dajakan, Gebangan, Gletak, Pandowan, Koelwaroe, Kalidengen
		Art	Djantoeran, Tajuban

Source: Data Analysis Result, 2022

Meanwhile, not all meanings of geographical names were obtained from dictionaries or history books. The unknown geographical names are assumed to be derived from a foreign language or a combination of regional activities that occurred in the past and were not recorded in history books. The origins and meaning of these geographic names were not found, such as Promasan, Pengos, Kleben, Krijan, Ngipikredjo, Semaken, Tempel, Senden, Toebin, Toerip, Sewonan, Temanggang, Bandjarardjo, Gegoeloe, Kasijan, Lemoeroe, and Imorenggo.

Geographic name grouping was carried out by referring to the results obtained from deciphering their meanings. Furthermore, geographical names with complete information related to meanings from dictionaries and history books tend to show how they are related to spatial data (Figure 2.). The analysis of the spatial associations between the geographical names and the regional characteristics that underlie each geographical name of the village is grouped into two, namely physical and non-physical categories.

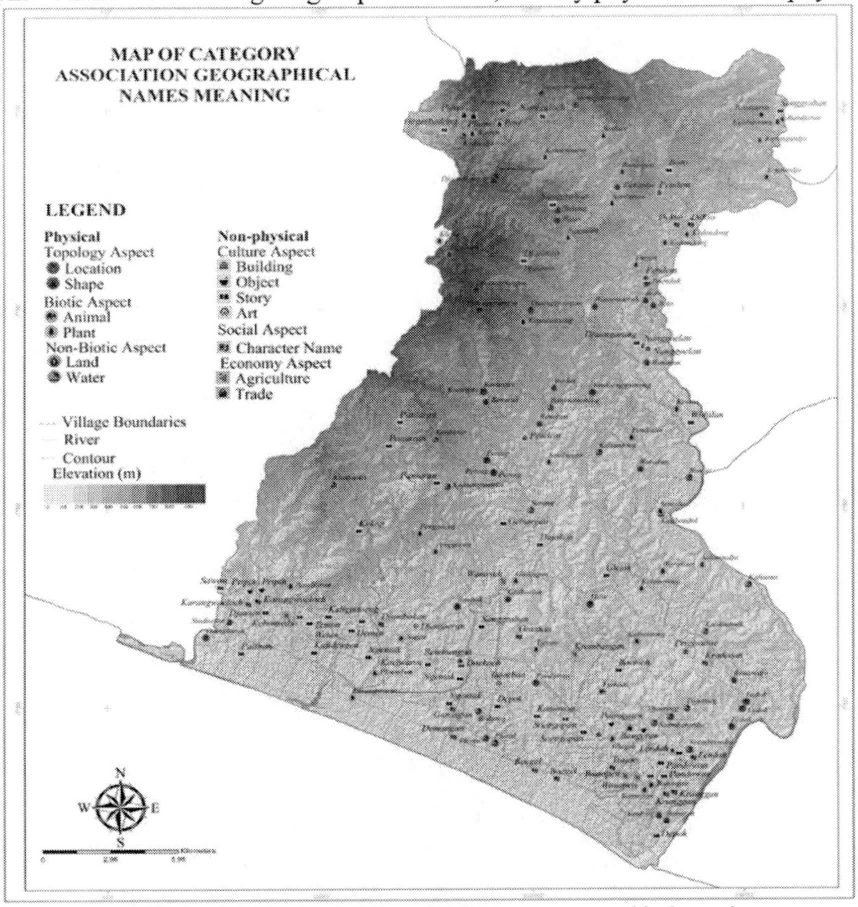

Figure 2: Map of category association name geographical meaning
Source: Data Analysis Result, 2022

3.1 Physical Category

The physical category includes 78 geographical names or 56% of the total known village geographical names. These are categorized into biotic, non-biotic, and topological aspects. Meanwhile, the biotic feature consists of plant and animal sub-aspects. The plant sub-aspect group consists of 35 villages or 45% of the total geographical names grouped under the physical category, namely Klepoe, Karangwoeni, Bandjaran, Sogan, Kembang, Penggoeng, Ploembon, Sambiroto, Pendjalin, Kemiri, Kamal, Kepoendoeng, Salamrejo, Tandjoeng, Glagah, Barongan, Degan, Kalinongko, Kedondong, Nogosari, Petet, Sambiroto, Soeloer, Soeren, Kalibawang, Karangardjo, Tegalardjo, Gadingan, Kalimenur, Kaliwinong, Tjerme, Sentolo, Worawari, Kemiriombo, and Djatiroto. According to its meaning in Javanese and its historical origin, the village is similar to plants. In Kalurahan Pendjalin, it means rattan, based on the Javanese-Indonesian Dictionary. Historically, Kalurahan Kalibawang originated from an area where onion plants were cultivated in large quantities. Generally, the relationship plants have with the formation of geographical names in the Kulon Progo Regency is listed specifically concerning the type, which is not explicitly symbolized on the map. The animal sub-aspects group consist of two or three% of the total geographical names categorized as physical. The villages are Kalurahan Poetjoeng and Kalurahan Tjabean. Based on the Javanese-Indonesian Dictionary, Poetjoeng means the name of an animal, 'pucung'. At the same time, Tjabean, or its pronunciation, namely Chilian, is based on the Javanese dictionary, namely the name of a small bird that eats caterpillars.

Geographical names included in the non-biotic features are categorized in the water and soil sub-aspects. The water sub-aspect comprises 15 villages or 19% of the geographical names categorized as physical, namely Kaliredjo, Kaliandong, Djoeroeg, Banjoemeneng, Kalikoetoek, Kalisono, Kaliwiroe, Kaliboeka, Bodjong, Plered, Soemberredjo, Bantar, Kedoengtangkil, Kalibondol, and Serang. According to its meaning in Javanese, the village has similar meanings to river. Bantar in the Javanese language dictionary means rapids or waterfalls. If seen on the map, it is located near Kali Progo, directly adjacent to the river in the east, as shown in Figure 3. Another example is the Kalurahan Serang, whose name was derived from Kali Serang because it passes the village, as shown in Figure 4. This fairly large river plays an important role in balancing ecosystems and regional development in the DIY Province.

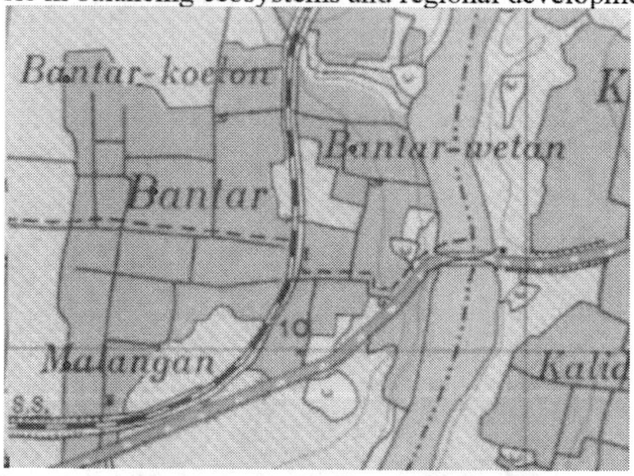

Figure 3: Kalurahan Bantar is directly adjacent to Kali Progo [14].

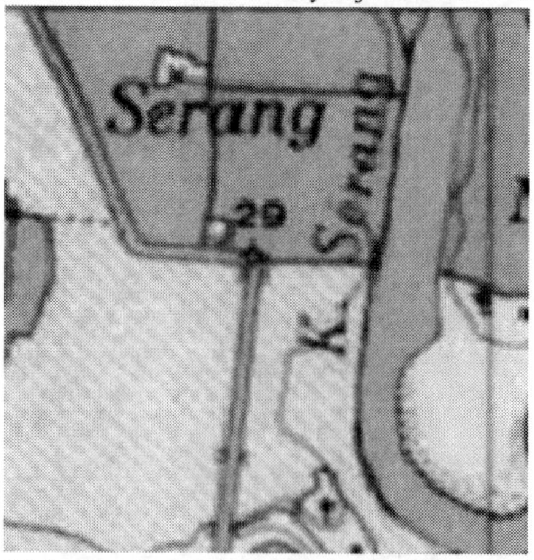

Figure 4: Kalurahan Serang is associated with Serang River [15].

Generally, the relationship between several geographical names and rivers or springs is evident in the word 'Kali'. Although, this needs to be reviewed by studying previous maps to determine whether or not the area is near a river. This is because the naming of times is also used in several geographical names that are not directly related to rivers or springs. The land sub-aspect consists of 11 villages or 14% of the total geographical names categorized as physical, namely Tegalsari, Banaran, Bonoredjo, Kenteng, Boto, Toekardjo, Watoemoerah, Sindoetan, Paras, Grahoelan, and Wadas. According to the meaning in Javanese or historical stories, the village has certain meanings related to soil or stone. Toekardjo consists of two words, namely 'toek', which means to obtain in the Javanese dictionary, and the terms 'ardjo or harja', meaning fertile. Toekardjo is a fertility place, and this is closely related to the soil. From the map, soil fertility in this village is evidenced by the discovery of several plants in the area, as shown in Figure 5.

Figure 5: Toekardjo indicates soil fertility marked by the number of plants [16].

Geographical names included in the sub-aspect of soil have a similar meaning as a stone and indicate a specific type of rock, such as Kalurahan Boto, which depicts red stone. Furthermore, Kalurahan Paras and Kalurahan Kenteng imply rock-solid and stone, respectively. The interpretation of the meaning of geographical names related to stones cannot be directly proven on the map because they do not show a specific symbolization.

Meanwhile, geographical names included in the topological features are categorized based on their form and location sub-aspects. The form sub-aspect group comprises two villages or 3% of the total geographical names categorized as physical, namely Balong and Djoemblangan. According to its meaning in Javanese, the village is related to the shape of the area itself. The Javanese-Indonesian dictionary stated that Balong means curvy and has a lot of water. When viewed from the map, Kalurahan Balong is situated in an area with curvy contours, meaning it forms a valley and mountains. It is also fed by several tributaries, meaning the area has a lot of water, as shown in Figure 6.

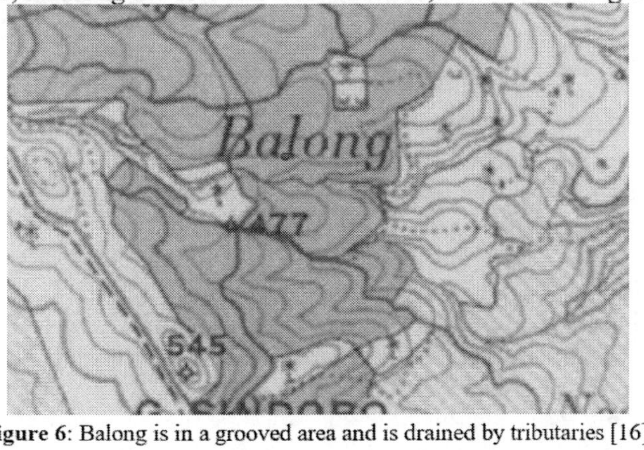

Figure 6: Balong is in a grooved area and is drained by tributaries [16].

The location sub-group comprises 13 villages or approximately 17% of the total geographical names categorized as physical, namely Tambak, Wates, Pereng, Tandjoenggoenoeng, Koetogiri, Djoenggrangan, Menggermalang, Ngargowahono, Poendak, Djangkaran, Ledok, Milir, and Pandjatan. The meaning of village in Javanese is related to location, such as the altitude and distance to another place. Djoenggrangan means 'high' based on the Javanese-Indonesian dictionary, depicting a high place. Kalurahan Djoenggrangan is located with dense contours, as shown in Figure 7. The closer a contour, the more steep the area. A vertical region implies the area is located on higher ground.

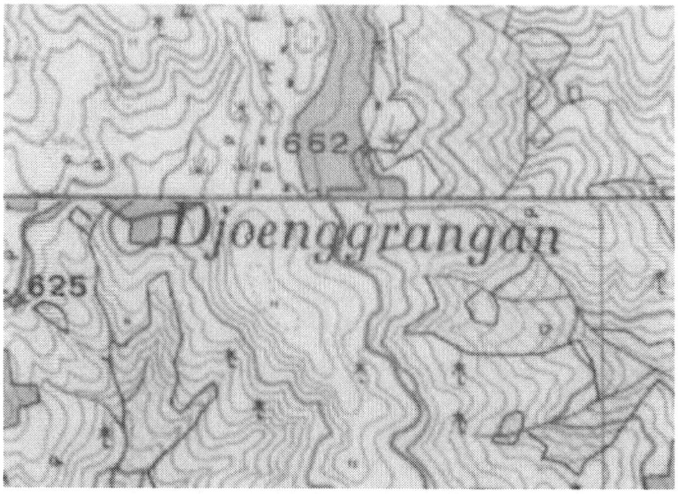

Figure 7: Djoenggrangan shows the height of the place seen from the density of the contour [17]

3.2 Non-physical Category

The non-physical group comprises 61 geographical names or 44%, categorized in social, economic, and cultural aspects. The social feature only consists of the sub-aspect of the character's name, with 14 villages or 23% of the total geographical names categorized as non-physical, Djombokan, Koeloer, Kradenan, Boegel, Kranggan, Lendah, Trajoe, Sermo, Dekso, Samigaloeh, Karangwoeloeh, Garongan, Demangan, and Gotakan. The meaning of the figurehead's name is known from the history or origin of the village. This is generally adopted from the name of the big figure who influenced the establishment of the village. Like the geographical name in Kalurahan Dekso, the name Dekso comes from the word Daksya, or Rakai Daksya, a King of Mataram Shiva who it was believed to have lived temporarily in the village on the outskirts of Progo [12]. Meanwhile, from a local story, it was reported that the name Dekso was also coined out from the name of a great figure in the Tumenggung Joyopidekso [13]. Based on this opinion, it was interpreted that Dekso is identical to the name of a big figure who has a connection with the village. Another example of a geographical name in the sub-aspect of a character's name is Garongan, proven by the fact that a leader gives orders and is quite loud on the street while going around all the residents to ensure that the command can be heard and this is known as Ki Lurah Garong.

The economic features consist of sub-aspects of trade and agriculture. Moreover, the trade sub-aspect group comprises two villages or relatively 4% of the total geographical names categorized as non-physical, namely Plono and Bangeran. It was included in this activity because it is recorded that trade, namely the market, has a large influence on the village. Kalurahan Bangeran is related to the existence of Bangeran Market, which has existed since the Dutch colonial era. Kalurahan plono,'plono' literally means 'lapak', a place that is often associated with buying and selling activities. Similar to Kalurahan Bangeran, Kalurahan Plono also has a market.

The agriculture sub-aspect group comprises four villages or 7% of the total geographical names categorized as non-physical, namely Kebonredjo, Wonopeti, Kembangan, and Wanasidi. The village has a linguistic meaning related to agriculture and forestry. In Javanese, Kebonredjo is a geographic name, although its specific name is derived from a combination of 'Kebon', meaning field or land, and 'redjo', depicting fertile. All plants planted in Kebonredjo land usually thrive. Kebonredjo has seen fertile plants, such as bananas, coconuts, rice, etc. Even during the dry season, many people actively cultivate crops, including watermelon, melon, chilli, shallots, etc. The map shows that Kebonredjo is surrounded by features symbolized as agriculture or rice fields, as shown in Figure 8. Similarly, Wonopeti village is also a specific combined name. In Javanese, it comes from' Wono', which means forest, and 'Peti', depicting quiet, therefore it is interpreted as a quiet forest. However, when viewed from the map, Wonopeti is not symbolized as a forest but as a settlement. It is assumed that this decision was made long before the settlement, leading to a deserted forest that was rarely touched. This is proof that some of the geographical names in Kulon Progo Regency were formed long before the map was generated, thereby depicting that their meaning, name and appearance on the map are sometimes irrelevant, therefore, supporting data is needed.

(a) (b)

Figure 8: (a) Kebonredjo surrounded by agriculture [18], (b) Wonopeti name that don't show forest [19].

The cultural aspect consists of buildings, objects, stories, and arts sub-aspects. The building sub-aspect group comprises six villages or 10% of the total geographical names categorized as non-physical, namely Bendoengan, Nanggoelan, Prigiwatoe, Maesan, Pendem, and Sewoegaloer. In Javanese, village is related to buildings. Bendoengan and nanggoelan mean 'dam' and fortress, respectively. Geographical names with broad implications for buildings are not directly seen on the map because the scale is not detailed. Few geographical names can be directly related to the map, such as Kalurahan Pendem, which means cemetery (Figure 9.). The object sub-aspect group comprises three villages or 5% of the total geographical names categorized as non-physical, namely Pripih, Potrogaten, and Dalangan. In Javanese, the village is related to objects. Geographical names related to objects cannot be directly seen on the map but needs to be proven directly in the field.

The story sub-aspect group comprises 29 villages or 49% of the total geographical names categorized as non-physical, namely Gegerbadjing, Kalidengen, Temon Wetan, Djanten, Temon Koelon, Ngentak, Doekoeh, Kedoendang, Kaligintoeng, Demen, Kokap, Djatingarang, Pantaran, Depok, Kanoman, Soengapan, Brosot, Kempleng, Boro, Sanggrahan, Palihan, Sawon, Widjilan, Boetoeh, Dajakan, Gebangan, Gletak, Pandowan, and Koelwaroe. The grouping of these villages is based on their historical origins or formation, which is a fairly long story. One example is Kalurahan Brosot, in which the term 'Brosot' is derived from the word 'brojol' or 'mbresotan', related to the character Ki Truno Bongso who often leaves without saying goodbye after saving Kalurahan from every threat of danger [13]. The art sub-aspect group consists of two villages or 3% of the total geographical names categorized as non-physical, namely Djanturan and Tajoeban. Djanturan means a story or explanation told by an insider during a wayang art performance. At the same time, Tajoeban is the art of tayuban or tayub traditional dance, which originated in Central Java and the Old Javanese Kingdom.

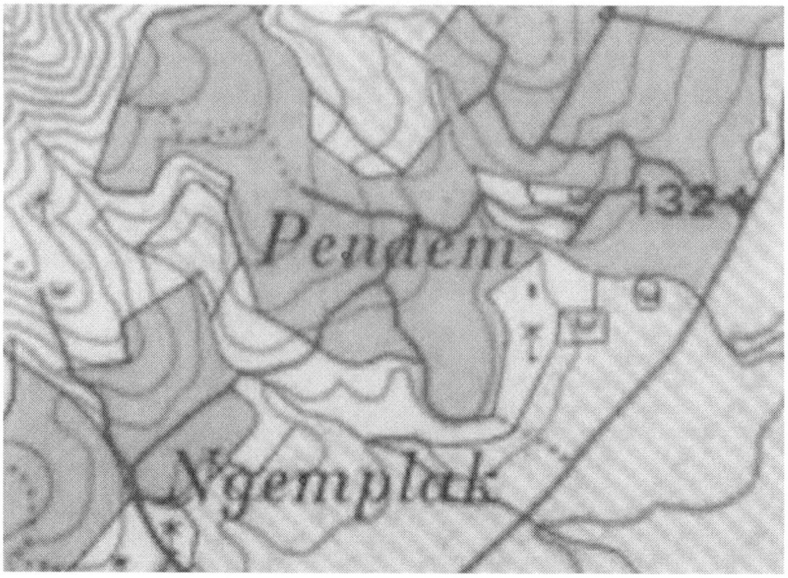

Figure 9: Geographical name Pendem is related to graves [20]

4 Conclusions and Recommendations

In conclusion, the inventory of geographical names effectively reveals past events utilizing multitemporal historical maps. The interpretation of the meaning of village names in the Kulon Progo Regency can be obtained through the Javanese – Indonesian Dictionaries, Old Javanese Dictionaries, and Kulon Progo Literatures, including historical books. In accordance with the interpretation results, it was discovered that physical geographical names dominated the name of villages in Kulon Progo Regency, with as many as 78 villages (54%), comprised of 45% biotic, 35% non-biotic, and 20% topological aspects. Meanwhile, non-physical geographical names comprised 61 villages, 26% social, 8% economic and 10% cultural aspects.

5 Acknowledgements

The authors are grateful to the Leiden University Library for the opportunity and permission to access their collection of maps and the Indonesia Geospatial Information Agency for the support. This research was funded by several research schemes, such as The Faculty of Geography Lecturer Research Grant (*Hibah Dosen Fakultas Geografi*), which supported data collection and field survey, as well as the RTA (*Rekognisi Tugas Akhir*) UGM Grant for conference publication. The authors are also grateful to the government of Kulon Progo from village to regency level for their assistance in collecting secondary data.

6 References

[1] F. Wang, L. Zhang, G. Zhang, and H. Zhang, "Mapping and spatial analysis of multiethnic toponyms in Yunnan, China," *Cartogr. Geogr. Inf. Sci.*, vol. 41, no. 1, pp. 86–99, 2014, doi: 10.1080/15230406.2013.831529.

[2] Presiden Republik Indonesia, "Peraturan Pemerintah Republik Indonesia Nomor 45 Tahun 2021 tentang Penyelenggaraan Informasi Geospasial." 2021.

[3] A. Apriadsa, H. S. Anggoro, A. Cahyono, and R. Apriadna, "Unveil the 'lost toponyms' in the northern part of the Menoreh Mountains, Java, Indonesia," *Int. J. Cartogr.*, May 2019, DOI: 10.1080/23729333.2019.1611025.

[4] P. A. Longley, M. F. Goodchild, D. J. Maguire, and D. W. Rhind, *Geographic Information*

Science and Systems, 4th ed. New Jersey, United States: John Wiley & Sons, 2015.
[5] D. F. Lach, "Asia in the eyes of Europe: sixteenth through eighteenth centuries," 1991. https://www.lib.uchicago.edu/e/su/southasia/lach.html (accessed Nov. 30, 2017).
[6] F. Ormeling, "Colonial Cartography of the Netherlands East Indies 1816-1942," 2005.
[7] A. Cahyono, "Studi nama geografi melalui layanan pemetaan urundaya di Desa Giripurwo, Purwosari, Gunungkidul D.I. Yogyakarta (Geographical names study through crowdsourced mapping services in Giripurwo Village, Purwosari, Gunungkidul D.I. Yogyakarta)," *Spatial*, vol. 18, no. 2, pp. 105–114, 2018.
[8] M.-J. Kraak and F. Ormeling, *Cartography : Visualization of Geospatial Data*, 4th ed. Florida, United States: Taylor & Francis, 2021.
[9] Pemerintah Kabupaten Kulon Progo, "Kondisi Umum," 2020. https://kulonprogokab.go.id/v31/detil/7672/kondisi-umum (accessed Mar. 12, 2021). Pemerintah Kabupaten Kulon Progo, "Sejarah," 2020.
[10] Pemerintah Kabupaten Kulon Progo, "Sejarah," 2020. https://kulonprogokab.go.id/v31/detil/7671/sejarah (accessed Mar. 12, 2021).
[11] A. Apriadsa, H. S. Anggoro, A. Cahyono, and R. Apriadna, "Unveil the 'lost toponyms' in the northern part of the Menoreh Mountains, Java, Indonesia," *Int. J. Cartogr.*, May 2019, DOI: 10.1080/23729333.2019.1611025.
[12] Soeradjiman, P, *Sendangsono: Promasan Kalibawang*, Muntilan: Kanisius, 1968.
[13] Athoillah, A, *Desa Mawa Carita Sejarah Desa dan Kota di Kulon Progo*, Yogyakarta: Ombak, 2021.
[14] Topografische Dienst. 1935. Java Gouvt Jogjakarta, 1:25,000. Batavia:Topografische Dienst.
[15] Topografische Dienst. 1935. Java Gouvt Jogjakarta:Wates, 1:25,000. Batavia:Topografische Dienst.
[16] Topografische Dienst. 1927. Java Res Jogjakarta:Balong, 1:25,000. Batavia:Topografische Dienst Weltevreden.
[17] Topografische Dienst. 1935. Java Res Jogjakarta:Nanggoelan, 1:25,000. Batavia:Topografische Dienst.
[18] Topografische Dienst. 1935. Gouvt Jogjakarta en Midden-Java:Krijan , 1:25,000. Batavia:Topografische Dienst.
[19] Topografische Dienst. 1934. Java Gouvt Jogjakarta:Sewoegaloer, 1:25,000. Batavia:Topografische Dienst.
[20] Topografische Dienst. 1935. Java Res Jogjakarta:Nanggoelan, 1:25,000. Batavia:Topografische Dienst.

Formal-Informal Employment and Poverty Dynamics: A Case Study of Mentawai Islands Regency, 2019-2021

R F Putri[1*], M K A Sarajwati[1], F A N Aini[1], H K Setiacahyandari[1], I Dairina[1], A K Dewi[1]

[1]Department of Geography and Environmental Science, Faculty of Geography, Gadjah Mada University, Yogyakarta, Indonesia

*Corresponding author: ratihfitria.putri@ugm.ac.id

Abstract. Mentawai Islands Regency is part of the frontier, remote, and underdeveloped regions in West Sumatera. Participation in the labor force is often considered a key to ensuring a good income and resources, which can help people escape or avoid poverty. The absorption of labor shown by the Labor Force Participation Rate (LFPR) in this regency shows the highest value within West Sumatera i.e 82,57% in 2021. However, its poverty rate was 14,84%, whereas West Sumatera's rate was only 6,63%. This study aimed to analyze the dynamics of poverty and employment in Mentawai Islands Regency using secondary data from BPS Indonesia with descriptive and comparative approaches. The results show that 79,4% of the working population and 67.8% of the poor are involved in the informal sector. The high proportion of the poor workers in the informal sector can be influenced by numerous variables, including the lack of education. Even though the sector utilizes many workers, informal workers' average monthly net income in Mentawai Islands Regency is among the lowest in West Sumatera. Therefore, despite the high number of LFPRs, not all forms of work are equally sufficient to avoid poverty, and employment does not guarantee a way out of poverty.

Keywords: poverty, employment, formal-informal sectors, Mentawai Islands Regency, West Sumatera

1. Introduction

One of the labor market's processes, the absorption of labor, shows the number of jobs that have been filled as indicated by the high labor force participation rate. The absorption of workers is caused by the demand for labor from employment and is influenced by the rate of population growth and the labor force [1]. Increased economic growth has the potential to enhance new job opportunities. The proportion of the working-age population participating in economic activity can be determined using the Labor Force Participation Rate (LFPR). Mentawai Islands is a regency that has the highest LFPR value in West Sumatera Province. Although the regency's relatively small economic growth, the LFPR of the Mentawai Islands Regency reached 82.57% in 2021. This rate is higher than other West

Sumatera regencies which are further along in terms of economic growth, such as Padang, which had only 63.78% for its LFPR.

The other labor absorption parameter is employment status which consists of formal and informal employees [2]. Formal employees include businesses assisted by permanent workers and all official workers. Meanwhile, informal employees include self-employment, trying to be assisted by precarious/unpaid workers, casual workers in agriculture, casual workers in non-agriculture, and family/unpaid workers. The Mentawai Islands Regency's rural-like socioeconomic conditions lead to the majority of the laborers working in the informal sector. In addition, the income of workers in the informal sector, which is frequently lower than those in the formal sector, also contributes to the lower level of welfare in the Mentawai Islands Regency than in other regencies of West Sumatera. The majority of people who work in the informal economy have low incomes, hence, there is a link between informal employment and poverty [3].

Participation in the labor force is often considered a key to ensuring a good income and resources, which can help people escape or avoid poverty. Poverty has always been a complex social issue that the government must take seriously in various regions in Indonesia. The inability of an individual or family to provide for their basic needs, such as clothing, food, housing, and other social requirements, in order to lead a minimally decent life is an indication of poverty [4]. A person who is unable to meet the needs of a decent standard of living is then referred to as a poor person. According to Damayanti and Ratnasari's definition of poverty, a person with a per capita monthly spending that is at or below the poverty line [5].

Rowntree's theory states that an increase in economic growth has a positive effect on the growth of employment which believed that employment is the best way out of poverty [6]. However, the Mentawai Islands Regency has contrasting conditions due to its high LFPR; in recent years, this region has had the highest proportion of poor people in West Sumatera Province. Badan Pusat Statistik (BPS) recorded that at least 14.84% of the Mentawai Islands Regency's population was classified as poor people in 2021. This high poverty rate can be influenced by various factors, such as the economic development and employment condition in this regency. Therefore, this study aimed to analyze the dynamics of poverty and employment in Mentawai Islands Regency in 2019-2021.

2. Methodology

This study analyzes the dynamics of employment and poverty in the Mentawai Islands Regency. The study area, Mentawai Islands Regency, comprises 92 islands and is subdivided into ten sub-districts (**Figure 1**). Descriptive and comparative analysis methods are used in this study. The descriptive method aims to provide an overview of a society or group of people and also a description of the condition or connection between two or more indicators. Whilst, the comparative method is a technique of comparing an object with another object, either in a different location or at a different time. The data collection method that is used in this study is a literature study using secondary data. Literature study is a method for creating a theoretical framework and suitable analysis approaches to problem-solving through the research for the collecting of relevant literature material [7]. The secondary data were obtained from Badan Pusat Statistik (BPS), including total workers by employment status, employment income, percentage, and the total of poor people in 2019-2021.

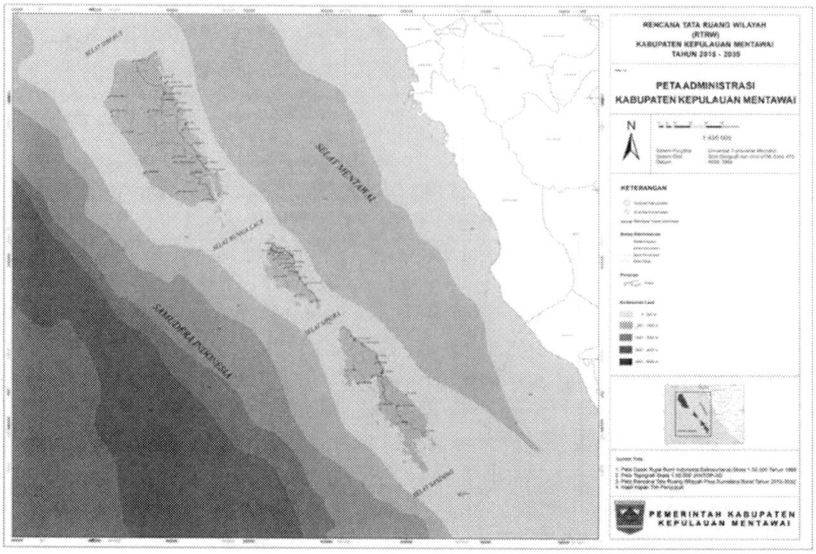

Figure 1. Administrative Map of Research Location
Source : Mentawai Islands Regency Spatial Plans 2015-2035

3. Results and Discussion

3.1. Employment

Labor Force Participation Rate (LFPR), which provides a comparative indicator of the workforce available to engage in the production of goods and services, can be understood as the percentage of the working-age population that is actively participating in the labor market, either by working or looking for work (15-64 years) [8]. The proportion of the working population illustrates the condition of labor absorption for the production of goods and services as well as economic growth in an area. Therefore, it is considered that the labor force participation rate can be used to determine economic conditions in a region. The Mentawai Islands Regency has one of the highest LFPRs among the regencies in West Sumatera Province. Compared to the LFPR of the West Sumatera, which is only 67.72%, this regency's LFPR in 2021 is relatively high at 82.57%.

The Mentawai Islands Regency's fluctuating LFPR patterns from 2019 to 2021 are depicted in **Figure 2**. The Mentawai Islands Regency's LFPR in 2020 decreased from 81.65% in 2019 to 82.77%. Moreover, this workforce participation rate increased to 82.57% in 2021. The decline of LFPR in 2020 was caused by the Covid-19 pandemic, which started the first phase that year. The corona virus outbreak in Indonesia had an effect on various industries, including the employment sector. Covid-19 pandemic has caused several businesses to suffer financial difficulties, which led to the layoff of their employees in the employment sector [9] thus affecting the number of unemployed people and reducing working hours [10]. In 2021 the labor force participation rate begin to increase as a result of various policies that have been implemented by the government so that economic activity slowly begins to increase.

Labor Force Participation Rate (LFPR) can be divided by gender i.e male LFPR and female LFPR. Figure 1 shows the fluctuation of LFPR by gender in the Mentawai Islands Regency. In 2020, the LFPR decreased due to Covid-19 pandemic and it increased in 2021. Men consistently participate in the labor force at a higher rate than women do, moreover, in 2019 its LFPR exceeded 90%. This issue is caused by the local perception that in a family, a man has an obligation to work. Besides that, the LFPR for women is relatively high (more than 70%) as well in 2019–2021. The possibility that a woman will choose to work depends on her age, education, salary, and husband's income [11], nevertheless, having dependents, being married, and having greater education all increase the probability that a woman will work [12].

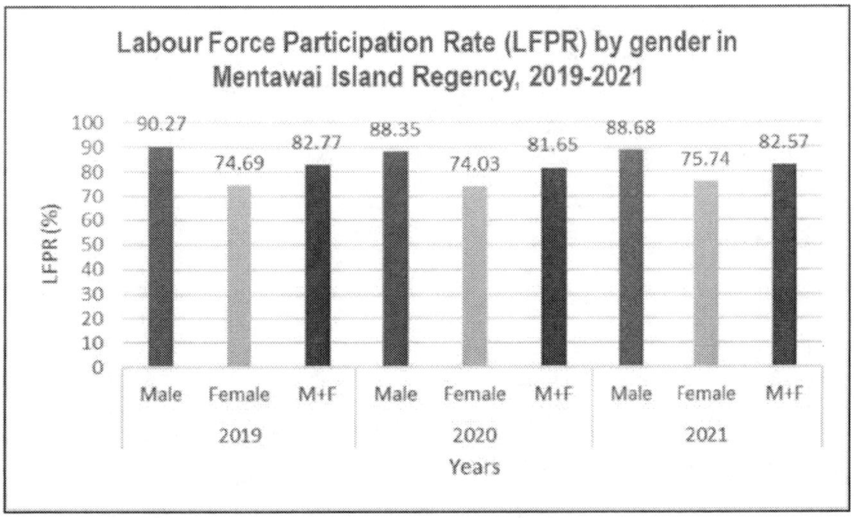

Figure 2. Labor Force Participation Rate (LFPR) by Gender in Mentawai Island Regency, 2019-2021
Source : Badan Pusat Statistik (2019, 2020, and 2021)

Developing countries such as Indonesia typically have more informal workers than the formal [13]. One Indonesian region that contributes to the high number of informal workers is the Mentawai Islands Regency. **Figure 3** illustrates the huge contrast between formal and informal workers in the Mentawai Islands Regency. The number of informal workers is significantly higher than the number of formal workers. In 2021, the number of informal workers has reached 41.050 people or 79,4% of the workforce, whilst the formal workers only reach 10.618 people or 20,6%. That vast difference may be a result of the Mentawai Islands Regency's social and physical circumstances. The social conditions that could have an impact consist of education, gender, age, and place of residence [14]. According to School Enrollment data [15] particularly for the 16 to 18-year-old age group, education in the Mentawai Islands Regency is still rated as inadequate. Mentawai Islands Regency's School Enrollment in 2021 for 16 to 18-year-olds was only 74,91%, whereas West Sumatera's School Enrollment was 84.07%. The lack of education in the Mentawai Island Regency impacts the proportion of informal workers. This condition is also affected by the physical characteristic of the Mentawai Island Regency as well. Archipelagic areas with hilly topography make this area difficult to access. Other than the topography, its main land use, forestry, also makes accessibility hard. Forestry (85.10%) and agricultural land (14.35%) are the two main land uses in the Mentawai Islands Regency [16]. That land use supports agricultural activities, which may have an impact on the proportion of farmers who are classified as informal workers.

From 2019 to 2021, the Mentawai Islands Regency had various changes in the dynamics of formal and informal workers. The trend of formal workers has decreased in 2019-2020, while the number of informal workers has increased (**Figure 3**). This condition was caused by the Covid-19 pandemic that happen throughout the world. A significant number of employees have been dropped due to layoffs, notably in the formal sector, as a result of the fall in demand for products and services in 2020 [17]. According to [18], the Covid-19 pandemic has a negative impact on almost all formal industries. However, the increase in informal workers in 2020 coincided with a fall in formal workers, indicating that some formal workers in the Mentawai Islands Regency transitioned to informal workers.

The formal workers' trend in 2020-2021 has increased, while informal workers have decreased. Conforming to [19], a decline in the proportion of informal workers indicates an improvement in economic and labor force productivity. The community's condition, which had started to adapt to Covid-19, was what led to this improvement in economic conditions. Economic activities in 2021 have started to recommence, albeit with some safety protocols to prevent the Covid-19 viruses spread. The Human Development Index (HDI), which measures the quality of living, rises in response to improvements in the Mentawai Islands Regency's economy in 2021 [20]. The HDI for the Mentawai Islands Regency increased from 61,09 to 61,35 in 2021. Hence, the Mentawai Islands Regency's workforce dynamics are important for enhancing both the economy and the welfare of the society.

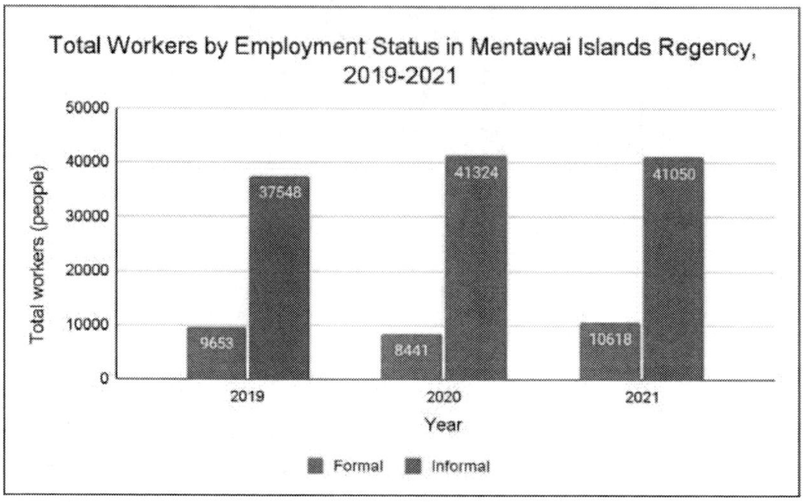

Figure 3. Total Workers by Employment Status in Mentawai Islands Regency, 2019-2021
Source : Badan Pusat Statistik (2018, 2019, and 2020)

3.2. Formal-Informal Sector within The Poor

Poverty indicates a situation of complete deprivation that is not desired by poor people. Poverty is generally viewed from a variety of perspectives or multidimensional, not just from an economic one. According to Chambers [21], poverty is an integrated concept consisting of five dimensions, namely 1) poverty, 2) powerlessness, 3) vulnerability to emergency situations, 4) dependence, and 5) alienation both geographically and sociologically. The characteristics of poor people consist of 1) on average they do not have their own production factors such as land, capital, work equipment, and skills, 2) have a low level of education, 3) mostly work or are self-employed and are small businesses (informal sector).), underemployed or unemployed (not working), 4) mostly in rural or certain urban areas (slum areas), and 5) lack of opportunity to obtain basic necessities, clothing, housing, health facilities, drinking water, education, transportation, communication facilities, and other social welfare in sufficient quantities [22]. According to the World Bank, a person is classified as poor if a person's income is less than US$1 per day (equivalent to IDR 8,500 per day).

The Mentawai Islands Regency is a 3T area (underdeveloped, border, and outermost), which has a high rate of poverty as one of its characteristics. Based on **Figure 4**, 14% of the population is classified as poor. This poverty rate increased from 14.43% in 2019 to 14.89% in 2021. Lack of physical access to infrastructure and public services like road networks, electricity, and communication in rural areas is one of the causes of poverty, especially in rural areas [23]. The Mentawai Islands Regency has natural poverty, which is characterized by a lack of public infrastructure and natural resources as well as parched soil conditions. This is due to its geographical location, which is 174 km far from the capital of West Sumatera Province and is bounded by the ocean. The geographic condition of the islands makes it difficult for the government to construct road infrastructure in this regency. According to Sagalak in [24], the Mentawai Islands' poverty is caused by a lack of supporting infrastructure, particularly a lack of road infrastructure. We acknowledge that the costs of development in the Mentawai are relatively high in comparison to West Sumatera's mainland.

Alongside geographic limitations, the socioeconomic circumstances of the archipelagic people also contribute to their persistent poverty. The Head of Mentawai DSP3A, Nicolaus Sorot Ogok, claims that social and cultural, specifically a lack of resilience and mindset to control production elements, are also factors contributing to poverty in the Mentawai Islands [25]. The high rate of early marriage is another social aspect that contributes to this regency's poverty rate which is substantially greater than those of West Sumatera Province.

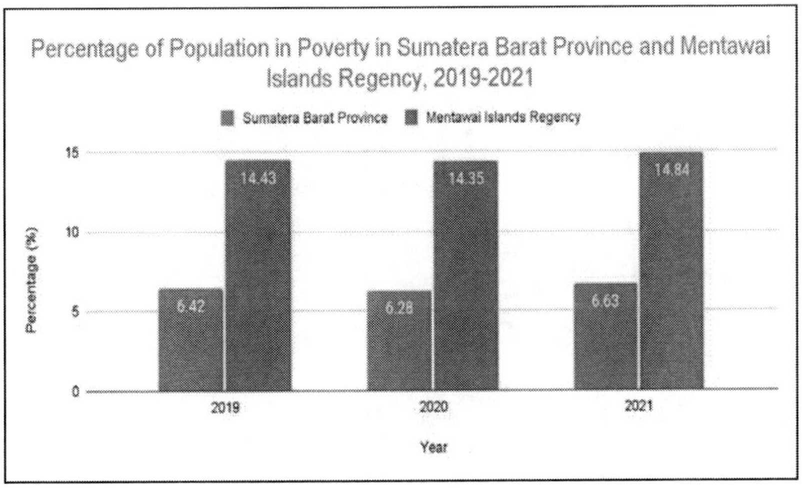

Figure 4. Percentage of Population in Poverty in Sumatera Barat Province and Mentawai Island Regency, 2019-2021
Source : Badan Pusat Statistik (2019, 2020, and 2021)

There are two groups of poor people in the labor force i.e employed and unemployed people. Based on their employment status, the Mentawai Islands Regency's poor workers can be divided into two groups which include those who work in the formal sector and those who do so informally. **Figure 5** illustrates the proportion of poor people in the Mentawai Islands by employment status from 2019 to 2021 based on publicly available data from the Badan Pusat Statistik (BPS). In this regency, the percentage of poor people who work in the formal economy is the lowest, at only about 3%. Subsequently, this percentage dropped to only 1.56% in 2021. In contrast, the Mentawai Islands Regency's poor workers in the informal sector show the highest proportion, with a rate of over 65 percent, and this value appears to have been rising over the past three years. The high proportion of the poor workers in the informal sector can be influenced by numerous variables, including the lack of education. It is widely known that poor people usually difficult to pursue higher education due to the inability to pay the comparatively high tuition costs. This is in accordance with the results of Nariswari's research [26] that one of the factors influencing workers' choices to work in the informal sector is their level of education. Many of the poor people in the Mentawai Islands Regency choose to work in the informal sector due to its traits that may absorb a lot of workers, particularly the young laborers who have low levels of education and skills.

The disparity between the poor's percentage in the formal and informal sectors emphasizes the hypothesis that the Mentawai Islands Regency's poor workers in the informal sector may be a contributing factor to the regency's poverty. This is in line with [27] statement that the informal sector has a small but significant impact on poverty. The informal sector proportion represents a lack of development and economy in an area [28]. Due to its geographical characteristics, which are part of the 3T (Frontier, Remote, and Least Developed) area, the Mentawai Islands Regency is now lagging behind other adjacent areas in terms of regional development and economic growth. This is the reason why there are more jobs in the informal sector than in the formal sector in the Mentawai Islands Regency. The informal sector is related to population welfare either. It shows that many people who work in the informal sector are unable to obtain their basic demands because of the flexible work hours and low salaries. Hence, more people in the Mentawai Islands Regency will be classified as poor as a result of the increasing number of informal sector employees who are unable to achieve their basic needs.

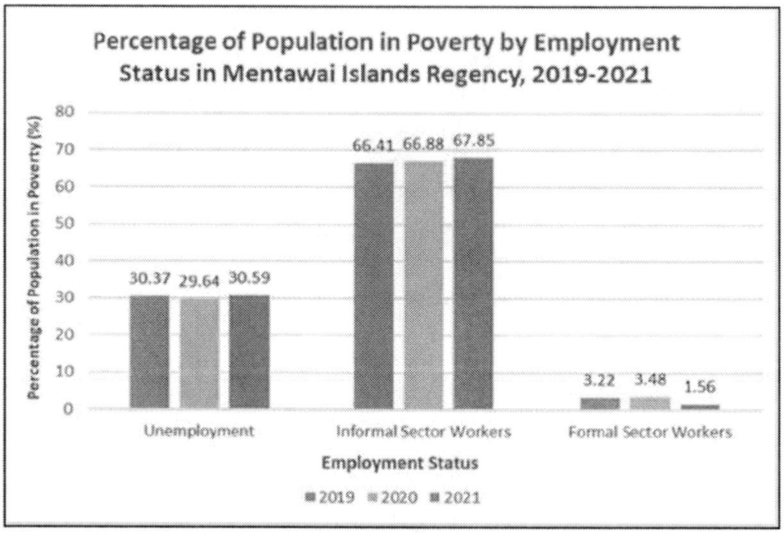

Figure 5. Percentage of Population in Poverty by Employment Status in Mentawai Islands Regency, 2019-2021
Source : Badan Pusat Statistik (2019, 2020, and 2021)

Figure 6 shows that in 2021, Mentawai Islands Regency's average monthly income or net salary for both official and informal workers is lower than in West Sumatra Province. This indicates that the average monthly net income of workers in Mentawai Islands Regency is lower than the majority of other regencies in West Sumatra Province. In contrast, the average monthly income or net salary for formal workers in Mentawai Islands Regency is higher than the informal workers, with an average difference of IDR 1,100,000 in 2019-2021 (**Figure 7**). It is obvious that there has been income inequality in this regency between the formal and informal sectors.

The Regional Minimum Wage (RMW) in 2021 for West Sumatra Province including Mentawai Islands Regency is IDR 2,484,041 per month or equal to 174 USD. However, in the province scope, only the formal sector workers have earned this minimum wage, as shown in **Figure 7**. Meanwhile, the average monthly net income or salary of formal sector workers in Mentawai Islands Regency is quite close to the RMW. In comparison, the average monthly net income of informal sector workers is significantly below West Sumatera's RMW. This occurs because there is no law requiring the payment of informal sector workers in accordance with RMW. For instance, Micro, Small, and Medium Enterprises (MSMEs), which dominate 99% of business units in Indonesia are able to absorb approximately 97% of the workforce and contribute 61% to the national Gross Domestic Product (GDP) or worth to IDR 8,573.89 trillion [29]. However, 64.13 million of Indonesia's 64.19 million

MSMEs remain in the informal sector [30]. The terms for calculating the amount of salaries for individual companies, micro, and small businesses are based on a mutual agreement between the company and the workers, according to Government Regulation of the Republic of Indonesia Number 36 Article 36 of 2021 about Wages. The salaries must fulfill the demands, one of which is that they are at least 25% over the provincial poverty line. Therefore, if the poverty line in West Sumatra Province, which is IDR 568,703 [31], is used as a benchmark, the monthly minimum salary for informal workers in Mentawai Island Regency will only be IDR 710,878 in 2021 or 49.81 USD. This implies that business in the informal sector can pay their employees less than the RMW, whereas RMW is established to ensure that employees/workers could live a decent life.

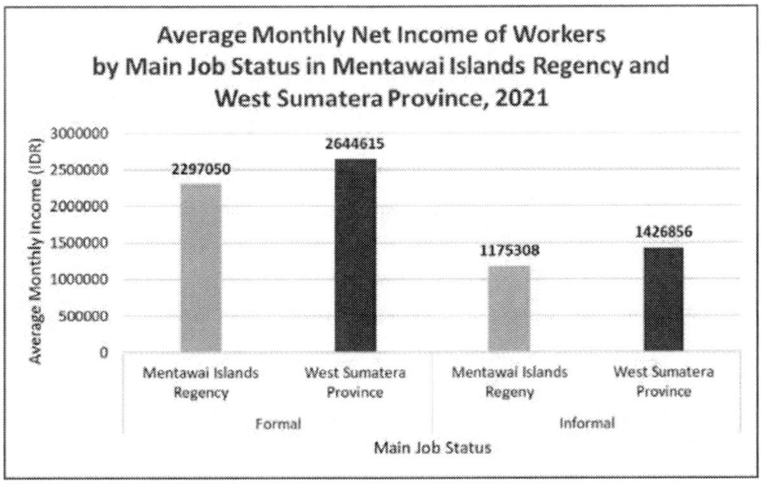

Figure 6. Average Monthly Net Income of Workers by Main Job Status in Mentawai Islands Regency and West Sumatera Province, 2021
Source : Badan Pusat Statistik (2021)

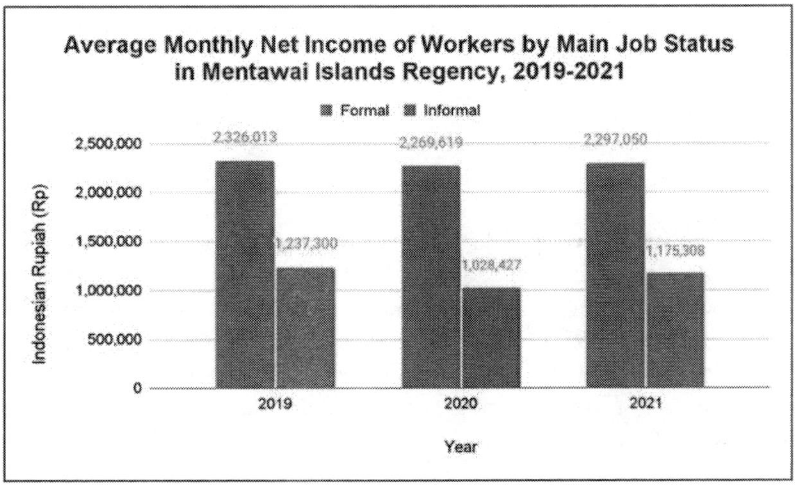

Figure 7. The Average Monthly Net Income of Workers by Main Job Status in Mentawai Islands Regency, 2019-2021
Source : Badan Pusat Statistik (2019, 2020, and 2021)

Workers in the informal sector are more vulnerable to a decline in income as a result of this situation, especially when experiencing an economic shock. This is obvious that, in comparison to the formal sector, the income or net salary of workers in the informal sector is less stable throughout the Covid-19 pandemic in 2020. Despite the net salary that was already low, Mentawai Islands Regency informal sector workers' average net income fell by IDR 208,873 in 2020, whereas formal sector workers' net salaries only fell by IDR 56,395. The population's per capita expenditure actually increased from 2019 to 2020, in contrast to the decline in the informal sector workers' net income to IDR 1,028,427 per month. In 2019, only 19% of the population in Mentawai Islands Regency had a monthly expenditure of more than IDR 1,000,000, however, in 2020, 35% of the population did [32]. The net income of workers in the informal sector is extremely low when compared to per capita expenditure, especially if one person is responsible for one family. Particularly in 2020, with a population percentage of 33.6 %, the per capita expenditure range of IDR 500,000–749,000 has the highest percentage. The number of wages for workers in the informal sector with a calculation of at least 25% above the poverty line at the provincial level, which is IDR 710,878, still cannot meet the needs of all 33.6 %, even if it is based on Government Regulation of the Republic of Indonesia Number 36 Article 36 of 2021 concerning Wages. The net income of workers in the informal sector has increased in 2021 during the Covid-19 pandemic recovery phase, but it hasn't been able to reach its well as the pre-pandemic.

Given the high number of persons employed in the informal sector, the disadvantages faced by those workers undoubtedly have a significant impact on the economic situation of the population in Mentawai Islands Regency. Moreover, in 2021, 67% of the poor people worked in the informal sector, making it difficult to increase welfare levels as long as there are numerous individuals employed in the informal economy sector. Referring to [33] who examined the dynamics of poverty and informality in Argentina, they examined whether engaging in informal work in the past leads to future poverty, and whether experiencing poverty leads to future informality. There are a number of factors such as residential segregation, spatial labor mismatch, and labor discrimination, coupled with the condition of poverty that influence and finally lead to a situation in which it's harder for someone to find a job, even if they have the qualifications. Positive spillover effects are discovered, from past poverty to current informal employment and from past informal employment to current poverty status. Therefore, this corroborates the view that the two processes are shaped by interconnected dynamics. Hence, despite the high number of LFPR, not all forms of work are equally sufficient to avoid poverty and employment does not guarantee moving out of poverty.

4. Conclusion

Based on the explanation above, it can be concluded that:

1. The Mentawai Islands Regency is a district that has the highest LFPR compared to other districts in West Sumatra Province. The value of male LFPR is always higher than female LFPR due to the perception that men are the breadwinners. The number of informal workers is significantly higher than the number of formal workers. The difference in the number of formal and informal workers can be influenced by physical and social conditions in the Mentawai Islands Regency. The trend in LFPR and the number of formal-informal workers in this district tend to fluctuate and decreased in 2020 as a result of the Covid-19 pandemic.

2. The Mentawai Islands Regency is an area that is categorized as a 3T area (underdeveloped, foremost, and outermost) which has the highest poverty rate in West Sumatra reaching more than 14% and continues to increase from 2019-2021. The reason for the high level of poverty in the Mentawai Islands is that the majority of the population are informal workers. The low income of informal workers compared to formal workers and irregular working hours have caused many

residents to be unable to meet their needs. The growth of the informal sector in the Mentawai Islands Regency is due to the condition of its territory in the form of an archipelago so that economic activities that can develop are in the form of agriculture and fisheries. In brief, despite the high rate of LFPR, it does not necessarily ensure that the residents in Mentawai Islands Regency are not poor.

References

[1] Effendi R 2014 Analisis Penyerapan Tenaga Kerja pada Sembilan Sektor Ekonomi di Sumatera Selatan *Jurnal Penelitian dan Pengembangan Akuntansi* Vol. 8 No. 1 hal. 25-52
[2] Badan Pusat Statistik (BPS) 2021 *Booklet SAKERNAS Februari 2021* (Jakarta : BPS-Statistics Indonesia)
[3] Nazier H and Ramadan R 2019 Informality and Poverty : A Causality Dilemma with Application to Egypt *The Economic Research Forum* No. 895 hal. 1-29
[4] Iqram M 2020 Pengaruh PDRB, Pendidikan, Kesempatan Kerja, dan Persentase Penduduk Sektor Informal Terhadap Kemiskinan di Provinsi Bali *E-Jurnal EP Unud* Vol. 9 No. 6 hal. 1200-1229
[5] Damayanti Y and Ratnasari V 2013 Pemodelan Penduduk Miskin di Jawa Timur Menggunakan Metode Geographically Weighted Regression (GWR) *Jurnal Sains dan Seni Pomits* Vol. 2 No. 2 hal. 329-334
[6] Rowntree B S 1918 *The Human Needs of Labour* (London : Nelson)
[7] Syahputra R 2017 Analisis Faktor-Faktor Yang Mempengaruhi Pertumbuhan Ekonomi Di Indonesia *Jurnal Samudra Ekonomika* Vol. 1 No. 2 hal. 183-191
[8] Badan Pusat Statistik 2012 *Indikator Pasar Tenaga Kerja Indonesia Februari 2012* (Jakarta : BPS)
[9] Randi Y 2020 Pandemi Corona Sebagai Alasan pemutusan Hubungan Kerja Pekerja Oleh Perusahaan Dikaitkan dengan Undang-Undang Ketenagakerjaan *Jurnal Universitas Islam Malang* Vol. 3 No. 2 hal. 119-136
[10] Ayupijaya M 2021 Dampak Pandemi Covid-19 terhadap Ketenagakerjaan di Provinsi Bali *Jurnal Bali Membangun Bali* Vol. 2 No. 3 hal. 151-168
[11] Herlina E 2016 Faktor-faktor yang Mempengaruhi Kesempatan Kerja Wanita dan Implikasinya terhadap Kesejahteraan Keluarga di Kabupaten Cirebon *JURNAL EKONOMI* Vol. 18 No. 2 hal 172-207
[12] Nurkumala R D 2020 Faktor-Faktor Yang Mempengaruhi Pengambilan Keputusan Tenaga Kerja Wanita DI Desa Wonoasri Kecamatan Tempurejo Kabupaten Jember Untuk Bekerja Kembali ke Luar Negeri *Laporan Penelitian* Fakultas Ekonomi dan Bisnis Universitas Jember
[13] Dewi M M et al 2020 Dampak Pandemi Covid-19 terhadap Tenaga Kerja Formal di Indonesia *Populasi* Vol. 28 No. 2 hal. 32-53
[14] Sharma B P and Adhikari D B 2020 Informal Economy and Poverty Dynamics: A Review *Quest Journal of Management and Social Sciences* Vol. 2 No. 1 hal. 168-180
[15] Badan Pusat Statistik (BPS) 2021 Angka Partisipasi Sekolah (APS) Penduduk Berumur 5-18 Tahun Menurut Kabupaten/Kota dan Kelompok Umur 2019-2021 Available in https://sumbar.bps.go.id/indicator/28/117/1/angka-partisipasi-sekolah-aps-penduduk-berumur-5-18-tahun-menurut-kabupaten-kota-dan-kelompok-umur.html (accessed on 24th June 2022)
[16] Badan Perencanaan Pembangunan Daerah 2018 *Rencana Pembangunan Daerah (RKPD) Kabupaten Kepulauan Mentawai Tahun 2019* (Tuapejat : Badan Perencanaan Pembangunan

Daerah Kabupaten Kepulauan Mentawai)
[17] Bank Indonesia Provinsi Sumatera Barat 2021 *Laporan Perekonomian Provinsi Sumatera Barat Februari 2021* (Padang : Kantor Perwakilan Bank Indonesia Provinsi Sumatera Barat)
[18] Ozili P K and Arun T 2020 Spillover of COVID-19: Impact on the Global Economy Available in https://papers.ssrn.com/sol3/papers.cfm?abstract_id=3562570 (accessed on 24th June 2022)
[19] Sari N P 2016 Transformasi Pekerja Informal ke Arah Formal: Analisis Deskriptif dan Regresi Logistik *Jurnal Ekonomi Kuantitatif Terapan* Vol. 9 No. 1 hal. 28-36
[20] Bank Indonesia Provinsi Sumatera Barat 2022 *Laporan Perekonomian Provinsi Sumatera Barat Februari 2022* (Padang : Kantor Perwakilan Bank Indonesia Provinsi Sumatera Barat)
[21] Nasikun 2001 *Isu dan Kebijakan Penanggulangan Kemiskinan* Diktat Mata Kuliah (Yogyakarta : Magister Administrasi Publik, Universitas Gadjah Mada)
[22] Salim E 1980 *Pembangunan Ekonomi dan Pemerataan* (Jakarta : Idayu)
[23] Suryawati, Chrisdarwani 2005 Memahami Kemiskinan Secara Multidimensional *JMPK* Vol. 8 No. 3
[24] Haluan.com 2020 Infrastruktur Masih Kurang, Penyebab Angka Kemiskinan Tinggi https://www.harianhaluan.com/sumbar/pr-10244987/infrastruktur-di-mentawai-masih-kurang-penyebab-angka-kemiskinan-tinggi
[25] Mentawaikab.go.id 2020 Tekanan Angka Kemiskinan, Ini Yang Dilakukan Pemkab Mentawai http://mentawaikab.go.id/berita/detail/tekan-angka-kemiskinan-ini-yang-dilakukan-pemkab-mentawai (accessed on 25th June 2022)
[26] Nariswari R 2020 Analisis Peranan Sektor Informal Terhadap Kemiskinan di Jawa Timur *Jurnal Ilmiah Mahasiswa FEB Universitas Brawijaya* Vol. 9 No. 1 hal. 1-14
[27] Paramita D A M 2012 Sektor Informal, Pengangguran dan Kemiskinan di Provinsi Bali Tahun 2004 - 2010 *E-Jurnal Ekonomi Pembangunan Universitas Udayana* Vol. 2 No. 1 hal. 29-38
[28] Manning, Chris and Tadjuddin Noer Effendi 1996 *Urbanisasi, Pengangguran dan Sektor Informal di Kota* Edisi Ketiga (Yogyakarta : Yayasan Obor Indonesia)
[29] Rizal N Y 2021 Pandemi Covid-19 Mengakibatkan Melemahnya Usaha Mikro, Kecil, dan Menengah di Wilayah Kejeran, Kota Surabaya *Jurnal Inovasi Penelitian* Vol. 1 No. 8 hal. 1553-1558
[30] Bahtiar R A 2021 Dampak Pandemi Covid-19 terhadap Sektor Usaha Mikro, Kecil, dan Menengah serta Solusinya *Pusat Penelitian Badan Keahlian DPR RI* Vol. 8 No. 10 hal. 19-24
[31] Badan Pusat Statistik (BPS) 2021 Garis Kemiskinan Sumatera Barat https://sumbar.bps.go.id/indicator/23/33/1/garis-kemiskinan-menurut-kabupaten-kota-di-sumatera-barat.html
[32] Badan Pusat Statistik (BPS) 2021 Persentase Penduduk Menurut Kabupaten/Kota dan Golongan Pengeluaran per Kapita Sebulan (Persen), 2019-2021 Available in https://sumbar.bps.go.id/indicator/5/601/1/persentase-pengeluaran-per-kapita-sebulan-menurut-kelompok-komoditas-dan-daerah-tempat-tinggal.html (accessed on 25th June 2022)
[33] Devicienti F, Groisman F, and Poggi A 2009 Informality and poverty: Are these processes dynamically interrelated? Evidence from Argentina *The Society for the Study of Economic Inequality (ECINEQ)* doi :10.1108/S1049-2585(2010)0000018007

Service Area Analysis of Primary Health Facilities in Yogyakarta Urban Area

Mohammad Isnaini Sadali[1], Raden Rijanta[2], Lutfi Mutaali[2], Andri Kurniawan[2], and Dewi Setianingsih[2]

[1] Doctoral Program, Faculty of Geography, Universitas Gadjah Mada, Yogyakarta, Indonesia
[2] Department of Development Geography, Faculty of Geography, Universitas Gadjah Mada, Yogyakarta, Indonesia
m.isnaini.s@ugm.ac.id

Abstract. The National Health Insurance System in Indonesia was started in 2014 with the hope that the community would have equal access rights (equity) to health care facilities. The distribution of health facilities in Yogyakarta Urban Area administratively has met the minimum standards specified by SNI 03-1733-2004 and Government Regulation no. 47 of 2016, i.e. one sub-district has at least one Puskesmas. However, the location of existing health facilities needs to be analyzed spatially, because they have various ranges (distance and service area), so that they can affect access and utilization. The purpose of this research is to map the distribution of Primary Health Facilities and analyze the range of services. This study uses quantitative research methods, with secondary data and spatial data as the main data. Quantitative analysis in this study consists of crosstab analysis of attribute data and spatial modeling. The analysis was carried out descriptively to explain each research objective. The results of the study indicate that the distribution of Primary Health Facilities in Yogyakarta Urban Area tends to cluster in Yogyakarta City and Sleman Regency. However, the range of services from health facilities has been able to cover the Yogyakarta Urban Area.

1. Introduction

The health insurance system in Indonesia is formed by legislation which is described in the National Health Insurance System (SJSN) Law of 2004 and Social Security Administration Agency (BPJS) Law of 2011. Along with the development of the Health Insurance System The national program was started in 2014 with the hope that the community would have equal access rights (equity) to health care facilities. On the other hand, the distribution of health facilities and human resources or health workers in Indonesia is also still experiencing obstacles. Basically, the availability of health service facilities can affect the level of public health in the area, region, or country. The definition of health facilities according to the law, namely Undang-Undang No. 36 of 2009 concerning Health is a tool and/or place used to organize health service activities, whether promotive, preventive, curative, and rehabilitative, carried out in synergy by the central government, local government, and also community participation.

The hope shown by the Indonesian government is a commitment to providing more adequate health services for the community. Currently, the Government of Indonesia is trying to complete 3 (three) basic plans, namely the improvement and development of medical (health) personnel from 2011 to 2025, the improvement of the Social Security Administering Body system which is currently in the process of development, and the addition and expansion of health infrastructure. There are 3 (three) indicators in an effort to strengthen primary health services through the implementation of puskesmas (community health centers) as set out in the 2015-2019 National Medium-Term Development Plan and the 2015-2019 Ministry of Health Strategic Plan, these indicators are: (1) each sub-district has at least 1 unit of puskesmas that accredited; (2) availability of inpatient puskesmas and non-inpatient puskesmas with health services that meet existing standards; (3) the availability of health centers that have cooperation, both with the Blood Transfusion Unit and with hospitals to support blood services in an effort to reduce the value of the Maternal Mortality Rate (MMR).

The Yogyakarta Special Province Health Office is carrying out its duties, principals, and functions to improve the level of public health by improving basic/primary and referral/advanced health services, where primary health care facilities are held by community health centers (puskesmas) and other similar health facilities, while for advanced health services held hospital. As an initial illustration, the number of puskesmas in the Yogyakarta Special Province has 121 (one hundred and twenty-one) puskesmas units, including 45 nursing health centers and 76 (seventy-six) non-maintenance health centers where all existing health centers have accreditation. As for referral health services carried out by hospitals in the Yogyakarta Special Province has 76 (seventy-six) units, which include 54 (fifty-four) general hospitals and 22 (twenty-two) special hospitals. The availability of adequate health facilities in Yogyakarta Urban Area needs to be supported by regional policies that also favor the ease of accessing health services. A good example is shown by the Yogyakarta City Government with its efforts to succeed in the "Universal Health Insurance Coverage" program in the Yogyakarta City, manifested by the Yogyakarta City Government will register Yogyakarta City residents to Health BPJS starting January 2019 [1].

Data obtained from the 2019 Indonesian Health Profile shows that the ratio of puskesmas per sub-district in Yogyakarta Province is 1.55. This shows that the availability of health facilities in Yogyakarta Urban Area has met the minimum standard number of health facilities that must be provided in each sub-district. However, the distribution of health facilities in a region tends to be concentrated in the city or regional center [2]. The distribution of health facilities and population in Indonesia and in Yogyakarta is spatially uneven [3], so that access to health facilities varies [4]. Therefore, a more comprehensive study is needed regarding the reach of health care facilities in Yogyakarta Urban Area, especially Primary Health Facilities so that people can access health facilities equally. However, the location of each health facility in each sub-district does not reflect equity and equal access (reach of service). This research begins with identifying Primary Health Facilities and mapping the existence of Primary Health Facilities in Yogyakarta Urban Area. The spatial distribution of the population that is the users and the Primary Health Facilities in Yogyakarta Urban Area needs to be analyzed further to determine the distance and reach of health facility services. This study seeks to provide an explanation regarding the distribution of Primary Health Facilities in the Yogyakarta Urban Area and their services in order to ensure equitable access for users.

2. Methods

The study was conducted using quantitative research methods. The type of data used in this study is secondary data, both in the form of statistical data and spatial data obtained from official publications. The analysis was carried out descriptively to explain each objective. The scope of the research area and location is in the Yogyakarta Urban Area, with the basis of determining the Yogyakarta Urban Area according to Regional Regulations of the Yogyakarta Special Province no. 5 of 2019 concerning the RTRW of the Yogyakarta Special Province Year 2019-2029. Yogyakarta Urban Area includes the following administrative areas:

- The whole area of Yogyakarta City
- Parts of Sleman Regency: Gamping Sub-District, Godean Sub-District, Mlati Sub-District, Depok Sub-District, Ngemplak Sub-District, and Ngaglik Sub-District.
- Bantul Regency: Kasihan Sub-District, Banguntapan Sub-District, and Sewon Sub-District.

Secondary data in this study were obtained from competent agencies in the form of documents containing data and official publications of data sourced from agencies in accordance with their fields.

The official publication of the data needed in this research is accessed through the official website (web) from trusted agencies (such as the Central Statistics Agency, Regency/City Governments, and the Geospatial Information Agency). The spatial data on the distribution of health facilities was also cross-checked with data from the Social Security Administration (BPJS) and its spatial distribution through location plotting on Google Earth.

Analysis of the coverage of health facilities in Yogyakarta Urban Area using network analyst facilities in ArcGIS. In the analysis, there is also a service area analysis facility which is a method for creating or calculating the scope of an object's area [5]. One solution to analyze service coverage using the Service Area Analysis method is to utilize the Network Analyst facility in ArcGIS, which is the pioneer of software for this analysis [6] [7]. Analysis of the scope of this area uses the travel time component (associated with distance) to the object by passing the road network that connects the objects to be analyzed. In this case, the connected objects are health facilities and residential areas.

3. Results and Discussion

3.1. Distribution of Primary Health Facilities in Yogyakarta Urban Area

Data on the number and types of health facilities in this case using Primary Health Facilities are secondary data obtained through several parties or agencies such as the Health Office, the Central Statistics Agency (BPS), and Social Security Administration Agency (BPJS) in Yogyakarta Urban Area covering 3 (three) regencies/cities (Yogyakarta City, Sleman Regency, and Bantul Regency). In detail, the study unit of this research was carried out in the Yogyakarta Urban Area, which was "Perda D.I. Yogyakarta" No. 5 of 2019 concerning the RTRW of Yogyakarta Province Year 2019-2029. The Yogyakarta Urban Area includes Gamping Sub-District, Godean Sub-District, Mlati Sub-District, Ngaglik Sub-District, Depok Sub-District, and Ngemplak Sub-District in Sleman Regency. Banguntapan Sub-District, Sewon Sub-District, and Kasihan Sub-District in Bantul Regency, as well as all sub-districts in Yogyakarta City. In the context of the spatial system and spatial integration in the distribution of urban areas, the Yogyakarta Urban Area is a real phenomenon of a regional interaction that forms a unified spatial system [8] [9] [10].

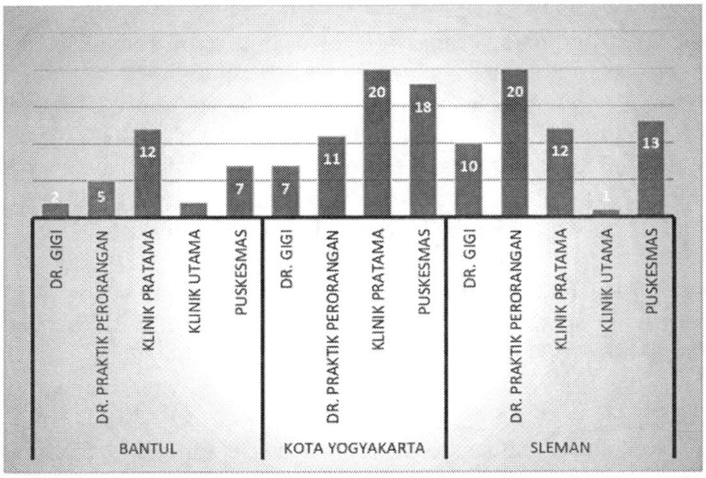

Figure 1. Graph of Number of Primary Health Facilities by Type in Yogyakarta Urban Area
Source: Secondary Data Analysis, 2022

There are a total of 28 Primary Health Facilities in Bantul Regency with details of 2 dental practice units, 5 individual practice doctor units, 12 pratama clinic units, 2 main clinic units, and 7 puskesmas units. As for the Primary Health Facilities in Yogyakarta City, there are 56 units with details of 7 dentists, 11 individual practice doctors, 20 pratama clinics, and 18 puskesmas units. There is no main clinic in Yogyakarta City. The total number of Primary Health Facilities in Sleman Regency is 56 units with details of 10 dentists, 20 individual practice doctors, 12 pratama clinics, 1 main clinic, and 13 puskesmas units. From Figure 1, it can be seen that the highest number of dentists and individual practice doctors is in Sleman Regency. Meanwhile, the highest number of pratama clinics and puskesmas are in Yogyakarta City. The largest number of main clinics is in Bantul Regency. Based on a total of 140 units of Primary Health Facilities distributed in Yogyakarta Urban Area, if sorted from the most types are primary clinics (44 units), puskesmas (38 units), individual practice doctors (36 units), dentists (19 units), and main clinics (3 units).

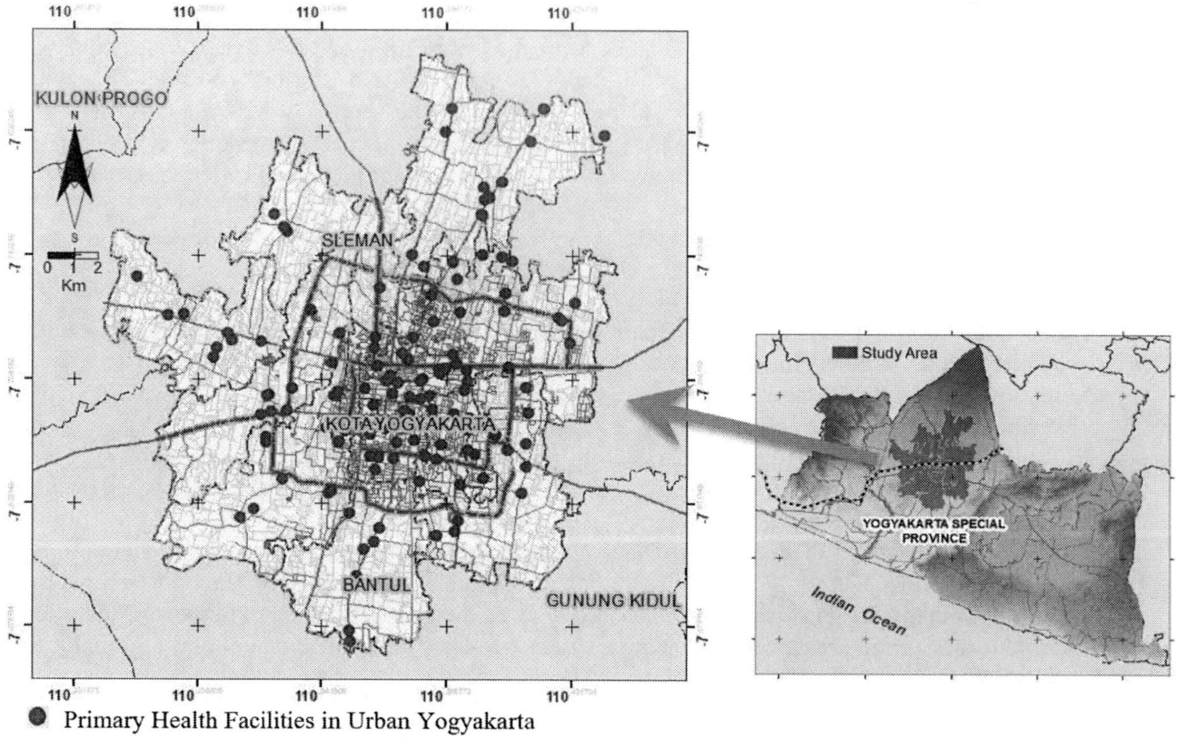

Figure 2. Distribution of Primary Health Facilities in Yogyakarta Urban Area
Source: Spatial Analysis Results, 2022

Based on the spatial analysis that has been carried out, the distribution of primary health facilities in Yogyakarta Urban Area shows that there is a higher density level in the Yogyakarta City compared to the surrounding areas (Sleman Regency and Bantul Regency). This is also supported by the results of previous studies which showed that the concentration index of health facilities in the Yogyakarta City was much higher than in other areas in the Yogyakarta Special Province.

3.2. Coverage of Primary Health Facilities in Yogyakarta Urban Area

In Figure 2. it can be seen the spatial distribution of Primary Health Facilities locations in the Yogyakarta Urban Area. The distribution of Primary Health Facilities in Yogyakarta City tends to have a higher density, this is because the area of Yogyakarta City is smaller when compared to the area of Bantul Regency and Sleman Regency. When viewed from the number of FKTPs between Sleman Regency and Yogyakarta City, the number of Primary Health Facilities is similar, but the area of Sleman Regency is much wider than the area of Yogyakarta City. This is because the function of Yogyakarta City as the capital of the Yogyakarta Special Province which is also the center of economic, government, and educational activities which are hierarchically higher than the surrounding area [11] and also the population density of Yogyakarta City is also high so that the need for health facilities must also be able to meet the needs of the community.

The number of Primary Health Facilities in Bantul Regency is the least between Sleman Regency and Yogyakarta City, which are only 28 units or it can be said that only half of the number of Primary

IOP Conf. Series: Earth and Environmental Science 1233 (2023) 012035 doi:10.1088/1755-1315/1233/1/012035

Health Facilities units in Sleman Regency and Yogyakarta City. In terms of population, Bantul Regency is the area with the second largest population after Sleman Regency, but the number of Primary Health Facilities in Bantul Regency is much less than in Sleman Regency. This is because geographically, the distance between Sleman Regency and the center of Yogyakarta City is relatively closer so that there are many new growth centers that are influenced by the location of the campus. This will directly affect the number of health facilities that should be available so that the community's needs for health facilities can be met. There are several factors identified as the cause of inequality that also occurs in Yogyakarta Urban Area, such as the spatial distribution of residential areas, spatial allocation of health facilities, road networks, and topographic conditions [12] [13].

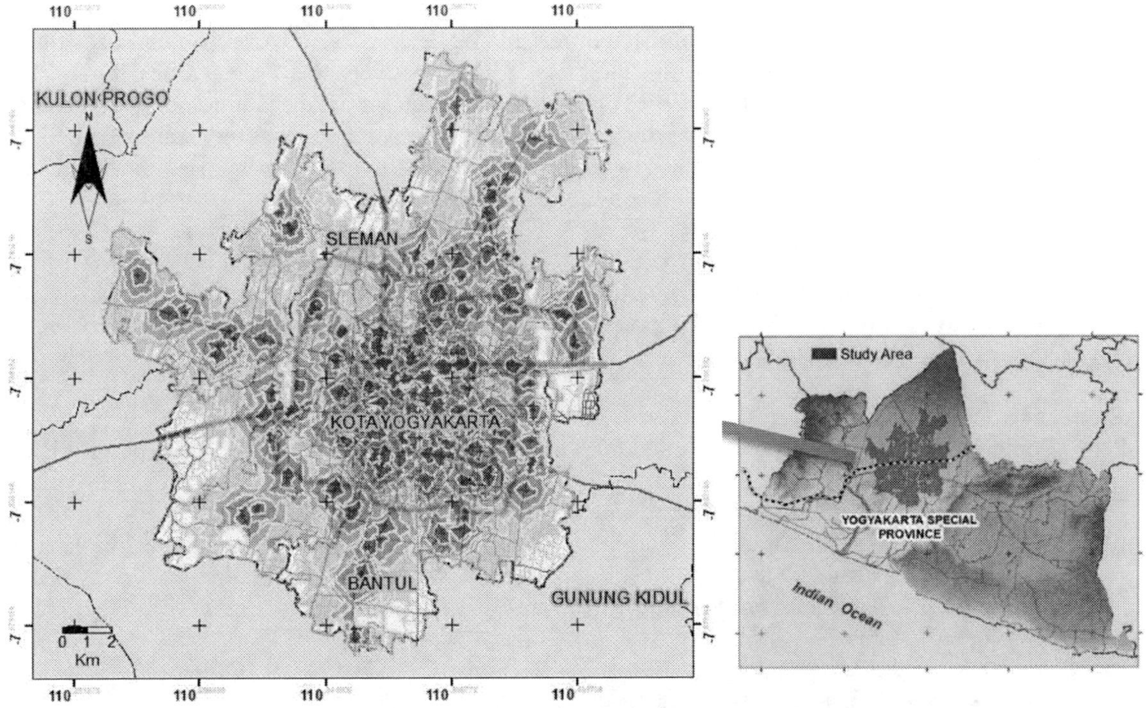

Figure 3. Distribution of Primary Health Facilities Locations in Yogyakarta Urban Area

Source: Spatial Analysis Results, 2022

Health care planning is a process that assesses overall health needs by geographic area or population and determines how these needs can be met in the most effective way through the allocation of existing and future resources [14]. In other words, planning is done by identifying the population's needs for health care facilities and then determining the best way to meet those needs. Some of the covering principles in the analysis of service coverage that have been carried out by previous researchers are the location set covering model [15] [16] and the maximal covering location model [6] [17]. The coverage of Primary Health Facilities services in Yogyakarta Urban Area is classified according to the range (distance) and access (road) that connects it, with the provisions of the hierarchy of service coverage as follows:

I : 0 – 500 meters
II : 500 – 1.000 meters
III : 1.000 – 1.500 meters
IV : 1.500 – 3.000 meters

Hierarchy I (high) indicates that health facilities have optimal services in terms of coverage or service area. In other words, the higher of the hierarchy, it represents the optimal range of health facility services. Based on the spatial analysis of the distribution and service coverage of health care facilities in Yogyakarta Urban Area, it can be seen that the city of Yogyakarta has optimal health facilities (Primary

Health Facilities) in providing access to services to users (residents). Meanwhile, health facilities located around the city of Yogyakarta (Sleman Regency and Bantul Regency) have a fairly adequate range of services and are almost affordable spatially, although there are some areas that have not been included in the service area. Based on this hierarchical analysis of service areas, it can be concluded that the development of primary health facilities in urban Yogyakarta needs to be directed at areas (areas) that have not been covered by health facilities, by placing health facilities in the middle of settlements to make them more optimal (affordable) [18]. A small part of the area that is not covered by the hierarchical area of health service areas in Bantul Regency and Sleman Regency can be categorized as a blank spot area or fragile area [12] [13].

4. Conclusion

The spatial distribution of Primary Health Facilities in Yogyakarta Urban Area is evenly distributed. However, in terms of the number of health facilities, administratively the Yogyakarta City still has a large and complete number of health facilities. The relatively even spatial distribution of health facilities in Yogyakarta Urban Area has been able to cover all administrative areas in Yogyakarta City, Sleman Regency, and Bantul Regency. In other words, the availability of Primary Health Facilities in Yogyakarta Urban Area has been able to reach the general population.

Acknowledgments

This research is part of the Doctoral Dissertation of Geography Study Program, Universitas Gadjah Mada. Thanks to the promoter team who have provided input for this research, and also everyone who have supported this research.

References

[1] Pemerintah Kota Yogyakarta, "Penduduk yang didaftarkan oleh Pemda sebagai peserta BPJS," Pemerintah Kota Yogyakarta, 17 Desember 2018. [Online]. Available: https://warta.jogjakota.go.id/detail/index/7679. [Accessed 29 July 2022]

[2] M. I. Sadali, M. A. F. Alfana, Z. Hadijah, E. L. Rosewidiadari dan R. Andika, "Dominasi kota sebagai konsentrasi fasilitas kesehatan (Studi kasus: Daerah Istimewa Yogyakarta)," *REGION: Jurnal Pembangunan Wilayah dan Perencanaan Partisipatif*, vol. **17**, no. 1, pp. 136-150, 2020.

[3] A. Muawanah, "Blogspot," 26 Januari 2012. [Online]. Available: http://annisamuawanah.blogspot.com.

[4] Susilowati, "Ketidakmerataan Akses Pelayanan Kesehatan Rawat Jalan di Indonesia," Universitas Gadjah Mada, Yogyakarta, 2004.

[5] Esri, "Service area analysis," 2020. [Online]. Available: https://desktop.arcgis.com/. [Accessed 08 April 2020].

[6] R. Church and F. Medrano, "Location-allocation Modeling and GIS," in *The Geographic Information Science & Technology Body of Knowledge (3rd Quarter 2018 Edition)*, J. P. Wilson, Ed., Washington, DC, University Consortium for Geographic Information Science, 2018, p. 9.

[7] Vinan, "Network Analysis / Analisis Jaringan," 24 Agustus 2018. [Online]. Available: https://savinotes.wordpress.com/.

[8] R. Bintarto and S. Hadisumarno, Metode Analisa Geografi, Jakarta: Lembaga Penelitian Pendidikan dan Penerangan Ekonomi dan Sosial, 1979.

[9] N. Sumaatmadja, Studi Geografi Pendekatan dan Analisa Keruangan, Bandung: Alumni, 1988.

[10] H. S. Yunus, Struktur Tata Ruang Kota, Yogyakarta: Pustaka Pelajar, 2002.

[11] M. I. Sadali, R. Rijanta, L. Mutaali and A. Kurniawan, "Study of the service functions of health facilities in Yogyakarta Special Province," in *ICST 2021 – The 2nd Geoscience and Environmental Management Symposium*, Yogyakarta, 2021.

[12] A. Geofana, "GIS Application for Health Facility Coverage Mapping in Temanggung Regency,

IOP Conf. Series: Earth and Environmental Science 1233 (2023) 012035 doi:10.1088/1755-1315/1233/1/012035

Central Java Province," *Journal of Regional and Rural Development Planning*, vol. **5**, no. 3, pp. 160-173, 2021.

[13] V. R. Verma dan U. Dash, "Geographical accessibility and spatial coverage modelling of public health care network in rural and remote India," *PLOS ONE*, vol. **15**, no. 10, pp. 1-23, 2020.

[14] R. K. Thomas, Health Services Planning, New York: Kluwer Academic/Plenum Publishers, 2003.

[15] G. Sahin and H. Süral, "A review of hierarchical facility location models," *Computers & Operations Research,* vol. **34**, pp. 2310-2331, 2007.

[16] J. Baray and G. Cliquet, "Optimizing locations through a maximum covering/p-median hierarchical model: Maternity hospitals in France," *Journal of Business Research,* vol. **66**, pp. 127-132, 2013.

[17] S. M. Algharib, "Distance and Coverage: An Assessment of Location-Allocation Models for Fire Stations in Kuwait City, Kuwait," Kent State University, Ohio, 2011.

[18] U. F. Kurniawati, S. Nurlaela, C. Susetyo dan F. Firmansyah, "Spatial analysis of health facility service coverage in handling of COVID-19 patients in the area Surabaya City Settlement," in *IOP Conference Series: Earth and Environmental Science*, Surabaya, 2022.

[19] D. DiBiase, M. DeMers, A. Johnson, K. Kemp, A. T. Luck, B. Plewe and E. Wentz, "Location-allocation modeling and p-median problems.," in *The Geographic Information Science & Technology Body of Knowledge*, Washington, DC, 2006.

[20] R. Church and C. ReVelle, "The maximal covering location problem," *Regional Science Association,* vol. **32**, no. 1, pp. 101-118, 1974.

IOP Conf. Series: Earth and Environmental Science 1233 (2023) 012036 doi:10.1088/1755-1315/1233/1/012036

Study of Regional Economic Resources Potential at the Former Residency of Banyumas

R Setiawan[1]

[1]Department of Environmental Geography, Faculty of Geography, Gadjah Mada University, Yogyakarta, Indonesia

Abstract. Resources are assets as well as important capital owned by each region. Resources can be a determinant in the success of the development. The former Banyumas Residency is an area that includes the Regencies of Banyumas, Banjarnegara, Cilacap, and Purbalingga. Every year the value of GRDP for each regency always increases even though there is still inequality. The GRDP of the Cilacap Regency is much higher than other regencies because of the high value-added and dependence on the manufacturing sector. Development in the former Banyumas Residency requires efforts to focus on sectors that can develop. This study aims to examine the leading sectors that become potential economic resources and the pattern and structure of economic growth in each regency from 2015 to 2019. The method used is the Sectoral Contribution Index (IKS), Location Quotient (LQ), Classic Shift Share, and Klassen Typology. The leading sectors in each regency are different. Construction is the leading sector in Banyumas Regency; agriculture, forestry, and fishery sectors in Banjarnegara and Purbalingga regencies; and the manufacturing sector in Cilacap Regency. The pattern and structure of sectoral growth in each regency also differ based on the Klassen typology.

1. Introduction

Resources are assets as well as important authorized capital owned by each region. Natural resources can provide raw materials, especially in various production activities, while human resources can be a driving force in utilizing existing natural resources through their processing technology. Both of these resources can be a determinant in the success of development, where if one can use them optimally, it will increase prosperity. The maximum utilization capacity of a region's resources is an indicator of the success of economic development because it describes the amount of potential income the area receives [1]. Furthermore, investment in both natural resources and human resources in the long term participates in determining economic growth [2].

Economic development is a strategic issue often discussed in addition to social development. Economic development in developing countries tends to have relatively higher inequality or disparities. Income disparities between provinces can be found in provinces in Indonesia [3]. However, the economic growth of each region is much different, where an increase in economic growth has not been followed by an even distribution of income [4]. The emergence of this inequality is strengthened due to the differences in regional characteristics, human resources, and the provision of infrastructure facilities that have not been evenly distributed. The progress of a region is also much influenced by geographical conditions [5], which gives each region the potential for resources and advantages of different economic sectors. We can also find this disparity in Central Java Province, where economic growth tends to stagnate and is in the second-lowest category among other provinces in Java [6]. This is because there is still high dependence on the manufacturing sector, so it does not focus on different sectors. There is a lack of sustainable economic growth sources, low human resources, and limited mobility of people's savings [7].

The success of national economic development can be seen from the success of economic development in the regions because for developing countries like Indonesia; regional economic development is an important factor in increasing and strengthening national development. There are various indicators to measure the success of economic development. The hands that can be seen are economic growth, economic structure, and the smaller or reduced value of income inequality per capita, between regions, and between sectors [8]. An indicator often used to see the success of a regional government's performance in regional economic development efforts is to know its economic growth. Analysis of economic growth in an area can help local governments make better decisions and solve problems [9]. One of them is in terms of identifying the advantages of the region.

Talking about economic development and regional economic growth in Indonesia is inseparable from the regional autonomy policy. One form of opportunity in the era of regional autonomy is the need to sharpen the orientation of development based on the potential of regional economic resources [10]. It is also the biggest challenge for local governments because they must identify and understand the region's potential. Regions that can identify and understand the potential of their region will be able to utilize it optimally for the sake of the sustainability of the regional economy and development will not only rely on one particular sector but other sectors that can develop.

The former Banyumas residency was one of the administrative areas during the Dutch East Indies era. The former residence of Banyumas includes the regencies of Banyumas, Banjarnegara, Cilacap, and Purbalingga. The former Banyumas residency area, included in the Barlingmascakeb Region, is strategic in Central Java because it has the second-largest average economic growth rate in Central Java [11]. The high rate of growth in the region is still found to have disparities in economic growth seen from the GRDP value of each regency. This indicates that the high economic growth in the cooperation region is not necessarily followed by equitable development in each regency.

Every year from 2015 to 2019, the total GRDP value for each region also increased even though there was still inequality where the GRDP of Cilacap Regency was much higher than other regions (Figure 1). The value of GRDP in 2019 based on constant prices in 2010 shows that Cilacap Regency's GRDP is the highest among the other three regencies, which is 100,445,727.17 million rupiahs, Banyumas Regency is 39,779,320.86 million rupiahs, Purbalingga Regency is 17,387,941.80 millions of rupiah, and Banjarnegara Regency of 15,246,870.00 million rupiahs. The highest GRDP value in 2019 for Banyumas Regency, Cilacap Regency, and Purbalingga Regency was in the manufacturing sector because in these areas, there were quite a lot of companies engaged in the manufacturing sector, while for Banjarnegara Regency, the highest GRDP value was in the agricultural, forestry, and fisheries sector because the area is dominated by mountains and is close to Wonosobo Regency, so the agricultural sector is more developed in these areas.

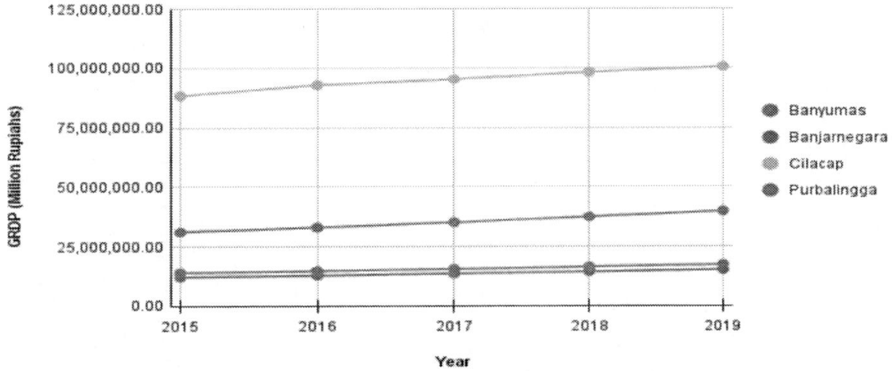

Figure 1. Total GRDP of each regency in the Former Residency of Banyumas in 2015-2019

IOP Conf. Series: Earth and Environmental Science 1233 (2023) 012036 doi:10.1088/1755-1315/1233/1/012036

Regional development in the former Banyumas residency requires efforts to focus on sectors that can develop. In addition, it can also be done to determine the leading sector because each region must have a different superior sector. Therefore, the purpose of this study is to examine the region's top sectors, potential economic resources, and the pattern and structure of economic growth in each district in the former Banyumas residency in 2015 – 2019. It is hoped that with the identification of the leading sectors in each of these areas, it can be developed to support regional economic development and encourage other sectors to develop.

2. Method

This conducted the research in one of the former residencies in Central Java Province, the former Banyumas Residency. The former Banyumas Residency area consists of four regencies, namely Banyumas, Banjarnegara, Cilacap, and Purbalingga Regencies (Figure 2). The area is 542,855,119 hectares, with the largest district being Cilacap Regency.

The data used in this study is secondary data sourced from the Central Bureau of Statistics (BPS) publications for each regency and local government, involving data on Gross Regional Domestic Product (GRDP). The research uses a quantitative approach with the basic method of descriptive analysis. The data processing method is carried out using the method of calculating the Sectoral Contribution Index (IKS), Location Quotient (LQ), and Classic Shift Share to determine the leading sector that becomes the economic resource potential of each district, as well as the Klassen Typology to determine the pattern and structure of sectoral economic growth in each district in 2015 – 2019. The data used is GRDP data by a business field which consists of 17 sectors. The 17 sectors are sector A (agriculture, forestry, and fisheries); B (mining and quarrying); C (manufacturing); D (electricity and gas); E (water supply, waste management, and recycling); F (construction); G (wholesale and retail, car, and motorcycle reparation); H (transportation and warehousing); I (accommodation and food and beverage); J (information and communication); K (financial services and insurance); L (real estate); MN (business services); O (government administration, defense, and social assurance); P (education services); Q (health services and social activity); RSTU (other services).

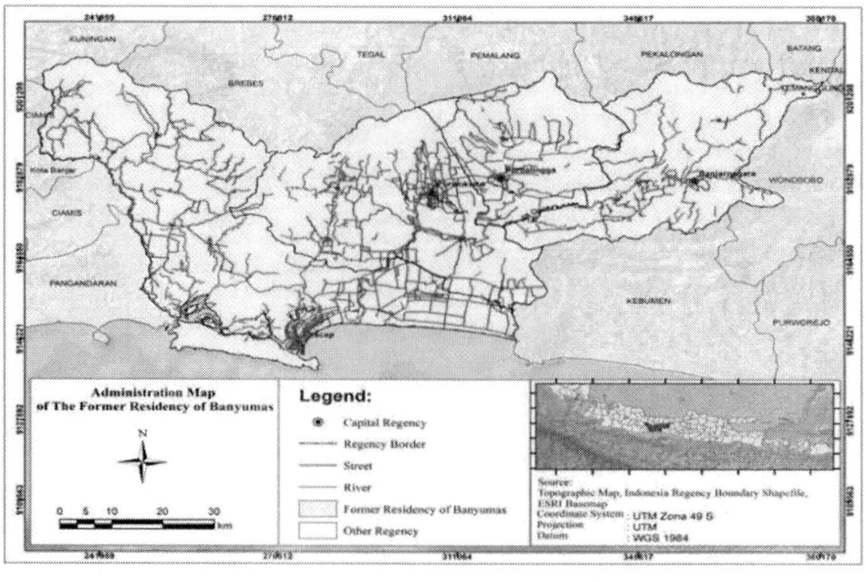

Figure 2. Administrative Map of Former Residency of Banyumas

The novelty of this study compared to previous studies can be seen in the research analysis unit. Research using former residence units of analysis is also rare because research on determining the economic resource potential of a region is mostly carried out at provincial and national units of study. The novelty of the method is the use of mapping data to present the results of the calculation of the leading sectors.

2.1. Sectoral Contribution Index

The sectoral contribution index determines each economic sector's contribution to each district's total regional income in the former Banyumas residency in a certain year. The formula for calculating IKS is as follows [12]:

$$IKS = \frac{GRDPsi}{GRDPt} \quad (1)$$

Description
GRDPsi : GRDP sector i regency in the year of analysis
GRDPt : GRDP total all sectors regency in the year of analysis

The range of IKS calculation values is from 0 to 1. An index value closer to 1 means that the sector has a greater or more dominant contribution, so the more likely the sector will become the leading sector in a region. The trend of the sectoral contribution index can be seen when presented in time series. Determining the index value trend in this study using the free hand method is to determine if the index value increases or decreases when compared to the index value of the previous year [13].

2.2. Location Quotient (LQ)

Location Quotient analysis can explain the differences in specialization from one region to another [14]. The difference in the occupation of an area or region is also known as comparative advantage. According to neoclassical growth theory, comparative advantage arises due to differences in regional productivity related to the intrinsic factors of the region, such as human resources, natural resources, and the physical characteristics of the region [15]. Comparative advantage can be a characteristic that describes the area so that it distinguishes it from other regions. The formula for calculating LQ is as follows [12]:

$$LQ = \frac{\frac{psi}{pl}}{\frac{PSi}{PL}} \quad (2)$$

Description
psi : GRDP sector i in regency
pl : GRDP total all sectors in regency
PSi : GRDP sector i in former residency
PL : GRDP total all sectors in former residency

The calculation results have several possibilities and different interpretations, as shown in Table 1 below [12]:

Table 1. Interpretation of LQ Value

LQ Value	Sector Classification	Specialization	Market Share
LQ > 1	Base and leading sector	Specialized sector	Can serve market needs within and outside the region (export)
LQ = 1	The sector is balanced with the reference area	The sector is the same as a reference area	Only able to meet domestic needs region (non-export)
LQ < 1	Non-base, non-leading,	Unspecialized	Not yet able to meet market

IOP Conf. Series: Earth and Environmental Science 1233 (2023) 012036 doi:10.1088/1755-1315/1233/1/012036

	and non-potential sector	sector	needs outside the region (non-export)

2.3. Classic Shift Share

Shift share is an analytical tool used to see the criteria for economic growth. This analysis will divide the added value or sectoral economic growth into three interrelated components, namely national share (Nij), proportional shift (Mij), and differential shift (Cij) [16]. The three main components can explain in more detail what elements most influence economic growth in the study area. The details in the shift-share analysis method are as follows [16]:

a. The real impact of district economic growth (Dij):

$$Dij = Nij + Mij + Cij \quad (3)$$

Then if the calculation is applied using GRDP, then:

$$Dij = Yij^{*} - Yij \quad (4)$$

b. Effect of national economic growth or reference area (national share) (Nij):

$$Nij = Yij\ x\ rn \quad (5)$$

c. Proportional shift (Mij):

$$Mij = Yij(rin - rn) \quad (6)$$

d. Effect of competitive advantage (differential shift) (Cij):

$$Cij = Yij(rij - rin) \quad (7)$$

Description:
Yij : GRDP sector i regency in the early year of analysis
rij : GRDP growth of sector i in regency
rin : GRDP growth of sector i in former residency
rn : Total GRDP growth in all sectors in former residency
superscript mark (*) showing GRDP at the end of the analysis year

2.4. Klassen Typology

Development activities in each region will be different according to the region's characteristics. These developments will certainly cause changes, especially in the economic structure. The analysis that can be used to determine changes in the economic structure is the Klassen typology. This method is the basis for describing the pattern and structure of regional sectoral economic growth. Classification in the Klassen typology is listed in Table 2 below [12]:

Table 2. Klassen Typology Matrix

	y region > y reference	**y region < y reference**
r region > r reference	**Quadrant I** Advanced and fast-growing sector	**Quadrant II** Potential sector or can still grow rapidly
r region < r reference	**Quadrant III** Advanced sector but depressed	**Quadrant IV** The sector is relatively lagging behind

Description
y region : Sectoral GRDP contribution in regency
y reference : Sectoral GRDP contribution in former residency

r region : GRDP growth of sector i in regency
r reference : GRDP growth of sector i in former residency

Grouping based on Klassen typology is divided into four, namely: advanced and fast-growing sectors, fast-growing sectors, advanced sectors but depressed, and the sector is relatively lagging.

3. Result and Discussion

3.1. Leading Sectors: Potential Regional Economic Resources

3.1.1. Sectoral Contribution Index

The Sectoral Contribution Index (IKS) is one of the analytical tools to see the contribution aspect of a business sector or GRDP. The IKS value shows the size of the added value of the business sector to the total GRDP of a region. Sectors with a high IKS value also have the potential to become the leading sector of the region.

The results of the IKS calculation show values that vary in each regency. The highest IKS scores in each regency have similarities in the agriculture, forestry, and fishery sector, manufacturing; construction; and wholesale and retail repair of cars and motorcycles (Figure 3). Although the agriculture, forestry, and fisheries sectors in each district are among the sectors with a high contribution to the formation of total GRDP, when viewed in time series from 2015 to 2019, it shows that the IKS value continues to decline.

The decline in the value-added contribution of the agricultural, forestry, and fishery sectors during the analysis period was caused by two main factors, namely the reduction in the area of agricultural land due to changes in land function for non-agricultural activities and the increase in prices for the products of the business field itself which tended to be slow compared to products from other business sectors. In addition, there are also natural factors where this sector is a business sector with the highest dependence on nature, so its economic growth is the most unstable. The existence of a prolonged dry season in the analysis year, such as in 2019, made the production of agricultural products very reduced, which ultimately affected the added value [17].

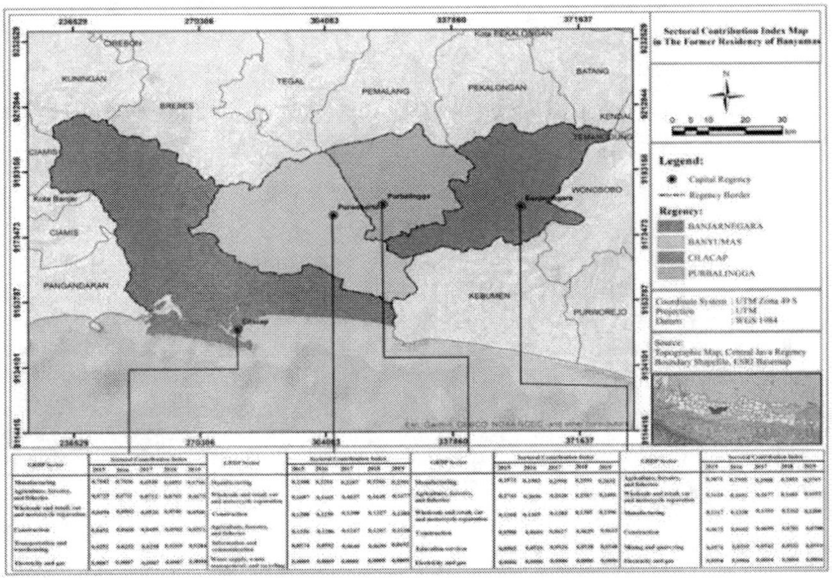

Figure 3. Map of Highest and Lowest IKS Values in the Former Residency of Banyumas

Different contributing factors cause the contribution of the business sector in each district. Banyumas Regency has the highest IKS value in the manufacturing sector because it is driven by the increase in micro and small industries and the added value contribution from the food and beverage industry sub-sector as well as the chemical, pharmaceutical, and traditional medicine sub-sectors, both of which are important sub-sectors that are needed by the community every day. The construction sector also has a high contribution index compared to other sectors due to many infrastructure development projects.

Banjarnegara Regency has the highest contribution index value in the agriculture, forestry, and fishery sectors, although experiencing a downward trend in index values. The high index in the agriculture, forestry, and fishery sectors are because the horticultural crop sub-sector contributes the largest added value, followed by the food crops, plantation crops, and livestock sub-sectors. The wholesale, retail, car, and motorcycle reparations also have a high IKS value. They tend to increase due to the added value contribution from the wholesale and retail trade sub-sector, which dominates.

The highest contribution index value in Cilacap Regency is in the processing industry because it is home to Indonesia's largest oil processing industry. Manufacturing, especially oil and gas processing, is the main foundation in the total GRDP of Cilacap Regency because the added value is very large compared to other sectors even though the index value has decreased. The decline occurred due to the relatively low world oil price in 2019, which was the impact of the trade war issue, so oil production in Cilacap Regency also decreased, ultimately affecting the sector's added value that year.

The contribution of GRDP seen from the IKS value in Purbalingga Regency is the highest in the manufacturing sector and tends to be regular yearly. The high index value in this sector results from the increase in new industries from the micro, small to medium, and large levels, as well as increased production in existing industries. The next highest sector in Purbalingga Regency is the agricultural, forestry, and fishery sector, although the value is experiencing a downward trend. The downward trend is due to unfavorable climatic factors such as a prolonged dry season.

3.1.2. Location Quotient (LQ)

Location Quotient (LQ) analysis is one of the methods in the economic base theory model [18]. The LQ method is used to identify potential economic resources by dividing them into basic and non-basic sectors. This division is the result of relative comparison of the study area to the reference area [19].

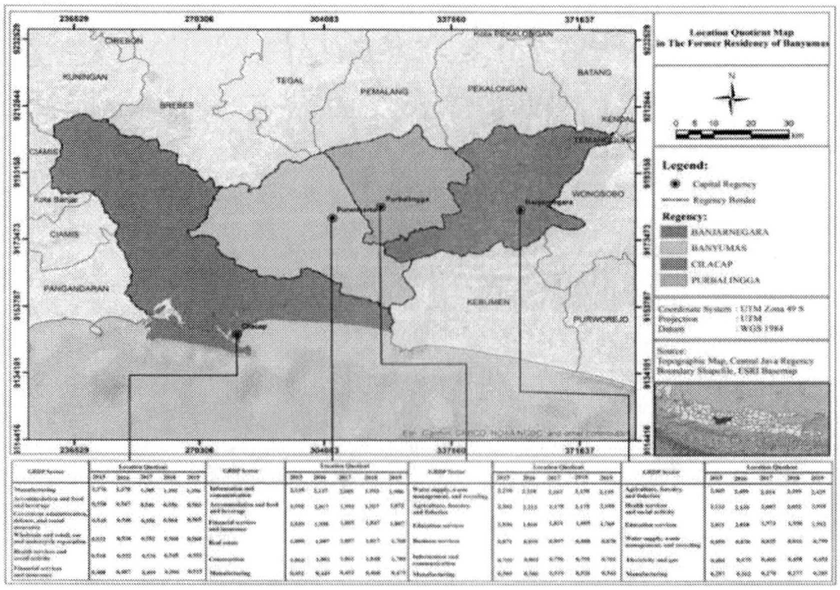

Figure 4. Map of Highest and Lowest LQ Values in the Former Residency of Banyumas

The results of the LQ calculation show that the average LQ value varies for each sector of the business field (Figure 4, Table 3). The highest average LQ values from 2015 to 2019 in Banyumas Regency are in the information and communication sector, the accommodation, food, and beverage sector, and the construction sector; Banjarnegara Regency has the highest LQ values in the agriculture, forestry, and fishery sectors, the health services, and social activity sector, and the education service sector; Purbalingga Regency with the highest LQ scores is in the agriculture, forestry, and fishery sectors, the water supply sector, and the education service sector; and lastly, Cilacap Regency in the manufacturing sector, where the added value of the GRDP compilers is also the highest among other sectors.

Banyumas Regency has a business sector, almost all of which have an average LQ value > 1 except for the manufacturing sector (Table 3). The sector with the highest LQ value both in time series and on average is the information and communication sector due to several factors, including this sector having a relatively stable role in total GRDP because it is a supporter of all economic activities and cannot be separated from the efforts made by telecommunications companies in Indonesia. Banyumas Regency in improving facilities and service speed. The sector of accommodation and food and beverage related to tourism in Banyumas Regency also has a high LQ value due to the increasing number of tourist visits coming in the analysis year period, encouraging the development of various lodging places such as hotels and food and beverage (FnB) around tourist destinations.

Table 3. Average Value of LQ in the Former Residency of Banyumas

Code	Sector	Average Value of LQ (2015-2019)			
		Banyumas	Banjarnegara	Cilacap	Purbalingga
A	Agriculture, forestry, and fisheries	1.031	2.432	0.587	2.184
B	Mining and quarrying	1.432	1.546	0.730	1.196
C	Manufacturing	0.458	0.270	1.385	0.519
D	Electricity and gas	1.408	0.467	0.955	0.827
E	Water supply, waste management, and recycling	1.540	0.836	0.630	2.178
F	Construction	1.849	1.000	0.704	0.892
G	Wholesale and retail, car and motorcycle reparation	1.737	1.739	0.548	1.441
H	Transportation and warehousing	1.233	1.366	0.834	1.165
I	Accommodation and food and beverage	1.960	1.163	0.554	1.407
J	Information and communication	2.051	1.222	0.614	0.790
K	Financial services and insurance	1.883	1.773	0.499	1.383
L	Real estate	1.846	1.331	0.652	0.925
MN	Business services	1.438	1.912	0.726	0.886
O	Government administration, defense, and social assurance	1.638	1.908	0.556	1.472
P	Education services	1.638	1.973	0.489	1.820
Q	Health services and social activity	1.521	2.085	0.534	1.713
RSTU	Other services	1.480	1.947	0.577	1.661

Banjarnegara Regency has various LQ values (Table 3). Sectors with the highest LQ scores include agriculture, forestry, fishery, health services, social activities, and education services. The sectors with a low LQ value of less than one are the water supply, electricity and gas, and manufacturing sectors. The agricultural, forestry, and fisheries sectors are the specialties of Banjarnegara Regency because they have a comparative advantage where their production in the year of analysis can meet the needs within their territory and the needs of other regions. The agricultural sector is well developed.

Cilacap Regency in the analysis year had the highest LQ value only in the manufacturing sector, and other sectors had an LQ value of less than one (Table 3). This is because there is a very large dependence on this sector. After all, it plays a role of 60% of the total GRDP of Cilacap, and other sectors tend to lose in terms of added value. The high LQ value in the manufacturing sector in Cilacap Regency is reinforced by the coal industry and oil and gas refineries which account for almost 70% of the sector's total GRDP. Oil and gas refining industry activities in Cilacap Regency continue to grow and supply nearly 30% of national needs. This means that the production of the oil and gas processing industry, apart from meeting the needs of its region, can also meet the needs of other regions in Indonesia.

Purbalingga Regency has various LQ values (Table 3). The highest LQ values in the analysis year included the agriculture, forestry, and fishery sectors, the water supply, and the education services sectors. The agriculture, forestry, and fisheries sectors are the sectors that have the strongest comparative advantage, as is Banjarnegara Regency, although they are experiencing a downward trend. This is due to a long dry season, especially in 2019, due to El Nino and climate change. Climate change affects agriculture, impacting crop production growth, development, and yield because of its high dependence on weather and climatic conditions [20].

3.1.3. *Classic Shift Share*

Analysis shift-share is an analytical tool used to see the criteria for economic growth. This analysis divides sectoral economic growth into three interrelated components, namely national share (Nij), proportional shift (Mij), and differential shift (Cij). National share or Nij can be used to see the effect of the economic growth of a wider reference area, namely the former residency, on the economic growth of the study area, namely the district; proportional shift (Mij) is useful for looking at sectors that are the concentration or center of the regional economy because they grow faster than growth the reference area economy as well as to see whether there are external factors that affect sectoral economic growth [21], and the last is differential shift or Cij which is used to explain which sectors have competitiveness or competitive advantage compared to the same sector. On a wider regional scale, namely the former residency.

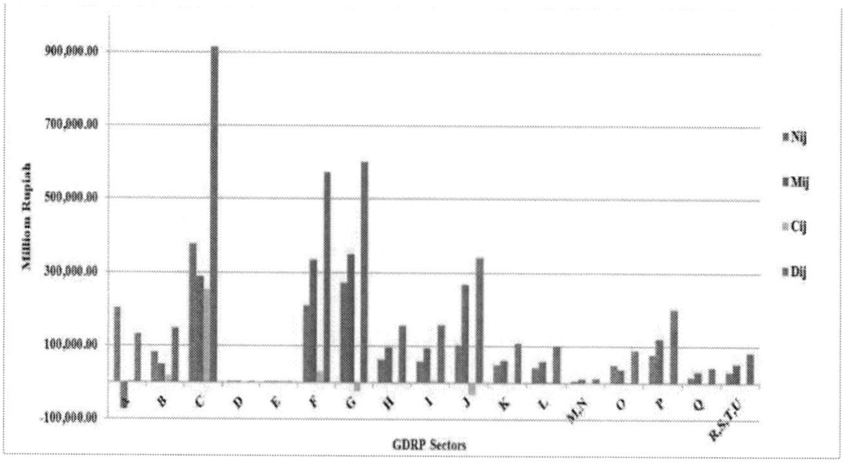

Figure 5. Graph of Shift Share Component Value in Banyumas

IOP Conf. Series: Earth and Environmental Science 1233 (2023) 012036 doi:10.1088/1755-1315/1233/1/012036

The results of the classic shift-share in Banyumas Regency show various values (Figure 5). Economic growth (Dij) in Banyumas Regency for all business sectors gives a positive value with the manufacturing sector as the sector with the highest growth, which is 916,342.34 million rupiahs, followed by wholesale and retail trade, repair of cars and motorcycles amounted to 600,583.67 million rupiahs, the third was the construction sector with a growth of 572,177.18 million rupiahs. The sectors with the highest growth values are all secondary and tertiary economic sectors which indicate that there has been a shift in the economic sector from the primary economic sector to the secondary and tertiary economic sectors. If we dissect the Dij component, we see that the highest Nij component is in the manufacturing sector. In that case, the Mij component in all sectors has a positive value, and the highest is in the wholesale and retail trade and car and motorcycle repairs. The highest Cij component is in the manufacturing sector.

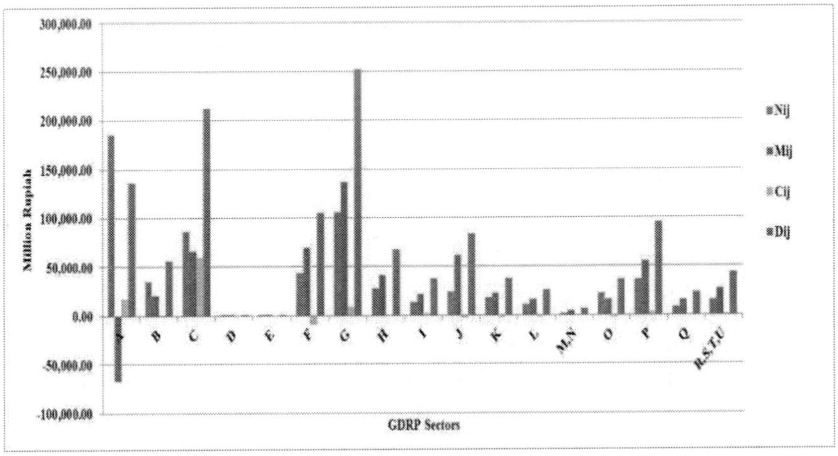

Figure 6. Graph of Shift Share Component Value in Banjarnegara

Banjarnegara Regency has positive economic growth (Dij) for all business sectors (Figure 6). The highest economic growth was in the wholesale and retail trade, car and motorcycle reparation sector with a growth value of 251,726.91 million rupiahs, followed by the manufacturing sector, which grew by 211,915.60 million rupiahs, the agriculture, forestry, and fishery sectors with a growth value at 135,973.4 million rupiahs, and the construction sector with a growth value of 104,720.57 million rupiahs. This shows that the primary sector, namely the agriculture, forestry, and fishery sectors, is still the sector with the highest economic growth compared to other sectors. If the Dij is dissected, it will be seen that the highest Nij component is in the agriculture, forestry, and fishery sectors, the Mij component of all sectors is positive except for the agriculture, forestry, and fishery sectors which have negative values due to slowing growth. Then the highest Cij component is the manufacturing sector.

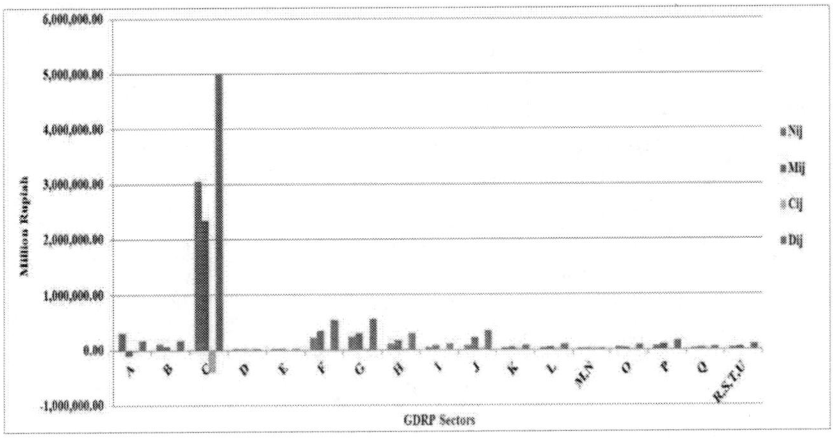

Figure 7. Graph of Shift Share Component Value in Cilacap

IOP Conf. Series: Earth and Environmental Science 1233 (2023) 012036 doi:10.1088/1755-1315/1233/1/012036

Analysis of classical shift-share in Cilacap Regency shows variable sectoral economic growth (Dij). The highest growth value or Dij is the manufacturing sector, which has a very high growth value compared to other sectors, which is 4,993,411.07 million rupiahs, then the wholesale and retail trade, car and motorcycle reparation sector, which is 556,723.34 million rupiahs, and the construction sector with a growth value of 538,853.60 million rupiahs (Figure 7). Let's look at this sectoral growth into three components. We will see that the highest Nij component is in the manufacturing sector, the Mij component is only in the agriculture, forestry, and fisheries sectors with negative values , and the highest Mij value is in the manufacturing sector. Still, the highest Cij component is information and communication. The Cij value for the manufacturing sector is quite negative because the processing industry in Cilacap Regency only relies on the oil and gas processing industry, which each region does not always have, so it does not have a competitive advantage compared to the other areas.

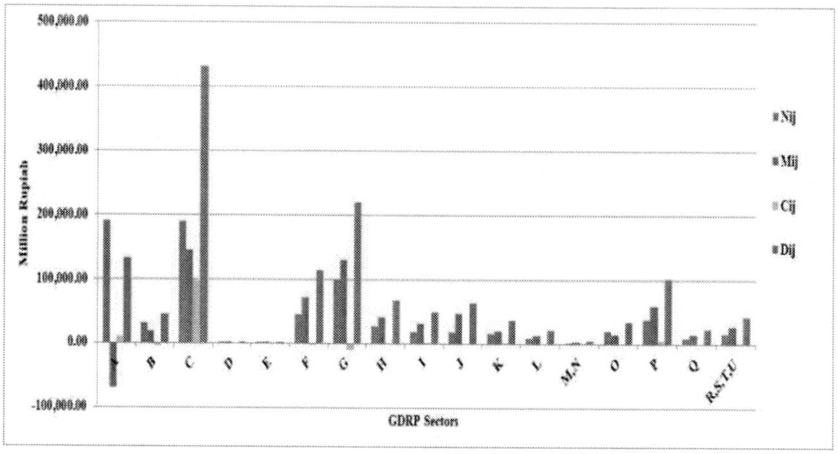

Figure 8. Graph of Shift Share Component Value in Purbalingga

Purbalingga Regency has a positive economic growth value (Dij) in all business sectors, and its dominated by several sectors (Figure 8). The sectors with the highest Dij value include the manufacturing sector with a growth value of 430,969.33 million rupiahs, followed by wholesale and retail trade, car and motorcycle reparation sector at 220,063.56 million rupiahs, the agriculture, forestry, and fishery sectors of 132,284.24 million rupiahs, and the construction sector amounting to 113,566.23 million rupiahs. As seen from each component, the highest Nij component is the agriculture, forestry, and fishery sector, and the highest Mij component is the manufacturing sector. In contrast, the agriculture, forestry, and fishery sectors have a negative Mij value because growth is slower than the former residency, and the manufacturing sector's highest Cij component.

3.2. Pattern and Structure of the Regional Economy

Analysis Klassen typology aims to see an overview of each regency's economic patterns and structures in the former Banyumas residency. This analysis is an analysis with combined criteria which uses growth and contribution indicators from each sector of the business field. Klassen typology will group 17 business sectors into four categories. The 17 sectors are Sector A (agriculture, forestry, and fisheries); B (mining and quarrying); C (manufacturing); D (electricity and gas); E (water supply, waste management, and recycling); F (construction); G (wholesale and retail, car, and motorcycle reparation); H (transportation and warehousing); I (accommodation and food and beverage); J (information and communication); K (financial services and insurance); L (real estate); MN (business services); O (government administration, defense, and social assurance); P (education services); Q (health services and social activity); RSTU (other services).

Table 4. Classification Results of Klassen Typology in Banyumas Regency

	y region > y reference	y region < y reference
r region > r reference	**Quadrant I** ***Advanced and fast-growing sector***	**Quadrant II** ***Potential sector or can still grow rapidly***
	Sector B, E, F, H, I, K, L, O	Sector C
r region < r reference	**Quadrant III** ***Advanced sector but depressed***	**Quadrant IV** ***The sector is relatively lagging***
	Sector A, D, G, J, MN, P, Q, RSTU	-

The results of the typology classification analysis in Banyumas Regency, as shown in Table 4 above, show that several sectors are divided into different quadrant categories. We can see that no sector occupies quadrant IV as a lagging sector. The sectors that occupy quadrant I are advanced and fast-growing sectors. There are eight business sectors: the mining and quarrying sector, the construction sector, transportation, warehousing, and the provision of accommodation and food and beverage. This follows the LQ analysis, where these sectors are included in the leading sectors in the analysis year. In addition, access to Trans Java bus transportation to several tourist attractions in Banyumas can be an opportunity to improve the accommodation and food and beverage provision sector in the future. The sectors that occupy quadrant III or the developed sector but are depressed because their growth tends to fluctuate or weaken include the agriculture, forestry, and fishery sectors, information and communication, and education services.

Table 5. Classification Results of Klassen Typology in Banjarnegara Regency

	y region > y reference	y region < y reference
r region > r reference	**Quadrant I** ***Advanced and fast-growing sector***	**Quadrant II** ***Potential sector or can still grow rapidly***
	Sector A, G, H, I, RSTU	Sector C
r region < r reference	**Quadrant III** ***Advanced sector but depressed***	**Quadrant IV** ***The sector is relatively lagging***
	Sector B, J, K, L, MN, O, P, Q	Sector D,E,F

Banjarnegara Regency, as shown in Table 5, has an economic pattern and structure spread in several typology classifications. Five sectors are considered advanced and growing fast: agriculture, forestry, and fishery; wholesale and retail trade; car and motorcycle reparation; and accommodation and food and beverage sectors. This sector occupies quadrant I because of its higher growth and contribution than in the former residency. In addition, the agricultural, forestry, and fisheries sectors are the leading sectors in terms of LQ values, but they need consistent and sustainable management. The manufacturing sector, similar to Banyumas Regency, is included in the category of a fast-growing sector, which means this sector has faster growth than the aggregate growth in the former residency. However, its contribution value is still less than the contribution of the same sector in the former residency.

Table 6. Classification Results of Klassen Typology in Cilacap Regency

	y region > y reference	y region < y reference
r region > r reference	**Quadrant I** ***Advanced and fast-growing region***	**Quadrant II** ***Potential sector or can still grow rapidly***
	-	Sector D, G, J, K, L, MN, P, Q
r region < r reference	**Quadrant III** ***Advanced sector but depressed***	**Quadrant IV** ***The sector is relatively lagging***
	Sector C	Sector A, B, E, F, H, I, O, RSTU

IOP Conf. Series: Earth and Environmental Science 1233 (2023) 012036 doi:10.1088/1755-1315/1233/1/012036

Analysis of Klassen typology in Cilacap Regency, as shown in Table 6 above, divides the business sector into three quadrant categories. No business sector occupies quadrant I as a developed and fast-growing sector because the value of growth and contribution is lower than the former residency. In addition, the concentration of added value only focused on the manufacturing sector also makes other sectors crash. Seven sectors occupy quadrant II as fast-growing sectors such as wholesale and retail trade, car and motorcycle reparation; information and communication, financial services, and insurance. The sector in quadrant III as a developed but depressed sector is the manufacturing sector because of its weak growth value compared to the growth of the former residency.

Table 7. Classification Results of Klassen Typology in Purbalingga Regency

	y region > y reference	y region < y reference
r region > r reference	**Quadrant I** ***Advanced and fast-growing region***	**Quadrant II** ***Potential sector or can still grow rapidly***
	Sector A, E, H, I, K, O, P, RSTU	Sector C, MN
r region < r reference	**Quadrant III** ***Advanced sector but depressed***	**Quadrant IV** ***The sector is relatively lagging***
	Sector B, G, Q	Sector D, F, J, L

The Klassen typology in Purbalingga Regency, as shown in Table 7 above, shows that the district's existing economic patterns and structures are evenly divided into four categories. The agriculture, forestry, fishery, transportation, warehousing, accommodation, food, and beverage sectors are three of the eight business fields that are advanced and fast-growing sectors in Purbalingga. According to the LQ value, the agriculture, forestry, and fishery sectors are indeed the leading sectors. The transportation and warehousing sector, the accommodation and food and beverage sector, and the agricultural sector is the leading sector seen from the LQ value. With new transportation facilities such as airports and Trans Java buses, the Purwokerto-Purbalingga corridor can be an opportunity to develop these two sectors in the future. The sector that is growing or in quadrant II is the manufacturing sector and corporate services because its growth is higher in the residency, but its contribution is still inferior to the same sector in the residency.

4. Conclusion

The sectors that have the potential to become leading sectors in each district are different. The leading sectors in 2015-2019 in Banyumas Regency are in all sectors except the manufacturing sector; the leading sectors in Banjarnegara Regency are almost all sectors except the manufacturing sector, the electricity, and gas sector as well as the water supply sector, waste management, and recycling; the leading sector in Cilacap Regency is the manufacturing sector; and finally, the leading sectors in Purbalingga Regency are almost all sectors except the manufacturing sector, the electricity and gas sector, the construction sector, the information, and communication sector, the real estate sector, and the business services sector.

The pattern and structure of the economy of each district have several sectors. Eight business sectors in Banyumas Regency are developed and growing fast, including the mining and quarrying; construction; and accommodation, food, and beverage sectors. There are five developed and fast-growing sectors in Banjarnegara Regency: agriculture, forestry, and fisheries; wholesale and retail, car and motorcycle reparation; and the sector of providing accommodation and food and beverage. Furthermore, in Cilacap Regency, seven sectors are growing/fast-growing, such as wholesale and retail, car and motorcycle reparation; information and communication; and the financial services and insurance sectors, while the manufacturing sector is a developed but depressed sector. Finally, in Purbalingga Regency, the agriculture, forestry, and fishery sectors, the transportation and warehousing sector, as well as the accommodation and food and beverage sectors are three of the eight business fields that are developed and fast-growing sectors.

Acknowledgment
The author expresses his deepest gratitude to RTA-UGM 2022, which has funded this research. Dr. Rika Harini, S.Si, M.P as the thesis supervisor. And also really appreciate reviewers for suggestions to make this article better. Without all of these supports, this research would not be successful. Hopefully, this article will be useful for advancing science and future research.

References
[1] Oki K K and Pangastuti M D 2020 Peran Sumberdaya pada Pertumbuhan Ekonomi Daerah Perbatasan Timor Leste *e-J. Ekonomi Bisnis dan Akuntansi* **VII**(2) 98-103.
[2] Mankiw N G 2003 *Teori Makroekonomi Edisi Kelima Terjemahan* (Jakarta: Erlangga).
[3] Zainuri A and Jamal Abd 2017 Disparitas Pembangunan Ekonomi di Indonesia *J. Ilmiah Mahasiswa (JIM) Ekonomi Pembangunan* **2**(1) 1-10.
[4] Syaifullah 2021 Pengaruh Pertumbuhan Ekonomi terhadap Disparitas Pendapatan di Indonesia J. Ekonomi-Qu **11**(1) 21-36.
[5] Rahmah I M M 2020 Dikotomi Keruangan Wilayah: Karakteristik Wilayah dan Daya Saing Daerah Provinsi Jawa Timur *J. of Architecture and Urban Research* **3**(2) 154-165.
[6] Fahrizal N S Sudati J and Whinarko 2019 Analisis Ketimpangan Ekonomi Provinsi Jawa Tengah Tahun 2008 – 2017 *DINAMIC: Directory Journal of Economic* **1**(4) 399-417.
[7] Badan Perencanaan Pembangunan Nasional 2015 *Seri Analisis Pembangunan Wilayah Provinsi Jawa Tengah 2015* simreg.bappenas tahun 2015 online: https://simreg.bappenas.go.id/assets/temaalus/document/Publikasi/DokPub/03.%20Analisis%20Provinsi%20Jawa%20Tengah%202015_ok.pdf.
[8] Santosa S H 2015 Disparitas Pertumbuhan Ekonomi dan Pembangunan Wilayah di Satuan Wilayah Pembangunan IV Provinsi Jawa Timur *Media Trend* **10**(2) 116-128.
[9] Herath J Gebremedhin T and Maumbe B M 2011 A Dynamic Shift Share Analysis of Economic Growth in West Virginia *J. of Rural and Community Development* **6**(2) 155-169.
[10] Prawoto N 2010 Pengembangan Potensi Unggulan Sektor Pertanian *J. Ekonomi dan Studi Pembangunan* **11**(1) 1-9.
[11] Sutrisno A 2012 Analisis Ketimpangan Pendapatan dan Pengembangan Sektor Unggulan di kabupaten dalam kawasan Barlingmascakeb Tahun 2007-2010 *Economic Development Analysis Journal* **1**(1) 42 – 49.
[12] Muta'ali L 2015 *Teknik Analisis Regional untuk Perencanaan Wilayah Tata Ruang dan Lingkungan* (Yogyakarta: Badan Penerbit Fakultas Geografi).
[13] Faradiba 2020 *Analisis Data Berkala* (Jakarta: Universitas Kristen Indonesia).
[14] Guimarães P Figueiredo O and Woodward D 2009, Dartboard tests for the location quotient *Regional Science and Urban Economics* **39** 360–364.
[15] Soebagiyo D 2014 Implications and Competitiveness of Regions on Regional Development of Central Java *J. Ekonomi Pembangunan* **15**(2) 158–171.
[16] Abidin Z 2015 An Application of the Shift Share Analysis for Transformation of the Agricultural Sector in Economic Areas at South East Sulawesi *J. Informatika Pertanian* **24**(2) 163-178.
[17] Badan Pusat Statistik 2020 *Produk Domestik Regional Bruto Kabupaten Banyumas Menurut Lapangan Usaha 2015-2019* (Banyumas: BPS Banyumas).
[18] Setiawan R Insani A A Indrastuti A N Purwanto D Rohmah H N Suci R P Ayu M M Omar R C and Putri R F 2022 Dynamics of Agricultural Land and Sustainability of the Agricultural Sector in the Province of Bali 2015-2020 *Atlantis Press: Biological Science Research, Proceedings of the 2nd International Conference on Smart and Innovative Agriculture (ICoSIA 2021)* **19** 46-55.
[19] Jumiyanti K R 2018 Analisis Location Quotient dalam Penentuan Sektor Basis dan Non Basis di Kabupaten Gorontalo *Gorontalo Development Review* **1**(1) 29-43.
[20] Ruminta Handoko and Nurmala T 2018 Indikasi perubahan iklim dan dampaknya terhadap produksi padi di Indonesia (Studi Kasus: Sumatera Selatan dan Malang Raya) *J. Agro* **18**(2) 83-93.
[21] Amalia F 2012 Penentuan Sektor Unggulan Perekonomian Wilayah Kabupaten Bone Bolango

dengan Pendekatan Sektor Pembentuk PDRB *J. Etikonomi* **11**(2) 196 – 207.

IOP Conf. Series: Earth and Environmental Science 1233 (2023) 012037 doi:10.1088/1755-1315/1233/1/012037

Analysis of Social and Environmental Vulnerability Against Landslides Using Scoring and Weighting Methods at Probolinggo District

Cherie Bhekti Pribadi[1], Farhan Himawan[1]

[1]Institut Teknologi Sepuluh Nopember, Surabaya, Indonesia
cherie_b@geodesy.its.ac.id

Abstract. Landslides phenomena majority occur in area which is adjacent with mountain or hill where the soil and rock structure move down the slopes. It is caused by land instability such as loss of plants or trees in highland which have function of binding soil grains while maintaining pores of soil beneath. Based on data from the Regional Disaster Management Agency (BPBD), one of districts/cities in East Java is categorized as prone against landslides is at Probolinggo District where 108 disaster occurred over 2021. Landslide disasters certainly have an impact on the environment and social aspects in vicinity. Therefore, this research aim to assess areas which has prone level against landslides in terms of social and environmental vulnerabilities through analysis of potential for landslide disasters in Probolinggo District. The research using weighting parameter methods in order to obtain the level of social and environmental vulnerability. It is divided into four categories namely low, medium, high, and very high. Maron sub-district become the largest area with very high level of social vulnerability about 33.7% of the area of one sub-district. Meanwhile, in area of environmental vulnerability which is dominated by very high levels of vulnerability is at Krucil sub-district about 88.5% of the area of one sub-district.

1. Introduction

Natural disasters as one of natural phenomena can occur along the time as well as can cause harm people and their activity. However, natural disasters almost occur when there was significant increase of rain intensity namely landslides, about 45% of land area in Indonesia is mountain land which is prone against landslide and erosion [5]. Minister of Public Works No. 22/PRT/M/2007 states that geographically most of the territory of the Unitary State of the Republic of Indonesia in prone area against natural disaster and one of natural disasters often occurs namely landslides [7].

Landslide is one of natural disasters which majority occur in highland areas or those adjacent against mountain or hill where the soil and rock structure move down the slopes, either in large number or only partially. The natural disaster itself can generate property loss as well as loss of life and cause damage to other facility and infrastructure which have an impact on economic and social condition [4]. Generally, landslide is caused by land instability such as loss of plants or trees in highland which have function of binding soil grains while maintaining pores of soil beneath [6]. Vulnerability assessment can be conducted by analyzing condition as well as characteristics of local people as well as location of their livelihood. It can be determined by assessing the security aspects of their livelihoods or the condition caused by physical, socio-economic, as well as environmental factors or processes which can increase the vulnerability of citizens against the threats and impacts of disasters [2]

Based on data from the East Java Regional Disaster Management Agency (BPBD), there is about 20 area are categorized as prone level against landslides in three classes such as high, medium, and low categories [3]. Probolinggo District is one of the areas that have the potential for landslides. It can happen due to geographical conditions of Probolinggo District, especially Sukapura District, which mostly have mountain area and has a fairly steep slope. Based on results of the Disaster Event Data Recap, the Center for Disaster Management Operations Control (Pusdalops PB), there were 108 disaster events from January 1, 2021 to December 31, 2021, including 106 natural disasters, 1 accident, and 1 non-natural disaster. The occurrence of landslides dominates about 49 events [2].

Landslide disaster certainly has impact on environment and social aspects in vicinity. Therefore, this research aim to assess area which has prone level against landslides in terms of social and environmental vulnerabilities through analysis of potential for landslide disasters in Probolinggo District using weighting parameter methods.

2. Methodology

2.1. Study Area

Geographically, Probolinggo District is located at 7°40' South Latitude - 8°10' South Latitude and 112°50' East Longitude - 113°30' East Longitude. Probolinggo District has area about 169,616.65 Ha or 1,696.17 km^2. Administratively, Probolinggo District is bordered on north side by the Madura Strait, on the east side by Situbondo District and Jember District, on the west side with Pasuruan District, and on the south side with Lumajang District and Malang District. Meanwhile, to the north of the center there is an Autonomous Region, namely the City of Probolinggo [3].

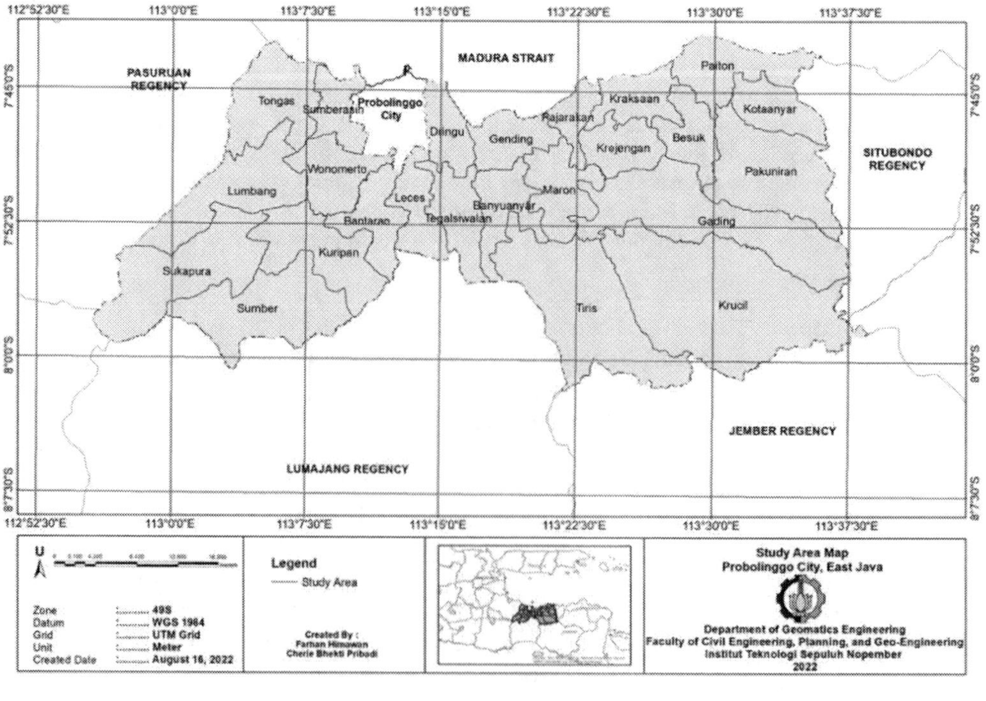

Figure 1. Study Area

2.2. Data and Software

The data used in this study are Probolinggo District population density data, gender ratio data, vulnerable group ratio data, poor population ratio data, disabled population ratio data which is those data was obtained from Central Agency of Statistics (BPS), land cover map of Probolinggo District, and vector data in shapefile format of Probolinggo District administrative boundary from Geospatial Information Agency (BIG).

2.3. *Research Method*

2.3.1. Social Vulnerability First of all, in order to gain social vulnerability map, we have conducted the overlay methods of several data such as population density data, gender ratio data, vulnerable group ratio data, poor population ratio data, disabled population ratio data. Social vulnerability consist of parameter of population density and vulnerable aspect in each area (village or sub-district).

Then, these data is calculated based on weight and class as follows. Spatially, each parameter value is distributed in the area of settlement per district. Processing was conducted by giving scores and weights to form levels of social vulnerability through the classification process.

Table 1. Social Vulnerability Estimating and Scoring Parameters [1]

Parameter	Weight (%)	Class		
		Low	Moderate	High
Population Density	60	<500 person/km^2	500-1000 person/km^2	>1000 person/km^2
Vulnerable Group	40			
Gender Ratio (10%)		>40%	20-40%	<20%
Vulnerable Group Ratio (10%)				
Poor Population Ratio (10%)		<20%	20-40%	>40%
Disabled Population Ratio (10%)				
Scoring		1	2	3

Next step, we have conducted reclassify method in order to classify each village into several classes based on social vulnerability level and it can be generated map of social vulnerability in Probolinggo District.

2.3.2. Environmental Vulnerability. First step, environmental vulnerability Environmental vulnerability consists of parameters of forest, shrubs, and swamps as listed in the following table. Environmental vulnerabilities obtained from land cover data and respective weights itself as shown in the following table.

Table 2. Environmental Vulnerability Estimating and Scoring Parameters [1]

Parameter	Weight (%)	Class		
		Low	Moderate	High
Forest	80	<20 ha	20 – 50 ha	>75 ha
Mangrove	10	<10 ha	10 – 30 ha	>30 ha
Shrubs	10	<10 ha	10 – 30 ha	>30 ha
Scoring		1	2	3

The Processing data similar to social vulnerability. Then two vulnerability maps are combined with the hazard map by spatial analysis (overlay method) in order to make a map of social and environmental vulnerability to landslides.

Furthermore, we also have conducted reclassify method in order to classify each village into several classes based on social vulnerability level and it can be generated map of social vulnerability in Probolinggo District.

In order to determine the social and environmental susceptibility against erosion based on the landslide hazard map using overlay method in two maps by considering the vulnerability risk analysis matrix. The data analysis technique uses a risk matrix presented by the Australian Geomechanics Society and Sub-committee on Landslide Risk Management in Concepts and Guidelines for Landslide Risk Management (Qualitative Risk Analysis) for vulnerability and in the Regulation of the Head of the National Disaster Management Agency (BNPB). which is adjusted according to the vulnerability class from the hazard and vulnerability map as shown in table 3 below.

Table 3. Risk Analysis Matrix [1]

Risk Matrix		**Vulnerability**		
		High	Moderate	Low
Hazard	Very High			
	High			
	Moderate			
	Low			

From the risk matrix table 3, we decide four level of vulnerability, namely low (green), moderate (yellow), high (orange), and very high (red).

3. Result and Discussion

3.1. *Social Vulnerability*

Population data obtained from BPS calculated according to the class and weight in the table 1. The results of weighting of each parameter then are summed and social vulnerability index is obtained in which determination of low, middle, and high classes which is based on class length formula. The results of class calculations and weighting above can be shown in table 2.

Table 4. Social Vulnerability Index Calculation

No	Sub-district	Social Vulnerability Index	Class
1	Sukapura	1,2	Low
2	Sumber	1,3	Low
3	Kuripan	1,3	Low
4	Bantaran	2,5	High
5	Leces	2,5	High
6	Tegalsiwalan	1,9	Moderate
7	Banyuanyar	2,5	High
8	Tiris	1,3	Low
9	Krucil	1,3	Low
10	Gading	1,3	Low
11	Pakuniran	1,3	Low
12	Kotaanyar	1,9	Moderate
13	Paiton	2,4	High
14	Besuk	2,5	High
15	Kraksaan	2,5	High
16	Krejengan	2,5	High
17	Pajarakan	2,5	High
18	Maron	2,5	High
19	Gending	2,5	High
20	Dringu	2,4	High
21	Wonomerto	1,8	Moderate
22	Lumbang	1,3	Low
23	Tongas	1,9	Moderate
24	Sumberasih	2,5	High

To figure out social vulnerability against landslide based on a landslide hazard map, using overlay method of two maps by considering vulnerability risk analysis matrix. The data analysis technique uses a risk matrix as presented by the Australian Geomechanics Society and the Sub-committee Landslide Risk Management in Landslide Risk Management Concepts and Guidelines (Qualitative Risk Analysis) for vulnerabilities which are then adjusted according to the classes of the hazard and vulnerability map as shown in table 3.

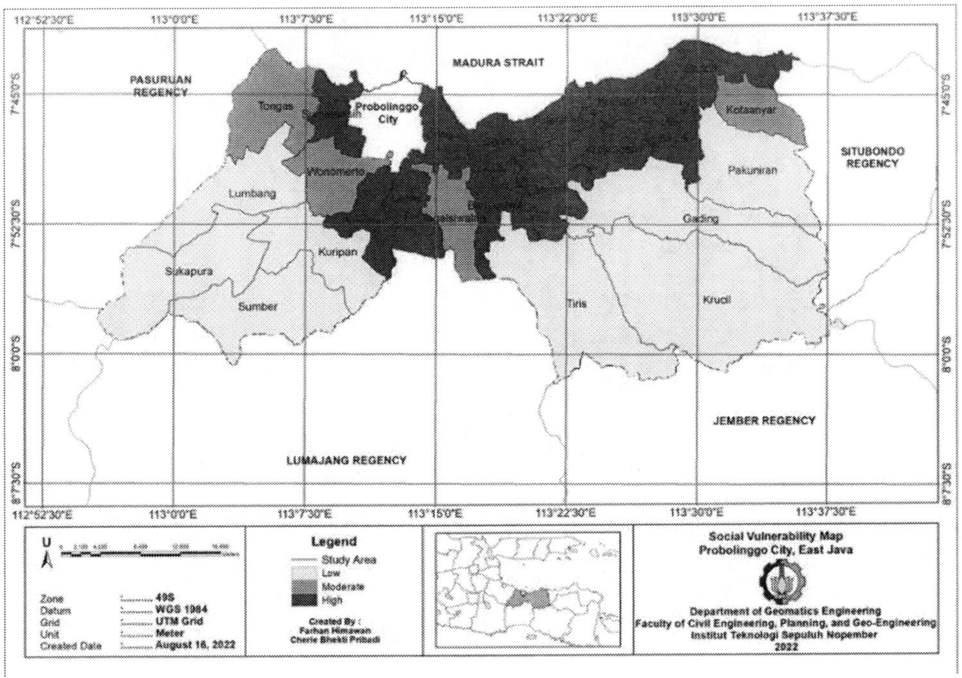

Figure 2. Social Vulnerability Map

Figure 2 shows a high level of vulnerability in densely populated sub-districts, which consists of 12 sub-districts such as Sumberasih, Bantaran, Leces, Dringu, Gending, Banyuanyar, Maron, Krejengan, Kraksaan, Besuki, Paiton, and Pajarakan. Meanwhile, the low level distribute in sub-districts which has low population density in 8 sub-districts such as Lumbang, Sukapura, Sumber, Kuripan, Tiris, Krucil, Gading, and Pakumian. Moreover, medium-level vulnerabilities spread only in 4 sub-districts such as Tongas, Wonomerto, Tegalsiwalan, and Kotaanyar. The population index for vulnerable group will also adjust to the number or population density in a sub-district.

Based on risk analysis matrix which can be seen in table 3, we get four levels or classes of vulnerability, namely low, moderate, high, and very high. A map of social vulnerability to landslides is obtained as shown in figure 3.

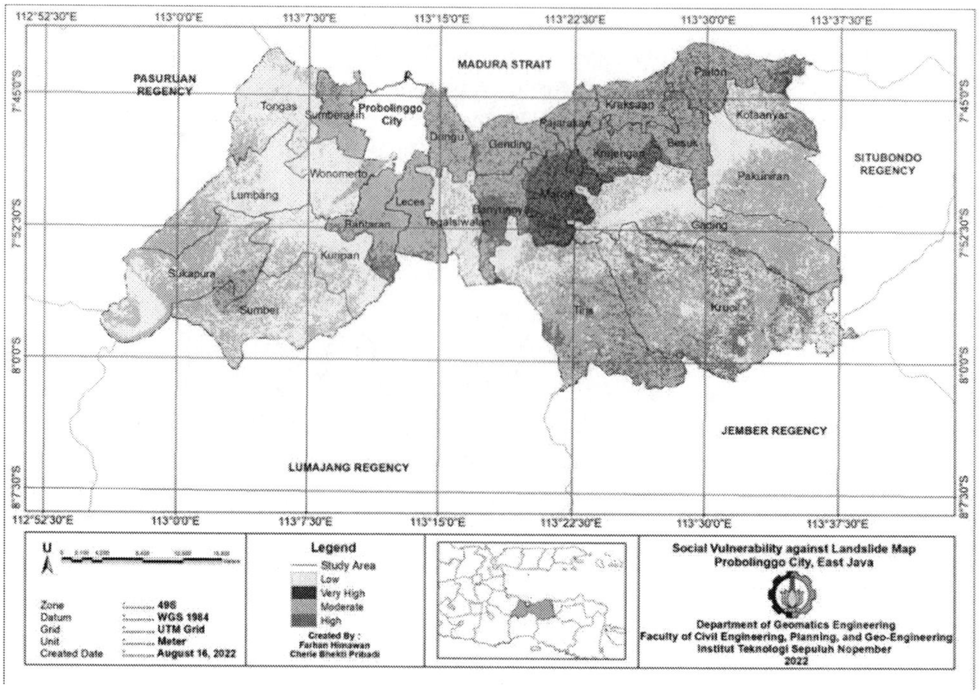

Figure 3. Social Vulnerability against Landslides Map

Figure 3 describes the largest area with a very high level of social vulnerability is dominated by Maron District about 33.7% of the area of one sub-district. While the low level of vulnerability is distributed in Wonomerto District about 86.2% of the area of one sub-district. It can be said that the area which classified as safe area from social vulnerability against landslides is in Wonomerto District.

3.2. *Environmental Vulnerability*

Similar to social vulnerability, the results of weighting of each parameter are then summed and environmental vulnerability index obtained in which the determination of low, middle, and high classes based on class length formula. The results of the class calculations and weighting can be seen in table 5 below.

Table 5. Environmental Vulnerability Index Calculation

No	Sub-district	Environmental Vulnerability Index	Class
1	Sukapura	2,8	High
2	Sumber	2,8	High
3	Kuripan	2,8	High
4	Bantaran	2,6	High
5	Leces	1,8	Moderate
6	Tegalsiwalan	1,8	Moderate
7	Banyuanyar	1,8	Moderate
8	Tiris	2,8	High
9	Krucil	2,8	High
10	Gading	2,8	High
11	Pakuniran	2,8	High
12	Kotaanyar	2,6	High
13	Paiton	2	Moderate

IOP Conf. Series: Earth and Environmental Science 1233 (2023) 012037 doi:10.1088/1755-1315/1233/1/012037

No	Sub-district	Environmental Vulnerability Index	Class
14	Besuk	1	Low
15	Kraksaan	1,2	Low
16	Krejengan	1	Low
17	Pajarakan	1,2	Low
18	Maron	1	Low
19	Gending	1,2	Low
20	Dringu	1,2	Low
21	Wonomerto	2,6	High
22	Lumbang	2,8	High
23	Tongas	2	Moderate
24	Sumberasih	1,2	Low

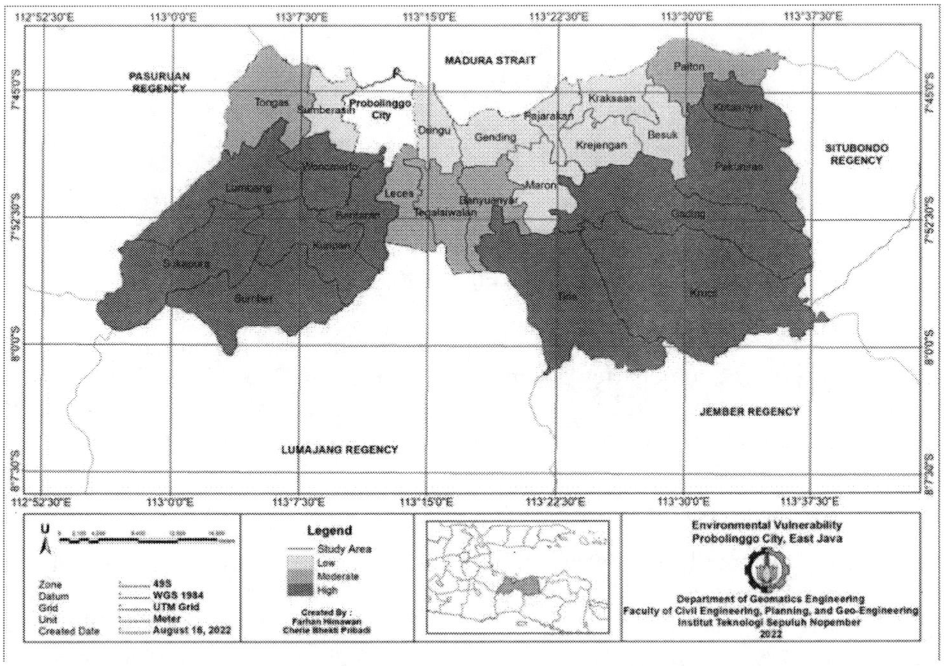

Figure 4. Environmental Vulnerability Map

Figure 4 shows low level of environmental vulnerability in 5 sub-districts where type of vulnerable land cover slightly found in these sub-districts. Area with moderate environmental vulnerability located in the north side of Probolinggo District, which is spread over 8 sub-districts. Meanwhile, the high level of environmental vulnerability is distributed over 11 sub-districts located in the south side of Probolinggo district.

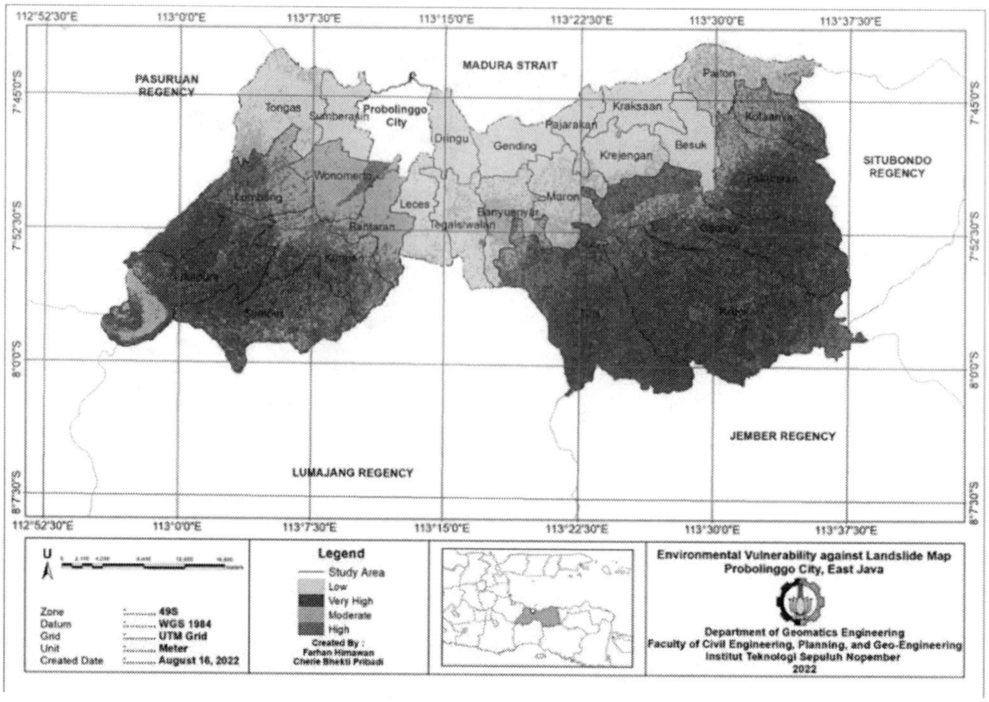

Figure 5. Environmental Vulnerability against Landslides Map

Figure 5 illustrates a very high level of environmental vulnerability is dominated by Krucil District, about 88.5% of area of one sub-district. Meanwhile, Sumberasih Sub-district is area with 100% of its territory is a low level of environmental vulnerability against landslides. It can be said that the area classified as safe from social vulnerability against landslide is Wonomerto District.

4. Conclusion

Probolinggo District has four level of social and environmental vulnerability against landslides, namely low, medium, high, and very high. The largest area with a very high level of social [8]vulnerability is dominated by Maron District about 33.7% of the area of one sub-district. While the low level of vulnerability is distributed in Wonomerto District about 86.2% of the area of one sub-district. Social vulnerability influenced by population density as well as number of vulnerable people in the area. The environmental vulnerability is dominated by a very high level of vulnerability is Krucil District, about 88.5% of the area of one sub-district. Meanwhile, Sumberasih Sub-district is an area with 100% of its territory is a low level of environmental vulnerability against landslides.

References

[1] Badan Nasional Penanggulangan Bencana 2012 *Peraturan Kepala Badan Nasional Penanggulangan Bencana No 02 Tahun 2012 tentang Pedoman Umum Pengkajian Resiko Bencana* Jakarta: Badan Nasional Penanggulangan Bencana.

[2] Badan Penanggulangan Bencana Daerah 2018 *Dokumen Kajian Risiko Bencana Kabupaten Probolinggo 2019-2023* Kabupaten Probolinggo: Badan Penanggulangan Bencana Daerah.

[3] Badan Pusat Statistik 2021 *Kabupaten Probolinggo dalam Angka 2022* Kabupaten Probolinggo: Badan Pusat Statistik.

[4] J A Nugroho B M Sukojo and I L Sari 2009 *Pemetaan Daerah Rawan Longsor dengan Penginderaan jauh dan Sistem Informasi Geografis* Surabaya: ITS.

[5] J Susilo 2008 *Pengembangan Model SIG Penentuan Kawasan Rawan Longsor sebagai Masukan Rencana Tata Ruang (Studi Kasus: Kab. Tegal)* Semarang: Universitas Diponegoro.

[6] K D Priyono and Y Priyana 2006 *Analisis Tingkat Bahaya Longsor Tanah Di Kecamatan Banjarmangu Kabupaten Banjarnegara* Forum Geografi vol 20 pp 175-189.

[7] P. Prasindya T Hariyanto and A Kurniawan 2020 *Analisis Potensi Tanah Longsor Menggunakan Sistem Informasi Geografis dan Analytical Hierarchy Process (AHP) (Studi Kasus: Kecamatan Songgon, Kabupaten Banyuwangi)* Geoid vol 16 no 1 pp 19-27.

IOP Conf. Series: Earth and Environmental Science 1233 (2023) 012038 doi:10.1088/1755-1315/1233/1/012038

Land transportation influence on the spatial distribution of Lead (Pb) in urban soils of Yogyakarta, Indonesia

J L Magno[1], W Budianta[1*]

[1]Department of Geological Engineering, Universitas Gadjah Mada, Jl. Grafika 2 Kampus UGM, Yogyakarta, Indonesia

*Corresponding author: wbudianta@ugm.ac.id

Abstract. Land transportation is a vital component of urbanization to promote economic activities in the society. But land transportation is also one of the major contributors of Lead (Pb) contamination in urban soils that might affect the health of the population. This study aims to look at the perspective of land transportation influence to spatial distribution of Pb in metropolitan Yogyakarta. We obtained 168 unique samples in grid manner and quantified its Pb concentration using ICP-AES. In relation, Inverse distance weighted (IDW) geostatistical technique is utilized to produce a spatial distribution map of Pb concentration and correlated to road density, traffic points, and transportation points perspective as a context of influence. The findings show that Yogyakarta City as highly urbanized and more land transportation activities compared to Sleman and Bantul regencies is a place of Pb contamination hotspots. Traffic areas, and Transport areas (e.g., fuel station, terminals, stopover hubs) are amplifying components of land transportation in the context of influence. Thus, land transportation sustainable urban planning is critical to address its growing influence in Pb spatial distribution in urban that might lead to population health issues upon circulating through food, air-breathing, or drinking.

1. Introduction

The global population is continuously growing that directly promotes urbanization. Occurrence of large masses of people is becoming a challenge in many cities transportation system [1]. In relation, there is a notion of accommodating movement of people from one place to the next (i.e., mobility). Building transportation infrastructure is a decent solution to improve mobility and steer economic activities in a society, although, it can have substantial environmental impacts if planning is not in sustainable perspective. Indeed, in urban areas, transportation activities support increasing mobility demands of goods/cargo and passengers; conversely, transportation activities implied more high levels of environmental outwardness. This come to a point where transportation is a dominant source of pollutants via emission, which has a wide range of environmental consequences [2]. Undeniably, the relationship of transportation system and the environment is multi-faceted.

Transportation is becoming an important facet in the concept of sustainability, which is projected to be the primary focus of transport activities in the near future [3], ranging from infra-planning, vehicle emission to management practices of green supply chain (e.g., Transpo-digital integration, EV cars). Southeast Asia is the best scenario to study and address transportation sustainability, particularly Indonesia, having the largest population and economy in the region [4]. A country with growing economy and large population, its transportation activities have resulted in higher amount of motorization and promoting congestion. Hence, consideration should be taken as the important impacts of transportation on the environment come down to air quality, noise, water quality, climate change, land take-biodiversity, and soil quality.

Transportation system is divided into three land transport, marine, and aviation. Majority of the environmental impact felt is in the land, in such, studies in Europe shows that 72% of emissions is from land transport, while 14% from marine and 13% from aviation transport [5]. In Southeast Asia, a regional strategy framework in sustainable land transport was established to act as a guideline to address intensifying mobility and emission risk [6]. Land transport in urban areas is given emphasis as motor transport and railway transport emits not only Lead (Pb) pollution (i.e., primary element emitted from fossil fuel burning and study of interest), but also with other heavy metals such as Copper (Cu), Zinc (Zn), Cadmium (Cd), etc. and absorb in soils (i.e., Urban soils). It is necessary to note that soil pollution by heavy metals is dangerous to health, and metal transfer happens into various environment (e.g., animals, man, water, plants) through leaching, food circuits, and organisms influence [7]. Consequently, tackling land transport sustainability is crucial for a sustainable urban development especially in developing countries (i.e., growing economy and population) such as Indonesia.

Yogyakarta City incorporating Sleman and Bantul regencies is considered as fastest growing urban area or as one of the next metropolitans in Indonesia [8]. Hence, as metropolitan Yogyakarta develops, it is important to fill the gap of understanding on the heavy metal lead (Pb) spatial distribution in its urban soil – as it is majorly emitted by transportation sector. This study intends to assess the relative importance of land transportation in influencing Lead (Pb) concentrations (i.e., dominant element of concern for land transportation) in urban soils of metropolitan Yogyakarta. Results contribute to the improvement of land transportation system planning in terms of source distribution approaches (i.e., Pb hotspots) and provide a basis for more cost-effective urban soil pollution/contamination remediation strategies.

2. Methods and Materials

This study determined Lead (Pb) concentration in the urban soil's samples collected from Yogyakarta City, Sleman, and Bantul regencies using Inductively Coupled Plasma – Atomic Emission Spectrometry (ICP-AES). Pb distribution geostatistical analysis and spatial processing of open-source transportation data are analyzed utilizing ArcGIS Pro 2.9.3.

2.1. Study Area

The special region of Yogyakarta is composed of five regencies, namely: (1) Yogyakarta City, the center of region development; (2) Kulon progo regency in the west, as aviation mobility center of the region (i.e., New Yogyakarta International Airport); (3) Slemen regency located north of Yogyakarta City; (4) Bantul regency located south of Yogyakarta city; and (5) Gunung Kidul in the east of region. The 168 unique samples were collected in areas of Sleman, Bantul, and Yogyakarta City (*Figure 1*), as per urbanization concept that Yogyakarta City is still expanding and integrating the nearest regencies (e.g., Sleman and Bantul). Thus, the idea streamlines into Yogyakarta City becoming more urbanized, transportation mobility intensifies affecting nearby regencies.

2.2. Sampling and Elemental Analysis

A grid method (1.5 km x 1.5 km) is used to determine the 168 unique sampling locations (*Figure 1*). At a depth of 10 cm, samples were obtained as it represents the topsoil layer using hand drill (i.e., hand auger). For laboratory processing, 500g of soil samples at each sampling sites were collected and placed in sealed plastic bag. In the lab, samples were dried for two weeks at room temperature. Soil samples were crushed using agate mortar and pestle then filtered via sieving to obtain fine size of 2 millimeters. After, soil samples undergo digestion using triacid (HF, HCl, and HNO_3) method in line with EPA – Method 3050B [9]. Triacid were mixed and put into conical flask where samples emplaced then putted in microwave digestion. Following digestion, the solution form samples were analyzed using ICP-AES to quantify Lead (Pb) concentration of soil samples.

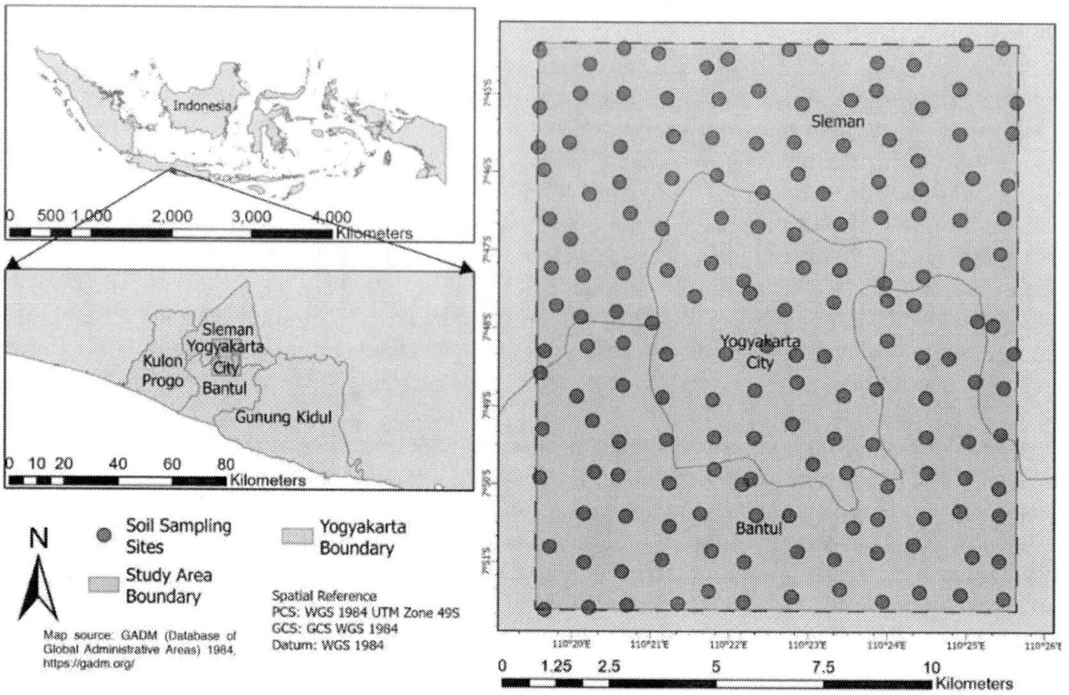

Figure 1. Location of Samples in Yogyakarta Metropolitan, Indonesia.

2.3. *Geostatistical Analysis*

Inverse distance weighted (IDW) geostatistical technique is used in this study to characterize spatial variability and interpolating between samples points. IDW clearly assumes that samples that are close to one another are more alike in value than those that are farther apart [10]. In relation, IDW act as complement to grid method sampling of urban soils as contamination. Also, IDW utilized measured values surrounding the identified location to predict a value for any unmeasured location. Hence, IDW is an effective geostatistical technique to study soil contaminants spatial distribution. In this study, the analysis was accomplished by utilizing ArcGIS Pro 2.9.3 – Lead (Pb) concentration key in as z-field values in IDW map produced.

2.4. *Spatial Digitizing*

In order to visualize effectively and interpret spatial distribution of Lead (Pb) in relation to land transportation system of metropolitan Yogyakarta, digitizing of relevant data was conducted (e.g., railway and roads). Road network class data included were residential, primary, service, streets, and major highways. In addition, traffic points and transport points (e.g., hubs, terminals, fuel station, stop points) were identified to act as supplemental data for hotspot spatial assessment. All road network class is combined to produce a road density map covering all sampling locations using ArcGIS Pro 2.9.3. OpenStreetMap is the primary source of road network, traffic points, and transport points data under the license of Open Data Commons Open Database License (ODbL) [11]. Datasets were verified through field & using blended Landsat 9 imagery dated April 24, 2022 (i.e., LC09-20220424) then produced plots via ArcGIS Pro 2.9.3.

IOP Conf. Series: Earth and Environmental Science 1233 (2023) 012038 doi:10.1088/1755-1315/1233/1/012038

3. Results and Discussion

3.1. Analysis of Pb concentrations in urban soil

Figure 1 shows the IDW spatial distribution of Lead (Pb) concentration of urban soils for metropolitan Yogyakarta. In position, the smallest concentration value is 18 mg/kg, and the largest concentration value is 129 mg/kg (*see Table 1 – range values*). Statistical summary of Table 1 indicates that the whole 168 unique soil samples have an average Pb concentration of 70.04 and it exceeds the background values (i.e., natural concentration of Pb without anthropogenic influence). In relation, skewness and kurtosis value of the samples is -0.21 and -1.10, respectively, implying existence of extremity samples. As evidently shown in the Pb concentrations map, there are extremely high values (i.e., contamination hotspots) and there are low values (cold spots) that are below background value. Majority of the Pb hotspots coincide in inner metropolitan Yogyakarta (i.e., Yogyakarta City), spreading contamination throughout Sleman and Bantul regencies - especially near borders with Yogyakarta City portion.

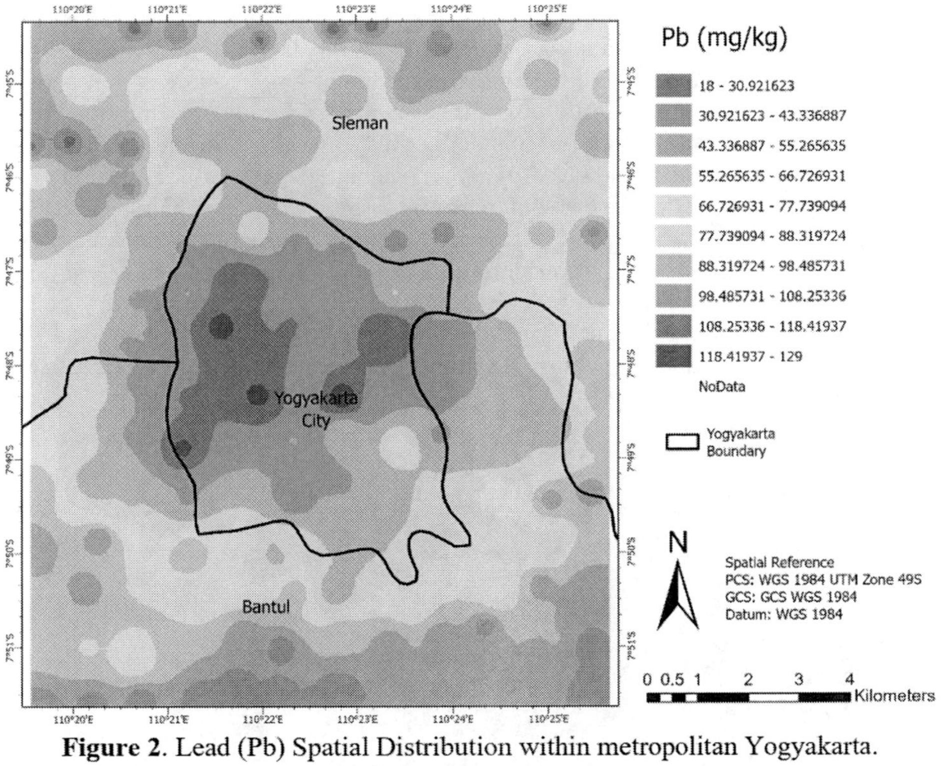

Figure 2. Lead (Pb) Spatial Distribution within metropolitan Yogyakarta.

Table 1. Pb concentration statistical summary (mg/kg) in urban soils.

	Pb
Background values	25.5
Mean	70.04
Range	18-129
SD	29.26
Skewness	-0.21
Kurtosis	-1.10

In urban areas, there are a lot of factors that affect soil contamination, but dominantly caused by anthropogenic activities[12]. Sleman and Bantul regencies are becoming highly contaminated with Lead (Pb), yet there are places still within low values (i.e., most outermost portion of the study area). In this study, the resulting factor is land transportation. Land transportation is the major contributor of Lead (Pb) emission as it burns up fossil fuels.

3.2. Land transportation a key factor in urbanization

Land transportation is the backbone of society as it promotes mobility of goods and people. A place like metropolitan Yogyakarta requires a robust land transportation system as it caters growing economic activity [8]. Consolidating all road network data class resulted into a road density map (*Figure 3*). Figure 3 shows the road density map that represents the ratio of the length of metropolitan Yogyakarta's total road network to the land area (kilometers per 100 square kilometers). The study area has high road density values act as a proof that it is an urbanized area, with a maximum value of 38 km/km^2. Lowest values (white) are found in the outer portions - < 4 km/km^2, and some pockets of low values in metropolitan as there are presence of open spaces such as parks and occupied space such as buildings. To develop further the economic status of metropolitan Yogyakarta, mobility access is built more through road networks. It is understood, the higher the value of your road density it is more expose or promotes land transportation mobility, in return, there's a lot of Lead (Pb) emission from burning of fossil fuels. The pattern of road density map is congruous to spatial distribution of Pb concentration. Notice the upper right of Figure 3 road density map, it has high density values, meaning more presence of vehicles mobility. In connection, upper right portion of Figure 2, in Sleman regency, starting to have hotspots pockets. The cover up pattern of road density map in metropolitan Yogyakarta complements and follows the spatial distribution of Pb concentration. Evident enough that road network is an influence on soil contamination spatial distribution. This scenario calls on to put necessary attention to road network system and underlying amplifying factors such as traffic system and transportation hubs/terminals, for example, during mobility there are leakages of fuels that amplify Pb contamination in soils.

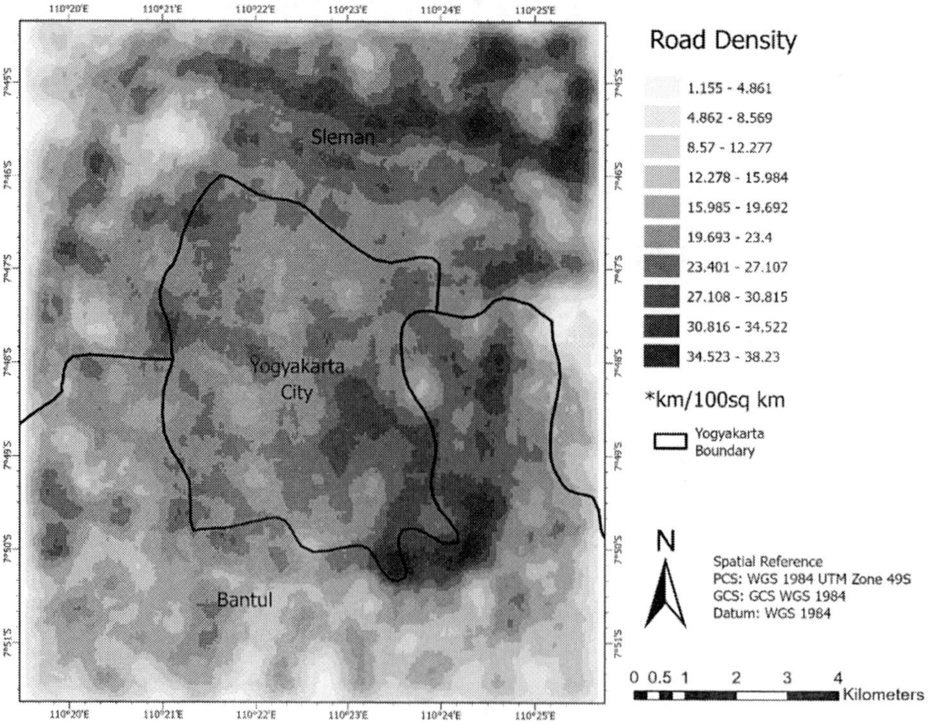

Figure 3. Road Density Map within metropolitan Yogyakarta.

3.3. Land transportation correlation to Lead (Pb) distribution

The system of land transportation as an influence into soil contamination is complex. In this study, traffic points and transport points are considered amplifying context of land transportation system contribution to contamination. As mentioned earlier, road density is congruous or in-match with the spatial distribution of Pb concentration in metropolitan Yogyakarta. Cementing the idea that vehicle mobility is influential to Pb contamination spreading. But land transportation mobility is not only about continuous movement in one place, it also add-in the stopover occurrences. Figure 4 shows the location of traffic areas, and transport areas such as terminals, fuel station, unloading/loading passenger areas. The existence of many traffic areas and transport areas in inner metropolitan Yogyakarta suggest high volume of land transportation activity that promotes more emission of Pb contaminants. In Figure 4, it is identified that intense traffic areas also happened to be within areas where there are hotspots of Pb contamination. Majority of traffic and transportation points are located in Yogyakarta City and nearby borders Sleman & Bantul regencies, in which high Pb concentration values are present. In relation, less traffic and transportation points in the outermost of study area (far-border Sleman and Bantul Regencies), there is less contamination. This finding agrees to the past assessment study of Pb contamination in Yogyakarta City [13][14]. Hence, land transportation influences the distribution of Pb contaminant in metropolitan Yogyakarta.

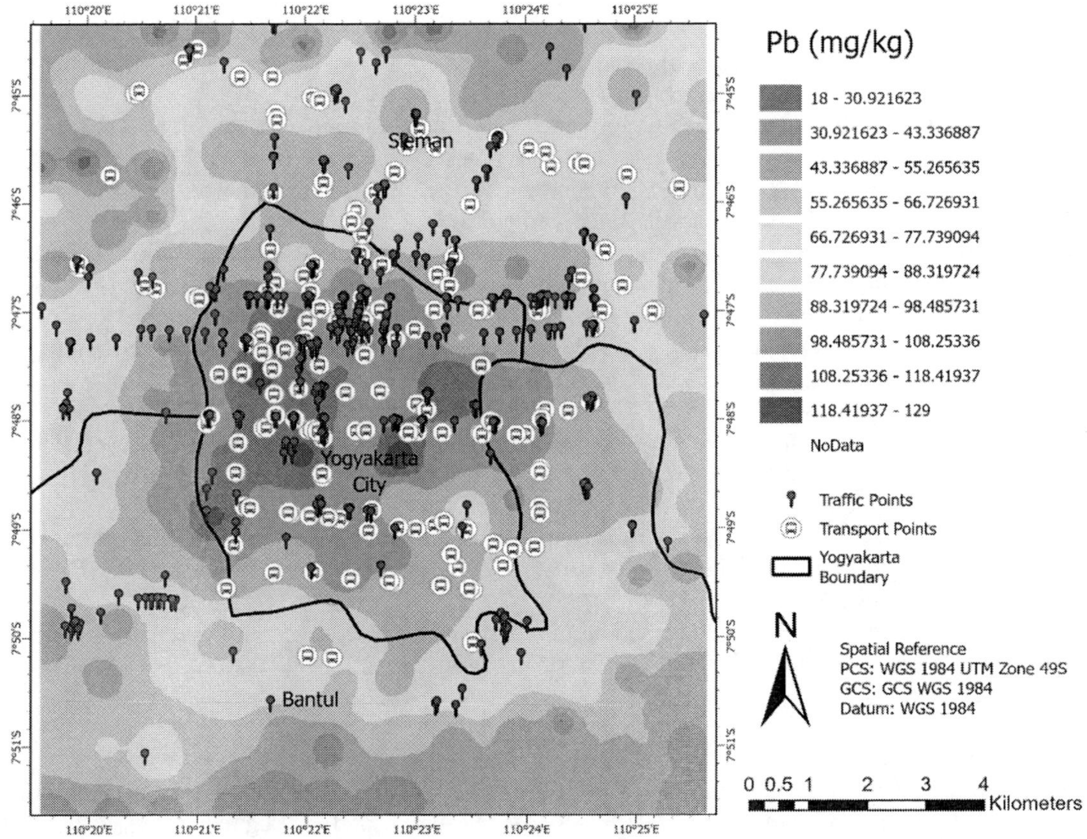

Figure 4. Traffic and transportation hubs plots within metropolitan Yogyakarta (*spatial distribution of Lead (Pb) as based for correlation*) – raw data source out through OSM Open database [11].

4. Conclusion

To attain sustainable land transportation development, it is crucial to put attention to systematic urban planning especially the subject that it might influence in positive and negative way. In this study, through analyzing the Lead (Pb) concentration of urban soils in metropolitan Yogyakarta, Indonesia, it found out that land transportation influences the Pb spatial distribution. A highly urbanized area where land transportation activities is more intense such as Yogyakarta City accumulates more Pb contaminant compared to Sleman and Bantul regencies as having less land transportation activities (especially the outermost part of the study area). In order to strengthen the urban planning and mitigation measures, it is advisable to look into industrial perspective influence on Pb distribution as there are key components in industrial system that contributes to Pb emissions. Overall, aligning regulations and policies in relation to growing economic activity and urbanization expansion is important to address problems like Pb contamination that may cause health problems to the society in a long run.

References

[1] Goetz A R 2019 Transport challenges in rapidly growing cities: is there a magic bullet? *Transp Rev* **39** 701–5.
[2] Aminzadegan S, Shahriari M, Mehranfar F and Abramović B 2022 Factors affecting the emission of pollutants in different types of transportation: A literature review *Energy Reports* **8** 2508–29.
[3] Langeland O, Andersson M and Flotve B L 2021 Changes and Challenges in Future Transport Drivers and Trends.
[4] Asian Development Bank 2019 *The Evolution of Indonesia's Participation in Global Value Chains:* (Manila, Philippines).
[5] EEA 2020 *Transport - EEA Report No 19/2020.*
[6] ASEAN 2019 ASEAN Regional Strategy on sustainable land transport. *ASEAN Book's Collection.*
[7] Briffa J, Sinagra E and Blundell R 2020 Heavy metal pollution in the environment and their toxicological effects on humans *Heliyon* **6** e04691.
[8] Pradoto W 2011 Dynamics of Peri-urbanization and Socioeconomic Transportation: Case of Metropolitan Yogyakarta, Indonesia *International Journal of Arts & Sciences 19-20.*
[9] U.S. EPA 1996 *Method 3050B: Acid Digestion of Sediments, Sludges, and Soils.*
[10] Gotway C A, Ferguson R B, Hergert G W and Peterson T A 1996 Comparison of Kriging and Inverse-Distance Methods for Mapping Soil Parameters *Soil Science Society of America Journal* **60** 1237–47.
[11] OSM 2022 Export | OpenStreetMap - Open Data Commons Open Database License (ODbL).
[12] Garbuio F J, Howard J L and dos Santos L M 2012 Impact of Human Activities on Soil Contamination *Appl Environ Soil Sci* **2012** 1–2.
[13] Sekarningsih A T, Budianta W, Warmada I W and Hinode H 2021 Assessing the spatial distribution of urban soil lead contamination in Yogyakarta City, Indonesia *E3S Web of Conferences* **325** 03001.
[14] Magno J L and Budianta W 2022 Spatial distribution and pattern of heavy metals in urban soils of Yogyakarta, Indonesia *IOP Conf Ser Earth Environ Sci* **1071** 012032.

Acknowledgments

This research work is financially supported by JICA Project for AUN/SEED-Net in collaboration with Universitas Gadjah Mada for their CUBE-ASEAN Program and Rekognisi Tugas Akhir (RTA) Program. Authors would also like to thank Tokyo Institute of Technology, Japan for their kind facilitation of sample analysis.

THE UTILIZATION OF GEOGRAPHIC INFORMATION SYSTEMS FOR SETTLEMENT LOCATION DETERMINATION IN SUPPORTING SUSTAINABLE DEVELOPMENT: STUDY CASE IN SEDAYU SUB-DISTRICT, BANTUL REGENCY, DIY

N. Nur Awalia[1], D. Luklu Chusnia[1], S. Fajar Nugroho[1], A. Bayu Sekaranom[1]

Department of Environmental Geography, Faculty of Geography, Universitas Gadjah Mada, Indonesia

Corresponding author: andung.geo@ugm.ac.id

Abstract. The phenomenon of unsuitable settlement becomes a major issue in environmental problems in big cities. The purpose of this study is to provide innovation design recommendations for determining the location of appropriate settlements to realize the implementation of smart cities. The resulting output of this study is an innovation in settlement location determination. The innovation using a geographic information system equipped with a variety of spatially processed information with overlay techniques. The results of the geographic information system process are then extracted into a mobile application design. Study case in Sedayu District show that the area consists of 36.9% of the area is not suitable for analysis; 36.4% of the area is not suitable for use, and 26.7% of the area is suitable for use. Mobile application design in this study called Smart Application for Smart Resident with two main features, My Contribution (KOKU) and Recommendations and Settlement Locations feature. This innovation can help the community by providing information and recommendations for suitable, less suitable, and unsuitable settlement locations. Thus, the existence of the result study can be a consideration of the community in determining the location of settlements by aspects of sustainable development.

Keywords: geographic information system, mobile application, smart city, utilizing, settlement

1. Introduction

Land needs for settlement development should consider certain criteria. The physical condition of nature and existing social conditions are the important factors that must be considered in determining land for settlement [7]. The characteristics of settlements land are important to note so that the construction of these settlements does not cause problems for the surrounding environment. Various cases of loss and damage can occur due to the unsuitable settlement land that exceeds the limits of its land capabilities [4]. Therefore, land suitability analysis for settlements needs to be carried out so that the negative impact of settlements on the environment can be minimized.

Land suitability analysis provides an assessment of the level of land suitability by identifying and comparing various existing land uses [8]. Analysis of the suitability of settlement land is useful for identifying land characteristics and providing information for determining settlement locations [5]. This is suitable as done by previous research on Land Suitability Analysis for Housing in Pesisir Selatan Regency, which states that diverse environmental conditions make it impossible to develop housing in every location. Spatial analysis is therefore useful in determining land suitability for housing development so that environmental problems are avoided [11].

Data on the physical and social condition of the land can be processed and analyzed using geographic information systems. Geographic information systems can be used in land suitability analysis for settlements because they have the principle of rapid digital spatial data processing, large data storage capacity, and can produce results that are following the needs of analysis [12]. Geographic information systems can be used in the analysis because they have the principle of rapid digital spatial data processing, large data storage capacity, and can produce results that are following the needs of analysis [12].

The overlay technique is one of the methods in geographical information systems that can be used for the analysis of the suitability of settlement land. The principle of this technique is to combine spatial information from two or more maps into a piece of new information. Through the use of geographical information, systems will be obtained data output that can produce new information that is precise and accurate [6]. The overlay technique is one of the methods in geographical information systems that can be used for the analysis of the suitability of settlement land [1]. The principle of this techniqueis to combine spatial information from two or more maps into a piece of new information. With geographical information, systems will obtain data output that can produce new information that is precise and accurate.

The high rate of population growth can have an impact on increasing the needs of settlement land. This caused many settlements to be built not based on the suitability of existing land for settlement [10]. The use of land and social characteristic data for the selection of settlement locations is expected to provide appropriate settlement location recommendations. Based on these conditions, this study does an analysis of the suitable land for settlement in the form of a settlement location determination system. This system will be put in the form of a mobile application. This study aims to understand the potential utilization of the application design in the era of industrial revolution 4.0 in determining the location of settlements. In addition, it is expected to be a consideration for the community in determining the location of settlements that are in accordance with aspects of sustainable development.

2. Method

The type of data collected in this study was secondary data relating to indicators of settlement development. The type of data collected in the study is secondary data related to development indicators.

TABLE 1. Data Secondary Source

Data	Data Resources	Description
Trash Bins Distance	Geographic Information System Laboratory, Universitas Gadjah Mada	Scale 1:40.000
River Distance	Geographic Information System Laboratory, Universitas Gadjah Mada	Scale 1:40.000
Road Network	Geographic Information System Laboratory, Universitas Gadjah Mada	Scale 1:40.000
Vegetation Density	Geographic Information System Laboratory, Universitas Gadjah Mada	Scale 1:40.000
Land Use	Geographic Information System Laboratory, Universitas Gadjah Mada	Scale 1:40.000
Soil Texture	Geographic Information System Laboratory, Universitas Gadjah Mada	Scale 1:40.000
River Distance	Geographic Information System Laboratory, Universitas Gadjah Mada	Scale 1:40.000

IOP Conf. Series: Earth and Environmental Science 1233 (2023) 012039 doi:10.1088/1755-1315/1233/1/012039

Population Density	Central Bureau of Statistics	-

The data processing process in this study was carried out by applying the principles of geographic information systems. Data processing was assisted by using ArcMap software version 10.8. The indicator data was presented at the location that has been obtained, then in the form of overlaid thematic maps. Before carrying out the technical overlay, each thematic map showing the selected location indicator had a weight or value in the form of scoring. The basis used to determine the score or weight for each indicator was the difference in its role in determining the selected location [2].

After the score was obtained, then the arithmetic summation was carried out on each thematic map and then analyzed ised the standard deviation method to determine the classification of a good settlement recommendation location. Recommendations for settlement locations were obtained based on the results of overlays and scoring results with a tiered quantitative approach. The overlay results was processed used a tiered quantitative approach and are weighted with a score of 8-20. The scores were then processed based on the principle of standard deviation, so the result was a classification of settlement recommendations.

TABLE 2. Classification of Recommendations Settlement

Classification of Recommendations Settlement	Score
Unsuitable	8-13
Less Suitable	14-15
Suitable	16-20

As an example of study location, Sedayu District, Bantul Regency, was used as a sample analysis. This sub-district consists of four villages, such as Argodadi, Argorejo, Argosari, and Argomulyo. Sedayu District was chosen as a sample location because it was included as a suburban area that has a relatively high growth rate of irregular settlements.

3. Result and Discussion

3.1. Analysis of Sedayu Distric as Study Location

The indicator data for determining the location of settlements that have been obtained are then presented in the form of thematic maps. Before the overlaying, each thematic map showing indicators for determining the location of settlements is given weight or value in scoring. The basis used to determine the score or weight for each indicator is the difference in its role in determining the location of settlements. Based on the results of the overlay of the seven indicator maps above, a map is obtained according to the land to be used in Sedayu District.

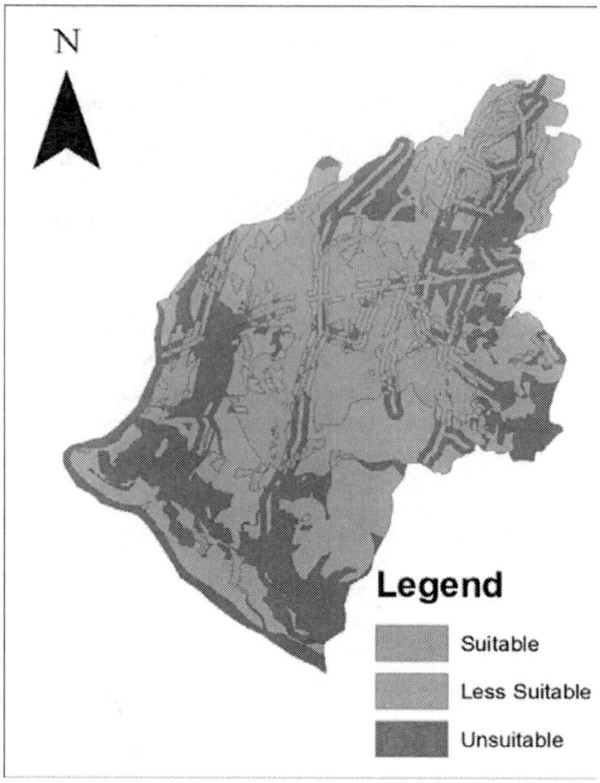

FIGURE 1. Thematic Map Overlay Process in Sedayu District (Source: data processing, 2022)

The land is considered suitable for development if it has a high score on weighting. The higher the score obtained, the more suitable the land for land use. The results that have been carried out on a sample of research locations, Sedayu District, show that 36.9% of the area is not suitable for analysis; 36.4% of the area is not suitable for use, and 26.7% of the area is suitable for use. Based on that proportion, the area that is suitable to be seen in Sedayu District is relatively small. This can bea warning for the government and local communities to use the existing land wisely and as well as possible. Suitable area for settlement located close to the road network. Among the four villages in Sedayu District, the most suitable location for this is Argorejo Village. It can be seen that Argorejo Village has a high potential to be used as a location that utilizes.

3.2. How the System Works

In the design of the SANFOR (Smart Application for Smart Resident) application, there are five features on the main menu, such as the two main features in the form of Recommendations and Settlement Locations and the KOKU (My Contribution) feature. In addition to the main features, there are supporting features in the form of Housing and Settlement Guidelines, Application Usage Instructions, and User Profiles. When entering the application, users are directed to be able to create anaccount first on the Profile feature by completing personal data and verifying the account. After the user has an account, the user can take advantage of the main features, such as Recommendations and Residential Locations to find out recommendations for the location of the house which are entered in the location section.

The recommendations used to refer to seven indicators covering physical and social aspects, such as land use, vegetation density, road network, soil texture, disposal to trash cans, river distance, and population density. The main function in the system is mapping and scoring on each indicator to determine the feasibility level of a location. In the mapping process, SANFOR has used digital maps and integrated them with the Geographic Information System on the Map of Sedayu District, DI Yogyakarta. In classifying process, this application design will be stacking (overlay) the seven indicators information as determinant for deciding area recommendation for settlement [9].

In addition to the main features, there are supporting features in the form of Housing and Settlement Guidelines, Application Usage Instructions, and User Profiles. When entering the application, users are directed to be able to create an account first on the Profile feature by completing personal data and verifying the account. After the user has an account, the user can take advantage of the main features, namely Recommendations and Residential Locations to find out recommendations for the location of the house which are entered in the location section. The recommendations used to refer to seven indicators covering physical aspects and social, namely land use, vegetation density, road network, soil texture, the distance of trash cans, the distance of rivers, and population density.

The SANFOR application design will classify the seven indicators and provide conclusions as recommendations from the locations that have been entered. The results of the recommendations are classified into three categories, such as appropriate, less appropriate, and inappropriate. In addition to functioning as a virtual assistant in providing recommendations, this application also functions as a distributor of information related to the physical and social conditions of the surrounding environment which includes the seven indicators above from fellow users. This is because users can take advantage of the KOKU (My Contribution) feature by providing information and descriptions that will be useful for fellow application users to make decisions regarding the location of their homes. This information will be displayed on SANFOR (Smart Application for Smart Resident) application design below.

3.3. App View

FIGURE 1 shows the main page of the SANFOR application design which consists of news pages related to smart cities, the urgency of settlement suitability, and a login menu. At the bottom, there is information about SANFOR which includes social media accounts, such as Instagram and email. After the user creates an account and performs account verification, the user will enter a username and password to log in. Next, the user will be directed to the main menu section to select the features offered by SANFOR. There is a profile feature on SANFOR which consists of some information in the form of personal data that must be completed by the user as a condition to be able to create an account in this application. Data requirements are only used for database management and will be kept confidential.

To make it easier for users to be more effective and efficient, SANFOR also provides short tutorials related to application usage instructions that are packaged in the form of animated videos and equipped with a short description at the bottom. The video is expected to help users understand how to use this application. KOKU features that can be used by application users to contribute by providing information and descriptions related to environmental conditions, which include physical and social conditions. The description given can include seven predetermined indicators, namely land use, vegetation density, road network, soil texture, the distance of place garbage, river distance, and population density. The user will be directed to the google form when pressing the KOKU option on the page. In this section, the user is required to enter the location first. After the location is entered, the system will automatically display general recommendations in the column on the right. In addition, in more detail, users can see recommendations on the other seven aspects by pressing the column for these aspects at the bottom. More detailed information and recommendations on each indicator will appear if the user presses the indicator column. The information presented is in the form of a description of the indicators and the results of their recommendations.

IOP Conf. Series: Earth and Environmental Science 1233 (2023) 012039 doi:10.1088/1755-1315/1233/1/012039

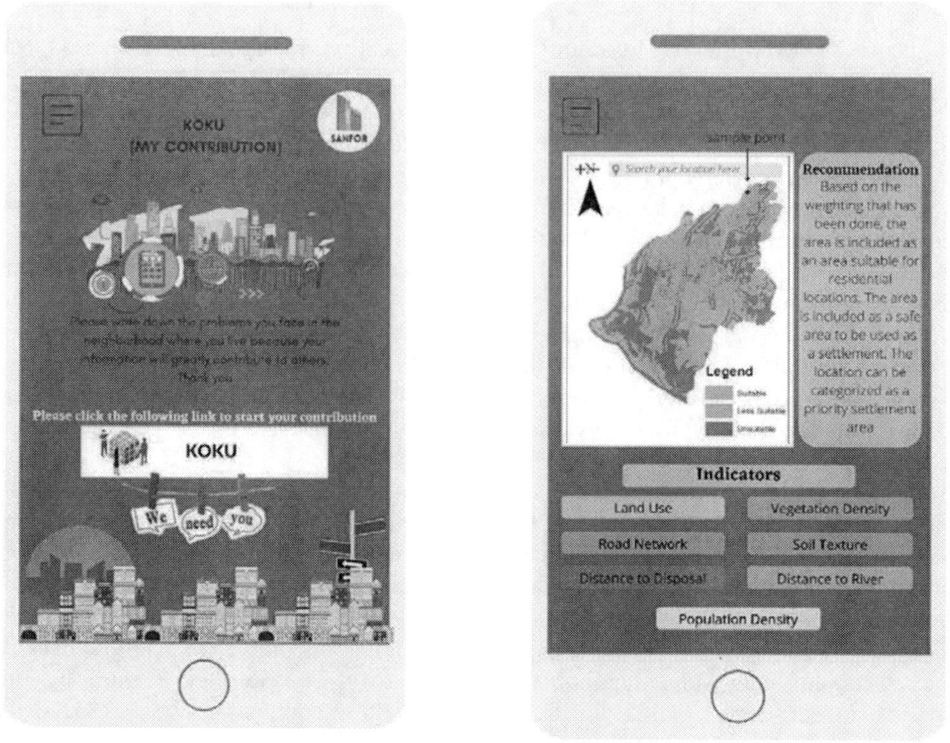

FIGURE 2. Design of SANFOR Application

Based on research conducted by Haryana et al, 2013 there are similarities in utilizing Remote Sensing and Geographic Information System for settlement land use classification planning. Both use the overlay technique of the various parameters for settlement classification. The difference in the SANFOR application is that the Earthquake Hazard Map is not used, this is because SANFOR focuses more on settlement classification of land feasibility in terms of various parameters such as trash bins distance, river distance, road network, vegetation density, land use soil texture, and population density. Based on these parameters, this application does not pay attention to disaster parameters such as earthquakes. Meanwhile, previous research conducted by Haryana et al. only focused on my disaster aspect, namely earthquakes, so it did not pay attention to parameters such as vegetation, roads, and water conditions for housing classification. Another similarity with previous research is the location of the study, namely Bantul Regency, DIY. So that both use the Topographic Map of Bantul as the basic data.

4. Conclusion

Determination of the location is helpful for knowing an area that will be used as a suitable settlement location. Determination of location recommendations can be obtained by overlaying all settlement indicators using a tiered quantitative method. In the future, it is necessary to reconstruct the weighting of each indicator to know the role of each indicator in the settlement location. SANFOR still needs development, so it would be better if in the future this application is able to provide analysis and information related to disaster conditions in an area. So that the housing recommendations provided will be more inclusive for users. The method used in this study is the overlay technique on several maps based on predetermined parameters, then provides scoring to then provide recommendations to the user. SANFOR can be an alternative solution for the community to determine a good and appropriate housing location based on various parameters used by utilizing the Geographic Information System.

5. Acknowledgments
This research is supported by grant of "Desa Binaan" from "Direktorat Pengabdian Masyarakat (DPKM) Universitas Gadjah Mada 2022" with Dr. Andung Bayu Sekaranom as principal investigator (Contract number: 461/UN1/DPM/YANMAS/PM/2022)

References
[1] Amir, A. K., Wunas, S. & Arifin, M., 2020. Settlement development based on land suitability. Earth and Environmental Science, Volume 419, pp. 1755-1315.
[2] Becker, William, et al. "Weights and importance in composite indicators: Closing the gap." Ecological indicators 80 (2017): 12-22.
[3] Haryana, Inneke. K., Fikriyaha, Vidya. N., Yulianti, Nur. Vita., 2013. Application of Remote Sensing and Geographic Information System for settlement land use classification planning in Bantul based on Earthquake Disaster Mitigation (Case Study: Bantul earthquake, May 27th 2006). Procedia Environmental Sciences, pp. 434 – 443.
[4] Khaerani, R., Sitorus, S. R. dan Rusdiana, O., 2018. Analisis Penyimpangan Penggunaan Lahan Berdasarkan Rencana Tata Ruang Wilayah Kabupaten Sumedang. (Tataloka, Semarang, 2018),pp. 399—409.
[5] Luhukay, M. R., Sela, R. L., dan Franklin, P. J., Analisis Kesesuaian Penggunaan Lahan Permukiman Berbasis (SIG) Sistem Informasi Geografi di Kecamatan Mapanget Kota Manado.(*Spacial Journal*, 2019), pp. 271—281.
[6] Makalalag, D. H., Rindengan, Y. D. Y., dan Sengkey, R., 2016. Sistem Informasi Geografis Monitoring dan Evaluasi Proyek Dinas Pekerjaan Umum Kota Kotamobagu. E-journal Teknik Informatika. Volume 13, pp. 1—9.
[7] Missah, R. E., Sela, R. L., dan Takumansang, E. D., 2019. Analisis Kesesuaian Lahan Permukiman Berdasarkan Rencana Tata Ruang Wilayah (RTRW) di Kabupaten Minahasa Tenggara (Studi Kasus: Kecamatan Ratahan). (*Spacial Journal*, 2019), pp. 247— 258.
[8] Ratnawati, H., dan Djojomartono, P. N., 2020. Analisis Kesesuaian Lahan Permukiman di Kecamatan Playen Kabupaten Gunungkidul menggunakan Pendekatan Analytical Hierarchy Process. (Journal of Geospatial Information Science and Engineering, 2020), pp. 123—132.
[9] Tanjung, N. A., Rustiadia, E. & Widiatmakaa, 2021. Suitable and available land for settlement development in Cianjur Regency. Journal of Natural Resources and Environmental Management, Volume 11, pp. 550-566
[10] Tekouabou, S. C. K., Diop, E. B., Azmi, R., Jaligot, R., & Chenal, J. 2021. "Reviewing the application of machine learning methods to model urban form indicators in planning decision support systems: Potential, issues and challenges." Journal of King Saud University-Computer and Information Sciences.
[11] Widiyastuti, D., Ermawati, H., Septiawan, L. & Kumara, I. S. W., 2019. Land Suitability Analysis for Housing in Pesisir Selatan Regency, West Sumatra, Indonesia. ASEAN Journal on Science & Technology for Development, Volume 36, pp. 45-52.
[12] Yuliyanto, W. A., Nugraha, A. L., & Hani'ah, 2016. Desain Pengembangan Aplikasi Pelayanan Air Bersih Desa Sarirejo Kecamatan Guntur Kabupaten Demak Berbasis GIS Desktop. Jurnal Geodesi Undip. Volume 5, pp. 151—164.

The Linkage of Rainfall Anomaly due to El Nino Southern Oscillation (ENSO) on Cassava Productivity in Central Java Province

Anugrah Jorgi Firmansyah[1,2 a)], and Emilya Nurjani[1]

[1]*Department of Environmental Geography, Faculty of Geography, Universitas Gadjah Mada, Indonesia*
[2]*Center for Natural Disaster Studies, Universitas Gadjah Mada, Indonesia*

[a)] Corresponding author: anugrahjorgi@mail.ugm.ac.id

Abstract. El Nino Southern Oscillation (ENSO) is a global climate phenomenon that affects the climate of a region and causing climate anomalies. ENSO can cause rainfall anomalies that have an impact on the agricultural sector. Cassava is one of the plants known as climate-resistant plants. Cassava has high productivity and is almost evenly distributed in Central Java Province. This study aims to determine the relationship between rainfall anomalies due to ENSO and cassava productivity in Central Java from 1990-2019 at each different ENSO phase. El Nino in weak, moderate, and strong phases triggers a decrease in annual rainfall intensity with an even distribution from the north and becomes softer towards the south of the province. When La Nina begins, the intensity of rainfall increases with a strong intensity starting from the south side and weakening towards the province's north. The increase in rainfall during the La Nina phase tends to increase cassava productivity and the El Nino phase tends to decrease cassava productivity, although there are case findings in certain areas.

1. Introduction

El Nino Southern Oscillation (ENSO) is one of the global climate phenomena that can affect the climate of a region or country so that it changes climate patterns and causes climate anomalies. ENSO is a non-periodic Global Climate System. This ENSO phenomenon affects marine conditions in Indonesia. During El Nino, Indonesian sea waters will be colder and when La Nina will be warmer [1]. The impact of El Nino in 1997/1998 caused the rainy season to be delayed by 2 - 3 months. This resulted in the delay in the 1998 dry season planting time. In the 1997/1998 period, rice production decreased by 6.5 percent and rice imports increased by 3 million tons [2]. This climate change also affects secondary crops such as cassava and corn, significantly affecting planting time and production yields [3].

The effect of ENSO on annual rainfall is 16.36% higher than the effect of El Nino (11.28%) and La Nina (12.80%). The effect of ENSO and El Nino on seasonal rainfall in Indonesia takes place in the middle of the year. La Nina does not have a significant effect on the rainfall of early dry season, peak dry season, early rainy season and peak rainy season in Indonesia [4]. El Nino events had more of an impact on secondary crops, the decline in productivity of secondary crops and tubers due to El Nino only occurred in 2002. On the other hand, El Nino in 2015 had a positive impact on several commodity

IOP Conf. Series: Earth and Environmental Science 1233 (2023) 012040 doi:10.1088/1755-1315/1233/1/012040

productivities. La Nina in 2000, 2008, and 2010 caused a decrease in palawija productivity in several sub-districts, but did not affect the productivity of tubers [5].

Cassava is a multipurpose plant because the tubers, stems, and leaves are helpful [6]. Cassava is used as food (58%), industrial raw materials (28%), feed ingredients (2%), and exported in the form of cassava (8%). The marketing chain for cassava in Java is relatively simple. Generally, farmers sell to traders who then distribute to the cassava processing industry group. In Central Java Province alone, there are 36 cassava processing industrial companies, more than East Java (15 companies) and West Java (26 companies) [7]. Research on the relationship between climate elements and the productivity of cassava (Manihot esculenta Crantz) in Malang Regency shows that rainfall climate elements affect cassava production by 80% and the other 20% do not feel the effect [8].

Cassava is a climate-resistant crop with a growing period of 8 to 12 months. However, it turns out that cassava requires unique criteria for climatic conditions, especially rainfall for its growth [9][10]. The climatic conditions in Indonesia are undoubtedly inseparable from the impact of global climate phenomena both in Indonesian waters and waters in other parts of the world, one of which is ENSO. This is one of the exciting factors in studying whether ENSO has a relationship with cassava productivity in Central Java. The aim of this research is to identify the linkage of rainfall anomaly due to El Nino Southern Oscillation (ENSO) on cassava productivity in Central Java Province. The map of the research area can be seen in Figure 1.

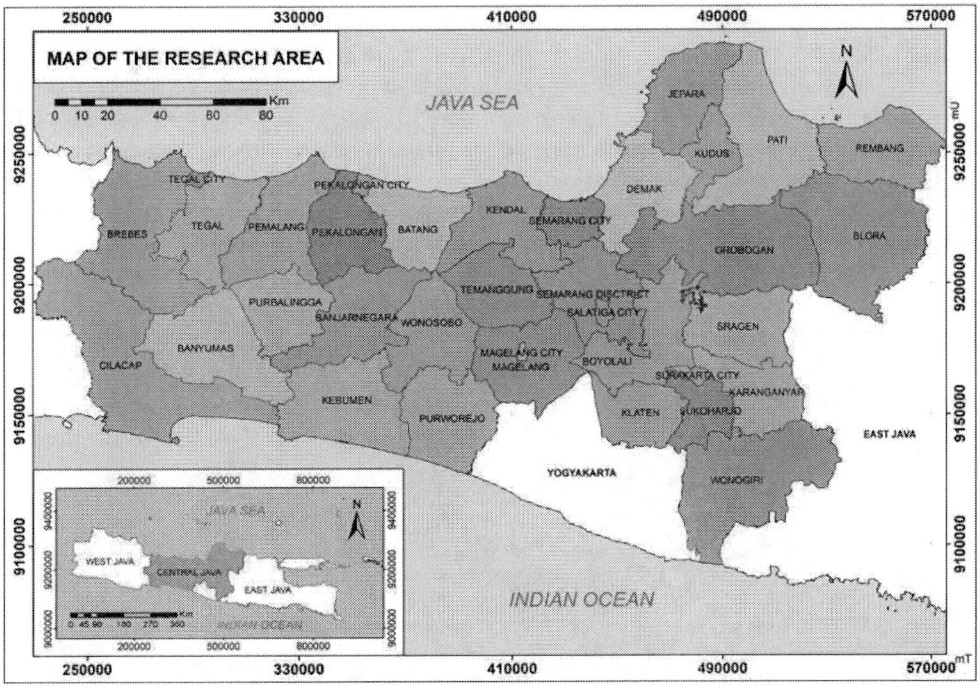

Figure 1. Map of the research area in Central Java, Indonesia

2. Data and Methodology

We use monthly rainfall data CHIRPS (Climate Hazard Group Infrared Precipitation with Station Data), which is reanalyzed rainfall data from the combination of satellite data, models, and surface observations [11]. This CHIRPS was obtained from the Climate Hazard Center website of the University of California, Santa Barbara (https://data.chc.ucsb.edu/products/CHIRPS-2.0/). Central Java Province cassava productivity data for 1990-2019 was obtained from the Central Java Provincial Statistics Agency (https://jateng.bps.go.id/).

The relationship between rainfall anomalies caused by ENSO and cassava productivity in Central Java Province was carried out by comparing rainfall anomalies with cassava productivity anomalies.

The anomaly of cassava productivity was calculated using the moving average method. Moving average of order three is used to find the average value of the three years before the El Nino/La Nina event, which will calculate the productivity of cassava. The trend of moving average of order 3 is the most suitable model to estimate the variables of palawija production in Maluku Province [12].

$$\acute{Y}_t = \frac{(Y_{t-1} + Y_{t-2} + Y_{t-3})}{3} \ldots. (1)$$

where:

$\acute{Y}_t$ = Expected value in t-period

$Y_{t-1,dst.}$ = Productivity value (year)-1, and so on

This classification of productivity anomalies is classified as multiples of 50 quintal/ha in each class. The anomaly is neutral if the change in cassava productivity during the ENSO phase in that year is between -25 to 25 quintal/ha and it symbolized by a gray color. Negative anomalies are said when there is a decrease in productivity starting from -25 quintal/ha with multiples of -50 quintal/ha, symbolized by colors ranging from light yellow to dark orange. Positive anomalies are said if productivity increases starting from 25 quintal/ha with multiples of 50 quintal/ha, which is symbolized by light blue to dark blue.

3. RESULT AND DISCUSSION

The observations of rainfall anomalies with cassava productivity during a weak El Nino can be seen in Figure 2. In the area of climate type B, there is an anomaly of evenly distributed rainfall in the area. The decline in cassava productivity occurred only in a few districts such as Cilacap, Banjarnegara, and Purbalingga with low intensity. In regions with climate type C, decreased rainfall occurred in several districts/cities bordering climate types B and D, but cassava productivity decreased on the southern side of the area, namely in Magelang, Klaten, Boyolali, and Sukoharjo Regencies. Climate type D, which is in the northeast and mixed in the northwest of the province when El Nino is weak, gets a high decrease in rainfall. At the same time, the calculation of the anomaly in cassava productivity is normal, neither increases nor decreases, there is only one district that experienced a rise in productivity cassava at Pemalang Regency.

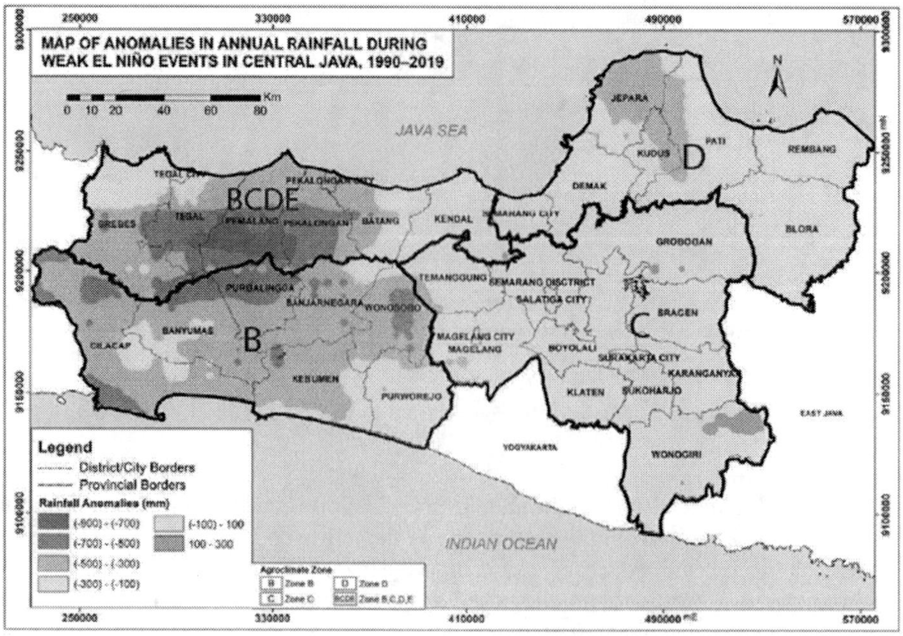

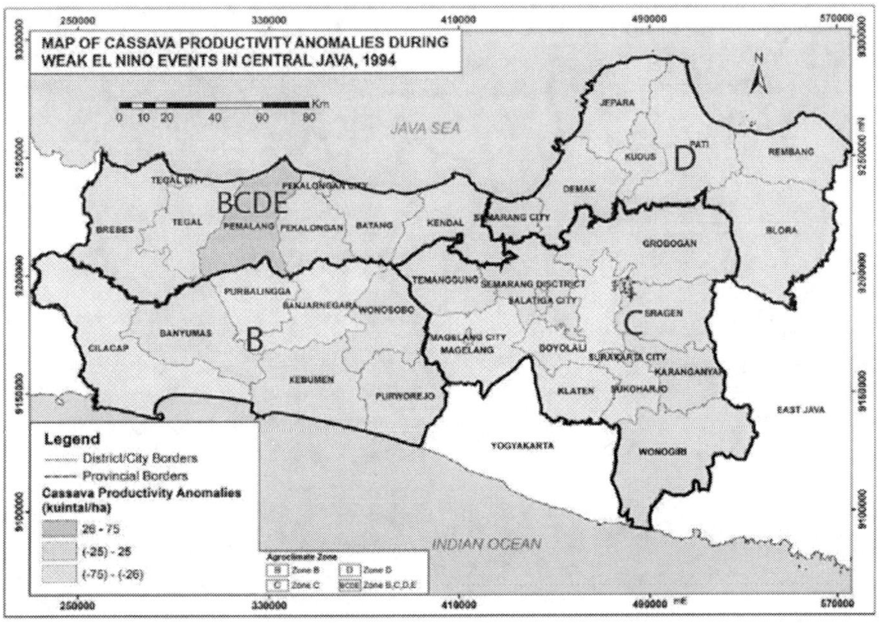

Figure 2. Comparative Map of Rainfall Anomaly and Cassava Productivity during the 1994 Weak El Nino in Central Java Province

The results from observations during a moderate El Nino can be seen in Figure 3. In areas with climate type B, the decrease in rainfall was uneven and only spread in the western part. However, the increase in cassava productivity occurred in Purbalingga and Kebumen Regencies, which experienced a relative decrease low in rainfall. In the climate type C area, most productivity decreases in the districts bordering climate type D with a reduction in light rainfall, but there is also an increase in productivity in the Klaten and Wonogiri districts followed by normal rainfall anomalies. The climate type D area experienced decreased cassava productivity on average during a moderate El Nino, except for Pati Regency, which increased to 95 quintal/ha. In the mixed climate type area (BCDE), the regencies/cities located on the east side did not experience anomalies, while the west side experienced a mild decrease in cassava productivity, namely in Brebes, Tegal, Pemalang, and Tegal Regencies.

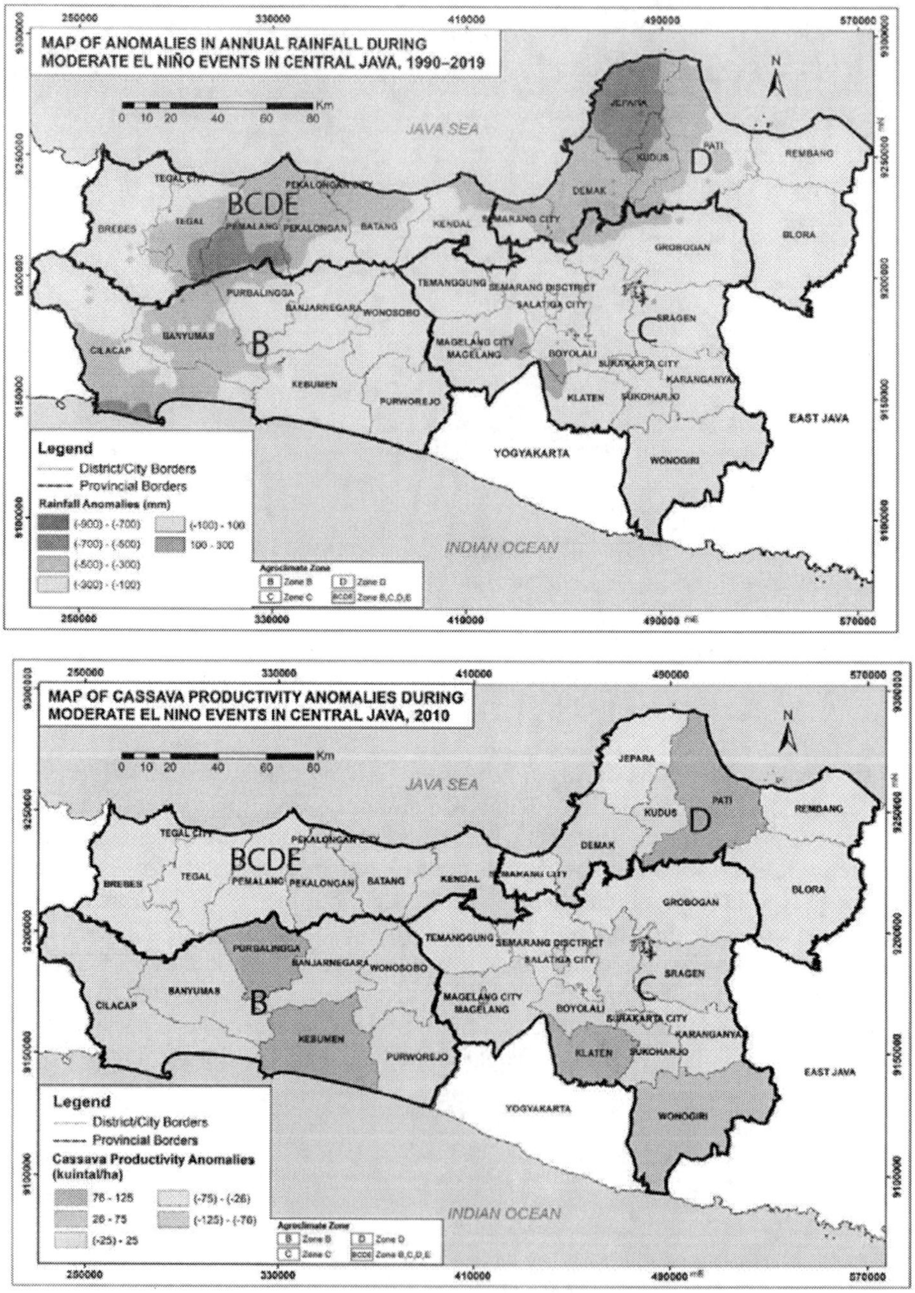

Figure 3. Comparative Map of Rainfall and Cassava Productivity during the 2010 Moderate El Nino in Central Java Province

The comparison of rainfall anomalies and cassava productivity in Central Java Province during a strong El Nino is shown in Figure 4. The decrease in annual rainfall intensity is evenly distributed when there is a strong El Nino in this province. In climate type B, two districts experienced a decline in cassava productivity, namely Cilacap Regency with -118 quintal/ha and Banyumas Regency with -32 quintal/ha. In climate type C, the anomaly of cassava productivity is heterogeneous. Three districts experienced a decline in productivity, namely Temanggung Regency (-32 quintal/ha), Magelang (-144 quintal/ha), and Wonogiri (-28 quintal/ha). Magelang Regency was affected by the most significant decline in

IOP Conf. Series: Earth and Environmental Science 1233 (2023) 012040 doi:10.1088/1755-1315/1233/1/012040

productivity during a strong El Nino in 2015. This happened because in the three years before this strong El Nino event, Magelang Regency had always experienced an increase in cassava productivity which peaked in 2014 reaching 413 quintal/ha, and decreased rapidly in 2015. There are also districts/cities that experienced an increase in productivity when El Nino was strong in climate type C areas, namely Klaten (+93 quintal/ha), Boyolali (+44 quintal/ha), Karanganyar (+61 quintal/ha) /ha), and Salatiga City (+38 quintal/ha). The area of climate type D and mixed climate type (BCDE) experienced a decrease in annual rainfall when a strong El Nino occurred in Central Java Province in the last 30 years. The climate type D area, which experienced a reduction in cassava productivity was only in Rembang Regency with -107 quintal/ha, while other regencies/cities tended to be normal. In diverse climate type areas, most regencies/cities are normal and not affected by rainfall anomalies. Only Batang Regency experienced an increase in productivity of 55 quintal/ha during the strong El Nino in 2015.

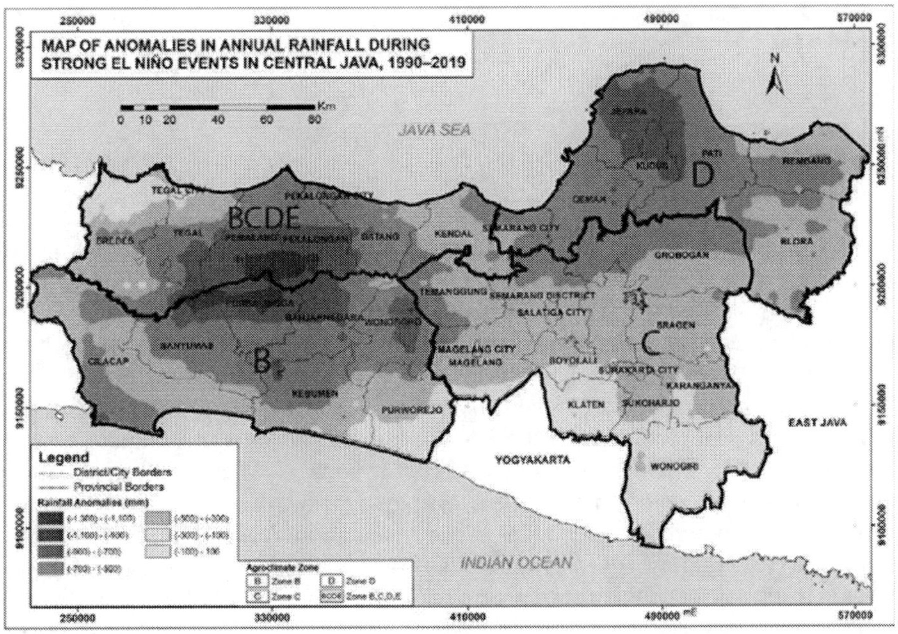

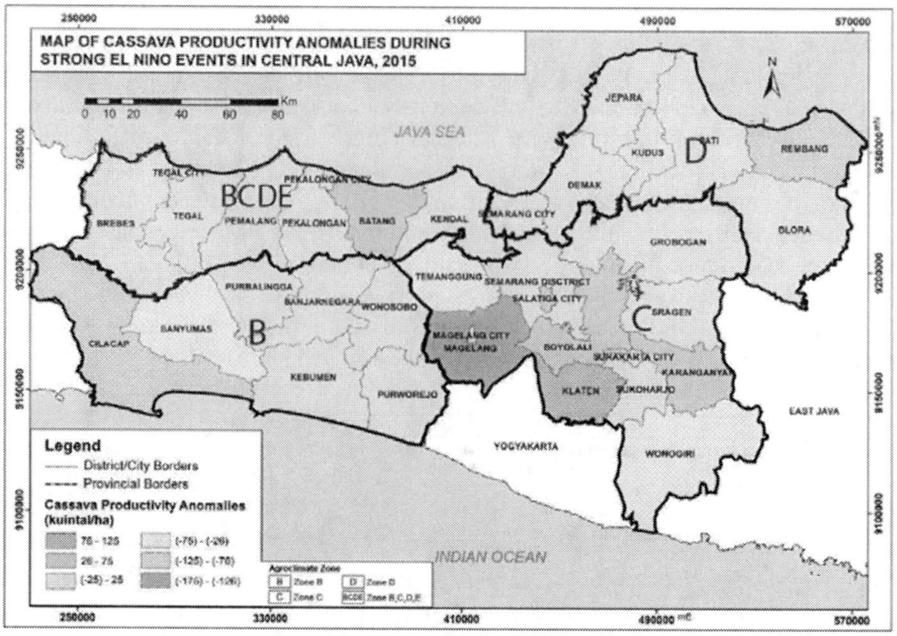

Figure 4. Comparison Map of Rainfall Anomaly and Cassava Productivity during Strong El Nino in 2015 in Central Java Province

The relationship between rainfall anomaly and cassava productivity anomaly when La Nina is weak in Central Java Province can be seen spatially in Figure 5. Areas with climate type B experience an increase in light rainfall and are spread unevenly when linked to a map of the cassava productivity anomaly during this phase. According to the ENSO, all districts/cities in climate type B tend to be normal and not affected by rainfall anomalies when La Nina is weak. In the climate type C area, productivity anomalies only occur in urban areas with a slight increase in productivity and other areas tend to be normal. Areas in climate type C that experienced an increase in productivity were Magelang City (+39 quintal/ha), Surakarta City (+44 quintal/ha), and Salatiga City (+57 quintal/ha). Areas with climate type D when a weak La Nina experienced a decrease in light rainfall in the northern areas around Jepara and Kudus regencies. However, this decrease in rainfall has no impact on the existing cassava productivity anomaly; it can be seen that all districts/cities in climate type D areas tend to be normal. Areas with a mixed climate type experienced an increase in rainfall on the southeast and southwest sides and a slight decrease in rainfall around Pekalongan City. The anomaly of cassava productivity tends to be normal in the districts in the region; however, there are two cities, namely Tegal City (-97 quintal/ha) and Pekalongan City (-82 quintal/ha), which experienced a decrease in cassava productivity. During the weak La Nina 1996, the two cities did not publish data on cassava productivity, resulting in a moderate decline compared to the previous three years.

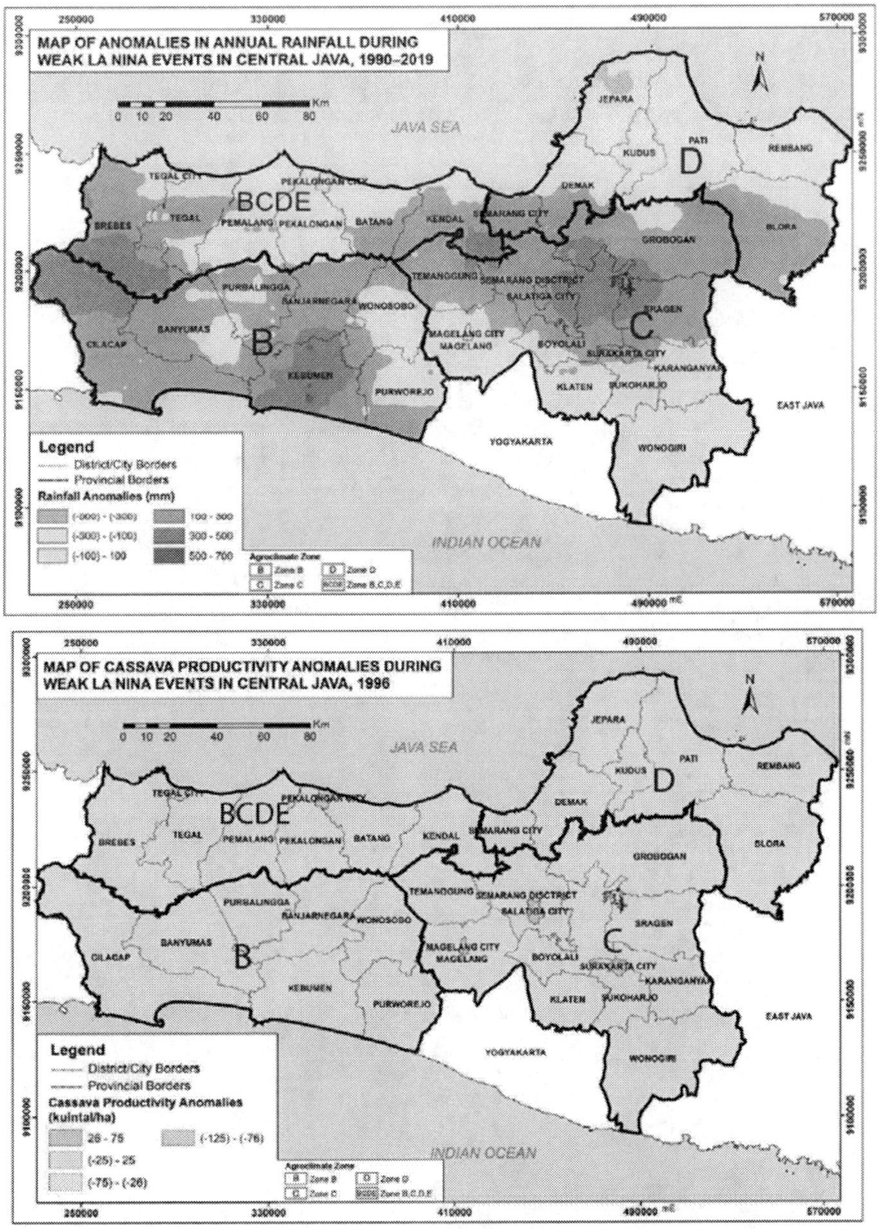

Figure 5. Comparison Map of Rainfall Anomaly and Cassava Productivity during Weak La Nina 1996 in Central Java Province

The comparison of rainfall anomaly with cassava productivity anomaly during moderate La Nina in Central Java Province spatially can be seen in Figure 6. In the climate type B area, the positive anomaly of rainfall is unevenly distributed and there is only an increase in light rainfall on the southeast and north sides bordering the mixed climate type area. The positive anomaly of cassava productivity occurs in almost all districts in climate type B. Only Cilacap and Banyumas Regencies are in normal condition. In the climate type C area, the rainfall anomaly is not much different from the climate type B, but some mild negative anomalies on the northeast side bordering the climate type D area. The productivity of cassava tends to increase slightly to moderately in Semarang Regency with an increase of 81 quintal/ha compared to the previous three years' average. Areas with climate type D experience a negative rainfall anomaly with a decrease in light rainfall during moderate La Nina for 30 years in Central Java Province.

However, when La Nina was in 2009, most districts/cities in the region continued to experience a slight increase. Only Kudus Regency did not experience growth and was included in normal conditions. In diverse climate-type areas, positive anomalies of cassava productivity occur in almost all districts/cities. The most significant increase was in the City of Tegal with +190 quintal/ha. Only Pekalongan City was included in normal conditions when La Nina was in 2009.

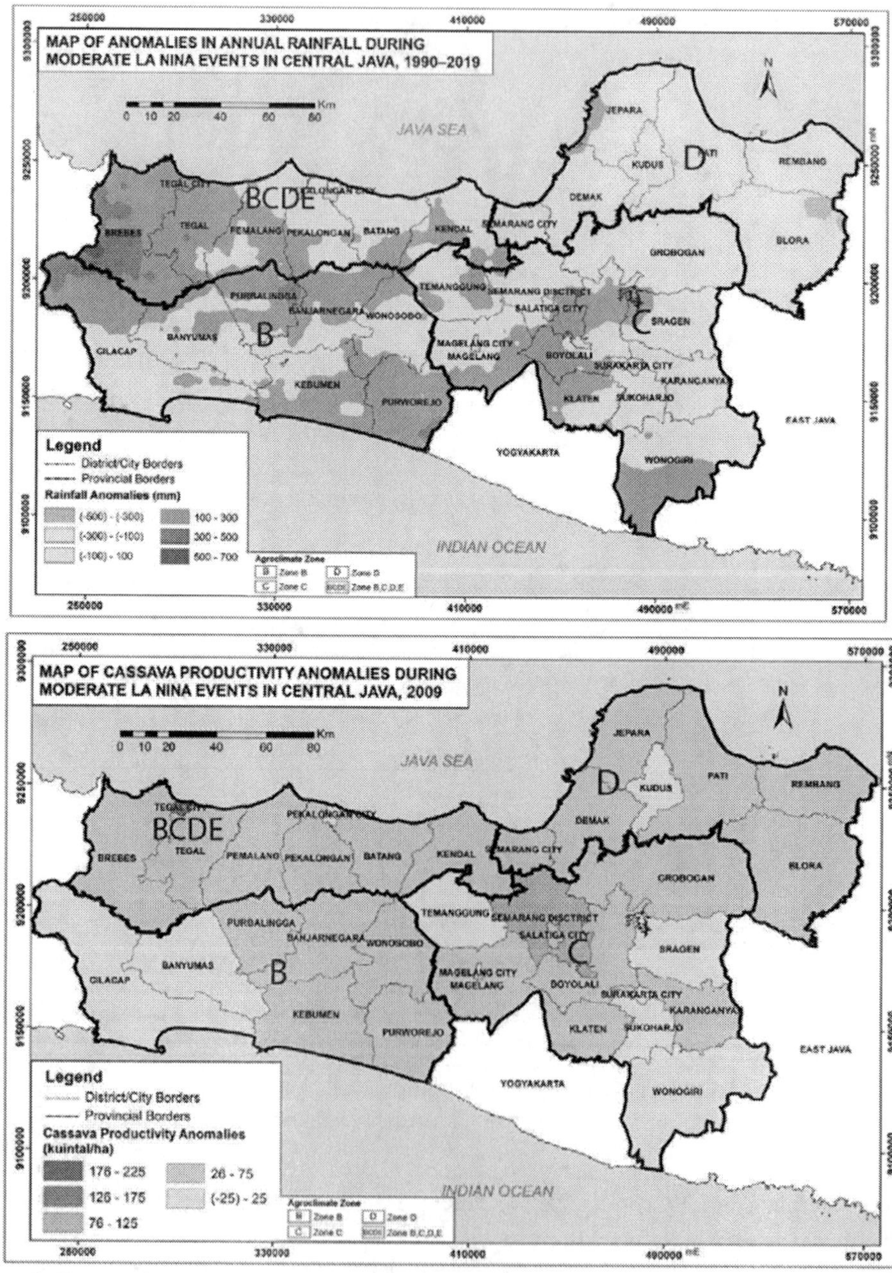

Figure 6. Comparison Map of Rainfall Anomaly and Cassava Productivity during Moderate La Nina 2009 in Central Java Province

The comparison of rainfall anomaly with cassava productivity anomaly when La Nina is strong in Central Java Province spatially can be seen in Figure 7. The increase in annual rainfall is evenly

IOP Conf. Series: Earth and Environmental Science 1233 (2023) 012040 doi:10.1088/1755-1315/1233/1/012040

distributed in almost all areas of Central Java Province, only Jepara Regency which tends to be normal when La Nina has been strong in the last 30 years. Areas with climate type B experience positive rainfall anomalies evenly distributed when this strong La Nina occurs. However, in 2012 productivity, not all regions in climate type B experienced positive anomalies. Only Cilacap (+193 quintal/ha), Banyumas (+58 quintal/ha), and Wonosobo (+40 quintal/ha) districts experienced an increase in cassava productivity. There is a decline in productivity in the climate type B area, namely in Banjarnegara Regency with -27 quintal/ha.

Areas with climate type C experienced an increase in cassava productivity in the districts of Temanggung (+47 quintal/ha), Semarang (+44 quintal/ha), Klaten (+49 quintal/ha), and Grobogan (+30 quintal/ha). There was also a decline in productivity in Boyolali (-26 quintal/ha) and Sragen (-31 quintal/ha) districts. Other regencies/cities in climate type C areas have normal productivity.

The climate type D area does not experience an anomaly of positive rainfall that is flat when there is a strong La Nina. However, when La Nina was strong in 2012, the district that did not experience an increase in rainfall increased its productivity was Jepara Regency (+48 quintal/ha). Other regencies that experienced increased productivity in this climate type area were Pati (+96 quintal/ha) and Rembang (+94 quintal/ha). Meanwhile, other regencies/cities tend to have normal productivity anomalies.

Areas of mixed climate type where La Nina is strong experienced most declines in cassava productivity. This decrease occurred in Pemalang Regency (-47 quintal/ha), Batang (-39 quintal/ha), Kendal (-57 quintal/ha), and City of Tegal (-135 quintal/ha). Meanwhile, the other four regencies/cities, Brebes, Tegal, Pekalongan, and Pekalongan city, are in normal condition when La Nina is strong in this climate type area.

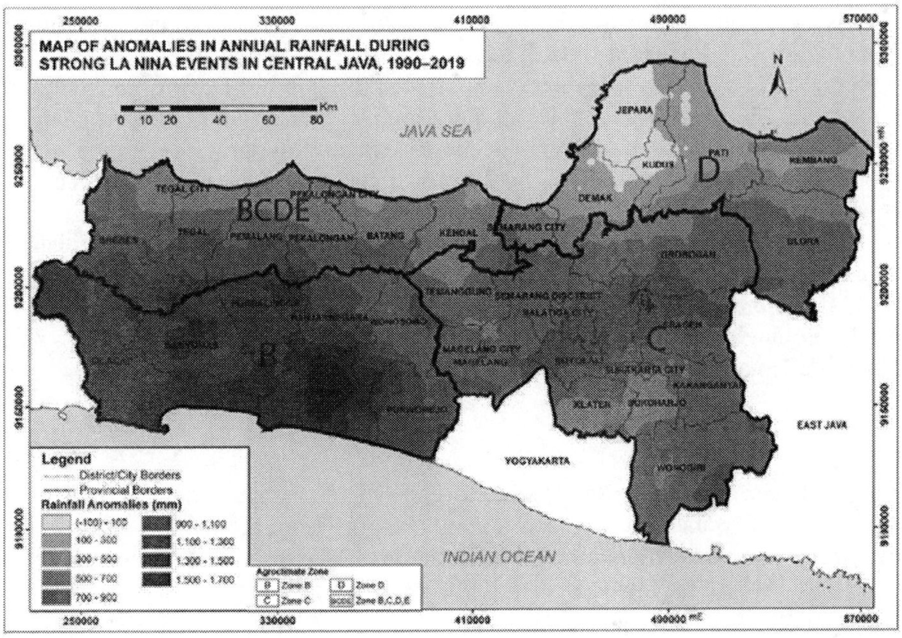

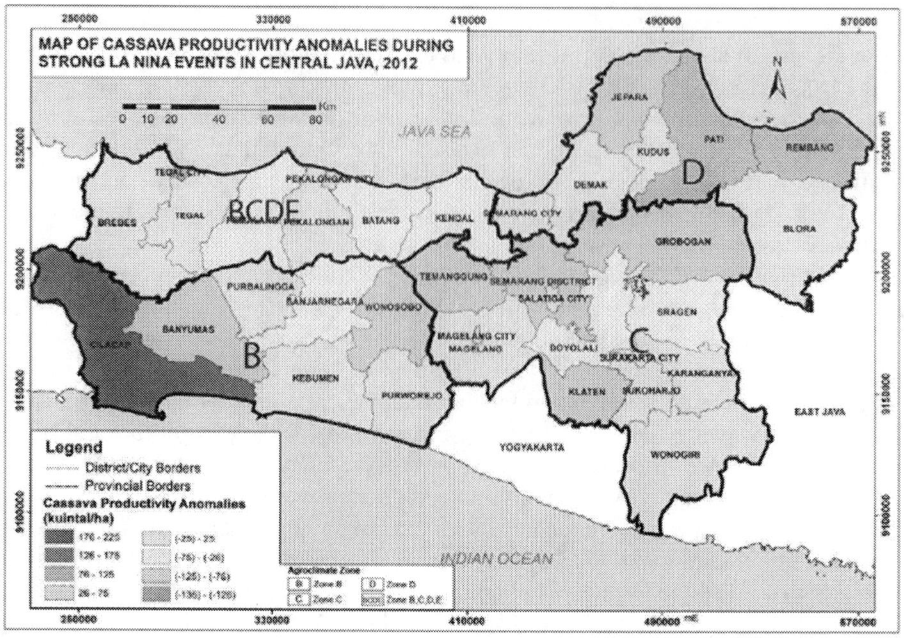

FIGURE 7. Comparison Map of Rainfall Anomaly and Cassava Productivity during Strong La Nina 2012 in Central Java Province

Cilacap, Banyumas, and Rembang Regencies during the 2015 El Nino experienced negative productivity anomalies, but during La Nina, it turned out to have positive anomalies. It can be concluded that cassava productivity in these three districts is affected by rainfall. During La Nina, the increase in rainfall in Cilacap and Banyumas regencies is quite high, reaching 1.000 mm/year, and in Rembang Regency, it is increasing up to 700 mm/year. The increase in rainfall during La Nina is one of the factors in increasing the productivity of cassava in the three districts, apart from other technical factors such as cropping patterns to the treatment of plants by farmers.

The effect of ENSO on rainfall anomalies in Central Java Province impacts some districts/cities that rely on rainfall as one of their main water sources. This can be seen when there is a strong El Nino or La Nina. It can affect the amount of agricultural productivity, especially cassava in the region. Most agriculture in developing countries still depends on climatic conditions, so phenomena such as ENSO, both El Nino and La Nina can affect agricultural productivity in these countries [13].

4. CONCLUSION

The relationship between rainfall anomalies due to ENSO and cassava productivity is evident, but the intensity is not always the same in every district/city. Judging from the pattern when El Nino is weak, moderate, or strong, several districts/cities in Central Java Province tend to experience a decline in cassava productivity. When La Nina was in 2009, the increase in cassava productivity was evenly distributed in Central Java Province. When La Nina is strong, some districts experience high productivity increases, but there is also a decline in productivity in some districts. This can be due to the treatment and changes in cropping patterns from farmers who adjust their respective climatic and geographical conditions to maintain agricultural conditions and maintain their economy.

Planting cassava during La Nina has a positive effect on farmers. However, when an El Nino occurs, it also does not have a negative impact as long as the water needs for plants are still fulfilled. Suggestions from the authors for the study of the effect of ENSO on cassava in Central Java Province, it is necessary to add more detailed climate parameters for each district/city to see whether other climate parameters affect cassava productivity. A detailed study of the effect of ENSO on cassava must also determine the treatment and cropping patterns of farmers in each district/city to suit the agricultural characteristics of the area.

Acknowledgments
This research is supported by the grant of "Rekognisi Tugas Akhir (RTA) Universitas Gadjah Mada 2022" with Dr. Andung Bayu Sekaranom as principal investigator (Contract number: 3550/UN1.P.III/Dit-Lit/PT.01.05/2022 on May 20th, 2022)

References

[1] Aldrian, E. *Meteorologi Laut Indonesia*. Jakarta: Badan Meteorologi dan Geofisika. Retrieved from https://www.researchgate.net/publication/305809658_Meteorologi_Laut_Indonesia (2008).
[2] Apriyana, Y., & Kailaku, T. E. Variabilitas Iklim dan Dinamika Waktu Tanam Padi di Wilayah Pola Hujan Monsunal dan Equatorial. *Seminar Nasional Masyarakat Biodiversiti Indonesia*, *1*(2), 366-372 (2015).
[3] Herlina, N., & Prasetyorini, A. Effect of Climate Change on Planting Season and Productivity of Maize (Zea mays L.) in Malang Regency. *Jurnal Ilmu Pertanian Indonesia*, *25*(1), 118-128. https://doi.org/10.18343/jipi.25.1.118 (2020).
[4] Sitompul, Z., & Nurjani, E. Pengaruh El Niño Southern Oscillation (ENSO) Terhadap Curah Hujan Musiman Dan Tahunan Di Indonesia. *Jurnal Bumi Indonesia*, *2*(1), 11–18 (2013).
[5] Anggara, J. *Pergeseran Pola Musim Hujan dan Musim Kemarau Kabupaten Jember Tahun 1999-2014 Serta Pengaruhnya Terhadap Produktivitas Tanaman Palawija dan Umbi-Umbian Pada Berbagai Jenis Tanah Utama*. Skripsi. Universitas Jember (2017).
[6] Saleh, N., Taufiq, A., Widodo, Y., Sundari, T., Gusyana, D., Rajagukguk, R. P., & Suseno, S. A. *Pedoman Budi Daya Ubi Kayu di Indonesia* (A. Taufiq, N. Saleh, & D. Gusyana, Eds.). Jakarta: Indonesian Agency for Agricultural Research and Development (IAARD) Press. Retrieved from https://balitkabi.litbang.pertanian.go.id/monograf/pedoman-budi-daya-ubi-kayu-di-indonesia-2016/ (2016).
[7] Direktorat Budidaya Aneka Kacang dan Umbi. Laporan Tahunan Direktorat Budidaya Aneka Kacang dan Umbi Tahun 2010. In *Kementerian Pertanian* (2010).
[8] Maulana, A. R., & Herlina, N. (2020). Hubungan Unsur Iklim Terhadap Produktivitas Tanaman Ubi Kayu (Manihot esculenta Crantz) di Kabupaten Malang. *PLANTROPICA: Journal of Agricultural Science*, *5*(2), 118–128. https://doi.org/10.21776/ub.jpt.2020.005.2.3
[9] Suharno, Djasmin, Rubiyo, & Dasiran. Budi Daya Ubi Kayu. Kendari: Badan Peneliti dan Pengembangan Pertanian (1999).
[10] Wargiono, J., Hasanuddin, A., & Suyanto. Teknologi Produksi Ubikayu Mendukung Industri Bioethanol. Bogor (2006).
[11] Funk, C. C., Peterson, P. J., Landsfeld, M. F., Pedreros, D. H., Verdin, J. P., Rowland, J. D., Romero, B. E., Husak, G.J., Michaelsen, J. C., Verdin, A. P. A quasi-global precipitation time series for drought monitoring. *U.S. Geological Survey Data Series 832*, *832* (2014).
[12] Santoso, A. B. Pengaruh Perubahan Iklim terhadap Produksi Tanaman Pangan di Provinsi Maluku. *Jurnal Penelitian Pertanian Tanaman Pangan*, *35*(1), 29. https://doi.org/10.21082/jpptp.v35n1.2016.p29-38 (2016).
[13] Limsakul, A. Impacts of el niño-southern oscillation (ENSO) on rice production in Thailand during 1961-2016. *Environment and Natural Resources Journal*, *17*(4), 30-42. https://doi.org/10.32526/ennrj.17.4.2019.29 (2019).

IOP Conf. Series: Earth and Environmental Science 1233 (2023) 012041 doi:10.1088/1755-1315/1233/1/012041

Determination of River Border Width at the Manikin River in Kupang Regency, Nusa Tenggara Timur Province

Christiani Chandra Manubulu[1*], Yulius Patrisius Kau Suni[1], Don Gaspar Da Costa[1]

[1]Faculty of Engineering, Universitas Katolik Widya Mandira Kupang

*Corresponding author: christianichandra@gmail.com

Abstract. River Border Lines was made and determined so that there some efforts were made for the protection, use and control of existing resources in the river. One of the river segments that are prioritized for determining river border lines is a river section adjacent to or within a developing area, one of which is the Manikin River in Kupang Regency, which is currently under construction of the Tefmo Dam which is targeted to be completed by the end of 2022. Therefore, in the future, of course, area around the weir will experience great pressure on land use which can have an impact on the function of the river due to developing community activities. The purpose of this study is to determine the width of the river border on the Manikin River. The length of the river segment where the width of the border line will be determined is 10 km from the Manikin weir to the downstream of the river. The results of this study are to obtain the width of the Manikin river border and get an overview of the location around the Manikin River about land conditions and functions, facilities and utilities that are included in the border area.

1. Introduction

Manikin River was located in Kupang Regency with 23.1 km main river length[1]. The upstream is in Oben Village, Nekamese District and the downstream is in Tarus Village, Central Kupang District. To fulfill the community's agricultural irrigation needs, the construction of the Manikin weir was beginning to build in 2019 and planned to be completed by the end of 2022. The Manikin weir will be the main water source that encourages agricultural productivity in that region, it will certainly trigger human activities around the Manikin river[2]. Therefore, it is necessary to determine the width of the river border before the area around the Manikin weir becomes densely populated, so that it can minimize losses due to scouring of river cliffs which can threaten buildings or other public facilities due to eroded river currents. With the establishment of river boundaries, the riverbed becomes more stable, the air quality is better, the life of flora and fauna increases, the aesthetics are also more attractive because there is a harmonious life among the elements of nature, including humans in it.

The Manikin River Border Study activity in Kupang Regency was carried out for 10 km with the zero point taken from the Manikin Dam Location (0 + 7.92 km) which is located in Kuaklalo Village, Taebenu District and the 10 km point (0 + 17.92 km) taken from the zero point towards downstream of the river which is located in Oelnasi Village, Central Kupang District. The location of the Manikin River Border Study Activity in Kupang Regency passes through 2 Districts, including:

- Taebenu District

Study of the Manikin River Border in Taebenu District passing through Kuaklalo Village, Bokong Village, East Baumata Village and North Baumata Village

- Central Kupang District

  The study of the Manikin River Border in Central Kupang District only passes through Oelnasi Village.

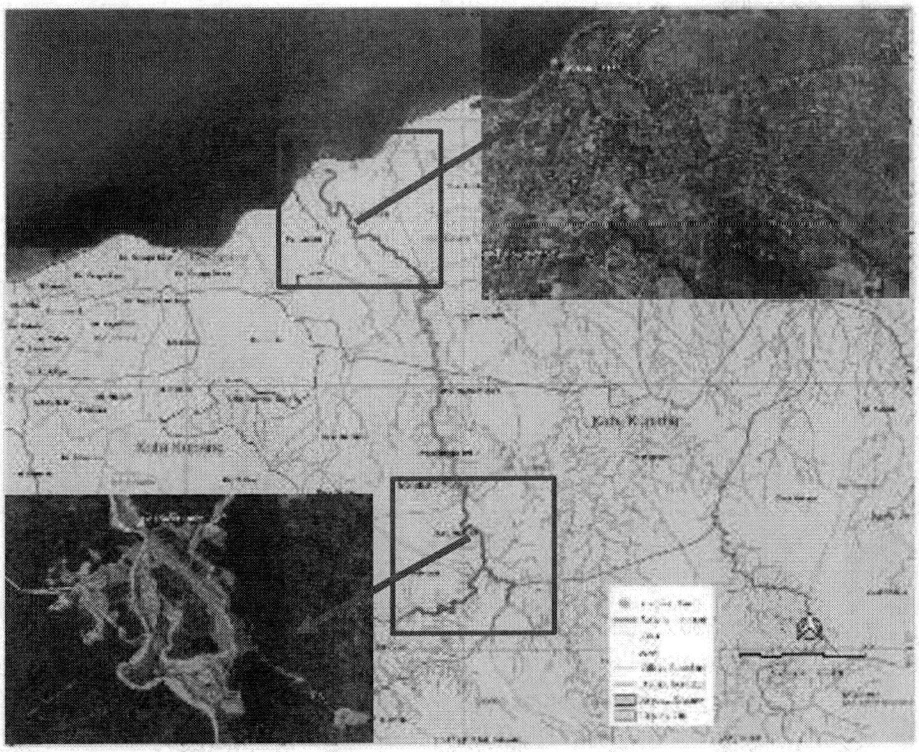

Figure 1. Manikin River in Kupang Regency[3]

The river border is a virtual line on the left and right of the riverbed which is designated as a river protection boundary. River Border Line is made and determined in order for the protection, use and control of the resources that exist in the river. While the purpose of determining the river border based on Minister Regulation No. 28 / PRT / M / 2015 is as follows:

- The function of river and lakes is not disturbed by human activities around the rivers and lakes;
- Utilization activities and efforts to increase the value of benefits of existing resources in rivers and lakes can provide optimal results while preserving the functions of rivers and lakes; and
- The destructive water presseure from river and lake to the environment can be limited.

2. Methodology

In this study several stages of the work were carried out including:

1. Preparation phase

   At this stage, secondary data collection related to the manikin river is carried out in the form of the length of the river, the large area of the Manikin Watershed, and information about activities around the manikin river.
2. Field survey

   a. Location inventory

IOP Conf. Series: Earth and Environmental Science 1233 (2023) 012041 doi:10.1088/1755-1315/1233/1/012041

The purpose of the site inventory is to obtain information on the condition and function of land, facilities and utilities around the location of the Manikin river border activities.

b. Topographic survey
At this stage, topographic measurements are carried out to obtain detailed topographic data regarding the depth and width of the river as well as details of the facilities and infrastructure at the measurement location.

3. Data analysis
At this stage, topographic data analysis and technical planning for border line determination.

The Criteria and Area Width of the Rivers outside the urban areas can be seen in Table 1[4].

Table 1. Criteria and River Border Areas in rural areas

No	Type of River	Criteria	Width (m)
1.	Embankment river (measured from the outer rigid embankment)	-	5 m
2.	Un-embankment river (measured from the riverside)	Big river (large of DPS > 500 km^2)	100 m
		Small river (large of DPS < 500 km^2)	50 m
3.	Influential rivers high tide (from the river)	-	100 m

3. Result and Discussion

The starting point of river border activities on the Manikin River is 7.92 km from the upstream of the river which is the location of the Manikin weir[5]. The length of the activity location is 10 km. Inventory, identification and topographic measurements were carried out from the position of the Manikin weir downstream along 10 km. The inventory and identification carried out aims to obtain data on buildings along the left and right of the river in the form of infrastructure around the river, public facilities and also community settlements around the study location of the Manikin River border. Based on the results of the inventory, identification and topographic measurements carried out at the study site of the Manikin River border, data obtained that the minimum width of the Manikin River is 18.11 m and the maximum width of the Manikin River is 85.47 m. The minimum depth of the Manikin River is 0.58 m and the maximum river depth is 13.38 m. The depth and width of the Manikin River from topographic measurements can be seen in Table 2.

Table 2. The depth and width of the Manikin River

	River width (m)		River depth (m)	
KM	Minimum	Maximum	Minimum	Maximum
0 - 1	31,4	59,06	2,01	4,83
1 - 2	34,36	85,47	1,50	5,78
2 - 3	33,38	71,03	0,97	4,64
3 - 4	39,6	61,24	1,22	9,32
4 - 5	23,37	78,51	1,04	4,38
5 - 6	22,89	56,21	0,91	6,2
6 - 7	19,51	57,17	0,81	13,38
7 - 8	22,49	44,38	0,82	4,01
8 - 9	18,11	61,83	0,58	6,85
9 - 10	24,77	63,97	1,42	6,57

The width of Manikin River border in Kupang Regency is taken 50 m from the left and right sides of the river adjusted to the area of the watershed that is smaller than 500 (five hundred) km^2 which is 49,31 km^2 on Minister Regulation No. 28 / PRT / M / 2015. Based on the width of the river border that

IOP Conf. Series: Earth and Environmental Science 1233 (2023) 012041 doi:10.1088/1755-1315/1233/1/012041

has been obtained, and then the study traced the conditions, land functions, facilities and utilities that exist in the river border area. The depiction of the conditions carried out on the left and right of the river for every 1 km. The results of tracking infrastructure included in the Manikin River border can be seen in Table 3.

Table 3. Condition, Landuse, Facilities and Utilities of the Left and Right Sides of the Manikin River.

km	Condition and Land Function	Facilities and Utilities
0 - 1	Left side (0 – 1) km Open field, Trees, Shrubs, and Public Gardens	Left side (0 – 1) km Footpath
	Right side (0 – 1) km Open field, Trees, Shrubs, and Public Gardens	Right side (0 – 1) km Footpath
1 - 2	Left side (1 – 2) km Open field, Trees and Shrubs	Left side (1 – 2) km Footpath
	Right side (1 – 2) km Open field, Trees and Shrubs	Right side (1 – 2) km Footpath
2 - 3	Left side (2 – 3) km Open field, Trees, Shrubs, and Public Gardens	Left side (2 – 3) km Footpath
	Right side (2 – 3) km Open field, Trees, Shrubs, and Public Gardens	Right side (2 – 3) km Footpath
3 - 4	Left side (3 – 4) km 2 units of residential home, Open field, Trees, Shrubs, Public Gardens and Public Rice Field	Left side (3 – 4) km Environmental road, Footpath, and Bridge Oprit
	Right side (3 – 4) km Open field, Trees, Shrubs, Public Gardens and Public Rice Field	Right side (3 – 4) km Environmental road, Footpath, Bridge Oprit and Baumata Timur Weir
4 - 5	Left side (4 – 5) km Open field, Trees, Shrubs, Public Gardens and Public Rice Field	Left side (4 – 5) km Footpath and Electricity Pole
	Right side (4 – 5) km Open field, Trees, Shrubs, Public Gardens and Public Rice Field	Right side (4 – 5) km Footpath
5 - 6	Left side (5 – 6) km Open field, Trees, Shrubs, Public Gardens and Public Rice Field	Left side (5 – 6) km Footpath
	Right side (5 – 6) km Open field, Trees, Shrubs, Public Gardens and Public Rice Field	Right side (5 – 6) km Footpath
6 - 7	Left side (6 – 7) km Open field, Trees and Shrubs	Left side (6 – 7) km Footpath
	Right side (6 – 7) km Open field, Trees and Shrubs	Right side (6 – 7) km Footpath
7 - 8	Left side (7 – 8) km Open field, Trees and Shrubs	Left side (7 – 8) km Environmental road, Footpath and Oelnasi Weir
	Right side (7 – 8) km Open field, Trees and Shrubs	Right side (7 – 8) km

km	Condition and Land Function	Facilities and Utilities
		Environmental road, Footpath and Oelnasi Weir
8 - 9	Left side (8 – 9) km Trees, Shrubs, Public Gardens and Public Rice Field	Left side (8 – 9) km Environmental road and Footpath
	Bagian Kanan (8 – 9) km Open field, Trees and Shrubs	Left side (8 – 9) km Footpath
9 - 10	Left side (9 – 10) km Trees, Shrubs, Public Gardens and Public Rice Field	Left side (9 – 10) km Footpath
	Right side (9 – 10) km Open field, Trees and Shrubs	Left side (9 – 10) km Environmental road and Footpath

4. Conclusions and Recommendations

4.1 Conclusions

Based on the results of the analysis and discussion in the previous chapter, several conclusions can be drawn as follows:

1. The width of the Manikin River border is set at 50 km in Kupang Regency
2. The survey results also obtained data on land use in the border area of the river that is 2 residential units, tree areas, shrubs, gardens and community fields. And facilities and utilities included in the border area are village roads, footpaths, bridges oprit, and Weir.
3. The minimum width of the Manikin River is 18.11 m and the maximum width of the Manikin River is 85.47 m. While the minimum depth of the Manikin River is 0.58 m and the maximum river depth is 13.38 m.

4.2 Recommendations

The following are some recommendations based on the results of the study:

1. The importance of the purpose of determining the river border, it is expected that the determination of the river border can be done for all rivers, especially rivers in urban areas or outside urban areas that will become developing areas.
2. The results of river border activities should also be disseminated to all regional management units so that the boundaries of the river border are known. Thus the policies issued by the authorities can be in line with the determination of river borders that have been carried out at river locations that already have river border boundaries.

References

[1] E. Hunggurami, "Keandalan Metode Haspers dan Weduwen pada Das Manikin," vol. VII, no. 2, pp. 193–204, 2018.
[2] M. Ningtyas, "Bab III - Metode Penelitian Metode Penelitian," *Metod. Penelit.*, pp. 32–41, 20014.
[3] Badan Informasi Geospasial 2006, *Peta Rupa Bumi Indonesia (RBI) dan Peta Kontur*, Bogor.
[4] Permen PUPR No 28 2015, *Peraturan Menteri Pekerjaan Umum dan Perumahan Rakyak, 2015, Nomor 28/PTR/M/2015, Tentang Penetapan Garis Sempadan Sungai dan Garis Seempadan Danau, Jakarta, Lampiran halaman 3, 4, 5, 6, 7, dan 8.*
[5] Y. Falo, D. Djunaedi, and A. Nama, "Analisis Debit Air Limpasan Permukaan (Run Off) Di Daerah Aliran Sungai (Das) Manikin Kabupaten Kupang Provinsi Nusa Tenggara Timur," *JUTEKS J. Tek. Sipil*, vol. 4, no. 2, p. 45, 2019, doi: 10.32511/juteks.v4i2.299.

IOP Conf. Series: Earth and Environmental Science 1233 (2023) 012042 doi:10.1088/1755-1315/1233/1/012042

Symptoms of Gentrification in the Yogyakarta Urban Area and Its Implications for the Sustainability of Yogyakarta City

F T Nugrahaningtyas[1] and D Marwasta[1*]
[1]Department of Environmental Geography, Faculty of Geography, University of Gadjah Mada, Indonesia
*Corresponding author: jakamar@ugm.ac.id

Abstract. One of the most common symptoms in developing country cities today is gentrification. Gentrification is the recessive of residence of economically weak groups by strong investors. This study tries to examine gentrification in Yogyakarta urban area and its implications for the sustainability of Yogyakarta city. The spatial approach is used in this study considering that space is an integral part of a city which is a place for various urban activities and functions. In general, this research aims to: (1) Identifying the symptoms of gentrification in the Yogyakarta Urban Area; and (2) Assessing the implications of gentrification on urban sustainability. This research is a study in the field of geography with the main focus on the spatial approach. The data analysis framework in this study consisted of two kinds, namely spatial analysis and descriptive statistical analysis. The results showed that (1) the symptoms of gentrification in the Yogyakarta urban area had occurred in several places, and (2) the gentrification that occurred had implications for the sustainability of the city, both positive and negative.

1. Introduction

The high level of urbanization that occurs causes competition in meeting the need for land for housing and business places to increase in a high escalation [1] [2]. As a consequence of this, the bargaining position of the lower middle class which have settled in the city for such a long time becomes weak and vulnerable to being invaded by the upper middle class and rich people who have large and strong capital. The urgency of housing for economically weak groups of people by strong investors is referred to as gentrification.

The term of gentrification was firstly stated by Ruth Glass in 1964, a sociologist from England. Glass in his publication describes the socio-economic transformation that occurred in London City [3]. This phenomenon of gentrification then began to be widely discussed since the 1970s until this phenomenon also occurs in many cities in many countries. Gentrification is the process by which high-income residents replace low-income residents [4]. The Ministry of Public Works and Public Housing defines gentrification as change in social stratification; stratification of lower-level city dwellers from slum areas in the middle of the city which later became luxurious settlements complete with all the living facilities for the urban middle class or the upper class.

The rapid population growth in most urban in developing countries has resulted in tighter competition for land tenure. The consequence is, competition to earn land for housing or other urban functions [5]. This symptom also take place a lot in Indonesia, especially on the Java island. One of the areas experiencing symptoms of gentrification is Yogyakarta Urban Area.

Gentrification has two opposite sides, positive and negative. This phenomenon has a positive impact on environmental quality due to efforts to revitalize and improve urban conditions [6]. The condition of the area that was previously an unhealthy and unfriendly area has become increasingly better with the presence of gentrification. Atkinson also mentions that gentrification causes other positive impacts, such as an increase in city tax revenues, an increase in property values, an increase in social inclusion, and an improvement in local services and the physical environment [6]. But on the other hand, gentrification also raises social inequality [7]. The existence of development to meet the needs of the middle class will further eliminate marginal groups. Research related to the phenomenon of gentrification in various locations also reveals the positive and negative impacts of this gentrification.

This study tries to examine gentrification in the Yogyakarta urban area and its implications for the sustainability of Yogyakarta city. The spatial approach employed in this research considering that space is a substantially part of a city, a place for numerous urban functions and activities. This research is also tied to the agenda of sustainable city that has been proclaimed by the United Nations [8]. This research emphasize more on socio-spatial aspects, particularly the analysis of spatial processes, and their implications for physical and administrative spaces.

Gentrification is a dilemma faced by cities in developing countries. The contestation and competition for very limited land in urban areas to fulfill housing needs with urban functions other than settlements is a dilemma. Both were important for the city's development, but one had to be the losing side in the competition. Settlements that have the lowest bargaining power among other urban functions are often the losers.

Symptoms of gentrification vary from place to place, as well as the process and its implications [9]. Yogyakarta urban area is an interesting area due to study the phenomenon of gentrification. Several assumptions that underlie the existence of gentrification symptoms in the Yogyakarta Urban Area include: (1) as a tourist area, so that the level of hotel demand is very high; (2) as a cultural city, the social ties of the indigenous people are very high; (3) as a city of education (especially universities) causing investment in the provision of dormitories is very prospective; and (4) as one of the economic centers in Java, so that investment comes in very large. With these pre-conditions, the Yogyakarta Urban Area becomes a very attractive area for investors (especially hotels, apartments, shops, and so on).

The development of urban areas coupled with the presence of newcomers in the Yogyakarta Urban Area causes physical and social changes. From year to year the number of hotels, shopping centers, and apartments is increasing. This increase in the number of hotels can be assumed to be caused by the gentrification phenomenon [10]. The increasing number of hotels also creates other problems, triggering a rise of land prices in the vicinity. This, of course, results in the reduced accessibility of residents in buying land to build decent settlements. Another problem that arises is the increasing recessive of low-class economic community settlements, because they are often considered as disturbing the aesthetics of the city.

In addition to causing negative impacts, the phenomenon of gentrification in the Yogyakarta urban area also has positive implications for the community and the region. With the increase in population due to immigrants from various other regions, economic growth is encouraged to follow the increase, and in the end will also encourage the construction of other facilities, such as housing, commercial areas, and infrastructure. This will improve the aesthetic identity of the City of Yogyakarta, and has the implication of becoming an attraction for outsiders to come. The phenomenon of gentrification that gives rise to positive and negative implications in the Yogyakarta urban area is interesting to be studied more comprehensively.

In general, this research aims to (1) Identifying the symptoms of gentrification in the Yogyakarta Urban Area; and (2) Assessing the implications of gentrification on urban sustainability.

2. Methods

Build upon the research objectives to be achieved, this study is included on exploratory-descriptive research because it aims to identify the symptoms of gentrification and its implications for environmental conditions. The analytical technique used is a deductive approach. It is started with the study of general phenomena,which is hereinafter used as a basis for understanding more specific phenomena.

Regarding the object of studyh, this research used a survey method, observing and measuring various variables used to examine the process and impact of gentrification. The data collection uses a combination of secondary data (SUSENAS 2021) and primary data (structured interviews) at selected locations by using purposive sampling. The data analysis used is dominated by spatial analysis.

Regarding to the analytical method, this research is include in a quantitative research. The location of the study is Yogyakarta Urban Area. This location was chosen based on the consideration that the Yogyakarta Urban Area has potential conditions for gentrification. This condition is quite suitable for the purpose of research that wants to examine the process and implications of gentrification.

In order to accurately assess the gentrification process, administrative and morphological boundaries are used simultantly. The morphological city boundaries is determined by refering to Pryor [11], which is an area that has more than 75% land occupied by urban function. In addition, this study also uses the types of urban physical growth as proposed by Ellman [12], Antrop [13], and Wilson [14] as a reference to determine the process of urban development in the research area.

This research is a study in the field of geography with the main focus on the spatial approach. The main object of study is the gentrification phenomenon with an emphasis on the assessment of the symptoms of taking over low-class residential land by investors to be transformed into new residential land intended for the upper class or into commercial urban functions. The identification of the gentrification process is carried out by tracing the symptoms of taking over residential land in the last 10 years (2011-2021). Multitemporal satellite imagery is used as a data source. Identification of this process is intended to obtain an overview of the gentrification that occurs.

In this research, the activities carried out include interpretation of digital images in the laboratory, secondary data analysis, observation, and physical field measurements. As the first stage, activities in the laboratory include making maps, overlapping maps, compiling a list of field observations and searching for secondary data, compiling a list of questions for informants, and determining sample areas.

The data used in this study are primary and secondary data. Some of the primary data collected in this study are maps that derived from interpretation of satellite imagery, data taken from field observations, as well as data from interviews with selected respondents. Some secondary data collected in this study include thematic maps, tables and/or graphs (especially population data) published from various agencies. The variables and indicators in this sutdy show in Table 1.

Table 1. Variables and Indicators

Related to research aim	Symptoms/Factors	Variables	Indicators
- Identifying the symptoms of gentrification	- Initiale stage of gentrification	- House Renter	- Total household of house renter
		- Accesibility	- Time travel rate to work place
		- Land	- Land and building price
	- Advance stage of gentrification	- House Rent	- House Rent Price
		- Invader	- Total number of middle-High income newcomer
		- Function Changes	- Total number of new café or gallery etc
- Iimplications of gentrification on urban sustainability	- Economics	- Tax	- Tax increase
	- Social	- Neighborhood	- Social bonding
	- Ecology	- Quality	- Comfortability

The data analysis framework in this study consisted of two kinds, namely spatial analysis and descriptive statistical analysis. Spatial analysis is intended to answer the first objective of the study, namely the identification of the gentrification process. Descriptive statistical analysis was used to elaborate the second research objective, namely to examine the implications of gentrification on the environment and urban sustainability. Framework flow diagram of this study is presented in Figure 1.

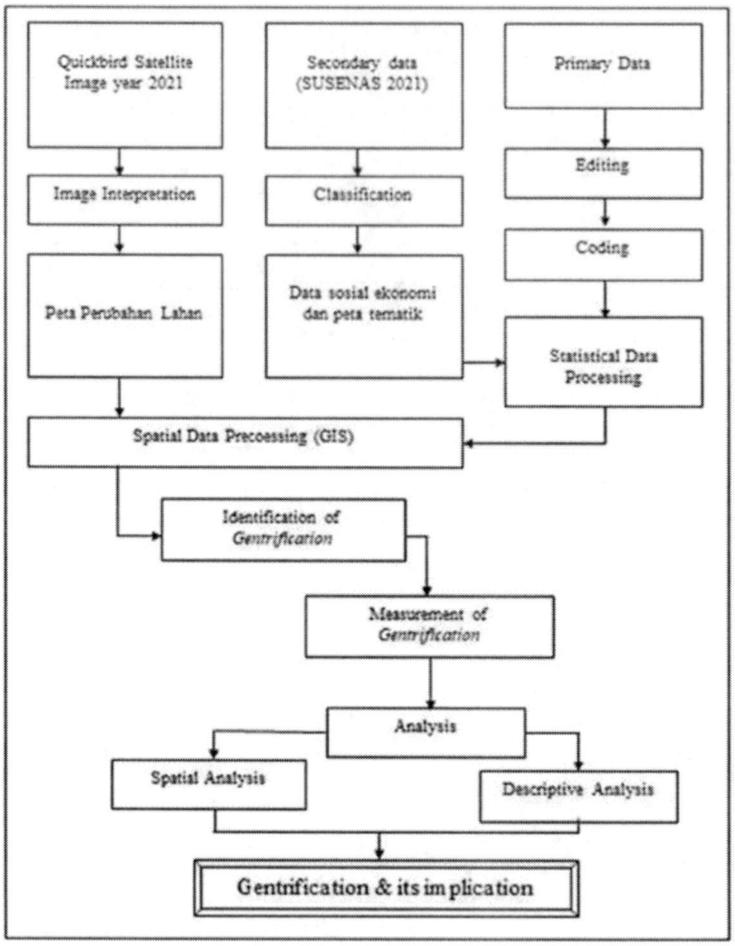

Fig 1. Research framework flow diagram

3. Results and discussions

The world's population increases year by year continuely. United Nations stated, the earth's human population has doubled in the 50 years (between 1950 to 2000), from 3 billion to 6 billion [15]. The rapid growth of world's population is mostly contributed by population growth in developing countries.

Especially in developing countries, the rapid population growth in urban areas is more dominantly caused by the uncontrolled rate of urbanization. The city as a center of socio-economic and cultural activities has experienced growth in line with population growth, both non-physical (social, economic, political, and cultural) as well as physical growth. However, many cities that experience high population growth have a bad impact on the physical and socio-cultural environment in and around the city [16]. The expectation as a safe, comfortable, and prosperous place for urban residents did not materialized because the growth of the city, especially population growth and building density, did not match what was expected. The symptoms of gentrification are closely related to the distribution of slums. Figure 2 presents the distribution of slum settlements in Yogyakarta City.

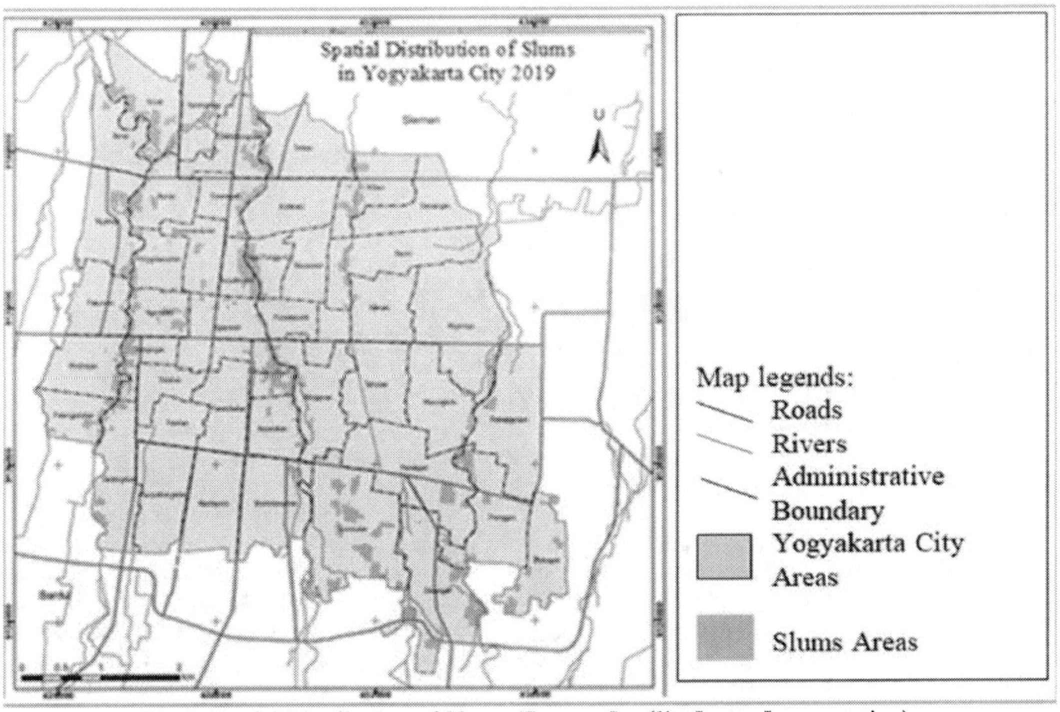

Fig 2. Spatial Distribution of Slums (Source: Satellite Image Interpretation)

Several urban researchers have tried to explain the phenomenon of gentrification in various perspectives, based on the symptoms of urban restructuring [17]. Some researchers still see gentrification as only related to the spatio-temporal context of certain regions (especially London and New York), and not a phenomenon that can occur in many places [18]. This view causes the study of gentrification to stagnate. Several other researchers have argued that gentrification is a micro-economic process entangle formal property rights and playing out in the real estate market [19]. The views show a tendency (1) to assume cities in developing countries as qualitatively distinct and detached from the capitalist accumulation process; (2) assume squatters and slums as different urban spaces and accumulation of capital cannot change them; (3) ignoring the formal and informal processes that coalesce in the global economy; and (4) downplaying the role of speculators and landlords in informal settlements [20]. This view also reduces the phenomenon of gentrification in Latin America and Asia from the 1970s. The forced transfer of informal housing in many pieces of the world is an integral part of the gentrification process [21].

From the results of in-depth interviews supported by secondary data, it is obtained the fact that there has been a process of gentrification in Yogyakarta. Early indications of gentrification in Yogyakarta city include: (1) The high average tenant, (2) easy access to work centers (highways, public transportation, shuttle, new sub-terminal), (3) Traffic jams, (4) High architectural value, and (5) The low comparison of house building values.

The indications of gentrification is taking place in the Yogyakarta city include: (1) Changes from rented houses to owned houses, (2) Increased down payment ratio, decreased home asset financing, (3) Influx of households and individuals interested in urban comfort and culture (artists, young professionals, schoolars), (4) An influx of conveniences catering to the rich, for example music cafes and galleries. Some of the variables identified as the drivers of gentrification are: (1) Working opportunity growth, (2) Proprety market dynamic, (3) Preference for living in the city, and (4)

Increasing of Congestion. The analysis of various data sources can be summarized in a diagram as presented in Figure 3.

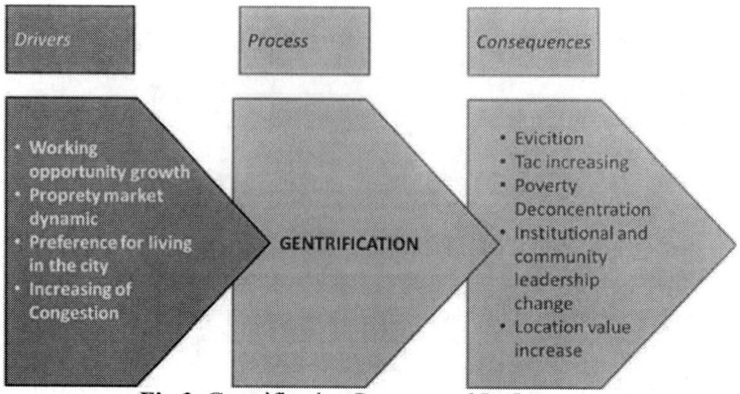

Fig 3. Gentrification Process and Its Impacts

Gentrification is an urban phenomenon that arises from the social aspect, which cannot be solved by a planning approach that only pays attention to the physical aspect. However, gentrification has not been widely discussed in most cities in Indonesia, although its impact is indirectly felt by people who have experienced it. From a physical point of view, gentrification may not appear to be a problem, because its negative impact is actually only accepted by the community. In addition, the consequence of the planning approach is that there is no action that can provide solutions to the community to overcome a situation that may have the opportunity or even have occurred in certain areas.

In both local and global perspectives, cities must have experienced disinvestment, and at other times experienced revival. Urban revival appears in the form of new housing, business centers, and the migration of the upper middle class from the suburbs to the city center, known as gentrification. Subjectively, gentrification benefits homeowners because they can sell their homes at a higher price [6], while tenants, and lower middle class individuals face the negative effects of gentrification. Some tenants can no longer afford to pay rent and will move elsewhere. From an educational perspective, low-income families are left behind by having to move their children's schools [22].

Various studies have found that people who migrate to gentrified environments are mainly high- and very high-income, professionals and nuclear families, while those displaced are low-class, elderly, and unemployed [23]. Gentrification displaces poor people and minorities from their social communities [24]. Unfavorable impacts, such as displacement of indigenous inhabitants, local enterpreneur, and cultural facilities; housing inaccessibility, socio-economic diversity; homelessness, as well as increasing land/housing speculation [25] have prompted analyzes aimed at identifying gentrification and the factors that characterize it.

The decline in the ability to obtain affordable housing, racial and economic segregation, public investment that accelerate gentrification and public dissatisfaction on environmental changes, make gentrification take place more intensively. Identifying environments prone to gentrification phenomena can help mitigate the triggers or factors that can exacerbate gentrification [26], [27], [28].

When referring to Freeman's research, there are two ways of measuring displacement; i.e all types of housing mobility as proxies for displacement, and all respondents providing reasons for moving [29]. Freeman found from his study that displacement and high mobility did not play a most significance role in gentrification. The environment will be gentrified but no displacement occur. Freeman argues that gentrification can increase investment and attract upper-middle-class households into an underinvested environment as well. Furthermore, Freeman states that the good impacts of gentrification have the potential to increase the tax base and even escalate socio-economic integration

[29]. Freeman's view ignores the social aspects of marginalized communities as groups that are very negatively affected by the genrification process.

4. Conclusion

By virtue of the study results, it can conclude as follow: (1) the symptoms of gentrification was take place in Yogyakarta Urban Area, identified by (1) changes from rented houses to owned houses, (b) increased down payment ratio, decreased home asset financing, (c) influx of households and individuals interested in urban comfort and culture (artists, young professionals, schoolars), (d) an influx of conveniences catering to the rich, for example music cafes and galleries.; and (2) the implications of gentrification on urban sustainability can be positive (i.e. increasing tax revenue, increasing land value, and increasing city aestethic) and negative (ie. eviction of low income people, poverty deconsentration, and institutional and community leadership change).

Acknowledgement

Very high commendation to the the Faculty of Geography which provide an opportunity to conduct this research, and to all of the informants involved in this research as well.

References

[1] Yunus H S 2006 *Megapolitan: Konsep, Problematika, dan Prospek* (Yogyakarta: Pustaka Pelajar)
[2] Yunus H S 2008 *Dinamika Wilayah Peri Urban: Determinan Masa Depan Kota* (Yogyakarta: Pustaka Pelajar)
[3] Glass R 1964 *London : Aspects of Change* (London: MacGibbon & Kee)
[4] Kennedy M and Leonard P 2001 *Dealing with neighborhood change: A Primer on Gentrification and Policy Changes* (Washington DC: The Brookings Institution on Urban and Metropolitan Policy)
[5] Yunus H S 2000 Perubahan Penggunaan Lahan di Daerah Pinggiran Kota: Kasus di Pinggiran Kota Yogyakarta; *Disertasi* (*unpublished*) Yogyakarta: Universitas Gadjah Mada, Yogyakarta
[6] Atkinson R 2002 *Does gentrification help or harm urban neighborhoods?* An assessment of the evidence-base in the context of the new urban agenda (*unpublished*). ESRC Center for Neighborhood Research
[7] Harvey D 2009 The new imperialism: accumulation by dispossession, in Panitch L and Leys C (eds) *Social Register 2004: The New Imperial Challenge*, London: Merlin
[8] Yunus H S 2004 *Pembangunan Kota Berkelanjutan: Permasalahan dan Strategi Pencapaiannya*, (*unpublished*), Yogyakarta: Universitas Gadjah Mada, Yogyakarta
[9] Sullivan D 2007 Reassessing Gentrification Measuring Resident Opinion using Survey Data *Urban Affairs Review* 42(4)
[10] Vidal M 2019 *Gentrification in Porto: will the city turn into a hotel*? Open Democrazy, 9 Januari, Available at https:www.opendemocrazy.net/en/can-europe-make-it/gentrification-in-porto-will-city-turn-into-hotel/
[11] Pryor R J 1968 Defining The Rural-Urban Fringe. *Social Force*. Vol. 47. Issue 2
[12] Ellman T 1997 Infill: the cure for sprawl? *Arizona Issue Analysis 146*. [online] URL:http://www.urbanfutures.org/p82897.html
[13] Antrop M 2000 Changing Patterns In The Urbanized Countryside Of Western Europe, *Landscape Ecology*, Vol 15
[14] Wilson E H 2002 Development Of A Model To Quantify and Map Urban Growth, *Thesis* (*unpublished*), University of Connecticut
[15] United Nations 2001 *World Urbanization Prospects The 2001 revision Data Tables and*

Highlights. ESA/P/WP.173.http://www.un.org/esa/population/publications/wup2001/wup2001dh.pdf
[16] Parker S 2004 *Urban Theory and the Urban Experience: Encountering the City* (London: Routledge)
[17] Lees L and Phillips M 2018 *Handbook of gentrification studies* (Cheltenham-UK: Edward Elgar Publishing Limited)
[18] Maloutas T 2012 Contextual diversity in gentrification research *Critical Sociology*, 38(1)
[19] Ghertner D A 2015 *Rule by Aesthetics: World-class City Making in Delhi* (Oxford: Oxford University Press)
[20] Lees L 2014 The urban injustices of New Labour's "New Urban Renewal: The case of the Aylesbury Estate in London *Antipode*
[21] Lees L, Shin H and López-Morales E 2016 *Planetary Gentrification* (Cambridge: Polity Press)
[22] Hankins K B 2007 The Final Frontier: Charter Schools as New Community Institution of Gentrification. *Paper on Seminar (unpublished)*
[23] Atkinson R 2000 Measuring gentrification and displacement in greater London *Urban Studies* 37(1)
[24] Kirkland E 2008 What's Race Got to Do with it? Looking for the Racial Dimensions of Gentrification. *The Western Journal of Black Studies*, 32(2), 18-30. from https://www.researchgate.net/publication/287842782_What's_race_got_to_do_with_it_Looking_for_the_racial_dimensions_of_gentrification
[25] Atkinson R and Bridge G 2005 *Gentrification in a Global Context: The New Urban Colonialism*. (London: Routledge)
[26] Chapple K 2009 *Mapping Susceptibility to Gentrification: The Early Warning Toolkit*. August. Center for Community Innovation, UC Berkeley. Available at: https://communityinnovation.berkeley.edu/reports/Gentrification-Report.pdf
[27] Chapple K, Loukaitou-Sideris A and Waddell P 2017 *Developing a new methodology for analyzing potential displacement*. Available at:https://trid.trb.org/view/1464806
[28] Pollack S, Bluestone B, and Billingham C 2010 *Maintaining Diversity in America's Transit-Rich Neighborhoods: Tools for Equitable Neighborhood Change*. Boston, MA: Northeastern University and Dukakis Center for Urban and Regional Policy
[29] Freeman L 2005 Displacement or succession? Residential mobility in gentrifying neighborhoods. *Urban Affairs Review* 40(4)

IOP Conf. Series: Earth and Environmental Science 1233 (2023) 012043 doi:10.1088/1755-1315/1233/1/012043

The water balance analysis of several land uses in Tinalah sub-watershed, Kulon Progo

M Amalia[1] and A Kusumandari[2]

[1]Alumni of the Faculty of Forestry, Universitas Gadjah Mada
[2]Senior lecturer at the Faculty of Forestry, Universitas Gadjah Mada
*Corresponding author: ambar_kusumandari@ugm.ac.id

Abstract. Tinalah sub-watershed has vulnerable to the threat of erosion and landslides. Erosion in the upstream watershed can trigger catastrophic flooding in the area below it. In addition, the problems that occur every year in the Tinalah sub-watershed are drought during the dry season and flooding during the rainy season. The water balance was analyzed by using Thornthwaite and Matter. The result shows that the dry index increased from June and the highest is in August. The highest runoff is in June (0.93). Mostly the land uses have deficit water during 4 months that are June to September, except the mixed planting has 3 months of deficit, whereas the area of the settlement has a longer deficit until 5 months. The simulation of land use in an agroforestry system shows the decreasing value of the deficit which then it will affect the value of the Drought Index (Ia) which also decreases.

1. Introduction

The Tinalah sub-watershed located in Kulon Progo Regency is part of the Progo Watershed. The Tinalah sub-watershed has an important role, namely as a source of water for the surrounding area. Water originating from the Tinalah sub-watershed is the main source for the existing irrigation network. The upstream part of the Tinalah sub-watershed is mostly in the form of dry field. It is estimated that according to the results of calculations carried out by [1] the area of the Tinalah sub-watershed is 44.43 km2, so that the Tinalah sub-watershed is included in the broad sub-watershed category (more than 10 km2). The Tinalah sub-watershed has a round shape and has a fast flood flow velocity, so it only takes a relatively short time to reach the peak of the flood.

The upstream part of the Tinalah sub-watershed is an area that is vulnerable to the threat of erosion and landslides [1]. Erosion in the upstream watershed can trigger catastrophic flooding for the area below it. In addition, the problems that occur every year in the Tinalah sub-watershed are drought during the dry season and flooding during the rainy season. Samigaluh and Kalibawang sub-districts, which are in the Tinalah sub-watershed area, are included in the high level of drought hazard according to data from the Kulon Progo BPBD. If the situation of shortage and excess of water occurs drastically, it can result in a disaster that is detrimental to many parties. It is necessary to manage water resources based on detailed and measurable water availability by considering the natural factors that influence it, including aspects of conservation and land use [2]. One of the water resources management that is appropriate is by calculating the amount of water using the water balance model.

The water balance is a description of the relationship between inflow and outflow in a certain period of the water circulation process [3]). According to [3] the water balance is detailed information about changes in all inputs, outputs, and storage of water in the soil to determine the water content in the soil and describe the water gain (surplus or deficit) over time. The availability of water resources can be influenced by several factors, namely climatic conditions, topography, soil type, land cover, and the geological structure of an area [4] . Water balance is calculated by using the Thornthwaite method because of the simplicity of the calculation method and the input data required.

Calculation of the water balance needs to find out the actual picture of water availability in a watershed. Based on the understanding of the water balance we can predict floods and droughts, so that we can formulate the prevention activities that can be carried out. Therefore, it is important to calculate the water balance in the Tinalah sub-watershed. This research aims of analyzing the condition of the water balance of several land uses in the Tinalah sub-watershed, Kulon Progo Regency.

IOP Conf. Series: Earth and Environmental Science 1233 (2023) 012043 doi:10.1088/1755-1315/1233/1/012043

2. Material and methods

2.1. Research time and location

The research location is the Tinalah sub-watershed (Figure 1) which is located in Samigaluh and Kalibawang Districts, Kulon Progo Regency. The Tinalah watershed is part of the Progo watershed. The upstream area of the Tinalah sub-watershed is located in the Menoreh structural hill complex. Based on watershed morphometry calculations carried out by [1] the Tinalah sub-watershed area is 44.43 km^2 with a main river length of 13.02 km.

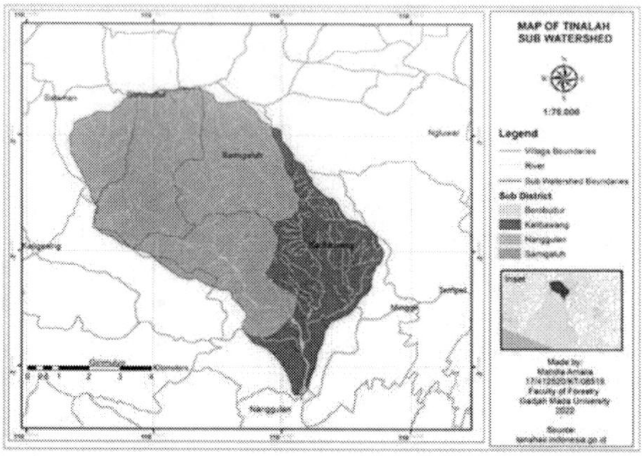

Figure 1. Research location map (source: Indonesia Geo Spatial)

2.2. Material and tools

Tools used during this research consists of:

Table 1. Equipment used for this research.

No	Name	Function
1	*Software* ArcGIS	to make maps and process spatial data
2	*Software* Microsoft Excel	to process and analyze data
3	Stationaries	to record the data obtained
4	*Global Positioning System* (GPS) and *Software Avenza*	to find and mark location coordinates
5	Scoop	to take soil samples in the field
6	Plastic bag	to store soil samples obtained in the field.
7	Label	to write a description of the soil sample

The materials used are monthly rainfall data for the Tinalah sub-watershed for 2011-2020 and monthly air temperature data for the Tinalah sub-watershed 2011-2020. Some map supports this research includes map of the administrative area of the research site, soil type map, land use map, slope map, and contour map. Besides, we need coordinate data of rain stations in the research location and elevation data.

2.3. Data collection methods

The data consists of two groups that are primary and secondary data. The two data sources will determine the data collection process. Primary data is data collected based on direct interaction between researchers and data sources. While secondary data includes data collected from printed sources of previous research [5].

2.3.1. Primary data. Field data measurements uses purposive sampling technique to take soil samples for each type of land use. Purposive sampling or can also be referred to as judgment sampling chooses sample members intentionally by researchers because of their qualities [6]. Soil samples consists of disturbed soil samples or more commonly known as disturbed soil samples, are soil samples taken using a hoe,

IOP Conf. Series: Earth and Environmental Science 1233 (2023) 012043 doi:10.1088/1755-1315/1233/1/012043

shovel, or soil drill from a certain depth of 1-2 kg [7]. The water content analysis uses the disturbed soil samples. Soil samples were taken at a depth of 15 cm in a clean condition of roots and gravel. We packed the soil samples in plastic bags and then labeled with information on the location, type of land use, date of collection, and soil depth.

2.3.2. Secondary data. The source of secondary data consists of data obtained from a literature study in the form of literature, written sources that have relevance to this research. We obtained the secondary data from the Geospatial Information Agency (BIG) and the Yogyakarta Meteorology, Climatology and Geophysics Agency (BMKG) includes watershed administration maps, land use maps, slope maps, contour maps, soil type maps, rain station coordinates, rainfall data during 10 years from 2011 to 2020 and air temperature data.

2.4. Data analysis

This research location was in Samigaluh and Kalibawang Districts, Kulon Progo Regency. The method used in the water balance analysis is the method of Thornthwaite and Mather [8]. This method has been widely used to analyze the water balance because it is easy to use and the parameters are readily available. The Thornthwaite-Mather water balance equation in [9] uses soil moisture capacity to estimate water supply. The parameters used in this method are as follows.

Potential evapotranspiration is the potential of the atmosphere to remove water from the soil surface. We use the Thornthwaite method. The next step, we subtract the rainfall (CH) for a given month from the potential evapotranspiration (ETP) in the same month. We use the APWL to determine the potential for water loss in dry months, by adding up the consecutive negative CH-ETP values until the last negative CH-ETP value. APWL increases during the dry season and decreases during the wet months due to soil moisture recharge. The APWL value can be zero (0) when the soil moisture is equal to the soil's available water holding capacity [9]. Thornthwaite and Mather [8] proposed the AWC value determination uses data of land use, soil texture, and root depth by establishing a water holding capacity (WHC) table [9] in different soil and vegetation combinations. The water content in the soil / soil moisture describes the amount of water that can be stored by the soil. If the CH-ETP is positive, then the KAT value is the same as the AWC. Meanwhile, if KAT is negative, it means that water released for evapotranspiration [9].

The difference between actual evapotranspiration (ETA) and potential evapotranspiration (ETP) lies in their relationship to KAT. In ETP, water loss only occurs at the soil surface due to atmospheric events (heat), while in ETA there is a change in KTA at the soil surface. When rainfall (CH) is higher than ETP, it means that KAT is still saturated with excess rainwater. Therefore, ETA is the same as ETP in that there is no change in KAT. When CH is lower than ETP it means that there is a change in KAT. Thus, ETA equals CH minus the change in KAT (ΔKAT), because all CH and (ΔKAT) will be evapotranspiration.

The deficit illustrates that there is a difference between ETA and ETP (Thornthwaite and Mather, 1957 in Nugroho et al., 2019). When soil moisture reaches the maximum soil moisture capacity, namely AWC, the excess rainfall will become a surplus value, and then the surplus is equal to CH-ETP [9].

$$D = ETP - ETA \tag{1}$$

$$S = CH - ETP \tag{2}$$

Thornthwaite and Mather suggest that only 50% of the surplus water in the watershed will run off each month. The remaining 50% is assumed to be in the form of runoff in the following month [9].

$$R_i = 50\%R_i + 50\%R_{i-1} \tag{3}$$

For the month of January, R_{i-1} was calculated by using 50% of surplus in December [10].

3. Result and discussion

3.1. Water Balance Concept

This research uses the Thiessen polygon method for calculating the rainfall average. The water balance is calculated for each land use based on the distribution of the rainfall area coverage. We analyze the water balance to determine the relationship between research variables and the condition of the water balance at each land use so that we can get data about the water capacity stored in the soil. These results is useful as a basis for simulating land use design in the Tinalah sub-watershed to improve water balance conditions.

IOP Conf. Series: Earth and Environmental Science 1233 (2023) 012043 doi:10.1088/1755-1315/1233/1/012043

Table 2 shows the largest land use is a mixed planting with an area of 17.08 km^2. The next largest land use is the dry and field of 12.75 km^2. The paddy fields covers an area of 8.07 km^2. The settlement has an area of 7.52 km^2, whereas the shrubs covers an area of 3.48 km^2. The smallest land use is grass area of 0.04 km^2.

Table 2. Land use area of Tinalah sub Watershed

No	Land use type	Area (km^2)	Percentage (%)
1	Paddy field	8.69	16.3
2	Dry land	12.75	25.8
3	Mixed planting	17.08	34.5
4	Shrubs	3.48	7.0
5	Grass	0.04	0.1
6	Settlements	7.52	15.2
7	River	0.55	1.1
	Total	49.50	100

The amount of water holding capacity (WHC) in the Tinalah sub-watershed is shown in Table 3 According to [8, factors influencing the WHC value is soil texture and type of ground cover vegetation. Soil texture will affect the ability of the soil to infiltrate and hold water. The same type of vegetation will have different root depths if it grows on different soil types, so that it will affect the different WHC or Available Water Capacity (AWC) values. To obtain the AWC value or soil moisture capacity, we multiply the available water value by the root and the percentage of land use area.

The highest water holding capacity of the soil occurred in the use of garden land at 69.37 mm. While the lowest WHC value is in the grassland of 0.12 mm. This happens because in the garden land use there are more trees than other land uses. Types of trees that exist include *sengon*, mahogany, cloves, and *jabon*. Other influencing factor is soil texture. Clay on mixed planting land has many micro pores (small) or not porous because the smaller the size of the separate means the greater the number and the greater the surface area so that the particles are denser. The less porous the soil, the more difficult it is for water and air to circulate or in other word we say that the drainage and aeration are poor, but the existing water is not easily lost from the soil [11]. The root zone on the garden area is only 0.67 m because the roots are difficult to penetrate due to the non-porous clay [11].

Table 3. Estimation of Water Holding Capacity (WHC) on different combinations of soil and vegetation in the Tinalah sub-watershed

Land use	Soil types	Water available (mm/m)	Root zones (m)	% Area	AWC (mm)
Paddy field	silty clay loam	250	0.8	16.3	32.61
Dry land field	silty clay	150	1.67	25.8	64.53
Mixed planting	clay	300	0.67	34.5	69.37
Bush	clay	300	0.67	7.0	14.14
Grass	silty clay	150	1	0.1	0.12
Settlement	clay	300	0.1	15.2	4.56
	Total				185.33

A. Paddy field B. Dry land field C. Mixed planting

D. Bush E. Grass F. Settlement

Figure 2. Several land uses in Tinalah Sub-watershed

3.2. Water balance of several land uses

3.2.1. Paddy fields. The paddy fields mostly dominates the Banjararum and Banjarsari areas as well as some parts of the Gerbosari, Purwoharjo, and Sidoharjo areas. The area has a slope of 0 - 8% and 8 - 15% i.e. flat to sloping. The height of the place varies between 64.07 – 197.82 m. The types of agricultural commodities planted by the community are rice and corn. The soil type in paddy fields is Latosol; it has a dusty clay-clay texture.

The graph of the water balance in the paddy fields area (Figure 3) shows that based on rainfall data from 2011-2020 the Tinalah sub-watershed experiences a dry season in the middle of the year, namely June – September. August is the driest month with an average rainfall of only 7.90 mm. The rainy season occurs at the beginning and end of the year with the highest average rainfall in December of 414.65 mm. The potential for water loss in dry months in paddy fields is 118.42 mm. The value of soil moisture at field capacity (ST) dependences on the type of land use, soil texture, and root depth. The highest ST value occurred in the wet month (32.61 mm). Runoff decreases in the dry months (June – September) and will increase again when rainfall increases in wet months. The highest runoff value occurred in January at 296.88 mm and the lowest in October at 4.25 mm.

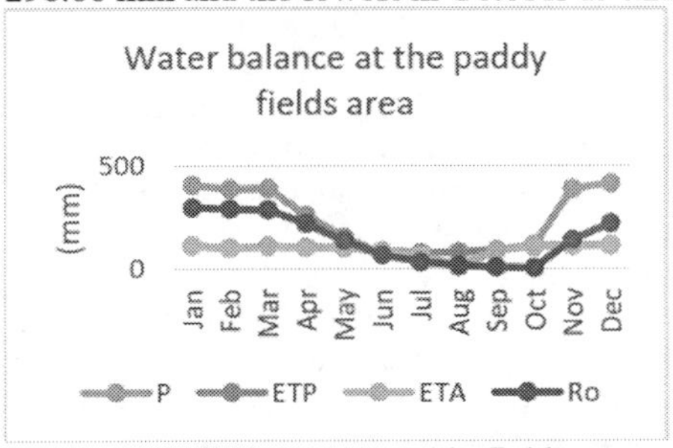

Fig 3. Water balance on paddy field area

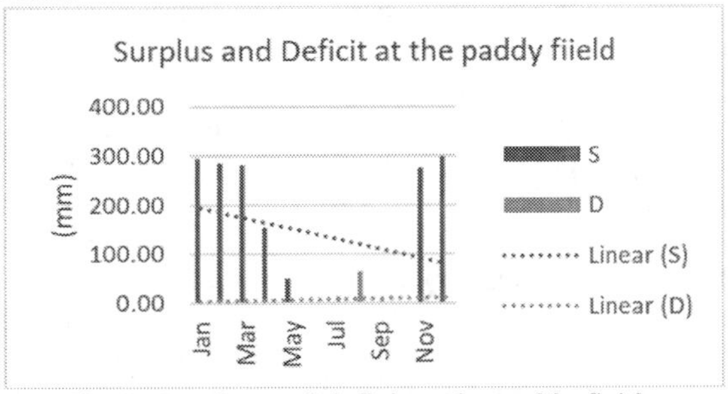

Fig 4. Surplus and deficit at the paddy field.

In the Thornthwaite-Mather method, the surplus value will be runoff. Figure 4. Shows the surplus occurs in the wet months, namely the beginning and end of the year with the highest surplus value in December of 299.31 mm. The deficit occurs in the dry months with the highest value occurring in August of 65.17 mm. At the end of the year, the surplus trend decreased, while the deficit trend increased.

3.2.2. Dry land fields. The dry land fields spreads in almost all areas of the Tinalah sub-watershed. The slope varies between 15 – 25% (slightly steep), 25 – 45% (steep), and > 45% (very steep). The altitude of the location of the fields ranges from 215.33 m – 965.17 m. The soil type is latosol with a sandy loam texture. Most of the land use in the fields is located on steep land so that the most common types of vegetation are tree species. The types of trees planted by the community are mahogany, jabon, and sengon. In addition to trees, there are also understory plants used by the community as animal feed.

Figure 5 shows the water balance of the dry land fields in Tinalah sub-watershed uses data of the average monthly rainfall for the years 2011-2020. Rainfall acts as an input to the water balance. December has the highest rainfall input of 414.65 mm. Meanwhile, the lowest rainfall input occurred in August at 7.9 mm. The dry season in the Tinalah sub-watershed occurs in June – September. While the rainy season occurs in January - May and October - December. Rainfall influences the evapotranspiration process. The potential for water loss in the dry season in field land use is 118.42 mm. In addition to sufficient evapotranspiration, rainfall is stored as a water reserve. If the water reserves are sufficient, the rainfall will become a surplus of water. According to Thornthwaite-Mather only 50% of the surplus becomes runoff in each month and the remaining 50% becomes runoff in the following month. In field land use, the largest runoff occurred in January and the smallest runoff occurred in October.

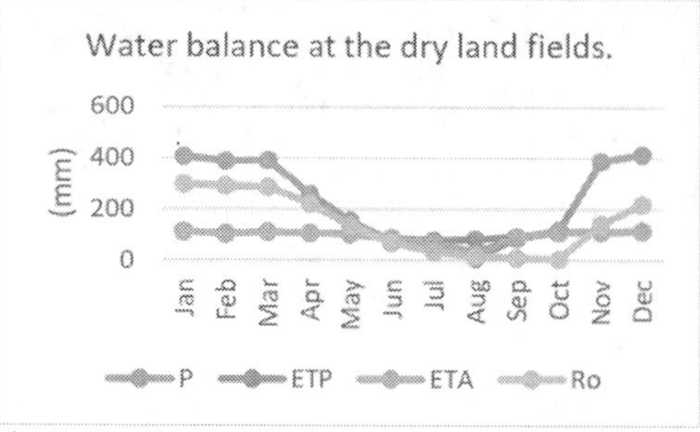

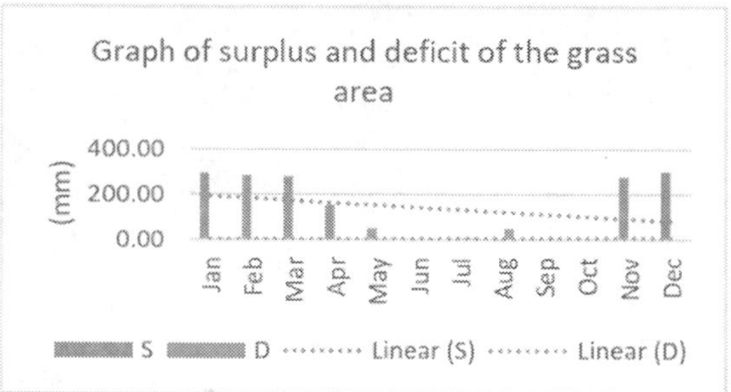

Fig 5. Graph of water balance on dry land field Fig 6. Surplus and deficit graph at the dry land field.

Figure 6. shows the surplus and deficit conditions in the dry land field. When the water storage has been covered by rainfall, the remaining rainfall becomes a surplus value. At this dry land field, the highest surplus occurred in December of 299.31 mm. The deficit describes the difference between potential evapotranspiration and actual evapotranspiration. The highest deficit occurred in August of 50.01 mm. At the end of the year, the surplus value trend will decrease and the deficit trend will increase.

3.2.3. Mixed planting. The land use of mixed planting spreads throughout the Tinalah sub-watershed area. The garden area has a slope of 15 – 25% (slightly steep), 25 – 45% (steep), and > 45% (very steep). The height of the place is between 215 – 965.17 m. The type of soil in the garden area is latosol with a loamy texture. In the use of garden land, the community applies an intercropping system. The types of trees planted are sengon, mahogany, clove, jabon, and coconut. Plants under cultivation such as *empon-empon*, cassava, *lompong*, grass, and there are banana herbs.

Figure 7 shows the water balance of mixed planting in the Tinalah sub-watershed. The component that acts as an input is rainfall. While the components that act as output are evapotranspiration and runoff. Rainfall uses data on the average monthly rainfall for the years 2011-2020. Heavy rains occur at the beginning and end of the year with the highest rainfall occurring in December of 414.65 mm. The lowest rainfall occurs in August at 7.90 mm. Evapotranspiration is the opposite of rain events, namely the loss of water from the earth to the atmosphere. If rainfall is reduced by negative potential evapotranspiration, it is called a dry month and vice versa if it is positive it is called a wet month. Wet months occur in January-May and October-December and dry months occur in June-September. The potential for water loss in dry months in garden land use is 118.42 mm. After sufficient storage of water reserves in the soil, the remaining rainfall will become a surplus. Only 50% of the surplus will be runoff in each month and the rest will be runoff in the following month. In garden land use, the largest runoff occurred in January of 296.88 mm and the lowest runoff occurred in October of 4.25 mm.

Figure 8. shows the surplus and deficit on mixed planting land use. The high surpluses occur at the beginning and end of the year when rainfall is also high. The highest surplus value occurred in December of 299.31 mm. while the water deficit occurs in the middle of the year when rainfall is low. The highest deficit value occurred in August of 48.14 mm. At the end of the year, the surplus value rend will decrease and the deficit trend will increase.

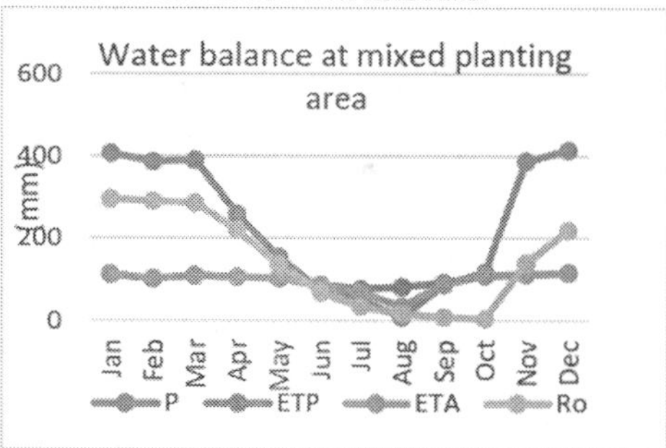

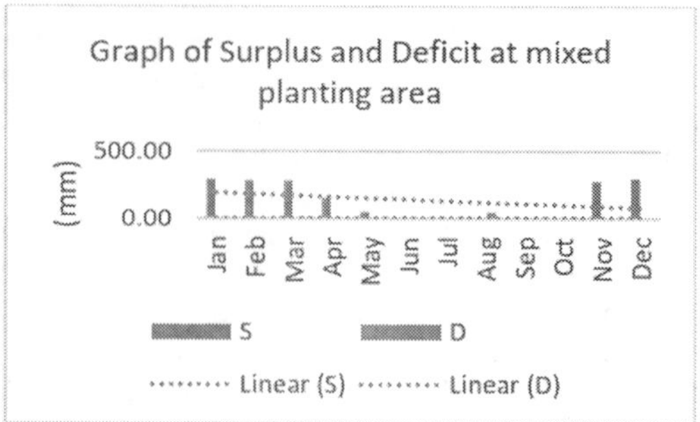

Fig 7. Water balance at the mixed planting area. Fig 8. Surplus and deficit at the mixed planting area.

IOP Conf. Series: Earth and Environmental Science 1233 (2023) 012043 doi:10.1088/1755-1315/1233/1/012043

3.2.4. Shrubs land use. In the Tinalah sub-watershed, bush land use covers the area of Gerbosari, Sidoharjo, Purwoharjo, and Banjarsari. The slope of the land varies between 15 – 25% (slightly steep), 25 – 45% (steep), and > 45% (very steep). Bush land is located at an altitude between 317.50 m – 965.17 m. The soil type is Latosol with a clay texture. The plants found were bamboo, *melinjo, gamal*, teak, *porang* (*Amorphophallus muelleri*), *and sembukan* (*Paederia foetida*).

Figure 9 reflects the water balance on bushland land use in the Tinalah sub-watershed. Heavy rainfall occurs at the beginning and end of the year. In the middle of the year, the rainfall decreases. The highest rainfall occurred in December of 411.65 mm. Meanwhile, the lowest rainfall occurred in August at 7.90 mm. January – May and October – December are wet months. The dry months occur in June – September. In the wet month, there is the potential to experience a water surplus and the dry month the potential to experience a water deficit. The potential for water loss in the dry month in bush land use is 118.42 mm. Water surplus of 50% will be runoff every month and the remaining 50% will be runoff in the following month. The highest runoff on bushland land use occurred in January of 296.88 mm. In contrast, the lowest runoff occurred in October at 4.25 mm.

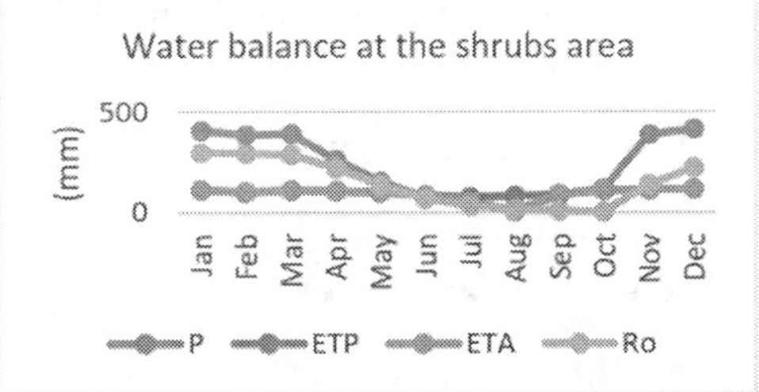

Fig 9. Graph of water balance on bushland

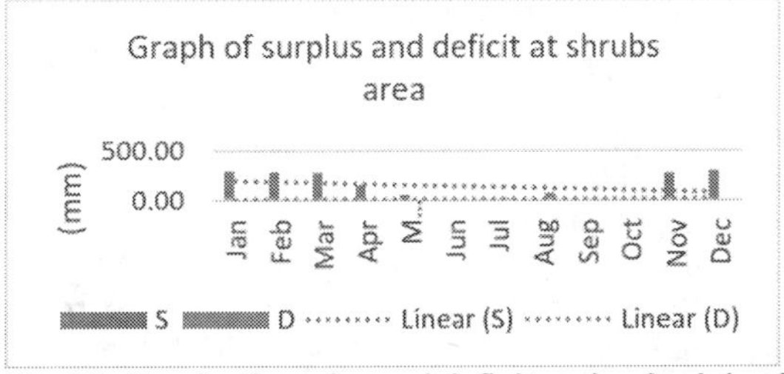

Fig 10. Graph of surplus and deficit at the shrub land

Figure 10 shows the surplus and deficit on land use in the Tinalah sub-watershed. Based on the results of the analysis of the water balance of the scrubland, there was a surplus in January – May and November – December. Meanwhile, the deficit occurred in June – October. Surplus occurs when the rainfall that falls exceeds the value of water loss through evapotranspiration and infiltration. On the other hand, a deficit condition occurs when there is less rainwater for evapotranspiration. The highest surplus value occurred in December of 299.31 mm. Meanwhile, the highest deficit value occurred in August of 73.02 mm. At the end of the year, the surplus value trend will decrease and the deficit trend will increase.

3.2.5. Grass area. The grassland covers the area of Purwoharjo. Reeds live in riverside areas has land slopes ranging from 0 – 8% (flat) and 8 – 15% (sloping). Grassland is located at an altitude between 64.07 – 317.50 m asl. The soil type is latosol with a sandy loam texture. The plants found were Bahia grass (*Paspalum notatum*).

Figure 11 draws the water balance of grassland area in the Tinalah sub-watershed. The water balance analysis of the Tinalah sub-watershed uses data on average rainfall for 10 years (2011-2020). The lowest average rainfall occurs in August at 7.90 mm and the peak rainfall occurs in December at 414.65 mm. The highest potential evapotranspiration occurred in December at 115.34 mm with an average temperature of 24.43°C. Meanwhile, the lowest potential evapotranspiration value occurred in July at 77.10 mm with an average temperature of 21.71°C. Differences in evapotranspiration values influenced by climatic factors and soil conditions [4]. In the grass area, the water balance in January – May and October – December are called wet months because the rainfall is higher than the potential evapotranspiration. These months have the potential to experience a surplus of water. Meanwhile, June-September is the dry month because the potential evapotranspiration is higher than rainfall, so it has the potential to experience a water deficit. As much as 50% of the surplus will be runoff in each month and the other 50% will be runoff in the following month. The highest runoff occurred in January at 296.88 mm and the lowest runoff occurred in October at 8.16 mm.

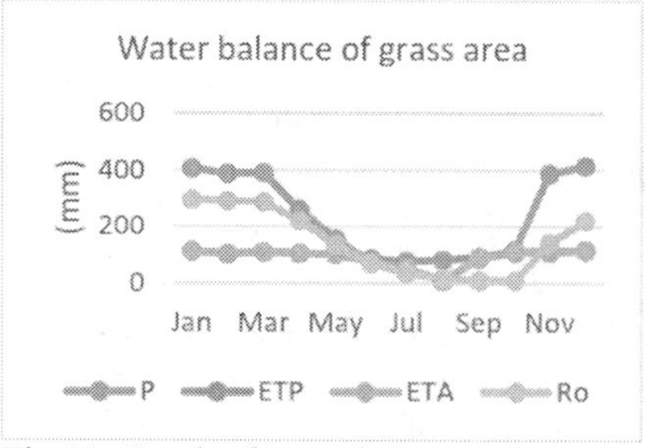

Fig 11. Graph of water balance of grass area

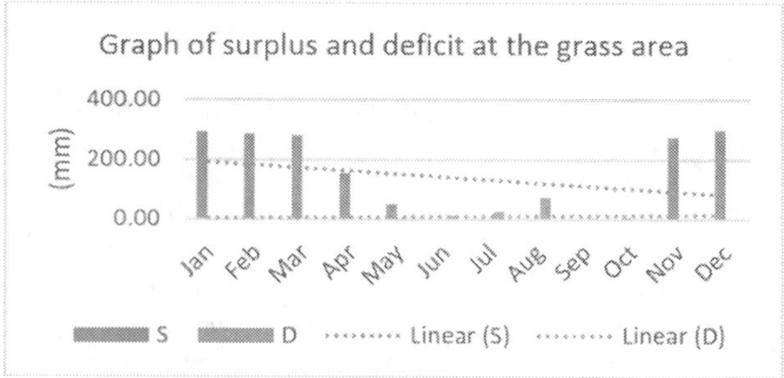

Fig. 12. Graph of surplus and deficit at the grass area

Figure 12 shows the results of the water balance analysis, the surplus and deficit values. The highest surplus occurred in December of 299.31 mm. The highest deficit occurred in August of 73.89 mm. The surplus value trend will decrease and the deficit trend will increase at the end of the year.

3.2.6. Settlements. The settlement spreads throughout the Tinalah sub-watershed. People live on land that has a slope of 0 – 8% (flat), 8 – 15 % (sloping), and 15 – 25% (rather steep) with an altitude between 64.07 – 722.29 m asl. The type of soil in the residential area is latosol with clay texture. Many people plant plants in their yard. The types of plants are banana herbs, ornamental plants, and fruit trees. On land with a lot of overgrown with grass.

Figure 13 shows the water balance of the settlements area. The average monthly rainfall was calculated using rainfall data of the years 2011 – 2020. The peak of rainfall occurred in December of 414.65 mm and the lowest rainfall occurred in August of 7.90 mm. The highest potential evapotranspiration occurred in December at 115.34 mm with an average temperature of 24.43°C. Meanwhile, the lowest potential evapotranspiration value occurred in July at 77.10 mm with an average temperature of 21.71°C. January – May and October – December are the wet months because the rainfall is greater than the potential evapotranspiration. The dry months occur in June – September because the rainfall is lower than the potential evapotranspiration. Rainfall reduced due to evapotranspiration and infiltration will become runoff on the soil surface. The highest runoff on residential land occurred in January of 296.88 mm and the lowest runoff occurred in October of 5.94 mm.

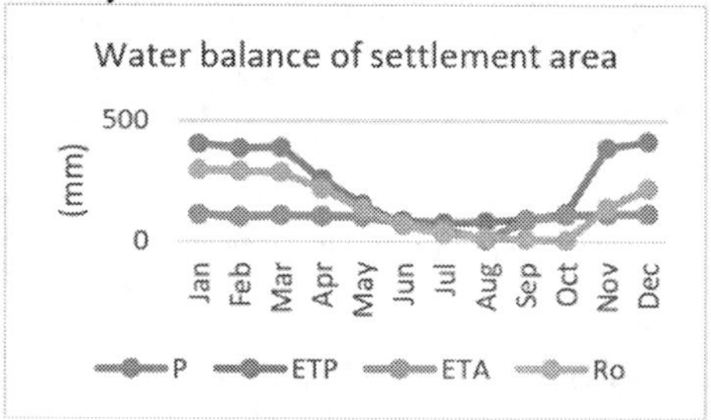

Fig 13. Water balance on settlement land use

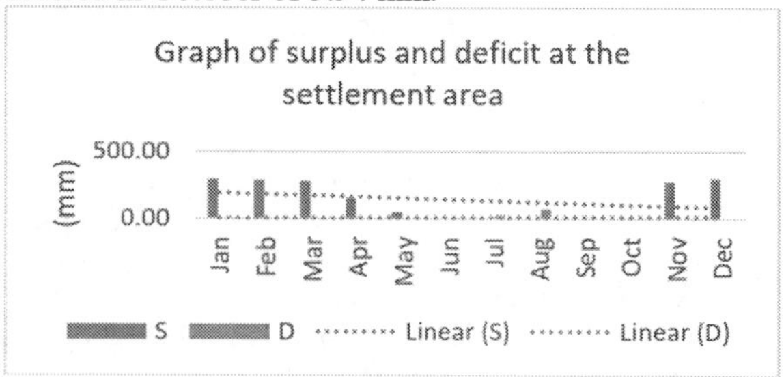

Fig 14. Graph of surplus and deficit at the settlement area

Figure 14 reflects the water balance analysis, the surplus and deficit values. The highest surplus occurred in December of 299.31 mm. The highest deficit occurred in August of 73.89 mm. At the end of the year, the surplus value trend will decrease and the deficit trend will increase.

3.3. Drought index

The drought index is useful to detect, monitor, and evaluate drought events in an area. Figure 15 shows an overview of the drought index in various forms of land use in the Tinalah sub-watershed. During the wet month, which is a condition where the rainfall is greater than the potential evapotranspiration, there is no drought (Ia = 0). On the other hand, during the dry month period where the potential evapotranspiration is higher than rainfall, a drought will occur. During the dry month period, the amount of rainfall that falls is not able to meet the water needs for plants, causing a lack of soil moisture (moisture deficiency). The relationship between water deficit and potential evapotranspiration will produce a drought index value.

The Tinalah sub-watershed experiences drought in June – September with the peak of drought in August. In June, the grassland drought index had the highest drought index value compared to other land uses, which was 15.47, which means a little lack of water. Meanwhile, the mixed planting land use has the lowest drought index of 1.42, which means it also experiences little or no water shortage. In July, all types of land use drought index increased. Imperata land use and settlements have the highest drought index, which is 33.36, which means they are experiencing severe water shortages. Meanwhile, the lowest drought occurred in the mixed planting land use with a drought index of 10.54, which means that it experienced a slight shortage of water. In August, the Tinalah sub-watershed faced a peak of drought. Imperata and settlements have the highest drought index of 90.34, which means there is a severe water shortage. While the lowest drought index occurred on garden land of 58.86, which means severe water shortages. In September, the drought began to decline. All types of land use almost have the same drought index. The highest drought index happens at the land use of shrubs, grassland, and settlements of 5.54, which means that there is a slight shortage of water. While the lowest drought index occurred in the dry land field at 4.62, which means that, there was a slight shortage of water.

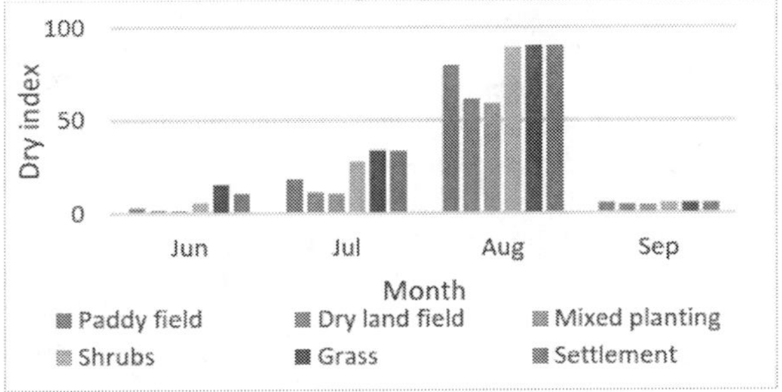

Fig 15. Dry index in Tinalah sub watershed

3.4. Runoff coefficient

The runoff coefficient (C) shows the magnitude of the ratio between the runoff rate and the amount of rainfall. The value of the runoff coefficient ranges from zero to one. If the value of C is equal to zero it means that there is no runoff in a watershed. If the value of C is equal to one, it indicates that the watershed is experiencing large runoff or rainfall that falls entirely becomes runoff. This shows that there is damage to the watershed area because it does not function properly as a water reservoir.

Based on the results of the analysis of the runoff coefficient in January for all types of land use, it was 0.73. This value illustrates that the Tinalah sub-watershed experiences a high surface runoff. According to the Director General of RLPS [12] means that 73% of the rainfall becomes runoff. In February and March, the runoff coefficient values spreads between 0.75 and 0.73, which means that the surface runoff is quite high. In April and May, the Tinalah sub-watershed experienced higher surface runoff than in the previous months with runoff coefficient values of 0.85 and 0.88, respectively. In June, the Tinalah sub-watershed experienced the highest runoff coefficient of 0.93. Starting in July the runoff begins to decrease. The runoff coefficient in July was 0.66, which means that there was moderate runoff. In September and October, surface runoff in the Tinalah sub-watershed is at its lowest. The value of the runoff coefficient in September is 0.10. Whereas in October each land use has a different coefficient value. The runoff coefficient value on the use of paddy fields, fields, gardens, and shrubs is 0.04, which means that there is almost no surface runoff. Meanwhile, the runoff coefficient values for Imperata land and settlements are 0.07 and 0.05. In November and December, the Tinalah sub-watershed began to experience an increase in surface runoff with runoff coefficients of 0.36 and 0.53, which means that there was moderate runoff. Factors that can affect the runoff coefficient value are soil infiltration rate, land slope, land cover, and rainfall intensity. In addition, there are other factors such as groundwater, the degree of soil density, soil porosity, and deposit depression [13].

3.5. Land Use Simulation

The land use simulation aims to achieve conditions where there is an increase in the ability of the soil to store water (water holding capacity) and a decrease in surface runoff (runoff). The formulation of the propose land uses the Kulon Progo Regency Spatial Plan and considers information from key informants as people with an interest in the Tinalah Sub-watershed. Simulation of land use type's uses land use types such as dry fields and shrubs as a pattern for designing agroforestry.

The Tinalah sub-watershed is located in the northern area of Kulon Progo Regency, which is the Menoreh hills plateau with an altitude between 500-1,000 meters above sea level so that the area is a conservation cultivation area and is a landslide-prone area. The Kulon Progo Regency government's efforts to improve the quality of the environment are carried out through tree planting activities (vegetative) and the construction of technical civil buildings for land and water conservation [14]. Reforestation efforts can be carried out by applying agroforestry, namely a mixture of food crops, seasonal crops, and other woody plants planted between trees or forest plants so that they get double benefits. Land uses such as gardens, fields, and shrubs in the Tinalah sub-watershed are mostly on steep and even very steep slopes with a slope of >45% so that they require the right form of land use, one of which is by applying agroforestry.

Vegetation plays an important role in strengthening the soil, especially on land with steep slopes that have the potential to experience erosion and shallow landslides. Tree roots have the ability to strengthen soil grip and help reduce soil saturation by rainwater, increase infiltration capacity, and increase water holding capacity. Hydrologically beneficial root length density will also increase soil aggregation thereby strengthening slopes [15]. Trees act as ground cover and preservatives for groundwater when it rains in the rainy season, they will be accommodated and stored in the soil so that the danger of erosion or flooding can be reduced. Meanwhile, when the dry season arrives, water reserves can be used to reduce the risk of water shortages.

Table 4 shows the results of the simulation of land use design in the Tinalah sub-watershed. At the paddy fields area, hardwood is no trees because it can shade agricultural crops which results in the development of agricultural crops not being optimal. Farmers plant trees around the rice fields that can act as guardrails and selected species that have economic value. In dry fields, mixed planting, and shrubs because they are on a steep slope, they should select the appropriate type of vegetation. The selected plant species should such a tree that has the function of interception, evapotranspiration, and root stability. The ideal type of tree for steep slopes is a fast growing species because it will quickly form a canopy and roots. The canopy will play a role in reducing the kinetic energy of rainwater. Tree species that have a high evapotranspiration rate is a good tree chosen because they can reduce the load force by water. In addition, trees with deep roots and wide root distribution good [15]. Examples of tree species suitable for planting on steep slopes include *sengon*, acacia, mahogany, jackfruit, *rambutan*, and mango. In grassland, an agroforestry system cannot be applied because the area is found along the riverbank. On residential land, greening is suitable for planting yard plants. The results of the simulation of land use design show an increase in the value of the capacity of the availability of ground water to increase to 239.84 mm, which was previously 185.33 mm after modification of land use into an agroforestry system.

Table 4. Estimation of Water Holding Capacity after Simulation of Land Use Design in the Tinalah Sub-watershed.

Land use	Soil types	Available water (mm/m)	Root zone (m)	% Area	AWC (mm)
Paddy field	silty clay loam	250	1.5	16.3	61.13
Dry land field	silty clay	150	1.67	25.8	64.53
Mixed planting	clay	300	0.67	34.5	69.37
Bush	clay	300	0.67	7.0	14.14
Grass	silty clay	150	1	0.1	0.12
Settlement	clay	300	0.67	15.2	30.53
		Total			239.84

Runoff due to high slopes can be overcome by applying conservation techniques either mechanically, vegetative, or both. Mechanically it can be applied by building terraces, water and soil collection channels (SPAT) and drainage channels. Erosion and shallow erosion found in the Tinalah sub-watershed resulted by the absence of a water path in the soil for water to flow so that the soil becomes saturated with water. Vegetative practices, the runoff control consists planting terrace-strengthening plants, setting cropping patterns, and planting in strips.

In Table 5, we can see the calculation of actual evapotranspiration, runoff, surplus, deficit, and drought index after the simulation. The actual evapotranspiration value increases in June – August (dry months) due to an increase in groundwater reserves. The value of the water deficit during the dry months (June - September) decreased indicating that there was sufficient water storage due to the role of vegetation. Because of the decreasing value of the deficit, it will affect the value of the Drought Index (Ia) which also decreases.

Table 5. Estimation of ETA, Ro, S, D, and Ia Values after Simulation of Land Use Design in Tinalah Sub-watershed

Com-ponent	Month											
	Jan	Feb	Mar	Apr	May	Jun	Jul	Aug	Sep	Oct	Nov	Dec
Actual Evapotranspiration												
Before	114.08	103.39	111.32	106.80	104.17	85.76	68.95	49.39	72.16	111.28	114.21	117.30
After	114.08	103.39	111.32	106.80	104.17	86.21	70.84	55.37	75.46	111.28	114.21	117.30
Runoff.												
Before	234.63	219.32	224.46	175.80	93.53	46.76	23.38	11.69	5.85	2.92	110.65	175.44
After	234.63	219.32	224.46	175.80	93.53	46.76	23.38	11.69	5.85	2.92	110.65	175.44
Surplus												
Before	229.02	204.01	229.60	127.13	11.26	0	0	0	0	0	218.37	240.24
After	229.02	204.01	229.60	127.13	11.26	0	0	0	0	0	218.37	240.24
Deficit												
Before	0	0	0	0	0	2.07	9.26	33.60	21.07	0	0	0
After	0	0	0	0	0	1.62	7.38	27.62	17.77	0	0	0
Drought Index (I												
Before	0	0	0	0	0	2.36	11.84	40.49	22.60	0	0	0
After	0	0	0	0	0	1.84	9.43	33.28	19.06	0	0	0

4. Conclusion

It concludes that the dry index increased from Jun and the highest is in August. The highest runoff is in June. Mostly the land uses has the deficit water during 4 months, that are Jun to September, except the mixed planting has 3 months of deficit, whereas the settlements area has a longer deficit until 5 months. The simulation of land use into an agroforestry system show the decreasing value of the deficit, it will affect the value of the Drought Index (Ia) which also decreases.

Acknowledgement

We would like to thanks for the funding from Final Project Recognition (RTA) Universitas Gadjah Mada year 2022 in supporting this research.

References

[1] Vienastra, S. (2018). Watershed Geomorphology and Morphometry of Tinalah in Kulon Progo Regency, Yogyakarta Special Province. *Journal of Teknologi Technoscientia*, *11*(1), 21–28.

[2] Narulita, I. D. A. (2017). Spatial Water Balance Estimation for Evaluation of Water Resources Availability Case Study: Cerucuk Watershed, Belitung Island. *Journal of Teknologi Lingkungan*, *18* (January), 120–129.

[3] Paski, J. A. I., S L Faski, G. I., Handoyo, M. F., & Sekar Pertiwi, D. A. (2017). The Water Balance Analysis at Land for Paddy and Corn in Bengkulu. *Journal Ilmu Lingkungan*, *15*(2), 83. https://doi.org/10.14710/jil.15.2.83-89

[4] Tufaila, M., Mpia, L., & Karim, J. (2017). Water Balance Analysis at the Soil Developed in Karst Area in Parigi, Muna Regency, South East Sulawesi. *Agritech*, *37*(2), 215. https://doi.org/10.22146/agritech.16747

[5] Wibisono, D. (2003). *Business Research Manual for Practician and Academicians*. Jakarta: Gramedia Pustaka Utama.

[6] Etikan, I. (2016). Comparison of Convenience Sampling and Purposive Sampling. *American Journal of Theoretical and Applied Statistics*, *5*(1), 1. https://doi.org/10.11648/j.ajtas.20160501.11
[7] Suganda, H., Rachman, A., & Sutono, S. (2006). Manual of Soil Sampling. *Bogor: Balai Besar Litbang Sumberdaya Lahan Pertanian*.
[8] Thornthwaite, C.W. and Matter, J.R. (1957) Instructions and Tables for Computing Potential Evaporation and the Water Balance. Climatology, 10, 311.
[9] Nugroho, A. R., Tamagawa, I., Riandraswari, A., & Febrianti, T. (2019). Thornthwaite-Mather water balance analysis in Tambakbayan watershed, Yogyakarta, Indonesia. *MATEC Web of Conferences*, *280*, 5007.
[10] Widiyono, M. G., & Hariyanto, B. (2016). Fulfill of Domestic Water Needs in Dry Area of Trowulan, Mojokerto Regency. *Swara Bhumi*, *1*(1), 10–17.
[11] Hanafiah, K. A. (2014). *Introduction of Soil Science* (1st Ed.). Surabaya: Rajawali Press.
[12] Paramitha, T. A., & Rauf, A. (2018). Runoff Coefficient (C) of Taripa sub-watershed in Toaya, Donggala Regency. *Journal of Warta Rimba*, *6*(1), 102–108.
[13] Handayani, F., Hardiyenti, R., Agus, I., Sadtim, & Hartati (2020). The Study of Run off Coefficient of Batang Arau Watershed in Padang. *Journal of Ilmiah Poli Rekayasa*, *15*(2), 55. https://doi.org/10.30630/jipr.15.2.167
[14] Dinas Kehutanan Kulon Progo (2021). Area Status of Kulon Progo Regency Environment Year 2020.
[15] Sittadewi, E. H. (2020). Function of Interception, Evapotranspiration and Root Reinforcement of Plant on Slope Stabilization. *Journal of Sains Dan Teknologi Mitigasi Bencana*, *15*(1), 19–26. https://doi.org/10.29122/jstmb.v15i1.4116

Dynamics of Land Use Change and The Relationship with Food Sufficiency in Pacitan Regency

Revia Ardyaning Pramesti[1*], Rika Harini[2]

[1]Department of Environmental Geography, Faculty of Geography, Gadjah Mada University, Indonesia
[2]Department of Environmental Geography, Faculty of Geography, Gadjah Mada University, Indonesia

*Corresponding author: revia.a@mail.ugm.ac.id

Abstract. Infrastructure development in an area affects changes in existing land use, one of them to meet demand in development. Agricultural land use change can affect agricultural production and productivity. This study aims to determine the conditions of land use change and food sufficiency in Pacitan Regency and the relationship between the two variables. The data used in the analysis are data on land use and food production from 2009-2017 to calculate food sufficiency. The research method is quantitative descriptive with Pearson correlation statistics. The results showed that agricultural land use in Pacitan Regency was dominated by dry land agriculture. Paddy fields and dry land agricultural have decreased, while non-agricultural land has increased. The food sufficiency of the four food commodities, namely rice, corn, sweet potatoes, and cassava, has fluctuated in recent years. However, the overall condition of food sufficiency from 2009-2017 decreased from 6.48 in 2009 to 5.04 in 2017. Based on the Pearson correlation test results in Pacitan Regency, there was no connection and significant relationship between land use change and food sufficiency. This result can occur due to an increase in population, changes in the cultivation of food crops into woody plants, changes, and a decrease in land productivity.

1. Introduction

A land is a place for every human activity, for example, agriculture, animal husbandry, industry, settlements, recreation areas, and roads for transportation. Development demands can encourage an increase in land demand, resulting in land conversion, primarily agricultural land converted into built-up land [1]. Theoretically, in the axis theory, the existence of a transportation axis in an area can affect regional development, which impacts development and encourages land use change [2]. The increase in land demand is not only triggered by development but also by the increasing number of people who play a role in it [3]. However, agricultural land is essential in supporting the implementation of food security. Reduced agricultural land area decreased agricultural production and productivity and threatened regional food security [4].

Pacitan Regency is one of the areas traversed by the 89-kilometer National Southern Cross Road, known as Jalan Lintas Selatan (JLS), that crosses this area. The construction of the JLS was carried out from 2010 to 2017. This development is one of the efforts of the Ministry of Public Works to increase the availability of infrastructure in the southern part of Java. As an area that has tourism potential to be developed, road infrastructure development is an important thing. In addition, the construction of the JLS also impacts the development of education, trade, and industry. The ease of access to transportation also attracts investors to build industrial facilities in this area [5]. However, on the other hand, the process and impact of the construction of JLS can increase the demand for land. In addition, the population of Pacitan Regency, which continued to increase in 2009, was 565,827 people to 592,060 people in 2017. Population increase has also influenced the increase in the need for land and also for food [6].

One of the keys to achieving food security is maintaining land existing farms not be converted [7]. The higher demand for land can have an impact on decreasing food availability in the future. On the other hand, the population's food needs are also increasing. Therefore, this study was conducted to determine the conditions of land use change and food sufficiency conditions in Pacitan Regency during the period 2009-2017 and to determine the relationship between land use changes and food sufficiency.

2. Methods

2.1 Research Locations

This research was carried out in all sub-districts included in the administrative area of Pacitan Regency. There are 12 sub-districts in the study area including Donorojo, Punung, Pringkuku, Pacitan, Kebonagung, Arjosari, Nawangan, Bandar, Tegalombo, Tulakan, Ngadirojo, and Sudimoro. Four of them are crossed by the Jalan Lintas Selatan, namely Pacitan, Kebonagung, Tulakan, and Ngadirojo. Figure 1 shows the study area in the study.

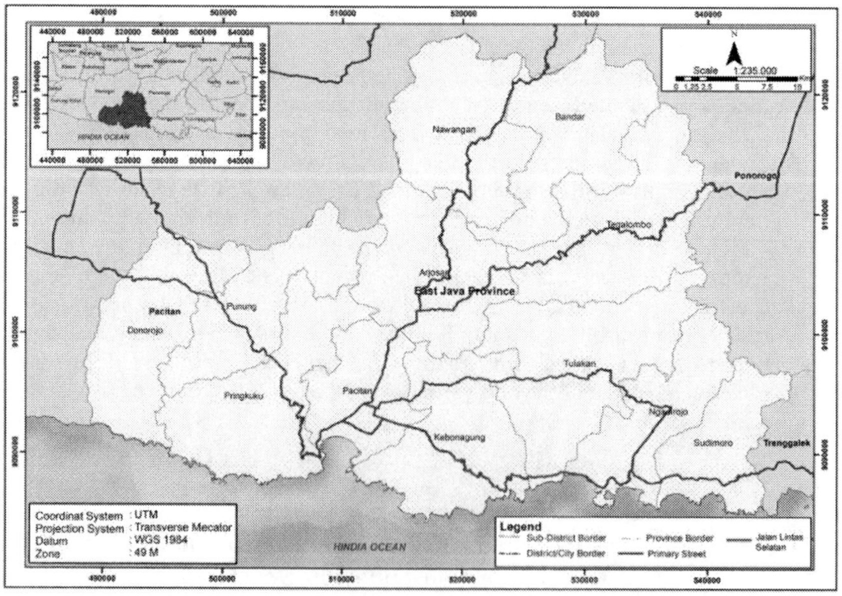

Figure 1. Pacitan Regency administration map.

2.2 Data

This research uses secondary data from 2009-2017. The data used in this study consist of secondary data published by the Bureau of Statistics (BPS) and the Pacitan Regency Agriculture Service. Some of the data include data on land use area, data on the amount of food production in 4 commodities, namely rice, corn, sweet potatoes, and cassava, and also data on the mid-year population per sub-district in Pacitan Regency.

2.3 Data Processing

This research is quantitative research was conducted by processing and analyzing secondary data, which aims to determine the conditions of land use change and food sufficiency in Pacitan Regency and examine the relationship between the two variables. The primary analysis method used descriptive

IOP Conf. Series: Earth and Environmental Science 1233 (2023) 012044 doi:10.1088/1755-1315/1233/1/012044

quantitative to describe each data in the form of numbers from the calculation results in tables and graphs.

2.3.1 Food Sufficiency. The calculation method used to determine the value of food sufficiency is as follows [8]:

Calculation of rice food production:

$$Rnet = (P \times (1-(S+F+W))) \times C \tag{1}$$

Description:
Rnet : Net production of rice (ton/year)
P : Rice production (ton/year)
S : Seeds (0.9 %)
F : Feed (0.44%)
W : Scattered (5.4%)
C : Conversion of rice to rice (62.74 %)

Calculation of corn, cassava and sweet potato food production:

$$Mnet = P \times (1-(S+F+W)) \tag{2}$$

$$Cnet = P \times (1-(S+F+W)) \tag{3}$$

$$SPnet = P \times (1-(S+F+W)) \tag{4}$$

Description:
Mnet : Net production of corn (ton/year)
Cnet : Net production of cassava (ton/year)
SPnet: Net production of sweet potato (ton/year)
P : Production of each commodity (ton/year)
S : Seeds (maize: 0.9%; cassava: 0%; sweet potato: 0%)
F : Feed (maize: 6%; cassava: 2%; sweet potato: 2%)
W : Scattered (maize: 5%; cassava: 2.13%; sweet potato: 10%)

Calculation of potato food production:

$$Tnet = 1/3 \times (Cnet + SPnet) \tag{5}$$

Description:
Tnet : Net production of cassava and sweet potato (ton/year)
Cnet : Net production of cassava (ton/year)
SPnet : Net production of sweet potato (ton/year)

Cereal food production:

$$Pfood = Rnet + Mnet + Tnet \tag{6}$$

Description:
Pfood : Cereal food production (ton/year)
Rnet : Net production of rice (ton/year)
Mnet : Net production of corn (ton/year)
Tnet : Net production of cassava and sweet potato (ton/year)

Availability of cereal food:

$$F = Pfood/(tpop \times 365) \tag{7}$$

Description:
F : Availability of cereal food (ton/capita/day)
Pfood : Cereal food production (ton/year)
tpop : Total mid-year population (people)

Food availability ratio:

$$Iav = C\ normative/F \tag{8}$$

Description:
Iav : Food Availability Ratio
C normative : Normative consumption (gr/capita/day)
F : Cereal Food Availability (gr/capita/day)

The results of the calculation of food sufficiency through the ratio of food availability and normative consumption are classified as follows (Table 1) [9]:

Table 1. Classification of food availability ratio

Food Availability Ratio	Category
< 0.90	Deficit
0.90 – 1.14	Enough
> 1.14	Surplus

2.3.2 *Pearson Correlation.* This research also used Pearson correlation to determine the relationship between variables, land use change (X) and food sufficiency (Y). The reason for choosing the Pearson correlation method is that the two variables used are ratio data. Pearson correlation has a value ranging from -1 to 1, with the following formula:

$$\boldsymbol{r} = \frac{\boldsymbol{xy} - \frac{(\Sigma x)(\Sigma y)}{n}}{\sqrt{(\Sigma x^2 - \frac{(\Sigma x)^2}{n})(\Sigma y^2 - \frac{(\Sigma y)^2}{n})}} \tag{9}$$

The results of the Pearson correlation are described by the notation r with the criteria in Table 2 below [10]:

Table 2. Level of relationship based on correlation strength (r)

Interval Coefficient	Level of Relationship
0.00 - 0.199	Very low
0.20 - 0.399	Low
0.40 - 0.599	Moderate
0.60 -0.799	Strong
0.80 - 1.00	Very strong

3. Results and Discussion

3.1 Land Use Conditions in Pacitan Regency

Land use is divided into two types: agricultural land and non-agricultural land. The use of agricultural land consists of paddy fields or wetland and dry land agriculture. Based on data from the Bureau of Statistics (BPS), the land use area from 2009 to 2017 did not experience significant changes (Figure 2).

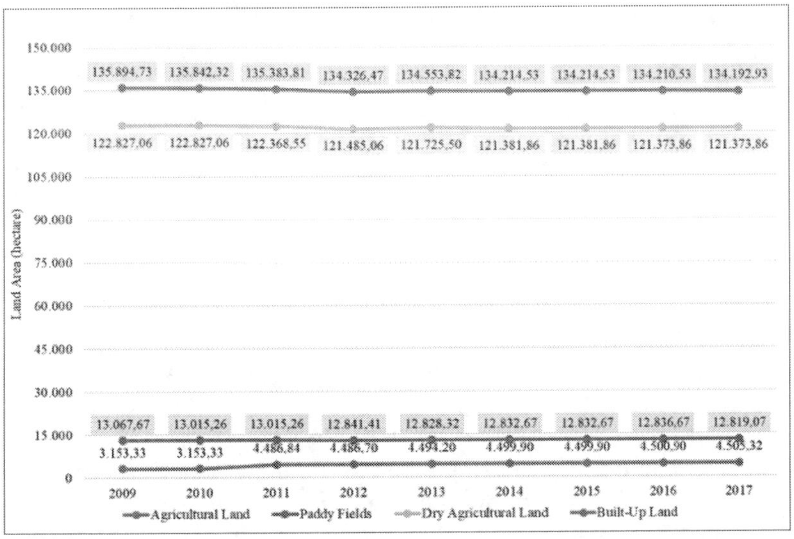

Figure 2. Trends in land use areas in Pacitan Regency, 2009-2017.

Overall, the area of paddy fields in Pacitan Regency has decreased from 2009-2017. The area of paddy fields in 2009 was 13,067.67 hectares, and in 2017 it became 12,819.07 hectares, or there was a decrease in paddy fields by 248.60 hectares over nine years. This decrease is due to the increasing demand for land for the population. Wetland agricultural land is often the target of land conversion because it is often considered to have less favorable economic value when compared to other economic activities such as trade [11]. The paddy fields in Pacitan Regency are indeed not large enough considering the physical condition of Pacitan Regency, which is dominated by dry land. In addition, several areas in the Pacitan sub-district are included in the karst area, including Donorojo, Punung, and Pringkuku sub-district. One of the influencing factors is that paddy fields require abundant surface water resources. In contrast, the condition of surface water resources in karst areas is less than that of subsurface water resources [12].

Dry land agriculture in Pacitan Regency has a larger area than rice fields. Dry land agricultural in Pacitan Regency reaches 85% of all agricultural land, including house yards [13]. Forests are the dominant land use in almost all areas in Pacitan Regency. Land use data in 2017 shows that forest land use is 73,177.86 hectares [14]. The dominance of dry land agriculture in Pacitan Regency is due to the condition of the existing landform in some areas in the form of karst landform, so agriculture suitable for cultivation is dry land agriculture [15].

Dry land agriculture in Pacitan Regency has decreased but experienced fluctuations in 2012-2014. The dry land agricultural area increased in 2012-2013 by 240.44 hectares, then shrunk again in 2013-2014 by 343.64 hectares. The increase in the area of dry land agricultural is partly due to changes in paddy fields to other uses, such as fields and moor. The total area of agricultural land converted for nine years (2009-2017) is 1,453.20 hectares of converted land.

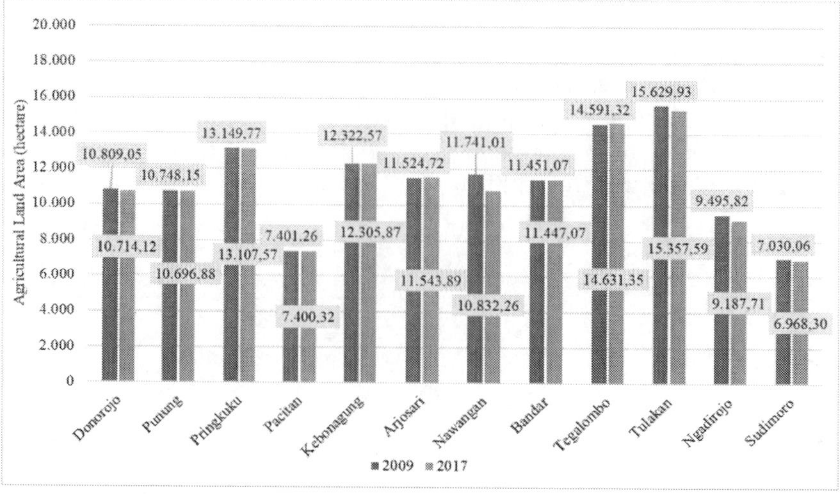

Figure 3. Comparison of agricultural land areas in Pacitan Regency, 2009 and 2017.

Figure 3 shows a graph of the agricultural land area in the form of the total area of paddy fields and dry land agricultural in 2009 and 2017 in all sub-districts in Pacitan Regency. Based on the graph, the largest agricultural land area is in the sub-district of Tulakan, Tegalombo, and Pringkuku. This condition is caused by climatic conditions that are suitable for agricultural land. The lowest area of agricultural land is in the Sudimoro sub-district, followed by Pacitan sub-district. Sudimoro sub-district has a low area of agricultural land because this area is the sub-district with the smallest area. However, Pacitan sub-district has a lower agricultural land area than other sub-district. This condition is caused by Pacitan sub-district, the capital of Pacitan Regency, so most of the population is concentrated in Pacitan sub-district, causing the available land to be widely used for built-up land for economic activities and settlements.

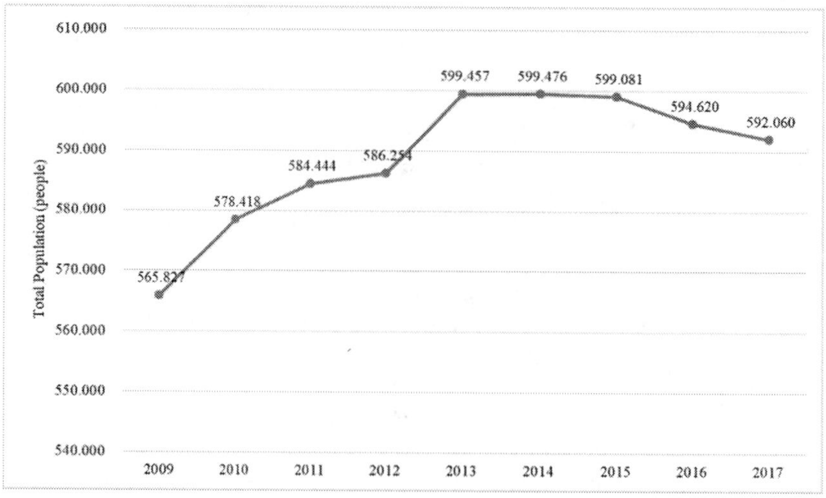

Figure 4. Total population in Pacitan Regency, 2009-2017.

Overall, based on the data, although the area of agricultural land, especially paddy fields, there is no significant decrease in land area. However, the conditions in the field show many changes in paddy fields into built-up land. This condition is due to an increase in the population. Figure 4 shows an increase in the population in Pacitan Regency from 2009-2017. The increasing population does not rule out the possibility of causing an increase in non-agricultural land. And this condition can continue to occur in the coming years. In addition, in Pacitan Regency, Jalan Lintas Selatan (JLS) is constructed by converting agricultural land, including paddy fields and dry land agriculture, into built-up land for roads. Data on the agricultural land area, which tends to be stagnant and different from the conditions in the field, is because the standard area of land under the authority of the National Land Agency (BPN) has not been updated.

3.2 Food Sufficiency Conditions in Pacitan Regency
Food sufficiency is characterized by the sufficient availability of food needed for each and takes place sustainably [16]. Measuring food sufficiency cannot be separated from the aspect of food needs using the ratio of food availability and normative consumption. Normative consumption is defined as the amount of food a person consumes daily to obtain energy [17]. Food crops that produce carbohydrates as an energy source are rice, corn, cassava, and sweet potatoes, called cereal food crops. The four cereal crops are used as parameters to determine food sufficiency. This parameter is based on the fact that Indonesia's food consumption pattern shows that almost 50 percent of the total caloric needs come from cereal crops [18].

Table 3. Food sufficiency in pacitan regency based on food availability ratio, 2009-2017

Year	Land Area (ha)		Food Production (ton/year)				Food Availability (gr/capita/day)	Food Sufficiency	Desc.
	Paddy Fields	Dry Land Agriculture	Rice	Corn	Cassava	Sweet Potato			
2009	13.067,67	122.827,06	153.814	121.015	636.245	746	1.943,46	6,48	Surplus
2010	13.015,26	122.827,06	166.922	123.595	511.735	781	1.765,62	5,89	Surplus
2011	13.015,26	122.368,55	170.257	107.449	564.430	801	1.758,75	5,86	Surplus
2012	12.841,41	121.485,06	179.040	138.297	499.730	1.355	1.801,82	6,01	Surplus
2013	12.828,32	121.725,50	198.169	101.683	361.432	1.094	1.476,22	4,92	Surplus
2014	12.832,67	121.381,86	180.662	111.640	539.420	550	1.713,21	5,71	Surplus
2015	12.832,67	121.381,86	202.619	115.512	451.265	254	1.657,92	5,53	Surplus
2016	12.836,67	121.373,86	223.946	122.639	288.162	185	1.510,47	5,03	Surplus
2017	12.819,07	121.373,86	223.860	110.126	318.234	196	1.512,66	5,04	Surplus
Total	116.089	1.096.745	1.699.289	1.051.956	4.170.653	5.962	15.140,12	50,47	-
Avrg.	12.898,78	121.860,52	188.809,89	116.884,00	463.405,89	662,44	1.682,24	5,61	-

Based on Table 3, the value of food sufficiency fluctuated in specific years, namely in 2011 by 5.86, increasing to 6.01 in 2012, then decreasing again in 2013 to 4.92, and in 2015 increasing to 5.71, while in other years it decreased. Despite the decline, food sufficiency in Pacitan Regency from 2009-2017 is still in a food surplus condition with an average food sufficiency value of 5.61. This food sufficiency is reviewed through the value of the ratio of food availability to normative consumption, with an average food availability ratio of 1,682.24 grams/capita/day. This condition shows that the food needs of the population in the Pacitan Regency are still fulfilled through domestic production from this region. Table 3 also illustrates that although the area of rice fields in Pacitan Regency is not large enough, the food sufficiency is still fulfilled. Food production in the Pacitan Regency comes not only from rice commodities but also from corn and sweet potatoes. Corn and cassava commodities have a reasonably large production yearly, so the value of cereal food sufficiency is still classified as a surplus.

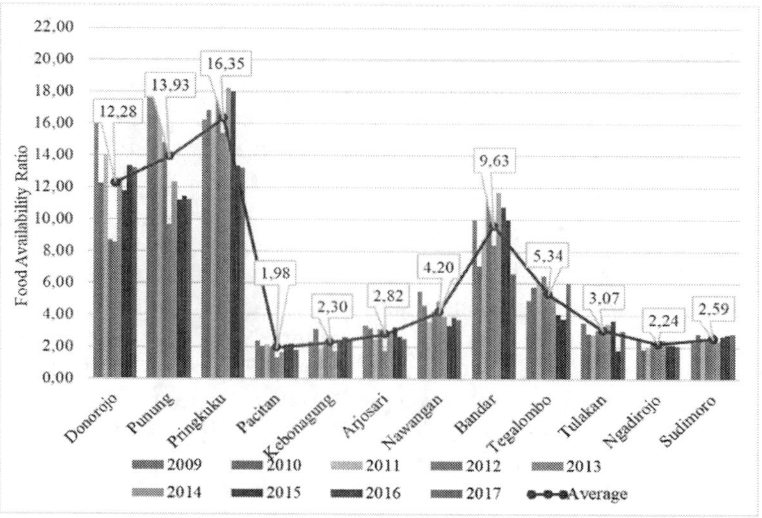

Figure 5. Food sufficiency level of each sub-district in Pacitan Regency, 2009-2017.

Figure 5 shows the condition of food sufficiency for each sub-district in Pacitan Regency, with the highest level of food sufficiency in Pringkuku sub-district, then Punung and Donorojo sub-districts with an average of 16.35 respectively; 13.93; and 12.28. The lowest food sufficiency levels are Pacitan and Ngadirojo sub-districts, with an average of 1.98 and 2.24. The area with the highest food sufficiency (Figure 5) is not in the sub-district with the largest area of agricultural land (Figure 3). Tulakan and Tegalombo sub-districts have the largest agricultural land area, but the highest average food sufficiency is in Pringkuku, Punung, and Donorojo sub-districts.

The low food sufficiency in Pacitan sub-district is influenced by the low food production, considering that this area is the capital of Pacitan Regency. Land use in Pacitan sub-district is widely used for central government, economic activities, and housing. In addition, there are factors in the area's physical condition, where the Pacitan sub-district has a flat topography, which causes the concentration of the Pacitan Regency population to live in this area. The condition is different in the Ngadirojo sub-district, which has a lower food availability ratio than other sub-districts in Pacitan Regency. This difference is influenced by the physical condition of Ngadirojo sub-district, where almost 90% of its area is dry land in the form of forests and dry fields [19].

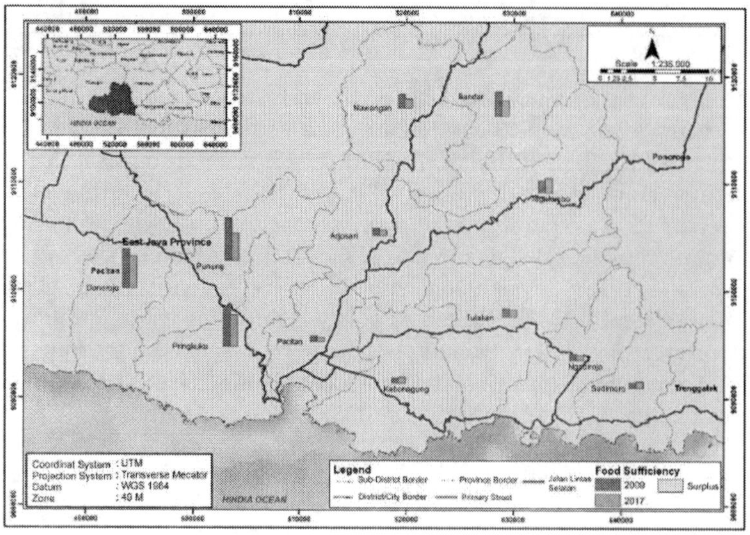

Figure 6. Comparison of food sufficiency map of Pacitan Regency in 2009 and 2017.

Figure 6 compares food sufficiency between 2009 and 2017 in all sub-districts of Pacitan Regency. Spatially, food sufficiency in the western region (Sub-districts of Donorojo, Punung, and Pringkuku) is higher than in the eastern and northern regions. The high food sufficiency in the western region is due to higher food production than in other districts. Not only in the production of corn, sweet potato, and cassava commodities but also rice commodities. It should be noted that the three sub-districts are classified as having a smaller area of rice fields compared to other areas but having a larger rice production. This condition is caused by rice grown in the western region dominated by upland rice or rice grown on dry agricultural land. Therefore, although the western region is dominated by dry agricultural land, rice production is quite large compared to areas with larger paddy fields. This condition is supported by the availability of production facilities and infrastructure, one of which is the existence of reservoirs in the area.

3.3 The Relationship between Changes in Land Use and Food Sufficiency in Pacitan Regency

The description in the previous section shows that the sub-district with the highest average food sufficiency is not the sub-district with the highest agricultural land area. Therefore, this study also examines the relationship between land use and food sufficiency through the Pearson correlation. The variables used are the extent of changes in agricultural land use as the independent variable and the value of food sufficiency as the dependent variable.

Table 4. The results of the pearson correlation test between the area of change in land use and food in Pacitan Regency in 2009-2017

Regency	Correlation	Significance
Donorojo	0,390	0,300
Punung	-0,041	0,917
Pringkuku	-0,077	0,844
Pacitan	0,385	0,306
Kebonagung	-0,494	0,177
Arjosari	-0,323	0,396
Nawangan	0,148	0,704
Bandar	-0,066	0,865
Tegalombo	0,225	0,561
Tulakan	-0,090	0,818
Ngadirojo	-0,254	0,510
Sudimoro	.790*	0,011
Kabupaten Pacitan	-0,417	0,264

*. Correlation is significant at the 0.05 level (2-tailed).

Based on the Pearson correlation test results in Table 5.11, the area agricultural land use change and food sufficiency in Pacitan Regency has a correlation value of -0.417 and a significance value of 0.264. However, the significance value is > 0.05, so H0 is rejected. Therefore, it can be concluded that there is no relationship between land use change and food sufficiency in Pacitan Regency. Likewise, the 11 sub-districts in this region do not have a strong and significant relationship with food sufficiency because the significance value is > 0.05. However, one sub-district has a strong positive relationship, namely Sudimoro sub-district, with a correlation value of 0.790 and a significance level of 0.011. If the significance value is <0.05, then H0 is accepted. This result means that changes in agricultural land use that occur in Sudimoro sub-district are related to the level of food sufficiency. This positive relationship shows that the conversion of agricultural land in Sudimoro Sub-district has experienced smaller or near negative fluctuations in the area of change and is related to food sufficiency, which has also decreased.

The results of the Pearson correlation test in almost all sub-districts in Pacitan Regency do not have a strong and significant correlation between changes in land use and food sufficiency. This condition is certainly different from the general theory and previous studies that changes in agricultural land use will cause a decrease in food production [20]. However, the conditions that occur are the area of agricultural land that has not experienced a large enough change, and food sufficiency has decreased even though it is still classified as a food surplus. This condition can occur due to various factors such as the increasing number of people, changes in the cultivation of food crops into woody plants, and climate change which causes limited planting time and decreased land productivity. The decline in food sufficiency yearly does not rule out the possibility of a decline in the next few years to a deficit condition. Therefore, there needs to be more attention from the Pacitan Regency government to maintain the condition of food sufficiency in the region. The hope is that food sufficiency will remain in surplus, one of which is through increasing land productivity so that the production of food commodities is maintained and the population's food supply remains fulfilled.

4. Conclusions

Agricultural land use in Pacitan Regency is dominated by dry land agricultural. Paddy fields and dry land agricultural from 2009-2017 have decreased, with a rice field area of 248.60 hectares and a dry land agricultural area of 1,453.20 hectares. On the other hand, non-agricultural land has increased by 1,351.99 hectares over nine years. The results of the calculation of food sufficiency in 2009-2017 show that the food sufficiency of Pacitan Regency is still classified as a food surplus. However, it fluctuates and tends to decrease, so more attention is needed by the government related to efforts to increase local food production. Based on the Pearson correlation, it was found that there is no strong and significant correlation between agricultural land use change and food sufficiency in Pacitan Regency. The conditions indicate the area of agricultural land that tends to stagnate and fluctuates in food sufficiency. It can be caused by population growth, changes in the cultivation of food crops into woody plants, climate change which causes limited planting time, and decreased land productivity.

References

[1] L. K. B. Sembanyang, "Analisis Keterkaitan Ketersediaan Infrastruktur Dengan Pertumbuhan Ekonomi Di Indonesia: Pendekatan analisis Granger Causality," JEJAK: Jurnal Ekonomi dan Kebijakan, 4(1). Mountain Research and Development, vol. 26, no. 3, 2011.

[2] H. S. Yunus, Struktur Tata Ruang Kota. Yogyakarta: Pustaka Pelajar, 2000.

[3] D. Djoni, S. Suprianto, and E. Cahrial, "Kajian alih fungsi lahan pertanian pangan di Kota Tasikmalaya," Mimbar Agribisnis: Jurnal Pemikiran Masyarakat Ilmiah Berwawasan Agribisnis, vol. 1, no. 3, pp. 233-244, 2018.

[4] W. A. Saputro and Y. Fidayani, "Faktor-faktor yang mempengaruhi ketahanan pangan rumah tangga petani di Kabupaten Klaten," In Agrica (Jurnal Agribisnis Sumatera Utara), vol. 13, no. 2, 2020.

[5] M. Fashidullisan, Mukodi, and S. Sugiyono, Pacitan dalam Badai Perubahan Analisis Dampak Pembangunan Jalan Lintas Selatan. Yogyakarta: Dialektika, 2018.

[6] Sunanto and A. W. Rauf, "Respon petani terhadap pelaksanaan display padi gogo VUB pada lahan sub optimal," In Jurnal Sosial Ekonomi Pertanian, vol. 14, no. 2, 2018.

[7] A. Mulyani, and F. Agus, "Kebutuhan dan ketersediaan lahan cadangan untuk mewujudkan cita-cita Indonesia sebagai lumbung pangan dunia tahun 2045," Analisis Kebijakan Pertanian, vol. 15, no. 1, 2017.

[8] Badan Ketahanan Pangan, Sistem Kewaspadaan Pangan dan Gizi Tingkat Kabupaten/Kota. Jakarta: BKP Kementerian Pertanian RI, 2014.

[9] R. Harini, H. N. Handayani, and F. R. Ramdani, "Food security assessment in the coastal area of Demak Regency," In IOP Conference Series: Earth and Environmental Science," vol. 148, no. 1, 2018.

[10] F. I. Rahmatullah and L. Muta'ali, "Analisis swasembada beras di Provinsi Jawa Tengah tahun 2005–2014," Jurnal Bumi Indonesia, vol. 6, no. 1, 2017.

[11] S. W. Ani, "Tren alih fungsi lahan pertanian di Kabupaten Klaten," Sepa: Jurnal Sosial Ekonomi Pertanian dan Agribisnis, vol. 8, no. 2, 2012.

[12] S. Sulastoro, "Karakteristik sumberdaya air di daerah karst (studi kasus daerah Pracimantoro)," Journal of Rural and Development, vol. 4, no. 1, 2013.

[13] H. Noviyanto, "Evaluasi dampak program kawasan rumah pangan lestari Kabupaten Pacitan," JI, vol. 10, no. 1, 2021.

[14] BPS, Kabupaten Pacitan dalam Angka. Pacitan: BPS, 2018.

[15] E. Antriyandarti, R. U. Fajarningsih, S. Marwanti, S. Supardi, J. Sutrisno, M. Ferichani, and R. Khairiyakh, "Poverty alleviation system of dryland farm community in karst mountains Gunungkidul, Indonesia," In IOP Conference Series: Earth and Environmental Science, vol. 200, no. 1, 2018.

[16] M. Gardjito, A. Djuwardi, and E. Harmayani, Pangan Nusantara: Karakteristik dan Prospek untuk Percepatan Diversifikasi Pangan. Jakarta: Kencana, 2013.

[17] D. T. Kurniawan, "Ketersediaan pangan sebagai indikator peta ketahanan dan kerentanan pangan," Journal of Research and Technology, vol. 1, no. 1, pp. 45-51, 2015.

[18] M. Rusliyadi, B. Hj. M. J. Azahraini, and R. T. Kumalasari, "Analysis of household food security policy: case of food security village programme, Indonesia," Asian Journal of Agriculture and Rural Development, vol. 9, no. 1, pp. 19-32, 2019.

[19] A. Jauhari, "Pemanfaatan sig untuk pemetaan kawasan produksi komoditas unggulan tanaman pangan di Kabupaten Pacitan," In Journal of Regional and Rural Development Planning (Jurnal Perencanaan Pembangunan Wilayah dan Perdesaan), vol. 4, no. 3, 2020.

[20] I. M. Y. Prasada, and T. A. Rosa. "Dampak alih fungsi lahan sawah terhadap ketahanan pangan di Daerah Istimewa Yogyakarta," Jurnal Sosial Ekonomi Pertanian, vol. 14, no. 3, 210, 2018.

IOP Conf. Series: Earth and Environmental Science 1233 (2023) 012045 doi:10.1088/1755-1315/1233/1/012045

Contribution of the Agricultural Sector in Green Space Sleman Regency

Rika Harini[1]*, Isnaini Dairina[1], Khrisna Wasista Widantara[1], Ragil Setiawan[1], Marinda Mustika Ayu[1], Dimas Maulana Ichsan [2]

[1] Department of Environmental Geography, Faculty of Geography, Gadjah Mada University, Indonesia
[2] Department of Geographic Information Science, Faculty of Geography, Gadjah Mada University, Indonesia

*Corresponding author: rikaharini@mail.ugm.ac.id

Abstract. Sleman Regency is the largest producer of agricultural commodities in Yogyakarta Province. This study aims to examine the development of the agricultural sector's contribution, identify the spatial distribution of the agricultural sector as part of green space, and analyze the trend of the agricultural sector's contribution to the area of green space in Sleman Regency. The data used is secondary data in the form of GRDP data from BPS and land use area from BAPPEDA. The results show that income from the agricultural sector has increased. Still, its contribution to GRDP has decreased, as evidenced by the analysis of location quotients (LQ) as a non-basic sector. The area of agricultural land, including shrubland, vacant land, irrigated rice fields, rainfed rice fields, gardens, forest, and moor area, has decreased to 58.51% of the total area. This change has been caused by population growth and massive urbanization over the last few years and shows that the role of the agricultural sector in green space has decreased but still does not exceed the limit set by law.

Keywords: agriculture sectoral, green space, location quotients, GDRP, land use area

1. Introduction

The agricultural sector is one of the economic sectors in Indonesia, which is still a source of livelihood for the community. However, the agricultural sector is one of the sectors that is negatively affected by economic development [1]. This is because indirectly land use will be related to socio-economic activities. Depreciation of agricultural land area in addition to having a direct impact on the decline in food production will also have an impact on reducing green space in an area. The decrease in the area of green spaces can cause severe problems for the condition of regional spatial planning and environmental quality. The existence of green space is essential because it has various benefits for environmental sustainability and is a place for social interaction in the community. Green space also has the potential as a green infrastructure (GI) to address urban challenges such as climate change and contribute to sustainable development and increasing urban resilience [2][3].

Sleman Regency is one of the regencies in the DIY Province. Administratively, Sleman Regency consists of 17 sub-districts, most of which are rural areas where many agricultural activities are carried out. Sub-districts bordering the city of Yogyakarta, such as Depok, for agricultural activities are relatively low compared to other economic activities (home industry, trade, and services). In addition,

Ngaglik and Mlati sub-districts have a tendency to significantly reduce agricultural land area compared to other sub-districts in Sleman Regency. The limited availability of land in the city center and high prices will cause urban development to tend to expand from suburban areas to rural areas, followed by various urban functions and activities.

Regional development in Sleman Regency has increased quite rapidly and is directly proportional to the increase in population. Population growth in Sleman Regency has increased yearly because Sleman Regency has an attraction that causes urbanization. This high population density results in high demand for land for settlements and the construction of infrastructure and facilities that can cause environmental problems [4]. The availability of land that cannot be increased causes forests and agricultural land to be converted into built-up land (city) [5]. The occurrence of land use change can be made possible due to a lack of intensive attention to the agricultural sector by the local government so that people turn to other sectors such as industry, trade, or services[6].

The development of an area has positive and negative impacts on various aspects, one of which is the environmental aspect. Population growth that affects the need for land will directly affect the availability of green space in an area. The importance of the need for green space in an area is emphasized in Law Number 26 of 2007 concerning Spatial Planning to realize a safe, comfortable, productive, and sustainable urban area. The existence of green space is essential because it has various benefits for environmental sustainability and is a place for social interaction in the community. Green space also has the potential as green infrastructure (GI) to address urban challenges such as climate change and contribute to sustainable development and increasing urban resilience [7][8][9][10].

The availability of the agricultural sector as a part of green space in Sleman Regency needs to be considered to maintain ecological balance. The conversion of agricultural land is important to be controlled by Law Number 41 of 2009 concerning the Protection of Sustainable Food Agricultural Land (LP2B). The protection of agricultural land in Sleman Regency is stipulated in the Sleman Regency Regional Spatial Plan (RTRW) for 2011-2031, in which the level of availability, both quality, and quantity, must always be taken into account in the urban planning process. This is because green spaces have many ecological benefits by providing ecosystem services and contributing to human well-being, thereby improving the quality of life [11]. This issue has become one of the strategic issues for Sleman Regency to harmonize spatial planning, development, and environmental conservation.

This study describes the spatial distribution of agricultural land as part of green space along with the economic contribution analysis using LQ analysis as a representation of agricultural land productivity. This information can serve as a reference in spatial planning and agriculture. The availability of green space, especially in the sector in Sleman Regency, needs to be studied to maintain ecological balance. In addition, this study can also support the evaluation and monitoring of the achievement of SDGs in goals 2 (zero hunger) and 15 (life on land). From the description of the existing problems, the objectives of this study can be formulated, namely to examine the development of the contribution of the agricultural sector, determine the spatial distribution of the agricultural sector as part of green space, and analyze the trend of the contribution of the agricultural sector to the area of green space in Sleman Regency.

2. Method

2.1. *Data collection*

The method used in this study is quantitative research, which is defined as a research method based on the philosophy of positivism, used to examine specific populations or samples, data collection using research instruments, and quantitative/statistical data analysis, with the aim of testing hypotheses that have been developed set [12]. Meanwhile, the determination of the research area is carried out using a purposive method, namely, the selection of research areas is carried out according to specific considerations where Sleman Regency is one of the regencies in the Special Region of Yogyakarta which is experiencing regional development which will have a negative impact on the environment, especially in the agricultural sector. The percentage of quantity and quality of the environment is not considered with the widespread conversion of agricultural land into built-up land.

The main data used in this study is secondary data. Secondary data were obtained from related institutions such as the Central Statistics Agency (BPS) and the Regional Development Planning Agency (BAPPEDA). The types and sources of data collected can be seen in Table 1.

Table 1. Types and Sources of Data

No.	Data	Data source
1	Geographical Condition	Central Bureau of Statistics
2	Residential Condition	Central Bureau of Statistics
3	Land Use	Development Planning Agency at Sub-National Level

Source: Author, 2022

2.2. Data Analysis

2.2.1. Location Quotient. Location Quotient (LQ) method aims to identify leading sectors. The LQ. technique is widely used to discuss economic conditions, leading to the identification of specialization of economic activity or measuring the relative concentration of economic activity to obtain an overview in determining the leading sector as the leading sector of industrial economic activity. The LQ calculation formula is as follow s [13]:

$$LQ = \frac{psi/pl}{PSi/PL} \quad \ldots (1)$$

Information:
psi = GRDP value of sector i district
pl = total value of district GRDP
PSi = GRDP value of sector i of the region at the top level
PL = total value of regional GRDP at the top level

Table 2. Definition of LQ Score

LQ value	Classification of Base and Non-Based Sectors	Specialization	Market share
LQ >1	Base and leading sectors	Specialized sectors	Can serve the needs of the market inside and outside the region (export)
LQ = 1	Sector balanced by reference area	Sector is the same as the reference area	Only able to meet the needs within the region (non-export)
LQ<1	Non-basic, non-seeded, and non-potential sectors	Unspecialized sector	Not able to meet market needs outside the region (non-export)

Source: Muta'ali, 2015

2.2.2. Land use change. Land area was calculated by Development Planning Agency at Sub-National Level. Secondary data was generated from image interpretation in 2010 and 2020. Land area was obtained through geometric calculations to produce land area figures in hectares. Changes in the land area are calculated by subtracting the 2020 and 2010 land areas based on the type of land use.

IOP Conf. Series: Earth and Environmental Science 1233 (2023) 012045 doi:10.1088/1755-1315/1233/1/012045

3. Results and Discussion

3.1. Description of Research Site

3.1.1. Physical condition. Geomorphologically, Sleman Regency is located in the form of volcanic and fluvial processes. The northern part is an area with highlands and mountains, while the southern part is a low-lying area traversed by many rivers [14], such as the Progo River, Krasak River, Sempor River, Nyoho River, and Boyong River. The abundant water supply apart from rivers that pass through Sleman Regency is also supported by a wet tropical climate with average rainfall in 10 years (2010-2020) of 2028 mm/year. The level of rainfall is very supportive of crop production [15]. The type of regosol soil that dominates most of the Sleman Regency makes the agricultural sector has a high potential as a supplier of agricultural products in the DI Yogyakarta.

Sleman Regency has an area of 57,482 ha divided into 17 districts (Figure 1). The length of productive land (paddy fields and fields) is 28,434.81 ha or 49.5% of the total land area in 2018 [16]. However, the land has decreased from 2010, where productive land in Sleman Regency has an area of up to 29,091.93 ha or 50.6% of the total land area [17]. This shows a change in the function of productive land to other land uses caused by various factors. Factors that can cause land use change include population, number of industries, and efforts to increase Gross Regional Domestic Product (GRDP) [18]. An increase in population will certainly make the need for residential areas higher so that residential areas will continue to increase.

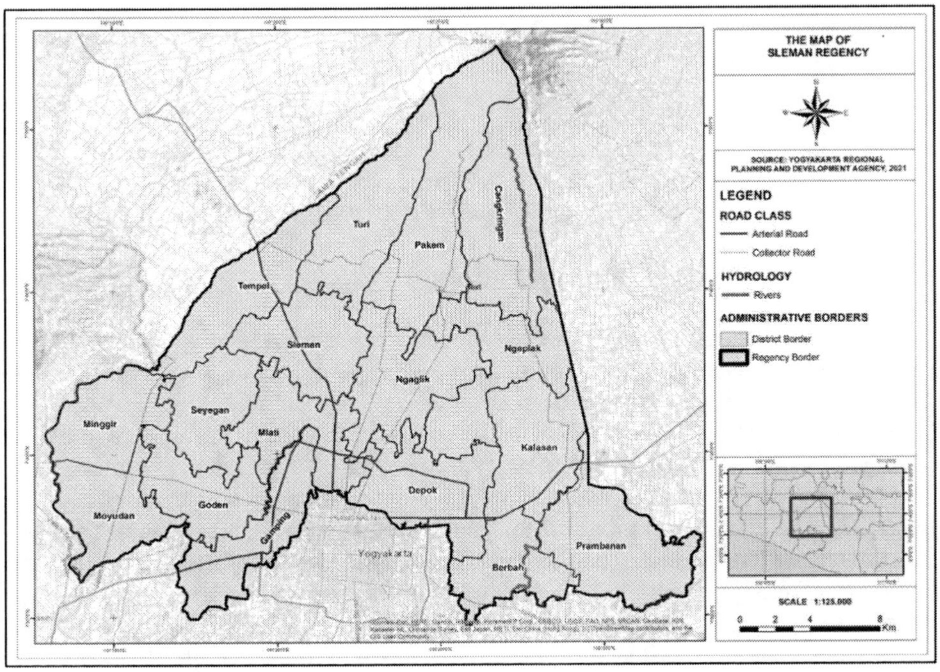

Figure 1. Map of Research Location
Source: Yogyakarta Regional Planning and Development Agency, 2021

3.1.2. Socio-Economic and Population Conditions. Sleman Regency has a population of 1,093,110 people in 2010 and increased to 1,125,804 people. The population in Sleman Regency has continued to increase by +3% [19] over the last ten years. This increase is quite low, and this is due to the total fertility rate (TFR) value of Sleman Regency, which was only 1.7 in 2019

[20]. This increase in population has become the capital as well as an actor in economic development.

Population growth is not only an opportunity for a demographic transition but also a transition to the dominant labor sector in Sleman Regency. Based on Figure 2, from 2010-2020, there was a transition from primary to secondary and tertiary sectors. The primary sector includes the agricultural and mining sectors. Meanwhile, the secondary sector, which consists of water, electricity, and gas processing and supply, has increased. The primary sector is experiencing a downward trend. One of the factors is that many workers switch to the secondary and tertiary sectors. The causes of this shift are, among others, income outside the non-agricultural sector which is greater than the agricultural sector, a negative image of agriculture, increased education, narrow ownership of agricultural land, and rural convenience accessibility [21].

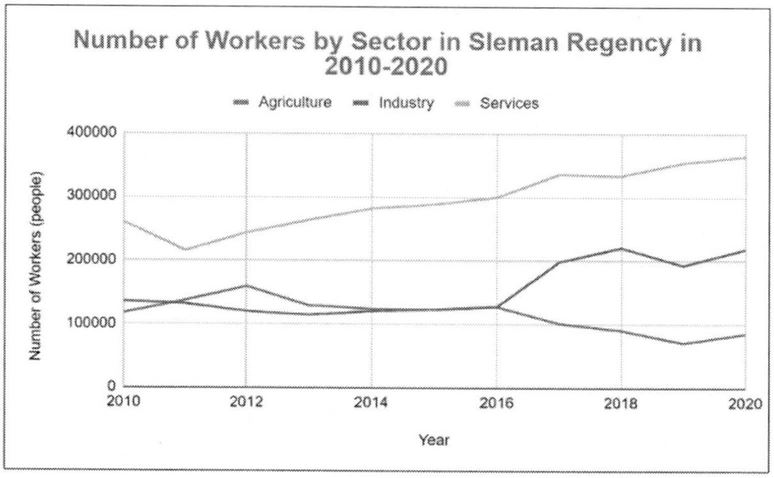

Figure 2. Number of Workers by Sectors in 2010-2020
Source: BPS Sleman Regency 2010-2021, Data Processing (2022)

3.2. Spatial distribution of the existing agricultural sector as part of green space in Sleman Regency

Agricultural land in Sleman Regency experienced a decrease in the area of agricultural land due to land conversion. Massive land use change in Sleman Regency occurs due to urbanization. Urbanization does not only mean changes in land use from rural areas to urban areas but is also related to changes in livelihoods and patterns of population activities, such as from farming to other livelihoods [22]. Population pressure in Sleman Regency can occur due to the many attractions such as housing destinations, tourist destinations, and cultural developments, as well as campus centers and educational facilities [23]. The high rate of urbanization is a consequence of the location of Sleman Regency, which borders the city of Yogyakarta. The Sleman Regency area, directly adjacent to the city of Yogyakarta, tends to be an alternative residential land with relatively more affordable prices and is still close to the city center.

Changes in land use in Sleman Regency currently occur as a result of regional developments that occur. This can be seen by comparing land use patterns in 2010 and 2020, as shown in Figure 3 and Figure 4. Based on Table 3, shrubs, open land, irrigated rice fields, rainfed rice fields, and grasslands experienced a decrease in the area, with an average reduction of 11.75%. The most significant decrease in the area occurred in irrigated rice fields, namely 5.94%. The construction of the Yogyakarta "Ring Road" (an Arterial Road), which has been fully utilized starting in 2010, provides access to lands that were formerly irrigated agricultural land.

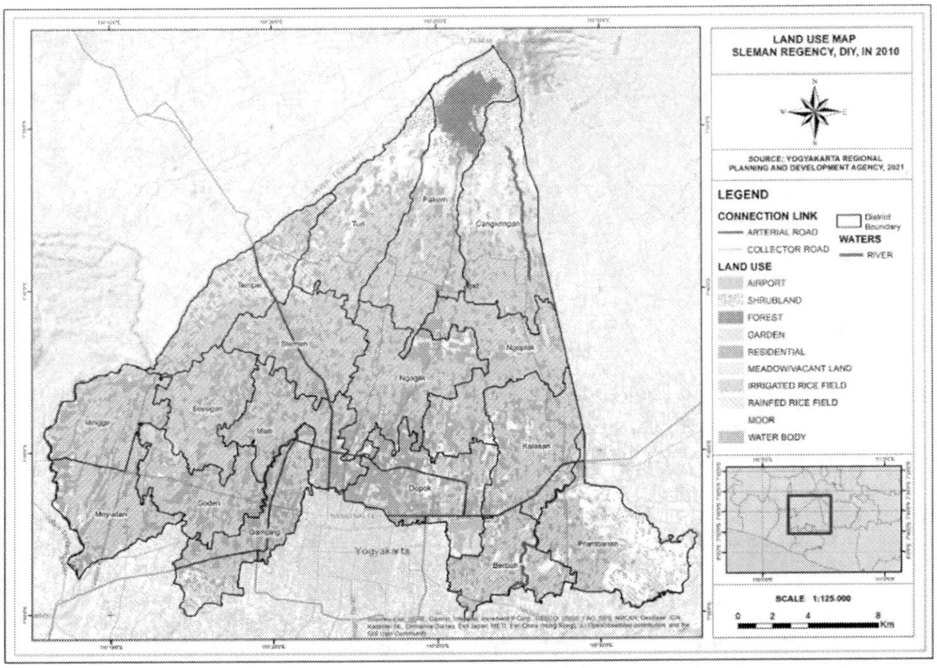

Figure 3. Land Use Map of Sleman Regency in 2010
Source: Regional Development Planning Agency (Bappeda), 2010 (Processed, 2022)

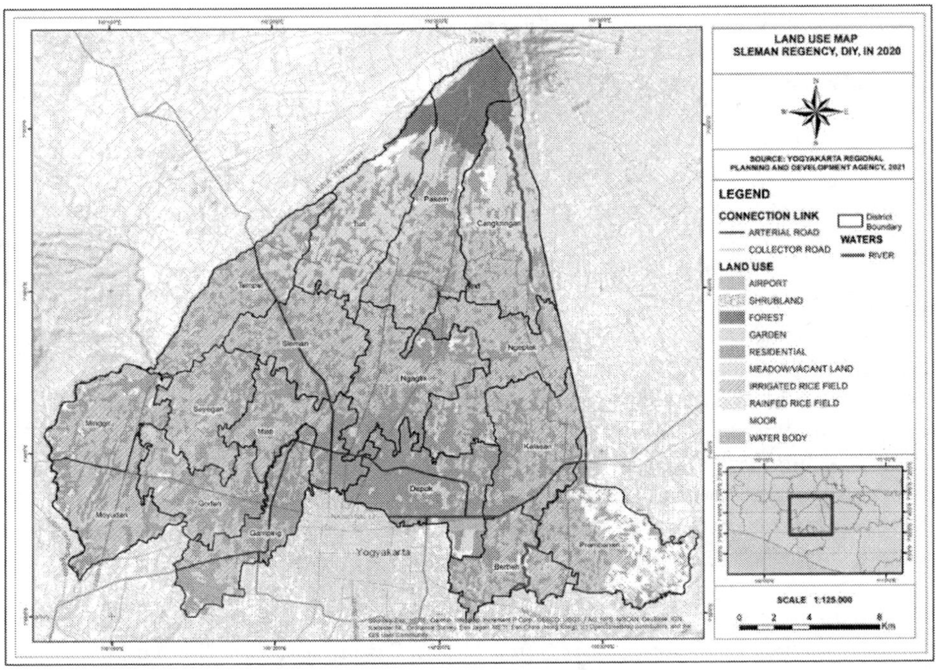

Figure 4. Land Use Map of Sleman Regency in 2020
Source: Regional Development Planning Agency (Bappeda), 2020 (Processed, 2022)

Table 3. Land Use Change of Sleman Regency in 2010 and 2020

Land Use	Land Area (ha)		Percentage (%)		Change
	2010	2020	2010	2020	
Airport	184	137	0.32	0.24	-0.08
Shurbland	1200	114	2.1	0.2	-1.9
Forest	933	1907	1.63	3.33	1.69
Garden	5172	6359	9.06	11.09	2.03
Residential	16630	23169	29.13	40.41	11.28
Meadow/Vacant Land	819	210	1.43	0.37	-1.07
Irrigated Rice Field	25504	22210	44.68	38.74	-5.94
Rainfed Rice Field	1571	276	2.75	0.48	-2.27
Moor	4907	2472	8.6	4.31	-4.28
Water Body	166	483	0.29	0.84	0.55

Source: Author, 2022

Other land uses experiencing rapid changes are agricultural and residential areas. Agricultural areas consist of paddy fields (irrigated rice fields, rainfed rice fields) and non-rice fields, which include gardens, dry meadows, community forests, temporarily uncultivated land, and other non-paddy agricultural lands [24]. Overall, there is a tendency to change the land use from agricultural to residential land. Agricultural land is evenly distributed in most areas of Sleman Regency to the area bordering the TNGM area in the north [25]. Areas still dominated by agricultural land in 2020 include Turi District, Pakem District, Cangkringan District, Tempel District, Sleman District, Seyegan District, Minggir District, Moyudan District, Ngeplak District, Kalasan District, Berbah District, and Prambanan District.

Meanwhile, other sub-district areas tend to be dominated by residential land. Residential lands are spread mainly along roads, with the most densely populated regions in areas directly adjacent to the city of Yogyakarta, such as Depok District, Mlati District, and Gamping District. Based on previous research [26], the development of universities, hospitals, government offices, shopping malls, and residential buildings was built in this decade. This accelerates the spread of urban development (urban sprawl) in the Sleman Regency. Areas traversed by the ring road tend to develop into denser residential areas, such as Ngaglik District.

The decrease in agricultural land due to land changes brings various impacts that can arise. The decreasing area of agricultural land causes a reduction in the area of green space. In fact, green space plays an essential role in maintaining and improving the urban landscape as well as environmental sustainability from high population density [27]. Green space can also provide ecological services such as preventing flooding, reducing air pollution, and maintaining ambient air temperature [28]. The lack of agricultural area, especially in densely populated areas such as Depok District and Mlati District, causes the minimum area of green space to be found. This is a spatial planning problem due to urbanization in suburban areas, which is weak in its implementation.

3.3. Development of Agricultural Sector Contribution in Development in Sleman Regency

The agricultural sector is a strategic sector that has an essential role in the national economy and the community's survival, especially in contributing to GDP, employment, and domestic food [29][30]. The agricultural sector is the sector that has proven to be the most capable of supporting the national economy in critical conditions [31]. Likewise, what was explained by the Sleman Regent in 2021 was that until now, Sleman was able to contribute 40% of the local rice supply. During the Covid-19 pandemic, the agricultural sector was sector that was able to survive compared to other sectors. Figure 5 shows that the income generated from the agricultural sector has an increasing trend in the last decade. Three agricultural sub-sectors have become prominent sectors and can export part of their production for seven years, namely horticultural crops, livestock, and agricultural and hunting services [31].

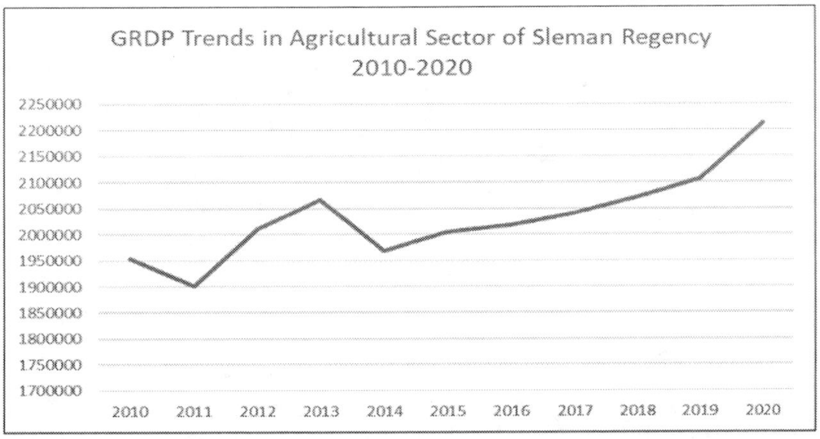

Figure 5. GRDP Trends in Agricultural Sector of Sleman Regency, 2010-2020
Source: Central Bureau of Statistics of Sleman Regency, 2011-2021

This study further discusses the magnitude of the role of the agricultural sector in economic development in the Sleman Regency. The theory of economic basis suggests that the determinants of economic growth in a region are closely related to the demand for goods and services from outside the area or the magnitude of the increase in exports from the region [32]. The economic sector is grouped into two, namely the basic and non-basic sectors. The Location Quotients (LQ) approach identifies the two sectors by calculating the GRDP. A sector includes the base sector when LQ >1 and the non-basic sector when the value is LQ <1. The sector that can meet the needs of its region and other regions is called the base sector. Meanwhile, sectors that cannot meet domestic needs, let alone non-domestic ones, are called non-basic sectors.

The dynamics of the development of the economic sector during 2010-2020 in Sleman Regency are shown in Figure 6. The agriculture, forestry, and fisheries sectors have been non-basic for the last decade. The LQ value of the sector is ranked the lowest in the bottom three. This shows that based on the location quotients approach, the agricultural sector has not been able to meet the needs of Sleman Regency, as evidenced by the low income of the district's GRDP. This lower agricultural land will also affect our food security, so in 2039 it is estimated that DIY will experience a food security crisis.

Location quotients (LQ) analysis shows that the role of the agricultural sector in Sleman Regency during the last decade has decreased. Broadly speaking, the trend of the LQ value of the agricultural sector in Sleman Regency is decreasing. The decline in the agricultural sector's contribution to economic development is caused by the change in livelihoods and activities of the local population. Population growth and changes in the structure of the economy cause land downloads from rice fields to non-rice fields (BPS, 2018). Sleman Regency has experienced an economic transformation in the last decade from the agricultural sector to the service sector, such as companies, construction, and education. The share of the contribution declines as one moves up from the lowest to the highest income per capita. [33]. This is evidenced by the more significant contribution of the GRDP contribution from the industrial and service sectors compared to the contribution from the agricultural sector. Based on the vision of the DIY Province in the RPJPD DIY 2005-2025, one of the visions to be achieved by the DIY province is to become a leading education center in Southeast Asia. The presence of 41 universities in Sleman Regency encourages property investment and the establishment of companies in this district.

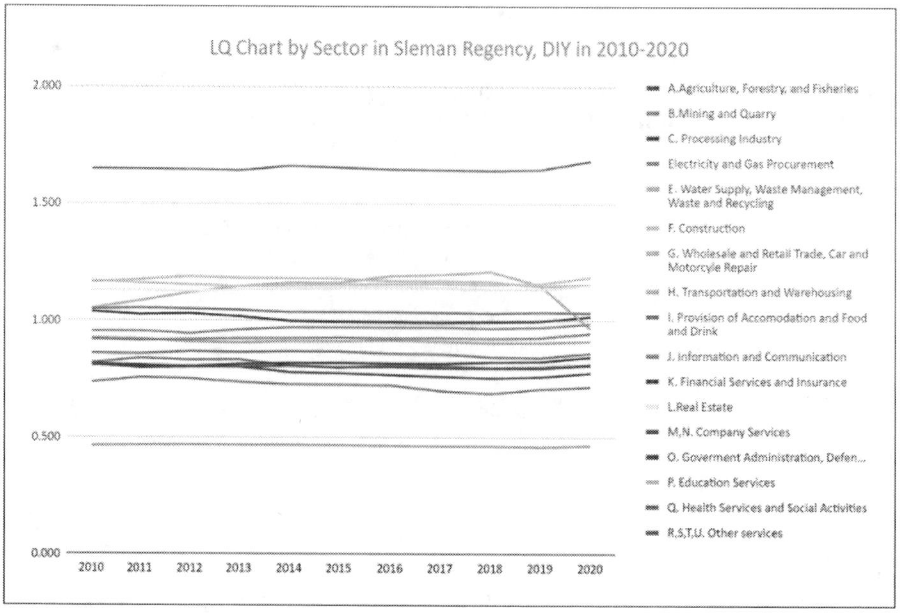

Figure 6. LQ Chart Trends by Sector in Sleman Regency, 2010-2020 Year
Source: Central Bureau of Statistics of Sleman Regency, 2011-2021 (Processed, 2022)

Although the role of the agricultural sector in economic development is less than that of the secondary and service sectors, the agricultural sector in Sleman Regency still has good prospects in the future. Population growth originating from birth and migration every year is followed by increasing demand for food. The increasing global demand for agricultural products is an excellent opportunity to enhance the sector's contribution to GDP and raise incomes, especially for the rural population [34]. Moreover, the increase in the economy in Sleman Regency leads to an increase in revenue which will increase public consumption. Income growth, taste preferences, and population increase will significantly impact demand [35].

3.4. The dynamics of the contribution of the agricultural sector to the area of green space in Sleman Regency

Green space is an elongated area/lane and/or clustered area whose use is more open, where plants grow, both naturally and intentionally planted. Green space has several functions: social and cultural, economic, and aesthetic. The existence of green space has an essential role in maintaining the quality of human life [36][37]. Based on Law no. 26 of 2007 concerning Spatial Planning, the proportion of green space is at least 30% of the total area. The minimum proportion of public green space is 20% of the total area. There are two types of green space based on ownership, namely public green space and private green space. A private green space is a garden or yard, or building owned by the community that is planted with plants, one of which is agricultural land. Based on Table 4, green space in Sleman Regency is 58.51% of the total land area. Meanwhile, the built-up area is 40.95%, and non-green space, namely water bodies, is 0.84%. This area calculation shows that the area of green space in Sleman Regency is still considered sufficient from the minimum limit determined by law.

Table 4. Length of Green space

Area	**Percentage (%)**	
	2010	**2020**
Green space	70.26	58.51
Non Green space	0.29	0.84

Build Land	29.45	40.65

Source: Regional Planning & Development Agency Sleman Regency, 2011-2021 (Processed, 2022)

Agricultural sector land contributes to private green space, which has an economic function. The agricultural sector is included in green economic activities. According to the United Nations Environment Program (UNEP), a green economy is an economy that produces low carbon, is resource-efficient, and is socially inclusive. Green GDP is an indicator of economic growth, with the environmental impact on that growth factored into the traditional GDP [38]. According to the OECD, the agricultural sector has a role in the green economy, such as efficient use of resources, reducing environmental impacts, focusing on the economy, and encouraging research and technology development activities to improve the productivity of resource use.

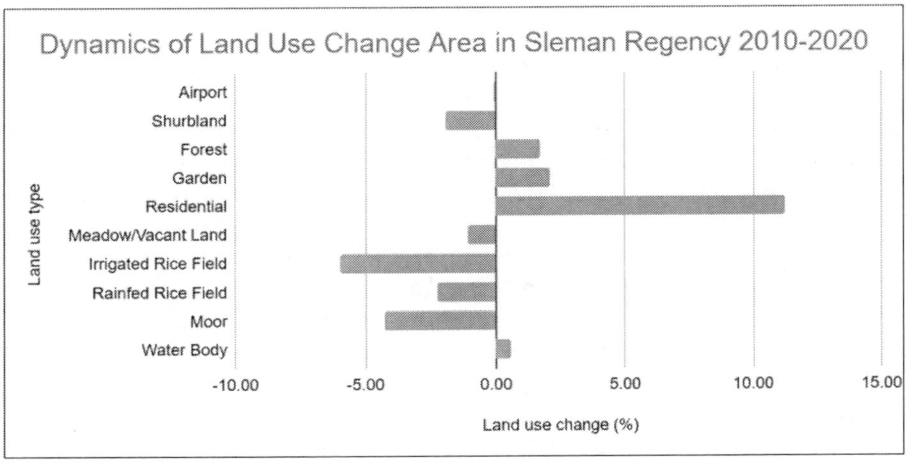

Figure 7. Dynamics of Land Use Change in Sleman Regency, 2010-2020
Source: Regional Planning & Development Agency Sleman Regency, 2011-2021 (Processed, 2022)

Figure 7 shows that the area of agricultural land in Sleman Regency has decreased by 1.9% in shrubland; 1.07% in vacant land; 5.94% in irrigated rice fields, 2.27% in rainfed rice fields, and 4.28% in moor areas over ten years. Otherwise, the forest area was increased by 1.69%, and the garden was increased by 2.03%. This shrinkage was accompanied by a massive increase in residential land area of 11%. Shrinking agricultural land not only reduces food production but also eliminates environmental services whose value is incalculable because if the environment is damaged, human economic activities will stop. Based on the Environmental Kuznets Curve (EKC) hypothesis, which was adapted to the environmental degradation relationship and obtained [39], an increase in per capita income will initially increase environmental degradation until it reaches a turning point. This can be applied to economic development in Sleman Regency, Yogyakarta. The income per capita of Sleman Regency in 2010 was Rp. 12,452,292, and in 2020 it tripled to Rp. 37,180,000 in one year [17][19]. However, during this period, environmental degradation occurred, one of which was reflected in the massive development of built-up land by 13% and the reduction of agricultural land by 8%. Therefore, agricultural growth must be accelerated even though it has limited land and water resources to achieve food security.

Depreciation of agricultural land in Sleman Regency in the last ten years will lead to food security, economy, and ecology problems. The agricultural sector supports the village economy and is a source of livelihood for the poor [40][41]. Meanwhile, the ecological problem is the reduction of green space, which has benefits as an absorber of carbon dioxide (CO_2), oxygen enhancer, lowering the environmental temperature, absorbing rainwater, and reducing noise. Moreover, Sleman Regency is located in the upstream area, so the existence of green space as a rainwater absorber is needed to maintain the stability of the ecosystem in the middle and downstream areas. For example, flooding occurred in the city of Yogyakarta and its surroundings due to extreme weather accompanied by a lot of water runoff from upstream areas due to the lack of rainwater infiltration. Responding to this phenomenon of shrinkage of agricultural land, the government issued Presidential Regulation No. 59/2019 concerning the Control of the Transfer of Rice Fields. This policy is in the form of assisting agricultural facilities and infrastructure, irrigation facilities and infrastructure, acceleration of land certification, and/or other forms by the provisions of laws and regulations.

Thus, the contribution of green space from the agricultural sector in Sleman Regency in 2010-2020 has decreased. Changes also followed this change in the number of workers and production in the agricultural sector. The agricultural sector's contribution to economic development every year decreases even though the amount of income continues to grow. This indicates that other sectors, such as the manufacturing and service sectors, grow faster and dominate than the agricultural sector in Sleman Regency.

4. Conclusion

The area of green space (rice fields and non-rice fields) in Sleman Regency from 2010 to 2020 decreased. It was 70.26% to 58.51% of the total land area. Overall, the area of green space still meets the standards that have been set. Most of the Sleman Regency area is still dominated by agricultural land, except for Depok District, Mlati District, and Gamping District, which are dominated by built-up land. An increasing trend occurred in the GRDP of the agricultural sector in Sleman Regency for a decade (2010-2020). However, based on Location Quotients (LQ) analysis, the agricultural sector is included in the non-basic sector, which means that it has not been able to meet domestic needs. The area of green space in the agricultural sector has decreased by around 11.75% of the total area of Sleman Regency from 2010-2020. This study's results suggest an increase in GRDP in Sleman, but it is followed by a reduction in green space, which results in a decrease in environmental quality. This is in line with the Kuznets Environmental Curve Hypothesis (EKC). This is due to the conversion of agricultural land, especially to build land. This change was also triggered by income outside the non-agricultural sector which is greater than the agricultural sector, a negative image of agriculture, increased education, narrow ownership of agricultural land, and rural convenience accessibility.

Acknowledgments

The authors would like to thank the Research Directorate of Gadjah Mada University, Indonesia, for financially supporting this research through the geography lecturer research grant of 2022 year.

References

[1] Creutzig, Felix, Christopher Bren d'Amour, Ulf Weddige, Sabine Fuss, Tim Beringer, Anne Gläser, Matthias Kalkuhl, Jan Christoph Steckel, Alexander Radebach, and Ottmar Edenhofer. "Assessing human and environmental pressures of global land-use change 2000–2010." *Global Sustainability* **2** (2019). DOI: https://doi.org/10.1017/sus.2018.15

[2] Toth A., & Damyanovic D. 2019. Planning and Implementation of Green Infrastructure in Austrian Cities, University of Natural Resources and Life Sciences, Vienna, Vienna.

[3] Hansen R., Rall E., Chapman E., Rolf W., & Pauleit S. 2017. Urban Green Infrastructure Planning: A Guide for Practitioners. *Green Surge*. Retrieved April 1, (2022) https://www.e-pages.dk/ku/1340/html5/

[4] Ratnasari, Amalia, Santun RP Sitorus, and Boedi Tjahjono. "Perencanaan Kota Hijau Yogyakarta Berdasarkan Penggunaan Lahan Dan Kecukupan Rth." *TATALOKA* **17**, no. 4 (2015): 196-208. Retrieved by: http://ejournal2.undip.ac.id/index.php/tataloka

[5] Dadashpoor, Hashem, Parviz Azizi, and Mahdis Moghadasi. "Land use change, urbanization, and change in landscape pattern in a metropolitan area." *Science of the Total Environment* **655** (2019): 707-719. DOI:10.1016/j.scitotenv.2018.11.267

[6] Ngadi, Erastus D., Alfetri NP Lango, and Lika Bernadina. "Analisis Faktor–Faktor Yang Mempengaruhi Alih Fungsi Lahan Pertanian Ke Lahan Non Pertanian Di Kota Kupang." *Buletin Ilmiah IMPAS* **22**, no. 2 (2021): 162-170.

[7] Toth A., & Damyanovic D. Planning and Implementation of Green Infrastructure in Austrian Cities . Vienna: University of Natural Resources and Life Sciences, (2019).

[8] Hansen, Rieke, Emily Rall, Eleanor Chapman, Werner Rolf, and Stephan Pauleit. "Urban green infrastructure planning: A guide for practitioners." *Green Surge* (2017). https://www.e-pages.dk/ku/1340/html5/

[9] Galan, J. "Landscape in Green Infrastructures & Interscalar Planning. Aalto, Finland: Aalto University." (2015). Retrieved from https://aaltodoc.aalto.fi/handle/123456789/17239

[10] Austin, G. Green Infrastructure for Landscape Planning, Integrating human and natural systems . London: Routledge, (2014).

[11] Yu et al. View-based greenery: a three-dimensional assessment of city buildings' green visibility using Floor Green View Index. *Landscape and Urban Planning*, **152**, 13-26, (2016) DOI: 10.1016/j.landurbplan.2016.04.004

[12] Sugiyono, Dr. "Statistika untuk Penelitian (Cetakan ke-30)." *Bandung: Cv Alfabeta* (2019).

[13] Muta'Ali, Lutfi. "Teknik analisis regional untuk perencanaan wilayah, tata ruang dan lingkungan." *Yogyakarta: Badan Penerbit Fakultas Geografi (BPFG)* (2015).

[14] Santosa, Langgeng Wahyu. Keistimewaan Yogyakarta dari sudut pandang geomorfologi. *UGM PRESS.* (2016).

[15] Tasong, Selsus Lorgen, Dyah Ully Parwati, and Abdul Mu'in. "Kajian Curah Hujan Di Kabupaten Sleman Diy Terhadap Tingkat Produksi Padi Sawah." *JURNAL AGROMAST* **1**, no. 2 (2019).

[16] Sleman Regency in Figures 2018, BPS Sleman. Sleman (2019)

[17] Sleman Regency in Figures 2010, BPS Sleman. Sleman (2011)

[18] Rosidah, Umi, Hadi Sasana, and Gentur Jalunggono. "Analisis Faktor-Faktor Yang Mempengaruhi Alih Fungsi Lahan Pertanian Di Kabupaten Sleman Tahun 1998-2017." *DINAMIC: Directory Journal of Economic* **1**, no. 3 (2019): 315-324.

[19] Sleman Regency in Figures 2020, BPS Sleman. Sleman, (2021)

[20] Alfana, Muhammad Arif Fahrudin, Luthfia Adlina, and Hidayati Nur Rohmah. "Transisi Demografi di Kabupaten Sleman: Proses dan Analisis Perubahan Struktur Penduduk Dalam Satu Dasawarsa Terakhir." *Geomedia: Majalah Ilmiah dan Informasi Kegeografian* **17**, no. 2 (2019).

[21] Nugroho, Agus Dwi, and Lestari Rahayu Waluyati. "Upaya memikat generasi muda bekerja pada sektor pertanian di Daerah Istimewa Yogyakarta." *JPPUMA: Jurnal Ilmu Pemerintahan dan Sosial Politik UMA (Journal of Governance and Political Social UMA)* **6**, no. 1 (2018): 76-95.

[22] Nia, K. Pontoh, and Iwan Kustiawan. "Pengantar perencanaan perkotaan." *ITB. Bandung* (2009).

[23] Astuti, Farida Afriani, and Herwin Lukito. "Perubahan Penggunaan Lahan di Kawasan Keamanan dan Ketahanan Pangan di Kabupaten Sleman." *Jurnal Geografi: Media Informasi Pengembangan dan Profesi Kegeografian* **17**, no. 1 (2020): 1-6.

[24] Sarjana, I. Made, Listia Dewi, and Ida Ayu. "Faktor-Faktor Pendorong Alihfungsi Lahan Sawah Menjadi Lahan Non-Pertanian (Kasus: Subak Kerdung, Kecamatan Denpasar Selatan)." *Jurnal Manajemen Agribisnis* **3**, no. 2 (2015): 26303.

[25] Marhaento, Hero, and Asep N. Kurnia. "Refleksi 5 tahun paska erupsi gunung Merapi 2010: menaksir kerugian ekologis di kawasan Taman Nasional Gunung Merapi." *Geoplanning: Journal of Geomatics and Planning* **2**, no. 2 (2015): 69-81.

[26] Giyarsih, Sri Rum. "Factors That Affect Urban Sprawl Symptoms in Sub Urban Areas of Yogyakarta." (2017).

[27] Xue, Fei, Zhonghua Gou, and Stephen Lau. "The green space development model and associated use behaviors in dense urban settings: Lessons from Hong Kong and Singapore." *Urban Design International* **22**, no. 4 (2017): 287-302.

[28] Siahaan, Esra Oktavia, Slamet Santoso, and Dyah Lituhayu. "Evaluasi Dampak Pera Nomor 7 Tahun 2010 Kota Tentang Penataan Ruang Terbuka Hijau Kota Semarang (Studi di Kecamatan Candisari Kota Semarang)." *Journal of Public Policy and Management Review* **7**, no. 1 (2018): 94-103.

[29] Latumaerissa, Julius R. *Perekonomian Indonesia dan dinamika ekonomi global*. Mitra Wacana Media, (2015).

[30] Zhang, Xin, Guolin Yao, Srishti Vishwakarma, Carole Dalin, Adam M. Komarek, David R. Kanter, Kyle Frankel Davis et al. "Quantitative assessment of agricultural sustainability reveals divergent priorities among nations." *One Earth* **4**, no. 9 (2021): 1262-1277. DOI: https://doi.org/10.1016/j.oneear.2021.08.015 ,

[31] Lubis, Wahyu Wardana, Listiyani Listiyani, and Danang Manumono. "Analisis Tipologi dan Sektor Unggulan Pertanian Kabupaten Sleman." *Jurnal Masepi* **3**, no. 2 (2018).

[32] Muta'Ali, Lutfi. "Teknik analisis regional untuk perencanaan wilayah, tata ruang dan lingkungan." *Yogyakarta: Badan Penerbit Fakultas Geografi (BPFG)* (2015).

[33] Siwar, Chamhuri, Rospidah Ghazali, and Mohammed Ibrahim. "Agriculture and Food Security." In *Malaysia's Leap Into the Future*, pp. 117-139. Springer, Singapore, (2022).

[34] Meuwissen, Miranda PM, Peter H. Feindt, Alisa Spiegel, Catrien JAM Termeer, Erik Mathijs, Yann De Mey, Robert Finger et al. "A framework to assess the resilience of farming systems." *Agricultural Systems* **176** (2019): 102656. DOI: https://doi.org/10.1016/j.agsy.2019.102656

[35] Artmann, Martina, Luis Inostroza, and Peilei Fan. "Urban sprawl, compact urban development and green cities. How much do we know, how much do we agree?." *Ecological indicators* **96** (2019): 3-9. DOI: 10.1016/j.ecolind.2018.10.05

[36] Pauleit, Stephan, Bianca Ambrose-Oji, Erik Andersson, Barbara Anton, Arjen Buijs, Dagmar Haase, Birgit Elands et al. "Advancing urban green infrastructure in Europe: Outcomes and reflections from the GREEN SURGE project." *Urban Forestry & Urban Greening* **40** (2019): 4-16. DOI: 10.1016/j.ufug.2018.10.006

[37] Yu, Siyi, Bailang Yu, Wei Song, Bin Wu, Jianhua Zhou, Yan Huang, Jianping Wu, Feng Zhao, and Weiqing Mao. "View-based greenery: A three-dimensional assessment of city buildings' green visibility using Floor Green View Index." *Landscape and Urban Planning* **152** (2016): 13-26. DOI: 10.1016/j.landurbplan.2016.04.004

[38] Kunanuntakij, Kultida, Viganda Varabuntoonvit, Natanee Vorayos, Chanin Panjapornpon, and Thumrongrut Mungcharoen. "Thailand Green GDP assessment based on environmentally extended input-output model." *Journal of Cleaner Production* **167** (2017): 970-977. DOI: https://doi.org/10.1016/j.jclepro.2017.02.106

[39] Grossman, Gene M., and Alan B. Krueger. "Environmental impacts of a North American free trade agreement." (1991).

[40] Rao, Ch Srinivasa, K. A. Gopinath, J. V. N. S. Prasad, and A. K. Singh. "Climate resilient villages for sustainable food security in tropical India: concept, process, technologies, institutions, and impacts." *Advances in Agronomy* **140** (2016): 101-214.

[41] Bournaris, Th, Ch Moulogianni, S. Arampatzis, F. Kiomourtzi, D. M. Wascher, and B. Manos. "A knowledge brokerage approach for assessing the impacts of the setting up young farmers policy measure in Greece." *Environmental Impact Assessment Review* **57** (2016): 159-166.

Use of Sustainable Transport System Approach in Tourism Zone Development at Kupang City

Don Gaspar Noesaku da Costa[1]*, Mauritius Ildo Rivendi Naikofi[2], Christiani Chandra Manubulu[2]

[1]Transportation Studies Centre, Faculty of Engineering, Universitas Katolik Widya Mandira, Jl. A. Yani 50-52, Kupang, East Nusa Tenggara, Indonesia
[2]Study Program of Civil Engineering, Faculty of Engineering, Universitas Katolik Widya Mandira, Jl. A. Yani 50-52, Kupang, East Nusa Tenggara, Indonesia

*Corresponding author: dnoesaku@gmail.com

Abstract. The augmentation of high scale social-economic activities usually triggers a great number of small-scale derived activities. However, the presence of such activities has been ignored in traffic impact analysis. The crowded traffic produced surround it then became a laten issue. This study aim is to determine the effect of roadside friction to road capacity and to provide adequate parking zone. Subsequently, this research also discusses the possible implemented solution based on land use availability and/or road environmental condition. Therefore, the traffic volume and geometrical survey have been carried out, and the road segment capacity was calculated using Indonesian Highway Capacity Manual, whereas the effect of varying roadside friction level to road capacity were simulated under 3 conditions, i.e.: 1) without on street parking, 2) with one side on street parking, and 3) with two side of on street parking. It was found that: 1) high level of roadside friction due to the existing on-street parking vehicle and pedestrian as well as entry-exit vehicles produced strongly influenced road capacity, degree of saturation 2) since only one side on street parking is allowed, then the authority should provide additional parking zone by convert potential land into off-street parking zone.

Keywords: parking index; parking zone; road ride friction; transport system; zoning regulation

1. Introduction

Development of civil engineering infrastructure that is not integrated with land use characteristics, as well as socio-economic and environmental characteristics can have an impact on decreasing the quality of the natural and artificial environment around it, because increasingly going beyond their core function [1]. In this case, the impact of high travel-attractive infrastructure development, such as a culinary tourism center built on the Kupang City (the capital city of East Nusa Tenggara Province, Indonesia) arterial road, is indicated to have an impact on decreasing road performance and even decreasing the quality of the surrounding environment. The laws and regulations in Indonesia expressly state that disturbances to the smoothness and safety of travel along arterial roads must be effectively limited [2]–[4]. However, the results of field observations indicate that the activity of roadside parking vehicles occurs in many urban arterial road segments, including in this study location as can be seen in Fig.1. However, it should be noted that such on-street parking vehicle occurs due to unplanned and/or unintentional socio-economic activities (such as street vendors), because the main activity planned is beachside restaurant. Therefore, since the road segment performance is getting worst so that the aim of this study is to determine the effect of roadside friction to road capacity, and to provide adequate parking space based on the value of the existing parking index as well as the predicted attraction and trip generation due to the operation of the restaurant in the near future. The culinary tourism area which is used as the object of this research is built on a relatively limited land so that it is not equipped with a parking area. Subsequently, this research also discusses the possible implemented solution based on land availability and/or road environmental conditions. So far, there have been no research results that specifically report the impact of the construction of public facilities that are not equipped with adequate parking spaces on the capacity and service level of the surrounding road sections. The non-integration of tourism infrastructure planning products with the carrying capacity and capacity of the land is the gap in this research.

Figure 1. Characteristics of Traffic and Land Use at Study Location
Source: private documentation, 2022

Thus, the research focus is directed at evaluating the impact of roadside parking control on road performance. Similar situation also occurs Thessaloniki, Greece [5]. Road performance in this study is assessed not only based on the degree of saturation and travel speed but also based on the level of accident risk that can arise due to pedestrian crossing activities and/or entering-exiting vehicles.

2. Study Design

2.1. Research Variables

Since the aim of this study is to evaluate the effect of roadside friction level change to Road Capacity (C) and Degree of Saturation (DS) as well as Level of Service (LoS) values then these are some basic steps needed to perform this study, i.e.:

a. Determination of roadside friction level due to existing on-street parking, entering-exiting vehicles, pedestrian crossing as well as slow motion vehicle movements. The roadside obstacle level is determined based on the frequency value of the weighted occurrence of the volume of pedestrian movement across the road, the number of roadside parking vehicles and vehicles entering and leaving the parcel and the number of slow vehicle movements [6].

b. Measuring travel speed along the road segment which is intended to identify variations in the speed of vehicles crossing the road segment during periods of heavy, medium or light traffic.
c. Calculating traffic volume (Q) during the peak hour period. The highest traffic volume value obtained is then used to calculate the degree of saturation, while the composition of traffic flow for each direction of movement is used to calculate the capacity value of the road segment [6].
d. Determination of Road Capacity (C) and Degree of Saturation (DS) as well as Level of Service (LoS).
e. Simulating the effect of roadside friction level change to road capacity, DS and LoS

2.1.1. Roadside Friction Level. The roadside obstacle class is determined based on the frequency value of weighted events, namely the product of the multiplication of the number of parking vehicle movements, incoming and outgoing vehicles, pedestrians / pedestrians and slow vehicle movements with each weighting criteria.

2.1.2. Travel Speed. Vehicle speed is the distance traveled by the vehicle in a certain unit of time, which in this case is assumed to be the result of the influence of variations in roadside obstacle classes.

2.1.3. Traffic Volume. The number of vehicles that pass through a section of the road is greatly influenced by not only the width of the road lane and the number of directions of movement (whether 1-way or 2-way) but also the condition of the roadside barriers around it

2.1.4. Road Capacity. According to Indonesian Highway Capacity Manual [6], road capacity could be calculate using Eq.1 as follow:

$$C = FC_{cw} x\ FC_{sf}\ x\ FC_{sp}\ x\ FC_{cs} \quad (1)$$

Where is:

$FC_{cw} = Road\ Capacity\ Adjustment\ Factor\ due\ to\ Effective\ Width\ of\ road\ Lane$
$FC_{sf} = Road\ Capacity\ Adjustment\ Factor\ due\ to\ Roadside\ Friction\ Level$
$FC_{sp} = Road\ Capacity\ Adjustment\ Factor\ due\ to\ Split\ Direction$
$FC_{cs} = Road\ Capacity\ Adjustment\ Factor\ due\ to\ City\ Size$

2.1.5. Level of Service. Road performance indicators such as degree of saturation and travel speed as well as accident risk level were used to describe the effect of parking on-street restriction schemes. Therefore, the following criteria was used to evaluate it:

a. Degree of Saturation/DS i.e., the ratio between traffic volume and road capacity. Indonesian Highway Capacity Manual [6] stated that if DS $\geq$ 0.75 then it means such road segment has started experiencing capacity constraints, visually indicated through a progressive decrease in vehicle speed, in line with the increasing in the DS value.
b. Travel Speed i.e., the distance travelled by a vehicle in a given unit of time. In this particularly case, such speed was determined based on spot speed value of a number of sample vehicle when passing through the road segment in front of Aston Hotel and surrounding culinary area.
c. Free flow speed i.e., traffic flow speed prediction for various road conditions and environments. In this study, the change in free flow speed is predicted through 3 restriction schemes namely 1) no vehicle parking at all 2) vehicle parking only allowed in one side 3) two side on-street parking scheme. Noted, the 2nd scheme corresponds to or refers to the existing parking conditions at study location.

3. Result and Discussion

3.1. Effect of on-street Parking to Road Capacity, Degree of Saturation and Level of Service

Table 1 below clearly shows that the existing effect of one side parking still could be accepted (LoS D). However, since the Timor Raya Road is the main route between regencies, it is highly recommended that only motorcycles are allowed to park in 1 side position on the road section so that the effective width of the road lane is still around 4.5 m. The width of the road is very necessary to pass the type of heavy vehicle, such as containers. Furthermore, if it is in line with the operation of the main activity in the area, namely restaurants, it is estimated that vehicle parking will occur on both sides of the road. If that happens, LoS will decrease and the vehicle speed is estimated to decrease to E and 25 km/hour. This condition is not desirable. Thus, efforts are needed to provide off-street parking spaces so that the impact of attraction and trip generation on the performance of inter-city arterial roads, where the culinary tourism area is built, does not decrease.

Table 1. Effect of Various Level of Roadside Friction to Road Performance

No	Scheme	Capacity (pcu/h)	Volume (pcu/h)	Degree of Saturation	% DS	Free Flow Speed (km/h)	% Free Flow Speed	Leve of Service	Traffic Performance
1	No Parking	3063,88	1455	0,47	7,06%	42,32	6,52%	C	busy traffic conditions, speed is quite limited but obstacles from other vehicles parked on the road and from pedestrians are starting to be felt
2	on side parking	2861,87	1455	0,51		39,56		D	traffic conditions are saturated, vehicle speeds start low and obstacles from other vehicles parked on the road and from pedestrians are getting bigger
3	two side parking	1905,30	1455	0,76	50,21%	29,87	24,49%	E	traffic jam conditions, the class of roadside barriers is getting higher so that the speed and volume of traffic into traffic is getting lower

Source: result analysis, 2022

These findings strongly indicate that each building should be provided with adequate parking zone. Therefore, it is required to develop the building licensing mechanism by adding the design parking index based on minimum parking lot needs as a mandatory criterion. In South Korea, the government regulates aspects of development permits for a number of activities without parking facilities, but this is only allowed in certain areas [7].

3.2. Implication of Study Results

Because the culinary tourism area does not provide parking space, the potential land around it must be converted. From the results of observations and field measurements, it is known that there are 2 lands that can be converted into the intended parking lot. The first parking area is along the coastal road that stretches from the rear of the Bundo Kanduang restaurant to the Hotel "On the Rock" and Hotel "M-Swiss bell court" (approximately 410 m, 5-6 m wide). this road is a dead end, and so far, it has only been used as a park/public space to enjoy sea and/or sunset views. Consequently, if this zone is used, it is necessary to provide a number of turning areas (at least every 50 m) so that light vehicles can access and maneuver comfortably. the second is a small plot of land used as a family burial site (limited) which is less than 200 m^2 in size.

This first option is much more strategic than the second one because it does not require land acquisition and parking building costs but only costs for the construction of a vehicle turning area. Furthermore, it is estimated that the parking zone can accommodate about 40 light vehicles and 200 motorcycles. In addition, this option can support the goal of sustainable development, namely to improve not only the quality of the natural but also the artificial environment, because this parking zone can also be combined with a food court and/or park area, according to the objectives of managing the transportation system [8].

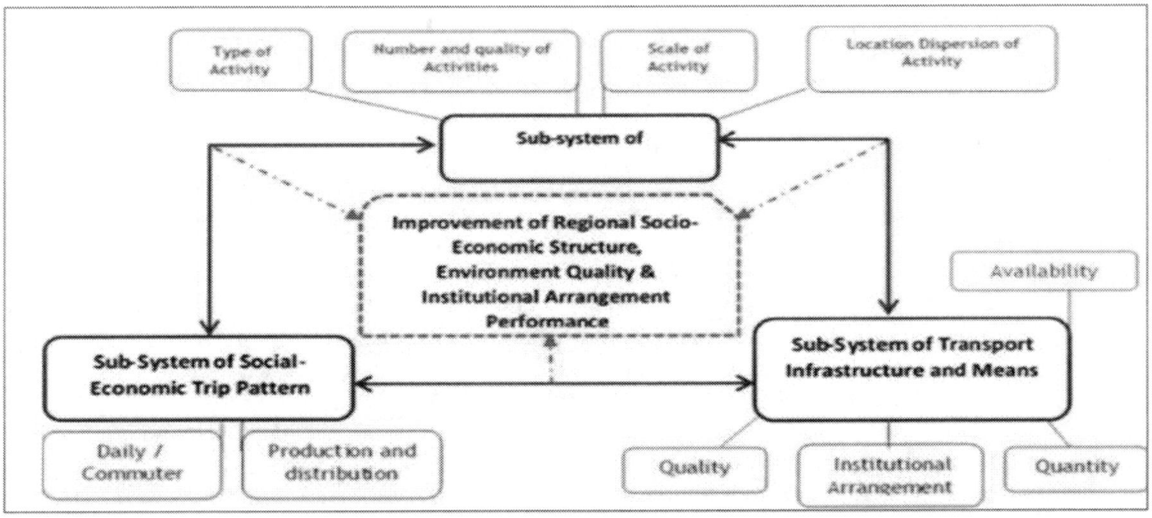

Figure 2. Estuary of interaction between the components of the transportation system [8]

Thus, the implication of the results of this study is that the government must improve the performance of infrastructure development licensing, especially through the preparation of a traffic impact analysis document, including evaluating the availability of minimum parking spaces which is a mandatory requirement for the construction of large-scale activity centers [4]. Therefore, because this corridor is a fast-growing and developing area, the government must immediately prepare zoning regulations so that the type, amount, scale and density of activities in the area can be efficiently limited. These restrictions are needed so that the carrying capacity and capacity of the area are not exceeded [9].

4. Conclusion

Based on the results of analysis and discussion it can be concluded that: 1) roadside parking policy can be given as long as it does not have a negative impact on capacity, degree of saturation, travel delays and queue opportunities caused by vehicle density, vehicle maneuvers in and out of the parking lot or due to road crossing activities, 2) in order to minimize the negative impact of culinary center visitor vehicles, efforts are needed to convert coastal road land into the main parking area. This concept is in line with the results of previous research which states that land use changes have an impact on changes in travel patterns (travel attraction and generation) to a center of activity [10] 3) the non-integration of development policies and the carrying capacity of land is indicated to have an impact on the decline in the quality of the artificial and natural environment. The development of the center of the culinary tourism area is proven to trigger the growth of the type, number and scale as well as the density of associated activities so that the cumulative impact must be anticipated. Warnings about this side effect, among others, have been published in a number of previous studies [1], [7].

References

[1] N. Bobylev and I. F. Jefferson, "Integration of Urban Physical Infrastructure Into Land Use," *Annu. World Bank Conf. L. Poverty*, no. March, pp. 3–6, 2014, doi: 10.13140/2.1.1563.3603.

[2] Pemerintah Republik Indonesia, "Undang Undang Republik Indonesia No 22 Tahun 2009 tentang Lalu Lintas dan Angkutan Jalan." Jakrta, 2009.

[3] Pemerintah Republik Indonesia, "Peraturan Pemerintah Republik Indonesia Nomor 32 Tahun 2011 tentang Manajemen dan Rekaysa,Analisis Dampak Lalu Lintas serta Manajemen Kebutuhan Lalu Lintas." Jakarta, 2011.

[4] Kementrian Perhubungan, "Peraturan Mentri Perhubungan Republik Indonesia Nomor PM 75 Tahun 2015 tentang Penyelenggaraan Analisis Dampak Lalu Lintas." 2015.
[5] A. Tsakalidis and P. Tsoleridis, "The Impacts of Illegal Parking on the Urban Areas' Traffic and Environmental Conditions: The Case of The City of Thessaloniki," *SPATIUM*, vol., no. 33, pp. 41–46, 2015, doi: 10.2298/SPAT1533041T.
[6] Direktorat Jenderal Bina Marga, "Manual Kapasitas Jalan Indonesia." Jakarta, p. 573, 1997.
[7] J. Jeong, M.-J. Cho, and M.-H. Lee, "An Analysis of the Effect of Non-Parking Facilities in Parking-Only Buildings on the Traffic Inducement Rate," *Sustainability*, vol. 14, no. 4250, 2022.
[8] D. G. N. da Costa, R. M. Rayawulan, K. R. Bela, M. I. Naikofi, E. N. B. Seran, and R. Hendrikus, "Use of Train Transport to Bridge Social-Economic Activities Between Indonesia and Timor Leste Along the Northern Coast of the Island of Timor: A Preliminary Study," *Adv. Eng. Res.*, vol. 199, no. ICoSITEA 2020, pp. 6–10, 2021, doi: https://doi.org/10.14716/ijtech.v9i4.941.
[9] D. G. N. da Costa, R. M. Rayawulan, and S. S. L. M. F. Seran, "Penentuan Jenis, Jumlah, Skala dan Kerapatan Aktivitas di Kawasan Perkotaan berdasarkan Tarikan dan Bangkitan Perjalanan," in *Prosiding Simposium Forum Studi Transportasi antar Perguruan Tinggi ke-22*, 2019, pp. 1–10.
[10] W. Salim and U. Faoziyah, "The Effect of Transport Infrastructure on Land-use Change: The Case of Toll Road and High-Speed Railway Development in West Java," *J. Reg. City Plan.*, vol. 33, no. 1, pp. 48–65, 2022, doi: 10.5614/jpwk.2022.33.1.3.

IOP Conf. Series: Earth and Environmental Science 1233 (2023) 012047 doi:10.1088/1755-1315/1233/1/012047

An Ecological Study of Karst Landscapes Around the Jalur Jalan Lintas Selatan (JJLS) of Gunungkidul Regency

B I D Astuti [1*], E Haryono [1]

[1] Department of Environmental Geography, Faculty of Geography, Universitas Gadjah Mada, Indonesia

*Corresponding author : Bernadettaindri99@mail.ugm.ac.id

Abstract. JJLS is a newly constructed highway connecting the southern area of Java. Along the Gunungkidul Regency trace, the road crosses protected karst landscape of Gunungsewu. This study aims to analyze landscape change during 2005-2020 period after the construction of the highway. Landscape change was analyzed by using Spatial Landscape Impact Assessment (SLIA). The data used for this research were Landsat 7 ETM+ images, google earth images, and Sentinel 2A images . The result study shows that the construction of JJLS has changed the landscape structure within the radius of 100 road sides. The change is mostly land use and landforms duty cut and fill activity. JJLS Road construction produces a very high impact on environmental sustainability. The trickle down effect must be controlled in the future development in order to ensure not to deteriorate environmental and water resources.

1. Introduction

The Jalur Jalan Lintas Selatan (JJLS) is one of the national roads that was built as a form of government policy to improve the economy of the people in the southern region of Java. This road was built in 2005 and is estimated to have a length of 127,000 km from West Java to East Java [1]. Gunungkidul Regency is one of the regencies passed by JJLS. The JJLS construction in Gunungkidul Regency overall passes through the Gunungsewu Karst landscape. Karst is one of the landscapes in a tropical climate. The existence of karst is characterized by the presence of limestone layers of CaCO3 [2].

The infrastructure development of the Jalur Jalan Lintas Selatan (JJLS) has had positive and negative impacts on the surrounding environment. The positive impact of JJLS development can be seen in the increasing number of manufacturing and service industries along the way [3]. Meanwhile, the negative impact on the environment caused by the construction of JJLS is a change in the ecological structure of the physical landscape. Landform changes are marked by the activity of cutting karst hills that decreases the volume of limestone as the main material for forming karst landforms and by land fragmentation. The karst hill cutting activity will impact the hydrological and ecological function of the karst area. The hydrological function is related to the karst's ability to store water in the karst hills [4]. Meanwhile, the ecological function is related to the karst's function as a place for living creatures to live. Another negative impact is the potency increase in the groundwater contamination around JJLS [5].

Landscape ecology is a science that focuses on the spatial pattern of an area. Landscape ecology studies can be applied in various fields and objects on the earth's surface, including analyzing changes in landscapes and living habitat conditions. Spatial and ecological approaches that are formal aspects of geography are very needed. The development of geography related to remote sensing and geographic information systems (GIS) is also helpful in identifying landscape patterns and structures to be used as the basis for environmental sustainability-based development planning [6]. The current development of landscape ecological studies has led to sustainable landform planning or in this case it can also be interpreted as sustainable environmental planning. The principle of sustainable environmental planning is professionalism and decision making related to ecological science and research [7].

Landscape ecology studies are concerned with issues of conservation, habitat, land fragmentation, and land use change. The four studies will be related to changes in landform structure consisting of matrices, patches, and corridors. The many issues of environmental degradation due to human actions to exploit resources have caused the study of landscape ecology to be very necessary to ensure the sustainability of landforms as a space for interaction of living things [8]. This study aims to analyze changes in the abiotic ecological aspects of the karst landscape that occurred due to the construction of the Jalur Jalan Lintas Selatan in Gunungkidul Regency in 2005-2020.

2. Methods

2.1. Location

The research location is around the Jalur Jalan Lintas Selatan (JJLS) of Gunungkidul Regency. The area is geomorphologically located in the Gunungsewu Karst Landscape. The unit of observation and unit of analysis is in a radius of 100 meters around JJLS.

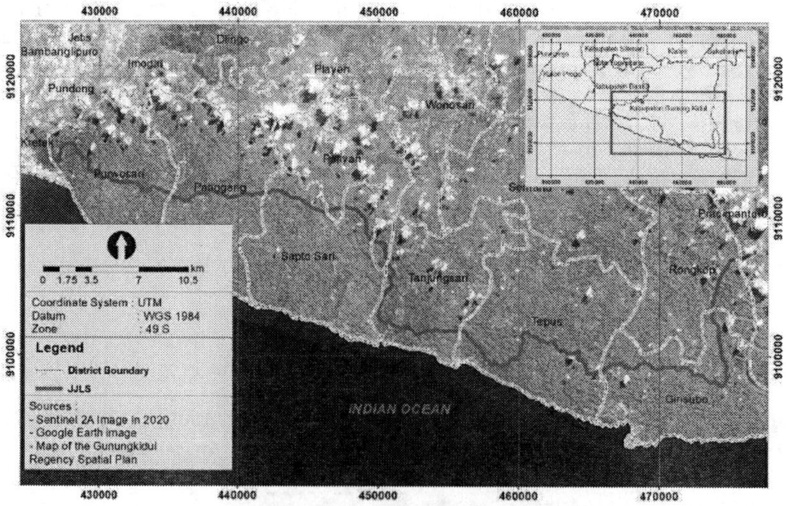

Figure 1. Map of Study Area

2.2. Data and Population

The secondary data used in this study are satellite images in 2005 and 2020. The satellite images consist of Landsat 7 ETM+ images, 2A sentinel images, and google earth images. Google Earth images were used to determine JJLS 2020 and validate land use maps in 2005. Sentinel 2A images were used to find out land use in 2020 and Landsat 7 ETM+ images to determine land use in 2005. The accuracy testing of image interpretation results in 2020 is performed through direct measurements and observations in the field. The population of this research is the use of existing land in a radius of 100 m around JJLS in Gunungkidul Regency which is obtained from satellite image processing. The sampling method used in this study is the census method. This method is a technique for determining data samples if the entire population is used as a sample [9].

IOP Conf. Series: Earth and Environmental Science 1233 (2023) 012047 doi:10.1088/1755-1315/1233/1/012047

2.3. Data Processing

Data processing is done using Landsat 7 ETM+ imagery, 2A sentinel imagery, and google earth imagery. The image was processed using software QGIS, ArcGIS, and SNAP to obtain land use map data in 2005 and 2020. Land use map derivatives were used to calculate the variables of abiotic ecological aspects change of the karst landscape, namely surface area, land fragmentation, and significant reduction in ecological areas. The surface area variable is the total area used in the construction of JJLS. This variable was obtained through the process of digitizing the sentinel 2A image in 2020. The variable of important ecological area reduction was obtained through the overlay method between the 2005 land use map and the JJLS map to determine the reduced vegetation area due to the JJLS construction and the development of the built-up area around it. The variable of land fragmentation was calculated based on the number of patches formed as a result of the JJLS construction. The three variables are categorized based on the SLIA rating scale presented in table 1. The Spatial Landscape Impact Assessment (SLIA) method is a weighting method used to assess the impact of human activities on landscape ecological changes in terms of landscape surface area, reduction of important ecosystem areas, and landscape fragmentation [10].

Table 1. SLIA Rating Scale based on Ostman (2015)

Aspects of Landscape Change	**Numerical Influence rating**	**The Area of Surface Change**	**Impact Description**
Surface Area	0	0 m^2	No impact
	1	1-449,000 m^2	Very low impact
	4	450,000-649,000 m^2	Low impact
	8	650,000-849,000 m^2	Significant impact
	13	850,000-1,049,000 m^2	High impact
	20	≥1,050,000 m^2	Very high impact
Reduction of important ecological areas	0	0 km^2	No impact
	1	0.1 – 1.0 km^2	Very low impact
	4	1.1 – 1.5 km^2	Low impact
	8	1.6 – 2.0 km^2	Significant impact
	13	2.1 – 2.5 km^2	High impact
	20	≥2.5 km^2	Very high impact
Land fragmentation	0	0 patch	No impact
	1	1 – 74 *patches*	Very low impact
	4	75 – 99 *patches*	Low impact
	8	100 – 124 *patches*	Significant impact
	13	125 – 149 *patches*	High impact
	20	≥150 *patches*	Very high impact

2.4. *Data Analysis*
The data analysis process in the form of a table is carried out using a descriptive approach supported by quantitative data. This approach was chosen because qualitative research is useful for understanding a phenomenon experienced by research subjects that emphasizes the process that occurs from a phenomenon [11]. The data is presented in the form of a ratio scale and categorized from no impact to very high impact based on the Spatial Landscape Impact Assessment (SLIA) rating scale. The classification process is carried out by referring to the SLIA rating scale (table 1). The analysis process is performed by explaining the level of impact of the JJLS construction on the ecological conditions of the surrounding physical landscape. The points of analysis are focused on how big the impact level of the JJLS construction is based on the results of data processing.

3. Results and Discussions
3.1. *Surface Area*
The surface area parameter is an important parameter to consider in analyzing the ecological changes of the karst landscape in the Gunungkidul Regency. This parameter shows the area used to build the Jalur Jalan Lintas Selatan (JJLS) in Gunungkidul Regency (table 2). The area used in the construction of this JJLS is 221.94 ha. The total surface area used in the construction of JJLS is rated 20 and is categorized as the very high impact class [10].

Table 2. The Area of Surface Change dan SLIA Rating

Assessment components	Change plan components	Area (m^2)	Area (ha)
Surface Area	Built-up area of JJLS	2,219,388.18	221.94
Total		2,219,388.18	221.94
SLIA Rating		20	

The area of land use that experienced the highest change to JJLS was rainfed rice fields (table 3). The total area of reduced rainfed rice fields for the construction of JJLS reached 68.213 ha, equivalent to 30.7349% of JJLS built on rainfed rice fields. The lowest area used for JJLS construction is a building with an area of 0.001 ha. It is equivalent to 0.0004% of the total area of JJLS built in 2020.

Table 3. Area of Land use used for JJLS Construction

No	Land use Change	Area (ha)	Area (%)
1	Buildings changed to JJLS	0.001	0.0004%
2	Fresh water turned into JJLS	0.231	0.1041%
3	Scrub/bush turned into JJLS	44.429	20.0184%
4	Plantation turned into JJLS	12.709	5.7264%
5	Settlement turned into JJLS	37.224	16.7722%
6	Grassland turns into JJLS	1.071	0.4826%
7	Rainfed rice fields turn into JJLS	68.213	30.7349%
8	Moor changed to JJLS	58.062	26.1611%
	Total	221.939	100%

3.2. *Land Fragmentation*
Fragmentation can occur due to natural processes or human activities, such as the opening of JJLS in the Gunungkidul Regency. Land fragmentation impacts changes in vegetation structure, the composition of plant species, the abundance of sources, and microclimate conditions [12]. In the JJLS construction in the Gunungkidul Regency, one of the areas experiencing land fragmentation is Girisekar and Girisubo District. Figure 2 shows that the construction of JJLS causes land fragmentation in shrubs/bush, dry fields, and rainfed rice fields. Land fragmentation in Girisekar District is caused by the opening of a new road corridor, namely JJLS.

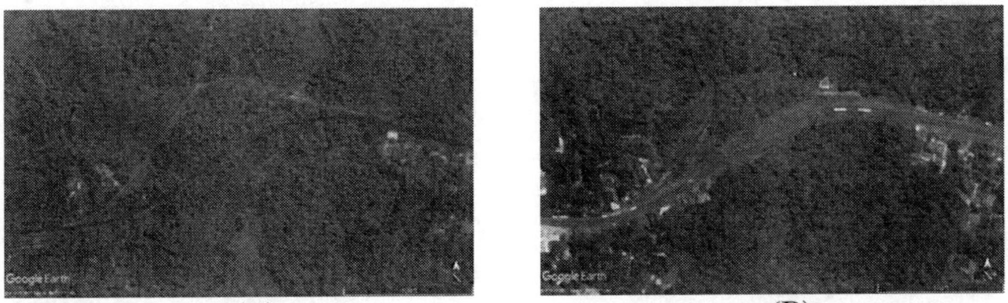

(A) (B)

Figure 2. Land Fragmentation in (A) Girisekar District on 2010, (B) Girisekar District on 2020

Land fragmentation around the Jalur Jalan Lintas Selatan in Gunungkidul Regency is marked by the appearance of patches, matrices, and corridors as a result of disturbances in the form of road construction around JJLS. Land fragmentation around JJLS increases the vulnerability to disturbances such as drought. Table 4. shows that the rating from the SLIA classification for the land fragmentation parameter is 20, which means that the JJLS development in Gunungkidul Regency has a very high impact on changes in the ecology of the surrounding landscape. It can be seen based on the number of patches formed, reaching a number of 150 patches. Land fragmentation that causes the formation of patches in the form of roads occurs in areas of use of scrub and dry land, while road patches are slightly formed in areas of use of freshwater land and buildings.

Table 4. Land Fragmentation and SLIA Rating

Assessment components	Land use	Number of patches
Land fragmentation	Freshwater	3
	Scrub	87
	Building	1
	Plantation	20
	Settlement	82
	Grassland	3
	Rainfed rice field	58
	Moor	45
Total		299
SLIA Rating		20

3.3. Reduction of Important Ecological Areas

The reduction of important ecological areas is a parameter that shows the important ecological areas that are disturbed and turned into the Jalur Jalan Lintas Selatan corridor and patches in the form of residential areas in the Gunungkidul Regency. The land use parameters included in important ecological areas consisted of fresh water, shrubs, plantation, grassland, rainfed rice fields, and dry fields. Table 5 shows the rating scale for the reduction of important ecological areas reached 20. It is because the total area reduced is 349,481 ha.

The construction of JJLS also affects the development of the built-up area around it. Built-up areas that develop around JJLS in Gunungkidul Regency include settlements and buildings. Table 5 shows that the area of freshwater used for JJLS construction is 0.105 ha. The development of residential areas around JJLS also decreases the land use area in the form of freshwater by 0.272 ha. Other land uses that also decrease in the area are shrubs, plantations, grassland, and rainfed rice fields. Based on the data

IOP Conf. Series: Earth and Environmental Science 1233 (2023) 012047 doi:10.1088/1755-1315/1233/1/012047

processing results (table 5), it can be seen that the highest land use used for road construction is rainfed rice fields.

Table 5. Reduction of Important Ecological Areas and SLIA Rating

2005	2020	Area (m^2)	Area (ha)
Freshwater	Street	1,050.2	0.105
Freshwater	Settlement	2,721.9	0.272
Scrub	Building	934.7	0.093
Scrub	Street	336,832.4	33.683
Scrub	Settlement	609,351.9	60.935
Plantation	Street	115,882.3	11.588
Plantation	Settlement	56,336.95	5.634
Grassland	Street	9,730.156	0.973
Rainfed Rice Fields	Street	568,675.4	56.868
Rainfed Rice Fields	Settlement	353,077.8	35.308
Moor	Street	350,290	35.029
Moor	Settlement	1,056,789	105.679
Moor	Building	33,137.7	3.314
Total		3,494,810	349.481
SLIA Rating		20	

The reduction of important ecological areas around JJLS Gunungkidul Regency is due to the activity of opening lanes and widening roads. In Figure 3, it can be seen that the Tepus District experienced a reduction in important ecological areas in the form of shrubs/bush and dry fields. In 2020, it can be seen that one of the segments in the Tepus District experienced karst cone cutting activities for the widening and opening of the Jalur Jalan Lintas Selatan.

(A)

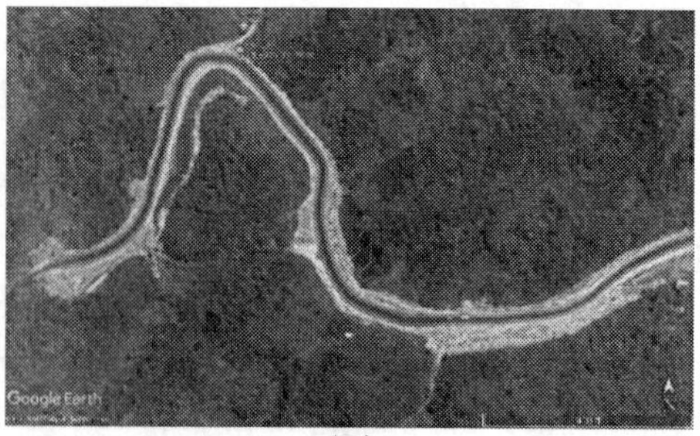

(B)

Figure 3. Reduction of Important Ecological Areas on Tepus District (A) in 2010 and (B) 2020

4. Conclusion

JJLS construction in Gunungkidul Regency has a very high impact on changes in the abiotic environment around a 100 meter radius from the road based on SLIA methods. This can be seen from the increase in the area of JJLS built, land fragmentation, and reduction of important ecological areas. This condition is caused by the cut and fill activity. Based on these results, the three parameters used are surface area, land fragmentation, and ecological area reduction, respectively worth 221.939 ha, 299 patches, and 349.481 ha.

References

[1] F. Lustina, "Analisis Pengaruh Pembangunan Jalur Jalan Lintas Selatan Terhadap Nilai Tanah Di Kabupaten Gunungkidul. Studi Pada Kecamatan Purwosari Tahun 2016," Universitas Gadjah Mada, 2017.

[2] E. Haryono and T. N. Adji, *Geomorfologi Dan Hidrologi Karst*. 2017.

[3] B. P. Semedikun, "Tinjauan Kondisi Bentang Lahan Kawasan Tembalang Akibat Perubahan Tata Guna Lahan (Land Use)," *Presipitasi Media Komun. dan Pengemb. Tek. Lingkung.*, vol. 6, no. 1, pp. 40–46, 2009, doi: https://doi.org/10.14710/presipitasi.

[4] A. Cahyadi, "Sumberdaya Lahan Kawasan Karst Gunungsewu Ahmad," *Cycle*, 2017, doi: 10.31227/osf.io/4dbr5.

[5] A. Cahyadi, E. Haryono, T. N. Adji, and M. Widyastuti, "Why are Allogenic Watersheds not Protected (Again)?," *Soc. Humanit. Educ. Stud. Conf. Ser.*, vol. 3, no. 1, 2020, doi: 10.20961/shes.v3i1.45054.

[6] J. Wu, "Key concepts and research topics in landscape ecology revisited: 30 years after the Allerton Park workshop," *Landsc. Ecol.*, vol. 28, no. 1, pp. 1–11, 2013, doi: 10.1007/s10980-012-9836-y.

[7] A. Botequilha Leitão and J. Ahern, "Applying landscape ecological concepts and metrics in sustainable landscape planning," *Landsc. Urban Plan.*, vol. 59, no. 2, pp. 65–93, 2002, doi: 10.1016/S0169-2046(02)00005-1.

[8] F. Ardian and E. Haryono, "Karakteristik Struktur Ekologi Bentanglahan untuk Kesesuaian Habitat Kukang Jawa (Nycticebus Javanicus) di Kabupaten Temanggung," *Fak. Kehutanan. Univ. Gadjah Mada. Yogyakarta*, vol. 2, pp. 1–10, 2018.

[9] L. Nurmala, S. Soetoro, and Z. Noormansyah, "Analisis Biaya, Pendapatan dan R/C Usaha Tani Kubis (Brassica Oleraceal) (Suatu Kasus di Desa Cibeureum Kecamatan Sukamantri Kabupaten Ciamis)," *J. Ilm. Mhs. Agroinfo Galuh*, vol. 2, no. 2, p. 97, Apr. 2017, doi: 10.25157/jimag.v2i2.64.

[10] D. Ostman, "A New Approach for Assessing Landscape Impacts of Geothermal Power Plants: A Case Study of Hellisheiði," University of Iceland, 2015.

[11] M. Lexy J., *Metodologi penelitian kualitatif Edisi Revisi*, 30th ed. Bandung: PT Remaja Rosdakarya, 2018.

[12] L. B. Prasetyo, *Hand Out Ekologi Lanskap*. Bandung: Departemen Konservasi Sumberdaya Hutan dan Ekowisata Fakultas Kehutanan. Institut Pertanian Bogor, 2006.

IOP Conf. Series: Earth and Environmental Science 1233 (2023) 012048 doi:10.1088/1755-1315/1233/1/012048

Differences in soil-water characteristics of monoculture oil-palm plantations, agroforestry oil-palm plantations and natural forest

M. J. van Oosterhout[1], Hero Marhaento[2*], Martijn J Booij[1], Darmawati Ridho[2]

[1]Faculty of Engineering University of Twente, Enschede, The Netherlands
[2]Faculty of Forestry, Universitas Gadjah Mada, Yogyakarta 55281, Indonesia

*Corresponding author: marhaento@ugm.ac.id

Abstract. This study aims to determine the soil water retention curves and infiltration rates for three different land use types located in Jambi province, Sumatra, Indonesia: natural forest, monoculture and agroforestry oil palm plantations (divided into new agroforestry (2,5 years) and old agroforestry (over 10 years)). Data on the infiltration rate of the soil for the different land use types was collected by using a double infiltrometer. A Horton curve was fitted to these measurements. The results on the soil water characteristics were compared between all land use types. This concluded that there is a significant difference in constant infiltration rate between the value of the natural forest and the new agroforestry and the value of the other land use types. The forest also has a significant smaller porosity. The results of the study concluded that plantations have a bad influence on the soil characteristics investigated, but in general agroforestry has a less bad impact than monoculture. However, this is highly dependent on the features of the locations investigated.

1. Introduction

Due to a growing population, more stress is put on the agriculture all over the world [1]. This causes the need for more agricultural land, and more productive agriculture. The first need can result into deforestation all over the world. While the second needs research into how more productive agriculture could be reached.

Some areas in the world where a lot of deforestation takes place due to agriculture are Borneo and Sumatra, both located in Indonesia [2]. The major cause for deforestation in Indonesia is agriculture with mainly oil palm plantations as driver [3]. The agricultural sector plays a vital role in the economy of Indonesia [4]. This makes sustainable agriculture even more important in these areas. It is important to have the agricultural production as high as possible and keep the ground suitable for agriculture for a long time period. Mostly deforested areas are transformed in monoculture plantations, which have a negative impact on the climate and ecosystem of an area. Therefore, more sustainable options should be found for this problem.

University Gadjah Mada (UGM) works together with the Kehati Foundation on a project called "Strategi Jangka Benah". The aim of the project is to transform monoculture oil palm plantations in Jambi province, Sumatra, into agroforestry plantations which are expected to have a better influence on the climate and ecosystem. However, this has not been proven yet. Therefore, this research focusses on soil water characteristic, namely water retention curves and infiltration rates. These are two of the important factors which can help provide more insight in the impact of the change of monocultural plantations to agroforestry plantations, but also the changes regarding to natural forest. The impact of these different land uses on the soil-water characteristics can then be linked to the growth of vegetation, as the two characteristics indicate the water availability for plants and the flow into the soil which can transport nutrients into the ground which are beneficial for plant growth.

In the research of Dislich et al. [5] and Tarigan et al. [6] there is only focussed on the difference in infiltration between monoculture oil palm plantation and the natural forest. This means knowledge on the agroforestry land use and the differences of this land use with the other two land use types, is still missing as well as the knowledge on water retention curves for these land use types. This study aims to investigate the differences in soil-water characteristics of monoculture oil-palm plantations, agroforestry oil-palm plantations and natural forest. The results of this study can be beneficial for agricultural decisions and specifically for the "Strategi Jangka Benah" project to transform more monoculture into agroforestry plantations, when the conclusion can be made that the agroforestry land use type has better soil water characteristics.

2. Method

First, data was gathered on mostly visual features of the different areas like soil colour, vegetation density, organic material at the surface layer and the bulk density. This was done to make a fair comparison between the different land use types and explain the differences between them. Most features were determined in the field. The bulk density of the soil however was determined in the lab after oven drying the samples for 24 hours at a temperature of 80°C by using equation 1 in which $M_{s\,(od)}$ is the mass of the oven dried soil (g) and V_s is the volume of the soil sample (cm^3).

$$D = \frac{M_{s\,(od)}}{V_s} \qquad \text{(Equation 1)}$$

Data on the water retention curves was collected by collecting 15 undisturbed samples in the field in cylindrical rings (with a height of 5,4 cm and a diameter of 4,9 cm). First the porosity of the soil was determined for one sample (see equation 2). For this one sample of each land use type was soaked in water for 48 hours. This determined the range in water content for the filter paper method. All samples were given a specific water content by adding water to the samples. The filter paper method was executed on the samples as described by Almeida et al. [7]. After 7 days an equilibrium has been set between the soil and the filter paper. By determining the weight of the wet filter paper, the water content of the filter paper was determined. By using equation 3 and 4 for the calibrated Whatman filter paper the suction corresponding with the water content was determined [8]. Both were plotted in a graph.

$$\varphi = \left(1 - \frac{\rho_s}{\rho_p}\right) = \frac{V_p}{V_t} \qquad \text{(Equation 2)}$$

$$For\ wc_{fp} > 45{,}26\% \qquad log_{10}\,Suction = 5{,}327 - 0{,}0779\,(wc_{fp}) \qquad \text{(Equation 3)}$$

$$For\ wc_{fp} \leq 45{,}26\% \qquad log_{10}\,Suction = 2{,}412 - 0{,}0135\,(wc_{fp}) \qquad \text{(Equation 4)}$$

In the field the data on the infiltration rates was gathered by using a double ring infiltrometer. At each land use type at least 6 measurements were done, which should be a sufficient number following to get representative results on the infiltration rates [9]. As the aim was to collect the infiltration rate of the soil, the organic matter at the surface layer was removed before placing the double ring infiltrometer. The decrease in water level was measured at constant time steps. This data was plotted into a graph and the Horton equation (see equation 5) [10] was fitted to the measured data. In equation 5 f_c is the final constant infiltration capacity at saturation (cm/h), f_0 is the initial infiltration capacity (cm/h), k is soil specific decay constant depending on soil and vegetation (-) and t is the time (h)

$$f_p = f_c + (f_0 - f_c)e^{-kt} \quad \text{(Equation 5)}$$

Finally, the data collected on the soil water characteristics was compared (by an independent sample t-test) and explained by making use of the found features of the different locations. For further detail information about the equations used in this study, we suggest readers to refer the primary references.

3. Results and Discussion

3.1. Features of the land use types

First the features of the research locations were determined. In **Error! Reference source not found.** the locations of the research areas are shown.

shows a picture of land use types to illustrate the differences between the locations. Based on the research there was found that the vegetation density of the forest is the highest followed by the new agroforestry plot. The old agroforestry and monoculture both had a really low vegetation density. This corresponded with the thickness of the organic matter layer which was the largest for the natural forest. The bulk density was investigated in this part and this resulted in an average bulk density of 1,15 g/cm^3 for the natural forest and an average bulk density of around 1,56 g/cm^3 for the other land use types. This difference can be explained by the compaction of plantation plots, while preparing or working on the plots.

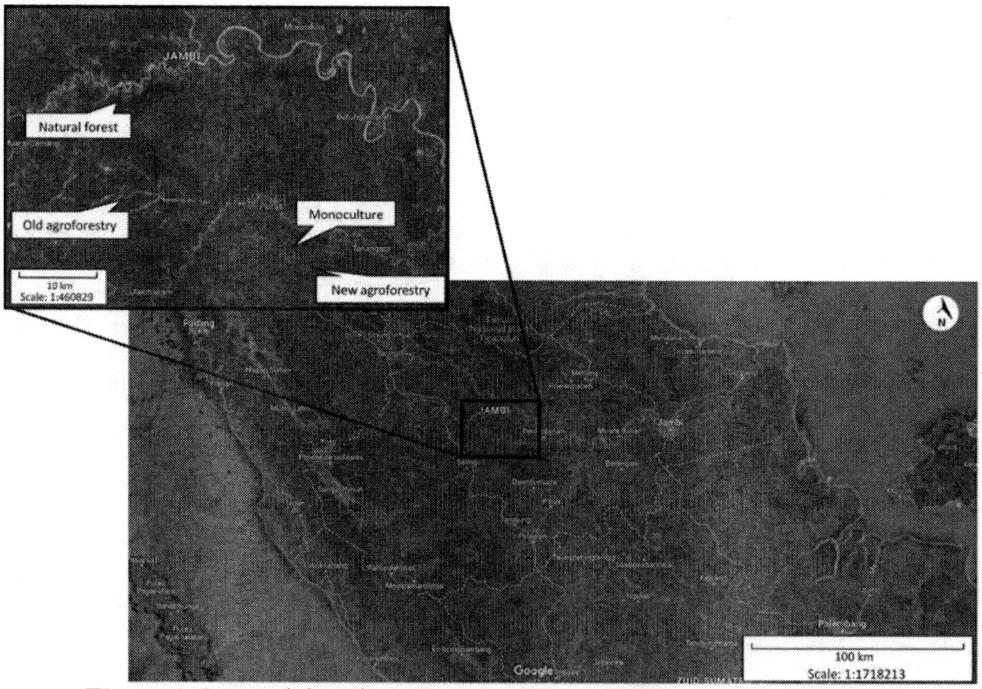

Figure 1. Research locations Sumatra, Indonesia (Source: Google Maps)

IOP Conf. Series: Earth and Environmental Science 1233 (2023) 012048 doi:10.1088/1755-1315/1233/1/012048

Figure 2. Different land use types of the study area

3.2. *Water retention curves*

The results from the filter paper method could unfortunately not be used. The results only showed the suction of the filter paper at two suction values (see figure 3), which is not representable for the actual water retention curve. This failure was most likely caused by the use of not certified filter paper, which was not known before using. Due to time limits the method could not be executed again. However, the porosity of the soils can be used. This gave that both agroforestry plots had the lowest porosity (porosity new agroforestry = 0,28, porosity old agroforestry = 0,29). The monoculture plot had a similar porosity of 0,31, while the forest had a way higher porosity of 0,44. This could be partly explained by the bulk density found.

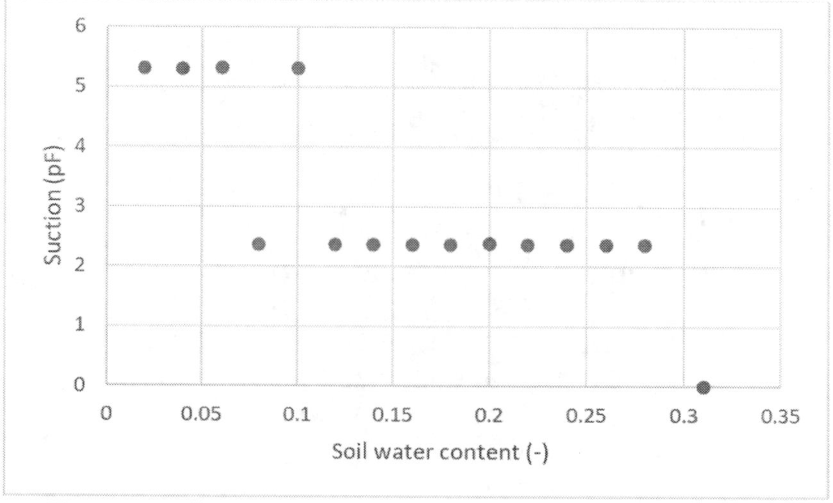

Figure 3. Soil water retention curve of the monoculture plantation

3.3. *Infiltration rates*

For the results on the infiltration rates, it was found that there was a lot of variability within all land us types, but this variability was hard to explain. The Horton equation fitted to the measurement points gave for most measurement locations a high correlation (0,77 or higher), four measurements had a correlation below this value, but still showed a correlation with a correlation value between 0,65 and 0,70. The range in parameter values from the Horton equation is given in table 1.

Table 1. Infiltration parameters

Land use type	f_0 min – max (cm/h)	f_c min – max (cm/h)	k (t^{-1})
Natural forest	3.27 – 131.90	0.90 – 39.60	0.99 – 19.13
Old agroforestry	0.30 – 4.77	0.06 – 1.50	1.37 – 5.38
New agroforestry	8.87 – 86.28	2.40 – 10.95	0.81 – 1.85
Natural forest	0.16 – 3.02	0.075 – 0.90	1.02 – 2.51

The values for the constant infiltration rate for all land use types are shown in figure 4. it turned out there is a significant difference between the values of the forest and new agroforestry and the values of the old agroforestry and monoculture, despite the big variability. Both the natural forest and new agroforestry has significantly higher constant infiltration rates compared to the other two land use types. The differences between the old agroforestry and monoculture are not significant. This is the same case for the new agroforestry and natural forest which differences in constant infiltration rate are also not significant. The decay factor gained from the Horton equation was also compared between land use types, this is shown in figure 5. For this there was only a significant difference found between the new agroforestry plot and the other three land use types. The decay factor for the agroforestry is lower, which means it takes longer for the soil to reach a constant infiltration rate.

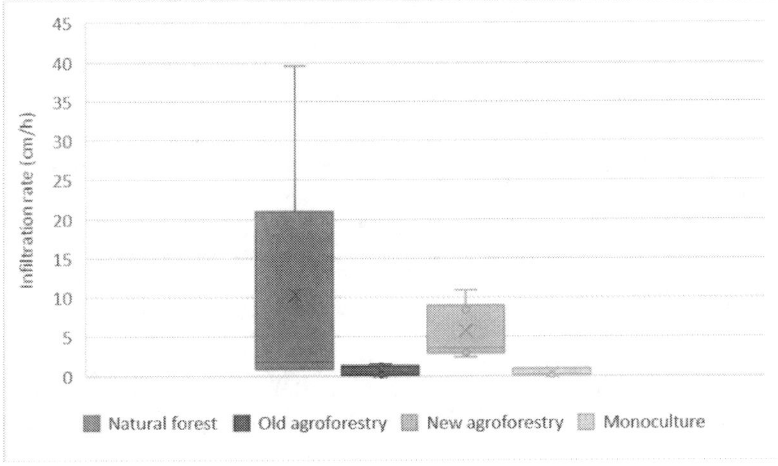

Figure 4. Constant infiltration rate different land use types

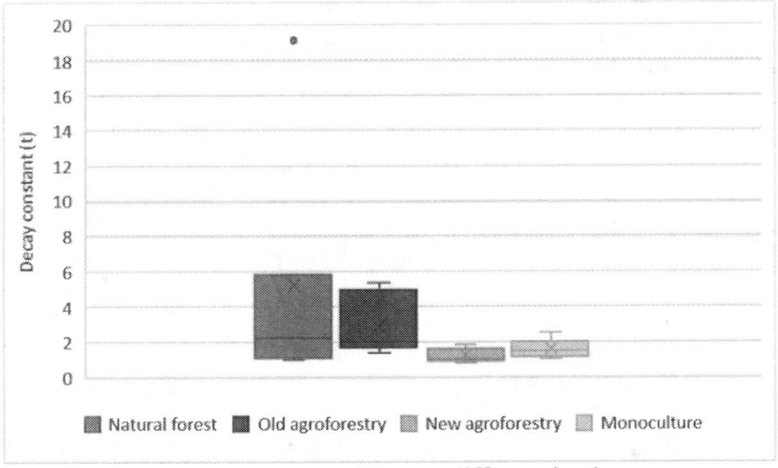

Figure 5. Decay constant Horton different land use types

3.4. Discussion

The methods chosen and choices made during the project could have influenced the outcomes of the research. One of the points up for discussion is the available data. There was almost no data available which made it hard to explain the results and differences between the land use types. Besides, the precision of the double infiltrometer and its set up could have given slight deviations in the actual infiltration rates, due to human factors as well by factors caused by the procedure.

4. Conclusion

Based on the research, generally the conclusion can be made that the natural forest has the most optimal soil water characteristics, namely the highest constant infiltration rate and the highest porosity, followed by agroforestry. Monoculture as a negative effect on the soil water characteristics investigated. This is in line with the research of Suprayogo [11] and Dislich [2]. However, the soil water characteristics are highly dependent on the features of the locations investigated which can indicate that the statement made above is not always the case, this can be seen in this research in the difference between the old agroforestry and new agroforestry. Finally, the results of this study can support the application of the

"Strategi Jangka Benah" project to transform monoculture oil palm into oil palm agroforestry inside the state forest since the oil palm agroforestry has resulted in better soil water characteristics.

Acknowledgement
The first author would like to express appreciation to the Strategy Jangka Benah (SJB) team from Faculty of Forestry UGM for allowing the first author to do the research in their demonstration plot. All author thanks Universitas Gadjah Mada for partly funding this research through *Rekognisi Tugas Akhir* (RTA) 2021 program.

References

[1] Calicioglu, O., Flammini, A., Bracco, S., Bellù, L., & Sims, R. (2019). The Future Challenges of Food and Agriculture: An Integrated Analysis of Trends and Solutions. *Sustainability 2019, Volume 11, article 222.*
[2] Global Forest Watch. (n.d.). *Global.* Retrieved from Global Forest Watch: https://www.globalforestwatch.org/. accessed July 1st 2022
[3] Austin, K. G., Schwantes, A., Gu, Y., & Kasibhatla, P. S. (2019). What causes deforestation in Indonesia? *Environmental Research Letter 14.*
[4] Bashir, A., Suhel, Azwardi, Atiyatna, D. P., Hamidi, I., & Adnan, N. (2019). The Causality Between Agriculture, Industry, and Economic Growth: Evidence from Indonesia. *Etikonomi, Volume 18,* 155-168.
[5] Dislich, C., Keyel, A. C., Salecker, J., Kisel, Y., Meyer, K. M., Auliya, M., . . . Töll, M. H. (2017). A review of the ecosystem functions in oil palm plantations, using forests as a reference system. *Biological reviews,* 1539-1569.
[6] Tarigan, S. D., Sunarti, Wiegand, K., Dislich, C., Slamet, B., Heinonen, J., & Meyer, K. (2016). Mitigation options for improving the ecosystem function of water flow regulation in a watershed with rapid expansion of oil palm plantations. *Sustainability of Water Quality and Ecology, Volume 8,* 4-13.
[7] de Almeida, E. L., Teixeira, A., da Silva Filho, F. C., de Assis Junior, R. N., & de Oliveira Leão, R. A. (2015). Filter paper method for the determination of the soil water retention curve. *R. Bras. Ci. Solo, 39,* 1344-1352.
[8] Kim, H., Prezzi, M., & Salgado, R. (2016). Calibration of Whatman Grade 42 filter paper for soil suction measurement. *Canadian Journal of Soil Science 97(2),* 93-98.
[9] Burgy, R. H., & Luthin, J. N. (1956). A test of the single- and double-ring types of infiltrometers. *Eos, volume 37 (2),* 189-192.
[10] Gikonyo, E., Zaharah, A., Hanafi, M., & Anuar, R. (2010). Phosphorus Leaching in an Acid Tropical Soil "Recapitalized" with Phosphate Rock and Triple Superphosphate. The Scientific World Journal 10, 1498–1508.
[11] Suprayogo, D., van Noordwijk, M., Hairiah, K., Meilasari, N., Rabbani, A. L., Ishaq, R. M., & Widianto, W. (2020). Infiltration-Friendly Agroforestry Land Uses on Volcanic Slopes in the Rejoso Watershed, East Java, Indonesia. *Land, 9(8), 240.*

Characteristics of entrenched meanders in catchment landscapes: a case study of Java, Indonesia

Puguh Dwi Raharjo[1,3*], Eko Haryono[2], M. Anggri Setiawan[2,4], Haryadi Permana[1], Ardhan Farisan[1]

[1] Research Center for Geological Resources, National Research and Innovation Agency, Bandung 40135, Indonesia
[2] Department of Environmental Geography, Gadjah Mada University, Yogyakarta, Indonesia
[3] Doctoral Program in Geography, Gadjah Mada University, Yogyakarta, Indonesia.
[4] Center for Disaster Study, Gadjah Mada University, Yogyakarta, Indonesia.

*Corresponding author: puguh.draharjo@gmail.com; puguh.dwi.raharjo@brin.go.id

Abstract. The diversity of physiography on the island of Java reflects the existence of different landforms that have properties and characteristics in the interaction of river formation. Rivers respond to longitudinal profile deformation, one of which can form river meanders. Meander rivers in Java are formed on alluvial plains (free meanders) and in hilly areas (entrenched meanders). The meander river formation factor is the high flow volume, which can be seen from the morphometric conditions. This study aimed to determine the morphometric characteristics of water catchment in entrenched meander rivers on the island of Java, as well as its relationship with the geodynamic conditions. The method used is the identification of entrenched meander rivers on Java Island, and morphometric calculations are carried out; besides that, it is also associated with morphostructural conditions to determine the process of its formation. The results obtained that the catchment area of the entrenched meander river has a tendency for small volumes of water and relatively longer floods; this is different from the nature of the free meander. The morphostructure much controls the process of forming entrenched meander rivers.

1. Introduction

River dynamics is the interaction of water flow with the surface over a short to a long period [1]–[3]. The river will adapt to an area's shape, dimensions, and profile [4]. Rivers respond to longitudinal profile deformation due to deflection, aggradation, and degradation [5]. Changes in slope cause the river to experience differences in the speed of water flow, affecting the discharge and erosion. The river's slope is one of the main factors affecting the shape of the river channel [6].

Topography along the river has a role in forming river meanders [7], [8]. The speed of water flow is influenced by the condition of the slope of the surface. Steep slopes provide high water flow velocities, and a reduced slope will decrease the rate. The existence of a topographic gradient that minimizes water flow energy makes the adaptation of rivers on the surface [9], [10]. Rivers adapt by lengthening the flow. Erosion is a process that occurs in the display of streams and forms the curvature of the river.

Rivers with a curvature of 1.5 (the ratio of river length to wavelength) are called meander rivers [1], [11]–[14]. Slope gradient, discharge, and erosion factors are the main elements in the formation of river meanders [15]–[23]. The longitudinal profile of the river is one of the main factors affecting the shape of the river [6]. Changes in the longitudinal profile will affect the flow of the river [5]. Meander river mechanics is concerned with the distribution of energy losses and causes river curvature. Differences in slope and flow velocity affect flow turbulence which causes river erosion and the formation of meander patterns [16], [20], [24].

Meander rivers can form on rock and soil substrates [1], distributed in alluvial plains and hilly areas [25]–[27]. The location of the formation of a meandering river can be categorized as a free meander if it is formed on alluvial plains and an entrenched meander if the river is in the form of bedrock. Entrenched meanders in bedrock are autogenic and not inherited and not due to river regeneration [28], [29], but Charlton [1] argues that meander channels in this bedrock inherit the previous alluvial phase. The last river meander was rejuvenated, which became an entrenched meander [30]. An entrenched meander affects the river material in the form of bedrock is often covered by sedimentation [31]. After a decrease in sediment input and thinning alluvial, the bedrock material will be eroded [32].

The availability of flow is an essential factor that can form the accumulation of flow in rivers [33]. The flow accumulation is caused by the volume of water from upstream. The characteristics of the water catchment much influence the Flow accumulation. The Characteristics, especially from the area and shape and the level of drainage density. The water catchment in the upstream area, which tends to have a rounded shape, has a high flow accumulation and is prone to flooding in the area around its outlet [34]. The level of drainage density in the upstream area also influences the volume of flow; the higher the drainage density in the upstream area, the higher the volume. Knowledge of this river system evolution related to geomorphic processes in the catchment can be used to assess the system's response to disturbance events. A landscape unit is a topographical feature easily identified by a characteristic pattern of landforms. Landform configuration affects the spatial characteristics and biophysical patterns represented in a water catchment area. A catchment area provides a topographical and hydrological entity described as a landscape unit. In addition, topographical forms (elevation, relief-dissection) also reflect rock types and morphological characteristics [35]. Catchment variability provides the character and behaviour of rivers determined by the landscape unit's configuration.

Water catchment is essential to know the volume of water that collects to become a flow that passes through a meandering river. In principle, this water catchment is a unit that limits the amount of rainwater that falls on the surface so that the energy from the contribution of water is known in making flow accumulation and forming meander rivers. In theory, meander rivers can form when the flow is high. This high flow can be determined by the morphometric characteristics of the watershed, namely the area of the watershed, the drainage density, and the shape of the watershed. The entrenched meander river in Java Island is geomorphologically spread over areas with structural origins of land formation, so the river is closely related to the structural conditions and genesis of the area.

The fundamental question in this study is, does entrenched meander river also occur in landscapes with catchment areas with high volume characteristics and catchment in free meander rivers? They have apparent differences. This study aimed to determine the morphometric characteristics of water catchment in entrenched meander rivers on the island of Java, as well as its relationship with the geodynamic conditions. The entrenched meander river is a natural phenomenon with unique characteristics and is not found in all locations. The research results are expected to be helpful for the development of science, especially in terms of finding the landscape characteristics of the entrenched meander river catchment area. The study's results can be used as reference material, especially for entrenched meander rivers related to structural processes in hilly areas. In addition, this research also has practical benefits that are expected to provide input to the government in anticipating environmental problems, especially flooding and river bank erosion in hilly areas with an entrenched meander river phenomenon.

2. Method

2.1. Study area

This research was conducted on a meander river object located in the hills on the island of Java. The entrenched meander river is administratively spread over the provinces of West Java, Central Java, the Special Region of Yogyakarta, and East Java. Sampling is carried out based on the representation of the physiographic zone. The appearance of the meandering river can be detected using remote sensing imagery with good spatial resolution. Google earth is a remote sensing image that can be used on a global scale and can also be used on a local scale [36]. Satellite image data from google earth is used to determine the type of entrenched meander river in the bedrock. This magnification using google earth images is carried out up to a scale of 1: 5000. DEM image data (demnas) is used to ensure that this entrenched meander river is in a hilly area and not in an alluvial area. Secondary data such as the Java Island Physiographic Map [37], Geological Map (ESDM), RBI Map (BIG), Indonesian Island Tectonic Map (USGS, 1991); and Tectonic Maps of Java Island [38] are needed to determine the surface and geodynamic conditions associated with entrenched meander rivers.

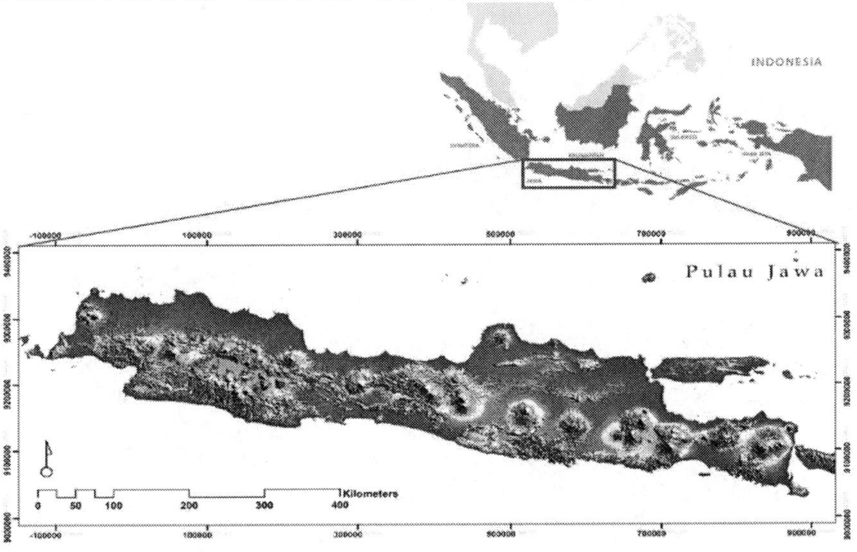

Figure 1. Research area Java Island, Indonesia

The data in this study's limiting water catchment uses a dem with data processing using a geographic information system (GIS). The delineation process for water catchment and river flow uses the automatic function of GIS; this process can provide surface hydrological parameters and components quickly and accurately. The grid on the DEM has a surface height value that can be used for flow direction analysis, and the flow will flow following the lower surface and form a river channel [39].

2.2. The area and shape of the catchment

The shape of a water catchment can be categorized into several forms: elongated, oval, circular, and fan-like [40]. Catchment areas with a rounded shape tend to have a concentration time to become a maximum discharge faster than catchment areas with an elongated shape [34]. The elongated form has tributaries that flow into the main river at larger intervals of time and space. In this study, the watershed's shape is expressed qualitatively using the circularity ratio (Rc) approach. The equation can derive the circularity ratio [41].

In this study, water catchment forms are classified into rounded and elongated shapes, such as the classification by Suwarno [42]. In the classification, it is stated that with the Rc value > 0.5, the shape of the catchment area is rounded, and with the Rc value <0.5, the shape of the catchment area is elongated. Water catchment areas can also be divided into three based on their size: small, medium, and

IOP Conf. Series: Earth and Environmental Science 1233 (2023) 012049 doi:10.1088/1755-1315/1233/1/012049

large. Each phase has different storage characteristics. Singh distinguishes the catchment area based on its size and a catchment area is said to be small if it has a size of fewer than 250 km2, medium with a size of 250 km2-2500 km2, and a large catchment area if the area is more than 2500 km2 [43].

2.3. Drainage density

Drainage density (Dd) measures the water balance of a catchment area at a time that varies based on hydrometeorological, geomorphic, and pedomorphic conditions [44]. This process relates the runoff characteristics and the topographical or geomorphic characteristics of the catchment [45]. Geomorphically, Dd is an essential indicator of landform and is related to the potential for runoff and topographic texture ([46], [47]. The study of drainage density is an excellent parameter to predict the volume of water and is related to the release of flow in a catchment area. A high drainage density value indicates a high current density and a fast response to precipitation events [48], [49]. drainage density is the ratio of the total flow length of all orders (ΣNu) with the area of the watershed (A) [50].

Shankar & Mohan [51] classified drainage density into four classes, namely very low (<0.1), low (1.0-2.0), moderate (2.0-3.5), and high (>3,5), while Linsley [52]also classified into four classes based on runoff conditions which were divided into four classes, namely, Dd <1.0; Dd 1.0 to 2.0; Dd 2.0 to 5.0; and Dd is more significant than 5.0. In this study, the catchment area is an upstream area which is a source of flow that affects the formation of meander rivers.

According to Leopold & Langbein [53]meandering in rivers considers two contexts. The first involves the entire profile of the upstream tributary and the main river, i.e., the longitudinal profile of the river system. Changes in the slope gradient that result in the minimization of flow energy are measured regarding the height difference and the distance of the slope difference. Increasing the gradient (less than 1°) can result in significant river changes, including tributaries' development due to slope changes [54]. In addition, it can also cause vertical erosion to the front [50].

2.4. Morphostructure

Active morphostructure is an endogenous dynamic process (tectonic) in the form of lifting, folding, faulting, and jointing. The tectonic setting of Java as a reference in this study has been carried out by Hall et, al [55]. Java Island also has a structure with a west-east physiographic zone broadly following the structural zone. The distribution pattern of entrenched meanders in Java Island was analyzed based on its relationship to the regional physiographic development of Java Island. This regional analysis uses physiographic map data as the basis. Tectonic maps of the island of Java, geological maps, and literature studies of areas that have entrenched meander river phenomena are used to determine the development of morphostructure and geomorphology.

3. Results and Discussion

Identification of entrenched meander river curvature in hilly areas on Java Island was detected in 1786 locations. The curvature of the entrenched meander river on the hills in West Java Province is in Sukabumi Regency, Ciancur Regency, and Pangandaran Regency. Central Java province has a river curvature in the hills located in the northern part of Kebumen Regency, the western part of Wonosobo Regency, and the western part of Wonogiri Regency. The curvature of the river on the hills in Wonogiri Regency has the same river segment as the curvature of the river in Gunungkidul Regency, Special Region of Yogyakarta Province. East Java Province is an area that has the most distribution of river curvature in the hills. The distribution of the river curvature in the hills is in the northern part of the Nganjuk Regency, the western part of the Jombang Regency, and the southern part of the Bojonegoro Regency. The entrenched meander rivers on Java are all located in hilly areas with elevations above 100 meters above sea level.

In the West Java river segment, there is also a branching river with a meander pattern, but most of the meander river segments are single segments. River segments with a series of meanders on their tributaries and main rivers are also found in several areas in East Java. The meander river segments in Central Java are all single segments, and there is no curvature of the meandering river, which is in a series of river systems. This area has three segments of the meandering river in the hilly area. In Gunungkidul Regency, there are three meander river segments. Meander rivers have a single segment, and meander segments a series of river systems.

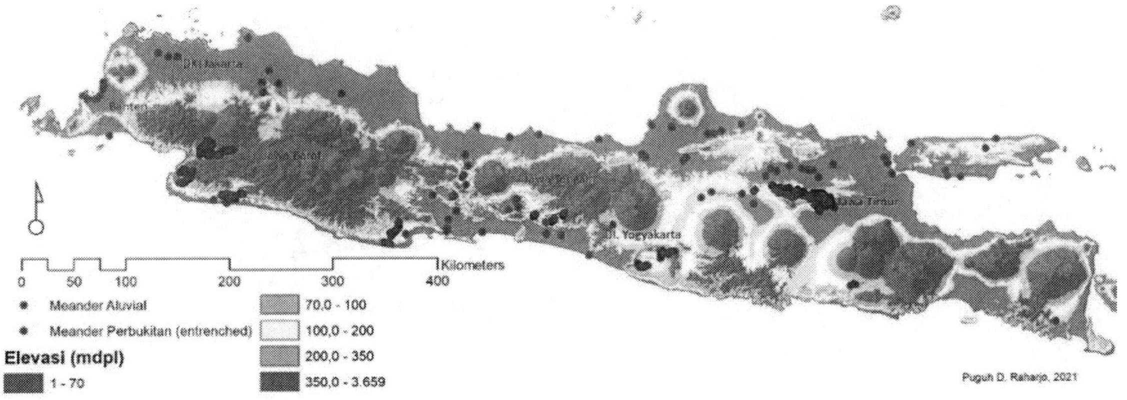

Figure 2. Distribution of meander rivers based on topography in Java, Indonesia

In this study, the morphometry measured in the catchment area where there is an entrenched meander is area and shape, drainage density, and river segments in the catchment area. Entrenched meander rivers in the hills on the island of Java are located in 68 water catchment areas. The distribution includes, namely: 11 water catchment areas in West Java Province, three catchment areas in Central Java Province, three water catchment areas in the Special Region of Yogyakarta Province, and 51 water catchment areas in East Java Province. The catchment area with an entrenched meander river object in Java Island has a varied area, from a small catchment area to a large catchment area. Most of the entrenched meanders on Java Island are in water catchment areas with an area of fewer than 100 km2. Even more than 50% (37 catchment areas) are in an area below 10 km2.

Most of the catchment areas that have a high area are in West Java Province. The Agribinta-Cianjur, Cigugur-Pangandaran, Bantargadung-Sukabumi areas are areas with water catchments in West Java with an area of more than 500 km2. The largest water catchment area on the island of Java, where there is an entrenched meander river, is in the Agribinta-Cianjur area, with an area of 1188.99 km^2. The catchment area, which has an area of less than 10 km^2, is mainly located in East Java Province.

In East Java Province, the entrenched meander river is located in a small catchment area. However, there is one catchment area that has a significant difference in the area from other water catchment areas, which is about 154.43 km2 (water catchment area Jombang-Plandaan 1). In this water catchment area, the entrenched meander occurs not in the main river but also the sub-catchments in its tributaries. In the catchment area in Jombang-Plandaan 1, 10 sub-catchments have entrenched meander rivers. Entrenched meander rivers like this also occur in catchment areas such as in Bojonegoro-Bubulan 3, Bojonegoro-Bubulan 4, Bojonegoro-Kaligandong 1, Bojonegoro-Kaligandong 1b, Bojonegoro-Kaligandong 5, Bojonegoro-Kaligandong 8, Bojonegoro-Kedungadem 3, Bojonegoro- Ngasem 1, Bojonegoro-Ngasem 3, Bojonegoro-Tambakrejo 1, Jombang-Plandaan 5, Jombang-Plandaan 6, and Nganjuk-Gondang 1. Entrenched meander rivers, which have tributaries in the form of entrenched meanders, are also all located in East Java Province. The size of the smallest catchment area is 0.74 km2 which is located in the Kaligandong-Bojonegoro 6 area. This location is also located on an entrenched meander river which is a tributary with the main river being entrenched as well.

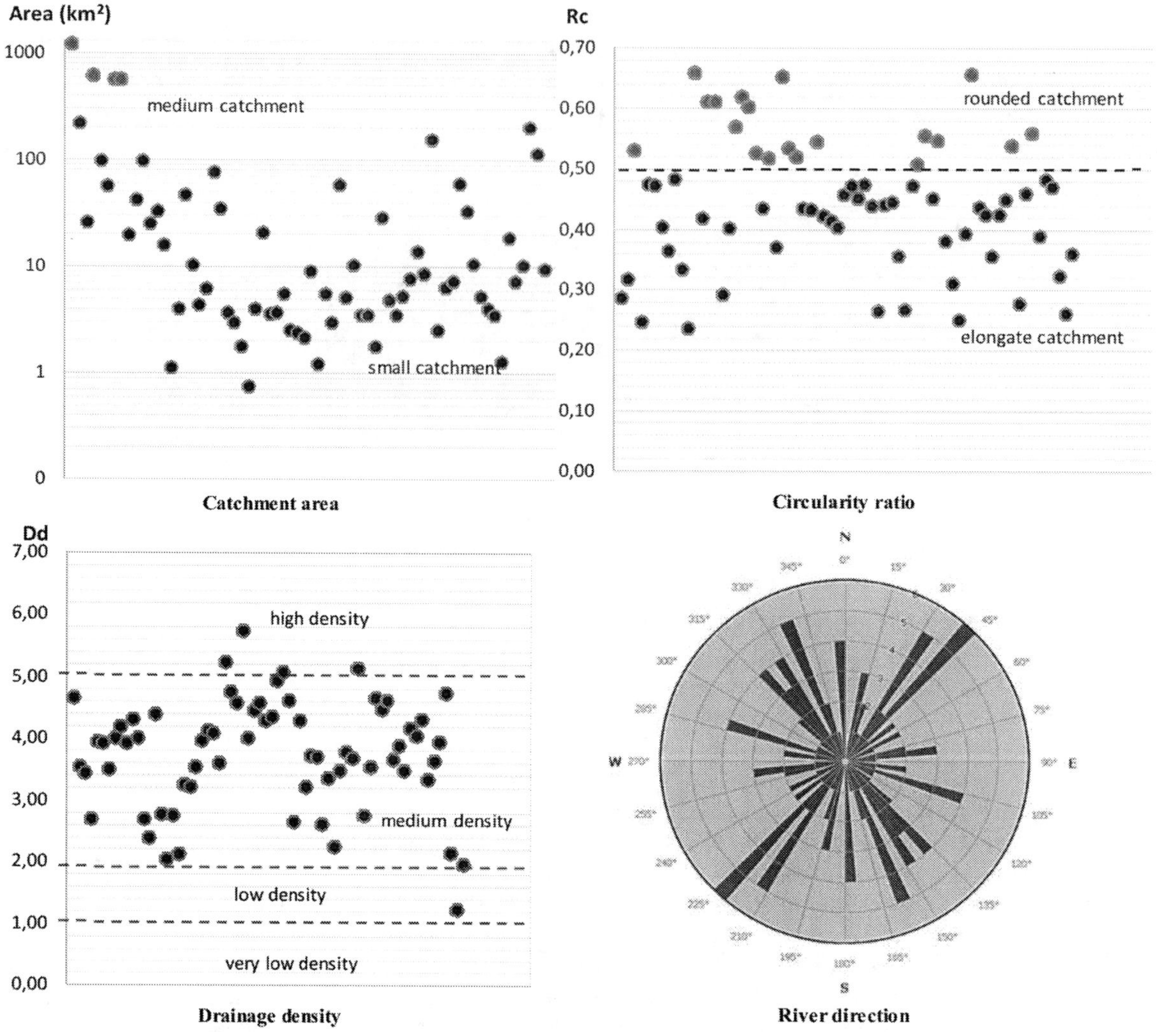

Figure 3. Morphometric classification of entrenched meander catchments in Java, Indonesia

According to the classification of catchment area size by Singh (1994), the entrenched meander rivers on the island of Java are primarily located in small catchment areas. The percentage of entrenched meander rivers in small catchments is around 94%, and those in medium catchments are around 6%. The size of the catchment area in the medium category is all in West Java Province. There are no entrenched meander rivers on Java Island, located in large catchment areas. The wider the catchment area, the larger the volume and can quickly form a meander, but most of this entrenched river on the Java island is in a small catchment area.

Most of the entrenched meander rivers on the island of Java are located in catchment areas with an elongated form of the catchment. The shape of this catchment area is based on the circularity ratio (Rc) of Miller [41]. The highest Rc value is 0.66, and the lowest is 0.24. [42] classifies the shape of the catchment area based on the value of the circularity ratio. Rc values more than 0.5 catchment areas have a rounded shape, while Rc values less than 0.5 catchment areas have an elongated shape. Of the 68 catchments in Java, most of the catchments are elongated (49 catchments), and 19 are rounded. Seyhan [34] explained the difference in the shape of the rounded and elongated catchment from the side of the runoff. Catchment areas with rounded shapes tend to have a concentration-time to reach maximum discharge faster than catchment areas with elongated shapes. The rounded shape will be more prone to flooding. The distribution of water catchment areas with a rounded shape in West Java Province has one catchment area. In Central Java Province, there are two catchment areas, and in East Java Province, there are 16 catchment areas. Most of the water catchment areas in Java Island have an elongated shape (72%). Most of the entrenched meander rivers on the island of Java occur in the form of an elongated catchment.

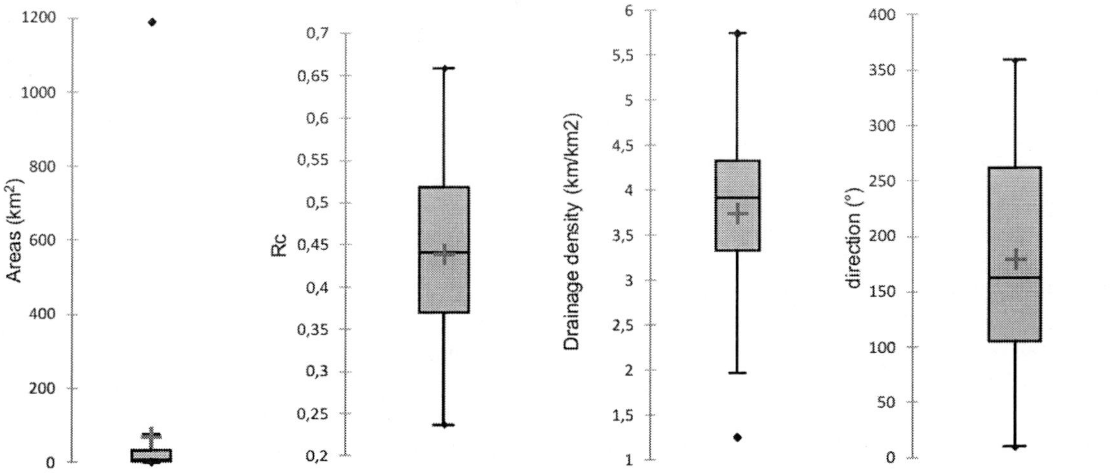

Figure 4. Catchment morphometry box plot

Drainage density in catchment areas with an entrenched meander river on Java Island shows varying values. The values range from 1.26 km/km2 to 5.75 km/km2. The difference in Dd is not only caused by differences in lithology and pedology [44]. It is also a factor that affects the volume of the main river. A high Dd indicates a high current density, and a low Dd indicates a low flow density. Drainage density correlates with an increase in discharge. The Dd value using the classification from Linsley et al., (1949)shows that there is no shallow drainage density class (Dd less than 1) in the entrenched meander river catchment area on Java Island, while low Dd is only found in 2 (3%) areas. Water catchment. The catchment area with a value of 1-2 Dd is in the catchment area in Yogyakarta Province, namely in the water catchment area in Panggang-Kaliyo 1 and Panggang-Kalioyo 2. This catchment class is in the Wonosari Formation and Kepek Formation, which are dominated by carbonate rocks. Medium density drainage (Dd, 2-5) has the highest number, 62 (91%) catchment areas. Drainage density in this catchment mostly indicates that the drainage system is quite good, surface storage is low, and drainage is good [56], [57]. High drainage density (Dd more than 5) in 4 (6%) catchment areas.

Meander rivers occur at high water volumes, so the drainage density should also be high. The higher the drainage density, the shorter the distance that runoff has to travel to the river, so the time needed to reach the peak of the flood is faster, but this does not happen to entrenched meanders on Java Island because the flow that occurs is primarily moderate. This moderate drainage density illustrates that entrenched meander rivers mainly occur on the type of surface where the flow passes through rocks that tend to be soft [42]. Drainage density is the most crucial element of landform characteristics [45], so the entrenched meander river in Java is influenced by surface characteristics in the catchment area. Horton [58]stated that this characteristic condition is related to its geological structure.

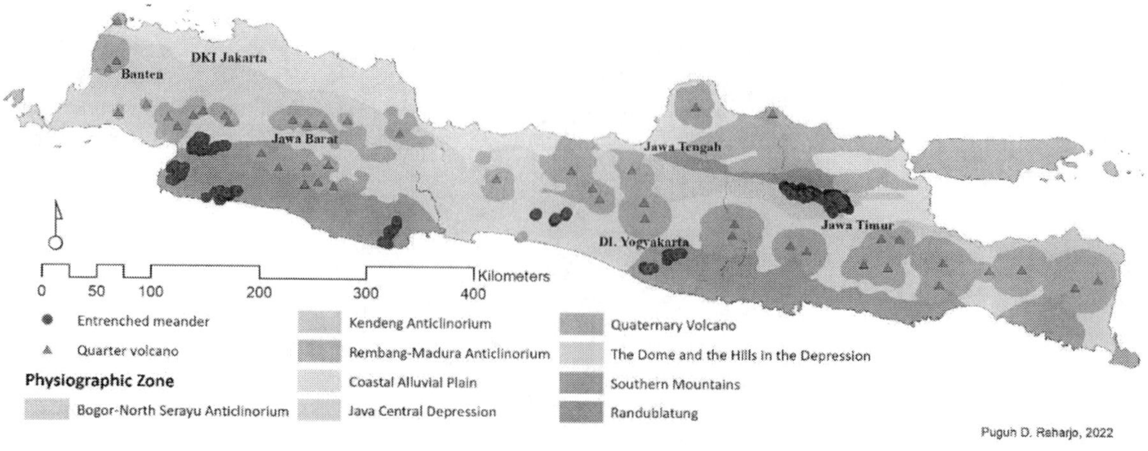

Figure 5. Distribution of entrenched meandering based on Physiographic Map of Java Island [37]

The most significant percentage of the entrenched meander river segment on the island of Java leads to the Northeast - Southwest, with a percentage of 28%. Rivers leading to the Southeast – Northwest total 17 river segments (25%), and those leading to the North-South and West-East have the same percentage of 24% or totalling 16 river segments. The diversity of relief that forms a slope is the primary determinant of the setting of the valley in which the river is formed [59], so it can be seen that the direction of the entrenched meander river is primarily determined by the position of the slope on the hills with specific characteristics. The entrenched meander is related to the morphostructure in each landscape, especially in certain areas related to structural landforms. This structural landform is strongly influenced by geodynamics, which is part of the orogenic belt on the island of Java. It is known that Java Island has an active geodynamic history and has undergone several phases of tectonic development. Most of the meander rivers in the hills on the island of Java are incised symmetrically (entrenched). The catchment area on structural landforms (folds and faults) affects the formation of entrenched meander rivers in the hills. The presence of structural landforms that result in differences in topographic gradients (deflection) provides a specific morphology that can disrupt flow stability.

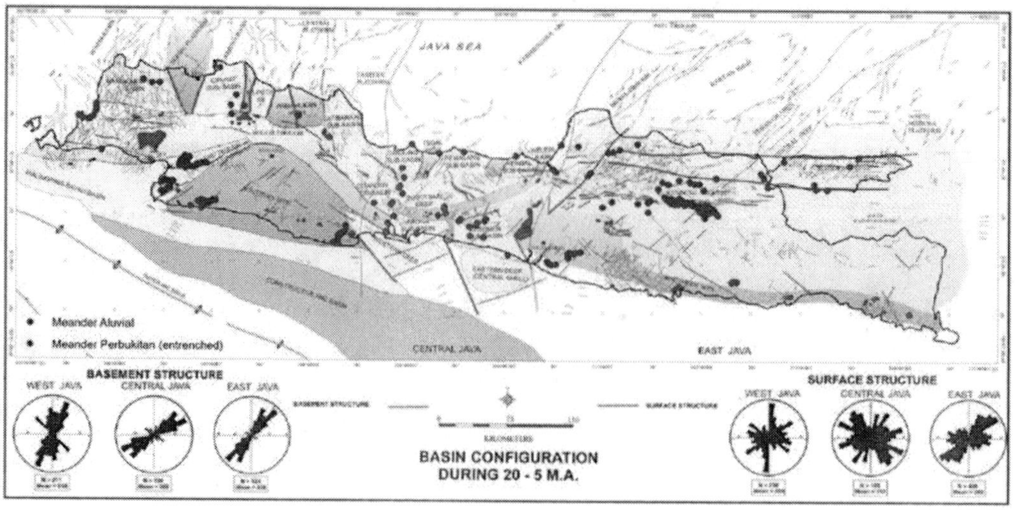

Figure 6. Distribution of entrenched meandering based on Tectonic Framework of Java Island at 20 – 5 Ma [60]

A southward shift of the Indian-Australian Plate boundary took place, followed by continuous magmatic activities occupying almost the whole of the island of Java. In the northern part, back-arc basins developed, subdivided into several sub-basins, which were separated by basement highs, controlled by basement block faulting. In northern West Java the orientation of basement faulting remains north-south, whereas in the Southwest and Central Java a significant NE-SW and NW-SE fault system seems more pronounced (Figure 6). These patterns may have been initiated as strike-slip pair faulting during the process of northward subduction along the southern part of Java [60].

It can be concluded that entrenched meandering, based on Sribudiyani et al. (2003), on the island of Java is statistically located in the zone controlled by structures with high intensity, especially in the transpressional zone. The configuration of the surface structure plays an essential role in the control of entrenched meandering. The main directions of the main structure of West Java, Central Java, and East Java have quite visible differences. The main directions of the structure of western Java are N-S with a mean value of N059^{0}E, Central Java with a trend towards NW-SE and SE-NW with a mean of N112^{0}E, and eastern Java with a NE-SW direction with a mean of N042^{0}E.

4. Conclusion

The island of Java has many entrenched meander rivers with an even distribution. The distribution is in hilly topography, especially in structural landforms. The high flow volume is an important factor in forming a meandering river. It can be seen from the level of morphometry of the water catchment. Entrenched meander rivers in Java Island tend to be in small catchment areas, relatively elongated catchments, and have medium drainage density. The morphometric characteristics are the nature of water catchment which tends to have a small volume of water, and floods that occur relatively longer. The catchment of the entrenched meander river tends to have a different character from the catch of the free meander (alluvial).

The main rivers in the entrenched meander in West Java and Central Java tend to have a direction to the south, southwest, and west. This is different from the direction of the entrenched meander river in East Java. Structural landforms due to morphostructure in the Ciletuh-Jampang Kulon area (West Java), Karangsambung area (Central Java), Gunung Kidul area (DI Yogyakarta), and the Kendeng Mountains area (East Java) greatly affect the formation of entrenched meander rivers and their catchment characteristics.

References

[1] R. Charlton, *Fundamentals of Fluvial*. New York: Taylor & Francis e-Library, 2007.
[2] D. J. Furbish, "Using the Dynamically Coupled Behavior of Land-Surface Geometry and Soil Thickness in Developing and Testing Hillslope Evolution Models," in *Prediction in Geomorphology*, P. R. Wilcock and R. M. Iverson, Eds. Washington, D.C., 2003, pp. 169–182.
[3] B. I. Gartsman and L. M. Korytny, "A new fundamental work on stream channel studies," *Geogr. Nat. Resour.*, vol. 29, no. 3, pp. 301–302, 2008, doi: 10.1016/j.gnr.2008.09.018.
[4] G. Ioana-Toroimac, L. Zaharia, G. Minea, R. Zarea, and M. Borcan, "Channel Typology Based on Stability Criteria. Case Study: Ialomiţa and Buzău Watersheds (Romania)," *Procedia Environ. Sci.*, vol. 14, pp. 177–187, 2012, doi: 10.1016/j.proenv.2012.03.017.
[5] J. Holbrook and S. A. Schumm, "Geomorphic and sedimentary response of rivers to tectonic deformation: A brief review and critique of a tool for recognizing subtle epeirogenic deformation in modern and ancient settings," *Tectonophysics*, vol. 305, no. 1–3, pp. 287–306, 1999, doi: 10.1016/S0040-1951(99)00011-6.
[6] E. W. Lane, *A Study of the shape of channels formed by natural streams flowing in erodible material*. U.S. Army Engineer Division, Missouri River, 1957.
[7] R. C. Moore, "Significance of Inclosed Meanders In The Physiographic History of The Colorado Plateau Country," *J. Geol.*, vol. 34, no. 2, pp. 29–57, 1926, doi: doi.org/10.1086/623284.
[8] A. Winslow, "The Osage River and Its Meanders," *Adv. Sci.*, vol. 22, no. 546, pp. 31–32, 1893,

doi: 10.1126/science.ns-22.546.31.
[9] W. B. Langbein and L. B. Leopold, "River meanders - Theory of minimum variance. Physiographic and hydraulic studies of rivers," *Geol. Surv. Prof. Pap.*, no. 422-H, pp. 1–15, 1966, [Online]. Available: https://pubs.usgs.gov/pp/0422h/report.pdf.
[10] B. L. Rhoads and M. R. Welford, "Initiation of river meandering," *Prog. Phys. Geogr.*, vol. 15, no. 2, pp. 127–156, 1991, doi: 10.1177/030913339101500201.
[11] D. W. Burbank and R. S. Anderson, *Tectonic Geomorphology*. Australia: Blackwell Science Ltd, 2001.
[12] A. Gupta, *Tropical Geomorphology*. New York: Cambridge University Press, 2011.
[13] L. B. Leopold, M. G. Wolman, and J. P. Miller, *Fluvial Processes in Geomorphology*. San Francisco: Freeman and Company, 1964.
[14] S. A. Schumm, "Patterns of Alluvial Rivers," *Ann. Rev. Earth Planet. Sci.*, vol. 13, pp. 5–28, 1985, doi: 10.1146/annurev.ea.13.050185.000253.
[15] M. L. Assine, E. R. Merino, F. Do Nascimento Pupim, H. De Azevedo Macedo, and M. G. M. Dos Santos, "The Quaternary alluvial systems tract of the Pantanal Basin," *Brazilian J. Geol.*, vol. 45, no. 3, pp. 475–489, 2015, doi: 10.1590/2317-4889201520150014.
[16] J. R. Barbour, "The origin and significance of sinuosity along incising bedrock rivers," Columbia University, 2008.
[17] W. . M. . Davis, "Meandering Valleys and Underfit," *Ann. Assoc. Am. Geogr.*, vol. 3, pp. 3–28, 1913, doi: 10.1080/00045601309356993.
[18] L. B. Leopold and M. G. Wolman, "River Channel Patterns: Braided, Meandering, and Straight," *Geol. Surv. Prof. Pap.*, no. 282-B, p. 50, 1957, doi: 10.3133/pp282B.
[19] D. R. Montgomery and K. B. Gran, "Downstream variations in the width of bedrock channels," *Water Resour. Res.*, vol. 37, no. 6, pp. 1841–1846, 2001, doi: 10.1029/2000WR900393.
[20] J. Petrovszki and G. Timár, "Channel sinuosity of the Körös River system, Hungary/Romania, as possible indicator of the neotectonic activity," *Geomorphology*, vol. 122, no. 3–4, pp. 223–230, 2010, doi: 10.1016/j.geomorph.2009.11.009.
[21] J. D. Stock and D. R. Montgomery, "Geologic constraints on bedrock river incision using the stream power law," *J. Geophys. Res. Solid Earth*, vol. 104, no. B3, pp. 4983–4993, 1999, doi: 10.1029/98jb02139.
[22] W. A. Tarr, "Intrenched and Incised Meanders of Some Streams on the Northern Slope of the Ozark Plateau in Missouri Author (s): W . A . Tarr Source : The Journal of Geology , Vol . 32 , No . 7 (Oct . - Nov ., 1924), pp . 583-600 Published by : The University of Ch," *J. Geol.*, vol. 32, no. 7, pp. 583–600, 1924, doi: doi.org/10.1086/623142.
[23] K. J. Tinkler, "Active Valley Meanders," *Area*, vol. 5, no. 1, pp. 41–43, 1973, [Online]. Available: https://www.jstor.org/stable/20000014.
[24] P. F. Macar, "Effects of Cut-Off Meanders on the Longitudinal Profiles of Rivers," *J. Geol.*, vol. 42, no. 5, pp. 523–536, 1934, doi: 10.1086/624197.
[25] A. J. Hartley, A. Owen, G. S. Weissmann, and L. Scuderi, "Modern and ancient amalgamated sandy meander-belt deposits: recognition and controls on development," in *Other publications Number 48 of the International Association of Sedimentologists*, 2019, pp. 349–384.
[26] M. Kirkby, "Alluvial and Non-Alluvial Meanders," *Area*, vol. 4, no. 4, pp. 284–288, 1972, [Online]. Available: https://www.jstor.org/stable/20000712.
[27] B. Zhang, N. S. Ai, Z. W. Huang, C. B. Yi, and F. C. Qin, "Meanders of the Jialing River in China: Morphology and formation," *Chinese Sci. Bull.*, vol. 53, no. 2, pp. 267–281, 2008, doi: 10.1007/s11434-007-0516-2.
[28] J. Challinor, "The 'Incised Meanders' near Pont-erwyd, Cardiganshire," vol. 70, no. 2, pp. 90–92, 1933, doi: doi.org/10.1017/S0016756800091287.
[29] C. R. Twidale, *Structural Landforms*. Canberra: Australian National University Press, 1971.
[30] W. . S. Cole, "The Interpretation of Intrenched Meanders," *J. Geol.*, vol. 38, no. 5, pp. 423–436,

1930, [Online]. Available: http://www.jstor.org/stable/30067328.

[31] G. S. Hancock and R. S. Anderson, "climate Numerical modeling of fluvial strath-terrace formation in response to oscillating climate," *Geol. Soc. Am. Bull.*, vol. 114, pp. 1131–1142, 2002, doi: 10.1130/0016-7606(2002)114<1131.

[32] R. S. Anderson and S. P. Anderson, *Geomorphology: The Mechanics and Chemistry of Landscapes*. Cambridge, UK: Cambridge University Press, 2010.

[33] P. D. Raharjo, S. Winduhutomo, K. Widayanto, and E. Puswanto, "Analisa Hidrologi Permukaan dalam Hubungannya dengan Debit Banjir DAS Lukulo Hulu dengan Menggunakan Data Penginderaan Jauh," *J. Geogr.*, vol. 13, no. 2, pp. 163–178, 2016, doi: https://doi.org/10.15294/jg.v13i2.7974.

[34] E. Seyhan, *Fundamentals of Hydrology*. Geografisch Instituut der Rijksuniversiteit, 1977.

[35] A. M. Gurnell *et al.*, "A multi-scale hierarchical framework for developing understanding of river behaviour to support river management," *Aquat. Sci.*, 2015, doi: 10.1007/s00027-015-0424-5.

[36] A. R. G. Large and D. J. Gilvear, "Google Earth as a virtual globe tool for Earth science applications at the global scale: progress and perspectives," *Int. J. Remote Sens.*, vol. 33, no. 12, pp. 3966–3986, 2012, doi: 10.1002/rra.

[37] V. Bemmelen, *The Geology of Indonesia*, I.A. Government Printing Office, 1949.

[38] R. Hall, B. Clements, H. R. Smyth, and M. A. Cottam, "A new interpretation of Java's structure," in *Proceedings 31st Annual Convention*, 2007, no. May, pp. 63–86, doi: 10.29118/ipa.1077.07.g.035.

[39] P. D. Raharjo, "Pemodelan Hidrologi untuk Identifikasi Daerah Rawan Banjir di Sebagian Wilayah Surakarta Menggunakan SIG (Sistem Informasi Geografi)," *Limnotek Perair. Darat Trop. di Indones.*, vol. XVI, no. 1, pp. 1–11, 2009.

[40] I. Zavoianu, *Morphometry of Drainage Basins*, 1st ed. Amsterdam, The Netherlands: Elsevier Science Publisher, 1985.

[41] V. C. Miller, "A quantitative geomorphic study of drainage basin characteristics in the clinch mountain area virginia and tennessee," 1953.

[42] Suwarno, *Hidrologi: Pengukuran dan Pengolahan Data Aliran Sungai, (Hidrometri)*. Bandung: Nova, 1991.

[43] V. P. Singh, *Elementary Hydrology*. New Delhi: Prentice Hall, 1994.

[44] D. G. Day, "Drainage density variability and drainage basin outputs," *J. Hydrol. (New Zealand)*, vol. 22, no. 1, pp. 3–17, 1983.

[45] C. W. Carlston, "Drainage Density and Streamflow," *Geol. Surv. Prof. Pap.*, no. 422–C, 1963.

[46] M. Mark, "Geomorphometric parameters: a review and evaluation," *Geogr. Ann. Ser. A, Phys. Geogr.*, vol. 57, no. 3, pp. 165–177, 1975, doi: 10.1080/04353676.1975.11879913.

[47] A. Strahler, "Quantitative Analysis of Watershed Geomorphology," *Trans. Am. Geophys. Union*, vol. 38, no. 6, pp. 913–920, 1957, doi: 10.1029/TR038i006p00913.

[48] S. L. Dingman, "Drainage Density and Streamflow : A Closer Look," *Water Resour. Res.*, vol. 14, no. 6, pp. 1183–1187, 1978, doi: 10.1029/WR014i006p01183.

[49] J. A. Prieto-Amparán *et al.*, "A Multivariate Geomorphometric Approach to Prioritize Erosion-Prone Watersheds," *Sustainability*, vol. 11, no. 5140, pp. 1–21, 2019, doi: doi:10.3390/su11118514.

[50] R. E. Horton, "Erosional Development of Streams and Their Drainage Basins ; Hydrophysical Approach to Quantitative Morphology," *Bull. Geol. Soc. Am.*, vol. 56, no. 3, pp. 275–370, 1945, doi: 10.1130/0016-7606(1945)56.

[51] M. N. R. Shankar and G. Mohan, "Assessment of the groundwater potential and quality in Bhatsa and Kalu river basins of Thane district, western Deccan Volcanic Province of India," *Environ. Geol.*, vol. 49, no. 7, pp. 990–998, 2006, doi: 10.1007/s00254-005-0137-5.

[52] R. K. Linsley, M. A. Kohler, and J. L. H. Paulhus, *Applied hydrology*, 1st ed. New York: McGraw-Hill, 1949.

IOP Conf. Series: Earth and Environmental Science 1233 (2023) 012049 doi:10.1088/1755-1315/1233/1/012049

[53] L. B. Leopold and W. B. Langbein, "River Meanders," *Sci. Am.*, vol. 214, no. 6, pp. 60–73, 1966, [Online]. Available: jstor.org/stable/24930965.
[54] R. Raj, "Journal of Asian Earth Sciences Active tectonics of NE Gujarat (India) by morphometric and morphostructural studies of Vatrak River basin," *J. Asian Earth Sci.*, vol. 50, pp. 66–78, 2012, doi: 10.1016/j.jseaes.2012.01.010.
[55] R. Hall, "Cenozoic geological and plate tectonic evolution of SE Asia and the SW Pacific: computer-based reconstructions, model and animations," *J. Asian Earth Sci.*, vol. 20, no. 4, pp. 353–431, 2002, doi: doi.org/10.1016/S1367-9120(01)00069-4.
[56] V. Te Chow, D. R. Maidment, and L. W. Mays, *Applied hydrology*. New York: McGraw Hill, 1988.
[57] T. Gunawan, "Penerapan Teknik Penginderaan Jauh untuk Menduga Debit Puncak Menggunakan Karakteristik Lingkungan Fisik DAS, Studi Kasus di Daerah Aliran Sungai Bengawan Solo Hulu Jawa Tengah," Institute Pertanian Bogor, 1991.
[58] R. E. Horton, "Drainage-basin characteristics," *Eos, Trans. Am. Geophys. Union*, vol. 13, no. 1, pp. 350–361, 1932, doi: 10.1029/TR013i001p00350.
[59] G. Brierley and K. Fryirs, *Geomorphology and River Management, Applications of the River Styles Framework*. Oxford, UK: Blackwell Publishing, 2005.
[60] Sribudiyani *et al.*, "The Collision of the East Java Microplate and Its Implication for Hydrocarbon Occurrences in the East Java Basin," 2003, doi: 10.29118/ipa.1530.03.g.085.

Groundwater quality analysis using water quality index in Wonosari basin, Gunungkidul regency-Indonesia

R A Bagaskara[1] **and M Widyastuti**[1*]

[1]Department of Environmental Geography, Faculty of Geography, Universitas Gadjah Mada, 55281 Special Region of Yogyakarta, Indonesia

*Corresponding author: mwiwik@ugm.ac.id

Abstract. Wonosari Basin is a physiographical unit in Gunungkidul Regency, Indonesia, that has enormous groundwater potential. Limestone inside the basin and dense population can harm the environment. The research aimed to analyze the groundwater quality in the Wonosari Basin based on the applicable quality standards, determine the water condition using the Water Quality Index (WQI), and determine the spatial pattern of the WQI grade. The total sample used in this research is 21 samples. Total Dissolved Solid (TDS), pH, Iron (Fe), Manganese (Mn), Sulfate (SO4), Nitrate (NO3), Hardness (CaCO3), and Total Coliform values of the groundwater samples were inputted to the weighted arithmetic WQI using Microsoft Excell to analyze the water quality. The spatial analysis was carried out using the cross-tabulation method with Microsoft Excell. The water quality standard used in this research is Regulation of the Minister of Health of the Republic of Indonesia number 23 of 2017. The research result showed that total coliform is the only parameter that exceeds its maximum acceptable presence, WQI values indicate that most of the groundwater samples have good quality, and the spatial pattern of water quality follows the population density.

1. Introduction

Water is a type of resource essential to sustain the lives of living organisms. The hydrological cycle creates a variety of extractable water resources such as rain, surface water, and groundwater [1]. Groundwater is freshwater with an abundant quantity, amounting to 1.3% of the total water on earth, produced through the infiltration process in the hydrological cycle [2]. Gunungkidul Regency is among the regions in Indonesia abounding with groundwater potential. However, with bioclastic limestones forming most of the aquifer [3], the calcium element is dissolved into the water, rendering the water quality less acceptable for varying purposes [4]. The Gunungkidul population is concentrated in the Wonosari Basin with Wonosari District is the most populated district in Gunungkidul Regency based on Central Bureau of Statistics Indonesia data (2021), where the gently sloped terrain suits and furthers settlement developments and human activities. Densely populated settlements can indicate the most current and potential economic activities of sizable intensities that generate large amounts of domestic waste, including land conversion into agricultural fields, infrastructures, and even mining that also changes the water quality in the area [5].

The combination of limestone aquifers and high population density conclusively modifies the groundwater quality. The denser the population, the worse the quality of the surrounding environment, therefore the quality of the water will be polluted as well, this is in accordance with the Environmental Kuznet's Curve theory [6]. Limestone causes high hardness, domestic waste also pollutes, while the water demand in the Basin is quite high, absolutely it will cause problems. The Wonosari Basin is also an area with minimal groundwater quality studies, although it has quite unique area characteristics.

The problem's above warrants the need for water quality analyses in the Wonosari Basin area, especially those concerning the possibility of contamination that may disturb the water's quality and irreplaceable function in human life. Water Quality Index (WQI) is one of many water quality analysis techniques that indicate suitability for consumption in index values [7]. In addition, the index aids the local populations in understanding the quality of the water they use, thus ultimately facilitating communication between experts or professionals and the public [8].

The research is conducted in the Wonosari Basin (Figure 1) to analyze the groundwater quality for domestic use based on applicable standards of quality, measure the WQI, and determine and interpret the spatial pattern of the water quality grades. Besides suitability to meet local domestic water usage, the results also describe the correlation between the WQI and its influencing factors. Examples of these factors are population density and rock formation.

2. Data and method

2.1. Data type and acquisition

Wonosari Basin was selected as the research area based on several considerations. The basin is the center of the Gunungkidul population concentration, thus creating more sources of pollution than its neighboring areas. Furthermore, the local communities rely heavily on the enormous groundwater potential in the basin morphology to meet their daily and consumption needs. To analyze the water quality, a set of parameters were tested: total dissolved solids (TDS), hardness ($CaCO_3$), pH, sulfate, nitrate, manganese (Mn), iron (Fe), and total coliform, and then used to calculate the WQI. Other parameters were measured but not included in the WQI calculation, namely electrical conductivity, odor, color, and taste.

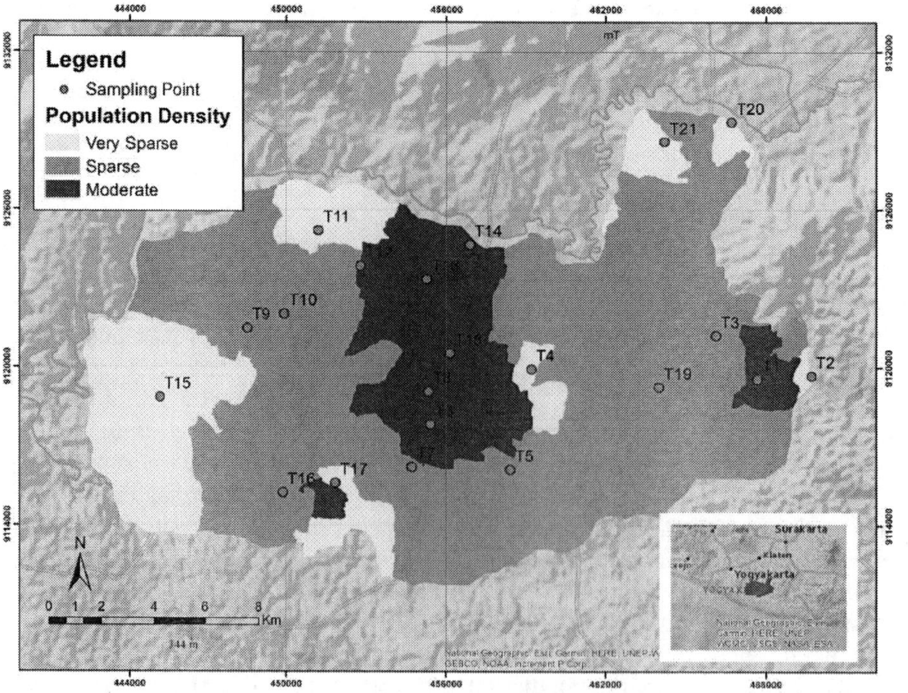

Figure. 1. Map of the Research Location and Groundwater Sampling Points (classification of density refers to Munawir (2005))
Source: Data processing (2022)

For the groundwater quality analyses, primary and secondary data were used. The primary data included water quality parameter values that were recorded or measured directly in the field during the rainy season in November 2021 using stratified random sampling based on population density (resulting in 21 samples, see Figure 1) and were obtained from the laboratory analysis results of the collected water samples. The secondary data were as follows:

1. Population density per subdistrict as the unit of analysis in deciding the sampling points.
2. Rock formation to characterize the study area.

2.2. Data analysis technique

2.2.1. Water Quality Index (WQI) Calculation. The gathered data were analysed using the Weighted Arithmetic Water Quality Index [9] and the water quality standards issued by the Minister of Health (MoH) in Regulation No. 23 of 2017 as reference. The calculation consisted of a series of steps. The first one used equation (1) below:

$$\mathrm{WQI} = \frac{\Sigma \mathrm{Qi} \times \mathrm{Wi}}{\Sigma \mathrm{Wi}} \tag{1}$$

where Qi indicates the quality rating scale and Wi the weight value of all the parameters used. Qi was obtained using equation (2):

$$\mathrm{Qi} = 100 \times \frac{(\mathrm{Vi}-\mathrm{Vo})}{Si-Vo} \tag{2}$$

where Vi indicates the measured value of parameter i, Vo the ideal value of parameter i in water, and Si the standard value of parameter i. Vo can be considered zero (0), except when the water pH is 7. Wi in the WQI equation was obtained using equation (3):

$$\mathrm{Wi} = \frac{\mathrm{K}}{\mathrm{Si}} \tag{3}$$

where Si denotes the standard value of parameter i and K is a constant value obtained from equation (4):

$$\mathrm{K} = \frac{1}{\Sigma(1/\mathrm{Si})} \tag{4}$$

The calculation above is operated using Microsoft Excel. The value of WQI must be classify according to table 1.

Table 1. Water Quality Rating and Grading

WQI	Rating of Water Quality	*Grading*
0-25	*Excellent Water Quality*	A
26-50	*Good Water Quality*	B
51-75	*Poor Water Quality*	C
76-100	*Very Poor Water Quality*	D
>100	*Unsuitable for Drinking Purpose*	E

Source: Tyagi (2013)

2.2.2. Spatial analyst. Spatial analysis was conducted using the quantitative descriptive method. The data observed were the derived WQI values and their spatial distribution and correlation with factors that influenced the quality. The factor linked to water quality was population density, and their correlation was tested using the cross-tabulation method with Microsoft Excel.

3. Result and discussion

3.1. Water quality for domestic use

3.1.1. Odor, taste, and color. These parameters were measured qualitatively with the researcher's senses. The observation results showed that all the 21 groundwater samples (see the distribution in Figure 1) had no odor, taste, or color that might indicate pollution. Generally, the groundwater in the Wonosari Basin is odorless, tasteless, and colorless or transparent, meaning that organoleptically it is not contaminated.

3.1.2. Electrical Conductivity (EC). This parameter describes the ability of water to conduct electricity. This parameter is closely linked to chemical properties, i.e., the abundance of ions dissolved in water, which can be used to indicate the relative quality of groundwater [10]. Furthermore, according to the definition, total dissolved solids (TDS) also include the concentration of dissolved ions, meaning that EC can indicate TDS levels in the water [11].

Based on the measurement results, the EC of the groundwater samples ranged from 348 mhos/cm (T12) to 796 mhos/cm (T16) (Figure 2). The smallest EC was found at the wells in areas with moderate population density, whereas the largest was in sparsely populated areas. These findings show that EC is not related to population density but may be dependent on the concentration of dissolved ions such as nitrate and sulfate. High nitrate content in the water is attributable to the decomposition of organic matter [12], and the concentration of sulfate, the main anion in water, can increase because of human activities. Nevertheless, other ions also regulate electrical conductivity, particularly those sensitive to rock formations (aquifer's composition), more than human activities. Nitrates and sulfates are merely a few of the many ions present in water.

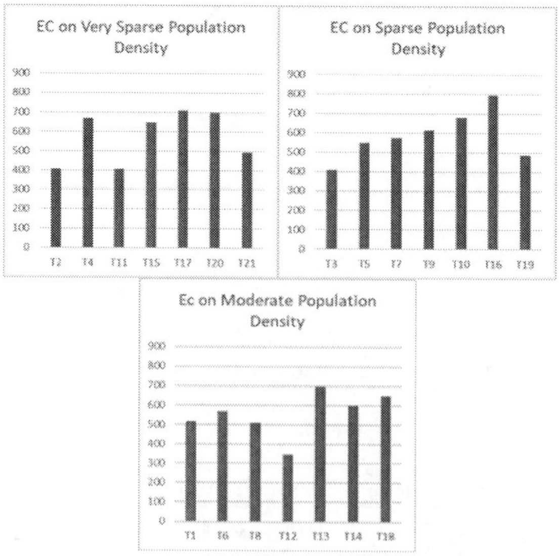

Figure. 2. Electrical Conductivity Graph of Groundwater Samples by Population Density
Source: Data processing (2022)

The electrical conductivity of groundwater in natural waters varies typically from 20 to 1500 mhos/cm [13]. The EC levels measured from 21 groundwater samples in the study area are within the normal range (Figure 2). This result shows that the water in the Wonosari Basin is of relatively good quality.

3.1.3. Total Dissolved Solids (TDS). Figure 3 shows the total dissolved solids (TDS) contained in the groundwater samples. The highest TDS was found at the well located in sparsely populated areas (564 mg/l, well T16), whereas the lowest coincided with moderate population density (247 mg/l, well T12). Based on the information depicted in this figure, it can be inferred that TDS is randomly distributed and shows no relationship with population density. However, also based on Figure 3, the graphic pattern of the TDS levels is similar to that of the electrical conductivity, suggesting a relationship between the two parameters. In addition, the spatial pattern of the TDS may be influenced more by the aquifer's composition than human activities. The aquifer of the Wonosari Basin mainly consists of bioclastic limestones, which are bordered by tuffs to the north and carbonate limestones to the south, according to the geological map sheets of Surakarta Giritontro [14].

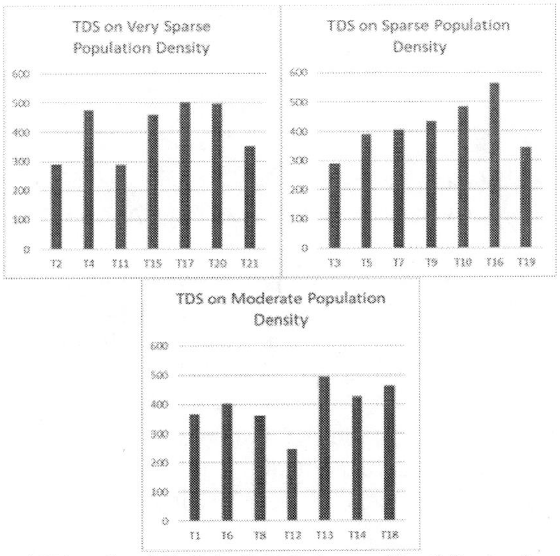

Figure. 3. Graph of the Total Dissolved Solids (TDS, in mg/l) of Groundwater Samples by Population Density Source: Data processing (2022)

To understand the water quality from this parameter, the standards issued in the MoH Regulation No. 32 of 2017 were used. TDS contents of all the 21 groundwater samples (564–247 mg/l) are still below the threshold value for hygiene and sanitation purposes, i.e., 1000 mg/l. Therefore, the groundwater in the basin is generally safe for domestic use by the local populations.

3.1.4. pH. Figure 4 shows the measurement results of the groundwater pH. Generally, all the samples had a PH value close to neutral (7), except for well T2. This well had a pH of higher than 8 and was located in areas with very sparse population density.

pH is linked to the dissociation of water molecules into hydrogen ions, making the water acidic, and hydroxide ions, contributing to basic properties [11]. In other words, the presence of dissolved acids and bases affects water pH. A low pH indicates corrosive water, whereas a high pH shows alkaline water.

The MoH Regulation No. 32 of 2017 suggests a range of 6.5 to 8.5 for a safe freshwater pH. The upper limit is 8.5 because alkaline water is safer for consumption than water with acidic properties. All the groundwater samples had a relatively similar pH (close to 7 or neutral); therefore, according to the spatial pattern formed in the graph (Figure 4), there is no relationship between pH and population density. The bar chart also shows that the groundwater observed is safe for domestic use because the pH values are within the regulated upper and lower bounds.

IOP Conf. Series: Earth and Environmental Science 1233 (2023) 012050 doi:10.1088/1755-1315/1233/1/012050

Figure. 4. pH Graph of Groundwater Samples by Population Density
Source: Data processing (2022)

3.1.5. Nitrate (NO_3). Figure 5 shows the distribution of the nitrate content of 21 groundwater samples according to population density. Based on the measurement results, nitrate appears to be relatively high in locations with very sparse population density, while the low nitrate levels tend to coincide with moderate population density, except for well T6 (3.9 mg/l). Pollution by organic matter, such as plant litter and feces and urine produced by living organisms, is responsible for the accumulation of nitrates in water [11]. Plantations and farms are the main sources of nitrates [15] due to the use of organic matter to fertilize crops; therefore, Figure 5 illustrates that agricultural land in areas with sparse and very sparse population density determines the nitrate distribution.

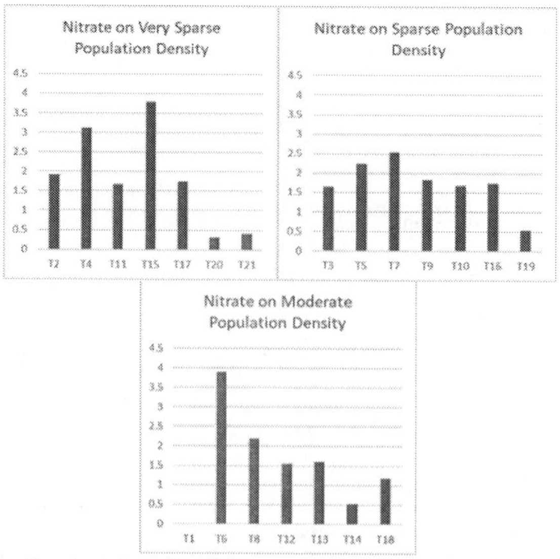

Fig. 5. Graph of the Nitrate Content (mg/l) of Groundwater Samples According to Population Density
Source: Data processing (2022)

The MoH water quality standard for nitrate contents in freshwater suggests a maximum of 10 mg/l. However, as seen in Figure 5, the groundwater samples contained 0.0132 mg/l (T1) to 3.9 mg/l (T6) of nitrates. Therefore, the specified threshold is not exceeded at this range, meaning that the groundwater in the Wonosari Basin is of good quality and safe for daily use.

3.1.6. Sulfate (SO_4). Figure 6 shows the sulfate content of 21 groundwater samples in different population densities. Based on the measurement results, sulfate tends to be lower in highly populated areas and vice versa. Sulfate is a naturally occurring dominant anion in waters. Still, it does not exclude the possibility that some sulfates are the byproducts of human activities [16], including runoff from croplands that apply chemical fertilizers. The sulfate distribution pattern is similar to nitrate, indicating the influence of agricultural land in areas where populations are sparse and very sparse.

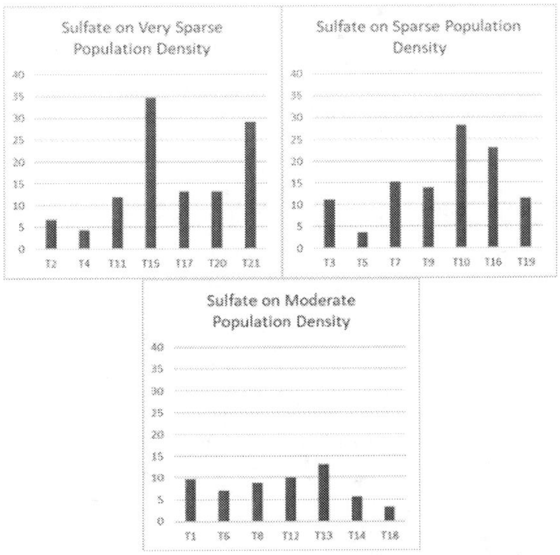

Figure. 6. Graph of the Sulfate Content (mg/l) of Groundwater Samples According to Population Density
Source: Data processing (2022)

According to Regulation No. 32 of 2017, freshwater should not contain more than 400 mg/l of sulfate. Based on the measurement results presented in Figure 6, the sulfate contents were in the range of 3.2 mg/l (T18) to 34.7 mg/l (T15) or far below the threshold. Therefore, the groundwater in the Wonosari Basin can be used to meet the local domestic needs.

3.1.7. Hardness ($CaCO_3$). Figure 7 shows that the measured hardness forms a random spatial pattern and is unrelated to population density. It confirms the theory that hardness results from contacts between water and rock [13]. Limestones that compose the rock formation contribute significantly to hardness because the Ca elements are easily dissolved in water [4]. For this reason, the hardness level in the Wonosari Basin is generally higher.

The hardness level of the groundwater samples was between 156 mg/l (T12) and 356 mg/l (T17). Although this range is higher than the hardness in non-limestone areas, it is still below the threshold. In Regulation No. 32 of 2017, freshwater's hardness should not exceed 500 mg/l. It means that the groundwater samples meet the water quality standard for domestic water. However, observations in the field found that the groundwater extracted through the wells had to be boiled and filtered to remove the lime.

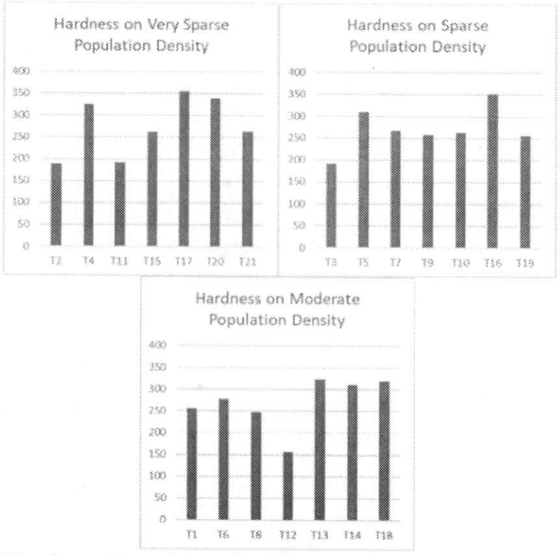

Figure. 7. Graph of the Hardness Level (mg/l) of Groundwater Samples According to Population Density
Source: Data processing (2022)

3.1.8. Iron (Fe). Figure 8 shows that the measured iron contents form a particular spatial pattern indicating a relationship with population density. Locations with moderate population density tend to have higher iron levels than those with sparse and very sparse populations. Human activities that generate industrial, domestic, and agricultural wastes substantially affect the presence of iron in water [17]. Large amounts of heavy metals leached out of landfills into the environment contaminate underground or downstream water systems [18].

Per Regulation No. 32 of 2017, the water quality standard for iron concentration is set at 1 mg/l. Based on the measurement results shown in Figure 8, all the groundwater samples contained small amounts of iron, with the highest, 0.289 mg/l, found at well T12 (moderate population density). Therefore, all the groundwater samples are safe for domestic use.

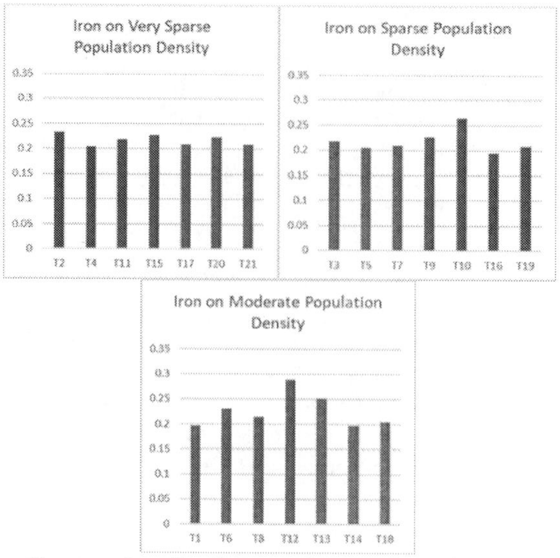

Figure. 8. Graph of the Iron Content (mg/l) of Groundwater Samples According to Population Density
Source: Data processing (2022)

3.1.9. Manganese (Mn). Figure 9 shows that the spatial pattern of the manganese distribution in the study area is similar to that of iron, i.e., a more densely populated area has a higher manganese level. Also, like iron, manganese is a heavy metal originating from industrial, domestic, and agricultural waste, signifying the contribution of human activities to manganese content in waters. According to the MoH water quality standard, the maximum allowable presence of this heavy metal is 0.5 mg/l. Therefore, the sampled groundwater is safe for domestic use.

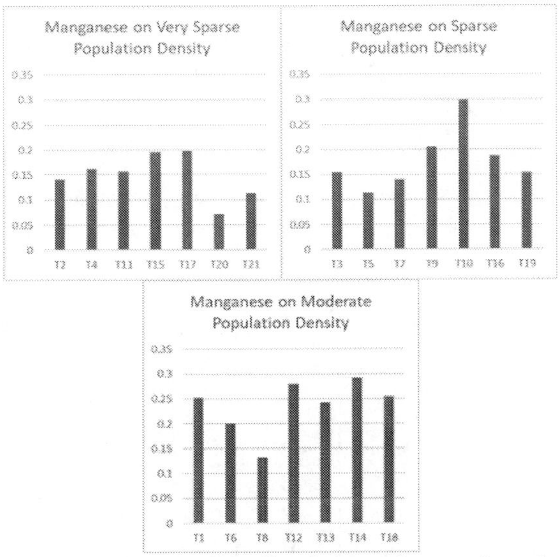

Figure. 9. Graph of the Manganese Content (mg/l) of Groundwater Samples According to Population Density
Source: Data processing (2022)

3.1.10. Total Coliform. The spatial pattern of the total coliform counts in Figure 10 indicates a relationship with population density: a more densely populated area has a higher total coliform concentration. Total coliforms measure groups of coliform bacteria found in water [11], which may come from human excreta. Human activities determine if coliform bacteria are present and are found in substantial amounts that cause pollution.

Per the MoH regulation, total coliform counts in freshwater should not exceed 50 MPN/100 ml. However, total coliforms in all the samples were above the standard; some were even higher than 1600 MPN/100 ml (too many to count by the instrument used), except for well T17 (total coliform = 1.8 MPN/100 ml). Therefore, it is not recommended to use the groundwater to meet daily domestic needs.

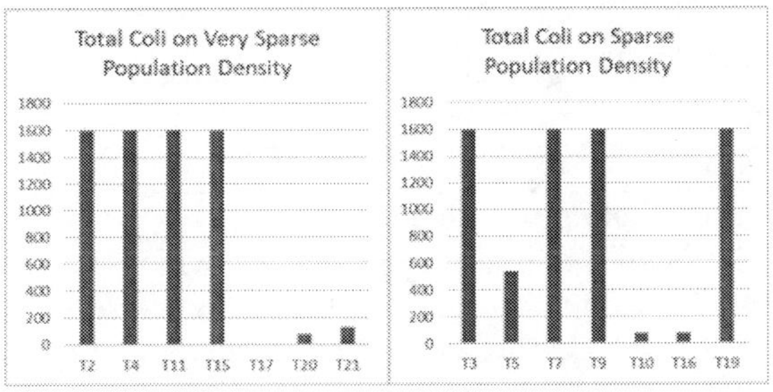

IOP Conf. Series: Earth and Environmental Science 1233 (2023) 012050 doi:10.1088/1755-1315/1233/1/012050

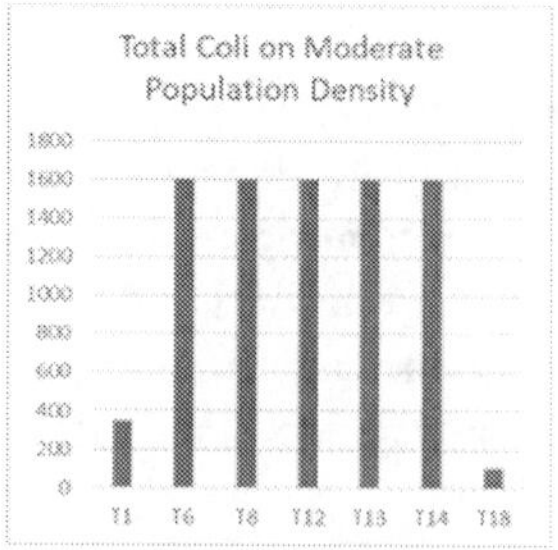

Figure. 10. Graph of the Total Coliform (MPN/100 ml) of Groundwater Samples According to Population Density
Source: Data processing (2022)

3.2. Water Quality Index (WQI)

The Water Quality Index was calculated from the derived parameter values. Table 2 shows that most of the groundwater samples are of good quality, but some are relatively poor. Wells T20 and T21 are groundwater samples with excellent or grade-A quality.

Table 2. Water Quality Index (WQI) Values and Their Classification

No	WQI Value	Grade	Rate of water quality
T1	42.46	B	Good
T2	46.63	B	Good
T3	46.84	B	Good
T4	46.78	B	Good
T5	27.59	B	Good
T6	53.25	C	Poor
T7	44.91	B	Good
T8	43.99	B	Good
T9	52.36	C	Poor
T10	47.48	B	Good
T11	47.20	B	Good
T12	64.78	C	Poor
T13	57.56	C	Poor
T14	61.79	C	Poor
T15	53.29	C	Poor
T16	31.07	B	Good
T17	31.08	B	Good
T18	38.82	B	Good
T19	46.56	B	Good
T20	16.36	A	Excellent
T21	24.82	A	Excellent

Source: Data processing (2022)

The WQI calculation involves several stages, one of which is weighting. At this stage, each parameter is assigned a weight value according to their respective impacts on water quality, which determines the outcome to a very great degree. Table 3 shows the weight value of each parameter, with the highest given to manganese (Mn) and then iron (Fe). Accordingly, both parameters provide the most contribution to the derived WQI.

The WQI analysis revealed that 13 of the 21 groundwater samples had good quality and were suitable for domestic use. Two others had excellent quality—the best groundwater condition observed in the Wonosari Basin—because their parameter values meet the standards for consumption. On the contrary, six samples showed poor quality, a condition that demands carefully planned treatment strategies before the local populations can extract and use the groundwater to meet their daily water requirements.

Table 3. Weight Values of WQI Parameters

Parameter	Weight
Hardness	0.00062
Sulfate	0.00077
Nitrate	0.03083
Iron	0.30834
Manganese	0.61668
pH	0.03628
TDS	0.00031
Total Coliform	0.00617

Source: Data processing (2022)

3.3. Spatial pattern of the WQI grade

The spatial pattern of the Water Quality Index (WQI) grade was analyzed by considering two underlying factors: rock formation and population density. Wonosari Basin is mainly composed of clastic limestones and, consequently, has homogenous materials across the rock formation. On the other hand, the population density varies from very sparse to moderate, and the settlements are concentrated at the center of the basin area. Therefore, spatial variations in the water quality, if any, are visible across different population densities.

The cross-tabulation results (Table 4) showed a relationship between population density and the WQI grade. It is because population density represents the existence of human activities as a determinant of groundwater pollution. In addition, the spatial pattern of the WQI grade further confirms that a more densely populated area has a poorer environmental condition.

Table 4. Cross Tabulation between the Water Quality Index (WQI) Grade and Population Density

Pop. Density / WQI Grade	Very Sparse	Sparse	Moderate
A (excellent)	2	0	0
B (good)	4	6	3
C (poor)	1	1	4

Source: Data processing (2022)

4. Conclusion

In the majority, the groundwater in the Wonosari Basin does not meet the water quality standards for domestic use based on the Regulation of the Minister of Health No. 32 of 2017 because one of the parameters is present in excessively high concentrations in 20 groundwater samples, that is, total coliforms. Only the remaining sample contains total coliforms below the threshold, i.e., well T17. On the contrary, the parameters other than total coliform indicate that the groundwater quality of the Wonosari Basin meets the specified standards. For this reason, it is highly recommended to boil the extracted groundwater to remove the bacteria before consumption.

The groundwater quality in the basin has three grades according to the Water Quality Index (WQI): A or excellent (two groundwater samples or wells), B or good (13), and C or poor (6). To calculate WQI, each of the parameters used in the research has been assigned a weight value depending on their degree of impact on water quality based on the applicable quality standards. Iron (weight = 0.3) and manganese (0.6) are the most influencing parameters; both heavy metals can substantially harm human health when present in drinking/consumed water. Iron and manganese are believed to be responsible for grade-C or poor water quality in the study area.

The spatial pattern of the calculated index values demonstrates a relationship between WQI and population density. Population density is a surrogate parameter for human activities as a source of groundwater pollution; a higher population density means more intensive human activities. In addition, it has been identified that iron, manganese, and total coliform appear to have higher concentrations in locations with moderate population density. Overall, the derived WQI grades and rates are directly proportional to population density.

Acknowledgements

The authors would like to thank Universitas Gadjah Mada for funding the research with the Final Project Recognition grant (RTA) for the fiscal year 2022. Gratitude is also extended to the Department of Environmental Geography, Faculty of Geography, for providing the opportunity and permission to conduct the research.

References

[1] Musy A and Higy C 2011 *Hydrology: A Science of Nature* (Enfield: Science Publishers).
[2] Davie T 2008 *Fundamentals of Hydrology* (London: Routledge Taylor and Francis).
[3] Setiawan T and Asgaf N M A 2016 *Jurnal Lingkungan dan Bencana Geologi.* **7** 155-167.
[4] Haryono E and Adji T N 2004 *Geomorfologi dan Hidrologi Karst* (Yogyakarta: Kelompok Studi Karst Fakultas Geografi Universitas Gadjah Mada).
[5] Liyanage C P and Yamada K 2017 *Sustainability.* **9** 1405.
[6] Stern D I 2004 *World Development.* **32**, 1419-1439.
[7] Tyagi S, Sharma B, Singh P and Dobhal R 2013 *American Journal of Water Resources.* **1** 34-38.
[8] Brown R M, McClelland, Nina I, Deininger R A and Tozer R G 1970 *Water Sewage Works.* **117** 339-343.
[9] Brown R M, McClelland N J, Deininger R A and O'Connor M F A 1972 *Proceedings in International Conference on water pollution Research, Jerusalem.* **6** 787-797.
[10] Santosa L W and Tjahyo N A 2014 *Karakteristik Akuifer dan Potensi Airtanah Graben Bantul* (Yogyakarta: Gadjah Mada University Press).
[11] Todd D K and Mays L W 2005 *Groundwater Hydrology Third Edition* (US: John Wiley and Sons Inc).
[12] Mustofa A 2015*Jurnal DISPROTEK.* **6** 13-19.
[13] Effendi H 2003 *Telaah Kualitas Air, Bagi Pengelolaan Sumber Daya dan Lingkungan Perairan* (Yogyakarta: Penerbit Kanisius).
[14] Surono, Toha B and Sudarno I 1992 *Peta Geologi Lembar Surakarta–Giritontro* (Bandung: Pusat Penelitian dan Pengembangan Geologi).
[15] Rusydi A F, Naily W and Lestiana H 2015 *RISET Geologi dan Pertambangan.* **25** 87-97.
[16] Ananda M S 2019 *AMINA.* **1** 35-38.
[17] Putra A Y and Mairizki F 2020 *Jurnal Katalisator.* **5** 47-43.
[18] Fatmawinir, Suyani H and Alif A 2015 *J. Ris. Kim.* **8** 101-107.

IOP Conf. Series: Earth and Environmental Science 1233 (2023) 012051 doi:10.1088/1755-1315/1233/1/012051

An Isotopic Approach to Investigate Seepage in Dam: Case Study Ciawi and Sukamahi Dry Dams, West Java, Indonesia

F K Dewi[1,2], W Wilopo[1*], I Azwartika[2]

[1]Departement of Geological Engineering, Gadjah Mada University, Yogyakarta, Indonesia
[2]Ministry of Public Works and Housing, Indonesia

*Corresponding author: wilopo_w@ugm.ac.id

Abstract. Seepage control in the dam is vital in maintaining dam safety. It is important to determine the source of seepage water to be controlled so that there is no excessive seepage that threatens dam stability. This study used Physico-chemical parameters and stable isotopes (δ^2H and $\delta^{18}O$) to determine groundwater characteristics, the groundwater origin around the dam, and the connectivity of the Ciawi and Sukamahi Dry Dam's seepage with surrounding water sources. This research shows that the groundwater in the study area includes freshwater with TDS values varying from 62 ppm to189 ppm and EC values from 62 µS/cm to 414 µS/cm. The results of stable isotope analysis showed that the composition of seepage and groundwater sources around the dams are around the local meteoric water line (LMWL); this indicates that groundwater is more influenced by rainwater. Comparing the composition of δ^2H and $\delta^{18}O$ showed that the seepage water from the two dams has the same genesis as springs in Sukakarya Village (MA02) and dug wells in Gadog Village (SG13). This information is beneficial in determining the handling of seepage in the Ciawi dan Sukamahi Dry Dams.

1. Introduction

Ciawi and Sukamahi Dams were built as part of flood control in Jakarta. These dams are located in the upper reaches of the Ciliwung Watershed in Bogor Regency. The construction of these two dams is expected to reduce the maximum flood discharge heading to Jakarta. Ciawi and Sukamahi Dams are national strategic projects and are the first dry dams in Indonesia. The effective storage volume of the Ciawi Dam is $\pm 5.03x10^6$ m^3c, and the effective storage volume of the Sukamahi Dam is $\pm 1.32x10^6$ m^3[2].

Ciawi and Sukamahi dams are located in the Bogor groundwater basin with a large relative groundwater discharge; this is influenced by a large amount of groundwater recharge from rainfall in the Bogor groundwater basin recharge area [3]. The location of the two dams in the productive aquifer [4] is feared to have problems such as stability and seepage.

Seepage control is essential to maintain dam safety. Seepage through the dam body is planned and drained by a filter mounted on the dam[5]. Excessive seepage can cause structural failure in the dam. Finding the causes of a dam leak's leaky water is essential for fixing the dam, restoring it, and maintaining it [6]. Therefore, seepage in the dam needs to be controlled not to disturb the dam's stability. Prevention and handling of seepage in dams can be done with a hydrogeological approach [7]. One method that can be used to track water seepage in dams is the isotope method. Several authors have used methods to determine seepage in the dam. In China, a combination of ^{222}Rn, hydrogen and oxygen isotopes, anion, and temperature uses to identify the water leakage source in the Pushihe pumped storage power station [6]. Hydrological and isotopic analyses to identify sources and causes of leaks in earth dams[5] were also carried out in Taiwan. Piezometric and isotopic analyses were used to determine the leaks in Gazelle dams, Algeria [7]. In Indonesia, several studies using isotopes to identify seepage and the origin of water around the dam were carried out on the Ngancar dam [8][9]. Several studies using isotopes to detect seepage in dams have been carried out on dams with water in the reservoir. At the same time, at the Ciawi and Sukamahi Dry Dams, there is no water in the reservoir, and there is no use of the stable isotopes approach to identify the source of seepage in these two dams.

In the study presented here, a novel approach of applying physico-chemical properties and stable isotope composition of the water is used for assessing seepage water and water sources around the dam. This study aimed to determine the groundwater characteristics, the groundwater origin, and the connectivity of the Ciawi and Sukamahi Dry Dam's seepage with surrounding water sources. Determine the characteristics of the water around the dam based on Physico-chemical properties. In this study, as many as seventeen samples of seepage, springs, dug wells, and rivers, measured physical and chemical properties consisting of total dissolved solids (TDS), electrical conductivity (EC), temperature, and pH. Stable isotope (δ^2H and $\delta^{18}O$) analysis of the seventeen samples was carried out to determine the groundwater origin. Comparison of stable isotope compositions can provide information about seepage connectivity with groundwater sources around the dam. This approach can provide a scientific reference for handling seepage and drainage problems using engineering technology, improving the dam's operational efficiency and safety.

2. Study area

2.1. Location

This research was conducted in Ciawi and Sukamahi Dams, Bogor Regency, West Java Province, Indonesia, as shown in Figure 1. This dam is located in the upper Ciliwung Sub-Watershed. Like its function, this dry dam only holds water when there is an enormous flood discharge on the Ciliwung River Upstream and Sukabirus River. It releases storage water when the river discharge returns to normal. The distance between the two dams is ± 5km.

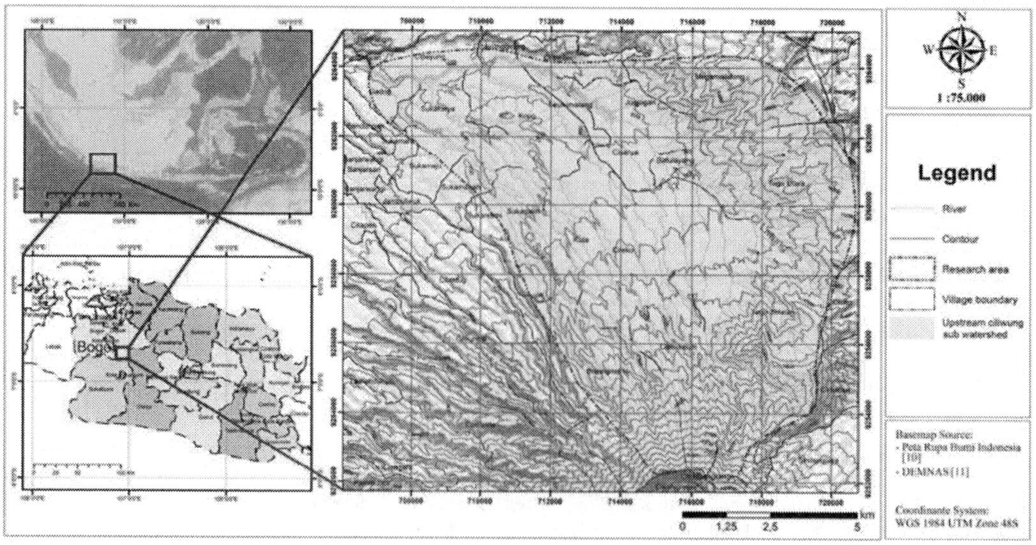

Figure 1. Research area.

The research location is in a tropical climate, with the rainy season starting in September/October and the dry season starting in April/May. The average annual rainfall in this area is 3,178 mm/year, with a maximum temperature of 24.2 °C and a minimum temperature of 20 °C [12].

2.2. Geological and hydrogeological conditions

The research location is in the old quarter volcanic rock formations consisting of lava, lava, andesite breccia to basalt and pumice tuff [13]. From old to young, the constituent rocks in the dam consist of volcanic breccia, andesite lava, tuffaceous sandstone, and colluvial and alluvial deposits [1].

Figure 2. (a) Seepage of the left pedestal of the Sukamahi Dam (RS02); **(b)** Sukagalih Spring (MA11).

Ciawi and Sukamahi Dams are located in extensive and moderately productive aquifers. These aquifers vary widely in transmitivity, and the depth of unconfined groundwater is generally deep[4]. Unconfined aquifers in the study area include pore aquifers consisting of weathered breccia and tuffaceous sandstone. Figure 2(a) shows the seepage from the weathered andesite breccia. Likewise, Sukagalih Village's spring (Figure 2(b)) andesite breccia rocks are still found.

3. Material and methods

3.1. Sample collection

Water samples were taken for isotope testing, as many as seventeen samples at the dam location and the upstream part of the watershed of the two dams. Sampling was carried out on dam seepage (S), river water (RS), springs (MA), and dug wells (SG) around the dam location, which were thought to have a similar genesis as dam seepage water. Samples of river water, springs, and dug wells have a higher elevation than the dam site.

Water sources are measured in situ using a pH meter to determine water characteristics based on physico-chemical properties, including pH, temperature, EC, and TDS. The groundwater table is measured using a roll-meter for dug wells and GPS for their location. Then processed using GIS software to obtain groundwater flow patterns and direction

The isotopic composition of seepage water was analyzed along with water samples from water sources around the dams to identify the source of seepage in the dam. For this purpose, a water sample was collected from March 15, 2022, until March 25, 2022. A sample at each location was taken as much as 100 ml (for δ^2H and $\delta^{18}O$ analysis) using airtight glass bottles. The bottle was filled with sample water by bringing it closer to the water source or placing it in a water bucket. The bottle was filled to the brim, and no bubbles were trapped to avoid evaporation and condensation. Analysis of δ^2H and $\delta^{18}O$ was carried out at the Hydrogeology and Hydrochemistry Laboratory, Bandung Institute of Technology, using a liquid water stable isotope analyzer, Picarro L-2130i.

Table 1 shows seepage samples taken from six seepage locations at each dam with the sample code RS. River samples taken on the Ciliwung River and Sukabirus River, dammed by the two dams, are coded S. The four spring samples are coded MA, and five dug wells are coded SG. The samples are scattered in the upstream Ciliwung Sub-Watershed and Sukabirus Sub-Watershed, as shown in Figure 3.

Table 1. Sample coordinate and elevation

Sample Code	Location	Coordinate		Elevation (m asl)
		X (m)	Y(m)	
RS 01	V Notch Ciawi Dam	707713	9263657	505
RS 02	Sukamahi Dam	708144	9262193	619
RS 03	Sukamahi Dam	708161	9262398	585
RS 04	V Notch B. Sukamahi	707925	9262446	586
RS 05	Ciawi Dam	707777	9263765	523
RS 06	Ciawi Dam	707640	9263472	529
S 02	Sukabirus River	710287	9260457	696
S 03	Ciliwung River	715098	9261437	840
MA 02	Sukakarya Village	709812	9262728	715
MA 06	Cisarua Village	714341	9260952	855
MA 03	Kuta Village	710315	9261944	720
MA 11	Sukagalih Village	710571	9260237	768
SG 04	Citeko Village	713756	9259598	963
SG 08	Kopo Village	712450	9261537	810
SG 09	Jogjogan Village	713321	9263169	749
SG 11	Tugu Selatan Village	717320	9258745	1097
SG 13	Gadog Village	707770	9263313	721

3.2. *Deuterium and $\delta^{18}O$ isotopic approach*

Stable isotope analysis is one method to determine the origin of groundwater, recharge areas, and aquifer potential [14]. The stable isotopes δ^2H and $\delta^{18}O$ are the main constituents of water, and δ^2H and $\delta^{18}O$ have relatively more minor abundances than hydrogen (^1H) and Oxygen-16 (^{16}O). To find the ratio of isotopes with the same element ($^2H/^1H$ and $^{18}O/^{16}O$) expressed relative to an international standard, the Mean Ocean Water Standard (SMOW). The isotopic composition of water is calculated using mass spectrometry and is represented by the notation (delta) in units per mil. The values of δ^2H‰ (D‰) and $\delta^{18}O$‰ are obtained from equation[15]:

$$\delta D‰ = \frac{(^D/_H)\text{sample}-(^D/_H)\text{SMOW}}{(^D/_H)\text{SMOW}} \text{x } 1000 \quad (1)$$

$$\delta^{18}O\ ‰ = \frac{(^{18}O/^{16}O)\text{sample}-(^{18}O/^{16}O)\text{SMOW}}{(^{18}O/^{16}O)\text{SMOW}} \text{x } 1000 \quad (2)$$

A positive value indicates that the sample's isotopic composition is larger than the standard. In contrast, a negative value indicates that the sample's isotopic composition is smaller than the standard. There is a close relationship between $\delta^{18}O$ and $\delta\,^2H$ in global freshwater. The equation "global meteoric water line" (GMWL) is as follows [16]:

$$\delta D = 8\,\delta^{18}O + 10\ ‰ \quad (3)$$

BATAN researchers have calculated the composition of the ratio of $\delta^{18}O$ and $\delta^{2}H$ isotopes contained in local rainwater in Bogor and Jakarta. Based on the data, the local meteoric water line (LMWL) equation for the study area is as follows[17]:

$$\delta D = 7.78\,\delta^{18}O + 12.93\ ‰ \quad (4)$$

The ratios of $\delta^{2}H$ and $\delta^{18}O$ isotopes were plotted on a graph and compared with GMWL/LMWL to determine the groundwater genesis. If there are similarities in the values of $\delta^{2}H$ and $\delta^{18}O$, it can be concluded that groundwater comes from the same source (same genesis) or in the same flow pattern [18].

4. Results and discussion

4.1. The Physico-chemical properties

The Physico-chemical properties measured in this study included pH, TDS, EC, and temperature. The measurement results are shown in Table 2. TDS values in seepage samples ranged from 62 ppm to 189 ppm (99.67 ppm average); this value was higher than the value of other groundwater sources with an average of 59.54 ppm. The TDS values of the tested samples can be grouped into two groups; group 1 has a TDS value < 100 ppm, and group 2 has a TDS > 100 ppm. Table 2 shows that only three samples were included in group 2 (TDS>100 ppm), which are RS01, RS04, and MA03. However, the TDS value of the whole sample was freshwater because the TDS value is < 1000 ppm [19]. A low TDS value can also indicate a relatively short time of contact between water and rocks [18].

EC value can be used as a parameter to determine water quality. From Table 2, EC values range from 62 µS/cm to 414 µS/cm. Six samples have an EC value of less than 100 µS/cm, S02, MA02, MA11, SG04, SG11, and SG13. All water samples are freshwater because they range between 30 µS/cm - 2000 µS/cm[20].

PH is the negative logarithm of the hydrogen ion concentration in water; neutral pH values are 6.5 to 8.5 [20]. The measurement results in the field obtained a pH value of 5 to 8.2 with an average value of 6.6. An acidic pH was below <6.5[16], so eight samples are acidic. The temperature shown in Tabel 2 ranges from 25.1°C to 30.7°C, with an average value of 28.4°C.

Table 2. The measurement result of the physico-chemical properties of water

Sample Code	TDS (ppm)	EC (µs/cm)	pH	Temperature (°C)	Note
RS 01	189	389	7.4	27.5	perennial
RS 02	63	126	6.5	29.4	perennial
RS 03	62	118	6.5	29.4	perennial
RS 04	122	244	6.9	28.1	perennial
RS 05	79	158	6.0	29.4	perennial
RS 06	83	166	8.2	30.4	perennial
S 02	46	92	7.8	26.8	perennial
S 03	50	100	7.8	30.7	perennial
MA 02	31	62	5.0	30.5	perennial
MA 03	210	414	6.5	30.5	perennial
MA 06	54	118	7.2	25.7	perennial
MA 11	49	98	6.4	27.9	perennial
SG 04	36	66	5.8	28.1	perennial
SG 08	52	104	5.8	29.3	perennial
SG 09	51	104	6.2	28.7	perennial
SG 11	33	78	6.2	25.1	perennial
SG 13	43	88	5.7	25.6	perennial

IOP Conf. Series: Earth and Environmental Science 1233 (2023) 012051 doi:10.1088/1755-1315/1233/1/012051

4.2. Characteristic distribution of stable isotopes

The isotope analysis results of the seepage samples from the dam and around the dam are shown in Table 3. The groundwater level analysis estimated groundwater flows from south to northwest (Figure 3). The increase of the $\delta^{18}O$ isotope concentration was observed in the direction of groundwater flow from upstream to around the dam (high to low elevation). Figure 4 shows that the isotopic composition of the study area is fractionated due to evaporation and the effect of altitude. The lower the altitude, the isotopic composition will be heavier.

Table 3. δ^2H and $\delta\ ^{18}O$ isotope analysis results

Sample Code	$\delta^{18}O$ (‰)	SD	δD (‰)	SD
RS 01	-6.106	0.007	-35.467	0.081
RS 02	-6.679	0.048	-39.070	0.089
RS 03	-6.591	0.066	-39.100	0.069
RS 04	-6.615	0.034	-38.720	0.087
RS 05	-6.593	0.037	-39.893	0.010
RS 06	-6.322	0.038	-37.271	0.035
S 02	-7.732	0.014	-45.402	0.009
S 03	-7.734	0.078	-45.894	0.007
MA 02	-6.634	0.022	-38.357	0.060
MA 06	-7.929	0.033	-47.699	0.146
MA 03	-7.029	0.013	-42.369	0.150
MA 11	-6.917	0.038	-40.533	0.023
SG 04	-7.545	0.025	-43.757	0.086
SG 08	-7.195	0.033	-41.691	0.037
SG 09	-7.138	0.247	-43.026	0.057
SG 11	-7.840	0.035	-46.236	0.016
SG 13	-6.772	0.008	-39.514	0.028

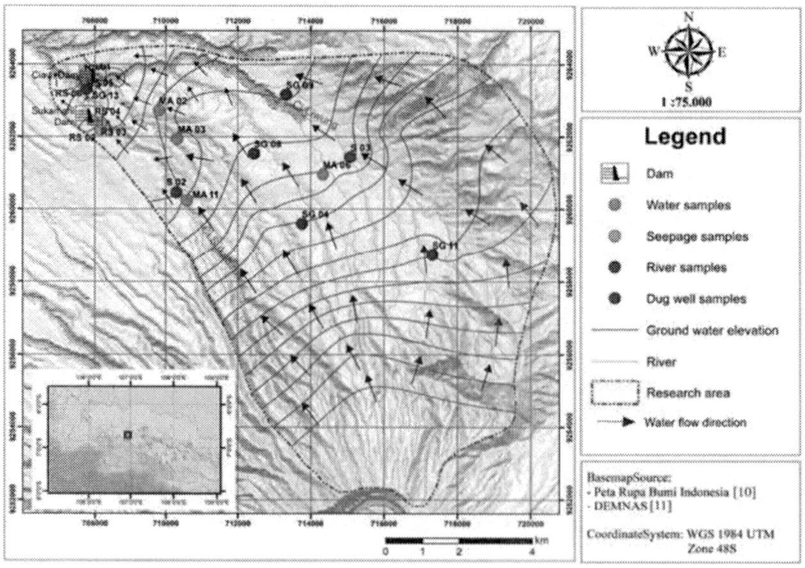

Figure 3. Groundwater flow patterns in the study area

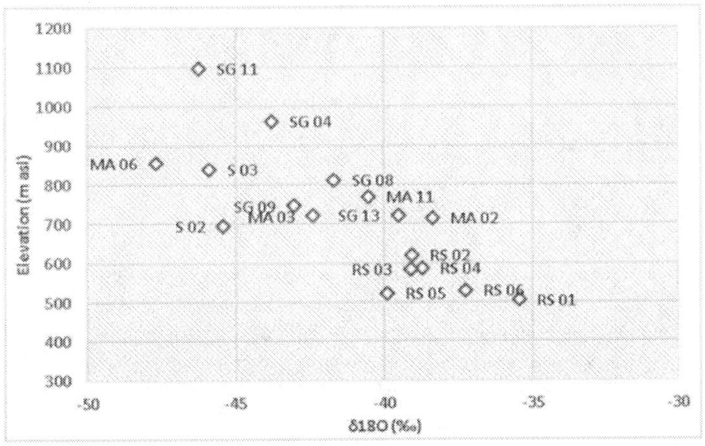

Figure 4. Relationship between $\delta^{18}O$ and elevation

4.3. Groundwater Origin

The relationship between $\delta\,^{2}H$ and $\delta^{18}O$ with LMWL(4) and GMWL(3) can be seen in Figure 5. It can be seen that the isotopic composition of water samples tends to be close to the local meteoric line (LMWL). It shows that during the rainy season (when sampling), some of the shallow aquifers around the dam are dominantly derived from rainwater. All groundwater samples have a lower slope than the slope of GMWL; this indicates that isotope fractionation has occurred, resulting in changes in the molecular composition of light and heavy isotopes. The slope value, which is less than the slope of GMWL, indicates that the humidity in the recharge area is relatively smaller than that of GMWL [14]. The stable isotopes of seepage samples in both dams were from -6.679‰ to -6.106‰ for $\delta^{18}O$ with an average of -6.484‰. Deuterium is in the range of -39.893‰ to -35.467‰ with an average of -38.254‰.

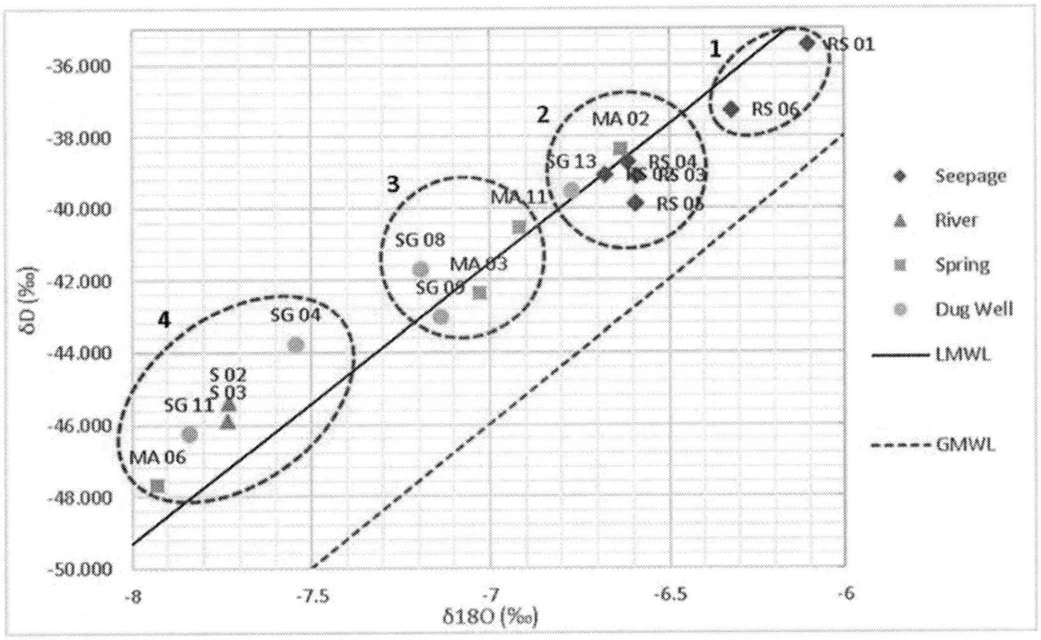

Figure 5. Relationship between $\delta^{2}H$ and $\delta^{18}O$

IOP Conf. Series: Earth and Environmental Science 1233 (2023) 012051 doi:10.1088/1755-1315/1233/1/012051

Based on the relationship between δ^2H, $\delta^{18}O$, and LMWL (Figure 5), groundwater origin around the two dams is grouped into four as follows:

a. Group 1 consists of RS01 and RS02, both of which were Ciawi Dam's seepage samples. This group is heaviest in δ^2H and $\delta^{18}O$ isotopic compositions than the other groups. Deuterium compositions value between -35,467‰ to -37.271‰ and δ ^{18}O compositions between -6.106‰ to -6.322‰. Both samples deviated from the local meteoric line so that both have experienced evaporation from the recharge area. The heavier isotopic composition can also be caused by the lower elevation of these two samples than the other samples [21].
b. Group 2 consists of RS02, RS03, RS04, RS05, MA02, and SG13. Group 2 mainly contains seepage samples from dams, where four out of six samples in this group are seepage samples from dams. Sample RS05 deviated from LMWL; it indicates RS05 has evaporated from the recharge area. The isotopic composition in group 2 is lighter than in group 1; this shows that the elevation of the catchment area of this group is higher than that of group 1. The range values of δ^2H and $\delta^{18}O$ points RS03, RS04, RS05, MA02, and SG13, show the same genesis. However, RS05 and MA02 are assumed to have different groundwater flows from the other samples. Groundwater sources have lighter isotopic composition at higher elevations, but RS02 has lighter isotopic composition even though they are at lower elevations. Likewise, MA02 has heavier isotopic composition even though it is at a higher elevation. It shows that RS05 and MA02 have different groundwater flow and recharge areas, although they have the same genesis as other samples.
c. Group 3 consists of MA03, MA11, SG08, and SG09. The value of isotopic composition in this group is lighter than in group 1 and group 2. This group's water source comes from rainwater because δ^2H and δ ^{18}O are close to the LMWL value. The sample in this group has a range of deuterium values between -43.026‰ to -40,533‰ and δ ^{18}O between -7.195‰ to -6.917‰. From the range of sample values in group 3, they have the same genesis but do not have the same groundwater flow.
d. Group 4 consists of S02, S03, SG04, SG11, and MA06. In this group, there are water samples from the Ciliwung River and Sukabirus River, which are the Ciawi and Sukamahi Dams. The value of the isotopic composition of this group is the lightest among other groups; this shows that the catchment area of this group is higher than the other groups; this can also be seen from the elevation of the location of the water source, which is higher than the other samples. Isotope analysis shows that seepage water in the dam has a different genesis from river water.

5. Conclusions

The characteristics of water around Ciawi and Sukamahi dams based on Physico-chemical properties are freshwater, with TDS values varying between 31 ppm to 210 ppm and EC from 62 μS/cm to 414 μS/cm. The analysis results of δ^2H and δ ^{18}O, dam's seepage have isotopic compositions around the local meteoric line (LMWL). Therefore it can be concluded that the seepage source is more influenced by rainwater. Most seepage in the Ciawi and Sukamahi Dry Dams is in Group 2. The analysis results show that the river is not a source of seepage. The comparison of stable isotope compositions shows that seepage sources' genesis is similar to springs in Sukakarya Village (MA02) and dug wells in Gadog Village (SG13). However, the springs in Sukakarya Village do not have similar connectivity and recharge area to seepage water. Its indicated by the MA02 isotopic composition value, which is heavier than the other samples even though the elevation is higher. The study found that rainwater significantly affects seepage water, and there is a similarity in the genesis of seepage water with the other two water sources. These results can be used as one of the considerations to determine the proper treatment to reduce seepage in the Ciawi and Sukamahi Dry Dams. One of the methods used to control seepage is a sub-drain to lower the groundwater level.

Acknowledgments
The authors are grateful to the River Basin Organization for Ciliwung-Cisadane (BBWS) for permit and support data; and the Human Resource Development Agency of the Ministry of Public Works and Housing Indonesia for financial support for this research.

References

[1] PT. Indra Karya (Persero) 2015 Laporan Akhir Bendungan Ciawi (Jakarta: Balai Besar Wilayah Sungai Ciliwung Cisade) 471p.
[2] PT. Indra Karya (Persero) 2015 Laporan Akhir Bendungan (Jakarta: SukamahiBalai Besar Wilayah Sungai Ciliwung Cisade) 350p.
[3] Sutopo M F 2011 Pengembangan Kebijakan Pembayaran Jasa Lingkungan Dalam Pengelolaan Air Minum (Studi Kasus DAS Cisadane Hulu) Unpublish dissertation (Bogor: Institute Pertanian Bogor) 327p.
[4] Poespowardoyo R S 1986 Peta Hidrogeologi Indonesia Lembar Jakarta (Jawa) (Bandung: Direktorat Geologi Tata Lingkungan)
[5] Peng T R and Wang C H 2008 Identification of sources and causes of leakage on a zoned earth dam in northern Taiwan: Hydrological and isotopic evidence *Applied Geochemistry* **23** 2438-2451
[6] Yi P, Yang J, Wang Y, Mugwanezal V de P, Chen L and Aldahan A 2018 Detecting the leakage source of a reservoir using isotopes *J. Environ. Radioact.* **187** 106–14
[7] Ratiat A, Khettal T and Meddi M 2020 The piezometric and isotopic analysis of leaks in earth dams: the case of the fountain of Gazelle dam, Biskra, Algeria *Environ. Earth Sci.* **79** 1–10
[8] Wibagiyo, Indroyono, Alip, Pratikno B, and Haryono 1998 Penentuan Lokasi Rembesan Pada Dasar Bendungan Dengan Teknik Radioisotop di Bendungan Ngancar, Wonogiri *Penelitian dan Pengembangan Aplikasi Isotop dan Radiasi* 229-232
[9] Sidauruk P, Indrojono, Wibagyo, Pratikno B I, and Evarista Ristin P 2000 Penelitian Asal-Usul Berbagai Sumber Air di Sekitar Bendungan NGancar Wonogiri, Jawa Tengah dengan Metode Isotop Alam *Risalah Pertemuan Ilmiah Penelitian dan Pengembangan Teknologi Isotop dan Radiasi* 195-200
[10] Badan Informasi Geospatial Republik Indonesia 2019 Peta Rupa Bumi Indonesia Bogor Jawa Barat accessed from https://tanahair.indonesia.go.id/portal-web/.
[11] Badan Informasi Geospatial Republik Indonesia 2019 DEM Nasional Bogor Jawa Barat accessed from https://tanahair.indonesia.go.id/demnas/#/demnas.
[12] Azizah M, Subiyanto A, Triutomo S, and Wahyuni D 2022 Pengaruh Perubahan Iklim Terhadap Bencana Hidrometeorologi di Kecamatan Cisarua-Kabupaten Bogor *PENDIPA Journal of Science Education* **6**(2) 541-546
[13] Effendi A and Hermanto B 1998 Peta Geologi Lembar Bogor Jawa (Bandung.: Pusat Penelitian dan Pengembangan Geologi)
[14] Kharisma H L, Wijatna B A, and Wilopo W 2015 Aplikasi Isotop Alam untuk Mengetahui Asal-Usul Air Umbul Cokro, Kecamatan Tulung, Kabupaten Klaten *Forum Teknik* **36**(1) 73-80
[15] Mazor E 2004 Chemical and Isotopic Groundwater Hydrology Third Edition (New York: Macer Dekker, Inc.) 453 p
[16] Clark I 2015 Groundwater Geochemistry and Isotopes (Boca Raton: Taylor & Francis Group) 421 p
[17] Pratikno B and Satrio 2014 Studi Garis Air Meteorik (Meteoric Water Line) Beberapa Wilayah di Indonesia Dengan Metode Isotop Stabil (Bandung, Fakultas Teknik Geologi, Universitas Padjdjaran) 30
[18] Siftianida I I, Wijatna B A, and Pratikno B 2016 Aplikasi Isotop Alam untuk Pendugaan Daerah Resapan Air Mata Air di Kecamatan Cijeruk, Kabupaten Bogor, Jawa Barat *Jurnal Ilmiah Aplikasi Isotop dan Radiasi* **12** 97-106

[19] Griffioen J 2004 Groundwater Quality In: V. S. Kovalevsky, G. P. Kruseman & K. R. Rushton, eds. Groundwater studies: An international guide for hydrogeological investigations (Paris: United Nations) 43-66.
[20] Mandel S and Shiftan Z L 1981 Groundwater Resources: Investigation and Development (New York: Academic Press, Inc.) 269p
[21] Aggarwal P K, Froehlich K, Kulkarni K M and Garcia-Agudo E 2004 Isotopes Techniques in Groundwater Investigations. In: V. S. Kovalevsky, G. P. Kruseman & K. R. Rushton, eds. Groundwater studies: An international guide for hydrogeological investigation (Paris: United Nations) 239-270

IOP Conf. Series: Earth and Environmental Science 1233 (2023) 012052 doi:10.1088/1755-1315/1233/1/012052

Analysis of Rainfall Variability and Shifts in Season in Relation to the Rice Cultivation Period in Sleman Regency

L B Firdauzi[1,2], U Suarma[1]

[1]Department of Environmental Geography, Faculty of Geography, Universitas Gadjah Mada, Yogyakarta, Indonesia
[2]CV Greenterra Mapindo, Yogyakarta, Indonesia

utiasuarma@ugm.ac.id

Abstract. The research aimed to determine seasonal rainfall variability in Sleman Regency to determine shifts in season in Sleman Regency, and the relationship between the shift in season and rice cultivation period in Sleman Regency. The method was the coefficient of variation to determine seasonal rainfall variability. The shift in season was analysed in 10-year periods. The relationship between the shift in season and the rice cultivation period was analysed using the Chi-Square test. The result of this research showed that the rainfall variability in the dry season was higher than that in the rainy season in Sleman Regency from 1990-2019. The dry season in 2000-2009 came 10 days earlier in 6 subdistricts. The dry season in 2010-2019 was delayed in a period of 10-50 days. The rainy season in 2000-2009 came 20 days earlier in 7 subdistricts, delayed in a period of 10 days in 3 subdistricts. The rainy season in 2010-2019 was delayed in a period of 10 days in 8 subdistricts. The Chi-Square count was greater than the Chi-Square table, meaning that there was a relationship between shift in season and rice cultivation period in Sleman Regency.

1. Introduction

Indonesia is a maritime continent that is passed by some global wind circulation [1]. Wind circulation influences the rainfall in Indonesia, eventually affecting the season in this country. The season in Indonesia is affected by the monsoon which is generally found periodically, but the season does not always start at the same time every year due to various disturbances such as ENSO and IOD which influence the seasonal and inter-seasonal rainfall variability in Indonesia [2].

Rainfall variability affected by climate-driven disturbances is characterized by shift in season because rainfall variability has two components that can be affected: spatial and temporal rainfall variability [3]. The temporal rainfall variability is seasonal, bringing an impact on various sectors of community life; one of the most vulnerable sectors is the agricultural sector [4]. The most vulnerable agricultural sector is the rice farming sector because rice plants require optimal water for their optimal growth approximately 200 mm/month [5].

There have been many studies on the impact of a change in rainfall variability on shift in season, which also has an impact on a change in the rice cultivation period. Changes in rainfall variability caused by ENSO and IOD led to a change in the beginning of the rice cultivation period in West Java due to the shift in the beginning of the rainy season [6]. In addition, rainfall variability was higher in a lower area, and the rice cultivation period in Kebumen from 1981-2010 changed [3]. This shows a relationship between rainfall variability and shift in season with the rice cultivation period in Sleman Regency.

Sleman Regency had large rice field with an area reaching 27,725.3 ha in 2019, around 49% of the entire area of Sleman Regency [7]. In addition, this regency had the second largest harvest area in the Special Region of Yogyakarta in 2019 around 24,465.85 ha [8] which produced 115,540.59 tons [9]. This causes the agricultural sector in Sleman Regency to be vulnerable when there are changes in rainfall variability and shift in season. Based on this description, the present research aimed to (1) analyse the seasonal rainfall variability in Sleman Regency from 1990-2019, (2) analyse the shift in season in Sleman Regency from 1990-2019, and (3) analyse the relationship between the shift in season and the rice cultivation period in Sleman Regency.

2. Method

Geographically, Sleman Regency is located at 110°33′00″ to 110°13′00″ East Longitude and 7°34′51″ to 7°47′30″ South Latitude. The research area is shown in Figure 1.

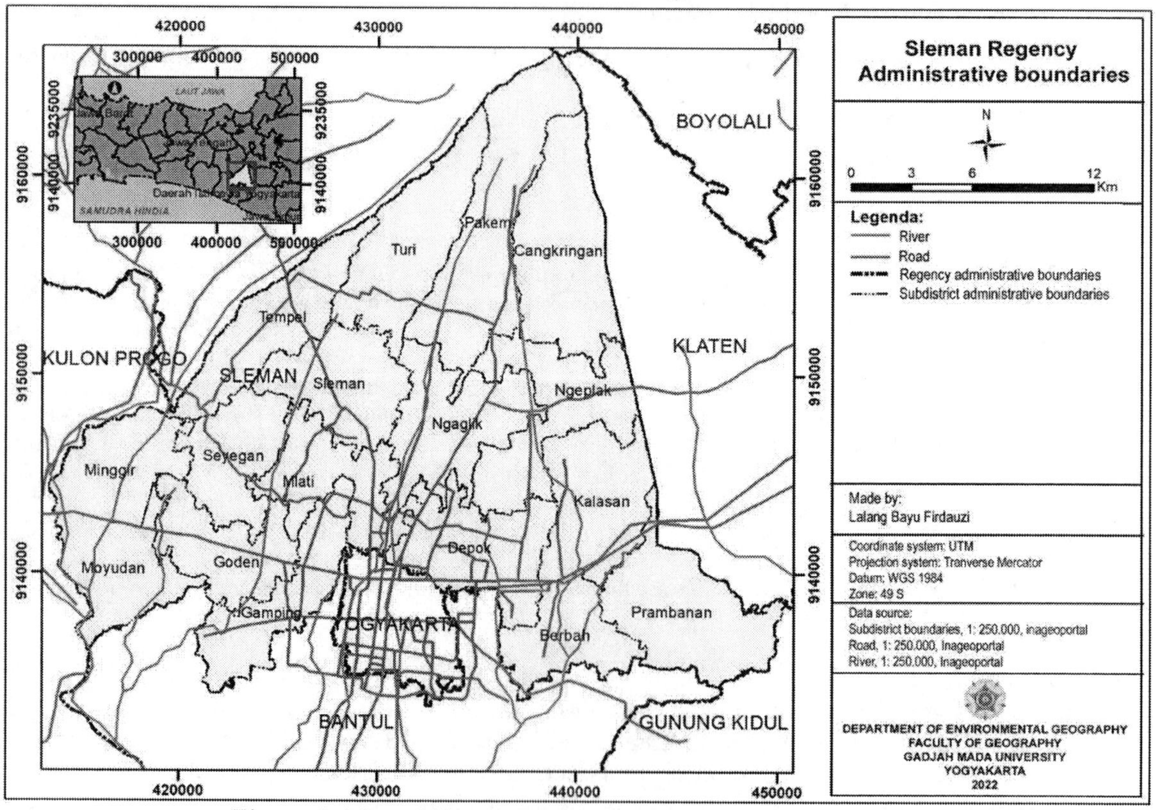

Figure 1. Map of Sleman Regency Administrative Boundary

This research used secondary data obtained from BBWS Serayu-Opak, CHIRPS rainfall data, SRTM, and the Integrated Planting Calendar of Sleman Regency 2011-2019.

2.1. Selection of rain gauge

The selection of rain gauge was done because several rain gauges had incomplete data. According to the records from BBWS Serayu-Opak, there were 11 rain gauges in Sleman Regency with more than 10 years of data. The rainfall data from 11 rainfall gauges from 1990-2019 were first completed based on the observation rainfall data using the Inverse Square Distance. The consistency of the data from the observed rain gauges and CHIRPS were tested using the Rescaled Adjusted Partial Sums (RAPS) [10].

The consistency of RAPS test data was determined by the value of $Q/\sqrt{n}$. The value of $Q/\sqrt{n}$ shows consistent data if the value of $Q/\sqrt{n}$ count is smaller than the value of $Q/\sqrt{n}$ table or the critical value of the appropriate confidence level [10]. The values of $Q/\sqrt{n}$ table and $R/\sqrt{n}$ table are shown in Table 1.

Table 1. Values of Q/√n table and R/√n table

n	Q/√n			R/√n		
	90%	**95%**	**99%**	**90%**	**95%**	**99%**
10	1.05	1.14	1.29	1.21	1.28	1.38
20	1.10	1.22	1.42	1.34	1.43	1.60
30	1.12	1.24	1.46	1.40	1.50	1.70
40	1.13	1.26	1.50	1.42	1.53	1.74
50	1.14	1.27	1.52	1.44	1.55	1.78
100	1.17	1.29	1.55	1.50	1.62	1.86
∞	1.22	1.36	1.63	1.62	1.75	2.00

2.2. *Validation of CHIRPS rainfall data*
CHIRPS rainfall data were validated using Pearson's correlation test and the error rate was tested using RMSE and MAE. Pearson's correlation test was used to determine the correlation between CHIRPS rainfall data and observation rainfall data [11] [12]. The seasons were divided into rainy and dry seasons based on BMKG criteria. The seasonal rainfall variability was analysed using the coefficient of variation method [7]. Shift in season analysis was carried out using criteria as determined by BMKG. The beginning of the rainy season is characterised as rainfall higher than 50 mm in the first ten days, followed by the next ten days, while the beginning of the dry season is the opposite. The shift in season was analyzed in 10-year periods.

2.3. *Characteristics of planting patterns in Sleman Regency*
The data of the planting pattern characteristics in Sleman Regency was processed using the Integrated Planting Calendar of Sleman Regency and Oldeman classification. The Oldeman classification was obtained from a literature study and calculation using rainfall data from 1990-2019. The Integrated Planting Calendar was used to determine the planting pattern in Sleman Regency based on the prediction of planting rotation shown in the Integrated Planting Calendar of Sleman Regency 2011-2019.

2.4. *The relationship between shift in season and rice cultivation period*
The relationship between shift in season and rice cultivation period was determined by comparing the shift in the beginning of the season with the beginning of the rice cultivation. The relationship was tested using the Chi-Square test [13]. The Chi-Square test was determined using 2 criteria, if the Chi-Square count is greater than the Chi-Square table, then H_0 is rejected and H_1 is accepted; if the Chi-Square count is smaller than the Chi-Square table, then H_0 is accepted and H_1 is rejected.

3. Discussion
3.1. *Selection of rain gauge*
The selection of rain gauges with more than 10-year data aimed to perform correlation and validation tests on the CHIRPS rainfall data because the measurement used 30-year data from 1990-2019. In fact, 30-year rainfall data are the minimum period of data required to calculate normal rainfall in a region [14]. The selected rain gauges represented 1-3 subdistricts as the scope of the analysis. Before the rain gauges were used for the analysis, the quality of the data was first tested by a consistency test. Based on calculation, all rain gauges in Sleman Regency were consistent on both observation rainfall data and CHIRPS rainfall data.

3.2. *Validation of CHIRPS rainfall data*
The CHIRPS rainfall data were validated using Pearson's correlation test and error rate tests using MAE and RMSE. Based on the correlation value and error rate of the CHIRPS rainfall data calculation, the highest value was found at Godean and Bronggang station with a correlation value of 0.71, while the

lowest were at Kemput, Prumpung, and Tanjungtirto station with a correlation value of 0.66. The highest error rate was found at Kemput rainfall station, while the lowest at Tanjungtirto rainfall station.

3.3. Seasonal rainfall variability

Seasonal rainfall variability was calculated in the time span of 1990-2019 and divided into three periods, i.e., 1990-1999, 2000-2009, and 2010-2019. The seasonal rainfall variability in Sleman Regency shown in Table 2.

Table 2. Seasonal Rainfall Variability 1990-2019 in Sleman Regency

Nr.	Rainfall Station Name	Subdistrict	Elevation (msl)	Dry Season (%)			Rainy Season (%)		
				1990-1999	2000-2009	2010-2019	1990-1999	2000-2009	2010-2019
1.	Kemput	Pakem	523	55.45	76.52	48.96	24.80	23.89	31.17
2.	Prumpung	Ngaglik	276	57.53	78.88	50.83	25.61	24.05	26.49
3.	Angin-angin	Turi	330	56.79	78.88	50.83	25.48	24.05	26.49
4.	Beran	Sleman	206	71.04	77.78	53.29	23.36	27.34	27.00
5.	Tanjungtirto	Berbah, Prambanan	114	72.05	84.17	59.71	22.56	27.48	26.57
6.	Santan	Depok	115	74.73	86.05	61.67	26.82	32.11	25.35
7.	Plataran	Ngemplak, Kalasan	291	66.11	78.58	54.21	19.88	24.18	23.80
8.	Gemawang	Mlati	147	68.14	83.07	59.60	31.20	31.39	25.05
9.	Bronggang	Cangkringan	370	66.21	78.58	54.21	20.22	24.18	23.80
10.	Seyegan	Seyegan, Tempel, Minggir	179	61.96	78.63	58.64	26.77	21.91	30.38
11.	Godean	Godean, Gamping, Moyudan	149	60.80	76.33	54.63	24.49	24.18	26.09
Average				**64.62**	**79.77**	**55.14**	**24.66**	**25.89**	**26.56**

Source: Data analysis, 2022

Based on Table 2, the rainfall variability in the dry season was higher than the rainfall variability in the rainy season. The rainfall variability in the dry season had an average of 66.51% while that in the rainy season had an average of 25.70%. The region with the highest dry season rainfall variability in 1990-2019 was Depok Subdistrict. The lowest dry season rainfall variation in 1990-1999 was in Pakem Subdistrict. The lowest dry season rainfall variation in 2000-2009 was in Godean, Gamping, and Moyudan Subdistricts. Meanwhile, the region with the lowest dry season rainfall variation in 2010-2019 was Pakem Subdistrict.

The region with the highest rainy season rainfall variation in 1990-1999 was Mlati Subdistrict, while the lowest rainy season rainfall variation in Sleman Regency in 1990-1999 was in Ngemplak Subdistrict. The highest rainy season rainfall variation in 2000-2009 was found in Depok Subdistrict, while the lowest rainy season rainfall variation in 2000-2009 was found in Seyegan, Tempel, and Minggir Subdistricts. The highest rainy season rainfall variation in 2010-2019 was in Pakem Subdistrict, while the region with the lowest rainy season rainfall variation was Cangkringan and Ngemplak Subdistricts.

Based on Table 4, elevation also affects seasonal rainfall variability. The seasonal rainfall variability in Cangkringan Subdistrict of which the elevation was 370 meters above sea level, was different from the rainfall variability of Depok Subdistrict, which was 115 meters above sea level. The rainfall variability of the dry season was, in general, lower in higher areas, i.e., in the western part of Sleman Regency, while the rainfall variability of the rainy season in Sleman Regency was in general lower in higher areas, i.e., in the eastern part of Sleman Regency.

IOP Conf. Series: Earth and Environmental Science 1233 (2023) 012052 doi:10.1088/1755-1315/1233/1/012052

3.4. *Shift in season*

The beginning of both the rainy season and the dry season is important for farmers in determining the planting period, especially for rice plants that require abundant water for their growth stage of approximately 60 days [15]. The changes in the beginning of the rainy season can be seen in Table 3.

Table 3. Shifting Season during 1990-2019 in Sleman Regency

Nr.	Rainfall Station Name	Subdistrict	Dry Season Change (10-Days Rainfall)			Rainy Season Change (10-Days Rainfall)		
			1990-1999	2000-2009	2010-2019	1990-1999	2000-2009	2010-2019
1	Kemput	Pakem	May II	May I	June III	Oct III	Oct III	Oct III
2	Prumpung	Ngaglik	May II	May I	June I	Oct III	Oct III	Oct III
3	Angin-angin	Turi	May II	May I	June I	Oct III	Oct III	Oct III
4	Beran	Sleman	May I	May I	June I	Nov II	Oct III	Nov I
5	Tanjungtirto	Berbah, Prambanan	April III	April III	June I	Nov II	Oct III	Nov I
6	Santan	Depok	April III	April III	May I	Nov II	Oct III	Nov I
7	Plataran	Ngemplak, Kalasan	May I	May I	June I	Nov II	Oct III	Nov I
8	Gemawang	Mlati	April III	April III	May I	Oct III	Oct III	Nov I
9	Bronggang	Cangkringan	May I	May I	June I	Nov II	Oct III	Nov I
10	Seyegan	Seyegan, Tempel, Minggir	May II	May I	June III	Oct II	Oct III	Oct III
11	Godean	Godean, Gamping, Moyudan	May I	May I	June I	Oct III	Okt III	Oct III

Source: Data analysis, 2022

Based on Table 3, the beginning of the dry season in 2000-2009 came 10 days earlier in Pakem, Ngaglik, Turi, Seyegan, Tempel, and Minggir Subdistricts. The beginning of the dry season in 2010-2019 was delayed 10-50 days. The beginning of the rainy season in 2000-2009 came 20 days earlier in Cangkringan, Ngemplak, Kalasan, Depok, Berbah, Prambanan, and Sleman Subdistricts; the beginning of the rainy season in Seyegan, Tempel, and Minggir Subdistricts was delayed 10 days. The beginning of the rainy season in 2010-2019 was delayed 10 days in Cangkringan, Ngemplak, Kalasan, Depok, Berbah, Prambanan, Mlati, and Sleman Subdistricts.

The beginning of the rainy season in Sleman Regency had a direction from west to east due to topography and wind direction. Topography is influential because the western part of Central Java and Special Region of Yogyakarta have more mountainous areas so orographic rainfall can occur in those areas [16]. Wind direction also has an impact on the beginning of the rainy season which came earlier in the western part of Sleman [17]. The change in the monsoon leads to changes in rainfall in Indonesia [1]. In addition to the monsoon, the shift in the beginning of the season was also due to changes in rainfall variability caused by ENSO and IOD. ENSO and IOD generally has an impact on seasonal rainfall variability, causing a shift in the beginning of the season in Indonesia.

Shift in the beginning of the season in Sleman Regency has an impact on agricultural sector. In fact, the shifts in the beginning of the season in Sleman Regency had an impact on rice production. Rice production in Sleman Regency in 2010, when a strong La Nina hit Indonesia, increased to 410,352 tons compared to that in 2005 with a rice production of around 306,201 tons because a weak El Nino hit Indonesia in that year [18].

IOP Conf. Series: Earth and Environmental Science 1233 (2023) 012052 doi:10.1088/1755-1315/1233/1/012052

3.5. Characteristics of crop patterns in Sleman Regency

In fact, it is crucial to arrange the crop patterns on a land to obtain maximum agricultural products and reduce the risk of harvest failure by taking into account the climate and land conditions [19]. The crop patterns in Sleman Regency can be identified from various things, especially in terms of the rainfall that affects farmers' crop pattern planning. Based on rainfall, the crop patterns in Sleman Regency can be identified based on the Oldeman classification. Sleman regency had a climate types B2 and C2 in 1975 based on Oldeman Classification Map in Java Island. In the last decade, there was a change in the climate type in Sleman Regency, i.e., the climate in Sleman Regency was drier [20] and the climate in Sleman Regency in 2014 was dominated by climate type C3. Based on the rainfall in 1990-2019, the climate type in Sleman Regency throughout its subdistricts was climate type C3 based on Oldeman classification, meaning that rice crops can be planted once and *palawija* crops also once, but the *palawija* crops should not be planted in the dry seasons.

The crop patterns in Sleman Regency can also be identified from the Integrated Planting Calendar obtained from BPTP Yogyakarta. The Integrated Crop Pattern Calendar (KATAM) presents spatial and tabular information regarding the beginning of planting, and crop patterns based on predictions of variability and climate change [21]. Based on KATAM, the normal crop patterns in Sleman Regency were divided into 2, namely rice-rice-rice, and rice-rice-*palawija*.

3.6. Relationship between shift in season and rice cultivation period

The relationship between the shift in the beginning of the rainy season and the rice cultivation period was determined using the Chi-Square test. In this research, the beginning of the rainy season and the rice cultivation period were considered the nominal data. The Chi-Square analysis was done to the data of the occurrence of shifts in the beginning of the rainy season and the rice cultivation period in 2012-2019 in 17 subdistricts.

The Chi-Square analysis was used to test the hypothesis of a relationship between shift in season and rice cultivation period. The hypothesis was divided into 2, namely there is no relationship between the shift in season and the rice cultivation period as H_0 and there is a relationship between the shift in season and the rice cultivation period as H_1. Based on calculation, the Chi-Square count was 24.804, while the Chi-Square table was 9.488. The result showed that the Chi-Square count was greater than the Chi-Square table. This indicates that in Sleman Regency there was a significant relationship between the shift in season and the rice cultivation period.

4. Conclusion

The conclusions of this study are as follows:

1. The rainfall variability in the dry season was 65.51%, higher than that in the rainy season 25.70% in Sleman Regency from 1990-2019, so the dry season had higher rainfall variability than the rainy season.
2. The dry season in 2010-2019 was delayed in a period of 10-50 days. The rainy season in 2000-2009 came 20 days earlier in 7 subdistricts, delayed in a period of 10 days in 3 subdistricts. The rainy season in 2010-2019 was delayed in a period of 10 days in 8 subdistricts.
3. The Chi-Square count of 24.804 was greater than the Chi-Square table of 9.488, meaning that there was a relationship between shift in season and rice cultivation period in Sleman Regency.

Acknowledgement

This publication is generate based on the final bachelor thesis work and the authors would like to convey gratitude to Balai Besar Wilayah Sungai (BBWS) Serayu-Opak and Balai Pengkajian Teknologi Pertanian (BPTP) Yogyakarta who granted permission to access secondary data and have assisted in collecting research data needed.

IOP Conf. Series: Earth and Environmental Science 1233 (2023) 012052 doi:10.1088/1755-1315/1233/1/012052

References

[1] Hermawan, E. 2015. *Indeks Monsun Asia-Australia dan Aplikasinya*. Jakarta: LIPI Press.

[2] Narulita, I. 2017. Pengaruh ENSO dan IOD pada Variabilitas Curah Hujan di DAS Cerucuk, Pulau Belitung. *J. Tanah dan Iklim*, **41**(1): 45–60.

[3] Indratmoko, S., Harmantyo, D., and Kusratmoko, E. 2017. Variabilitas curah Hujan di Kabupaten Kebumen. *J. Geogr. Lingkung. Trop.* **1**(1): 29–40.

[4] Sekaranom, A. B., Nurjani, E., and Nucifera, F. 2021. Agricultural Climate Change Adaptation in Kebumen, Central Java, Indonesia. *Sustainability*. **13**(13):1–16.

[5] Yulianto and Sudibyakto. 2012. Kajian Dampak Variabilitas Curah Hujan terhadap Produktivitas Padi Sawah Tadah Hujan di Kabupaten Magelang. *J. Bumi Indones.***1**(1).

[6] Apriyana, Y. 2015. Strategi Adaptasi Kalender Tanam Terhadap Variabilitas Iklim Pada Sentra Produksi Padi di Wilayah Monsunal dan Equatorial. *Semin. Nas. Masy. Biodiversiti Indones.***1**(7): 1725–1734.

[7] Ambarwatik, R. 2020. Analisis Kesesuaian Penggunaan Lahan terhadap Fungsi Kawasan di Kabupaten Sleman Tahun 2019. *Skripsi*. Universitas Muhammadiyah Surakarta, Jawa Tengah.

[8] BPS Daerah Istimewa Yogyakarta. 2020. *Provinsi Daerah Istimewa Yogyakarta Dalam Angka 2020*. Daerah Istimewa Yogyakarta: Badan Pusat Statistik Daerah Istimewa Yogyakarta.

[9] BPS Kabupaten Sleman. 2020. *Kabupaten Sleman Dalam Angka 2020*. Sleman: Badan Pusat Statistik Kabupaten Sleman.

[10] Lashari, Kusumawardani, R., and Prakasa, F. 2017. Analisa Distribusi Curah Hujan di Area Merapi menggunakan Metode Aritmatika dan Poligon. *J. Tek. Sipil Perenc.* **19**(1): 39–48.

[11] Jarwanti, D. P., Suhartanto, E., and Fidari, J. S. 2021. Validasi Data Curah Hujan Satelit TRMM (Tropical Rainfall Measuring Mission) dengan Data Pos Penakar Hujan di DAS Grindulu, Kabupaten Pacitan, Jawa Timur. *Jurnal Teknologi dan Rekayasa Sumber Daya Air*. **1**(2): 772–785.

[12] Sharifi, E., Steinacker, R., and Saghafian, B. 2016. Assessment of GPM-IMERG and Other Precipitation Products against Gauge Data under Different Topographic and Climatic Conditions in Iran: Preliminary Results. *Remote Sensing*. **8**(2).

[13] Bunganaen, W., Krisnayanti, D. S., and Klau, Y. A. 2013. Analisis Hubungan Tebal Hujan dan Durasi Hujan Pada Stasiun Klimatologi Lasiana Kota Kupang. *J. Tek. Sipil*. **2**(2):181–190.

[14] Molle, B. A., and Larasati, A. F. 2020. Analisis Anomali Pola Curah Hujan Bulanan Tahun 2019 terhadap Normal Curah Hujan (30 Tahun) di Kota Manado dan Sekitarnya. *J. Meteorol. Klimatologi dan Geofis*. **7**(1):1–8.

[15] Subagyono, K., Dariah, A., Surmaini, E., and Kurnia, U. 2004. *Tanah Sawah dan Teknologi Pengelolaannya*. Bogor: Pusat Penelitian dan Pengembangan Tanah dan Agroklimat.

[16] Wredaningrum, I. and Sudibyakto. 2014. Analisis Perubahan Zona Agroklimat Daerah Istimewa Yogyakarta ditinjau dari Klasifikasi iklim menurut Oldeman. *J. Bumi Indones.* **3**(4).

[17] Fadhika, U., Rifai, A., and Rochaddi, B. 2014. Arah dan Kecepatan Angin Musiman serta Kaitannya dengan Sebaran Suhu Permukaan Laut di Selatan Pangandaran, Jawa Barat. *J. Oseanografi*. 3(3): 429–437.

[18] Tasong, S. L., Parwati, D. U., and Mu'in, A. 2016. Kajian Curah Hujan di Kabupaten Sleman DIY Terhadap Tingkat Produksi Padi Sawah. *J. Agromast*. **1**(2).

[19] Nganji, M. U. and Simanjuntak, B. H. 2020. Penentuan Pola Tanam Tanaman Pangan Berdasarkan Neraca Keseimbangan Air di Kecamatan Umbu Ratu Nggay Barat, Kabupaten Sumba Tengah, Provinsi Nusa Tenggara Timur. *J. Ilm. Teknol. Pertan. Agrotechno*. 5(2).

[20] Harmoni, K. 2014. Analisis Persebaran Iklim Klasifikasi Oldeman di Provinsi Daerah Istimewa Yogyakarta. *Skripsi*. Universitas Muhammadiyah Surakarta, Jawa Tengah.

[21] Balitbang Pertanian. 2021. *Kalender Tanam Tanaman Padi Musim Hujan, Oktober-Maret 2021/2022, Kabupaten Sleman, Provinsi Yogyakarta*. Sleman: Badan Penelitian dan Pengembangan Pertanian.

IOP Conf. Series: Earth and Environmental Science 1233 (2023) 012053 doi:10.1088/1755-1315/1233/1/012053

Analysis of sanitation system and coliform number in groundwater in Wonosari District, Gunungkidul Regency, Indonesia

Afifudin[1] and M. Widyastuti[1]

Department of Environmental Geography, Faculty of Geography, Universitas Gadjah Mada, Indonesia
*Corresponding author: afifudin@mail.ugm.ac.id

Abstract. Sanitation visualizes the environment's health status, comprising housing, sewage disposal, and clean water provision. In general, sewerage can have on-site and offsite sanitation facilities. The research was conducted to characterize the sanitation system for sewage disposal, determine the water quality from biological parameters, and enumerate coliform bacteria in groundwater for every sanitation system observed in Wonosari District, Gunungkidul Regency, Indonesia. These sets of information can help detect groundwater contamination by coliform that reflects the sanitary characteristics of the sewage disposal to be further considered in sanitation management and treatment policies. The research used purposive sampling, field surveys, interviews with a questionnaire, laboratory analyses, and a descriptive-comparative design. The results showed that the sanitation system for sewage disposal at the study site consisted of individual septic tanks and wastewater treatment plants (WWTPs). In terms of quality, the coliform counts exceeded the existing standards, with the highest presence found in groundwater downstream of improperly maintained WWTPs and substandard septic tanks close to other sources of pollutants.

1. Introduction

Sanitation is a subject of concern in water resource management. When management is lacking, sanitation facilities can increase the likelihood of bacterial contamination in groundwater. It is common for many communities in rural areas and low-income urban settlements to use on-site sanitation facilities, which, in and of itself, can be a major source of pollutants. On-site sanitation collects and stores fecal matter on the same plot where it is deposited without introducing further treatments [1]. Poor sanitation systems, open defecation, and improperly maintained wastewater disposal and treatment facilities are sources of pollutants that present coliforms in groundwater. With the increasing human activities that inevitably generate more domestic waste, it is imperative to identify the characteristics of a sanitation system to determine potential groundwater contamination by bacteria.

Wonosari is the most densely populated district in Gunungkidul Regency (1158.18 people/km^2) according to BPS-Statistics of Indonesia [2]. It occupies a basin, which is a bowl-like landform enclosed by karst hills (Gunungsewu) in the south and structural hills (Batur Agung) in the west, north, and east [3]. This basin morphology forms a place where water accumulates, contributing to the abundant water resources in the district. Nevertheless, inherently, it is susceptible to water pollution because the surrounding karst hills, which make up most of the physical environment, have a thin layer of soil through which harmful substances can easily enter and taint the groundwater.

Groundwater contamination by coliforms is most likely to occur in Wonosari. Data from the Regional Environmental Status of Gunungkidul Regency in 2014 showed that all five samples of the local well water tested positive for coliforms. The groundwater samples collected in Sumbermulyo RT 01/02 (Kepek Village) and Besari RT 01/04 (Siraman Village) had the highest fecal coliform count and total coliform amounting to $\geq$1898/100 ml (relatively high presence).

In addition to human activities, variations in the adopted sanitation systems also regulate groundwater vulnerability to coliform contamination. Based on the data from the Public Works Service, the sanitation system in Wonosari District varies in service coverage, year of construction, and maintenance. In addition, potential or possible contamination differences may result from other characteristics, such as structural maintenance and distance between wells and sanitation facilities. Public Works has established sanitation development procedures based on the Indonesia National Standard SNI 2398 [4] issued by the Ministry of Public Works, including septic tank construction and a safe distance of at least 10 m from the septic tank to wells acting as clean water sources and 5 m to infiltration wells. To determine the extent to which variations in the sanitation system affect coliform contamination in groundwater, it is necessary that each type of the system is characterized and analyzed for such effect. Thus, the kind of sanitation to best prevent groundwater pollution by coliforms can be identified.

2. Data and Method

2.1. Data

The data used in the study were gathered through field surveys, sample testing in the laboratory, literature studies, and relevant agencies. The secondary data collected for the analysis included WWTPs in Gunungkidul Regency from the Department of Public Works and Housing, water quality in Wonosari District, population density from the Indonesian Statistics Agency, and sources of pollutants. The primary data consisted of information on the sanitation system's characteristics, water-table elevation at the sampled wells, and coliform count in the groundwater (through laboratory tests).

2.2. Method

A questionnaire was used in structured interviews as the research instrument to identify the characteristics of the sanitation system. In addition, other equipment like Avenza maps, field checklists, and a set of sampling tools consisting of a measuring tape, an icebox, water sample bottles, and a water checker was used. Furthermore, the Indonesia Topographic Map (RBI) of Gunungkidul at a scale of 1:25,000 was used to determine the physical condition of the study site, including roads, rivers, and administrative boundaries. Also, data from the Public Works were consulted to determine the number and location of communal WWTPs.

The sanitation system for sewage disposal was characterized using the field survey and interview results. Laboratory analysis was conducted to enumerate the presence of coliform in groundwater by testing fecal and total coliform parameters. The test results were then compared with water quality standards for sanitary hygiene in Regulation No. 32 of 2017 and drinking water quality standards in Regulation No. 492 of 2010 issued by the Minister of Health (MoH). Further, the test results of the water samples collected from the dug wells of sanitation system users were intercompared to determine which one contained the smallest to largest coliform to confirm fecal contamination in the groundwater.

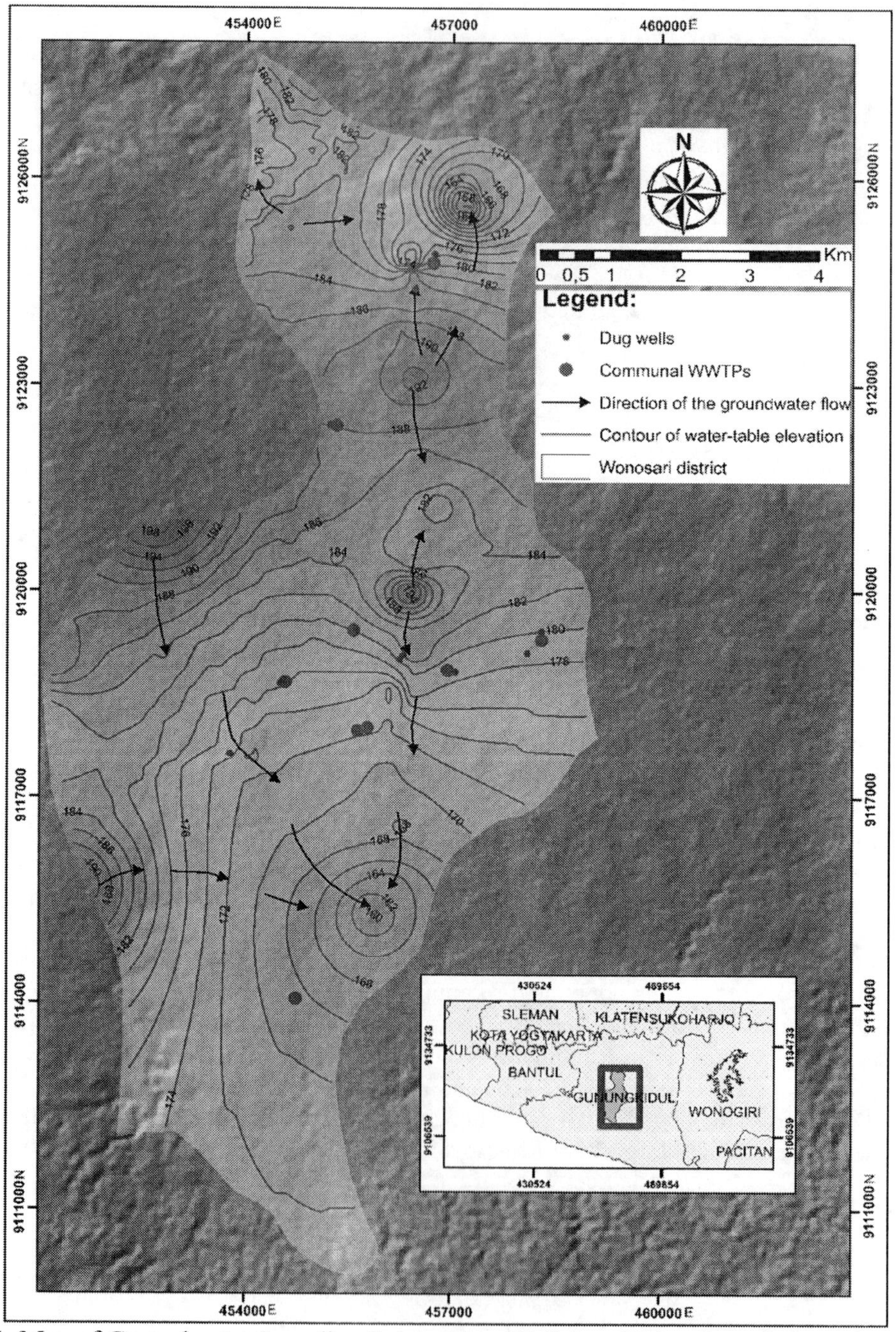

Figure 1. Map of Groundwater Sampling Points (Dug Wells) and Direction of Groundwater Flow

The purposive sampling technique determined the samples, which collects data by applying specific criteria. In this case, the criterion was as follows: the sample should represent each existing sanitation system across the district: communal, individual, and local. Therefore, groundwater samples were collected at dug wells downstream of each system (i.e., sampling unit), as shown by the direction of the groundwater flow (Figure 1). Based on the data on elevation and depth of the groundwater table, the value of the groundwater level can be seen. The data is used to map groundwater flow so that the direction of groundwater flow can be known. This process was intended to choose dug wells that accurately represented the sanitation system and enumerate coliforms present in the groundwater.

3. Results and Discussion

3.1. Sanitation Systems for Sewage Disposal in Wonosari District
Sanitation systems for sewage disposal in Wonosari District consist of various types. Generally, divided into a centralized (offsite) sanitation system and a decentralized (on-site) sanitation system.

3.1.1. Centralized or Offsite Sanitation. Centralized or offsite sanitation is a system that disposes of wastewater flows from several households into a collection channel and then a centralized treatment plant [5]. Based on the data from the Public Works, the centralized system can be found in nine WWTPs located in some of the villages in Wonosari, all of which were included as the object of observation in the research.

Further observations revealed that these WWTPs were introduced or maintained through different programs: community-based environmental sanitation (*Sanitasi Lingkungan Berbasis Masyarakat* – SLBM), community-based sanitation (*Sanitasi Berbasis Masyarakat* – SANIMAS), and Communal WWTPs. These three centralized systems had different characteristics, including system capacity, and were installed at different times (Table 1). Based on the Public Works data—supported by information from the WWTP management, the plant installation was funded and implemented by the Public Works and the Working Unit of Environmental Sanitation Development. SLBM and SANIMAS were government initiatives aiming to increase community access to proper sanitation [6].

Table 1. Sanitation system, capacity, year of construction, and maintenance funding of nine WWTPs in Wonosari District

No	Sewage Disposal System (WWTPs)	Sanitation Type	Year of Installation	System Capacity (households)	Maintenance Funding
1	WWTP Karangrejek	SLBM	2013	150	Contribution funds
2	WWTP Rusunawa Karangrejek (Flat)	Communal	2019	350	Flat management
3	WWTP Siraman	SANIMAS	2010	42	Contribution funds
4	WWTP Mulo	SLBM	2014	180	Contribution funds
5	WWTP Rejosari, Baleharjo	SANIMAS	2016	150	Contribution funds
6	WWTP Kepek I	SLBM	2013	80	Contribution funds
7	Domestic WWTP Selang V	Communal	2018	44	Contribution funds
8	WWTP Piyaman	SANIMAS	2016	150	Contribution funds
9	WWTP Karangtengah	Communal	2019	51	Contribution funds

Sources: Field Survey, 2021, and Public Works Service, 2022

Construction is an essential aspect of a building. Based on the field survey results and information from the WWTP management, the nine plants observed had a similar structure, i.e., a combination of iron frames and concrete cast. Experts were consulted in the careful planning, design, and construction of the plants, guaranteeing the physical quality of their structure more than it would be without professional involvement. In addition, construction is an aspect that significantly determines possible contamination associated with management and maintenance during the structure's operation. The district's wastewater management technology mainly combined an anaerobic baffled reactor (ABR) with an anaerobic filter (AF). The treatment chamber of ABRs consists of a sedimentation basin connected to reactors that are compartmentalized by a series of baffles to create upflows, whereas AFs are packed with layers of rocks that are larger at the bottom and increasingly smaller toward the top of the filtration chamber [7].

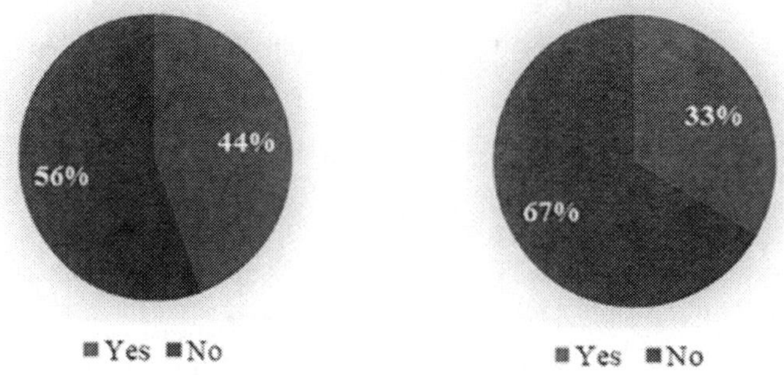

Figure 2. Percentages of water quality monitoring at the WWTP outlet (left) and routine inspection on WWTP condition (right)

Wastewater is treated at WWTPs to remove contaminants and discharge the remaining water, called effluent, back into water bodies or the environment without causing pollution [8]. It is thereby imperative to monitor water quality at the plant outlet. Observations and additional information from the WWTP management revealed that such monitoring had been conducted at four of the nine plants (44%) but not at the remaining five (56%) (Figure 2-left). In addition, only three (33%) were routinely checked for their most current conditions by responsible agencies (Figure 2-right). From these facts, it can be inferred that damages such as leaks and clogged drains may not be detected early. More often than not, reports are made to the office when the affected community is experiencing unpleasant odor and puddles are observed around the plant, which then prompts an inspection. Monitoring of water quality at the outlet measures to what extent the installed plant successfully treats wastewater and, thus, ensures that the effluent can be returned to water bodies or the environment without harming environmental sustainability.

Using a centralized sanitation system to treat wastewater before its disposal in the environment can reduce the potential for water pollution, provided that proper management and maintenance are implemented. The maintenance in question can include desludging, draining, repairing pipes, and monitoring water quality at the WWTP outlets. According to the detailed technical planning guidelines for sludge treatment facilities, the treatment tank must be desludged and drained at least once in three years [9]. However, field observations and interviews with the management found that the WWTPs in the district were not desludged and drained regularly. Moreover, some had never been desludged. It was also found that only three of the nine sampled plants were routinely desludged and drained, namely WWTPs Rusunawa Karangrejek and Kepek I (once a year) and Mulo (every three years). In conclusion, because of the required routine desludging, all the observed WWTPs in Wonosari District do not comply with the guidelines, except for Rusunawa Karangrejek, Kepek I, and Mulo.

Irregular desludging and lack of proper, periodic maintenance can cause problems, e.g., drain leaks, clogging, and unpleasant odor. Leaks are the leading cause of groundwater pollution [10]. Furthermore, poor maintenance can disrupt the plant's capacity to treat wastewater, rendering this function less than optimal and potentially increasing the risk of bacterial contamination in the water and the environment.

3.1.2. Decentralized or On-site Sanitation. Decentralized or on-site sanitation uses a disposal and treatment system located in the vicinity of where the wastewater is generated without conveying it to centralized treatment sites or receiving water bodies [5]. An individual septic tank is an example of on-site sanitation adopted in Wonosari. The working principle of a septic tank involves a very slow wastewater flow that allows solids like fecal matter to settle in the tank, thus increasing the opportunity for anaerobic microbes to decompose the organic matter [11]. This naturally occurring process can separate solids in the form of stable mud from liquids.

Proper and healthy sanitation contributes to public health and environmental sustainability. A sanitation facility is considered adequate if it meets these health requirements: the facility is used by one household or together with certain other households, equipped with toilets with a built-in gooseneck trap, and connected to a final disposal site of fecal matter such as a septic tank or WWTP [12]. Based on the field survey findings, the inlets of all the eight observed septic tanks were toilets with gooseneck traps, complying with the standards for proper sanitation (Table 2).

Table 2. Toilet types and septic tank constructions

No	Locations	Toilet Designs	Septic Tank Constructions
1	Karangrejek Subvillage	Gooseneck traps	Stone, cast top, with filters
2	Wareng Village	Gooseneck traps	Stone, cast top
3	Rejosari, Baleharjo	Gooseneck traps	Stone, cast, filters
4	Ringinsari, Wonosari	Gooseneck traps	Brick
5	Wonosari Village	Sitting toilets	Stone, filters made from stiff Arenga fibers
6	Selang	Gooseneck traps	Stone, cast top
7	Karangtengah	Gooseneck traps	Stone, cast top
8	Karangtengah	Gooseneck traps	Stone, cast top

Source: Field Survey, 2021

In addition to toilet design, septic tank construction is another aspect contributing to the properness of a sanitation system. The septic tanks in the district were mostly made from stones, bricks, and a cast top. Some were only used to dispose of and store sewage without removing the solids from liquids because they did not have filters or filter beds. Only three of the eight septic tanks observed (37%) were equipped with advanced processing, i.e., filtration. Based on information from the residents, filters were not added to the tank to minimize construction costs. An effluent still contains several organic parameters [11]; therefore, without the needed further processing, it increases the potential for water pollution, primarily due to seepage from the septic tank. This explains why offsite sanitation (i.e., WWTPs) technically poses a lower risk to the environment than on-site sanitation, such as septic tanks [1], mainly because the former is equipped with wastewater treatment technology, and related agencies directly supervise its construction.

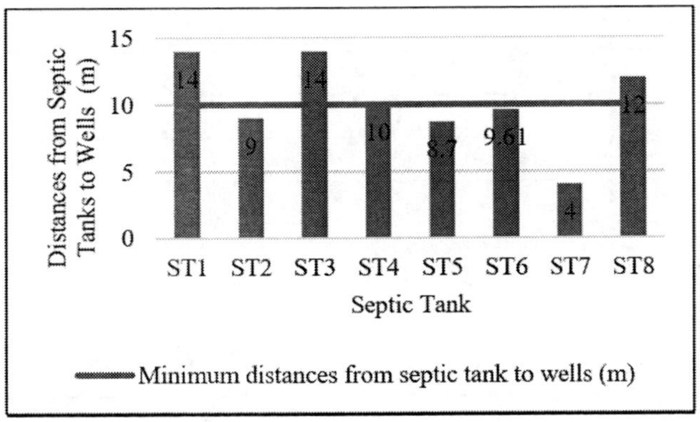

Figure 3. Distances from the Eight Septic Tanks Observed to the Nearest Wells

Per the SNI 2398 [4], the Public Works established standardized sanitation construction procedures that include septic tank construction and a safe distance of at least 10 m from water wells and 5 m from infiltration wells. During the field survey, it was found that the septic tanks were installed at varying distances from the dug wells used to meet the local domestic water needs (Figure 3). Also, only 37% of the eight sampled septic tanks were installed more than 10 m or at a safe distance from the nearest dug wells. Therefore, around 63% do not meet the above provision, i.e., > 10 m to wells. The distance between septic tanks and the nearest buildings, wells for clean water sources, and rain infiltration wells should be carefully considered in planning the construction [13] because the closer the septic tank is installed to a well, the greater risk of contamination it poses, especially by coliform bacteria [14].

Based on the survey results, all the septic tanks observed were never drained or desludged. In addition, interviews with the residents revealed no routine inspection of their condition and water quality monitoring at the surrounding wells. This situation can have a negative impact on the future. According to the construction procedures in SNI 2398, a septic tank should be drained or desludged once every 2–5 years, assuming a family of five members uses it with sludge sediment forming at 30–40 L/person/year.

Draining or desludging reduces the potential for leaks, clogging, and explosion due to the decomposition by-products, i.e., gas [15]. All the septic tanks in the district are never drained or desludged; thus, it is highly likely that bacteria-containing water and mud from tanks with permeable bottoms and without wastewater filtration have seeped into the ground. Septic tanks with permeable bottoms can contaminate waters in downstream wells, especially if both are located less than 10 meters from each other [16].

3.2. Groundwater Quality Based on Coliform Count in Wonosari District, Gunungkidul Regency

Coliforms, including fecal coliform and total coliform, are the biological parameters of water quality. In addition, based on the MoH Regulation No. 32 of 2017, fecal coliform or *E. coli* and total coliform are mandatory biological parameters to check to determine whether or not a water body is suitable for sanitation and hygiene purposes and, in another regulation (No. 492 of 2010), drinking water. Furthermore, per Regulation No. 32 of 2017, sanitation and hygiene purposes include using water to maintain personal hygiene, such as bathing, teeth brushing, and washing food ingredients, eating and cooking utensils, and clothes. Water with this level of quality can also be used as raw water for drinking water.

Table 3. Coliform Count of the Well Water Samples in Wonosari District

No	Sample Codes	Fecal Coliforms (MPN/100 mL)	Total Coliforms (MPN/100 mL)
1	S1	4.5	49
2	S2	540	>1600
3	S3	11	130
4	S4	350	1600
5	S5	11	110
6	S6	33	79
7	S7	110	>1600
8	S8	920	>1600
9	S9	170	>1600
10	S10	7.8	49
11	S11	>1600	>1600
12	S12	540	>1600
13	S13	17	49

Source: Laboratory Test Results, 2021

Based on the laboratory analysis, all the water samples collected from the dug wells were contaminated with high coliform levels (Table 3). Per Regulation No. 32 of 2017, clean water for daily use or sanitation and hygiene purposes can contain a total coliform of <50 MPN/100 mL and 0 MPN/100 mL fecal coliform. With a range of 4.5 to >1600 MPN/100 mL, the fecal coliform levels of all the test samples far exceeded the water quality standard. In addition, three samples had a total coliform of 49 MPN/100 mL or only one point lower than its maximum acceptable presence (50 MPN/100 mL). The high coliform count is attributed to the physical condition of the area surrounded by karst hills with a thin layer of soil, which allows contaminants to easily seep into and pollute groundwater [3]. Moreover, the presence of pollutant sources, e.g., cattle pens, near the wells exacerbates the risk of contamination. In addition, improperly managed domestic wastewater can also increase the number of coliforms in groundwater.

Per the MoH Regulation No. 492 of 2010, water is considered suitable for drinking if it contains no coliforms (fecal and total coliform = 0 MPN/100 mL). Because the laboratory test returned with high coliform counts (Table 3), all the samples did not meet the drinking water requirements. In conclusion, based on the coliform parameter, the groundwater in the district has poor quality, which corresponds to the Regional Environmental Status of Gunungkidul Regency in 2014.

3.3. Groundwater-Contaminating Coliform Counts in Each Sanitation System in Wonosari District, Gunungkidul Regency

The waste disposal system in Wonosari District consists of a centralized sanitation system such as WWTPs and decentralized or offsite sanitation in the form of septic tanks. The presence of coliforms in groundwater in each sanitation system was enumerated from the samples collected at the dug wells of the system users. The test results showed varying concentrations of coliforms (Table 4). The highest total coliforms, >1600 MPN/100 mL, were detected in the dug wells downstream of WWTPs Piyaman, Karangrejek, and Selang. The samples with the highest concentrations of fecal coliforms were the dug wells downstream of WWTPs Piyaman (>1600 MPN/100 mL), Karangrejek (540 MPN/100 mL), and Selang V (170 MPN/100 mL). The high coliform levels in the well water are caused by the physical environmental condition and poor maintenance of the sanitation system.

Table 4. Coliform Counts in the Groundwater Samples of the WWTP Users (in MPN/100 mL)

Dug Well Locations	Sanitation System	Fecal Coliform	Total Coliform
Karangrejek	WWTP Karangrejek	540	>1600
Siraman	WWTP Siraman	11	130
Kepek	WWTP Kepek RT I	33	79
Selang	WWTP Selang V	170	>1600
Piyaman	WWTP Piyaman	>1600	>1600

Source: Analysis Results, 2022

WWTP Piyaman was built in 2016 to accommodate 150 households, but this large system capacity and construction made from sturdy iron frames and cast were not complemented by sufficient management and maintenance. Also, the survey results showed that this plant had never been drained or desludged since its construction. According to the detailed technical planning guidelines for sludge treatment facilities, the treatment tank should be drained and desludged at least once in three years [9]. However, this minimum frequency of necessary maintenance was not met at WWTP Piyaman. In addition, the water quality and sanitation system were not monitored regularly. During the interviews, it was revealed that the residents around this plant often experienced unpleasant odor, clogged drains, and leaks and saw lots of mosquitoes.

The laboratory test results showed that the well water downstream of WWTP Karangrejek had a high number of coliforms: fecal coliform of 540 MPN/100 mL and total coliform of > 1600 MPN/100 mL. The tank had been drained only once since its construction in 2013. As a result, some problems that often occurred were pipe leaks, unpleasant odor, and clogging. Also, there had been no routine checks and monitoring of water quality and no assistance from related agencies. The observed well was located only 6 m from or very close to WWTP Karangrejek. Wastewater seepage from the leaking tank and the poorly maintained plant are believed responsible for the high coliform presence in the well water.

The laboratory test results showed the lowest coliform count at the dug well downstream of WWTP Kepek I. The fecal coliform was 33 MPN/100 mL, and the total coliform was 79 MPN/100 mL. WWTP Kepek I was among the plants receiving the best care and management in the district. Based on the observation and survey findings, the tank was drained or desludged once a year, complying with the WWTP treatment guidelines. This plant was built in 2013 to accommodate 80 households and was routinely inspected (once a year) by related governmental services. Records showed that no pipe leak was recorded since its construction. The sewage treatment system in this plant was also monitored regularly, during which a solution consisting of molasses, yeast, and EM4 as nutrients for the decomposing bacteria was added. As a result of this maintenance, the sewage treatment operates optimally and potentially reduces the risk of groundwater contamination. It can be inferred from these findings that a relatively low coliform count is associated with a properly maintained WWTP.

Compared to WWTPs, individual septic tanks (decentralized or on-site sanitation system) are more common in the district. The laboratory test results showed varying concentrations of coliforms in the water samples (Table 5). Fairly high coliform counts were detected in Wonosari, Karangtengah 1, and Wareng Villages, whereas relatively low coliform levels were found in Karangrejek, Baleharjo, Selang, and Karangtengah 2. These tanks were installed at varying distances to the sampled wells.

Table 5. Coliform Counts in the Groundwater Samples of the Individual Septic Tank Users

Well Locations	Sanitation System	Distance from Septic Tank to Well (m)	Fecal Coliform (MPN/100 mL)	Total Coliform (MPN/100 mL)
Karangrejek	Individual Septic Tank	14	4.5	49
Wareng	Individual Septic Tank	9	350	1600
Baleharjo	Individual Septic Tank	14	11	110
Wonosari 1	Individual Septic Tank	10	110	>1600
Wonosari 2	Individual Septic Tank	8.7	920	>1600
Selang	Individual Septic Tank	9.61	7.8	49
Karangtengah 1	Individual Septic Tank	4	540	>1600
Karangtengah 2	Individual Septic Tank	12	17	49

Source: Analysis Results, 2022

Based on the data from the laboratory test (Table 5), the well water with the least coliform presence, i.e., in Karangkejek Village, was located 14 m from the septic tank. The fecal coliform was 4.5 MPN/100 mL, and the total coliform was as high as 49 MPN/100 mL. According to SNI 2398, apart from a distance, the number of coliforms in the well is also influenced by well and septic tank constructions. The well at Karangkejek had watertight walls, and the septic tank was made from stone and cast top and equipped with a filter composed of fibers and stones. In addition, no other pollutant sources affected this well.

The well with the highest coliform count was found in Wonosari Village. It was downstream of several pollutant sources, i.e., WWTP at the Wonosari Market and other domestic wastes, potentially contaminating the groundwater observed easily. In addition, Wonosari Village is the center of activities in Gunungkidul and has a high population density. For these reasons, a large amount of waste is generated.

The presence of other pollutant sources can increase the potential for groundwater pollution. For instance, the well sample in Karangtengah 1 Village contained 540 MPN/100 mL of fecal coliforms, and the total coliform amounted to >1600 MPN/100 mL, indicating bacterial contamination. In this location, the distance between the well and the septic tank was only 4 m or far from the minimum distance set in the SNI (10 m). Moreover, the septic tank used was never drained or desludged and checked regularly, and some sewerage was clogged. Cattle pens located about 1 m from the well also contributed to the high contamination by coliforms.

WWTPs for waste disposal and treatment are technically better in minimizing water pollution than individual septic tanks. The reason is that the plants treat the wastewater with sufficient technology and system so as not to pollute the environment [17]. In contrast, on-site sanitation, which generally accommodates and stores fecal matter at the place where it is generated without applying further processing like filtration [1], poses a higher risk of bacterial contamination. Nevertheless, installing WWTPs does not guarantee low pollution. This is evident in the laboratory test results, which suggest that some of the plants in the district have not been able to reduce bacterial contamination in groundwater. The main factor precluding the optimal operation of a WWTP is poor management and treatment after its construction. Poor sanitation systems, open defecation, and improperly maintained wastewater disposal and treatment facilities are some sources of coliforms that contaminate groundwater [1].

4. Conclusion

The sanitation systems for sewage disposal in Wonosari District consist of offsite sanitation such as communal wastewater treatment plants (WWTPs) and on-site sanitation in the form of individual septic tanks. These systems have different characteristics, including post-construction management and maintenance. Based on the field survey findings, only one-third of the WWTPs are inspected and drained or desludged regularly, and water quality monitoring is conducted at only four of the nine plants observed. In addition, the coliform counts in the district's groundwater have exceeded their maximum allowable presence, as specified in the Minister of Health Regulations No. 32 of 2017 and No. 492 of 2010. Groundwater samples with high coliform counts are found in the dug wells downstream of poorly maintained WWTPs, very close to substandard individual septic tanks with poor management, and near other sources of pollutants such as livestock pens.

References

[1] Lawrence A R and Macdonald D 2001 *Guidelines for Assessing the Risk to Groundwater from On-Site Sanitation* (UK: British Geological Survey).

[2] BPS-Statistics of Gunungkidul Regency 2021 *Kabupaten Gunungkidul dalam Angka 2020* (Jakarta: Badan Pusat Statistik).

[3] Kurniawan A and Sadali M I 2015 *Keistimewaan Lingkungan Daerah Istimewa Yogyakarta* (Yogyakarta: UGM Press).

[4] BSN-National Standardization Agency of Indonesia 2017 Standar Nasional Indonesia (SNI) 2398:2017 tentang Tata Cara Perencanaan Tangki Septik dengan Pengolahan Lanjutan (Jakarta: Badan Standardisasi Nasional).

[5] Fajarwati A 2000 Perencanaan Sistem Penyaluran Air Buangan Domestik Kota Palembang (Studi Kasus: Kecamatan Ilir Timur I dan Kecamatan Ilir Timur II), *Thesis* ITB Bandung.

[6] Directorate General of Human Settlements 2015 *Rencana Strategis Direktorat Jenderal Cipta Karya Tahun 2015-2019* (Jakarta: Ditjen Cipta Karya Kementerian PUPR).

[7] Kurnianingtyas E Prasetya A and Yuliansyah A T 2020 Kajian Kinerja Sistem Instalasi Pengolahan Air Limbah (IPAL) komunal (Studi Kasus: IPAL Komunal Kalisong, Kelurahan Sembung, Kecamatan Tulungagung, Kabupaten Tulungagung, Jawa Timur) *Media Ilmiah Teknik Lingkungan* (5) 62-70.

[8] Pranajaya I 2019 *Apa Itu IPAL Komunal dan Apa Manfaatnya* (Aceh: Sistem Informasi Database Infrastruktur Kabupaten Aceh Tamiang).

[9] Directorate of Environmental Sanitation Development (PPLP) 2019 *Pedoman Perencanaan Teknik Terinci Instalasi Pengolahan Lumpur Tinja* (Jakarta: Direktorat PPLP Kementerian PUPR).

[10] Sasmito B 2014 Sistem Sanitasi dan Kualitas Air Sumur Dangkal di Kabupaten Sumenep

Rekayasa Teknik Sipil (3) 166-172.
[11] Sudarmadji 2014 Tangki Septik dan Peresapannya sebagai Sistem Pembuangan Air kotor di Permukiman Rumah Tinggal Keluarga *Jurnal Teknik Sipil* 9(2) ISSN 1907-6975.
[12] BPS-Statistics of Gunungkidul Regency 2020 Persentase Rumah Tangga yang Memiliki Akses terhadap Layanan Sanitasi Layak (Jakarta: Badan Pusat Statistik).
[13] Nurhidayat A 2009 Strategi Pengelolaan Air Limbah Domestik Dengan Sistim Sanitasi Skala Lingkungan Berbasis Masyarakat di Kota Batu Jawa Timur *Prosiding* Seminar Nasional Manajemen Teknologi X.
[14] Sapulete M 2010 Hubungan Antara Jarak Septic Tank ke Sumur Gali dan Kandungan Escherichia Coli dalam Air Sumur Gali di Kelurahan Tuminting, Kecamatan Tuminting, Kota Manado *Jurnal Biomedik* (2) 179-186.
[15] Pujari N 2007 Sistem Sanitasi dan Kualitas Air Sumur Dangkal di Kabupaten Sumenep *Rekayasa Teknik Sipil* (3) 166-172.
[16] Mardikanto A 2020 *Hati-Hati Tangki Septik Meledak* (Jakarta: National Water & Sanitation Information Services).
[17] Hermawan T 2010 Pengaruh Sanitasi Air Terhadap Kesehatan Lingkungan Ditinjau Secara Bakteriologis: Kasus Penyediaan Air Bersih Di Kecamatan Guguk Panjang Kota Bukittinggi Propinsi Sumatera Barat *Master Thesis* Pascasarjana Universitas Gadjah Mada.

Acknowledgment
The authors would like to thank Universitas Gadjah Mada for funding this research through the Thesis Recognition Grant (RTA) scheme in the fiscal year 2022 with a Letter of Assignment Number is 3550/UN1.P.III/Dit-Lit/PT.01.05/2022. Gratitude is also extended to the Faculty of Geography, Universitas Gadjah Mada, for the opportunity to conduct this research.

Analysis of the Amount of Filling and Emptying Time Communal Rainwater Harvesting Tank (RWH)

A Kurniawan[1,3], A Maryono[1], P T S Sembada[1], F N M Jayatri[2], R T Onerio[1], T Abieza[2]

[1] Department of Civil Engineering, Vocational College, Universitas Gadjah Mada, Yogyakarta 55281, Indonesia
[2]Environmental Science Student, Post Graduated, Universitas Gadjah Mada, Yogyakarta 55281, Indonesia

[3] Email: adhy_kurniawan@ugm.ac.id

Abstract. Utilization of stagnant water storage tanks in Banjararum Village, Kalibawang District, Kulon Progo Regency with a total of 23 units needs to be optimized by using them as communal rainwater harvesting (RWH). The amount of time for filling and emptying the tank is an important aspect to be able to optimize rainwater used for daily needs. The aims of this research was to determine the effectiveness of the use of Communal PAHs applied in the community. This is analyzed through several stages, including determining the planning location unit, analyzing hydrological characteristics and analyzing the flow rate and duration of filling and emptying the tank using the Prandtl-Colebrook formula and the Bernoulli formula. Researchers took samples in Dusun Kisik by optimizing 2 roofs that can distribute rainwater of 27.09 m^3/week for roof 1 and 27.19 m^3/week for roof 2. Assuming 6 people live around the tank who use water storage every day, then the communal PAH can supply 3,000 liters within 52 minutes while the emptying process for distribution to other media takes 69 minutes. This difference is due to the different dimensions of the inlet pipe and outlet pipe. So that the effectiveness of the use of Communal PAHs applied in the community can be one of the recommendations in the provision of sustainable clean water.

Keywords: Communal Rainwater Harvesting Tank, Drought Disaster, Emptying Time, Filling Time

1. Introduction

The drought disaster in 2018 occurred in the Special Region of Yogyakarta, especially Kulon Progo Regency in 13 villages, one of which was Banjararum Village, Kalibawang District [1]. This makes it difficult for the availability of clean water to be used to meet the daily needs of the community. On the other hand, Banjararum Village which is located in the Menoreh Hills area and adjacent to the Progo River is always hit by drought during the long dry season, this is very different when the rainy season comes, where some areas experience a surplus of water and even often cause landslides. So there is a need for education to deal with drought disasters during the dry season and disasters during the rainy season. One of them is taking advantage of the untapped potential of rainwater in the area.

In line with the potential for rainwater that has not been utilized, Banjararum Village has the infrastructure to supply water needs which until now has not been utilized properly due to the failure of its planning. This resulted in 23 abandoned water storage tanks with a capacity of 3,000 liters or the equivalent of 3 m^3. The position of the water storage tank is in a residential area, so it is unfortunate if it is not used properly even in the long dry season, the tank is used to receive clean water from drop-off assistance from various aid providers. Therefore, along with the increasing number of people who are struggling to take advantage of the potential of rainwater as an alternative to water to meet their daily needs and mitigate drought and flood disasters, Banjararum Village is right if it has to optimize the potential of rainwater by utilizing abandoned water tanks. The design used to collect rainwater in Banjararum Village uses a communal rainwater harvesting design. This was chosen because of the availability of large water storage infrastructure owned by Banjararum Village compared to the community's water tank which is only 1,000 liters (simple house scale).

Figure 1. A stalled water storage tank in Banjararum Village

Rainwater harvesting is one of the efforts to support the achievement of SDGs (Sustainable Development Goals) on indicator 6, namely clean water and sanitation, indicator 13, namely climate action, and indicator 15 for life on land. Improving the quality of life and the economy will also be obtained when the potential for rainwater is used to meet the daily needs of the surrounding community, because the quality of rainwater is better than the quality of well water and river water which is widely used by the people of Banjararum Village, Kalibawang District, Kulon Progo Regency.

Utilization of rainwater storage by using storage media in the form of tanks, ponds, and others, which in developed countries individually is one aspect of stability in several countries in the world. Countries that require regulations such as Catalonia, Spain for new buildings of a certain size, in addition, Belgium is required to accommodate rainwater with a roof area of more than 100 m^2 [2]. In Australia, 26% of households use rainwater tanks as a water source [3]. Meanwhile, in 2013, households in Queensland Australia were required to meet a water-saving target of 70 Kilo liters per year through the installation of a connected 5 Kilo Liter tank with a roof area of 100 m or half of the available roof area, less under Queensland Development regulations [4].

Progress in optimizing rainwater potential in developed countries is one step that can be duplicated in various regions in Indonesia. Although harvesting rainwater is a culture that has existed for a long time, it is necessary to develop the use of technology that can provide good services, especially regarding the quality of water that will be consumed by the community. The development of the implementation of Rainwater Harvesting (RWH) activities in Indonesia has been running since 2011 with positive public acceptance. Indicators of community acceptance with a case study of the Special Region of Yogyakarta used as a sample of user communities include the use of technology, rainwater quality, usability, and several other indicators. The results presented in the article show that 75% feel happy and useful with the Rainwater Harvesting [5]. Thus, the application of RWH activities can be carried out in Banjararum Village but with a communal design. This is the development of RWH technology, which currently uses a lot of individual methods.

Communal Rainwater Harvesting Technology has advantages and disadvantages in several aspects. Aspects of the advantages of communal RWH can be seen from the equipment maintenance process. Individual RWH technology is usually neglected because equipment owners pay less attention to its maintenance, while communal RWH has more controlled maintenance because it consists of several roofs and the process of its use is divided into several houses. On the other hand, the weakness of communal RWH is the process of managing water entering and leaving the tank. This is due to the analysis of the equations used for the distribution process must follow the needs or volume owned. Thus, the distribution of the amount of discharge (m^3/s) of clean water can be obtained evenly. This is one of the responsibilities of the communal tank manager. This study aims to determine the effectiveness of the use of Communal PAHs applied in the community.

2. Method

The research method to calculate the amount of time required for filling and emptying communal rainwater harvesting tanks (RWH) utilizes several stages of analysis, including determining the planning location of a total of 23 unused tanks with a capacity of 3,000 liters in Banjararum Village, Kalibawang District, Kulon Regency. Progo. Next, analyze the hydrological characteristics to determine the potential volume of rainwater that can be accommodated. The monthly high rainfall data (mm) used utilizes data from 2006 - 2015 with sources from BPS. Then do the planning and drawing of the location units that have been determined already as well as the plan for the placement of communal RWH. First, the equation used in the calculation and planning of the communal RWH design is the rational formula to get the volume (m^3) of rainwater that can be accommodated [6].

$$V = I\,x\,T\,x\,A\,x\,\alpha\,x\,\beta \tag{1}$$

or

$$V = h\,x\,A\,x\,\alpha\,x\,\beta \tag{2}$$

where:

I	:	Rain Intensity
T	:	Rain Duration (hours)
A	:	Roof Area (m^2)
α	:	Run Off Coefficient (0,9)
β	:	Rain Distribution Coefficient (1,0)
h	:	Rain Height (mm)
V	:	The volume of rainwater accommodated (m^3)

Furthermore, the equations used to analyze the flow rate and duration of filling and emptying the tank use the equations of the Prandtl-Colebrook formula and the Bernoulli formula [7, 8]. Analysis of the calculation of the discharge that can enter the rainwater tank.

$$Q = \left(\frac{1}{4}\pi D^2\right)x(\sqrt{2gh} \tag{3}$$

where:

Q	:	Pipeline discharge (m^3/s)
D	:	Pipe diameter (m)
G	:	gravity (9,81 m/s^2)
H	:	Height Difference (m)

$$Q = \frac{V}{t} \tag{4}$$

where:

Q	:	Pipeline discharge (m^3/s)
V	:	Volume of rainwater (m^3)
t	:	duration of rain (s)

Pipe cross-sectional area formula:

$$a = \frac{1}{4}\, x\, \pi\, x\, D^2 \tag{5}$$

$$Q = Cd\, x\, a\, x\, \sqrt{2\, g\, H} \tag{6}$$

To calculate constant:

$$Q = Cd\, x\, a\, x\, \sqrt{2\, g\, H}$$

$$q = Cd\, x\, a\, x\, \sqrt{2\, g\, H} \tag{7}$$

$$q = K\, \sqrt{2\, g\, H} \tag{8}$$

where:

Q : Pipeline discharge (m^3/s)
Cd : Coefficient
a : Pipe cross-sectional area (m^2)
g : Gravity (9,81 m/s^2)
H : Height Difference (m)
K : Constant

The calculation to find out the equation of filling and emptying time is as follows the derivative results after knowing the amount of discharge (m^3/s) distributed in rainwater harvesting (RWH):

$$t = \frac{2\, x\, A}{K^2}\, [\, Q \ln\left(\frac{Q - K\sqrt{H1}}{Q - K\sqrt{H2}}\right) + K\left(\sqrt{H1} - \sqrt{H2}\right)] \tag{9}$$

where:

T : Time (s)
Q : Pipeline Discharge (m3/s)
A : Pipe Diameter Area (m2)
H1 : Height Difference (m)
H2 : Height Difference (m)
K : Constant

3. Results and Discussions

The condition of stagnant water tanks in Banjararum Village totals 24 units with 1 unit which is no longer usable. So that the total tanks which are usually used to receive water supply during the dry season are 23 units spread across 16 hamlets. Each hamlet has 1 to 3 tanks with the distribution units shown in Table 1.

After knowing each water tank location unit, then it is analyzed to select sample locations in planning units that can be used for communal rainwater harvesting (RWH). The parameters used as assessment indicators are the condition of the area in terms of water sufficiency, community water needs, the distance of the tank to the nearest house, ease of access to the location, the elevation of the location roof, and aspects of supporting infrastructure. So from these 6 parameters, 3 priority locations were obtained for applying communal RWH. The selected locations include Kisik Hamlet, Kedondong 1 Hamlet, and Kedondong Hamlet 2. The values that are highly considered are community water needs, access, elevation, tank distance, and supporting infrastructure such as the location of places of worship around the unit.

Table 1. Location of Tank Units in each Hamlet in Banjararum Village

Dusun	Number of Tanks	Dusun	Number of Tanks
Kanoman 1	1	Kagongan 2	2
Kanoman 2	1	Kedondong 1	2
Sayangan	2	Kedondong 2	3
Kisik	1	Semaken 1	1
Jogobayan	2	Semaken 2	1
Ngipikrejo 1	1	Semaken 3	2
Ngipikrejo 2	1	Kemesu	1
Kriyan	1	Degan 1	1

The next stage is to analyze the volume (m^3) of rainwater that can be accommodated. The aspect or component to be considered is the high rainfall data (mm) in Banjararum Village. Data were accessed from 2006 to 2015 through the BPS website and is presented in Table 2.

Table 2. Rainfall Height Data (mm) Banjararum Village

Years/ Month	Jan	Feb	Mar	Apr	Mei	Jun	Jul	Ags	Sep	Okt	Nov	Des
2015	331	346	259	312	145	105	80	0	0	0	227	516
2014	563	344	203	385	211	100	128	0	0	173	240	427
2013	365	365	308	62	128	-	0	0	0	98	508	551
2012	414	320	340	314	304	0	0	0	0	38	510	207
2011	474	231	361	214	185	94	45	127	360	347	290	464
2010	401	248	217	144	176	55	6	0	0	17	399	359
2009	104	416	234	540	36	0	0	0	0	13	246	442
2008	104	427	244	536	36	0	0	0	0	13	254	455
2007	486	383	289	327	182	34	0	0	0	0	0	478
2006	486	383	289	327	182	34	0	0	0	0	0	478
Average	372,8	346,3	274,4	316,1	158,5	46,8	25,9	12,7	36	69,9	267,4	437,7

Based on Table 2, it can be concluded that the season that has a high potential for rainfall occurs from January to April, and continues in November and December. This is because during this period it was the rainy season. Meanwhile, from May to October, rainfall tends to decrease due to the ongoing dry season at that time. The maximum average rainfall in the period 2006 to 2015 occurred in December, which was 437.7 mm. Meanwhile, the minimum average rainfall occurs in August, which is 12.7 mm.

After knowing the character and average of high rainfall (mm) in the Banjararum Village area, then an analysis of the availability of rainwater is carried out that can be used at the three predetermined units. The availability of rainwater is then put into a stalled water tank which will be redesigned to become a communal RWH. Before calculating the availability of water using equation 2, the water needs of the 3 hamlets were selected according to SNI 03-7065-2005 where each person needs 120 liters of water per day. liters/day with a population of 466 people, and Dusun Kedondong 2 with a population of 46,320 liters/day with a population of 386 people. The water needs of the 3 hamlets will be met with the potential for rainwater in addition to using wells and PAMDES from the Progo River springs. The hope is to reduce the burden on the Banjararum Village Government.

Availability of rainwater (m^3) which has been calculated by equation 2 in Kisik Hamlet (from 3 units, 1 is chosen which is more priority, although the other 2 units are also calculated for RWH capacity until the implementation activity) produces a magnitude of 27.09 m^3/week for roof 1 and 27.19 m^3/week for roof 2. The amount of availability of these 2 roofs will then be related to the needs of the community in the area which is assumed to be 6 people with a need of 120 liters/day of 5.04 m^3/week. Then obtained tables and graphs of the relationship between the availability of rainwater that can be accommodated in Kisik Hamlet with the needs of the community's use of rainwater around the unit.

Table 3. Relationship between Rainwater Availability (m^3) and Water Demand (m^3)

Month	Rainfall Height (mm) Weekly Average	Availability of Rainwater (m^3) Communal/week	Community Water Needs (m^3)/week
Januari	93,2	8,56	5,04
Februari	86,6	7,95	5,04
Maret	68,6	6,30	5,04
April	79	7,26	5,04
Mei	39	3,64	5,04
Juni	11,7	1,08	5,04
Juli	6,5	0,59	5,04
Agustus	3,2	0,29	5,04
September	9	0,83	5,04
Oktober	17,5	1,60	5,04
November	66,9	6,14	5,04
Desember	109,4	10,05	5,04

In Figure 2, it can be seen that in the rainy season, the availability of water is abundant every day, but it is different from the dry season when the need for water is greater than the availability of water. Thus, the potential for rainwater in the rainy season will be optimized by maximizing the water reservoir that has been converted into a communal RWH with a capacity of 3,000 liters.

The use of a 3-dimensional pipe in the communal RWH installation structure makes the process of distributing rainwater into water reservoirs take time, where the amount of time for filling and emptying the tank is an important aspect to be able to optimize rainwater used for daily needs. The equation used is based on equations 5-9. The following is the result of the analysis of the time of filling the water storage tank which is used as a communal RWH.

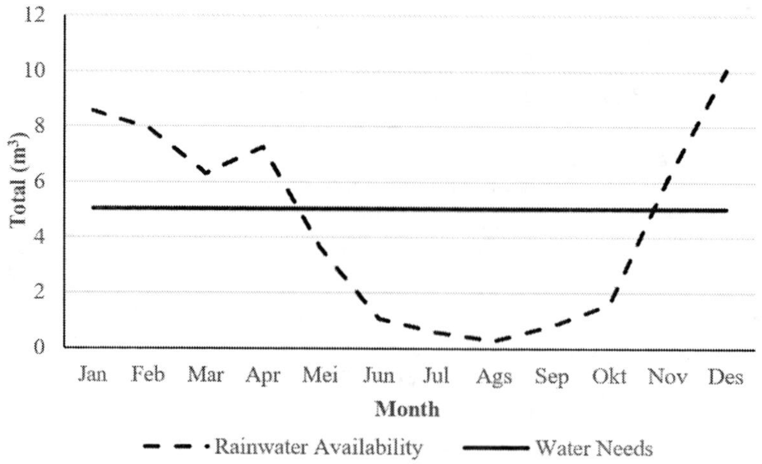

Figure 2. Graph of Relationship between Rainwater Availability (m^3) and Water Demand (m^3)

Based on the analysis of the calculation results using equations 1-8, the pipe cross-sectional area (a) with dimensions of 3 inches is 0.0045 m^2, the amount of discharge generated when filling RWH (Q) is 0.0167 m^3/s and the constant (K) which is used as the basis for calculating knowing the charging time is 0.0125. Thus, the amount of time it takes to fill rainwater into the RWH is $t_1 = 3127.78$ s = 52 minutes. With the same formula, the amount of time for emptying the tank is longer than the tank filling process because the dimensions of the pipe for emptying the tank are 1 inch with 2 external holes, where the pipe dimensions are smaller than the pipe that distributes rainwater into the tank. The time required is $t_2 = 4170{,}37$ s = 69 minutes.

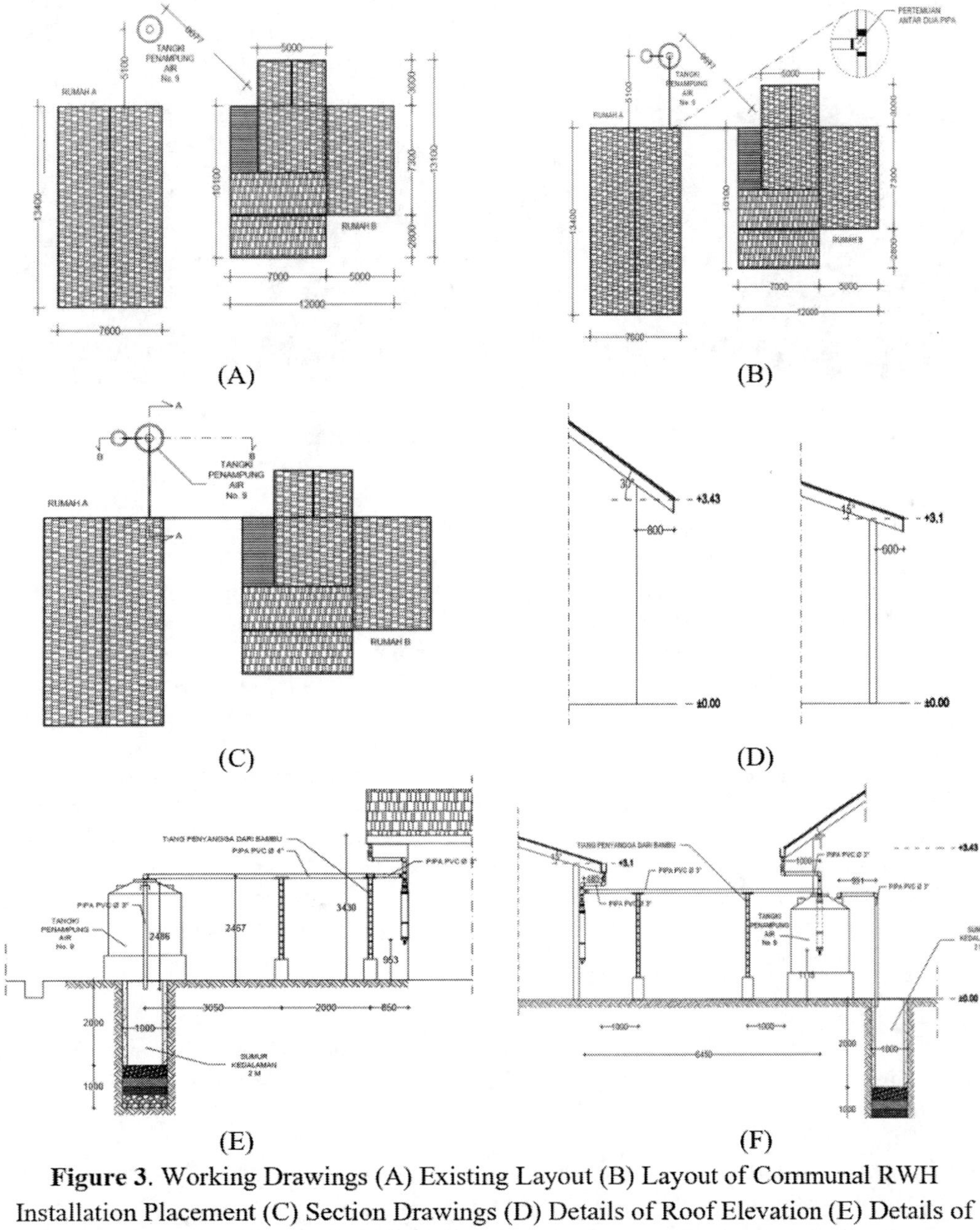

Figure 3. Working Drawings (A) Existing Layout (B) Layout of Communal RWH Installation Placement (C) Section Drawings (D) Details of Roof Elevation (E) Details of Communal RWHs Sections A-A and (F) Details of Communal RWHs Sections B-B

After knowing the amount of time for filling and emptying the tank, then an implementation design is planned in the form of working drawings to change the water storage tank into a communal RWH.

This drawing process presents the layout design of the building and tank, details of the roof area and/or top view of the building, roof elevation of 2 buildings, and details of communal RWH work. A working picture of communal RWH in Kisik Hamlet, Banjararum Village, Kalibawang District is presented in Figure 3.

The activity of making communal RWH in Banjararum Village was carried out in 2018 by involving various multi-stakeholders such as village administrators, communities, and other groups. Good community acceptance makes this program still running and sustainable. In the future, the development of Communal PAHs will enter the stage of product refinement to the adoption of electronic devices to measure the amount of water entering and leaving so that long voids do not occur.

Figure 4. Communal RWH Realization Result

4. Conclusion

Utilization of stalled water storage tanks in Banjararum Village, Kalibawang District, Kulon Progo Regency with a total of 23 units used as communal rainwater harvesting (RWH) (3 priority units were selected from the results of the assessment based on several parameters). Banjararum Village has a short rainy season in the vulnerable period from January to April, with the community's water needs, especially in Kisik Hamlet, which is 30,120 liters/person per day causing the potential for rainwater to be an alternative source to meet the needs of the community. In this study, samples were taken in Kisik Hamlet by optimizing 2 roofs that can distribute rainwater (m^3) of 27.09 m^3/week for roof 1 and 27.19 m^3/week for roof 2. Assuming 6 people live around the communal RWH tank can supply 3,000 liters within 52 minutes, while the emptying process for distribution to other media takes 69 minutes. This difference is due to the different dimensions of the inlet pipe and outlet pipe. The time required for the process of distributing rainwater is the right time for the community to optimize alternative water resources for their daily needs. Communities who live around the unit where the water tank turned into a communal RWH welcome this process until now. So that the effectiveness of the use of Communal PAHs applied in the community can be one of the recommendations in the provision of sustainable clean water.

5. Acknowledgments

This research was supported by Vocational School and Directorate of Research Universitas Gadjah Mada with funding for research, under a UGM Final Project Reconstruction (RTA) 2019.

6. References :

[1] https://yogya.inews.id/berita/bencana-kekeringan-di-kulonprogo-meluas-13-desa-krisis-air-bersih diakses pada tanggl 4 Juni 2022 pukul 17.52

[2] Domènech L and Saurí D 2011 A comparative appraisal of the use of rainwater harvesting in single and multifamily buildings of the Metropolitan Area of Barcelona (Spain): social experience, drinking water savings and economic cost *Journal of Cleaner Production* 19 (6-7), 59 – 60

[3] Australian Bureau of Statistics (ABS) 2010 Environmental Issues: Water use and conservation.

IOP Conf. Series: Earth and Environmental Science 1233 (2023) 012054 doi:10.1088/1755-1315/1233/1/012054

Mar 2010 *Available at: http://www.abs.gov.au/ausstats/abs@.nsf/mf/4602.0.55.003*, [Last Accessed: May 2012
[4] Department of Infrastructure and Planning (DIP) 2008 *Water Saving Targets. For Councils, Plumbers, Builders and Developers: A Guide to the Queensland Development Code Part MP 4.2* (Queensland Government)
[5] Maryono A, Nuranto S, Sembada P T S, and Petrus H T B M 2022 GAMA-RainFilter: a modified rainwater harvesting technique to meet the demand for clean water in Indonesia *International Journal of Hydrology Science and Technology* 13(1):1–22
[6] Maryono 2016 *Memanen Air Hujan* (Yogyakata: Gadjah Mada University Press)
[7] Triatmodjo 1994 *Soal Penyelesaian Hidraulika 1* (Yogyakarta: Beta Offset)
[8] Maryono A, Muth W, and Eisenhauer N 2003 *Hidrolika Terapan* (Yogyakarta: Pradnya Paramita)

Analysis of the Effect of Land Cover Changes on the Runoff Dynamics Using the HEC-HMS Model in Gajahwong Watershed, Yogyakarta

Nur Ayumi[1] and Nugroho Christanto[1,*]

[1]Department of Environmental Geography, Faculty of Geography, Universitas Gadjah Mada, Sekip Utara, Bulaksumur, Yogyakarta 5528, Indonesia

*Corresponding author: *n.christanto@ugm.ac.id*

Abstract. The Gajahwong watershed is part of the Opak watershed. The development of urban areas in the Opak watershed is very massive. The development of the urban regions in the Gajahwong watershed is the occurrence of land conversion from non-built land to built-up land. The land cover change will result in changes in the hydrological response of the watershed. The results showed that in the Gajahwong watershed, there was a decrease in agricultural land in the Gajahwong watershed, which was initially 30,209 km^2 in 2016, down to 20,204 km^2 in 2020. Agricultural land declines due to the conversion of functions to built-up land. Built-up land in the Gajahwong watershed increased by 4,483 km^2. This condition resulted in an increase in the CN value in the Gajahwong watershed from the original in 2016 having a CN II of 73 and then increasing to 74. The results show that in 2016, the Gajahwong watershed had a peak discharge of 14.9 m^3/s; in 2020, the peak discharge in the Gajahwong watershed decreased to 11.8 m^3/s. The volume runoff in the Gajahwong watershed also reduced from 2880.21 mm in 2016 and then in 2020 to 2011,39 mm.

1. Introduction

Land cover plays an important role in regulating the hydrological function of the watershed. Therefore, changes in land cover will result in changes in the hydrological function in the watershed in the form of a decrease in the rate of soil infiltration. The impact of decreasing soil infiltration rate is an increase in erosion and sedimentation, runoff, and baseflow reduction [1]. An increase in runoff will cause an increase in the potential for flooding in a watershed.

Coefficient runoff or curve number (CN) is used to assess the watershed's hydrological condition. One of the determinants of the CN value is the land cover condition. The CN value varies greatly, from 0 to 100 [2]. Therefore, the greater the value of CN, the greater the volume of rainwater that is not absorbed by the soil. This phenomenon is caused because most of the land cover in the watershed is in the form of built-up land, so the land quality in the watershed is very poor.

The Gajahwong Watershed has experienced massive urban development [3]. The impact of the development of urban areas in the Gajahwong watershed is the conversion of land from non-built land to built-up land. The land use change will result in a difference in the hydrological response in the form of an increase in runoff.

This study aims to: (1) examine the types and changes in land cover in the Gajahwong watershed in 2016 and 2020; (2) examine the characteristics of the CN value in the Gajahwong watershed in 2016 and 2020 because of the influence of land cover changes; (3) assessing the characteristics of runoff in the Gajahwong watershed in 2016 and 2020 based on the results of modeling in HEC-HMS. The HEC-HMS software version 4.4 was chosen for daily flow simulation with parameters that affect the hydrological response of the Gajahwong watershed. This research is important because it remembers that the existence of a watershed in an ecosystem area has an essential role and function in supporting its various lives [4].

2. Data and method

2.1. Study Area

The research activities are in the Gajahwong watershed, with the review outlet located at the Wonokromo Automatic Water Level Recorder (AWLR) in Pleret District, Bantul Regency. This watershed covers an area of 69,274 km^2, with a river length of 31,409 km. The Gajahwong watershed is located at coordinates 9129211 – 9155146 mU and 431431 – 436533 mT. The research location is shown in figure 1.

2.2. Research Data

The data for this research activity comes from secondary data and primary data. Secondary data is obtained by conducting surveys in various sources and related agencies. Meanwhile, primary data was obtained based on direct sampling in the field. The data used in this study consisted of:

a) DEMNAS data
b) River Shapefile
c) Outlet Point (AWLR Wonokromo)
d) Sentinel-2 images of 2016 and 2020
e) Daily rainfall data for 2016 and 2020
f) AWLR Wonokromo daily debit data for 2016, 2017 and 2020
g) Soil texture data

This research uses tools such as:

a) ArcGIS 10.3.1 Software
b) HEC-GeoHMS Extension
c) HEC-HMS 4.4 software
d) ENVI 5.3 software
e) Google Earth Pro software
f) Shovel and Soil Drill
g) Plastic Sample
h) GPS
i) Permeability Ring

2.3. Data Analysis

2.3.1. Analysis of Land Cover Types in the Gajahwong Watershed. The research was carried out quantitatively to examine the direction of land cover in the Gajahwong watershed in 2016 and 2020. The direction provides an overview of changes in land cover types in the Gajahwong watershed that occurred in 2016 and 2020.

2.3.2. Characteristic analysis of CN values in the Gajahwong watershed. Analysis of the characteristics of CN values was carried out quantitatively and presented statistically inferentially in the form of a table of the distribution of CN values in the Gajahwong watershed. The CN value distribution table is then analyzed to determine the spatial distribution of CN values to see which areas have the highest and lowest CN values.

2.3.3. Analysis of Runoff Characteristics in the Gajahwong Watershed. The research was carried out quantitatively by presenting inferential statistical data in the form of a table from the simulation results of runoff at HEC-HMS to know the characteristics of runoff in the Gajahwong watershed in 2016 and 2020.

2.3.4. Analysis of the Effect of Land Cover Change on Volume Runoff in the Gajahwong Watershed. The research was carried out quantitatively to determine how much influence changes in land cover had on changes in runoff components such as peak discharge (Qp) and volume runoff.

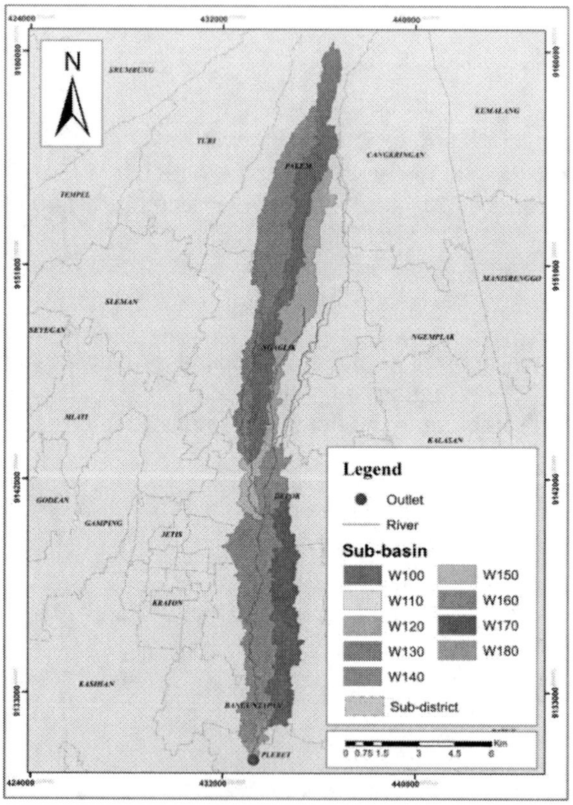

Figure 1. Gajahwong Watershed
Source: Data Processing, 2022

3. Result and discussion

3.1. Hydrologic Soils Group in the Gajahwong Watershed

Hydrologic Soils Group (HSG) is based on the ability of the soil to infiltrate rainwater. The HSG group was divided into four classes, namely classes A, B, C, and D; the characteristics of the grouping were based on soil texture and minimum infiltration rate. The Gajahwong watershed consists of only 2 HSG groups: class A and class B. Table 1. shows the percentage of HSG area in the Gajahwong watershed. The HSG class, A group, has the characteristics of a low potential for runoff due to its very high ability to pass water. The soil texture in the Gajwahwong watershed, belonging to the HSG class A group, is loamy sand with a minimum infiltration rate of 210 mm/hour [5]. The percentage of HSG class A in the Gajahwong watershed is 1,509%. The HSG class, A group, is only located in the upstream part of the Gajahwong watershed. The upstream watershed is still strongly influenced by volcanic activity from the Merapi Volcano, whose material tends to be sand or loamy sand.

Meanwhile, the HSG class B group is the most dominant in the Gajahwong watershed. The HSG class B group has an area percentage of 98.491%. The characteristic of the HSG class B group is that the potential for runoff is rather low, and the rate of water passing tends to be slow. The HSG class B group is evenly distributed from upstream to downstream of the Gajahwong watershed with a soil texture of sandy loam or sandy loam. Soil texture in the form of sandy loam generally has a minimum infiltration rate of 26 mm/hour [5].

Table 1. Percentage HSG Area in the Gajahwong Watershed

Group	Area (Km^2)	Percentage (%)
A	1,045	1,509
B	68,229	98,491

Source: Data Processing, 2022

3.2. Type Analysis Land Cover in the Gajahwong Watershed

Based on the results interpretation of land cover by automatic using Sentinel-2 Image (table 2), we get results that domination land cover in the Gajahwong watershed 2016 in the form of land agriculture of 30,209 km^2 or about 43.608% of the total area of Gajahwong watershed. Meanwhile, 2020 is in the form of land awakened with an area as big as 34,368 km^2 or about 49.612% of the total area of the Gajahwong watershed. Decrease agriculture land area caused as existence over function land Becomes land built up. Appendix 1. shows that the land cover map of Gajahwong Watershed in 2016 and 2020.

Table 2. Area Land Cover in the Gajahwong Watershed

Land cover	2016		2020	
	Area (Km^2)	%	Area (Km^2)	%
Vegetated land	7,894	11,395	7,829	11,302
Agricultural land	30,209	43,608	20,204	29,165
Built-up land	29,885	43,140	34,368	49,612
Open field	0,403	0,582	3,751	5,415
Non-agricultural land	0,883	1,275	3,122	4,507
Total	69,274	100	69,274	100

Source: Data Processing, 2022

3.3. Characteristics of CN Value in the Gajahwong Watershed

The CN value is used to assess a watershed's hydrological conditions ranging from 0 to 100 (Tassew et al., 2019; Arsyad, 2010). CN value is determined based on land cover conditions, soil moisture conditions (AMC), and soil hydrological conditions (HSG). Land cover with good hydrological conditions (HSG A) will have a smaller CN value. The smaller the CN value indicates, the greater the volume of rainwater absorbed by the soil [6].

Based on the calculation of the results CN value of the area in the Gajahwong watershed in 2016 earned the CN II value is 73. The CN value means that falling rain to surface soil is assumed by 100%, then 73% will become runoff, and the rest will be absorbed into in soil or undergoes an evaporation process. In 2020, the value of CN II in the Gajahwong watershed experience enhancement will become 74. Occurrence enhancement CN value is due to existing non-built land over built-up land. More specifically, the distribution CN value at subbasin Gajahwong is served in table 3.

Table 3. CN Value in sub-basin Gajahwong 2016 and 2020

Sub basin	2016			2020		
	CN I	CN II	CN III	CN I	CN II	CN III
W520	52	72	86	54	73	86
W530	52	72	85	54	73	86
W630	57	75	87	55	74	87
W640	52	72	85	53	73	86
W690	55	74	87	56	75	87
W710	55	74	87	56	75	87
W730	54	74	87	55	74	87
W850	55	74	87	55	75	87
W860	54	73	86	54	74	87

Source: Data Processing, 2022

3.4. HEC-HMS Hydrological Model

3.4.1. HEC-HMS Input Model Parameters

3.4.2. Selection of models and methods estimation runoff customized with completeness and availability of data and components that can represent score runoff such as peak flow and volume runoff (table 4). Modeling hydrology in the Gajahwong watershed using the HEC-HMS software is carried out by making components of the subbasin, junction, reach, and outlet shown in Figure 2.

Table 4. Model Input Parameters in HEC-HMS

No	Model	Methods	Parameter
1	*Precipitation*	*Gauge Weight*	*Precipitation (mm)*
2	*Volume Runoff*	*SCS-Curve Number*	*Curve Number*
			Impervious (%)
			Initial Abstraction (mm)
3	*Direct Runoff*	*SCS Unit Hydrograph*	*Lag Time (MIN)*
4	*Baseflow*	*Constant monthly*	*Baseflow (m^3/s)*
5	*Routing*	*Lag*	*Lag Time (MIN)*

Source: Data Processing, 2022

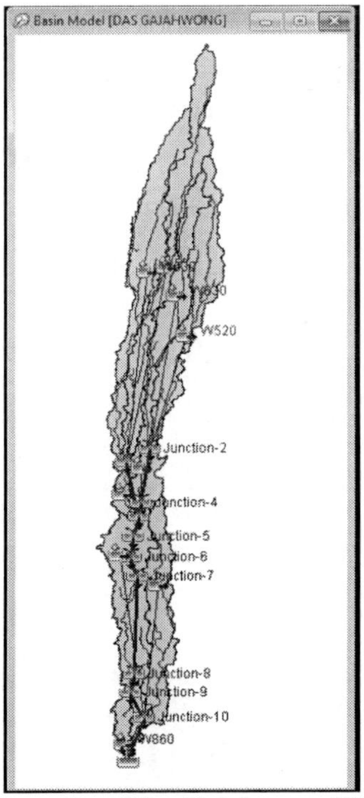

Figure 2. Schematic of the Gajahwong Watershed Model at HEC-HMS
Source: Data Processing, 2022

3.4.3. Watershed Morphometry Results HEC-GeoHMS. The process of determining morphometry in the Gajahwong watershed is carried out automatically with the HEC-GeoHMS extension in ArcGIS so that the Gajahwong watershed is divided into nine subbasins (table 5). The division of the subbasin is based on the characteristics of the river branch and the ease of analysis.

Table 5. Gajahwong sub-watershed morphology.

Sub basin	Area (km)	Slope (%)
W520	10,198	3,848
W530	12,552	4,207
W630	7,637	4,105
W640	12,268	4,513
W690	1,905	3,804
W710	2,657	3,775
W730	8,620	3,573
W850	12,624	3,888
W860	0,812	3,751

Source: Data Processing, 2022

3.4.4. Determination of Volume Runoff SCS-CN Method. Determining volume runoff using the SCS-CN method requires three parameters: initial Abstraction (Ia) value, CN value, and percent impervious area (table 6).

Table 6. Initial Parameter Values for the Loss Method Process with the SCS CN Method in 2016 and 2020.

Sub basin	2016			2020		
	Initial Abstraction (mm)	CN	Impervious Area (%)	Initial Abstraction (mm)	CN	Impervious Area (%)
W520	19,475	72	1,689	18,417	73	2,182
W530	19,964	72	2,365	18,492	73	2,115
W630	16,687	75	3,532	17,858	74	6,047
W640	19,915	72	3,256	18,717	73	6,575
W690	17,446	74	20,353	17,119	75	18,388
W710	17,566	74	20,795	17,040	75	19,366
W730	17,896	74	15,257	17,401	74	14,507
W850	17,684	74	15,054	17,299	75	14,770
W860	18,544	73	2,060	17,983	74	2,313

Source: Data Processing, 2022

3.4.5. Determination of Direct Runoff Method of SCS Unit Hydrograph. Determination of direct runoff is used to determine the runoff value in the Gajahwong watershed. The parameters needed in deciding the transform method require input in the form of CN value, river length (feet), and watershed slope so that tlag will be obtained (table 7).

Table 7. Initial Parameter Value of Gajahwong Watershed Tlag for Transform Process with SCS UH Method in 2016 and 2020

Sub basin	River length (feet)	Slope (%)	2016		2020	
			CN	Tlag (min)	CN	tlag (min)
W520	13.343,195	3,848	72	96,818	73	93,880
W530	15.114,086	4,207	72	103.727	73	99,420
W630	11.217,768	4,105	75	74.994	74	77,795
W640	14.907,877	4,513	72	98.921	73	95,578
W690	4.875,910	3,804	74	40,975	75	40,558
W710	5.953,525	3,775	74	48,433	75	47,643
W730	12.063,152	3,573	74	88,473	74	87,134
W850	15.166,057	3,888	74	101,208	75	100,006
W860	2.922,802	3,751	73	28,326	74	27,853

Source: Data Processing, 2022

3.4.6. Determination of Baseflow Constant Monthly Method. Baseflow determination in this study is estimated automatically using the *https://sephydro.hydrotools.tech/* software using the TR – 55 methods. The baseflow value of each sub-basin in the Gajahwong watershed is shown in table 8.

Table 8. Calculation of Baseflow Constant Monthly Gajahwong Watershed in 2016 and 2020

Sub basin	Baseflow (m^3/s)	
	2016	2020
W520	0,696	0,465
W530	0,857	0,573
W630	0,521	0,349
W640	0,838	0,560
W690	0,130	0,087
W710	0,181	0,121
W730	0,589	0,393
W850	0,862	0,576
W860	0,055	0,037

Source: Data Processing, 2022

3.4.7. Determination of Routing Method Lag. The Gajahwong watershed is divided into ten types of reach named reach 1, reach 2, reach 3, reach 4, reach 5, reach 6, reach 7, reach 8, reach 9, and reach 10. The routing values for each reach in The Gajahwong watershed are presented in table 9.

Table 9. Initial Parameter Value of Gajahwong Watershed Tlag for the Lag Method Routing Process in 2016 and 2020.

Reach	Lag (min)	
	2016	2020
Reach 1	115,163	110,566
Reach 2	104,021	99,869
Reach 3	35,151	33,748
Reach 4	47,596	45,696
Reach 5	51,003	48,967
Reach 6	46,603	44,743
Reach 7	180,892	173,671
Reach 8	47,473	45,578
Reach 9	68,428	65,696
Reach 10	88,980	85,475

Source: Data Processing, 2022

3.5. Running Model HEC-HMS

Running the HEC-HMS model was carried out in 2016 using observation discharge data from the same period. The purpose of running the model is to determine the characteristics of the model produced by HEC-HMS before the calibration and validation process is carried out. Based on table 10, the peak flow from the simulation results is 16.2 m3/s, and the observed peak flow is 45.1 m^3/s with a percent difference of -64.08%. Meanwhile, the simulated volume runoff is 3051.70 mm, and the observed volume runoff is 2565.83 mm, with a percent difference of 18.94%. Furthermore, the results of the 2016 modeling will be calibrated to obtain more representative results with the study area.

Table 10. Results of the 2016 Gajahwong Watershed HEC-HMS Model

Result	Unit	*Simulated*	*Observed*	*Difference*	*Percent Difference* (%)
Peak Flow (m^3/s)	m^3/s	16,2	45,1	-28,9	-64,08
Volume (mm)	mm	3.051,70	2.565,83	485,87	18,94

Source: Data Processing, 2022

3.6. Calibration and Validation Stage of HEC-HMS Model Results
The calibration process uses 2016 data, while the validation or verification process is carried out with 2017 and 2020 data.

3.6.1. HEC-HMS Model Result Calibration Process. The calibration process is carried out by retrying (trial and error). Furthermore, the calibration process is complete if it has R^2, NSE, and PBIAS values, which explains that the model can describe the observation conditions. The parameters calibrated in this study were the values of CN and Ia. Based on table 11, the calibration results in 2016 are obtained in the form of a simulation result peak discharge of 14.9 m^3/s and an observation peak discharge of 45.1 m^3/s with a percent difference of -66,96%. The volume runoff value from the simulation results is 2880.21 mm, and the observed volume runoff is 2565,83 mm, with a percent difference of 12.25%.

Table 11. Calibration Results of the 2016 Gajahwong Watershed HEC-HMS Model

Result	Unit	Simulated	Observed	Difference	Percent Difference (%)
Peak Flow (m^3/s)	m^3/s	14,9	45,1	-30,2	-66,96
Volume (mm)	mm	2.880,21	2.565,83	314,38	12,25

Source: Data Processing, 2022

If viewed based on the statistical test of percent bias (PBIAS), the coefficient of determination (R^2), and the value of the objective function presented in table 12, from the value of the statistical test results, the model can be judged to be good or acceptable because it describes according to the study area. Based on the Nash-Sutcliffe Efficiency (NSE) statistical test, the model is considered bad because it has an NSE value of 0.251. The three statistical tests indicate that the HEC-HMS modeling accurately determines runoff characteristics in the Gajahwong watershed. Furthermore, the statistical test R^2 obtained results of 0.735, which means that 73.5% of the model can describe the effects of observations.

Validation of HEC-HMS Model Results. Validation of the HEC-HMS model was carried out to test the calibration results parameters with data in different year periods so that it could be known whether the calibration parameter values were following the characteristics of the Gajahwong watershed or not. The HEC-HMS model results were validated with 2017 and 2020 data. Based on table 13, the validation results for 2017 are obtained in the form of a simulation result peak discharge of 25.8 m3/s and an observation peak discharge of 46.5 m3/s with a percent difference of -44.52%. The volume runoff value from the simulation results is 2702.18 mm, and the observed volume runoff is 2395.94 mm, with a percent difference of 12.78%.

Table 12. Statistical Test Results for Modeling of Volume Runoff in the Gajahwong Watershed in 2016

Statistic test	Score	Description
objective function	6,07	Accurate model
Nash-Sutcliffe Efficiency (NSE)	0,251	Bad
percent bias (PBIAS)	12,20 %	Good
koefisien determinasi (R^2)	0,735	Models are acceptable

Source: Data Processing, 2022

Table 13. Validation Results of the 2017 Gajahwong Watershed HEC-HMS Model

Result	Unit	Simulated	Observed	Difference	Percent Difference (%)
Peak Flow (m^3/s)	m^3/s	25,8	46,5	-20,7	-44,52
Volume (mm)	mm	2.702,18	2.395,94	306,24	12,78

Source: Data Processing, 2022

Based on the results of the statistical test of percent bias (PBIAS), the coefficient of determination (R^2), and the value of the objective function, which are presented in table 14, from the value of the statistical test results, the model can be judged to be good or acceptable because it describes according to the study area. Based on the Nash-Sutcliffe Efficiency (NSE) statistical test, the model is considered bad because it has an NSE value of 0.449. The three statistical tests indicate that the HEC-HMS modeling accurately determines the runoff characteristics in the Gajahwong watershed. Furthermore, the statistical test of R^2 obtained a result of 0.703, which means that 70.3% of the model can describe the effects of observations.

Table 14. Statistical Test Results of the Gajahwong Watershed Volume Runoff Modeling in 2017

Statistic test	Score	Description
Nash-Sutcliffe Efficiency (NSE)	0,449	Bad
percent bias (PBIAS)	12,93 %	Good
koefisien determinasi (R^2)	0,703	Models are acceptable

Source: Data Processing, 2022

Based on table 15, the validation results for 2020 are obtained in the form of a simulation result peak discharge of 11.8 m3/s and an observation peak discharge of 21.7 m3/s with a percent difference of -45.62%. The volume runoff value from the simulation results obtained is 2011.39 mm, and the observed volume runoff is 1924.59 mm, with a percent difference of 4.51%. Based on the results of the Nash-Sutcliffe Efficiency (NSE) statistical test, percent bias (PBIAS), the coefficient of determination (R^2), and the objective function values are presented in table 16; from the value of the statistical test results, the model can be judged to be good or acceptable because it already describes according to the study area. The four statistical tests indicate that the HEC-HMS modeling accurately determines the runoff characteristics in the Gajahwong watershed. Furthermore, the statistical test R^2 obtained results of 0.759, which means that 75.9% of the model can describe the effects of observations.

Table 15. Results of Validation of the Gajahwong Watershed HEC-HMS Model in 2020

Result	Unit	Simulated	Observed	Difference	Percent Difference (%)
Peak Flow (m^3/s)	m^3/s	11,8	21,7	-9,9	-45,62
Volume (mm)	mm	2.011,39	1.924,59	86,8	4,51

Source: Data Processing, 2022

Table 16. Statistical Test Results for Gajahwong Watershed Volume Runoff Modeling in 2020

Statistic test	Score	Description
Nash-Sutcliffe Efficiency (NSE)	0,541	Good
percent bias (PBIAS)	4,54 %	Very good
koefisien determinasi (R^2)	0,759	Models are acceptable

Source: Data Processing, 2022

3.7. Analysis Characteristics volume runoff in the Gajahwong watershed

The characteristics of runoff in the Gajahwong watershed in 2016 and 2020 are represented based on the peak discharge parameters and the volume runoff presented in table 17. In 2016, the Gajahwong watershed had a peak discharge of 14.9 m3/s; in 2020, the peak discharge in the Gajahwong watershed decreased to 11.8 m3/s. The volume runoff in the Gajahwong watershed also reduced from 2880.21 mm in 2016 and then in 2020 to 2011.39 mm.

Suppose surface runoff characteristics are seen based on the sub-watershed aspect. In that case, the W630 sub-watershed with a catchment area of 7,637 km^2 has the largest volume runoff in 2016 and 2020, 885.39 mm and 695.72 mm, respectively. Based on its location, the W630 sub-watershed is located upstream in the middle of the Gajahwong watershed. Based on table 4.4, the W630 sub-watershed has normal CN values in 2016 and 2020 of 75 and 74. The dominance of land cover in the W630 sub-watershed in 2016 was agricultural land, while in 2020, it was dominated by built-up land with an area of 3,997 km^2 and 3,464 km^2. The smallest volume runoff is in the W860 sub-watershed at 185.75 mm and 350.97 mm, with a catchment area of 0.812 km^2, respectively. When viewed from its location, the W860 sub-watershed is located downstream of the Gajahwong watershed with CN values of normal conditions in 2016 and 2020 of 73 and 74. The dominance of land cover in the W860 Sub-watershed in 2016 was agricultural land with an area of 0.338 km^2, while in 2020 dominated by land cover in the form of built-up land with an area of 0.379 km2. Furthermore, the sub-watershed with the highest peak discharge in 2016 was the W850 sub-watershed of 8.8 m^3/s with a catchment area of 12.624 km^2, while in 2020, the highest peak discharge was 3.6 m3/s and was in the W520 sub-watershed with an area of the catchment area of 10.198 km^2. The W850 sub-watershed in 2016 had a normal condition CN value of 74 with the dominance of land cover in the form of built-up land with an area of 9,389 km^2. Meanwhile, the W520 sub-watershed in 2020 has a normal condition CN value of 73, which is dominated by land cover in the form of built-up land with an area of 4,649 km^2. In 2016 and 2020, the

IOP Conf. Series: Earth and Environmental Science 1233 (2023) 012055 doi:10.1088/1755-1315/1233/1/012055

smallest peak discharge was in the W860 subwatershed, with peak discharge values of 0.4 m^3/s and 0.2 m^3/s.

Table 17. Characteristics of Runoff in the Gajahwong Watershed in 2016 and 2020

Sub-basin	Area (km^2)	2016		2020	
		Peak Flow (m^3/s)	*Volume* (mm)	*Peak Flow* (m^3/s)	*Volume* (mm)
W520	10,198	3,3	843,71	3,6	666,35
W530	12,552	4,4	844,92	3,4	608,55
W630	7,637	2,5	885,39	2,8	695,72
W640	12,268	4,7	669,03	3,4	614,82
W690	1,905	1,8	675,38	0,6	574,37
W710	2,657	2,5	670,19	0,8	559,01
W730	8,620	6	462,28	2,4	473,33
W850	12,624	8,8	466,01	3,5	473,99
W860	0,812	0,4	185,75	0,2	350,97
Gajahwong watershed	69,274	14,9	2.880,21	11, 8	2.011,39

Source: Data Processing, 2022

Based on these conditions, it can be concluded that the characteristics of runoff in each sub-basin are highly dependent on the area of the catchment area, the type of land cover, and the CN value. The smaller the catchment area, the smaller the peak discharge value and the volume runoff. The wider the watershed, the greater the capacity, resulting in a greater volume of water that can be stored and contributed by the watershed [7]. The type of land cover will affect the volume of runoff produced. Under conditions in the Gajahwong watershed, land cover in the form of agricultural land has the largest tendency to make runoff compared to land cover in the form of settlements. Other factors affect surface runoff characteristics in the Gajahwong watershed, namely the rainfall conditions. According to [8], rain influences runoff characteristics, including the duration, intensity, type, and distribution of rain, which will determine the size of runoff in a watershed. The volume of runoff and peak discharge in a watershed is closely related to the intensity and duration of rain that occurs in the watershed [9]. Peak discharge and volume runoff in the Gajahwong watershed decreased. This condition is caused by the total rainfall in 2016 being greater than in 2020, which is 3,717 mm and 3,006 mm.

3.8. Effect Analysis Change Land Cover against volume runoff in the Gajahwong watershed

Changes in land cover affect changes in runoff in the Gajahwong watershed. The pattern of land cover changes in the Gajahwong watershed from agricultural land to built-up land. If seen based on table 18, peak discharge and runoff volume in 2020 decreased compared to 2016. This condition was caused because, in 2020, agricultural land fell and changed its function to built-up land.

In 2016, the Gajahwong watershed had an agricultural land area of 30,209 km^2 with a percentage of 43,608%. Agricultural land, in this case, rice fields, has a plow pan layer [10]. This layer is very useful to support water availability for rice plants because of its ability to withstand the movement of percolation water, thus facilitating the occurrence of puddles on the soil surface. Therefore, the plow pan has a soil moisture condition that tends to be saturated and a soil layer impervious to water, so falling water tends to run off into volume runoff. This condition resulted in the Gajahwong watershed in 2016 having a peak discharge of 14.9 m^3/s and a volume runoff of 2880.21 mm.

In contrast to 2020, agricultural land in the Gajahwong watershed has decreased to 20,204 km^2 and has been converted into built-up land. An increase in built-up land will increase the potential for volume runoff. However, this condition does not apply in the Gajahwong watershed; the increase in built-up land results in a decrease in peak discharge and volume runoff in the Gajahwong watershed [11]. This condition is because the development of built-up land requires the construction of rainwater infiltration wells and biopore infiltration holes that reduce volume runoff and prevent flooding.

Infiltration wells and biopore infiltration holes must be constructed in areas with less than 30% vegetation cover, densely populated settlements, and areas that have a flat topography [12] [13].

Table 18. Land Cover Area and Runoff Characteristics in the Gajahwong Watershed in 2016 and 2020

Land cover	2016		2020	
	Area (Km²)	**%**	**Area (Km²)**	**%**
Vegetated land	7,894	11,395	7,829	11,302
Non-agricultural land	0,883	1,275	3,122	4,508
Agricultural land	30,209	43,608	20,204	29,165
Built-up land	29,885	43,140	34,368	49,612
Open field	0,403	0,582	3,751	5,415
Total	69,274	100	69,274	100
Runoff Characteristics of Gajahwong Watershed				
Peak flow (m^3/s)	14,9		11,8	
Volume (mm)	2.880,21		2.011,39	

Source: Data Processing, 2022

Furthermore, the falling water will enter the infiltration well or drainage system. Water that enters the drainage system will flow into the river branches. The Gajahwong watershed has a characteristic elongated shape. The elongated shape of the watershed will cause the travel time of volume runoff from the upstream to the outlet to be relatively longer. Runoff channeled by river branches causes this condition, and runoff on the ground tends to be concentrated for longer due to the long distance from the main river. Concentrated water tends to experience an infiltration process, and this condition is supported by HSG in the Gajahwong watershed, with a percentage of 98.690% belonging to class B HSG. The class B HSG category in the Gajahwong watershed has soil texture characteristics in the form of sandy loam or sandy loam with a minimum infiltration rate of 26 mm/hour. The soil texture in the form of sandy loam or sandy loam is classified as good at absorbing water into the soil, and this indicates that the soil texture in the Gajahwong watershed has good potential to withstand the volume runoff rate so that it will result in a decrease in the volume runoff rate.

4. Conclusion

Gajahwong watershed in 2016 and 2020 is classified as a watershed with critical conditions because it only has a vegetated land area of less than 30%, or 11.395% and 11.301%, respectively. In 2016, land cover in the Gajahwong watershed was dominated by agricultural land of 30,209 km², or about 43.608% of the total area of the Gajahwong watershed. Meanwhile, in 2020 it will be built up land with an area of 34,368 km² or around 49.612% of the total area of the Gajahwong watershed. In 2016, land cover in the form of open land had the smallest area of 0.403 km² or about 0.582%. In 2020, land cover in non-agricultural land had an area of at least 3,112 km² or about 4.507% of the total area of the Gajahwong watershed. The Gajahwong watershed experienced an increase in the CN value in 2016 and 2020 by 73 and 74, respectively. The increase in the CN value in the Gajahwong watershed was caused by the conversion of land functions from non-built land to built-up land. Characteristics of runoff in the Gajahwong watershed in 2016 and 2020 are represented based on peak discharge and volume runoff parameters. In 2016, the Gajahwong watershed had a peak discharge of 14.9 m3/s; in 2020, the peak discharge in the Gajahwong watershed decreased to 11.8 m3/s. Volume runoff in the Gajahwong watershed also decreased from 2,880.21 mm in 2016 to 2,011.39 mm in 2020. The decrease in volume runoff characteristics is caused by changes in land cover, watershed morphometry, and rainfall.

IOP Conf. Series: Earth and Environmental Science 1233 (2023) 012055 doi:10.1088/1755-1315/1233/1/012055

References

[1] Utami, N. D., & Suprayogi, S. 2014. Kajian debit banjir akibat perubahan penggunaan lahan di Sub DAS Belik, Daerah Istimewa Yogyakarta. Jurnal Bumi Indonesia, 3(3).
[2] Arsyad, S. 2010. Konservasi Tanah dan Air (Edisi Kedua) (H. Siregar (ed.); Edisi Kedu). IPB Press.
[3] Keputusan Menteri Pekerjaan Umum Nomor 590 tentang Pola Pengelolaan Sumber Daya Air Wilayah Sungai Progo-Opak-Serang.
[4] Ramadhani, E., Suprayogi, S., & Hadi, M. P. 2021. Pengaruh Perubahan Penggunaan Lahan Terhadap Limpasan Menggunakan Multidata Iklim Satelit di Sub DAS Samin. 22(1), 31–50.
[5] Triatmodjo, B. (2010). Hidrologi Terapan (Cetakan Kedua). Betta Offset Yogyakarta.
[6] Kartika. (2013). Pengaruh Perubahan Penggunaan Lahan Terhadap Limpasan Permukaan di DAS Code, Yogyakarta (Skripsi, Universitas Gadjah Mada).
[7] Asdak, C. 2010. Hidrologi dan Pengelolaan Daerah Aliran Sungai. Gadjah Mada University Press.
[8] Cahyono, H. K. 2008. Pengelolaan Daerah Aliran Sungai (DAS) Terpadu (Volume 3) Bandung: Prisma.
[9] Fauzi, R. G. N., Utomo, D. H., & Taryana, D. 2018. Pengaruh perubahan penggunaan lahan terhadap debit puncak di sub DAS Penggung Kabupaten Jember. Jurnal Pendidikan Geografi, 23(1), 50-61.
[10] Agus, F. 2004. Tanah sawah dan teknologi Pengelolaannya. Bogor: Pusat Penelitian dan Pengembangan Tanah dan Agroklimat (Puslitbangtanak).
[11] Nasjono, J. K., Utomo, S., Marawali, U. D. B., Debit, P. K., & Number, S. C. S. 2018. Keandalan Metode Soil Conservation Services-Curve Number untuk Perhitungan Debit Puncak DAS Manikin. VII(2), 183–192.
[12] Peraturan Menteri Negara Lingkungan Hidup Nomor 12 Tahun 2009 tentang Pemanfaatan Air Hujan.
[13] Peraturan Menteri Pekerjaan Umum dan Perumahan Rakyat Nomor 11/PRT/M/2014 tentang Pengelolaan Air Hujan pada Bangunan Gedung dan Persilnya.

IOP Conf. Series: Earth and Environmental Science 1233 (2023) 012055 doi:10.1088/1755-1315/1233/1/012055

Appendix 1. Land Cover Map of Gajahwong Watershed for 2016 and 2020

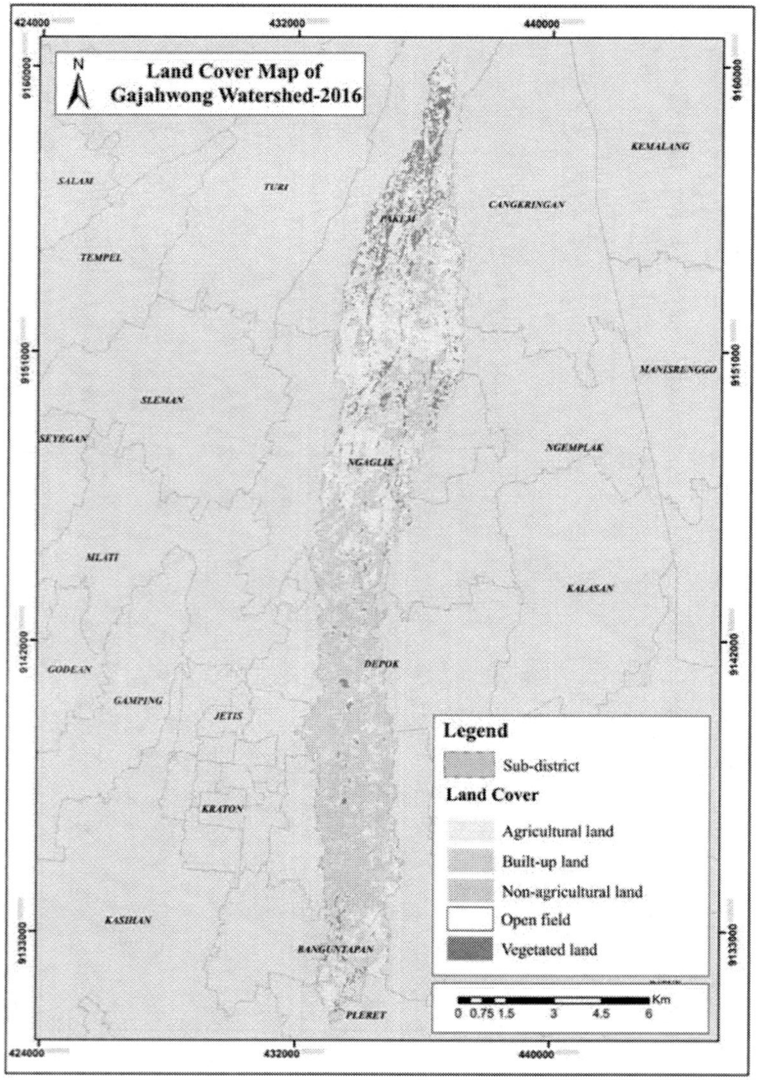

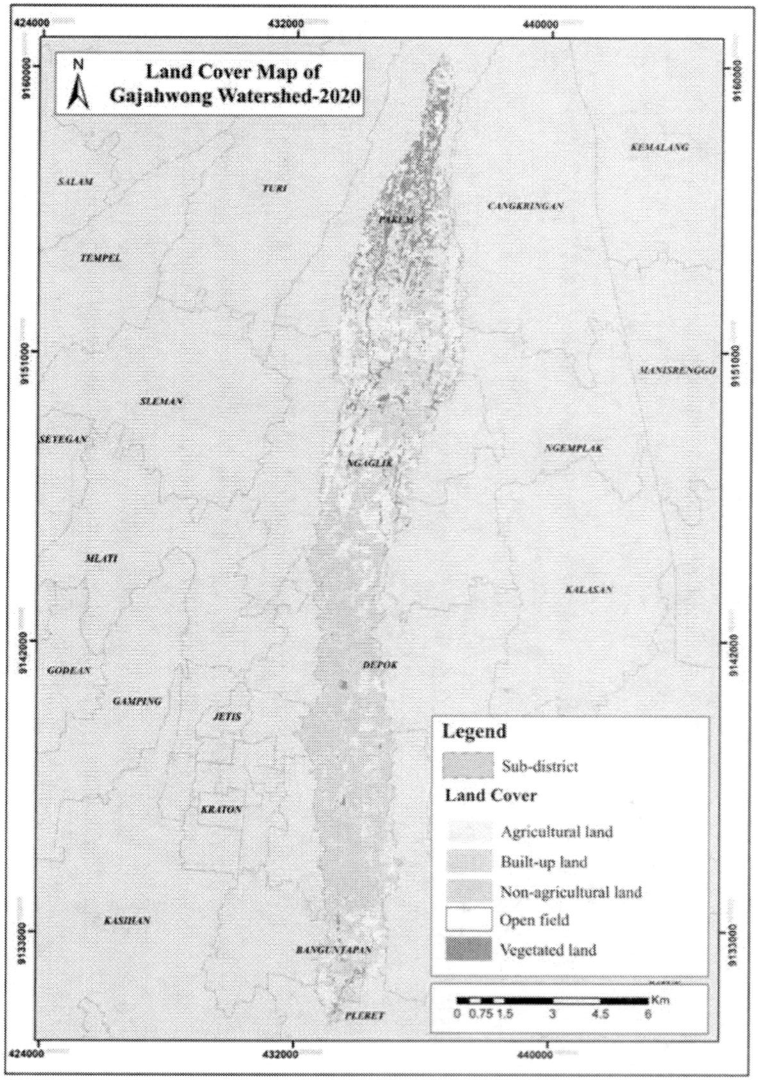

IOP Conf. Series: Earth and Environmental Science 1233 (2023) 012056 doi:10.1088/1755-1315/1233/1/012056

Analysis of Total Maximum Daily Loads of Winongo River, Special Region of Yogyakarta, Using Water Quality Analysis Simulation Program (WASP)

A N Indrastuti, M Widyastuti* and L N Fadlillah

Department of Environmental Geography, Faculty of Geography, Universitas Gadjah Mada, Daerah Istimewa Yogyakarta, Indonesia
*Corresponding author email: mwiwik@ugm.ac.id

Abstract. A total maximum daily load (TMDL) is the amount of a pollutant load that a waterbody can accommodate without exceeding the water quality standard for that particular pollutant. The research aimed to measure pollutant loads in the Winongo River, TMDLs, and pollutant reduction needed if pollutant loads were higher than TMDLs. The Water Quality Analysis Simulation Program (WASP) model was used to determine water quality responses to the potential amounts of pollutants discharged into the river, resulting in information on existing pollutant loads. After testing the water quality parameters, it was found that pH and BOD at points 2–7 and TDS at all the sampling points meet the standards for Class 1 water per the Governor Regulation No. 20 of 2008. The WASP model showed a total pollutant load of 51,78 kg/day and a TMDL of 139,08 kg/day. Segment 1 had the lowest pollutant load of 2.57 kg/day, then the TMDL was 4.56 kg/day. The highest pollutant load measured was found in segment 6 of 20.11 kg/day, then the TMDL was 79.13 kg/day. Based on the results, the Winongo River can accommodate the pollutant loads while maintaining the water quality.

1. Introduction

Water resources are at the top of what living organisms fundamentally need, including humans. Most human activities use water, such as drinking, bathing, washing, and irrigating agricultural land. It also plays a part in industrial activities and water transportation to support economic growth. To a large extent, its usage pays attention to not only quantity but also quality. As a result of economic growth and urbanization, water consumption increases, and clean water has become increasingly less available [1]. In this context, clean water quality deteriorates because of several factors, including the growing human population, expanding urban functions, and increasing discharge of industrial and agricultural wastes.

Water resources consist of surface water and groundwater. The former is more easily polluted than the latter because it is more exposed to and creates direct access for industrial and agricultural waste disposal [2]. A pollutant load can contaminate a river, an example of surface water, if it exceeds the total maximum daily load (TMDL) for this water body. According to Government Regulation No. 82 of 2001, a pollutant load is defined as the amount of a polluting element of waste that is discharged into a water body, and TMDL indicates the maximum amount of a pollutant load that a water body can accommodate while maintaining water quality according to applicable standards. Therefore, it is imperative to monitor pollutant loads and TMDLs for sustainable use of good-quality water resources and proper waste disposal management.

Declining water quality has been at the core of many water resources and pollution management decisions worldwide. An example includes the Winongo River, part of the Opak Watershed in the Special Region of Yogyakarta, Indonesia. Per the Governor Regulation No. 22 of 2007, this river has Class 1 water quality or is suitable for the raw water of drinking water and other designated uses that require the same standards. However, the characteristics needed to

IOP Conf. Series: Earth and Environmental Science 1233 (2023) 012056 doi:10.1088/1755-1315/1233/1/012056

support this level of quality have been diminishing because of factors that cause disturbances in the ecosystem, e.g., urbanization, improper agricultural practices, deforestation, and pollution [3]. Monitoring by the Environmental Service in 2020 revealed that Winongo was the third-most polluted river in the region after Code and Belik because of high biochemical or biological oxygen demand (BOD). Built-up land covers the largest proportion of the watershed area, including residential buildings on the riverbanks. Also, because Winongo traverses densely occupied urban areas, there is a high potential for water pollution.

Based on these problems, calculating the pollutant loads and TMDLs of various segments of Winongo River becomes necessary. Water Quality Analysis Simulation Program (WASP) has been widely used to model contamination or potential pollutant loads and how they are transported in surface water by applying the concept of unsteady-state mass balance [4]. Previous research conducted an analysis of the pollutant load capacity using the mass balance concept but did not consider the speed of advection and dispersion which influenced by water discharge [5]. In this study, BOD is observed and modeled in the program that has an algorithm that considers the process so that it can produce different conditions from previous research. It is considered indicative of water quality and sensitive to the addition of pollutant loads. Should the latter happen, BOD will represent the increased amount of oxygen that microorganisms need to decompose organic matter, in which case the TMDL is potentially exceeded [6].

2. Study Area

The research was conducted in the Winongo Watershed, which starts from the slope of Merapi Volcano in the north and then flows downstream, forming a confluence with the Opak River, before emptying into the IndianOcean. Figure 1 shows the research area of the Winongo catchment area. The main channel of the Winongo River traverses three administrative units: Sleman Regency, Yogyakarta City, and Bantul Regency. According to Governor Regulation No. 22 of 2007, the Winongo River segments from Mraen (a sub-village in Sendangadi, Mlati District, Sleman Regency) downstream to Gadingdaton (a sub-village in Donotirto, Kretek District, Bantul Regency) are designated for the raw water supplies for drinking water or other purposes with the exact requirements (Class 1 water quality). Figure 2 shows the land use map of the Winongo Watershed. It also flows through urban areas where residential buildings are dominant, indicating intensive human activities that can induce pollution due to the disposal of both solid and liquid waste into the river.

IOP Conf. Series: Earth and Environmental Science 1233 (2023) 012056 doi:10.1088/1755-1315/1233/1/012056

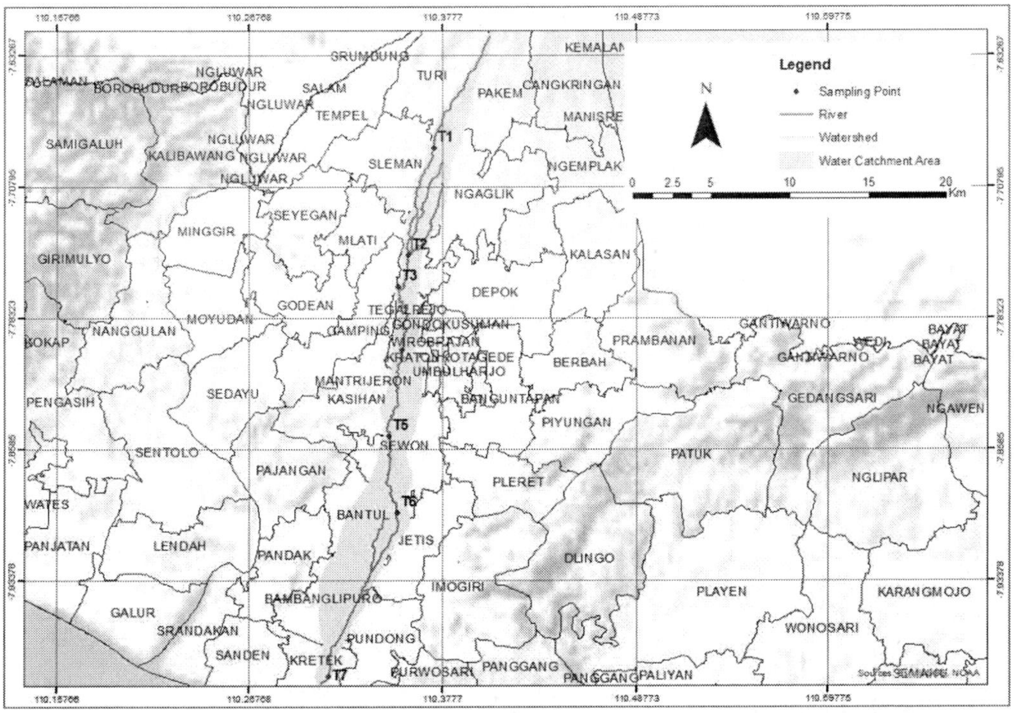

Figure 1. Research Area and Distribution of Water Quality Sampling Points

(Data Processing, 2022)

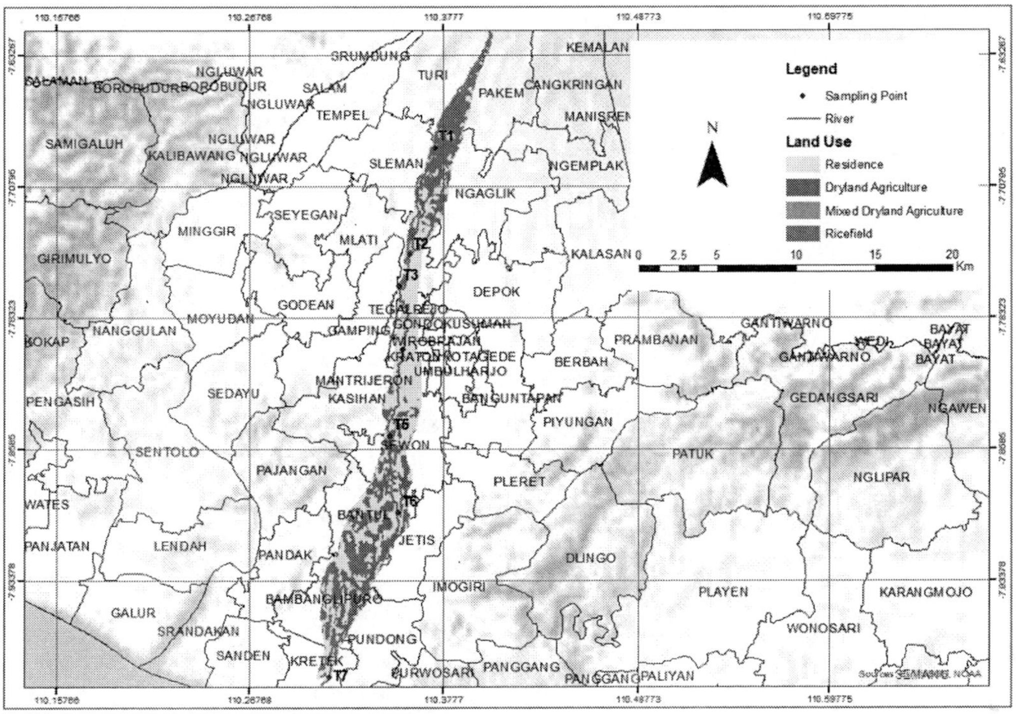

Figure 2. Land Use Map of Winongo Watershed

(Data Processing, 2022)

3. Method

3.1. Data Acquisition Method

3.1.1 River Water Sampling. The research divided the entire length of Winongo River into six segments with a total of seven sampling points, as shown in Figure 2. The segment division was based on the type of waste- generating land use. Then, the river water samples were tested in the laboratory to measure the BOD and determine the water quality of the river segments at the time of data collection.

3.1.2 River Discharge Measurement. Details on river discharge and morphology were collected during the water sampling. First, the discharge was measured using the velocity-area method with a current meter. Then, it was calculated using the equation below.

$$V = aN+B \tag{1}$$

$$A = \sum_{x=1}^{n} \frac{b_{(x+1)} - b_{(x-1)}}{2} dx \tag{2}$$

$$Q = \frac{V}{A} \tag{3}$$

3.2 Calculation of Pollutant Loads and Total Maximum Daily Loads

Pollutant loads were determined by estimating the number of pollutants discharged from point and nonpoint sources based on the Regulation of the Minister of Environment No. 01/2010. Nonpoint-source pollution included domestic activities, agricultural practices, livestock farms, hotels, and hospitals. Pollutant loads from these sources were estimated from the weighted average of population size, the number of livestock, and the area of agricultural land. This estimation requires the BOD emission factor of each source. Domestic pollutant loads with a focus on BOD levels (PL_{BOD}) were calculated according to the urban category (equivalent ratio, see Table 1) and the distance of settlements from the river (α, Table 2) using equation 4 below.

$$Domestic\ PL_{BOD} = 0.04\ x\ Population\ size\ x\ Eq.\ ratio\ x\ \alpha \tag{4}$$

Table 1. City Equivalent (Eq.) Ratio

Regional Category	Eq. Ratio
Cities	1
Towns and suburbs	0.8125
Rural areas or hinterlands	0.625

Source: Ministry of Environment and Forestry, 2017

Table 2. Load Transfer Coefficients (α)

Distance to River (m)	α
0–100	25
100–500	20
>500	20

Source: Regulation of the Minister of Public Works No. 03/PRT/M/2013

PL_{BOD} from livestock farms were calculated with equation 5 according to types of livestock and their respective emission factors (Table 3). The average pollutant load from this source is 20%.

$$Livestock\ PL_{BOD} = Number\ of\ livestock\ x\ BODemission\ factor\ x\ 20\% \quad (5)$$

Agricultural land area and BOD emission factors (Table 4) were inputted to equations 6–8 to calculate agricultural pollutant loads per cropping season. Two types of agricultural land were observed: (1) rice fields and (2) other plantations, non-irrigated croplands, or mixed-species plantations, which, in Indonesia, producean average of 10% and 1% pollutant loads, respectively.

$$Rice\ fields = Area\ x\ BOD\ emission\ factor\ x\ 10\% \quad (6)$$

$$Non\text{-}irrigated\ croplands = Area\ x\ BOD\ emission\ factor\ x\ 1\% \quad (7)$$

$$Agricultural\ PLBOD\ (kg/day) = pollutant\ loadper\ cropping\ season/number\ of\ days\ in\ the\ cropping\ season \quad (8)$$

Table 3. BOD Emission Factors of Different Types of Livestock

Livestock	BOD Emission Factor (kg/day)
Cow	0.292
Buffalo	0.207
Horse	0.226
Pig	0.128
Sheep	0.056
Goat	0.034
Chicken	0.0024
Swan	0.0025
Duck	0.0009

Source: Ministry of Environment and Forestry, 2017

Table 4. BOD Emission Factors of Agricultural Lands

Agricultural Land	BOD Emission Factor (kg/ha/cropping season)
Rice fields	0.292
Other plantations/non-irrigated croplands/mixed-species plantations	0.207

Source: Ministry of Environment and Forestry, 2017

The hospital pollutant load was calculated from the number of beds, while the hotel pollutant load was from the number of rooms. Equation 9 was used to calculate the pollutant loads from hospitals and hotels with emission factors shown in Table 5.

$$Hospital,\ Hotel\ PL_{BOD} = Number\ of\ beds,\ roomsx\ BOD\ emission\ factor \quad (9)$$

IOP Conf. Series: Earth and Environmental Science 1233 (2023) 012056 doi:10.1088/1755-1315/1233/1/012056

Table 5. BOD Emission Factors of Hospitals and Hotels

Source of Pollution	BOD Emission Factor(kg/day)
Hospitals (per bed)	0.123
Hotels (per room)	0.055

Source: Ministry of Environment and Forestry, 2017

To calculate total maximum daily loads (TMDLs), several data were used to construct the WASP model, namely water discharge, river morphology, water quality, and pollutant load. This model helps determine if the simulated river flow and water quality are close to or represent the data obtained in the field. It was designed in two stages: (1) model simulation from the field-obtained data for pollutant load estimation and (2) model simulation based on Class 1 water quality standards to estimate the TMDL.

To determine the validity of the model outcome, a reliability test was conducted using the relative bias formula below.

$$rB = \frac{R_m - R_o}{S_o} \quad (10)$$

$$F = \frac{{S_m}^2}{{S_o}^2} \quad (11)$$

rB shows whether or not the model is accepted to simulate the TMDL. If -0.5 < rB < 0.5 and 0.5 < F < 1.5, the model is accepted. On the contrary, the model is rejected if rB < -0.5 or rB > 0.5 and F < 0.5 or F > 1.5.

4. Results and Discussion

4.1 River Water Quality

Water pH is naturally in the 6.5–9 range [7]. pH is closely linked to alkalinity: pH will increase because of the carbon and oxygen ions present in the water if alkalinity increases. Figure 3 (a) shows that the water pH measured at all seven sampling points varied from 7.57 to 8.57. The highest pH was found at point 1, which can be attributed to a large amount of agricultural waste, such as excess fertilizer that runoffs into the river.

Electrical conductivity (EC) is a parameter that represents the concentrationof ions in water [8]. More dissolved residues or solids (i.e., total dissolved solids, TDS) can lead to a higher EC value. Figure 3 (b) shows that the EC measured at point 1 was 205 μS/cm and then continuously increased to383 μS/cm at point 5. However, it had substantially decreased to 108 μS/cm at point 6 before rising again to 448 μS/cm at point 7. An instrument error can cause the drop at point 6 during the EC reading. The increasingtrend is attributed to the accumulation of anthropogenic wastes or residues transported by runoff into the river. The highest EC observed at point 7 may result from the high agricultural activities around the last segment that produce waste containing minerals or ions.

Dissolved residues refer to the concentration of dissolved ions in water [9]; this indicates that total dissolved solids (TDS) are related to EC. Figures 3 (b) and (c) show that the EC values and TDS generally increased downstream. In other words, the lowest TDS was found at point 1 (146 mg/l), whereas the highest was at point 7 (317 mg/l). Based on this range of values, the water at all the sampling points meetsClass 1 quality standards per the Governor Regulation No. 20 of 2008, i.e., 1000 mg/l.

Like TDS, salinity or degree of saltiness is also linked to EC because it represents the number of salt ions contained in the water. A higher concentration of salt ions is manifested in a higher salinity value like the water in karst regions. Figure 3 (d) shows that the salinity at point 1 was the lowest (100 ppm). It increased toward the lower part of the watershed and reached the highest value at point 7 (215 ppm). This pattern is positively correlated with TDS and EC values.

BOD shows the level of oxygen needed by microorganisms in a water body to decompose organic matter [10]. It is sensitive to the amount of agricultural, residential, livestock, hotel, and hospital wastes released into the water. Also, it is a key parameter in determining TMDLs. Based on Figure 3 (e), BOD fluctuated across the sampling points. Observation BOD values have increased and decreased due to several factors, including pollutant load from organic matter, flow velocity, turbulence, and river morphology [11]. The highest (2.17 mg/l) was detected at point 1, believed to result from the accumulation of waste from nearby agricultural land and some settlement clusters. This level exceeds the Class 1 quality standard, i.e., 2 mg/l. The BOD continued to lower until point 5 (0.73 mg/l) before increasing to 1.72 mg/L at point 6. The last sampling point had the lowest BOD, 0.51 mg/l. According to [12], running water can re-purify itself throughout the length of the water body (i.e., river) before appearing with a better quality downstream because any organic matter has been decomposed into nitrate, sulfate, and carbon dioxide.

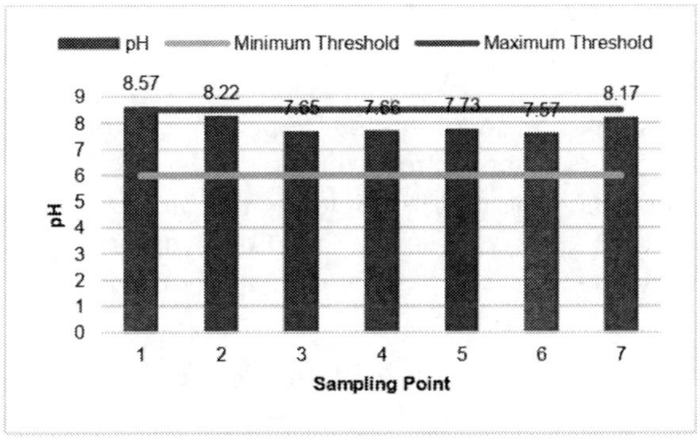

(a)

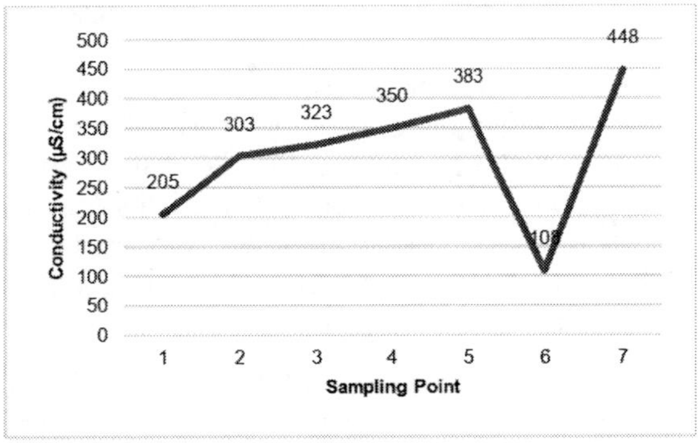

(b)

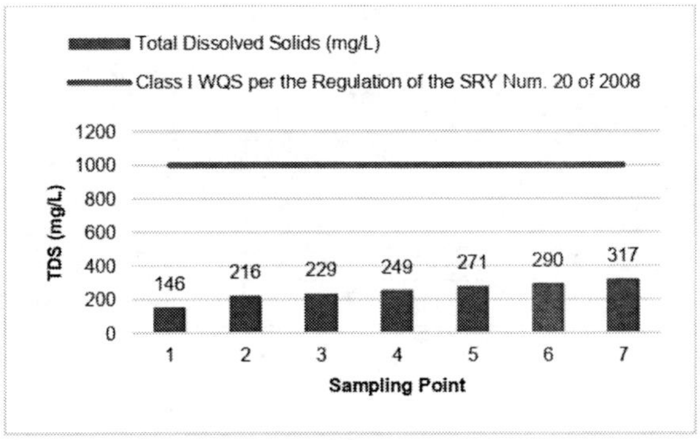

(c)

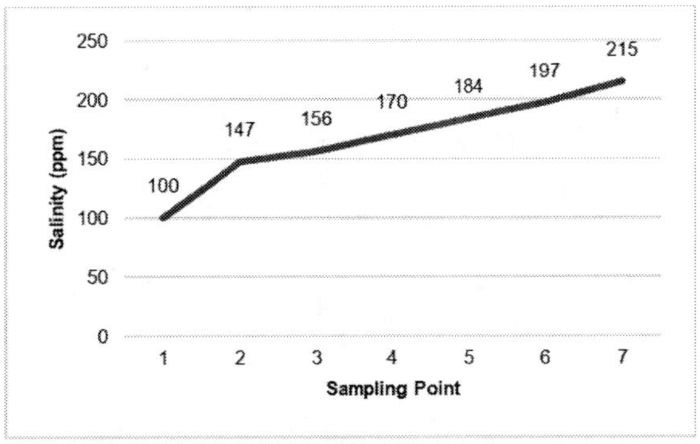

(d)

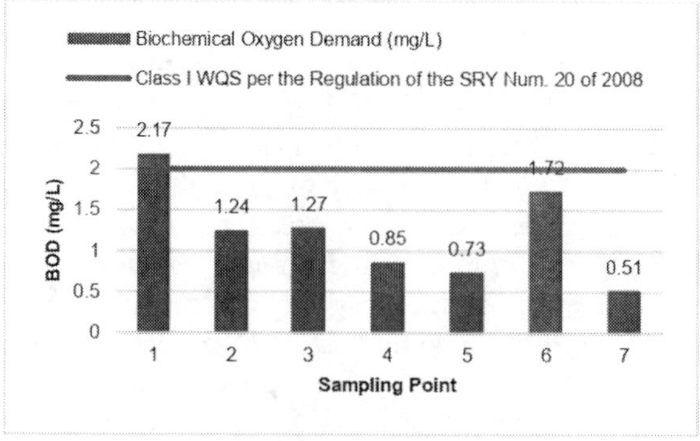

(e)

Figure 3. Water Quality of Winongo River During the Rainy Season: (a) pH, (b) Electrical Conductivity, (c) Total dissolved Solids (TDS), (d) Salinity, (e) Biochemical Oxygen Demand (BOD)

(Data Processing, 2022)

4.2 Pollutant Loads Inventory

The pollutant loads were inventoried by type: point-source or nonpoint-source pollution. Nonpoint-source pollution consisted of domestic activities, livestock farms, agricultural lands, hotels, and hospitals. Pollutant sources in the Winongo watershed and potential to discharge into the river were calculated using secondary data obtained from the relevant agencies. The results of the inventory and calculations represent the estimated pollutant loads. Table 6 shows the estimated and measured pollutant loads.

Table 6. Estimated Pollutant Loads (PL_{BOD}) According to Segments and Sources of Pollution

Segment	PL_{BOD} (kg/day)						
	Domestic	Agricultural	Livestock	Hotel	Hospital	Point-source	Total
1	508.18	0.02	55.93	49.50	0	1.27	614.89
2	229.85	0.001	8.17	1623.60	110.70	14.61	1986.93
3	1874.18	0	5.62	361.35	350.55	17.87	2609.57
4	1595.26	0.003	70.98	153.45	313.65	13.73	2147.07
5	365.22	0.01	154.67	56.93	3370.20	63.61	4010.63
6	490.93	0.02	450.22	0	313.65	8.31	1263.13

Source: Data Processing, 2022

4.3 Total Maximum Daily Loads (TMDLs)

4.3.1 BOD Simulation. Due to the limited availability of relevant details, the pollutant loads were measured using less accurate data. Therefore, a model was created in the research to determine the amounts of pollutants entering the river by identifying and understanding the water quality response. The model outcome, i.e., estimated BOD, was calibrated with a trial-and-error technique to obtain the closest value to the measured BOD (derived from laboratory analysis). The results of the statistical test of the relative bias method (Table 7), the values of rB and F were -0.03 and 0.999, respectively. Based on these results, statistically the model can be accepted. Figure 4 shows the estimated and measured BOD for all sampling points.

Table 7. Reliability Test Using Relative Bias Method

Segment	Mod BOD (mg/l)	Obs BOD (mg/l)	Xi-$X_{average}$ Mod	Xi-$X_{average}$ Obs	(Xi-$X_{average}$ Mod) ^{2}	(Xi-$X_{average}$ Obs) ^{2}
1	1.13	1.24	0.09	0.19	0.01	0.03
2	1.29	1.27	0.25	0.22	0.06	0.05
3	0.85	0.85	-0.20	-0.20	0.04	0.04
4	0.73	0.73	-0.31	-0.32	0.09	0.10
5	1.73	1.72	0.69	0.67	0.48	0.44
6	0.51	0.51	-0.53	-0.54	0.28	0.30
Sum	6.25	6.32			0.97	0.97
Rm	1.04				0.16	0.16
Ro	1.05					
Sm	0.40					
So	0.40			rB =	-0.029	
Sm²	0.16			F =	0.999	
So²	0.16					
The model is accepted if -0,5 < rB < 0,5 and 0,5 < F < 1,5 Conclusion: Accepted						

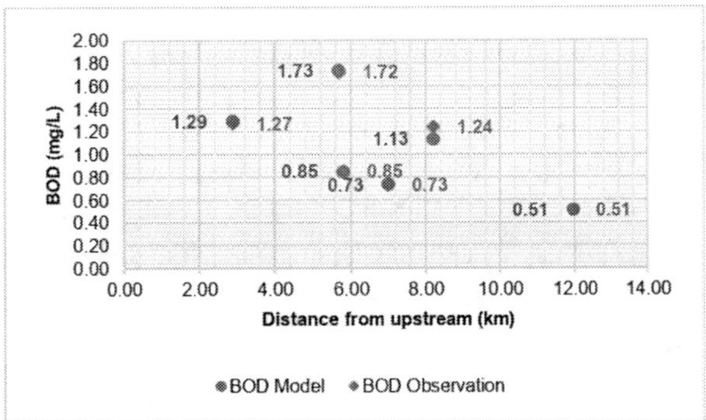

Figure 4. Estimated and Measured BOD Levels (in the picture: BOD Model and BOD Observation, respectively)

(Data Processing, 2022)

4.3.2 Existing Pollutant Loads. An existing pollutant loads is the value of the loads that discharge into the river after modelling. Pollutant loads in the six river segments (expressed in kg/day) are the sum of the pollutant loads from point and nonpoint sources of pollution. Table 8 shows the pollutant loads derived from the estimation and the model. It was found that the estimated loads were significantly different from the measured ones. The reason for such difference is that not the entire estimated pollutant loads are discharged into the river. Another cause is the calculation of the weighted average for population size, areas of residential and agricultural land, and the number of livestock.

IOP Conf. Series: Earth and Environmental Science 1233 (2023) 012056 doi:10.1088/1755-1315/1233/1/012056

Table 8. Estimated and Existing Pollutant Loads by River Segment

Segment	Estimated PollutantLoad (kg/day)	Existing Pollutant Loads (kg/day)
1	614,89	2,57
2	1986,93	5,33
3	2609,57	3,43
4	2147,07	9,87
5	4010,63	10,47
6	1263,13	20,11
Total	12632,23	51,78

Source: Data Processing, 2022

Based on the results presented in Table 8, the pollutant loads increased from segment 1 to 6 but decreased in segment 3. Segment 1 receive the lowest pollutant loads than the other segments (2.57 kg/day, BOD = 1.24mg/l). This segment is in the upstream of the watershed so that it has a higher slope than the middle and downstream so that the waste is directly carried away by the river flow. Residential land of a smaller area in this segment is believed to cause the lower BOD levels. Agricultural practices, both rice fields and dry land agriculture, influence BOD more than residential activities.

The highest pollutant load measured was found in segment 6 (20.11 kg/day, BOD = 0.51 mg/l) because it is in the downstream of the river where the pollutant load accumulates. The pollutant loads in this segment derived from the previous segment and the pollutant sources around the river in segment 6. Segments 5, 4, and2 had pollutant loads of 10.47; 9.87; and 5.33 kg/day, respectively. The lowest pollutant loads after segment 1 is segment 3 (3.43 kg/day, BOD = 0.85 mg/l). Residential buildings are believed to cause the BOD levels because cover all the neighboring land. Pollutant loads can be derived from the domestic, hotel, hospital, and road runoff.

4.3.3 Total Maximum Daily Loads (TMDLs). The modelling of the pollutant load capacity was carried out using a BOD class I standards of Water Quality Index (2 mg/l). Figure 5 shows the model and water quality standards BOD.

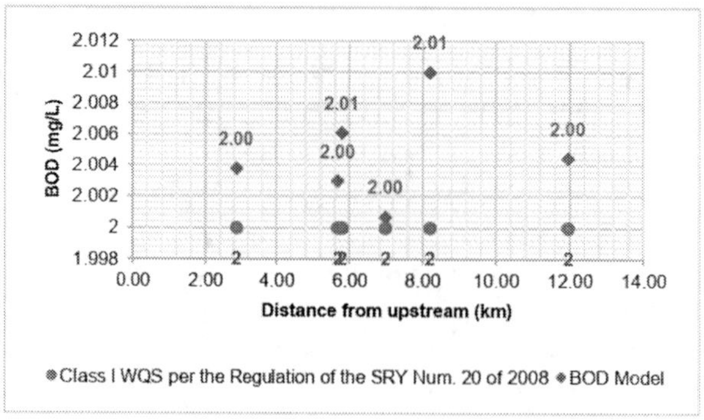

Figure 5. Water Quality Standards and Model BOD Levels (in the picture: Class I WQS and BOD Model, respectively)

(Data Processing, 2022)

Based on the TMDL calculation results, all segments of river can accommodate the loads. The pollutant loads discharge into the river is less than the TMDL. The highest TMDL is in segment 6 of 79.13 kg/day, while the lowest is in segment 1 of 4.56 kg/day. Based on each segment and the total number, the Winongo River can still accommodate the pollutant loads for the sampling period in December or the rainy season. Table 9 presents the measured pollutant, TMDLs, and pollutant reduction targets, if necessary.

Table 9. Existing Pollutant Loads, Total Maximum Daily Loads, and Pollutant Reduction Targets of Each Segment

Segment	Existing Pollutant Loads (kg/day)	TMDLs (kg/day)	Difference(kg/day)	Pollutant Loads to Reduce
1	2,57	4,56	1,99	No reduction necessary
2	5,33	8,27	2,94	
3	3,43	8,13	4,70	
4	9,87	26,90	17,03	
5	10,47	12,09	1,62	
6	20,11	79,13	59,02	
Total	51,78	139,08	87,30	

Source: Data Processing, 2022

5. Conclusion

The research has measured and estimated the pollutant loads and total maximum daily loads (TMDLs) and determined pollutant reduction targets for segments whose TMDLs are exceeded. Winongo River (Winongo Watershed, part of Opak Watershed in the Special Region of Yogyakarta) receives a pollutant load of 51.78 kg/day, and the maximum number of pollutants allowed to be discharged into this river is 139.08 kg/day. These results demonstrate that the river can accommodate the pollutant loads while maintaining the water quality. The lowest pollutant loads are in the segment 1 of 2.57 kg/day, then the TMDL is 4.56 kg/day. The highest pollutant loads are in segment 6 of 20.11 kg/day, then the TMDL is 79.13 kg/day. Each segment had a pollutant loads lower than the TMDL so that it is still able to accept the pollutant loads.

Further research needs to be carried out in more detail so that the results of the calculation of the pollutant load entering the river are more accurate. Emission factors used in the calculation of nonpoint pollutant sources should be re-measured. To support further research, the government needs to improve the availability and good data management so that there is no missing or empty data. In addition, the sampling period is better carried out in two periods, namely the dry and rainy seasons, to determine the condition of river water at different times and seasons.

6. References

[1] Asian Development Bank 2016 *Indonesia, Country Water Assessment* (Manila: ADB Avenue)

[2] Liqin D, Jinming S, Yayan X, Xuegang L, and Ying Z 2010 *Journal of Hazardous Materials,* **183** 1 p 155-164

[3] WWAP (United Nations World Water Assessment Programme) 2015 *The United Nations World Water Development Report 2015: Water for a Sustainable World* (Paris: UNESCO)

[4] Robert B A and Tim A W 2017 *WASP8 Stream Transport-Model Theory and User's Guide, Supplement to Water Quality Analysis Simulation Program (WASP) User Documentation.* (Washington: US Environmental Protection Agency)

[5] Liangliang G and Daoliang L 2014 *Front. Agr. Sci. Eng.* **1** 4 p 267-276

[6] Winardi D N and Lintang C 2007 *Jurnal Presipitasi* **3** 2 p 93-101

[7] Claude E B 2019 *Water quality: an introduction*. Springer Nature

[8] Mariely H B D, Alexandra A M, Marcelo C B, Jean P H B O, Plinio B D C, Alex V K, Maria V B, Reynaldo L V and Luiz A M 2002 *Water, Air, and Soil Pollution* **136** 1 p 189-206

[9] Anna F R 2018 *IOP Conf. Ser.: Earth Environ. Sci.* **118** 012019 vol 118 (Bandung: IOP Publishing)

[10] V. Darsono 2013 *Panduan Pengelolaan Green Industry* (Yogyakarta: Cahaya Atma Pustaka)

[11] Rahmat R A, Winardi D N and Sudarno 2015 *Jurnal Teknik Lingkungan* **4** 3 p 1-15

[12] Frank R S 1996 *Stream Ecology and Self-Purification, an Introduction for Wastewater and Water Specialist* (United States of America: Technomic Publishing Company)

[13] *Peraturan Gubernur Daerah Istimewa Yogyakarta Nomor 22 Tahun 2007 tentang Penetapan Kelas Air Sungai di Provinsi Daerah Istimewa Yogyakarta*. Yogyakarta, (2007) (in Indonesian)

[14] Pergub DIY. *Peraturan Gubernur Daerah Istimewa Yogyakarta Nomor 20 Tahun 2008 tentang Baku Mutu Air di Provinsi Daerah Istimewa Yogyakarta*. Yogyakarta, (2008) (in Indonesian)

Acknowledgement

This article is preliminary research. The authors would like to thank Universitas Gadjah Mada for funding the research with the Thesis Recognition Grant (RTA) with a Letter of Assignment Number is 3550/UN1.P.III/Dit-Lit/PT.01.05/2022. Gratitude is also extended to the Department of Environmental Geography, Faculty of Geography, for providing the opportunity and permission to conduct the research.

IOP Conf. Series: Earth and Environmental Science 1233 (2023) 012057 doi:10.1088/1755-1315/1233/1/012057

Application of MODIS land surface temperature data on ENSO-based analysis in Kalimantan

R A Zahra[1]

[1]Department of Environmental Geography, Faculty of Geography, Universitas Gadjah Mada, Indonesia

Corresponding author: rahma.aulia.zahra@mail.ugm.ac.id

Abstract. El Nino Southern Oscillation as climate anomaly has been impacted Land Surface Temperature (LST) in Indonesia. Higher temperature during El Nino has been considered to give more risk in fire hazards in Kalimantan, thus, analysis of spatial and temporal distribution of LST became crucial. This study aims to analyze the spatial and temporal distribution of land surface temperature in Kalimantan based on the ENSO phases from 2014 to 2020. MODIS MOD11A1 images and ONI index were used as analysis materials. Image processing was conducted by selecting pixels using the QC band and then processing the LST values using a scale using RStudio software. The result shows the distribution of higher LST values on the southern side of Kalimantan, influenced by low elevation, peat soil, and land use change. The temporal distribution shows a tendency for higher temperatures in the dry months of August to October, with an increase of 4.5 percent during El Nino and a decrease of 0.9 percent during La Nina from 2014 to 2020. This study could contribute to management and mitigation for fire risk in certain area and period as ENSO occurred in Kalimantan.

1. Background

Climate change effect on increasing temperature has widely impacted various regions in the world as it causes many drawbacks, such as fire hazard as one of the most hazardous effects. The increasing temperature has known to give more contribution on fire risk across all climatic zone, such as in China and California. In China, temperature positively affected the fire size severity during spring and summer season [1]. The historical trend analysis uncovered that weather-driven autumn wildfire risk in California has been increased as temperature rose for +0.30°C/decade [2].

In the tropical area, fire risk in Indonesia also surged during the dry months due to increasing rate of evapotranspiration [3]. The climate anomaly El Nino Southern Oscillation (ENSO) strongly associated with drought event in Indonesia, and it also underpinned the fire occurrences massively in 2015 [4]. Increasing rate of strong El Nino occurred affected the higher land surface temperature (LST) across Kalimantan [5]. However, it becomes crucial to assess LST based on El Nino Southern Oscillation in spatial and temporal distribution in Kalimantan by using remote sensing data.

The development of technology for atmospheric modeling in remote sensing plays a role in studying phenomena on the earth's surface. Many satellite-based remote sensing products are used for climate and fire monitoring. One of the advantages of satellite-based remote sensing compared to climate monitoring stations is that it can evenly record spatial variations of rainfall [1]. In addition, remote sensing products also have a database that facilitates the computation and analysis of various phenomena on the earth's surface, including LST.

LST data is crucial in climate and environment analysis [2]. Several remote sensors that currently provide LST products are Earth Observing System/Moderate Resolution Imaging Spectroradiometer (EOS/MODIS), National Oceanic and Atmospheric Administration/Advanced Very High Resolution Radiometer (NOAA/AVHRR), and Feng Yun/Visible and Infrared Spin Scan Radiometer (FY/VISSR) [3]. Moderate Resolution Imaging Spectroradiometer (MODIS) is one of the remote sensing images used in drought studies with a high spatial resolution [4]. MODIS images are also one of the data sources used by the Indonesian Ministry of Environment and Forestry to monitor fire incidents [5]. This image is commonly used to observe sea surface temperature, soil surface temperature, and vegetation by assessing evaporation rates. The wide usage of MODIS LST has been implemented as an evaluation to determine the accuracy of the product using in-situ measurements [6].

MODIS image processing technique in LST analysis requires Quality Control (QC) processing. MODIS image LST data has a spatial resolution of 1 km, which is susceptible to cloud disturbances. Due to cloud sizes smaller than 5 km, aerosols can interfere with the pixel values in LST. LST pixel removal is usually done by identifying the QC band. Some previous studies have already conducted an analysis of pixel removal by analyzing the QC band in MODIS LST [7] in several areas, such as the US [8], Tibetan Plateau [9], and Portugal [10].

This research implemented MODIS image processing in Kalimantan as the most significant part of the Indonesian archipelago. El Nino Southern Oscillation was compared to the results of the LST, as ENSO becomes one of the most significant tropical climate anomaly that influences on weather variability in Indonesia. Kalimantan was chosen as a large coverage area to compare the spatial and temporal distribution of LST during El Nino, La Nina, and neutral conditions.

Several studies have also contributed in analyzing Kalimantan's temperature [5,16,17]. However, most of them used the Sea Surface Temperature (SST) when analyzing its impacts based on ENSO and fire hazards. In addition, only limited research focused on the specific spatial and temporal distribution of LST during ENSO as conducted in this study. This study aimed to analyze the spatial and temporal distribution of land surface temperature in Kalimantan based on the ENSO phases from 2014 to 2020 by using MODIS and RStudio. The research findings could be analyzed as the basic conditions of how the changes of LST differs across Kalimantan during El Nino, La Nina, and neutral condition. Furthermore, it could be used as the basic strategy of peatland management as well as fire hazard mitigation in Kalimantan.

2. Materials and methods

2.1. Research area

The research was conducted in Kalimantan as part of the island of Borneo which covers the administrative area of Indonesia. This region has an area of 743,330 km^2 and consists of West Kalimantan, Central Kalimantan, South Kalimantan, East Kalimantan, and North Kalimantan. Kalimantan has the largest tropical forest biodiversity across Asia, yet it also experienced fire hazard almost every year. One of the physical factors regarding to this related to the peat soil which is prone to the soil moisture. During the dry months, several fire hotspots occurred in the peat soil because of the decreasing rainfall and massive evapotranspiration [16]. Evaporation rate is commonly calculated based on the land surface temperature, and its analysis became crucial to obtain the spatial and temporal differences across Kalimantan.

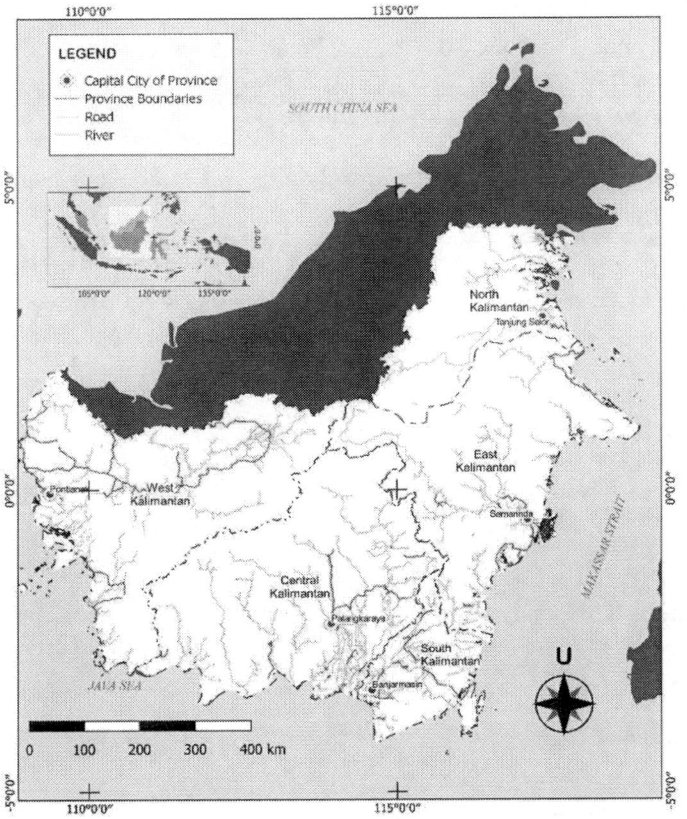

Figure 1. Study area (Source: Data processing, 2021. Base map obtained from Indonesian Geospatial Agency (BIG) 2020)

2.2. *Data collection*

The data collection was carried out by obtaining secondary data, as shown in Table 1. This study used ONI and MODIS Land Surface Temperature images from 2014 to 2020.

Table 1. Data

No	Data	Materials	Organization	Sources
1.	LST	MODIS (MOD11A1.006) 2014-2020	NASA	https://lpdaac.usgs.gov/tools/appeears/
2.	Oceanic Nino Index (ONI)	ONI 2014-2020	NOAA	https://origin.cpc.ncep.noaa.gov/products/analysis_monitoring/ensostuff/ONI_v5.php

2.3. *Data processing*

2.3.1. ONI index. The ONI index was identified by events based on phase. El Nino occurs when ONI increases >0.5°C, while La Nina occurs when ONI decreases <-0.5°C. ONI was identified monthly from 2014 to 2020.

2.3.2. MODIS LST. MODIS LST image's risk to cloud disturbances is caused by its 5 km spatial resolution. Due to the cloud sizes smaller than 5 km, aerosols can interfere with the pixel values in

LST. LST grids were selected by identifying the Quality Control channel (QC band). Pixels considered inappropriate were removed from the analysis by a clip process with an LST raster. The QC band has several values; they are "0," "1," "2," and "3" flags, associated with the average errors <1 K, 1–2 K, 2–3 K, and >3 K, respectively [9]. LSTs with average errors of 2–3 K and >3 K (QC flag =2 and 3) were removed in RStudio for this study.

2.3.3. MODIS LST scale factor. MODIS LST pixel value was scaled into a specific temperature value [18] as shown in the equation (1) below:

$$\text{Value} = \text{Pixel Value} \times 0.02 \quad (1)$$

The raster value was converted from Kelvin into a Celsius unit by subtracting the raster value with 273. These computations were conducted in RStudio with the 'raster' package.

2.4. Data analysis

2.4.1. Descriptive analysis. Descriptive analysis was carried out by describing the data and research results in the form of tables, diagrams, and graphs. Descriptive analysis through graphs presented ONI data and LST. The ENSO phases were described through ONI to identify El Nino, La Nina, and normal conditions.

2.4.2. Spatial analysis. This analysis was conducted to know the spatial perspective through maps based on the ENSO phase. Maps also depicted the distribution of land surface temperature based on the ENSO phase.

3. Result and discussion

3.1. ENSO phase

ENSO consists of three phases, namely El Nino (ONI > 0.5), La Nina (ONI < -0.5) and a neutral phase (-0.5 < ONI < 0.5) [11]. There are methods to identify ENSO, one of which is through the Oceanic Nino Index (ONI). ONI is an index that measures changes in SST in the Nino 3.4 region as a feature of ENSO.

ONI value variability from 2014 to 2017 is shown in Figure 1. At the beginning, the value was still neutral, rose in the mid-2014, then sloped and continued to grow until the end of 2014. It passed the value of 0.5, which indicates the El Nino condition. Then, the ONI value continued to increase throughout 2015. The ONI value decreased in 2016 until it experienced a neutral state in the middle of the year. This value declined to -0.5 at the end of the year. In 2017, the ONI value was generally neutral, then became below -0.5 at the end of the year. This condition continued in early 2018 then increased to 0.5 from the end of the year to mid-2019. The temporal pattern of ONI tended to fluctuate until 2020. After mid-2020, the ONI value declined below -0.5, which indicates La Nina conditions.

3.2. Land surface temperature variability

The distribution of LST data based on the ENSO phases presented in Figure 2 indicates some differences in LST values during the El Nino, La Nina, and neutral phase. During El Nino, LST tended to be higher than La Nina and impartial, with an average value of 27.3 °C. However, the difference in average LST was not very significant under neutral conditions. The average value under neutral conditions was in the range of 27.1-27.2°C. A significant difference was found in the LST curve based on the La Nina phase, which had values of 26.5-26.8°C.

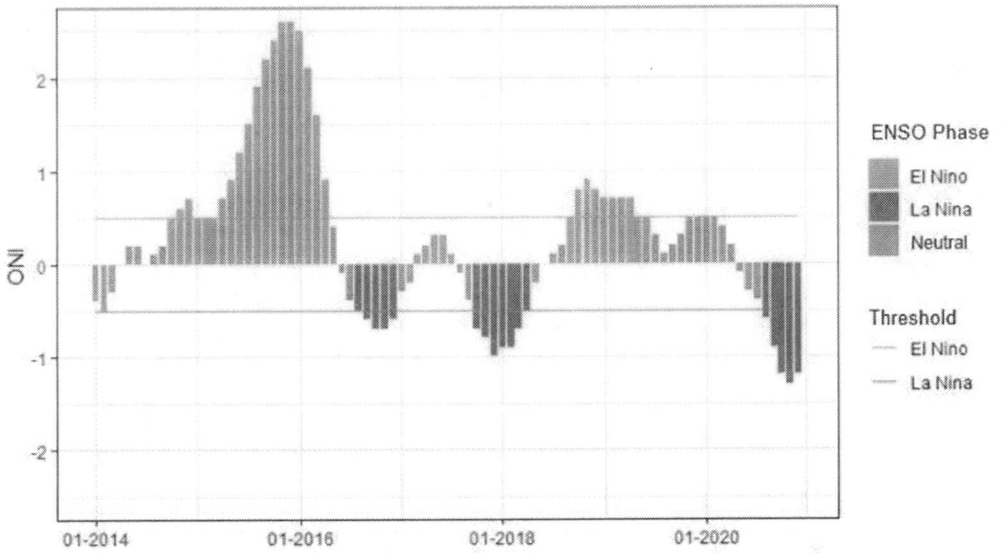

Figure 2. ONI 2014-2020 (Source: Data processing, 2021)

LST during El Nino conditions had many variations in data distribution compared to other ENSO phases. The minimum value was 24°C, and the maximum value reached 32°C in several months. The mean value was 27.2 - 27.3°C, indicated by the central peak of the density curve. However, several data trend segments were based on peaks at values of 26°C and 30-31°C.

LST during La Nina also had variations, but the values were more concentrated in the range of 26.5-26.8°C. The minimum value was 24°C and the maximum value tends to be lower than the other phases at 29°C. Several segments of the trend of surface temperature data during La Nina are based on other peaks in the range of 25.0-25.5°C and 28.0-28.2°C.

Analysis of the LST at neutral conditions was assumed to be a normal LST condition in Kalimantan. Temperature data in neutral conditions were more concentrated at 27.1-27.2°C. The minimum data value was in the range of 25°C, while the maximum value was 30°C. Several segments of LST data trend on neutral accumulated at values of 26.5-26.7 °C and 28.5 °C. However, these two peaks were not very significant as the data was more concentrated on the average, which had a value of 27.1-27.2°C. This average value had nearly the same when El Nino, with an average of 27.2 - 27.3 °C. In addition, the most significant difference between these two conditions was the maximum value during the year. LST during El Nino had a maximum data distribution that far exceeded the maximum value for neutral conditions. It implied a considerable influence on the maximum value of LST during El Nino in Kalimantan from 2014 to 2020.

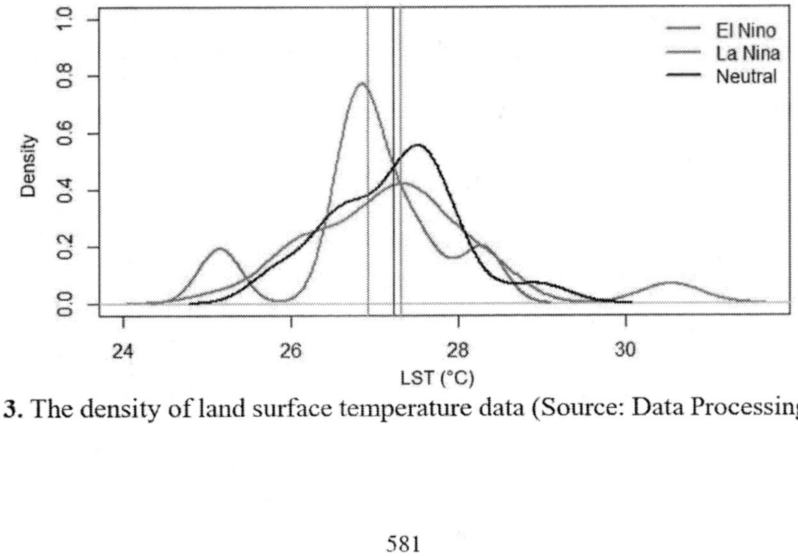

Figure 3. The density of land surface temperature data (Source: Data Processing, 2021)

3.3. Temporal distribution of land surface temperature

The temporal pattern of Kalimantan's LST in 2014-2020 in Figure 3 shows the average monthly LST was in the range of 26-31°C. The graph shows a peak in the average monthly LST in 2014, 2015, and 2019 with temperatures above 29°C. All highest temperature occurred in August-October. The highest LST occurred in September 2015 at 30.6°C. High-temperature values also occurred in October 2014 at 29.6°C and September 2019 at 28.8°C. Based on the graph, there is a tendency for ONI to increase when the LST rises. Meanwhile, LST conditions in other years tend to be evenly distributed every month. There was a significant decrease in February 2017, December 2019, and December 2020s associated with La Nina conditions (ONI < -0.5). The end of 2020 was the time with the lowest surface temperature throughout 2014-2020 at 25°C.

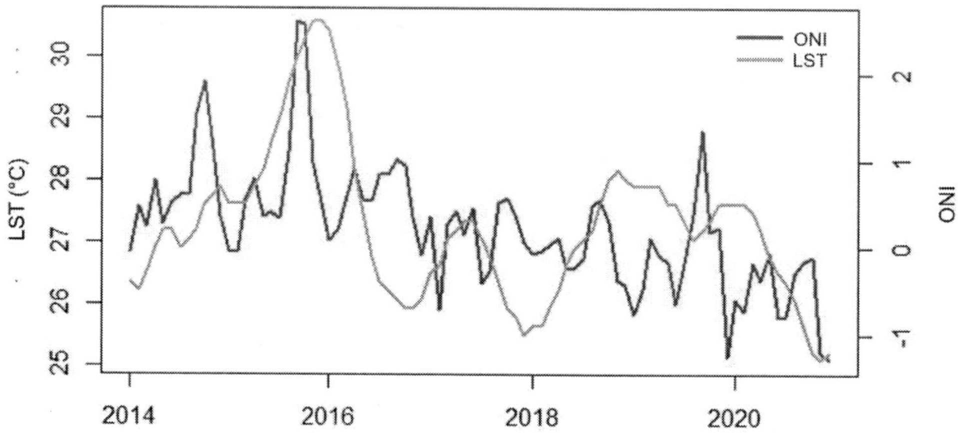

Figure 4. The average land surface temperature Kalimantan 2014-2020 (Source: Data processing, 2021)

The high temperature in 2015 was also influenced by the rise of ONI in these months. Since May 2015, there had been a strong El Nino affecting the rise in LST. The presence of ONI more than 2.0 began to occur from September to December 2015, created significant increase of LST reached to 30.6°C. In addition to ONI influence, there was an anomaly from the Indian Ocean Dipole (IOD), which had the highest number of 1.17 in September 2015 [12] and affected the higher LST.

3.4. Spatial distribution of land surface temperature

The spatial distribution of Kalimantan's LST in 2014-2020 in Figure 4 focused on dry months, especially in August, September, and October as this period affected ENSO significantly [13]. LST values tend to be higher in the southern part of Kalimantan, both in El Nino, La Nina, and neutral phases. The lower temperature values tend to spread over the northern and central sides of Kalimantan.

LST in 2014 generally shows a higher value in southern and western Kalimantan. The neutral phase occurred in August and September when the average LST was 27.8°C and 29.1°C. El Nino phenomenon also influenced the high increase in average LST in September in October with an ONI value of 0.5 (weak El Nino). The rise in surface temperature also continued in October, which can be observed in several areas in southern Kalimantan.

LST in 2015 shows the dominance of high v in most areas of Kalimantan. From August to October, the average LST values were 28.5 °C, 30.6 °C, and 30.5 °C, respectively. The LST values in this period were the highest condition throughout 2014-2020 in the same month. The intense El Nino influenced this increased value in August (ONI=1.9), September (ONI=2.2), and October (ONI=2.4). The most significant impact can be observed in the southern part of Kalimantan, where the surface temperature was more than 36°C in these months.

In contrast to 2015, in 2016, the LST conditions of Kalimantan tended to be lower due to the La Nina phases. The spatial distribution of LST also had a similar pattern, lower in the north and higher in the south of Kalimantan. LST values were 28.1°C, 28.3°C, and 28.2°C, respectively.

The distribution of LST in 2017 had almost similar conditions from August to October. The LST in a neutral phase for August and September commonly had the highest values in the southern part, while in some conditions, the value was 32-36°C. Meanwhile, there was a La Nina event in October, which decreased LST in the southern part compared to the previous month.

LST in August-October 2018 had the mean values of 27.6°C, 27.7°C, and 27.3°C. The existence of El Nino conditions in September and October did not significantly impact increasing LST in these two months. In the southern region, the higher LST was incredibly closed to the peat area around Palangka Raya City.

LST fluctuated in 2019 as the values were 27.5°C, 28.8°C, and 27.2°C. The difference in LST values can be observed in the western and southern parts of Kalimantan. LST in 2020 tended to be lower than in the previous years. Figure 4.6 shows that LST ranged from 24-28 °C in most areas. LST have higher values in the south but are still between 28 and 36°C. This condition was associated with a vital La Nina phase in these months.

LST of Kalimantan tended to be higher in the south and lower in the middle and north as analyzed by previous study [5]. A significant increase of LST was found during El Nino 2015, while a decrease can be found during La Nina phase such as in 2020. In general, northern area's land cover was dominated by tropical rainforest vegetation, while southern area was more affected by human activities. Studies predicted the high rate of deforestation and more extreme El Nino events will have far reaching impacts on the environment, agricultural productivity, and human livelihoods in Kalimantan in coming decades [16]. This impact is shown in the area located around the urban area of Palangka Raya, whose higher LST than other areas. Physical factor of topography also influenced LST spatial distribution. In the southern region of Kalimantan, the morphology of the basin and plains had a lower LST than the topographic conditions in the northern and central parts of Kalimantan.

3.5. Analysis of land surface temperature during ENSO phases

LST condition in Figure 5 shows differences in each ENSO phase. The average LST was 27.75 °C in neutral condition, 29.03 °C in El Nino, and 27.48 °C in La Nina. These values indicate a significant change in El Nino conditions as LST rose up to 4.6 percent from normal conditions in August to September. Meanwhile, in the La Nina phase, the difference in LST was not too significant compared to neutral conditions, as the LST decreased to only 0.99%.

The spatial distribution of LST in the neutral phase was indicated by higher LST at lower elevations in Kalimantan. The average value in August to October was 27.75°C. The northern and central sides of Kalimantan were still dominated by tropical rainforests with an average LST < 24°C. The highest LST distribution reaching more than 36°C was distributed on the south and west sides of Kalimantan, associated with the coast and agricultural land use dominated by human activities. The higher LST value in this area was influenced by deforestation as previous study found a correlation between forest decline, daily temperature increase, and precipitation reduction in Kalimantan [14].

The linkage of El Nino with LST generally occurred in south-eastern Kalimantan, associated explicitly with peatlands and high intensity of agriculture. LST increased up to 4.5% in the El Nino phase from August to October. The Sea Surface Temperature anomaly influenced this due to the El Nino anomaly. A study implied that projecting the location of El Nino events might be more important than projecting their strength for fire management in Borneo [4]. Figure 6 shows an increase in the area with a LST impacted by El Nino of more than 36 °C in the southeast, south, and some parts of the West Kalimantan coast. Supporting this findings, previous study related to land cover influence on temperature found that stronger temperature in the southern area increased for almost 1°C because of massive deforestation [5].

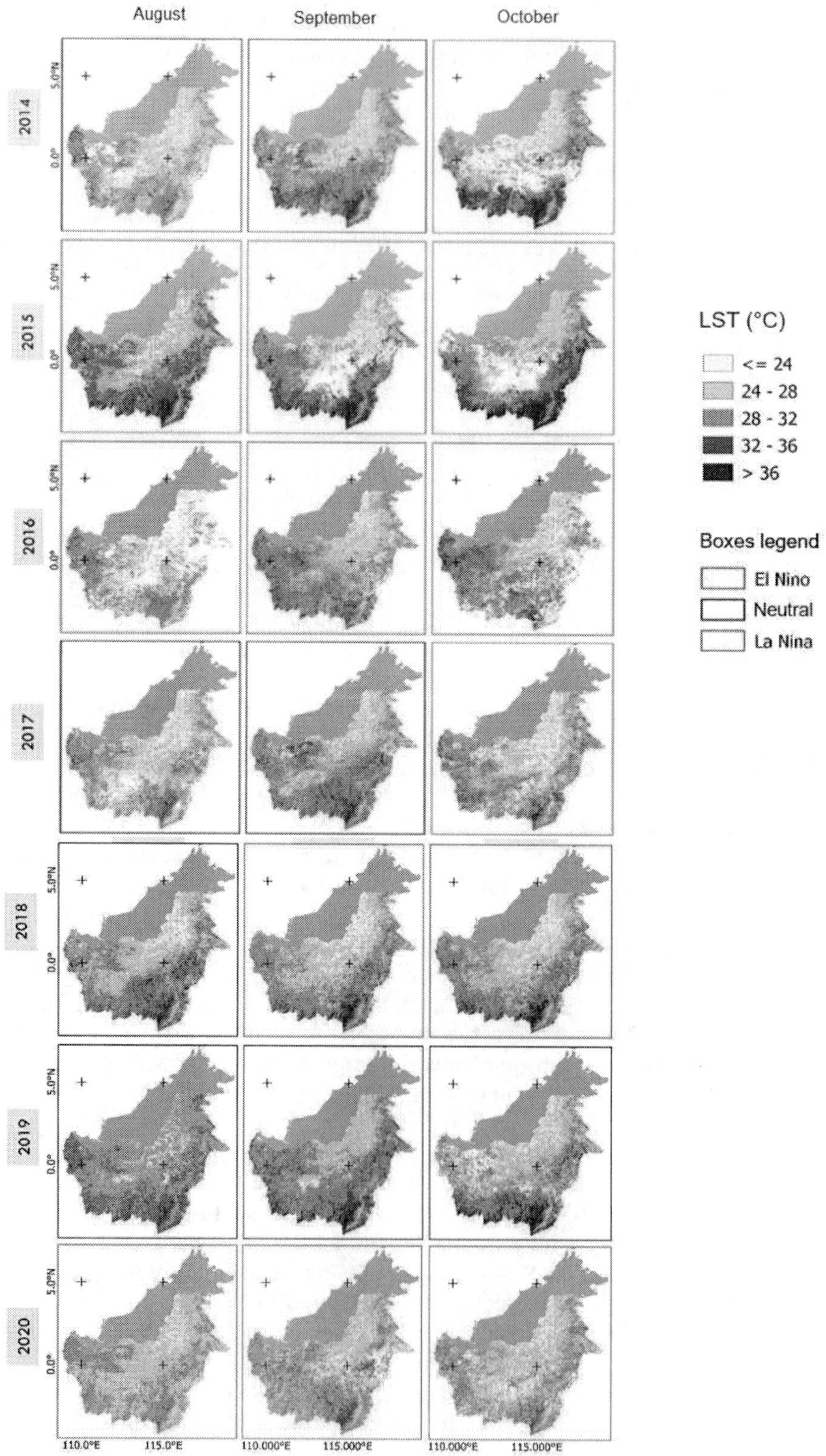

Figure 5. Spatial distribution of LST in Kalimantan (Source: Data processing, 2021)

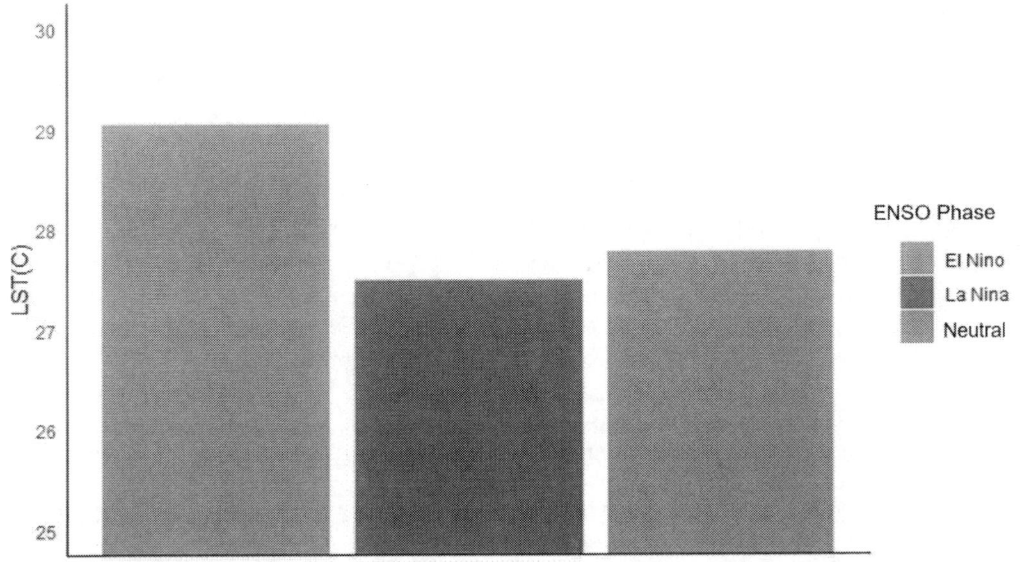

Figure 6. Average of LST based on ENSO phases (Source: Data processing, 2021)

La Nina also contributed to the decrease in the average LST of Kalimantan. The reduction in LST occurred to 0.9 percent from August to October during La Nina. The most significant declined in the south-eastern side of Kalimantan, associated with agricultural land and peatlands whose experienced massive land-use conversions. The declining LST in this area indirectly affected the intensity of fires that may occur during the dry month.

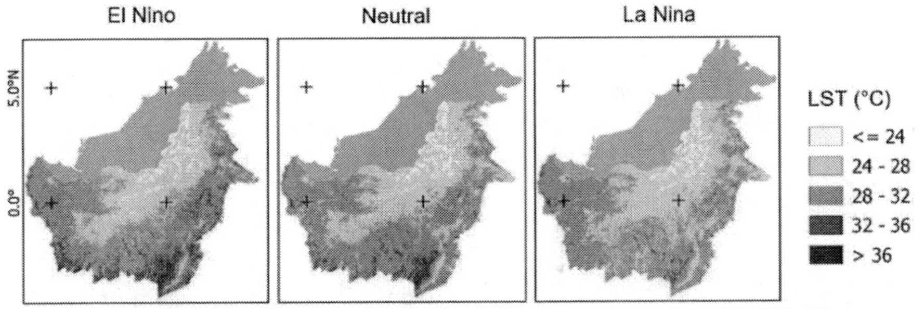

Figure 7. Spatial distribution of LST in Kalimantan based on ENSO (Source: Data processing, 2021)

The research findings relevant with previous studies related to the LST in Kalimantan. The effect of El Nino was generally stronger in higher LST as the consequence of reduced evaporative cooling, especially in southern Kalimantan which experienced stronger LST [5]. Several studies also found that August-October also became the driest period and potentially increased the occurrence of fires during El Niño years [4,10,17,20]. However, this research has limitation on achieving the field LST data. In the long run, it is essential to develop this study to analyze the spatial distribution of the future LST prediction. The implication of this study benefits management in land and water for disaster management and food security in Kalimantan.

4. Conclusion

LST distribution from MODIS had a significant value with the spatial and temporal conditions in Kalimantan during ENSO. The LST was higher in the southern and western Kalimantan, associated with the lower elevation, peat soil distribution, and deforestation. During El Nino, the average LST in southern Kalimantan could reach more than 36°C specifically near Palangkaraya and peatland areas with massive agriculture and deforestation activities.

August to October became the period with highest significance with ENSO condition as well as the driest period in Kalimantan. LST increased to 4.5 percent during El Nino and a decrease to 0.9 percent during La Nina. This study could significantly contribute to resources management in cropland area as well as mitigation for fire risk in certain area and period as ENSO occurred in Kalimantan.

Acknowledgements
This research was supported by the grant of "Rekognisi Tugas Akhir (RTA) Universitas Gadjah Mada 2022" with Dr. Andung Bayu Sekaranom as principal investigator (Contract number: 3550/UN1.P.III/Dit-Lit/PT.01.05/2022 on May 20th, 2022). The author would like to sincerely thank Dr. Andung Bayu Sekaranom and Dr. Emilya Nurjani as supervisors and climate experts from Universitas Gadjah Mada who had contributed to this research improvement.

References
[1] Fan Q, Wang C, Zhang D and Zang S 2017 Environmental influences on forest fire regime in the Greater Hinggan Mountains, Northeast China *Forests* **8 (10)** 1–16
[2] Goss M, Swain D L, Abatzoglou J T, Sarhadi A, Kolden C and Williams AP 2020 Climate change is increasing the likelihood of extreme autumn wildfire conditions across California *Environ Res Lett.* **15(9)**
[3] Sloan S, Locatelli B, Wooster M J, Gaveau D L A 2017 Fire activity in Borneo driven by industrial land conversion and drought during El Niño periods, 1982–2010 *Glob Environ Chang.* **47** 95-109
[4] Chen CC, Lin H W, Yu J Y and Lo M H 2016 The 2015 Borneo fires: What have we learned from the 1997 and 2006 El Niños? *Environ Res Lett.* **11(10)** 1–7
[5] Chapman S, Syktus J, Trancoso R, Salazar A, Thatcher M and Watson J E M 2020 Compounding impact of deforestation on Borneo's climate during El Nino events. *Environ Res Lett.* **15(8)**
[6] Sekaranom A B, Suarma U and Nurjani E 2020 Climate extremes over the maritime continent and their associations with Climate extremes over the maritime continent and their associations with Madden-Jullian Oscillation *OP Conf Ser Earth Environ Sci.* **451**
[7] Shi S, Helman D and Lensky I M 2021 Worldwide continuous gap-filled MODIS land surface temperature dataset *Sci Data* 1–10
[8] Yu P, Zhao T, Shi J, Ran Y, Jia L and Ji D 2022 Data Description tor global spatiotemporally continuous MODIS land surface temperature dataset. *Sci Data* **9(143)** 1–15
[9] West H, Quinn N and Horswell M 2019 Remote sensing for drought monitoring & impact assessment: Progress, past challenges and future opportunities *Remote Sens Environ* **232(7)**
[10] Larasati B, Kanzaki M, Purwanto R H and Sadono R 2019 Fire regime in a peatland restoration area: lesson from Central Kalimantan. *J Ilmu Kehutan* **13(2)** 210
[11] Duan S B, Li Z L, Li H, Göttsche F M, Wu H, Zhao W, et al. 2019 Validation of Collection 6 MODIS land surface temperature product using in situ measurements *Remote Sens Environ.* **225(2)** 16–29
[12] Phan T N and Kappas M 2018 Application of MODIS land surface temperature data: a systematic literature review and analysis *J Appl Remote Sens.* **12(04)** 1
[13] Zeng L, Wardlow B D, Tadesse T, Shan J, Hayes M J, Li D and Xiang D 2015 Estimation of daily air temperature based on MODIS land surface temperature poducts over the corn belt in the US. *Remote Sens.* **7** 951-70
[14] Zhang H, Zhang F, Ye M, Che T and Zhang G 2016 Estimating daily air temperatures over the Tibetan Plateau by dynamically integrating MODIS LST data. *J Geophys Res Atmos.* **14** 11425-41
[15] Benali A, Carvalho A C, Nunes J P, Carvalhais N, and Santos A 2012 Remote sensing of environment estimating air surface temperature in Portugal using MODIS LST data. *Remote Sens Environ.* **124** 108–121

[16] Mcalpine C A, Johnson A, Salazar A, Syktus J, Wilson K and Meijaard E, et al. 2018 Forest loss and Borneo's climate. *Environ Res Lett.* **13** 044009
[17] McPhaden M J, Santoso A and Cai W 2020 *Introduction to El Niño Southern Oscillation in a changing climate* (Washington: American Geophysical Union and John Wiley & Sons) p 1-19
[18] Wan Z 2007 Collection-5 MODIS land surface temperature products users' guide. (Santa Barbara: ICESS, University of California)
[19] Yananto A and Dewi S 2016 Analysis of the 2015 El Nino event and its influence on the increase of hotspots in Sumatera and Kalimantan *J Sains Tekno Modif Cuaca* **17(1)** 11–9
[20] Santika T, Budiharta S, Law E A, Dennis R A, Dohong A and Struebig M J, et al. 2020 Interannual climate variation, land type and village livelihood effects on fires in Kalimantan, Indonesia. *Glob Environ Chang.* **64** 1–11

IOP Conf. Series: Earth and Environmental Science 1233 (2023) 012058 doi:10.1088/1755-1315/1233/1/012058

Flood characteristics analyses in the Jratunseluna river basin due to the impact of Kedungsepur national strategic area development

N H Hanifah[1], H G Mawandha[1*], F Abiy[1], C Setyawan[1], L M Kesuma[2] and D S Rulisyani[3]

[1]Departmen of Agricultural Engineering and Biosystem, Faculty of Agricultrural Technology
Universitas Gadjah Mada, Yogyakarta, 55281, Indonesia
[2]Engineer Professional Study Program, Universitas Gadjah Mada, Yogyakarta, 55281, Indonesia
[3]Central River Region Pemali Juana, Ministry of Public Works and Public Housing, Semarang, 50191, Indonesia
*Email: hanggar.g.m@ugm.ac.id

Abstract. The area of the Jratunseluna river basin is categorized as a high-risk flood vulnerability. The urban development plan can worsen this condition through Kedungsepur national strategic area. The objective of this study is to identify flood characteristics in Jratunseluna river basin caused by the alteration of land use due to Kedungsepur national strategic area development through rainfall-runoff 2D simulation. The RRI 2D model was chosen because it could simulate flood inundation comprehensively in all river systems while the other models such as HEC-HMS, TOPMODEL, and Tank models have a limitation on it. The input data used for the RRI model were rainfall data, topography (DEM), and land use maps. The rain data used in the simulation was the Global Precipitation Measurement (GPM) from 2010 to 2020. The study results indicate that the flood risk level would increase during the implementation of the Kedungsepur national strategic area development. The flood duration and peak discharge increase mainly characterize the increase in flood risk. The most significant increase in peak discharge among the four outlets is obtained in the Garang watershed. In the case of a 2-year flood, the peak discharge increased by 44.91% from 262.18 m3/s to 379.92 m3/s and for the case of a 50-year flood, the peak discharge increased by 54.34% from 331.63 m3/s to 511.83 m3/.

1. Introduction

The area of the Jratunseluna river basin is categorized as a high-risk flood vulnerability. With a large river basin coverage area, which is about 9,896 km^2 and dominated by the urban area, the flood impact may affect many people living in the area. In 2014, flood occurrence sharply increased along the region of the north coast of Central Java, followed by a dramatic drop in 2015 and a steady increase afterward in 2018 [1]. This condition makes a flood analysis in the Jratunseluna river basin essential to carry out to prevent further damage.

Land use change can increase flood occurrence. The built-up area in the Jratunseluna River Basin has expanded from 1222 to 1581 km^2 from 2009 to 2018. Most of the expansion occurred in the urban

IOP Conf. Series: Earth and Environmental Science 1233 (2023) 012058 doi:10.1088/1755-1315/1233/1/012058

area at approximately 137%. The development of urbanization promotes flooding due to the increase in total impervious regions, leading to excessive rainfall [1]. The study of flood characteristics analysis shows that land use changes, urbanization from 1976 to 1997, can be associated with an increase in peak flow and a decrease in travel time [2].

The main issue of the water management strategic plan in the Jratunseluna river basin is the development of the Kedungsepur National Strategic Area. Kedungsepur (Kendal-Demak-Ungaran-Salatiga-Semarang-Purwodadi) will be developed into an urban area. The Kedungsepur Strategic National Area plan is prioritized to develop a relatively advanced area and lead to equitable development. The Kedungsepur area is expected to be able to compete in the social, economic, cultural, and political fields in the era of globalization.

In the hydrological aspect, the Kedungsepur area covers the main rivers in the Jratunseluna which have the potential used as a raw water source and irrigation water. The land use changes due to the development of the Kedungsepur national strategic plan may affect the hydrological condition. Analysis of the flood hydrograph in the jratunseluna river basin is essential to mitigate disasters occurred due to flood risk. Flood analysis is possibly done with rainfall-runoff modeling and inundation analysis. Two models usually perform the inundation simulation: the rainfall-runoff model for predicting river discharge and the hydraulic model for water propagation [3]. Many rainfall-runoff models, such as HEC-HMS, TOPMODEL, Tank model, and others, may simulate 1D rainfall-runoff within a single catchment, while the RRI model developed by Sayama et al. (2012) could simultaneously compute flood inundation, runoff, and extent areas in the river system [4]. In addition, The RRI model has been successfully applied to different areas around the world, such as the Yasu River basin in Japan [5], the Kabul River basin in Pakistan [2], the deltaic area of Myanmar [4], and Citarum watershed in Indonesia [3]. Therefore, using the RRI model is applicable to identifying the impact of land use change on flood risk.

The study on the Kedungsepur national strategic area development effect on flood risk remains less evidence. The study that has been published mainly focused on the social and economic aspects. Therefore, this study aims to identify flood characteristics in the Jratunseluna river basin caused by the alteration of land use due to Kedungsepur national strategic area development through rainfall-runoff 2D simulation. Moreover, the study can provide appropriate flood mitigation in the future.

2. Study Area

Jratunseluna river basin covers an urban area that consists of eleven regencies and two cities, as shown in Fig.1. Jratunseluna river basin is adjacent to five other river basins that consist of Bodri Kuto in the west, Progo Opak Serang in the southwest, Serayu Bogowonto in the south, Bengawan Solo in the southeast to the east, and Wiso Gelis in the north. The focus of this study area is validated at Garang, Babon, Dolok, and Tuntang watersheds as there are affected by Kedungsepur National Strategic Development.

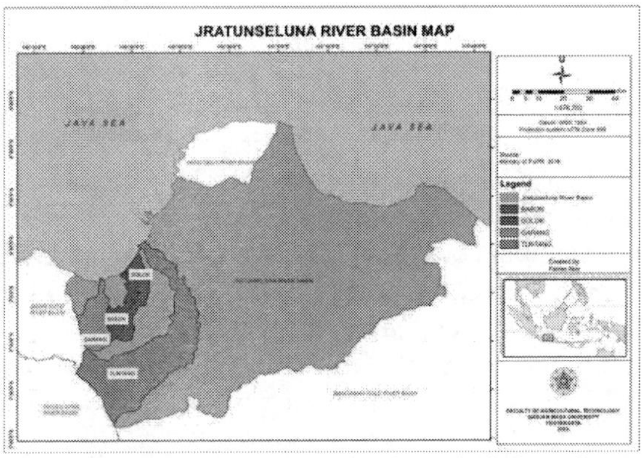

Figure 1. Jratunseluna River Basin Administration Map

The Kedungsepur national strategic area, as the abbreviation of the region included in the strategic plan, i.e., Kendal-Demak-Ungaran-Salatiga-Semarang-Purwodadi, is in Central Java Province (Fig. 2). The site is situated between 109°10' – 111°25' east longitude and 6°43'26" – 7°32' latitude. The total area of the Kedungsepur area is 5,256.53 km^2 or about 16.25% of the total area of Central Java Province.

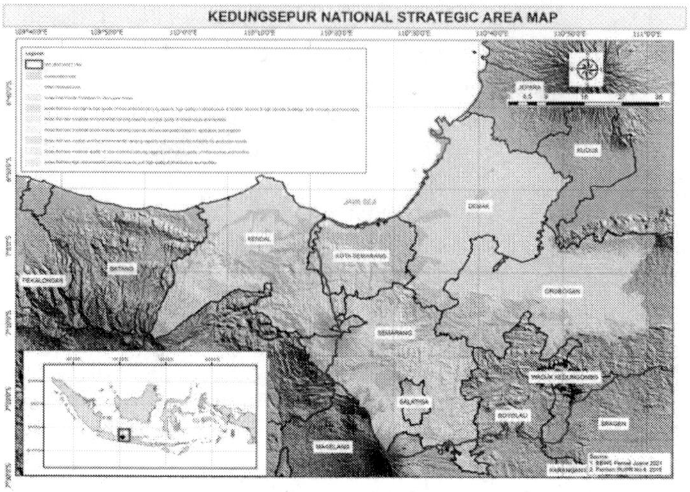

Figure 2. Kedungsepur National Strategic Area Map
(Source:PUSDATARU Provinsi Jawa Tengah, 2021)

3. Method

A data quality test was carried out for rain data and measured discharge data. This test is conducted since the possibility of data being corrupted due to several factors. The data quality test carried out is the persistence test and the outlier test. The persistence test equation is displayed on eq. 1 and eq. 2, y_H is the upper limit, and y_L is a lower limit. Meanwhile, the outlier test equation is shown on eq. 3 and eq. 4. For eq. 1 and eq. 2, For eq. 3 and eq. 4, KS is Spearman correlation coefficient, m is the number of data, di is the difference between rank to X_i and X_{i-1}, and t is t test count value.

$$y_H = \bar{y} + K_n . s_y \tag{1}$$

$$y_L = \bar{y} + K_n . s_y \tag{2}$$

$$KS = 1 - \frac{6 \sum_{i=1}^{m}(di)^2}{m^3 - m} \tag{3}$$

$$t = KS \left[\frac{m-2}{1-KS^2}\right]^{0.5} \tag{4}$$

The rainfall-runoff simulation is conducted by using the RRI model. The RRI model is a two-dimensional (2D) distributed-based hydrological model [2]. The model is applicable to capture the natural mechanism of rainfall-runoff transformation through a watershed into river discharge, river depth, and slope water depth simultaneously [4]. There are three outputs from the RRI model: the slope water depth (m), the river depth (m), and the river discharge (m^3/s). In this case, rainfall intensity and distribution are the most crucial input parameters for the RRI model.

All the datasets used in this study are listed in Table 1. The rainfall data used is from 2010 to 2020. The daily rainfall data from meteorological stations are often unavailable through the years. Thus, daily precipitation data from the Global Precipitation Measurement satellite is used. Using satellite data, the model parameter for flood simulation was calibrated with the observed data. The calibrated model parameters were then used for simulation at 2, 5, 10, 20, and 50-year floods. The topography data used for RRI input consisted of a digital elevation model, accumulation flow, and direction flow map obtained from HydroSHEDS USGS. A soil map is used to get soil texture data for RRI input.

Two simulation conditions are used in this case. The first simulation uses existing land cover; in the second simulation, Kedungsepur land cover data is used. Land cover is classified into seven classes: water body, cropland, plantation area, forest, swamp, building, and bare land. The output simulation with current land use and Kedungsepur land use are compared to capture the effect of land use change.

Table 1. Datasets input for the RRI model

Dataset	Date	Sources	Notes
Daily Precipitation	2010 - 2020	GPM (Global Precipitation Measurement) satellite	Resolution: 10 km (approximately)
Daily Precipitation	2019	Central River Region Pemali Juana	35 rain gauge station
Observed discharge	2019	Central River Region Pemali Juana	Serang-Muncar river
Digital Elevation Model, Accumulation Flow, and Direction Flow		HydroSHEDS USGS	Resolution : 0.5 km
Soil Map	2007	Food and Agriculture Organization	Digital Soil Map Of The World resolution: 50 km
Land Cover Map	2019	Ministry of Environment and Forestry Indonesia	Resolution : 2.5 km
Kedungsepur land-use map	2021	Central River Region Pemali Juana	

The simulation results are compared based on the flood hydrograph characteristics (Fig. 3). A flood hydrograph has several components such as peak flow, inflection point, recession limb, and time to peak [7]. In this study, the parameter compared are peak flow, time to peak, and recession limb, with the flood simulation duration set to 24 hours.

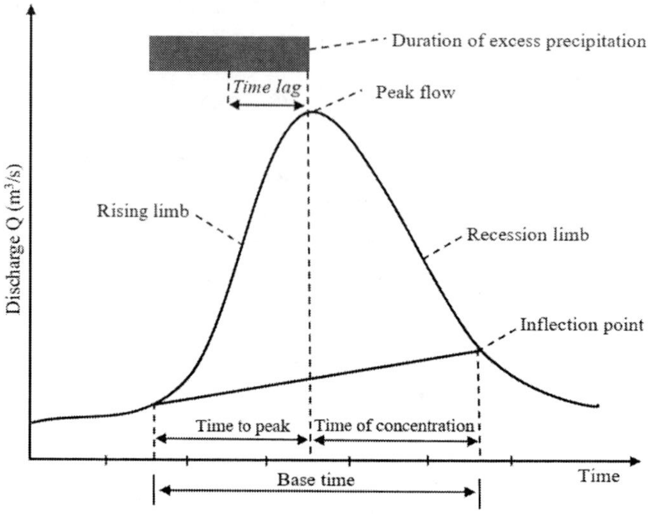

Figure 3. Typical flood hydrograph

(redrawn)

4. Results and Discussion

Quality data test for rainfall is needed to determine whether the data can be used for further analysis or not. The quality data test was applied to the observed data. The qualified observed data is then used for calibration. Table 2 shows the appropriateness of the observed discharge and rainfall data used in the analysis. It is known that there are several data identified as outlier data. Outlier data will not be used for further analysis. The persistence test result shows that discharge and rainfall depend on each other, meaning the trend of rainfall and discharge data should be equivalent.

Table 2. Data quality test

No	Data	Persistence Test	Outlier Test
1	Observed discharge	Dependent	Outlier
2	Observed rainfall	Dependent	Outlier

The distributed-based hydrological model consists of watershed parameters that influence the rainfall with the output discharge connecting the river and the grid slope. The calibrated parameters used in the RRI model consist of land use, manning's roughness on the slope (ns_slope), and saturated subsurface flow hydraulic conductivity (ka). Each parameter is used with different combinations and scenarios to determine the best-fit runoff prediction. The detail of the parameters used is presented in Table 3. The land cover map of 2019 was used for flood simulation using the RRI model in the existing condition, and the observed discharge data were obtained from the Central River Region Pemali Juana, Ministry of Public Works and Public Housing. Table 4 shows the comparison among the calibrated parameters.

The comparison of the calculated discharge using the RRI model at Serang-Muncar river with the observed discharge for 1 March 2019 to 30 April 2019 flood events is used to calibrate the RRI parameters as shown in Fig. 4. The best parameter to estimate the observed data based on the two statistical indicators, which consist of root mean squared error (RMSE) and correlation factor (R), is obtained by the 5th scenario. In this case, the results of the calculated discharge at the Serang-Muncar river could resemble the observed data even though the best-fit condition was not reached. Thus, the model parameters can be used for flood simulation at 2, 5, 10, 20, and 50-year-flood return periods.

Table 3. Parameter used in the simulation

Parameter	Classification of landuse						
	Waterbody	Cropland	Plantation area	Forest	Swamp	Building	Bare land
ns slope	0.035	0.3228	0.3228	0.32	0.035	0.0404	0.0113
soil depth	1	1	1	1	1	1	1
gamma	0.437	0.398	0.398	0.464	0.437	0.476	0.453
kV	0	0	0	0	0	0	0
sf	0.0495	0.2185	0.2185	0.2088	0.0495	0.3163	0.1101
ka	0	0.1	0.1	0.1	0	0	0
beta	8.d0	8.d0	8.d0	8.d0	8.d0	8.d0	8.d0

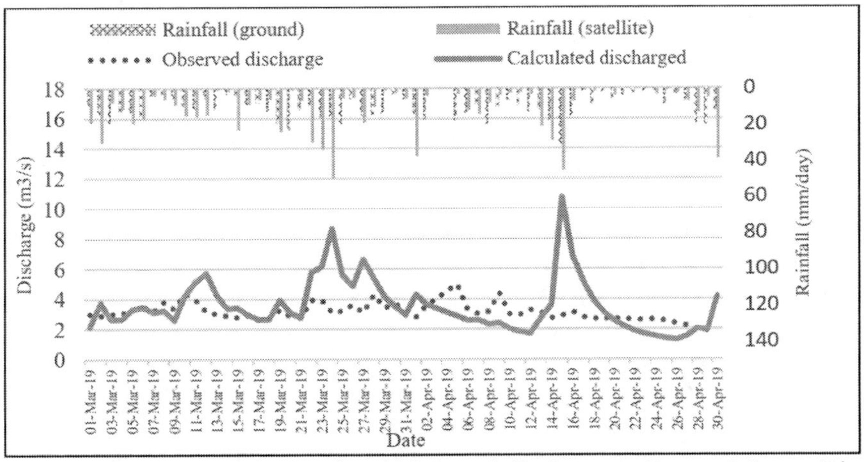

Figure 4. Comparison of the calculated and observed discharge at Serang-Muncar River from 1 March 2019 to 30 April 2019.

Table 4. Comparison of the calibrated parameter

Scenario	Parameter			RMSE	R
	Landuse	ns_slope	ka		
1	1	0.035	0-0.1	10.68	0.317
2	1	0.1	0-0.1	1.75	0.275
3	7	0.1	0-0.1	2.14	-0.210
4	7	0.1	0	1.19	0.244
5	7	0.035-0.3228	0-0.1	1.75	0.275
6	7	0.08	0	1.19	0.252

The maximum 24-hour rainfall from the 2010-2020 flood ranged from around 46.34-302.20, with an average of 126.49 mm. The maximum 24-hour rainfall for a 2-year flood, 5-year flood, 10-year flood, 20-year flood, and 50-year return period to capture an extreme rainfall event was calculated based on the Normal, Log-Normal, Log-Pearson III, and Gumbel distributions was about 122.33 mm, 154.73 mm, 174.88 mm, 193.64 mm, 217.60 mm, 235.62 mm, respectively.

Fig. 5 and 6 show the condition of flood in the case of 2-year flood, 5-year flood, 10-year flood, 20-year flood, and 50-year return periods under current and Kedungsepur land cover conditions, respectively. The inundation maps represent slope water depth concentrated during the flood return period. An increase in excessive return period shows that flood inundation is increasing in the southern area. Moreover, the comparison between existing and Kedungsepur land cover shows that the inundated areas spread more spacious and followed by the increase of inundation depth. The flood-risk increase may be due to the expansion of building areas. The excessive development of the urban regions dominated by buildings and roads may reduce infiltration rates due to an impervious area increase [6].

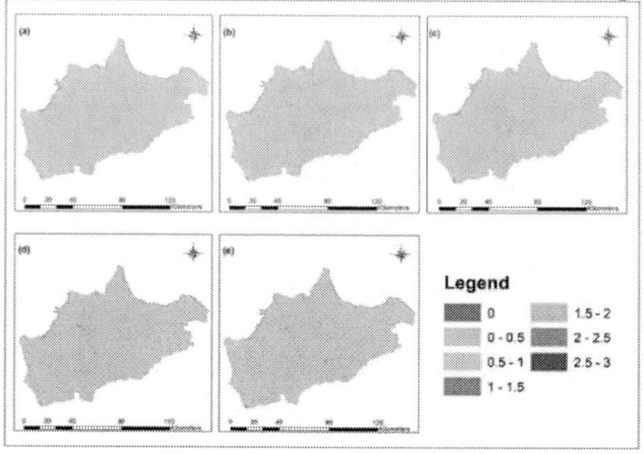

Figure 5. Inundation map under the existing condition (a) 2-year flood (b) 5-year flood (c) 10-year flood (d) 20-year flood (e) 50-year flood

Fig. 5 show the inundation map in the Jratunseluna river basin under existing condition. The map shows that the inundation mainly occurs in the center of the Jratunseluna river basin. The maximum inundation height is 2.78 meters which may happen in a 50-year flood. The average inundation height in all cases of year-flood is 0.5-1.5 meters.

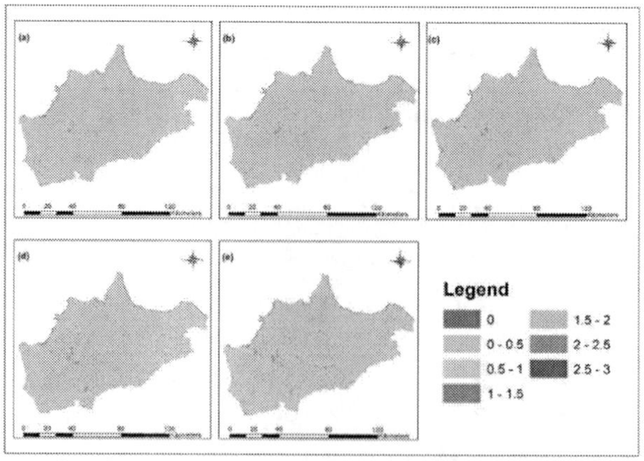

Figure 6. Inundation map under the Kedungsepur national strategic area development (a) 2-year flood (b) 5-year flood (c) 10-year flood (d) 20-year flood (e) 50-year flood

The results of flood hydrograph at different watersheds correspond to the Kedungsepur coverage area, namely Garang, Babon, Dolok, and Tuntang watersheds, to sample the difference in flood responses. The results show that the flood discharge may increase in the future due to the development of the Kedungsepur national strategic area. Land use changes can be associated with an increase in peak flow and a decrease in travel time. Moreover, an increase in the building and cropland areas decreases has elevated the flood inundation depth and inundated areas.

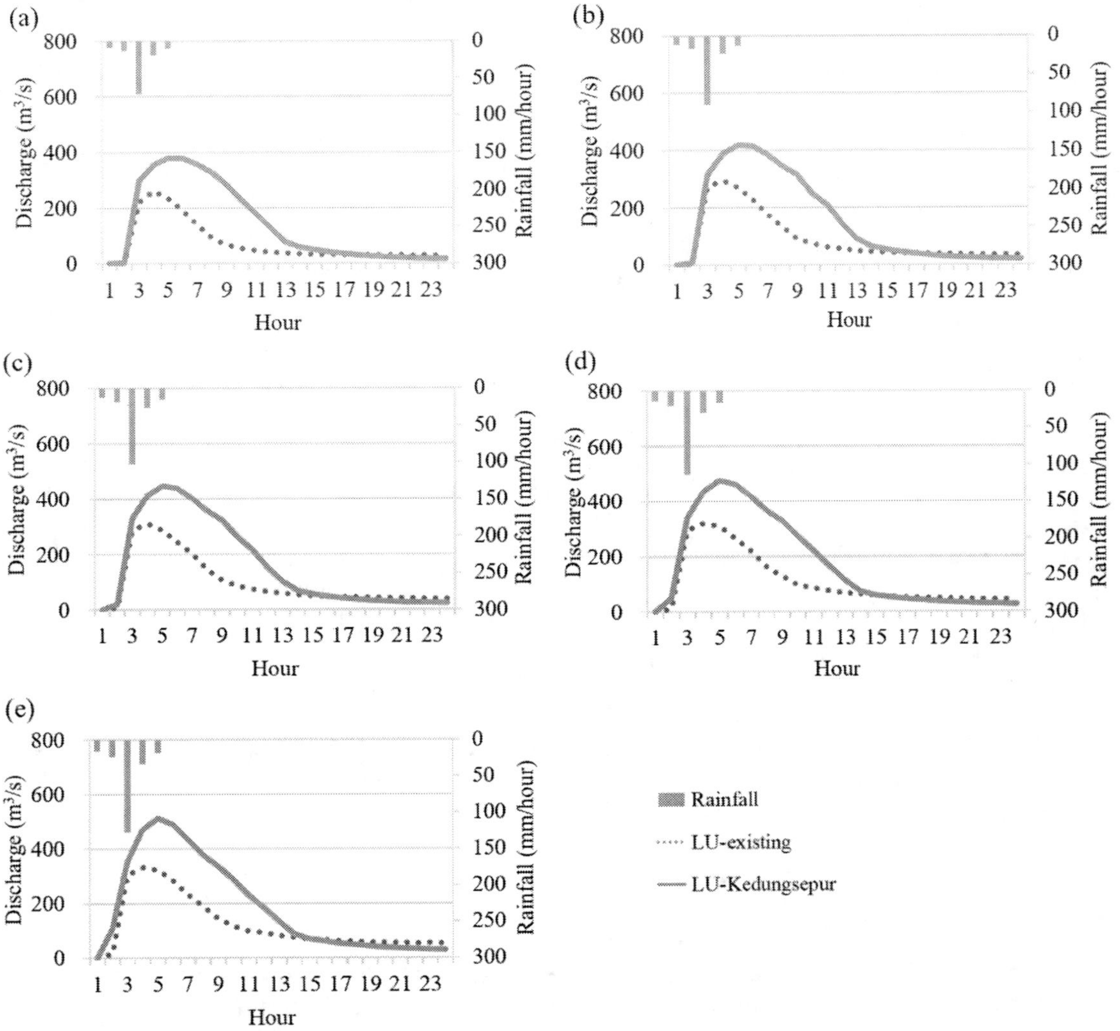

Figure 7. Comparison of the calculated flood discharge with different land cover at the Garang watershed for various flood return periods (a) 2-year flood (b) 5-year flood (c) 10-year flood (d) 20-year flood (e) 50-year flood.

The results between flood simulation with existing land cover and Kedungsepur at the Garang watershed, as shown in Fig. 7, indicate a significant increase in peak discharge and flood duration but a slower time to peak obtained. The peak discharge increased by 117.74 m^3/s, 124.28 m^3/s, 137.66 m^3/s, 156.09 m^3/s, and 180.20 m^3/s for 2-year flood, 5-year flood, 10-year flood, 20-year flood, and 50-year flood, respectively. The former peak time reaches the 3rd hour while under development of national strategic area reaches the 4th hour.

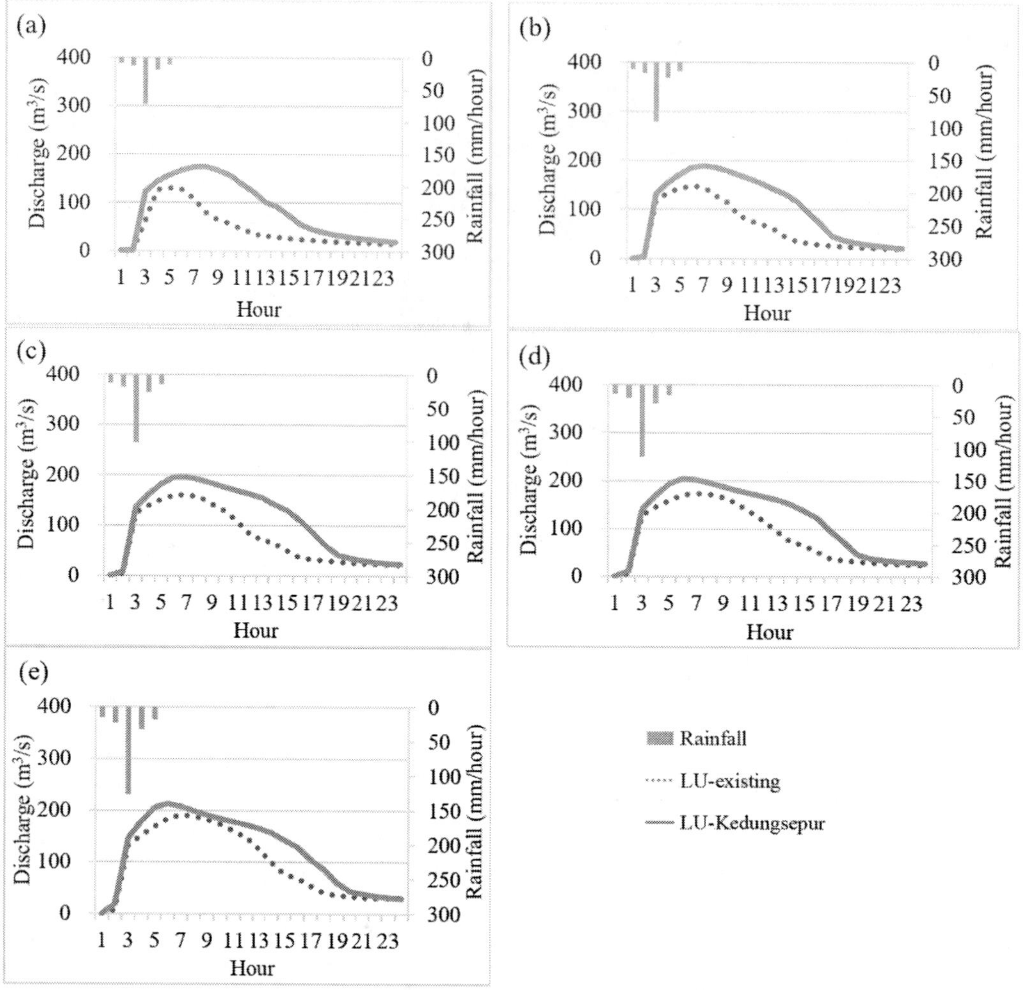

Figure 8. Comparison of the calculated flow discharge with different land cover at the Babon watershed for various flood return periods (a) 2-year flood (b) 5-year flood (c) 10-year flood (d) 20-year flood (e) 50-year flood.

The results of the flood hydrograph at the Babon watershed, as shown in Fig. 8, indicate a significant increase in peak discharge and time to peak. The peak discharges increased by 44.13 m^3/s, 41.18 m^3/s, 35.72 m^3/s, 30.58 m^3/s, and 21.72 m^3/s for 2-year flood, 5-year flood, 10-year flood, 20-year flood, and 50-year flood, respectively. The former peak time reaches the 7th hour while under the development of a national strategic area reaches the 6th hour.

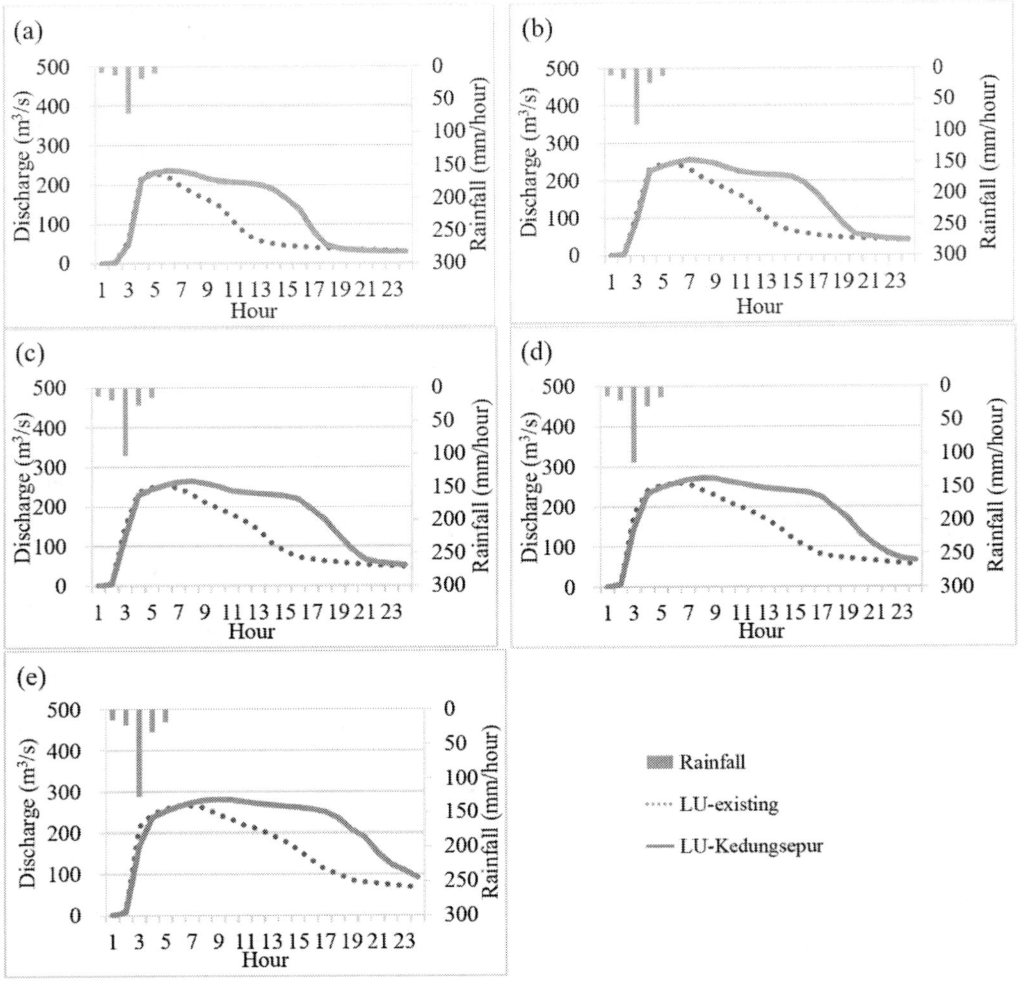

Figure 9. Comparison of the calculated flow discharge with different land cover at the Dolok watershed for various flood return periods (a) 2-year flood (b) 5-year flood (c) 10-year flood (d) 20-year flood (e) 50-year flood

The results between flood simulation with existing land cover and Kedungsepur at the Dolok watershed, as shown in Fig. 9, indicate a significant increase in flood duration but a less significant increase in peak discharge and time to peak obtained. The peak discharge increased by 6.39 m^3/s, 10.38 m^3/s, 11.57 m^3/s, 12.32 m^3/s, and 15.16 m^3/s for 2-year flood, 5-year flood, 10-year flood, 20-year flood, and 50-year flood, respectively. The peak time reaches the 6th hour. Moreover, the recession limb of the existing condition is steeper than that under development of a national strategic area. This condition indicates that flood duration will be much longer than the former condition, which causes a higher potential loss due to flood.

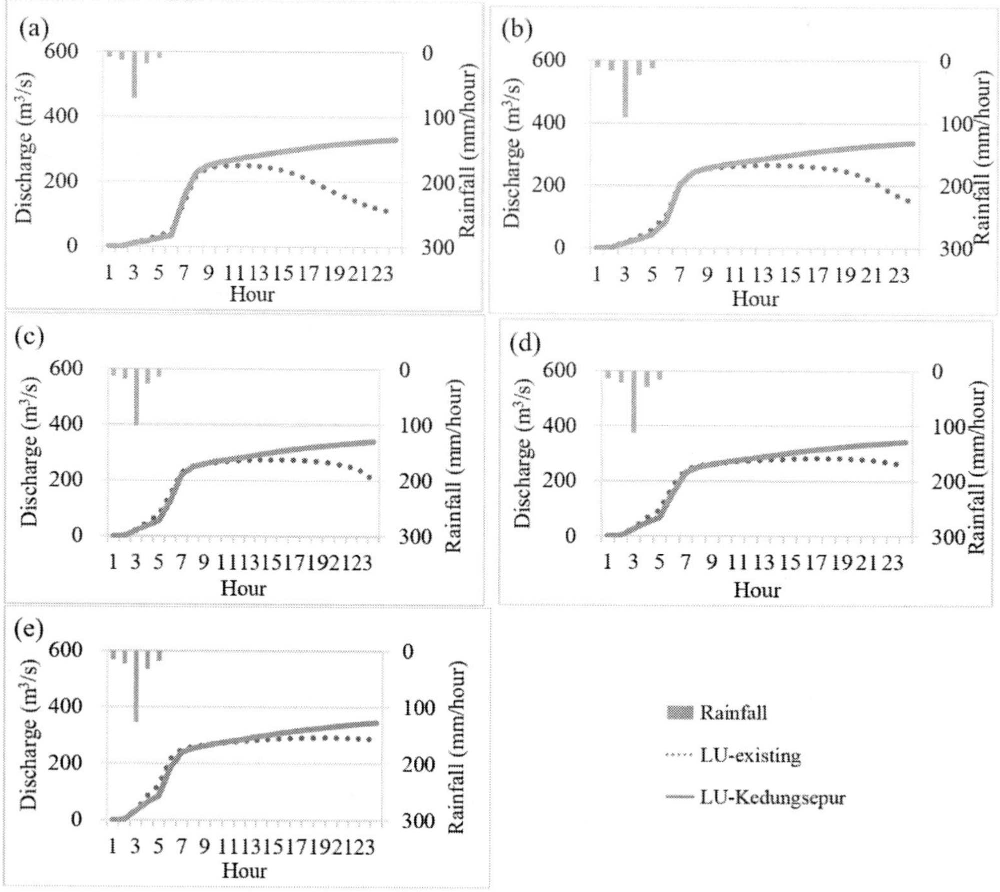

Figure 10. Comparison of the calculated flow discharge with different land cover at the Tuntang watershed for various flood return periods (a) 2-year flood (b) 5-year flood (c) 10-year flood (d) 20-year flood (e) 50-year flood

The results between flood simulation with existing land cover and Kedungsepur at the Tuntang watershed, as shown in Fig. 10, indicate a significant increase in flood duration but a less significant increase in peak discharge and time to peak obtained. The peak discharge increased by 80.15 m^3/s, 71.06 m^3/s, 65.28 m^3/s, 59.71 m^3/s, and 53.22 m^3/s for 2-year flood, 5-year flood, 10-year flood, 20-year flood, and 50-year flood, respectively. The former peak time reaches the 6th hour, while under development of a national strategic area is later than that. There is a significant difference in peak discharge, time to peak, and recession limb in the case of a 2-year flood but less in the case of a 50-year flood. This condition is similar to the Dolok watershed. It indicates that flood duration will be much longer than the former condition, which causes a higher potential loss due to flood.

5. Conclusion

The flood risk level would increase during the implementation of the Kedungsepur national strategic area development, which makes a higher potential loss in the urban area. The increase in flood risk is mainly indicated by the flood duration and peak discharge increase. The flood hydrograph obtained at the Dolok and Tuntang watersheds shows the most vulnerable area to flood risk characterized by a longer flood duration and a higher peak discharge. However, the most significant increase in peak discharge among the four outlets is obtained in the Garang watershed. In the case of a 2-year flood, the peak discharge increased by 44.91% from 262.18 m^3/s to 379.92 m^3/s and for the case of a 50-year flood, the peak discharge increased by 54.34% from 331.63 m^3/s to 511.83 m^3/s.

Acknowledgment

Our special thanks are presented to the Laboratory of Land and Water Resource Engineering, Department of Agricultural and Biosystem Engineering, Faculty of Agricultural Technology, Universitas Gadjah Mada for facilitating this research. We also appreciate the collaborative work with the Central River Region Pemali Juana to make available the necessary data for analysis. Last but not least, the authors are grateful to Universitas Gadjah Mada for the support of funds to finish this research through the 2022 Thesis Recognition Grant (RTA) with a Letter of Assignment Number is 3550/UN1.P.III/Dit-Lit/PT.01.05/2022.

References

[1] W. Handayani, U.E. Chigbu, I. Rudiarto, I.H.S. Putri, Urbanization and Increasing Flood Risk in the Northern Coast of Central Java-Indonesia: An Assessment towards Better Land Use Policy and Flood Management, Land 9 (2020)

[2] T. Sayama, Ozawa, G, T. Kawakami, S. Nabesaka, K. Fukami. Rainfall-runoff-inundation analysis of the 2010 Pakistan flood in the Kabul River basin. Hydrological Science 57, 298–312 (2012)

[3] K.D. Nastiti, K. Yeonsu, J. Kwansue, and A. Hyunuk. The Application of Rainfall-Runoff-inundation (RRI) Model for Inundation Case in Upper Citarum Watershed, West Java-Indonesia. Procedia Engineering 125, 166-172 (2015)

[4] S.S. Bhagabati and A. Kawasaki, Consideration of the rainfall-runoff-inundation (RRI) model for flood mapping in deltaic area of Myanmar. Hydrological Research Letter 3, 155-160 (2017)

[5] T.A. Kimaro, Y. Tachikawa, and K. Takara, Distributed hydrologic simulations to analyze the impacts of land use changes on flood characteristics in the Yasu River basin in Japan. Japan *Natural Disaster Science* 27, 85-94 (2005)

[6] B.B. Shresta, Approach for Analysis of Land-Cover Changes and Their Impact on Flooding Regime. Quartenary 27 (2019)

[7] P. Ganguli and M.J. Reddy (2012) "Probabilistic assessment of flood risks using trivariate copulas," in *Theoritical and Applied Climatology*. DOI: 10.1007/s00704-012-0664-4

IOP Conf. Series: Earth and Environmental Science 1233 (2023) 012059 doi:10.1088/1755-1315/1233/1/012059

GIS Based Monitoring of Critical Area Upstream Watershed of PB. Soedirman Reservoir, Central Java, Indonesia

A. R. Khidzir[1], **C. Setyawan**[1*], **S. Susanto**[1], **M. K. Zaki**[1]

[1]Department of Agricultural and Biosystems Engineering, Faculty of Agricultural Technology, Universitas Gadjah Mada, Indonesia.

*Corresponding author: chandra.tep@ugm.ac.id

Abstract. Upstream watershed has an important role as a conservation area to protect environmental functions for life. The damage of upstream watersheds causes various environmental problems such as soil erosion, landslides, river sedimentation, flooding, and others. Monitoring of upstream watersheds is important to control potential environmental damage. This study used a geographic information system (GIS) for monitoring the critical area upstream of the study site. The aim of this study is to identify critical areas in the watershed by using GIS. Six indicators presented in raster maps such as annual rainfall, soil type, land use type, land slope, population density, and water quality were used to determine the critical area. These indicators were then quantitatively analyzed by using spatial analysis tools of Arc GIS 10.6. Five classes of the critical category were used in this study namely very low, low, moderate, high, and very high. The result showed that 61.74% of the study site was covered by a low class of critical areas. Meanwhile, 37.89% and 0.36% of the study site were covered by moderate and high class critical areas, respectively. Areas with moderate and high class critical areas were found in the eastern part of the study site. The findings of this study provide a reference for determining priority areas for conservation planning.

1. Introduction

In a watershed, there is an ecosystem consisting of various land uses, geomorphological forms, flora and fauna, physical buildings, humans, and their activities that compose each other in the unity of the ecosystem [1]. The damage to the watershed will also affect all sectors that depend on it, including agriculture. Currently, Indonesia has around 458 watersheds where 60 watersheds were in severely critical condition, 222 and 176 were in critical and potentially critical condition, respectively [2]. This condition has triggered various disasters downstream areas such as floods in the capital city of Jakarta due to the damage to the Ciliwung watershed. Material and economic losses due to natural disasters were very large [3]. Repairing environmental damage due to natural disasters takes a long time and is quite expensive. Prevention efforts are one of the best solutions to reduce various disasters, especially in watershed areas.

The impact of watershed damage can be prevented by maintaining the condition of the watershed. A modeling to determine the distribution of critical areas in the watershed is required as a basis for conservation planning. Studies on critical watershed assessments have been carried out, but mostly those previous studies still use a single parameter and have not clearly considered spatial aspects. In this study, critical areas were identified based on six indicator reflecting important aspects of the watershed system including socio-economic factor which was not considered in previous studies.

2. Method

This study was performed in the upstream watershed of Panglima Besar (PB) Reservoir (total area: 97,839 Ha) Upstream of Serayu River administratively located in Central Java Province of Indonesia (Figure 1). Six indicators were used for critical area determination such as annual rainfall, soil type, land use type, land slope, population density, and water quality. All indicators were presented as a raster map with 30-meter resolution (1: 250,000 scale) and obtained from various sources for the period of 2021. The critical level of watershed areas was determined based on the total score which was obtained from the sum of indicator values (map 1 + map 2 + map 3 + map 4 + map 5 + map 6) overlaid by using a raster calculator of Arc GIS.

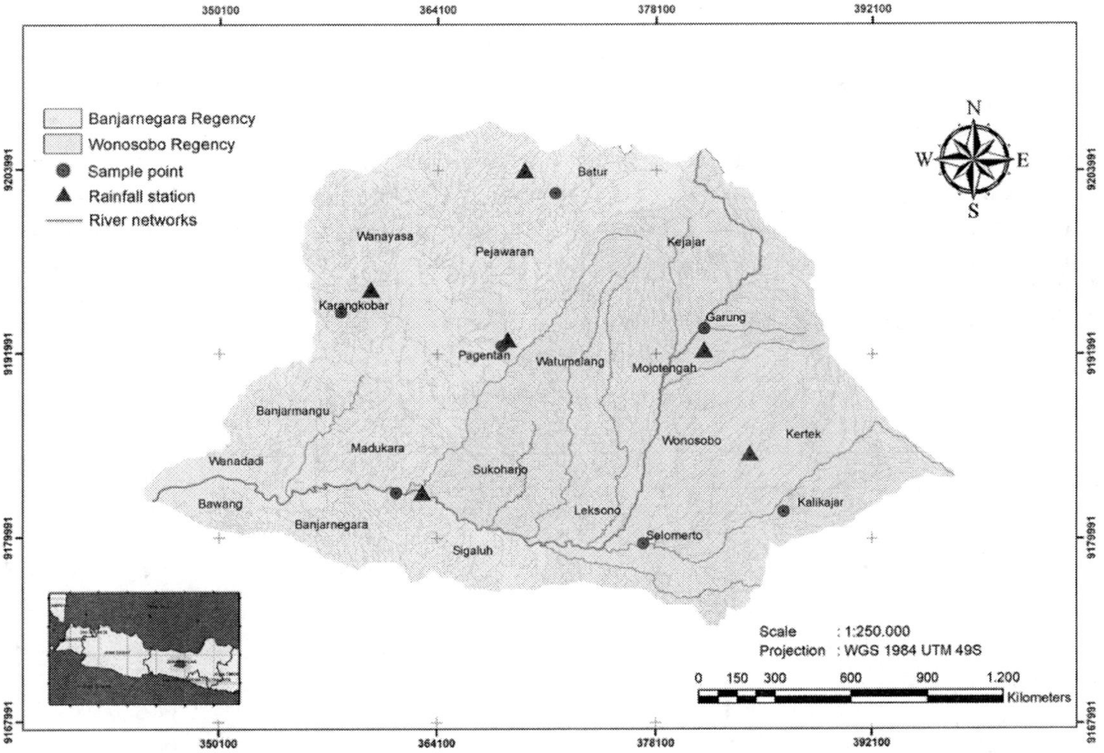

Figure 1. Map of Upstream Watershed of PB. Soedirman Reservoir

Refers to Setyawan et.al., 2019 [4], each indicator has three scores and categories namely good (score 1), moderate (score 3) and bad (score 5) as shown in Table 1.

Table 1. Category and Score for Indicator of Critical Area Assessment

Indicators	Good (score: 1)	Moderate (score: 3)	Bad (score: 5)
Annual Rainfall	> 2000 mm	1500-2000 mm	< 1500
Land Slope	0-15%	15-40%	>40%
Soil Type	Latosol	Lithosol/Andosol	Regosol/Grumusol
Land Use Type	Vegetation	Settlement	Farmland
Population Density	0-250 people/km^2	250-750 people/km^2	>750 people/km^2
Water Quality (COD)	<100 mg/l	100-200 mg/l	>200 mg/l

Annual rainfall data were collected from six stations in the study site (see Figure 1) for the period of 2021 and analyzed by using Inverse Distance Weighting (IDW) method in Arc GIS 10.3 for the resulting rainfall map. Land slope and land use maps were generated from Geospatial Information Agency (Badan

Informasi Geospasial) in 2021. Soil type was obtained from the main office of serayu opak river, the ministry of public works in 2021. The population density was obtained from Statistics Indonesia (Badan Pusat Statistik), while COD data were collected from seven sampling points in the study site in 2021. The total score of the indicator was then classified into seven classes of criticality levels (Table 2) using reclassify tools in Arc GIS.

Table 2. Critical Level of Watershed Area Based on Total Score

Total Score	Critical Level Category
0-6	Very Low
7-12	Low
13-18	Moderate
19-24	High
25-30	Very High

3. Results and Discussion

3.1. Annual Rainfall and Land Slope

Annual rainfall data were obtained from the Meteorological, Climatological, and Geophysical Agency of Indonesia or Badan Meteorologi, Klimatologi, dan Geofisika (BMKG) for six rainfall stations in the study site namely Sigaluh, Batur, Pagentan, Karangkobar, Garung, dan Kertek. The rainfall data record from those rainfall stations indicates an annual rainfall value of more than 2000 mm (Figure 2a), hence the category for annual rainfall indicator was good (score: 1). Rainfall data from six stations are shown in Table 3.

Table 3. Rainfall Data in Study Site (mm)

Month	Name of Station					
	Sigaluh	Batur	Pagentan	Karangkobar	Garung	Kertek
January	491	552	495	459	540	554
February	521	328	531	467	530	392
March	551	472	529	362	525	532
April	318	365	389	291	379	322
May	242	250	289	208	271	324
June	147	155	192	128	148	155
July	99	106	98	108	66	131
August	46	70	50	46	66	79
September	172	237	230	251	151	206
Oktober	207	333	361	401	241	289
November	393	507	522	567	491	563
December	517	561	581	549	583	607
Total	3704	4209	4266	3836	3993	4152

Rainfall is one of the determining factors for the rate of soil erosion, landslides, and surface runoff that causes watersheds to be damaged [5]. The difference in annual rainfall values from the six stations indicates a high variability of rain in Indonesia [6]. For the land slope indicator, the land slope map was generated from the *Digital Elevation Model* (DEM) and analyzed by using the spatial analysis tool of Arc GIS 10.3. The DEM was issued by Geospatial Information Agency or Badan Informasi Geospasial (BIG), Indonesia. Land slope distribution in the study site can be seen in Figure 2b. The study site was dominated by an area with a land slope between 0-15% (score: 1) covering about 79.25% of the total area. While areas with land slope between 15-40 % (score: 3) and > 40% (score 5) covered about 20.6%

and 0.16% of the total area. Areas with high slopes have a high risk of damage [7], hence this area should be used as a conservation area.

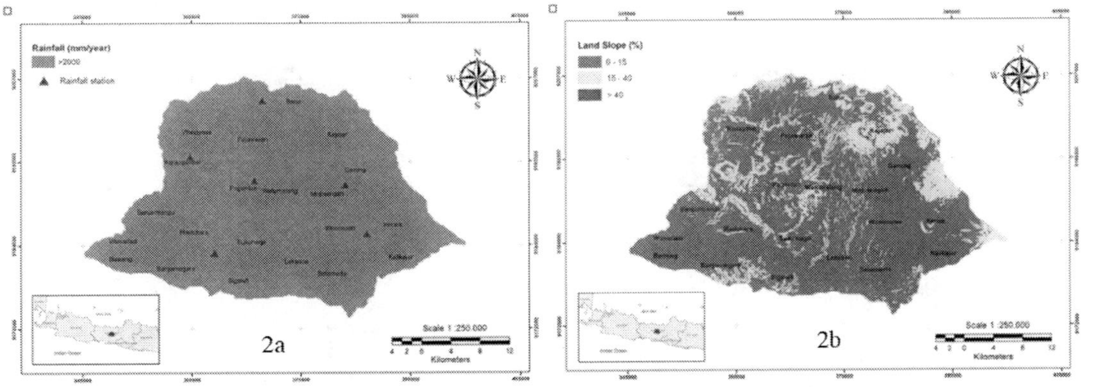

Figure 2. Map of Rainfall (a) and Land Slope (b) in the Study Site

3.2. Soil Type and Land Use

Soil type map was obtained from the Main Office of Serayu Opak River, the Ministry of Public Work Indonesia. Four types of soil (Figure 3b) were identified at the study site such as alluvial (score: 3), grumusol (score: 5), latosol (score: 1) and regosol (score: 1). Latosol covered about 87.59% of the total area, followed by grumusol (7.18% of total area), regosol (3.28% of total area) and alluvial (1,95% of total area). Erodible soil such as grumusol and alluvial will increase the criticality value of the watershed [8]. The land use map, it was obtained from BIG. The land use map indicated that the study site was dominated by farmland (score: 5) which covered about 80.21 % of the total area as shown in Figure 3b. In general, agricultural practices in upstream areas tend to cause a lot of environmental damage such as soil erosion [9,10].

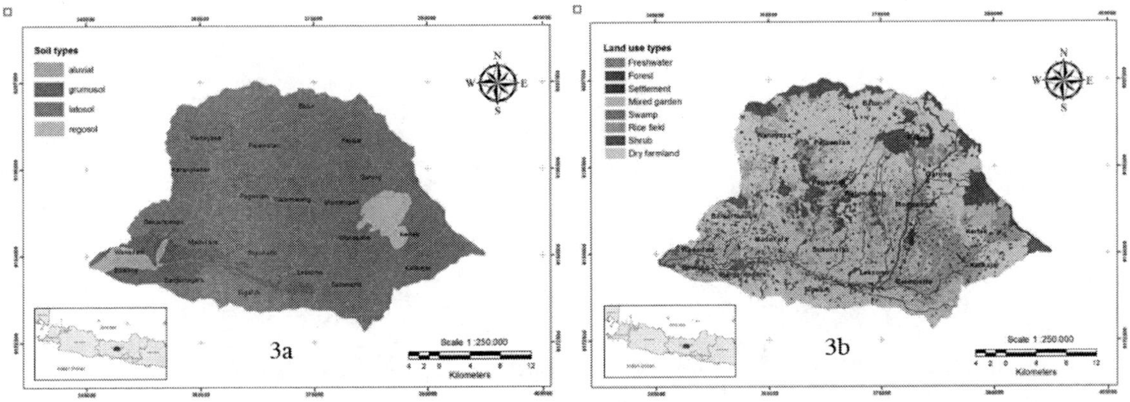

Figure 3. Soil Type (a) and Land Use Type (b) in the Study Site

3.3. Population Density and Water Quality

Population density map was obtained based on statistical data of the population in sub-districts. The study site was covered by areas with a population density between 250-750 people/km^2 (35.1% of the total area) and >750 people/km^2 (64.9% of the total area) as shown in Figure 4a. Population density >750 people/km^2 indicates a high risk of environmental damage, especially in developing countries [4]. A higher population number increases the potential for human activities that exploit nature and the environment. While for Chemical Oxygen Demand (COD), the map of COD (Figure 4b) was obtained based on water sample analysis from seven points in the study site. The result indicated that COD content was < 100 mg/l (score: 1).

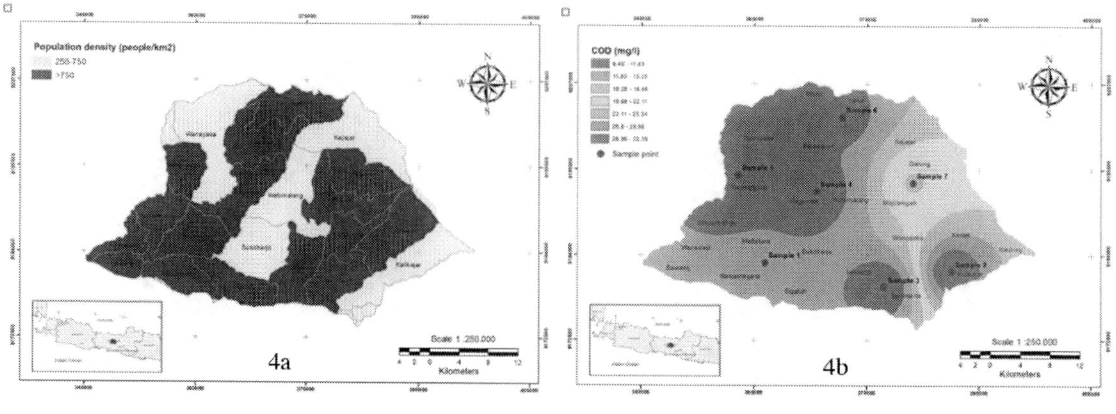

Figure 4. Population Density (a) and COD Content in the Study Site

3.4. Critical Area of Watershed

Spatial analysis by using scoring revealed three types of critical areas in the study site such as low (37.89% of the total area), moderate (61.74% of the total area) and high (0.36% of the total area) as shown in Figure 5. The area with a high category of the critical area was covered by farmland with high land slopes and high population density. Farming activities in high land slope areas were reported increasing the risk of environmental damage [11,12].

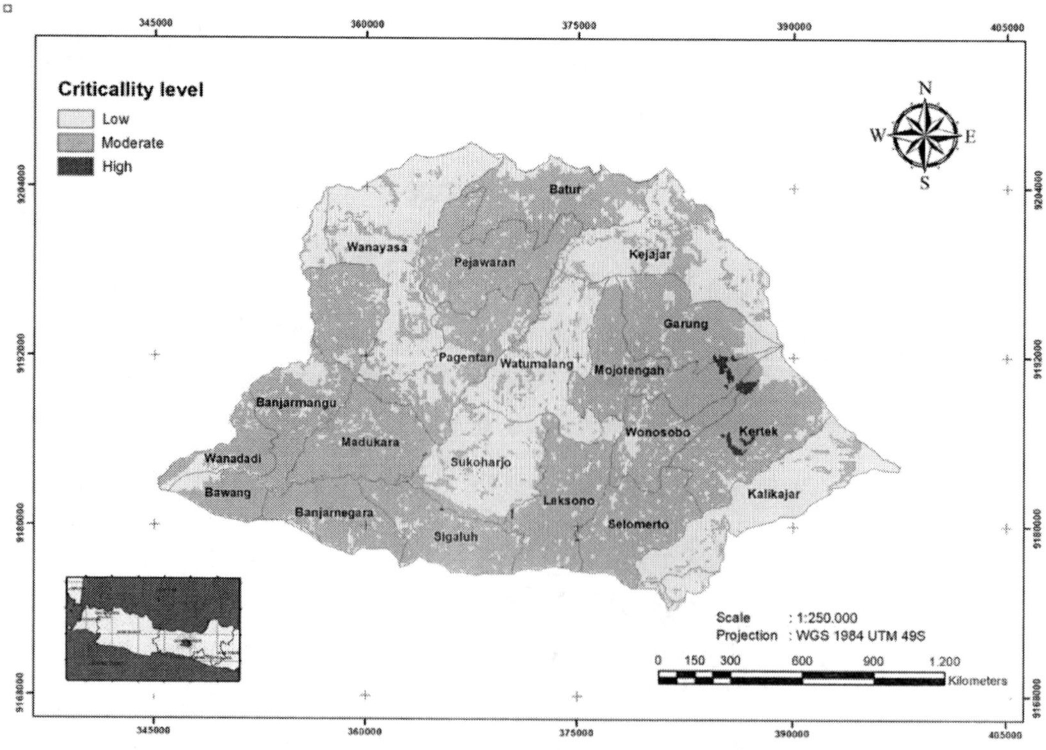

Figure 5. Distribution of Critical Area in Study Site

Even though it is only 0.36% of the total watershed area, the area with a high criticality potential can contribute or trigger a large-scale environmental problem such as soil erosion [13,14]. Hence, areas with a high critical level need to be a priority for conservation programs. Meanwhile, areas with moderate critical levels indicate potential environmental problems. Indicators that have a bad category value need special attention as a basis for problem solving. In this study, the type of land use in the form of dry farmland and the slope of the land contributed significantly to the bad score of those indicators. This study uses more than one indicator to determine the criticality level of the watershed so that it can detect several potential problems based on the indicator values obtained [13]. Some previous studies have only focused on assessing one indicator to determine problems in the watershed [15,16]. The use of GIS enables more specific spatial mapping of critical areas in watersheds. The use of more indicators can identify problems in the watershed more specifically.

4. Conclusions

GIS based monitoring revealed that the upstream watershed of PB. Soedirman Reservoir was dominated by an area with a moderate category of critical area. Areas with a high category of critical area were found in the site covered by farmland with high land slope and high population density particularly in the eastern part of the study site. The use of GIS enables more specific spatial mapping of critical areas in watersheds.

Acknowledgments

The authors would like to thank the Department of Agricultural and Biosystems Engineering Faculty of Agricultural Technology, Universitas Gadjah Mada which provided research facilities and the Directorate of Research Universitas Gadjah Mada which provided research funds through RTA Research Grant 2022 (letter of assignment number: 3550/UN1.P.III/Dit-Lit/PT.01.05/2022).

References

[1] Sodikin, A. 2012. Kinerja Daerah Aliran Sungai Berdasarkan Indikator Penggunaan Lahan pada DAS Padang Guci Bengkulu. *Naturalis – Jurnal Penelitian Pengelolaan Sumberdaya Alam dan Lingkungan* 1(2):105-112.

[2] Ishak, I.P., Asman, A.I., dan Ahmad, D.N.A. 2016. Pemanfaatan Teknologi Spasial dalam Pengelolaan Daerah Aliran Sungai (DAS) Binana Lumbua Kabupaten Jeneponto Sulawesi Selatan. *Jurnal Geomatika* 22(1):1-8.

[3] Sumadilaga, D.H. 2020. Bencana Alam Buat Indonesia Rugi Rp22,8 Triliun Setiap Tahun. URL: https://www.cnnindonesia.com/ekonomi/20200924134122- 92-550410/bencana-alam-buat-indonesia-rugi-rp228-triliun-setiap-tahun.

[4] Setyawan, C., Susanto, S., and Lee, C.Y. 2019. Spatial modelling of watershed health assessment by using GIS. *IOP Conf. Series: Earth and Environmental Science* 355.

[5] Triastuti, A. 2017. *Analisis Kekritisan Lahan di Sub DAS Samin dengan Pemanfaatan Sistem Informasi Geografi*. Skripsi. Geografi. Universitas Muhammadiyah Surakarta. Surakarta.

[6] Lee, H.S. 2015. General rainfall patterns in Indonesia and the potential impacts of local season rainfall intensity. *Water* 7: 1751-1768. http://dx.doi.org/10.3390/w7041751.

[7] Ochoa, P.A., Fries, A., Mejía, D., Burneo, J.I., Ruíz-Sinoga, J.D., Cerdà, A. 2016. Effects of climate, land cover and topography on soil erosion risk in a semiarid basin of the Andes. *Catena* 140: 31-42. https://doi.org/10.1016/j.catena.2016.01.011.

[8] Andriyanto, C., Sudarto, Suprayogo, D. 2015. Estimation of soil erosion for a sustainable land use planning: RUSLE model validation by remote sensing data utilization in the Kalikonto watershed. J of Degraded and Min Lands Manag 3 (1): 459- 468. http://dx.doi.org/ 10.15243/jdmlm.2015.031.459.

[9] Borrelli, P., Paustian, K., Panagos, P., Jones, A., Schütt, B., Lugato, E. 2016. Effect of good

IOP Conf. Series: Earth and Environmental Science 1233 (2023) 012059 doi:10.1088/1755-1315/1233/1/012059

agricultural and environmental conditions on erosion and soil organic carbon balance: A national case study. Land Use Policy 50: 408-421. http://dx.doi.org/10.1016/j.landusepol.2015.09.033.

[10] Seika, A.N., Setyawan, C., Ngadisih, Tirtalistyani, R. 2021. Soil Erosion Mapping using GIS Based Model in Agricultural Area of Progo Watershed, Central Java, Indonesia. IOP Conf. Series: Earth and Environmental Science 686. https://doi.org /10.1088/1755-1315/686/1/012024

[11] Mondal, A., Khare, D., Kundu, S. 2016. Impact assessment of climate change on future soil erosion and SOC loss. Nat Hazards 82: 1515. https://doi.org/10.1007/s11069-016-2255-7.

[12] Setyawan, C., Susanto, S., Lee, C.Y. 2019. Land and water conservation practices in tropical agricultural watershed. Journal of Degraded and Mining Lands Management 6 (2),1697-1704.

[13] Setyawan, C., Susanto, S., dan Lee, C.Y. 2019. Spatial Modelling of Watershed Health Assessment by Using GIS. *IOP Conf. Series and Enviromental Science, The 3rd International Symposium on Agricultural and Biosystem Engineering*. 6-8 Agustus 2019, Sulawesi Selatan, Indonesia. Vol 355.

[14] Yudhistira, A., Setyawan, C., Ngadisih, Tirtalistyani, R. 2021. IOP Conf. Series: Earth and Environmental Science 686. https://doi.org /10.1088/1755-1315/686/1/012033.

[15] Gebretsadik, Z.M. 2014. Watershed degradation and the growing risk of erosion in Hawassa-Zuria District, Southern Ethiopia. *J. Flood Risk Manag*. 7, 118-127.

[16] Thakkar, A.K., Desai, V.R., Patel, A., Potdar, M.B. 2017. Impact assessment of watershed management programmes on land use/land cover dynamics using remote sensing and GIS. *Remote Sens. Appl.: Soc. and Environ*. 5, 1-15.

IOP Conf. Series: Earth and Environmental Science 1233 (2023) 012060 doi:10.1088/1755-1315/1233/1/012060

Groundwater Conditions and Springs Potency for Domestic Requirement in Klaten and Banggai Kepulauan Regency, Indonesia

Krisman Tuyu

Environmental Science Master Programme, Universitas Gadjah Mada, Indonesia

Email : Krismantuyu@gmail.ugm.ac.id

Abstract. At present, sustainable management of water resources has become a major issue for the world community. In water management efforts, potency assessment becomes an important thing. The purpose of this study is to analyze the availability of ground water and water from springs both qualitatively and quantitatively and evaluate their potency for domestic . To determine the groundwater conditions in the research area, a qualitative descriptive analysis was carried out based on geological and hydrogeological data. To determine the potency of the springs, a quantitative descriptive analysis was carried out with the help of tables and diagrams. The potency for springs is calculated based on the discharge data of all springs contained in each district in relation to water for domestic purposes. The results showed that Kebonarum District has very good groundwater availability because it is located in an area of unconsolidated rock between mountains. In South Tinangkung District, groundwater conditions are not good. Groundwater is generally found in cracks or rock fractures. From the calculation results, it is known that the total spring discharge in Kebonarum District is 9,870,768 m^3/year, while the spring discharge in South Tinangkung District is 64,144,224 m^3/year. With the water in Kebonarum District is 915,858 m^3/year and in South Tinangkung Sub District is 355,393 m^3/year, the potency of water from springs in these two districts is sufficient to be used as a water source for domestic.

Keywords: groundwater, springs, Klaten, Banggai Kepulauan

1. Introduction

Water resources are one part of natural resources whose current existence is increasingly worrying, even though water is the primary source of life needed for all human activities such as domestic, agricultural and industrial requirement [1 ; 2]. To anticipate more severe consequences, various efforts have now been made to manage this resource so that its existence remains sustainable with an always sufficient quantity and quality that meets quality standards [3].

In the management of water resources, determining the presence of water is something that must be done first, and the presence of groundwater is no exception. The existence of water on earth follows a process called the hydrologic cycle [4 ; 5 ; 6]. Disruption of this process will result in various problems in the distribution of water on earth.

Groundwater is part of the most important water resources [7]. Groundwater is part of water resources whose existence is also strongly influenced by this cycle, because the main source of groundwater is from rainfall that falls on the earth.

In addition to being influenced by the size of the rainfall, the presence of groundwater is also very dependent on the presence of rock layers that can accommodate, store and drain water called aquifers. An area composed of sand and gravel material has great potency to contain groundwater. Another aspect that also plays an important role in relation to the presence of groundwater is vegetation and slopes. An area with dense vegetation and with a flat topography, has the potency to be able to absorb rainwater that falls on its surface as groundwater input.

Springs are one of the most important types of water resources, especially for drinking water and irrigation. In fact, most or all of the bottled drinking water we consume comes from springs. Therefore, this resource inventory becomes important for the development of its use in the future.

Springs can be said to be everywhere and appear in various ways. However, the distribution is uneven, depending on things that affect the availability of springs. A symptom of the emergence of specific springs can be seen in strato volcanoes. At certain heights, spring belts are found, which are related to their orohydrological properties. In addition, this path of appearance is also related to slope changes caused by changes in the constituent material.

At present, sustainable management of water resources (including groundwater and springs) has become a major issue for the world community [8]. In water management efforts, potency assessment becomes an important thing [9]. This should also be done in the Districts of Kebonarum and South Tinangkung. These two sub-districts are two quite important sub-districts in Klaten Regency and Banggai Islands Regency. Based on this background, the objectives of this study are (1) to analyze the availability of ground water and spring water both qualitatively and quantitatively and (2) to evaluate its potency for domestic .

2. Research Methods

To determine the groundwater conditions in the research area, a qualitative descriptive analysis was carried out based on geological and hydrogeological data, while to determine the potency of the springs a quantitative descriptive analysis was carried out with the help of tables and diagrams. From the two analyzes, an integrated analysis of the potency for groundwater and springs can be carried out in the research area.

The potency for springs is calculated based on the discharge data of all springs contained in each sub-district, in relation to water requirement for domestic purposes. The population will determine the amount of domestic water requirement.

In this study, the use of water that is taken into account is the use of water for domestic purposes because the sector uses springs. The use of water for domestic purposes is determined according to the size of the population and the amount of water needed per person per day. According to Mangku Sitepoe [10], water requirement in big cities are generally >150 liters/capita/day, in medium cities 80-150 liters/capita/day, sub-district cities 60-80 liters/capita/ day and villages ranged from 30-60 liters/capita/day. Based on this criterion, because Kebonarum and South Tinangkung are medium-sized cities, the water usage is determined to be between 120 liters/person/day.

3. Results and Discussion

3.1. Location and area

Kebonarum District is one of the districts in Klaten Regency. This district is divided into 7 villages, namely Basin Village, Malangjiwan Village, Gondang Village, Menden Village, Pluneng Village, Karangduren Village and Ngrundul Village. Kebonarum District has an area of 9.07 km^2.

South Tinangkung is a district in the Banggai Kepulauan Regency, Central Sulawesi, Indonesia. This district is the result of the division of the Tinangkung District. South Tinangkung consists of 9 villages i.e. Tobing Village, Paisumosoni Village, Tobungin Village, Mansamat A Village, Mansamat B Village and Kampung Baru Village. South Tinangkung has an area of 439.12 km^2 which consists of 187.89 km^2 of land and 251.23 km^2 of sea.

3.2. *Geological Condition*
Kebonarum District is located on the eastern slope of the foot of the Merapi Volcano. The area is an area that has a slope ranging from 0-8% to 8-15%, with the classification of the slope being gentle to flat. The formation of this area is strongly influenced by river flows that carry sediment resulting from volcanic eruptions which are then deposited in flat areas. This formation is composed of volcanic breccia, lava, and tuff as a result of the lava deposits of Mount Merapi which are still active today.

The rocks that make up the South Tinangkung sub-district are metamorphic rocks (skis, gneiss and quartzites), granite and volcanic rocks. The geological structure is controlled by faults and folds. The fault structure consists of a descending fault, an ascending fault and a containment fault, while the fold structure consists of a weak fold with a gentle slope.

3.3. *Rainfall*
The climate condition in Kebonarum District cannot be separated from the climatic conditions of Java Island as a whole, which is influenced by monsoons, namely global winds that blow due to the influence of the location of the sun, so that it will always change direction every half year. During October to April when the sun is in the southern hemisphere, mainland Australia is the center of low pressure, so the wind blows from mainland Asia which is the center of high pressure. When passing through the equator, the direction of this wind is deflected from the east of the ocean (North-Easterly) to the west of the ocean (North-Westerly). When passing through Java, this wet monsoon becomes a westerly wind, so it is often called a wet west monsoon (Wet-West Monsoon). During this period, Java Island in general and Klaten District in particular experienced the rainy season. On the other hand, from May to October, mainland Asia becomes the center of low pressure and Australia becomes the center of high pressure. The monsoon wind blows from Australia and does not carry water vapor (East-dry monsoon), so parts of Java experience a dry season.

Base on the rainfall, Kebonarum District has an annual rainfall of 2,640 mm with the highest rainfall occurring in February with a total rainfall of 416 mm and the lowest rainfall occurring in August with a total rainfall of 36 mm. The climatic conditions in South Tinangkung District are classified as local climates with an annual rainfall of 1,074 mm. The highest rainfall occurs in June with rainfall of 233 mm and the lowest rainfall occurs in September with rainfall of 2.2 mm (Figure 1).

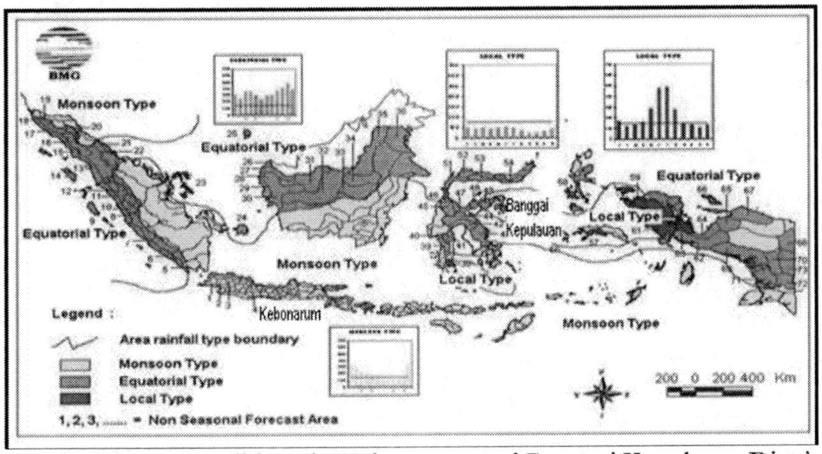

Figure 1. Climate Conditions in Kebonarum and Banggai Kepulauan Districts
(Sources : Indonesian Meteorology Climatology and Geophysics Council)

IOP Conf. Series: Earth and Environmental Science 1233 (2023) 012060 doi:10.1088/1755-1315/1233/1/012060

3.4. Groundwater Potency

Groundwater is water that is in the pores of rocks below the soil surface in a water-saturated zone, with the hydrostatic pressure of water equal to atmospheric pressure when it is in a free aquifer or greater than atmospheric pressure when it is in a confined aquifer. Like other types of water found on land, the source of groundwater is rainwater that seeps into the ground. 90% of groundwater occupies the rock layers, especially those that contain alluvial deposits.

Groundwater is mostly found in rock layers called aquifers which are generally found on alluvial plains. The presence of groundwater in alluvial deposits can still be divided into 4 categories with varying potency, namely watersheds, buried valleys, plains and valleys between mountains. Watersheds are located on the right and left of the river, with the constituent material in the form of river deposits. For areas with lower groundwater levels than river water, the potency for groundwater will be very high, because in addition to getting water supply from rainwater that falls in the area, the supply also comes from river water infiltration.

The buried valley is an area that was previously a river channel. However, due to changes in the trajectory of the river flow, the area turned into a valley. Although it is not as high as the watershed area, the buried valley is an area with quite high groundwater potency. The thing that causes this area to have a high groundwater content is because this area is composed of loose material, so that rainwater that falls on the ground surface can easily experience infiltration and percolation. Therefore, the buried valley is a good aquifer and contains a lot of water, especially in the rainy season. It can be said that the nature of this area is similar to that of a watershed, only that the added water is smaller because it no longer gets water supply from the river.

The plain area is a large area with rock material in the form of sand and gravel. Because it is composed of material that has not been compressed, rainwater that falls on the ground can easily seep into the ground. Therefore, the potency for groundwater in the plains is also quite high.

An intermountain valley is a valley between mountains. Just like other areas that include alluvial deposits, the material making up this area is also in the form of sand and gravel. Besides coming from rainwater that falls in this area, the source of recharge also comes from rainwater that falls on the surrounding mountains that flows into the valley. Therefore, the groundwater potency in this area is also quite high, and it is even possible to find confined aquifers.

Beside alluvial deposits, although most have lower potency, several other rock formations can also function as water reservoirs such as limestone, volcanic rock, sandstone and igneous and metamorphic rocks. Limestone is a type of rock that has unique properties. If in other rocks, the older the age the more compressed the granules, then in the older limestone, the more pores are the result of dissolving by water. Therefore, the characteristics of limestone (density, porosity and permeability) as water storage also vary greatly, especially depending on its age. At different ages, the degree of compression and fracture development is usually different. In young limestone, the holes formed are usually microscopic in size, while in old limestone caves and rivers will form underground, so that in old limestone it is possible to have high groundwater potency.

There are various types of volcanic rocks, both of which have aphanitic and paneric sizes, both of which contain low or high ferromagnesian elements. Some examples of volcanic rocks are granite, syenite, diorite, trachyte, andesite and basalt. Of the several examples of these volcanic rocks, only basalt rock can function as a good aquifer. This is because basalt can have a lot of gas vents. In addition, basalt rock also often experiences many cracks during its formation. The former gas holes and cracks will be occupied by groundwater originating from rainwater that falls on the basalt rocky area.

Sandstone is rock formed by cemented sand. Due to the presence of cement, the sandstone is in a compressed state. The same is true for conglomerate rocks. Conglomerate rocks are formed by cemented rounded gravel and gravel. However, if the rock has experienced many cracks, it will also be able to store water even in a limited amount.

IOP Conf. Series: Earth and Environmental Science 1233 (2023) 012060 doi:10.1088/1755-1315/1233/1/012060

Hydrogeologically, Klaten Regency, Central Java Province is located in the Karanganyar-Boyolali Groundwater Basin (CAT), which has high groundwater potency because it is located in an unconsolidated area between mountains. Referring to the Groundwater Basin Map of the Province of Central Java and the Special Region of Yogykarta published by the Directorate General of Geology and Mineral Resources on a scale of 1: 250,000, the discharge of the free aquifer in the Karanganyar-Boyolali CAT is 1330 million m3/year in the free aquifer and 21 million m3/year in the aquifer. stressed.

According to [11], the Mount Merapi Rock Formation originates from the Holocene which was formed during the Quaternary Period. This area has a high potency of groundwater because it is an area of unconsolidated sediment that is water-passing. This is also supported by the relatively wet climatic conditions. Climatic conditions in general are closely related to rainfall because they indirectly determine the amount of water availability in an area.

Banggai Kepulauan Regency has many rivers, lakes and springs that spread. The existence of surface water and rivers not only serves as a source of drinking water, but is also used as a source for agricultural cultivation activities, while the utilization of surface water and rivers for fisheries has not been optimally achieved. Areas where islands require clean water are obtained by utilizing shallow groundwater fluctuations which are strongly influenced by rainwater. Groundwater is stored in the aquifer in the form of fractures or solid rock fissures and is found in a shallow position. Water resources in Banggai Kepulauan Regency are ground water such as dug wells located in coastal and lowland areas with a height of 0-15 meters above sea level, shallow groundwater levels ranging from 0.5 to 1.0 meters from the ground water level.

3.5. Springs Potency

The appearance of springs in Kebonarum District is caused by the orohydrological nature of this area which is located on the eastern slope of Mount Merapi. As is known, there is a phenomenon of the appearance of unique springs in strato volcanoes, namely the presence of a spring belt at certain heights and the presence of spring belts at certain heights. changes in slope due to changes in the constituent materials. There are 5 springs in this district, namely Brondong, Tirtomulyono, Tortomulyani, Brintik and Bayanan springs.

Brondong spring is a large tourist spot and there are 4 pools for children and adults. There is a watercress vegetable farm, where visitors can sit and enjoy the green vegetables in the gazebo. Brintik Spring is located in Malangjiwan Village, Kebonnarum. It is used for health therapy because there are lots of water sources under it. Tirtomulyono and Tirtomulyani springs are located in Pluneng Village. Tirtomulyono Spring is for bathing men, while Tirtomulyani is for bathing women.

Concerning from the way it occurs, there are two forces that cause the appearance of groundwater to the surface or springs, namely non-gravitational energy and gravitational power (Bryan in Todd Mays, 2005). Springs that occur due to non-gravitational forces include volcanic springs and fissure springs, which are usually hot springs. For springs whose appearance is caused by gravity, they are divided into 5 types, namely (1) depression springs, namely springs caused by the cutting of the groundwater table due to sharp slope changes, (2) contact springs, namely springs that appear on contact area between permeable and impermeable rock, (3) artesian springs, namely springs whose water comes from compressed groundwater, (4) springs in impervious rock (impervious rock springs), namely springs that occur in channels or cracks in rock. impermeability and (5) tubular or fracture springs, namely springs that occur from lava pipes, dissolution or rock cracks associated with groundwater. Based on this classification, the springs in Kebonarum District occur due to gravity and are classified as depression springs.

In South Tinangkung District, one spring was found, namely the Paisu Musoni Spring [12]. The genesis of these springs is caused by gravitational forces and is classified as tubular or fracture springs, namely springs that occur from lava pipes, dissolving or rock cracks associated with groundwater. This spring appears in marl limestone. Concerning from the potency of the spring, the

Paisu Musoni spring has a very large discharge rate of 2,034 liters/sec, while the Tirtomulyani spring has the smallest flow rate of 41 liters/sec (Table 1).

Table 1. Springs potency in Kebonarum and South Tinangkung Districts

Springs Name	Discharge (l/det)	Flow Properties	Classification
Brondong	70	Perennial	Depresion Spring
Tirtomulyono	52	Perennial	Depresion Spring
Tirtomulyani	41	Perennial	Depresion Spring
Brintik	84	Perennial	Depresion Spring
Geneng	28	Perennial	Depresion Spring
Bayanan	38	Perennial	Depresion Spring
Paisu Musoni	2,034	Perennial	Fracture Spring

3.6. Domestic Water Requirement

The amount of domestic water requirement in an area is known by multiplying the average or assumed water use of each person with the population of the study area, so that the amount of water requirement will be directly proportional to the population. The population in Kebonarum District is 20,910 people, while in South Tinangkung District it is 8,114 people. With the water requirement per person of 120 liters/day, the water requirement in Kebonarum District is 915,858,000 liters/year or 915,858 m^3/year, while the water requirement in South Tinangkung District is 355,393,200 liters/year or 355,393 m^3/year.

Referring to Table 1, the total spring discharge in Kebonarum District is 313 liters/sec or 9,870,768 m^3/year, while the spring discharge in South Tinangkung District is 2034 liters/sec or 64,144,224 m^3/year. Based on this data, the water potency from the springs in these two sub-districts is sufficient to be used as a water source for domestic .

Table 2. Water requirement and availability of spring water for domestic

Districs	Number of Population (people)	Water (m^3/year)	Water Availability from Springs (m^3/year)
Kebonarum	20.910	915.858	9.870.768
Tinangkung Selatan	8.114	355.393	64.144.244

4. Conclusion

1. Based on groundwater conditions, Kebonarum District is included in the Karanganyar - Boyolali groundwater basin where the influence of Merapi's activity is quite affecting the ground water condition. The research area is located in the Mount Merapi Formation, which has very good potency for water availability. This is also supported by the relatively wet climatic conditions. In South Tinangkung District, groundwater conditions are not good. Groundwater is generally found in cracks or rock fractures.
2. From the calculation results, it is known that the total spring discharge in Kebonarum District is 9,870,768 m^3/year, while the spring discharge in South Tinangkung District is 64,144,224 m^3/year. With the water requirement in Kebonarum sub-district is 915,858 m^3/year, while the water requirement in South Tinangkung District is 355,393 m^3/year, the water potency from springs in these two sub-districts is sufficient to be used as a water source for domestic requirement.

Acknowledgement

This article is the result of a study financed by the Final Assignment Recognition (Rekognisi Tugas Akhir/RTA) Programme Universitas Gadjah Mada Fiscal Year 2022.

References

[1] Purnama I L S., Marfai, M.A. 2012. Saline water intrusion toward groundwater : issues and its control. Journal of Natural Resources and Development. 2012 (02) : 25-32.

[2] Fenta A A & Kifle A. (2014). Spatial Analysis of Groundwater Potential Using Remote Sensing and GIS-Based Multi-Criteria Evaluation in Raya Valley, Nothern Ethiopia. Hydrogeology Journal 23 (2015) : 195-206.

[3] Moench, A.F and Barlow, P.M. (2000). Aquifer Response to Stream-Stage and Recharge Variations. I. Analytical Step-Response Functions. Journal of Hydrology 230 (2000) : 192-2010.

[4] Todd, D.K. And L.W. Mays. 2005. *Groundwater Hydrology*. John Wiley & Sons, New York.

[5] Fetter, C.W. 1997. *Applied Hydrogeology*. Macmillan Publishing Company, New York.

[6] Wannielista, M., R. Kersten and R. Eaglen. 1997. *Hydrology : Water Quantity and Quality Control*. New York : John Wiley and Sons Inc.

[7] Naghibi S A & Pourghasemi H A. (2015). GIS-Based Groundwater Potential Mapping Using Boodted Regression Tree, Classification and Regression Tree and Random Forest Machine Learning Models in Iran. Environ Monit Asses 188 (44) : 1-25.

[8] Al-Salamah I.S., Ghazaw, Y.M. and Ghumman, A.R. 2011. Groundwater Modelling of Saq Aquifer Buraydah Al Qassim for Better Water Management Strategies. *Environ Monit Assess (2011) 173 : 851-860.*

[9] Lubis, R.J., Sakura, Y and Delinom, R. 2008. Groundwater Recharge and Discharge Processes in The Jakarta Groundwater Basin, Indonesia. *Hydrogeology Journal (2008) 16 : 927-938.*

[10] Priyana Y dan D. Safriningsih, 2005. Sistem Penyediaan Air Bersih Penduduk di Kecamatan Musuk Dalam Menghadapi Musim Kemarau. *Forum Geografi* 19 (1) : 81-87.

[11] Riastika,M .2011. Pengelolaan Air Tanah Berbasis Konservasi di Recharge Area Boyolali Studi Kasus Recharge Area Cepogo Boyolali Jawa Tengah. Jurnal Ilmu Lingkungan Program Pascasarjana UNDIP. 9(2) :86-97.

[12] Magister Pengelolaan Lingkungan Sekolah Pascasarjana Universitas Gadjah Mada. 2013. Kajian Potensi Mataair Dan Pengelolaannya Kabupaten Banggai Kepulauan. Badan Pengelolaan Lingkungan Hidup Kabupaten Banggai Kepulauan, Salakan.

IOP Conf. Series: Earth and Environmental Science 1233 (2023) 012061 doi:10.1088/1755-1315/1233/1/012061

Hydrological Potential Mapping to Generate Micro-Hydro Power in Code River

M P Hadi[1,2], U Suarma[1], A A K Sumunar[1] and M R B Girsang[1]

[1]Faculty of Geography, Universitas Gadjah Mada, Yogyakarta, Indonesia
[2]Center for Environmental Studies, Universitas Gadjah Mada, Yogyakarta, Indonesia
*Corresponding author: mphadi@ugm.ac.id

Abstract. Fossil fuel energy utilization in Indonesia is consider high consisting of 65% coal sources and 20% fuel, while only 15% are generate from new and renewable energy sources. This number is critical due to the commitment to reduce greenhouse gases by 29% or 41% as declared Glasgow in 2021. Alternatives source of energy could be obtained from solar, wind, and water. The energy generate from those sources are consider intermittent, however, water utilization can be optimized by construct several weirs (small-scale reservoirs) to accumulate the rain water as a kinetic energy source to generate electricity. Only some part of the world, especially tropical Indonesia archipelago, has enough rain to generate energy. This research intends to measure the potency of Code River, representing small river in generating micro-hydro power. The Code River is representing a small river on the island of Java as part of the archipelago in Indonesia which has the characteristics of a river, a small water catchment area, and has high discharge input from rainfall because it is in the tropics. The weir is equipped with a penstock pipe, as water intake to the generator. The pipe inner diameter made of high-density polyethylene material can be 200, 300, 400, 500, 700, 800, 1,000 and 1,600 mm. Each pipe size can produce discharges of 0.42, 0.75, 1.62, 2.1, 3.1, 3.5, 6.0 and 24.0 m^3/s respectively. The combination of discharge and head height will produce electrical energy (in watts). Based on hydrological modelling consisting of 52 sub-watersheds, the highest electricity production in 2021 is 8,359,218.3 kWh with 800 mm pipe size (3.5 m^3/sec discharge). The electricity production is based on reliable discharge between 7-96% which is different for each weir location. In conclusion, the electricity supply for the whole year (100%) is inadequate

1. Introduction

Climate change known as a major threat to global prosperity and development. One that supports climate change is because of carbon dioxide emissions that increase from year to year in the world, including in Indonesia [1]. Burning fossil fuels is one of the causes of the use of energy sources for carbon dioxide emissions and supports the increased risk of climate change [2]. One percent increase in the use of non-fossil fuels in electricity generation can reduce per capita carbon dioxide emissions from power generation by about 0.82% [3]. Hence, global cooperation and collaboration are considered significant to successful climate change mitigation and adaptation. At the World Leaders Summit on Climate Change (COP26) in 2021, Indonesia is committed to running a green economy to tackle climate change issues. In 2022, Indonesia is the host for G20 Presidency and invite many countries to jointly reduce the trend of increasing the earth's temperature by 1.5°C due to the production of greenhouse gases (GHG).

In running a green economy, Indonesia will implement climate finance and others innovative financing such as mixed financing, green partnership or green *sukuk*. Programs to increase the area of mangrove forests, peat restoration, reforestation of forest areas, forest protection from fires, are significant steps that have been taken. On the other hand, Indonesia currently still uses 20% of fuel (fuel oil) and natural gas and 65% of coal-based energy, so that the total fossil-based energy is 85%. Indonesia currently only utilizes 15% of New Renewable Energy (NRE) such as water, wind and solar energy. In fact, Indonesia has implemented a 5-10% mix program for fossil energy, however, it is considered inadequate. Based on Government Regulation 79/2014 concerning National Energy Policy, the target for NRE is 23% in 2025 and 31% in 2050. Nevertheless, NRE source for 2022 is currently only 11.5%.

Indonesia has high amount of rainfall because of its geographical location in the tropics. Hence, it has potency as renewable energy source to be converted into kinetic energy which could operate a generator to produce electrical energy. Even though the rainfall distribution is uneven, utilization of water energy for electricity is a strategic choice since it will lower the CO_2 emissions [4].With regard to the rainfall-runoff process, narrow watershed will produce less energy potential. Narrow watershed is an implication that Indonesia is an archipelagic country. Climatologically, rainfall in Indonesia categorize as monsoon, equatorial and local types (Figure 1) [5]. The monsoon type is known for its dry season and rainy season in Java and Nusa Tenggara, parts of Sumatera and parts of southern Kalimantan. Equatorial type characterized by rainfall throughout the year without any dry season. Areas with equatorial rainfall types will benefit the abundance of water supply while the monsoon type with dry period needs to consider additional infrastructure to conserve water.

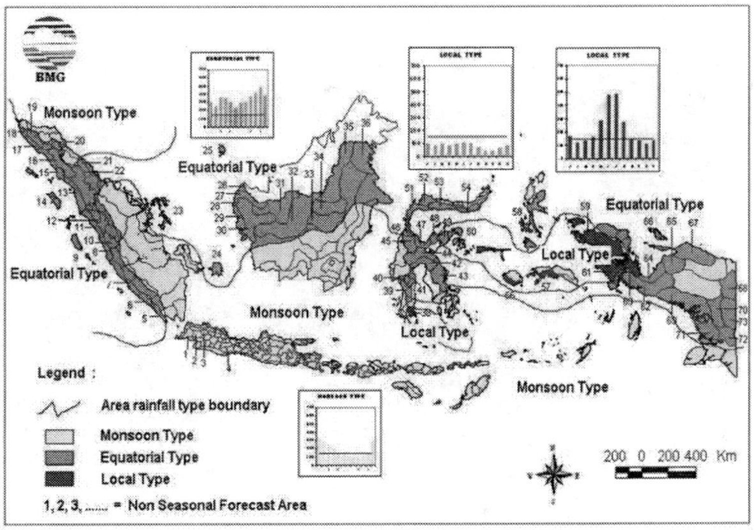

Figure 1. Rainfall type in Indonesia.

Besides, Indonesia as an archipelagic country means that the percentages of ocean cover is higher than the continent. As consequences, river's length are tend to be shorter than where it dominates by continent such as countries in central Asia. Therefore, the resources are more vulnerable when compared to continental regions [6]. Each island have its own water resources potency, however, because most of the island separate by the ocean, hence this potency tend to be used locally. This condition could trigger high vulnerability of water resources. Java Island with population of more than 137 million have limited water supply per capita, even lead to water scarcity [7]. By assuming the water demand is 1,800 m^3/person/year [8], almost 50% of water source has been exploit by the community.

Infrastructure and human development which tend to be concentrated in Java Island resulted in the disparities. The existence of Java-Bali connectivity network to encourage energy supply efficiency is one of the evidence that disparities does occurs. Support by gross domestic product per capita < USD 10,000 and demographic bonus by 2025-2045, essentially important to improve the country's wellbeing. Several studies projected that by 2050; Indonesia's GDB is estimated to reach USD 35,000 per capita [5]. In order to boost the economic growth, large amount of investment and energy is required. One of the strategies is to process raw materials (*minerba*) into finished or semi-finished goods.

The provision of low-carbon energy is an option that must be initiated. Mapping the potential electrical energy from rivers in Java Island is predominance and the first step should be initiated. Series of weirs construction on these rivers will be an enormous advantage since it will increase the water reserves (storage) especially during the dry season. Slope characteristic will be crucial to model weirs location and density. Converting rainfall data into discharge data could visualize the potential for river flow. The combination of flow rate and weir head height is a potential electrical energy that can be generated. Therefore, the use of hydropower plants can produce renewable energy without carbon emissions while cutting electricity production costs. This research can help show how much cost advantage can be generated when electrical energy is generated by utilizing river flows such as the Code River. This can help people, especially those who live around Code River, to take advantage of renewable electrical energy at a much lower cost or even free of charge. If this innovation is implemented in the Code River, it can also provide a source of electrical energy that is closer to the community so that it can support the formation of other productive activities that require electrical energy. One of the example from Chinese government (particularly Guangdong Province) which has the potential to installed hydropower plants of 62 GW, consisting of 45,000 units of hydropower installed with average production per unit is 1.35 MW. River system with larger watershed area is beneficial to generate this potency. It means there are high potency of energy generates from rainfall (Figure 2).

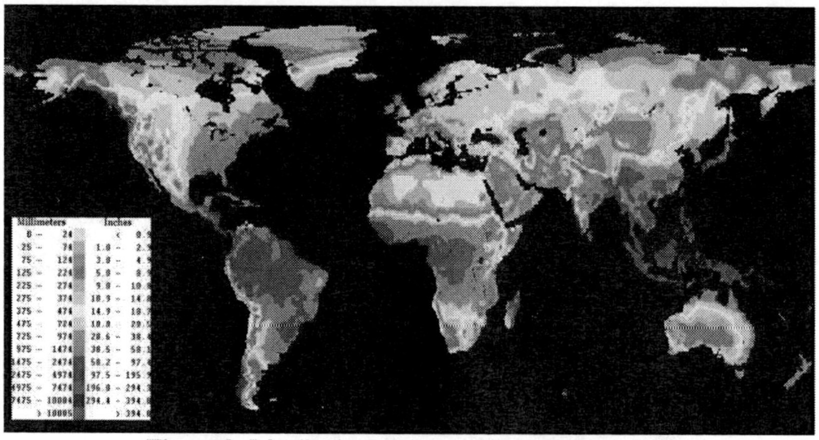

Figure 2. Distribution of the world's annual rainfall.

An environmentally friendly energy source is obligatory considering the growing energy demand and responsibility in reducing greenhouse gases at the same time. Rough topography (mountain) produces steep slopes, thus it has a large energy potential to generate electrical energy. In principle, DEM (Digital elevation model) is beneficial to calculate and analyze inundation volume [9]. Series of weirs could be structured in order to produce several heads/falls and eligible as Micro-Hydro Power (MHP) [10]. Code Watershed is one of the representations in the tropics area that have high rainfall and has a diversity of land uses and varied regional characteristics. Hence, the method and system micro-hydro power can be applied in other watersheds that are still included in the tropics. To conclude, the research objectives are: 1) Assessing the hydrograph characteristics of Code River flow, and 2) Mapping the potential for flow and head flow for electricity generation in Code River.

2. Methods

2.1. Site Selection

Weir construction is often carried out with the intention of producing puddles as water reserves for fresh water, irrigation, fisheries, hydropower energy, as well as flood controlling function. Weirs are large in size and often produce extensive inundation. Practically, to construct a weir will be costly since there's a need for land acquisition own by the local communities. Therefore, construction of a small-scale weir could be beneficial because the puddle would entirely fills the rivers. Construct a dams resulted in a height difference (head) and provides potential plunge to

IOP Conf. Series: Earth and Environmental Science 1233 (2023) 012061 doi:10.1088/1755-1315/1233/1/012061

produce electrical energy. It will be beneficial as NRE source presuming that the weirs construct in several location in the entire watershed. Code River could represent rivers with intermittent flow which tend to be higher in the rainy season and less to none in the dry season. In addition, the topography ranging from mountainous to plain also contributes (Figure 3). This location has the characteristics of a monsoon type with dry and rainy seasons.

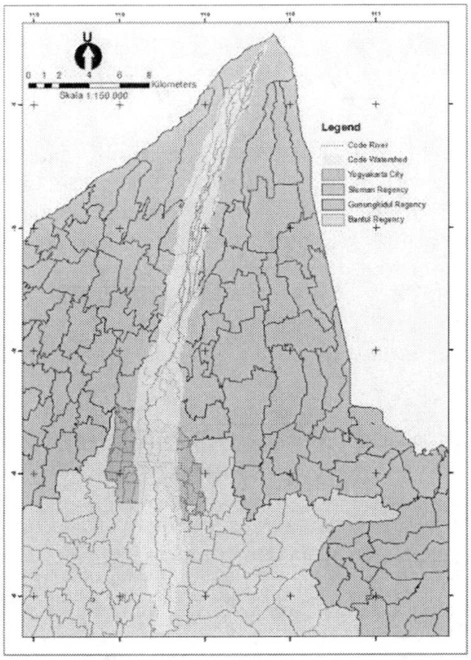

Figure 3. Code Watershed as research area and its administrative zone map.

2.2. *Data collection and analysis*

Digital Terrain Model (DTM) with an adequate scale is necessary to obtain contours which could describe the cross-sectional profile of the river section. Watershed and sub watershed boundary data are required for hydrological analysis using HEC-HMS software. Land use information and soil data (hydrologic soil group) are mandatory in order to assess Curve Number (CN) value using the SCS-CN triangular hydrograph concept and obtain runoff potential value. Daily observed and radar based rainfall data for several stations are used to determine the temporal and spatial distribution as well as to generate flow hydrographs. Base flow which is important for calculating potential electrical energy will be measure based on the flow hydrograph on each river's section. By using Geographic Information System (GIS) tool, 3D analysis can be prepare to calculate volume of the applied weir design. The inundation area designed specifically in safe distance from other cultivated land uses.

Discharge is calculated from the applied hydrological modelling. Therefore, a time lag parameter (T_L) which is calculated based on the concentration time (T_c) is needed. Tc is calculated according to the formula1 [12].

$$T_c = 0.066(L/\sqrt{S})*0.64 \text{..........} (1)$$

Where T_c is the time of concentration (minutes), L is the length of the runoff path (meters), and S is the slope of the sub-watershed.

Runoff volume was calculated based on the CN model. The CN method is an empirical approach to estimate runoff from the relationship between rainfalls; land cover and soil hydrological groups (Cover complex classification). Formula 2 and 3 are used for calculating runoff volume.

$$Q = (P\text{-}0.2S)*2/\ ((P+0.8S)) \quad (2)$$

$$S=25400/CN\text{-}254 \quad (3)$$

Where Q = runoff thickness (mm), P = rainfall (mm), S = maximum potential water retention or maximum storage capacity (saturated) following runoff or water which infiltrates to the soil (mm).

Hydraulic potential is the energy potential generated by water pressure due to earth's gravity [13]. The potential of micro-hydro energy available in nature is a hydraulic potential in the form of potential energy, the amount of which is influenced by the magnitude of the discharge and the height/head. To obtain hydraulic power (P_h) which is a potential energy resource of water, formula 4 is used.

$$P_h=Q\,H\,g \quad (4)$$

Where P_h is the hydraulic power potential, Q is the flow rate, H is the head difference, and g is the earth's gravity (9.8).

The generating power capacity (potential energy produced) can be calculated with formula 5.

$$P\sim.=9.81\,\rho\,Q\,H\,n \quad (5)$$

Where $P\sim.$ = energy produced (watts), ρ = density of water (kg/m^3), Q = flow rate m^3/second (mainstream discharge), H = head, height difference (m), n = 0.56 (efficiency)

Electrical energy estimate from each weir is then tabulated in data series. Discharge data simulated based on the outlet pipe (orifice) with the size of the discharge capacity of 0.42, 0.75, 1.62, 2.1, 3.1, 3.5, 6.0 and 24.0 m^3/sec, and data for the height difference (head) of each weir to calculate the potential electrical energy (watts). Furthermore, total number of electrical energy data from the simulation results in one year is added to produce accumulative energy (watt hours). The maximum total energy produced per year based on certain discharge criteria is an option to be recommended in the development of MHP in Code River.

3. Result and Discussion

Code Watershed has a fairly complex land use consisting of forests, rice fields, dry fields, livestock areas, rural settlements, urban settlements, offices, and other forms of combined land use [14] [15]. The watershed has an elongated shape starting from the slopes of Mount Merapi which has a peak height of 2,930 (mean sea level), thus topographically it has a high flow slope. Orographic rainfall often occurs in the southern slope of Mount Merapi and cause this area tends to have more rainfall.

The CHRS rainfall data of Persiann-CCS which has a spatial resolution of 4 km x 4 km is used. In this study, the location of CHRS rainfall station was selected based on the correlation value between observed rainfall data (institutional) and CHRS rainfall data at the same station. There are three rainfall stations in the surrounding area of Code Watershed namely Gemawang (7.7579440°S, 110.370055°E), Prumpung (7.70675 °S, 110.3926944 °E) and Kemput (7.6438 °S, 110.4020 °E). Based on the regression analysis of daily rainfall data (2021) at the three stations, the correlation coefficient (r) values were 0.35059, 0.36406, and 0.54381, respectively. It appears that Kemput station has the highest correlation value and statistically has significant correlation. Consequently, CHRS hourly rainfall data at the Kemput station then used as the input for hydrological modelling. In Figure 4, the distribution of the CHRS Portal hourly rainfall in 2021 is presented.

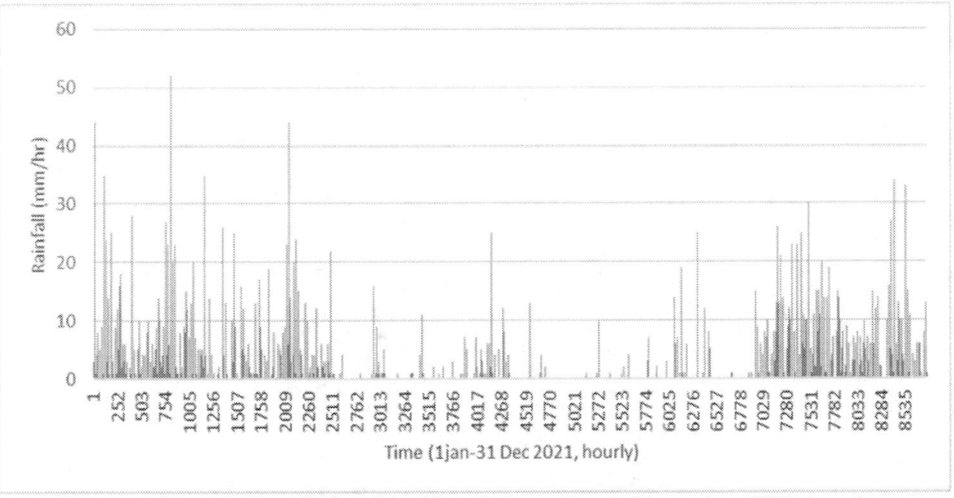

Figure 4. Distribution of rainfall in 2021 (1 January – 31 December 2021) based on CHRS Portal data (at 7.6438°S; 110.4020°E).

3.1. *Sub-watershed division and selection of weir location*

The selection of potential locations as MHP is based on DTM data, river width, river network, and the flexibility of settlements and or other infrastructure buildings, as well as upstream-downstream height differences or can be represented by river slope data. The tentative locations were determined based on visual observations via Google Earth data, and several of them were verified directly in the field. As a result, fifty weir locations and sub-watershed outlets successfully determined (Figure 5). Furthermore, weirs are designed with outlets and overflows (dam top). Weir outlet designed in the form of an orifice (pipe) with a certain diameter. From this outlet, the flow discharge subsequently channelled to drive the turbine and turn a generator to produce electrical energy. The dam top functions as a spillway if the flow rate exceeds the weir. In Figure 6, a plot of the weir height against the distance from the highest point upstream of the Code Watershed (y axis with vertical aggregation) is presented, which provides an overview of the kinetic energy potential of each weir.

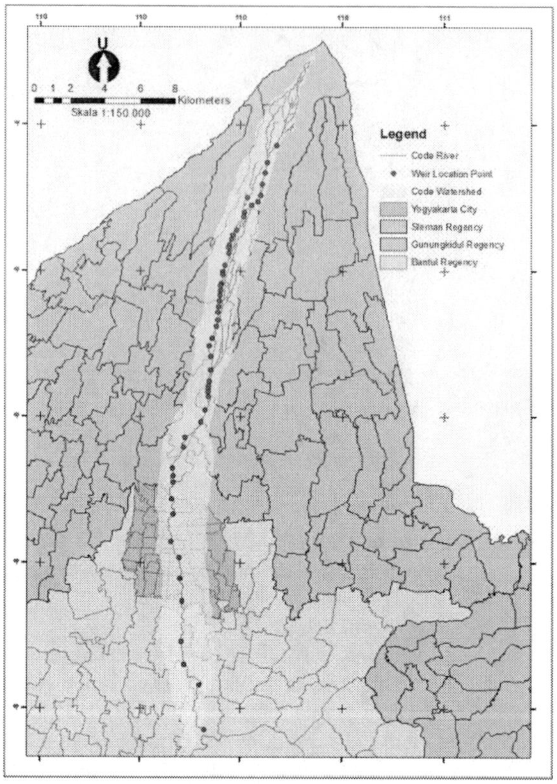

Figure 5. Weir location map along Code River.

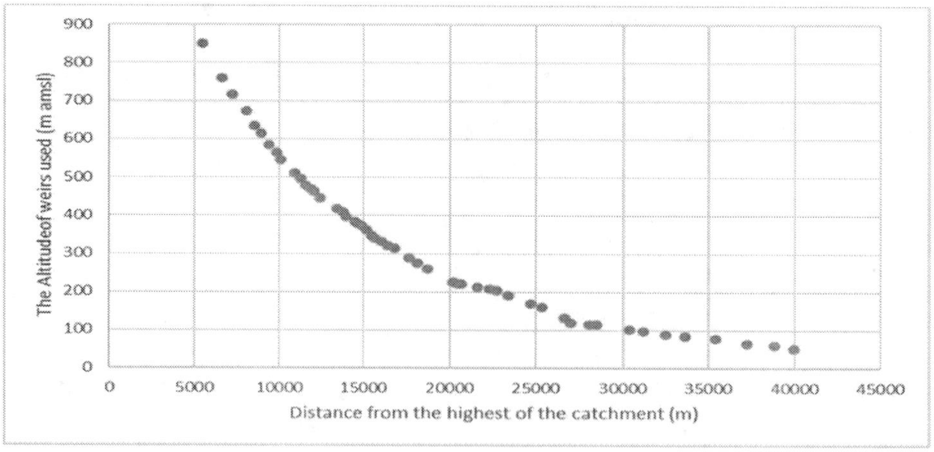

Figure 6. The altitude of weirs used in the modelling and the distance from the highest point of the catchment.

3.2. *Hydrological modelling*

The flow rate is calculated based on two stages, namely determining the effective rainfall and generating runoff. Effective rainfall is determined by the CN method, while runoff generation is based on the SCS-CN (soil conservation service). Runoff discharge calculation is carried out in each sub-watershed to determine the potential flow. The morphometric data for the sub-watershed and the weighted CN value of the calculation results are presented in Figure A1. Runoff generation by SCS method requires time lag data. Time lag is a delay between the times a rainfall event begins over a watershed until runoff reaches its maximum peak. The lag of a watershed may be thought of as a weighted time of concentration. The value 1.67 was used for the lag time (T_L) defined as the time difference between the center of mass rainfall excess and the peak discharge (formula 6) [16][17].

$$T_c = 1.67\ T_L \text{ or } T_L = 0.6T_c \text{.......... (6)}$$

Based on the rainfall data, CN value and T_L value, the flow rate for each sub-watershed can be calculated. In Figure 7 an example of a flood hydrograph is presented in the D01 sub-watershed (upstream) and in D55 (middle part) resulting from modelling for 2021. As a result, it appears that there are periods of low flow discharge, in the form of base flow. In sub-watershed Z01 (area 2.626 km^2) there is no base flow (zero) measure, especially in August (scarce rain), while at D55 weir location with watershed accumulation area of 38.91 km^2 and base flow of 1.3 m^3/s. This amount of flow is seepage from springs along the river.

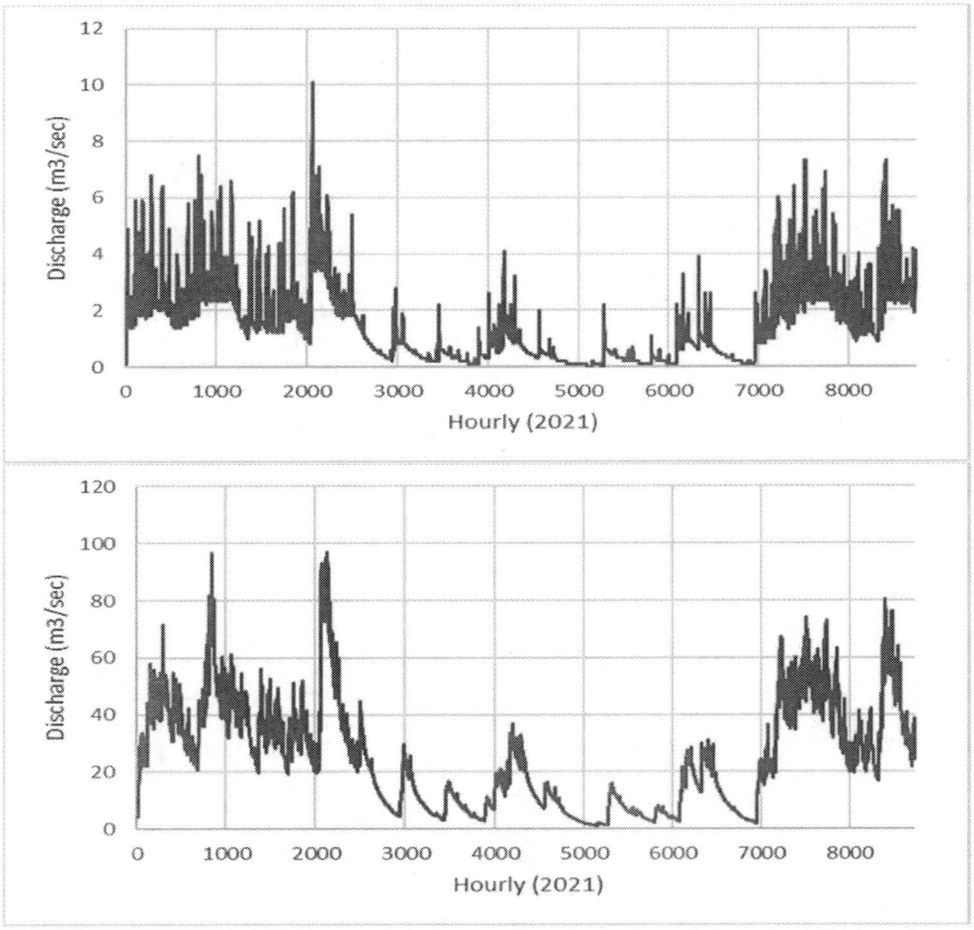

Figure 7. Total flood hydrographs for D01 (upper) and D55 (lower) dams for 2021 based on modelling results as an examples.

Micro-hydro distributed generations (PLTMH) have been utilized as a renewable energy source for electricity generation. The major advantage of these PLTMH is that there are no needs for the construction of big dams and reservoirs. However, these electricity system suffer from large variations in generation due to variable water flow caused by a seasonal rainfall. Figure 8 shows the schematic operation diagram for PLTMH. It can be seen that the fundamental components of PLTMH include the weir, penstock, turbine, generator, and supporting equipment. The head and flow rate (discharge) of available water resources are two major components for PLTMH, which need careful attention before designing PLTMH. The head is the vertical distance through which the water falls, while the flow rate is the volume of the water per unit of time. Electricity generation from PLTMH depends on the combination of suitable head and flow rates by using efficient and well-installed equipment. Generally, the efficiency of the PLTMH is about 50% due to frictional losses and imperfection of turbines.

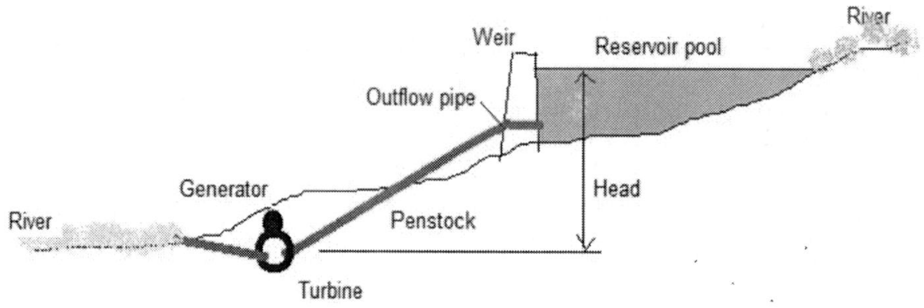

Figure 8. Concept of weir and intake for PLTMH.

The hydrograph characteristics of each single weir flow outlet are strongly influenced by the weir's height. Even though the weir outlet prepare with matching size during the simulation, discharge will be affected by the inundation height due to runoff accumulation. Outlet pipe identical size is intended to accommodate the procurement of materials on the market, including selection of the turbine and generator type and capacity. In Figure 9, a comparison of the flow discharge from weirs D01 and D55 outlets is presented with different characteristics. The D55 weir tends to have a more stable discharge, even though the head is small (3 meters), compared to the D01 weir (15 meters). Weir D01 has 10 m^3/s peak discharges, while weir D55 is 4.5 m^3/s.

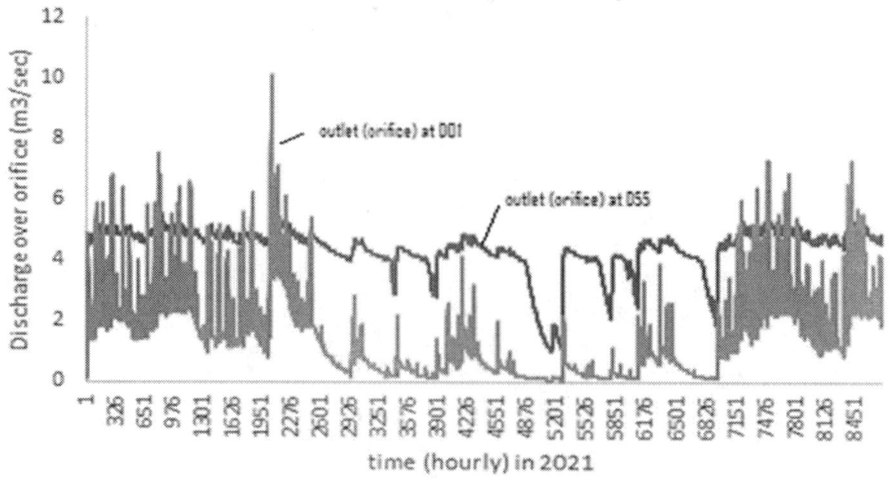

Figure 9. Outlet flow discharge comparison of weirs D01 and D55 in 2021.

The characteristics of peak discharge and runoff volume moreover show a lower trend in the larger sub-watershed area, as shown in Figure 10 and Figure 11. The distance from the highest peak in Code Watershed actually describes area of the sub-watershed from Z01 (the smallest is 2.623 km^2), to watershed outlets (80 km^2 area). The specific runoff volume is higher in the upstream due to river slope factors, sub-watershed area factors, and slope length factors. The shorter the runoff path, the less infiltration, the greater the portion of the water that becomes runoff, hence the specific runoff volume is higher.

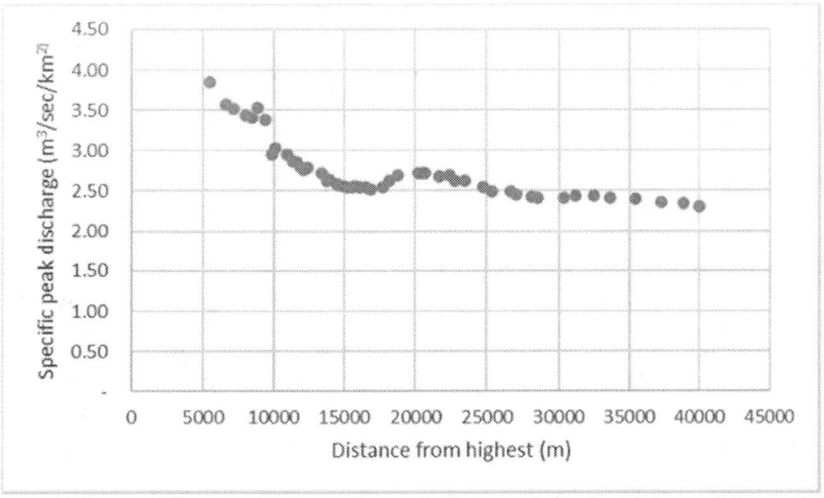

Figure 10. The specific peak discharge for each distance from the highest of the catchment.

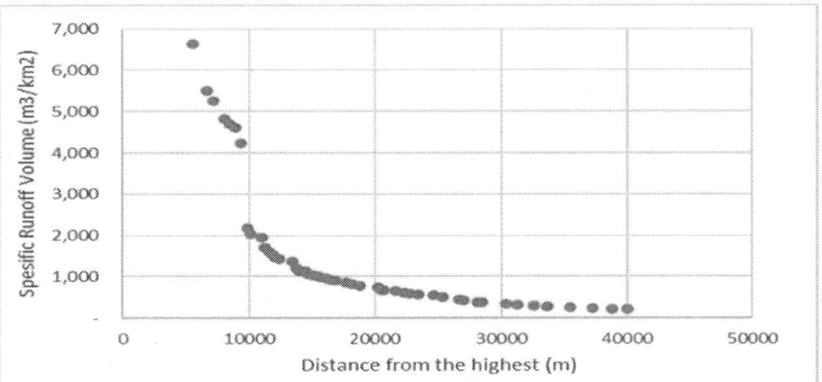

Figure 11. The specific runoff volume for each distance from the highest of the catchment.

Conclusion

Based on hourly rainfall data from the CHRS Portal (2021) and calibration of modelling parameters for several sub-watershed locations carried out in 2018, the hourly flood hydrograph for Code River can be calculated properly. The daily average discharge data reported for discharge measurement point (SPAS) at Pogung (D55) are incompatible for calibration due to lack of detail hydrograph ordinate to describe the flood hydrograph. Therefore these set of data unable to be used for modelling purposes, especially for hourly interval. Potential flow rate in generating micro-hydro power is calculated based on hydrological modelling and elevation data design for the weirs. There are fifty two locations equipped with an intake pipe which could generate total 8,359,218.3 kWh during 2021 with the pipe discharge measure at 3.5 m^3/second and the inner pipe size of 800 mm. The total energy that can be produced has a large enough value when calculated at the current price of electricity per kWh, which is 1,200 rupiah. This explains that the required electricity production can reach around 10 billion rupiahs. Therefore, the use of renewable energy through micro hydropower plants can save up to 10 billion rupiahs each year, which is very profitable while reducing carbon emissions. This electricity production is based on reliable discharge between 7-96% with median as much as 74.2% which is different for each weir location. In conclusion, the electricity supply for the whole year (100%) is inadequate. It's mean that the intermittent factor unable to be overcome using small reservoir such as weir.

Acknowledgement
The authors would like to express sincere gratitude to the Faculty of Geography UGM as the funding institution through Independent Grant for Lecturers 2022 scheme.

References

[1] Sasana H and Putri A E 2018 The Increase of Energy Consumption and Carbon Dioxide (CO2) Emission in Indonesia *E3S Web of Conferences* p 31
[2] Andres R J Boden T A and Higdon D 2014 A new evaluation of the uncertainty associated with CDIAC estimates of fossil fuel carbon dioxide emission *Chemical and Physical Meteorology* **66** (1)
[3] Liddle B and Sadorsky P 2017 How much does increasing non-fossil fuels in electricity generation reduce carbon dioxide emissions *Applied Energy* **197** pp 212-221
[4] Purwanto T Ernawati I Dwiastuti and Wiranta S 2017 *Pembangkit Listrik Mikrohidro (PLTMH) Sebuah Pilihan: Belajar dari Koperasi Mekar Sari Subang* LIPI Press Jakarta
[5] Aldrian E and Susanto R D 2003 Identification of Three Dominant Rainfall Regions within Indonesia and Their Relationship to Sea Surface Temperature *Int. J. Climatol* **23** pp 1435-1452
[6] Hadi M P 2000 The method for estimating water resources on small island *Proc. Int. Conf. on The Management for Coastal and Small Island Ecosystem* Faculty of Geography Universitas Gadjah Mada Yogyakarta
[7] Rijsberman F R 2006 Water scarcity: Fact or Fiction *Agricultural Water Management* **80** pp 5-22
[8] Falkenmark M 1990 Rapid Population Growth and Water Scarcity: The Predicament of Tomorrow's Africa *Population and Development Review* (*Population Council*) **16** 81-94
[9] Irvem A 2011 Application of GIS to Determine Storage Volume and Surface Area of Reservoirs: The Case Study of Buyuk Karacay Dam *International Journal of Natural and Engineering Sciences* **5**(1) pp 39-43
[10] Salim S Hadi M P and Berahim H 2012 Modeling of river water resource potential development in supporting the supply of electrical energy in the Bone watershed, Gorontalo *Unpublished report* Faculty of Geography Universitas Gadjah Mada Yogyakarta.
[11] PricewaterhouseCooper 2017 *Indonesian Banking Survey 2017* http://www.pwc.com/id
[12] Sandoval-Erazo W Toulkeridis T Rodríguez-Espinosa F and Mora M J M 2018 Velocity and time of concentration of a basin – A renewed approach applied in the Rio Grande Basin, Ecuador *IOP Conf. Series: Earth and Environmental Science* **191**
[13] Tobi M D and VanHarling V N 2017 Studi Perencanaan Pembangunan PLTMH di Kampung Sasnek Distrik Sawiat Kabupaten Sorong Selatan Provinsi Papua Barat *Electro Luceat* **3**(1) pp 32–43
[14] Prasena A and Shrestha D B P 2013 Assessing the effect of Land Use Change on Run Off in Bedog Sub Watershed Yogyakarta *Indonesian Journal of Geography* **45**(1) pp 48-61
[15] Saragih M P 2019 Kajian Respon Hidrologi Sungai Boyong di Atas Bendung Lempongsari Menggunakan Model HEC-HMS *Skripsi* Departemen Geografi Lingkungan Fakultas Geografi Universitas Gadjah Mada Yogyakarta
[16] McCuen R H Wong S L and Rawls W J 1984 Estimating urban time of concentration *J. Hydraul. Eng* **110** (7) pp 887–904

[17] McCuen R H 1998 *Hydrologic analysis and design* 2nd Ed. Prentice-Hall Upper Saddle River N J
[18] Raju P E S N and Jain T 2019 Chapter 2: Distributed energy resources and control, Editor(s): Chauhan R K and Chauhan K *Distributed Energy Resources in Microgrids* Academic Press pp 33-56

IOP Conf. Series: Earth and Environmental Science 1233 (2023) 012061 doi:10.1088/1755-1315/1233/1/012061

Figure A1. Characteristics of morphometric and CN values of sub-watersheds in Code Watershed.

No	Sub Catchment code	Outlet altitude (m amsl)	Sub Catchment area (km^2)	Length of runoff (m)	Hydraulic gradient	Time of concentration (min)	Head potential of weir (m)	CN
1	Z01	851	2.626	5500	0.1725	559	15	61
2	Z02	761	0.842	1099	0.0819	162	10	64
3	Z03	716	0.31	585	0.0769	89	15	61
4	Z04	674	0.44	832	0.0505	156	10	60
5	Z05	635	0.27	290	0.0833	42	6	65
6	Z06	614	0.213	250	0.0532	46	8	66
7	Z07	584	0.506	822	0.0636	138	10	66
8	Z08	615	3.76	4800	0.08	717	9	66
9	Z09	566	0.99	850	0.1021	112	10	66
10	Z10	545	0.503	934	0.0745	145	10	65
11	Z11	512	1.675	1084	0.0398	230	6	70
12	Z12	495	0.716	1065	0.0511	199	5	71
13	Z13	479	0.633	1046	0.0533	191	5	72
14	Z14	470	0.631	1016	0.0294	250	5	73
15	Z15	463	0.352	1020	0.0405	214	4	78
16	Z16	447	0.658	1038	0.05	196	6	76
17	Z17	419	2.105	1047	0.0277	265	5	70
18	Z18	409	0.778	1000	0.0263	260	2	68
19	Z19	397	0.317	957	0.0741	149	5	69
20	Z20	384	0.985	937	0.025	250	2	78
21	Z21	378	0.298	900	0.037	198	5	76
22	Z22	372	0.505	890	0.0213	258	3	77
23	Z23	362	0.485	875	0.0364	194	3	78
24	Z24	347	0.586	880	0.0449	175	4	77
25	Z25	340	0.269	855	0.0452	170	3	77
27	Z27	332	0.73	900	0.0192	274	2	75
28	Z28	320	0.666	960	0.0335	221	2	75
30	Z30	314	0.846	1040	0.0142	369	3	77
33	Z33	289	1.693	1060	0.031	254	3	76
34	Z34	275	1.038	1067	0.0287	266	3	76
37	Z37	261	1.188	926	0.0235	255	3	75
43	Z43	225	2.676	871	0.0242	237	3	70
44	Z44	222	0.474	873	0.011	351	2	76
45	Z45	220	0.264	890	0.0133	326	2	76
46	Z46	213	1.744	895	0.0072	447	2	75
48	Z48	209	1.306	907	0.0055	516	2	74
50	Z50	204	0.715	950	0.013	352	3	75
51	Z51	191	1.276	955	0.0194	290	3	82
53	Z53	169	2.844	1250	0.0171	404	3	82
55	Z55	160	1.678	1444	0.0144	507	3	85
57	Z57	133	4.011	1516	0.0199	454	3	85
58	Z58	120	1.022	1516	0.0386	326	3	83
61	Z61	115	3.25	1516	0.0047	938	3	79
62	Z62	114	1.179	1516	0.0026	1,263	2	78
67	Z67	102	5.529	1400	0.0063	743	3	70
69	Z69	98	2.331	1375	0.0048	842	3	70
71	Z71	88	3.541	1345	0.0077	648	3	78
72	Z72	84	2.938	1309	0.0036	920	0.6	88
73	Z73	78	4.607	1300	0.0034	942	3	75
74	Z74	66	4.853	1295	0.0064	683	3	78
75	Z75	60	4.063	1293	0.0038	883	3	90
76	Z76	52	2.606	950	0.0069	484	3	87
78	OYO [illegible]	40	1.46	3013	0.004	2,017		

IOP Conf. Series: Earth and Environmental Science 1233 (2023) 012061 doi:10.1088/1755-1315/1233/1/012061

Figure A2. The percentage of discharge for each discharge category (orifice pipe diameter) on each weir location during 2021 (no data (-) means zero).

Wier Code	Discharge category (m^3/sec)							
	24.00	6.00	3.50	3.10	2.10	1.62	0.75	0.42
D01	-	3,659	17,709	22,710	39,092	42,995	32,397	21,420
D02	-	14,981	47,473	55,024	64,149	59,977	37,823	24,338
D03	-	25,117	62,903	69,177	72,576	64,394	39,338	25,255
D04	-	39,653	79,487	83,507	80,243	68,336	41,068	26,123
D05	-	53,200	92,725	94,006	84,655	71,421	42,588	26,746
D06	-	67,786	101,493	101,568	87,943	73,939	43,639	27,203
D07	-	93,644	115,539	114,162	92,148	76,972	45,005	27,868
D08	-	39,653	79,487	83,507	80,243	68,336	41,068	26,123
D09	-	209,141	155,830	144,509	109,920	89,985	49,572	29,384
D10	-	219,623	160,272	147,677	111,236	91,844	50,196	29,519
D11	-	241,229	169,501	155,698	116,842	95,786	51,494	29,799
D12	-	643	173,048	159,300	119,576	97,079	51,859	29,913
D13	-	-	176,192	161,574	121,930	98,278	52,143	30,014
D14	-	40,048	178,874	163,260	123,297	99,518	52,316	30,093
D15	-	-	180,172	164,103	124,006	100,151	52,378	30,145
D16	-	264,566	182,941	166,172	125,477	101,232	52,508	30,193
D17	-	1,730	188,853	172,405	129,233	103,658	52,860	30,287
D18	-	-	93,129	142,619	130,617	104,343	53,021	30,311
D19	-	63,484	191,940	176,390	131,361	104,792	53,101	30,311
D20	-	-	18,228	122,694	133,195	105,794	53,274	30,311
D21	-	47,959	195,804	180,247	133,593	106,136	53,330	30,311
D22	-	-	197,477	181,524	134,251	106,664	53,423	30,311
D23	-	-	199,698	182,776	135,099	107,217	53,534	30,311
D24	-	-	201,803	184,564	135,930	107,850	53,645	30,311
D25	-	-	202,755	185,433	136,206	108,140	53,682	30,311
D27	-	-	51,540	134,266	136,795	108,707	53,762	30,311
D28	-	-	28,582	128,569	137,435	109,208	53,837	30,311
D30	-	-	207,773	189,750	137,989	109,604	53,874	30,311
D33	-	-	211,927	193,173	139,442	110,474	53,960	30,311
D34	-	-	213,801	194,706	140,446	110,725	54,010	30,311
D37	-	-	216,368	196,545	141,571	111,015	54,065	30,311
D43	-	-	220,954	199,866	143,561	111,634	54,127	30,311
D44	-	-	124,537	161,267	143,803	111,687	54,127	30,311
D45	-	-	142,794	166,683	144,011	111,740	54,127	30,311
D46	-	-	112,107	158,508	144,651	112,004	54,127	30,311
D48	-	-	99,503	154,370	145,049	112,188	54,127	30,311
D50	-	-	224,992	203,161	145,257	112,320	54,127	30,311
D51	-	-	226,261	204,490	145,603	112,544	54,127	30,311
D55	-	-	228,972	206,431	146,122	112,913	54,127	30,311
D57	-	-	232,491	210,059	146,762	113,599	54,127	30,311
D58	-	359,061	233,356	210,876	146,866	113,718	54,127	30,311
D61	-	-	235,259	212,025	147,212	113,942	54,127	30,311
D62	-	-	157,416	180,503	147,333	114,021	54,127	30,311
D67	-	22,496	238,028	213,916	147,904	114,350	54,127	30,311
D69	-	27,342	238,864	214,299	148,147	114,495	54,127	30,311
D71	-	35,005	240,422	214,912	148,597	114,733	54,127	30,311
D72	-	-	38,013	58,601	76,955	66,213	35,500	20,399
D73	-	59,084	241,951	215,806	149,099	114,944	54,127	30,311
D74	-	71,790	242,787	216,317	149,427	115,076	54,127	30,311
D75	-	80,987	243,306	216,649	149,670	115,181	54,127	30,311
D76	-	388,815	243,883	215,981	149,912	115,260	54,127	30,311
Total (kWh)	-	2,470,686.20	8,359,218.30	8,287,335.70	6,452,438.40	5,137,131.30	2,602,685.00	1,495,089.70

IOP Conf. Series: Earth and Environmental Science 1233 (2023) 012061 doi:10.1088/1755-1315/1233/1/012061

Figure A3. The power generated (kWh) for each discharge category (pipe diameter) on each weir location for 2021.

Wier Code	Head (m)	Discharge (m^3/sec) [pipe inner diameter (mm)]							
		24.00 [1600]	6.00 [1000]	3.50 [800]	3.10 [700]	2.10 [500]	1.62 [400]	0.75 [300]	0.42 [200]
D01	15	-	0.8	7.0	10.1	25.8	37.2	59.8	70.7
D02	10	-	3.5	18.8	24.6	42.3	51.9	69.9	80.3
D03	15	-	5.8	24.9	30.9	47.9	55.8	72.7	83.3
D04	10	-	9.2	31.5	37.3	52.9	59.2	75.9	86.2
D05	6	-	12.3	36.7	42.0	55.8	61.8	78.7	88.2
D06	8	-	15.7	40.2	45.4	58.0	64.0	80.6	89.7
D07	10	-	21.6	45.7	51.0	60.8	66.6	83.1	91.9
D08	0	-	9.2	31.5	37.3	52.9	59.2	75.9	86.2
D09	10	-	48.3	61.7	64.6	72.5	77.9	91.6	96.9
D10	10	-	50.7	63.4	66.0	73.4	79.5	92.7	97.4
D11	6	-	55.7	67.1	69.6	77.1	82.9	95.1	98.3
D12	5	-	0.1	68.5	71.2	78.9	84.1	95.8	98.7
D13	5	-	-	69.7	72.2	80.4	85.1	96.3	99.0
D14	5	-	9.2	70.8	73.0	81.3	86.2	96.6	99.3
D15	4	-	-	71.3	73.3	81.8	86.7	96.7	99.4
D16	6	-	61.1	72.4	74.3	82.8	87.6	97.0	99.6
D17	5	-	0.4	74.7	77.0	85.3	89.7	97.6	99.9
D18	2	-	-	36.9	63.7	86.2	90.3	97.9	100.0
D19	5	-	14.7	76.0	78.8	86.7	90.7	98.1	100.0
D20	2	-	-	7.2	54.8	87.9	91.6	98.4	100.0
D21	5	-	11.1	77.5	80.5	88.1	91.9	98.5	100.0
D22	3	-	-	78.2	81.1	88.6	92.4	98.7	100.0
D23	3	-	-	79.0	81.7	89.1	92.8	98.9	100.0
D24	4	-	-	79.9	82.5	89.7	93.4	99.1	100.0
D25	3	-	-	80.3	82.9	89.9	93.6	99.2	100.0
D27	2	-	-	20.4	60.0	90.2	94.1	99.3	100.0
D28	2	-	-	11.3	57.5	90.7	94.6	99.4	100.0
D30	3	-	-	82.2	84.8	91.0	94.9	99.5	100.0
D33	3	-	-	83.9	86.3	92.0	95.7	99.7	100.0
D34	3	-	-	84.6	87.0	92.6	95.9	99.8	100.0
D37	3	-	-	85.6	87.8	93.4	96.1	99.9	100.0
D43	3	-	-	87.5	89.3	94.7	96.7	100.0	100.0
D44	2	-	-	49.3	72.1	94.9	96.7	100.0	100.0
D45	2	-	-	56.5	74.5	95.0	96.7	100.0	100.0
D46	2	-	-	44.4	70.8	95.4	97.0	100.0	100.0
D48	2	-	-	39.4	69.0	95.7	97.1	100.0	100.0
D50	3	-	-	89.1	90.8	95.8	97.2	100.0	100.0
D51	3	-	-	89.6	91.4	96.1	97.4	100.0	100.0
D55	3	-	-	90.6	92.2	96.4	97.8	100.0	100.0
D57	3	-	-	92.0	93.9	96.8	98.4	100.0	100.0
D58	3	-	82.9	92.4	94.2	96.9	98.5	100.0	100.0
D61	3	-	-	93.1	94.7	97.1	98.7	100.0	100.0
D62	2	-	-	62.3	80.7	97.2	98.7	100.0	100.0
D67	3	-	5.2	94.2	95.6	97.6	99.0	100.0	100.0
D69	3	-	6.3	94.5	95.8	97.7	99.1	100.0	100.0
D71	3	-	8.1	95.2	96.0	98.0	99.3	100.0	100.0
D72	0.6	-	-	15.0	26.2	50.8	57.3	65.6	67.3
D73	3	-	13.6	95.8	96.4	98.4	99.5	100.0	100.0
D74	3	-	16.6	96.1	96.7	98.6	99.6	100.0	100.0
D75	3	-	18.7	96.3	96.8	98.7	99.7	100.0	100.0
D76	3	-	89.8	96.5	97.0	98.9	99.8	100.0	100.0

IOP Conf. Series: Earth and Environmental Science 1233 (2023) 012062 doi:10.1088/1755-1315/1233/1/012062

Macro Fungi in Urban Forest of Universitas Gadjah Mada and Their Potential Uses as Medicinal Fungi

G Anggara[1], S Rahayu[1]*, and H H Nurjanto[1]

[1]Faculty of Forestry Universitas Gadjah Mada, Indonesia
*Corresponding author: sri.rahayu2013@ugm.ac.id

Abstract. Macro fungi are important ecosystem component, containing carbohydrates, fibers, proteins, vitamins, and have enormous medicinal purposes. Previous study stated some medicinal macro fungi were found on Lawu Mountain. These species can become alternative source of healthy food to increase human's immunity. Another study found macro fungi growing in the urban forest of UGM, but studies on their medical benefits have not been conducted yet. This research aimed to understand macro fungi which has potential uses as medicinal fungi in urban forest of UGM. Literature review was done to identify the medicinal purposes of macro fungi in urban forest of UGM. The result shows 15 fungus in urban forest of UGM are able to be natural source of medicine, Agaricus, Auricularia, Ganoderma, Mycena, Earliella, Microporus, Poria, Pycnoporus, Trametes, Schizophyllum, Stereum, Tremella, Pleurotus, Tricholoma, and Xylaria. According to this information, urban forest of UGM has potential to be source of medicinal macro fungi which can be developed as a source of high-quality fungal genetics through pharmaceutical research. Moreover, this research can be used as consideration in making management policies of urban forest of UGM so that the medicinal macro fungi can be used as source of food and for advanced research purposes.

1. Introduction

Macro fungi are important component of ecosystem because they play vital role in the ecosystem functions [16]. Just like another living things, the growth of macro fungi also affected by several micro climate such as light intensity, temperature, air humidity, and soil humidity [7]. Previous study about medicinal macro fungi on Lawu Mountain found some species of macro fungi that contain several chemical compunds that be able to boost the human immunity in Covid-19 pandemic [22].

Macro fungi contain carbohydrates, fibers, proteins, vitamins, and have enormous medicinal purposes. They are rich in protein, contain essential amino acids, also essential minerals [6]. The chemical compound of macro fungi also has been reported by several studies. Medicinal macro fungi has theurapeutic benefits, anti-oxidant activity, anti-tumor activity, anti-viral activity, anti-microbial activity, anti-malarial agents, anti-Alzheimer's disease, hypocholesterolemic activity, and anti-diabetic activity [8].

The ecosystem of urban forest of UGM consists of various species of trees with an average age of 15 years with different trees density for each location depending on the utilize of the location. In addition, on some locations there also shurbs, grass, and other small plants [19]. The urban forest of UGM be able to promote the growth of macro fungi either directly and indirectly. It has suitable environmental requirements to support the growth of macro fungi such as various substrate, humidity, and sunlight condition. According to Anggara [2], there were 25 genera belonged to 15 families from 6 orders in the urban forest of UGM. They came in various shapes, colors, and sizes. The macro fungi lived as a saprophyte, decomposer, mycorrhiza, and pathogen on the various substrate such as soil, litter fall, degraded wood, and living trees.

Since the ecosystem campus of UGM through the urban forest ecosystem was very suitable for growing and developing many macro fungi, thus the aim of this research is to understand the macro fungi which has potential uses as medicinal fungi in urban forest of UGM. This information indicates that urban forest of UGM has potential as a source of germplasm of medicinal macro fungi which can be developed as a source of high-quality fungal genetics through pharmaceutical research.

2. Materials and Methods/Methodology

2.1. Description of research location

The research was conducted in the urban forest of UGM which divided into three types of location according to the light intensity (**Table 1)** on June 2020.

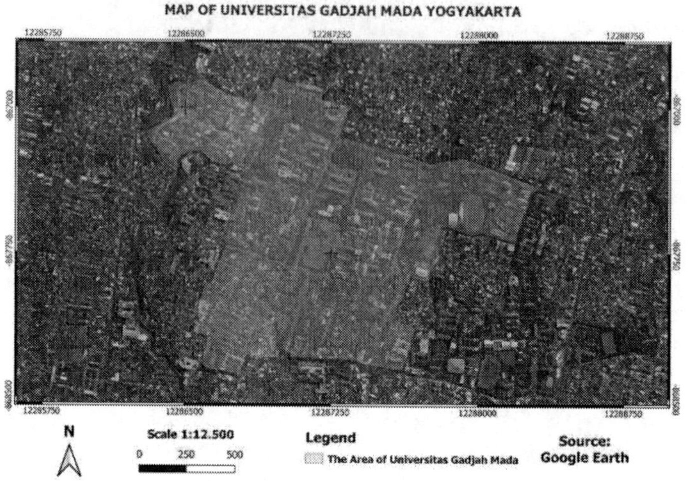

Figure 1. Map of research location (Source: Google Earth, 2022 [11])

Table 1. Determination of research location

No	Light Habitats	Environmental Conditions
1	Low-light intensity area	- Relative Light Intensity ranged between 3%-5% - Dominated with trees - Trees density was 414 N/Ha
2	Medium-light intensity area	- Relative Light Intensity ranged between 17%-23% - Dominated with trees - Trees density was 117 N/Ha
3	High-light intensity area	- Relative Light Intensity ranged between 40%-47% - Dominated with shrub, grass, and other small plant - Trees density was 63 N/Ha

Source: Modification from Anggara [2].

2.2. Sample record

The specimens of macro fungi in urban forest of UGM were observed by Complete Survey [25]. The observation was done by walking along the site and record the detail morphological features of macro fungi included the substrate, distribution, and specimen condition. The morphological features recorded were cap size, shape, colour, surface texture, gill colour, attachment, stem size, stem shape, and stem type.

2.3. Identification the genera and medicinal potential uses

The identification was done according to the recorded characteristic features of the specimens. The literature used in identification were A field Guide to the Larger Fungi of FRIM by S.S. Lee (2017)

IOP Conf. Series: Earth and Environmental Science 1233 (2023) 012062 doi:10.1088/1755-1315/1233/1/012062

[12], A guidebook to the macrofungi of Tasik Bera by Zainuddin et al. [26] , and A guidebook to the macrofungi of Fraser's Hill by Thi et al. [24]. The most literature used were from Malaysia, with consideration that Indonesia and Malaysia are tropical countries, which have high ecological similarities including the species of macro fungi. The identification of medicinal potential uses was done by available literature.

3. Results and discussion

3.1. Distribution of medicinal potential macro fungi

Each location has different number of specimens of macro fungi; low-light intensity area was 4305; medium-light intensity area was 2013; and high-light intensity area was 2960. These specimens belonged to 25 genera, 15 families, and 6 orders which were Agaricales, Auriculariales, Polyporales, Russulales, Tremellales, and Xylariales. According to the literature, the potential medicinal macro fungi in urban forest of UGM belonged 10 families. The distribution of potential medicinal macro fungi presents in **Table 2.** The fruiting bodies of potential medicial macro fungi was quite high which indicates that urban forest of UGM is the suitable habitat of them. Auricularia, Schizophyllum, and Xylaria are the most common fungus found in urban forest of UGM. In addition to their role as a saprophyte, the high number of these fungi is affected by suitable habitat and the amount of substrate.

Table 2. Distribution of potential medicinal macro fungi in urban forest of UGM

Genera	Family	Number of Fruiting Bodies at Ecosystem With					
		Low-light Intensity		Medium-light intensity		High-light intensity	
		Presence	Number	Presence	Number	Presence	Number
Agaricus	Agaricaceae	-	0	-	0	v	6
Auricularia	Auriculariaceae	v	1142	v	324	v	203
Ganoderma	Ganodermataceae	v	109	v	31	v	39
Mycena	Mycenaceae	v	16	-	0	v	1
Earliella	Polyporaceae	v	251	-	0	v	61
Microporus		v	115	v	2	v	31
Poria		-	0	-	0	v	4
Pycnoporus		-	0	-	0	v	1
Trametes		v	496	v	96	v	188
Schizophyllum	Schizophyllaceae	v	96	v	1250	v	1945
Stereum	Stereaceae	-	0	v	109	-	0
Tremella	Tremellaceae	v	3	-	0	-	0
Pleurotus	Tricholomataceae	v	6	v	22	-	0
Tricholoma		v	6	-	0	-	0
Xylaria	Xylariaceae	v	1930	v	50	v	23

3.2. The potential medicinal purpose of macro fungi

According to the literature cited, the potential medicinal purpose of macro fungi in urban forest of UGM were various. Medicinal fungi in urban forest of UGM consist of mushroom (umbrella like fruiting body), pore fungi, jelly fungi, and coral fungi. Most of them have potential to be natural source of antioxidant. They were found grew on the various substrate such as soil, litter fall, decaying wood, and living trees. In detail they are present in **Table 3.**

Table 3. The medicinal potential of medicinal macro fungi in urban forest of UGM

No	Genera	Family	Substrate	Medicinal Potential	Selected References
1	Agaricus	Agaricaceae	Soil	Theraputic potential, Antimicrobial, Antioxidant	[9]; [16]; [21]
2	Auricularia	Auriculariaceae	Litter fall, decaying wood, living trees	Medicine	[22]
3	Earliella	Polyporaceae	Litter fall, decaying wood, living trees	Antimicrobial, Antioxidant	[15]
4	Ganoderma	Ganodermataceae	Decaying wood, living trees	Antimicrobial, Antiviral, Antioxidant, Anti-cancer	[15]; [10]; [8]
5	Microporus	Polyporaceae	Litter fall, decaying wood	Antimicrobial, Antioxidant	[15]
6	Mycena	Mycenaceae	Litter fall, soil, decaying wood, living trees	Antioxidant	[14]
7	Pleurotus	Tricholomataceae	Litter fall, soil, decaying wood, living trees	Medicine	[22]
8	Poria	Polyporaceae	Litter fall	Antitumor, Anti-inflammatory	[17]
9	Pycnoporus	Polyporaceae	Litter fall	Antioxidant	[20]; [4]
10	Schizophyllum	Schizophyllaceae	Litter fall, soil, decaying wood, living trees	Medicine	[22]
11	Stereum	Stereaceae	Decaying wood	Medicine	[22]
12	Trametes	Polyporaceae	Litter fall, decaying wood, living trees	Medicine	[22]
13	Tremella	Tremellaceae	Decaying wood	Anti-aging, Anti-inflamatory	[23]
14	Tricholoma	Tricholomataceae	Soil	Antioxidant, Anti-cancer, Anti-aging	[3]; [4]
15	Xylaria	Xylariaceae	Litter fall, soil, decaying wood, living trees	Anti-bacterial, Anti-cancer, Antioxidant	[1]

Auricularia (**Figure 2 a., b.**) is a jelly fungus and becomes source of food especially in Asian countries. It has potential as natural immunomodulator source [22]. Other jelly fungus in urban forest of UGM was Tremella (**Figure 2 c.**). Tremela has been popular as Chinese cuisine ingredient for centuries as it contains rich dietary fibers, protein, minerals, antioxidant, and high of vitamin D. Tremella consists of Tremella polysaccharides which it becomes the major component and activity unit of Tremella. *Tremella fuciformis* potentially utilized as precautionary in neurodegenerative disease like Alzheimer disease [21].

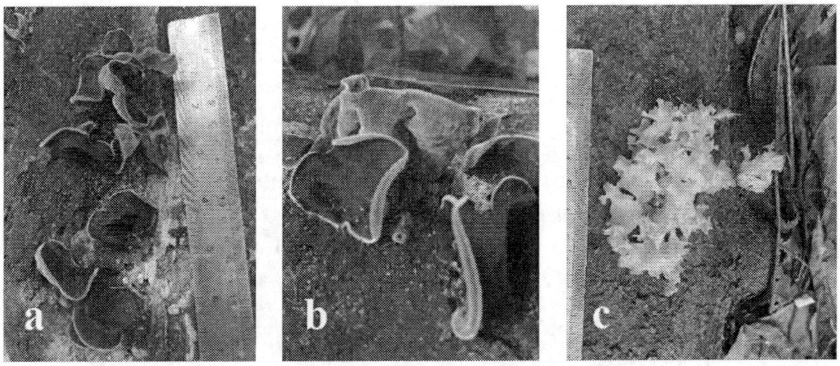

Figure 2. The fruiting body of **a.**, **b.** Auricularia and **c.** Tremella

Schizophyllum (**Figure 3**) fungi are characterized with split gill on the underside of basidiocarp. In the urban forest of UGM, Schizophyllum were abundantly growing on litter, degraded wood, and living trees. It also popular as source of food. *S. commune* are reported have polysaccharides which shows structural elucidation and determines immune stimulation [22].

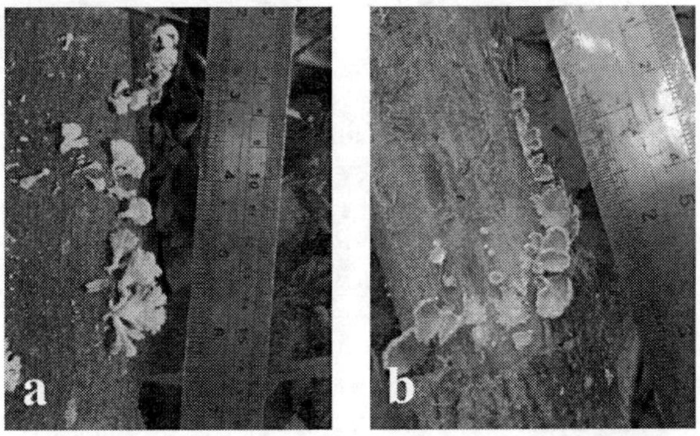

Figure 3. The fruiting body of Schizophyllum **a.** Upperside, **b.** Underside

The only Ascomycota fungi in urban forest of UGM is Xylaria (**Figure 4**). It has black color and coral-like fruiting body. One of species of Xylaria, *Xylaria primorskensis* contains xylaranic acid which was obtained form its terpenoids. It showed an antibacterial activity, antioxidant potential against DPPH & H_2O_2 radicals, and anticancer activity against human lung cancer cells [1].

Figure 4. a., b. The variation of fruiting body of Xylaria

Agaricus (**Figure 5**) has been proven to have many health benefits. Beside being source of food, Agaricus also have antitumor activity. De Miranda et al [9] reported that consuming *A. brasiliensis* supplemented diet be able to inhibit the growth of Sarcoma 180 tumor on mice. The extract of *A. blazet* was identified has a greater inhibition for the Gram-positive bacteria (compared with Gram-negative bacteria), such as *Salmnonela typhy, Streptococcus pneumontae, Streptococcus mutans, Streptococcus sobrtnus, Staphylococcus aureus, B. cereus, L. monocytogenes* and coliforms. Furthermore, *A. blazet* contains ergosterol, other types of fatty acids, polysaccharides, and alkaline which important in sygernistic antimicrobial action [16]. Study by Orsine et al [4] reported that alcoholic extract of *A. sylvaticus* has a great antioxidant potential (74,6%) compared with aqueous and ether extract (14,6% each).

Figure 5. The fruiting body of Agaricus **a.** Upperside **b.** Side Surface

Mycena (**Figure 6 a., b., c., d.**) is a saprophyte [2] that usually found solitary on litter, soil (covered by litter), living trees, and degraded wood. Study by Liang et al [14] reported that two polysaccharides (MDPs) that purified form *Mycena dendrobii* showed that both MDPs showed a different antioxidant activities. MDP-2 showed an execellent superoxide radical activity better than BHT (butylated hydroxytoluene). MDP-1 composed of mannose, xylose, and galactose. While, MDP-2 consits of falacturonic acid, galactose, and rhamnose.

Figure 6. The fruiting body of Mycena **a.** Cap surface, **b.** Gill, **c., d.** Variation of striate cap

Pleurotus (**Figure 7 a., b.**) has white fleshy fruiting body. Pleurotus are popular in the Asian countries as food. It has antitumor and immunodeficiency activities also becomes co-adjuvant in chemotherapy [22].

IOP Conf. Series: Earth and Environmental Science 1233 (2023) 012062 doi:10.1088/1755-1315/1233/1/012062

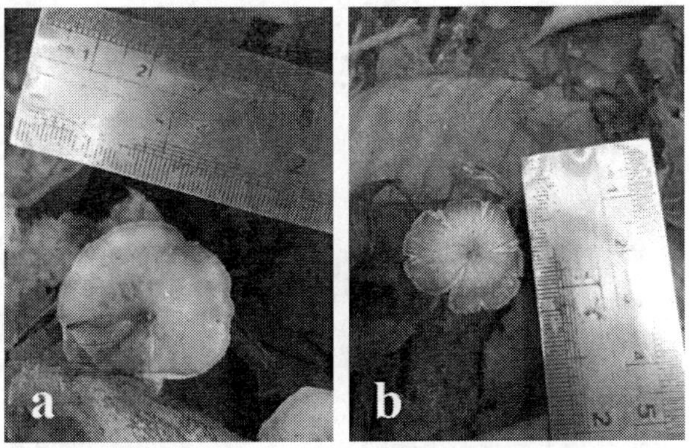

Figure 7. a., b. The fruiting body of Pleurotus

Tricholoma (**Figure 8**) found grew on soil and has white color. It is a mycorrhiza [2]. Bal [3] reported that Total Antioxidant Status (TAS) of *Tricholoma imbricatum* and *T. virgatum* were higher than *Suillus granulatus, Infundibulicybe geotropa, Lactifluus rugatus, Cerioporus varius,* and *Lentinus trigrinus* with result 3,474 and 3,745, respectively. TAS value is an indicator of all antioxidant compounds in the living organisms. Other species of Tricholoma, *Tricholoma matsutake* has been popular had anticancer, immunomodulatory, anti-oxication, and anti-aging activities [16].

Figure 8. The fruiting body of Tricholoma **a.** Upperside, **b.** Side surface

Earliella (**Figure 9**), Ganoderma (**Figure 10**), and Microporus (**Figure 11**) are polypore fungi which have pores on the underside of the basidiocarp. Some studies reported those fungus have antimicrobial, antiviral, antioxidant, and anti-cancer activities. Study by Liew et al [15] reported that *Ganoderma australe* demonstrated the highest DPPH scavenging activity of 89,56%, while *Earliella scabrosa* and *Microporus xanthopus* were 21,27% and 6,48%, respectively. Those fungus are classified as strong inhibitors as the Minimum Inhibitory Concetration (MIC) values recorded were <500 $\mu g\ mL^{-1}$. Other species of Ganoderma, *Ganoderma lucidum*, reported contains 18 kinds of amino acids and the most abundant amino acid was leucine, which has strong hypoglycemic and antioxidant activities [10]. The polysaccharides, triterpenes, glycopeptides, peptidoglycans, and triterpenods of *G. lucidum* reported have anti-tumor and anti-cancer activities [5]. El Sheikha et al [10] reported *G. lucidum* has been used as alternative medicine to support medicine of diabetes, hepatitis, and cancer.

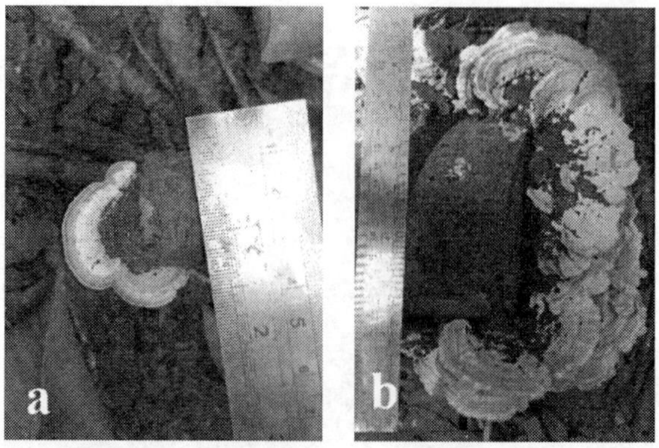

Figure 9. The fruiting body of Earliella **a., b.** Variation of concentric zones

Figure 10. The variation of fruiting body of Ganoderma **a.** Young fruiting body, **b., c., d.** Variation of fruiting body, **e.** Underside

Figure 11. The variation of fruiting body of Microporus **a**. Upperside, **b**. Underside with stem, **c**. Fruiting body without stem (resupinate)

Poria (**Figure 12**) is a bracket fungi which attach directly to the substrate and had tough fruiting body. Some studies found the compounds of triterpenes and polysaccharides in Poria had antitumor activity to coloncancer cells, lung adenocarcinoma cells, kidney cancer cells, human prostate cancer cells, cervical cancer cells, and human breast cancer cells [17]. Other bracket fungi is Pycnoporus (**Figure 13**). Study by Orango Bourdette et al [20] reported *Pycnoporus sanguineus* had very high antioxidant acitivity in the aqueous extract (IAA = 3,13) compared with the water-ethanol extract and ethanol extract (IAA = 0,90 and 0,22, respectively). The fungal extract of *Pycnoporus sanguineus* contains phenolic compunds, flavonoids, and proanthocyanidines. Also, it can become a natural source of antioxidant since it presented a satisfactory antioxidant potential compared with *Ramaria flava, Rhizopohon roseolus*, and *Russula delica* [4].

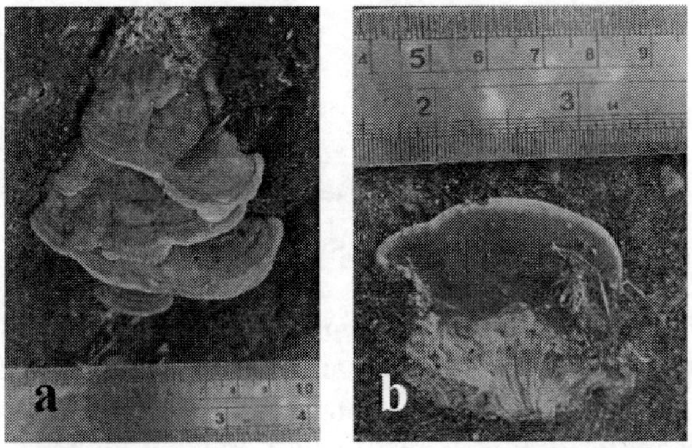

Figure 12. The fruiting body of Poria **a**. Upperside, **b**. Underside

Figure 13. The fruiting body of Pycnoporus **a**. Upperside, **b**. Underside

Trametes (**Figure 14**) is a saprophyte and has brown concentric zones on the upper surface of basidiocarp. Report by Rahayu et al [22] explained *Trametes versicolor* has strong immunomodulatory effect due to a polysaccharide called Krestin (PSK). Stereum (**Figure 15**) was found grew in group on living trees in urban forest of UGM. Dry extract of *Stereum hirsutum* had significant inhibitory effect on thrombin. Furthermore, the extract of *S. hirsutum* be able to increase the antioxidant and biomodulatory properties (synergistic effect) [22].

Figure 14. a., b. The variation of fruiting body of Trametes

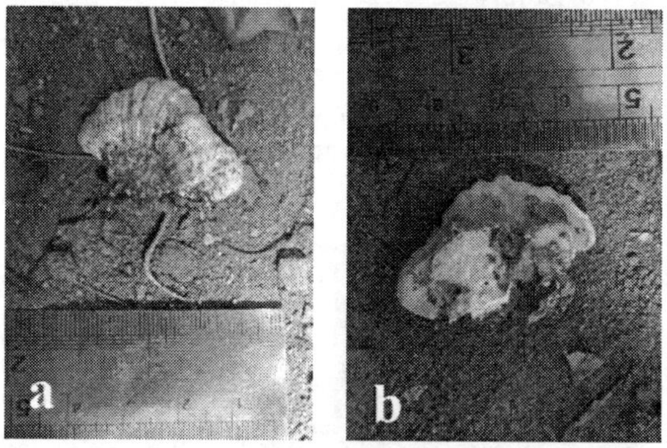

Figure 15. The fruiting body of Stereum **a**. Upperside, **b**. Underside

Several compunds that has been identified in medicinal macro fungi have various therapeutic effect, including a high antioxidant compunds. Some studies reported that the chemical compounds of medicinal fungi can become an alternative of human immune booster. The advanced studies should be conducted considering that urban forest of UGM can become the suitable habitat for medicinal macro fungi which has a high potential as natural medicine source and food source.

4. Conclusion

From total 25 genera that found in urban forest of UGM, according to the literature 15 genera have a high potential of natural medicine source. They are Agaricus, Auricularia, Earliella, Ganoderma, Microporus, Mycena, Pleurotus, Poria, Pycnoporus, Schizophyllum, Stereum, Trametes, Tremella, Tricholoma, and Xylaria. The polysaccharides compound in those fungus showed antioxidant, antimicrobial, anti-infalmmatory, and therapeutic potential activities. The most common medicinal fungus in urban forest of UGM, Auricularia, Schizophyllum, and Xylaria, need to be cultivated more intensively on their natural habitat.

References

[1] Adnan M, Patel M, Reddy M N and Alshammari E 2018 Formulation, evaluation and bioactive potential of bioactive potential of Xylaria primorskensis terpenoid nanoparticles from its major compound xylaranic acid *Scientific Reports* **8**

[2] Anggara G 2021 The Diversity of Macro Fungi on the Urban Forest of Univesitas Gadjah Mada During Lockdown of University *Undergraduate Thesis*

[3] Bal C 2021 Antioxidant Potential and Element Contents of Wild Mushroom *Tricholoma imbricatum KSU J. Agric Nat* 24 **(1)**: 196-199, 2021

[4] Borderes J, Costa A, Guedes A and Tavares L B B 2011 Antioxidant Activity of the Extracts from Pycnoporus sanguineus Mycelium *Brazilian Archives of Biology and Technology* Vol. **54**, n. 6: pp 1167-1174

[5] Bulam S, Üstün N and Peksen A 2019 Health Benefits of Ganoderma lucidum as a Medicinal Mushroom *Turkish Journal of Agriculture – Food Science and Technology*, **7** (sp1): 84-93

[6] Chaturvedi V K, Argawal S, Gupta K K, Ramteke P W and Singh M P P 2018 Medicinal mushroom: boon for theurapeutic applications *Biotech* **8** 334

[7] Chen Y, Yuan Z, Bil S, Wang X, Ye Y and Svenning Y 2018 Macrofungal species distributions depend on habitat partitioning of topography, light, and vegetation in a temperate mountain forest *Scientific Reports Journal* **8**:13589

[8] De Silva D D, Rapior S, Sudarman E, Stadler M, Xu J, Alias S A and Hyde K D 2013 Bioactive metabolites from macrofungi: ethnopharmacology, biological activities and chemistry *Fungal Diversity* (2013) **62**:1-40

[9] De Miranda A M 2022 Agaricus brasiliensis (Sun mushroom) and its therapeutic potential: A review *Arch Food Nutr Sci.* 2022; **6**: 006-015

[10] El Sheikha A F 2022 Nutritional Profile and Health Benefits of Ganoderma lucidum "Lingzhi, Reishi, or Mannentake" as Functional Foods: Current Scenario and Future Perspective *Foods 2022*, **11**, 1030

[11] Google Earth 2022 Map of Universitas Gadjah Mada Yogyakarta

[12] Lee S S 2017 A Field Guie to the Larger Fungi of FRIM *Forest Research Institute Malaysia*

[13] Li M, Zhang Y, Lu Q, Gao Y, Ye Ting, Wang C and Xing D 2022 Structure, bioactives and applications of the polysaccharides from Tricholoma Matsutake: a review *Food Sci. Technol, Campinas,* **42**, e44922, 2022

[14] Liang X X, Gao Y Y, Pan Y, Zou Y F, He M, He C L, Li L X, Yin Z Q and Lv C 2019 Purification, chemical characterization and antioxidant activities of polysaccharides isolated from *Mycena dendrobii Carbohydate Polymers* **203** (2019) 45-51

[15] Liew G M , Khong H Y and Kutoi C J 2015 Phytochemical Screening, Antimicrobial and Antioxidant Activities of Selected Fungi from Mount Singai, Sarawak, Malaysia *International Journal of Research Studies in Biosciences* Vol. 3, Issue **1**, January 2015 PP 191-197

[16] Lima C U J O, Gris E F and Karnikowski M G O 2016 Antimicrobial properties of the mushroom Agaricus blazei – integrative review *Revista Brasileira de Farmacognosia* **26** (2016) 780-786

[17] Lu J, Tian J, Zhou L, Meng L, Chen S, Ma C, Wang J, Liu Z, Li C and Kang W 2021 Phytochemistry and Biological Activities of Poria *Journal of Chemistry* Vol **2021** Article ID 6659775

[18] Mueller G M, Bills G F and Foster M S 2004 Biodiversity of Fungi: Inventory and Monitoring Methods *Elsevier Academic Press* pp 1

[19] Na'iem M, Adriyanti D and Musyafa 2014 Pedoman Pengelolaan Vegetasi di Lingkungan Universitas Gadjah Mada *Direktorat Pengelolaan dan Pemeliharaan Aset Universitas Gadjah Mada* pp 3-10

[20] Orango Bourdette J O, Calixte E, Ngoua Meye Misso R L, Sima Obiang C, Prudence Y, Ondo, J P, Roger N and Obame Engonga L 2019 Phytochemical Screening, Antioxidant and

Antiangiogenic activities of Daedaleopsis nitida, Pycnoporus sanguineus and Phellinus gilvus Medicinal Mushrooms from Gabon *The Pharmaceutical and Chemical Journal* **6**(2). 71-80

[21] Orsine J V C, da Costa R V and Novaes R C G 2012 Mushrooms of the genus Agaricus as functional foods *Nutr Hosp.* 2012; **27** (4): 1017-1024

[22] Rahayu S, Wardana A, Nurjanto H, Anggara G and Mahayani N 2021 Macrofungi diversity in Lawu Mountain Forests and their potential uses as medicinal mushroom for inducing immunity against Covid-19 *IOP Conf. Series: Earth and Environmental Science* **914** (2021) 012077 (pp. 1-9). IOP Publishing

[23] Shahrajabian M H, Sun W, Shen H and Cheng Q 2020 Chemical compounds and health benefits of Tremella, a valued mushroom as both cuisine and medicine in ancient China and modern era *Amaz. Jour. Of Plant Resear*. **4**(3):692-697, 2020

[24] Thi B K, Lee S S, Zainuddin N and Chan H T 2011 A guidebook to the macrofungi of Fraser's Hill *Forest Research Insitute Malaysia*

[25] Van Norman K and Huff R 2012 Survey & Manage Category B Fungi Equivalent-Effort Survey Protocol pp 1-22

[26] Zainuddin N, Lee S S, Chan H T and Thi B K 2010 A guidebook to the macrofungi of Tasik Bera *Forest Research Institute Malaysia*

IOP Conf. Series: Earth and Environmental Science 1233 (2023) 012063 doi:10.1088/1755-1315/1233/1/012063

Management of Springs in Pluneng Village, Kebonarum, Klaten

N Khasanah[1] and S Purnama[2]
[1)]Environmental Science Master Programme, Universitas Gadjah Mada, Indonesia
[2)]Faculty of Geography, Universitas Gadjah Mada, Indonesia
*Corresponding author: nurilkhasanah95@mail.ugm.ac.id

Abstract. Springs are one of the most important natural resources for some people. Springs are often used as the main water source to meet household needs. In Pluneng Village, Kebonarum District, Klaten Regency, there are two springs that are very important for the surrounding community, namely the Tirtomulyono Spring and the Tirtomulyani Spring. The purpose of this research is to formulate a management strategy for the two springs that is in accordance with the conditions of the area. This research is analytical descriptive using survey method. Collecting data by interviewing local community leaders and field observations. The spring management strategy is to define and analyze the spring protection zone, namely the spring point protection zone, the spring protection zone from contaminants in the form of pathogenic bacteria, and the water catchment area protection zone. In addition, if the two springs will be used for domestic water requirement, treatment should be carried out first because the total coliform bacteria for both springs do not meet the quality standards.

Keywords: springs, management, Pluneng, Klaten Regency

1. Introduction

Water is a very important resource for human life. It can be said that water is a blessing because its presence affects the quality of life of every human being. According to Moench and Barlow (2000), the increasing need for water related to population growth raises awareness from the community to manage the existence of limited water resources.

Once the importance of this water resource in human life, it is often the cause of conflicts of interest. For example, there are still some conflicts between farmers who use water because they are fighting over water for irrigation. Then referring to reports from several newspapers, in Klaten in 2014 there was a dispute between mineral water companies and local residents regarding the use of water resources from springs.

Unfortunately, although it is known the importance of groundwater and springs as a source of daily water, the saving aspect is still often neglected [1 ; 2]. Whereas the amount of groundwater and springs is not unlimited because it is very dependent on rainwater that seeps into the ground through the mechanism of the hydrological cycle. Currently, there are indications that the infiltration of rainwater into the soil is decreasing due to land conversion [3 ; 4]. On the other hand, the need for living facilities and infrastructure is increasing, including the need for clean water [5 ; 6]. For this reason, management efforts must begin to be considered as early as possible before problems related to these two water sources occur in the future.

The purpose of water resource management is the realization of optimal hydrological conditions, namely obtaining water yields in accordance with the needs and requirements [7]. The problem in water resource management is how to make water available where it is needed in sufficient quantity, good quality, and available at all times. Therefore, water resource problems can be divided into three main issues, namely quantity problems, quality problems, and water distribution problems, both distribution according to space and time.

At present, integrated and sustainable management of water resources (including groundwater) has become a major issue for the world community. In an effort to manage water resources, an assessment of the potential of groundwater and springs is important. Management of groundwater and springs cannot be separated from overall water management. Good groundwater management will have a good impact on overall water management. Base on this background, the purpose of this research is to formulate a management strategy for the two springs in the research area i.e. the Tirtomulyono Spring and the Tirtomulyani Spring that is in accordance with the conditions of the area.

2. Research Methods

This research was conducted in Pluneng Village, Klaten Regency, where the land use is in the form of settlements and irrigated rice fields (Figure 1). This research is analytical descriptive using survey method. Collecting data by interviewing local community leaders and field observations. The form of questions used is a combination of closed and open questions. The instrument used is an interview guide to find out the use of land around the springs, the type of distribution and utilization of the springs, the users of the springs and other questions that can support information about the springs.

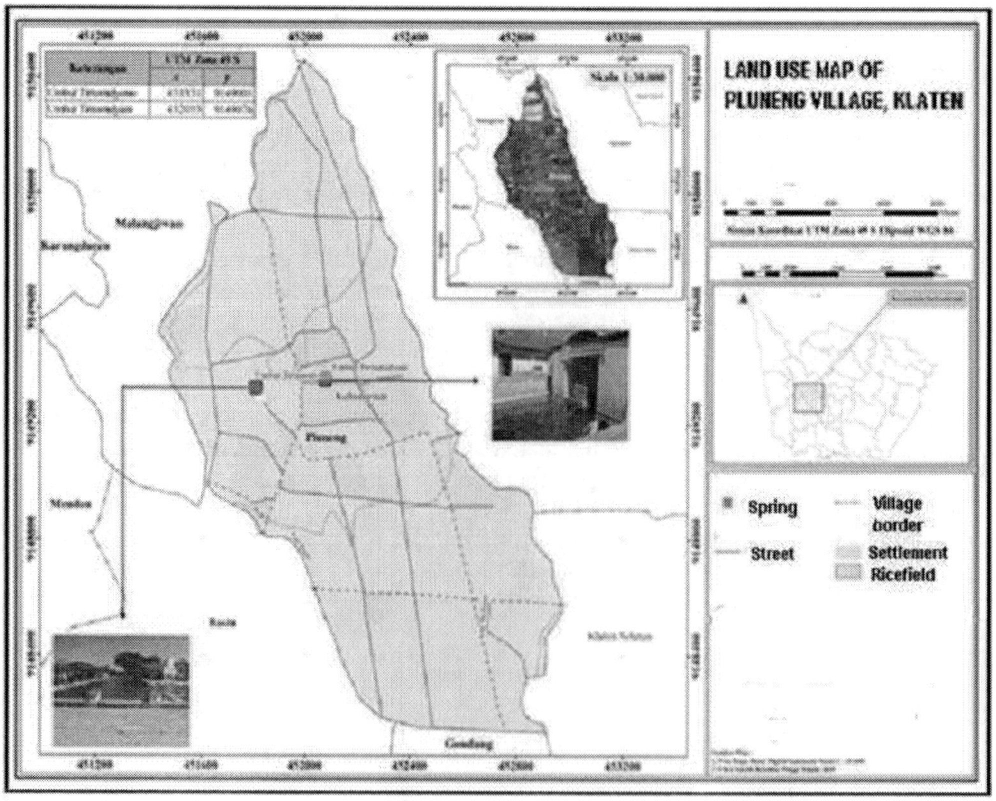

Figure 1. Land Use in Pluneng Village, Klaten

The interview technique used is an in-depth interview, which is conducted face to face with the 27 informant. The selection of one informant to the next is carried out sequentially using the snowball technique. The selection of informants ends if there has been repetition of information, meaning that by interviewing the next informant, it can be said that there is no additional meaningful new information or information saturation. Informants selected in this study were actors involved in the utilization and management of springs in Pluneng Village, namely village officials, water managers and community leaders. The method used to develop a management strategy is to study literature from various books, legislation or government regulations, similar research, field observations in the research area and water quality data from laboratory tests so that the strategies formulated and produced can be useful, focused and effective. useful but still in accordance with the conditions and situations in the research area.

3. Results and Discussion

3.1. Community-Based Water Supply and Sanitation Programme

The management of clean water supply in the community can be analyzed using several criteria. The criteria used are in accordance with those stated in Government Regulation number 16 of 2005. These criteria include operational, institutional, financing, regulatory and community participation aspects.

Operational aspects relate to water quality, quantity, and continuity, distribution, maintenance and operational infrastructure techniques [8]. The Tirtomulyono spring has been used by the residents of Pluneng Village for generations. Tirtomulyono spring is estimated to be tens of years or even hundreds of years old. According to local people, this banner was previously surrounded by rocks that fenced off the springs. Then around the 1970s a barrier wall was built between the spring area and the rice fields around the spring. After the existence of a clear spring management coupled with the formation of Tirta Sejahtera Village Owned Enterprises (BUMDes) as the manager of the Tirtomulyono Spring, the facilities were repaired to become pools to support water tourism needs such as learning to swim, playing in water and even believed to be a place of therapy to cure several diseases.

The Tirtomulyono spring is also used to meet the raw water needs of the people of Pluneng Village. Prior to the Community-Based Water Supply and Sanitation Program (PAMSIMAS), the community previously only relied on local residents' wells. Due to the presence of several residents' wells which were dirty and several wells suspected of being contaminated, PAMSIMAS was formed to drain water from the springs to the houses of residents in Pluneng Village. The selection of the Tirtomulyono spring water source which is used as the raw water source for PAMSIMAS is due to the quality, quantity and continuity that can be said to meet the standards to meet the needs of the community. However, some communities with adequate quality and quantity of well water still use well water sources and add springs. The use of PAMSIMAS is also due to easy access and saving time and energy because they no longer need to draw well water for their daily needs.

The clean water supply system in Pluneng Village includes raw water units, distribution units to consumers as clean water recipients. The raw water unit has the most important part, namely the spring catcher building.

Water sources are the most important part of the clean water network. Springs are usually easier to protect because the water that comes out of the ground is directly accommodated by the Spring Catcher, which is usually a reservoir and is given a fence around it for security. Ponds of sufficient size are made to raise water so that it can be easily drawn or flowed through the intake pipe. In general, water treatment plants are a very important component of the clean water network. However, in the raw water supply system in Pluneng Village, this installation is not needed because water from springs is considered to meet drinking water requirements. Distribution and service units cover the distribution of clean water to consumers. In order for water continuity to work well, the clean water distribution system in Pluneng Village uses gravity and pumping. The pump is connected to the distribution pipe as well as the taller sump. When the water requirement is low, the water is stored or flows into the reservoir. However, when the demand for water increases, the flow of water in this distribution is supplied from the pump and from the reservoir itself by utilizing the height of the place.

According to [9], a reservoir is a place to store clean water. Reservoir types can be divided based on the shape, function, height of the reservoir to the surrounding soil surface and construction materials. Based on the height of the reservoir relative to the surrounding land surface, reservoir types can be divided into surface reservoirs and tower reservoirs. The type of reservoir contained in the clean water supply system of Pluneng Village is a tower reservoir.

A tower reservoir is a reservoir in which the entire reservoir reservoir is located higher than the surrounding ground level. Based on the construction material, the reservoir in Pluneng Village is a cast concrete reservoir. The advantages of a cast concrete reservoir are that it is watertight and does not leak easily. The main function of the reservoir is to balance the production discharge and water usage discharge, because often the clean water production discharge is not the same as the water consumption. When the amount of clean water production is greater than the amount of water used, the excess water is stored in a reservoir, and is used to meet water shortages when the amount of clean water production is less than the amount of water used. In addition to these main functions, the reservoir also has a function to increase water pressure in the distribution pipeline network so that the water pressure in the distribution pipeline network is relatively stable, regulates water pressure and overcomes other emergencies.

The type of pipe used for the distribution of clean water in Pluneng Village is PVC. The advantage of using this type of pipe is that it is light in weight, making it easy to transport and install and does not react with water. According to [10], the type of pipe has a very large influence on network services, durability, and investment and operating costs. In addition, the type of pipe will also determine the water pressure in the pipe that can be held. In general, the selection of the type of pipe is very dependent on the desired water pressure, water quality, chemical reactions, the influence of sunlight, theft and security against water pressure from outside.

The operation of the service unit includes house connections and the existence of a public faucet with a water meter. Currently, every house connection in Pluneng Village is equipped with a water meter. However, until now there has been no fee charged to the community as PAMSIMAS water customers. It is planned that in 2022 there will be a retribution of Rp. 1,400/m^3 of water.

3.2. Institutional aspects

After years of individual management of the Tirtomulyono and Tirtomulyani springs using an auction system, on November 5, 2015, the Tirta Sejahtera BUMDes was established. Since January 1, 2018 until now the Tirtomulyono and Tirtomulyani springs, especially for tourism activities, have been managed professionally by BUMDes Tirta Sejahtera. In addition to several improvements and additional facilities, an expansion of the Umbul Tirtomulyono area has also begun to reach an area of 4,000 m2. In addition to developing the potential for water tourism, the Tirtomulyono Spring is also developed as a source of raw water through the Pamsimas program which can provide clean water and the community to meet domestic water needs.

The management of the PAMSIMAS program is also planned to be transferred to BUMDes. The existence of BUMDes is expected to strengthen the existing PAMSIMAS operational activities. In addition, the plan to add employees is also expected to assist operational activities and maintenance of the existing clean water supply system.

The people of Pluneng Village also have unwritten rules regarding water sources from springs that every citizen must obey. This regulation is a form of local wisdom for the preservation of springs. Efforts to conserve springs that are applied in the community include not littering near springs, keeping springs clean and not using soap or other chemicals near springs.

3.3. Aspects of community participation

BUMDes Tirta Sejahtera as the manager of the Tirtomulyono and Tirtomulyani springs has recruited employees from Pluneng Village residents. The recruitment is indirectly a form of community

IOP Conf. Series: Earth and Environmental Science 1233 (2023) 012063 doi:10.1088/1755-1315/1233/1/012063

representation in the active involvement of the management of the springs. Currently, the management of BUMDes Tirta Sejahtera is still related to the development of water potential as a tourist destination. The management of cleanliness around the springs has become the responsibility of BUMDes as an effort to protect the springs. In addition, the village government and the community also play an active role in this effort through PAMSIMAS, while operationalization and services are the responsibility of the village as the institution responsible for running the Pluneng Village PAMSIMAS program.

There is a banyu thanksgiving tradition in Pluneng Village. This annual tradition began in 2014 and was last held in 2018. The series of banyu thanksgiving activities usually begins with tree planting and cleaning in the river with the culmination of a cultural carnival and wayang kulit performance. There are 18 tumpeng mountains originating from all RWs in Pluneng Village and other villages in the Kebonarum District. The mountains are paraded from the mountain of arts and culture of Pluneng Village to Umbul Tirtomulyono. This banyu celebration is held as a form of gratitude and an effort to preserve and keep the water that is enjoyed always flowing and can be felt for generations to come.

3.4. Strategy of spring management in Pluneng Village

Along with population growth, it is estimated that the population's need for clean water in Pluneng Village in 2035 will also increase. The development and utilization of water resources should be a serious concern for the local government, by not focusing on efforts to develop the tourism potential of water resources but prioritizing the utilization and development of the potential of springs as raw water for the needs of the local community. From a physical point of view, there is no local residents' concerns about water shortages or decreasing quality. However, it should be noted that there may be impacts that will occur in the future if development in the tourism sector is prioritized without paying attention to environmental conditions.

Aspects of maintenance at the location of the springs need to be done by keeping the plants or trees around the springs. According to information from local residents, previously at the location of the Tirtomulyono spring there was a banyan tree, but now the tree can no longer be found, replaced by densely populated residential areas and areas that have become permanent facilities such as bathrooms for visitors to the springs. The lack of public interest in replanting spring protection plants such as banyan, is due to the assumption that the amount of water resources is still sufficient to meet the daily needs of the people of Pluneng Village. In the current condition, the impact of the loss of spring protection plants may not be felt, but in the long term it is feared that it will result in a decrease in water resources along with the increase in population and the increase in clean water.

It is important to maintain plants as part of the landscape aspect that will affect the condition of the quantity and quality of groundwater. An example is the decrease in bamboo plants that switch functions to built-up land which results in a decrease in the quantity and quality of water from springs and groundwater [11].

According to [12], the concept of spring protection has a wider spectrum and is not just the area around the springs, but covers the entire water catchment area (DTA) of the springs. Therefore, it is important to apply spatial zoning for the protection of springs. Based on laboratory tests, the condition of the Tirtomulyono and Tirtomulyani springs did not meet the quality standards in terms of total coliform, which is the reason for the importance of zoning the protection of the springs so that there will be no deterioration in quality as well as from a chemical and physical point of view (Table 1).

Table 1. Management Strategy of Tirtomulyono and Tirtomulyani Spring

Springs Name	Problems	Management Strategy

Tirtomulyono	• The demand for water in 2035 is estimated to increase as well as the need for water for PAMSIMAS as well as the development of water tourism activities, requiring the availability of sufficient water. Tourism development is also expected to pay attention to the spring environment so that the availability of spring water can always be met. • There are no regular measurements so that data on spring discharge cannot be monitored for its availability. • There are parameters that do not meet the required quality standards PerMenKes RI No. 492/MENKES/PER/IV/2010, namely the total coliform bacteria of > 1600 MPN/ 100 ml.	• The need for the role of the government and related agencies regarding monitoring and controlling activities that can reduce, change or eliminate the protected function of an area. Even if necessary, prohibition can be carried out. • Create spring protection zoning and spring area analysis. • Provide treatment to spring water.
Tirtomulyani	• The need for spring water is increasing from year to year. • Varied rainfall and fluctuating discharge • Total coliform > 1600 MPN/100 ml sample that does not meet the required quality standards. • Based on direct observations in the field, local communities carry out domestic activities such as washing and bathing directly in the spring area which should be protected from chemicals, such as laundry soap and shampoo.	• Protect water catchment areas. • Create spring protection zoning and analysis of spring zone areas.

Spring protection zones can refer to the area of raw water sources carried out in Europe [13]. The zoning system allows it to be applied in Indonesia, especially in the case of spring protection in the Pluneng Village area. One of the concrete actions to maintain water quality so that it can be used as a source of raw water for the community, is to protect the source from the pollution process. The enactment of water source protection zones is the first step in protecting the quality of the raw water sources used by the community.

In simple terms, the division of protection zones consists of at least the location of the springs, the area around the springs, and the water catchment area of the springs. Each protection zone has specific characteristics and protection objectives, which will form the basis for determining the protection strategies that can be implemented. The catchment area protection zone needs special attention because the condition of its existence greatly affects the sustainability of the function of water sources in an area. One of the catchment areas that requires protection due to sand extraction activities is the Kemalang area, Klaten Regency, which is one of the water catchment areas in Klaten Regency. Based on the study of the environmental status of the Klaten Regency, sand extraction activities in protected areas are mostly carried out on the upper slopes to the middle of Mount Merapi, which are included in the Kemalang and Prambanan Districts. In the long term, it is feared that the impact of the damage will affect the water absorption capacity which will result in a decrease in the quality and quantity of water sources. For this reason, it is necessary to carry out further studies and determine environmental strategies in handling the protected area of the water catchment area.

So that the community's need for raw water sources can be met at all times, the government and all authorized institutions as well as local communities are expected to play an active role in efforts to protect springs. Activities that can reduce, change or eliminate the protected function of an area should be controlled, and even if necessary, prohibition can be carried out. Water tourism activities that have existed so far can be evaluated by taking into account aspects of the environmental conditions of the springs so that they do not cause negative impacts in the future.

4. Conclusion

Management strategies that can be carried out regarding the two springs include the following:

1. Define and analyze the spring protection zone, namely the spring point protection zone, the spring protection zone from contaminants in the form of pathogenic bacteria, and the water catchment area protection zone.
2. Tirtomulyono and Tirtomulyani springs can be used to domestic water requirement, especially for drinking water. However, its use must be processed first because the results of laboratory tests state that the total coliform for both springs based on the Minister of Health Regulation No.492/MENKES/PERIV/2010 does not meet the quality standards.

Acknowledgement

This article is part of a study for thesis entitled "Analysis of Springs Potency to Fulfill Domestic Water Requirement in Pluneng Village, Kebonarum District, Klaten" financed by Final Assignment Recognition (Rekognisi Tugas Akhir/RTA) Programme Universitas Gadjah Mada year 2022

References

[1] Moench, A.F and Barlow, P.M. (2000). Aquifer response to stream-stage and recharge variations. analytical step-response functions. *Journal of Hydrology* 230 : 192-2010.
[2] Agarwal R. & Garg P. K. (2016). Remote sensing and GIS based groundwater Potential and recharge zones mapping using multi-criteria decision making technique. *Water resour Manage* 30 : 243-260.
[3] Mishra N., Khare D., Gupta K.K., Shukla R. (2014). Impact of land use change on groundwater-a review. *Advances in Water resource and Protection* 2 : 28-41.Wanielista M, Kersten R and Eaglin R. 1997. Hydrology : Water Quantity and Quality Control. John Wiley and Sons Inc, New York.
[4] Tam V.Y & Nga T.T.V. 2018. Assesment of urbanization impact on groundwater resources in Hanoi, Vietnam. *Journal of Environmental Management* 227 : 107-116.
[5] Jhariya, D. C., Shrivastava S., Nage S. K., Chateerjee R. S. (2015). Integrated remote sensing and GIS approach to groundwater potential delineation in the Doon Valley, Uttarakhand, India. *SAARSC International Journal of Geo Science and Geoinformatics* 2 (3) : 1-11.
[6] Machival D., Rangi., N., Sharma A. (2015). Integrated knowlewdge and data diven approaches for groundwater potential zoning using GIS and multi-criteria decision making techniques on hard-rock terrain of Ahar Catchment, Rajasthan, India. *Environ Earth Sci* 73 : 1871-1892.
[7] Fenta A. A. & Kifle A. (2015). Spatial analysis of groundwater potential using remote sensing and GIS based multi-criteria evaluation in Raya Valley, Northern Ethiopia. *Hydrology Journal* 23 : 195-206.
[8] Akhmadi. 2011. Pola Pemanfaatan Mata air Tuk Babon Dan Tuk Pakis Oleh Masyarakat Lokal Di Kawasan Taman Nasional Gunung Merbabu. *Thesis*. Semarang: Program Pasca Sarjana Universitas Diponegoro.
[9] Human Resources Development Agency of the Ministry of Public Works. 2018. Perencanaan Jaringan Pipa Transmisi Dan Distribusi Air Minum. (Online).

https://bpsdm.pu.go.id/center/pelatihan/uploads/edok/2018/11/aee05_3._Fungsi_Ground_Reservoir.docx.pdf. Diakses pada tanggal 30 Oktober 2021.

[10] Kaunang, C.D dan Halim L.K.F. 2015.Pengembangan Sistem Penyediaan Air Bersih di Desa Maliambao Kecamatan Likupang Barat Kabupaten Minahasa Utara. *Jurnal Sipil Statistik*. 3 (6): 361-372.

[11] Suharjo, Musiyam M., dan Arozaq M. 2008. Agihan Tanaman Bambu dan Implikasinnya Terhadap Potensi Air Tanah Daerah Kabupaten Klaten Jawa Tengah.*The 5th Urecol Proceeding*.Yogyakarta: Universitas Ahmad Dahlan.

[12] Setiawan, O. 2020. Perlindungan Mata Air di Tatanan Normal Baru: Sebuah Tinjauan Praktis. (Online) *https://www.menlhk.go.id/site/single-post/3164/perlindungan-mata-air-di-tatanan-normal-baru-sebuah-tinjauan-praktis*. Diakses pada tanggal 4 November 2021.

[13] Ekklessia, R., Sukandarrumidi dan Hendrayana, H. 2005. Penentuan Zona Perlindungan Sumber Air Baku Pada Sumur Bor Mojosongo, Kadipiro dan Jebres Kota Solo Propinsi Jawa Tengah. *Jurnal Manusia dan Lingkungan*. 12 (2) : 88-103.

Manganese prospecting using time domain induced polarisation (TDIP), case study: Karangnunggal district, Tasikmalaya regency

R M Antosia[1], R Yolandika[1] and W W Parnadi[2*]

[1] Department of Geophysical Engineering, Institut Teknologi Sumatera
[2] Department of Geophysical Engineering, Institut Teknologi Bandung

*Corresponding author: wahyudi@gf.itb.ac.id

Abstract. Indonesia has good manganese potential, but its utilization has not been optimal. It has been unable to meet domestic needs because of the low production level. Thus, it needs to conduct more exploration activities. Karangnunggal district, Tasikmalaya regency, is one of the manganese minerals located. The method to prospect the manganese distribution in the area was applying a geophysical approach. The measurement employed the time domain induced polarisation (TDIP) method in 17 lines. In this study, the process deployed the Wenner array. Each line has a length of about 117.5 m with an electrode spacing of 2.5 m. The outcome shows that a manganese mineral has a chargeability of more than 5 ms. Based on the 3D visualization, it can be identified that most deposits spread to the south and the east while partially located in the north and the west of the study area. The 3D chargeability modeling generates the estimating manganese volume of about 2.4 million cubic meters.

1. Introduction

Manganese is an abundant natural resource in Indonesia that is widely used by industry, but its utilization has not been optimal [1]. Especially with the development of manganese, consumption in the world is increasing daily, where manganese has a significant role in various fields such as the iron/steel industry, food, agriculture, and chemical [2,3,4]. The steel industry still consumes 90 % of manganese production [3,5]. Then, with the development of the manganese-based battery industry [4,5], for example, Nickel Manganese Cobalt Oxide and Lithium Manganese Oxide [4], the demand for manganese minerals will also increase. Indonesia's manganese potential is spread over several islands, such as Sumatra, Java, Sulawesi, Nusa Tenggara, and Maluku [4,6]. Because the production level is still low, it has not been able to meet domestic needs, so more imports from other countries [1,4]. Therefore, it is necessary to increase exploration activities as one of the steps to increase Indonesia's manganese production.

One of the manganese-producing areas on the island of Java is Karangnunggal [6,7]. Karangnunggal district, Tasikmalaya regency, West Java province, is where manganese minerals were first discovered in Indonesia [6,7]. However, the exploitation of manganese in the area has been discontinued since 2005 [6]. Perhaps this happened because the mining was only carried out based on the visible presence of manganese on the surface. Geologically, the region has hilly geomorphology with relatively parallel and partially irregular ridges. It is because the area is a subduction area of the India-Australia plate with the Eurasian plate [8]. The area during the Middle Miocene experienced an uplift, causing the formation of faults and folding and a shift in volcanism from the north [9]. With these conditions, the area still has good potential for manganese reserves. It is supported by data on reserves on the island of Java, reaching around 12.5 million tons of manganese ore [4].

In mineral exploration, applying the science of geology and geophysics is necessary to obtain subsurface information easier. A geological/ geophysical survey is very effective because it can be carried out quickly and cheaply in addition to being environmentally friendly [10]. Previously, [11,12] studied manganese in the area from a geological point of view. But, there has been no study in terms of geophysics to determine the estimated amount of manganese in the study area. One of the geophysical methods that can map the earth's subsurface is the time domain induced polarization (TDIP) method. The parameter acquired is chargeability; it initially describes the ability to store a material's electrical energy and is sensitive to a conductive material content [10,13-16]. A high chargeability value indicates the presence of metallic mineral zones in the study area [16-19]. Many studies have implemented the induced polarization method for minerals prospecting [14-22]. Even the electrical resistivity methods only have also been applied [23,24]. Some focused on manganese minerals with similar results in interpreting from chargeabiliy data [14-20]. But, some combine it with other methods, such as resistivity [16-19] or electromagnetics [20]. Therefore, because there was no scientific study related to the distribution of the mineral in the Karangnunggal region, in this study, the authors will apply the same approach of the TDIP method to approximate the manganese distribution.

2. Material and Method

According to the geological map of the Karangnunggal [8], the research area (figure 1) in the Karangnunggal district, Tasikmalaya regency, is part of the *Anggota Genteng-Jampang* Formation (Tmjg). In the formation, planktonic foraminifera showed the age of the Lower Miocene-Middle Miocene. In addition, the tuff rocks are interspersed with dacitic breccias and limestone inserts, where this unit contains black minerals and quartz as mixing minerals [8]. The tuff rock is similar in composition to igneous rock but physically like sedimentary rock and has porosity. It is a type of pyroclastic rock formed by volcanic eruptions that release magma, rock debris, dust, and other materials.

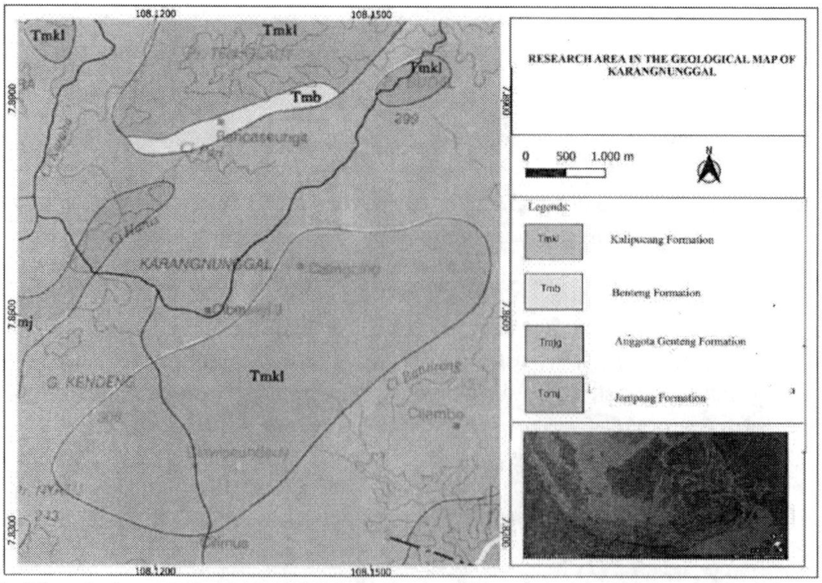

Figure 1. The study area in the geological map of Karangnunggal (modified from [8,25]).

In this study, there are 17 lines of 2D TDIP in an area of about 270 thousand square meters. Each line has a 117.5 m length with an electrode spacing of 2.5 m. A total of 6 lines spread from south to north, and 11 stretches from north to south. In detail, the design of the TDIP measurement trajectory can be seen in figure 2. The potential difference change due to the polarization in the time domain measurement (TDIP). The polarization effect is a condition where when the current injection is stopped, the measured potential difference does not immediately become zero but will slowly decay to zero. So that the time domain will have a primary potential difference (V_P) measured when a current injection is turned on and a second potential difference (V_S) is measured when the current injection shuts down. Thus, the parameter will be obtained in the form of a ratio of V_P/V_S, called chargeability (ms) [10,13].

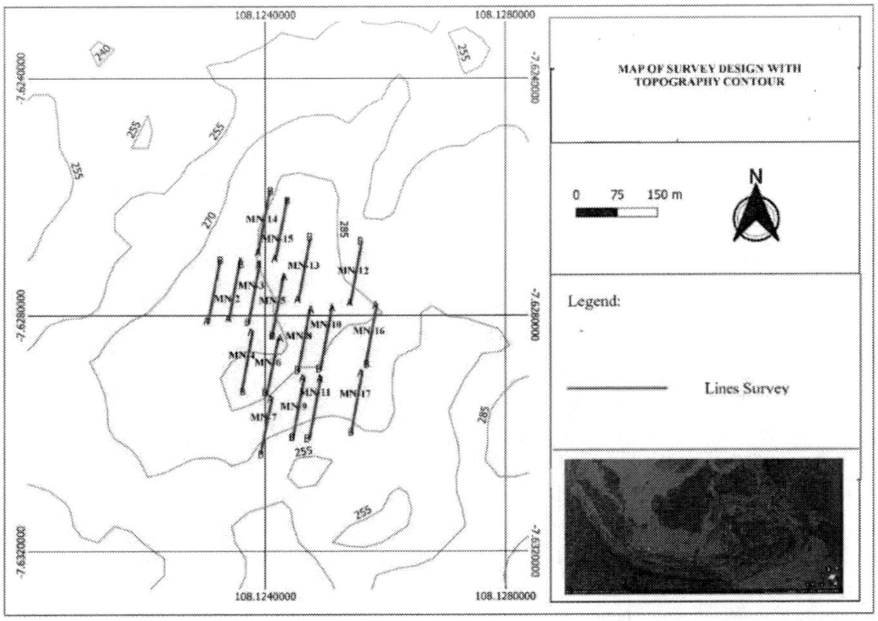

Figure 2. The 2D TDIP survey design with topography contour (modified from [25]).

Figure 3 illustrates the TDIP method when the current is injected, and the potential difference shows a gradual decrease. This condition describes the measurement of the primary potential difference (V_P) and secondary potential difference (V_S). The figure also shows the relationship between potential difference and time (the time will increase as the potential slowly reaches zero value). $V_S(t_1)$ is the measured potential at t_1 and $V_S(t_2)$ is the measured potential at t_2. The shaded part is the measured chargeability value and can be calculated using the following equation [13],

$$M = \frac{1}{V_P}\int_{t_1}^{t_2} V_S(t)dt \quad (1)$$

where, M = chargeabiliy (ms), V_P = primary potential (V), and $V_S(t)$ = secondary potential concerning time (V).

The Wenner configuration is one of the configurations in the electrical resistivity method. This configuration is often used for profiling purposes. The profiling result is a 2D pseudo-section [19]. The arrangement of the electrodes in the Wenner configuration is symmetrical in a line and has the same electrode spacing ($C_1P_1 = P_1P_2 = P_2C_2 = a$). The array has a good vertical resolution and fairly horizontal resolution [26]. The illustration can be viewed in figure 4. Although the geometrical factor value is not used for the TDIP method, it also applies the same array. The process only involves electrode positions of the configuration and the chargeability value of the datum point. So, the output will also generate a 2D pseudo-section.

IOP Conf. Series: Earth and Environmental Science 1233 (2023) 012064 doi:10.1088/1755-1315/1233/1/012064

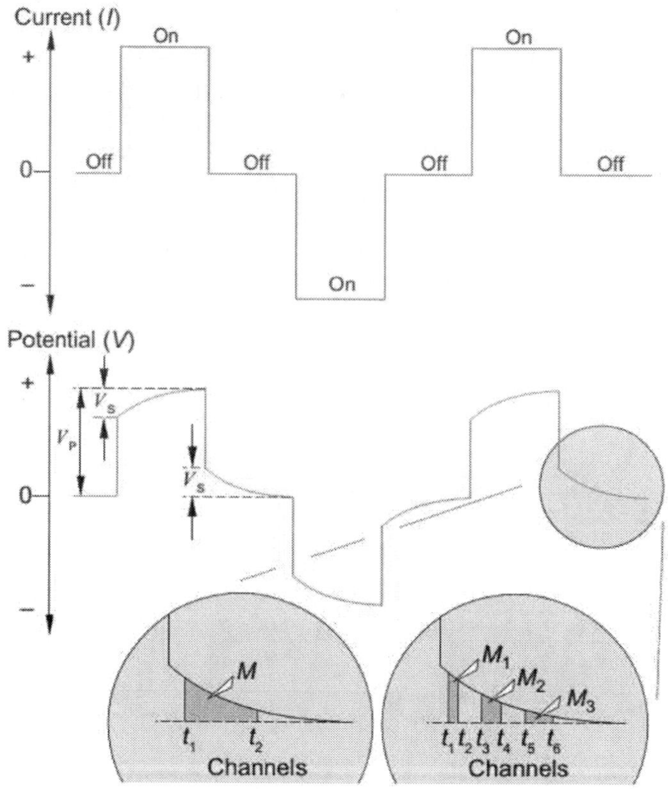

Figure 3. The signal waveforms used in TDIP measurements [13], the time domain bipolar square wave signal, and the distorted wave measured by the receiver. Polarisation effects produce the slow rise and decay in the received signal. The time intervals used to measure chargeability (M) are shown shaded.

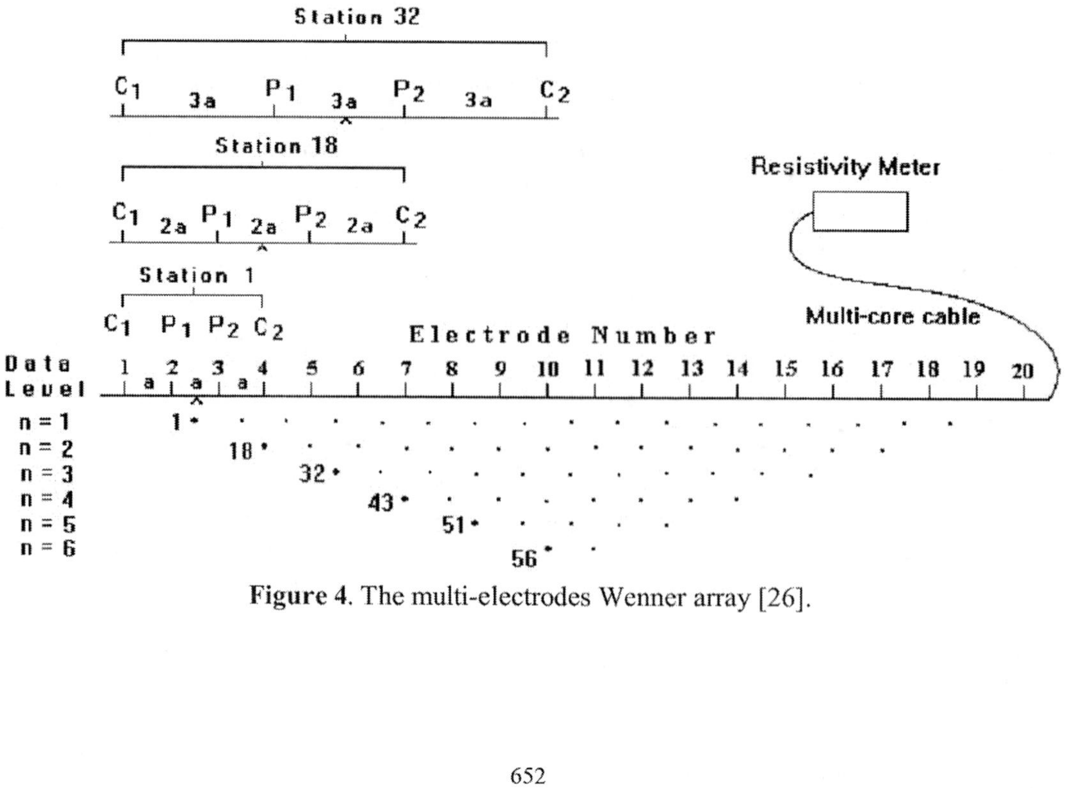

Figure 4. The multi-electrodes Wenner array [26].

3. Result and Discussion

After summarizing some studies, the manganese chargeability values close to the study area are about 9-20 ms [15-18]. The authors select the minimum cut-off level for the manganese mineral as 5 ms because, in some cases, the value of 5 ms is estimated as a conductive mineral. The area's 2D profiles of the TDIP measurement are shown in figures 5-8. Figure 5 presents no conductive material in the subsurface area. It consists of lines 3, 5, 6, and 14 (see figure 2 to view the survey map). The profiles have a value distribution below 5 ms. Although there are spotted anomaly zone in lines 3 (top right corner) and 14 (bottom middle), it still has low chargeability.

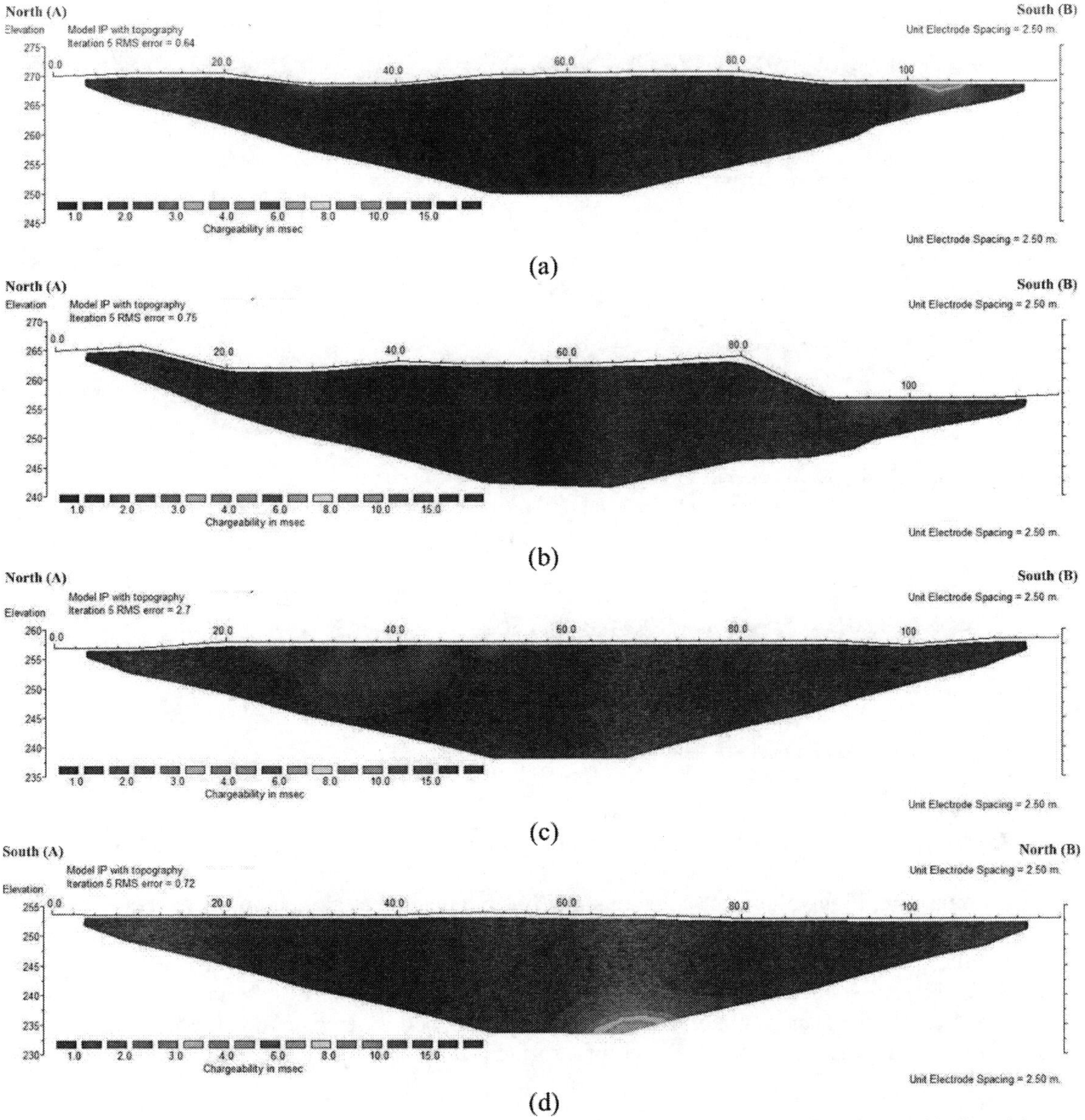

Figure 5. The 2D chargeability pseudo-section (a) line 3, (b) line 5, (c) line 6, and (d) line 14.

IOP Conf. Series: Earth and Environmental Science 1233 (2023) 012064 doi:10.1088/1755-1315/1233/1/012064

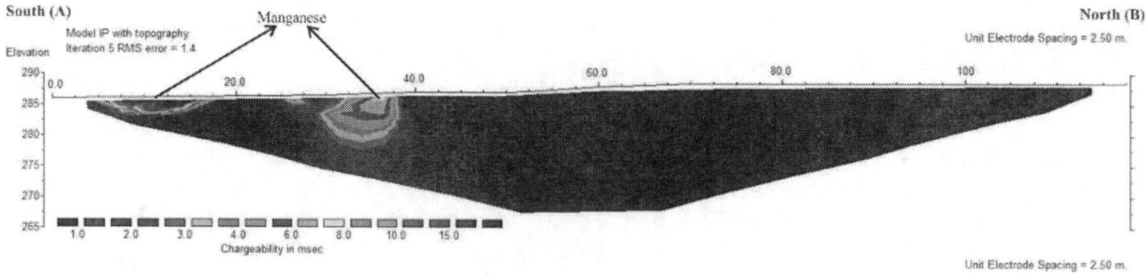

(a)

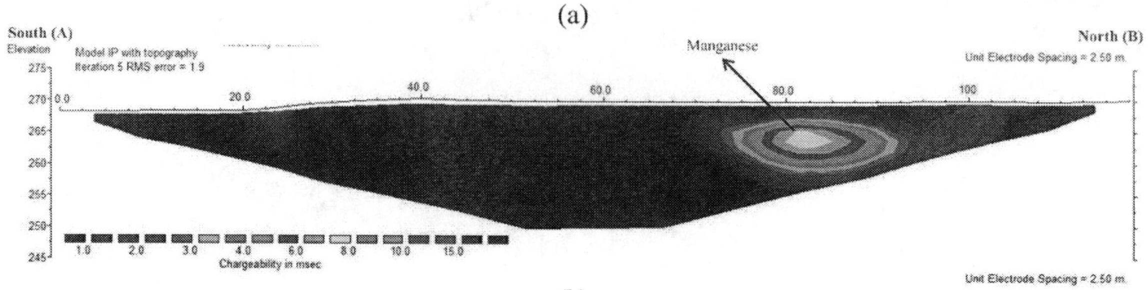

(b)

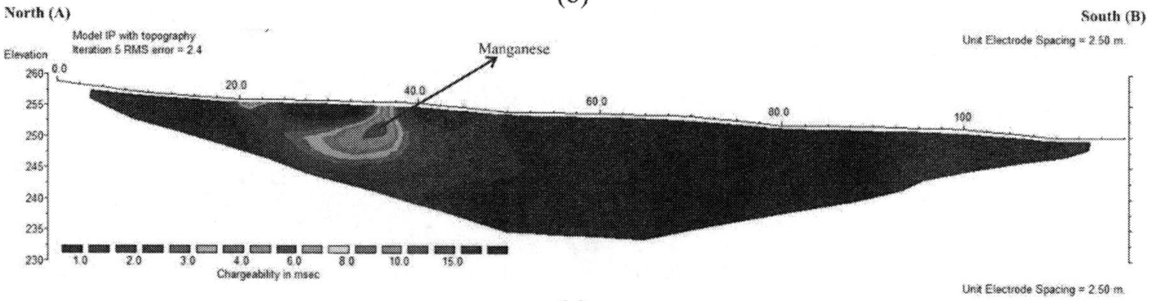

(c)

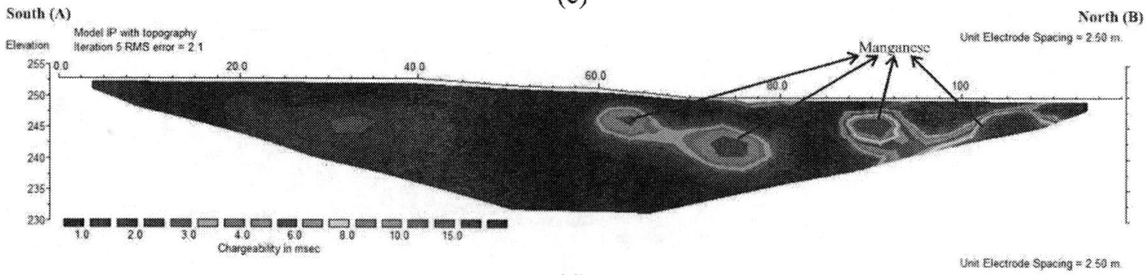

(d)

Figure 6. The 2D chargeability pseudo-section (a) line 1, (b) line 2, (c) line 9, and (d) line 12.

In comparison, high-value anomalies in some lines indicate manganese boulders. They are lines 1, 2, 9, and 12 in figure 6. It has small boulders with approximate diameters of 5 m up to 10 m. Meanwhile, giant boulders have a diameter of about 10-20 m. They can be seen in figure 7, which consists of lines 4, 7, 13, and 17. The anomaly is marked with a sign arrow for each profile (figures 6-8). Moreover, the manganese minerals are like a spreading bed in lines 8, 10, 11, 15, and 16. They are almost filled along the profile under the ground surface with a depth of more than 5 m.

IOP Conf. Series: Earth and Environmental Science 1233 (2023) 012064 doi:10.1088/1755-1315/1233/1/012064

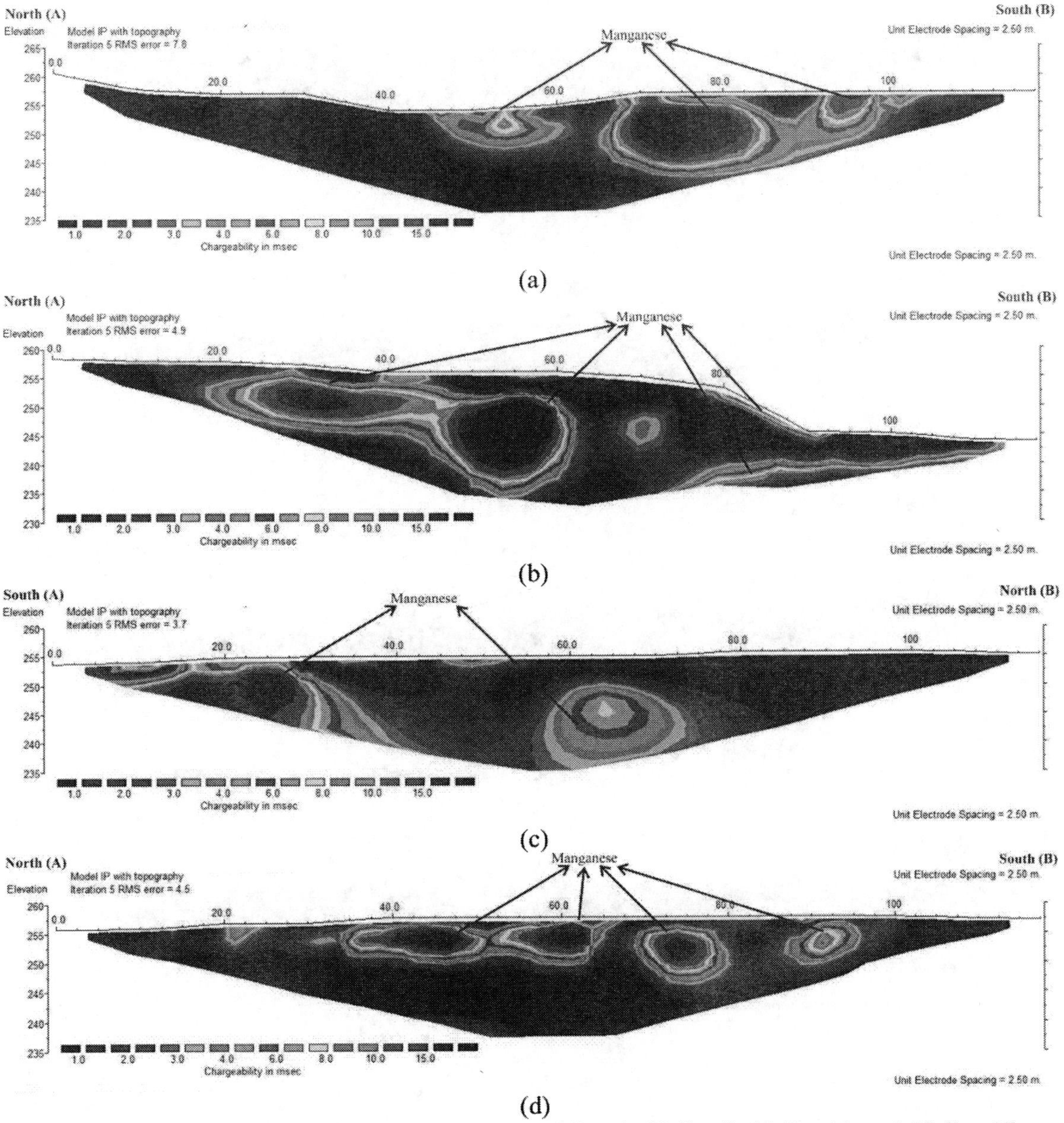

Figure 7. The 2D chargeability pseudo-section (a) line 4, (b) line 7, (c) line 13, and (d) line 17.

According to [11], the constituent rocks on each line can be grouped into three lithologies: tuff stone, limestone tuff, and limestone. The presence of manganese in the area is associated with limestone tuff and limestone. Based on figure 9, the manganese deposit is distributed in the south and the east. It is identified as a primary deposit. The formation is due to hydrothermal influence, the area's geological conditions in the form of folds and fractures to control the manganese formation in the study area [12]. Some deposits are partial as a boulder in the north and the west, as secondary manganese deposits, which are thought to have formed from the deposition of primary sediment [12]. The 3D chargeability modelling (figure 9) shows that the manganese volume is approximately 2.4 million cubic meters.

IOP Conf. Series: Earth and Environmental Science 1233 (2023) 012064 doi:10.1088/1755-1315/1233/1/012064

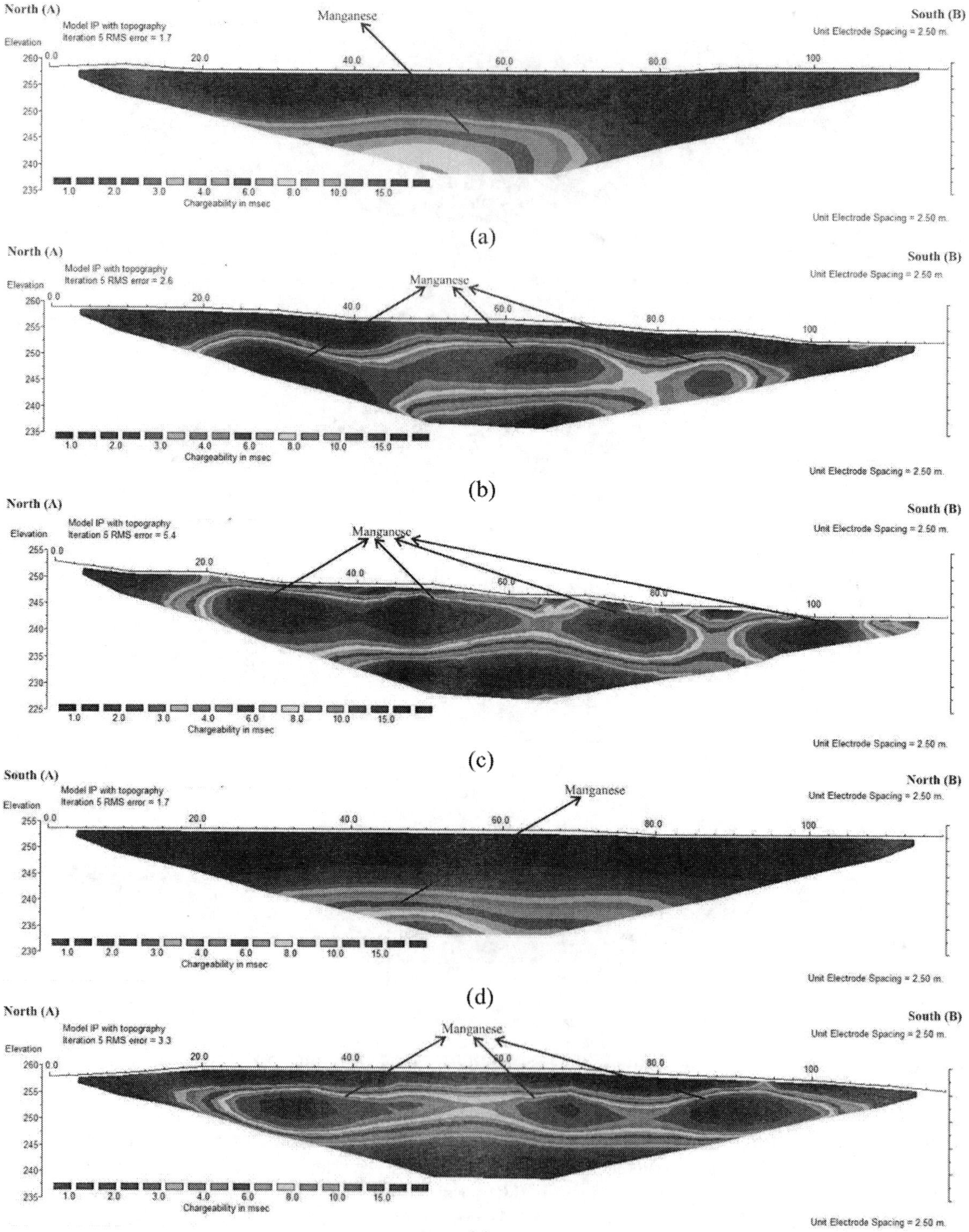

(e)

Figure 8. The 2D chargeability pseudo-section (a) line 8, (b) line 10, (c) line 11, (d) line 15, and (e) line 16.

Estimating a manganese reserve (in tons) in the study area can not be determined from the result. It needs further chemical study to get information about the effective manganese mass of a meter cubic to calculate the true manganese reserve.

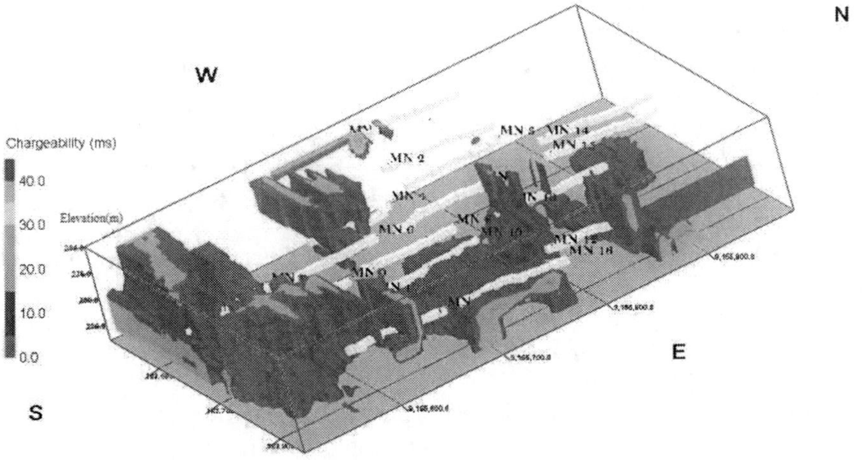

Figure 9. The 3D chargeability visualization.

4. Conclusion

The study area has good potential as one of the manganese producers in Indonesia. From the time domain induced polarisation (TDIP) method results, the manganese mineral anomaly has a chargeability value above 5 ms up to 20 ms. Then, the 3D modeling result shows that the primary deposit spreads to the south and the east. Meanwhile, the partial mineral is in the area's north and west. It has a significant volume of 2.4 million cubic meters. It is necessary to conduct further research to support the results, such as a geochemical study to measure the effective mass to get the estimated manganese reserve (in tons). In addition, it needs to expand the similar study in the region to get the big picture of the distribution of the mineral reserve (to increase manganese production in the future).

References

[1] Panjaitan R R 2011 Kajian Pemanfaatan Batu Mangan / Senyawa Mangan dalam Industri *Berita Litbang Industri* (http://lib.kemenperin.go.id/neo/download_artikel.php?id=744. [Accessed: 17-Aug-2022])

[2] Mufakhir F R and Sumardi S 2015 Kajian Pengolahan Bijih Mangan dan Prospek Aplikasinya di Indonesia *Lembaga Ilmu Pengetahuan Indonesia* (http://lipi.go.id/publikasi/kajian-pengolahan-bijih-mangan-dan-prospek-aplikasinya-di-indonesia/2950. [Accessed: 18-Aug-2022])

[3] Vale 2016 Mangan dan campuran besi *Pertambangan* (http://www.vale.com/indonesia/BH/business/mining/manganese/Pages/default.aspx#:~:text=Hampir%2090%25%20hasil%20produksi%20mangan,%2C%20pakan%20hewan%2C%20dan%20mobil. [Accessed: 17-Aug-2022])

[4] Yasin C M, Ramanda Y, Wardhana A I, Ferlianta W, Praditya A and Asdriargo A 2021 *Peluang Investasi Mangan Indonesia* (Jakarta: Direktorat Jenderal Mineral dan Batubara, Kementerian Energi dan Sumber Daya Mineral)

[5] ANS PT 2022 Kegunaan Batu Mangan *Batu Mangan PT. Anugrah Nusantara Sejahtera* (http://www.batumangan.com/batu-mangan/. [Accessed: 17-Aug-2022])

[6] Supriadi A, Sunarti, Kencono A W, Kurniasih T N, Prasetyo B E, Kurniawan F, Kurniadi C B, Alwendra Y, Aprilia R, Rabbani Q, Setiadi I and Anggreani D 2017 *Kajian Dampak Hilirisasi Mineral Mangan Terhadap Perekonomian Regional*, Cetakan pertama (Jakarta: Pusat Data dan Teknologi Informasi Energi dan Sumber Daya Mineral, Kementerian Energi dan Sumber Daya Mineral, Kementerian Energi dan Sumber Daya Mineral)

[7] Sulistiyono E, Sufiandi D and Suharyanto A 2015 Pemanfaatan Bjih Mangan Dari Karangnunggal Tasikmalaya Sebagai Bahan Baku Baterai Lithium *Lembaga Ilmu Pengetahuan Indonesia* (http://lipi.go.id/publikasi/pemanfaatan-bjih-mangan-dari-karangnunggal-tasikmalaya-sebagai-bahan-baku-baterai-lithium/3132. [Accessed: 31-Jul-2022])

[8] Supriatna S, Sarmili L, Sudana D and Koswara A 2005 *Peta Geologi Lembar Karangnunggal*, Edisi Kedua (Bandung: Pusat Penelitian dan Pengembangan Geologi)

[9] Van Bemmelen R W 1949 *The Geology of Indonesia*, vol. 1A (The Hague: Government Printing Office)

[10] Reynolds J M 2011 *An Introduction to Applied and Environmental Geophysics* (Chichester: Wiley-Blackwell)

[11] Sudarsono S and Setiawan I 2009 Genesa Cebakan Mangan Karangnunggal Tasikmalaya: Pendekatan secara Petrografi dan Mineragrafi *Prosiding Pemaparan Hasil Penelitian Puslit Geoteknologi-LIPI* **2009** 1–8.

[12] Sata M 2002 Geologi Endapan Mangan Daerah Karangnunggal Jawa Barat *Perpustakaan Digital ITB* (https://digilib.itb.ac.id/index.php/gdl/view/13974. [Accessed: 01-Aug-2022])

[13] Dentith M C and Mudge S T 2018 *Geophysics for the mineral exploration geoscientist* (Cambridge: Cambridge University Press)

[14] Yatini Y, Santoso D, Laesanpura A, and Sulistijo B 2016 Effects of changes in metallic mineral to time domain induced polarization (TDIP) response on physical modeling fieldscale *AIP Conference Proceedings* **1746** 020010

[15] Vieira L B, Moreira C A, Côrtes A R P and Luvizotto G L 2016 Geophysical modeling of the manganese deposit for Induced Polarization method in Itapira (Brazil) *Geofísica Internacional* **55** 107–117

[16] Srigutomo W, Trimadona T and Pratomo P M 2016 2D Resistivity and Induced Polarization Measurement for Manganese Ore Exploration *Journal of Physics: Conference Series* **739** 012138

[17] Bahri A S, Hilyah A and Putra A N 2021 Pengoptimalan Dugaan Mineral Mangan (MN) Dengan Korelasi Metode Resistivitas 2D Dan Polarisasi Terinduksi 2D Di Lapangan 'X', Kabupaten Timor Tengah Utara, Nusa Tenggara Timur *Jurnal Geosaintek*, **7** 103–110

[18] Setiarini A, Linuwih S and Khumaedi K 2018 Identifikasi Mineral Mangan Menggunakan Metode Polarisasi Terinduksi Di Desa Pucung Kecamatan Eromoko Kabupaten Wonogiri," *Dinamika Rekayasa* **14** 119–124

[19] Anaperta Y M and Gusman M 2016 Survey Geofisika Induksi Polarisasi (IP) Mangan di Nagari Jambak Kecamatan Lubuk Sikaping Kabupaten Pasaman *Jurnal Teknologi Informasi dan Pendidikan* **9** 122–128

[20] Zakaria M F 2018 Identification of distribution and calculation of manganese reserves from VLF and IP method in Pengasih, Yogyakarta *EAGE-HAGI 1st Asia Pacific Meeting on Near Surface Geoscience and Engineering* **2018** 1–5

[21] Nasruddin N M, Saputra A V, Nafisah M D and Sehah S 2020 Identifikasi Lapisan pembawa Mineral Emas Dengan Metode resistivitas Dan Induksi Polarisasi di Desa paningkaban kecamatan gumelar kabupaten banyumas jawa tengah *Jurnal Fisika Flux* **17** 20–29

[22] Truffert C, Gance J, Leite O and Texier B 2018 Full-3D electrical resistivity tomography and induced polarization for enhancing deep mineral exploration *2nd Conference on Geophysics for Mineral Exploration and Mining* **2018** 1–5

[23] Seran R and Edi E 2021 Kajian Geofisika Dan Geokimia Mangan di Desa Oetalus Kabupaten

TTU *Indonesian Journal of Applied Physics* **11** 40–50

[24] Antosia R M, Mustika, Putri I A, Rasimeng S and Dinata O 2021 Andesite prospect at West Sungkai of North Lampung: Its distribution based on electrical resistivity tomography *IOP Conference Series: Earth and Environmental Science* **882** 012086

[25] Earth G. 2022 -7.628 108.1200 *Google Earth* (https://earth.google.com/web/search/-7.628+108.1200/@-7.628,108.12,247.60281214a,1046.08182638d,35y,0h,45t,0r/data=ClMaKRIjGU-Nl24Sgx7AIUfhehSuB1tAKg8tNy42MjggMTA4LjEyMDAYAiABIiYKJAkH91EuNTA3QBHlRw0TFkkpwBlZDAuLcNVjQCGjD3ykSfpRQA. [Accessed: 18-Aug-2022])

[26] Loke M H 2022 Tutorial: 2-D and 3-D electrical imaging surveys *Geotomo software – download* (https://www.geotomosoft.com/downloads.php. [Accessed: 01-Aug-2022])

Physico-chemical properties and stable isotope composition of springwater in Leuwisari subdistrict and surrounding, Tasikmalaya, West Java, Indonesia

S Fadillah[1,2], D A Kuncoro[2], F Priandani[2]

[1]Department of Geological Engineering, Faculty of Engineering, Universitas Gadjah Mada, Indonesia
[2]Directorate General of Water Resources, Ministry of Public Works and Housing, Indonesia

Abstract. Springs are a crucial source of water supply in quarternary volcanic environments, such as Galunggung Volcano in Tasikmalaya Regency, yet their quality and relationships remain poorly defined. Physicochemical properties and stable isotopes of hydrogen (δ^2H) and oxygen ($\delta^{18}O$) were used to examine springwater's eligibility as drinking water sources, assess the connectivity among springs and surface water, and evaluate the origin of groundwater resources. Compared with WHO standards, the springwater samples met the criteria for drinking water eligibility. Eight springs assessed are expected to originate from an unconfined aquifer actively recharging. The isotopic data indicate little or no influence of direct recharge from the rivers on this aquifer. These isotope data, integrated with available information on physical hydrogeology and water chemistry, will be used to identify and define the spring recharge and protection zones.

Keywords: groundwater, spring, Tasikmalaya, stable isotopes

1. Introduction

Water is essential to life, and the world's most extensive freshwater supply is underground [1]. Groundwater, which accounts for more than 95 percent of the world's unfrozen freshwater reserves, is primarily a renewable resource and, if properly managed, guarantees a long-term supply that helps meet growing demand [2]. While natural water supplies are generally stable, their demand rises with the human population [3].

In the past few decades, the rate of population expansion in Tasikmalaya Regency and Tasikmalaya City, West Java, Indonesia, has been exceptional. This region's population increased from 2,300,000 in 2008 to 2,600,000 in 2021 [4], [5]. The population of Tasikmalaya is anticipated to expand by 0.97 percent every year until 2045, when it will reach 3,300,000 people. People in this region depend on groundwater, particularly springwater, as their primary potable water supply.

Springs are an essential water source for various uses in the Tasikmalaya region. They provide good quality water for domestic use, freshwater fishing, and paddy irrigation. However, the springwater's quality and the relationships among the springs of the reclaimed area are still lacking. In order to inform, manage, and conserve future water resources, it is necessary to collect new information on the water cycle and the linkages between precipitation, surface waters, and groundwaters.

Spring is an intense discharge of groundwater that emerges as a stream of flowing water at the land surface [1]. The amount of water collected by aquifer recharge, groundwater extraction, and spring capture can change significantly. These rates of change can be important information for developing aquifer and spring management strategies. The groundwater system is dynamic; changing one part of the system will ultimately affect other factors [6]. The vast majority of groundwater systems are dynamically linked with surface water [6]. Spring research is a liminal field because it signifies the transition between groundwater and surface water [7]. It may also investigate groundwater dynamics in areas with few boreholes or wells [8].

Stable isotopes of hydrogen and oxygen have long been utilized in water cycle research as naturally occurring tracers [9], [10]. The significance of precipitation, the sources and flow of water to and from surface water within watersheds, and the origin and age of groundwater resources could be determined by analyzing the isotopic data [6], [9], [10]. When integrated with available information on physical hydrogeology and water chemistry, the isotopic data of springwater aid in understanding the source and process of recharge, groundwater movement and sustainability, groundwater travel time distribution, hydraulic connectivity, groundwater origin, and evolution [11]. This isotopic data may apply mainly due to water-rock interaction, aquifer depth, recharge area temperature, and the source and outcome of groundwater contamination [11]. The use of these isotopes emerges from their predictable fractionation as water flows through the hydrologic cycle, depending on the physical and chemical mechanisms involved [12].

Research on the application of stable isotopes to determine the origin of groundwater has been carried out in various geological and climatological settings across the globe. In Canada, exhaustive isotopic precipitation, surface water, and groundwater analyses were conducted in the Okanagan Valley [13]. In Indonesia, a study to determine the origin of groundwater in Umbul Cokro, the giant spring on the eastern foot slope of Merapi Volcano, using isotope carbon-13, deuterium, and oxygen-18, has been conducted [14]. Using a similar method, another study identified that Umbulan Spring (Pasuruan, East Java) originates from an unconfined aquifer [15]. No isotopic assessments of springwater in the study area have been conducted. This lack of information denotes a significant lack of knowledge concerning the hydrogeology of the Tasikmalaya Groundwater Basin.

This study used the physicochemical properties and stable isotopes of water to achieve the following: (1) examine the springwater's eligibility based on physicochemical parameters in comparison to the WHO drinking water standard, (2) compare the isotopic composition and physicochemical properties of springwater and surface water in the study area to determine the relationship among them, and (3) evaluate the isotopic composition and physicochemical properties of springwater to precipitation in the study area to determine the origin of groundwater. When integrated, the stable isotope data from all critical hydrological components will provide the basis for future water supply evaluation in the research area. This study is the initial step in further scientific and management discussion to implement conservation strategies for the Tasikmalaya region. Furthermore, it is essential to adapt existing approaches, such as isotope hydrology, and to enhance their reliability in this humid tropical environment with a wide variety of altitudes.

2. Study Area

Leuwisari subdistrict is one of the subdistricts in Tasikmalaya Regency, West Java. Leuwisari is located on the southern side of the Galunggung Volcano, one of about 129 active volcanoes in Indonesia [16]. Galunggung Volcano is situated at around 7°15’ S and 108°03’ E, including a part of Garut and Tasikmalaya Regency, West Java. This volcano was recorded to have erupted in 1822, 1894, 1918, and 1982 [17]. Young deposits were formed due to these eruptions on the ruins of older deposits from eruptions before 1822 [18]. Volcanic deposits resulting from quarternary volcano activity are productive aquifers, marked by the appearance of springs [19].

IOP Conf. Series: Earth and Environmental Science 1233 (2023) 012065 doi:10.1088/1755-1315/1233/1/012065

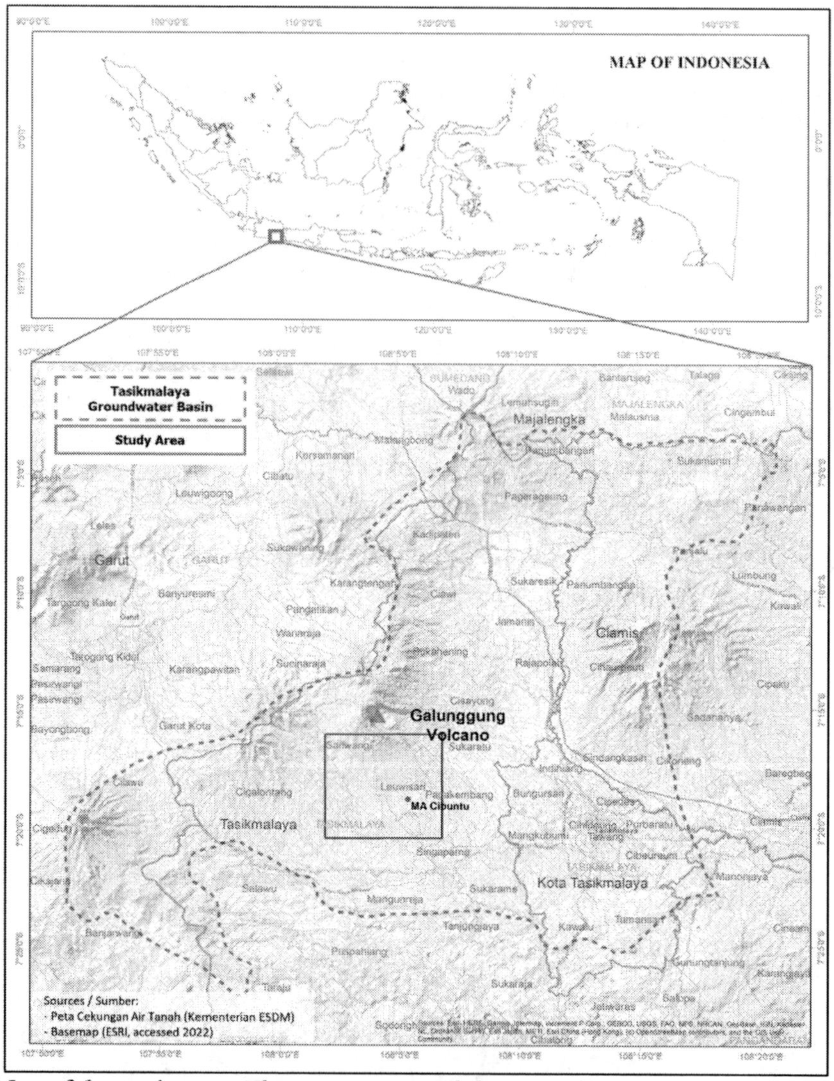

Figure 1. Map of the study area. There are twenty-eight groundwater basins in West Java, one of which is the Tasikmalaya Groundwater Basin, bordered by the red dashed line. The study location, Leuwisari subdistrict and surrounding, is in the Tasikmalaya Groundwater Basin, marked by a red square. [20]

According to regional geological maps [21], shown in Figure 2, rocks in the study area consist of 3 formations: (1) Younger Volcanics/Qvg1, containing volcanic breccia, lahars, and andesitic to basaltic tuffs from Galunggung; (2) Galunggung Volcanic Breccia/Qvb4, weakly consolidated volcanic breccias, containing of blocks andesitic lava forming isolated blocks; and (3) Older Volcanics/Qtvk, containing volcanic breccia, flow breccia, tuff, andesitic lava from Mt. Kukus.

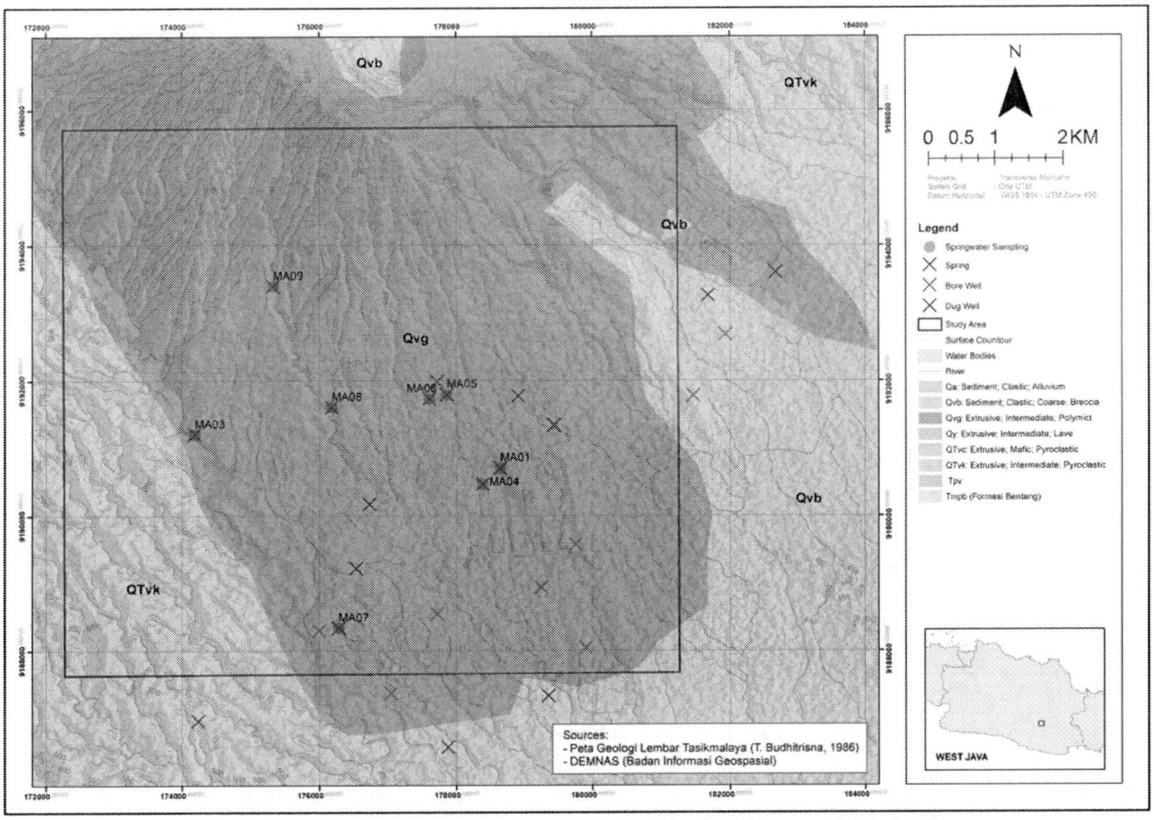

Figure 2. The regional geologic map of the study area. [21], [22]

Based on the updated Köppen-Geiger climate classification [23], the climate of the study area is classified as tropical rainforest. Rainfall almost occurs over the whole year. Overall, the precipitation amount in the basin is high, 3,500~5,000 mm/year, and minimum precipitation occurs in June, July, and August. The average annual temperature area is approximately 24.9°C, and the temperature is relatively uniform throughout the year, with a variation that never exceeds 7°C. The average monthly humidity is 84.1%, with a minimum value of 67.1% and a maximum of 92.5% [4].

Groundwater is the primary source of water in the Tasikmalaya region. More than 90% of the water provided by Tasikmalaya Municipal Waterworks (i.e., PDAM Tasikmalaya) originates from springs, or approximately 480 liters per second [24].

There are three main drainage patterns: Cimanuk to the north, Citanduy to the east, and Ciwulan to the south. The study area is in the Ciwulan watershed. The land use of the study area is mainly irrigated rice fields, particularly on the northern and eastern of Galunggung's foot slope.

3. Material and Methods

3.1. Water table and physicochemical properties measurement

The physicochemical parameters are pH, electrical conductivity (EC), Total Dissolved Solids (TDS), and temperature were measured in situ using HANNA instrument HI9812-5. Before use, the HANNA instrument is calibrated by dipping the probe in calibration solution and adjusting the knob to match the solution's value. The instrument's probe was dipped into the cup, and readings were recorded after stabilization. Water-table elevations are measured using a roll-meter in a network of dug wells in the study area's unconfined aquifer (Figure 3). The Topo to Raster method in GIS Software interpolates data and spatially computes values between measurements for the water table mapping.

IOP Conf. Series: Earth and Environmental Science 1233 (2023) 012065 doi:10.1088/1755-1315/1233/1/012065

3.2. Springwater samples

The springwater samples were collected once in March 2020. The springwater sampled in the Leuwisari subdistrict is shown in Figure 3. Eight of the eleven springs identified were chosen to test, and all were screened in younger volcanic formations. A list of springwater sampling locations is provided in Table 1.

Table 1. Springwater sample locations in the study area.

No.	**ID**	**Spring Name**	**UTM WGS 1984 49S X (m)**	**UTM WGS 1984 49S Y (m)**	**Altitude (m asl)**
1	MA01	Cibuntu	178646	9190709	554
2	MA03	Cipondok	174190	9191208	632
3	MA04	Citerewes	178385	9190469	536
4	MA05	Kalipicung	177858	9191796	638
5	MA06	Manggung	177602	9191732	629
6	MA07	Cijambe	176278	9188341	508
7	MA08	Cipiit	176176	9191609	652
8	MA09	Cipager	175324	9193399	857

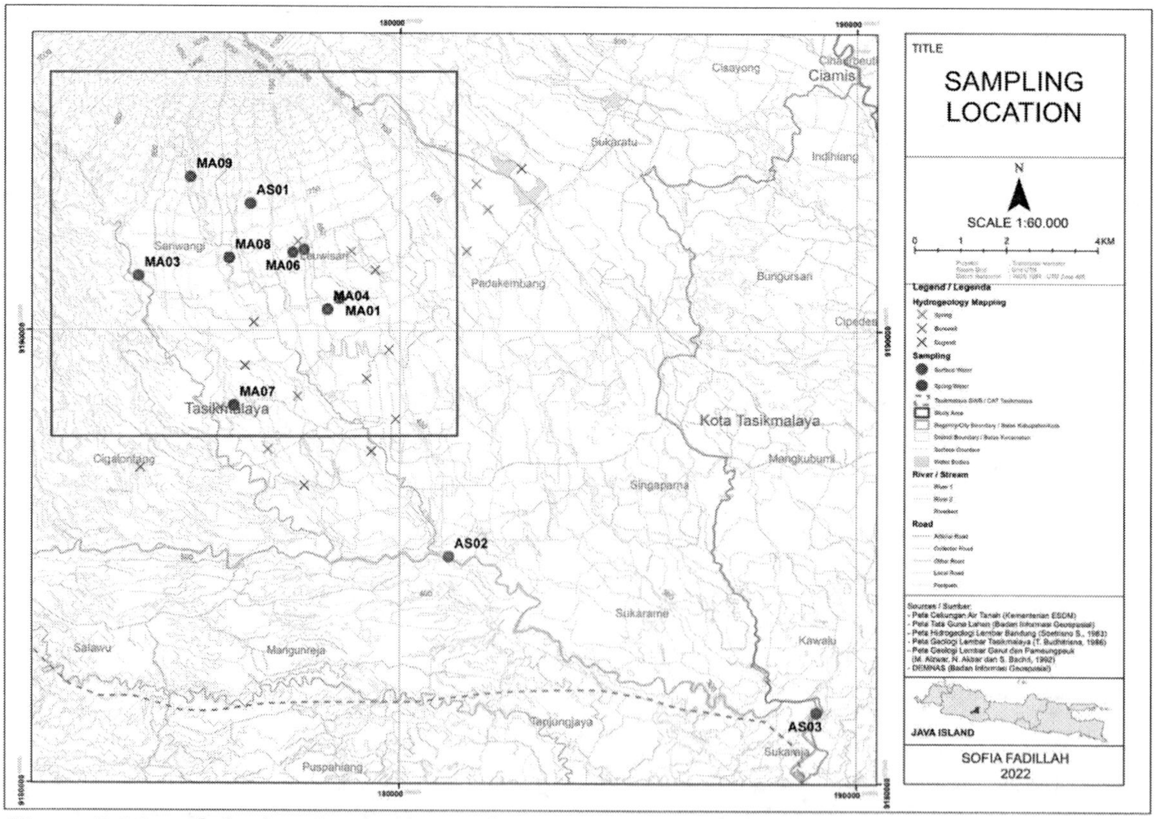

Figure 3. Map of physico-chemical properties measurement and sampling locations. The cross marks are the locations of physico-chemical measurement, and the blue and purple cross marks are the locations of water level measurement. The blue dots are spring water sampling locations, and the magenta dots are surface water sampling locations.[20], [25]

3.3. Surface water samples

The surface water samples were collected once in March 2022 at 3 locations representing the upstream, midstream, and downstream in Tasikmalaya Groundwater Basin, as shown in Figure 3. Table 2 provides a list of surface water places.

Table 2. Surface water sample locations in the study area.

No.	ID	Object Name	UTM WGS 1984 49S X (m)	UTM WGS 1984 49S Y (m)	Altitude (m asl)
1	AS01	River (Upstream)	176645	9192812	773
2	AS02	River (Midstream)	181080	9184997	393
3	AS03	River (Downstream)	187605	9188048	264

3.4. Water sampling methods

Unfiltered groundwater and surface water samples were collected using 100 ml amber glass bottles. Before being filled, the glass bottles were washed with sample water. The bottle is filled until there are no air bubbles and then sealed with the inner and outer closures immediately to avoid exposure to air. Springwater is collected as close as possible to a visible outlet of spring. In easy access points, river water samples were taken approximately 10 cm below the surface. During collection (handling and preservation), samples are refrigerated in a cooler box to avoid evaporation fractionation.

3.5. Stable isotopic analysis

Water samples were analyzed for δ^2H and $\delta^{18}O$ at the Laboratory of Hydrogeology and Hydrogeochemistry, Institut Teknologi Bandung, Bandung, West Java, Indonesia. Picarro Water Stable Isotope Analyzer for Water Isotope used to conduct isotopic analysis.

Generally, the isotopic composition of water is determined by comparing it to a selected sample of ocean water, called Standard Mean Ocean Water (SMOW) [26]. The stable isotopic composition is expressed in terms of the relative value δ.

$$\delta_x = \left(\frac{R_x}{R_{SMOW}} - 1\right) \times 1000 \quad (1)$$

Where δ is expressed in parts per thousand (‰), R_x is the isotope ratio (e.g., $^2H/^1H$ or $^{18}O/^{16}O$) in the substance x, and R_{SMOW} is the isotope ratio in the corresponding international standard substance.

3.6. Precipitation isotopic data

The first measurements of $\delta^{18}O$ and δ^2H in freshwaters worldwide were done in 1961 [26]. This measurement showed a high correlation between $\delta^{18}O$ and δ^2H. Therefore, it is possible to calculate the recharge input for specific places using the characteristic Meteoric Water Lines (MWLs). Globally, the oxygen and hydrogen ratios in precipitation (GMWL) are given by the equation [26]:

$$\delta^2H = 8(\delta^{18}O) + 10 \quad (2)$$

Rainwater isotope data is needed to estimate the Local Meteorite Water Line (LMWL). In this study, LMWL was taken from previous research conducted in some areas in Indonesia, including in Garut (Garut MWL) and other areas on Java Island. The Garut MWL, which will be used in this study, is described by the equation [27]:

$$\delta^2H = 8.06(\delta^{18}O) + 16.48 \quad (3)$$

IOP Conf. Series: Earth and Environmental Science 1233 (2023) 012065 doi:10.1088/1755-1315/1233/1/012065

4. Results and Discussion

4.1. Water table and physicochemical properties

Water table measurements are critical for management purposes and establishing flow directions. Interpreting groundwater flow patterns using data collected from observations and measurements of groundwater levels is essential for understanding the hydrogeology of the study area.

The altitudes of spring heads and river channels obtained from topographic maps also give essential information on the water table's level since they reflect the points at which the saturated zone meets the ground surface [28]. The slope of the water table is determined by the drainage system's base, the land surface's topography, and the presence of impermeable rock beds [10]. Groundwater discharge points, such as wells and springs, can be observed to gather information on the depth and elevation of the groundwater level. In this study, groundwater level measurements were carried out at 11 springs, and 3 dug wells. The acquired data is then processed to create a contour map of the height of the groundwater level by interpolating the data from the observation points.

Groundwater flow patterns can be determined from the contour map. Contours are interpolated between the measured water-table elevations using the topo to raster method in GIS. In addition, the lack of primary and supporting data in the northern part of the study area means that the drawing of the groundwater level contour lines can only be understood in light of the surrounding topography contours and morphological features. A dashed line in the northern part of the study area represents these equipotential lines. Then, the flow lines are drawn at the equipotential lines perpendicularly, as shown in Figure 4.

According to the groundwater flow pattern in the research area, groundwater generally moves from northeast to southeast, following the existing hydrogeological map [29]. The equipotential lines in the northern study area tend to be denser than the southern part. This pattern suggests a decrease in both the topographic gradient and the hydraulic gradient towards the south of the study area.

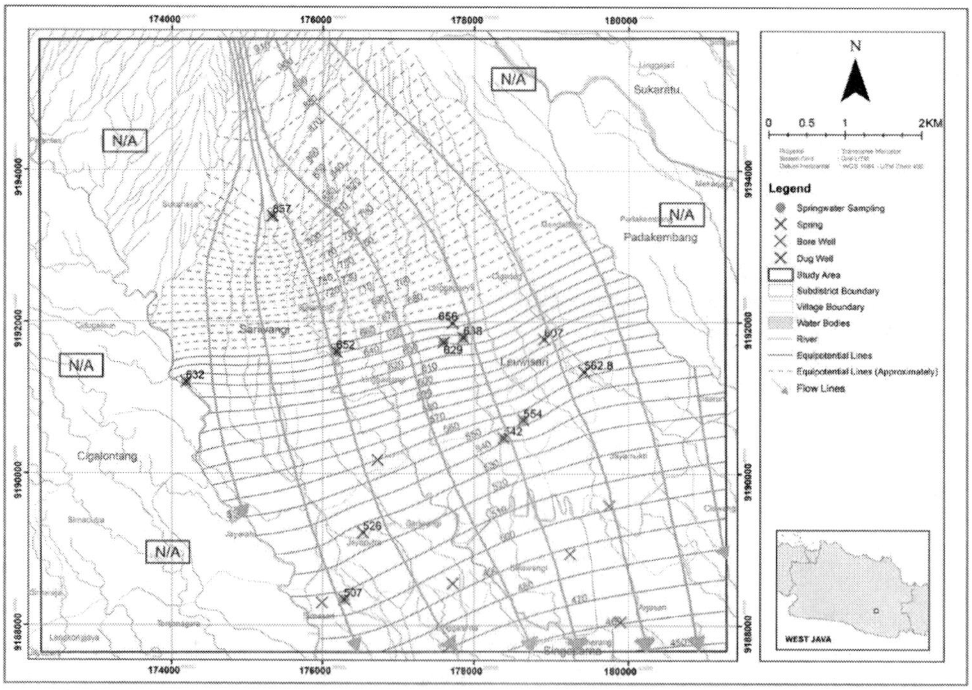

Figure 4. The locations of water table measurement; the equipotential lines based on the location of water table measurement; The arrows represent the predicted groundwater flow directions of an unconfined aquifer in the study area.

Temperature data is highly correlated with hydrologic conditions. Groundwater with a temperature near the local mean annual surface temperature is classified as a shallow active water cycle, with movement restricted to 100 m and, on rare occasions, 200 m [10]. The springwater's temperature fluctuates from 22°C to 26°C (Table 3), near the average annual surface temperature of 24.9°C, indicating that the observed springs come from a shallow active water cycle. The slight discrepancy between them is presumed to indicate an unconfined aquifer. Based on the categorization of water temperature [30], all identified springs are normal (Figure 5.a).

Table 3. The physicochemical properties of springwater in the Leuwisari subdistrict, Tasikmalaya Groundwater Basin.

No.	ID	Object Name	Altitude (m asl)	Physico-chemical Properties			
				T (°C)	pH	TDS (mg/L)	EC (µS/cm)
1	MA01	Cibuntu Spring	554	26	6.8	90	190
2	MA03	Cipondok Spring	632	22	6.9	87	180
3	MA04	Citerewes Spring	536	24	6.7	80	163
4	MA05	Kalipicung Spring	638	23	6.5	70	150
5	MA06	Manggung Spring	629	22	6.7	73	160
6	MA07	Cijambe Spring	508	26	7.0	133	280
7	MA08	Cipiit Spring	652	24	7.2	120	250
8	MA09	Cipager Spring	857	22	7.1	70	150
			Maximum	26	7.2	133	280
			Minimum	22	6.5	70	150
			Average	24	6.9	90	190

The pH scale specifies the composition of water: a value of 7 represents neutral water, a higher value suggests basic water, and a lower value suggests acid water [10]. Depending on the composition of the water and the type of construction materials used in the distribution network, the optimal pH needed will vary between 6.5 and 8.5 for various water sources [31]. The pH values of springwaters sampled in this study range from 6.5 to 7.2, with an average value of 6.9 (Table 3). Thus, according to pH classification [32], all springwaters are considered neutral.

Total Dissolved Solids (TDS) refer to the total concentration of dissolved material in groundwater [3]. According to WHO guidelines [31], TDS up to 600 mg/l is considered good, while 1,000 mg/l is the maximum allowed level; anything over that will be unpalatable. The TDS value of the springwater samples ranges from 70 mg/l to 133 mg/l (Table 3), much lower than the WHO standard as shown in Figure 5.c. According to the TDS classification [33], all observed spring water is classified as freshwater (TDS<1000 mg/L).

Electrical conductivity (EC) is the ability of 1 cm^3 of water to transmit electrical current [3]. EC generally correlates with dissolved solids' presence and gives an indirect method for determining water salinity [34]. The EC value for drinking water must not exceed 400 µS/cm [35]. The EC values of observed springs vary from 163 µS/cm to 280 µS/cm (Table 3). As per EC classification [34], seven springwaters are rated excellent for human consumption or irrigation purposes, while one spring water is deemed good (Figure 5.d).

Based on the values of pH, TDS, and EC, compared with WHO standards and recommendations, it can be concluded that all spring water samples met the criteria for drinking water eligibility. Apart from being an indicator of the feasibility of drinking water, this physicochemical parameter can also support the analysis of the origin of groundwater. The origin of springwater that comes from rainwater and belongs to the meteoric water cycle can be indicated by the low EC, TDS, and similar chemical properties to rainwater [6], [19], [36]. As observed from the previous data presentation, the springwater in this research is sourced from the current water cycle.

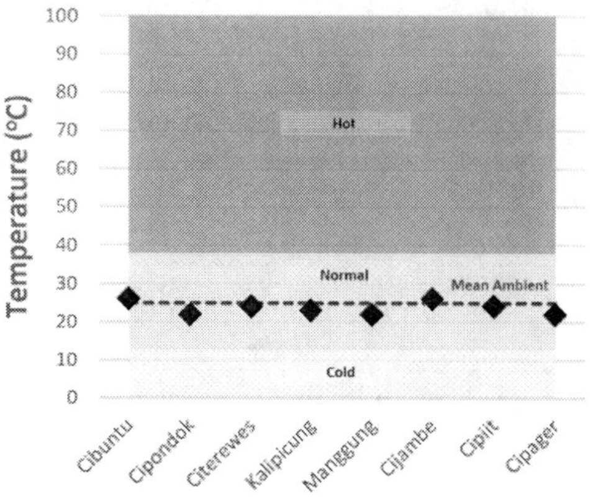

(a)
The results of temperature measurements in springwater and classification of springs based on mean ambient temperature.

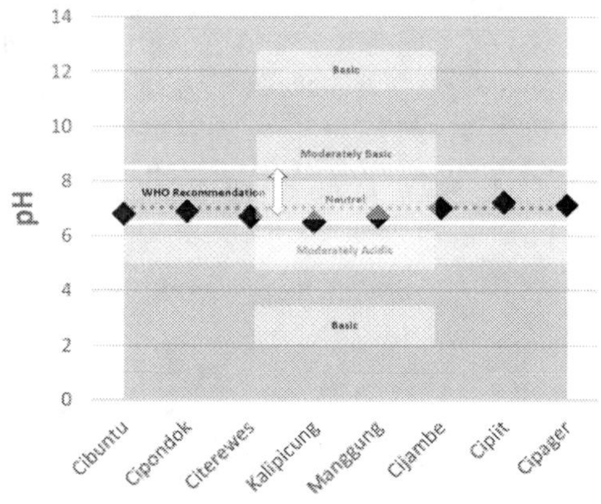

(b)
The results of measuring pH in spring water and classification of springwater based on pH.

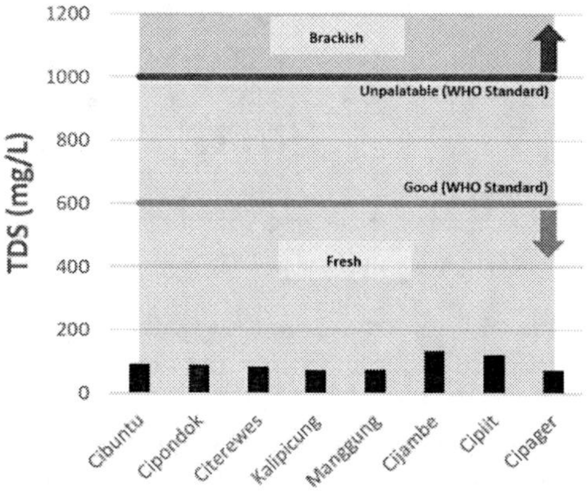

(c)
The results of the TDS measurement in spring water are compared to the WHO standard, and the classification of springwater based on the TDS value

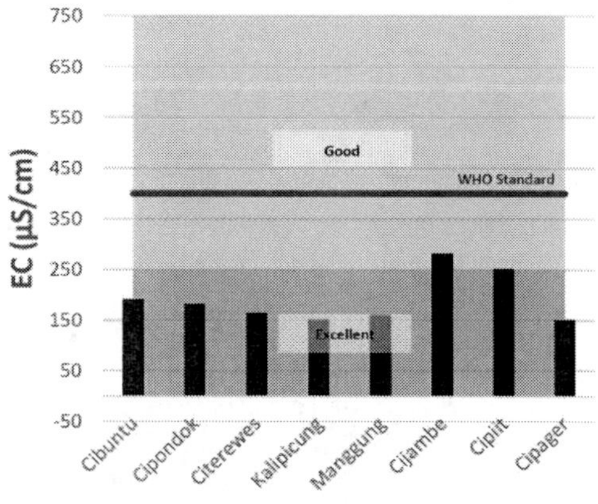

(d)
The results of the EC measurement in spring water are compared to the WHO standard and the classification of springwater based on the EC value.

Figure 5. Physico-chemical properties of springwater samples in Leuwisari subdistrict and surrounding compared to WHO Standard and other classification

4.2. Stable isotopic composition

The springwater sampled in this study has $\delta^{18}O$ and $\delta^{2}H$ ranging from -6.681 ± 0.038 ‰ to -6.134 ± 0.030 ‰ and -36.197 ± 0.089 ‰ to -34.596 ± 0.065 ‰, respectively (Table 4 and Figure 6). Samples plot on, slightly above, or slightly below the meteoric water line, suggesting an origin of precipitation, with or without evaporation prior to infiltration. The small isotopic values indicate that the recharge process has impacted springwater under similar climatic conditions. This range of values is comparable to those anticipated for Garut MWL recharge.

Table 4. The isotopic springwater and surface water composition in the Leuwisari subdistrict, Tasikmalaya Groundwater Basin.

No.	ID	Object Name	Sampling Date	Altitude (m asl)	$\delta^{18}O$ max (‰)	$\delta^{18}O$ min (‰)	δ^2H max (‰)	δ^2H min (‰)
1	MA01	Cibuntu Spring	09/03/2022	554	-6.375	-6.473	-35.256	-35.434
2	MA03	Cipondok Spring	09/03/2022	632	-6.643	-6.719	-35.556	-35.714
3	MA04	Citerewes Spring	15/03/2022	536	-6.418	-6.550	-35.806	-35.858
4	MA05	Kalipicung Spring	15/03/2022	638	-6.545	-6.683	-36.108	-36.286
5	MA06	Manggung Spring	15/03/2022	629	-6.274	-6.438	-34.688	-34.842
6	MA07	Cijambe Spring	15/03/2022	508	-6.194	-6.278	-35.173	-35.413
7	MA08	Cipiit Spring	13/03/2022	652	-6.181	-6.231	-34.952	-35.070
8	MA09	Cipager Spring	13/03/2022	857	-6.104	-6.164	-34.531	-34.661
9	AS01	River (Upstream)	16/03/2022	773	-7.099	-7.159	-41.279	-41.469
10	AS02	River (Midstream)	15/03/2022	393	-7.086	-7.116	-43.491	-43.675
11	AS03	River (Downstream)	15/03/2022	264	-7.158	-7.246	-45.240	-45.272

The δ^2H and $\delta^{18}O$ values of the study area were plotted on a δ^2H- $\delta^{18}O$ diagram and compared to GMWL, Indonesia MWL, and Garut MWL (Figure 6). The δ^2H and $\delta^{18}O$ values of the surface water ranged from -41.374 ± 0.095 ‰ to -45.256 ± 0.016 ‰ and -7.101 ± 0.015 ‰ to -7.202 ± 0.044 ‰., respectively. The slightly lower isotope levels of surface water samples depicted on the meteoric water line probably indicate rainwater mixing. The minimal spatial differences in δ^2H and $\delta^{18}O$ ratios of surface water samples from the Ciwulan River Basin revealed that the streams function as a single hydrologic unit.

In the study of the origin of the Umbul Cokro spring (216 masl) in Klaten, Central Java [14], the δ^2H value was -52.00‰, and $\delta^{18}O$ was -9.16‰. Meanwhile, Umbulan Spring (33 AMSL) in Pasuruan, East Java, has a δ^2H value of -47.10‰ and $\delta^{18}O$ of -7.23‰ [15]. In general, springs in this Tasikmalaya study area have greater positive isotope values than the Umbul Cokro Spring and Umbulan Spring. This is likely due to the Tasikmalaya research area's higher altitude than the Klaten and Pasuruan study areas. Moreover, in these two previous studies, the recharge location of each spring can be determined by using a graph of the isotopic value of rainwater against altitude. However, in this Tasikmalaya spring study, the estimated location of the recharge springs cannot be known due to the absence of elevation data and rainwater isotope values.

The stable isotopic composition of shallow groundwater distinct from the river indicates zero or minor contributions from the river (Figure 6). It is observed that the springwaters are substantially lighter than the adjacent rivers, suggesting that little or no mixing occurs.

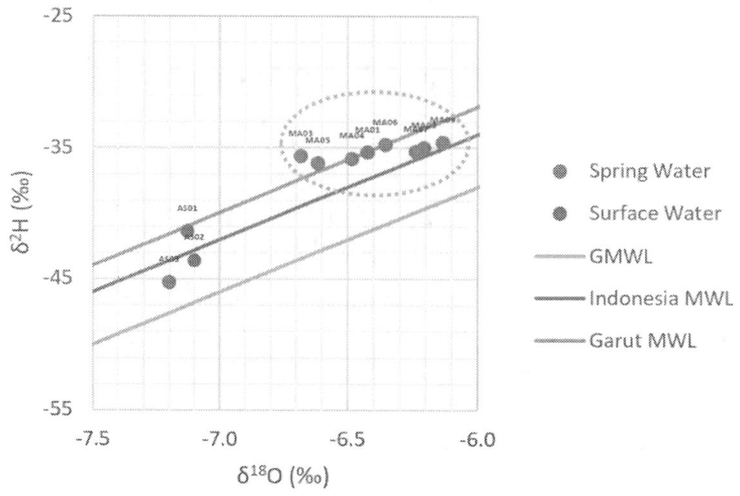

Figure 6. Plot of δ2H and δ18O of springwater and surface water in Leuwisari subdistrict, Tasikmalaya Groundwater Basin, compared to the Garut MWL, Indonesia MWL, and GMWL.

5. Conclusions

Based on the physicochemical properties in the present study, it can be concluded that the springwaters in the Leuwisari subdistrict and surrounding met WHO's standard for drinking water. Interpretation of isotope analysis reveals that the springwater can be categorized into one group with similar isotopic signatures. These springwaters originate from an unconfined aquifer, recharging under present conditions. The isotopic data suggest that direct river recharging has little or no influence on this unconfined aquifer. Combined with the other physical hydrogeology and water chemistry information, these isotope data will provide the basis for future water supply assessments in the research area.

More regular sampling (— for example, monthly) and investigation of the groundwater and surface water in the basin would enable the monitoring of seasonal fluctuations and the evaluation of whether seasonality influences the isotopic composition of groundwater and surface water. Additional samples must be taken from springs, bore wells, dug wells, rivers, lakes, and other water sources in the basin to increase sampling density and the study's accuracy. The δ^2H and $\delta^{18}O$ of precipitation in the basin may be utilized to provide a more reliable data set. Using tritium to determine the age of groundwater would improve the development of models that estimate the time scale of water origin and movement.

Acknowledgment

This endeavor would not have been possible without Dr.rer.nat. Ir. Doni Prakarsa Eka Putra, ST., MT., IPM. and Dr.Eng. Ir. Wawan Budianta, ST., M.Sc., IPM. from Department of Geological Engineering of Universitas Gadjah Mada. We extremely grateful for their help with data interpretation and analysis. We also thankful to Kementerian PUPR (Ministry of Public Works and Housing) for their financial support during the study.

References

[1] Todd D K and Mays L W, 2005 *Groundwater Hydrology* 3rd Editio Hoboken: John Wiley and Sons, Inc.

[2] Ragone S E and Llamas R, 2006 The Alicante Declaration: Steps along the Pathway to a Sustainable Future *Ground Water* **44**, 4 p. 500–503.

[3] Fotouhi F and Kresic N, 2010, Springwater Treatment, in *Groundwater Hydrology of Springs*, 1st Editio, (Butterworth-Heinemann: Elsevier Inc.), p. 269–304.

[4] BPS Kota Tasikmalaya, 2022, Kota Tasikmalaya dalam Angka (Tasikmalaya Municipality in

Figures) 2022, Kota Tasikmalaya.
[5] BPS Kabupaten Tasikmalaya, 2022, Kabupaten Tasikmalaya dalam Angka 2022, Kabupaten Tasikmalaya.
[6] Kresic N and Stevanovic Z, 2010 *Groundwater Hydrology of Springs* Oxford: Elsevier Inc.
[7] Brune G, 1975 Major and Historical Springs of Texas *Texas Water Dev. Board* March p. 94.
[8] Manga M, 2001 Using Springs to Study Groundwater Flow and Active Geologic Processes *Annu. Rev. Earth Planet Sci.* **29**.
[9] Clark I, 2015 *Groundwater Geochemistry and Isotopes* 1st Editio Boca Raton: CRC Press.
[10] Mazor E, 2004 *Chemical and Isotopic Groundwater Hydrology* Third Edit, **23**, 2 New York: Marcel Dekker.
[11] Trček B and Zojer H, 2010, Recharge of Springs, in *Groundwater Hydrology of Springs*, 1st Editio, (Butterworth-Heinemann: Elsevier Inc.), p. 87–127.
[12] Criss R E, 1999 *Principles of Stable Isotope Distribution* New York: Oxford University Press.
[13] Wassenaar L I Athanasopoulos P and Hendry M J, 2011 Isotope hydrology of precipitation, surface and ground waters in the Okanagan Valley, British Columbia, Canada *J. Hydrol.* **411**, 1–2 p. 37–48.
[14] Kharisma H L Budhie A and Wilopo W, 2015, Aplikasi Isotop Alam untuk Mengetahui Asal-Usul Air Umbul Cokro, Kecamatan Tulung, Kabupaten Klaten, *Forum Teknik*, **36**, 1. p. 73–80.
[15] Seizarwati W, 2013, Penyebab Penurunan Debit Mataair Umbulan, Institut Teknologi Bandung.
[16] PRATOMO I, 2006 Klasifikasi gunung api aktif Indonesia, studi kasus dari beberapa letusan gunung api dalam sejarah *Indones. J. Geosci.* **1**, 4 p. 209–227.
[17] Bronto S, 1989, Volcanic geology of Galunggung, West Java, Indonesia, University of Canterbury.
[18] Mulyanto B, 1999 Morphological, Physical, and Chemical Characteristic of Some Volcanic Soils of Mt. Galunggung *J. Ilmu Tanah dan Lingkung.* **2**, 1 p. 25–32.
[19] Setiawan O, 2019, Zonasi Kawasan Konservasi Mataair Berbasis Morfologi dan Material Permukaan di Lereng Selatan Gunungapi Rinjani, Lombok, Universitas Gadjah Mada.
[20] Menteri Energi dan Sumber Daya Mineral Republik Indonesia, 2017 *Peraturan Menteri Energi dan Sumber Daya Mineral Republik Indonesia No. 2 Tahun 2017 tentang Cekungan Air Tanah di Indonesia* Jakarta.
[21] Budhitrisna, 1986, Peta Geologi Lembar Tasikmalaya Skala 1:100.000. Pusat Penelitian dan Pengembangan Geologi, Bandung.
[22] Badan Informasi Geospasial Republik Indonesia, 2018 *Digital Elevation Model Nasional (DEMNAS)* V1.0 Bogor, Jawa Barat: Badan Informasi Geospasial (BIG).
[23] Peel M C Finlayson B L and McMahon T A, 2007 Updated World Map of The Köppen-Geiger Climate Classification *Hydrol. Earth Syst. Sci.* **11**, 11 October 2007 p. 1633–1644.
[24] Pemerintah Kabupaten Tasikmalaya, 2019, Laporan Akhir Pemutakhiran Dokumen Rencana Induk Sistem Penyediaan Air Minum (RISPAM) Kabupaten Tasikmalaya, Kabupaten Tasikmalaya.
[25] Badan Informasi Geospasial Republik Indonesia, 2019 *Peta Rupabumi Digital Indonesia* Bogor, Jawa-Barat.
[26] Craig H, 1961 Isotopic variations in meteoric waters *Science (80-.).* **133**, 3465 p. 1702–1703.
[27] Pratikno B and Satrio, 2014 Studi Garis Air Meteorik (Meteoric Water Line) Beberapa Wilayah di Indonesia dengan Metode Isotop Stabil in *Seminar Nasional Geologi untuk Meningkatkan Kesejahteraan Masyarakat.*
[28] Dunne T and Leopold L B, 1978 *Water in Environmental Planning* San Fransisco: W. H. Freeman.
[29] Soetrisno, 1983, Peta Hidrogeologi Indonesia Lembar Bandung Skala 1:250.000. Direktorat Geologi Tata Lingkungan, Bandung.
[30] Alfaro C and Wallace M, 1994 Origin and classification of springs and historical review with

current applications *Environ. Geol.* **24**, 2 p. 112–124.
[31] World Health Organization, 2011 *Guidelines for Drinking-water Quality, Fourth Edition* Malta: Gutenberg.
[32] Springer A E *et al.*, 2004 A comprehensive springs classification system: integrating geomorphic, hydrogeochemical and ecological criteria *Arid L. springs North Am. Ecol. Conserv.* p. 49–75.
[33] Fetter C W, 1980 *Applied Hydrogeology 4th Edition* New Jersey: Prentice-Hall.
[34] Şen Z, 2014 *Practical and Applied Hydrogeology* Amsterdam, The Netherlands: Elsevier.
[35] WHO, 2004 *Guidelines for Drinking Water Quality, Third Edition, Vol. 1. Recommendations* Third Edit, **1** Geneva: World Health Organization.
[36] Healy R W, 2010 *Estimating Groundwater Recharge* Camb: Cambridge University Press.

Preliminary study of isotope characteristics of groundwater in karst hydrogeology in Pacitan Regency, East Java

T Anggraeni [1,2], D P E Putra[1], A Taufiq[2]

[1]Department of Geological Engineering, Universitas Gadjah Mada, Indonesia
[2]Ministry of Public Works dan Housing Republic of Indonesia

Email: trianggraeni1995@mail.ugm.ac.id

Abstract. The Pacitan Karst region is well known to be one of drought area in the Pacitan Regency during the dry season. Consequently, during the dry season, springs from karst system are required as a water source for the daily life. The appearance of these springs are a result and advantage of the karst terrain, which contains potential groundwater resources. The purpose of this study was to investigate the origin and characteristics of spring's water in the Pacitan karst region based on its isotope concentrations. This research was conducted by collecting groundwater samples from springs, dug wells, bore wells, and rivers in Pacitan Regency and analyzing their ^{18}O and ^{2}H concentrations. Information on the source of groundwater is thought to be necessary for sustainable groundwater protection planning (conservation). According to the results of the investigation, there are four categories of isotope ratio distribution. The pattern of ^{18}O and ^{2}H concentrations in the samples around the Local Meteoric Water Line (LMWL) shows that groundwater originates from the same elevation of the catchment area from where rainwater falls and has the same characteristics as rainwater or meteoric water.

Keywords, groundwater, origin, stable isotope, karst, Pacitan.

1. Introduction

Pacitan Regency is an area in East Java Province that has administrative boundaries in the west is Wonogiri Regency, Wonogiri and Ponorogo Regency in the north, Trenggalek Regency in the east, and in the South, the Indonesian Ocean [1]. Pacitan Regency is dominated by the highlands of the Southern Mountains, with the Wonosari formation being the underlying geology.

The Wonosari Formation has identical characteristics, with limestone being the primary rock lithology and the appearance of the karst phenomenon, the emergence of springs or rivers connected to the forming of subsurface rivers. [2], [3]. The limestone that composes the Wonosari Formation has a porous nature and is less able to hold water, so the absorbed rainwater will be directly carried to the underground river [4], [5]. Therefore, it is quite difficult to make dug wells in the area. Dug wells are mainly rainfed wells, which hold water during the rainy season but dry up when the intensity of the rain decreases [6].

The strategic importance of karst terrain in Indonesia is reflected by the fact that it comprises over 20% of the country's area. One of the strategic values mentioned is the underground river system's role as a provider of water for domestic use [7]. Numerous research have been conducted in the Gunungkidul karst region, but fewer have been conducted in the Pacitan region.

As a result of hydrogeological mapping conducted by Prayogo in the Pacitan region, it was determined that the research area frequently experienced water problems due to the dominance of limestone lithology, and that it was necessary to manage water resources in an integrated and efficient way to maximize the utilization of the available potential [6]. The majority of residents in the Pacitan regency rely on PDAM to meet their daily needs, however as the dry season approaches, the PDAM water supply is insufficient to be supplied to all areas, so the community receives water allocations in turn [8]. At this time, the existence of springs is urgently required to increase the availability of clean water for communities [9], [10].

In regions with few infiltration zones, it is essential to identify the origin and source of infiltration in to protect (conserve) the sustainability of groundwater. As users of water resources, people must consider environmental sustainability in addition to engaging in exploitation. To sustain the function and use of the land, groundwater exploitation must be balanced with groundwater protection activities [11]. Infiltration of water into the Karst aquifer system can occur in two ways. Allogenic recharge is recharging that comes from other places that enter the karst system, through conduits or underground rivers. Autogenic recharge is recharge that comes from the karst area itself, so rainwater falls and seeps directly into the area [2]. Effective tracers that can be used to identify characteristics and investigate for the source of groundwater are the stable isotopes ^{18}O and ^{2}H [12].

Iffatul Izza Siftianida determined the source of springs in Cijeruk Subdistrict, Bogor Regency, West Java by determining the catchment area with the stable isotope method ^{18}O and ^{2}H. This results indicated that springs in Cijeruk Subdistrict, Bogor Regency, West Java are supplied by both rainwater and ground water [13]. Similar studies investigating the origin of groundwater with stable isotopes were also conducted by Kharisma et al in Umbul Cokro Klaten in 2015 [14], Satrio et al in Gunung Kidul [15], Yogyakarta in 2015, and Rusjan et al in Slovenia in 2019 [16].

The purpose of this study was to determine the characteristics of the isotopes ^{18}O and ^{2}H from spring's water including deep well and shallow well of groundwater but also surface water where the relationship leads to the determination of the origin of groundwater in the area. Origin of groundwater data can provide information and justification for groundwater planning and conservation. It can be used to identify the elevation of the area's groundwater recharge zone for future study. This research is expected to provide benefits for sustainable groundwater protection.

2. Study Area

The research location is the southern section of the Pacitan Regency, with a concentration on the Wonosari Formation, which extends from west of Pacitan City to the border of Central Java Province, covering Donorojo, Punung, and Pringkuku Subdistricts. For other areas, including the Arjosari Formation and lowlands, alluvial deposits provide as supplementary data. Donorojo Subdistrict has an administrative area of 109.09 km^2 with an altitude of 339 meters above sea level, Punung Subdistrict has an administrative area of 108.81 km^2 and an altitude of 389 meters above sea level, and Pringkuku Subdistrict has an area of 132.93 km² with an altitude of 325 meters above sea level [17], [18]. The research location is shown in Figure 1. below.

The research location is mostly composed of the Wonosari Formation, which consists of reef limestones, bedded limestones, fragmental limestones, sandy limestones, and marl from the Late Miocene to Pliocene ages. The Oyo Formation is composed of calcareous sandstone, tuffaceous sandstone, calcareous siltstone, tuffaceous limestone, sandy marl, and tuffaceous marl from the Late Miocene age. The Nampol Formation is composed of tuffaceous sandstone, siltstone, tuffaceous limestone, claystone, and lignite, with intercalation of conglomerates and breccias from the Middle Miocene age. Alluvium is composed of cobble, pebble, sand, silt, clay, and mud with a Quarter age, and the Arjosari Formation is composed of polymit conglomerate, sandstone, siltstone, limestone, claystone, sandy marl, pumice sandstone, intercalated by volcanic breccia, tuff, and lava from the Oligocene to Early Miocene age [19], [20].

The Wonosari Formation has an aquifer type that flows through fissures, fractures, and channels with moderate productivity. With the constituent lithology being layered and localized limestones, reef limestones, with varying levels of karst formation, have low to high graduations [21].

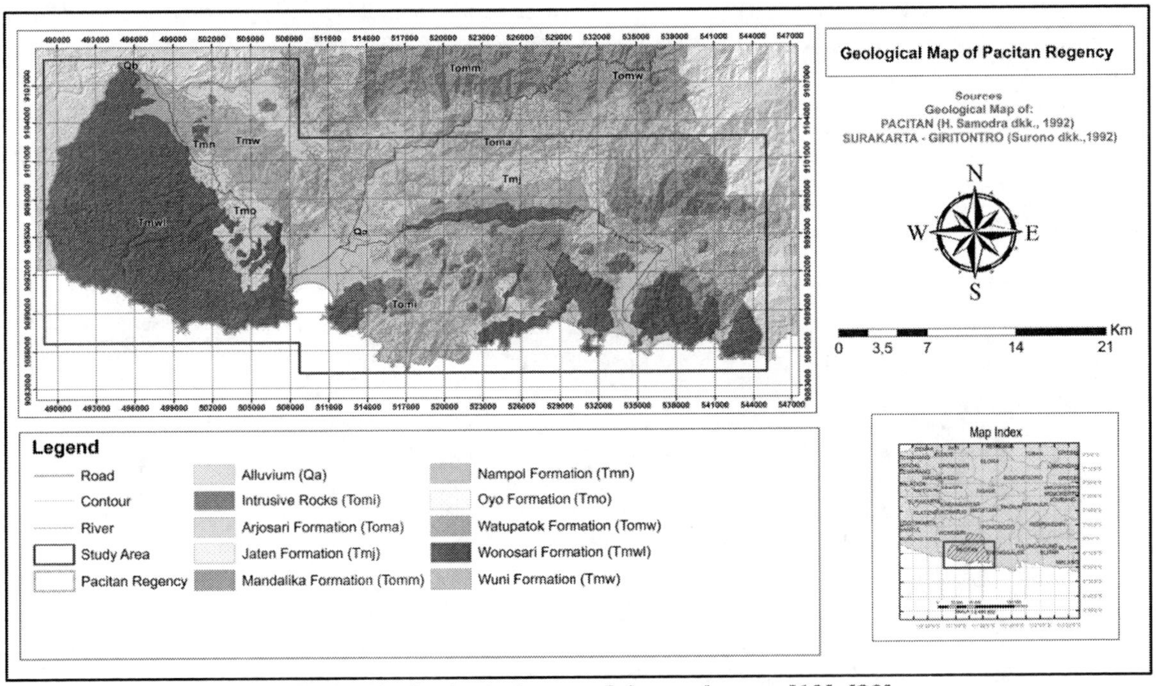

Figure 1. Geological map of the study area [19], [20]

3. Research Methodology

This research was conducted in three steps: determining the sampling location, collecting isotope samples, and sample testing and data analysis.

3.1. Determining the sampling location

The determination of the sampling location is based on the water source used and consumed by residents. Especially during the dry season, when the spring discharge continues to flow as long as possible in the Donorojo Subdistrict, Punung Subdistrict, and Pringkuku Subdistrict, where water is often difficult to find. The samples collected include rainwater samples and groundwater samples from springs, bore wells, and dug wells, and from the Maron River..

Table 1. Coordinates for isotope sampling in the study area

ID	Location		Location Coordinates		Altitude
	Village	Subdistrict	X	Y	(m asl)
AH 01	Dk. Dawung Kidul, Ds. Candi	Pringkuku	501184.00	9089430.00	114.11
AH 02	Dk. Kerok Wetan, Ds. Candi	Pringkuku	502781.79	9091408.79	272.28
AH 03	Dk. Mudah, Ds. Ngadirejan	Pringkuku	504995.00	9097094.00	387.61
Su 01	Sungai Maron, Ds. Punung	Punung	503459.00	9098671.00	295.47
SG 20	Dk. Kedawung, Ds. Mentoro	Pacitan	513757.32	9093947.01	7.25
SB 01	Dk. Jati, Ds. Cemeng	Donorojo	494256.21	9104338.05	377.90
MA 03	Dk. Kerok Wetan, Ds. Candi	Pringkuku	504089.80	9092558.76	201.35
MA 06	Dk. Barong Wetan, Ds. Candi	Pringkuku	499201.87	9088468.82	16.28
MA 13	Dk. Mudah, Ds. Ngadircjan	Pringkuku	504200.32	9097686.73	367.64
MA 14	Dk. Sooko, Ds. Sooka	Punung	502168.14	9099200.45	355.79
MA 17	Dk. Krajan, Ds. Sendang	Donorojo	493002.75	9094807.70	191.74
MA 20	Dk. Tumpakwatu, Ds. Widoro	Donorojo	491115.22	9093882.59	145.14
MA 30	Dk. Druju, Ds. Donorojo	Donorojo	498879.21	9104855.35	389.65
MA 37	Dk. Bolo, Ds. Kebonagung	Kebonagung	517277.70	9091125.79	51.78
MA 45	Dk. Salam, Ds. Bomo	Punung	497595.00	9097576.00	321.24

Source: field observation results

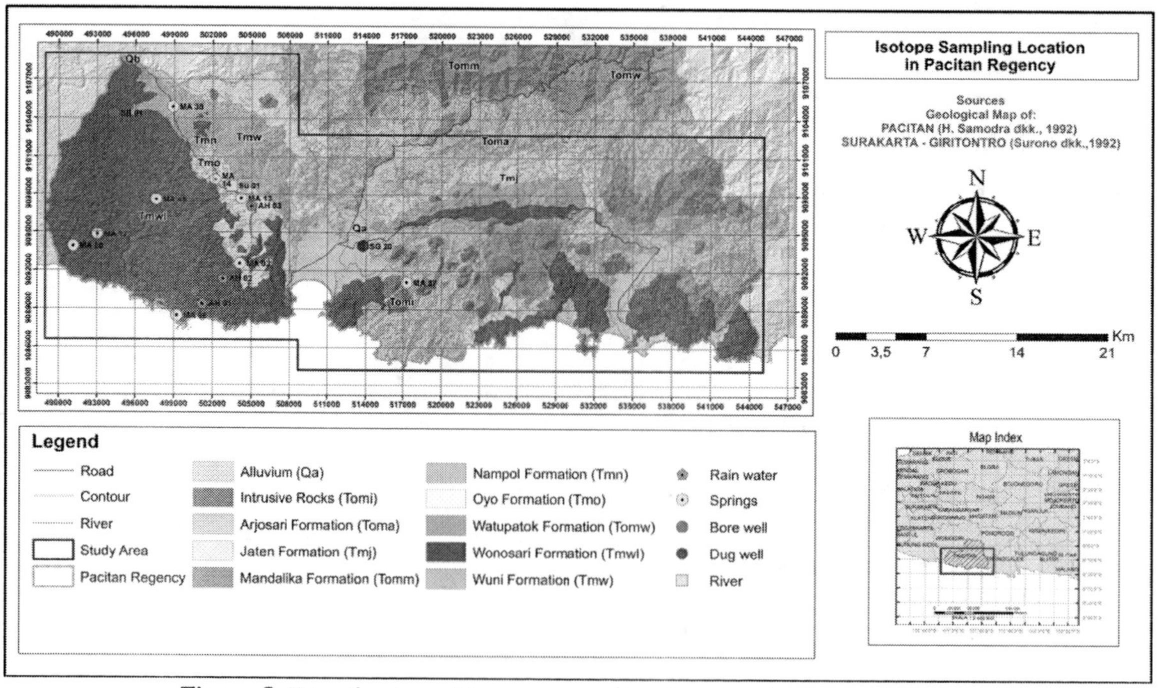

Figure 2. Distribution of isotope sampling location in the study area

3.2. Collecting Isotope Samples

The activities below are taken in March 2022 to collect the samples of water for ^{18}O and ^{2}H isotope analysis:

1. Prepare a glass bottle with that 100 mL capacity and an inner valve on the lid (double cap).
2. Rinse the bottle with sample water three times.
3. Insert a glass bottle into the water source and then place the sample water inside. This prevents air bubbles from entering the bottle.
4. After the bottle is completely filled, it is verified that there are no air bubbles, and it is closed tightly to prevent the entry of air.
5. Mark the sample code on the bottle to avoid confusion with other samples.

3.3. Sample Testing and Data Analysis

Testing of groundwater samples for stable isotopes ^{18}O and ^{2}H was carried out at the Hydrogeology and Hydrogeochemistry Laboratory, Faculty of Mining and Petroleum Engineering, Bandung Institute of Technology using the *Picarro L-2130i* instrument.

Isotope analysis to determine the origin of groundwater using the Local Meteoric Water Line (LMWL) equation, by correlating the ^{18}O with D from rainwater into a relationship graph to obtain an equation [12].

The measurement data for the ^{18}O and ^{2}H isotope ratios from the groundwater samples were then plotted in a ^{18}O vs D graph and compared with the Local Meteoric Water Line (LMWL) line equation that had been obtained.

To get an interpretation the genesis of groundwater can be seen from the groupings of groundwater to the local meteoric line. Groundwater that is in one line with the local meteoric water line (LMWL) so that it can be characterized that groundwater still has properties as meteoric water and gets recharge from rainwater [12].

4. Discussion

Isotopes are elements that occur in nature which have the same number of protons but different numbers of neutrons. Isotopes are divided into two groups, stable isotopes and radioactive isotopes. Stable isotopes are used for tracers, while radioactive isotopes are used to determine the age of groundwater. [22]

The study of stable isotopes is based on the tendency of several pairs of isotopes to fractionate or split into light and heavy fractions. This fractionation occurs in geological processes, such as evaporation or heating. The five elements used to study stable isotopes can be fractionated easily. This is because they have a relatively large difference in mass between the two isotopes and also exhibit one isotope being much more abundant than the other [23].

Determination of groundwater genesis using the ratio between heavy and light isotopes is determined by the difference in isotope ratios, which is $^{2}H/^{1}H$ and $^{18}O/^{16}O$. The ratio between heavy and light isotopes is denoted by (δ), which is the ratio between the ratio of the sample isotopes to the isotopic composition in seawater (Vienna Standard Mean Ocean Water, VSMOW), shown in the following equation:

$$\delta^{18}O\ sample = \left(\frac{(^{18}O\ /^{16}O)sample}{(^{18}O\ /^{16}O)VSMOW} - 1\right) x\ 1000\ ‰ \qquad (1)$$

$$\delta^2H\ sample = \left(\frac{(^2H\ /^1H)sample}{(^2H\ /^1H)VSMOW} - 1\right) x\ 1000\ ‰ \quad (2)$$

In 1961 Craig 1961 conducted research on ^{18}O and D for freshwater. Two principles were found, i.e., the ratio of ^{18}O and D in fresh water in areas with high temperatures will have a higher value and will decrease as the temperature of an area decreases, and it is found that there is a strong regression of the relationship between ^{18}O and D which is called the Global Meteoric Water Line (GMWL), which is the average of many different local or regional MWL due to the influence of climatic and geographical parameters of the region [12]. GMWL can be determined by the following equation:

$$\delta D = 8\,\delta^{18}O + 10\ ‰ \quad (3)$$

The ratio of D and ^{18}O isotopes in rainwater varies with a function of elevation, amount of rainfall and a certain distance. Variations in isotopic composition can be characterized by using the Local Meteoric Water Line (LMWL) in a region. Making it possible as a tracer of the catchment area.

Table 2. Isotope test results of rainwater samples in the study area

ID	$\delta^{18}O$ (‰)	δD (‰)
AH 01	-6.325 ± 0.003	-38.672 ± 0.080
AH 02	-9.962 ± 0.019	-63.988 ± 0.033
AH 03	-8.316 ± 0.088	-54.137 ± 0.018

Source: laboratory result

Table 2 above shows the results of testing and analysis of $\delta^{18}O$ vs δD isotopes from rainwater samples taken from the Karst area in Pacitan Regency with different elevations, which is at elevations of 114.11 masl, 272.28 masl, and 387.61 masl. These results show that the concentration of $\delta^{18}O$ isotope varies from -9.962‰ to -6.325‰ and the concentration of δD isotope ranges from -63.988‰ to -38.672‰. If the $\delta^{18}O$ and δD data are correlated in a relationship graph, it will look like in Figure 3. below:

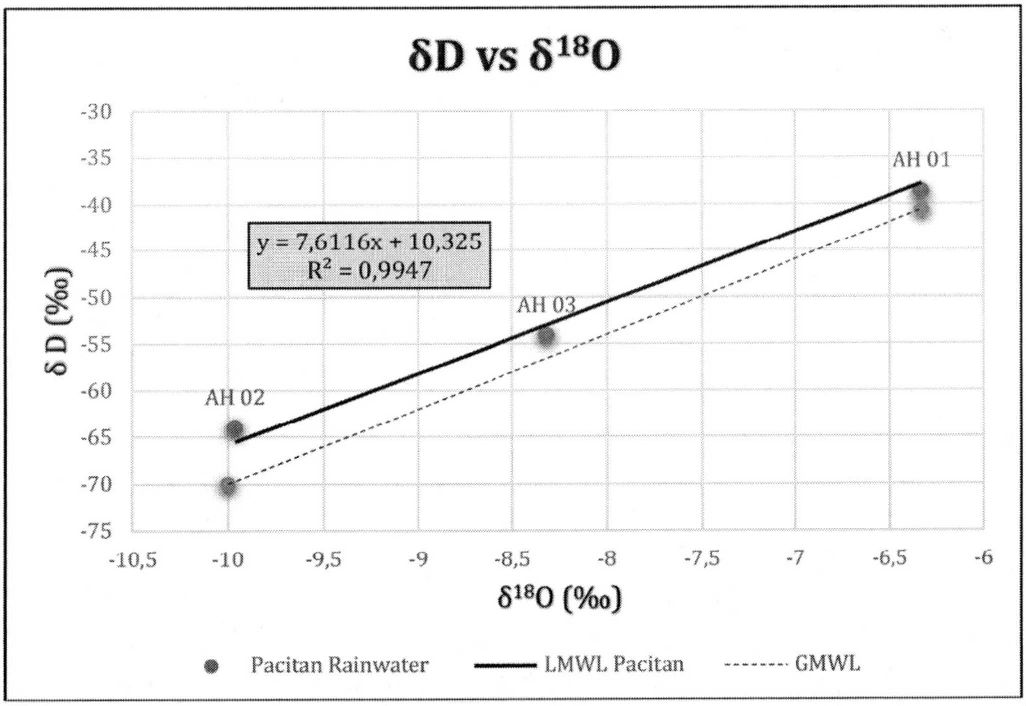

Figure 3. Graph of the relationship between $\delta^{18}O$ and δD concentrations of rainwater samples in the study area

The result of the correlation graph $\delta^{18}O$ and δD obtained from the Local Meteoric Water Line (LMWL) equation from the Pacitan Karst region is as follows:

$$\delta D = 7.6116\ \delta^{18}O + 10.325 \quad (4)$$

Determination of groundwater genesis can be done by correlating the concentrations of $\delta^{18}O$ and δD isotopes from groundwater samples into a graph and then combining it with Local Meteoric Water Lines that have been obtained from rainwater samples. Samples that have close isotopic values and form a group can be concluded that the groundwater comes from the same source.

Table 3. Isotope test results on groundwater samples in the study area

ID	δ18O (‰)	δD (‰)
Su 01	-7.030 ± 0.039	-43.369 ± 0.067
SG 20	-5.337 ± 0.033	-31.199 ± 0.041
SB 01	-6.912 ± 0.048	-41.414 ± 0.094
MA 03	-6.644 ± 0.009	-39.618 ± 0.096
MA 06	-7.081 ± 0.035	-42.103 ± 0.022
MA 13	-6.893 ± 0.054	-41.545 ± 0.033
MA 14	-6.808 ± 0.034	-40.090 ± 0.100
MA 17	-7.269 ± 0.027	-43.354 ± 0.055
MA 20	-6.920 ± 0.070	-41.626 ± 0.074
MA 30	-6.701 ± 0.012	-41.558 ± 0.088
MA 37	-6.374 ± 0.017	-36.445 ± 0.068
MA 45	-7.926 ± 0.010	-48.525 ± 0.065

Source: laboratory result

Table 3. above shows the results of testing and analysis of natural isotopes $\delta^{18}O$ vs δD from groundwater samples from the Karst area of Pacitan. From these data, it can be seen that the variation of stable isotope concentration for $\delta^{18}O$ ranges from -7.926‰ to -5.337‰ and the concentration of stable isotope δD ranges from -48.525‰ to -31.199‰. From the variation in the concentration of the stable isotope, when connected to a graph and combined with the Local Meteoric Water Line (LMWL) equation, can produce a relationship as shown in Figure 4. below.

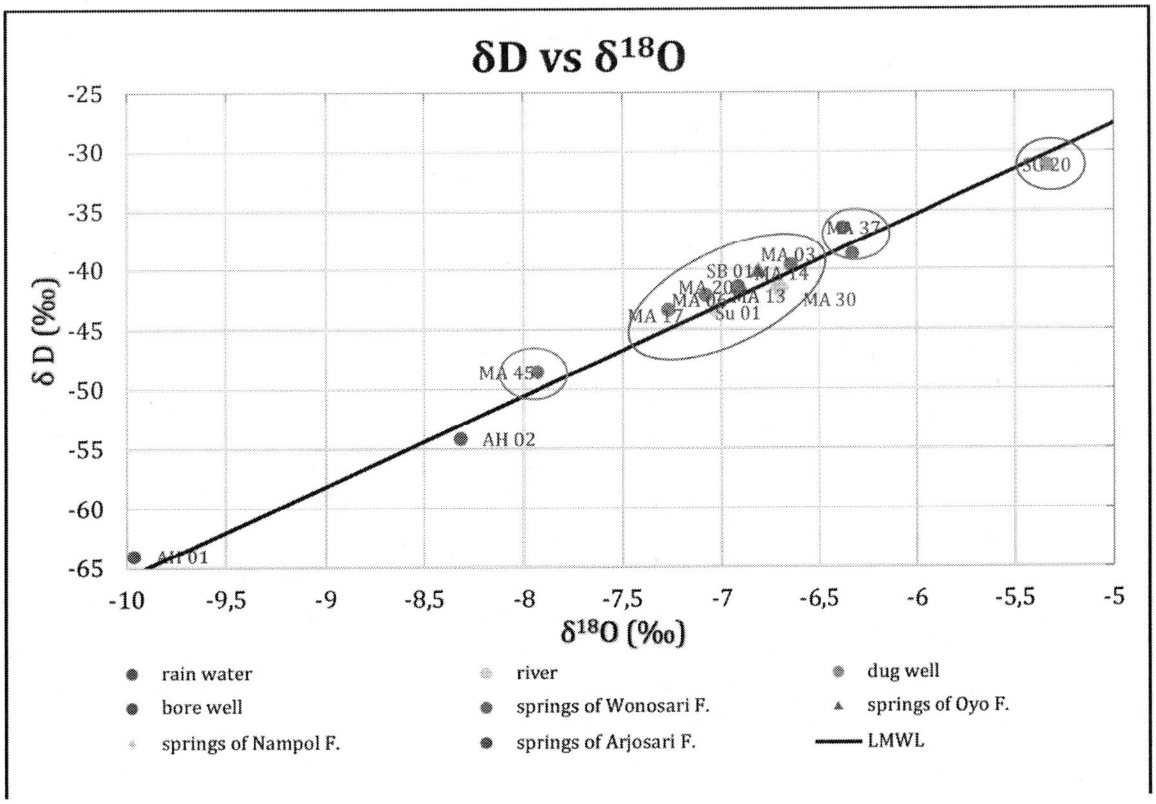

Figure 4. Graph of $\delta^{18}O$ vs δD relationship from groundwater samples in the study area

Based on the graph of the $\delta^{18}O$ vs δD relationship, the origin of groundwater in the Pacitan Karst area is divided into 4 groups, i.e.:

1. Group 1
   Groundwater in this group has a stable isotope concentration of $\delta^{18}O$ with a value of -5.337‰ and δD with a value of -31.199‰ and is the sample with a concentration of the richest $\delta^{18}O$ vs δD value compared to other groups. The groundwater is a sample of SG 20 and comes from a dug well located in an alluvial deposit area in Pacitan City. Based on the graph to the Local Meteoric Water Line (LMWL) equation, it is estimated that the groundwater comes from meteoric rain. The concentration of the SG 20 isotope is richer than the other groups, indicating that the elevation of the absorption area is lower than the other groups.
2. Group 2
   Groundwater in this group has a stable isotope concentration of $\delta^{18}O$ with a value of -6.374‰ and δD with a value of -36,445‰. The groundwater is a sample of MA 37 and comes from springs located in the Arjosari formation area in the Kebonagung Subdistrict, so it does not have the same genesis as water samples from the Wonosari Kast area. Based on the graph to the Local Meteoric Water Line (LMWL) equation, it is estimated that the groundwater comes from the same elevation of the catchment area as the rainwater falls and still has the same characteristics as meteoric water or rainwater.

3. Group 3
   Groundwater in this group has a stable isotope concentration of $\delta^{18}O$ with a range of -7,269‰ to -6,644‰ and δD with a value of -43.354‰ to -39,618‰. The groundwater is a sample of Su 01 which is a sample from the Maron River, a sample of SB 01 which is a sample from a bore well from the Wonosari Formation, the sample of MA 03, MA 06, MA 17, MA 20, and MA 30 which comes from springs located in the Wonosari Formation area. MA 13 and MA 14 samples were from springs in the Oyo Formation, and MA 30 is springs from the Nampol Formation. Based on this graph, it is estimated that the groundwater comes from the same genesis and the elevation of the catchment area is relatively the same. Samples Su 01 and MA 30 are samples that mostly come from rainwater because they have the same characteristics as meteoric water or rainwater.
4. Group 4
   Groundwater in this group has a stable isotope concentration of $\delta^{18}O$ with a value of -7,926‰ and δD with a value of -48,525‰ and is the groundwater sample with a concentration of $\delta^{18}O$ vs δD which is the depleted compared to the others. The groundwater is a sample of MA 45 and comes from Gong Cave which is located in the Wonosari Formation area. Based on the graph to the Local Meteoric Water Line (LMWL) equation, it is estimated that the groundwater comes from the elevation of the catchment area which is relatively the same and still has the same characteristics as meteoric water or rainwater. MA 45 has a depleted isotope concentration than the others because it has been evaporated or mixed with surface water from the dissolution of calcium carbonate minerals in the cave.

5. Conclusion

From the results of the research that has been carried out, it can be concluded that the characteristics of the origin of groundwater in the Pacitan Karst region are based on stable isotope concentrations of ^{18}O and ^{2}H. There are 4 distribution groups with $\delta^{18}O$ isotope concentration values ranging from -7.926‰ to -5.337‰ and δD isotope concentrations ranging from -48,525‰ to -31.199‰. Based on these results to the Local Meteoric Water Line (LMWL) equation, it is proved that groundwater still comes from the elevation of the catchment area which is the same as the fall of rainwater and still has the same characteristics as meteoric water or rainwater.

For further research, it is expected to be able to determine the elevation of the recharge zone related to the protection zone or groundwater conservation in the Pacitan Karst area.

6. Acknowledgment

The authors would like to thank the River Basin Organization for Bengawan Solo for the permission to conduct this research. The first author would also like to thank The Ministry of Public Works and Housing of Indonesia for providing financial support.

References

[1] Statistic of Pacitan Regency, 2014, Letak Geografis Kabupaten Pacitan. [Online]. Available: https://pacitankab.bps.go.id/statictable/2014/12/23/3/letak-geografis-kabupaten-pacitan.html.

[2] Ford D and Williams P, 2007 *Karst Hydrogeology and Geomorphology* West Sussex City: John Wiley & Sons, Ltd.

[3] Haryono E Barianto D H and Cahyadi A, 2017 Hidrogeologi Kawasan Karst Gunungsewu: Panduan Lapangan Fieldtrip PAAI 2017 *Pekan Ilm. Tah. Ahli Airtanah Indones.* December 2018 p. 1–33.
[4] Parno, 2018 *Gunung Sewu* Jakarta Timur: Badan Pengembangan dan Pembinaan Bahasa.
[5] Hikmah D S Dewi I K and Yatini, 2021 Identifikasi keberadaan rongga bawah permukaan dengan menggunakan metode geolistrik konfigurasi dipole-dipole daerah Karst Desa Monggol Kabupaten Gunungkidul in *Prosiding Seminar Nasional Ilmu Teknik dan Aplikasi Industri* 4 p. 5.
[6] Prayogo T, 2016 Pemetaan Geohidrologi Daerah Donorojo Kabupaten Pacitan in *Jurnal Teknologi Lingkungan* 10, 1 p. 54.
[7] Tjahyo Nugroho Adji, Eko Haryono S W, 1999 Kawasan Karst dan Prospek Pengembangannya di Indonesia in *Seminar PIT IGI* p. 10.
[8] Sihkusuma S R and Suhardono A, 2021 Debit Kebutuhan Pengembangan Pipa Air Bersih di Kecamatan Pringkuku in *Jurnal Online Skripsi* 2 p. 78–83.
[9] Haryono E, 2001 Nilai Hidrologis Bukit Karst in *Seminar Nasional Teknik SIpil UGM* p. 5.
[10] Nugroho J Zid M and Miarsyah M, 2020 Potensi sumber air dan kearifan masyarakat dalam menghadapi risiko kekeringan di wilayah karst (Kabupaten Gunung Kidul, Provinsi Yogyakarta) *J. Pengelolaan Lingkung. Berkelanjutan (Journal Environ. Sustain. Manag.* 4, 1 p. 438–447.
[11] Kılıç Z, 2020 The Importance of Water and Conscious Use of Water *Int. J. Hydrol.* 4, 5 p. 239–241.
[12] Clark I, 2015 *Groundwater Geochemistry and Isotopes* New York: CRC Press.
[13] Siftianida I I Wijatna A B and Pratikno B, 2017 Aplikasi Isotop Alam untuk Pendugaan Daerah Resapan Air Bagi Mataair Di Kecamatan Cijeruk, Kabupaten Bogor, Jawa Barat *J. Ilm. Apl. Isot. dan Radiasi* 12, 2 p. 97.
[14] Kharisma H L Budhie A and Wilopo W, 2015, Aplikasi Isotop Alam untuk Mengetahui Asal-Usul Air Umbur Cokro, Kecamatan Tulung, Kabupaten Klaten, *Forum Teknik*, 36, 1. p. 73–80.
[15] Satrio S and Sidauruk P, 2016 Studi Daerah Imbuh Sistem Air Sungai Bawah Tanah Gunungkidul–Yogyakarta Menggunakan Isotop Stabil δ18O dan δ2H *J. Ilm. Apl. Isot. dan Radiasi* 11, 2 p. 87.
[16] Rusjan S Sapač K Petrič M Lojen S and Bezak N, 2019 Identifying the Hydrological Behavior of a Complex Karst System Using Stable Isotopes *J. Hydrol.* 577, July p. 15.
[17] Statistic of Pacitan Regency, 2014, Luas Daerah dan Jumlah Pulau Menurut Kecamatan di Kabupaten Pacitan. [Online]. Available: https://pacitankab.bps.go.id/indicator/153/108/1/luas-daerah-dan-jumlah-pulau-menurut-kecamatan-di-kabupaten-pacitan-.html.
[18] Statistic of Pacitan Regency, 2019, Tinggi Wilayah dan Jarak ke Ibu kota Kecamatan di Wilayah Kabupaten Pacitan. [Online]. Available: https://pacitankab.bps.go.id/indicator/153/107/1/tinggi-wilayah-dan-jarak-ke-ibu-kota-kecamatan-di-wilayah-kabupaten-pacitan.html.
[19] Surono., Toha, B., Sudarno I, 1992 *1992. Peta Geologi Lembar Surakarta - Giritontro, Jawa* .
[20] Samodra, H., Gafoer, S., Tjokrosapoetro S, 1992 *Peta Geologi Lembar Pacitan, Jawa Timur* Pusat Penelitian dan Pengembangan Geologi.
[21] Djaeni A, 1982 *Peta Hidrogeologi Lembar Yogyakarta* Direktorat Geologi Tata Lingkungan.

IOP Conf. Series: Earth and Environmental Science 1233 (2023) 012066 doi:10.1088/1755-1315/1233/1/012066

[22] Mazor E and Dekker M, 1997 *Chemical and Isotopic Groundwater Hydrology* Third, 23, 2 New York: Marcel Dekker, Inc.
[23] Fetter C ., 2014 *Applied Hydrogeology, Fourth Edition* United States of America: Pearson.

IOP Conf. Series: Earth and Environmental Science 1233 (2023) 012067 doi:10.1088/1755-1315/1233/1/012067

The Study of Air Pressure System and Extreme Rain in The Southern Java Island

R W Manuhana[1], A B Sekaranom[1] and E Nurjani[1]

[1] Department of Environmental Geography, Faculty of Geography, Universitas Gadjah Mada, Indonesia
*Corresponding author: rinut.manuhana@gmail.com

Abstract. Extreme rain can trigger other natural phenomena that can interfere with human activities. Rain phenomenon extremes are common in tropical cyclones. This study aims to characterize the occurrence of extreme rain events concerning the air pressure system on Java Island. This study uses rainfall data from GSMAP MVK Product and Sea Level Pressure (SLP) from NCEP/NCAR Reanalysis. The identification of extreme rain is carried out with pixel detection of the GSMAP precipitation data from 2015 to 2019. The most extreme rain events were then ranked, and ten events with the highest rain rate were selected. The SLP data at the selected extreme rain event was analyzed and plotted as a map. The principal result shows that in the 10 highest extreme rain events, 4 of them occurred due to the influence of low-pressure systems and tropical cyclones. The rest are associated with other climatic phenomena. In conclusion, this study shows that extreme rainfall tends to be associated with low-pressure anomalies over the Indian ocean. However, not all of the low-pressure anomalies were intensified as a tropical cyclone. Therefore, extreme rainfall in Southern Java Island can be developed due to small air pressure differences.

1. Background

Tropical cyclones are meteorological phenomena characterized by strong winds moving around a center of low pressure. Tropical cyclones are generally accompanied by high-intensity rains and storms. Strong winds, extreme rain, and the presence of storm surge phenomena are often accompanied by tropical cyclones [1]. Tropical cyclones generally occur in warm waters or the oceans. The thermal energy from the ocean which supports the formation of cyclones. Tropical cyclones will also form when the Coriolis force exceeds the minimum, which is about 5° from the equator. Since the 1970s, tropical cyclones have gaining importance in the Indian Ocean and the Pacific Ocean [2].

The frequency of tropical cyclones that occur in the southern part of Indonesia has increased over the last 35 years [3]. At least 51 tropical cyclones in the southern region of Indonesia have recorded during the 1983-2017 period [4]. A few the above tropical cyclones resulted in heavy precipitation in several areas, especially in Java Island [5], [6]. The increase in rainfall due to the influence of tropical cyclones can reach 750% compared to the average precipitation [4]. In the study area, rain with a high intensity not only resulted in flooding but also trigger landslides that affect the dense population [7]. Currently, the Java Islands' population is reached 150 million, and most of them are prone to hydrometeorological hazards [6]. Therefore, studies related to the atmospheric process is important in increasing the resilience of the population.

Extreme rainfall events in the southern part of Java Island are often associated with tropical cyclones [3]. However, the contribution of tropical cyclones to the formation of extreme rain events in the area is still unknown. Thus, there is a gap in the study related to the relationship between tropical cyclones and extreme rainfall in the study area. One of the things associated with tropical cyclones is the emergence of low air pressure around the center of the cyclone [8]. The extreme rain events therefore can be related to the air pressure pattern. Based on the above condition, this study aims to characterize the occurrence of extreme rain events in relation to the air pressure system on Java Island. The results of this study are expected to give information about the influence of low-pressure systems on extreme rainfall in the study area.

2. Materials and Methods

Rainfall data from GSMAP MVK dataset V7 (https://sharaku.eorc.jaxa.jp/GSMaP/) from January 1st, 2015 to December 31th, 2019 was used in this analysis (5 years). The sea level pressure (SLP) data from NCEP/NCAR Reanalysis (https://psl.noaa.gov/data/gridded/data.ncep.reanalysis.html) was also utilized for the same period. Extreme rain was identified using the pixel detection method in the study area as shown in Figure 1. Each pixel in the data provides information on rainfall according to location coordinates. Each existing dataset has the same temporal resolution, i.e., every hour. Based on bulk data on rain and time of occurrence, pixel detection is carried out to identify the absence of extreme rain events in the entire region. This identification is done based on the 50 mm/hour threshold. In the first step, rain conditions with an intensity of less than 50 mm/hour will be eliminated. Furthermore, rain with an intensity of more than 50 mm/hour will be labelled as an extreme rain pixel. Cumulative number of extreme rain pixel across the study area was calculated. Then, the cumulative number of pixels was ranked from the highest to the smallest. From all the identified extreme rainfall in the dataset, the most extreme events were selected. The extreme rain and air pressure were then studied for their distribution both spatially and temporally. In this paper, only selected events that have distinct characteristics are presented in the result and discussion section.

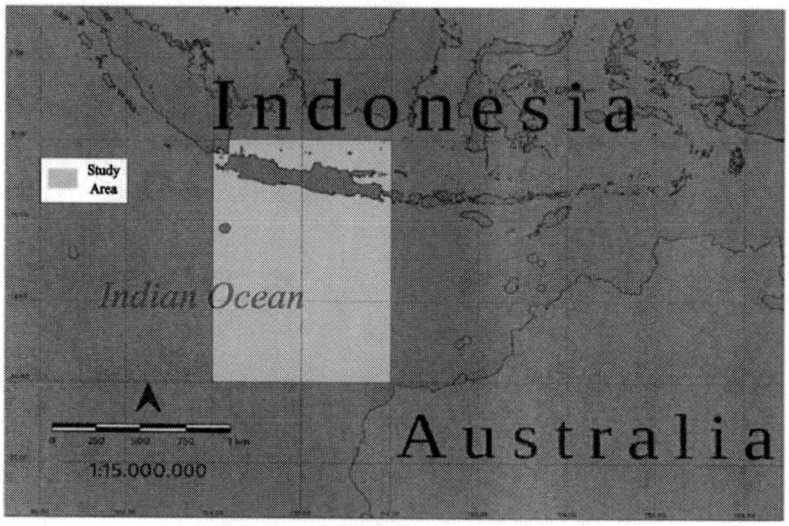

Figure 1. Study area location (Source: data processing, 2021; coastline data from Global Self-consistent, Hierarchical, High-resolution Shorelines (GSHHG) https://www.soest.hawaii.edu/pwessel/gshhg/)

3. Result and Discussion

Four of the ten most extreme rain events have a direct relationship with tropical cyclone phenomena. The tropical cyclone is Tropical Cyclone Cempaka-Dahlia in 2017, Tropical Cyclone Savannah in 2019, and Tropical Cyclone Riley in 2015. Among the three tropical cyclones that occurred, the Cempaka-Dahlia tropical cyclone brought the highest amount of extreme rain. Figure 2 indicates that the time distribution of extreme rain events on November 30th, 2017, is uneven. The rain occurred in the morning and reach its maximum on the afternoon and decreasing from the afternoon to the evening. The distribution of extreme rain that occurs appears to be relatively centered in the Indian Ocean by forming a circular pattern. The dominant extreme rain is mostly found on the ocean at the southern Java Island, followed by heavy rain that is spread more widely. The heavy rain not only occurred in the Indian Ocean, but also appeared to occur on the land surface. The longest duration of extreme rain occurred for 7 hours, starting from 4.00 am to 12.00 noon local time. This extreme event has been identified causing floods, landslides, and strong winds.

Air pressure data shows an anomaly in the low-pressure system adjacent to Java Island (Figure 3). On the previous days (28th and 29th November), low air pressure was in the central region of the Indian Ocean, although it had appeared to the west of Sumatra Island the day before. In the event of extreme rain on November 30th, 2017, there was an area of low air pressure surrounding the Java Island.

In addition to tropical cyclone Cempaka-Dahlia, tropical cyclone Savannah in 2019 also brought large amounts of extreme rain. Extreme rain events detected on March 15th, 2019, predominantly occurred during the day with a duration of up to 9 hours. The extreme rain associated with tropical cyclone Savannah causes flooding in the Gunungkidul area, which is located at the south part of Central Java [9]. The distribution of extreme rain that occurs during tropical cyclone Savannah is centered on two areas flanking the middle of the island of Java (Figure 4). This extreme rain flanked the north and south of the central part of Java Island. The distribution of extreme rain in the southern Java Island, is more dominant than extreme rain at the northern part.

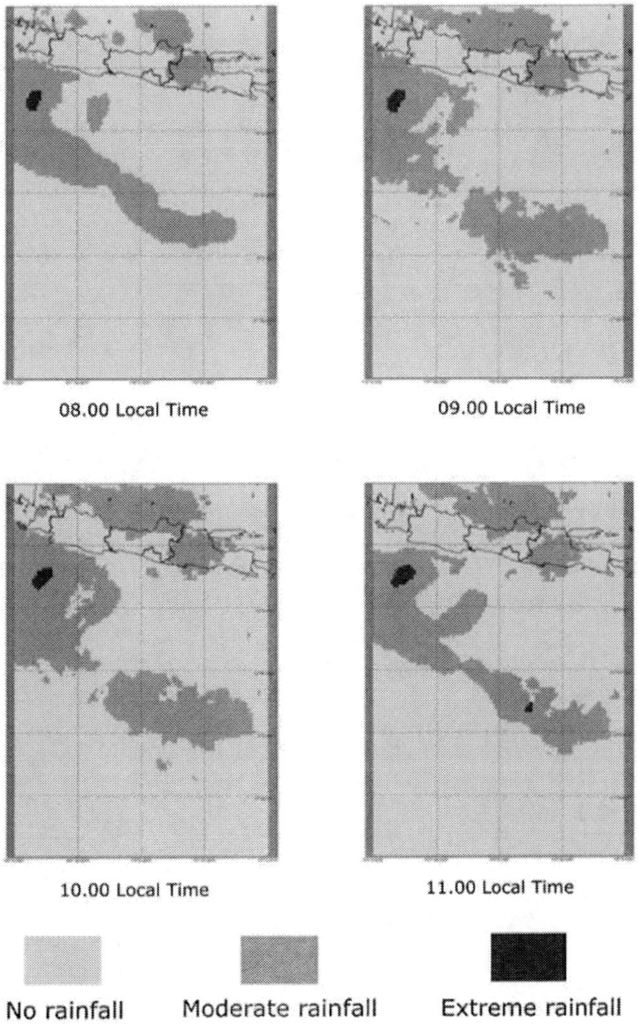

Figure 2. Rainfall distribution on November 30th 2017 (Source: Data processing, 2021)

IOP Conf. Series: Earth and Environmental Science 1233 (2023) 012067 doi:10.1088/1755-1315/1233/1/012067

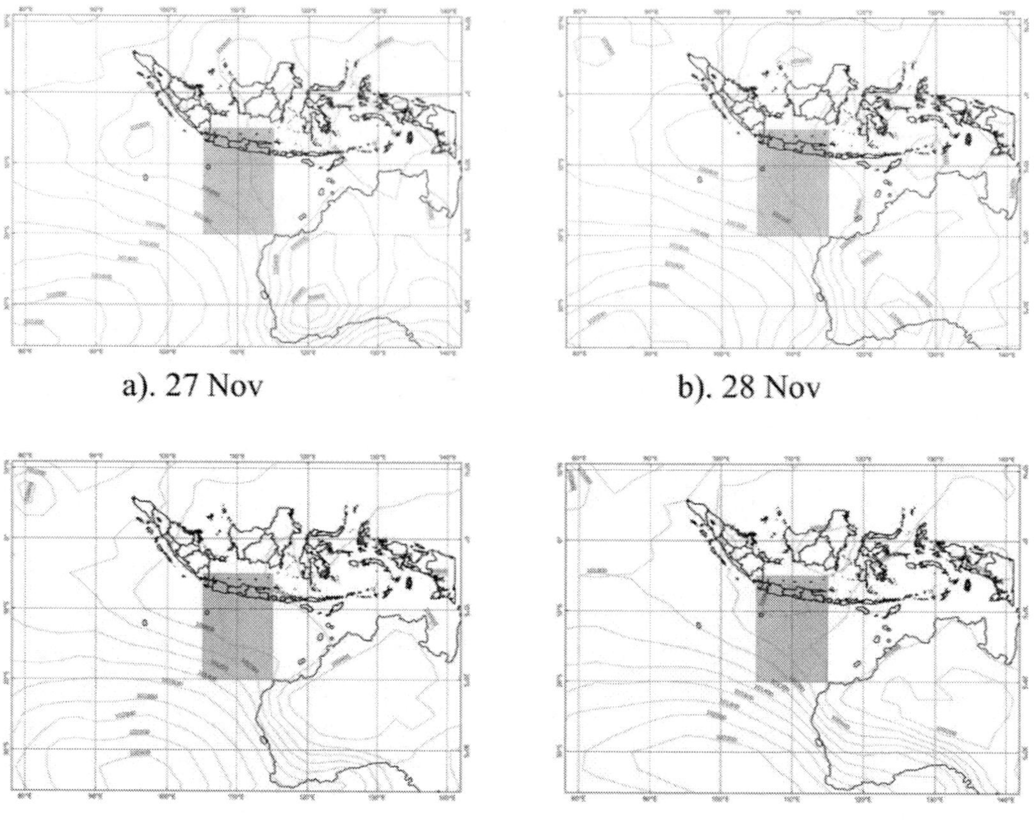

a). 27 Nov

b). 28 Nov

c). 29 Nov

d). 30 Nov

Figure 3. Sea Level pressure distribution during extreme rainfall on November 30th, 2017 (Source: Data processing, 2021)

The air pressure during extreme rains appears to be lower than the previous days (Figure 5). Low air pressure centered in the Australian area and spreads to the west close to the study area. The study area is flanked by two areas that have a low-pressure system from the west and east sides. The low-pressure system in the west can be seen in the Indian Ocean close to the Sumatra Island.

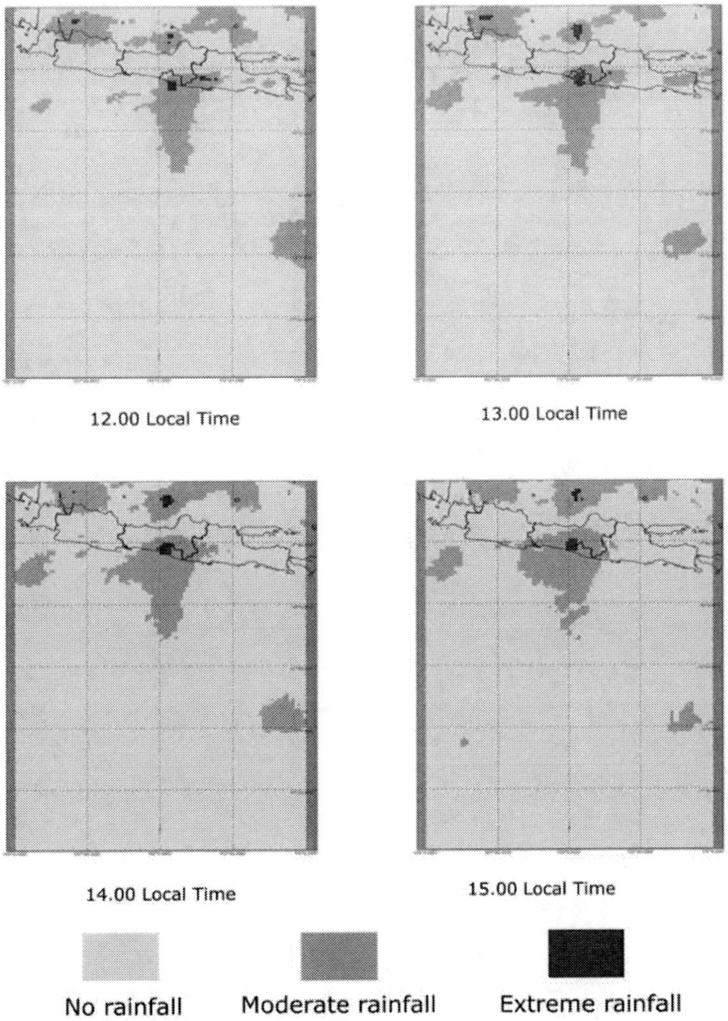

Figure 4. Rainfall distribution on March 15th 2019 (Source: Data processing, 2021)

IOP Conf. Series: Earth and Environmental Science 1233 (2023) 012067 doi:10.1088/1755-1315/1233/1/012067

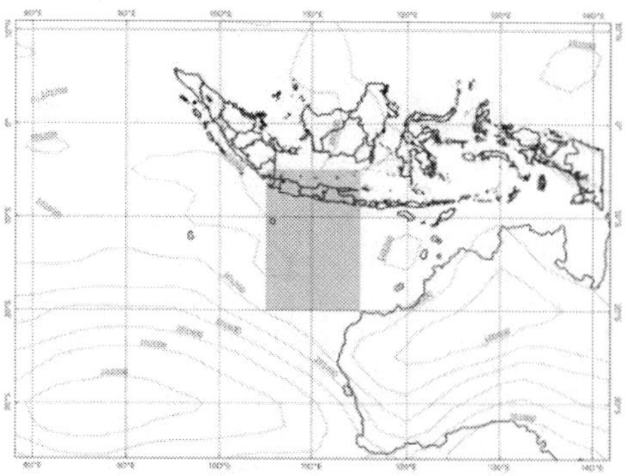

a). 12 Mar

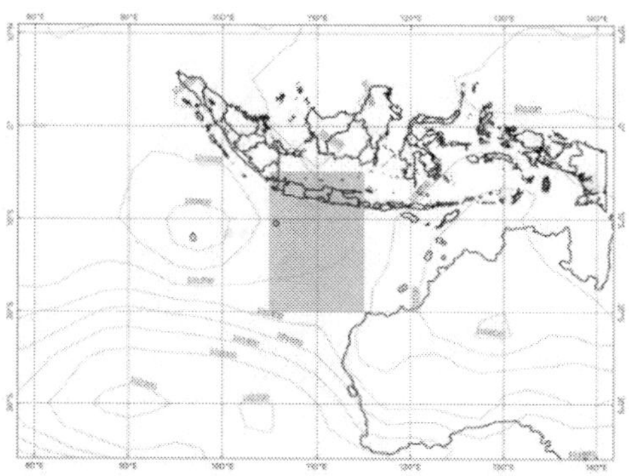

b). 13 Mar

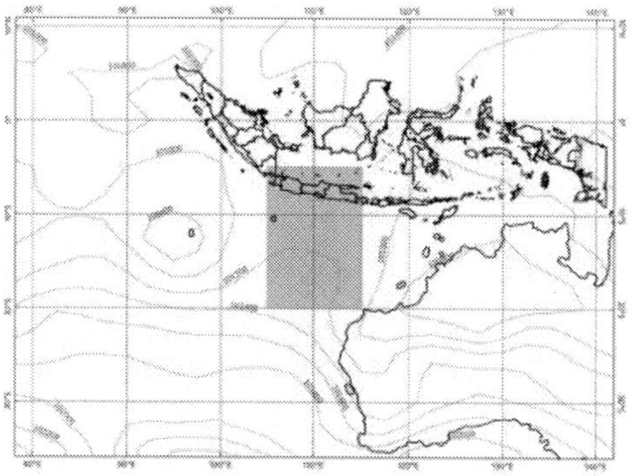

c). 14 Mar

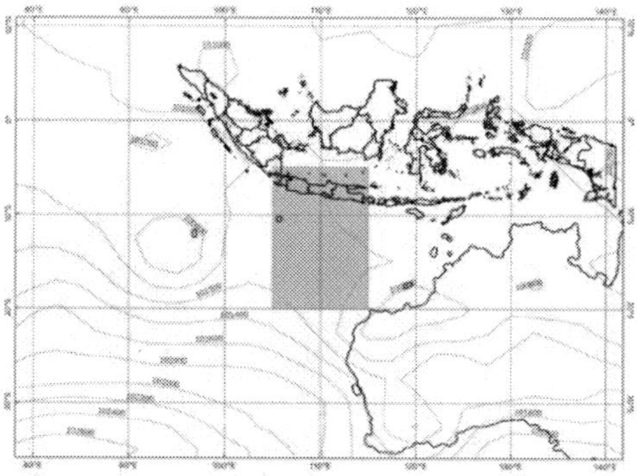

d). 15 Mar

Figure 5. Sea Level pressure distribution during extreme rainfall in March 15th 2019 (Source: Data processing, 2021)

The extreme rains detected in pixel detection are not always associated with tropical cyclones. An example is extreme rain events in September 2016 and January 2015. In September 2016, extreme rains were detected for four days, namely 16th, 24th, 17th, and 19th September. The most extreme rain was found on September 16th, 2016. Extreme rain occurred from the afternoon until the evening. The extreme rain that occurred on September 16th, 2016, predominantly occur on the land (Figure 6). The area that appears to be the center of extreme rain revolves around the border area between Central Java and West Java. In addition, the extreme rain that occurred was also accompanied by lower rain rate with wide coverage. The East Java area experiences almost no rain. Likewise, the Indian Ocean not experienced extreme rain events. The duration of the extreme rain that occurred reached 10 hours, starting from 2.00 pm to 11.00 pm. The peak of extreme rain seemed to occur at 17.00 in the afternoon.

IOP Conf. Series: Earth and Environmental Science 1233 (2023) 012067 doi:10.1088/1755-1315/1233/1/012067

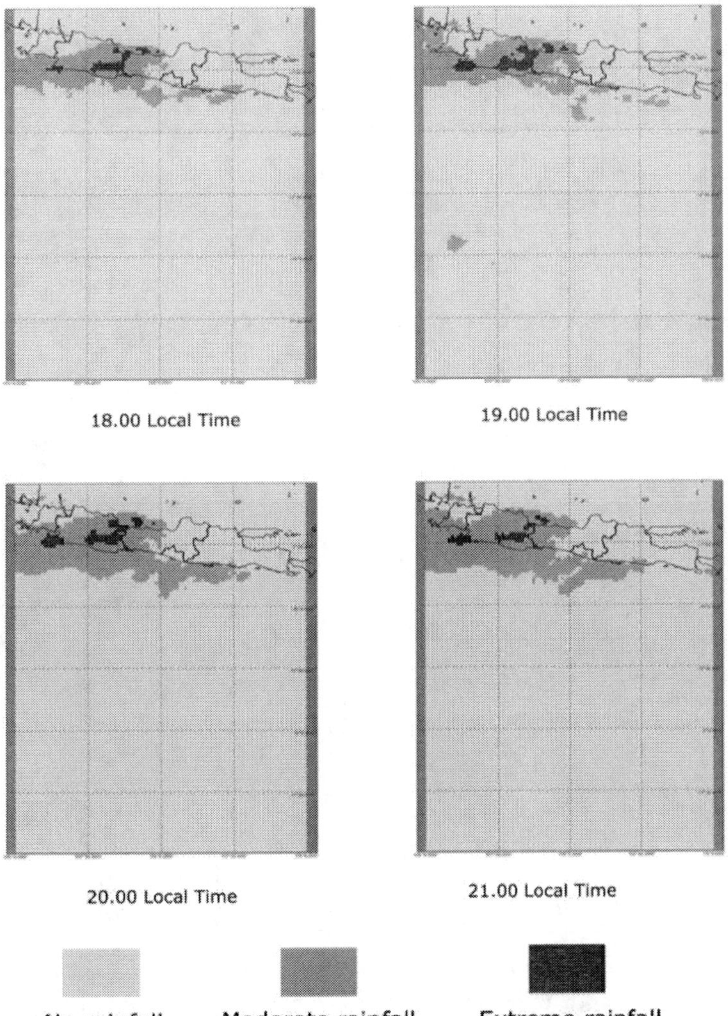

Figure 6. Rainfall distribution in September 16th 2016 (Source: Data processing, 2021)

The air pressure observed in the event of extreme rain on September 16th, 2016, appears to be dynamic. In Figure 7, the pattern of air pressure that occurs looks much different from the pattern of air pressure on the previous days. On September 13th and 14th, there was relatively minor low pressure to the west of Sumatra Island. On the 15th of September, the low pressure moved westward and continued to widen, but returned to a relatively similar position on the 16th. One thing that looks very different is the pressure pattern from the south side itself, where on September 16th, the air pressure centered from the mid-latitude region forming a clear circular pattern. The low pressure that previously appeared in the Australian region moved north and widened across the island of Java. During this period, it can also be seen that the area north of the equator has lower pressure than the area south of the equator. The meteorological phenomenon that is thought to be related to extreme rain and the description of this air pressure pattern is the Intertropical Convergence Zone (ITCZ) phenomenon. ITCZ appeared around the equator in early September 2016 [10].

IOP Conf. Series: Earth and Environmental Science 1233 (2023) 012067 doi:10.1088/1755-1315/1233/1/012067

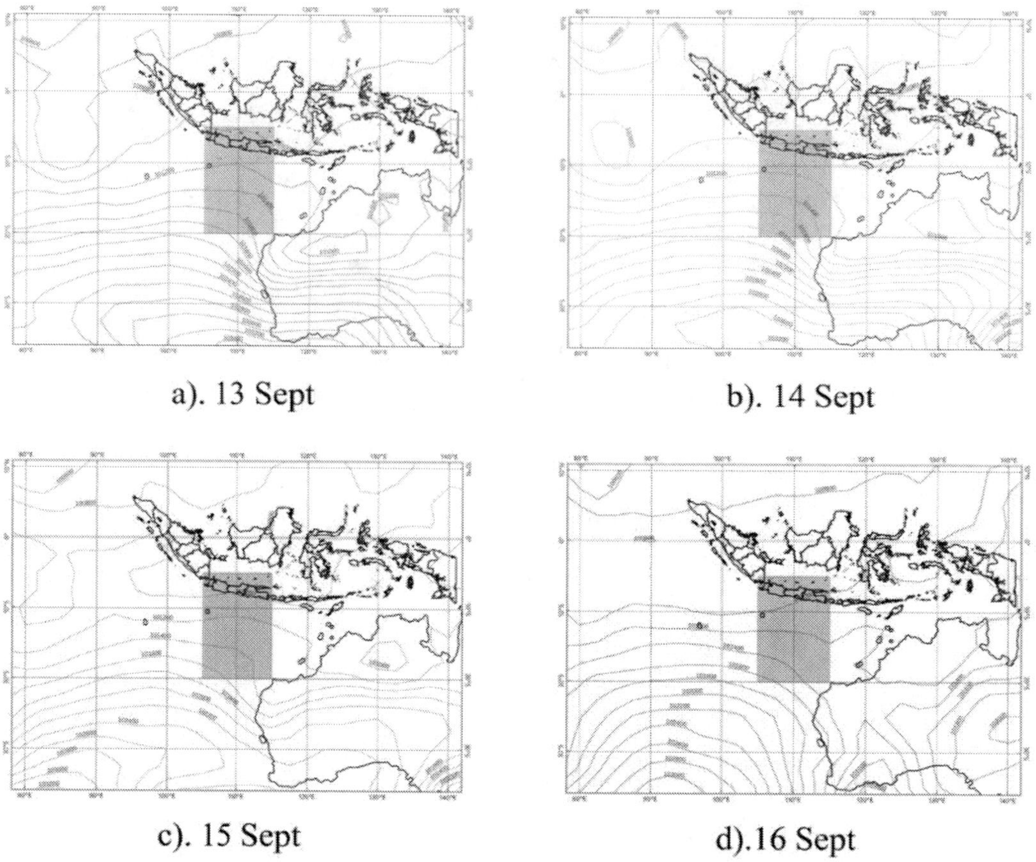

Figure 7. Sea Level pressure distribution during extreme rainfall September 16th, 2016 (Source: Data processing, 2021)

Another extreme rain event that is not related to tropical cyclones in the incident on January 25th, 2015. The extreme rain tends to be continuous. The peak of extreme rain appears to have occurred at 8 am, but the incident continued into the evening and again at night. The extreme rain that occurred on January 25th, 2015, appears to be randomly distributed (Figure 8). Based on the time sequence of rain events, the extreme rain that occurred appeared to have occurred in the western part of Java Island and then moved southeastward through the central part of Java Island until finally centered in the ocean. However, most areas of Java Island experience rain that is almost evenly distributed even though it is not classified as extreme. The peak of extreme rain at 8.00 was dominant in the central area of the island of Java, which is most likely to have a high elevation. After that, the extreme rain moves southeast towards the ocean surrounded by non-extreme rain events around it is forming a circular pattern.

The distribution of air pressure during extreme rain events is shown in Figure 9. There are two low-pressure areas located in the Indian Ocean and Australia as shown in the figure. The air pressure in the study area itself seems to be heavily influenced by the air pressure system in the Australian region. This extreme rain event is thought to be a phenomenon related to El Nino that occurred throughout 2015. The El-Nino phenomenon itself is the largest drought event in 2015. The formation of the El-Nino phenomenon does not occur instantly but is formed gradually. In January, the Indian Ocean Dipole (IOD) index was at a negative level which can contribute to heavy rainfall.

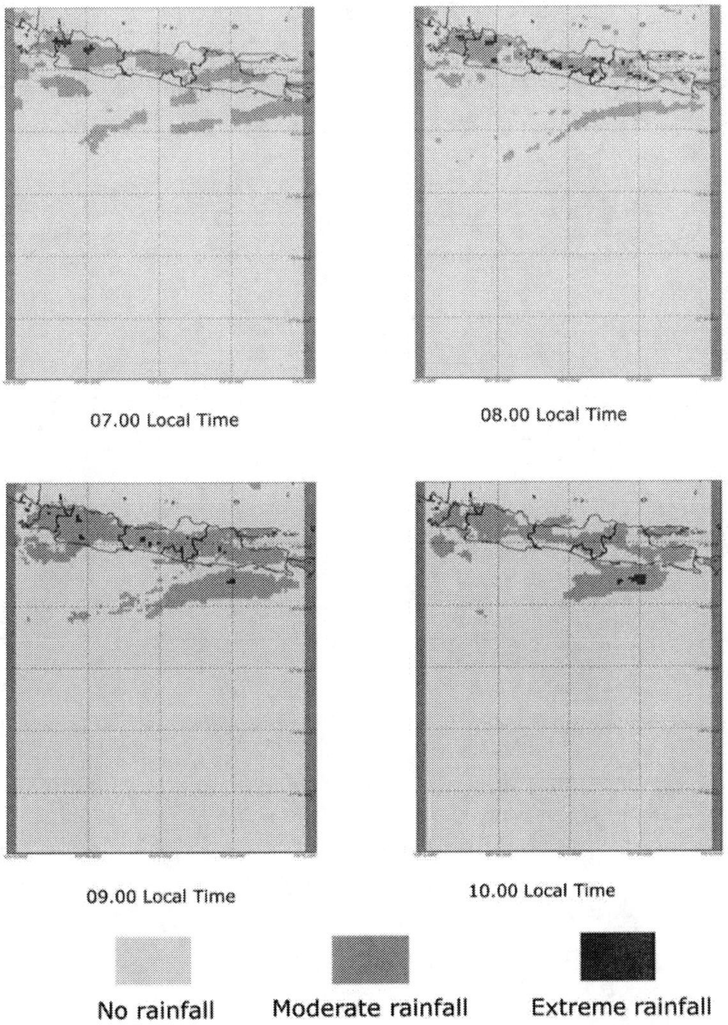

Figure 8. Rainfall distribution in January 25th 2015 (Source: Data processing, 2021)

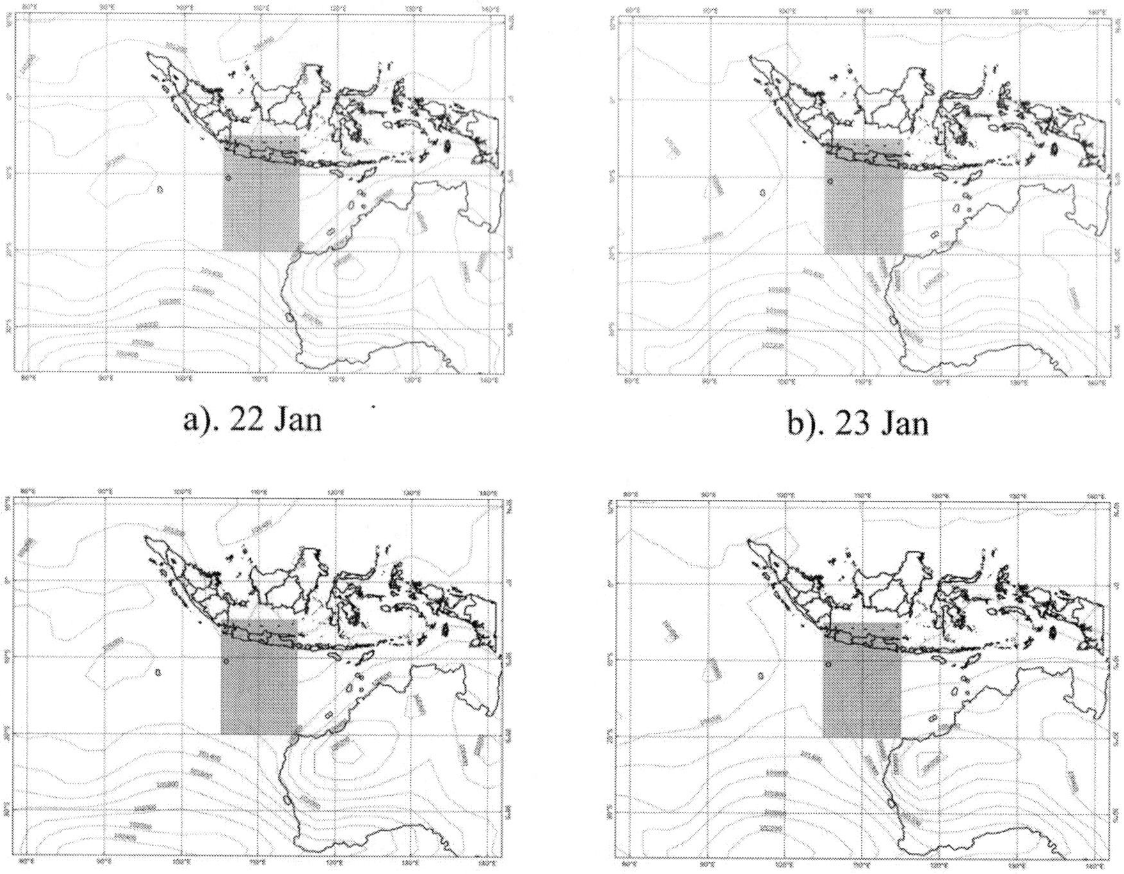

a). 22 Jan b). 23 Jan

c). 24 Jan d). 25 Jan

Figure 9. Sea Level pressure distribution during extreme rainfall in January 25th, 2015 (Source: Data processing, 2021)

The result obtained in this research shows that extreme rainfall can be triggered by various sources. Previous research shows that the Intertropical Convergence Zone (ITCZ), Madden-Jullian Oscillation (MJO), El Nino Southern Oscillation (ENSO), and Indian Ocean Dipole Mode (IOD) phenomena affect the occurrence of extreme rainfall in addition to tropical cyclone [11]– [13]. Similar research in the East Asian Monsoon region also indicated that the amount of precipitation due to tropical cyclone is less compared to precipitation generated by monsoon [14]. However, it should be noted that previous research also highlighted the increasing number of tropical cyclones in the study area [4]. Therefore, the contribution of precipitation due to tropical cyclones in the study area might increase in the future.

4. Conclusion

Extreme rain events tend to occur in tandem with the phenomenon of low-pressure systems. Four events out of the ten extreme events with the highest number in 2015-2019 are the influence of the tropical cyclone phenomenon. The remaining are related to the tropical climate dynamics. The most important part of the finding of this study is that heavy rainfall could occur even with small differences in the air pressure system. It means that to produce heavy rainfall, a low-pressure system does not require to be developed as a tropical cyclone. In general, tropical cyclone is often associated with a surface air pressure of less than 980 hPa. However, in the study area, the pressure can be higher than 1000 hPa. This means that a small tropical depression could trigger the occurrence of heavy rainfall events in the study area. One of the significances of this study is related to the development of heavy rainfall early warning systems. Heavy rainfall is often associated with a high surface temperature that triggers the convective process. However, previous research shows that in Indonesia, heavy rainfall could occur with less energy for the convective process [15]. Therefore, a heavy rainfall early warning system based on pressure might be less accurate.

Acknowledgments

Part is this research is supported by Grant of "Hibah Penelitian Dasar Unggulan Perguruan Tinggi (PDUPT)" in 2022 in the topics of precipitation estimation and cloud organization with Dr.Sc. Andung Bayu Sekaranom., M.Sc. as the principal investigator.

References

[1] Montgomery M T and Farrell B F 1993 Tropical cyclone formation *Journal of Atmospheric Sciences* 50(2) 285–310.
[2] Tjasyono B 1991 Tropical Storm Growth and Its Relationship with Warm Waters Around Indonesia (in Bahasa Indonesia) *Agromet* 7(2) 14–22.
[3] Hapsarini S A 2019 Analysis of Tropical Cyclone Rain and Flood Disasters - Cempaka and Dahlia Tropical Cyclone Case Study in 2017 (in Bahasa Indonesia) *Thesis* Yogyakarta: Universitas Gadjah Mada.
[4] Mulyana E Prayoga M B R Yananto A Wirahma S Aldrian E Harsoyo B Seto T H and Sunarya Y 2018 Tropical cyclones characteristic in southern Indonesia and the impact on extreme rainfall event *MATEC Web of Conferences* 229(1) 02007.
[5] Sekaranom A B Nurjani E and Pujiastuti I 2018 Cloud structure evolution of heavy rain events from the East-West Pacific Ocean: A combined global observation analysis *IOP Conference Series: Earth and Environmental Science* 148(1).
[6] Sekaranom A B Nurjani E Hadi M P, and Marfai MA 2018 Comparison of TRMM Precipitation Satellite Data over Central Java Region - Indonesia *Quaestiones Geographicae* 37(3) 97–114.
[7] Marfai M A Sekaranom A B and Ward P 2015 Community responses and adaptation strategies toward flood hazard in Jakarta, Indonesia *Natural Hazards* 75(2) 1127–1144.
[8] Trenberth K E and Caron J M 2000 The Southern Oscillation revisited: Sea level pressures, surface temperatures, and precipitation *Journal of Climate* 13(24) 4358–4365.
[9] Riyanto I A Cahyadi A Ramadhan F Naufal M Widyastuti M and Adji T N 2020 The Impact of Tropical Cyclone Savannah on the Kalinongko Karst Window, Gunungsewu Karst, Gunungkidul Regency, Indonesia (in Bahasa Indonesia) *Jurnal Geografi: Media Informasi Pengembangan dan Profesi Kegeografian* 17(1)7–14.
[10] Rohli R V and Vega A J 2017 *Climatology* Burlington: Jones & Bartlett Learning
[11] Sekaranom A B Nurjani E Harini R and Muttaqin A S 2020 Simulation of Daily Rainfall Data using Articulated Weather Generator Model for Seasonal Prediction of ENSO-Affected Zones in Indonesia *The Indonesian Journal of Geography* 52(2) 143–153.
[12] Izumo T Lengaigne M Vialard J Luo J J Yamagata T and Madec G 2014 Influence of Indian Ocean Dipole and Pacific recharge on following year's El Niño: Interdecadal robustness *Climate Dynamics* 42(1–2) 291–310.
[13] Tangang F T Juneng L Salimun E Vinayachandran P N Seng Y K Reason C J C Behera S K Yasunari T 2008 On the roles of the northeast cold surge, the Borneo vortex, the Madden-

Julian Oscillation, and the Indian Ocean Dipole during the extreme 2006/2007 flood in southern Peninsular Malaysia *Geophysical Research Letters* 35(14) 1–6.
[14] Chang C Lei Y Sui C, Lin X, and Ren F 2012 Tropical cyclone and extreme rainfall trends in East Asian summer monsoon since mid-20th century *Geophysical Research Letters* 39(18) 1-6.
[15] Sekaranom A B and Masunaga H 2019 Origins of heavy precipitation biases in the TRMM PR and TMI products assessed with cloudsat and reanalysis data *Journal of Applied Meteorology and Climatology* 58(1) 37–54.

IOP Conf. Series: Earth and Environmental Science 1233 (2023) 012068 doi:10.1088/1755-1315/1233/1/012068

Water Quality of Tirtomulyono and Tirtomulyani Springs, Klaten, Central Java

N Khasanah[1] and S Purnama[2]
[1] Environmental Science Master Programme, Universitas Gadjah Mada, Indonesia
[2] Faculty of Geography, Universitas Gadjah Mada, Indonesia
*Corresponding author: nurilkhasanah95@mail.ugm.ac.id

Abstract. Water quality is one aspect that is increasingly being considered in the management of water resources. In addition to the requirements regarding water quality that are more considered, currently there is a tendency for water pollution to occur in several water sources, including springs. The purpose of this study was to evaluate the water quality of the Tirtomulyono and Tirtomulyani springs in Kebonarum District, Klaten Regency, Central Java Province. To determine the condition of the water quality of the Tirtomulyono and Tirtomulyani springs, direct observations and water samples were taken. Direct observations in the field were carried out to determine physical parameters, while water samples were taken for analysis in the laboratory for chemical and biological parameters. The results showed that in terms of physical and chemical parameters, the quality of water in Tirtomulyono and Tirtomulyani springs was good because it met drinking water requirements. according to the Regulation of the Minister of Health No.492/Menkes/Per/IV/2010. However, based on the results of laboratory analysis, the Total Colliform value of the two springs is > 1600 MPN/100 ml even though according to the quality standard the Total Colliform value must be zero, so in terms of the Total Colliform value it does not meet the drinking water requirement.

1. Introduction

In the tropics there are three sources of water, namely rainwater, surface water and groundwater. Of the three water sources, most of the daily human water needs such as drinking, bathing and washing are sourced from groundwater [1][2][3][4][5]. This water source is usually obtained through dug wells or drilled wells. In Indonesia, the use of dug wells as a source of water for domestic needs has been carried out for a long time. However, drilling activities only started during the Dutch East Indies era with the construction of several artesian wells on the north coast of Java at the end of the 19th century. It is not surprising that most Indonesians use groundwater as a source of water for their daily life. The main reason is that this water is clearer or of better quality than river water or rainwater [6][7]. The availability of groundwater is also more sustainable, unlike rainwater which is highly dependent on the seasons.

Springs are one part of groundwater that occurs naturally on the ground surface [8]. Springs can be said to be everywhere and appear in various ways. However, the distribution is uneven depending on factors that affect the availability of springs, such as the level of rainfall, the hydrological characteristics of the soil surface (especially its permeability), topography, hydrological characteristics of aquifer formations and the geological structure of the area. In some places where groundwater is difficult to obtain through dug wells, springs are often used as an alternative source of water for the community's drinking water needs.

Water quality is one aspect that is increasingly being considered in the management of water resources. In addition to the requirements regarding water quality that are more considered, currently there is a tendency for water pollution to occur in several water sources, including springs. Overall, domestic and industrial waste is the largest source of water pollution which changes the nature of water which can affect the chemical properties of the water receiving the waste. For this reason, the purpose of this study was to evaluate the quality of water from the Tirtomulyono and Tirtomulyani springs in Kebonarum District, Klaten Regency, Central Java Province, Because these two springs are important water sources for people whose water quality is threatened by pollution.

2. Research Method

To determine the condition of the water quality of the Tirtomulyono and Tirtomulyani springs, direct observations and water samples were taken. Sampling of water from springs is carried out in the dry season in the early morning hours. Observations directly in the field were carried out to determine the physical parameters i.e. color, taste and odor.

For chemical and biological parameters, water samples were taken for analysis at the Yogyakarta Center for Environmental Health and Disease Control Engineering. The chemical and biological parameters analyzed were pH, Hardness as $CaCO_3$, Iron (Fe^{2+}), Manganese (Mn^{2+}), Sulfate (SO_4^{2-}), Nitrate (NO_3^-), Chloride (Cl^-) and Total Coliform. To evaluate the feasibility of the water quality for use as a source of drinking water, a comparison is made with the Government Regulation of The Republic of Indonesia Number 22 Year 2021 About Implementation of Protection and Management of The Environment.

3. Result and Dicussion

3.1. Physical properties of springs

This research was conducted in Tirtomulyono and Tirtomulyani springs in Kebonarum District. In terms of topography, the research area tends to have a flat to sloping topography (Figure 1).

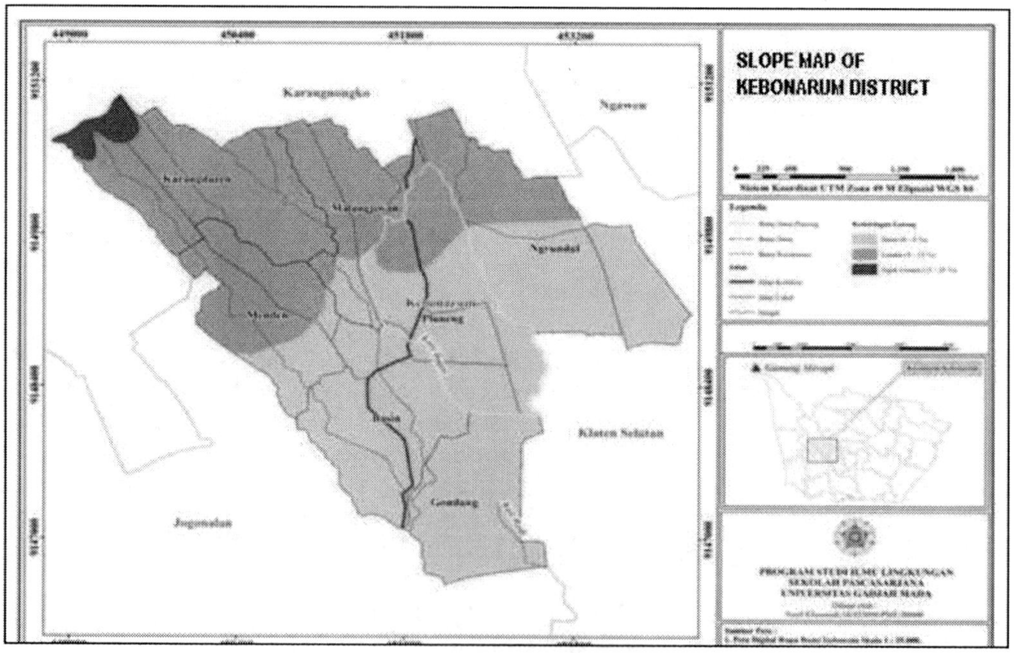

Figure 1. Topography of the research area

Physical parameters of water from springs are identified directly in the field using the five senses. In this study, the physical quality of spring water in terms of color, taste and odor parameters, the results of which are shown in Table 1. Observing Table 1, it can be seen that the physical properties of water from Tirtomulyono and Tirtomulyani springs are classified as good because they are colorless, tasteless and odorless.

Table 1. Results of Water Quality Analysis Based on Physical Properties

No.	Parameter	Maximum Standard	Tirtomulyono	Tirtomulyani
1.	Color	Colorless	Colorless	Colorless
2.	Taste	Tasteless	Tasteless	Tasteless

3.	Odor	Odorless	Odorless	Odorless

Sources : field work (2020).

3.1.1. Color

Naturally, the state of water is colorless or clear. The presence of color in water can be caused by the presence of iron ions, manganese ions, humus, plankton, aquatic plants and industrial waste. Industrial waste that has just been generated is generally gray in color. When organic compounds are decomposed by microorganisms, O_2 will be depleted due to the decomposition process and the waste will decompose marked by a change in color to black.

However, color is not an absolute indicator of pollution. For example, the brown color of peat soil is not the result of pollution but is caused by the presence of tannic acid and humic acid. In waters, colored water is not so desirable because it can reduce the photosynthetic activity of aquatic plants.

The results of field observations showed that the water from the Titomulyono and Tirtomulyani springs was colorless. Quality Standards Regulation of the Government Regulation of The Republic of Indonesia Number 22 Year 2021 concerning requirements for drinking water quality requires colorless water to be used as a source of drinking water, so that the water in both springs meets the quality standards.

3.1.2. Taste

Several types of water taste, among others, is salty, sour, bitter and bland. However, until now the taste parameters are still difficult to measure quantitatively, so they are still determined through the very subjective sense of taste. Good water is tasteless water. The results of field observations show that the water from the Tirtomulyono and Tirtomulyani springs is tasteless, so it is suitable for use as a source of drinking water.

3.1.3. Odor

Water that smells can be caused by the presence of gases from the decomposition of organic matter, such as an odor like rotten eggs can be caused by H_2S gas in an aerobic process. Odor in water can also be caused by dissolved organic matter such as phenol, chlorophenol and protein. Like taste, odor is also a parameter that cannot be quantified and good water is water that does not smell. Field observations show that water from Tirtomulyono and Tirtomulyani springs does not smell, so it is suitable for use as a source of drinking water.

3.2. Chemical properties of springs

Chemical parameters of spring water were identified through analysis of pH, hardness and elements of iron, manganese, sulfate, nitrite, nitrate and chloride. The results of the analysis are shown in Table 2.

Table 2. Results of Water Quality Analysis Based on Chemical Properties

No.	Parameter	Unit	Maximum Standard	Tirtomulyono	Tirtomulyani
1.	pH	-	6,5-8,5	6,9	6,8
2.	Hardness as $CaCO_3$	mg/l	500	72,36	80,40
3.	Iron (Fe^{2+})	mg/l	0,3	<0,0168	0,0212
4.	Manganese (Mn^{2+})	mg/l	0,4	<0,0066	<0,0066
5.	Sulfate (SO_4^{2-})	mg/l	250	20	17
6.	Nitrate (NO_3^-)	mg/l	50	3,66	5,25
7.	Nitrite (NO_2^-)	mg/l	3	<0,0314	<0,0314
8.	Chloride (Cl^-)	mg/l	250	<0,03	<0,03

Sources : result of laboratory analysis

3.2.1. pH

The pH value of a water indicates the balance between acid and base in the water by measuring the hydrogen ion concentration. Waters that have high carbonate, hydroxide and bicarbonate content will tend to be alkaline, while the presence of free mineral acids and carbonic acid will tend to make the water acidic. increase acidity. Under natural conditions, acidic waters are more common than alkaline waters. However, in general, freshwater pH ranges between 5 and 9, where in this pH range freshwater fish will still be able to live. The high and low pH value of the water will also be able to affect the type and composition of substances in the aquatic environment and affect the availability of nutrients and the level of toxicity of the trace elements.

Observing Table 2, Tirtomulyono springs and Tirtomulyani springs have pH values that are not much different, namely 6.9 and 6.8. In general, water with a pH of < 7 is found in areas with volcanic deposits, while water with a pH > 7 is found in areas with ultramafic rocks.

Good water to be used as drinking water According to the Government Regulation of The Republic of Indonesia Number 22 Year 2021 concerning requirements for drinking water quality, it is water with a pH value ranging from 6.0 to 9.0. Thus, the two springs meet the requirements to be used as drinking water when viewed from the pH value of the water.

The pH values of the two springs indicate that the acidity of the water is still in a neutral group. It can be said that the area where the springs appear which is still affected by volcanic activity has quite an effect on the pH of the springs. As is known, the tendency of the pH value of springs in volcanic areas is also often classified as neutral.

3.2.2. Hardness

Most of the causes of total hardness in water are Ca^{2+} and Mg^{2+} ions. However, Fe^{2+} and Mn^{2+} ions and all cations with a charge of 2 can also cause total hardness. In groundwater, high hardness is generally found in groundwater found in limestone areas, because it is a source of Ca^{2+} and Mg^{2+}.

Water that has a high level of hardness is not desirable to be used as a water source for household and industrial activities. Water with a hardness level of more than 350 mg/l will cause problems if the water is used for bathing purposes. The chemical relationship between the hardness ions and the soap molecules causes the detergent properties of the soap to be lost, making it difficult for soap to foam. As a result, there will be waste in the use of soap.

Hard water is also undesirable for industrial use. This is caused by the precipitation of Ca^{2+} and CO_3^{2+} which resulted in the formation of crust on the pipe wall by $CaCO_3$ deposition. As a result, the wet cross-sectional area of the pipe is reduced and the heating process in the boiler is prolonged.

Concerning to the hardness, the Tirtomulyono spring has a hardness value of 72.36 mg/l and the Tirtomulyani spring has a hardness value of 80.40 mg/l. This value is still below the specified quality standard limit of 500 mg/l. The level of hardness of spring water in the study area can be classified as low hardness (soft water).

3.2.3. Iron (Fe^{2+})

Iron is a chemical element that is often found in rock minerals [9], such as minerals that make up igneous rocks, namely pyroxene; amphibole; biotite, magnetite and olivine. Although most of the iron in igneous rocks is ferric, it can also be mixed with ferrous iron (Fe^{2+}), such as magnetite.

According to [8], in addition to these minerals, other sources of iron in igneous rocks are ferromagnesian minerals mica, ferrous sulfide and ferric sulfide or iron pyrite. Iron can also be sourced from sandstone minerals such as oxides (hematite), sulfides (pyrite) and iron clay minerals. In sedimentary rocks, iron is also found in ferrous form such as marcasite, siderite carbonate and in the form of mixed oxide magnetite which is an igneous or metamorphic rock mineral.

In surface water, iron with a concentration of more than 1 mg/l is rarely found, but in groundwater the concentration is often high which is generally dissolved as Fe^{2+} ions. ur is abundant in water, but iron is one of the important elements in groundwater. Because with a concentration of only a few milligrams of iron per liter, it is no longer suitable for several purposes. Therefore, in water quality studies often always include this elemental analysis. Iron is a water soluble substance that is highly undesirable for domestic use, because at concentrations greater than 0.30 mg/l/ it can cause a reddish-brown color like a rust tuft. High levels of iron also cause an unpleasant taste in drinking water and even in the long term will accumulate in the kidneys. In addition to having a bad effect on body health,

water with an iron content of 0.1 – 0.2 mg/l can also stimulate the growth of iron bacteria which will produce mucus which can cause blockages in water pipes.
Iron concentration in Tirtomulyono springs < 0.0168 mg/l and Tirtomulyani springs 0.0212 mg/l. Based on the quality standard of the Government Regulation of The Republic of Indonesia Number 22 Year 2021, the iron concentration has a maximum standard of 0.3 mg/l, so that when viewed from the iron concentration, the water from the two springs meets the requirements for source of drinking water.

3.2.4. Manganese (Mn^{2+})
In natural water, manganese comes from metamorphic rocks and sedimentary rocks. In addition, sedimentary rock minerals such as biotite mica and hornblende amphibole can also be contain large amounts of manganese [8]. Manganese concentrations in natural water are usually around 0.2 mg/l or even less, whereas acidic groundwater from mining areas may contain more than 10 mg. /l.
The presence of manganese in water will be disturbing even at low concentrations, so most water quality studies always include this ion in their analysis. According to the U.S. Public Services, water with manganese content of more than 0.05 mg/l should no longer be used as a daily water source.
The effect of manganese is almost the same as that of iron, although only at a lower concentration, namely because it can cause a brownish red color in washed clothes. Similarly, bathroom equipment can be covered by manganese rust mucus present in water with that concentration. In general, water with a manganese content of more than 0.1 mg/l should no longer be used as a water source for household purposes.
The permissible levels of Manganese according to the Government Regulation of The Republic of Indonesia Number 22 Year 2021 is 0.1 mg/l, then water that has manganese levels above this value does not meet the requirements while Tirtomulyono and Tirtomulyani springs have manganese levels that are far below the quality standard, which is less than 0.0066 mg /l.

3.2.5. Sulfate (SO_4^{2-})
Although not a major constituent of the Earth's crust, it is widely distributed in reduced form as metal sulfides in igneous and sedimentary rocks. According to [9], in igneous rocks, SO42- ions come from feldspathoid minerals, while in sedimentary rocks they come from evaporite rock minerals such as gypsum and anhydrite. Sulfates are also sourced from sulfide minerals, namely pyrite, FeS2, the detergent industry, domestic waste and acid rain.
The concentration of SO_4^{2-} ions in water is generally less than 400 mg/l. If the concentration reaches 500 mg/l the water will taste bitter, and if it reaches 1000 mg/l there will be a cathartic effect. The cathartic effect is nausea and vomiting due to the formation of $MgSO_4$ and Na_2SO_4 in the stomach. In addition to affecting health, excessive levels of SO_4^{2-} ions in water cause odors in waterways due to the reduction of SO_4^{2-} to H_2S [10]. For water to be used as a source of drinking water, it should have a sulfate content of less than 400 mg/l.
Sulfate concentration in Tirtomulyono springs are 20 mg/l and Tirtomulyani springs are 17 mg/l. Government Regulation of The Republic of Indonesia Number 22 Year 2021 concerning requirements for the quality of drinking water containing sulfate has a maximum standard of 300 mg/l, so that when viewed from the sulfate content, the water from the two springs meets the requirements for drinking water sources.

3.2.6. Nitrate (NO_3^-) and Nitrite (NO_2^-)
Nitrate is a form of nitrogen that is soluble in water. Nitrate from rock is generally small because most of the nitrate in groundwater comes from agricultural, household and industrial wastes [10]. Nitrate from agricultural waste comes from fertilizers used by farmers, while from households comes from animal and human feces contamination.
According to [8], nitrate is the most stable form of nitrogen compounds. Its concentration in natural water is generally less than 10 mg/l. NO3- levels that are more than 50 mg/l will be toxic to infants and ruminant animals, because in the digestive system, nitrate is reduced to nitrite which can bind to hemoglobin in the blood. As a result, blood transport is blocked and the baby's body turns blue, so it is

known as blue baby disease (methemoglobinemia). In addition, excessive nitrate levels in water can also cause the water to taste bad. If the water contains more than 100 mg/l of nitrate, it can cause a bitter taste on the tongue and a feeling of tension.

Nitrite is a reduced form of nitrate, so as with nitrate, sources of nitrite in water also come from agricultural, household and industrial wastes. As stated earlier, nitrite is a chemical substance that is toxic to infants, because it can inactivate hemoglobin. As a result there is a bottleneck in blood transport, known as blue baby disease. For this reason, drinking water should have a nitrite content of less than 1 mg/l.

Based on the results of laboratory analysis, the nitrate content of the Tirtomulyono spring was 3.66 mg/l and the Tirtomulyani spring was 5.25 mg/l. According to the Government Regulation of The Republic of Indonesia Number 22 Year 2021 a maximum of 10 mg/l, so that both springs can be used as a source of drinking water.

The nitrite concentration in Tirtomulyono and Tirtomulyani springs is < 0.0314 mg/l. This value is still far below the quality standard set by the Government Regulation of The Republic of Indonesia Number 22 Year 2021, because the maximum nitrite content especially for drinking water is 0.06 mg/l. Based on this value, it indicates that the two springs can be said to be safe and have not been contaminated by industrial waste or domestic waste and the use of nitrogen fertilizers in agricultural activities around water sources is still well controlled.

3.2.7. Chloride (Cl^-)

Chloride is found in the form of Cl- ions, which are sourced from sedimentary rocks that have not undergone diagenesis or sedimentary rocks that have not undergone consolidation such as evaporite rocks, especially the mineral halite, NaCl. According to [9], in igneous rocks, chloride ions come from feldspathoid sodalite minerals and phosphate minerals (apatite). If you find water with a high chloride ion content, it is certain that the water gets chloride ions from outside the rock [8].

Chloride concentration in water in humid areas are generally less than 10 mg/l, but in arid areas it can reach 1000 mg/l. According to [9], Cl- ions in water will increase if the mineral content increases. High concentrations of chloride are generally found around river mouths, groundwater and sea water, while in mountainous areas are generally low. water in natural waters can also be sourced from rainwater, tidal currents, seawater that enters the mainland due to the phenomenon of seawater intrusion and water pollution.Humans also produce chloride through ui urine output, which is about 6 grams / day. person. Excess Cl^- ions in water can cause an increase in water salinity. In water, chloride ions are readily soluble, and generally do not form strong complexes with metal ions in water. Actually, large amounts of chloride are not harmful to health, however, at a concentration of more than 250 mg/l this ion already gives a slightly salty taste, so it has a psychological effect. For this reason, the water to be used as a source of drinking water should have a chloride content of less than 250 mg/l. The results of the laboratory analysis showed that the chloride content in the water of the two springs was < 0.03 mg/l. This value is far from the maximum requirements of the quality standard according to Government Regulation of The Republic of Indonesia Number 22 Year 2021 which is 300 mg/l. Therefore, the water from both springs meets the requirements and is suitable for public use as drinking water.

3.3. Biological properties of springs

Coliform bacteria group includes facultative aerobic and anaerobic bacteria, gram negative rods and do not form spores. Coliform ferments lactose with acid and gas formation within 48 hours at 35°C. Concerning from the type, it is divided into Aerobacter coli which can come from soil or plants and Esherichia coli which is only and is always present in feces. Therefore, E. coli is the most efficient indicator bacteria in fecal contamination. Water to be used as a source of drinking water should not contain E. coli.

Based on the results of laboratory analysis, the total coliform bacteria of the two springs is > 1600 MPN/100 ml. According to the Government Regulation of The Republic of Indonesia Number 22 Year 2021, total coliform bacteria must be zero, so that if viewed from the Colliform quality standard, the water from the two springs does not meet the drinking water requirements. However, with proper processing before consumption, it is hoped that this water can be used for community needs. Coliform levels above the threshold could be due to the environment around the springs. The location of the two springs which are the community's raw water sources are very close to densely populated settlements.

4. Conclusion

Base on the physical and chemical parameters, the water quality in Tirtomulyono and Tirtomulyani springs is classified as good because it meets the drinking water requirements according to the Regulation of the Minister of Health No.492/Menkes/Per/IV/2010.
Based on the results of laboratory analysis, the total coliform value of the two springs is > 1600 MPN/100 ml. According to the Regulation of the Minister of Health No.492/Menkes/Per/IV/2010 the value of Total Colliform must be nil, so that in terms of the value of Total Colliform, it does not meet the drinking water requirements. The presence of coliforms in the two springs is probably due to the location of the two springs which are the community's raw water sources, which are very close to densely populated settlements.

Acknowledgements

This manuscript is part of a study for thesis entitled "Analysis of Springs Potency to Fulfill Domestic Water Requirement in Pluneng Village, Kebonarum District, Klaten" financed by Final Assignment Recognition (Rekognisi Tugas Akhir/RTA) Programme Universitas Gadjah Mada year 2022

References

[1] Belkhiri L, Tiri, A and Mouni L 2020 *Groundwater for Sustainable Development* **11** 100473
[2] Islam M B, Firoz A B M, Foglia L, Marandi A, Khan A R, Schüth C and Ribbe L 2017 *Hydrogeology Journal* **25** 3
[3] Khan A and Qureshi F R 2018 *Asian Journal of Water, Environment and* Pollution **15** 1
[4] Thyagarajan L P, Jeyanthi J and Kavitha D 2021 *Environmental Chemistry and Ecotoxicology 3*
[5] Vijakanth V, Sivakumar S S and Ratnaweera H C 2017 *International Journal of Scientific & Engineering Research* **8**
[6] Naghibi S A and Pourghasemi H A 2015 *Environ Monit Asses* **188** 44
[7] Patra S, Mishra P and Mahapatra S C 2018 *Journal of Cleaner Production* **172**
[8] Todd D K and Mays L W 2005 *Groundwater Hydrology* John Wiley and Sons.
[9] Hem J D 1970 *Study and Interpretation of The Chemical Characteristics of Natural Water*. United States Government Printing Office
[10] Sawyer C N and McCarty P L 1978 *Chemistry for Environmental Engineer* McGraw Hill Book Company

The effect of disaster knowledge on community preparedness in dealing with liquefaction disaster in Palu City, Central Sulawesi

D A Sekartaji[1] and E T W Mei[1]
[1]Department of Development Geography, Faculty of Geography, Universitas Gadjah Mada, Yogyakarta 55281, Indonesia
Corresponding authors: devyaulia99@mail.ugm.ac.id, estu.mei@ugm.ac.id

Abstract. Liquefaction occurred in Palu City on September 28, 2018 caused hundreds of casualties and thousands of buildings were damaged. The event became the Aha! Moment for many people that liquefaction disaster is real and people were not aware of the threat despite the mapping of liquefaction susceptibility in Palu had been published by Geological Agency in 2012. This study aims to analyse disaster knowledge and preparedness, as well as its influence on community preparedness in facing liquefaction. Quantitative method is used in this research by using questionnaire as primary data source to collect information regarding two variables namely disaster knowledge and community preparedness. Respondents were achieved by drawing the samples by accidental sampling. Quantitative descriptive analysis and Simple Linear Regression analysis were used to answer the objectives of this study. The results showed that disaster knowledge is in the high category generally, while community preparedness is in the medium category. Disaster knowledge has a positive and significant effect on disaster preparedness of the Palu City community in facing the threat of liquefaction, which is 43.8% with a regression coefficient of 5.013. This indicates that if disaster knowledge is higher, community preparedness will increase as well.
Keywords: Liquefaction, disaster knowledge, community preparedness

1. Introduction

Indonesia has a high potential for geological disasters threat. Geological disasters are natural disasters that occur because an area is in a zone of subduction of tectonic plates and active faults [1]. Palu City is one of the area prone to geological disasters, namely earthquake, this is shown by the multi-disasters phenomenon that occurred on September 28, 2018. The earthquake that occurred in 2018 in Palu City then triggered other disasters, namely tsunami and liquefaction which caused material losses and took the lives of hundreds of people. Of the three disasters, liquefaction is a disaster that has not received much attention from the community and stakeholders [2]. In contrast to liquefaction, earthquake, and tsunami are already widely known by the community and stakeholders, both from events in the area around Palu City and examples from incidents from other regions. However, liquefaction is not a new phenomenon that happened in Palu City. It has been recorded that some earthquakes had cause liquefaction in the past but the people at that time didn't give much attention to it because not many casualties and damage were inflicted [3]. This shows that liquefaction can happen anytime in the future and the community have to be ready to anticipate it, especially in reducing casualties and material loss.

Disaster knowledge can contribute to community's disaster resilience [4]. Therefore, it's necessary for Palu City community to improve their disaster knowledge to reduce liquefaction risk.

Liquefaction itself is a process of changing solid soil to become like liquid because the soil is in a saturated condition so that there is a movement of the ground surface which causes objects on the ground to move as well [5]. The liquefaction potential in Palu City is dominantly high to very high based on the Map of the Liquefaction Vulnerability Zone of Central Sulawesi Province [6]. The high potential for liquefaction is caused by the characteristics of the soil in Palu City in the form of young and loose deposits, and lots of sand. These characteristics cause unstable soil conditions in the event of an earthquake. This condition may be worsened in some areas with shallow groundwater levels and formed by alluvial soil prior to the earthquake shocks. Those area including Balaroa Village, Petobo Village, and Jono Oge Village, which are the areas that were impacted heavily by liquefaction in 2018 [7].

Liquefaction that happened in 2018 has become a shock and a momentum to improve community preparedness as well as Palu City preparedness in general. The phenomena caused more and more information being published regarding liquefaction that happened locally or internationally. The information can be used as tools to build community awareness in improving disaster preparedness since the threat can't be stopped, but the risk can be prevented by mitigation to reduce the potential damage and casualties caused by liquefaction. Although the publications related to liquefaction are increasing, most of it only discuss the disaster aspect of liquefaction. While the disaster aspect is important, social aspect regarding liquefaction needs more exposure. By discussing the social aspect such as community knowledge and preparedness, stakeholders can measure whether liquefaction in 2018 has been improving community awareness and preparedness or not and how does the community make us of the provided information. The approach on social aspect of liquefaction can be focused on measuring disaster knowledge and preparedness level as well as analysing the impact of disaster knowledge on preparedness.

Broadly speaking, the existing problems on the community level are related to the knowledge and preparedness of the people of Palu City in dealing with the liquefaction threat. The lack of public knowledge about the liquefaction in Palu City prior to the multi-disasters on 28 September 2018 had caused shock and unpreparedness of the community in dealing with disasters, as explained by Mei, *et al.* (2020) that a disaster can cause shock or trauma and also disrupt the course of education [8]. The lack of knowledge before the liquefaction phenomenon then had an impact on the disaster preparedness of the people of Palu City. Lack of knowledge then affects the decision of the people of Palu City to continue to live in areas with high liquefaction susceptibility such as settlements in Balaroa Village and Petobo Village. Whereas liquefaction susceptibility has been recognized by the local community and mapping of liquefaction disaster susceptibility in Palu City has been carried out since 2012 by the Geological Agency, Ministry of Energy and Mineral Resources.

The lack of disaster knowledge apparently was not a case in 1800s, local people at that time have known that some area such as Balaroa Village, Petobo Village, and Jono Oge Village were located in disaster prone area, namely liquefaction [9]. However, the local knowledge was slowly forgotten along the movement of migrants from other area to Palu. This affect urban sprawl and more lands were neede to build settlement which causing the growth of dense settlements in Balaroa Village, Petobo Village, and Jono Oge. The same statement was presented Sopaheluwakan (2006) where local people were familiar with the local term nalodo (submerged in mud) which indicates the occurrence of liquefaction in the past [10]. Unfortunately, this local wisdom fades along with the times and urban developments in this region. The lack of knowledge of the people of Palu City before the liquefaction disaster then had an impact on the disaster preparedness of the people of Palu City, causing the low preparedness in facing liquefaction. The low level of liquefaction preparedness resulted in 186 casualties and 3,455 house damage in Balaroa Village in West Palu and Petobo Village in South Palu [11].

By conducting this study, the researchers aim to identify community's disaster knowledge and preparedness relatively, as well as conducting analysis on disaster knowledge impact on preparedness. This study is expected to contributed to the current understanding of social aspect regarding liquefaction phenomenon in 2018 in Palu City, which can help local government and the community in improving disaster preparedness. This study can be used as reference by another researcher who focus on social aspects of disasters, especially liquefaction.

2. Method

The location of this research is Palu City which is in Central Sulawesi Province with coordinates 0°0'36" - 0°0'56" South Latitude and 119°45'0" - 120°0'0" East Longitude. There are two variables in this study, namely disaster knowledge as the independent variable and preparedness as the dependent variable. Knowledge is measured based on knowledge of liquefaction disaster threats and knowledge of disaster risk reduction efforts. Knowledge of disaster risk reduction is divided into pre-disaster, during disaster, and post-disaster knowledge [12]. Community preparedness is measured based on preparedness parameters by Sopaheluwakan, *et al.* (2006) [13].

This research was conducted using quantitative methods by utilizing the scoring results for disaster knowledge and community preparedness analysis. Simple Linear Regression analysis is used for analysing the effect of disaster knowledge on preparedness. Survey method is used as research design to gather primary data. The survey was conducted using a questionnaire instrument distributed online as an adjustment to the limited activities during the COVID-19 Pandemic.

The sampling technique (sample selection) used in this study is non-probability sampling with limited population who are the people of Palu City in 20-45 years age range. This sampling method is used, because not all residents in Palu City have the same opportunity to be sampled. The type of non-probability sampling used in this research is accidental sampling technique. Accidental sampling was used in this study because the sample was obtained freely, in the sense that there was no binding place or time so that the sample could come from anywhere as long as it met the respondent's criteria and within the data collection period determined by the researcher. The respondent criteria used by the researcher to filter the sample are domicile location and age. Research respondents are people who have a Palu City ID card and are currently domiciled in Palu City during the questionnaire filling period, while the age range used as the respondent's criteria is 20-45 years. The respondents are achieved by distributing questionnaire using Google Form through social media and based on the result, there are 100 respondents participating in this research. The number of respondents were determined by using Slovin formula.

Disaster knowledge is measured with liquefaction disaster knowledge factor and disaster risk reduction knowledge factor. Disaster preparedness is measured with disaster preparedness policy factor, emergency response plan factor, disaster early warning factor, and resource mobilization factor. Respondents' answers on disaster knowledge were given score 1 for each correct answer and 0 for each wrong answer, while respondents' answer on preparedness were given score 1 which indicate less preparedness to 5 which indicate more preparedness. Disaster knowledge is relatively low if the average score is less than 5, moderate if the average score is on 5-9 range, and relatively high if the average score is more than 9. Preparedness is relatively low if the average score is less than 42, moderate if the average score is on 42-108 range, and relatively high if the average score is more than 108.

The results of analysis in this study are presented in the form of statistical measures such as tables and graphs. In addition, data mapping is used to give better understanding about spatial aspect based on respondents' answers. Univariate analysis was conducted to see the level of knowledge and preparedness based on the total score, average, and percentage of respondents' answers. The level of knowledge and preparedness is determined using three categories according to Azwar (2017), namely low, medium, and high [14]. Bivariate analysis was performed with Simple Linear Regression. The classical assumption test on the data includes linearity test, normality test, and heteroscedasticity test were performed before conducting Simple Linear Regression analysis. The multicollinearity test was not carried out because there was only one independent variable in this study. The autocorrelation test was also not carried out because the data used in this study were not time series data.

3. Results and Discussions

Respondents' responses to disaster knowledge are presented in Table 1. Based on Table 1, respondents' knowledge of disaster is generally in the high/good category. Disaster knowledge, especially related to the relatively high liquefaction susceptibility, is expected to increase the preparedness of the people of Palu City in facing the potential for liquefaction which can occur at any time in the future. Disaster knowledge is a particularly crucial factor to support preparedness because disaster knowledge can play a role in increasing public understanding to protect themselves and their assets in the event of a liquefaction disaster. The elevated level of disaster knowledge of the people of Palu City can be influenced by the liquefaction phenomenon that occurred in 2018, so that the threat of liquefaction in Palu City gets more attention from the community and local government, as well as other related parties. As a result, there are more sources of information related to liquefaction, either through print media, electronic media, social media, or direct socialization about liquefaction disaster.

Table 1. Classification of community disaster knowledge (n=100).

Classification	Percentage
Low/Poor	10%
Medium/Moderate	27%
High/Good	63%

The lack of disaster knowledge causes low preparedness in dealing with disasters, especially for liquefaction disasters which cannot be predicted beforehand. The substantial number of fatalities and damage caused by the liquefaction disaster in 2018 showed that knowledge related to disasters was still lacking. The data/information related to the threat of liquefaction disasters have existed since 2012. However, the liquefaction disaster in Palu City has not received as much focus as the threat of flood, earthquake, and tsunami. The absence of focus on liquefaction in Palu City is also shown by planning documents such as the 2010-2030 Regional Spatial Plan (RTRW) of Palu City and the 2016-2021 Local Government Medium-term Development Plan (RPJMD) of Palu City which do not contain policies related to liquefaction. The lack of disaster knowledge in Palu City is explained in research by Josaphat and Santoso [15] which stated that the lack of knowledge and lack of awareness of the people of Palu City to learn basic science related to disasters causes the local community to have low preparedness, especially for sudden liquefaction disaster.

The average score for liquefaction disaster threats knowledge is relatively higher than disaster risk reduction knowledge showed on Figure 1. This shows that, even though the community's knowledge of disaster is included in the high category, knowledge of disaster risk reduction needs to be improved so that the knowledge is not only limited to the threat of liquefaction disasters but including the knowledge of liquefaction disaster management efforts as well. Disaster knowledge should also be in line with its implementation by the community, not only limited to knowing or understanding the threat of liquefaction disaster and efforts to reduce liquefaction disaster risk.

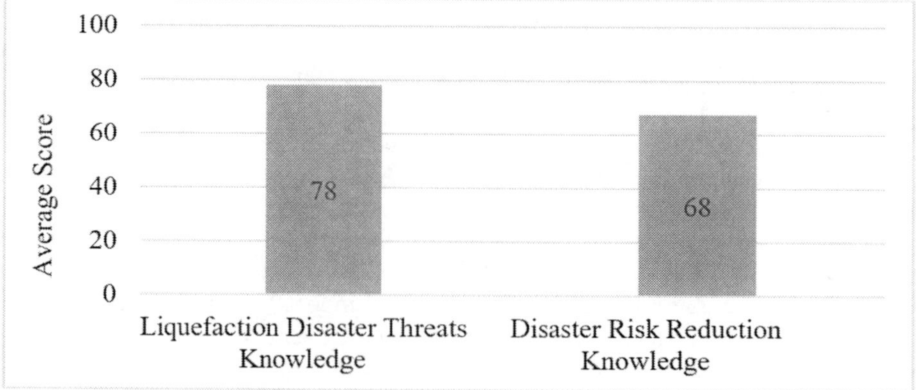

Figure 1. Average score on disaster knowledge factors (n=100).

Figure 2 shows the respondent's level of disaster knowledge based on their domicile location. In general, disaster knowledge in the high/good category spreads in agglomeration in the centre of Palu City, namely in the middle of Palu City, except for Ujuna Village because there is no data in that village. Although almost all villages in the centre of Palu City have relatively high/good disaster knowledge, there are villages that have moderate disaster knowledge, namely Baru Village and Kamonji Village. If it is associated with liquefaction events and liquefaction susceptibility, Balaroa Village and Petobo Village which were hit by liquefaction in 2018 have relatively high/good disaster knowledge. This shows that the occurrence of liquefaction can be a trigger for the community to increase disaster knowledge in the two villages, as well as other villages with high and medium levels of knowledge. However, there are villages with relatively low/poor levels of disaster knowledge, namely Palupi Village and Silae Village.

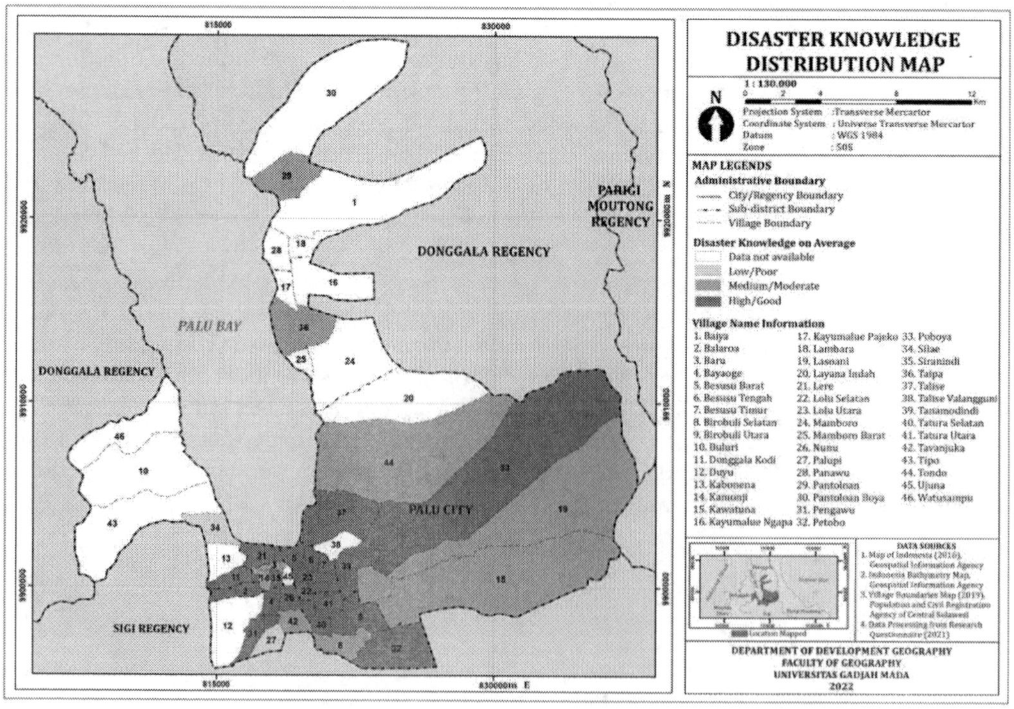

Figure 2. Distribution map of respondents' disaster knowledge (n=100).

IOP Conf. Series: Earth and Environmental Science 1233 (2023) 012069 doi:10.1088/1755-1315/1233/1/012069

Respondents' responses to disaster preparedness are presented in Table 2. Based on Table 2, the results of the classification of respondents' preparedness are generally in the moderate category. Disaster preparedness by the community, which is dominated by the moderate category, followed by preparedness in the good category in the facing the threat of liquefaction, is expected to minimize losses and casualties if liquefaction occurs again in Palu City. The existence of fatalities (186 people) and damage (3,455 housing units) that were not small due to liquefaction in Palu City in 2018 showed that community preparedness in dealing with liquefaction was still low/poor. The low preparedness of the Palu community is contained in the research of Samad, *et al.* [16] which is based on the results of a literature study and Forum Group Discussion (FGD), stated that the community does not yet have preparedness to face disasters in 2018.

Table 2. Classification of community preparedness (n=100).

Classification	Percentage
Low/Poor	9%
Medium/Moderate	54%
High/Good	37%

The occurrence of liquefaction in 2018 can be a momentum to improve disaster preparedness for the people of Palu City, considering that Palu City has a high liquefaction susceptibility. A similar momentum occurred in Aceh with the tsunami in 2004. Hidayati [17] in her research stated that awareness and concern for the importance of community preparedness in Aceh was only noticed and became important after the devastating tsunami in Aceh that attracted the attention of other countries in 2004. The average score for disaster preparedness policies and emergency response plans is relatively higher than disaster early warning and resource mobilization showed on Figure 3. This shows that, increasing preparedness can be focused on disaster early warning perspective and resource mobilization so that these two aspects can match preparedness policies and emergency response plans, while improving overall preparedness.

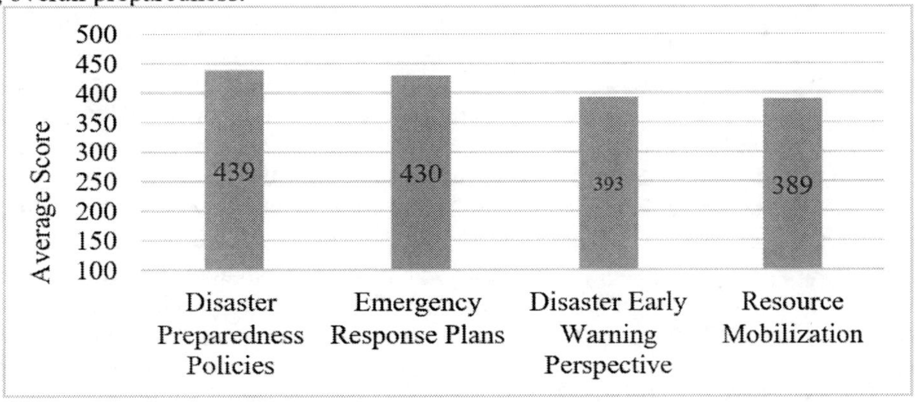

Figure 3. Average score on preparedness factors (n=100).

Figure 4 shows the level of preparedness of respondents based on their domicile location. Based on the mapping results, it can be seen that the total preparedness score of each respondent in Balaroa Village is not only higher than the average, Balaroa Village is also classified as having high/good preparedness overall. This is different from Petobo Village, where almost all respondents from Petobo Village have preparedness scores that are higher than the average, but Petobo Village as a whole has moderate preparedness.

Preparedness based on domicile location in Palu City is quite good overall, because preparedness is in the medium-high range. However, preparedness needs to be improved in order to optimally minimize the risk of liquefaction disasters. Preparedness in dealing with liquefaction is not only a challenge for the people of Palu City, but also for the Palu City Government. The City Local Government needs to be firm and needs to carefully consider future development plans, if necessary, restrictions on land use in the area can be made through Regional Spatial Plans, so that their use is focused on conservation functions or other activities other than cultivation functions.

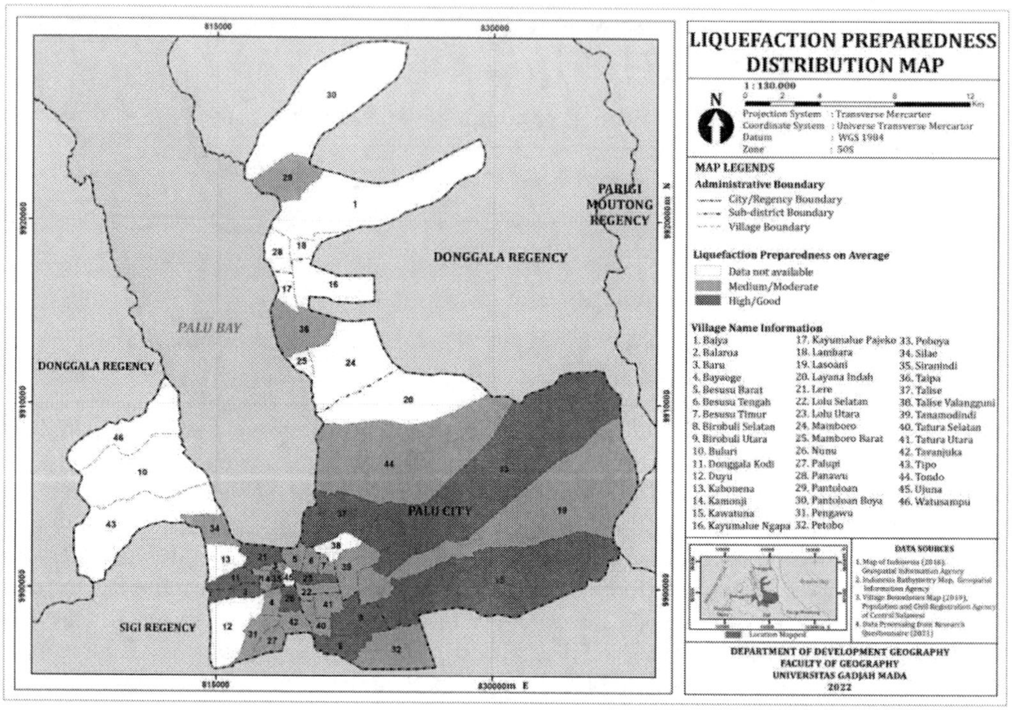

Figure 4. Distribution map of respondents' liquefaction preparedness (n=100).

Table 3, Table 4, and Table 5 present the analysis results of disaster knowledge effect on preparedness. Table 3 shows that disaster knowledge has a correlation with disaster preparedness that is 0.662, where the percentage of the effect of disaster knowledge on disaster preparedness is indicated by R Square which is 0.438. The R Square value indicates that disaster knowledge has an effect of 43.8% on disaster preparedness, while the remaining 56.2% is influenced by other variables outside the model or not analysed in this research. The R Square value of 0.75 can be categorized as having a large effect, 0.50 having a moderate effect, and 0.25 having a weak effect [18]. In addition, Moksony [19] suggests that a low R Square value, for example 10%, is acceptable for the field of arts and social humanities studies, because human attitudes cannot be predicted accurately. Meanwhile, a high R Square value, for example above 60%, is usually needed for studies in the pure science field, because the predictions can be made accurately. This shows that although the effect of knowledge on preparedness is relatively low, the R Square value obtained is acceptable in the context of research on social phenomena.

Table 3. Output model summary.

Model	R	R Square
1	0,662	0,438

Table 4 presents the regression results obtained from the ANOVA, namely the calculated F and the Sig value. Based on Table 1.4, the calculated F value is 16.707 or more than 2.760 (F table). Then, the significance value obtained is 0.000 or less than 0.100 (α used in this study), so it can be concluded that disaster knowledge has a significant effect on the disaster preparedness of the people of Palu City in facing liquefaction.

Table 4. Output anova.

Model 1	$F_{calculated}$	F_{table}*	Sig.
Regression	16,707	2,760	0,000

Based on Table 5, the calculated t value obtained is 3,447 or more than 1,660 (t table). Then, the significance value obtained is 0.001 or less than 0.100 (α which is used in this study), so it can be concluded that disaster knowledge has a partial significant effect on the disaster preparedness of the people of Palu City in facing liquefaction. The regression equation obtained in this study is Y= 72.316 + 5.013X or in other words the regression coefficient constant is 72.316 and the regression coefficient for disaster knowledge is 5.013. The regression coefficient in this study is positive or in other words, if disaster knowledge increases, preparedness will also increase.

Table 5. Output coefficients.

Model 1	Unstandardized Coefficients B	$T_{calculated}$	t_{table}*	Sig.
Constant	72,3	12,2	1,7	0,0
Disaster Knowledge	5,00	3,4	1,7	0,0

The results of the analysis in this study are in line with Noor's research [20] which stated that knowledge affects awareness and attitudes towards disaster preparedness, the more often someone acquires knowledge or information, the more preparedness attitudes are formed. The results of Chotimah's research [21] also suggest that community knowledge has a positive effect on preparedness to face landslides in Pasir Jaya Village, West Bogor District, Bogor City. Basic knowledge about landslides in accordance with the education and experience of each respondent can determine the ability to reduce disaster risk to minimize people who experience landslides or suffer losses due to landslides.

Table 6 shows that most of respondents who have medium disaster knowledge are categorized as respondents with high preparedness. The table also shows that not all respondents who have high disaster knowledge can be categorized as respondents with high preparedness, while most of respondents who have low disaster knowledge are categorized as respondents with low preparedness. This shows that despite not having 100% influence on preparedness, disaster knowledge still has important role in improving community preparedness. In other words, it is supporting the statement that disaster knowledge has partial significant effect on disaster preparedness.

Table 6. Cross table of disaster knowledge and community preparedness.

Community Preparedness	Disaster Knowledge			Total
	Low/Poor	Medium/Moderate	High/Good	
Low/Poor	6	4	0	10
Medium/Moderate	2	10	15	27
High/Good	1	40	22	63
Total	9	54	37	100

Although disaster knowledge has a noteworthy influence with a percentage of almost 50%, there are other variables that affect community disaster preparedness, such as disaster experience. In addition to disaster experience, community participation in disaster management programs can also be one of the influential factors such as preparedness. Mei, *et al.* [22] stated that community groups who are ready to face the threat of an eruption of Mount Merapi have experience in participating in disaster risk reduction programs so that most have awareness of the consequences that can be caused by the threat of an eruption of Mount Merapi. On the other hand, community groups who are not ready to face the threat of an eruption of Mount Merapi have minimal experience, resulting in a lack of preparation for evacuation, resulting in many victims from this community group. As for other factors affecting liquefaction preparedness, lack of simulation or practice can be one of the factors. Unlike earthquake, tsunami, flood, and other well-known disaster, simulation regarding liquefaction seldom to happen, thus it can affect community's perceptive of preparedness. Even though the community have high level of liquefaction knowledge, the impact will not be significant without proper simulation because people are not used to liquefaction phenomenon.

The results of Kurnianto's research [23] stated that the perception of disaster risk has a significant and positive effect on disaster preparedness in Kumur District, Pandeglang Regency. Research subjects with a high perception of disaster risk have the belief that their place of residence is at risk of natural disasters, so that, they feel responsible for reducing disaster risk by increasing preparedness. If it is associated with disaster knowledge in this study, high disaster knowledge needs to be accompanied by a high perception of disaster risk, because disaster knowledge alone is not enough to increase preparedness. Public awareness is needed so that disaster knowledge possessed can be useful for disaster preparedness in the future.

4. Conclusion

Disaster knowledge of the people of Palu City is in the high/good category. The high level of knowledge of liquefaction disaster in Palu City can be caused by the existence of sources of information in various forms of media such as print media, electronic media, social media and seminars or trainings that increase public knowledge regarding liquefaction as disaster. A high level of disaster knowledge is not enough if people only know or understand the concept of it. The community also needs to apply their disaster knowledge in order to realize the attitude or behaviour of liquefaction disaster preparedness.

Community preparedness in facing liquefaction is at moderate category. The level of community preparedness in the medium level indicates that liquefaction disaster preparedness needs to be further improved so that the community is better prepared to anticipate the potential for liquefaction that occurs in the future. Especially considering that Palu City has a high liquefaction susceptibility, the community needs to increase preparedness to reduce disaster risk so that casualties and losses can be minimized.

Disaster knowledge affects preparedness by 43.8%. The equation obtained from the regression model of this study shows that if the value of disaster knowledge does not exist, then the value of preparedness is 72,316 and if there is an increase in the value of disaster knowledge by one unit, then the value of preparedness increases by 5.013. Based on these results, it can be concluded that disaster knowledge has a positive and significant impact on the disaster preparedness of the Palu City community in facing the threat of liquefaction. If the community has high disaster knowledge, the community's preparedness in facing the threat of liquefaction disasters will also be higher. However, other aspects other than disaster knowledge still need to be improved so that community preparedness is even better.

This study has potential limitations because it only examines one independent variable, so that the influence of other variables that have a significant effect on preparedness cannot be analysed. Other independent variables were not studied because of limitations to conduct research in the midst of the COVID-19 Pandemic. Therefore, the variable independent research is only disaster knowledge because it's easier to fill out independently by the respondents. The using of non-probability sampling in this research is caused by the same condition when COVID-19 happened, non-probability sampling was chosen and not random sampling because of the difficulty to access entire population and give them equal opportunities in filling out online questionnaire. Reflecting on the limitation of this research, future researcher is expected to examine other variable other than disaster knowledge to obtain a more varied and in-depth discussion of the variables that affect preparedness and conducting the research using random sampling.

References

[1] National Earthquake Study Center Team 2019 *Investigasi Awal Longsor-Likuifaksi Geotechnical Extreme Events Reconnaisance (GEER) Akibat Gempa Palu 28 September 2018 (Geotechnical Report on 2018 Palu Donggala Earthquake)* (Bandung: Research Center for Housing and Settlement, Ministry of Public Works and Housing of the Republic Of Indonesia).

[2] Yanuarto T 2019 *Buku Saku: Tanggap Tangkas Menghadapi Bencana* (Jakarta: Data Center for Information and Public Relations, National Disaster Management Authority).

[3] Sopaheluwakan J 2006 *Kajian Kesiapsiagaan Masyarakat dalam Mengantisipasi Bencana Gempa Bumi & Tsunami* (Jakarta: LIPI-UNESCO/ISDR).

[4] Josaphat G and Santoso I 2020 fasilitas edukasi penanggulangan bencana di Kota Palu *J. eDimensi Arsitektur* **VIII**(1): 1081-1088.

[5] Samad M A, Tawil Y P and Kafiah M K A 2019 *Disaster Risk Management in Palu City (Case Study of Natural Disasters September 28 2018)* (Palu: Universitas Tadulako).

[6] Hidayati D 2008 Kesiapsiagaan masyarakat: paradigma baru pengelolaan bencana alam *J. Kependudukan Indonesia* **3**(1): 69-84.

[7] Mooi E and Sarstedt M 2011 *A Concise Guide to Market Research: The Proces, Data, and Methods Using IBM SPSS Statistics* (New York: Springer Publishing).

[8] Pahriannoor, Fauzan A and Hadi Z 2020 *Hubungan Pengetahuan dan Sikap Perawat dengan Upaya Pencegahan dan Penanggulangan Kebakaran di RSUD Ulin Banjarmasin Tahun 2020* (Banjarmasin: Universitas Islam Kalimantan Muhammad Arsyad).

[9] Chotimah A N 2019 Pengaruh pengetahuan dan sikap masyarakat terhadap Kesiapsiagaan menghadapi bencana longsor di Pasir Jaya Bogor *J. Manajemen Bencana (JMB)* **5**(2): 57-72.

[10] Kurnianto Y T 2019 *Pengaruh Persepsi Risiko Bencana terhadap Kesiapsiagaan Bencana pada Keluarga di Pesisir Pantai kecamatan Sumur* (Pandeglang Jakarta Timur: Universitas Negeri Jakarta).

[11] Sudibyakto, Retnowati A, Suryanti E D and Hizbaron D R 2012 Menuju Masyarakat Tangguh Bencana *Konstruksi Masyarakat tangguh Bencana: Kajian Integratif Ilmu, Agama, dan Budaya (Seri Agama dan Bencana* vol 3*)* ed A Indiyanto, A Baiquni and A Kuswanjono (Bandung: PT. Mizan Pustaka) pp 7-32.

[12] Rahman M A, Fathani T and Rifa'i A 2020 Analisis tingkat potensi likuifaksi di Kawasan Underpass Yogyakarta International Airport *J. Rekayasa Sipil (JRS-UNANDS)* **16**(2): 91-104.

[13] Mei E T W, Putri R F, Sadali M I, Febrita D, Yulandari D E, Anggriani M, Niam R A, Rasyida F Z O, Wardhani A G and Paramita T P 2020 Building volcanic disaster resilience community through school and education *IOP Conf. Series: Earth and Environmental Science* **451** (2020) 012014.

[14] Ministry of Energy and Mineral Resources 2019 *ATLAS Zona Likuefaksi Indonesia* (Bandung: The Geological Agency, Ministry of Energy and Mineral Resources).

[15] National Earthquake Study Center Team 2018 *Kajian Gempa Palu Provinsi Sulawesi Tengah 28 September 2018* (Bandung: Research Center for Housing and Settlement, Ministry of Public Works and Housing of the Republic of Indonesia).

[16] Azwar S 2017 *Penyusunan Skala Psikologi Jilid 2* (Yogyakarta: Pustaka Pelajar).

[17] Moksony F 1999 Small is beautiful: the use and interpretation of R2 in social research *Szociológiai Szemle* **9** (Special issue): pp 130-138.

[18] Mei E T W, Lavigne F, Picquot A, Bélizal Ed, Brunstein D, Grancher D, Sartohadi J, Cholik N and Vidal C 2013 Lessons learned from the 2010 evacuations at Merapi Volcano *J. Volcanology and Geothermal Research* 2013; **261**: pp 348-365.

[19] Lasimpo G 2019 *Mitigasi Bencana Berbasis Pengalaman Suku Kaili di Lembah Palu* (Sigi: Yayasan Lembaga Bantuan Hukum Asosiasi Perempuan Indonesia Keadilan Sulawesi Tengah).

[20] Rahardjo P P 2021 Study on the Phenomena of Liquefaction Induced Massive Landslide in 28 September 2018 Palu-Donggala Earthquake *Understanding and Reducing Landslide Disaster Risk Reduction* (*Catastrophic Landslides and Frontiers of Landslides Science* vol 5) ed Vilímek et al (The Capital Region of Denmark: Springer, Cham); pp 25-48.

[21] Jalil A, Fathani T F, Satyarno I and Wilopo W 2021 Liquefaction in Palu: the cause of massive mudflows *Geoenvironmental Disasters* 2021; **8**(21): pp 1-14.

[22] Prawoto E and Octavia L 2021 Disaster storytelling: extending the memory of the community toward disaster preparedness from myth, scientific explanation, and popular culture *J. Disaster Research*; **16**(2): pp 228-233.

[23] Bohari J A and Widiana I K 2021 Short training to Improve knowledge of disaster management *IOP Conf. Series: Earth and Environmental Science*; **708**: pp 1-7.

The effect of the eruption of mount Hunga-Tonga on density altitude, air pressure and flight operations in eastern Indonesian

Heri Ismanto*, Sabitul Hidayati, and Firman Setia Budi

Centre for Aeronautical Meteorology, Agency for Meteorology, Climatology, and Geophysics

*Corresponding author: heri.ismanto@bmkg.go.id

Abstract. There was an eruption of the Hunga Tonga volcano in the Pacific Ocean on January 15, 2022 and produced the phenomenon of Shock Waves in the atmosphere, in addition to sonic booms, and tsunami waves that spread to various parts of the world. Data from the automated weather observing system (AWOS) in several Airports in Indonesia were analyzed to assess the impact of the eruption of Mount Tonga on the Indonesian atmosphere. There are sudden changes in air pressure (QNH) at locations (Airports) including Sorong, Wamena, Ternate, Sumbawa, Biak, Geser, Lombok, Sabu, and Kupang, with a range between 0.6 to 1.9 Mb. These atmospheric waves propagate at a speed of about 1248 km/hour. This drastic change in air pressure is not followed by a significant change in density altitude which affects the effectiveness of the aircraft's lift, but a sudden change will cause a change in air pressure (QNH) in the altimeter which can cause a change in aircraft altitude. This sudden change has high probability to result in a plane crash due to a stall or undershoot or collision with aircraft.

1. Introduction

The aircraft in their critical period which are taking off and landing at the airport requires weather information as an important part of its operations. Atmospheric pressure changes, temperature changes, local weather made a lot of delays and cancelation especially in the mountaious area [10]. Meanwhile during in flight phase, significant weather such as: thunderstorm, turbulence and volcanic ash effect the efficiency and safety of flight operation [25, 26].

BMKG through its meteorological stations at each airport is obliged to provide weather information, one of which is used in aircraft operational activities at the airport, namely during the take-off and landing process. In this case, BMKG uses a weather data automation system through a digital automatic weather observation tool called the Automated Weather Observing System (AWOS). AWOS is automatic monitoring equipment for weather elements continuously for 24 hours which is used for aviation purposes. AWOS equipment is installed to determine the weather parameters on the Airport Runway. These weather parameters include pressure, density altitude (displayed on the TOWER/ATC display), wind direction and speed, air temperature and dew point, humidity, visibility, cloud conditions, precipitation, and lightning. AWOS equipment presents and displays processed data ranging from 10-20 seconds resolution, per minute, per hour, or day. The results of the AWOS output are METAR, SPECI, and MET REPORT/SPECIAL [15, 21]. In addition, AWOS can also provide a warning/alert if there is a significant change in weather elements. The results of AWOS data can be integrated or distributed to pilots and ATC using communication lines in the form of VHF radio,

NAVAID or ATIS, AFTN, and CMSS. In AWOS categories 1 or 2 or 3 there is Density Altitude (DA).

The eruption of Mount Tonga, located in the South Pacific Kingdom of Tonga, was observed on satellite with an intensive eruption level on January 15, 2022. This eruption had a direct and indirect impact on the surrounding area. The eruption of this mountain has been observed since December 2021, and the peak of the eruption on January 15 produced an atmospheric shock wave, sonic booms, and tsunami waves [19] that traveled the world and were heard as far away as Alaska [1]. Atmospheric shockwave and sonic boom made some radius of energy spreding, that as a function of Energy defided by contanta and atmospheric pressure [11]. Satellite-based lidar and limb sounder data indicated that the eruption was strong enough to send volcanic material well into the lower stratosphere (which generally begins 9 miles up in this part of the world) where, according to rawindsonde data from Fiji, easterly winds propelled the ash. Scientists watch closely when volcanic materials reach this relatively dry layer of the atmosphere because particles linger longer and travel farther than when they remain in the lower, wetter troposphere [22]. Volcanic eruption could produced atmospheric waves.

Figure 1. Volcanic Eruption of Mount Tonga 15 January 2022, 05.00 UTC [1, 16].

On Jan. 15, Hunga erupted and sent a towering plume of gas and particles billowing into the mesosphere, the third layer of the atmosphere above earth's surface. The plume reached 36 miles (58 km) tall at its highest point, making it the largest volcanic plume in the satellite record [17] and made a Concentric traveling Ionosphere Disturbences (CITDs) [18].

The indirect impact of the eruption on the air pressure, density altitude of Indonesian airports and flight operations is the basis for this research. Is there any changing of values in the daily trend of airport surface air pressure shortly after the eruption of Mount Tonga? more or less affects the density altitude value used in aircraft take-off and landing operations?. The purpose of this study is to find out how the effect of the eruption of Mount Tonga was on the value of air pressure (QNH) and density altitude (DA) and flight operations shortly after mount Tonga erupted.

2. Data and Method

2.1. Locations

The research locations are selected by purposive random sampling that will give the best estimation of the area of interest [8]. The eastern region and parts of the central region of Indonesia analyze how the influence of the eruption of Mount Tonga on the value of DA and changes in air pressure (QNH) in the territory of Indonesia. The selected locations consist of 9 airports with meteorological stations in charge of flight weather information. These locations include: Wamena, Biak, Lombok, Sentani, Kupang, Ternate, Sabu, Sorong, and Sumbawa (Fig.2)

2.2 Data and Method

Data on air pressure at the airport (QNH), temperature (T), and dew point (Td) are data obtained from airport AWOS observations with a data time resolution of 10 minutes. These data are then used to calculate and analyze whether there is a change in the value of DA by using the formula [9]:

$$h_{density} = 145366x(1 - \left(\frac{17.326xP_{inHg}}{Tv}\right)^{0.235} \tag{1}$$

where P inHg is the air pressure value converted from millibars (millibars) to inch Hg (P millibars x 0.02953) ,and Tv is the virtual temperature calculated using the formula:

$$Tv = \frac{T}{1-(\frac{e}{P_{mb}})x(1-0.622)} \tag{2}$$

where the value of vapor pressure (e) is calculated using the value of air temperature and dew point temperature (which has been converted to °K) with the formula:

$$e = 6.11x10^{(7.5xTd/237.7+Td)} \tag{3}$$

One of the important parameters in these periods is the density altitude. Density Altitude (DA) is formally defined as pressure altitude corrected for nonstandard temperature variations, which is determined by first finding pressure altitude, and then correcting this altitude for nonstandard temperature variations [4]. DA is such a simple way to give the pilot information on air density, this parameter being calculated such that the pilot can make a coarse assessment of aircraft performance in the current flight conditions [2]. DA is the vertical distance above sea level in the standard atmosphere at which a given density is to be found. The density of air has significant effects on the aircraft's performance because as air becomes less dense and will reduces: power of aircraft because the engine takes in less air, thrust of aircraft because a propeller is less efficient in thin air, and lift of aircraft because the thin air exerts less force on the airfoils [5]. Increasing DA value will likely result in more reduction in payloads [7]. QNH related to the altimetter setting of flight. Setting flight level using QNH value indicates of the aicraft altitude [13].

In analyzing and calculating the DA, the daily variational of the QNH value is taken into consideration to see any changes or disturbances in the pressure value trend of the shockwave that occurred due to the eruption of Mount Tonga. The existence of a disturbance in the pressure value trend is the basis for an analysis of the direct impact of the Tonga eruption on the DA value, which is the basis for pilot consideration when taking off or landing aircraft at the airport.
There are several things to watch out for when flying at high-density altitudes [4], including:

a. Horizontally: If it is going to land at one of the airports/airstrips in Papua with an altitude of 5000 feet, your TAS will increase by about 10%. So it requires a slightly wider downwind distance than usual. Because the speed of the aircraft is higher than that shown on the indicator. With a higher speed, the turning radius will also increase

b. Vertically: to calculate descent with profile 3' we usually use the rule of thumb → rate of descent = 5 x ground speed. So with an increase in TAS which also means an increase in ground speed, the rate of descent will increase quite significantly

c. Energy Management: At higher speeds due to the high-density altitude, the situation can be made worse if the aircraft's weight is close to its maximum. Because of the weight of this plane, it will take longer to decelerate. So try to stabilize with the speed to land (approach/landing speed) long before (low and slow).

d. Brake. The kinetic energy at the time of landing (touchdown) will be higher, so stronger braking is needed than usual. If there is a reverse or auto brake, then its use is highly recommended.

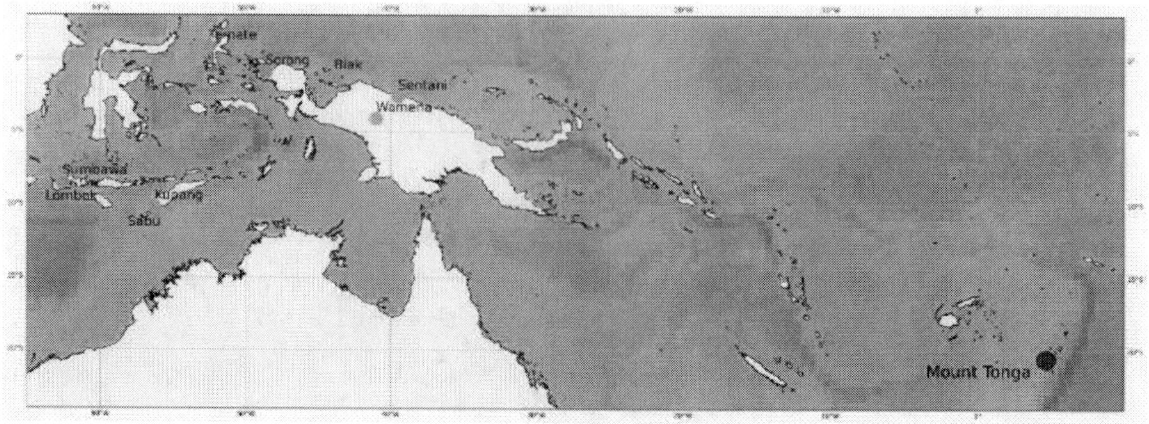

Figure 2. Research Area

3. Result and Discussion

Figure 3 - 5 shows the DA and pressure changes over 24 hours in Papua and Nusa Tenggara. There was a sudden change of pressure out of the diurnal cycle and daily variational of air pressure. Meanwhile the DA seem like unchanged/slight change during the atmospheric wave due to Hunga Tonga Mount Eruption.

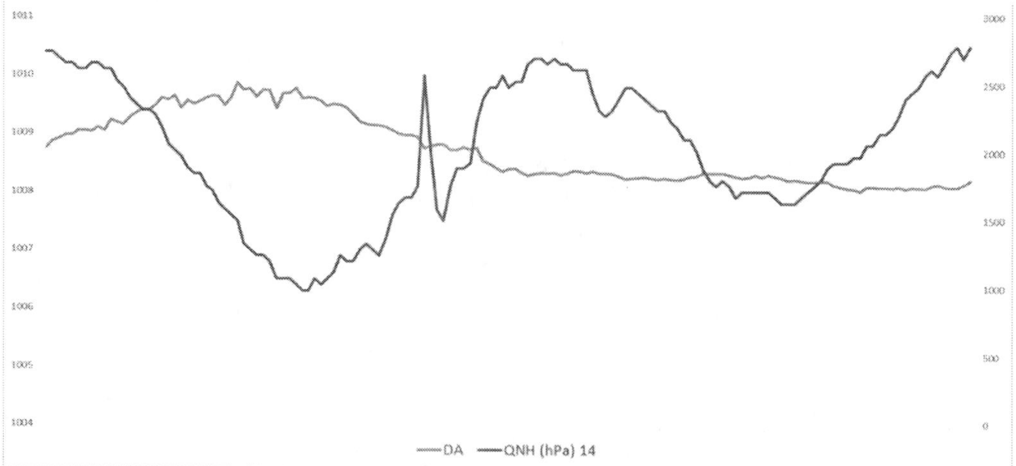

Figure 3. Pressure (QNH) and Density Altitude Changing over 24 Hours 15 January at El Tari Airport - Kupang, East Nusa Tenggara, Indonesia

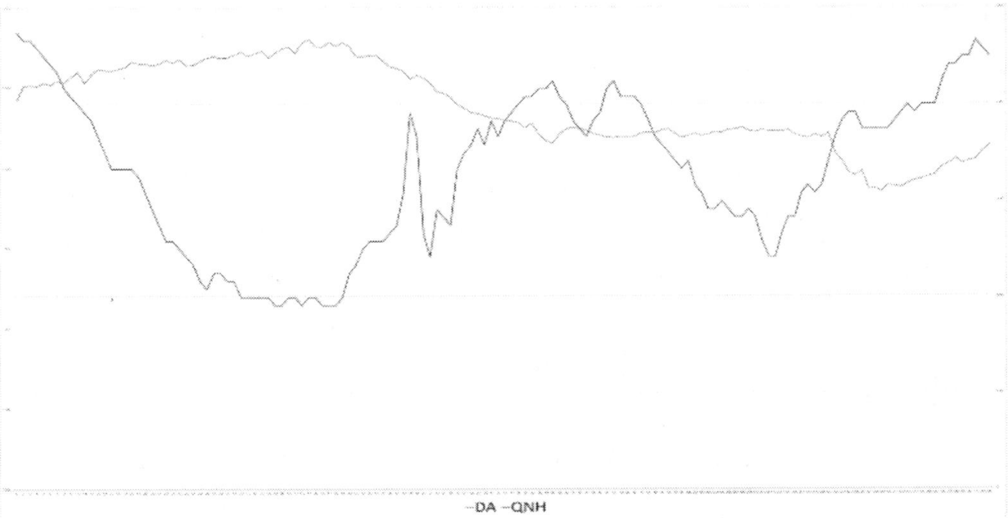

Figure 4. Pressure (QNH) and Density Altitude CHanging over 24 Hours 15 January at Biak Airport

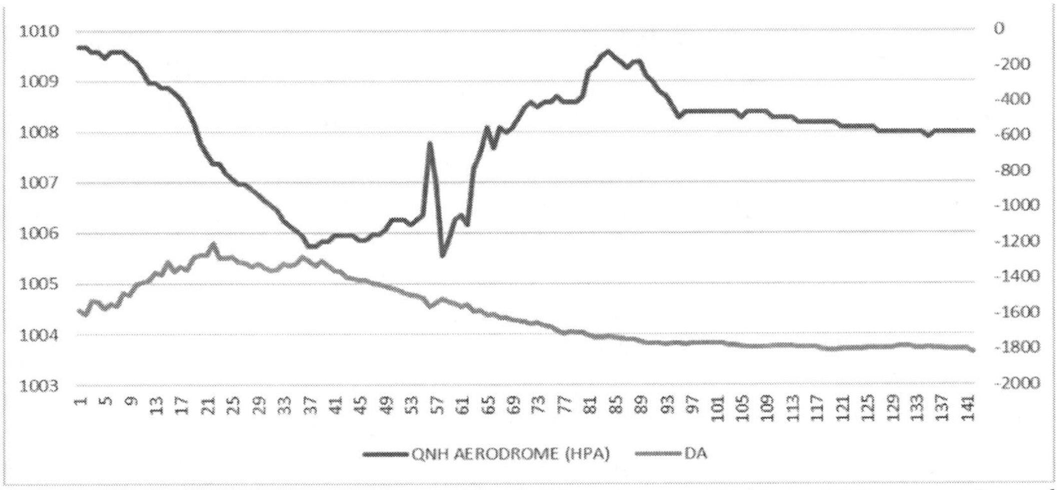

Figure 5. Pressure (QNH) and Density Altitude Changing over 24 Hours 15 January at Sentani Jayapura Airport

Density altitude change does not directly affect by the shock wave of mount Tonga eruption (Fig.3-5). It's changging mainly by the temperature deviation in the surface that follows the diurnal variation in the airports of the research area. DA could be significat during the long changing in the temperature during climate change [7]. Contrastly displayed with the air pressure (QNH) changing during the sonic boom of the Mount Tonga eruption. Most research related to volcano eruption and its impact on flight operation mainly discuses volcanic ash [12, 23, 24].

Volcano erupt explosively because trapped volatiles (mostly water, CO2, and SO2) rapidly come out of solution and undergo huge volume expansions as magma ascenf

The distance from Mount Hunga Tonga to the area of Indonesia is about 5,992 km, while atmospheric waves reach Indonesia in about 4-6 hours. So that the atmospheric waves spread with an average speed of 1248 km/hour. Comparing with Krakatao explosion [22], The Tonga mount eruption sonic boom pressure spread aproximately one thirds of krakatao exploid/eruption (pressure wave spreads reach 14.000 km).

The change in pressure due to this eruption is between 0.6 to 1.9 millibar or the equivalent of a change in height of 6 meters to 19 meters. In the world of aviation navigation, aircraft use location positions based on air pressure. If there is a sudden change that can cause an airplane to crash as a result of the pressure value suddenly dropping, the position of the aircraft recorded in the altimeter is in a much higher position, this difference will cause an overshoot (the aircraft lands further than the expected position). if the pressure suddenly rises then the position of the aircraft recorded in the altimeter is in a much lower position (closer to the runway surface), this sudden difference/change will cause an undershoot (aircraft lands shorter/closer than the expected touchdown position). Collition between other aircraft is possible because the pressure altitude impact in the vertical position could be unprecission, such previous incident between two aircraft in the heatrow airport that separated less than 1000 feet vertical due to altimeter setting error in low pressure condition [6].

Table 1. Air pressure (QNH) changing over eastern Indonesian airpot.

No	*Location*	*Time*	*Pressure Change (maximum)*	*Weather*
1	Mount Tonga (suspected source)	15 January 2022, 00.43-06.09 UTC		
2	Wamena (WAVV)	09.10-10.10 UTC	1,5 mb	Cloudy (SCT[b] cloud 1800 ft)
3	Biak (WABB)	09.30-10.50 UTC	1,2 mb	Cloudy (SCT[b] cloud 1800 ft)
4	Lombok (WADL)	10.51-11.22 UTC	1 mb	Rain, CB[a]
5	Kupang (WATT)	10.37-1.15 UTC	0,6 mb	Cloudy (SCT[b] cloud 1600 ft)
6	Sumbawa (WADS)	10.30-11.30 UTC	1,39 mb	Rain, CB[a]
7	Sabu (WATS)	09.20-11.30 UTC	0,7 mb	Rain
8	Sorong (WASS)	09.20-11.10 UTC	1,9 mb	Rain
9	Ternate (WAEE)	09.50-11.10 UTC	1,36 mb	Rain
10	Sentani (WAJJ)	09.10-10.10 UTC	1,42 mb	Cloudy (FEW[b] cloud 2000 ft)
Range		09.10-11.30 UTC	0,6 – 1,9 mb	Phenomea Cloudy to Rain

[a] Cb means Cumulonimbus Cloud

[b] SCT means Scatered 3-4 oktas cloud cover; FEW means Few 1-2 oktas cloud cover.

Oktaf (8/8) means full of disk sky fully by cloud.

Role of tumb based on Civil Aviation Safety Regulation (CASR) 91.121 (DGCA) and ICAO DOC 8168, Vol. 1, Part VI, Chapter 3 [14] related to flying aircraft separation between obstacle (such as mountain or others) and flight level shows that QNH (local air pressure at Airport/surface) is used to corrected the estimated real altitude/flight level of the flying aircraft. The corrected altimeter setting (to provide elevation/flight level information) is very crusial to avoid the crash between flying aircrafts and/or avoiding the obstacle. The QNH changing during sonic boom of mount Tonga possibly to make some flying aircraft accident/incident during flight operation (cruising/in the route of flying) due to their elevation/flight level changed.

The finding of sudden changes due to volcanic eruptions is very important as information for the world of air transportation, both from the operational side and related to the development of aircraft technology in the context of anticipating or preventive action from preventing aircraft accidents

4. Conclusion

The eruption of the Hunga Tonga volcano is a unique phenomenon because it has a far-reaching effect on the atmosphere of a volcanic eruption area. This volcano eruption produces atmospheric waves with a speed of about 1248 km/ hour. Several airport locations in the territory of Indonesia recorded sudden changes in air pressure. Recorded pressure changes a maximum of 1.9 millibars in a short period outside the diurnal cycle. These sudden changes in conditions can result in a hazard to flight because air navigation uses pressure as the altitude/flight level. It is to avoid collisions with other aircraft, or avoid mountains, or landing planes on the runway smoothly.

The sudden change in pressure due to this volcanic eruption is a finding of this paper and can be used as a caution regarding flight information services. This information is essential for aviation weather service providers, air traffic service providers, airlines, pilots, and stakeholders related to aviation.

A suggestion for further research is whether there is a relationship between the sudden change in pressure and the weather that has occurred or will occur. More detailed research is needed in this regard.

References

[1] Bennis, K., & Venzke, E. (2022). Report on Hunga Tonga-Hunga Ha'apai (Tonga). *Bulletin of the Global Volcanism Network*, 47:3. Global Volcanism Programe.

[2] Brandon, J. (2000). *Flight Teory Guide*. Retrieved from https://www.auf.asn.au.

[3] DGCA. (n.d.). Civil Aviation and Safety Regulation (CASR) part 91. Ministry of Transporation.

[4] FAA. (2008). Density Altitude. *FAA-P-8740-2.AFS-8*. Federal Aviation Administration.

[5] FAA. (2016). Pilot's Handbook of Aeronautical Knowledge. *FAA-H-8083-325*. Federal Aviation Administration.

[6] Foster, C., Plant, K., & Stanton, N. (2021). A very temporary operating instruction: Uncovering emergence and adaptation in air traffic control. *Reliability Engineering and System Safety*, doi.org/10.1016/j.ress.2020.107386.

[7] Goodman, C. J., & Griswold, J. D. (2018). Climate impacts on density altitude and aviation operations. *Journal Applied Meteorology and Climatology*, 517-523.

[8] Guarte, J., & Barrios, E. (2006). Estimation Under Purposive Sampling. *Communication in Statistics-Simulation and Computation*, 35 (2), 277-284. doi:10.1080/03610910600591610.

[9] Guinn, T. A., & Barry, R. J. (2016). Quantifying the Effects of Humidity on Density Altitude Calculations for Profesional Aviation Education. *International Journal of Aviation, Aeronautics, and Aerospace*, 3(3). https://doi.org/10.15394/ijaaa.2016.1124.

[10] Harris, T. (2021). AIR PRESSURE: Temporal Hierarchies in Nepali Aviation. *Cultural Antropology*, pp. 83-109.

[11] Henneton, M. P. (2012). Computational methods for evaluating meteorities as sources of sonic boom. *18th AIAA/CEAS Aeroacoustics Conference (33rd AIAA Aeroacoustic Conference.* Colorado Springs, CO.

[12] Hirtl, M., Arnold, D., Baro, R., Brenot, H., Coltelli, M., Eschbacher, K., . . . Rokitansky, C. (2020). A volcanic-hazard demonstration exercise to assess and mitigate the impact of

volcanic ash clouds on civil and military avaition. *Natural Hazards Earth System Science*, 1719-1739.

[13] ICAO. (2018). Meteorological Services for International Air Navigation. *ANNEX 3 Ammandement 79*. International Civil Aviation Organization.

[14] ICAO. (2018). Procedures For Air Navigation Services Aircraft Operation. *ICAO DOC 8168 sixth edition Vol 1 Flight Procedures Part VI, CHapter 3*. International Civil Aviation Regulation.

[15] ICAO. (2019). ICAO DOC 8896 Twelth Edition. *Manual of Aeronautical Meteorological Practices*. International Civil Aviation Organization.

[16] Kusky, T. M. (2022). Deja vu: Might Future Eruption of HUnga Tonga-Hunga Ha'apai Volcano be a repeat of the Devastating Eruption of Santorini, Greece (1650 BC)? *Journal of the Earth Science, Vol 33, No.2,*, pp.229-235; https://doi.org/10.1007/s12583-022-1624-2.

[17] Lanese, N. (2022, July 29). *Record-breaking Tonga volcano generated the fastest atmospheric waves ever seen*. Retrieved from https://www.livesscience.com/tonga-volcanic-eruption-fast-waves

[18] Lin, J.-T., Rajesh, P., Lin, C., Chou, M.-Y., Liu, J.-Y., Yue, J., . . . Kung, M.-M. (2022). Rapid Conjugate Appearance of the Giant Ionospheric Lamb Wave Signatures in the Northern Hemisphere After Hunga-Tonga Volcano Eruptions . *Geophysical Research Letter, 49, e2022GL098222*, https://doi.org/10.1029/2022GL098222.

[19] Manneela, S., & Kumar, S. (2022). Overview of the Tonga-Hunga Ha'apai Volcanic Eruption and Tsunami. *Geol. India*, 98:299-304. https://doi.org/10.1007/s12594-022-1980-7.

[20] NOAA. (2022, July 29). *The Hunga Tonga-Hunga Haápai Eruption, a Multi-Hazard Event*. Retrieved from https://www.nesdis.noaa.gov/

[21] Ratnasari, Y., Trilaksono, N., & Septiadi, D. (2022). Causes and impact of extreme Low Level Wind Shear (LLWS) event at SOekarno-Hatta International Airport. *IC-STCC-2022*. IOP Conference Series: Earth and Environmental Science.

[22] Reed, J. W. (1987). Air Pressure Waves From Mount St. Helens Eruption. *J. Geophysical Research*, pp. 11.979-11.992.

[23] Reichardt, U., Ulfarsson, G. F., & Petursdottir, G. (2019). Developing scenarios to explore impacts and weknesses in aviation response exercises for volcanic ash eruption in Europe. *Journal of Air Transport Management*, doi.org/10.1016/j.jairtraman.2019.101684.

[24] Schneider, D. J., Van Eaton, A., & Wallace, K. (2020). Satellite observation of the 2016-2017 eruption Bogoslof volcano: aviation and ash fallout hazard implications from a water-rich eruption. *Bulletin of Volcanology*.

[25] Takebayashi, M., Onishi, M., & Iguchi, M. (2021). Large volcanic eruptions and their influence on air transport: The case of Japan. *Air Transport Management*.

[26] Titos, M., Montenis, B., Barsotti, S., Sandri, L., Folch, A., Mingari, L., . . . Costa, A. (2021). Assessing potential impact of explosive volcanic eruptions from Jan Mayen Island (Norway) on aviation in the North Atlantic. *Natural Hazards and Earth System Sciences*, Preprint.

Urban Risk Assessment in Heritage Area for Developing Resilience Understanding

Gerarda Orbita Ida Cahyandari[1*], Suratman Worosuprojo[2], Djati Mardiatno[2]

[1]Department of Architecture, Universitas Atma Jaya Yogyakarta, Indonesia
[2]Department of Geography, Universitas Gadjah Mada Yogyakarta, Indonesia

[*]Corresponding author: ida.cahyandari@uajy.ac.id

Abstract. Heritage area is vulnerable to urban development. Risk and resilience were discussed separately, especially in a heritage area. The term urban risk describes the risk in an urban area. Main factors that link urban risk and resilience must be determined sequentially asa strategy to reduce urban risk and improve resilience. This paper was based on field observation and questionnaires distributed in Jeron Beteng Yogyakarta. The first questionnaires collected data from the owners or residents of heritage objects for addressing the basic risks assessment and social-community assessment. The second questionnaire on risk,resilience, and heritage was distributed to the surrounding communities of the heritage objects. The questionnaires were developed as an assessment for quantitative analysis. The scoring analysis revealed the essential roles of the government and the community in controlling urban risk and resilience. The multi-factors and multi-contexts have relations of interdependence and mutuality. Integrating urban risk and resilience must focus on government funds, professional assistance with renovation, and traditional authority. The social, economic, hazard-disaster mitigation, and building aspects are needed in inter-relationship with the factors. Governance factors must be prioritized in the strategy of preservation.

Keywords: Urban Risk, Resilience, Heritage, Yogyakarta Palace, Assessment, Community

1. Introduction

Risk and resilience emerge from different conditions. Risk is a possibility of adverse effect, loss, or destruction in a community, infrastructure, and service systems [1]. The definition of risk is commonly associated with hazards and disasters. Meanwhile resilience is the capability of impacted system, person, or community to anticipate, reduce, and adapt to the changes along with maintaining essential functions to recover [2][3]. Resilience comprises response, recovery, and mitigation [4][5]. If a city is developed to become a resilient city, the city is expected to achieve strong, prosperous, flexible, inclusive, integrated, and well-managed capacities [6].

The risk is often associated with various contexts, primarily physical and social contexts [7]. The contexts are included in basic problems faced by people, such as poverty, environmental degradation, dense population, and old buildings under specific conditions [8][9][10]. These vulnerabilities are mainly associated with urban risk. Resilience is the capacity of a community to adapt to vulnerabilities. Any kinds of density become an obstacle to achieving resilience [11].

The heritage area as the old part of the city is commonly located in the dense city center. In the heritage area, the urban risk approach is more appropriate for resolving basic and social problems, including disaster risk. High population density, housing density, and social-economic problems happen in the city center and can influence the heritage area and the environment. Heritage is a product of civilization, thinking, and behavior and must be preserved for the next generation [12][13]. The preservation requires integrated efforts amid urban development and urban risk.

Building is the most significant among many acknowledged tangible heritage objects [14]. Due to their aging condition, the under-maintenance heritage objects could endanger the surrounding neighborhood. Heritage buildings and the areas could pose risks to their surrounding environments and neighborhoods [15]. The continuous growth of urbanity and development in the city center could potentially threaten the existence of heritage objects [16].

The risk in an urban context starts from the everyday risk in urban daily life [1][17]. Numerous influences of urban risk urge redefinition of risk, especially urban risk in a heritage area. The definition leads to factors and clear parameters establishing the assessments and mitigation action. Therefore, risk assessment can also be used to assess resilience.

The factors of urban risk may also be the factors of resilience in a different way. Discussions on resilience began in the 1990s within the scope of urban planning and infrastructure improvement, and it further developed as part of risk management discussion from the 2000s onwards [18]. Therefore, resilience characteristics are frequently depicted in an adverse direction to vulnerability [19][20]. Risk and resilience have an opposte perspective in urban development and disaster occurrence. Although risk and resilience shall be treated distinctively, there is a tendency that the most vulnerable object is also often the least resilient one [21][22].

Development, risk, and resilience occur simultaneously in the urban heritage area. Risk and heritage are often separately discussed when addressing the impacts of developments in urban areas, even though consideration of multiple risks attempts to strengthen community resilience and safeguard [23]. Resilience rarely covers the heritage area. How to cope and adapt to the environmental changes can be learned from the heritage area and its community. The existence of heritage areas shows the capacity to accommodate social cohesion, socio-economic regeneration, poverty reduction, social well-being, creativity, and tourism benefits [24][25].

According to previous studies, heritage area is vulnerable to urban development. Urban risk and resilience occur simultaneously, but they were discussed separately, especially in a heritage area. The main factors that link urban risk and resilience must be determined sequentially as a strategy to reduce urban risk and improve the resilience of heritage area. Certain factors in specific contexts cause urban risk in the heritage area. It is essential to create an integrated strategy in relation to urban risk, resilience, and heritage area. This research aimed to identify the main factors that link risk and resilience to develop more effective and supportive strategies in a heritage area. The strategies can effectively reduce urban risk and improve the resilience of heritage area. The findings must be considered in the preservation efforts of heritage areas.

2. Materials and Methods

This paper was based on a literature study, field observation, and questionnaires. The history and contextual phenomenon established a literature review of the study area. The field observations comprised visual assessment and documentation of the heritage buildings, community, and environment. The documentation was conducted along with interviews to develop an understanding of the buildings and surroundings. The questionnaires were addressed to the owner or dweller of the heritage object and the surrounding community. The respondents represented the internal and external factors of urban risk from the perspective of heritage objects and the community. The questions were derived from the urban risk assessment [9], the definition of risk, resilience, and the influences of urban risk. The questionnaires were employed as tools for assessment.

2.1. Study Area

The research location is inside the traditional fort of Yogyakarta Palace, called Jeron Beteng (GPS: -7.803, 110.364). Jeron Beteng is part of the subdistrict of Kraton with three urban villages having traditional regulation and historical objects. It lies on a 1.4 sq. km area in the center of Yogyakarta, surrounded by walls with more than thirty objects categorized as cultural heritage objects (Figure 1). The settlements in Jeron Beteng are unique compared to other ordinary settlements in the city because of the historical and spatial relationships with the existence of the Sultan Palace.

Figure 1. The traditional fort and the area of research
Source: Google Earth

Urban development has been done in Jeron Beteng, although traditional unwritten regulation has been established. The population density was higher than the average density of Yogyakarta city in 2020. There are 15,646 people per km^2 compared to that of the city of Yogyakarta, which was 13,413 people per km^2. The recent socio-economic demands and the dynamics of the city have pushed Jeron Beteng and its inhabitants to accommodate modern and traditional lifestyle activities. These expandinguses and space alterations consequently changed the original form of dwelling and function. In Jeron Beteng, ordinary dwellings have been built within or around the walls of historical objects.

2.2. Data Collection and Analysis

Table 1 shows the scopes for the first questionnaire with 25 respondents. The subjects were basic risk and social community assessments (Table 1). The basic risk assessment and the social community assessment comprised questions from the residents' viewpoint about pertinent vulnerabilities and social development in the surrounding community.

Table 1. The Scopes of Questions for Approaching Basic Factors of Risk and Resilience

	The Questions	Assessments
1.	Generation of Inheritance (descendant)	Basic risk
2.	The Number of Families (privacy, psychological comfort, extended family)	Basic risk
3.	House Area and Number of Users (Density)	Basic risk
4.	Status of Land (Land Ownership)	Basic risk
5.	Frequency of Renovation (architecture and structure)	Basic risk

	The Questions	Assessments
6.	Function	Basic risk
7.	Visual Assessment (related to decent, safe, secured building)	Basic risk
8.	Functional Problems (uncomfortable to live in)	Basic risk
9.	Government Fund for Registered Heritage Building (availability and adequacy)	Basic risk
10.	Renovation Assistance (professional consultants for design and structure)	Basic risk
11.	Surrounding Buildings (physical attachment)	Social Community
12.	Sense of Community (sense of belonging and respect to the neighborhood)	Social Community
13.	Socio-cultural Participation (socio-cultural activities, formal-informal)	Social Community
14.	Peaceful Society (comfort, efficacy)	Social Community
15.	Security, Mitigation, & Risk-Reduction Participation	Social Community

Source: Preliminary Analysis

The second questionnaire was addressed to 15 heritage objects taken from 25 previous objects. The selection was based on the various areas and functions. The respondents were people who lived around the heritage object, inside or outside the walls. The number of respondents was from ten to twenty according to the area of each heritage object. The total was 185 respondents. The questions are derived from the Sendai Framework for Disaster Risk Reduction (2015-2030). The Sendai framework determines four sequential stages to achieve resilience [3][20][26]. Each stage consists of two resilience objects: health and well-being, social and economic, infrastructure and natural environments, and leadership and strategy [27]. Risk occurs when resilience stages fail to accomplish the resilience objects. The need to integrate the factors of risk, resilience, and heritage creates questions about the environment, building, social, cultural, economic, and hazard-disaster aspects. (Figure 1)

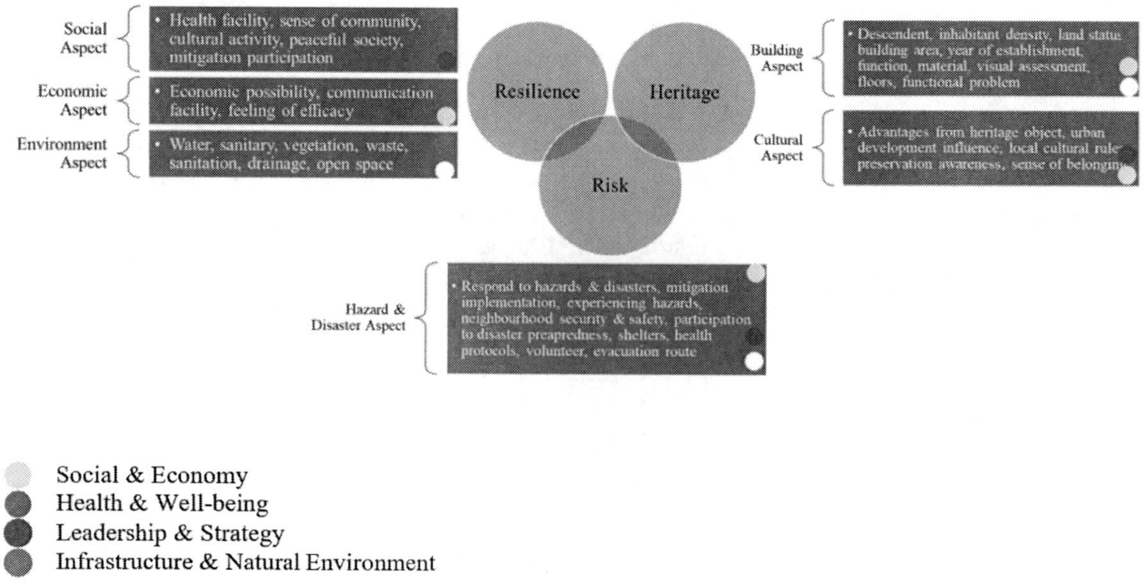

Figure 2. Aspects and Variables of Risk, Resilience, and Heritage
Source: Preliminary Analysis

The questionnaires were created using *Google form*. The results were automatically collected in excel file. The quantitative analysis of the first questionnaire used a score ranging from five to one, which was from high risk to low risk. The values were low, low-medium, medium, high-medium, and high-value. The quantitative analysis of the second questionnaire was presented in scores for each aspect. The score range was the same as the first questionnaire. The category was completed up and down to accommodate slight differences between scores. A low score means the risk is low, but the resilience is high, and vice versa. The high or low values are considered substantial risk or resilience factors.

3. Result

3.1. The First Questionnaire in Jeron Beteng

The first questionnaire reveals the internal perspective of the heritage objects by the residents. The basic risk assessment and social community assessment aimed to determine the internal influences of urban risk. The area of heritage object was possibly in a block or a single lot. The large heritage object is usually in a wide block area surrounded by houses. They occupy the land of the mansion. The singlelot means that the heritage object is in a smaller size of building and area. The surrounding community depends on the noblemen and still serves the noblemen and their mansions.

The high-risk value is indicated by the government funds, which is 4.04. It presumably influences the renovation frequency (3.76) and renovation assistance (3.92) which are included in high-medium risk. The renovation frequency also depends on the fund and professional assistance. There is an inter-relation between these three factors. Social community assessment conversely shows low-medium risk values for all the factors. (Figure 2)

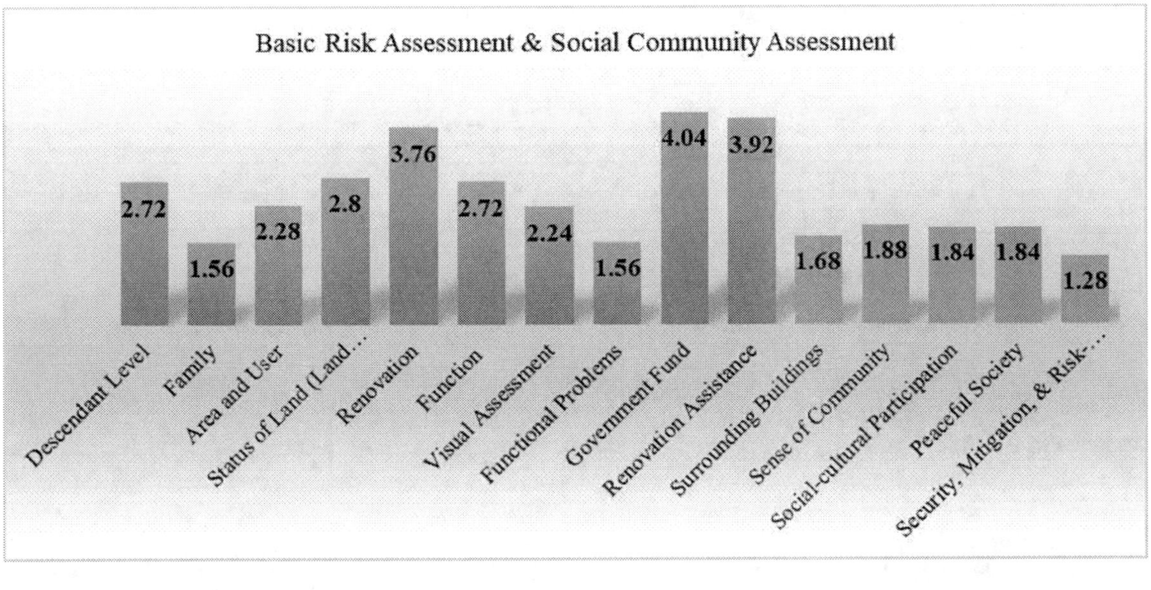

Figure 3. Basic risk assessment and social community assessment of 25 heritage objects
Source: Analysis

The result of each question in Table 2 illustrates the pattern of the score's level. The pattern is obtained from the highest and second scores consecutively (Table 2). The level of residents' understanding of their heritage object and the environment helps determine the significant factors of basic risk in the heritage area and in describing the priority of the heritage object.

IOP Conf. Series: Earth and Environmental Science 1233 (2023) 012071 doi:10.1088/1755-1315/1233/1/012071

Table 2. Category, value, and differences of internal perspective (basic risk and social community)

	Basic Risk Assess.	Category ((Amount out of 25); First and Second Percentage only)				
		Level 5 (High)	Level 4	Level 3	Level 2	Level 1 (Low)
1.	Descendant Level	not known (3)	4< (3)	4 (5; 20%)	3(12; 48%)	2 (2)
2.	Family	4< (0)	4 (0)	3 (1)	2(12; 48%)	1(12, 48%)
3.	Area and Users	very poor (1)	poor (1)	fair (7; 28%)	good (11; 44%)	very good (5)
4.	Status of Land	not known (0)	private property (10; 40%)	lease-use without a letter of agreement (0)	a usufructuary right given by the Sultan (15; 60%)	lease rights (under the Sultan agreement) (0)
5.	Renovation	never (2; 8%)	1-2 times (19; 76%)	3-4 times (1)	>4 (2; 8%)	anytime needed (1)
6.	Function	residential (5)	office (government, private) (3)	economic function (trade and service) (3)	social function (education, culture, society) (8; 32%)	mix-use (6; 24%)
7.	Visual Assessment	very poor (0)	poor (2)	fair (4; 16%)	good (17; 72%)	very good (2)
8.	Functional Problems	deterioration of construction and space (1)	a building needs maintenance and repairment (2)	insufficient space for users (1)	insufficient space for the function (2; 8%)	none (19; 76%)
9.	Government Fund	very poor (11; 44%)	poor (8; 32%)	fair (2)	good (4)	very good (0)
10.	Renovation Assistance	very poor (11; 44%)	poor (7; 28%)	fair (1)	good (6)	very good (0)
	Social Com. Assess.	Category ((Amount out of 25); First and Second Percentage only)				
		Level 5 (High)	Level 4	Level 3	Level 2	Level 1 (Low)
11.	Surrounding Buildings	other (0)	attaching inside and outside walls (3; 12%)	attaching inside walls (3; 12%)	attaching outside walls (2)	surrounding the walls at a certain distance (17; 68%)
12.	Sense of Community	very poor (0)	poor (0)	fair (0)	good (22; 88%)	very good (3; 12%)
13.	Socio-cultural Part	very poor (0)	poor (0)	fair (0)	good (21; 84%)	very good (4; 16%)
14.	Peaceful Society	very poor (0)	poor (0)	fair (0)	good (21; 84%)	very good (4; 16%)
15.	Security, Mitigation, & Risk-Reduction Participation	not known (0)	uninterested (0)	no time to participate (0)	moderate participation (7; 28%)	active participation (18; 72%)

Source: Analysis

There are three groups of answers (Table 3). The first group answered with more than 50% differences between the highest and lower levels. This group represents the substantial factors from the residents' perspective to measure basic risk in the heritage area. The second group gived consecutive answers without a gap between higher and lower levels. The factors in this group are needed for detailed assessment in a common area. The third group consists of answers with a gap between the highest and lower levels. This group is needed for contextual assessment to complete the other groups.

Table 3. Grouping analysis of the first questionnaire

Difference of more than 50%	Consecutive scores from the highest to lower	The gap between the highest and lower
	Descendant Level (M)	
	Family (L-M)	
	Area and Users (M)	
		Status of Land (M)
Renovation (H-M)	Renovation (H-M)	
	Function (M)	
	Visual Assessment (M)	
Functional problems (L-M)	Functional Problems (L-M)	
	Government Fund (H-M)	
	Renovation Assistance (H-M)	
Surrounding buildings (L-M)		Surrounding Buildings (L-M)
Sense of Community (L-M)	Sense of Community (L-M)	
Socio-cultural Part (L-M)	Socio-cultural Part (L-M)	
Peaceful Society (L-M)	Peaceful Society (L-M)	
Security, Mitigation, & Risk-Reduction Participation (L-M)	Security, Mitigation, & Risk-Reduction Participation (L-M)	

H-M= High Medium; M= Medium; L-M= Low medium
Source: Analysis

All of the highest scores are in the second group. This group is also completed with all value categories. The consecutive gradation of scores is presumably the objective perspective of the resident. If there is a gap in the score level between the highest and second, the respondents might have different contexts or backgrounds. It might happen as a unique condition or contextual factor. The factors are the status of land and surrounding buildings.

Status of land shows 60% for 'usufructuary right given by the Sultan' and 40% for 'privateproperty'. The second level for land status would be better if it were 'the lease rights under the Sultan agreement' or 'the lease-use without a letter of agreement'. Those categories are zero value.

The surrounding buildings keep a certain distance to the walls of heritage objects, which isincluded as low risk. The percentage is 68%, although there are still buildings which attach inside and outside the walls.

3.2. The Second Questionnaire in Jeron Beteng

The second questionnaire shows the results of an external perspective conducted in the surrounding community. The average was calculated from 185 respondents of 15 heritage objects. The 15 heritage objects were chosen from 25 heritage objects in the first questions. The groups of questions were about environment, building, economic, social, cultural, and hazard-disaster aspects. The environment consists of 9 questions, the building aspect 12 questions, the economic aspect four questions, the socialaspect six questions, the cultural aspect eight questions, and the hazard and disaster aspect 11 questions.

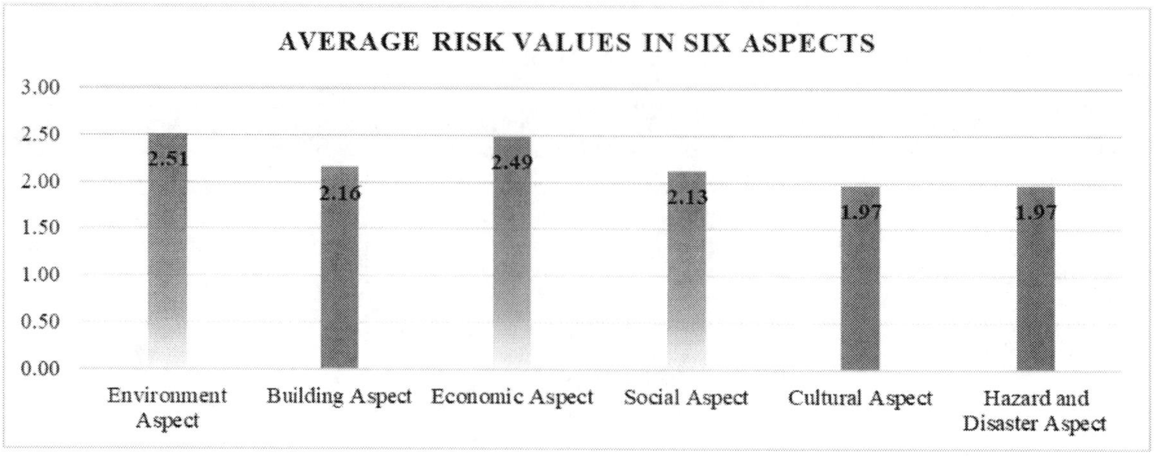

Figure 4. Average risk values in risk, resilience, and heritage aspects
Source: Analysis

The colors are related to the aspects related to risk (green), resilience (blue), and heritage (orange). The environment aspect has a medium-up value which is 2.51. The building aspect has a medium- down value, which is 2.16, together with the economic aspect (2.49) and social aspect (2.13). The cultural aspect (1.97) and hazard and disaster aspect (1.97) have low-medium-up values. All the differences between values are less than one (1) point. (Figure 3)

Consideration must be given to scores of questions in each aspect. The environment aspect has a high- medium up for 'open spaces for public activities beside square', which is 3.60, and 'waste management' (3.84). The 'role of vegetation in the property' has a high-medium down value of 3.33. The building aspect for 'status of land' (3.28) and 'building function' (3.18) are included in the high- medium down value. The hazard and disaster aspect have a high-medium down value for 'respond to fire hazard', which is 3.22, and a high-medium up value for 'center of joint secretariat volunteer in the subdistrict of Kraton', which is (3.68).

4. Discussion

4.1. Preliminary Assessments to Determine Factors of Urban Risk

The history of Kraton Yogyakarta is associated with the Sultan's authority and traditional regulation. The Sultan and the palace govern physical and nonphysical environments. Jeron Beteng originally accommodated Sultan Palace and mansions for the noble people. The Sultan gave the lands of the mansions under usufructuary rights or lease rights. The people who served the palace and the noble people also lived in the Jeron Beteng. The dwellings were built around the heritage objects inside or outside the walls.

Risk in the context of cultural heritage was introduced in 1972 because cultural heritage exemplified the integrated risk-management analysis [28]. Risk in this context was still about hazard and disaster. The heritage object proves that the multi-factors analysis helps to attain disaster risk reduction through good governance, preservation, and resilience efforts. Therefore, the vulnerability ofa building is added to the risk assessment [14]. The strength of heritage buildings lies in the construction, architectural, and historical values [29]. Architectural and historical values empower the preservation and resilience strategies.

Risk and resilience in urban spaces can be observed from the dwelling condition because dwellings represent integrated aspects of community [30]. It can be implied that dwelling is a reliable object to approach urban risk and resilience. According to the Sendai Platform, dwellings accommodate the aspects of resilience: health-well-being, social-economic, infrastructure-natural environments, and leadership-strategy. High-risk dwellings often have vulnerabilities that can worsen the consequences of disasters [31]. According to this fact, dwellings have multi contexts to draw risk and resilience [32].

Based on the assessments of Jeron Beteng, risk and resilience in heritage areas are essentially influenced by certain factors that lead to urban risk. The groups are governance, culture, building, owner/resident, and community. Multi-factors and multi-contexts in the heritage area are essential to identifying the urban risk level through assessment [7].

Table 4. Analysis assessments related to the group of factors

RANGE OF VALUE	CATEGORY		BASIC RISK ASSESSMENT	SOCIAL COMMUNITY ASSESSMENT
$0 \le X < 0.5$	Low	Down		
$0.5 \le X < 1$		Up		
$1 \le X < 1.5$	Low-Medium	Down		Security, Mitigation, & Risk-Reduction Participation
$1.5 \le X < 2$		Up	Fam. Number, Functional Problems	Surrounding Buildings, Sense of Community, Socio-cultural Participation, Peaceful Society
$2 \le X < 2.5$	Medium	Down	Area and User, Visual Assessment	
$2.5 \le X < 3$		Up	Function, Descendent Level, Status of Land,	
$3 \le X < 3.5$	High-Medium (HM)	Down		
$3.5 \le X < 4$		Up	Renovation, Renovation Assistance	
$4 \le X < 4.5$	High	Down	Government Fund	
$4.5 \le X < 5$		Up		

Governance	Culture	Building	Owner/Resident	Community

Source: Analysis

The low-medium risk deals with residents and community, and the high-medium risks are solely about governance. The medium risk is related to residents, buildings, and culture. The social community assessment results show that all the values are low-medium risks. It can be inferred that people who live in the heritage area have good responsibility for risk and resilience. The responsive community is excellent for reducing urban risk and creating resilience. (Table 4)

The contextual factors, which are the status of land and surrounding buildings, represent multi-context in the heritage area. The palace reserves the right to grant or decline permission to occupy the property. Urban risk and resilience in land ownership remain sensitive due to the interdependency between people and the owner/resident. Meanwhile, the physical distance between the surrounding dwellings and heritage objects implies respect and dependency on the heritage objects. The community gives high respect and less physical risk to the heritage object by keeping the distance of their houses or buildings from the inside and outside walls of the heritage objects.

4.2. Assessment of Community's Perspective for Understanding Resilience

Between 1855 and 1877, Sri Sultan Hamengkubuwono VI allowed people to occupy land in Jeron Beteng for safety [33]. This policy directly affected the densities of population and buildings in Jeron Beteng. Along the time, the development inside the Jeron Beteng was inevitably affected by social, political, and cultural activities. The community's opportunity occupied the land inside Jeron Beteng. The density also escalated when economic factors involved. The policy changed the community's activities but created an increased risk.

The densities worsened the consequences of disasters. The densities in buildings and population can be factors of urban risk and obstacles to achieving resilience [11]. In further analysis, demographicdata are needed. Density, including surrounding dwellings, is addressed as urban sprawl or slum area.

They may endanger heritage objects. Unexpectedly, the result of the first questionnaire was a low-medium risk for community matters.

The external perspective went deeper to resilience understanding because the respondents are the community. The aspects were chosen from urban communities regarding risk, resilience, and heritage. The resilience is established by social, economic, and environmental aspects, whereas the heritage is shaped by building and cultural aspects [7]. The risk is related to hazard and disaster aspect. Several variables specify each aspect. (Table 5)

Table 5. Category and value of external perspective (aspects of risk, resilience, and heritage)

RANGE OF RISK VALUE	CATEGORY		APPROACHES		
			Risk	Resilience	Heritage
$0 \leq X < 0.5$	Low	Down			
$0.5 \leq X < 1$		Up			
$1 \leq X < 1.5$	Low-Medium	Down			
$1.5 \leq X < 2$		Up	Hazard & Disaster		Cultural
$2 \leq X < 2.5$	Medium	Down		Economic, Social	Building
$2.5 \leq X < 3$		Up		Environment	
$3 \leq X < 3.5$	High-Medium (HM)	Down			
$3.5 \leq X < 4$		Up			
$4 \leq X < 4.5$	High	Down			
$4.5 \leq X < 5$		Up			

Source: Analysis

The second questionnaire strengthens the findings in the first assessments. In the second questionnaire, aspects of the community have low medium and medium values, especially the cultural and hazard-disaster mitigation aspects. It means that the aspects and factors related to the community in heritage areas are essential indicators for urban risk because the findings in the first and second questionnaires show similarity [14]. Although data in the heritage area are multi-factors and multi-contexts, the community and its derivatives always have low-risk values.

The integrated factors for urban risk and resilience in the heritage area start from the substantial factors. If the result tends to show high value, then the process is followed by a comprehensive assessment to understand potential high-risk factors and inter-relationship. The contextual assessment will be conducted if the area has unique characteristics. The topic questions can be changed according to the uniqueness. The last step for urban risk level needs demographic information, functional development, and public facilities to complete the scoring analysis. (Figure 4)

The results indicate that multi-factors and multi-contexts have pertinent interrelationships in the internal perspective and external perspective. The inter-relationships are interdependency and mutuality. A particular factor depends on the other factors and vice versa. This interdependency influences the risk value. Some factors mutually influence and exist because of each other. The inter- relationship within bad quality will increase risk values and decrease the capacity for resilience. (Figure 5)

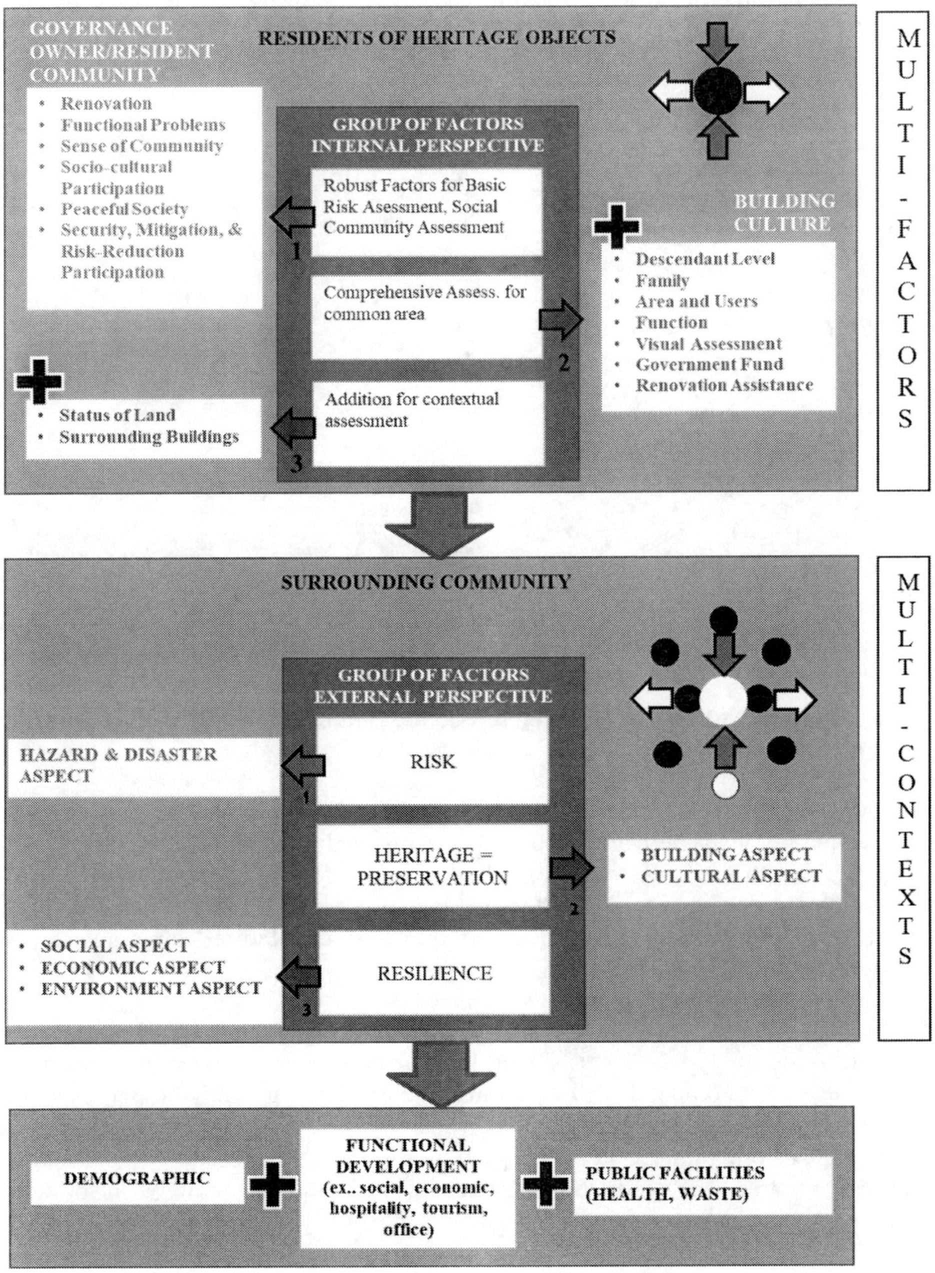

Figure 5. The integrated factors of urban risk and resilience: internal perspective, external perspective, and data of demography, functions, and facilities
Source: Analysis

Subsequently, the aspect is the context itself. Both still can be explored according to urban development. Building, economic, social, and hazard-disaster mitigation aspects are close to high risk and high-medium risk factors, which is governance. Building, socio-economic activities, and mitigation efforts should be under the regulation or good governance to implement urban risk reduction and resilience strategies. The cultural and environmental aspects prove the excellent existence of Jeron Beteng because those aspects and the factors of culture, owner/resident, andcommunity are in medium risk and low-medium risk values.

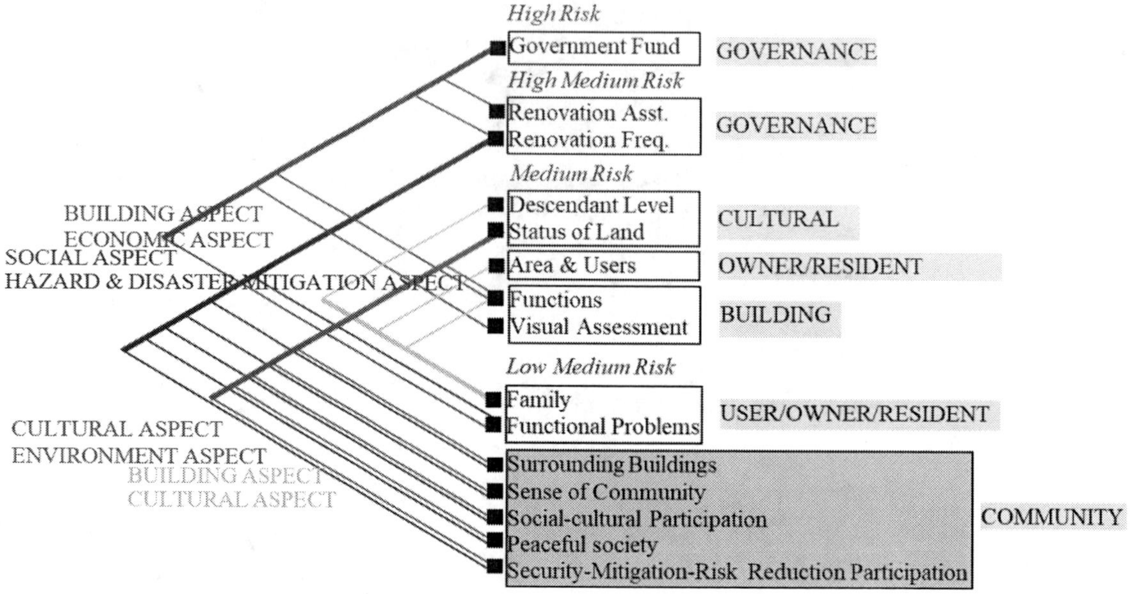

Figure 6. Multi-factors and multi-contexts in interdependency and mutuality
Source: Analysis

Without good governance, respective culture, building maintenance, supportive owner/resident, and responsive community, the heritage object and the surrounding community could possibly cause hazards or damage to each other. The inter-relation of factors shows the relationship between the aspects. When the community is at risk, the condition of the heritage object will also be affected. The community holds substantial sources of resilience, risk, and heritage. The surrounding community around the heritage object contributes to the existence of heritage objects.

5. Conclusion

Urban risk is a possibility of adverse effect, loss, or destruction in an urban area and its community system. Resilience is the capability of a person and community to apply proper responses and recovery to specific changes. Risk is associated with hazard and disaster, but the urban risk is associated with the urban scale and its community with the heritage values dan resilience.

Urban risk and resilience must be addressed as the integrated strategy in the heritage area. The strategy should consider multi-factors and multi-contexts [7]. The identification of factors starts from the community [33], and context identification starts from building or culture. Community is an essential substance for the risk and resilience strategies in the heritage area. These aspects emerged twice in the analysis. These aspects are vital for urban risk, resilience, and heritage.

Urban risk in the heritage area of Jeron Beteng occurs within the resilience of the palace, the heritage objects, and the community. The Jeron Beteng heritage area shows the capacity to accommodate urban risk and resilience within certain physical boundaries. The population and environment consist of ordinary and noble communities. The settlements in Jeron Beteng are considered physical and nonphysical elements because of the Sultan's and the palace's historical and spatial existences. Resilience is proved by the existence of heritage objects [32].

The high and low-risk scores can be seen as inversely proportion with the resilience values. The urban risk factors can be similar to resilience factors, but the contexts are different. For example, the hazard and disaster factors for urban risk and resilience are the same in the low-medium value, but they are different in context. Urban risk is about the occurrence of hazards and disasters, but the context for resilience is about the community's participation and response.

The basic risk assessment and social community assessment are the internal perspective of the owner/resident of the heritage object. The perspective resulted in main factors of urban risk in the heritage area: governance, culture, building, owner/resident, and community. The assessments were conducted on the owner/resident of each heritage object. The preliminary assessments show the consecutive risk factors from the highest to the lowest, from the governance factor to the community factor. The owner/resident assumed that the surrounding community became a supportive environment for the existence of the heritage objects.

The comprehensive assessments or external perspectives were conducted in the surrounding communities of each heritage object. The assessments consist of environmental, building, social, economic, culture, and hazard and disaster mitigation. The assessment aims to acquire the resilience factors for urban risk reduction. The values can be improved through data scoring analysis of demography, infrastructure, and facility based on administrative location.

Internal and external perspectives combine factors and contexts to deal with urban risk and resilience. The multi-factors and multi-contexts connect and adjust to each other in inter-relationship. The relations are interdependence and mutuality. For example, the community's loyalty resulted in opportunities to live nearby the noblemen's mansions. This interaction implies the social interdependence between people and the noblemen, which becomes a pattern of cultural and social resiliencies. It helps to protect the existing traditional supremacy. This condition subsequently supports the development of heritage resilience and strengthens the socio-cultural role.

Integrating urban risk and resilience in heritage areas focuses on government funds and regulation, professional assistance for renovation, and traditional authority. All is about governance factors in social, economic, hazard and disaster mitigation, and building aspects. According to the consecutive value, the owner/resident of a heritage building must increase the awareness of history. The history of building inheritance and family can build an understanding of safety, security, and comfort in the building, the land ownership, and the environment.

The Sendai framework in Figure 2 shows stages to achieve resilience in the multi-factors and multi-contexts. The findings are in accordance with the last stage of the framework, which is leadership-strategy. Governance factors must be prioritized in the preservation efforts of the heritage area. The preservation strategy must support the resilience establishment and reduce the impacts of urban risk.

Acknowledgment
We would like to thank Universitas Atma Jaya Yogyakarta and Universitas Gadjah Mada for funding the research. This research was supported by the community of Jeron Beteng Yogyakarta Palace as respondents and surveyors.

References
[1] International Federation of Red Cross 2011 *No Time for Doubt: Tackling Urban Risk* (Geneva: International Federation of Red Cross and Red Crescent Societies) www.ifrc.org, Accessed 20 January 2019 pp 12–13, 18–21
[2] Huck A, Monstadt J and Driessen P 2020 *Cities.* **98** 102573 pp 1–10
[3] International Federation of Red Cross and Red Crescent Societies 2017 *Building Urban Resilience: A Guide for Red Cross and Red Crescent Engagement and Contribution* (Geneva: International Federation of Red Cross and Red Crescent Societies) www.ifrc.org, Accessed 20 January 2019 pp 83–84
[4] Bizzotto M, Huseynova A and Estrada V V 2019 *Resilient Cities, Thriving Cities: The Evolution of Urban Resilience* (Bonn: ICLEI – Local Governments for Sustainability e.V.) p 8

[5] Larsen L N D, Howe P D, Brunson M, Yocom L, McAvoy D, Berry E H and Smith J W 2021 Risk Perceptions and Mitigation Behaviors of Residents following A Near-miss Wildfire *Landscape and Urban Planning* (Elsevier) **207** 104005 pp 1–11 10.1016/j.landurbplan.2020.104005, Accessed 4 February 2020
[6] Feng X, Xiu C, Bai L, Zhong Y and Wei Y 2020 Comprehensive Evaluation of Urban Resilience based on the Perspective of Landscape Pattern: A Case Study of Shenyang City *Cities* (Elsevier Ltd.) pp 1-15 doi.org/10.1016/j.cities.2020.102722, Accessed 20 March2021
[7] Romero-Lankao P, Hughes S, Qin H, Hardoy J, Rosas-Huerta A, Roxana Borquez R and Lampis A 2014 Scale, Urban Risk and Adaptation Capacity in Neighborhoods of Latin American Cities *Habitat International* (Elsevier) **42** pp 224–235 dx.doi.org/10.1016/j.habitatint.2013.12.008, Accessed 5 February 2019
[8] Castelluccio R 2018 Buildings as An Urban Risk *Techne* **15** pp 219–227
[9] Dickson E, Baker J L, Hoornweg D and Asmita Tiwari A 2012 *Urban Risk Assessments: Understanding Disaster and Climate Risk in Cities* (Washington: International Bank for Reconstruction and Development, The World Bank) p 41
[10] Samuelssona K, Colding J and Barthel S 2019 Urban Resilience at Eye Level: Spatial Analysis of Empirically Defined Experiential Landscapes *Landscape and Urban Planning* **187** pp 70–80
[11] Sharifi A 28 May 2019 Urban form resilience: A meso-scale analysis *Cities* **93** 2019 pp 238– 252
[12] Global Facility Disaster Reduction and Recovery (GFDRR) 2017 *Promoting Disaster-Resilient Cultural Heritage* (World Bank) p 2
[13] United Nations Educational, Scientific, and Cultural Organization (UNESCO) 2019 *Culture* en.unesco.org/fieldoffice/cairo/culture, Accessed 20 April 2020
[14] Maio R, Ferreira T M, Vicente R and Costa A 26 November 2018 Is the Use of Traditional Seismic Strengthening Strategies Economically Attractive in the Renovation of Urban Cultural Heritage Assets in Portugal? *Bulletin of Earthquake Engineering* pp 1–24
[15] Wu P-S, Hsieh C-M and Hsu M-F 2014 Using Heritage Risk Maps as an Approach to Estimating the Threat to Materials of Traditional Buildings in Tainan (Taiwan) *Journal of Cultural Heritage* (Elsevier Masson SAS) **15** pp 441–447 dx.doi.org/10.1016/j.culher.2013 10.005, Accessed 3 June 2018
[16] Zhang Y, Zhang H and Sun Z 16 May 2018 Effects of Urban Growth on Architectural Heritage: The Case of Buddhist Monasteries in the Qinghai-Tibet Plateau *Sustainability* **10** 1593 pp 1–18
[17] Satterthwaite D and Bartlett S 2017 Editorial: The full Spectrum of Risk in Urban Centres: Changing Perceptions, Changing Priorities *Environment & Urbanization* **29** 1 pp 3–14
[18] Heinzlefa C, Robert B, Hémond Y and Serre D 23 May 2020 Operating Urban Resilience Strategies to Face Climate Change and Associated Risks: Some Advances from Theory to Application in Canada and France *Cities* **104** 102762 pp 1–11
[19] Independent Evaluation Group (IEG) 2018 *The Urban Resilience Challenge* ieg.worldbankgroup.org/evaluations/urban-resilience, Accessed 25 January 2020
[20] United Nations International Strategy for Disaster Reduction 2012 *City Resilience in Africa: A 10 Essentials Pilot* (United Nations International Strategy for Disaster Reduction)
[21] Bergstrand K, Mayer B, Brumback B and Zhang Y 2015 Assessing the Relationship Between Social Vulnerability and Community Resilience to Hazards *Social Indicators Research* **122** pp 391–409
[22] Folke C December 2016 Resilience *Ecology and Society* **21** 4 p 30
[23] Ravankhah M, Schmidt M and Will T 2017 Multi-hazard Disaster Risk Identification for World Cultural Heritage Sites in Seismic Zones *Journal of Cultural Heritage Management and Sustainable Development* **7** 3 pp 272–289

[24] Labadi S, Giliberto, Rosetti I, Shetabi L and Yildirim E 2021 *Heritage and The Sustainable Goals: policy Guidance for Heritage and Development Actors* (Paris: International Council on Monuments and Sites – ICOMOS) p 135

[25] Lu H, Huang Y, Shi C and Yang X 2 April 2019 Exploring the Associations Between Urban Form and Neighborhood Vibrancy: A Case Study of Chengdu, China. *ISPRS International Journal of Geo-Information* **8** 165 pp 1–15

[26] UNDRR 2019 *Making Cities Resilient: My City is Getting Ready* www.unisdr.org/campaign/resilientcities/toolkit/article/the-ten-essentials-for-making-cities-resilient, Accessed 16 April 2020)

[27] City Resilience Index 2018 *City Resilience Index* www.cityresilienceindex.org/#/, Accessed 30 July 2020

[28] López P J M 2016 *Integrated Risk Assessment for Cultural Heritage Sites: A Holistic Support Tool for Decision-making* (Lucca: IMT School for Advanced Studies) p 21, 120

[29] Yu K K U, Oreta A W C, Ibabao R A and Hechanova N 2013 *Supporting Local Initiatives in Preserving Heritage Buildings in Iloilo City (Philippines) through Risk Assessment* (Bangkok: SEAMEO SPAFA with support from Japan Foundation) pp 1–19

[30] Espada Jr R, Apan A and McDougall K 19 February 2015 Vulnerability Assessment and Interdependency Analysis of Critical Infrastructures for Climate Adaptation and Flood Mitigation *International Journal of Disaster Resilience in the Built Environment* **6** 3 pp 313–346

[31] Flower B, Fortnam M, Kol L, Sasin P and Wood R G 2017 Using Participatory Methods to Uncover Interacting Urban Risks: A Case Study of Three Informal Settlements in Phnom Penh, Cambodia *Environment & Urbanization* (International Institute for Environment and Development (IIED)) **30** 1 pp 301–316 doi:10.1177/0956247817735481, Accessed 5 February 2019

[32] Collins A E 12 December 2018 Advancing the Disaster and Development Paradigm *International Journal of Disaster Risk Science* **9** pp 486–495

[33] Karaton Ngayogyakarta Hadiningrat 2017 *Tata Rakiting Wewangunan: Benteng Keraton Yogyakarta* www.kratonjogja.id/tata-rakiting-wewangunan/5/benteng-keraton-yogyakarta, Accessed 11 September 2019

IOP Conf. Series: Earth and Environmental Science 1233 (2023) 012072 doi:10.1088/1755-1315/1233/1/012072

Potential Carbon Footprint in Food System Based on Household Vegetable Consumption in Yogyakarta

A Musthofa[1], R Rijanta[2], D Widiyanto[2]

[1]Graduate School on Regional Development, Faculty of Geography, Universitas Gadjah Mada, Yogyakarta 55281, Indonesia
[2]Department of Development Geography, Faculty of Geography, Universitas Gadjah Mada, Yogyakarta 55281, Indonesia

azis.musthofa@mail.ugm.ac.id

Abstract. Consumption is one part of the food system in a region. Food is a basic need that cannot be separated from various activities. Food systems account for about a quarter of the global carbon footprint, indicating that every activity in it will impact the environment. This study aims to explain the potential carbon footprint resulting from the vegetable consumption commodities in Yogyakarta as part of the food system. The city of Yogyakarta became the research area with the consideration that this area is an urban area with a high population density, a high proportion of waste transported to the final landfill, and the composition of organic waste, which is more than half the total volume of waste. This study uses secondary data from the Socio-Economic Survey (Susenas) from the BPS-Statistics Indonesia and relevant literature. A quantitative descriptive analysis technique was used in this study to explain the vegetable food consumption of the people of Yogyakarta City. The potential for carbon footprint and food waste is calculated using emission factors that include production activities up to distribution. The results show that eggplant has the highest potential carbon footprint based on consumption data, with an average value of 0.25 kg CO2-eq/kg per household in one week. The average potential carbon footprint of vegetable consumption per household in one week in Yogyakarta City is 1.9 kg CO2-eq/kg produce. Potential food waste based on consumption activities has an average value of 0.15 kg per household in one week.

1. Introduction

The food system is a complex network of various activities in production, processing, transportation, and consumption [1,2]. The food system plays a role in the food security of a region. Each region has its characteristics. Physical conditions of the environment and culture will affect the food production of a region. Global and regional food trade have a big role in food distribution. The longer the distance will affect the length of the food system chain. In addition, the commodity type is also very influential in various processing, such as cooling, packaging, and various other processes.

Food systems contributed 19%-29% to anthropogenic global greenhouse emissions in 2008 [3]. This condition is caused by various factors such as agricultural production and land-use changes with regional variations. However, research on the impact of the food system on climate change still tends to be in the realm of agricultural production rather than discussing the food system as a whole. Some researchers have established a link between reduced distance travelled by food and lower energy consumption due to transportation. However, other researchers argue that the local food system also has poor energy consumption performance. These conditions are very dependent on the location [4].

The carbon footprint of food is carbon emissions or greenhouse gases resulting from growing, processing, transporting, storing, cooking, and consuming [5,6]. The carbon footprint is closely related to the food system. Every activity in the food system will create a carbon footprint resulting from energy consumption. Various processes and infrastructure are an integral part of the food system which becomes one from production and consumption activities.

Each commodity type of food has a varying amount of carbon emissions. Animal food products have a higher amount of emissions when compared to other commodities. However, it depends on how and where the food is produced. 25% of global emissions come from food. Based on several previous studies [7,8], dietary changes can significantly impact the environment.

Public access to food is closely related to the food system in an area. The food system is a complex network of activities involving production, processing, transportation, and consumption. Every activity in the food system plays a role in climate change. However, the research that has been carried out has explored more related to the agricultural sector [9]. The international food trade presents even greater challenges to climate change [10].

The city of Yogyakarta is an area with high cross-sectoral activities. Through the main market, it becomes the center of trading activity that connects various actors. The Giwangan Main Market is one of the markets that provide food for several areas at once. Traded vegetables are supplied to various locations that become producers around Yogyakarta [11]. As an urban area densely packed with tourism and education sectors, Yogyakarta requires high food supllies. In addition, the high intensity of activity in the food system will cause residues due to heigh energy consumption. Mitigation and efficiency of the food system's impact on climate change are not unlike policies in the energy sector [12]. Various issues such as food security, livelihoods, political interests, nutritional health, rural economic development, and biodiversity are why food system mitigation policies are far behind other sector policies.

The city of Yogyakarta is an urban area that functions as the center of regional activities. The urban characteristics with the main sectors in the form of services and trade, as well as high population density, will affect consumption patterns and food security. Urban communities will be more susceptible to food crises than rural communities, most of whom work as farmers. In addition, the potential carbon footprint resulting from the vegetable consumption commodities will also be different from that of rural communities. This study aims to explain the potential carbon footprint resulting from the vegetable consumption commodities in Yogyakarta as part of the food system.

2. Method

The city of Yogyakarta was the selection for the research area by considering the characteristics of the urban area (Figure 1). The area has a relatively high population with limited land availability. This condition makes urban areas have a higher proportion of built areas than protected areas. The dominant trade and service sector will impact the high demand and the resulting residue. The education and tourism sectors, the leading sectors in the city of Yogyakarta, also have a major impact on urban areas.In addition, the high potential for waste generation, the high proportion of waste transported to the Piyungan landfill, and the organic waste composition of more than 50% (Environmental Service, Yogyakarta City, 2020) were taken into consideration in the selection of Yogyakarta City as the research area.

The data used in the study were from micro data for the 2021 Socio-Economic Survey (Susenas) provided by Statistics Indonesia (BPS-Statistics Indonesia). 'Susenas' provides data on various social, economic, and living needs, such as clothing, food, housing, education, health, security, and job opportunities. The data used in this study focuses on consumption and expenditure. According to the data unit used, the household is the unit in this study. The number of household samples taken in the Susenas in Yogyakarta City is 722. Commodity consumption data used are 23 commodities.

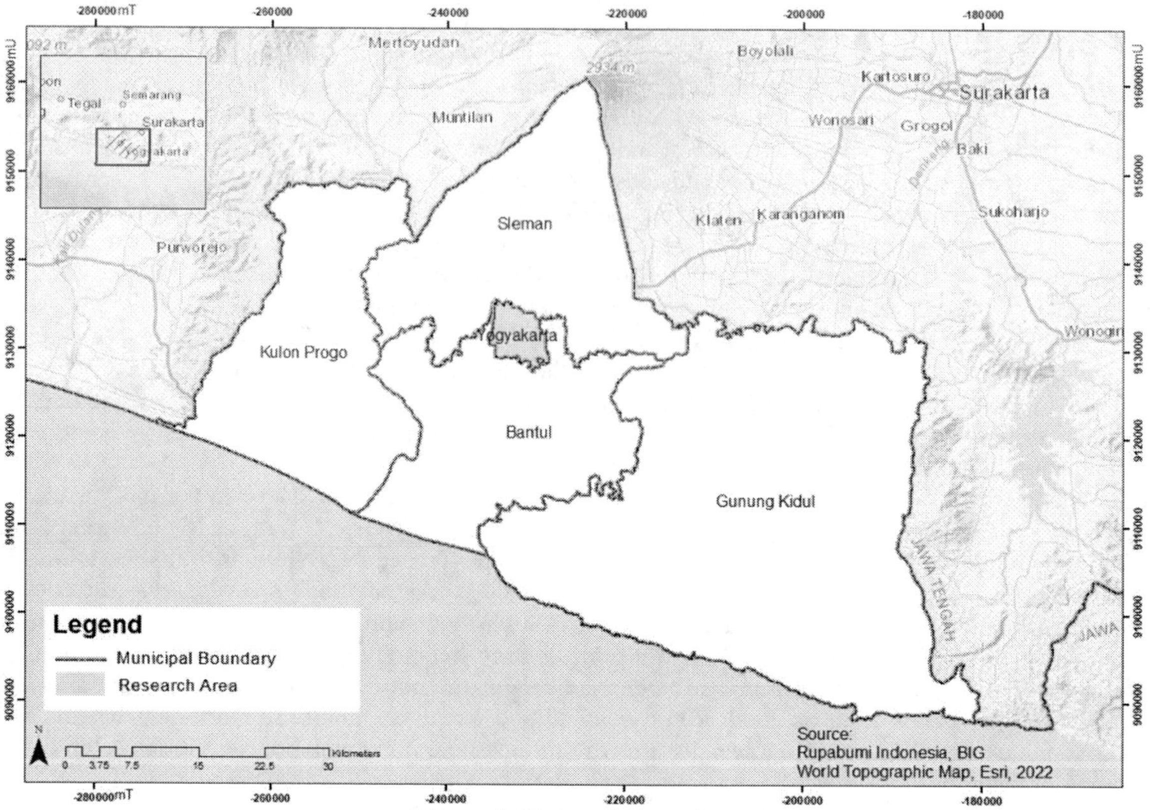

Figure 1. Research Location

Data analysis was done using descriptive analysis techniques by presenting consumption data using diagrams and graphs. The potential carbon footprint of vegetable consumption is calculated using the factor emission values for each vegetable commodity on a per capita scale [13,14]. The emission factor values are obtained based on a systematic literature study [15], represent the potential carbon footprint from the production process to distribution based on the type of food (Table 1). Although not all of the results of literature studies regarding the potential for carbon footprints or greenhouse gas emissions are most not carried out in Indonesia or Southeast Asia, the limited literature can be input for future research to measure the potential carbon footprints in Southeast Asian countries. To measure the potential for food waste in vegetable food consumption, it is calculated using the percentage by weight of food waste (Table 2) at the consumption stage [16].

$$\textit{Food emission} = \frac{(\textit{Total mass of food type (kg) x Emission factors for GHG } (CO_2e) \textit{ based on food type})}{\textit{Number of household}}$$

Table 1. Vegetable Factor Emission Value (kg CO2-eq/kg produce)

Vegetables	(kg CO2-eq/kg produce)
Spinach	0,54
Cabbage, brassica	0,23
Beans: green	0,51
Cowpea	0,48
Tomatoes	0,46
Carrot	0,22
Cucumber	0,33
Eggplant	1,35
Beansprout	0,62
Pumpkins	0,33
Papaya	0,34
Onion	0,18
Garlic	0,57
Other Vegetables	0,47
Other Fruits	0.5

$$\boldsymbol{Consumption\ waste} = \%\ \boldsymbol{weight\ consumption\ waste\ x\ consumption\ (kg)}$$

Table 2. Estimated/assumed waste percentages for fruits and vegetables group in step of food system for South and Southeast Asia

Food System	Fruits and Vegetables
Agricultural production	15%
Postharvest handling and storage	9%
Processing and packaging	25%
Distribution	10%
Consumption	7%

3. Results and Discussion

The residents' spending of Yogyakarta City based on BPS-Statistics Indonesia data in 2021 shows that the average spending percentage per capita in a month according to the goods group is predominantly by the non-food group with a 63.73% and the food group at 36.27%. The highest food expenditure was in the prepared food and beverage group, amounting to 17.02%. The vegetable group became the group with the second-highest percentage of expenditure, amounting to 2.62%, along with the grain group.

3.1. Household Vegetable Food Consumption

'Susenas' data divide consumption based on the source of origin, namely purchases and own production, gifts, and consumption of results other than buying. The total consumption of vegetable commodities in one week for the residents of Yogyakarta City shows that the percentage of vegetable consumption from purchases is 98%, and consumption from own production, gifts, and consumption results other than buying is 2%. The city of Yogyakarta, which is an urban area, has the primary sector in services and trade. This condition certainly impacts the population working in the secondary sector. Urbanization and the dominance of built-up land use in urban areas limit agricultural land.

The Giwangan Fruit and Vegetable Market in Yogyakarta City is a hub connecting smaller hierarchical markets and retailers. This market operates 24 hours and not only serves traders and the people of Yogyakarta City but also the surrounding area. Vegetable commodities entering the main market are mostly supplied from Central Jawa areas, such as Magelang, Wonosobo, and Boyolali. In contrast, fruit commodities are more diverse from various regions inside and outside Java Island.

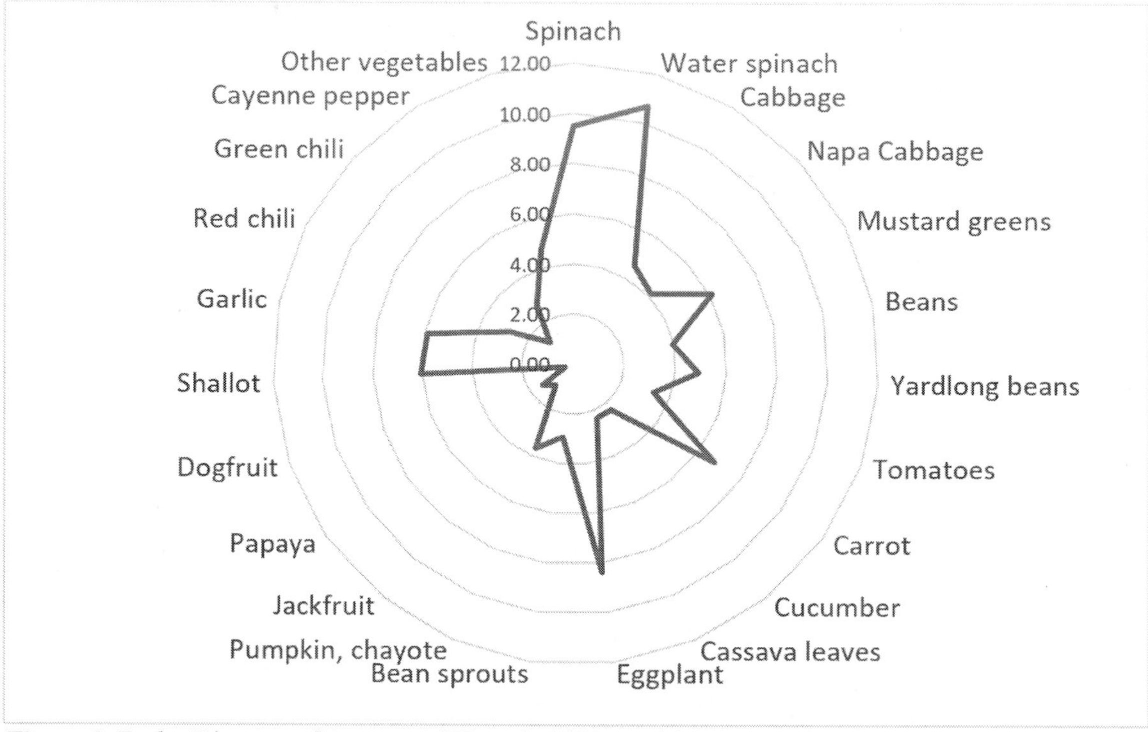

Figure 2. Radar Diagram of Amount of Household Vegetable Commodity Consumption in One Week in Yogyakarta City
Source: BPS-Statistics of Indonesia, 2021

There are 23 types of vegetable commodity categories consumed by residents of Yogyakarta City. In the Susenas data, there are commodities of young jackfruit, dog fruit, pumpkin/chayote, and papaya, which are included in the vegetable category. Residents of the City of Yogyakarta or Indonesia consume and process the fruit like a vegetable commodity. This study follows the categorization determined by BPS-Statistics Indonesia by including young jackfruit and papaya in the vegetable category. Water spinach is the commodity with the highest one-week consumption (Figure 2), reaching more than 10 kg for total consumption. In addition, spinach and eggplant commodities are consumed in large quantities of more than 8 kg in a week.

Dog fruit is the commodity with the lowest consumption. It is because the commodity has not yet been developed on a large scale. [17] and has only been developed on a household or individual scale. For other vegetable commodities consumed, it has been developed on a large scale in several highland areas in Central Java Province and distributed to Yogyakarta. Shallots, garlic, and chili are commodities that often experience price fluctuations based on availability. These commodities are included in the category of staple food commodities whose availability and price concern the government and other stakeholders.

3.2. Potential Carbon Footprint and Vegetable Food Waste

Potential carbon footprint or greenhouse gas emissions and food waste are measured based on the number of vegetable commodities consumed by households in Yogyakarta City. The amount of consumption is measured using Susenas microdata, which represents household consumption within a week. Based on the calculation method, the potential carbon footprint produced will represent the average carbon footprint value (kg CO2-eq/kg produce) based on household consumption in Yogyakarta City in one week. Food commodities will determine the value of the emission factor so that the value of the carbon footprint is very dependent on the type of commodity to be measured.

Food waste is measured by calculating the average consumption of vegetable food per household within a week. The total mass of the vegetable commodity will be multiplied by the estimated value or assumed percentage of food waste from population consumption activities. The factor value is based on a previous study, presenting a percentage value of 7% of the mass of vegetable and fruit food consumed specifically in the South Asian and Southeast Asian regions.

3.2.1. Potential Carbon Footprint.

The value of the potential carbon footprint based on vegetable food commodities per household in one week is presented in Figure 3. Based on the calculation results, garlic has the highest average potential carbon footprint with a value of more than 1 kg CO2-eq/kg produce per household. Eggplant commodity is the second commodity with the highest average carbon footprint. The number of commodities (kg) consumed and the factor values are proportional to the carbon footprint value produced by each commodity. The average carbon footprint value for others is below 1 kg CO2-eq/kg produce. However, some households have a carbon footprint value of more than 1 kg CO2-eq/kg produce, such as spinach, kale, cayenne pepper, and other vegetables.

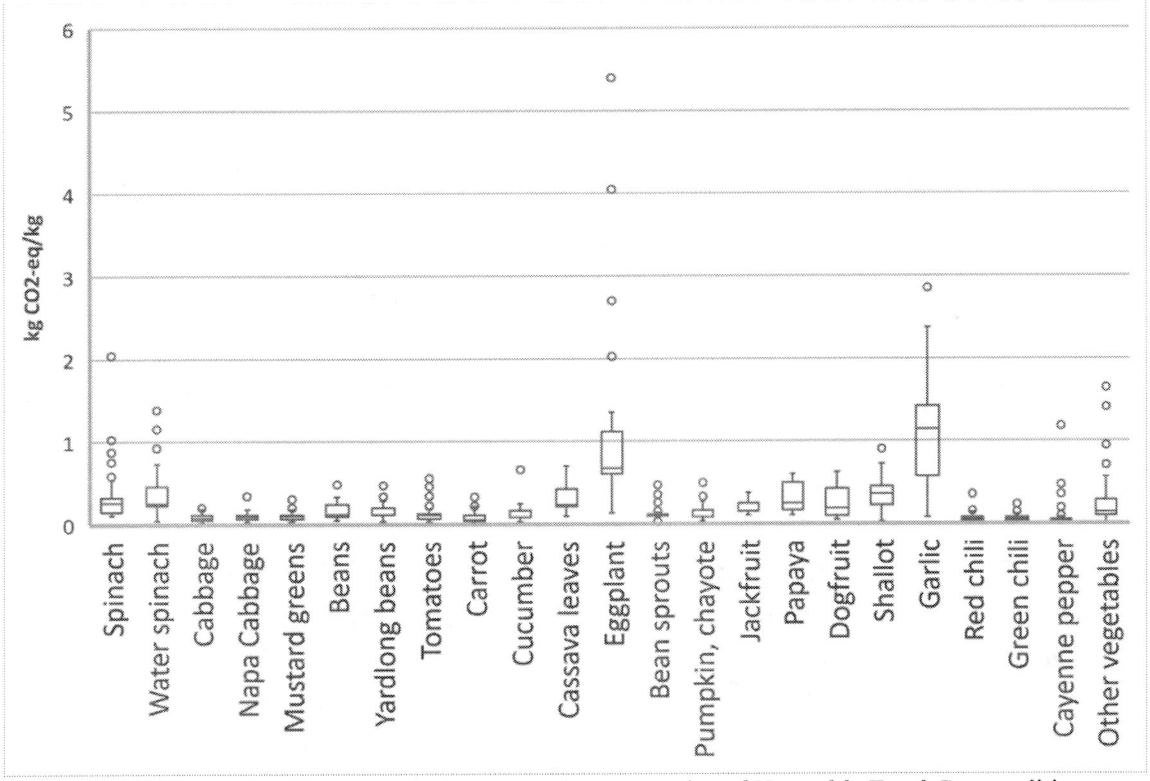

Figure 3. Box Plot Diagram of Potential Carbon Footprint of Vegetable Food Commodities per Household in One Week

The accumulated value of the carbon footprint produced by all household samples is presented in figure 4. Eggplant commodity has the highest carbon footprint value reaching 0.25 kg CO2-eq/kg produce. This value is very high when compared to other commodities. Referring to previous studies on the carbon footprint of eggplant commodities [18,19], eggplant commodities have a carbon value of 1.3 and 1.4 kg CO2-eq/kg produce, respectively. In the study, previous studies' average emission factor values were used as a reference for calculating the carbon footprint.. Another vegetable commodity with a high carbon footprint potential is spinach. Based on the studies [20], the average emission produced is 1.9 kg CO2-eq/kg produced per household.

Urban residents who work in the secondary sector or trade and services will be different from residents in rural areas. Rural communities, most of whom work as farmers, can consume their produce. When an economic crisis occurs, people in rural areas are not too affected. The city of Yogyakarta is the center of activity which cannot work alone in providing food. Although Yogyakarta city is one of the high food security categories along with other regions in the Special Region of Yogyakarta based on the Food Security and Vulnerability (FSVA) Map [21,22], rural urban linkages are an important factor for urban area. rural-urban linkages are a prominent factor for the urban area. Rural-urban linkages in food security see an area not limited to many administrative areas The relationship between the center and the hinterland is an interrelated system. If one of those components cannot work in the system well, it will impact other components Villages are as food producers from the primary agricultural sector, and cities are centers of activity with various economic activities that support processing activities as well as trade and services. There is a relationship between the center and the periphery. One of which is indicated by the food movement between villages and cities. The food movement can be the form of an interaction mechanism for surplus agricultural production from villages to cities.

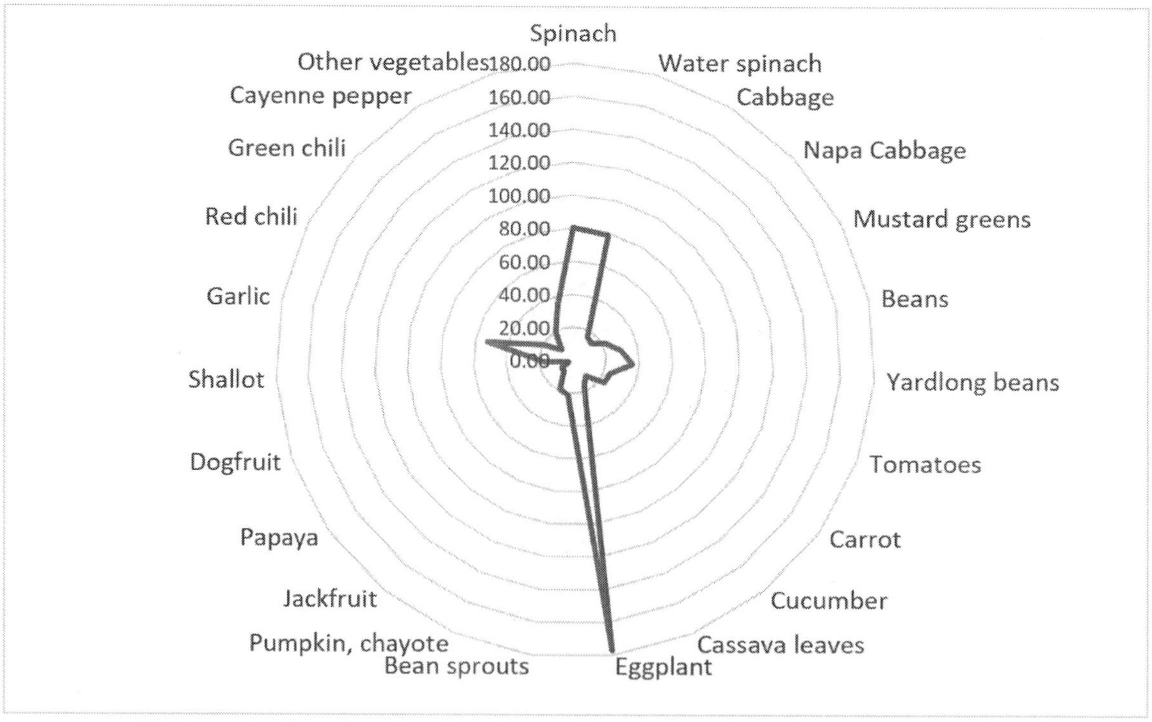

Figure 4. Radar Diagram of Potential Carbon Footprint Values by Commodity

3.2.2. Potential Waste of Food Consumption.

The calculation of the average potential household food waste at the consumption stage in Yogyakarta City uses an estimated assumption for Southeast Asian countries of 7% of the total consumption resulting in a value of 0.15 kg/household in a week. This value is generated based on data on the amount of consumption of vegetable commodities per household in one week. This amount will certainly depend on the number of residents in Yogyakarta. The potential for waste generation in Yogyakarta City based on the population has decreased, especially in 2021. The reduced population in the city of Yogyakarta is a benchmark for calculating the potential for waste generation, which also decreased from 361.96 tons/day in 2019 to 325.02 tons/day in 2021 (Figure 5). When referring to population data (BPS-Statistics Indonesia), the population in the city of Yogyakarta has always decreased. This phenomenon can be called suburbanization. The urbanization process occurs on the outskirts of urban areas can be caused by several attractions, such as universities, migration of people from suburban city centers and road corridors [23,24,25,26]. The population in urban areas has decreased and the surrounding areas have increased. This will have an impact on the formation of rural-urban areas and the increase in residential land as a result of the expansion of urban areas [27]. This condition also has an impact on the reduction of paddy fields that can produce food.

Land investment in urban areas makes the shift of urban settlements to the urban periphery. This situation makes the urbanization process on the outskirts of the city happen. Real estate development makes the city uncommercial and causes suburbanization. There are four areas in urban areas that need to be considered in relation to land investment, namely intra-city dynamics, peri-urban dynamics, the emergence of new cities and new infrastructure corridors [28]. Investments in urban areas such as various real estate, housing, and infrastructure projects will trigger urban renewal. The poor in urban areas will be the excluded party. The purchase of land will lead to gentrification and social segregation in urban areas.

These various conditions have an impact on reducing the amount of household waste produced. However, the development of various activity centers such as offices, education, and various supporting facilities will certainly produce a large amount of waste. Therefore, the reduced population does not necessarily reduce the amount of waste generated in the city of Yogyakarta.

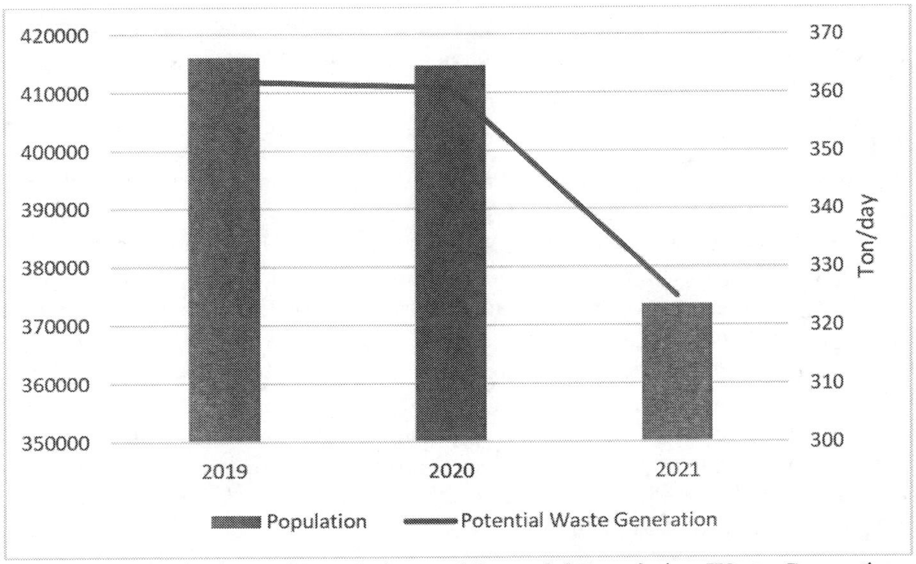

Figure 5. Diagram of Population and Potential Population Waste Generation
Source: Environmental Agency of Yogyakarta City, 2022

The percentage of waste types by type in Yogyakarta City is dominated by organic waste, reaching 53.3% in 2021 for residential waste (Yogyakarta City Environmental Service, 2022). Likewise, for non-residential waste, organic waste has the highest percentage with a value of 39.6%, and plastic has a percentage of 32.9%. Organic residential waste, of course, cannot be separated from food waste from consumption—both food waste and processing waste. Food losses are generated in the production, postharvest, and processing stages, while the final phases in the food system, namely sales and consumption, are called food waste [29]. In the food waste consumption stage, it is measured based on food waste generated from consumption on a household scale.

Mitigation and efficiency of the food system's impact on climate change are not like policy in the energy sector [30]. Various issues such as food security, livelihoods, political interests, nutritional health, rural economic development, and biodiversity are why food system mitigation policies are far behind other sector policies. Concerning regional and urban management, urban metabolism or urban metabolism discusses various systems that occur in an area. Food is one of the objects studied. Based on the previous explanation, activities in the food system significantly contribute to the carbon footprint and climate change. Resource efficient city is one idea that promotes sustainable production and consumption. At the city and regional scale, resource efficiency aims to improve the quality of life by minimizing resource extraction, energy consumption, waste generation, and ecosystem services. From a spatial perspective, Yogyakarta is an area whose needs are mostly met by the surrounding area. The agricultural area in Magelang Regency is a supplier of vegetable and fruit food needs for the Yogyakarta area. This food movement shows that the hinterland of Yogyakarta is outside its administrative area. The existence of Main Market is a commodity terminal that can be a means of distributing food from producers to consumers or producers to retailers. The higher the food supply that enters Yogyakarta, the higher the potential carbon footprint generated. Various innovations are needed in the production, packaging, distribution, and sale of food so that the energy released is more efficient and does not produce a large amount of residue or carbon footprint.

Previous research [31] conducted in 100 cities with a total of 132 food carbon footprints showed that the average carbon footprint in the cities reviewed was 2.1 tonnes of carbon dioxide per capita per year. This value is the result of calculations from a variety of foods. Meat and dairy are the main drivers of global warming. The results of the calculation of the carbon footprint of vegetable food in the city of Yogyakarta are limited to vegetable consumption in one week. Although the calculations are limited to vegetable commodities, this calculation is expected to be a consideration in avoiding food waste in urban areas. Research in different locations will show differences in the calculation results. This condition depends on how the rural and urban are connected in an area and how the food distribution flow moves.

4. Conclusion

The average potential carbon footprint of vegetable consumption per household in one week in Yogyakarta City is 1.9 kg CO2-eq/kg produce. The number of commodities (kg) and the type of commodity affect the potential value of the resulting carbon footprint. The potential for food waste in Yogyakarta City on average per household is 0.15 kg in one week. The population of Yogyakarta City, which is constantly decreasing, will impact reducing waste evocation from houses. However, overall, the evocation of non-household waste can increase along with the development of urbanization, investment, and areas with particular functions, such as tourism, offices, education, and various amenities.

The emission factor values used in the calculation of the study are sourced from studies that systematically review various sources of potential food carbon footprints in various countries. The unavailability of a complete study of food commodities in Southeast Asian countries is expected to be a research topic that can be developed in the future.

References

[1] Crippa, Monica; Solazzo, Efisio; Guizzardi, D.; Monforti, Fabio; Tubiello, Francesco & Leip, Adrian. (2021). Food systems are responsible for a third of global anthropogenic GHG emissions. *Nature Food. 2. 1-12. 10.1038/s43016-021-00225-9.*

[2] Francesco N Tubiello et al. (2021). *Environ. Res. Lett. 16 065007*

[3] Vermeulen, Sonja & Campbell, Bruce Morgan & Ingram, John. (2012). Climate Change and Food Systems. *Annual Review of Environment and Resources. 37. 195-222. 10.1146/annurev-environ-020411-130608.*

[4] Mundler, Patrick & Rumpus, Lucas. (2012). The energy efficiency of local food systems: A comparison between different modes of distribution. *Food Policy. 37. 609–615. 10.1016/j.foodpol.2012.07.006.*

[5] Garnett, Tara. (2011). Where are the best opportunities for reducing greenhouse gas emissions in the food system (including the food chain)?. *Food Policy. 36. 23-23. 10.1016/j.foodpol.2010.10.010.*

[6] Weber, Christopher & Matthews, H.. (2008). Food-Miles and The Relative Climate Impacts of Food Choices in the United States. *Environmental science & technology. 42. 3508-13. 10.1021/es702969f.*

[7] Kanyama, Annika & Nässén, Jonas & Benders, Rene. (2021). Shifting expenditure on food, holidays, and furnishings could lower greenhouse gas emissions by almost 40%. *Journal of Industrial Ecology. 10.1111/jiec.13176.*

[8] Vicente Vicente, José & Piorr, Annette. (2021). Can a shift to regional and organic diets reduce greenhouse gas emissions from the food system? A case study from Qatar. *Carbon Balance and Management. 16. 10.1186/s13021-020-00167-y.*

[9] Niles MT et al (2018). Climate change mitigation beyond agriculture: a review of food system opportunities and implications. *Renewable Agriculture and Food Systems 33, 297–308*

[10] Sandström, V., Valin, H. Lrisztin, T., Havlik, P., Herrero, M, d&Kastner, T. (2018). The role of trade in the greenhouse gas footprints of EU diets. *Global Food Security.*

[11] Maharani, N & Rijanta, R. (2018). Rantai Distribusi Sayuran Dalam Konteks Keterkaitan Desa Kota (Studi Kasus Wilayah Perdesaan Kabupaten Bantul Dan Kota Yogyakarta). *Jurnal Bumi Indonesia*

[12] Leahy, S.C., Clark, H., & Reisinger, A. (2020). Challenges and Prospects for Agricultural Greenhouse Gas Mitigation Pathways Consistent With the Paris Agreement. *Frontiers in Sustainable Food Systems.*

[13] Poore, J., and T. Nemecek. (2018). "Reducing Food's Environmental Impacts through Producers and Consumers." *Science 360* (6392): 987–92. https://doi.org/10.1126/science.aaq0216.

[14] Sari, D.; Rizki, M.; Nathania, B.; Ahmad, M.; Gunawam Gan, P.; Noor, N. (2021). Indonesia Zero Emissions Application (EMISI): Methodology for Calculating Individual Emissions from Food, Clothing, Electricity Consumption, and Solid Waste. *World Resour. Inst. 2021.*

[15] Clune S, Crossin E, Verghese K, Systematic review of greenhouse gas emissions for different fresh food categories, *Journal of Cleaner Production* (2016), doi: 10.1016/ j.jclepro.2016.04.082.

[16] Food and Agriculture Organization of The United Nations. (2011). *Global Food Losses and Food Waste. https://www.fao.org/fileadmin/user_upload/suistainability/pdf/Global_Food_Losses_and_Food_Waste.pdf*

[17] Maxiselly, Y., Ustari D., Ismail, A. & Karuniawan A. (2016). Pola Penyebaran Tanaman jengkol (Pithecellobium jiringa (Jack) Prain.) di Jawa Barat bagian selatan berdasarkan karakter morfologi. *Jurnal Kultivasi Vol. 15(1) Maret 2016. http://jurnal.unpad.ac.id/kultivasi/article/view/12007/5607*

[18] Audsley, E., M. Brander, J. Chatterton, D. Murphy-Bokern, C. Webster and A. Williams (2009).

How low can we go? An assessment of greenhouse gas emissions from the UK food system and the scope to reduce them by 2050. London: FCRN-WWF. Accessed: 17/02/2015 from: An assessment of greenhouse gas emissions from the UK food system end and the scope to reduce them by 2050 wwf.org.uk/downloads/how_low_report_1.pdf

[19] Stoessel, F., R. Juraske, S. Pfister and S. Hellweg (2012). "Life Cycle Inventory and Carbon and Water FoodPrint of Fruits and Vegetables: Application to a Swiss Retailer." *Environmental Science & Technology 46(6): 3253-3262.*

[20] Yoshikawa, N., K. Amano and K. Shimada (2008). Evaluation of Environmental Loads Related to Fruit and Vegetable Consumption Using the Hybrid LCA Method: Japanese Case Study. *Life Cycle Assessment VIII September 30-October 2, . Seattle, Washington.*

[21] BKP. (2020). *Panduan Penyusunan Peta Ketahanan dan Kerentanan Pangan (Food Security And Vulnerability Atlas/FSVA) Provinsi 2020*. Jakarta: Badan Ketahanan Pangan.

[22] Harini, Rika; Sukri, I; Ariani, R. D.; Faroh, E. P. I.; Nadia, H. & Kafafa, U. (2022). The Study of Food Security in the Special Region of Yogyakarta, Indonesia . *Forum Geografi.* DOI: 10.23917/forgeo.v35i2.15855

[23] Giyarsih, S. R. (1999). Mobilitas Penduduk Daerah Pinggiran Kota. *Majalah Geografi Indonesia 1999, XIII (2)*

[24] __________. (2017). Regional Management of Areas with Indications of Urban Sprawl in the Surrounding Areas of Universitas Muhammadiyah, Yogyakarta, Indonesia. *Indonesian Journal of Geography*

[25] ___________. (2001). Gejala Urban Sprawl Sebagai Pemicu Proses Densifikasi Permukiman di Daerah Pinggiran Kota (Urban Fringe Area) Kasus Pinggiran Kota Yogyakarta. *Jurnal of Regional and City Planning.*

[26] Rachmawati R., Rijanta, R. & Subanu, L. P. (2004). Peranan Kampus Sebagai Pemicu Urbanisasi Spasial Di Pinggiran Kota Yogyakarta. *Majalah Geografi Indonesia Vol 18, No. 1.*

[27] Bencardino, M. & Nestico, A. (2017). Demographic Changes and Real Estate Values. A Quantitative Model for Analyzing the Urban-Rural Linkage. *MDPI*

[28] Zoomers, Annelies &van Noorloos, Femke & Otsuki, Kei & Steel, Griet & van Westen, August. (2017). The Rush for Land in an Urbanizing World: From Land Grabbing Toward Developing Safe, Resilient, and Sustainable Cities and Landscapes. *World Development. 92. 10.1016/j.worlddev.2016.11.016.*

[29] Parfitt, J., Barthel, M. & Macnaughton, S. (2010). Food waste within food supply chains: quantification and potential for change to 2050, *Phil. Trans. R. Soc., vol. 365, pp. 3065-3081*

[30] Leahy, S.C., Clark, H., & Reisinger, A. (2020). Challenges and Prospects for Agricultural Greenhouse Gas Mitigation Pathways Consistent With the Paris Agreement. *Frontiers in Sustainable Food Systems.*

[31] Goldstein, Benjamin; Birkved, Morten; Fernández, John & Hauschild, Michael. (2016). Surveying the Environmental Footprint of Urban Food Consumption. *Journal of Industrial Ecology. 21. n/a-n/a. 10.1111/jiec.12384.*

Spatial Study of Population Pressure on the Carrying Capacity of Rice Field Agricultural Land in Bantul Regency

***Tri* Utami Setyawati**[1]

[1]Department of Environmental Geography, Faculty of Geography, Universitas Gadjah Mada, Yogyakarta, Indonesia

Corresponding author: triutami2018@mail.ugm.ac.id

Abstract. Bantul Regency has experienced high regional development and population growth. In the north, it is directly adjacent to the city of Yogyakarta, experiencing more rapid regional development, and has the potential to exceed the carrying capacity of agricultural land, especially rice fields. Based on this problem, this study's objectives are to determine the spatial variation of the carrying capacity of agricultural land and population pressure on agricultural land in the Bantul Regency. This research uses a quantitative descriptive method and secondary data from the Central Bureau of Statistics (2014-2019). The value of the carrying capacity of rice farming in the Bantul Regency is divided into Class I and Class III. Districts with a class I land carrying capacity can be self-sufficient in food and have no population pressure on agricultural land in eight sub-districts. On the other hand, there are seven sub-districts in class III that have not been able to be self-sufficient in food and have experienced population pressure on agricultural land, especially on the outskirts of Yogyakarta, including Banguntapan, Sewon, and Kasihan sub-districts. The value of the carrying capacity of agricultural land can affect the ability to be self-sufficient in rice food in Bantul Regency.

1. Introduction

Indonesia is an agrarian country where most of the population does agricultural work, either as farm laborers, small farmers, or working in agro-industrial plantations [1]. However, along with the increasing population growth in Indonesia, while the nature of the limited supply of land and land is a non-renewable natural resource, it creates intense land-use competition, which usually occurs between the agricultural and non-agricultural sectors. Another factor is that people think being a farmer is not their main livelihood because their income is uncertain, and they cannot meet their daily needs. Hence, people switch professions by opening shops and other businesses on agricultural land. This causes the conversion of agricultural land, which significantly reduces the availability of farmland and threatens the availability of food [2].

Bantul Regency experienced an increase in population from 968632 people in 2014 to 1022788 people in 2019. Along with the rise in population, the development that continues to be carried out in Bantul Regency cannot be avoided from changes in land use, especially agricultural land to non-agricultural. The area of lowland agricultural land from 2014 to 2019 decreased by 272.1 Ha, followed by a decrease in rice harvested land from 30160 Ha in 2014 to 28969 Ha in 2019 [3]. The use of agricultural land that is used for residential areas and economic businesses outside of agriculture has an impact on several aspects, one of which is the financial aspect where there is a decline in agricultural production (rice) which is a staple food so that it becomes one of the causes of the threat of food availability, especially rice in the area [4]. In addition, the increasingly rapid physical growth triggers an increase in land prices where land is considered to have a high commercial value making it attractive for landowners to sell their land, especially agricultural land.

Economically, agricultural land such as rice fields has a high selling price if the land is strategically located in the development area. Accessibility to infrastructure, geographical position, and the availability of resources are also the pull factors for residents outside the region to enter Bantul Regency both for commercial activities and to settle in the area concerned [5]. The existence of population pressure, while the availability of land resources is limited, can increase land conversion for settlement and carry the risk of growing food needs.

High population pressure followed by changes in land use, especially rice fields which cannot be avoided, has an impact on decreasing harvested land area and productivity of paddy fields. This causes the land carrying capacity threshold to be exceeded, so a spatial study of population pressure on the carrying capacity of agricultural land is needed to determine the distribution of areas experiencing population pressure and places with low land carrying capacity in Bantul Regency. The carrying capacity of agricultural land is the capacity or ability of the land to meet the community's food needs in an area. Determination of the carrying capacity value can be used as a basis for policies and evaluation of existing land use [6]. The analysis of the land's carrying capacity can provide input for the region in realizing food security and offer benefits to the population in the area. This research was conducted with a spatial approach to determine variations in population pressure on agricultural land and the carrying capacity of agricultural land in each sub-district in Bantul Regency.

2. Method

This research was conducted in Bantul Regency, Special Region of Yogyakarta Province, which consists of seventeen sub-districts, most of which experienced a reduction in paddy fields from 2014 to 2019. The research location is based on the conversion of paddy fields because the need for space for housing has increased, and developments such as housing and infrastructure trigger regional development. The analytical method used is descriptive quantitative, namely by performing calculations using mathematical formulas, then the results obtained are described and presented in the form of graphs, pictures, and tables.

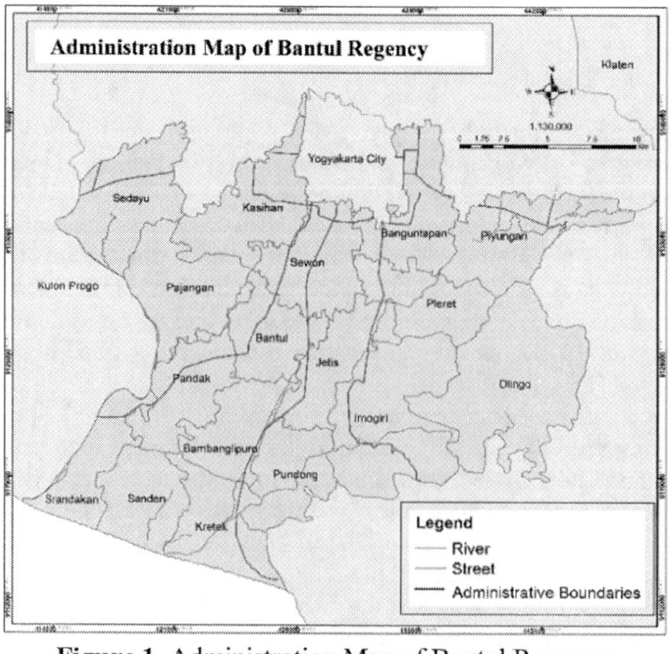

Figure 1. Administration Map of Bantul Regency

IOP Conf. Series: Earth and Environmental Science 1233 (2023) 012073 doi:10.1088/1755-1315/1233/1/012073

2.1 Data Analysis

This study uses secondary data indirectly published by the Central Bureau of Statistics and the Department of Agriculture Agency. The analysis of the carrying capacity of agricultural land uses the variables of population, harvested area, the average production of paddy fields per hectare, and Minimum Physical Needs (MPN) for each sub-district to determine the spatial distribution of the carrying capacity of agricultural land and population pressure in Bantul Regency.

2.1.1 Land Use Change. The average area of change in agricultural land use over a certain period is calculated using the formula for the rate of reduction of agricultural land to support the results of the analysis of the carrying capacity of agricultural land and population pressure on paddy fields. The formula used based on the concept of Yunus 2001 [7] to calculate the rate of change of agricultural land is as follows:

$$L = \frac{L1 - L2}{d}$$

Description:

L/year = Degradation rate of agricultural land/year (ha/year)
L_1 = Initial year area of agricultural land (ha)
L_2 = Final year agricultural land area (ha)
d = Time difference between the initial year and final year

2.1.2 Carrying Capacity. The carrying capacity of agricultural land (σ) in this study compares the area paddy rice harvesting area with a land area used for food self-sufficiency (rice). The space can be self-sufficient in food and meet the minimum physical needs of its population of 265 kg of rice/person/year, while a place that can provide a decent standard of living for the population is 650 kg of rice/person/year. The formula for calculating the carrying capacity of agricultural land using Odum et al. 1975 [8] concept.

$$\sigma = \frac{Lp/Pd}{MPN/Pr}$$

Description:

σ = Carrying capacity of paddy fields
Lp = Area of rice paddy harvested land (ha)
Pd = Total population (people)
MPN = Minimum physical requirement (kg/capita/year)
Pr = Average production of lowland agricultural land per hectare (kg/ha)

Based on the results of the calculation of the carrying capacity of the land, it is classified into:
$\sigma > 1$ means the area has a high carrying capacity of agricultural land and can be self-sufficient in food and the population in the area is below the optimal population.
$\sigma = 1$ means the carrying capacity of agricultural land in the optimal area, or one is the threshold for the land's carrying capacity.
$\sigma < 1$ means that the carrying capacity of agricultural land is low where the area has not been able to be self-sufficient in food, and the population in the area is above the optimal population.

2.1.3 Population Pressure. Calculation of population pressure uses the calculation results of the carrying capacity of agricultural land where the value of population pressure is inversely proportional to the value of the carrying capacity of agricultural land. The assumption is that the higher the value of population pressure, the lower the carrying capacity of agricultural land, which shows the ability of paddy fields to meet the population's food needs. The calculation formula used in calculating population pressure, namely:

$$\text{Population Pressure} = \frac{1}{\sigma}$$

Based on the calculation of population pressure, the assumption is
PP > 1 indicates a high population pressure so that it exceeds the land capacity limit
PP = 1 means moderate pressure population where optimal land use to land capability
PP < 1 indicates population pressure is still low where land capability is still not being utilized

3. Result and Discussion

3.1 Rate Change of Rice Fields

The City of Yogyakarta, as an urban center, has a fairly high population density which has resulted in the expansion of urban areas around the area, including those directly adjacent to the Bantul Regency. That is supported by the direction of the development of the northern part of the urban area of Yogyakarta City by building trade, education, and housing areas as well as other economic activities so that it leads to suburban areas where administratively the site is Bantul Regency and Sleman Regency [9]. The city's expansion was caused by several forces, one of which was physical strength, where the increase in population was the main factor. The development of the city causes the need for land to increase while the land area is limited so that it encourages the conversion of agricultural land to non-agricultural and threats that promote the protection of sustainable food agricultural land.

Table 1. The conversion rate of paddy fields in Bantul Regency in 2014-2019

District	Harvest Area		Lp/th
	2014	2019	
Srandakan	463	476	2.17
Sanden	988	970.4	-2.93
Kretek	888	915.1	4.52
Pundong	807	838.8	5.30
Bambanglipuro	1129	1046.1	-13.82
Pandak	957	868.5	-14.75
Bantul	1003	940	-10.50
Jetis	1127	1184.9	9.65
Imogiri	1108	1429.7	53.62
Dlingo	903	730	-28.83
Pleret	694	724	5.00
Piyungan	1209	1084	-20.83
Banguntapan	998	873.1	-20.82
Sewon	1177	1150	-4.50
Kasihan	563	543.1	-3.32
Pajangan	273	301.9	4.82
Sedayu	904	843.3	-10.12

The rate of conversion of paddy fields in Bantul Regency over the last six years can be seen in Table 1, where most paddy fields have decreased. The rate of land conversion has a negative value indicating that there is a conversion of paddy fields. In contrast, a positive value indicates an increase in the area of paddy fields. Rice field agricultural land with the highest rate of decline from 2014 to 2019 was in Dlingo District at a rate of 28.83 Ha/year, followed by Piyungan District at 20.83 Ha/year. Piyungan District, based on the RTRW, is classified as an industrial zone, so that the development of the industrial sector, as well as the housing sector, is quite rapid it encourages the conversion of rice fields to non-agricultural land [10].

Due to the shift of paddy rice commodities to other commodities, this happened in the Dlingo District, where there was the conversion of paddy fields into cocoa plantations due to the hilly topography with climatic conditions, temperature, soil, and altitude areas that were suitable for planting cocoa plants compared to rice plants. The existence a program carried out by the Bantul Regency Agriculture Service established Dlingo District as a cocoa center covering Terong Village, Mangunan Village, and Muntuk Village [11].

The conversion of paddy fields is also massively carried out in suburban areas in the Yogyakarta Urban Agglomeration area. Banguntapan District has decreased by 125 hectares, Sewon District has 27 hectares, and Kasihan District has an area of 20 hectares from 2014-2019. The site is prone to converting agricultural land where development towards the city's outskirts is growing due to limited land, so land use dynamics occur in suburban areas. The population's motivation to migrate to the outskirts of the city of Yogyakarta is because Bantul Regency has a location close to workplaces and education centers with a relatively affordable land value for housing, so migrants are interested in living in the area [12].

3.2 The Carrying Capacity of Agricultural Land

Based on Figure 2, the harvested area and average productivity of rice experience changes that tend to rise and fall every year. Bantul Regency had the largest rice harvest area in 2014, which reached 30160 Ha, while the smallest harvested area occurred in 2019, which decreased to 28611 Ha. The decrease in harvested area is one of the impacts of development that has implications for land conversion. The largest rice productivity was in 2015, reaching 6722 Kw/Ha; in 2019, it touched the smallest number, 6353 Kw/Ha.

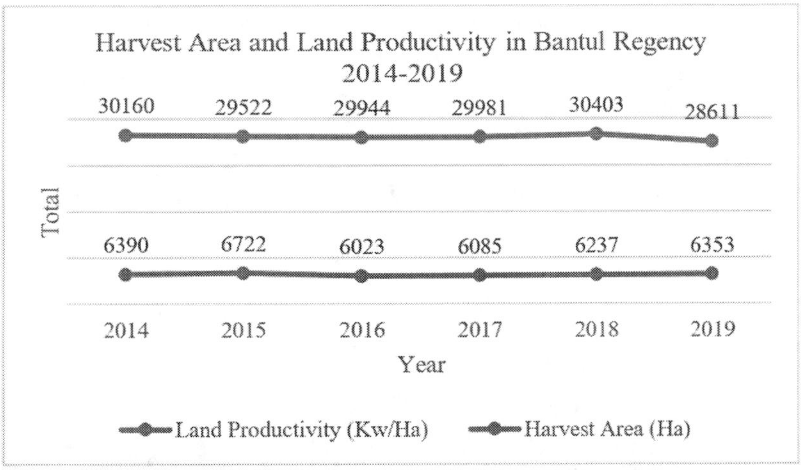

Figure 2. Harvest Area and Land Productivity in Bantul Regency 2014-2019

Changes also followed changes in rice productivity in the condition of total production in Bantul Regency, where in 2014 it was 192711 tons to 181767 tons or decreased by 5,6 percent. Pajangan District has the smallest total value of rice production among other sub-districts, which was 2530 tons in 2014 and continues to increase to 4304 tons in 2019. That is because Pajangan District has a relatively small area of rice fields and harvested land compared to the site of Pajangan in other districts. This condition can be influenced by topographic conditions where Pajangan District has a hilly topography of moderate relief so that the use of land in the form of rice fields requires extra effort for optimal results in terms of technical, cost, and labor.

On the other hand, Jetis Subdistrict, with a sloping topography, had the highest total rice production in 2014 of 23306 tons but decreased until 2019 to 21459 tons. That is because the area has a fairly large area of rice fields, so it affects the harvest area, which is quite high, supported by other factors such as irrigation systems, fertilizers, rainfall, and farmer experience. Agricultural production is not only influenced by the area of rice fields but also by rainy days, precipitation, harvested area, and fertilization [13]. In addition, rice production is also influenced by the variables of seeds, pesticides, phonska, and labor [14].

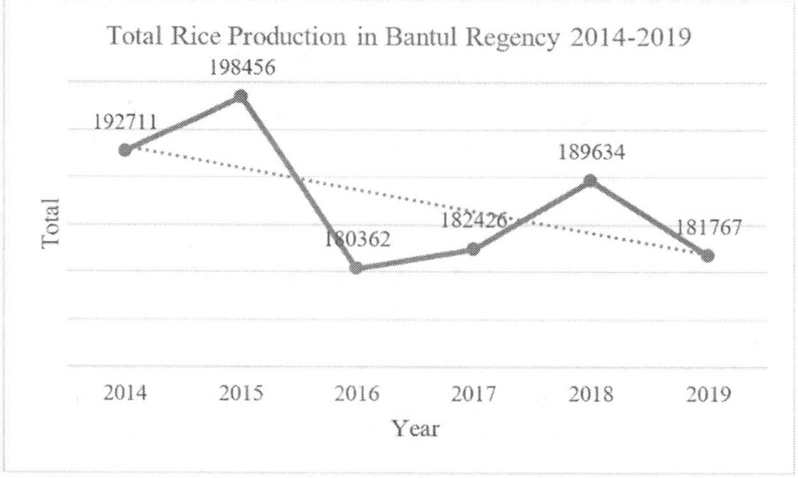

Figure 3. Total Rice Production in Bantul Regency 2014-2019

Land's carrying capacity indicates the ability to support human life, not only as agricultural land but also as land to live in or support other socio-economic activities [15]. The analysis of the carrying capacity of paddy fields in Bantul Regency carried out spatially-temporal can be shown from the results of calculations which are then classified based on the class of land carrying capacity in each sub-district from 2014 to 2019. The classification of rice fields' carrying capacity results shows that the sub-districts in Bantul Regency are in classes I and III.

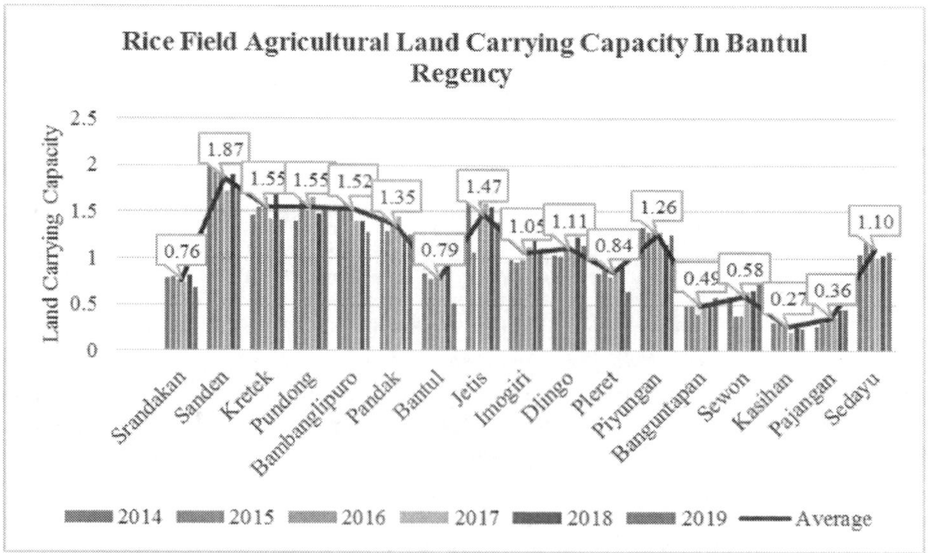

Figure 4. Agricultural Land Carrying Capacity In Bantul Regency

The highest land carrying capacity value was found in Sanden District at 1.99 in 2014 and decreased to 1.75 in 2019. Meanwhile, Imogiri District experienced a change in land carrying capacity where from 2014 to 2016, it was in class III with a value of 0.98 changed to class I with a value of 1.12 in 2017 and 1.07 in 2018. An increase in harvested area caused this change as a result of one of the agribusiness programs, namely increasing productivity, quality, and production of marketable agricultural commodities.

On the other hand, the sub-districts with the lowest carrying capacity of paddy fields in class III are located on the outskirts of Yogyakarta with an average value of 0.27 to 0.58, including Banguntapan

District and Sewon District, with the lowest score of 0.39 in 2016 and Kasihan District 0.20 in 2017. That shows that the area has not been able to be self-sufficient in rice food for its population or has exceeded the optimal population because the range of carrying capacity of rice fields is below one. Dominant factors can influence low land carrying capacity in urban areas to allow an increase in the rate of population growth [16].

The low carrying capacity of lowland agricultural land in class III from 2014 to 2019 is also found in areas with hilly topography. Pleret District had the lowest carrying capacity value of 0.64 in 2019 and Pajangan District 0.27 in 2014. Pajangan District has the lowest carrying capacity value because it has a topography dominated by structural hills of the Sentolo formation, composed of limestone and marl sandstone, so land use is dominated by plantations [17]. The condition of the administrative area and low paddy fields is one of the factors for the quiet harvest area and rice production followed by the increase in population, making Srandakan Subdistrict have a low carrying capacity of paddy fields. The distribution map of the land carrying capacity class can also be seen in Figure 5.

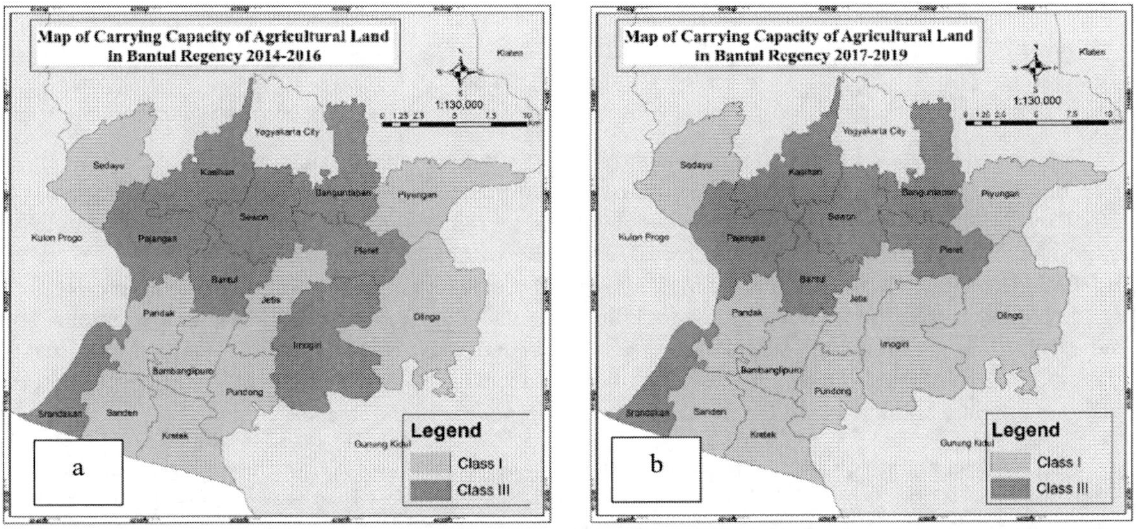

Figure 5. Map of carrying capacity of agricultural land in Bantul Regency **(a)** carrying capacity of rice fields in 2014-2016 **(b)** carrying capacity of rice fields in 2017-2019

3.3 *Population Pressure*

Based on Table 2 shows that from 2014 to 2019, each sub-district in Bantul Regency had varying values. The highest population pressure value of more than one is found in the suburbs of Yogyakarta City, namely Banguntapan District, Sewon District, and Kasihan District. That shows that the sub-district is experiencing population pressure, so the carrying capacity of paddy fields is exceeded. The characteristics of DIY, which has the predicate as a City of Students, City of Culture, and others, impact the high number of in-migration in this city so that the demand for land for housing, investment, industry, and tourism is getting higher. The population growth also leads to the city's outskirts, especially the Bantul Regency, which impacts the widening of urban activities towards the periphery and urges agricultural activities.

Table 2. Population Pressure on Agricultural Land in Bantul Regency 2014-2019

District	2014	2015	2016	2017	2018	2019	Population Pressure
Srandakan	1.28	1.26	1.31	1.37	1.24	1.46	PP > 1
Sanden	0.50	0.52	0.52	0.58	0.52	0.57	PP < 1

IOP Conf. Series: Earth and Environmental Science 1233 (2023) 012073 doi:10.1088/1755-1315/1233/1/012073

Kretek	0.69	0.64	0.60	0.71	0.55	0.71	PP < 1
Pundong	0.71	0.58	0.66	0.61	0.68	0.66	PP < 1
Bambanglipuro	0.66	0.50	0.64	0.72	0.72	0.78	PP < 1
Pandak	0.69	0.78	0.70	0.69	0.79	0.80	PP < 1
Bantul	1.19	1.30	1.24	1.25	1.02	1.94	PP > 1
Jetis	0.61	0.94	0.62	0.63	0.64	0.72	PP < 1
Imogiri	1.02	1.05	1.02	0.89	0.83	0.92	PP < 1
Dlingo	0.97	0.98	0.90	0.90	0.82	0.88	PP < 1
Pleret	1.19	1.01	1.25	1.12	1.11	1.57	PP > 1
Piyungan	0.75	0.79	0.78	0.78	0.87	0.98	PP < 1
Banguntapan	2.05	2.05	2.55	2.05	2.00	1.75	PP > 1
Sewon	1.69	2.57	2.57	1.54	1.52	1.20	PP > 1
Kasihan	3.30	3.33	3.40	4.97	4.00	4.09	PP > 1
Pajangan	3.66	3.24	3.07	2.66	2.23	2.22	PP > 1
Sedayu	0.95	0.77	0.88	0.99	0.97	0.94	PP < 1

Population pressure also occurs in Srandakan and Pajangan sub-districts, which have relatively low rice fields. That is because the Srandakan District experienced an increase in population from 2014 to 2019 by 2196 people with an area of 476 hectares of rice fields. Meanwhile, Pajangan Subdistrict experienced an increase in population by 1072 people, with a total area of 401.9 hectares of rice fields in 2019. The small rice fields and the high number of residents led to population pressure on rice fields, so the land was less able to support the population's needs for agricultural land. Rice in the region. Bantul sub-district is experiencing population pressure because it is the district's capital, so massive development is carried out where there has been a decrease in the area of rice fields by 63 hectares accompanied by an increase in the population of 2405 people in the last six years.

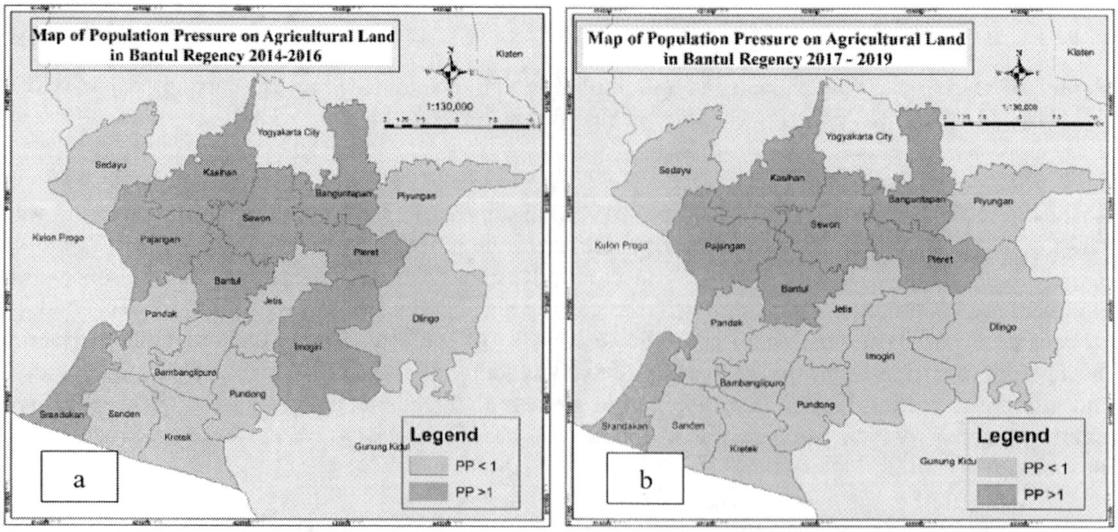

Figure 6. Map of population pressure on agricultural land in Bantul Regency **(a)** population pressure on paddy fields in 2014-2016 **(b)** population pressure on paddy fields in 2017-2019

On the other hand, several sub-districts that did not experience population pressure were found in Sanden District, Kretek District, Bambanglipuro District, Pundong District, Pandak District, Jetis District, Dlingo District, Piyungan District, and Sedayu District. That shows that the carrying capacity of paddy fields is still high because the area of paddy fields is relatively large so that it can support the needs of the population in the region despite the threat of massive conversion of agricultural land into non-agricultural land. Unlike the case with Imogiri District, which experienced changes in conditions where the value of population pressure from 2014 to 2016 experienced population pressure so that it exceeded the land capacity threshold. This condition changed the following year, from 2017 to 2019. The Imogiri District experienced a decrease in value, so there was no population pressure where agricultural land could still support the existing population. This was caused by an increase in the area of harvested land which in 2016 was 1734 Ha to 2055 Ha in 2019 so that paddy field agricultural production also increased.

4. Conclusion

The carrying capacity of paddy fields in each sub-district of Bantul Regency during 2014-2019 fluctuated and qualified in Class I, where the region could be self-sufficient in food and not experience population pressure. In class III, the area had not been able to be self-sufficient in food or experienced population pressure on land. Paddy farming, especially in the suburbs of Yogyakarta, which includes Kasihan, Banguntapan, and Sewon sub-districts, and other sub-districts have Bantul, Pajangan, and Sandakan sub-districts.

Acknowledgment

The author expresses his deepest gratitude to RTA-UGM 2022, which has funded this research. Dr. Rika Harini, S.Si, M.P as the thesis supervisor. And also really appreciate reviewers for suggestions to make this article better.

References

[1] Syuaib, M. F. Anthropometric study of farm workers on Java Island, Indonesia, andits implications for the design of farm tools and equipment. *Applied Ergonomics*, *51*, 222–235, (2015)

[2] Rondhi, M., Pratiwi, PA, Handini, VT, Sunartomo, AF, & Budiman, SA. Agricultural land conversion, land economic value, and sustainable agriculture: A case study in East Java, Indonesia. *Land*, *7*(4). (2018)

[3] Badan Pusat Statistik. *Kabupaten Bantul dalam Angka 2015*. Yogyakarta: Badan Pusat Statistik Kabupaten Bantul. (2019)

[4] M Chrisna Satriagasa, Rika Harini, Rina Dwi Ariani, dan S. Analisis Luas Lahan Pertanian terhadap Porduksi Padi di Kalimantan Utara. *Jurnal Kawistara*, *5415*, 15–27. (2019)

[5] Ruhimat, M. Tekanan Penduduk Terhadap Lahan Di Kecamatan Sukaraja Kabupaten Sukabumi. *Jurnal Geografi Gea*, *15*(2), 59–65. (2016)

[6] Setyati, S., Yusran, F. H., & Priatmadi, B. J. The Carrying Capacity Analysis of Land for Food Agriculture : Case Study in Banjar Regency , South Kalimantan. *International Journal of Research and Review*, *8*(March), 460–466. (2021)

[7] Yunus, H. S. Struktur Tata Ruang Kota. Yogyakarta: Fakultas Geografi, Universitas Gadjah Mada. (2000)

[8] Muta'ali, L. Daya Dukung Lingkungan untuk Perencanaan Pengembangan Wilayah. Yogyakarta: Universitas Gadjah Mada. (2012)

[9] Faizah, A. N., & Hendarto, M. Analisis difusi keruangan di sekitar kawasan perkotaan

yogyakarta. *Diponegoro Journal of Economics*, *2*(3), 1–9. (2013)

[10] Fadholi, H., Syamsiar, S., & Kismantoroadji, T. Dampak Konversi Lahan Pertanian Ke Nonpertanian Terhadap Sosial Ekonomi Masyarakat Di Desa Sitimulyo, Piyungan, Bantul. *Jurnal Dinamika Sosial Ekonomi*, *16*(2010), 15–26. (2015)

[11] Rizki, Yusnanda. Respon Petanian Terhadap Alih Fungsi Lahan Sawah Ke Perkebunanan Kakao di Desa Terong Kecamatan Dlingo Kabupaten Bantul. Skripsi. Universitas Muhammadyah Surakarta. (2020)

[12] Pradoto, W. Pola Pemanfaatan Lahan dan Faktor- faktor Perkembangan Wilayah Perkotaan di Kabupaten Sleman dan Kabupaten Bantul. *Conference on urban studies and development*, 207–220. (2015)

[13] Riang Enjelita Ndruru, Marihat Situmorang, G. T. Analisa Faktor-Faktor yang Mempengaruhi Hasil Produksi Padi di Deli Serdang. *Saintia Matematika*, *2*(1), 71–83.(2014)

[14] Onibala, A. G., Sondakh, M. L., Kaunang, R., & Mandei, J. Analisis Faktor-Faktor Yang Mempengaruhi Produksi Padi Sawah Di Kelurahan Koya, Kecamatan Tondano Selatan. *Agri-Sosioekonomi*, *13*(2A), 237. (2017)

[15] Putri, R. F., Naufal, M., Nandini, M., Dwiputra, D. S., Wibirama, S., & Sumantyo, J. T. S. The Impact of Population Pressure on Agricultural Land towards Food Sufficiency (Case in West Kalimantan Province, Indonesia). *IOP Conference Series: Earth and Environmental Science*, *256*(1). (2019)

[16] Prihatin, R. B. Alih Fungsi Lahan Di Perkotaan (Studi Kasus Di Kota Bandung Dan Yogyakarta). *Jurnal Aspirasi*, *6*(2), 105–118. (2016)

[17] Hertanto, I. R. *Pengembangan pariwisata berdasarkan ekosistem budaya di kecamatan pajangan kabupaten bantul*. *XI*(1), 1671–1686. (2012)

Typology of planned housing in the city of Yogyakarta

D Widiyasuti, S Y Adhitama, B O Ali, and B Aprillia

Faculty of Geography, Universitas Gadjah Mada, Yogyakarta, Indonesia

Email: s.yudha@mail.ugm.ac.id

Abstract. This paper examined the planned housing in Yogyakarta City distributed in eight sub-districts, namely Umbulharjo, Tegalrejo, Wirobrajan, Mergangsan, Mantrijeron, Kotagede, Gondokusuman, and Jetis. Planned housing in Yogyakarta City can be classified into a gated community and a non-gated community. They were commonly identified based on the similarity and regularity of the buildings and the road network. The data was taken from their Building Permit (IMB; Izin Mendirikan Bangunan) from the year 2000 – 2019, Google Street View, Pleiades Satellite Imagery of Yogyakarta City in 2018, Google Map, and ArcMap base-map. Data analysis was undertaken by using satellite imagery, map interpretation, and quantitative analysis by using scoring technique to describe the score of typology. Most of the planned housing in Yogyakarta City are classified into Typology II, which had a medium-strength gated community characteristic. The typology was composed based on the physical-geographical and social-economic indicators. Most of them were more than 10 years old and were provided with walls/fences and portals to restrict outsiders' access. The residents were classified into middle-class based on their social-economic status. Most of the planned housing in Yogyakarta have gated-community characteristics based on these features.

Keywords: planned housing, gated community, physical-geographical, social-economic, typology

1. Introduction

Housing is one of the basic needs of human life [1]. Housing segments are formed based on the income group of the population, location, and housing provision (formal and informal). The housing segment shows the targeted type of residents to secure the housing planning. The ideal housing type must be equipped with various facilities and infrastructure [2]. These facilities consist of public facilities and social facilities offered by the developer. It is essential to consider these following aspects: 1) The dimension of the neighborhood road, especially for two intersecting cars; 2) the feasibility of drainage gutters according to the criteria to reduce the stagnation of water and rainwater; 3) street lighting; 4) water facilities; 5) clean water; 6) telephone network, and; 7) shops or markets to fulfill daily life needs. Some social facilities also need consideration, including places of worship, sports fields, parks, areas open to the public, security posts, and meeting halls. The developers must be aware of security, health, comfort, compatibility aspects (beauty/harmony/regularity), flexibility, and affordability of distance and environment when deciding on the housing development location [3]. The ease of access, communication, and activities lead to comfort life.

Urban development will always increase, especially in terms of population and the rate of urbanization [4]. The availability of more permanent settlements and a dense population emphasize the characteristics of a city [5]. Various changes will appear, including new living environments formation, the rise of settlements, and the livelihood transformation from agricultural-based into non-agricultural-based. It proves that the development of cities will also influence the development of the settlements. Population development, housing needs for settlements, and limited land are parts of a dynamic nature, so it is necessary to observe the changes from time to time. A holistic plan is a better choice of anticipation by making a collaboration between the government, society, and academics [6].

The settlement development can lead to settlement transition patterns. The design of new housing often causes changes in the physical spatial arrangement because it does not pay attention to the existence of the environment. The imbalance of spatial changes and environmental design will create spatial segregation. It forms barriers to urban space and may cause inequalities [7].

The choice of settlement transition will result in planned and informal settlements, showing the origin of settlements formation [8]. Planned settlements include a collective center (mass shelter) and a planned camp. Meanwhile, informal settlements include host families, rural self-settlement, urban self-settlement, and self-settled camps. The types of settlements are related to the people's choice of living and movement patterns. Housing, as one of the planned settlements, is easily found in most cities [9]. Some of the characteristics that emerge are the regularity of physical form, including the road-facing direction, the building made from permanent materials, provided with walls, and equipped with electric lighting. Regularity is also shown by the arrangement of the road network, from the neighborhood-scale roads, local roads, to other larger road networks.

Housing in Yogyakarta City is divided into gated community and non-gated community housing. Gated community housing is a housing type with physical boundaries of walls or fences with one-gate access. It is equipped with public facilities, resulting in low social interaction between residents and the surrounding community.[10]. Residents live in gated community housing with the consideration of security factors, ease of access, availability of facilities, and economic affordability. [11]. The presence of impacts of gated community housing in suburban areas causes changes and challenges for housing residents. The residents may experience lifestyle, perspective, and home-style changes that reflect their social status and level. This gives the impression that there are apparent limitations and differences between the residents and the outsider community [12]. Nowadays, the rise of micro-scale gated community happened in many cities in Indonesia, e.g., Yogyakarta, Semarang, and Makassar with specific size and social interaction in dense area [13]–[15]. It leads the adaptation which is different with its origin concepts of gated community which is needed to be studied.

This research aimed to describe the characteristics of the planned housing in Yogyakarta City from a physical-geographical and social-economic context. This also aimed to compose the typology of planned housing in Yogyakarta City along with the rising trends of gated communities. This typology ought to be studied to understand both physical and social-economic characteristics correctly. A suitable database and planning may be developed from this research by considering the specific characteristics of planned housing in Yogyakarta City.

2. Methods

This research was conducted in the city of Yogyakarta, one of the medium-sized cities in Indonesia. This location was selected due to its rapid urban growth within the area of Special Region of Yogyakarta. Yogyakarta has a significant role as the center of services and economic development in this province, allowing people to come, live, and work. The primary driver of land-use changes and urban expansion is the choices of households in meeting their housing needs [16]. Residential development is one of the main drivers of land-use change. Rapid urbanization is characterized by unconsolidated lateral expansion of large cities that changed small towns and villages on mostly-agricultural lands. The cheap lands for housing development lead the urbanization process [17]. Rural land, primarily agricultural land, is the source of commercial housing. Rural lands usually have low costs for land acquisition [18]. Yogyakarta City has become more of an urban region since 1990. High accessibility of the major roads in Yogyakarta leads to the transformation of rural areas into urban areas. [19]. Every year, about 40 ha of farming land have been converted into housing in Yogyakarta City. It was marked by the small and dense housing of the non-agricultural population [20]. These facts indicated rapid urban growth in Yogyakarta City, triggering massive settlement spread, including planned housing type.

This research used secondary data from building development permits, also known as *Izin Mendirikan Bangunan* (IMB) for Housing in Yogyakarta City. The IMB data spans from the year 2000 to 2019, including the name of the housing area, location, and the year of construction. The data validation used Google Maps and housing profiles that can show the characteristics of the housing. In addition, spatial data processing used Yogyakarta City shapefiles obtained from Ina-Geoportal developed by the Indonesian Geospatial Information Agency (BIG; Badan Informasi Geospasial), Pleiades satellite imagery of Yogyakarta City in 2018, and ArcMap Basemap. The output was a map of the planned housing distribution in the city of Yogyakarta. This map and the database were essentially required for discussing better planning and management of planned housing development.

2.1. Population and Samples

The population of this research were the entire planned housing located in Yogyakarta City. The total number of this population was 102. This research used purposive sampling and proportionate stratified random sampling to obtain the samples. Some gated community features appear in the planned housing were used as criterias for purposive sampling, including the secondary data availability of each housing name. The samples were taken from housing with clear location and profile data.

There were 102 planned housing in Yogyakarta City, which were distributed in eight sub-districts, namely Umbulharjo, Tegalrejo, Wirobrajan, Mergangsan, Mantrijeron, Kotagede, Gondokusuman, and Jetis. Umbulharjo, the largest sub-district, had the most planned housing in Yogyakarta City. In addition, this sub-districts was a suburban area, so the planned housing was distributed along the periphery of Yogyakarta City. The proportionate stratified random sampling was used to maintan the ratio of sub-district representation within the samples so more samples were in Umbulharjo sub-district, followed by Tegalrejo,Wirobrajan, Mantrijeron, Kotagede, Gondokusuman, Mergangsan, and Jetis.

The total collected samples amounted to 32 housing areas from these eight sub-districts. These minimum samples were reasonable due to the limitations caused by the COVID-19 pandemic. Therefore, such limitation restricted this study to conduct a field survey to check the features of those housing areas. As a result, the use of secondary data was optimized.

2.2. Data Processing

The data was processed using ArcMap and Microsoft Excel. The ArcMap software processed the spatial data to create the necessary maps. Meanwhile, Microsoft Excel was employed to create a database and data visualization in the form of graphs and diagrams. The homogeneity illustrated the planned housing based on shape and size of the buildings and the road network layout. Data validation used IMB data, Google Maps, and Google Street View to prove the gated community characteristics tendency. Based on those data sources, the scoring process would be undertaken by using Microsoft Excel. Each parameters had specific score and weight to be calculated for composing typology.

2.3. Data Analysis

Data analysis was undertaken by using satellite imagery interpretation and quantitative analysis. The interpretation was carried out by identifying the shape, size, pattern, color, hue, and shade according to the characteristics of gated community housing. Quantitative analysis was used to display the characteristics of planned housing. Both physical-geographic and social-economic characteristics from a sample of planned housing in Yogyakarta City were described, respectively.

3. Results and Discussion

The planned housing in Yogyakarta City was identified using five features: homogeneity of rooftops, homogeneity of house size, regular agglomerated pattern, availability of walls/fences, and availability of a portal. All of the features can be interpreted based on satellite imagery. The keys of interpretation

IOP Conf. Series: Earth and Environmental Science 1233 (2023) 012074 doi:10.1088/1755-1315/1233/1/012074

lie on the form, size, pattern, and shadow [21]. The homogeneous rooftops and house size indicated that more than 90% are planned housing. They also had a regular pattern of agglomeration, which could be distinguished from the other buildings around them. Most of the planned housing have walls/fences and portals as borders. The availability of walls/fences and portals are a part of gated communities' characteristics to limit accessibility in planned housing.

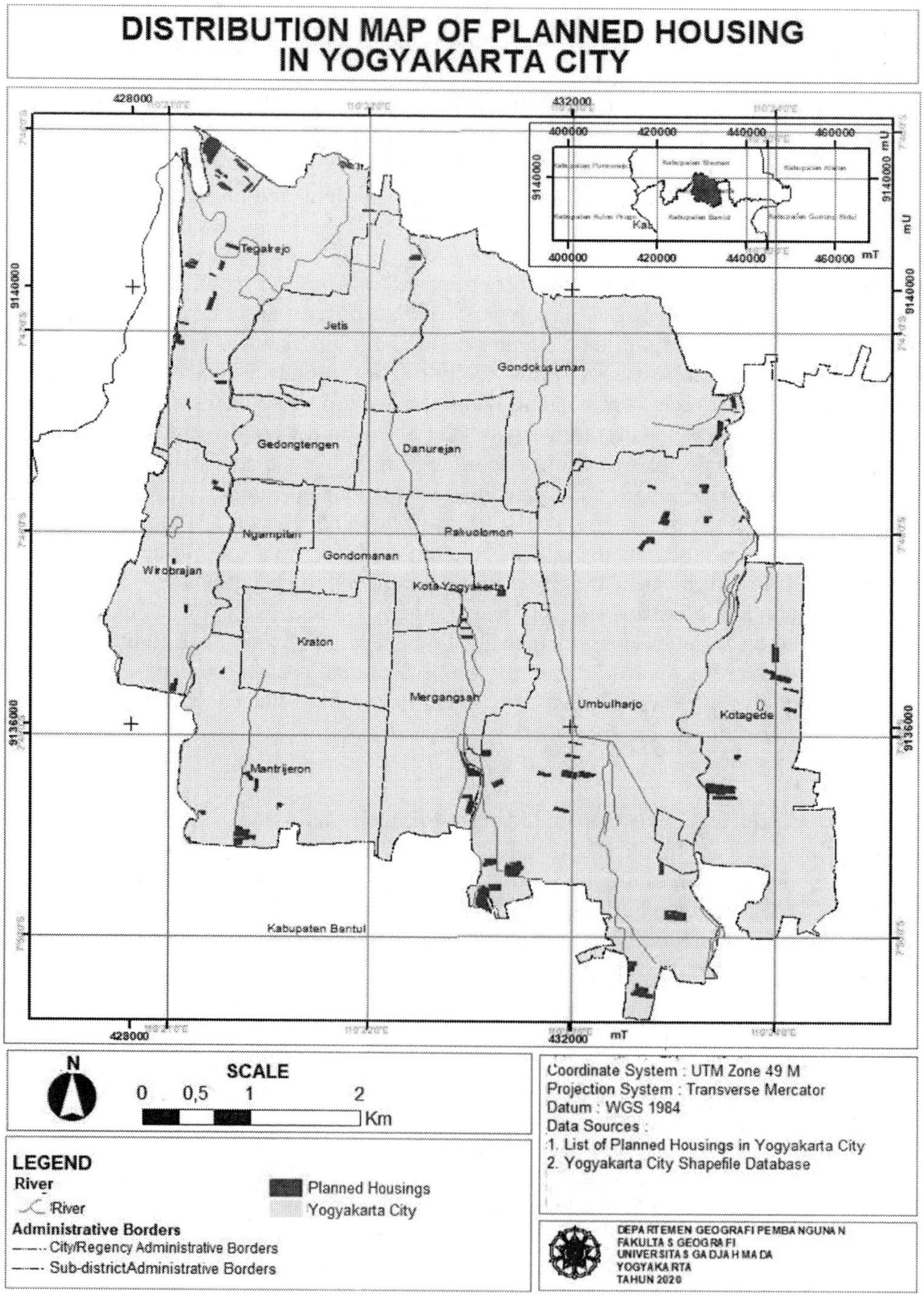

Figure 1. Map of Planned Housing in Yogyakarta City (processed from shapefile, 2021)

IOP Conf. Series: Earth and Environmental Science 1233 (2023) 012074 doi:10.1088/1755-1315/1233/1/012074

3.1. Characteristics of Planned Housing in Yogyakarta City

The characteristics of planned housing in Yogyakarta City show the tendency of settlement segregation. They can be interpreted based on the physical-geographic aspects and social-economic aspects. The physical-geographic characteristics were determined by the physical properties of housing, including estimated covered area, estimated total of houses, and access-restriction features. The other variables are the distance to geographical objects and public facilities around the housing. River and city center are some examples of geographical objects. It is essential to show the rivers since Yogyakarta City is crossed by the Code River, Gajahwong River, and Winongo River. Meanwhile, public facilities consisted of administrative, education, and social-economic facilities.

About 75% of planned housing areas in Yogyakarta City were more than ten years old, calculated as per 2020. Most of them were constructed in 2000–2010, so the planned housing growth in Yogyakarta City is not significant nowadays. It was assumed that limited availability of land and the dense area of Yogyakarta City contribute to the decreased growth. On the other hand, the spread of new planned housing takes place outside of Yogyakarta City, e.g., Sleman Regency, Bantul Regency, and Kulon Progo Regency, resulting from limited available areas for development in this city.

3.1.1. Physical-geographic characteristic. The majority of housing in Yogyakarta City covered less than one hectare area with its percentage reaching more than 84% of the total population. It can be argued that only about 3% of the population covers more than a two-hectare area. Housing size in the urban area needs concern to reduce urban sprawl [22]. The small size of housing can be understood to deal with urban sprawl tendency and limited land availability. This fact was supported by a smaller number of houses within the area. About 50% of planned housing in Yogyakarta City consisted of fewer than 30 houses. According to the Regulation of Community Housing Ministry Number 7 the Year 2013, housing must consist of at least 15 houses. It was common to find small clusters of planned housing in Yogyakarta City. The transformation of public space privatization appeared considerably in clusters, depicting the small-scale gated community. The small scale shows more gated residential enclaves, creating exclusivity from gates, walls, fences, or portals [23].

The majority of housing was completed with access-restriction features to secure the area by controlling everyone who enters the settlement area. These features show a strong tendency for more private and secure areas indicating gated community characteristics. About 97% of planned housing was occupied with portals and name plates to signify the housing area. It was accompanied by the presence of walls/fences and security posts, covering 87% of the population. Those restriction features may result in settlement segregation. However, the availability of CCTV covered only about 3% population due to their exclusivity and costly maintenance. Due to these physical characteristics, the market segment of planned housing in Yogyakarta City was likely targeted to small families to accommodate middle-class urban people. Most housing are close to many public facilities, located less than one kilometer from government offices, schools, rivers, local roads, markets, hospitals, and police offices. Theoretically, housing must consider components of affordability, legal security of tenure, habitability, and accessibility [24]. The availability of reachable facilities represents the consideration of accessibility.

3.1.2. Social-economic characteristic. This characteristic was represented by the size of houses and the availability of carport areas. Car is a standard transportation mode used to accommodate resident's mobility. However, the economic value of car ownership was higher than other common modes, like bicycles and motorcycles, so it represents high economic status. This research showed that the planned housing in Yogyakarta City provides a carport to each housing unit. The size of houses showed different results. About 56% of the population had medium-sized houses with 75–100 square meters in size. This unit small-medium in size was adapted with the limited size of the whole area of planned housings. Since this size was not too small, it still can be said that the houses were typical for middle-class residents. About 60% of the planned housing consisted of two-floor levels, resembling middle-class settlements. Both unit sizes, two-floor house design, and the carport availability suggest the combined

IOP Conf. Series: Earth and Environmental Science 1233 (2023) 012074 doi:10.1088/1755-1315/1233/1/012074

calculation of middle-class housing. Overall, this variable of middle-class housing covered about 68.75% of the total population. The growing upper-middle social-economy class has accelerated the spread of gated communities in several cities in Indonesia, including Jakarta and Yogyakarta [22]. This class commonly requested more secured and private settlements to be lived.

3.2. Typology of Planned Housing in Yogyakarta City

The planned housing in Yogyakarta City shared similar criteria with Secure Suburban Estates from Burke typology. This type is usually found in suburban areas with small houses [23]. However, they were not accompanied by proper internal facilities, so the residents still depended on outside public facilities to accommodate their needs of social and economic aspects. Therefore, the planned housing typology in Yogyakarta City is not exactly like the existing typology, so a local typology may be required to be devised to provide a more resembled explanation. The typology can be composed by considering housing profile, including area coverage, total units, or the type of units [24].

The typology of planned housing in Yogyakarta City was made using the identified physical-geographic and social-economic characteristics. However, the variables of distance radius to geographical objects and public facilities were eliminated since they shared variations too slightly to be composed for a typology. On the other hand, the social-economic characteristic was only represented by housing class since this variable resembled the combination of unit size, carport availability, and the two-floor design variables. Therefore, the variables were scored and weighted to compose the typology.

Table 1. The variables of composed typology and scoring method

Code	Indicators	Class		Score	Weight	
HAC	Housing Area Coverage	Small	< 1 hectare	10	15%	Housing area needs to adapt with limited land availability.
		Medium	1–2 hectare	20		
		Large	> 2 hectares	30		
NHU	Number of Housing Units	Small	< 30 units	10	15%	The number of house units needs to adapt with the existing housing areas and land availability.
		Medium	30–50 units	20		
		Large	> 50 units	30		
SGC	Strength of Gated Community by Access-Limitation Features	Weak	Have 1-2 Features	10	40%	Access limitation features act as central features of gated community tendency in planned housing.
		Medium	Have 3 Features	20		
		Strong	Have 4-5 Features	30		
HC	Housing Class by Social-Economy Class	Low	Low Class	10	30%	Housing classes characterize the residents' social-economic class.
		Medium	Medium Class	20		
		High	High Class	30		

Data processed from interpretation

The strength of gated community based on access restriction features has the highest weight value due to its principal characteristic of gated-community housing. Meanwhile, the area coverage and number of units shared the same value because they need to adapt with limited land availability. The calculation results were divided into three classes, namely Typology I (strong gated community features), Typology II (medium gated community features), and Typology III (weak gated community features). Each class describes the planned housing typology by the characteristics of a gated community. Scores were marked into 32 samples of planned housing in Yogyakarta City.

The scoring process showed that 69% of planned housing in Yogyakarta City is classified into the Typology II. Those strong gated community features were respectively indicated by secured and restricted access of the middle class of housing, resulting in exclusivity for their residents. However, the lowest score shown by the Typology III indicates that only a few of the planned housing did not share the gated community features. Thus, overall, most of the planned housing in Yogyakarta City share a similar characteristic of a secure and private area like a gated community. It may lead to the rise of inequality with surrounding area since there are tendency of exclusiveness [25], [27].

Figure 2. Typology 1; Kuantan Regency, Sorogenen (Google Street View, 2019)

Figure 3. Typology II; Agatama Regency, Giwangan (Google Street View, 2018)

Figure 4. Typology III; Griya Suryo Asri 2 (Google Street View, 2018)

4. Conclusion

The planned housing in Yogyakarta City is mostly constructed more than ten years ago with the typical characteristics of a gated community. Most of them cover less than one-hectare area with not more than 30 house units. The planned housing is indicated by the availability of walls/fences, portals, security guard post, name plates, and CCTV. Most planned housing are close to public facilities, like government offices, local roads, schools, places of worship, markets, hospitals, and police offices. The distances between the housing and those facilities are less than one kilometer. Each house in the planned housing regularly covers about 75–100 square meters including the carport. The residents living in the planned housing may develop low-social interaction as well as other types of urban settlement.

Most of the planned housing in Yogyakarta City belong to the Typology II. They are characterized by medium strength of gated community features. It is supported by few interactions between each resident and their surrounding environment. It can be said that most of the residents are in the middle class based on their economic and social status as urban people in Yogyakarta City. According to this finding, a specific strategy is required to deal with the common trend of gated-community tendency in the planned housing management in Yogyakarta City. This strategy must rely on the physical and social aspects of the rapid urban development and dense area. For further study, the effect of gated community trends in planned housing in Yogyakarta City can be analyzed since different characteristics of typology.

5. Acknowledgement

We would like to express our appreciation to the Faculty of Geography, Universitas Gadjah Mada, particularly for the Department of Development Geography, for encouraging this research publication. Furthermore, we appreciated the valuable supports from the Faculty of Geography, Universitas Gadjah Mada, during the planning, development, and funding of research. Thus, may this research enrich the geographical perspective of settlement development in urban areas, particularly the gated community.

References

[1] Komaruddin, 1997 *Menelusuri Pembangunan Perumahan dan Permukiman* Jakarta: Yayasan REI -Rakasindo.

[2] Asteriani F, 2011 Preferensi Penghuni Perumahan di Kota Pekanbaru dalam Menentukan Lokasi Perumahan *Jurnal Jurnal Ekonomi Pembangunan: Kajian Masalah Ekonomi dan Pembangunan* **12**, 1.

[3] Nasution A M, 2019 Analisis Permasalahan Perumahan dan Permukiman di Kota Medan *(Journal of Architecture and Urbanism Research* **3**, 1.

[4] Daldjoeni N, 1987 *Geografi kota dan desa* Bandung: Alumni.

[5] Payton N I and Taylor L, 1992 Housing: Symbol Structure Site *Journal of Architectural Education (1984-)* **46**, 2.

[6] Makarau vicky H, 2012 Penduduk, Perumahan Pemukiman Perkotaan dan Pendekatan Kebijakan *Jurnal Perencanaan Wilayah Kota* **3**, 1.

[7] Sofiyah Radliyatullah F and Dwisusanto Y B, 2020 Integrasi spasial dalam perancangan batas fisik perumahan kota, Studi kasus: Perumahan Setra Duta, Bandung, Jawa Barat *ARTEKS : Jurnal Teknik Arsitektur* **5**, 2.

[8] Corsellis T and Vitale A, 2005, Transitional settlement: displaced populations.

[9] Koestoer R H, 1997 *Perspektif lingkungan desa-kota : teori dan kasus* Jakarta: UI-Press.

[10] Osman W W Arifin M M. A T and Triahandayani A, 2018 Interaksi Sosial Gated Community Terhadap Masyarakat Sekitarnya p. A140–A147.

[11] Giawa F C R Silitonga S and Pakpahan R, Jun. 2018 Kajian Faktor Penarik (Pull Factor) Penghuni Bertempat Tinggal di Komunitas Berpagar (Gated Community) Studi Kasus Kecamatan Medan Selayang *ALUR :Jurnal Arsitektur* **1**, 1 p. 14–18.

[12] Supriadi E, 2021 Konstruksi Gated Community Perubahan dan Tantangan Masyarakat Perumahan (Studi di Perumahan BSB, Mijen Kota Semarang) **15**, 1 p. 107–128.

[13] Aulia D N and Suryani L, 2020 Gated community typology based on growth and development in Medan City, Indonesia in *IOP Conference Series: Earth and Environmental Science* **452**, 1.
[14] Susanti R Widjajanti R Wungo G L and Budiarti I, May 2020 Social Relationship Between Kampong Gendong Residents and Gated/Non-Gated Community in Sendangmulyo Village Tembalang District, Semarang *TATALOKA* **22**, 2 p. 261–270.
[15] Arifin M Rachman Rasyid A and Wahidah Osman W, Aug. 2019 The Vulnerable Socioeconomic Conflict on the Gate Community Neighborhood in Makassar *KnE Social Sciences*.
[16] Acheampong R A, 2018 Towards incorporating location choice into integrated land use and transport planning and policy: A multi-scale analysis of residential and job location choice behaviour *Land Use Policy* **78**.
[17] Liu Z and Robinson G M, 2016 Residential development in the peri-urban fringe: The example of Adelaide, South Australia *Land Use Policy* **57**.
[18] Yang Y Zhang D Meng Q Yu W and Yuan L, 2017 Stratified evolution of urban residential spatial structure in China through the transitional period: A case study of five categories of housings in Chengdu *Habitat International* **69** p. 78–93.
[19] Setyono J S Yunus H S and Giyarsih S R, Apr. 2016 The Spatial Pattern of Urbanization and Small Cities Development in Central Java: A Case Study of Semarang-Yogyakarta-Surakarta Region *Geoplanning: Journal of Geomatics and Planning* **3**, 1.
[20] Pramono R W D, 2018 Housing Development on the Urban Fringe and its Challenges to Sustainable Urban Growth in *IOP Conference Series: Earth and Environmental Science* **158**, 1.
[21] Sutanto, 1987 *Penginderaan Jauh Dasar Jilid 1* Yogyakarta: Gadjah Mada University Press.
[22] Remali A M Salama A M Wiedmann F and Ibrahim H G, 2016 A chronological exploration of the evolution of housing typologies in Gulf cities *City, Territory and Architecture* **3**, 1.
[23] Aulia D N and Marpaung B O, 2017 Transformation of gated community in the last three decades in Medan City, Indonesia *IPTEK Journal of Proceedings Series* 3.
[24] Paradis E, Jun. 2018, If You Build It, They Will Claim: Rights-Based Participation and Accountability in Canada's National Housing Strategy; Submission to the National Consultation on a Human Rights-Based Approach to Housing, Ottawa.
[25] Roitman S and Recio R B, May 2020 Understanding Indonesia's gated communities and their relationship with inequality *Housing Studies* **35**, 5 p. 795–819.
[26] Burke M, 2001 *The Pedestrian Behaviour of Residents in Gated Communities* .
[27] Hartanto I S and Rachmawati R, 2017 Assessing the spatial-temporal land use change and encroachment activities due to flood hazard in north coast of central Java, Indonesia *Indonesian Journal of Geography* **49**, 2.

AUTHOR INDEX